Study Guide with Student Solutions Manual

Algebra and Trigonometry

EIGHTH EDITION

Richard N. Aufmann

Richard D. Nation

Prepared by

Richard N. Aufmann

Richard D. Nation

Christine S. Verity

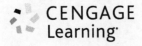

CENGAGE
Learning·

Australia • Brazil • Mexico • Singapore • United Kingdom • United States

For product information and technology assistance, contact us at **Cengage Learning Customer & Sales Support, 1-800-354-9706**.

For permission to use material from this text or product, submit all requests online at **www.cengage.com/permissions** Further permissions questions can be emailed to **permissionrequest@cengage.com**.

ISBN-13: 978-1-285-45112-1
ISBN-10: 1-285-45112-0

Cengage Learning
200 First Stamford Place, 4th Floor
Stamford, CT 06902
USA

Cengage Learning is a leading provider of customized learning solutions with office locations around the globe, including Singapore, the United Kingdom, Australia, Mexico, Brazil, and Japan. Locate your local office at: **www.cengage.com/global**.

Cengage Learning products are represented in Canada by Nelson Education, Ltd.

To learn more about Cengage Learning Solutions, visit **www.cengage.com**.

Purchase any of our products at your local college store or at our preferred online store **www.cengagebrain.com**.

Printed at CLDPC, USA, 09-23

Table of Contents

Study Guide

Student Solutions Manual

Chapter P: Preliminary Concepts

Word Search

```
L  W  I  S  P  Z  X  T  Z  C  J  X  V  C  I
H  A  F  M  Q  R  N  E  R  C  E  I  O  E  N
L  O  C  G  A  A  O  O  D  X  J  E  R  L  F
L  A  I  O  T  G  T  D  P  N  F  Q  A  B  I
A  H  N  S  R  A  I  O  U  F  I  Z  D  A  N
E  N  N  O  R  P  N  N  I  C  W  L  I  I  I
R  O  I  E  I  E  I  C  A  S  T  A  C  R  T
C  X  M  X  N  T  I  C  I  R  Q  I  A  A  Y
I  U  C  T  W  E  A  X  E  E  Y  M  L  V  U
N  Y  D  O  N  K  A  R  C  R  Z  O  G  V  J
R  S  V  T  F  A  C  T  O  R  I  N  G  H  H
R  O  T  A  N  I  M  O  N  E  D  Y  G  Q  C
L  A  V  R  E  T  N  I  L  D  I  L  F  R  B
I  N  T  E  G  E  R  Y  S  E  T  O  M  V  T
H  E  V  Y  M  U  O  G  X  E  L  P  M  O  C
```

AXIS	IMAGINARY	PRODUCT
COEFFICIENT	INDEX	RADICAL
COMPLEX	INFINITY	RATIONAL
CONSTANT	INTEGER	REAL
DENOMINATOR	INTERVAL	RECIPROCAL
EXPONENT	NUMERATOR	SET
FACTORING	POLYNOMIAL	VARIABLE

1

P.1: The Real Number System

The Real Numbers are the numbers we need to get along in our everyday lives – from just plain counting things, to representing numbers too big to count easily, to balancing the checkbook, to building houses, and more. We also need to be able to communicate about these numbers in order to get along in our everyday lives, so now we will talk a little about the ways we communicate about and classify these numbers.

There are a lot basic concepts in this section. Now is not the time to panic. Read and enjoy as much as you can!

Sets

In order to communicate about numbers, one of the first things we need to do is to distinguish between different kinds of numbers. So if we collect numbers together based on different attributes, we can call those collections "sets." Then we apply reasoning and set operations to them.

♣ composite number

A composite number is a positive integer that can be written as a product of other numbers besides one and itself. For example, 21 can be written as a product of 3 and 7, or 1, 3, 7, and 21 are its factors. In other words, a composite number is *not* a prime number. By the way, being able to represent numbers as a product is extremely useful. (When we *factor* polynomials later, for instance, we will use some of the same ideas.)

♣ negative integers

The counting or natural numbers all have opposites, created simply by putting a negative sign in front them. These would be the negative integers. The negative integers with the counting numbers (positive integers) and the number zero make up the set of integers.

✓ Questions to Ask Your Teacher

Sometimes a special symbol is used to symbolize the set of integers, Z. Ask if your teacher uses this notation.

★ **For Fun**

If you drew a number line with zero in the middle, placed some positive and negative numbers on it, could you come up with a workable meaning for the negative sign?

♣ **positive integers**

Of course, positive integers are the counting numbers, right? (They're also called the natural numbers.)

★ **For Fun**

Is zero a negative integer or a positive integer?

★ **For Fun**

What is the difference between the natural numbers and the whole numbers?

♣ **empty set, null set**

The empty set is actually the very subtle idea of a collection of absolutely nothing. (How could it be a collection, then?)

★ **For Fun**

Can you see a relation between the idea of the empty, or null, set and the idea of zero?

♣ **finite set**

One way to describe a finite set, or a collection of objects, is that it is conceivable that all the members of the set could be listed ... even if it took a lifetime or more to do so. For example, an easy finite set to list might look like $\{a,b,c\}$; on the other hand, the set of all integers between zero and two hundred million is a set that you can conceive of listing, but who would want to? The elements of a finite set can be counted.

♣ **infinite set**

An infinite set contains an infinite number of elements. Now, infinity is a hard concept to come to terms with, and that's what makes it different from finiteness. Counting until infinity is inconceivable. An infinite set is inconceivably large. Infinity is an idea you have to relax into and let your mind accept!

★ **For Fun**

But could an infinite set be *countable*? Don't answer too quickly.

♣ **set builder notation**

Set builder notation is another method to communicate about the members in a set. Since it's a notation, the notation holds all the information: $\{x \mid 0 < x \le 5\}$, for example. $\{x \mid 0 < x \le 5\}$ means "the set of all x's such that x is greater than zero and less than or equal to five." (Just a matter of translating from *math-ese* into English.)

✓ **Questions to Ask Your Teacher**

Maybe now would be a good time to go over the number line and greater than and less than if you have problems sorting out those symbols.

★ **For Fun**

Do you think that the set $\{x \mid 0 < x \le 5\}$ is finite or infinite? Again, don't answer too quickly!

Sets: Test Prep	
Specify those numbers which are integers, rational numbers, irrational numbers, prime, and reals: $$-\sqrt{3}, 0, 3\tfrac{4}{5}, 0.\overline{3}, \pi, -1, -\frac{100}{\sqrt{7}}$$	By inspection, we can sort these numbers accordingly. • integers: $0, -1$ • rationals: $0, 3\tfrac{4}{5}, 0.\overline{3}, -1$ • irrationals: $-\sqrt{3}, \pi, -\frac{100}{\sqrt{7}}$ • primes: none • reals: all
Try These	
Put the elements above into order from smallest to largest.	$$-\frac{100}{\sqrt{7}}, -\sqrt{3}, -1, 0, 0.\overline{3}, \pi, 3\tfrac{4}{5}$$

Find the exact decimal representation of $-\dfrac{4}{7}$	$0.\overline{571428}$

✓ **For Fun**

Where do the Whole Numbers fit in this Venn diagram?

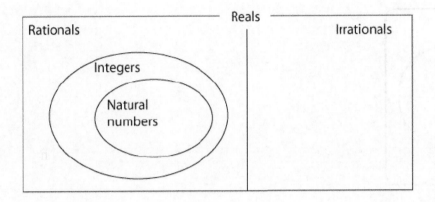

Set Builder Notation: Test Prep	
Write a description of this set in English: $\{x \mid 0 < x\}$	Noting that x's are strictly greater than zero, we can say, "The set of all x such that x is a positive real number."
Try These List the smallest three elements of the set $\{y \mid y = 2x, x \in \text{natural numbers}\}$	$2, 4, 6$
List the largest three elements of $\{x \mid x < 0,\ x \in Z\}$,	$-1,\ -2,\ -3$

Union and Intersection of Sets

Just like there are well-defined operations for numbers like addition and multiplication, sets have

operations that apply especially to them. The union of two or more sets is the operation that *includes* all the members those sets in one new set. The intersection is a more exclusive operation; it admits for members only those items that were already in all of the sets under consideration.

★ For Fun

Can you see the different sets, unions of sets, and intersections of sets in this Venn diagram?

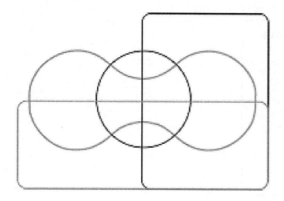

Union and Intersection of Sets: Test Prep	
Find $A \cap B$ given $A = \{p, e, a, c, h\}$ and $B = \{p, e, a, r\}$	By inspection, we see that the elements common to A and B are $\{p, e, a\}$
Try These	
What is the set formed by $\{0\} \cup \mathbf{Z}$	**W**, the set of whole numbers.
Find $A \cup B$ given A and B above.	$\{p, r, e, a, c, h\}$

Interval Notation

Dense sets that cannot be listed, like the rational numbers (fractions), require different symbols for describing them. One such tool is set-builder notation; another is interval notation.

♣ interval notation

Interval notation is a special kind of set notation. In interval notation, the set $\{x \mid 0 < x \le 5\}$ would be written $\left(0,\ 5\right]$. The parenthesis means that zero is *not* included in the set; the square bracket means five is included. And everything between those numbers is included, too.

♣ open interval

Intervals that include neither of the end-points are open intervals: (0, 5), for example.

♣ closed interval

Both end-points are included in a closed interval: [–3.5, 10], for example.

★ For Fun

Could you write [–3.5, 10] in set-builder notation? Could you signify this closed interval on a number line?

♣ half-open interval

You could just as well call this a half-closed interval, couldn't you? $\left(0,\ 5\right]$ is one.

♣ infinity symbol: ∞

In interval notation, "all real numbers" is symbolized by the open interval $(-\infty,\infty)$, "from negative infinity to positive infinity." Please notice that infinity, ∞, is *not* a number, but an idea, and so can never be an included end-point. (Could you ever actually reach infinity in order to include it?)

♣ negative infinity symbol: –∞

The number line goes forever in both directions. Negative infinity is where the arrowhead on the negative side of the number line is headed!

Interval Notation: Test Prep	
Graph this half-open interval on the number line: (–3, 7]	We note that –3 is not included but that 7 is: 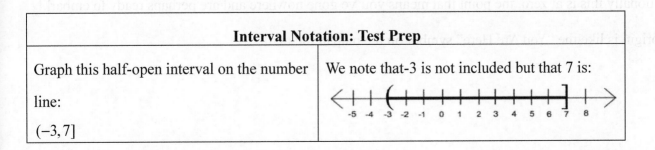

Try These	
Graph $(5, \infty)$ on the number	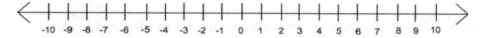
Write $\{x \mid -\infty < x \le 0\}$ in interval notation:	$(-\infty, 0]$

Absolute Value and Distance

Absolute Value is one of the easiest operations you learned in grade school. You know: just change it to a positive. It turns out, though, that the idea of absolute value is tied up with what it means to be a certain *distance* away. When it comes to numbers, the distance between them is very important, indeed. See if you can think differently about this easy operation.

❖ **coordinate axis**

An axis is a fixed line; the coordinates are the fixed values on that line. By the way, it's better to be consistent with distances between coordinates!

❖ **real number line**

The real number line is a coordinate axis that includes all real numbers: $(-\infty, \infty)$.

❖ **coordinate**

A coordinate is a member of a group numbers that specifies a position, like the numbers on the number line above!

❖ **origin**

Traditionally this is at zero, the point that means you've gone nowhere and are perhaps ready to embark! The origin is like the "You Are Here" symbol on the map at the mall.

♣ absolute value

Absolute value tells you have far you've gone, the distance you have travelled from the origin. And distance is always a positive notion, isn't it? Make sure you understand the formal definition of absolute value given on page 7 of your text book; the functional notation is important and understanding it will help tremendously in subsequent sections and math classes.

★ For Fun

Can you imagine a meaning for negative distance? How could you go negative three miles from where you are?

♣ exponential notation

Exponential notation is short hand for multiplication. 3^4 means $3 \cdot 3 \cdot 3 \cdot 3$.

♣ base

In the expression 3^4, the 3 is the base. In the expression x^4, what is the base?

♣ exponent

The presence of an exponent is what exponential notation is all about, right? In the expression 3^4, the 4 is the exponent! The exponent tells you how many times to multiply the base by itself.

Absolute Value: Test Prep	
Use the definition of absolute value to evaluate: 1. $\lvert 5 \rvert$ 2. $\lvert 0 \rvert$ 3. $\left\lvert -\dfrac{1}{7} \right\rvert$	1. Since $5 > 0, \lvert 5 \rvert = 5$ 2. Since $0 = 0, \lvert 0 \rvert = 0$ 3. Since $-\dfrac{1}{7} < 0, \left\lvert -\dfrac{1}{7} \right\rvert = -\left(-\dfrac{1}{7} \right) = \dfrac{1}{7}$

Try These	
Given $x = -3$, evaluate $|2x+1|$	5
Evaluate: $-|-5-1| + \left|-\dfrac{3}{4}\right|$	$-5\dfrac{1}{4}$ or $-\dfrac{21}{4}$

Distance: Test Prep	
Find the distance between -5 and 14.	$\begin{aligned} d(-5,\ 14) &= |-5-14| \\ &= |-19| = 19 \end{aligned}$
Try These	
Use absolute value notation to describe the distance between x and -9.	$|x+9|$
Use absolute value notation to describe that the distance between z and 2 is not more than 7.	$|z-2| \le 7$

Exponential Notation: Test Prep	
Evaluate $(-4^3)(-2)^2$	$(-4^3)(-2)^2 = -64 \cdot 4 = -256$
Try These	
Given $a = 3$ and $b = -2$, evaluate $\dfrac{b^2}{-a^2}$	$-\dfrac{4}{9}$
Given $x = -1$ and $y = 5$, evaluate $-x^2 + y$	4

Order of Operations Agreement

We have algebraic operations – addition, subtraction, multiplication, division – and now exponents, and we have algebraic expressions that use these operations between their parts. In order for the meaning in our mathematics to remain consistent and useful, we have to agree on how these symbols combine and what instruction is intended by the author. So we have the convention of "order of operations":

1. **Groups First**

   Use the following three steps to simplify any expressions inside of grouping symbols like brackets and parentheses, then use them to continue to simplify the expression as far as possible.

2. **Exponents**

   Simplify exponential expressions.

3. **Multiplication and Division**

   Carry out multiplication from left to right as those operations appear in the expression. Take as many steps as you need here to get the correct result!

4. **Addition and Subtraction**

   Carry out addition and subtraction from left to right as those operations appear in the expression. Take your time; don't be fooled by attractive, easy steps if they are not the next correct step as dictated by the order of operations agreement. And please, realize that just because you know an acronym does not mean you cannot be "fooled" into violating the agreement. Practice, and practice taking your time. Another very important aspect using the order of operations agreement: know how to input expressions in your calculator so that the order is observed when you evaluate in this manner, too.

♣ **variable term**

Variables are what algebra is all about, I like to think. It requires a lot of sophisticated thought to allow a letter to stand in for several possible choices for the idea of a number. Layers of abstraction. At any rate, a product that contains a letter standing in for a number is a variable term. For example, $-3x$, $12xy$, x^2, and so forth.

★ **For Fun**

Can you think of anything else besides letters to use as variables?

♣ constant term

A constant term does not have a variable in it. Constant terms have the appearance of ... [drum roll, please] ... numbers! For example, $-17, 2$, and the like.

♣ numerical coefficient

In a variable term, the product under question always has a constant, or numerical, factor. It's almost always written at the beginning of the term. For example, the numerical coefficient of $-3x$ is -3. It's the constant factor at the front.

★ For Fun

See how good you are at specifying the coefficients of these expressions: $17x^2y$, $-y^3$, xyz, $\dfrac{x}{3}$. The answers are, respectively: $17, -1, 1$, and $\dfrac{1}{3}$.

✓ Questions to Ask Your Teacher

If you're not clear about the last coefficient from the list just above, you might ask your teacher to go over fractions, and multiplication, and show a few more examples. The whole class might benefit from that clarification.

♣ variable part

Put simply: the part made up of letters, not the numerical coefficient, that's the variable part. Name the variable parts from the For Fun above. Easy.

♣ evaluate

To eVALUate means to find the VALUE of. By the way, values are numbers. So keep in mind that the answers to "evaluate" questions will be numbers; no variables left over! To evaluate an algebraic expression, you must replace each variable with its specified numerical value ... then simplify *carefully* according to the proper order of operations.

Order of Operations Agreement: Test Prep	
Evaluate $-2 \cdot 3^2 - \lvert -5 \rvert$	$\begin{aligned} -2 \cdot 3^2 - \lvert -5 \rvert &= -2 \cdot 3^2 - 5 \\ &= -2 \cdot 9 - 5 \\ &= -18 - 5 \\ &= -23 \end{aligned}$
Try These	
Evaluate $7 - 3(3-5)^2$	-5
Evaluate $-5 + [9 - (-1)]^2 \div 4$	20

Simplifying Variable Expressions

Simplifying means to make something simpler or easier to do or understand. In math something is usually "simpler" parenthesis are removed and operations are carried out wherever possible, like terms are combined, and, if applicable, ratios are reduced. In order to know what operations you can perform, and when and how, you have to know the operations, how to communicate about them, and how they are defined for use with the real numbers. Here follow some words about the operations, but be sure to thoroughly understand the rules for how these operations can be used with the real numbers; that is, what the properties real numbers are. These properties, and the properties of equality, are discussed on pages 11 through 13 in your text.

♣ addition

Addition is just a special word for "counting up all the units in order to find their total." It's a good thing we know how to add pretty quickly using addition tables and shortcuts (or even, gasp, calculators). Addition is where arithmetic starts, and so it's the basis for a lot of math. That means *counting* is the basis for a lot of math – Does that make sense?

♣ sum

A sum is the total you arrive at after performing an addition.

♣ terms

In an algebraic expression, like $3x + 2y + z$, three *terms* are separated by the plus sign. Please note that neither the numerical coefficients nor the variable parts make up a term in themselves!

♣ product

Like a sum is the result of addition, a product is the result of multiplication. In the above example, 6 is the product.

♣ factor

In the above example, the factors are 3 and 2. Note that if $3 \cdot 2 = 6$ then it's equally true that $6 = 3 \cdot 2$, and we've *factored* 6.

♣ additive inverse

The additive inverse is the number you can add to any number and obtain a sum of zero. Thus, $a + (-a) = 0$. Another word for "additive inverse is "opposite."

★ For Fun

By the way, zero is called the additive identity – it is the number you can add to any number and not change the *identity* of that number: $a + 0 = a$. Can you perform a little algebra on this equation to turn it back into the equation $a + (-a) = 0$?

♣ subtraction

And subtraction, again in terms of addition, is the addition of the opposite! For example: $10 - 3 = 10 + (-3) = 7$.

✓ Questions to Ask Your Teacher

Now is also a good time to review the difference between addition and subtraction using a number line. What does the minus sign really mean?

❖ difference

A difference is the result of subtraction. But it wouldn't hurt to also hang onto the idea what a difference means in English; you'll see how that meaning holds very soon in this course!

❖ multiplicative inverse

The multiplicative inverse is the number that you can *multiply* any number by and find the product to be one. Thus, $b \cdot \frac{1}{b} = 1$. Another word for "multiplicative inverse" is the reciprocal (keep reading).

❖ reciprocal

Same as the multiplicative inverse, you find the reciprocal by "turning a number upside down."

✓ For Fun

Find the reciprocals of $\frac{1}{3}, -\frac{8}{9}, 15$. Make sure when you multiply the reciprocal you specify with the original number that the product is 1.

❖ division

As soon as you write something like $\frac{1}{b}$ you're already talking about division. For instance, $b \cdot \frac{1}{b}$ means $b \div b$. What I'm saying is that division is multiplication by the reciprocal; maybe you've heard that before? In other words, we really know division as multiplication, just like subtraction is really defined as addition of the opposite. Note that you know $32 \div 8 = 4$ not because you really know or understand how many times 8 goes into 32; what you do know is $32 = 8 \cdot 4$.

❖ quotient

The special name we give to the result of a division operation is the quotient.

❖ numerator

Every division question can be written in the form of a fraction: $\frac{32}{8}$, for instance. The top part of a fraction is called the numerator. In a division problem, whether it's formatted as a fraction or not, the "numerator" can also be called the dividend.

★ **For Fun**

Is there a relation between this dividend and the dividend that is a positive return on an investment?

♣ **denominator**

The denominator is the bottom of a fraction, also called the divisor.

★ **For Fun**

Okay, this is important: what number can you never divide by?

On September 21, 1997, a divide by zero error on board the USS Yorktown (CG-48) Remote

Data Base Manager brought down all the machines on the network, causing the ship's

propulsion system to fail. "Sunk by Windows NT". *Wired News*. 1998-07-24.

Properties of Real Numbers: Test Prep	
Fill in the blank and name the property used: $-a+\_\_\_\_=0$	$-a+a=0$, Additive Inverse
Try These	
Name all the properties of real numbers used in $2u+4w=2(2w+u)$	Distributive Property of Multiplication Over Division and Commutative Property of Addition.
Simplify the variable expression: $5-7[2-(x+1)]-(x+10)$	$6x-12$

The commutative and associative laws do not hold for subtraction or division:

$$a - b \text{ is not equal to } b - a$$

$$\frac{a}{b} \text{ is not equal to } \frac{b}{a}$$

$$a - (b - c) \text{ is not equal to } (a - b) - c$$

$$a/(b/c) \text{ is not equal to } (a/b)/c$$

Try some examples with numbers and you will see that they do not work.

Properties of Equality: Test Prep	
Name the property of equality used If $a(b+c) = ab + ac$, then $ab + ac = a(b+c)$.	Since the left- and right-hand sides have traded places, this is the Symmetric Property of Equality.
Try These	
Name the Property of equality used: $2a = 2a$	Reflexive Property of Equality.
Name the Property of equality used: If $x = 3$, then $x + 5 = 8$	Substitution Property of Equality.

P.2: Integer and Rational Number Exponents

Remembering from last section that an exponent is shorthand for multiplication – that is,

$b^n = b \cdot b \cdot b \cdot \ldots \cdot b$ with b repeated n times in the product – we are now ready to explore all sorts of

interesting uses and properties of exponents. Be sure you have reviewed adding, subtracting, multiplying,

and dividing fractions.

Properties of Exponents

1. For any non-zero real number b, $b^0 = 1$.

2. If $b \neq 0$ and n is a natural number, then $b^{-n} = \dfrac{1}{b^n}$ and $\dfrac{1}{b^{-n}} = b^n$.

3. $b^n \cdot b^m = b^{n+m}$

4. $\dfrac{b^n}{b^m} = b^{n-m}$

5. $(b^m)^n = b^{mn}$

6. $(a^m b^n)^p = a^{mp} b^{np}$.

7. $\left(\dfrac{a^m}{b^n}\right)^p = \dfrac{a^{mp}}{b^{np}}$ $b \neq 0$

Know them, love them, practice them until they are no longer a stumbling block or a mystery.

✓ **For Fun**

You know what I'm going to suggest: prove a couple of these properties for yourself. Or you can more easily do a few examples with real numbers and small exponents!

Zero and Negative Exponents: Test Prep	
Evaluate $-\dfrac{1}{2^{-8}}$	$-\dfrac{1}{2^{-8}} = -2^8 = -256$
Try These	
Given non-zero variables, evaluate $(abc)^0$	1
Given non-zero variables, simplify $\dfrac{4^{-2}}{x^0}$	$\dfrac{1}{16}$

Properties of Exponents: Test Prep	
Use the quotient rule of exponents to Simplify: $\dfrac{a^2}{a^{-10}}$	$\dfrac{a^2}{a^{-10}} = a^{2-(-10)} = a^{12}$
Simplify: $\left(\dfrac{2}{x^2}\right)^5$	$\dfrac{32}{x^{10}}$

Operations with Rational Exponents: Test Prep	
Given non-zero variables, simplify $\left(\dfrac{x^0 y^{-10}}{3xy^2}\right)^{-2}$	$\left(\dfrac{x^0 y^{-10}}{3xy^2}\right)^{-2} = \dfrac{x^0 y^{20}}{3^{-2}x^{-2}y^{-4}}$ $= 3^2 x^{0-(-2)} y^{20-(-4)}$ $= 9x^2 y^{24}$
Try These	
Simplify: $(a \cdot a^4 \cdot b^3)^3$	$a^{15}b^9$
Simplify: $\left[\dfrac{1}{(-2)^3}\right]^{-\frac{5}{3}}$	-32

Scientific Notation

By now you know that mathematicians like to come up with shortcuts and shorthand for various

notations – exponential notation itself is one example. Scientific notation is another shorthand especially

used to specify extremely large and extremely small numbers.

★ scientific notation

In scientific notation, all numbers are written in the form $a \times 10^n$. Some examples:

decimal notation	scientific notation	on the calculator
2000	2×10^3	2 E3
$-6,720,000$	-6.72×10^6	-6.72 E6
0.000 012	1.2×10^{-5}	1.2 E-5

The exponential part of scientific notation obeys all the properties of scientific notation that you have become so familiar with!

★ For Fun

Who "invented" scientific notation: mathematicians or other scientists? What disciplines use it often?

★ For Fun

What is the difference between scientific notation and engineering notation? (How much do you know about the metric system?)

Scientific Notation: Test Prep	
Rewrite in scientific notation: 23,400,000	Moving the (understood) decimal point 7 places to the left so that $0 < 2.34 \le 10$, write 2.34×10^7. And check your answer!
Try These	
Rewrite in decimal notation: -5.6×10^{-3}	-0.0056
Rewrite in scientific notation: $\dfrac{3}{1000}$	3.0×10^{-3}

Rational Exponents and Radicals

Okay, this is a serious heads-up for any students going on to Calculus: you have to "get" this stuff! (And if you're not necessarily planning on Calculus, you need it for this course anyway.)

1. If n is an even positive integer and $b \geq 0$, then $b^{\frac{1}{n}}$ is the nonnegative real number such that

$$\left(b^{\frac{1}{n}} \right)^n = b .$$

2. If n is an odd positive integer, then $b^{\frac{1}{n}}$ is the real number such that $\left(b^{\frac{1}{n}} \right)^n = b$. Both these cases

basically say "$b^{\frac{1}{n}}$ is the nth root of b."

In other words, you can write $\sqrt{2} = 2^{\frac{1}{2}}$, or $\sqrt[4]{16} = 16^{\frac{1}{4}} = 2$, and so on and so forth! Now all these roots

are just *exponential expressions*, and you can use all the exponential rules to handle them. That's a really,

really good thing.

Simplify Radical Expressions

Once you know how to handle radical expressions in their exponential form, a lot of the work of

simplifying the radical can become easier ... especially if you have reviewed all your fractional operations.

Here come some properties radicals. I recommend that you translate the second item in particular into its

rational exponent version and just see how much sense it makes.

1. If n is a positive integer and b is a real number such that $b^{\frac{1}{n}}$ is a real number, then $b^{\frac{1}{n}} = \sqrt[n]{b}$.

2. For all positive integers n, all integers m, and all real numbers b such that $\sqrt[n]{b}$ is a real

number, $\sqrt[n]{b^m} = (\sqrt[n]{b})^m = b^{\frac{m}{n}}$.

3. If n is an even natural number and b is a real number, then $\sqrt[n]{b^n} = |b|$.

4. If n is an odd natural number and b is a real number, then $\sqrt[n]{b^n} = b$.

♣ **radicals**

Radicals are *roots*, like square roots, cube roots, 10th roots, you name it. "Radical" also refers to the

symbol: $\sqrt{}$.

♣ radicand

All the information *under* the radical symbol is the radicand. For example, $1-x$ is the radicand of the expression $\sqrt{1-x}$.

♣ index

The little integer nestled in the pocket in front of the radical symbol is the index: $\sqrt[9]{22}$, whose index is 9. The index tells you the denominator of the corresponding rational exponent version: $22^{\frac{1}{9}}$ in this case. So if $\sqrt{2}=2^{\frac{1}{2}}$, what is the "invisible" index? Square roots are understood to have an index of 2.

♣ principal square root, square root

The principal square root is the square root you get if you plug the radical into your calculator.

Technically we know that $\sqrt{4}$ is 2 or –2 because $2^2=4$ and $(-2)^2=4$. The principal square root is the *positive* square root. If you actually want *both* roots, you must specify $\pm\sqrt{4}=\pm 2$, or if you want the negative root: $-\sqrt{4}=-2$.

♣ simplest form of a radical

Here are the requirements for changing any radical into its simplest form:

1. The radicand contains only powers less than the index.

2. The index of the radical is as small as possible.

3. The denominator has been rationalized; that is, no radicals appear in the denominator.

4. No fractions occur under the radical sign (in the radicand).

♣ like radicals

You can add and subtract radicals just like you do with variable terms if they are *like* terms (next section, coming attractions!); that is, the variable parts are the same. Like radicals' radical parts are the *same*. Of course, you might have to *simplify* the radicals before you can tell if they are "like." For example,

$\sqrt{2}+\sqrt{200}=\sqrt{2}+10\sqrt{2}=11\sqrt{2}$

♣ rationalize the denominator

Step 3 above in the requirements for the simplest form of a radical makes this topic necessary.

Unfortunately not all radical expressions come automatically with with no radicals in the denominator.

The underlying idea involves multiplying the top and bottom of the ratio by something that will cause

the denominator to become a perfect root of the radical in question. For example:

$$\frac{1}{3-\sqrt{5}} = \frac{1}{3-\sqrt{5}} \cdot \frac{3+\sqrt{5}}{3+\sqrt{5}} = \frac{1(3+\sqrt{5})}{(3-\sqrt{5})(3+\sqrt{5})}$$

$$= \frac{3+\sqrt{5}}{9+3\sqrt{5}-3\sqrt{5}-\sqrt{25}} = \frac{3+\sqrt{5}}{9-5} == \frac{3+\sqrt{5}}{4}$$

★ For Fun

Which property real numbers are you using if you multiply the top and the bottom by the same value?

<table>
<tr><td colspan="2" align="center">Properties of Radicals: Test Prep</td></tr>
<tr>
<td>Rewrite the radical using

rational exponents: $\sqrt[3]{\dfrac{2^3}{x^2}}$</td>
<td>$\sqrt[3]{\dfrac{2^3}{x^2}} = \dfrac{\sqrt[3]{2^3}}{\sqrt[3]{x^2}} = \dfrac{(2^3)^{\frac{1}{3}}}{(x^2)^{\frac{1}{3}}}$

$= \dfrac{2}{x^{\frac{2}{3}}} = 2x^{-\frac{2}{3}}$</td>
</tr>
<tr><td colspan="2">Try These</td></tr>
<tr>
<td>Rewrite the radical using rational exponents:

$\sqrt[4]{a^2bc^3}$</td>
<td>$a^{\frac{1}{2}}b^{\frac{1}{4}}c^{\frac{3}{4}}$</td>
</tr>
<tr>
<td>Rewrite in radical notation: $(3x)^{\frac{2}{5}}$</td>
<td>$\sqrt[5]{9x^2}$</td>
</tr>
</table>

Simplifying Radicals: Test Prep	
Simplify the radical expression: $\sqrt{24x^2y}$	$\sqrt{24x^2y} = \sqrt{(4x^2)(6y)}$ $= \sqrt{4x^2} \cdot \sqrt{6x}$ $= 2x\sqrt{6x}$
Try These	
Simplify the radical expression: $2\sqrt{27} - 5\sqrt{3}$	$\sqrt{3}$
Simplify the rational exponent expression: $\left(2x^{\frac{1}{3}}y\right)\left(5x^{\frac{2}{3}}y^{\frac{1}{3}}\right)$	$10xy^{\frac{4}{3}}$

Rationalizing Denominators: Test Prep	
Simplify: $\dfrac{3}{\sqrt{x}-1}$	$\dfrac{3}{\sqrt{x}-1} \cdot \dfrac{\sqrt{x}+1}{\sqrt{x}+1} = \dfrac{3(\sqrt{x}+1)}{x-1} = \dfrac{3\sqrt{x}+3}{x-1}$
Try These	
Simplify the radical expression: $\dfrac{4}{\sqrt{32x}}$	$\dfrac{\sqrt{2x}}{2x}$
Simplify the rational exponent expression and rewrite in simplified radical form: $\dfrac{x^{\frac{1}{3}}y^{\frac{5}{6}}}{x^{\frac{2}{3}}y^{\frac{1}{6}}}$	$\dfrac{\sqrt[3]{x^2y^2}}{x}$

P.3:   Polynomials

Polynomials, like real numbers, are important mathematical objects that we can use to solve many problems. They are especially nice because, like real numbers, they have very specific properties that make them the next least complicated (yes, least complicated!) mathematical object to study.

Properties of Polynomials: Test Prep	
Determine the a) degree, b) leading coefficient, and c) rewrite the trinomial in descending order of degree: $-x-x^2+3$	a) 2 b) -1 c) $-x^2-x+3$
Try These	
Determine the degree of the polynomial: $3xy-5xy^2+11xyz$	3
Give an example of a fourth degree trinomial.	example: x^4-x+1 Answers will vary.

Operations on Polynomials

Now that we've reviewed a great deal about operations (adding, subtracting, etc.) with real numbers, it's time to learn about a new object, the polynomial, and how to perform operations on it. So first comes all the vocabulary for polynomials.

♣ degree of a monomial

First of all, a monomial is a single polynomial term. In order to be a term in a polynomial, the degree of the variable(s) has to be a whole number. (First, remember that the whole numbers are the counting numbers plus zero. Second, remember that radicals translate into rational exponents. Third, remember your negative exponent rules.) And the degree of a variable is its exponent. That's the degree of the monomial.

★ For Fun

Don't forget that $x = x^1$, so that the degree of x is 1. Having said that, I ask you, what is the degree of a constant?

★ For Fun

Now can you figure out what the degree of a term like $-3x^2yz^5$ is? First of all, do you agree that it's a monomial by definition?

❖ polynomial

A polynomial is a collection of one to many monomials that are added and subtracted from each other. Remember terms are separated by addition and subtraction, and so these *monomials* are *terms* in the polynomial. Also, it stands repeating from above, the degree of all these terms must be a whole number.

Some polynomial terms: -3, $13x^3$, $-2.2y$

Some *not* polynomial terms: $-\sqrt{42}$, $\dfrac{8}{x}$, $12x^{-2}$

❖ degree of a polynomial

The degree of the polynomial is the highest degree of one of the monomials that makes it up.

❖ like terms

Like terms have variable parts that match exactly.

Like terms: $-x$, $2x$, $\dfrac{1}{3}x$

Not like terms: $-xy$, $2x^2$, $\dfrac{1}{3}y^2$

❖ binomial

A binomial is a polynomial with exactly two terms: for example, $x + y$.

❖ trinomial

Naturally, that means that a polynomial with exactly three terms is a trinomial.

Polynomials with more than three terms don't get special names based on how many terms they have.

♣ constant polynomial

If you realized that the degree of a constant term is zero, and you know that zero is a whole number, then you know a constant term is a monomial. As such, it makes a perfectly good constant polynomial: for example, -12.5. That's a very simple and pretty boring polynomial, but it might be very important in the right context!

♣ leading coefficient

The leading coefficient is the numerical coefficient of the term with the highest degree (the term that gives its degree to the entire polynomial).

♣ constant term

The constant term is easy to recognize – it's the term with no variable. The degree of the constant term is zero because, for example, $2x^0 = 2 \cdot 1 = 2$. Remember the zero- exponent rule?

♣ FOIL method

FOIL is an acronym with stands for First, Outer, Inside, Last and, as such, gives an instruction about how to multiply two binomials:

$(a+b)(c+d) = ac + ad + bc + bd$. Be sure to include the $+/-$ signs! When you multiply polynomials, you're building a new polynomial whose terms are separated by ... plus or minus!

♣ evaluate a polynomial

To evaluate a polynomial, plug in the given values for the variables, follow the order of operations agreement to the letter, and, keeping in mind the *value* part of "evaluate," end up with a *number value*.

Operations on Polynomials: Test Prep	
Simplify: $(x^3 - x^2 + 3x) - (2x^2 - x + 10)$	$(x^3 - x^2 + 3x) - (2x^2 - x + 10)$ $= x^3 - x^2 + 3x - 2x^2 + x - 10$ $= x^2 - 3x^2 + 4x - 10$
Try These	
Multiply and determine the degree of the resulting polynomial: $(2x - 1)(x^2 + x + 5)$	$2x^3 + x^2 + 9x - 5$, degree 3
Multiply and simplify: $(5x + 2)^2 - (6x - 1)^2$	$-11x^2 + 32x + 3$

Special Products: Test Prep	
Simplify using FOIL and using a special product rule; show the results are identical: $(5x - 3)(5x + 3)$	1. $(5x - 3)(5x + 3) = 25x^2 + 15x - 15x - 9 = 25x^2 - 9$ 2. $(5x - 3)(5x + 3) = (5x)^2 - (3)^2 = 25x^2 - 9$ 3. $25x^2 - 9 = 25x^2 - 9$, identical!
Try These	
Use a special product rule to Simplify: $(x + 2y)^2$	$x^2 + 4xy + 4y^2$
Simplify: $(3a - 2)^2$	$9a^2 - 12a + 4$

Every polynomial in one variable can be written in the form

$$a_n x^n + a_{n-1} x^{n-1} + \cdots + a_2 x^2 + a_1 x + a_0 x$$

This form is sometimes taken as the definition of a polynomial in one variable.

Can you identify the degree, leading coefficient, and constant term?

Applications of Polynomials

The graph of $y = 3x + 1$ is a line and $3x + 1$; (Can you see that both of its terms, $3x$ and 1 satisfy the requirements for polynomial terms?). Everybody knows a straight line is the shortest distance between two points, right? Well, if a line (a polynomial) fails to describe how to get from one place to another, lots of times a higher-order polynomial might be able to do the job instead. (What is the degree of a linear polynomial? What do I mean by a higher-order polynomial? We've factored some of them, yes?) Of course, there are more complicated mathematical expressions that can be used to describe behaviors of systems, but beginning with lines and other polynomials will get you very far. Once you've found a nice polynomial, you'll need to evaluate it – see the previous section!

★ For Fun

Here are some obscure polynomial degree terms you can amaze your friends with:

Degree	Name	Example
$-\infty$	zero	0
0	(non-zero) constant	1
1	linear	$x + 1$
2	quadratic	$x^2 + 1$
3	cubic	$x^3 + 1$
4	quartic (or biquadratic)	$x^4 + 1$
5	quintic	$x^5 + 1$
6	sextic (or hexic)	$x^6 + 1$
7	septic (or heptic)	$x^7 + 1$
8	octic	$x^8 + 1$
9	nonic	$x^9 + 1$

NOW TRY THE MID-CHAPTER QUIZ

P.4: Factoring

In order to use mathematical ideas like The Zero-Factor Principle or The Factor Theorem, it's necessary to be able to find the *factors* first. Factoring is one of the first and most fundamental activities one undertakes in algebra. Although factoring in and of itself might be intellectually rewarding, there are very good reasons or applications for it, and they are coming!

♣ factoring

Factoring is generally recognized as the act of writing a polynomial as a product of polynomials, for example: $2x^3 - 2x^2 - 12x = 2x(x+2)(x-3)$. But remember composite numbers? Finding their prime factorization is factoring also, and many of the reasons for both kinds of factoring overlap.

★ For Fun

The factored polynomial above: Can you find the degree of each of its factors?

♣ factoring over the integers

Factoring over the integers makes life simpler for us algebra students. It means that you will be expected to learn how to recognize the factors of $(x^2 - 4) = (x+2)(x-2)$, but not necessarily how to factor this: $(x^2 - 2) = (x+\sqrt{2})(x-\sqrt{2})$. In the second case, we had to factor over the irrational numbers, which is much, much harder to do intuitively! Can you see the difference?

Greatest Common Factor

Finding the greatest common factor (GCF) of a polynomial is like using the distributive property in reverse. Instead of the usual $a(b+c) = ab + ac$, look at it the other way: $ab + ac = a(b+c)$. The a is the GCF of the expression on the left-hand side of the equation, and on the right-hand side it has been "factored out."

★ For Fun

What property of equality says that if $x = y$ then $y = x$? (Check out page 13 of your text if you cannot remember.)

❖ greatest common factor

The first step in factoring is to find the greatest common factor. To find the greatest common factor, look first for a common numerical factor for *all* the coefficients in the polynomial, then find the lowest power of the variable(s) common to *all* the terms. Use the distributive property to rewrite the polynomial with that GCF "factored out." Then check your factoring.

Greatest Common Factor: Test Prep	
Use the distributive property to factor out the GCF: $6a^2z - 2az + 14az^2$	$6a^2z - 2az + 14az^2 = 2az(3a - 1 + 7z)$ Check it!
Try These	
Factor out the GCF: $5x - 15$	$5(x - 3)$
Factor out the GCF: $2x^2 - 10x$	$2x(x - 5)$

Factoring Trinomials

After finding and factoring out the GCF, the next experiment is in factoring a trinomial. That's almost as far as this chapter goes, but in later chapters factoring larger and larger polynomials comes into play. So get good at the basics so that you can concentrate on the more difficult techniques later! There are basically two kinds of trinomials to learn to factor: $x^2 + bx + c$ or $ax^2 + bx + c$. In one case the *leading coefficient* is one, and these are usually the easiest to factor, they can be factored. More techniques, trials, and practice are needed to become proficient at factoring trinomials of the form $ax^2 + bx + c$.

♣ nonfactorable over the integers

Sometimes called a prime polynomial, but that's really misleading, so a trinomial is called "nonfactorable over the integers" if you cannot factor it by the means given in this section. This definition actually holds for polynomials of any degree, but you need more methods than those given in this section! So $(x^2 - 2) = (x + \sqrt{2})(x - \sqrt{2})$ is nonfactorable over the integers (but you may notice that it is *not* strictly "prime" – however that applies to polynomials – since I did factor it).

Factoring Quadratic Trinomials: Test Prep	
Factor: $b^2 - 7b + 12$	Noting that $-3 - 4 = -7;$ whereas, $(-3)(-4) = 12$, we factor the trinomial so: $(b-3)(b-4)$. And check it!
Try These	
Factor: $5x^2 + 14x - 3$	$(5x - 1)(x + 3)$
Factor the quadratic binomial: $4s^2 - 9$	$(2s - 3)(2s + 3)$

Special Factoring

More shortcuts! Again: this takes practice to recognize when you can use these shortcuts. Nice enough, though, if you cannot remember any shortcut that seems to apply, all the other methods this section will still work.

✓ Questions to Ask Your Teacher

Make sure you understand how thoroughly your teacher expects you to know, recognize, and use each of these shortcuts. Don't be penalized for using other methods if you can help it.

Factoring the Difference of Two Perfect Squares:

$$a^2 - b^2 = (a+b)(a-b)$$

♣ **perfect square**

A perfect square is a product of exactly two identical factors. $81 = 9 \cdot 9$ is a numerical perfect square. $x^4 = x^2 \cdot x^2$ is the perfect square of variables. And so on. Are you able to recognize a perfect square when you see it? You need to be able to find the perfect squares in a polynomial like $4x^2 - 81$ in order to be able to use the *Difference of Squares* factoring shortcut.

♣ **square root of a perfect square**

Likewise 9 is the square root of a perfect square, namely 81. Now you need to be able to find the square roots of the terms in $4x^2 - 81$ in order to factor it by this shortcut.

Factoring a Perfect Square Trinomial:

$$a^2 + 2ab + b^2 = (a+b)^2$$

$$a^2 - 2ab + b^2 = (a-b)^2$$

♣ **perfect square trinomial**

A perfect square trinomial is a trinomial that is in the form of the left-hand side of one of the two formulae above. These are harder to recognize than the difference of squares. Some examples are

$x^2 - 10x + 25$, $a^2 + 4a + 4$, $4x^2 - 4x + 1$.

Factoring a Sum/Difference of Cubes Binomial:

$$a^3 + b^3 = (a+b)(a^2 - ab + b^2)$$

$$a^3 - b^3 = (a-b)(a^2 + ab + b^2)$$

♣ **perfect cube**

Similarly, you need to recognize a perfect cube, a number or variable expression which is the product of three identical factors, for example 27 or $8x^3$.

♣ **cube root of a perfect cube**

Likewise the $2x$ is the cube root of the perfect cube $8x^3$.

Special Factoring Formulas: Test Prep	
Factor: $27a^3 + 8$	$27\,a^3 + 8 = (3a)^3 + (2)^3 = (3a+2)(9a^2 - 6a + 4)$
Try These	
Factor: $y^2 - 6y + 9$	$(y-3)^2$
Factor: $4x^2 + 81$	Nonfactorable over the integers.

♣ **quadratic in form**

Now everything you've learned about factoring quadratic expressions can be applied to any polynomial expression that is *quadratic in form* even though it may not actually be quadratic:

$$x^4 + 7x^2 + 12 \qquad\qquad 16x^4 + 8x^2 - 35 \qquad\qquad x^4 - 625 \qquad\qquad x^2 + 6\sqrt{x} + 5$$

Whereas these are important forms and become more and more important as you progress in algebra, they're hard to recognize. All I can say is "keep your eyes and mind open, look for patterns, and practice!"

Quadratic in Form: Test Prep	
Rewrite $z^4 - 5z^2 + 6$ in quadratic form, $au^2 + bu + c$, factor, and then rewrite with the original variables.	$\begin{aligned} z^4 - 5z^2 + 6 &= u^2 - 5u + 6 \qquad \text{Let } u = z^2. \text{ Then } u^2 = z^4. \\ &= (u-3)(u-2) \\ &= (z^2 - 3)(z^2 - 2) \end{aligned}$

Try These	
Factor: $y^4 + y^2 - 2$	$(y^2 + 2)(y^2 - 1)$
Factor: $a^2b^2 - ab - 6$	$(ab + 2)(ab - 3)$

Factor by Grouping

This method of factoring is one of my favorites. It requires that the polynomial be written in four terms, exactly the opportune four terms!

♣ **factored by grouping**

To factor by grouping, first you see the four terms as two groups of two (*grouping*), then factor the GCF out of each those groups, then see still another GCF in the remaining two terms. (You have to remember that *terms* are separated by + or –, no multiplication involved.)

✓ **Questions to Ask Your Teacher**

There's a way to use factoring by grouping to factor trinomials, too. See if you can get your teacher to show you that. It's fun and very useful!

Grouping: Test Prep	
Factor by grouping: $3x + 6a - ax - 2a^2$	$3x + 6a - ax - 2a^2$ $= 3(x + 2a) - a(x + 2a)$ $= (x + 2a)(3 - a)$
Try These	
Factor: $x^3 + x^2 - x - 1$ Factor: $x^2 - 4xy + 4y^2 - 16$	$(x - 1)(x + 1)^2$ $(x - 2y + 4)(x - 2y - 4)$

General Factoring

When it comes to factoring, you need to have as many tools as possible on hand to bring to bear on the problem. Review and practice all the methods of this section; look over a synopsis of them on page 47 of your textbook.

General Factoring: Test Prep	
Factor: $x^4 + x^2 - 20$	$x^4 + x^2 - 20 = (x^2 + 5)(x^2 - 4)$ $\qquad\qquad\qquad = (x^2 + 5)(x + 2)(x - 2)$
Try These	
Factor: $4x^4 - 2x^3 - 6x^2$	$2x^2(x + 1)(2x - 3)$
Factor: $2a^4 - 16a$	$2a(a - 2)(a^2 + 2a + 4)$

P.5: Rational Expressions

Rational expressions are ratios of polynomials, for example: $\dfrac{x^2 - 2x + 1}{x^2 - 4}$. Here's the definition:

♣ **rational expression**

A rational expression is a fraction in which the numerator and denominator are polynomials.

♣ **domain of a rational expression**

The domain of a rational expression is any set of numbers that can be used for replacement for the variable(s). Keep in mind that division by zero is not allowed!

★ **For Fun**

Find the domains these rational expressions:

1. $\dfrac{x + 2}{x - 5}$

2. $\dfrac{a^2 + 9}{a^2 - 64}$

3. $\dfrac{y - 5}{y}$

4. $\dfrac{x^2-4}{15}$

Simplifying Rational Expressions

Like simplifying fractions (also known as reducing fractions), you can *factor* the numerator and denominator of a rational expression and cancel common factors.

★ **simplify a rational expression**

Use the "equivalent expressions property" to eliminate factors common to both the numerator and denominator of the rational expression. Please be sure to factor first and then divide numerator and denominator by common factors.

✓ **Questions to Ask Your Teacher**

Have your teacher review dividing by common factors if you are unclear about this!

Simplify Rational Expressions: Test Prep	
Factor and simplify: $\dfrac{x^2+2x-3}{x^2-1}$	$\dfrac{x^2+2x-3}{x^2-1}=\dfrac{(x+3)(x-1)}{(x-1)(x+1)}=\dfrac{x+3}{x+1}$
Try These	
Describe the domain of the rational expression: $\dfrac{2x^2-x}{x^2+1}$	all real numbers
Describe the domain: $\dfrac{1}{x^2-16}$	$x\neq 4,-4$

Operations on Rational Expressions

Just like any other numbers, rational expressions can be added, subtracted, multiplied, and divided by each other. Since they are *fractions*, once again, you need to know how to do all these operations on fractions.

★ common denominator

When you add or subtract fractions, you must write all expressions involved on a common denominator, right? Same thing with rational expressions:

$$\frac{4x-2}{3x+12} - \frac{x-2}{x+4} = \frac{4x-2}{3(x+4)} - \frac{x-2}{x+4} \cdot \frac{3}{3} = \frac{4x-2-3(x-2)}{3(x+4)} = \frac{4x-2-3x+6}{3(x+4)} = \frac{x+4}{3(x+4)} = \frac{1}{3}$$

Be sure to simplify (reduce) your rational expression at the end, too.

✓ Questions to Ask Your Teacher

Review with your teacher the methods to find the least common denominator for fractions; see how this applies to rational expressions, too.

Determining the LCD of Rational Expressions

The least common denominator (LCD) of a rational expression is found in fundamentally the same way that the LCD is found for a group of fractions.

Consider again this example:

$$\frac{4x-2}{3x+12} - \frac{x-2}{x+4} = \frac{4x-2}{3(x+4)} - \frac{x-2}{x+4} \cdot \frac{3}{3} = \frac{4x-2-3(x-2)}{3(x+4)} = \frac{4x-2-3x+6}{3(x+4)} = \frac{x+4}{3(x+4)} = \frac{1}{3}$$

1. **Factor all the denominators.**

   In my example above, the two denominators factored thus $3(x+4)$ and $(x+4)$.

2. **Find the least common multiple of all those factors.**

   In the list of factors of the denominators, the least common multiple is the product of each factor in the list raised to the highest power that it occurs there. It's $3(x+4)$ in my example above.

3. **Multiply any rational expressions that are not already over the LCD by any factors missing from their denominators** – top and bottom because you can multiply by the number 1 and not change anything.

4. **Now combine all the numerators on their common denominator** – you're combining apples and apples now, instead of apples and oranges!

Operations on Rational Expressions: Test Prep	
Add: $\dfrac{2}{x+3}+\dfrac{x}{x+5}$	The LCD is $(x+3)(x+5)$. $\dfrac{2}{x+3}+\dfrac{x}{x+5}=\dfrac{2}{x+3}\cdot\dfrac{x+5}{x+5}+\dfrac{x}{x+5}\cdot\dfrac{x+3}{x+3}$ $\qquad=\dfrac{2x+10+x^2+3x}{(x+3)(x+5)}=\dfrac{x^2+5x+10}{x^2+8x+15}$
Try These	
Multiply: $\dfrac{x^2-2x+1}{x^2-4}\cdot\dfrac{x^2+4x+4}{x^2+x-2}$	$\dfrac{x-1}{x-2}$
Divide: $\dfrac{x^2-x-6}{x+4}\div\dfrac{x^2-9}{x^2+3x-4}$	$\dfrac{x^2+x-2}{x+3}$

Complex Fractions

❖ **complex fraction**

A complex fraction is a fraction whose numerator or denominator, or possibly both, also contain a fraction.

Complex fractions entail their own special challenges for simplifying. There are basically two methods:

Method One

1. Find the LCD of all the fractions internal to the complex fraction.

(For example, $\dfrac{\dfrac{1}{3}+\dfrac{3x}{2}}{2}$ has $3\cdot x\cdot 1=3x$ as its overall LCD.)

2. Multiply top and bottom of the complex fraction by this LCD.

(Continuing the example: $\dfrac{\dfrac{1}{3}+\dfrac{3}{x}}{2}\cdot\dfrac{3x}{3x}=\dfrac{x+9}{6x}$. *Be sure you use the distributive property to multiply the LCD correctly!*)

3. Simplify, if needed.

(The above example cannot be simplified, can it?)

Method Two

1. Simplify the numerator and denominator into a single fraction.

For example, $\dfrac{\dfrac{1}{3}+\dfrac{3}{x}}{2} = \dfrac{\dfrac{x+9}{3x}}{2}$

2. Multiply the numerator by the reciprocal of the denominator.

For example, $\dfrac{\dfrac{x+9}{3x}}{2} = \dfrac{x+9}{3x}\cdot\dfrac{1}{2}$

3. Multiply the fractions; simplify is necessary.

Continuing the example: $\dfrac{x+9}{3x}\cdot\dfrac{1}{2} = \dfrac{x+9}{6x}$

Complex Fractions: Test Prep	
Simplify: $\dfrac{1}{\dfrac{1}{S}+\dfrac{2}{R}}$	$\dfrac{1}{\dfrac{1}{S}+\dfrac{2}{R}}\cdot\dfrac{SR}{SR} = \dfrac{SR}{R+2S}$ The LCD of the entire complex fraction is SR.
Try These	
Simplify: $\dfrac{\dfrac{y}{x}+1}{\dfrac{x}{y}-\dfrac{y}{x}}$	$\dfrac{y}{x-y}$
Simplify: $\dfrac{4-\dfrac{1}{x}}{\dfrac{1}{y}-2}$	$\dfrac{4x-1}{1-2x}$

Applications of Rational Expressions

Rational expressions are one type of "more complicated" expression you might find used to describe a system. For example, capacitances in electronic circuits are summed up by their reciprocals:

$\frac{1}{c_1} + \frac{1}{c_2} + \frac{1}{c_3}$. Can you simplify this expression so that it shares a common denominator? Then, of course,

you might have to evaluate it for certain capacitances, or the capacitances might turn out to be described

by polynomials! Who knows?

P.6: Complex Numbers

Like negative numbers and irrational numbers, as mathematics progressed new numbers were required.

Imaginary numbers, which express the square roots of negative numbers, were a necessary evolution of

the number system. They are not real numbers as have been all the numbers we've considered so far this

chapter.

Complex numbers are numbers that have both real parts and imaginary parts.

★ **For Fun**

Can you illustrate the relationship between real numbers, imaginary numbers, and complex numbers in a

Venn diagram?

Introduction to Complex Numbers

Complex numbers are numbers that have both real and imaginary parts.

✤ **imaginary unit**

i is the imaginary unit and is equal to the square root of -1.

✤ **imaginary number**

An expression that is a multiple of i, the square root of -1, is an imaginary number. So $\sqrt{-3} = i\sqrt{3}$ is

an imaginary number.

✤ **complex number**

A complex number has a real part and an imaginary part, even if either part has zero magnitude. Even

zero is a complex number.

♣ **real part**

For the complex number $3-2i$, the real part is 3. The real part is the term that is not a multiple of i.

♣ **complex part**

For the complex number $3-2i$, the imaginary part is $-2i$. The imaginary part is the term that involves i.

♣ **standard form**

Complex numbers written in standard form appear like $a+bi$, where a is the real part and bi is the imaginary part. Sometimes it is more lovely to write the complex number in the form $a+ib$; as in the case: $5-i\sqrt{5}$ where the i should not appear as though it might be underneath the radical symbol.

Complex Numbers: Test Prep	
Write the complex number in standard form: $\sqrt{81}-\sqrt{-1}$	By definition of square root and the imaginary unit, $\sqrt{81}-\sqrt{-1}=9-i$
Try These	
Simplify and write in standard form: $\sqrt{-28}$	$2i\sqrt{7}$ (the result is purely imaginary)
Write in standard form: $\sqrt{24}-\sqrt{-6}$	$2\sqrt{6}-i\sqrt{6}$

Addition and Subtraction of Complex Numbers

If you think real parts to be *like* real parts and likewise for imaginary parts, then adding and subtracting complex numbers is exactly like combining like parts:

$4+3i-5+i=-1+4i$. Keep the sum or difference in standard form!

Multiplication of Complex Numbers

Proceed to multiply complex numbers just like you would polynomials, by the FOIL method, for instance. The difference is that $i^2 = -1$, which is a *real* number. So after the multiplication and simplification, you'll have another real term to combine according to the addition/subtraction algorithm above. For example:

$(2-i)(3+4i) = 6+8i-3i-4i^2 = 6+5i-4(-1) = 6+5i+4 = 10+5i$. Again the answer is given in standard form.

Division of Complex Numbers

Division of complex numbers is not exactly division at all. It's really more of a process like rationalizing denominators was. With complex numbers the idea is to turn the denominator into an entirely real number (you will need the complex conjugate for that) and then write the result in standard form.

♣ complex conjugates, conjugates

The complex conjugate of a complex number $a+bi$ is simply $a-bi$. Did you catch that? All you have to do is change the sign of the imaginary part. So that if you multiplied $(a+bi)(a-bi)$ you'd get $a^2 + b^2$. (Try it! It's just like the difference of squares, but that $i^2 = -1$ changes it into the *sum* of squares!

So here's how division works, for example:

$$\frac{3}{2-i} = \frac{3}{2-i} \cdot \frac{2+i}{2+i} = \frac{6+3i}{4+1}$$

$$= \frac{6+3i}{5} = \frac{6}{5} + \frac{3}{5}i \qquad \text{Divide by 5 so the result is in standard form.}$$

Operations on Complex Numbers: Test Prep	
Simplify $\sqrt{-10} \cdot \sqrt{-5}$ and write the result in standard form:	$\sqrt{-10} \cdot \sqrt{-5} = i \cdot \sqrt{10} \cdot i \cdot \sqrt{5} = -1 \cdot \sqrt{50}$ $= -5\sqrt{2}$
Try These	
Add: $(3+4i)+(4-5i)$	$7-i$
Divide the complex numbers, write the result in standard form: $\dfrac{2}{3-i}$	$\dfrac{3}{5} + \dfrac{1}{5}i$

Powers of i

If you investigate raising i to different powers, you can readily discover a pattern:

$i^0 = 1$	$i^3 = -i$	$i^6 = -1$
$i^1 = i$	$i^4 = 1$	$i^7 = $ _____
$i^2 = -1$	$i^5 = i$	$i^8 = $ _____

This repeat every four powers, yes? So we devise the Powers of i observation:

If n is a positive integer, then $i^n = i^r$, where r is the remainder of the division of n by 4. For example: $i^{14} = i^2 = -1$ because the remainder of dividing 14 by 4 is 2.

Powers of i: Test Prep	
Evaluate the power of $i: -i^{40}$.	Because 40/4 leaves a remainder of zero: $-i^{40} = -i^0 = -1$
Try These	
Evaluate: $\dfrac{-4 \pm \sqrt{4^2 - 4 \cdot 1 \cdot 6}}{2 \cdot 1}$	$-2 \pm i\sqrt{2}$
Simplify: $(2-i)^2$.	$3-4i$

NOW TRY THE CHAPTER P REVIEW EXERCISES AND THE

CHAPTER P PRACTICE TEST

Final Fun: Matching

FIND THE BEST MATCH. GOOD LUCK.

EMPTY SET $\qquad$ $a^2 + 2ab + b^2$

INFINITY $\qquad$ $b^0, b \neq 0$

INTERVAL NOTATION $\qquad$ $\sqrt{-1}$

ADDITIVE IDENTITY $\qquad$ $x + 5\sqrt{x} + 6$

SQUARE ROOT OF 3 $\qquad$ $a \times 10^n$

i $\qquad$ ZERO

PERFECT SQUARE TRINONIAL $\qquad$ $[-1, 8)$

QUADRATIC IN FORM $\qquad$ $18x$

GCF of $\{2x, 6, 18\}$ $\qquad$ ∞

LCM of $\{2x, 6, 18\}$ $\qquad$ 2

MULTIPLICATIVE IDENTITY $\qquad$ $3^{\frac{1}{2}}$

DEGREE OF A QUINTIC $\qquad$ 5

SCIENTIFIC NOTATION $\qquad$ $\varnothing$

Chapter 1: Equations and Inequalities

Word Search

```
W  E  Q  U  A  T  I  O  N  O  T  R  H  N  S
N  R  C  X  L  N  V  V  U  E  A  L  Y  O  O
Y  O  S  Q  U  A  R  E  S  E  A  O  P  I  L
F  R  I  Q  P  N  C  S  N  N  B  R  O  T  U
S  A  Z  T  U  I  E  I  O  N  E  E  T  A  T
I  D  U  U  C  M  L  I  T  L  A  Z  E  I  I
T  I  Z  K  V  I  T  P  N  I  V  I  N  R  O
A  C  I  T  A  R  D  A  U  Q  R  E  U  A  N
S  A  R  I  O  C  M  A  I  S  Q  C  S  V  E
Q  L  S  P  K  S  H  N  R  U  G  G  E  C  S
L  M  O  T  N  I  O  J  A  T  I  E  E  U  R
W  R  N  M  E  D  S  L  X  V  N  V  L  B  E
P  W  A  P  P  L  I  C  A  T  I  O  N  I  V
L  A  N  O  I  T  A  R  O  X  Q  N  C  C  N
I  O  D  C  Y  F  O  R  M  U  L  A  P  V  I
```

APPLICATION	INEQUALITY	RATIONAL
CONTRADICTION	INVERSE	SATISFY
CRITICAL	JOINT	SET
CUBIC	LEGS	SOLUTION
DISCRIMINANT	LINEAR	SOLVE
EQUATION	PROPORTIONAL	SQUARE
FORMULA	QUADRATIC	VARIATION
HYPOTENUSE	RADICAL	ZERO

1.1: Linear and Absolute Value Equations

From simplifying and otherwise manipulating expressions, in this chapter we progress to manipulating equations. Now we are working with mathematically complete thoughts, and we can begin to solve problems.

Linear Equations

There is a lot of very specific and meaningful information in the phrase "linear equation." First of all, the word equation lets you know that there must be an equals sign present, for example: $3 = 2x + 1$. What does the word "linear" tell you? Remember the degrees of polynomials discussed last chapter? "Linear" tells you the degree of the equation is *one*.

★ For Fun

Can you tell that the equation $3 = 2x + 1$ is linear?

♣ equation

An equation must contain equals sign in it. That makes an equation a complete mathematical sentence: the equals sign is the "verb." And this sentence or statement tells us about the absolute equivalence of the two side of the equals sign – the subject and the predicate, if you will.

♣ satisfy

To satisfy means something like "meeting the desires of," and satisfying an equation meets the particular and only desire of math – the truth. A value of a variable satisfies an equation if that value makes an equation true.

★ For Fun

Does the value $x = 1$ *satisfy* the equation $3 = 2x + 1$?

♣ solve

Solving an equation is the process of finding all the values of the variables that make the equation true. That set of values that satisfy the equation is called the solution set.

★ **For Fun**

Does the set $\{x \mid x = 1\}$ represent the *solution set* of the equation $3 = 2x + 1$?

♣ **equivalent equations**

Equivalent equations have exactly the same members in their solution sets. This is how you discover the basic rules of algebra that we can use to find solutions – Addition, Subtraction, Multiplication, and Division Rules of equality.

★ **For Fun**

See if you can figure out how my statement of the following properties is the same (and/or different?) as the statements of these properties in your textbook, page 76.

Addition and Subtraction Property of Equality

These are equivalent equations:

$$a = b$$
$$a + c = b + c$$
$$a - c = b - c$$

Multiplication and Division Property of Equality

These are equivalent equations:

$$a = b$$
$$a \cdot c = b \cdot c$$
$$a \div c = b \div c$$

♣ **linear equation**

So a linear equation is an equation of the form $ax + b = c$ where x is the variable and a, b, and c are constants, and $a \neq 0$.

★ **For Fun**

Can you tell why the equation $ax + b = c$ would no longer be "linear" if you allowed a to be equal to zero?

Linear or First-Degree Equations: Test Prep	
Solve and check your solution: $5x + 7 = x - 1$	$5x + 7 = x - 1$ $5x - x + 7 = x - x - 1$ $4x + 7 = -1$ $4x + 7 - 7 = -7 - 7$ $4x = -8$ $\dfrac{4x}{4} = \dfrac{-8}{4}$ $x = 2$ *check*: $5(-2) + 7 \overset{?}{=} (-2) - 1$ $-10 + 7 \overset{?}{=} -3$ $-3 = -3$
Try Them	
Solve: $6(5s - 11) - 12(2s + 5) = 0$	$s = 21$
Solve: $2(3 - 4x) = 6 - 7(2x + 4)$	$x = -\dfrac{14}{3}$

Related Comino Attractions

(*in other words*: *some of the topics you're preparing for now*)

- Graphing Linear Equations
- Linear Inequalities
- Graphing Linear Inequalities
- Systems of Linear Equations

Clearing Fractions: Test Prep	
Solve: $\dfrac{12+x}{-4} = \dfrac{5x-7}{3} + 2$	$\dfrac{12+x}{-4} = \dfrac{5x-7}{3} + 2$ $(-12)\left(\dfrac{12+x}{-4}\right) = (-12)\left(\dfrac{5x-7}{3} + 2\right)$ $3(12+x) = -4(5x-7) - 24$ $36 + 3x = -20x + 28 - 24$ $23x = -32$ $x = -\dfrac{32}{23}$
Try These	
Solve and check your solution: $\dfrac{x+3}{4} = 6$	$x = 21$
Solve and check your solution: $x - \dfrac{2}{3} = \dfrac{x}{6}$	$x = \dfrac{4}{5}$

Contradictions, Conditional Equations, and Identities

Now that we have equations, we start to organize them into different categories so that we know what we can, and cannot, accomplish *on* them and *with* them. Some might be for eliminating, some for solving, some for using as tools.

♣ contradiction

In English, a contradiction is a statement that can never be true because the parts are inconsistent with each other – they exclude the possibility of truth for the whole statement because they cannot both be true. In math, all it takes to be a contradiction is to exclude the possibility of truth; for example: $2 + 2 = 5$ or $a + 10 = a$.

♣ conditional equation

A conditional equation is an equation that is *true* under the right circumstances. For instance, $3 = 2x + 1$ is true only when $x = 1$. $x = 1$ is the *condition* that makes $3 = 2x + 1$ a true statement.

❖ identity

An identity is a statement that is always true, no conditions. Of course something like $2 = 2$ is an identity, but although this has value, it's pretty elementary. The identities that are really fruitful are the ones we turn into properties and put to use in, say, simplifying expressions: $a(b+c) = ab + ac$, the Distributive Property. This equation is always true, no matter what a, b, and c are.

★ For Fun

Can you come up with mathematical examples of contradictions, conditional equations, and identities on your own? Can you come up with examples *in English*, too?

Absolute Value Equations

As the title suggests, absolute value equations are equations that contain absolute value expressions in them. Here is a reminder of how absolute value is defined:

Absolute Value

$$|x| = \begin{cases} x, & x \geq 0 \\ -x, & x < 0 \end{cases}$$

Remember, if you will, that absolute value has everything to do with measuring distances, so an equation that includes an absolute value is frequently a question of some sort about how far away things are from each other. For example, the equation $|x - 1| = 2$ is a statement about how far away x is from 1; in fact, it is a statement that x is exactly 2 away from 1.

★ For Fun

Could you draw $|x - 1| = 2$ on the number line? Can you find the *conditions* that *satisfy* this equation, can you find the member(s) of the *solution set*?

★ For Fun

Would you classify the absolute value equation $|x| = -1$ as a contradiction, a conditional equation, or an identity?

✓ Questions to Ask Your Instructor

What happens if you try to use an expression of higher order than linear in an absolute value equation?

★ For Fun

Investigate the truth of this equation for different values of x: $\sqrt{x^2} = |x|$. Contradiction, conditional, or identity?

Linear Absolute Value Equations: Test Prep	
Solve: $|3x + 2| = 3$	$\begin{aligned}&|3x + 2| = 3\\ &3x + 2 = 3 \;\; or \;\; 3x + 2 = -3\\ &3x = 1 \;\; or \;\; 3x = -5\\ &x = \frac{1}{3} \;\; or \;\; x = -\frac{5}{3}\end{aligned}$ The solutions are $\frac{1}{3}$ and $-\frac{5}{3}$.
Try These	
Solve: $|4x - 1| = -12$	no solution
Solve: $3|x - 5| - 16 = 2$	$x = 11, -1$

Applications of Linear Equations

Applications of linear equations are frequent; anywhere in the real world that we can describe a behavior as following a line, we find linear equations applicable. Anywhere we are interested in distances, we might find absolute value equations useful too.

Pointers for working with word problems

1. Read the problem.
2. Read the problem carefully. (Yes, read the problem again!) And read the problem critically – what is the problem asking you for and what is it giving you to help you get there; what equations can you construct right away; what formulas do you need?
3. Pick (preferably meaningful) variables – like A for area or S for sum.
4. Are there any relationships between the variables? Can you write the relations in equations?
5. If it's even remotely helpful, Draw A Picture! Sketch A Diagram! Make A Table!
6. Can you write an equation that will give you the solution?
7. Check your answer. Don't simply check your answer in the equation(s) that you came up with, but really check your answer to see if it answers the question and if it makes sense!

1.2: Formulas and Applications

Since we now know how to evaluate all sorts of expressions – plug in values, right? – we can use them when they are in equations to actually get results. Formulas are equations with all sorts of variables waiting to be put to use.

Formulas

Formulas are like identities: $PV = nRT$, a statement of fact. Then they are like conditional equations if you need to find what P is when V, n, R, and T are given, for example. It takes someone a lot hours of experimentation to come up with some of these formulas. Just so that we can put them to use!

❖ formula

A formula is an *equation* that states how variables are related to each other under certain (usually physical) situations. Formulas range from the extremely simple, the area of a rectangle, $A = lw$; to completely inscrutable, the focal length,

$$\frac{1}{f} = (n-1)\left(\frac{1}{r_1} - \frac{1}{r_2}\right) + \frac{(n-1)^2}{n}\frac{t_c}{r_1 r_2}.$$

Evaluating a formula means plugging the appropriate value(s) into corresponding variable(s). Usually you wind up with one variable left over – the one you were looking for. Sometimes it's necessary to solve the formula for a particular variable – again, usually the one you're looking for!

Formulas: Test Prep	
Solve the formula for h: $A = 2\pi rh + 2\pi r^2$	$A = 2\pi rh + 2\pi r^2$ $A - 2\pi r^2 = 2\pi rh$ $\dfrac{A - 2\pi r^2}{2\pi r} = h$
Try These	
Solve for m: $y - y_1 = m(x - x_1)$	$m = \dfrac{y - y_1}{x - x_1}$
Solve for d: $40 = a + 21d$	$d = \dfrac{40 - a}{21}$

Applications

There are geometric formulas (area of a circle, volume of a rectangular solid, length of the hypotenuse of a right triangle), business formulas (compound interest, sales taxes, cost functions), physics formulas (motion, heat exchange, nuclear decay), chemical formulas (mixtures, stoichometry, titration), and many others (like coin collection, task time, and so forth). **Be sure you understand what each variable stands for.** Whether you name the variables and research your own formula or one is given to you, take the time to understand all the pieces of the puzzle. The careful work and understanding will pay off.

Some frequently used formulas for example:

- area of a triangle: $A = \frac{1}{2}bh$

- volume of a pyramid: $V = \frac{1}{3}bh$

- compound interest: $P = C\left(1 + \frac{r}{n}\right)^{nt}$

- quadratic formula: $x = \frac{-b \pm \sqrt{b^2 - 4ac}}{2a}$

Applications of Formulas: Test Prep	
Freda has 75 coins. Some are quarters, the rest are dimes. Altogether she has \$12.90. How many of each type of coin does she have?	Let x be the number of quarters and y be the number of dimes. $x + y = 75$ or $y = 75 - x$. Then $$0.25x + 0.10y = 12.90$$ $$25x + 10y = 1290 \quad \text{Multiply each side by 100.}$$ $$25x + 10(75 - x) = 1290 \quad \text{Replace } y \text{ by } 75 - x.$$ $$25x + 750 - 10x = 1290$$ $$15x + 750 = 1290$$ $$15x = 540$$ $$x = 36$$ and $$y = 75 - 36 = 39$$ So: 36 quarters and 39 dimes.
Try These	
The width of a rectangle is 3 yards more than twice the length. If the perimeter of the rectangle is 222 yards, find the length and width.	Length = 36 yd.; width = 75 yd.
A bicyclist traveling at 20 mph overtakes a in-line skater who is traveling at 10 mph and had a 0.5 hour head start. How far from the starting point did the bicyclist over take the in-line skater?	10 mi.

1.3: Quadratic Equations

Quadratic polynomials are polynomials of degree two. Linear polynomials are polynomials of degree one, and you now know that a linear equation is like a linear polynomial put into an equation. So, you know what quadratic equations are by extension. And since they're *equations* now, we can set about solving them! That higher degree, though, suddenly admits more solutions and more ways of finding those solutions.

Solving Quadratic Equations by Factoring

The first method is the safest, your quadratic is factorable. Remember first to put the equation into *standard quadratic form*, then factor the quadratic, and use the Zero Product Principle to set each factor equal to zero. Now you have linear equations to solve, and that's easy. It also yields the solution(s) to your quadratic equation.

The Zero Product Principle

If $A \cdot B = 0$, then either $A = 0$ or $B = 0$.

★ For Fun

Does the Zero Product Principle remind you of any of the properties of real numbers you studied?

♣ quadratic equation

A quadratic polynomial is a polynomial of degree 2. A quadratic equation is an *equation* of degree 2. For example: $3x^2 + x = 1$.

♣ standard quadratic equation

The standard form for a quadratic equation is $ax^2 + bx + c = 0$ where a cannot be equal to zero. We could convert the above example into standard form by subtracting 1 from both sides of the equation: $3x^2 + x - 1 = 0$.

♣ double solution, double root

A quadratic equation has two solutions or *roots*. Sometimes those two solutions are the same number. In this case, the solution is called a *double* solution or double root – it's sort a like the equation actually has two solutions but they just happen to be the same numbers.

Solving Quadratic Equations by Taking Square Roots	
Another method for solving some quadratic equations is by taking the square root of, say, x^2. Because $\sqrt{x^2} = \lvert x \rvert$ and $\lvert x \rvert$ equals either x or $-x$, we usually say $\sqrt{x^2} = \pm x$ and can use this to solve quadratic equations.	$x^2 - 49 = 0$ $x^2 = 49$ $\sqrt{x^2} = \sqrt{49}$ $\lvert x \rvert = 7$ $x = \pm 7$ The solutions are -7 and 7.

♣ square root procedure

The square root procedure involves

1. isolating the x^2 on one side of the equation

2. taking the square root of both sides, **being sure to take "both" square roots of the constant**

3. and simplifying the square root if possible.

Another example might look like

$x^2 - 24 = 0$

$\quad x^2 = 24$

$\sqrt{x^2} = \sqrt{24}$

$\quad \lvert x \rvert = \sqrt{24} = 2\sqrt{6}$

$\quad x = \pm 2\sqrt{6}$

This is an illustration of the square root of a complex number:

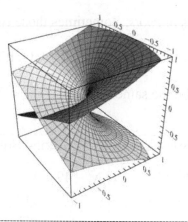

1. Aren't you glad we're not doing that? Isn't it lovely?
2. Doesn't math make lovely pictures sometimes?

3. **Solving Quadratic Equations by Completing the Square**

It's a natural progression to use the square root technique from above combined with our knowledge of factoring from last chapter to develop a technique to solve quadratic equations that depends on what is called "completing the square. If you can write $81 = 9^2$, then you have the concept required for completing the square.

First recall...

Factoring a Perfect Square Trinomial:

$$a^2 + 2ab + b^2 = (a+b)^2$$

$$a^2 - 2ab + b^2 = (a-b)^2$$

♣ **completing the square**

Completing the square is the method of rewriting a trinomial so that it is a perfect square trinomial.

For example, $x^2 - 6x$ would become a perfect square trinomial if I added 9 to it: $x^2 - 6x + 9 = (x-3)^2$

★ For Fun

Check it!

Completing the source is used in mathematics in these places, for example:

1. Solving quadratic equations (you're doing this now)

2. Graphing quadratic equations (next chapter)

3. Evaluating some integrals (that's Calculus, in case you're headed there)

4. Finding Laplace Transforms (and that's beyond Calculus)

   Completing the square is considered a basic operation in algebra.

Combining this with what we know about solving by taking square roots from above, lets you solve

more quadratic equations. For example:

$$x^2 - 2x = 2$$
$$x^2 - 2x + 1 = 2 + 1$$
$$(x-1)^2 = 3$$
$$\sqrt{(x-1)^2} = \sqrt{3}$$
$$|x-1| = \sqrt{3}$$
$$x - 1 = \pm\sqrt{3}$$
$$x = 1 \pm \sqrt{3}$$

Now, the BIG question is, "How did I know to add 1 to both sides of the equation?"

Here's how:

1. Find the coefficient of the linear term, then

   (*that's* −2 *in my example*)

2. divide *that* coefficient by 2,

   ($-2 \div 2 = -1$ *for me*)

3. and square that result – that will be the number to add **to both sides of the equation**

$((-1)^2 = 1$ *for my example*$)$

★ For Fun

Could you ever wind up having to *subtract* the same number from both sides of the equation?

✓ Questions to Ask Your Instructor

Make sure you know *how* your instructor will expect you to solve a given quadratic equation. (We are about to develop yet another method!)

> Since we are developing several methods to solve quadratic equations, get into the very, very good habit of *reading the instructions* given in the question. In other words, be prepared to use the method specified by your instructor in the problem.

Solving Quadratic Equations by Using the Quadratic Formula

By completing the square on the standard form of the quadratic equation, $ax^2 + bx + c = 0$, you can derive a very general solution to quadratic equations. We name it the quadratic formula (see? more formulas!).

> Given $ax^2 + bx + c = 0$, where $a \neq 0$, then the solution(s) are
> $$x = \frac{-b \pm \sqrt{b^2 - 4ac}}{2a}$$

★ For Fun

Can you see at least *two* reasons why a cannot be equal to zero in the above note?

Here are some considerations I would like you to take to heart when solving quadratic equations, by any method:

1. Solve them by factoring if you can – the method of factoring is slightly less error-prone than the quadratic formula.

2. Get into the habit, when using the quadratic formula, of actually writing the variable you are solving for, the "$x =$" in the above notation because

    -   it will help you remember that you are building the *solution set* of allowed variables, and

    -   it will help you make far fewer errors when your variables become more complicated (yes, that's a Coming Attractions bulletin).

3. Remember that the plus/minus symbol means $x = \dfrac{-b \pm \sqrt{b^2 - 4ac}}{2a}$ specifies **two** values.

Quadratic Equations: Test Prep	
Solve by completing the square: $x^2 - 6x + 7 = 0$	$x^2 - 6x + 7 = 0$ $x^2 - 6x + 9 = -7 + 9 \quad \left[\left(\dfrac{-6}{2}\right)^2 = 9\right]$ $(x-3)^2 = 2$ $x - 3 = \pm\sqrt{2}$ $x = 3 \pm \sqrt{2}$
Try These	
Solve by factoring: $3x^2 - 5x = 2$	$x = 2, -\dfrac{1}{3}$
Solve by using the quadratic formula: $x^2 - x + 2 = 0$	$x = \dfrac{1 \pm i\sqrt[3]{7}}{2}$

The Discriminant of a Quadratic Equation

Learning to be discriminating about your discriminant can help you discover information *about* the solution(s) to your quadratic equation ... before you ever have to solve it.

♣ **discriminant**

The discriminant of a quadratic equation is the small portion of the quadratic formula that resides under the radical, namely, $b^2 - 4ac$.

The Discriminant and the Solutions of a Quadratic Equation

- If $b^2 - 4ac > 0$, then there are two different real solutions

- If $b^2 - 4ac = 0$, then you have a *double solution*

- If $b^2 - 4ac < 0$, then there are no *real* solutions. (It so happens that there are exactly two *complex* solutions!)

The Discriminant: Test Prep	
Determine the discriminant and state the number of real roots: $8x^2 = 3x - 2$	$8x^2 = 3x - 2$ $8x^2 - 3x + 2 = 0$ $b^2 - 4ac = (-3)^2 - 4(8)(2)$ $= 9 - 64 = -55 < 0$ no real roots
Try These	
Determine the discriminant and state the number of real solutions: $9x^2 - 12x + 4 = 0$	0, one real solution
Determine the discriminant and state the number of real roots: $x^2 - 2x = 2$	12, two real roots

Applications of Quadratic Equations

Any system that can be described by a quadratic equation is a good candidate for exercising your quadratic-solving muscles. (Anything that involves an area, which is measured in *square* units, might be a candidate for the methods of quadratic equations whose dimensions are also squared – right?)

Simple motion in gravity is also a quadratic equation: $s(t) = -16t^2 + v_0 t + s_0$. s is called a position equation, and the variable is t, which denotes *time*. The position changes as time passes. Go ahead and get used to the notations v_0 and s_0. The subscript *zero* usually means *initial*, and *initial* means the

starting condition – in other words the condition when t was equal to zero.

The Pythagorean Theorem, frequently written $a^2 + b^2 = c^2$, couldn't look much more quadratic, could

it? So expect some applications having to do with the sides of right triangles. Here are some definitions

to go with those applications:

❖ **right triangle**

A right triangle is a triangle that contains one $90°$ angle.

★ **For Fun**

Is it possible for a triangle to contain more than one $90°$ angle?

❖ **hypotenuse**

The longest side of a right triangle is called the hypotenuse. It also happens to be the side that is

opposite, or across from, the $90°$ angle.

★ **For Fun**

Where does the word "hypotenuse" come from? Who coined it?

❖ **legs**

The other two sides of the triangle (not the hypotenuse) are called the legs. (There are *three* sides to a

triangle, right?)

Pythagorean Theorem: Test Prep	
An LCD television screen measures 72 in. on the diagonal, and its aspect ratio is 16 to 9. Find the width and height of the screen. Approximate your answers to the nearest tenth of an inch.	$(16x)^2 + (9x)^2 = 72^2$ $256x^2 + 81x^2 = 5184$ $x^2 = \dfrac{5184}{256 + 81} = \dfrac{5184}{337}$ $x = \sqrt{\dfrac{5184}{337}}$

	$\text{width} = 16x = 16\sqrt{\dfrac{5184}{337}} \approx 62.8 \ in.$
	$\text{height} = 9x = 9\sqrt{\dfrac{5184}{337}} \approx 35.3 \ in.$
Try These	
The length of each side of an equilateral triangle is 30 cm. Find the exact altitude of the triangle.	$15\sqrt{3}$
If the legs of a right triangle measure 8 in. and 17 in., find the length of the hypotenuse.	$\sqrt{353} \ in.$

★ For Fun

I've heard it said that Pythagoras' daughter was admitted into his mathematical and philosophical society. Apparently Pythagoras didn't practice sexual discrimination. What other off-beat tidbits can you find about Pythagoras? There are quite a few out there even though we actually know very little about him.

Can you write the equation of the Pythagorean Theorem that corresponds to this triangle?

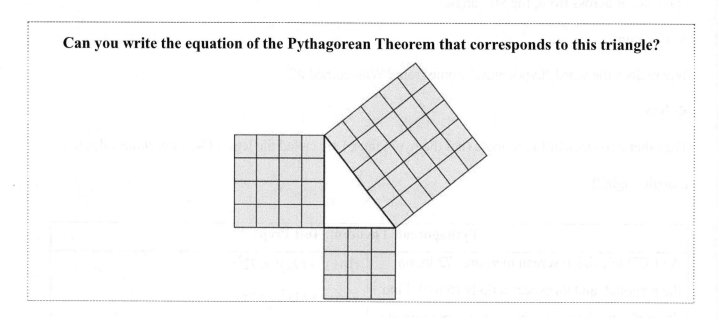

Applications of Quadratics: Test Prep

The height of a rock t seconds after being dropped from a 180 ft. tower is given by $h(t) = -16t^2 + 180$ How long does it take the rock to hit the ground, to the nearest hundredth of a second?	Height is zero at the ground, so $$0 = -16t^2 + 180$$ $$\frac{-180}{-16} = t^2$$ $t \approx 3.35$ sec.
Try These	
If the production cost of x number of pencil erasers is modeled by the equation $C(x) = 0.03x^2 + 2.5x + 150$ how many widgets can be manufactured for $4000?	319 erasers
The velocity in cm. per sec. of a fluid in a pipe goes according to the equation $v(r) = 16 - r^2$ where r is the radius of the pipe in cm. Find the radius of the pipe if the velocity of the fluid is found to be 7 cm. per sec.	3 cm.

NOW IS THE TIME TO TRY THE MID-CHAPTER 1 QUIZ.

1.4: Other Types of Equations

All the reasons we've investigated linear, absolute value, and quadratic equations now broaden out and include, well frankly, any kind of equation. Why not? In many ways, progress in algebra means progress in kinds of equations you can use to describe systems and methods you command to solve said equations.

Polynomial Equations

Certainly we already know that polynomials don't come in only *linear* and *quadratic* varieties. As you progress through this textbook, you will learn more and more about solving higher-order polynomials equations. For now, try what you do know: factoring.

♣ third-degree equation, cubic equation

A third-degree, or cubic, equation refers to a polynomial equation involving cubic polynomial(s). Inspect and see if you might be able to factor it – try grouping, try finding a GCF. Those are the tools you have now.

Polynomial Equations: Test Prep	
Solve; $2x^3 - x^2 - 4x + 2 = 0$	$2x^3 - x^2 - 4x + 2 = 0$ $x^2(2x-1) - 2(2x-1) = 0$ Factor by grouping. $(2x-1)(x^2-2) = 0$ $2x-1 = 0$ $x^2 - 2 = 0$ $\quad x = \dfrac{1}{2}$ or $x^2 = 2$ $\qquad\qquad\qquad x = \pm\sqrt{2}$ $x = \dfrac{1}{2}, \sqrt{2}, -\sqrt{2}$
Try These	
Solve: $x^3 = 4x$	$x = 0, 2, -2$
Solve: $16x^4 - 1 = 0$	$x = \pm\dfrac{1}{2}, \pm\dfrac{1}{2}i$

Rational Equations

You know what a rational expression is – a ratio of polynomials, right? Put one or more rational expressions into an equations, and, voila, you have a rational equation! Add the tools for solving these to your repertoire.

♣ rational equation

A rational equation is an equation that involves one or more rational expressions, which are ratios of polynomials.

★ For Fun

If the degree of an expression like $2xy$ is two, what is the degree, do you think, of an expression like $(x+1)(x-3)$ or $(x+1)^3$? Now what do you think you might be able to call the *degree* of a rational expression like, for instance: $\dfrac{x^2+1}{x-3}$?

★ For Fun

Remember how to find the domain of a rational expression? This domain question applies just as readily, if not more so, to rational equations. What is the domain of $\dfrac{x^2+1}{x-3}$? Would it be possible under any condition for $x = 3$ to be a solution to a rational equation like, say: $\dfrac{x^2+1}{x-3} = \dfrac{1}{2}$?

Rational Equations: Test Prep	
Solve and check your answer: $\dfrac{x+2}{x-3} + 2 = \dfrac{10}{2x-6}$	$\dfrac{x+2}{x-3} + 2 = \dfrac{10}{2x-6}$ $2(x-3)\left(\dfrac{x+2}{x-3} + 2\right) = \left(\dfrac{10}{2x-6}\right)2(x-3)$ $2x + 4 + 4(x-3) = 10$ $6x = 18$ $x = 3$ *answer?* $check: \dfrac{3+2}{3-3} + 2 \overset{?}{=} \dfrac{10}{2(3)-6}?$ Division by zero is not defined; therefore, there is no solution.
Try These	
Solve: $\dfrac{x}{x-1} = \dfrac{x+3}{x+1}$	$x = 3$
Solve: $3x - \dfrac{x+2}{x+3} = \dfrac{4x+13}{x+3}$	$x = \dfrac{5}{3}$

Radical Equations

The Power Principle makes it possible for you to raise both sides of an equation to a power and not upset the truth value of the equation, just like the Addition/Subtraction or Multiplication/Division Principles:

$$\text{If } P = Q, \text{ then } P^n = Q^n.$$

★ **For Fun**

Find examples this Power Principle for all types of n: positive and negative integers, radicals, etc.

★ **For Fun**

These principles always require that items like P and Q above must be "algebraic objects." Do you understand what an algebraic object is? (If not, maybe check with your instructor?) Can you think of something that is *not* an algebraic object that renders something like the Power Principle meaningless?

❖ **extraneous solution**

Like rational equations, radical equations may have extraneous solutions. The domain of a radical equation is important! Always check you answer(s)!

★ **For Fun**

Can you figure out why extraneous solutions arise when you raise both sides of an equation to an *even* power? (Hint: What happens to a negative number when you square it?)

Radical Equations: Test Prep	
Solve and check your answer: $\sqrt{2x-3}=7$	$\sqrt{2x-3}=7$ $\left(\sqrt{2x-3}\right)^2=7^2$ Square each side. $2x-3=49$ $2x=52$ $x=26$ *check*: $\sqrt{2(26)-3}\overset{?}{=}7$ $\sqrt{52-3}\overset{?}{=}7$ $\sqrt{49}=7$
Try These	
Solve: $\sqrt{2x}+\sqrt{x+1}=7$	$x=8$

Solve: $\sqrt{2x-7}-5=x+4$	No real solution.
Rational Exponent Equations If you can solve radical equations, then you can solve rational exponent equations ... because radicals can be written as rational exponents.	$x^{\frac{1}{3}}=-2$ $\left(x^{\frac{1}{3}}\right)^3=(-2)^3$ Raise each side to the third power, the number in the denominator of $\frac{1}{3}$. $x=-8$

★ **For Fun**

Does $x^{\frac{1}{2}}=-2$ remind you of *complex* numbers from last chapter?

Just as the Power Principle gives rise to extraneous solutions with radicals, it can give rise to the same

problems with rational exponent equations (they're really the same thing, after all) ... when you raise

both sides of the equation to an even power.

Equations With a Rational Exponent: Test Prep	
Solve: $2x^{\frac{3}{5}}+7=23$	$2x^{\frac{3}{5}}+7=23$ $2x^{\frac{3}{5}}=16$ $x^{\frac{3}{5}}=8$ $x=8^{\frac{5}{3}}=\left(8^{\frac{1}{3}}\right)^5=2^5=32$ $x=32$
Try These	
Solve: $x^{\frac{2}{3}}=25$ Solve: $(2x)^{\frac{3}{2}}+5=-3$	$x=125$ No solution

Equations That Are Quadratic in Form

If you can recognize when an equation that is not actually quadratic is, nonetheless, quadratic in form, you can probably make headway solving it.

♣ **quadratic in form**

Here are some examples of equations that are quadratic in form:

$$x^4 + 7x^2 + 12 = 0 \qquad 16x^4 + 8x^2 = 35 \qquad x^4 - 625 = 0 \qquad x + 6\sqrt{x} + 5 = 0$$

Could you perhaps use factoring to solve these? The quadratic formula? Can you imagine a strategy?

★ **For Fun**

How many solutions does an equation like the third one in the above list maybe have? It's not quadratic, after all.

★ **For Fun**

If you used the quadratic formula to try to solve the last equation in the above list, what would go on the left-hand side of the formula? In other words, what are you solving for with the quadratic formula ... because it's not x!

Equations That Are Quadratic in Form: Test Prep	
Solve: $x - 8\sqrt{x} + 15 = 0$	$x - 8\sqrt{x} + 15 = 0$ [*let* $u = \sqrt{x}$] $u^2 - 8u + 15 = 0$ $(u - 5)(u - 3) = 0$ $u = \sqrt{x} = 5 \ \text{or} \ u = \sqrt{x} = 3$ $x = 25 \ \text{or} \ x = 9$ $x = 25, 9$
Try These	
Solve using the quadratic formula: $2x^{\frac{2}{5}} - x^{\frac{1}{5}} - 6 = 0$	$x = -\dfrac{243}{32}, 32$
Solve: $x^4 - 3x^2 + 2 = 0$	$x = \pm 1, \pm\sqrt{2}$

Applications of Other Types of Equations

Some of the further applications of other types of equations explored in your textbook include motion problems, work problems, and distance problems. There are a lot of systems to describe, investigate, and solve. These are some of the reasons for studying algebra and building up a healthy set tools for working out these solutions.

Applications of Other Types of Equations: Test Prep	
If the time T in seconds between the instant you drop a stone into a well and the moment you hear its impact at the surface of the water is given by $T = \dfrac{\sqrt{s}}{4} + \dfrac{s}{1100}$ where s is measured in feet, find the depth to the foot of the water level if it takes 4 seconds to hear the impact.	$T = \dfrac{\sqrt{s}}{4} + \dfrac{s}{1100}$ $4 = \dfrac{\sqrt{s}}{4} + \dfrac{s}{1100}$ $4400(4) = 4400\left(\dfrac{\sqrt{s}}{4} + \dfrac{s}{1100} \right)$ $17600 = 1100\sqrt{s} + 4s$ $\sqrt{s} = \dfrac{-1100 \pm \sqrt{1100^2 - 4(4)(-17600)}}{2(4)}$ $= \dfrac{-1100 + \sqrt{1491600}}{8}$ Because $s \geq 0$, $s \approx 230$ ft.
Try These	
A ferry travels into a 24 mile deep bay at 10 mph and returns at 12 mph. The entire trip takes 5.5 hr. At what rate is the tide running out of the bay?	4 mph
Given the right triangle: where the ratio of the length of the hypotenuse to 7 is 4, find x. Round to two decimal places.	$x \approx 27.11$

1.5: Inequalities

For many of the sorts of equations we have studied already in this chapter, we can also investigate their behavior and solutions in inequalities instead. We will look at linear and other polynomial inequalities, rational inequalities, absolute value inequalities, etc. with the goal of increasing the tools we have to solve not only equations but also inequalities!

Properties of Inequalities

The next types of "equations" you can learn how to manipulate and solve are not equations. They are inequalities – mathematical sentences whose verb is not the equals sign but is instead "is greater than," "is less than," "is greater than or equal to," or "is less than or equal to." The symbols are well-known; they are, respectively, $>$, $<$, $\geq$, and $\leq$.

♣ solution set

A solution set is the set of all values of the variable that make the *inequality* time. Now this can admit a *lot* of values. For example, $x > 3$ is an infinitely large set of numbers; whereas, $x = 3$ is one number and one number only.

★ For Fun

Do you remember how to write $x > 3$ or other similar inequalities in set builder notation? in interval notation? graphed on a number line? (Now would be an especially good time to review all these.)

♣ equivalent inequalities

Like equations, equalities can be manipulated in order to find the solution set – that is, isolate the variable on one side of the inequality. These are some of the useful ways you can construct equivalent inequalities:

Properties of Inequalities

1. Addition/Subtraction Property:

If $a > b$, then $a + c > b + c$ or $a - c > b - c$. Likewise if $a < b$.

2. Multiplication/Division Property:

i.   If $c > 0$ and $a > b$, then $a \cdot c > b \cdot c$ or $a \div c > b \div c$. Likewise if $a < b$.

ii   If $c < 0$ and $a > b$, then $a \cdot c < b \cdot c$ or $a \div c < b \div c$. Likewise if $a < b$.

[Did you catch the difference between 2i and 2ii.?]

Linear Inequalities: Test Prep	
Solve: $-8x < 2x - 15$	$-8x < 2x - 15$ $-10x < -15$ $10x > 15$ $x > \dfrac{15}{10}$ $x > \dfrac{3}{2}$
Try These	
Solve and graph the solution set on the number line: $3x - 5 \geq 7$	$3x - 5 \geq 7$ $x \geq 4$
Solve: $5x + 6 < 2x + 1$	$x < -\dfrac{5}{3}$

Compound Inequalities

It isn't really possible to have a compound equation beyond something like $a = b = c$ for the very reason

that equality admits of only one possibility. In this case that possibility is that $a = c$.

★ For Fun

Can you create an interesting compound equality?

But because inequalities have infinite solutions, it's possible to write something interesting like

$-1 < 3x + 2 < 14$. After all, between -1 and 14 exist an infinity of numbers! And all the x's that define

those numbers can be discovered – the solution set.

♣ compound inequality

A compound inequality is a combination of inequalities that further defines the solution set of all of the inequalities.

There are two ways to combine inequalities:

1. AND

    And signifies the intersection $\cap$ of the solution sets – don't forget they are *sets* after all and subject to set operations. For example:

    $x - 5 > -2$ AND $2x > 10$

    $x > 3$ AND $x > 5$

    $\{x \mid x > 3\} \cap \{x \mid x > 5\} = \{x \mid x > 3\}$

    (Try graphing these sets on a number line if you have problems figuring out the result of the intersection.)

2. OR

    Or signifies the union $\cup$ of the solution sets. For example:

    $-3x > 30$ OR $x + 1 > 17$

    $x < -10$ OR $x > 16$

    $\{x \mid x < -10\} \cup \{x \mid x > 16\} = \{x \mid x < -10 \ \text{or} \ x > 16\}$

    (See if you can graph those on the number line? Can you think of a better way to describe the solution set?)

★ For Fun

Computers think with these ideas of AND and OR. Have you ever done a database search using these terms? Some search engines allow them – look under "Advanced Search Options" or something similar.

★ For Fun

Do you believe the compound inequality $-1 < 3x + 2 < 14$ is an AND or an OR case? Can you write this inequality as two separate inequalities with the correct set AND/OR connector?

Compound Inequalities: Test Prep	
Solve and graph the solution on a number line: $3x - 2 \geq 7$ OR $3x - 2 < -7$	$x \geq 3$ OR $x < -\dfrac{5}{3}$ number line from -3 to 4
Solve: $3 \leq 2x + 2 < 8$	$\dfrac{1}{2} \leq x < 3$

Absolute Value Inequalities

At this point, combining an absolute value expression with the notion of an inequality becomes interesting. Remember that absolute value has to do with distance. In an inequality absolute value can deal with ranges of distances, like ...

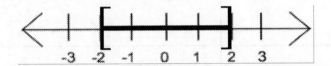

This graph is the set of all numbers that are 2 or less away from zero, right?

Learn these few simple "translation" rules for absolute value inequalities:

Properties of Absolute Value Inequalities
1. If $|E| < k$, then $-k < E < k$
2. If $|E| > k$, then $E > k$ OR $E < -k$ These properties hold for any E that is a algebraic expression and any k that is a real, non-negative number. They also hold if you replace the $<$ with $\leq$ and likewise for $>$.

★ For Fun

If k has to be non-negative, can it be zero? Why does it have to be non-negative?

★ For Fun

For the set graphed on the number line above, could you use the "translations" above to write an absolute value inequality that matched that set?

★ For Fun

Could you graph the set described by the absolute value inequality $|x-1| \leq 10$, for instance?

Absolute Value Inequalities: Test Prep	
Solve and graph the solution: $\|5-2x\| \leq 9$	$\|5-2x\| \leq 9$ $-9 \leq 5-2x \leq 9$ $-14 \leq -2x \leq 4$ $7 \geq x \geq -2$
Try These	
Solve and express the answer in interval notation: $\|3x+5\| < 7$	$\left(-4, \dfrac{2}{3}\right)$
Solve: $\|2x+1\| \geq 2$	$x \geq \dfrac{1}{2}$ OR $x \leq -\dfrac{3}{2}$

Polynomial Inequalities

The next algebraic object you can try out in an inequality is a polynomial. As you might expect, since higher-order polynomials can have more solutions than linear polynomials, the solution sets for polynomial inequalities can get more complicated. Learn the method and try it out. I offer some reminders:

1. Don't forget how to factor.

2. Use the number line.

3. Like the standard form for a quadratic equation, get one side of the inequality equal to zero.

> For all values of x between two consecutive real zeros, all values of the polynomial are positive or all values of the polynomial are negative.
>
> ("Sign Property of Polynomials" page 125 of your textbook)

♣ zero of a polynomial

Zeros of a polynomial are the values of the variable that cause the polynomial to be equal to zero. In other words, if $p(x)$ is a polynomial, then the zeros of $p(x)$ are the solution set of the particular case where $p(x) = 0$.

♣ critical values of the inequality

Real zeros of a polynomial in an inequality are the critical values of that inequality.

♣ test value

A test value is any value you care to plug into an algebraic expression to find its resultant value. In the case polynomial inequalities, you will need to pick test values that occur between the critical values of the polynomial in question. (By the way, picking good test values is a skill you will use extensively in calculus, too.)

Solving a Polynomial Inequality

1. If necessary, rearrange the inequality (using properties of equivalent inequalities) so that one side of the inequality is your polynomial and the other is zero.

2. Find all the real zeros of your polynomial-those are the critical values. Label them on a number line and list the factors in a column either above or below the number line.

3. In between the critical values, pick a test value. Find the sign of each factor for each test value. Determine the result of multiplication of these signs together (as the factors would be multiplied in order to build the original polynomial). Because of the Sign Property of Polynomials, everywhere in each interval the sign of the polynomial is that result.

4. Refer back to the polynomial inequality in step one in order to find the intervals that satisfy, then define the solution set of the inequality accordingly.

Polynomial Inequalities: Test Prep	
Solve: $x^2 - x - 2 < 0$	$x^2 - x - 2 < 0$ $(x-2)(x+1) < 0$ $$\begin{array}{lcccc} (x-2) & - & - & + \\ (x+1) & - & + & + & \textit{factors} \end{array}$$ $$\begin{array}{ccccc} & x=-2 & x=0 & x=3 & \textit{test} \\ \hline & & -1 \quad 2 & & \end{array}$$ $$x^2-x-2 \quad + \quad - \quad + \qquad \textit{polynomial}$$ The polynomial is negative between -1 and 2, so the solution set is $-1 < x < 2$
Try These	
Graph the solution of $9 \leq x^2$ on a number line.	 $$\begin{array}{cccccccccc} \leftarrow & + & + & + & + & + & + & + & + & \rightarrow \\ & -4 & -3 & -2 & -1 & 0 & 1 & 2 & 3 & 4 \end{array}$$
Solve the inequality $x^2 + 6 > 5x$ and express the answer in interval notation.	$(-\infty, 2) \cup (3, \infty)$

Rational Inequalities

Predictably we will next look at inequalities with rational expressions. We're investigating methods to find the solution set for these, too.

❖ **rational inequality**

Of course, a rational inequality is an inequality with a rational expression(s) included. For example: $\dfrac{x+1}{x-3} \geq \dfrac{2x}{x+5}$. Remember that a rational expression is a ratio of two polynomials.

❖ **critical value of a rational expression**

The critical values of a rational expression are all the zeros of both the numerators and denominators.

Solving a Rational Inequality

1. If necessary, rearrange the inequality (using properties of equivalent inequalities) so that one side of the inequality is zero. Also if necessary, rewrite the other side so that it is a single rational expression —this may involve all the manipulation of rational expressions that we talked about in the last chapter.

2. Find all the critical values of your rational expression. Label them on a number line and list the factors of both the numerator and denominator in a column either above or below the number line.

3. In between the critical values, pick a test value. Find the sign of each factor for each test value. Determine the result of multiplication and/or division of these signs together (as the factors would be multiplied in order to build the original polynomial).

4. Refer back to the rational inequality in step one in order to find the intervals that satisfy, then define the solution set of the inequality accordingly.

Rational Inequalities: Test Prep	
Solve: $\dfrac{x^2-11}{x+1} \leq 1$ Express your answer in interval notation.	$\dfrac{x^2-11}{x+1} \leq 1$ $\dfrac{x^2-11}{x+1} - 1 \leq 0$ $\dfrac{x^2-11-(x+1)}{x+1} \leq 0$ $\dfrac{x^2-x-12}{x+1} \leq 0$ $\dfrac{(x-4)(x+3)}{x+1} \leq 0$ The solution set is $(-\infty, -3) \cup (-1, 4]$
Try These	
Solve: $\dfrac{4}{x-1} \geq \dfrac{3}{x-4}$	$1 < x < 4 \ AND \ 13 \leq x < \infty$

Solve: $\dfrac{5x-9}{x-5} \le 3$	$[-3,5)$

Applications of Inequalities

As well as 3-D video games, statistics, geometry, and business are a few of the applications of inequalities. Careful reading with a firm basis of technique are the way to succeed with application problems. Be sure to name your variables meaningfully. And practice a lot!

Applications of Inequalities: Test Prep	
The sum of three consecutive odd integers is between 63 and 81. Find all possible sets of integers that satisfy these conditions.	3 consecutive odd integers could be expressed $x, x+2, x+4$. $$63 < x+x+2+x+4 < 81$$ $$63 < 3x+6 < 81$$ $$57 < 3x < 75$$ $$19 < x < 25$$ So x is either 21 or 23, and the two possible sets of integers are $\{21,23,25\}$ or $\{23,25,27\}$
Try These	
The average cost of producing x coin purses in dollar is $$\overline{C} = \dfrac{1.25x+2500}{x}$$ How many coin purses must be produced to bring the average cost below \$3.00?	more than 1429
A model rocket is launched from the top of a cliff 80 ft high. The equation $s(t) = -16t^2 + 64t + 80$ described the rocket's height in feet after t seconds. When will the rocket's height exceed the height of the cliff?	$0 < t < 4$ sec.

Here are some recent article titles from

The Journal of Inequalities and Applications:

- Refinements, Generalizations, and Applications of Jordan's Inequality and Related Problems
- Several Matrix Euclidean Norm Inequalities Involving Kantorovich Inequality
- Mixed Variational-Like Inequality for Fuzzy Mappings in Reflexive Banach Spaces
- Really! This is real math!

1.6: Variation and Applications

All those proportion problems you worked in math before college fall under the heading of variations: relations between variables that *vary* with each other in one way or another. Here come some general remarks, pointers, examples, and encouragements about these sorts of problems.

Direct Variation

Variation has to do with equations that relate variables – much like the formulas we have already talked about. Direct variation deals with equations of the form: $y = kx^n$.

❖ variation

In a lot ways a variation might seem like a formula; for example, $A = \pi r^2$ is a variation that relates the area of a circle to its radius. In other words, as the radius varies, so does the area.

❖ vary directly

If a relationship between variables, x and y for instance, can be written like $y = kx$, where k is a constant then you can say that y varies directly with x.

❖ directly proportional

$y = kx$ can also be expressed as "y is directly proportional to x."

❖ constant of proportionality, variation constant

And k gets the special label of constant of proportionality or variation constant when it plays this role in a variation equation.

❖ vary directly as the nth power

In my example above, $A = \pi r^2$, A varies directly the with second power of r.

★ For Fun

What is the constant of proportionality in $A = \pi r^2$?

Direct Variation: Test Prep	
Given $y = kx$, find k if y is 10 when x is 2.	$10 = k \cdot 2$ $\dfrac{10}{2} = k$ $5 = k$
Try These	
If y varies directly as x, and the constant of variation is $\dfrac{5}{3}$, what is y when $x = 9$?	15
If y varies directly as x, and $y = 8$ when $x = 12$, find k, write an equation that expresses this variation, and find y when $x = 5$	$k = \dfrac{2}{3}$ $y = kx$ $y = \dfrac{2}{3}x \qquad k = \dfrac{2}{3}.$ $y = \dfrac{2}{3}(5) \qquad x = 5.$ $y = \dfrac{10}{3}$

Direct Variation as the nth Power: Test Prep	
Given that weight varies with height according to the equation $w = kh^3$, and a man of 6 ft and 201 lb. wishes to meet a woman of similar proportions, how much would such a woman weigh, to the nearest pound, if she were 5.5 ft tall?	$201 = k \cdot 6^3$ $k = \dfrac{201}{216}$ *so* $w = \dfrac{201}{216} \cdot 5.5^3$ $\sim 155 lb$.
Try These	
The volume of a sphere is described by $V = \dfrac{4}{3}\pi r^3$. Approximate the volume of a sphere of radius 2.71 cm. to the nearest tenth of a cm.	83.4 c.c.
The distance that an object falls varies directly as the square of the time of the fall. If an object falls 256 ft. in 4 seconds, how long will it take to fall 64 ft?	2 sec.

Inverse Variation

Inverse variation has to do with relationships of the general form: $y = \dfrac{k}{x^n}$.

❖ **vary inversely, inversely proportional**

If a relationship can be written like this — $y = \dfrac{k}{x}$, where k is a constant – then you can say that y varies inversely with x or y is inversely proportional to x.

❖ **vary inversely as the nth power**

More generally you can say that y varies inversely with the nth power of x if $y = \dfrac{k}{x^n}$.

★ For Fun

Usually the *n* in these sorts of equations means *n* is a member of the natural numbers (the counting numbers). Does that restriction make sense here and in the direct variation equation?

<table>
<tr><td colspan="2" align="center">Inverse Variation: Test Prep</td></tr>
<tr>
<td>The water temperature of the Pacific Ocean varies inversely as the water's depth. At a depth of 1000m , the temperature is 4.4 degrees Celsius. What is the water temperature at a depth of 4000m ?</td>
<td>

$T = \dfrac{k}{d}$

$4.4 = \dfrac{k}{1000} \Rightarrow k = 4400$

and

$T = \dfrac{4400}{4000} = \dfrac{11}{10}$

$T \approx 1.1^\circ C$

</td>
</tr>
<tr><td colspan="2">Try These</td></tr>
<tr>
<td>If the number of potential buyers of a house varies inversely with the price and a house priced at \$100,000 attracts 250 potential buyers, find the variation constant for this market.</td>
<td>25,000,000</td>
</tr>
<tr>
<td>The time required to do a job varies inversely as the number of people working it. If it takes 100 days for 8 workers to complete a job, how long would it take 12 workers?</td>
<td>$66\dfrac{2}{3}$ *days*</td>
</tr>
<tr>
<td>Suppose *y* varies inversely with the square of *x*, and *y* = 4 when *x* = 5. Write the equation for this relationship and find *y* when *x* = 2.</td>
<td>

$y = \dfrac{k}{x^2}$ • *y* varies inversely as the square of *x*.

Replace *y* by 4 and *x* by 5. Then solve for *k*.

$4 = \dfrac{k}{25}$ or *k* = 100. Therefore, $y = \dfrac{100}{x^2}$

Replace *x* by 2 and find *y*.

$y = \dfrac{100}{2^2} = \dfrac{100}{4} = 25$

</td>
</tr>
<tr>
<td>The number of units of an item varies inversely with the square of the price, and 150 units sold</td>
<td>267</td>
</tr>
</table>

when the price was \$2. How many, to the nearest unit, would sell if the price were \$1.50?	
If y varies inversely as the square root of x, and $y = 17$ when $x = 49$, find the equation of variation.	$y = \dfrac{119}{\sqrt{x}}$

Joint and Combined Variations

Joint and combined variations mix up more variables – again a lot like some of the formulas you might have already seen. For example: $A = l \times w$, the area of a rectangle, which varies not with just the length or just the width of the rectangle, but both of them.

♣ vary jointly

If you can relate a variable z to more than one other variable like this – $Z = kxy$, where k is a constant again – then z varies jointly with x and y.

★ For Fun

This is the kind of equation my area of a rectangle falls under, right? Can you tell what the variation constant is for $A = l \times w$?

♣ combined variation

Combined variation means your variation may depend on more than one variable and the dependance may be direct or inverse or some of both. For example: $y = k\dfrac{xz}{w}$, where y varies directly with x and z and inversely with w.

Joint Variation: Test Prep	
y varies jointly as m and the square of n and inversely as p, and $y = 15$ when $m = 2$, $n = 1$, and $p = 6$. Find y when $m = 3$, $n = 4$, and $p = 10$.	$y = \dfrac{kmn^2}{p}$ so $15 = \dfrac{k \cdot 2 \cdot 1^2}{6} = \dfrac{k}{3}$

	$k = 45$
	and
	$y = \dfrac{45mn^2}{p}$
	then
	$y = \dfrac{45 \cdot 3 \cdot 4^2}{10} = 216$
Try These	
Simple interest varies jointly with the principal, rate and time in years: $I = prt$. Find the interest earned if $p = \$6000$, $r = 3.5\%$ after ten years.	$2100
The pressure of a gas depends on temperature and volume thus: $P = k\dfrac{T}{V}$. If $P = 2$ when $T = 298$ and $V = 9$, write the gas law.	$P = \dfrac{9}{149} \cdot \dfrac{T}{V}$

NOW TRY THE CHAPTER 1 REVIEW EXERCISES

AND THE CHAPTER 1 PRACTICE TEST

Final Fun: Matching

FIND THE BEST MATCH. GOOD LUCK.

$	x	$	PYTHAGOREAN THEOREM
DISCRIMINANT	$x + 3 = x$
$x = \dfrac{-b \pm \sqrt{b^2 - 4ac}}{2a}$	CIRCUMFERENCE OF A CIRCLE
STANDARD QUADRATIC	$\sqrt{x-1} = 17$
$x^2 + 10x + 25$	$x + 6\sqrt{x} + 5 = 0$
VARIATION	PERFECT SQUARE TRINOMIAL

CONTRADICTION

$$\begin{cases} x, x \geq 0 \\ -x, x < 0 \end{cases}$$

RADICAL EQUATION

$b^2 - 4ac$

$X^2 + Y^2 = R^2$

$z = kxy$

$C = \pi d$

QUADRATIC FORMLA

RATIONAL INEQUALITY

$\dfrac{x+1}{x-3} \geq \dfrac{2x}{x+5}$

QUADRATIC IN FORM

$ax^2 + bx + c = 0$

Word Search

```
D T N I O P D I M E D C D P N
N E N O M C E D T F O I E E O
A T T N I P I A O M M T C R I
I N N E O T N R P W A A R P S
S A S L R I A O C X I R E E S
E T S U D M S L I L N D A N E
T S C R I I I S S D E A S D R
R N O E T D S N E N P U I I P
A O N I N V A T A R A Q N C M
C C O U O T Z R A T G R G U O
I N T E R C E P T N I E T L C
N O I T A L E R R O C O R A O
I N C R E A S I N G L E N R Q
N O I T C E L F E R A E N I L
S N O I T C N U F O R I G I N
```

AXIS	DECREASING	ODD
CARTESIAN	DETERMINATION	ORDINATE
CENTER	DISTANCE	ORIGIN
CIRCLE	DOMAIN	PERPENDICULAR
COMPOSITION	FUNCTIONS	QUADRATIC
COMPRESSION	INCREASING	RADIUS

2.1: Two-Dimensional Coordinate System and Graphs

Since we're handling variables quite adequately now, let's learn how to handle two at a time and how to communicate about them in a graph.

Cartesian Coordinate Systems

In order to communicate in 2-dimensional geometry and algebra at the same time, we need to agree on *frame of reference*. In honor of René Descartes, who did much of the early work of combining these two topics, our agreed-upon frame of reference is called *Cartesian*. The Cartesian coordinate system breaks a 2-dimensional surface, such as the blackboard or a piece notebook paper, into 4 areas that everyone around the globe can navigate and communicate about.

♣ **coordinate**

A coordinate is a number that is associated with a position. A coordinate usually means that a scale is present so that the position has meaning. In algebra that scale is the number line.

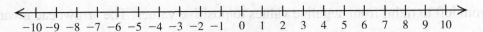

♣ **coordinate plane**

A coordinate plane is 2-dimensional. So instead of one scale, two scales are needed. Each scale is actually just a number line.

♣ **ordered pair**

An ordered pair of numbers, usually (x, y), is simply the coordinates required to specify a position in 2 dimensions. The order of the coordinates in an ordered pair is a *rigorous convention*!

✓ **Questions to Ask Your Instructor**

Make sure you understand when your instructor wants your answers in the form of an ordered pair. And then be sure to include all the elements of the notation: parentheses and comma.

♣ coordinates

See ordered pair. In this class coordinates will almost always come in ordered pairs. That means you will be working in two dimensions.

♣ *x*-coordinate, abscissa

Customarily the first coordinate in an ordered pair is the x-coordinate. The first coordinate, and remember that it's just a number, tells you the position along the scale, or number line, that refers to x-values. Abscissa is a slightly old-fashioned word (but still perfectly meaningful) for the x-coordinate. As your textbook points out, *abscissa* is related to the word *scissors*.

★ For Fun

Can you find any other words in English that are related to *abscissa* or *scissors*?

♣ *x*-axis

The x-axis is one of the number lines (scales) required to find positions on a coordinate plane. Usually the x-axis runs from left to right, from negative values to positive values. The arrowhead always points towards the positive end of the spectrum of x-values (or in the order in which values are increasing).

✓ Questions to Ask Your Instructor

How does your instructor want you to label your *x*-axis? (Or, for that matter, all your *axes*, which is the plural of *axis* even though it looks like the plural of a tool for chopping wood. Context is everything.)

♣ *y*-coordinate, ordinate

Like the *x*-coordinate, the *y*-coordinate is a number. This time it is the second number in an ordered pair, and it refers to the position on the scale of *y*-values.

★ For Fun

Look up *ordinate*. Find other meanings and see if they make any sense to you in the context of a 2-dimensional *co*-*ordinate* plane.

♣ *y*-axis

The scale of *y*-values is called the *y*-axis. Usually it runs from bottom to top, and negative values are at the bottom while positive values are at the top. It's another number line.

♣ origin

If you have two number lines running perpendicular to each other on a 2-dimensional plane, such as your paper or the blackboard, they *will* meet at a point (does that make sense?). We agree that they will meet at the position on each scale corresponding to a value of zero (or nothing). The ordered pair that describes this position on the coordinate plane is $(0, 0)$. This becomes the starting position against which all coordinates are measured: the origin. (The origin is like the "You Are Here" point on the maps at the mall. From "Here" you can get anywhere... given meaningful instructions!)

Here is a short list of possible coordinate systems in 2-dimensions:

Cartesian coordinate system

Polar coordinate system

Parabolic coordinate system

Bipolar coordinates

Hyperbolic coordinates

Elliptic coordinates

Descartes' version is simple. We'll stick with it until trigonometry!

♣ quadrant

The two crossing axes break a 2-dimensional surface up into four areas. Each of these areas is one quadrant. The quadrants are always numbered using Roman numerals (see picture).

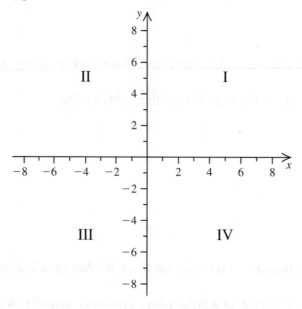

★ For Fun

Look up the derivation of the word *quadrant*. Think about it-obviously *quadrant* and *quadratic* are related

words. Why?

♣ Cartesian coordinate system

The *x*-axis, *y*-axis, and quadrants all together are frequently called a Cartesian coordinate system. A

standard 2-dimensional coordinate system is the same thing.

★ For Fun

Cartesian is named after René Descartes. Can you find out what the French words *"des cartes"* mean?

What else is René Descartes known for?

♣ plot a point

To plot is to place a point or points on the Cartesian coordinate plane. An actual *plot* is the picture that

results from plotting.

Distance and Midpoint Formulas

If we are navigating around a 2-dimensional surface, we might want to know how far we have to go to get from one point to another or where the halfway point lies. The distance and mid-point formulae (Latin plural of formula) enable us to completely specify these quantities.

The distance formula follows:

Given two points, (x_1, y_1) and (x_2, y_2), the distance between them is $d = \sqrt{(x_2 - x_1)^2 + (y_2 - y_1)^2}$

✓ **Questions to Ask Your Instructor**

Since the distance formula involves a square root and the mid-point coordinates involve division (that is, fractions!), you might make sure when your instructor wants an exact answer like $d = \sqrt{10}$ or an approximation like $d \cong 3.1623$. In the case of approximations, be sure you understand all about rounding.

★ **For Fun**

Do you readily recognize a similarity between the distance formula and the Pythagorean formula? Can you draw a picture that illustrates this relationship? Is it ever possible to have a negative value under the square root in the distance formula? Why not? Ask yourself, Why is distance always the positive square root?

♣ **midpoint**

A midpoint is the point exactly in the middle of two other points, or halfway between them.

Given two points, (x_1, y_1), (x_2, y_2), the midpoint between them is $\left(\dfrac{x_1 + x_2}{2}, \dfrac{y_1 + y_2}{2} \right)$

★ **For Fun**

Think about it: How is a midpoint like the average, or mean, of the two end points?

Distance Formula: Test Prep	
Find the distance between the points $A(2,1)$ and $B(4,-6)$.	$x_1 = 2,\ y_1 = 1,\ x_2 = 4,\ y_2 = -6$: $$d = \sqrt{(x_2 - x_1)^2 + (y_2 - y_1)^2}$$ $$= \sqrt{(4-2)^2 + (-6-1)^2}$$ $$= \sqrt{(2)^2 + (-7)^2}$$ $$= \sqrt{4 + 49}$$ $$d = \sqrt{53}$$
Try these	
Find the distance between the points $P(-3,-1)$ and $Q(0,2)$.	$d = \sqrt{18} = 3\sqrt{2}$
Find the distance between the points $P_1(-1, 16)$ and $P_2(4, 4)$.	$d = 13$

Midpoint Formula: Test Prep	
Find the midpoint of the line segment connecting the two points $P_1(-6,2)$ and $P_2(1,-3)$.	$\left(\dfrac{x_1 + x_2}{2}, \dfrac{y_1 + y_2}{2} \right)$ $\left(\dfrac{-6+1}{2}, \dfrac{2+(-3)}{2} \right) = \left(\dfrac{-5}{2}, \dfrac{-1}{2} \right)$
Try these	
Find the midpoint of the line segment connecting the two points $A(5, -3)$ and $B(-3, 4)$.	$\left(2, \dfrac{1}{2} \right)$
Find the midpoint of the line segment connecting the two points $P\left(-1, \dfrac{2}{3}\right)$ and $Q(2, 0)$.	$\left(\dfrac{1}{2}, \dfrac{1}{3} \right)$

Graph of an Equation

An equation is a mathematical proposition that frequently relates variables. In 2-dimensions, most equations will have 2 variables, for example:

$$y = 3x - 1$$

$$x^2 + y^2 = 16$$

$$y = |x + 3|$$

In order to "graph an equation," you need a coordinate system, and you need to understand what a graph of an equation means. Then you need to learn the methods to transfer your understanding onto your coordinate system.

♣ graph of an equation

The graph of an equation is a set of points that have been plotted on a Cartesian coordinate all of which satisfy the conditions of the equation – that means the points are in the solution set of the equation. Each ordered pair, (x, y), is a set of values that make the equation true. A graph of an equation is a *picture* of the solution set of the equation

Graph of an Equation: Test Prep	
Graph $y + x = 3$	$y = 3 - x$

Try these	
Graph $y = -x^2$	
Graph $y = -\lvert x+1 \rvert$	

Intercepts

In mathematics an intercept means a meeting point of two curves. (Is that anything like an interception in American football?) "Intercept" frequently refer to the meeting point between the graph of an equation and an axis on a Cartesian coordinate system. (However, later you will discover that *intercept* can also mean where the graphs of two equations meet.)

♣ intercept

An intercept is any place where two curves meet–like an intersection. For instance, the origin is the intercept of the x-axis and the y-axis.

♣ *x*-intercept

An x-intercept is the intersection of a graph of an equation with the x-axis. If a point lies *on* the x-axis, what must its y-coordinate be? All ordered pair representations of x-intercepts will have the form: $(x_0, 0)$. By the way, x_0 is just a way to say we are talking about an actual number on the x-axis instead of the variable x.

♣ *y*-intercept

Likewise, all ordered pair representations of *y* intercepts will have the form: $(0, y_0)$. If the point is plotted on the *y*-axis, what must its *x* coordinate be?

x-Intercepts and *y*-Intercepts: Test Prep	
Find the *x*- and *y*-intercepts of the equation $y = x^2 + 3x - 4$	To find the *y*-intercept, let $x = 0$. $$y = 0^2 + 3(0) - 4$$ $$= 0 + 0 - 4$$ $$y = -4$$ The *y*-intercept is $(0, -4)$. To find the *x*-intercept, let $y = 0$. $$0 = x^2 + 3x - 4$$ $$= (x + 4)(x - 1)$$ The solutions are –4, 1. The *x*-intercepts are $(-4, 0)$ and $(1, 0)$.
Try these	
Find the *x*- and *y*-intercepts of the equation $y - x = 9$	$(0, 9)$ and $(-9, 0)$
Find the *x*- and *y*-intercepts of the equation $x^2 = y^2 - 4$	$(0, 2)$, $(0, -2)$, and no *x*-intercepts

Circles, Their Equations, and Their Graphs

We all have a good idea what shape a circle is, but can you think of a good definition of that shape? The information contained in an equation that describes a circle depends on precise ideas about just what a circle is. Fortunately, because we know what a circle looks like, we have an advantage when it comes time to graph the equation that describes one – provided we recognize it and the information it contains.

♣ circle

A circle is a 2-dimensional round curve. In fact, it is the set of all points that are exactly the same distance from a central point. (The central point, however, is not part of the circle, is it?)

★ For Fun

If the definition for a circle is so "tied up" with the notion of distance, would you expect the distance formula to be involved? See if you can spot it ... or ask your instructor how they are related.

♣ radius

The radius of a circle is the length of one-half of its diameter, and a diameter is the length of a line that cuts a circle exactly in two. All points on the circle (on its circumference) are exactly one radius from the center of the circle. (That is, the "distance" mentioned in the definition for a circle above gets the special name of "radius.")

♣ center

Of course, the center of a circle is the point which lies exactly in the middle of the area defined by the circumference. Beware: the center of the circle is not part of the circle itself, which is the set points that make the equation for a circle true. The center is certainly a convenient way to understand more about where the circle is situated on a Cartesian coordinate system.

♣ standard form of the equation of a circle

The standard form of the equation of a circle looks like $(x-h)^2 + (y-k)^2 = r^2$ where the center is the point whose coordinates are (h, k) and whose radius is r.

♣ general form of the equation of a circle

The general form of the equation for a circle usually looks something like $x^2 + y^2 + Ax + By + C = 0$. When presented with the general form, you can no longer find the center coordinates or the radius by inspection.

> Review the method Completing the Square; you'll need it in order to coax the general form
> of the equation for a circle into the standard form.

Here's an example:

$x^2 + y^2 - 8x + 12y - 48 = 0$

$x^2 - 8x \qquad + y^2 + 12y = 48$

$x^2 - 8x + 16 + y^2 + 12y + 36 = 48 + 16 + 36 \qquad \left[\left(\dfrac{-8}{2} \right)^2 = 16 \right] \left[\left(\dfrac{12}{2} \right)^2 = 36 \right]$

$(x - 4)^2 + (y + 6)^2 = 100$

The center is $C(4, -6)$. The radius is $\sqrt{100} = 10$.

Equation of a Circle: Test Prep	
Find the radius and center of the circle described by this equation: $x^2 + y^2 + 6x + 8y - 11 = 0$	$x^2 + y^2 + 6x + 8y - 11 = 0$ $x^2 + 6x + y^2 + 8y = 11$ $(x^2 + 6x + 9) + (y^2 + 8y + 16) = 11 + 9 + 16$ $(x + 3)^2 + (y + 4)^2 = 36$ The center is $C(-3, -4)$. $r = \sqrt{36} = 6$
Try these	
Identify the center and radius of the circle described by this equation: $(x - 2)^2 + (y + 1)^2 = 50$	The center is $C(2, 1)$. $r = \sqrt{50} = 5\sqrt{2}$
Sketch the graph of this equation: $x^2 + y^2 - 6x + 2y - 15 = 0$	

2.2: Introduction to Functions

Functions are a cornerstone not only of algebra but of calculus and beyond. Work on a good foundation now. It'll pay off.

Relations

So the ordered pair is a clear way to relate two coordinates, and we're in the 2-dimensional coordinate system now, so we'll start investigating what ordered pairs mean in terms of what a relation is.

♣ relation

Simply put, any collection or set of ordered pairs forms a relationship. It specifies, if you will, how the x's are related to the y's.

For example:

$$\{(0,3),\ (2,-1),\ (-5,3),\ (0,-4.79),\ (3,0.25),\ (2,-2),\ (9,9)\},$$

or another less implicit example:

$$\{(x,y)\mid\quad y=-2x+3\}$$

Remember that an ordered pair carries with it the idea that you put an x into a relation "machine" and the machine spits out the y that goes with it.

Functions

Functions are special kinds of relations. We're still putting a coordinate in x and getting a second coordinate out y.

[Try diligently to keep in mind that x and y are by no means the only letters we can use for these variables, so you'll see them frequently referred to as the first and second coordinate ... and also as any number of other letter-variable-stand-ins. This requires some abstract thinking, but if you work at it, it will help your algebra (to say the least).]

♣ function

A function is actually just a special sort of relation – a subset of relations, if you will. A function is also a set of ordered pairs but with this important qualification: that no two ordered pairs have the same first coordinate while having different second coordinates.

Function

A function is a set of ordered pairs in which no two ordered pairs have the same first coordinate and different seconds coordinates.

(page 165 of your textbook)

If you read this "math-ese" carefully, you might notice that it would be okay to have the two identical ordered-pairs, (0,3) and (0,3) in a function, but, honestly, what does it add to a set to repeat a member? The important point in this definition is that you cannot have ordered pairs like $(4, 2)$ and $(4, -2)$ in a function (though it's perfectly okay for them to be in a relation). If two ordered pairs have the same first coordinate, then their second coordinate mustn't be different; that is, if they do *not* have the same first coordinate, everything is okay, and they do have the same first coordinate, they better be identical in first and second coordinate (and, shrug, then you simply have repetition of a point in your set).

★ For Fun

Keep this seemingly convoluted definition in mind for when we look at graphs of functions. Begin to consider what this restriction might mean graphically.

♣ domain

All the first coordinates (usually x's) make up the domain. Frequently when asked about the domain of a function, we are really trying to get at what first coordinates will result in *real* second coordinates coming out the other end of the function-machine.

♣ **range**

And those coordinates that come out of the function machine: they are the range, the output of the function.

★ **For Fun**

Can you come up with a definition of a function that doesn't even involve numbers, real or otherwise?

♣ **independent variable**

Variables that correspond to values in the domain are called independent variables.

♣ **dependent variable**

Variables that correspond to values in the range are dependent variables.

★ **For Fun**

Try to imagine why the coordinates are called *dependent* and *independent* like this. Keep it in mind for graphing time.

Definition of a Function: Test Prep	
Determine whether the equation states y as a function of x: $x + (y-2)^2 = 3$	$x + (y-2)^2 = 3$ $(y-2)^2 = 3 - x$ $y - 2 = \pm\sqrt{3-x}$ $y = 2 \pm \sqrt{3-x}$ Two possible values of y for one x-value means this is not a function.
Try these	
Determine whether the equation states y as a function of x: $x + 4y = 14$	
Determine whether the equation states y as a function of x: $x^2 + y = -1$	

Function Notation

Functions, like variables, generally are named with letters; the most common is f. f would be the name of the function, and it does nothing until you give it something to work on: $f(x)$. Since the x is what you put into the f-machine, the x-values are members of the domain of f. And the values that come out of $f(x)$, namely $f(x)$, make up the range of f. And y often stands in for $f(x)$, a la $y = f(x)$.

As soon as you see someone call an expression by $f(x)$ or $g(x)$ or some other similar name, one thing you know is that it is a function. Not only a relation.

♣ piecewise-defined functions

We've already met a piecewise-defined function – the absolute value function.

$$|x| = \begin{cases} x, & x \geq 0 \\ -x, & x < 0 \end{cases}$$

A piecewise-defined function is a function whose definition is broken up into "pieces," usually based on different parts of the function's domain.

★ For Fun

Can you *see* the domain in the definition above? Can you see (and name) the x- value where one piece of the definition changes into the other?

Evaluate a Function: Test Prep	
Given $Q(x) = -3$, find a. $Q(0)$ b. $Q\left(\dfrac{1}{4}\right)$ c. $Q(2)$	$Q(0) = -3,$ $Q\left(\dfrac{1}{4}\right) - 3$ $Q(2) = -3$ Note that this function is −3 all the time, regardless of the value of x. That's a *constant* function!

Try these	
Given $T(x) = \frac{1}{z}x - 4$, find a. $T(0)$ b. $T(5)$ c. $T(-5)$	$T(0) = -4$ $T(5) = -3$ $T(-5) = -5$
Profit is given by the equation $P = px$, where p is the price and x is the number of units sold. Find the profit when p is $63.50 and x is 226.	$14,351

Piecewise-defined Functions: Test Prep	
Given $$f(x) = \begin{cases} x^2 + 1 & x \geq 1 \\ x & 0 \leq x < 1 \\ -x^2 & x < 0 \end{cases}$$ evaluate a. $f(0)$ b. $f(10)$ c. $f(-5)$	a. $f(0) = 0$ (choose the middle definition of the function because $x = 0$) b. $f(10) = 10^2 + 1 = 101$ (choose the first definition of the function because $x > 1$) c. $f(-5) = -(-5)^2 = -25$ (choose the last definition of the function because $x < 0$)
Try these	
Use $$|x| = \begin{cases} x & x \geq 0 \\ -x & x < 0 \end{cases}$$ to evaluate $|5|$ and $|-5|$.	5, 5
Given $$f(x) = \begin{cases} 3x - 1 & x \geq 0 \\ 3 - x^2 & x < 0 \end{cases}$$ find a. $f(0)$ b. $f(-2)$	$f(0) = -1$ $f(-2) = -1$

★ **For Fun**

Try graphing, by plotting points, any of these piecewise-defined functions.

Domain and Range of a Function: Test Prep	
Find the domain of $f(x) = \sqrt{3-x}$	$3 - x \geq 0$ $-x \geq -3$ $x \leq 3$
Try these	
Find the domain of $f(x) = \dfrac{x}{x^2 - 1}$	$x \neq 1, -1$
Given the graph of f below, find the domain and range of f. 	domain: $[-4, 4]$ range: $[-4, 4]$

Graph of Functions

We already know that functions are nothing more than sets of ordered pairs – mind you, the set might very

well be infinite (and therefore hard to write in a set notation). But what are these ordered pairs?

You know when you evaluate a function that you replace the variable with the value given:

$g(x) = 2x^2 - x$
$g(-2) = 2(-2)^2 - (-2) = 8 + 2 = 10$

In this example, we have constructed the ordered pair $(-2, 10)$.

More generally, given a function $f(x)$ and an input value a, then we get the ordered pair $(a, f(a))$. And now that we have ordered pairs, we can plot them on our Cartesian coordinate system. At some point, we may then connect the dots so that we have a *picture* of the entire solution set of the function. (Before you connect the dots, though, be confident that you do have a very good idea of what the rest of the solution set behaves – what picture it makes.)

♣ graph of a function

The graph of a function is a plot of all the ordered pairs that belong to that function. For any x in the domain of f, the ordered pairs look like $(x, f(x))$. Now we have a real motivation for graphing functions. The picture will be much easier to get information out of than a description in, for instance, set builder notation, which might look something like ... $\{(x, f(x)) \mid x \in domain(f)\}$ or some other such nearly meaningless jargon. A graph is worth a thousand set notations!

♣ increasing

As your eye follows the graph on the page from left to right, if the curve of the graph is going up, then the graph (and by extension the function) is increasing in that region of the domain.

Sometimes a graph of a function will have arrowheads at the ends of the curve that depicts the function, like this:

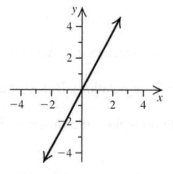

It's super important that you are still able to see that this graph is increasing as you look at it from left to right ... even though the left-hand end of the graph has an arrow pointing down which sometimes makes your eye see it as *decreasing*, but it's not. Do not let your eye fool your mind.

♣ **decreasing**

Similarly, if the graph is going down as your eye follows its curve from left to right, then the graph/function is decreasing in that part of the domain.

♣ **constant**

If the graph is flat, then the function is not changing – neither increasing nor decreasing – and it's constant. A constant function looks like, well, a constant. For instance: $f(x)=3$. (If you're looking at a piecewise-defined function, the function maybe be constant for only part of its domain. Can you read the domain information right out of a piecewise-defined function's description?)

The Zero of a Function

A zero of a function is a value in the domain of the function that results in an output of zero. You'll be working with zeros a lot in the next chapter and beyond.

Meanwhile, $(c,0)$ would be the point on a graph of a function where $f(c)=0$. As long as c is a real number that point will be an intercept. Which axis? Check back under in section 1 of this chapter under *Intercepts* if you're not sure.

One of the most important unsolved problems in mathematics concerns the location of the zeros of the Riemann zeta function.

Graph a Function: Test Prep		
Use a *T*-table to graph $y = 2 - x$	$\underline{x}$ $\quad$ $\underline{y}$ -1 $\quad$ 3 0 $\quad$ 2 1 $\quad$ 1	

Try these	
Use a T-table to graph $y = \sqrt{x+2}$	
Use a T-table to graph $y = -\lvert x \rvert + 3$	

Zero of a Function: Test Prep	
Find all the real zeros of $f(x) = x^2 - 144$	$0 = x^2 - 144$ $0 = (x-12)(x+12)$ $x = 12, -12$
Try these	
Find the zeros of $f(x) = 3x - 6$	$x = 2$
Find the zeros of $y = \lvert x - 2 \rvert - 5$	$x = 7, -3$

★ For Fun

If you remember the definition from the beginning of this chapter for a function in terms of ordered pairs, then you will remember that a set of ordered pairs is not a function if any two ordered pairs match in the first coordinate while their seconds coordinates differ. See if you can make that idea sync up with the visual test for a function as it is given below:

> **Vertical Line Test for a Function:** A graph is the graph of a function if and only if no vertical line intersects the graph at more than one point. (pg. 174 of your textbook)

★ For Fun

Which these six graphs represent functions?

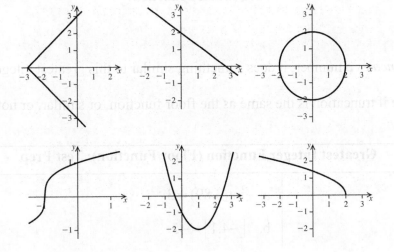

Greatest Integer Function (Floor Function)

A fun function to explore in order to strengthen your understanding of functional notation and graphs!

♣ greatest integer function, floor function, step function

Denoted variously by $\lfloor x \rfloor$, int(x), etc., the floor function puts out the greatest integer that is still less than or equal to x. For instance,

$$\left\lfloor \frac{1}{2} \right\rfloor = 0$$

$\lfloor -3.5 \rfloor = -4$

$\lfloor 2 \rfloor = 2$

And a graph of a floor function generally looks something like this:

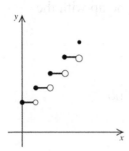

Can you make your understanding of the definition of a floor function match the graph above? Can you also see why this function is sometimes called the step function?

★ **For Fun**

When a computer *truncates* a number, it does something similar to this greatest integer function. Look up *truncation* online; see if truncation is the same as the floor function, or similar, or not at all?

Greatest Integer Function (Floor Function): Test Prep	
Evaluate a. int(3.89) b. $\lfloor -4.1 \rfloor$ c. $\lfloor -13 \rfloor$	a. $\text{int}(3.89) = 3$ b. $\lfloor -4.1 \rfloor = -5$ c. $\lfloor -13 \rfloor = -13$
Try these	
Evaluate a. $\lfloor -5 \rfloor$ b. $\lfloor 10.2 \rfloor$ c. $\lfloor \dfrac{3}{4} \rfloor$	a. $\lfloor -5 \rfloor = -5$ b. $\lfloor 10.2 \rfloor = 10$ c. $\lfloor \dfrac{3}{4} \rfloor = 0$

| Sketch $y = \dfrac{1}{2}\lfloor x \rfloor$ | |

Applications of Functions

Now we can add the challenge writing the *function* that satisfies the requirements of the application. Then

solving the problem! Start thinking "functions!"

2.3: Linear Functions

Lines are easy. Plus we use them in so many ways – the shortest distance between two points, huh? Lines

are basic, basic, basic to algebra, and since they're pretty easy, now is the best of possible times to learn

your way around a line from all directions. (If you're headed to calculus, you'll be covered up with the

ideas of lines there, too, so might as well master lines now.)

Slopes of Lines

The slope of a line can tell us a lot. It has universal and real-world meaning – everybody knows it's easier

to coast downhill than pedal uphill, right?

♣ **slope**

Aside from saying that the slope of a line is the ratio of the change in y to the change in x, like so

$$m = \frac{change(y)}{change(x)} = \frac{\Delta y}{\Delta x} = \frac{y_2 - y_1}{x_2 - x_1}$$

in algebra, slope also means all the things it means in English. The algebra is arranged so that the number,

m, tells you 1) direction (are you going up or down?) and 2) steepness (shallow or sharp?). The graph of a

line is increasing from left to right if the slope is positive, and decreasing if it's negative. Steepness is

harder to quantify and is really only a comparative measure. In general, a slope greater in absolute value than 1 will be getting pretty steep. A slope less than 1 in absolute value will appear pretty shallow.

Experience helps with this recognition.

★ For Fun

Consider: What would a slope of exactly one be like? How 'bout a slope of zero? What is the absolutely

steepest line you could envision?

The equation for a **horizontal** line has the form: $y = b$

The equation for a **vertical** line has the form: $x = a$

where a and b are constants.

★ For Fun

Which of the two cases above is actually a function? (One is definitely not.)

Find the slope of a line that passes through the points $(-2, -4)$ and $(2, 5)$.	$\dfrac{5-(-4)}{2-(-2)} = \dfrac{5+4}{2+2} = \dfrac{9}{4}$
Try these	
Find the slope of y from the graph:	$m = -\dfrac{4}{3}$
Find the slope of a line that passes through the points $(4, 3)$ and $(10, 3)$.	$m = 0$

Slope-Intercept Form

The slope-intercept form of an equation for a line is $y = mx + b$. The x and y are the variables in two dimensions. m is the slope, and the point $(0, b)$ is the y-intercept point for the line. (Does that intercept sync up with what you know about y-intercepts already?)

This equation is surely familiar. It's used frequently in graphing lines. For example, the equation $y = -\frac{1}{2}x + 3$ is in slope-intercept form. To graph this equation, first plot the y-intercept: $(0, 3)$. Then use the slope to plot one (or two?) more points:

$$-\frac{1}{2} = \frac{-1}{2} = \frac{change(y)}{change(x)}.$$

In other words, *from the point you already have* on the line, go -1 in the y-direction then $+2$ in the x-direction. Follow the process on this graph:

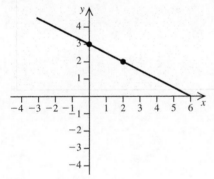

(Now is a good time to check our work. Did the line you just graphed actually turn out to have negative slope – is it decreasing? Is it pretty shallow since the slope is less in absolute value than 1?)

Slope-Intercept Form of the Equation of a Line: Test Prep	
Given a slope of $m = -\frac{2}{3}$ and a y-intercept of $(0, 2)$, write the equation of the corresponding line.	$m = -\frac{2}{3}$ and b is 2, so the equation can be written in slope-intercept form: $$y = -\frac{2}{3}x + 2$$

Try these	
Sketch the graph of $y = \dfrac{3}{4}x - 2$	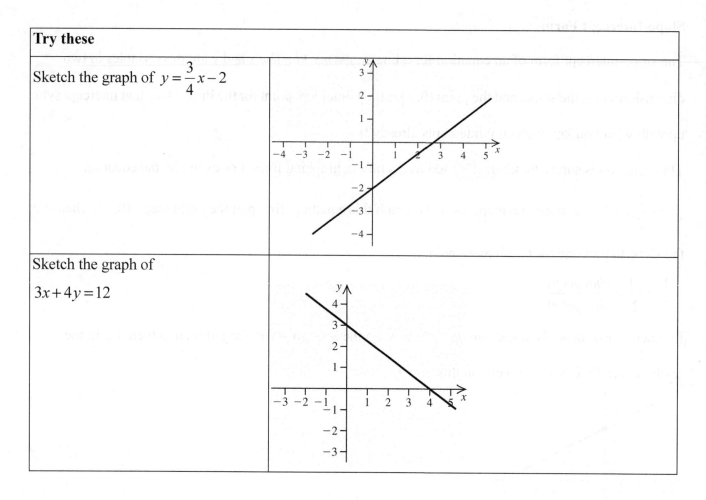
Sketch the graph of $3x + 4y = 12$	

General Form of a Linear Equation in Two Variables

The general form for a linear equation is

$$Ax + By = C$$

where A and B and C and real constants and neither A nor B are zero. (Why not?)

In many cases, you might want to change the general form into the point-slope form for the purpose of

graphing the line.

★ For Fun

Could you convert the general form into the slope-intercept form? What are the correspondences between

the constants in the two forms? (If you forget how to solve an equation for a particular variable, in this case

y, see Section 1.2 again.)

General Form of a Linear Equation in Two Variables: Test Prep	
Restate the equation of this line in slope-intercept form: $3x - y = -1$	$3x - y = -1$ $-y = -3x - 1$ $y = 3x + 1$
Try these	
Sketch the graph of this line: $x + y = 1$	
Sketch the graph of this line: $2x + 3y = 3$	

✓ Questions to Ask Your Instructor

There is another method that is sometimes popular for graphing a line whose equation is in standard form. This method completely skips solving the equation for y in order to put it into slope-intercept form and is sometimes called "the intercept method." If you're interested, you could ask your instructor to show you that method. (It's also illustrated in the alternative method to Example 3 in your text, page 187.)

Finding the Equation of a Line

Given an equation in either the general form or the slope-intercept form for a line in 2 dimensions, you should be able to graph it now. But what happens if you just have *information* about the line instead of its equation? What if someone tells you a point and the slope? Well, graphing that would be easy, right?

★ For Fun

Prove it. Graph the line that passes through the point $(-1, 3)$ and has a slope of $m = 2$.

Okay, now you have the graph, but what if you wanted to tell someone on the phone, as succinctly as possible, about this line? Or even more realistically, what if your instructor asks you to write the equation for this line on a quiz?

♣ point-slope form

Given a point (x_1, y_1) and the slope m of a line, you can write the equation for the line using the point-slope form of the equation of a line:

$$y - y_1 = m(x - x_1)$$

(And you may want to subsequently solve for y, thereby putting the equation into slope-intercept form.)

Point-Slope Formula: Test Prep	
Given $m = 3.5$ and the point $(-1, 2.5)$, write the equation for the line with this slope passing through this point, in slope-intercept form.	$y - 2.5 = 3.5(x + 1)$ $y = 3.5x + 3.5 + 2.5$ $y = 3.5x + 6$ $or\ y = \dfrac{7}{2}x + 6$
Try these	
Given a line that passes through the points $(-1, 4)$ and $(-2, -3)$, write the equation for this line in slope-intercept form.	$y = 7x + 11$
Given a line that passes through the points $\left(\dfrac{1}{3}, \dfrac{2}{5}\right)$ and $\left(-\dfrac{1}{3}, \dfrac{1}{5}\right)$, write the equation for this line in slope-intercept form.	$y = \dfrac{3}{10}x + \dfrac{3}{10}$

In two dimensions, you need two pieces of information about a line in order to graph it or to write its equation. Are two points sufficient to graph the line? Sure – plot two points and draw the line that passes through them. But how would you arrive at an equation for a line given only two points on the line?

1. Find the slope using the slope formula (Slope of Lines" above).

2. Use that slope and *one* of the points in the point-slope form.

3. Voilà!

Parallel and Perpendicular Lines

And now: a brief discussion of a couple of the special ways lines can be related to each other. Do you already have a clear idea in English what parallel (as in "parallel lines never meet") and perpendicular (meeting at ninety degrees) mean?

★ **For Fun**

The word *perpendicular* has a very interesting derivation. You wanna look it up? Might look up *parallel* too, while you're at it.

♣ **parallel**

Two lines are parallel if their slopes are equal.

★ **For Fun**

Hmm, could a line be parallel to itself?

♣ **perpendicular**

If their slopes are negative reciprocals of each other, then two lines are perpendicular.

★ For Fun

Fill in the blanks:

m	negative	reciprocal
$\dfrac{1}{2}$	-2	
3	_____	_____
$-\dfrac{5}{3}$	_____	_____
-1	_____	_____

Parallel Lines: Test Prep

Pick the parallel lines:	$y = 2x + 1$
$y = 2x + 1$	$y - 2x = 14$ $[y = 2x + 14]$
$x = -\dfrac{1}{2}y + 10$	$4x - 2y = 5$ $[y = 2x - \dfrac{5}{2}]$
$y = -\dfrac{1}{2}x + 1$	They all have a slope of 2.
$y - 2x = 14$	
$4x - 2y = 5$	
Try these	
Write an equation of a line parallel to $y = -2x + 3$ that passes through the point (1, 3).	$y = -2x + 5$
Write an example of an equation of a line parallel to $3x - y = 4$	example: $y = 3x + 1$

Perpendicular Lines: Test Prep

Find the equation of a line perpendicular to the line $y = -\dfrac{1}{3}x + 1$ and passing through the point (3, 0).	$y - 0 = 3(x - 3)$ $y = 3x - 9$
Try these	
Find a line perpendicular to $y = \dfrac{2}{5}x$	example: $y = -\dfrac{5}{2}x + 114$

Sketch a graph of the line perpendicular to $y = -x + 5$ and passing through the origin.

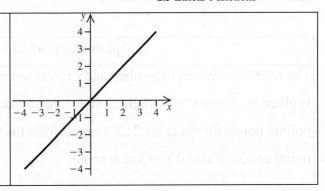

Applications of Linear Functions

As I said at the beginning this section, we use linear equations to describe lots of systems. Now we'll see a few of these occasions and use these equations to answer questions and draw graphs illustrating the system at hand.

♣ cost function

A function that models the cost of producing and selling x units: $C(x)$.

♣ revenue function

A function that describes the amount of money brought in by selling x units: $R(x) = xp$, where p is the price of the unit.

♣ profit function

A function that models the profit from selling x units. This is usually the difference between the cost $C(x)$ and the revenue $R(x)$: $R(x) - C(x) = P(x)$. (That makes sense, doesn't it?)

♣ break-even point

The break-even point is exactly where cost and revenue are equal: $C(x) = R(x)$. You can set two linear equations equal to each other and use this equation to solve or graph the system.

★ For Fun

What is the profit at the break-even point?

Applications of Linear Functions: Test Prep	
The relation between Fahrenheit and Celsius temperatures is given by $F = mC + n$ where m and n are constants. The boiling points for water are $212°F$ and $100°C$; the freezing points are $32°F$ and $0°C$. Find m and n.	The slope is $m = \dfrac{212 - 32}{100 - 0} = \dfrac{9}{5}$. When $C = 0$, $F = 32$. $$F = mC + n$$ $$32 = \frac{9}{5}(0) + n$$ $$32 = n$$
Try these	
Given a supply $S = \dfrac{1}{3}p$ and demand $D = -p + 200$, find the value of p at the break-even point.	$$\frac{1}{3}p = -p + 200$$ $$\frac{4}{3}p = 200$$ $$p = 150$$
Consider this table relating age and number of prescriptions. *age*　*prescriptions* 30　　1 40　　3 50　　6 55　　8 Write a linear equation based on the points at 40 and 50 years old. Use that equation to estimate the number of prescriptions for age 60.	$$y = \frac{3}{10}x - 9$$ A person 60 years old will have 9 prescriptions.

NOW IS THE TIME TO TRY THE MID-CHAPTER 2 QUIZ.

★ For Fun

Here's a piecewise-defined linear function. You might try graphing this! Good luck.

$$f(x) = \begin{cases} -x-3 & if & x \le -3 \\ x+3 & if & -3 < x \le 0 \\ -2x+3 & if & 0 \le x < 3 \\ x-3 & if & x \ge 3 \end{cases}$$

2.4: Quadratic Functions

The graphs of quadratic functions are parabolas (a symmetrical open plane curve formed by the intersection of a cone with a plane parallel to its side .... Okay, it's a big *U*-shaped curve!). The parabola is next degree up from linear and the next more complicated polynomial function. Quadratics functions, whose graphs are parabolas, also come with lots of applications. Here we go.

♣ quadratic function

A quadratic function is a *function* of degree 2:

$f(x) = ax^2 + bx + c$ where a is not zero.

★ For Fun

Beware, all that is degree two is not a function. Is $y^2 = x+1$ a quadratic function? You may have to go right back to the definition of a function to find out that, although it *is* quadratic, it's not a function.

♣ vertex of a parabola

The vertex of a parabola is the *turning point* of the parabola. It's the lowest point on a parabola that opens up and the highest on one that opens down:

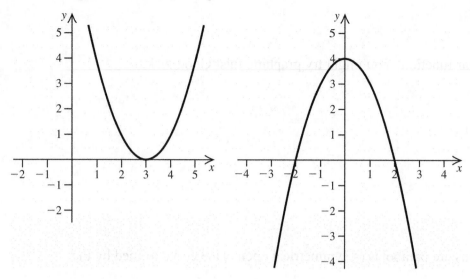

♣ symmetric with respect to a line

Symmetry with respect to a line means that the item is made up of exactly similar parts facing each other from opposite sides of an axis; the line is that axis of symmetry.

♣ axis of symmetry

The line that divides a symmetric object into its two similar parts is its axis of symmetry. (See above, okay?)

★ For Fun

Can you describe the axes of symmetry of the above two parabolas? Each have an axis that is a vertical line, so answer in both cases in the form of an equation for a vertical line. (Vertical lines were discussed in the last subsection.)

♣ standard form of a quadratic function

The standard form of a quadratic function is, as follows:

$f(x) = a(x-h)^2 + k$, where a is not zero.

This form is comparable to the standard form of an equation for a circle, and likewise it can be mined for information about the graph of the parabola. (Get ready to complete the square!)

1. Given a quadratic function in standard form, the vertex of the parabola is situated at the point (h, k).

2.   If a is positive, then the parabola opens up, and if a is negative, the parabola opens down.

3.   The axis of symmetry of the parabola is at the vertical line $x = h$.

Do you think that's enough information to be able to graph a good parabola from an equation in standard form?

By the way, don't forget you can still plot points, too. Don't ever forget you have that skill. It's an excellent way to make sure that your parabola has the right shape.

Quadratic Function: Test Prep	
Find the standard form of this quadratic function: $f(x) = x^2 - 4x - 1$	$\begin{aligned} f(x) &= x^2 - 4x - 1 \\ &= (x^2 - 4x) - 1 \\ &= (x^2 - 4x + 4) - 4 - 1 \\ f(x) &= (x - 2)^2 - 5 \end{aligned}$
Try these	
Identify the constants a, b, and c in this quadratic function: $f(x) = -x^2 + \dfrac{x}{3} - 12$	$a = -1$ $b = \dfrac{1}{3}$ $c = -12$
Sketch the parabola: $f(x) = -(x-2)^2 + 6$	

Vertex of a Parabola

If your quadratic function is in standard form, $f(x) = a(x - h)^2 + k$, then it's easy to pick out the coordinates of the vertex, namely (h, k). But even if the quadratic function is given in the form

$f(x) = ax^2 + bx + c$, you can find the vertex coordinates with those a, b, and c. The vertex is situated at,

$$\left(-\frac{b}{2a}, \frac{4ac - b^2}{4a}\right), \text{ or } \left(-\frac{b}{2a}, f\left(-\frac{b}{2a}\right)\right).$$

From the function, the sign of a will still tell you if the parabola opens up or down, and by extension, you can figure out if the vertex is the highest point (maximum) or lowest point (minimum) of the parabola. That may be enough, but if you want to graph the parabola, it's generally much easier if you have the function in the standard form: $f(x) = a(x - h)^2 + k$. So here's an example of using completing the square to rewrite a quadratic function into the standard form. This time, instead of adding the same thing to both sides of the equation, I'm going to *add zero* to one side. (I can add zero any time I want and not change the value of an expression, right? Zero, if you recall, is the additive identity.)

$$f(x) = x^2 - 10x + 22$$
$$= (x^2 - 10x + 25) - 25 + 22 \qquad \left[\left(\frac{-10}{2}\right)^2 = 25\right]$$
$$f(x) = (x - 5)^2 - 3$$

The vertex is at $(5, -3)$.

★ For Fun

Could you go ahead and graph that parabola? How 'bout this: could you find the x- and y-intercepts, if they exist? ... before you graph it? ... and verify them from your graph? Can you tell what the range is? ... before you graph it?

Parabola: Test Prep	
Find the standard form of the equation of this parabola and and its vertex: $f(x) = -x^2 - 2x + 7$	$f(x) = -x^2 - 2x + 7$ $= -(x^2 + 2x + 1 - 1) + 7$ $= -(x^2 + 2x + 1 - 1) + 7$ $f(x) = -(x + 1)^2 + 8$ *vertex* $(-1, 8)$

Try these	
Find the vertex of this parabola using the vertex formula: $f(x) = -x^2 - 4$	$(0, -4)$
Sketch the parabola $y = -2x^2 + 4$	 *(graph of parabola opening downward with vertex at $(0, 4)$)*

Maximum and Minimum of a Quadratic Function

Think of a parabola; for example:

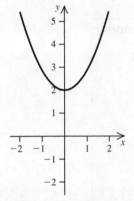

You can see from the graph that there are no coordinates on this curve with a y-value less than 2. In other words, since the vertex is at the point (0, 2) and the parabola opens up, the vertex is at the lowest point of the parabola. That means that the value 2 is the smallest value that this quadratic function will ever put out.

If you think about it, that tells you everything you need to know about the *range* of this particular quadratic. Seems like its range is $[2, \infty)$. What do you think?

♣ minimum value

From a quadratic function in standard form, the k is a minimum value of the function, or of its range, if the parameter a is positive.

♣ maximum value

Similarly, k is a maximum value if a is negative.

★ For Fun

What do you have if a is zero? (This is a trick question.)

Minimum or Maximum of a Quadratic Function: Test Prep	
Find the range of $f(x) = -(x-2)^2 - 5$	vertex: $(2, -5)$. The graph opens DOWN. Therefore, range is $(-\infty, -5]$.
Try these	
Find the range of $f(x) = x^2 + 7x + 2$	$\left[-\dfrac{41}{4}, \infty \right)$
Find the range of $f(x) = -2x^2 - 4x + 15$	$(-\infty, 17]$

There are lots of kinds of functions. So far we've looked at linear and quadratic functions. These are both algebraic functions. Here are some examples of functions that are not algebraic. Some you'll see later, most not!

- Hyperbolic functions
- Bessel functions
- Airy function
- Transcendental functions
- Theta function
- Hermite polynomials

Applications of Quadratic functions

Areas (*square* inches, *square* centimeters, *square* light-years) and trajectories (baseballs, rockets, thrown snowballs) are natural systems to be described by quadratic functions (they travel in parabolas). There are others, too (business, engineering, etc.). Now you can try out your expertise with quadratic functions on some applications.

Applications of Quadratic Functions: Test Prep	
Say the height h in feet of a grand slam baseball after t seconds is approximated by $h(t) = -16t^2 + 89t + 4$. When will the ball be caught be a fan 7 feet up in the stands, to the nearest tenth of a second?	$7 = -16t^2 + 89t + 4$ $0 = -16t^2 + 89t - 3$ $t = \dfrac{-89 \pm \sqrt{89^2 - 4 \cdot 16 \cdot 3}}{-32}$ approximately 5.5 seconds (disregarding the smaller solution because it will most likely be the second time the ball passes through 7 feet if its path is a parabola, right?)
Try these	
Given two rafter of 7.5 feet and roof-frame height of 1 foot, what will be the width of the ceiling, to the nearest foot? 7.5 ft. 7.5 ft. 1 ft.	15 feet
A dog breeder has 1200 meters of fence to enclose a rectangular pen with another fence dividing it in the middle as in the diagram: y $x \qquad x$ The breeder wishes to use all the fence and enclose the largest possible area. What must x and y be?	$x = 150$ m $y = 300$ m

2.5: Properties of Graphs

Now that we are graphing quite a bit, we can try to get some basic understanding about how graphs work and how to manipulate them. Or, perhaps, learn how manipulating the function changes the graph in predictable ways. Knowing this will become a tremendous time-saver, and, it goes without saying, increasing your comprehension of the meaning of displaying 2-dimensional systems as graphs is going to really pay off.

Symmetry

We investigated the idea of symmetry with respect to a line in the last section. That had to do with parabolas, but of course parabolas are not the only things that exhibit symmetry. In this section, we'll look into some special symmetries with an eye toward developing an aid for graphing. (Remember the graph is a picture of the solution set of the equation.)

♣ symmetric with respect to the *y*-axis

Symmetry with respect to (w.r.t.) the *y*-axis means that a graph includes all the similar points on both sides of the *y*-axis. In other words, if you folded the graph on the *y*-axis all points would coincide with their similar points on the other side of the axis.

Without a graph, you can test the equation of a function to see if it's symmetrical w.r.t. the *y*-axis by replacing *x* with-*x* and finding, after simplification, that the equation remains unaffected. In symbols: $f(-x) = f(x)$. Can you see that mathematical expression says that if you replace *x* with $-x$ it doesn't make a difference?

Visually, here's an example of symmetry w.r.t. the *y*-axis:

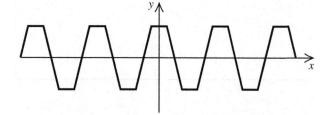

♣ symmetric with respect to the *x*-axis

Symmetry w.r.t. the *x*-axis is analogous to the above, but, of course, you would be folding the graph on the *x*-axis.

To test a function for this kind of symmetry, it's best to first rewrite every $f(x)$ with y; that is, $f(x) = y$.

Then replace y with $-y$, simplify, and see if the equation is unaffected. In other words: $y = -y$. [Think

about it: if a expression is symmetrical w.r.t. the x-axis, is it a function? If not, would I call it $f(x)$ then?]

An example:

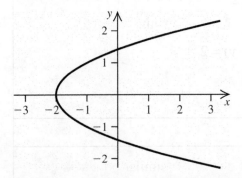

♣ symmetric with respect to a point

Symmetry w.r.t. an arbitrary point starts to sound much more technical: To be symmetrical about a point

Q, there must exist *every point* P its similar point P' so that Q is the midpoint of the line from P to

P'.

One special point for which is common to test for symmetry is the origin $(0, 0)$. To test an equation for

symmetry about the origin, first you replace *both* x with $-x$ and y with $-y$, simplify, and see if the

equation remains unaffected. Symbolically: $-f(-x) = f(x)$. [Symmetry about the origin does not

preclude function-hood (see graph below), and so I can name the expression a function $f(x)$ again.]

Example:

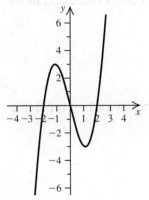

Symmetry of a Graph with Respect to ...: Test Prep	
Test the relation $xy = 2$ for symmetry w.r.t. ... *a.* x-axis *b.* y-axis *c.* origin	a. $x(-y) = 2$ $-xy = 2$ NO b. $(-x)y = 2$ $-xy = 2$ NO c. $(-x)(-y) = 2$ $xy = 2$ YES
Try these	
Test the relation $x^2 + y^2 = 4$ for symmetry w.r.t. ... *a.* x-axis *b.* y-axis *c.* origin	a. YES b. YES c. YES
Test the relation $xy = \lvert x \rvert$ for symmetry w.r.t. ... *a.* x-axis *b.* y-axis *c.* origin	a. NO b. NO c. YES

Even and Odd Functions

Like integers, some functions are even, some are odd. Unlike integers, some functions are neither even nor odd. This does have meaning for the graph of a function, but it's also an idea that becomes more and more important as you progress through trigonometry and calculus.

♣ **even function**

Like a function that is symmetrical w.r.t. the y-axis, an even function exhibits the behavior that replacing x with-x leaves the function unchanged.

$$f(-x) = f(x)$$

♣ odd function

Odd functions are harder to recognize. They behave like this: $f(-x) = -f(x)$. If you replace x with $-x$, then you get the function's *opposite* after simplifying. For example:

$f(x) = x^3$

$f(-x) = (-x)^3$

$= -x^3$

$= -f(x)$

Did you catch that? It takes some practice to see that one sometimes.

Even and Odd Functions: Test Prep	
Determine if this function is even or odd or neither: $f(x) = 3x + 1$	Neither. $f(-x) = 3(-x) + 1 = -3x + 1 = -(3x - 1)$ $f(-x) \neq f(x)$ $f(-x) \neq -f(x)$
Try these	
Determine if this function is even or odd or neither: $f(x) = |x| - 3$	Even
Determine if this function is even or odd or neither: $f(x) = 2x^3 - 4x$	Odd

Translations of Graphs

Now we can learn how to move graphs around the 2-dimensional plane. No, better still we'll learn to recognize when the function is a basic graph we already know but that has been moved horizontally or vertically or both to a new situation on the plane.

Vertical Translations

Assume that c is a positive constant, then

1. If $y = f(x) + c$, then y is $f(x)$ moved UP c units.

2. If $y = f(x) - c$, then y is $f(x)$ moved DOWN c units.

For example:

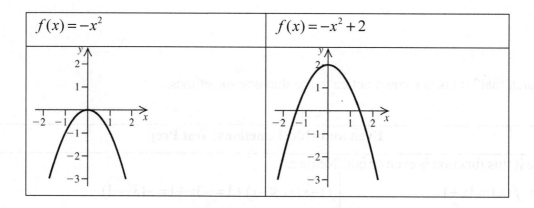

| $f(x) = -x^2$ | $f(x) = -x^2 + 2$ |

Horizontal Translations

Again assume that c is a positive constant, then

1. If $y = f(x + c)$, then y is $f(x)$ moved LEFT c units.

2. If $y = f(x - c)$, then y is $f(x)$ moved RIGHT c units.

For example:

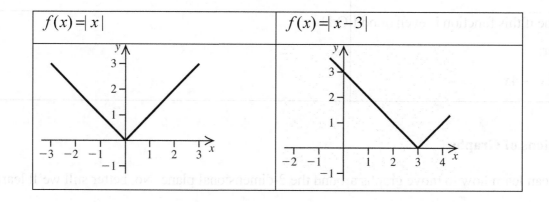

| $f(x) = |x|$ | $f(x) = |x - 3|$ |

Vertical Translation of a Graph: Test Prep

Given the graph *of* $f(x)$ below, graph $g(x) = f(x) - 2$	Move the graph *of* $f(x)$ DOWN two units:

Try these

Given the graph of $f(x)$ below, graph $f(x) + 3$	
Given $f(x) = x^2$, describe this translation: $g(x) = x^2 - 10$	$g(x) = x^2 - 10$ is the graph of $f(x)$ moved DOWN ten units.

Translation:

What does *translation* mean in English? Do any of its several possible meanings ring true for what we want to signify when we talk about translating graphs?

Horizontal Translation of a Graph: Test Prep

Given the graph *of* $f(x)$ below, graph $g(x) = f(x+2)$ 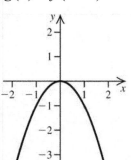	Move the graph *of* $f(x)$ LEFT two units: 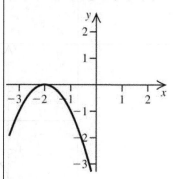

Try these

Given the graph of $f(x)$ below, graph $g(x) = f(x-3)$ 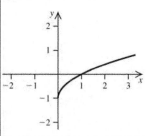	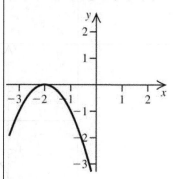
Given $f(x) = |x|$, describe this translation: $g(x) = |x-2|$	The graph of $g(x) = |x-2|$ is the graph of $f(x)$ moved RIGHT two units.

★ **For Fun**

Okay, now try doing both to one graph. Given the graph of *f* below, graph $y = f(x-2)+3$ 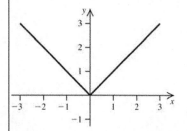	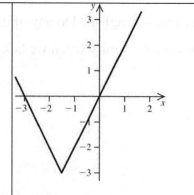

Reflections of Graphs

Besides moving graphs around, we can also "flip" them. The most common "flips" are reflections about the x- or y-axis.

Reflections

1. If $y = -f(x)$, then y is $f(x)$ reflected about the x-axis.

For example:

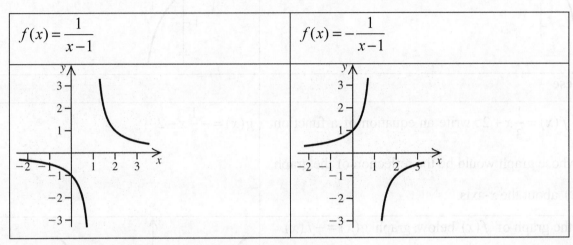

$f(x) = \dfrac{1}{x-1}$	$f(x) = -\dfrac{1}{x-1}$

2. If $y = f(-x)$, then y is $f(x)$ reflected about the y-axis.

For example:

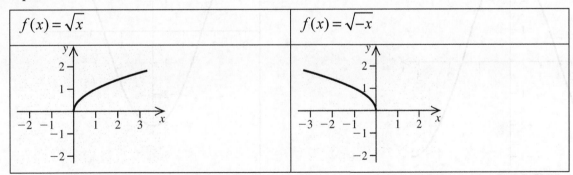

$f(x) = \sqrt{x}$	$f(x) = \sqrt{-x}$

Reflections of a Graph: Test Prep	
Given the graph of $f(x)$ below, graph $g(x) = f(-x)$. 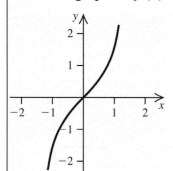	$g(x)$ is a reflection of $f(x)$ about the y-axis, thus: 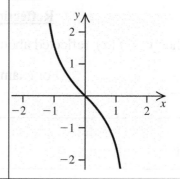
Try these	
Given $f(x) = \dfrac{1}{3}x + 2$, write an equation of a function $g(x)$ whose graph would be the reflection of the graph of $f(x)$ about the x-axis.	$g(x) = -\dfrac{1}{3}x - 2$
Given the graph of $f(x)$ below, graph $g(x) = -f(x)$. 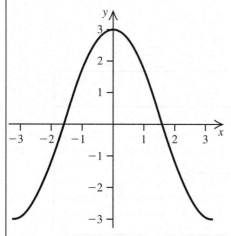	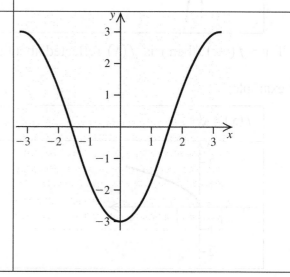

★ **For Fun**

Given that you know what the graph of $f(x) = x^2$ looks like, try graphing this combination of manipulations: $f(x) = -(x + 3)^2$.

Compressing and Stretching of Graphs

Compression and stretching of graphs is more subtle. But you'll know it when you see it! Basically, what is involved here is a *multiplicative* factor working on the function, as opposed to adding and subtracting in translations. (However, it's somewhat different from the multiplication by -1 that causes reflections.)

Vertical Stretching/Compression

Let c be a positive constant, and if $y = c \cdot f(x)$ then

1. if $c > 1$, y is $f(x)$ stretched vertically away from the x-axis by a factor of c

2. if $0 < c < 1$, y is $f(x)$ compressed towards the x-axis by a factor of c

For example:

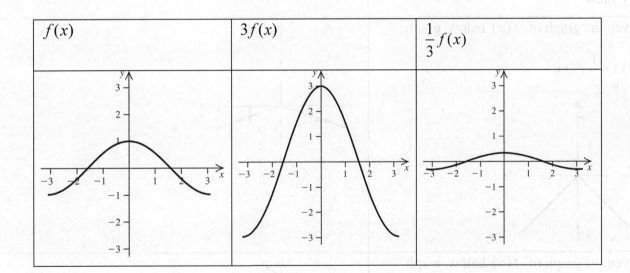

| $f(x)$ | $3f(x)$ | $\frac{1}{3}f(x)$ |

★ **For Fun**

Think about it: why is $c = 1$ not included in the above possibilities?

★ **For Fun**

What would happen if we allowed c to be negative? (Hint: see Reflections above.)

Vertical Stretching and Compressing of Graphs: Test Prep

Given the graph of $f(x)$ below, graph $g(x) = 3f(x)$ 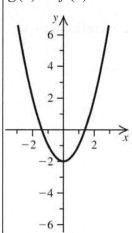	Every point will be 3 times bigger – the graph will "rise" 3 times faster than $f(x)$. This is a vertical stretching. 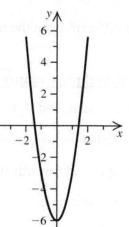

Try these

Given the graph of $f(x)$ below, graph $g(x) = \dfrac{1}{3}f(x)$. 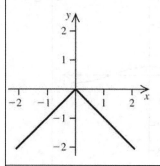	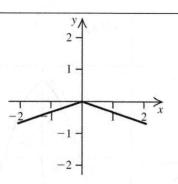
Given the graph of $f(x)$ below, graph $g(x) = 2f(x) + 1$. 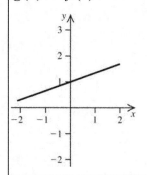	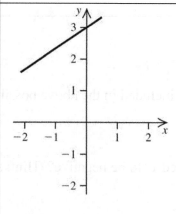

Horizontal Stretching/Compression

Let c be a positive constant, and if $y = f(c \cdot x)$ then

1. if $c > 1$, y is $f(x)$ compressed horizontally toward the y-axis by a factor of $\dfrac{1}{c}$.

    (Some people might see this as the same as a vertical stretching away from the x-axis.)

2. if $0 < c < 1$, y is $f(x)$ stretched horizontally away from the y-axis by a factor of $\dfrac{1}{c}$.

    (Some people will see this one as the same as compressing vertically towards the x-axis.)

For example:

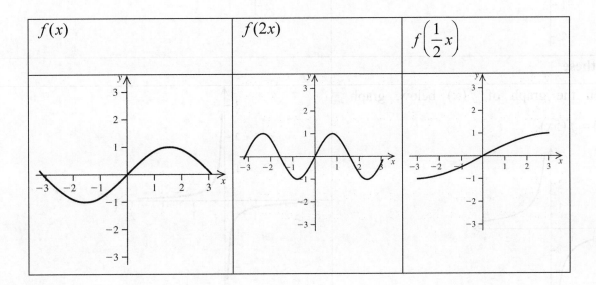

$f(x)$	$f(2x)$	$f\left(\dfrac{1}{2}x\right)$

★ **For Fun**

Combine what you know about the graphs of parabolas and all these modifications on functions to see if

you can match up all the translations, reflections, and compressions / stretchings in this equation:

$f(x) = -2(x - 4)^2 - 5$

You might try sketching this parabola and see if you can also visually find all the elements from this

section that apply.

Horizontal Stretching and Compressing of Graphs: Test Prep

Given the graph of $f(x)$ below, graph $g(x) = f\left(\dfrac{1}{4}x\right)$.

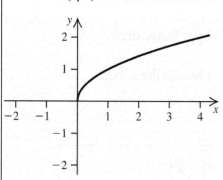

This graph is stretch horizontally away from the y-axis by a factor of four:

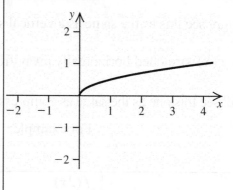

Try these

Given the graph of $f(x)$ below, graph $g(x) = f(5x)$.

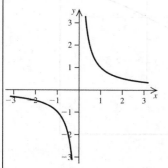

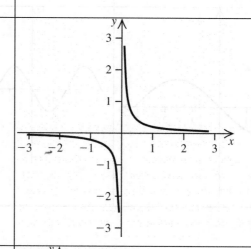

Consider $f(x) = x^2$ and use what you know about translations and stretching and compression to graph

$f(x) = 3(x-1)^2 - 2$

(Does this match what you expect from what you already know about parabolas?)

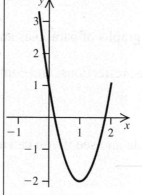

2.6: Algebra of Functions

The first part of this section is really easy – the sum, difference, product, and quotient of functions. The only fun part here is probably the reminder about *domains* involved in the quotient of two functions – you're still not allowed to divide by zero. The second part is *fundamental* and probably new to you: composition of functions.

Operations on Functions

This section has to do with how you add, subtract, multiply, and divide functions. I like to call it the "arithmetic of functions" myself. It's straightforward:

The Arithmetic of Functions

1. **Sum:** $(f+g)(x) = f(x) + g(x)$

   For example, given $f(x) = 3x$ $g(x) = 5x - 1$

   $(f+g)(x) = f(x) + g(x) = 3x + 5x - 1 = 8x - 1$

2. ***Difference:*** $(f-g)(x) = f(x) - g(x)$

   For example, given $f(x) = 3x$ $g(x) = 5x - 1$

   $(f-g)(x) = f(x) - g(x) = 3x - (5x - 1) = -2x + 1$

3. ***Product:*** $(f \cdot g)(x) = (f(x))(g(x))$

   For example, given $f(x) = 3x$ $g(x) = 5x - 1$

   $(f \cdot g)(x) = (f(x))(g(x)) = 3x(5x - 1) = 15x^2 - 3x$

4. ***Quotient:*** $\left(\dfrac{f}{g}\right)(x) = \dfrac{f(x)}{g(x)}$

   For example, given $f(x) = 3x$ $g(x) = 5x - 1$

   $\left(\dfrac{f}{g}\right)(x) \dfrac{f(x)}{g(x)} = \dfrac{3x}{5x - 1}, \quad x \neq \dfrac{1}{5}$

Operations on Functions: Test Prep	
Given $f(x)=x^2$ and $g(x)=x^2-1$, find a. $\left(\dfrac{f}{g}\right)(x)$ b. state its domain	a. $\left(\dfrac{f}{g}\right)(x)=\dfrac{f(x)}{g(x)}=\dfrac{x^2}{x^2-1}$ b. The denominator cannot be 0. Solve $x^2-1=0$ for x. $\qquad (x+1)(x-1)=0$ $\qquad x=-1,1$ The domain is $(-\infty,-1)\cup(-1,1)\cup(1,\infty)$.
Try these	
Given $f(x)=x+1$ and $g(x)=\dfrac{x}{2}$, find a. $(f+g)(x)$ b. $(f\cdot g)(x)$ c. $(f\cdot g)(4)$	a. $\dfrac{3}{2}x+1$ b. $\dfrac{x}{2}(x+1)=\dfrac{1}{2}x^2+\dfrac{1}{2}x$ c. 10
Given $f(x)=x$ and $g(x)=x^2-x-6$, find a. $\left(\dfrac{3f}{g}\right)(x)$ b. state its domain	a. $\dfrac{3x}{x^2-x-6}$ b. $x\neq 3,-2$

Difference Quotient

The difference quotient is another formula. It pops up lots of places – calculating averages, slopes, etc. – and so we practice on it a bit. By the way, knowing about the difference quotient is excellent preparation for the next section (and calculus even).

♣ difference quotient

This is the formula that is called the difference quotient (can you see a difference – subtraction – and a quotient – division?)

$$\frac{f(x+h)-f(x)}{h}$$

By the way, can you see why h cannot be equal to zero in this formula? (That's algebra thinking.)

The difference quotient is the slope of a secant line.

A secant line of a curve is a line that intersects two points on the curve. *Secant* comes from the Latin *secare*, to cut.

Secant functions are explored in trigonometry; the slope of the secant line (the difference quotient) comes around again in calculus.

How do you find $f(x+h)$?

Let $f(x) = x^2 + 1$.

Find these:

1. $f(3) = 3^2 + 1 = $ _____

2. $f(-1) = (-1)^2 + 1 = $ _____

3. $f(a) = $ _____

4. $f(b+1) = $ _____

   Did you get $f(b+1) = (b+1)^2 + 1 = b^2 + 2b + 1 + 1 = b^2 + 2b + 2$ for the last one?

★ **For Fun**

Look at the picture below. Could you pick out the coordinates of the two points where the curve and the line intersect? Could you use the slope formula to write an expression for the slope that line from those coordinates?

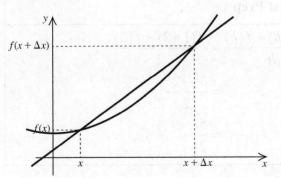

♣ average velocity

Average velocity is one those quantities that can be calculated with the difference quotient.

The formula is the same, but the variables and function have difference names:

$$\frac{s(a+\Delta t) - s(a)}{\Delta t}$$

where s is the position function (position is changing if a velocity is involved, right?), and a is the point in time when the beginning velocity is measured, and Δt is simply how long time runs.

So, for instance, say $s(t) = 6t^2$ measured in feet with t is measured in seconds. Let's find the average velocity for $2 \le t \le 3$:

Time is starting at 2 (seconds) so $a = 2$. Time ends at 3 (seconds), so $\Delta t = 3 - 2 = 1$.

$$\begin{aligned}
\frac{s(a+\Delta t) - s(a)}{\Delta t} &= \frac{s(2+1) - s(2)}{1} \\
&= s(3) - s(2) \\
&= 6(3)^2 - 6(2)^2 \\
&= 6 \cdot 9 - 6 \cdot 4 \\
&= 54 - 24 \\
&= 30
\end{aligned}$$

The average velocity is 30 feet per second.

★ For Fun

Can you see the units "feet per second" in the formula $\dfrac{s(a+\Delta t) - s(a)}{\Delta t}$ if the numerator is *position measured in feet* and the denominator is an interval of *time measured in seconds*?

Difference Quotient: Test Prep	
Given $f(x) = x^2$ and $x = 1$, and $h = 2$, find and simplify the difference quotient.	$\dfrac{f(x+h) - f(h)}{h} = \dfrac{f(1+2) - f(2)}{2}$ $= \dfrac{3^2 - 2^2}{2}$ $= \dfrac{5}{2}$

Try these	
Find the difference quotient for $g(x) = x^3$.	$3x^2 + 3xh + h^2$
Given $s(t) = -6t$, find the average velocity over the time interval $[1, 2]$.	-6

Composition of Functions

The composition of functions has to do with putting an x-value through two (or more) functions to see what you get out. That means that you take the result from one function and put that result into another function. Symbolically this is what that looks like:

$$f(g(x))$$

Do you recognize what's going on there? You're putting the result from running x through the g-function *into* the f-function. It's like any system that you move into another situation; the idea is very universal – it's like a human moving to a new city and seeing how he/she does.

Algebraically, you may find you need to practice this a bit, but it can definitely become very easy. Here's an example using the two functions that are defined like this:

$$f(x) = x + 1 \text{ and } g(x) = x^2$$

Let's find out what the result is if you run an x through g first and then through f, like so:

$$f(g(x)) = f(x^2) = x^2 + 1$$

Did you catch that? Everywhere there is an x in $f(x)$, you have to insert $g(x)$. That means wherever there is an x in $f(x)$, you have to insert x^2. And so you get the final result: $x^2 + 1$.

Now let's try $g(f(x))$ using the very same g- and f-functions.

$$g(f(x)) = g(x + 1) = (x + 1)^2 = x^2 + 2x + 1$$

Everywhere there was an x in $g(x)$, we had to insert $(x + 1)$. Then we simplified. *Are you catching how that works*? It's no big deal really – if you can plug, say, a into f and get $f(a)$, then you already have all the skills you need to do composition of functions.

While where here, it's important to notice that $f(g(x))$ was *not* equal to $g(f(x))$. In general, the composition of two functions is not *commutative*; that is, the order of the nesting makes a difference. (Later on we'll see a special case where the composition *is* commutative, and that case will have very special meaning.)

★ For Fun

You can try the "redundant" compositions of functions from my example *f*- and *g*- functions: $f(f(x))$ and $g(g(x))$. See what you get. I think that's all the possible combinations of those two functions … unless you want to try nesting them more than 2 deep. Can you imagine that?

There is a special *operator* that we use to denote the composition of functions. Now that you know how to "compose" two functions, all you have to know is what this operator means. Here's exactly how it translates into what you already know:

$(f \circ g)(x) = f(g(x))$ or vice versa $(g \circ f)(x) = g(f(x))$. First of all, can you see how the order of the nesting of the functions is specified by this $\circ$ operator?

Second of all, and this is important because students' eyes are sometimes fooled by this, you must be able to keep in mind that *f*(*x*) is *functional notation* and there is absolutely *nothing* about multiplying some "*f*" by some (*x*). There is also *no* multiplication going on between some $f \circ g$ and some *x*. $f \circ g$ is the name of some new function created by composing *f* and *g* – it's just the name of a function. You still have to give it an *x* to work on before it can actually do anything. And the work $f \circ g$ will do on its *x* is not that it multiplies itself by *x* – it does whatever the definition of the composition turns out to be.

It's like this: $\sqrt{}$ can't actually do anything until you put something inside it. Well, $(f \circ g)$ is just like that, and trying to multiply this $f \circ g$ times anything makes as much sense as this: $\sqrt{} \times 4$.

See if you can dissect these compositions of functions – in other words, can you see a function inside another function? (Like $(x+1)^2$ is $x+1$ inside a squaring function.)

$\sqrt{x+1}$

$(x-2)^3$

$\sqrt[3]{212x}$

$(3x)^5$

$\left(4\sqrt{x^5}\right)^3$

✓ Questions to Ask Your Instructor

How many layers of function-nesting can you spot in that last one? You might want to discuss this and other examples carefully and thoroughly in class. Make sure your "eye" is getting properly trained to spot these.

By the way, when you learned in the last subsection to do $f(x+h)$ for the difference quotient, that was a composition of functions, wasn't it? (Can you see that $x+h$ is a function inside of f?)

Composition of Functions: Test Prep	
Given $f(x)=2x-1$ and $g(x)=x^2$, find and simplify a. $(f \circ g)(x)$ b. $(g \circ f)(x)$	a. $(f \circ g)(x) = f(g(x))$ $\quad = f(x^2)$ $\quad = 2x^2 - 1$ b. $(g \circ f)(x) = g(f(x))$ $\quad = g(2x-1)$ $\quad = (2x-1)^2$ $\quad = 4x^2 - 4x + 1$
Try these	
Given $f(x) = \dfrac{1}{3}x^2$, find $(f \circ f)(3)$.	3
Given t the temperature in Fahrenheit, the temperature in Celsius is given by $C(t) = \dfrac{5}{9}(t-32)$ and the temperature Kelvin is given by $K(t) = C(t) + 273.15$. Convert $70°F$ to Kelvin. Round to the nearest hundredth.	$294.26°K$

2.7: Modeling Data Using Regression

After we learn so much about linear and quadratic functions and their graphs, how about trying to use these functions to *approximate* data that might fall roughly on a line or in a U-shape? Regression is a way to find functions that approximate graphs of data that you already have. (What we're doing is finding an equation from a graph instead of graphing from an equation like we have been doing throughout this chapter.)

Linear Regression Models

Just like a graph sometimes displays information in a way that minds can appreciate much more than a string a numbers, regression analysis allows us to construct curves on the cartesian coordinate system that our minds can grasp more meaningfully than a collection of dots.

♣ **regression analysis**

Regression analysis is the process we use to find a function (in this section, a line) that best describes the data (points on a 2-dimensional graph).

♣ **line of best fit, least-squares regression line**

The result of a linear regression analysis will be the line of best fit or the least squares regression line.

Although the equations for calculating the line of best fit are somewhat labor intensive and perhaps scary looking, this is something that you will likely have to learn to do in an introductory statistics course, or even sometimes in an introductory physics course. Don't let that scare you. The methods are actually straightforward – but a little bit beyond the scope, at this point, of precalculus algebra. Nonetheless, I assure you that it is definitely do-able with basic algebra skills... and time.

Having said that, here we will leave the analysis to a computer package. So get out your calculator guide and look up regression. Also Microsoft Excel and other spreadsheet applications can compute regression lines given (x, y) data points. Your textbook, page 238, also gives some steps for some TI calculators.

Linear Regression: Test Prep

For which set of data is the correlation coefficient closer to 1?	The set on the left because the points would cluster more closely to their regression line.
Try these	
The data show the distance in feet that a ball travels for various swing speeds in mph of a bat. bat speed (mph)　　distance (ft) 40　　　　　　　　200 45　　　　　　　　213 50　　　　　　　　242 60　　　　　　　　275 70　　　　　　　　297 75　　　　　　　　326 80　　　　　　　　335 a. Find the linear regression equation for these data. Keep 10 decimal places. b. What is the expected distance a ball will travel when the bat speed is 58 mph? Round to the nearest mph.	a. $y = 3.1410344282x + 65.09359606$ b. 263 ft.
Find the linear regression line for these ordered pairs: $(72, 5)$, $(75, 7)$, $(78, 8)$, $(81, 12)$, $(84, 15)$. Round to nearest hundredths.	$y = 0.83x - 55.60$

Correlation Coefficient

Now we introduce a way to quantify how good your "best-fit" line is.

♣ linear correlation coefficient *r*

Again, find out how to get the correlation out of your calculator when you give it data to find a best-fit line. The correlation coefficient, usually denoted by *r*, indicates how well that best-fit line fits your points. The correlation coefficient takes on values between –1 and 1. The closer it is to zero, the worse the fit. Both 1 and –1 are indicators of a good fit. Can you see those trends in these graphs and the values for their correlation coefficients?

♣ coefficient of determination *r²*

The square of the correlation coefficient is the coefficient of determination (see below).

> The coefficient of determination r^2 measures the proportion of the variation in the dependent variable that is explained by the regression equation. (page 240 in your text)

For example, if $r \cong 0.929$, then $r^2 \cong 0.86$ or something like 86%. That means that about 86% of the relation of the data to the best fit line is due to the way that your independent and dependent variables interact. It also means that there is a leftover approximate 15% that depends on something else.

★ For Fun

If a value closer to zero for the correlation coefficient means "not a good fit," what do you imagine that a coefficient of determination near zero would mean?

Quadratic Regression Models

Well, not everything fits a linear regression profile well. You can try a quadratic (power 2) regression instead; see how that goes. Instructions for quadratic regression for some TI calculators can be found in your textbook. Check out your calculator or spreadsheet application (like Excel).

This method will also furnish you with a coefficient of determination to aid you in deciding how much your independent and dependent variables really depend on each other. At least you can use it to decide if quadratic regression is a better tool than linear regression for your system.

Yes, of course, there are higher-order regressions. With a computer, regressions becomes pretty easy and very accessible.

Quadratic Regression: Test Prep	
The data show the oxygen consumption, in milliliters per minute, of a bird flying level at various speeds in kilometers per hour. speed consumption 20 32 25 27 28 22 35 21 42 26 50 34 a. Find a quadratic model for these data. Round to 5 decimal places. b. Use your model to find the speed to the nearest kph at which the bird has minimum oxygen consumption.	a. Enter the data in the table into your calculator or a spreadsheet application. Then find the quadratic regression model: $y = 0.05208x^2 - 3.56026x + 82.32999$ b. You can use the vertex formula to find the minimum or maximum of a quadratic function. In this case it will be a minimum: $x = -\dfrac{b}{2a}$ $= -\dfrac{-3.56026}{2(0.05208)}$ ≈ 34 Minimum consumption occurs at approximately 34 kph.

Try these	
Find the quadratic regression equation for these data: **Year** **AIDS cases** 1999 41356 2000 41267 **2001** 40833 2002 41289 2003 43171 Round to 4 decimal places.	$y = 345.1429x^2 - 1705.6571x + 42903.6000$ where x is years since 1998.
Find the quadratic regression equation for these data: $(-1, 6)$, $(0, -1)$, $(1, -3)$, $(2, -1.5)$, $(3, 5)$, $(4, 10)$. Round a, b, and c to two decimal places.	$y = 1.68x^2 - 3.91x - 0.23$

NOW TRY THE CHAPTER 2 REVIEW EXERCISES AND

THE CHAPTER 2 PRACTICE TEST

STAY UP ON YOUR GAME –

TRY THE CUMULATIVE REVIEW EXERCISES

Final Fun: Matching

FIND THE BEST MATCH. GOOD LUCK.

ORDERED PAIR

FUNCTION NOTATION

$x^2 + y^2 - 8x + 12y - 48 = 0$

QUADRANT

$(0, b)$

$3x + 2y = 12$

EQUATION OF A LINE

DISTANCE FORMULA

MIDPOINT FORMULA

$h(x)$

III

Y-INTERCEPT

$$\left(-\frac{b}{2a}, \frac{4ac-b^2}{4a}\right)$$

$f(g(x))$

DIFFERENCE QUOTIENT

$$\left(\frac{x_1+x_2}{2}, \frac{y_1+y_2}{2}\right)$$

$$d = \sqrt{(x_2-x_1)^2+(y_2-y_1)^2}$$

QUOTIENT OF FUNCTIONS

CONSTANT FUNCTION

$$m = \frac{y_2-y_1}{x_2-x_1}$$

FLOOR FUNCTION

(3, −2.1)

SLOPE FOMULA

COMPOSITION OF f and g

EQUATION OF A CIRCLE

$f(x) = -11$

$\lfloor x \rfloor$

$$\frac{x^2}{x-1}$$

$$\frac{f(x+h)-f(x)}{h}$$

VERTEX FORMULA

Chapter 3: Polynomial and Rational Functions

Word Search

```
D  S  D  N  U  O  B  X  M  E  B  R  L  V  R
Q  I  N  C  D  N  N  D  U  U  L  H  C  Q  E
U  M  V  S  I  L  F  B  L  O  M  W  P  C  D
O  U  T  I  U  B  M  G  T  R  Q  I  O  N  N
T  M  H  G  D  O  U  T  I  E  Y  E  X  I  I
I  I  E  V  F  E  U  C  P  Z  F  H  Q  A  A
E  N  O  G  D  G  N  N  L  F  P  R  S  X  M
N  I  R  T  Q  G  C  D  I  D  A  M  U  W  E
T  M  E  F  I  I  Q  C  C  T  O  N  D  O  R
U  Z  M  S  T  I  I  H  I  O  N  N  C  J  H
L  U  Z  R  P  E  L  O  T  I  A  O  X  L  S
Y  G  A  N  N  O  N  H  Y  L  T  E  C  D  L
Z  U  A  T  L  A  I  M  O  N  Y  L  O  P  K
Q  J  C  M  L  A  S  Y  M  P  T  O  T  E  X
K  R  O  S  I  V  I  D  F  A  C  T  O  R  S
```

ASYMPTOTE	MAXIMUM	SIGN
BOUNDS	MINIMUM	SMOOTH
COEFFICIENT	MULTIPLICITY	THEOREM
CONTINUOUS	POLYNOMIAL	ZERO
CUBIC	QUARTIC	
DIVIDEND	QUOTIENT	

3.1: Remainder Theorem and Factor Theorem

In Section 3.1, we need to get some basic skills in place. These are skills we'll use throughout the rest of the chapter (and you'll need them heavily in calculus, too). These are techniques for the most part. You can learn a technique – practice it – and we can see how it's used soon enough.

Division of Polynomials

Let's cover some basic vocabulary about division. These are all words you've heard before.

❖ **zeros**

A zero of a function is an x-value that causes the value of the function to become zero. In other words, if $x = c$ is a zero of $f(x)$, then $f(c) = 0$.

❖ **remainder**

Just like in grade school when you first learned about division, the remainder is still exactly the same item: what is left over when you carry out your division. If the remainder is zero, then the divisor divided the dividend exactly.

★ **For Fun**

When you write an improper fraction like $\frac{31}{9}$ as a mixed number, there's a remainder, right? What do you do with it in the mixed number? Once you're taking pre-calculus algebra, you no longer write the remainder like "R4." The pattern is more-or-less like the way you write the remainder for this mixed number.

❖ **quotient**

The quotient is the answer to a division problem which had no remainder.

❖ **dividend**

The dividend is the numerator when the division problem is written in fractional form; it's inside the bracket when the quotient is written out for long division. The dividend is the expression being divided *into*.

❖ divisor

The divisor is the denominator, the bottom of the fraction, outside the bracket in long division. It is the expression to be divided into the dividend.

★ For Fun

If $\dfrac{31}{9} = 3\dfrac{4}{9}$, can you name the quotient, divisor, dividend, and remainder?

Synthetic Division

First we'll discuss the long division of polynomials you've probably seen before. This method of division of polynomials is lengthy, and so you'll learn a new method: synthetic division.

Regardless of the method you use to divide polynomials, I assure you, wholeheartedly, that you need to practice these methods. Don't be fooled into thinking they look simple and sensible and so you know them. You won't remember the first step later, and then you're sunk. Plus they are simple and straightforward, so a little practice will go a long way and the speed you gain will pay off at test time.

❖ long division

Long division refers to the process (algorithm) you learned in grade school to carry out division of big numbers. Here we use it to divide polynomials. The algorithm is basically the same, but it looks somewhat more confusing:

$$
\begin{array}{r}
x+12 \\
x-3\overline{\smash{\big)}\,x^2+9x-16} \\
\underline{x^2-3x} \\
12x-16 \\
\underline{12x-36} \\
20
\end{array}
$$

The process starts when you ask yourself, "How many times does x go into x^2?" The answer to that question is x, and you write that on the top of the bracket *in the* x *place value column*:

$$
\begin{array}{r}
x \\
x-3\overline{\smash{\big)}\,x^2+9x-16}
\end{array}
$$

Then multiply that x by the divisor and write the result underneath the dividend, respecting the place value of the terms in the polynomial.

$$\begin{array}{r} x \\ x-3\overline{\smash{\big)}\ x^2+9x-16} \\ \underline{x^2-3x} \end{array}$$

Then subtract that result, bring down the next term from the dividend, and ask yourself if x goes into that result.

$$\begin{array}{r} x \\ x-3\overline{\smash{\big)}\ x^2+9x-16} \\ \underline{x^2-3x} \\ 12x-16 \end{array}$$

In our example, x goes into $12x$ twelve times. Write that on top of the bracket; include the operation of either addition or subtraction when you do because we're building a new polynomial up there.

$$\begin{array}{r} x+12 \\ x-3\overline{\smash{\big)}\ x^2+9x-16} \\ \underline{x^2-3x} \\ 12x-16 \end{array}$$

Multiply the new term in the quotient on top of the bracket by the divisor and write the result underneath the bottom row, respecting place value. Subtract and thereby find your remainder:

$$\begin{array}{r} x+12 \\ x-3\overline{\smash{\big)}\ x^2+9x-16} \\ \underline{x^2-3x} \\ 12x-16 \\ \underline{12x-36} \\ 20 \end{array}$$

So the division problem with its answer is

$$(x^2+9x-16) \div (x-3) = x+12+\frac{20}{x-3}$$

❖ synthetic division

Synthetic division is long division of polynomials written in a sort of shorthand.

First and foremost, in order to use synthetic division, the divisor must be able to be written in the form $x-c$. If the divisor cannot be written in this form, for example x^2-1 or $2x+3$, then you need to use good ol' long division.

Let's try the synthetic division for $(2x^3-x^2+3x-1)\div(x-3)$. First make a frame, with the value of x that makes the divisor zero on the outside (that's c of $x-c$). Across the top row, write the coefficients of the dividend. Copy the first coefficient to the result row under the frame:

$$3\underline{|2 \quad -1 \quad 3 \quad -1}$$
$$2$$

Multiply c (here it's 3) by that first coefficient that you copied into the bottom row. Write that result under the next place in the internal second row, add the result to the second coefficient in the top row, and write that result in the bottom, result row:

$$3\underline{|2 \quad -1 \quad 3 \quad -1}$$
$$\underline{ \quad 6}$$
$$2 \quad 5$$

Now multiply that number by c again (here, it's 3 times 5) and do the same thing with that result as you did in the last step, but this time everything is one more step to the right:

$$3\underline{|2 \quad -1 \quad 3 \quad -1}$$
$$\underline{ \quad 6 \quad 15}$$
$$2 \quad 5 \quad 18$$

Continue exactly like that until you run out of positions:

$$3\underline{|2 \quad -1 \quad 3 \quad -1}$$
$$\underline{ \quad 6 \quad 15 \quad 54}$$
$$2 \quad 5 \quad 18 \quad 53$$

The coefficients of the result and the remainder are in that bottom, result row. The quotient will be exactly one degree less than the dividend:

Now we can see the result: $(2x^3-x^2+3x-1)\div(x-3)=2x^2+5x+18+\dfrac{53}{x-3}$

★ For Fun

Now try this division $(x^2 + 9x - 16) \div (x - 3)$ by synthetic division. We already arrived at answer by long division for it. Pay very careful attention to all the ways that long division and synthetic division produce the same results, step-by-step. Also, please do make sure you get the right answer when you do it by a different method!

❖ fractional form

Fractional form includes writing the division in the form of numerator over denominator and the remainder over the divisor, thus:

$$\frac{2x^3 - x^2 + 3x - 1}{x - 3} = 2x^2 + 5x + 18 + \frac{53}{x - 3}$$

Synthetic Division: Test Prep	
Do the following division by both long division and synthetic division: $$\frac{6x^4 - 2x^3 - 3x^2 - x}{x - 5}$$	• long division: $x^2 - 1 = (x-1)(x+1) = 0$ $$\begin{array}{r} 6x^3 \quad +28x^2 \quad +137x \quad +684 \\ x=1 \text{ and } x-5 \overline{\smash{\big)}\, 6x^4 \quad -2x^3 \quad -3x^2 \quad -x \quad +0} \\ -\,(6x^4 \quad -30x^3) \\ \hline 28x^3 \quad -3x^2 \\ -(28x^3 \quad -140x^2) \\ \hline 137x^2 \quad -x \\ -(137x^2 \quad -685x) \\ \hline 684x \quad +0 \\ -\,(684x \quad -3420) \\ \hline 3420 \end{array}$$ • synthetic division: $$\begin{array}{r|rrrrr} 5 & 6 & -2 & -3 & -1 & 0 \\ & & 30 & 140 & 685 & 3420 \\ \hline & 6 & 28 & 137 & 684 & 3420 \end{array}$$

	• answer: $$6x^3 + 28x^2 + 137x + 684 + \frac{3420}{x-5}$$
Try Them	
Use synthetic division to carry out $(-x^3 + 3x^2 + 5x + 30) \div (x - 8)$	$-x^2 - 5x - 35 + \dfrac{-250}{x-8}$
Divide: $\dfrac{x^5 + 3x^4 - 2x^3 - 7x^2 - x + 4}{x^2 + 1}$	$x^3 + 3x^2 - 3x - 10 + \dfrac{2x+14}{x^2+1}$

Remainder Theorem

The Remainder Theorem is a nifty little item we'll use a lot:

<u>**The Remainder Theorem**</u>

Given

$$\frac{P(x)}{x-c} = Q(x) + \frac{R(x)}{x-c}$$

then $P(c) = R(x)$.

That means that if a polynomial $P(x)$ is divided by $x - c$, then the remainder of that division

is equal to P(c).

For example, if $\dfrac{2x^3 - x^2 + 3x - 1}{x - 3} = 2x^2 + 5x + 18 + \dfrac{53}{x-3}$ (from our last example), and if we name the

numerator like this $P(x) = 2x^3 - x^2 + 3x - 1$, then $P(3) = 53$. Did you catch that one? Try calculating

$P(3)$ the old-fashioned way. Is it 53?

If x gets to be anything remotely difficult, using synthetic division and finding the remainder that way

becomes must quicker and easier ... and less error-prone.

The Remainder Theorem: Test Prep	
Use synthetic division and the Remainder Theorem to find $P(-2)$ if $P(x) = 4x^3 - 5x^2 + 2x - 10$	$\begin{array}{r\|rrrr} -2 & 4 & -5 & 2 & -10 \\ & & -8 & 26 & -56 \\ \hline & 4 & -13 & 28 & -66 \end{array}$ $P(-2) = -66$
Try Them	
Given $P(x) = x^4 - 25x^2 + 144$, use synthetic division and the Remainder Theorem to ascertain if $c = 3$ is a zero.	Yes. $P(3) = 0$
Given $P(x) = x^4 - 2x^2 - 100x - 75$, find $P(-5)$.	$P(-5) = 1000$

Factor Theorem

Zeros and factors go together!

Factor Theorem

A polynomial $P(x)$ has a factor $(x - c)$ if and only if $P(c) = 0$. That is, $(x - c)$ is a factor of

$P(x)$ if and only if c is a zero of P.

(page 266 of your textbook)

Remember that being a factor means that said factor divides evenly, without a remainder, into the dividend. There you go – without a remainder. And by the Remainder Theorem, we know that the value of the polynomial for a specified x is given by its remainder. So the value of the polynomial must be zero if there is no remainder and no remainder means you've found a factor. At least it's all completely consistent!

★ For Fun

Consider $P(x) = x^4 - 8x^3 + 22x^2 - 29x + 24$. Determine whether $(x+1)$ and $(x-3)$ are factors. Now check to see if $x = -1$ and $x = 3$ are zeros. Use whatever method you prefer, but start to notice how the methods complement each other.

Take note: The Factor Theorem does not give you a method to *find* a factor, only to check and see if you have guessed one correctly.

<table>
<tr><td colspan="2" align="center">**The Factor Theorem: Test Prep**</td></tr>
<tr>
<td>Use synthetic division to determine if $(x+2)$ is a factor of $P(x) = x^3 + 8$.</td>
<td>

$$\begin{array}{r|rrrr} -2 & 1 & 0 & 0 & 8 \\ & & -2 & 4 & -8 \\ \hline & 1 & -2 & 4 & 0 \end{array}$$

$P(-2) = 0$; therefore, $x = -2$ is a zero of $P(x)$ and $(x+2)$ is a factor.
</td>
</tr>
<tr>
<td>**Try Them**</td>
<td></td>
</tr>
<tr>
<td>If the zeros of a polynomial $f(x)$ are given to be $x = -1, 1, 3$, list three distinct factors of $f(x)$.</td>
<td>$(x+1), (x-1), (x-3)$</td>
</tr>
<tr>
<td>Which is a factor of $2x^3 - 3x^2 - 11x + 6 -$ $(x+3)$ or $(x-3)$?</td>
<td>$(x-3)$</td>
</tr>
</table>

Reduced Polynomials

Just like we have to reduce fractions, it's possible to reduce polynomials. This is easy if you can completely factor the given polynomial. We've got the Remainder and Factor Theorems to help us check for factors now – that will help.

❖ **reduced polynomial, depressed polynomial**

A reduced polynomial (or depressed polynomial, but that sounds, well, depressing) is simply the quotient resulting from a division by a factor of that polynomial. Does that sound confusing? Don't panic: reducing polynomials is the same as reducing fractions. Find common factors and cancel them!

For example:

$$\frac{x^4 - 8x^3 + 22x^2 - 29x + 24}{x - 3}$$
$$= \frac{(x-3)(x^3 - 5x^2 + 7x - 8)}{x - 3}$$
$$= x^3 - 5x^2 + 7x - 8$$

It will be very important in coming sections to recognize these reduced polynomials.

★ **For Fun**

Think about this. Say you have a rational function,

$R(x) = \dfrac{x^4 - 8x^3 + 22x^2 - 29x + 24}{x - 3}$. Do you think that

$R(x) = x^3 - 5x^2 + 7x - 8$? Hint: think about domains.

Reduce these:

$$f(x) = \frac{2x^2 + x - 1}{x + 1}$$

$$g(x) = \frac{x^3 - 4x^2 + x - 4}{x - 4}$$

$$h(x) = \frac{x^3 + x^2 - 3x}{x^2 + 4x}$$

3.2: Polynomial Functions of Higher Degree

Now we embark on the great undertaking of learning about polynomials of degree higher than those we have already studied. We have studied degree 1 – lines – and degree 2 – parabolas.

❖ smooth continuous curves

There are rigorous criteria for smooth and continuous curves in calculus, but luckily they wind up meaning just about the same thing the words mean in English. A curve that is the graph of a function and doesn't have any pointy, sharp turns or gaps or holes or jumps of any kind is a smooth continuous curve. Polynomials are smooth continuous curves.

Far-Left and Far-Right Behavior

Since polynomials behave so nicely (for instance, they are smooth and continuous), there are a lot of generalizations we can make about their behavior and thereby their appearance on a graph.

General Form for a Polynomial

In general a polynomial can be defined by this expression:

$$P(x) = a_n x^n + a_{n-1} x^{n-1} + a_{n-2} x^{n-2} + \cdots + a_2 x^2 + a_1 x + a_0$$

This is the most general form. The degree is only specified as n, a natural number. All the coefficients a_n through a_0 are real numbers (remember they may be positive or negative or even zero in some cases). Ponder the form of this expression for a while; see if you can recognize the pieces in other polynomials you encounter. After all, math is about pattern recognition.

❖ leading term, dominant term

$a_n x^n$ is the leading term, or dominant term, of any polynomial. Recognize that this is the term of highest degree in the polynomial.

❖ leading coefficient

In the leading term, $a_n x^n$, a_n is the leading coefficient.

★ **For Fun**

Find the degree of the polynomial, leading term, and leading coefficient in these polynomials:

$$-x + 3x^2 - 2$$

$$x - 2x^4 + \frac{1}{3}x^3 - 12 + x^2$$

$$\frac{x^4}{4} + 2x^3 - x^2 + \frac{x}{2}$$

$$a_n x^n + a_{n-1} x^{n-1} + a_{n-2} x^{n-2} + \ldots + a_2 x^2 + a_1 x + a_0$$

While you're at it, make sure you can identify the *constant* term in each of them, too. Of course, you

noticed that there is no reason (except politeness) for writing a polynomial in descending order, is there?

Leading Term Test: Test Prep	
Identify the leading term of $P(x)$ and sketch its far-left and far-right behavior: $P(x) = 1 - 2x - 4x^3$	leading term: $-4x^3$
Try Them	
Describe the far-left and far-right behavior of $f(x) = -\dfrac{x^4}{4} + x - 16$	$f(x)$ goes down to both the far-left and the far-right.
Describe the far-left and far-right behavior of $T(x) = x^3 - x - 8$	$T(x)$ goes down to the far-left and up to the far-right.

❖ **far-left behavior**

The far-left behavior of a polynomial alludes to the value of the polynomial as x gets extremely small (goes towards $-\infty$). The value on the y-axis of any of these smooth and continuous (and unbounded) polynomials will always approach either ∞ or $-\infty$ (hence unbounded).

❖ **far-right behavior**

Similarly, the far-right behavior refers to the value on the y-axis of the polynomial as x approaches ∞.

★ **For Fun**

Describe the far-left and far-right behaviors of these polynomials:

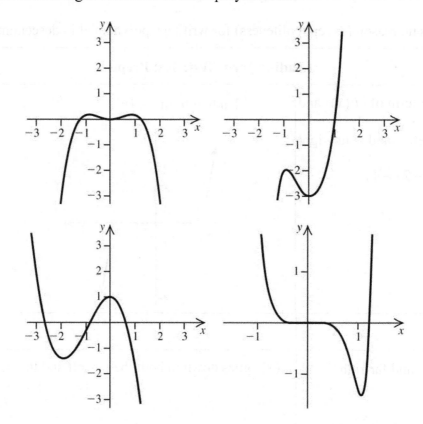

See the table on page 272 of your textbook to help you identify the sign of the leading coefficients and the even/odd nature of the degrees of these polynomial functions. By the way, this is all really good stuff to remember for calculus, if you're headed that way!

Maximum and Minimum Values

In between the far-left and far-right, a polynomial may "go up and down" quite a bit. One way to classify the behavior in these regions is to talk about where the value of the polynomial has a maximum or a minimum. We talked about the maximum and minimum of quadratic polynomials (parabolas) already.

❖ **turning points**

Turning points are points where the function changes from increasing to decreasing, where the graph of the function changes from going up to going down or vice versa. We talked about increasing and decreasing in Chapter 2.

A polynomial of degree n has at the very most $n-1$ turning points. So what is the maximum number of turning points of $f(x) = -x^4 + 3x^2 - x + 11$? Does knowing the possible number of turning points help you picture the graph of a polynomial?

❖ **absolute minimum**

The absolute minimum value of the function is the smallest value in the function's range.

❖ **absolute maximum**

Similarly, the absolute maximum value of a function, as read on the y-axis, is the maximum value of the function's range.

★ **For Fun**

Reason this one out: What are the absolute maximum and minimum values of a polynomial of odd degree?

❖ **relative minimum, local minimum**

Relative minima and maxima of functions occur at turning points. A relative minimum of a function will occur where the function changes from decreasing to increasing as you read a graph from left to right. In other words, a relative minimum occurs at the bottom of a "trough" in the polynomial curve. It may not be the absolute minimum but locally it is a minimum.

❖ **relative maximum, local maximum**

A relative maximum occurs at a turning point where the function changes from increasing to decreasing as you read from left to right. It may or may not also be an absolute maximum, but it is a maximum in the local vicinity of values of the polynomial.

★ **For Fun**

Can you point out absolute and relative min's and max's on the graphs from the previous page?

Definition of Relative Minimum and Relative Maximum: Test Prep	
Find the vertex of the parabola described by $f(x) = -(x-1)^2 - 3$. Is it a minimum or a maximum of $f(x)$?	By inspection (because the function is in the standard form for an equation of a parabola), the vertex is at $(1, -3)$ The vertex is a maximum because the leading coefficient is negative.
Try Them	
Mark the relative maximum values of $P(x)$ from the graph:	

Use the minimum and maximum value functions on a calculator or computer to estimate relative minima and maxima of $H(x) = -2x^3 - 3x^2 + 12x + 1$ Use what you know from the leading term test to name them relative or absolute extremes.	Relative maximum at $x \approx 1.0$ and $y \approx 8.0$; Relative minimum at $x \approx -2.0$ and $y \approx -19.0$

Real Zeros of a Polynomial Function

The *real* zeros (not imaginary zeros, those come later) of a polynomial are the self-same things as x's of the x-intercepts. Does that make sense? Isn't the value of the y-coordinate zero on the x-axis? Isn't where y equals zero exactly what we mean by the zero of the function?

At the very most, a polynomial of degree n can have n real zeros. Mind you, it may have fewer than n zeros, though.

Intermediate Value Theorem

Since a polynomial is continuous – no skips or jumps or holes – we can draw the very significant conclusion of the Intermediate Value Theorem, which basically says that in order to get from here to there without any jumping, you have to go through all the values in between the two places. Here it is more "officially":

The Intermediate Value Theorem

If P is a polynomial function and $P(a)$ is not equal to $P(b)$ for all $a < b$, then P takes on every value between $P(a)$ and $P(b)$ for all x's in the interval [a, b].

Try to visualize that every polynomial, whose domain is all real numbers, accepts each of the values in the interval [a, b] on the x -axis. Polynomials are continuous after all. This means that the polynomials also assign values on the y -axis between $P(a)$ and $P(b)$.

One of the most useful procedures that the Intermediate Value Theorem is used in is finding x -intercepts polynomial functions. Doesn't it make sense that if a polynomial is negative in one region:

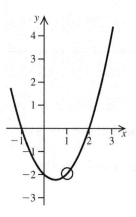

Then is positive in another region:

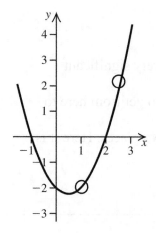

That there must be a point where it crosses the x -axis, and hence where there must be a zero of the function:

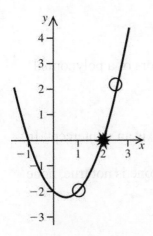

Helpful – now we know where to start looking for an *x*-intercept!

Intermediate Value Theorem: Test Prep	
Use the Intermediate Value Theorem to show $P(x) = -5x^3 + 11x^2 - 3x - 1$ has a zero between $a = 0$ and $b = 1$.	$P(0) = -1$ $P(1) = 2$ Since $P(0)$ and $P(1)$ have opposite signs, $P(x)$ has at least one real zero between $x = 0$ and $x = 1$.
Try Them	
Can you use the Intermediate Value Theorem to determine if $P(x) = -5x^3 + 11x^2 - 3x - 1$ has a zero between $a = -1$ and $b = 1$?	Inconclusive; no.
Use the Intermediate Value Theorem to show $P(x) = x^4 + 3x^3 - 2x^2 + 3$ has a zero between $a = -4$ and $b = -3$	$P(-4) = 35$ $P(-3) = -15$

Real Zeros, *x*-Intercepts, and Factors of a Polynomial Function

To review and collate important results regarding the zeros, x-intercepts, and factors of a polynomial function so far:

If $(x - c)$ is a factor of $P(x)$, then $x = c$ is a zero of $P(x)$, $P(c) = 0$, and $(c, 0)$ is an x-intercept. In fact all four conditions here are equivalent; that is, if one is true they all are, and if one is not true, none are.

Even and Odd Powers of $(x - c)$ Theorem

This topic deals with something that makes graphing a polynomial easier.

Say you have a polynomial higher order that can be factored thus:

$f(x) = (x-1)^2(x+2)^3$ Since $(x-1)$ is a factor, then $(1, 0)$ is an x-intercept. Likewise for $(-2,0)$. If you notice "how many times" $(x-1)$ is a factor $f(x)$, you are identifying the power of the factor $(x-1)$. Here it is 2, an even number, and the power of $(x+2)$ is 3, an odd number.

Here's the thing, if that power is even then the graph of the polynomial only touches the *x*-axis and turns around; whereas, if that power is odd, then the graph crosses the x-axis. Like this:

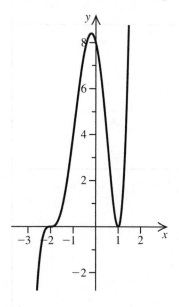

Even and Odd Powers of $(x - c)$ Theorem: Test Prep	
Given that $P(x) = x(x-6)^2(x+1)^2$, name its zeros and determine whether its graph crosses or only touches the x-axis at each zero.	• $x = 0$ – multiplicity 1, odd; therefore, crosses • $x = 6$ – multiplicity 2, even; therefore, only touches • $x = -1$ – multiplicity 2, even; therefore, only touches
Try Them	
Given that $P(x) = -(x-1.6)^3(x+11)$, name its zeros and determine whether its graph crosses or only touches the x-axis at each zero.	• $x = 1.6$ – multiplicity odd; therefore, crosses • $x = -11$ – multiplicity odd; therefore, crosses
Given that $P(x) = (x^2-1)^2$, name its zeros and determine whether its graph crosses or only touches the x-axis at each zero.	• $x = 1$ – multiplicity even; therefore, only touches • $x = -1$ – multiplicity even; therefore, only touches

Procedure for Graphing Polynomial Functions

Already we know a great deal about the behavior of a polynomial graph. Piece by piece you can put together a reasonable sketch. Here are some ideas to keep in mind while you work on graphing a polynomial function:

1. far-left and far-right behavior

2. y-intercept

3. x-intercept(s) and their powers

4. maximum number of turning points

5. symmetries

6. plot some more points using a T-table

★ **For Fun**

Why can there never be more than one y-intercept on the graph of a polynomial function?

	Matching:
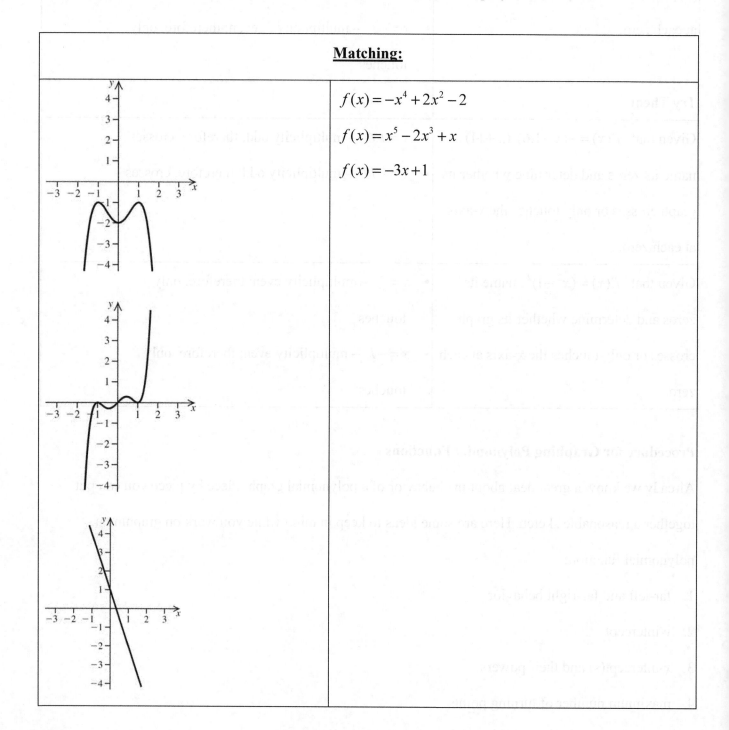	$f(x) = -x^4 + 2x^2 - 2$ $f(x) = x^5 - 2x^3 + x$ $f(x) = -3x + 1$

Procedure for Graphing Polynomial Functions: Test Prep

Identify which of the following is the graph of

$$y = -x^4 - 4x^3 - 3x^2 + 4x + 4:$$

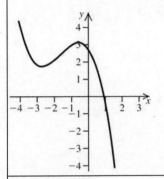

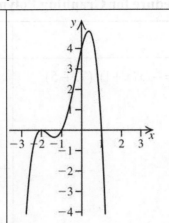

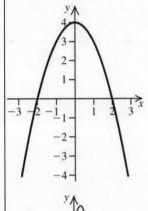

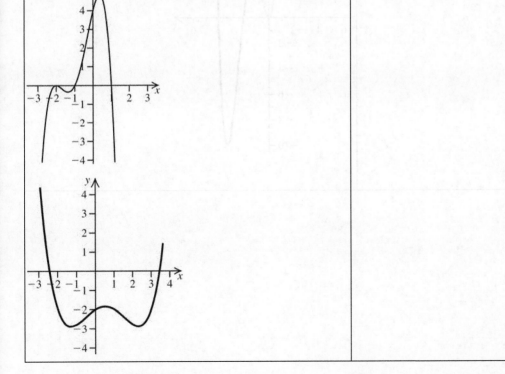

Procedure for Graphing Polynomial Functions: Test Prep

Try Them	
Sketch the graph of $P(x) = -2x(x+1)^2(7x-5)$	
Sketch the graph of $y = x^4 - 6x^3 + 8x^2$	

Cubic and Quartic Regression Models

Since we can now do something with polynomials of higher degree, we can also use regression of higher

degree. Once again, research how your graphing calculator or spreadsheet program handles this.

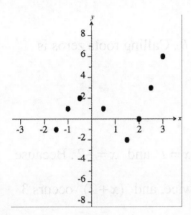

Have you developed the eye to recognize that perhaps a cubic regression might fit this data best? If you

don't see that yet, keep looking and thinking – you'll develop "the eye."

3.3: Zeros of Polynomial Functions

The skills of algebra and critical thinking (that you need to follow all the way through a problem of

finding the zeros of a higher degree polynomial) are historical ... and they will stand you in good stead in

future math courses and, I dare say, your life.

Of course, that's little comfort to you, and I have to admit that this requires a lot of mental work to keep all

the aspects of the problem organized in your head *and* on your paper. Practice is the only thing I know that

will get it for you. Or a graphing calculator, but your teacher is probably not so much interested in your

calculator skills as in your critical thinking skills.

Best foot forward!

Multiple Zeros of a Polynomial Function

A polynomial of degree n has at the very most n zeros. Sometimes, often even, it has fewer. Well, in fact, there is a way to look at it that counts these too-few zeros several times to make up the difference.

♣ **roots**

Root is just another word for zero of a function. If c is a root, then $P(c) = 0$. Calling roots zeros is slightly more popular now because, after all, it is more descriptive.

♣ **multiple zero**

A polynomial like we had last section, $f(x) = (x-1)^2 (x+2)^3$, has zeros at $x = 1$ and $x = -2$. Because the factors that go with these zeros occur more than once – $(x-1)$ occurs twice, and $(x+2)$ occurs 3 times – their zeros are multiple zeros. It's as if this polynomial had 5 zeros, but the zeros were of only two distinct values.

♣ **zero of multiplicity** k

Determining the multiplicity of a zero was what we did last section when we learned if the graph would show an x-intercept that crossed the x-axis or only touched it and turned around. The multiplicity of the zero $x = -2$ in the above polynomial is 3.

♣ **simple zero**

A simple zero has multiplicity of one.

★ **For Fun**

Would the graph cross or just touch the x-axis at a simple zero?

Rational Zero Theorem

Remember the general form for a polynomial function:

$$P(x) = a_n x^n + a_{n-1} x^{n-1} + a_{n-2} x^{n-2} + \cdots + a_2 x^2 + a_1 x + a_0$$

Have you learned to pick out the leading coefficient and the constant term? Good. Now we will use those numbers to list all the possible *rational zeros* of $P(x)$. Rational zeros will be only those zeros that can be written as a rational number – that means that irrational numbers like $\sqrt{2}$ will never appear on this list. Also, the list can contain many, many more rational numbers than there could possibly be zeros. That's okay; just remember there is a firm maximum number of zeros possible for a polynomial function of degree n. (And that max is n.)

Let's list all possible rational zeros for $f(x) = 3x^3 - x^2 + 5x - 6$.

1. First list all possible factors of -6, including both positive and negative possibilities: $\pm 1, \pm 2, \pm 3, \pm 6$. Call those p values.

2. Then list all the possible factors of the leading coefficient: $\pm 1, \pm 3$. Call those q values.

3. Now construct all the possible ratios of p's over q's:

$$\pm 1, \pm 2, \pm 3, \pm 6, \pm \frac{1}{3}, \pm \frac{2}{3}, \pm \frac{3}{3}, \pm \frac{6}{3}$$

But some of those can be reduced: $\pm 1, \pm 2, \pm 3, \pm 6, \pm \frac{1}{3}, \pm \frac{2}{3}, \pm 1, \pm 2$. And we can remove the duplicates so that we have this list finally:

$$\pm 1, \pm 2, \pm 3, \pm 6, \pm \frac{1}{3}, \pm \frac{2}{3}$$. Now that's a list with 12 elements. Our polynomial is only degree 3, and so it's impossible for all these to be zeros. In fact, it's possible for none of them to be zeros.

★ **For Fun**

Do you have an idea of how you might check to see if any of these values actually *are* zeros? Are you thinking "Remainder Theorem and synthetic division" like I am?

★ **For Fun**

How would this list be simpler to construct if the leading coefficient were 1?

Rational Zero Theorem: Test Prep	
List all possible rational zeros of $P(x) = -3x^3 - x + 1$	$p : \pm 1$ $q : \pm 1, \pm 3$ $\dfrac{p}{q} : \pm 1, \pm \dfrac{1}{3}$
Try Them	
List all possible rational zeros of $T(x) = -x^4 - 4x^3 - 3x^2 + 4x + 4$	$\dfrac{p}{q} : \pm 1, \pm \dfrac{1}{2}, \pm 2, \pm 4$
List all possible rational zeros of $f(x) = -2x^4 + 3x - 4$	$\dfrac{p}{q} : \pm 1, \pm \dfrac{1}{2}, \pm 2, \pm 4$

Upper and Lower Bounds for Real Zeros

There are some swell little procedures that help rule out some of the surplus possible rational zeros in your list. One of them is to discover when your x-value is either larger than the largest possible zero or when it's smaller than the smallest possible zero.

By the way, you need to be pretty good at synthetic division by now in order to proceed with these tests.

♣ upper bound

In the context of finding the zeros of polynomials, "upper bound" means a number past which there are no more zeros. Say if you know the polynomial value goes towards infinity by its far-right behavior. Isn't there a point on the x-axis past which (in the positive direction) there are not more x-intercepts (and hence no more zeros)?

How do you recognize one? Carry out synthetic division using your possible *positive* zero. If the bottom, result row has only positive values listed, then there cannot be any zeros larger than the one you are testing – whether or not it turned out to be a zero.

By the way, a zero on the bottom row can be counted as positive for this test. And remember this test only works for *positive* test values.

This synthetic division shows that 5 is an upper bound. All possible zeros larger than 5 can now be disregarded:

$$5 | \begin{array}{cccc} 1 & 0 & -19 & -18 \\ & 5 & 25 & 30 \\ \hline 1 & 5 & 6 & 2 \end{array}$$

Meaning and the number line:

If you discover that $x = 2$ is an upper bound for the zeros of this polynomial, is it still possible that $x = 1.5$ is also an upper bound?

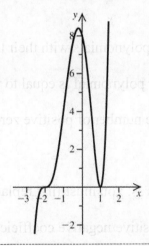

❖ lower bound

And a lower bound is an x-value past which (in the negative direction) there can be no more zeros.

This time the test is a tiny bit harder. The values in the bottom row of your synthetic division must alternate in sign in order for the test value to be a lower bound. If the values do alternate, you need not look for a zero smaller than your test.

Oh, yeah, this test only works for *negative* test values. Also, zero can be either positive or negative, but it cannot be both – that is, you cannot change its sign once you've fixed it.

Here's a result that shows –5 is a lower bound:

$$
\begin{array}{r|rrrr}
-5 & 1 & 0 & -19 & -28 \\
   &   & -5 & 25 & -30 \\
\hline
   & 1 & -5 & 6 & -58
\end{array}
$$

Descartes' Rule of Signs

Descartes' Rule of Signs is another way to discard possibilities from your list of potential rational zeros. It does require more algebra – nothing challenging, but a little bit time-consuming.

❖ Descartes' Rule of Signs

This rule has two parts, and it applies to polynomials with their terms listed in strictly descending order:

1. The number of *positive* zeros of your polynomial is equal to the number of sign changes of the coefficients of the polynomial. Or the number of positive zeros may be that number decreased by an even number.

2. To test for the number of *negative* zeros, you must first replace x with $-x$ in your polynomial, simplify, get a new distribution of positive/negative coefficients. The number of negative zeros is now equal to the number of sign changes in the new polynomial. Or the number of negative zeros may be decreased by an even number.

❖ variations in sign

Variations in sign refers to the number of sign changes in the coefficients of a polynomial when the polynomial is written in descending order of terms. We also assume all the coefficients are real.

★ For Fun

Count the number of sign changes in these polynomials:

$3x^3 - x^2 + 5x - 6$

$-x^4 + 3x^2 - x + 11$

$-x + 3x^2 - 2$

$\dfrac{x^4}{4} + 2x^3 - x^2 + \dfrac{x}{2}$

$x^3 - 5x^2 + 7x - 8$

Now change all the x's to $-x$'s, simplify, and count the sign changes again.

Descartes' Rule of Signs: Test Prep	
Use Descartes' Rule of Signs to state the number of possible positive and negative real zeros of $f(x) = 2x^4 - 19x^3 + 51x^2 - 31x + 5$	• There are four sign changes in $f(x)$, so 4, 2, or no positive real zeros. • $f(-x) = 2x^4 + 19x^3 + 51x^2 + 31x + 5$ has no sign changes, so there are no negative real zeros.
Try Them	
Use Descartes' Rule of Signs to state the number of possible positive and negative real zeros of $P(x) = 3x^3 + 11x^2 - 6x - 8$	• one positive real zero • two or no negative real zeros
Use Descartes' Rule of Signs to state the number of possible positive and negative real zeros of $P(x) = x^3 - 4x^2 - 3x$	• one positive real zero • one negative real zero

Zeros of a Polynomial Function

Here is a list of suggestions to consider when you begin to find all the zeros of a polynomial of higher degree:

- the degree of the polynomial

- the list of potential rational zeros: $\dfrac{p}{q}$

- Descartes' Rule of Signs and the possible number of positive or negative zeros

- upper and lower bounds on zeros

- reducing polynomials – the coefficients are in the bottom, result row once you have found a zero

- all your quadratic skills once the polynomial is reduced to second degree

This can be a slightly long procedure. Have a care and be patient. Finding the zeros of a higher order polynomial is really a reward in itself. Think of it like this: since time immemorial man has been working on this problem and you've joined those ranks!

Besides the quadratic formula, which is used to find the zeros of a second-degree polynomial, there are also general solutions up to quartic polynomials. However, it has been proven that polynomials of degree 5 or more do not produce zeros according to a formula in terms of their coefficients.

You know the quadratic formula. Can you find the "cubic formula" somewhere? One look and you'll understand why we don't memorize this one. It's easier and quicker to try the methods of this chapter!

Guidelines for Finding the Zeros of a Polynomial Function with Integer Coefficients: Test Prep	
Find the zeros of $P(x) = 2x^3 + 9x^2 - 2x - 9$ and state their multiplicities.	$p : \pm 1, \pm 3, \pm 9$ $q : \pm 1, \pm 2$ $\dfrac{p}{q} : \pm\dfrac{1}{2}, \pm 1, \pm\dfrac{3}{2}, \pm 3, \pm\dfrac{9}{2}, \pm 9$ [ordered] $\begin{array}{r\|rrr} 3 & 2 & 9 & -2 & -9 \\ & & 6 & 45 & 129 \\ \hline & 2 & 15 & 43 & 120 \end{array}$ Three is not a zero, but it is an upper bound. So $\dfrac{p}{q} : \pm\dfrac{1}{2}, \pm 1, \pm\dfrac{3}{2}, -3, -\dfrac{9}{2}, -9$ has fewer elements. $\begin{array}{r\|rrr} 1 & 2 & 9 & -2 & 9 \\ & & 2 & 11 & 9 \\ \hline & 2 & 11 & 9 & 0 \end{array}$ One is a zero. (Not only is it, too, an upper bound, but it is the only positive zero by Descartes' Rule.) Now find the zeros of the reduced polynomial, which can be factored: $2x^2 + 11x + 9 = (2x+9)(x+1)$ The zeros are $x = 1, -1, -\dfrac{9}{2}$
Try Them	
Determine the zeros, and their multiplicity, of $f(x) = (x-1)^2(x+2)(x+4)$	• $x = 1$, multiplicity 2 • $x = -2$, multiplicity 1 • $x = -4$, multiplicity 1
Find all the real zeros of $P(x) = 3x^5 + 10x^4 - 2x^3 - 28x^2 - 8x + 16$	$x = \dfrac{2}{3}, -2, \sqrt{2}, -\sqrt{2}$

Applications of Polynomial Functions

Now translate word problems into polynomial functions and solve. You're not restricted to degree one or two polynomials for these purposes any more!

NOW TRY THE MID-CHAPTER 3 QUIZ

3.4: Fundamental Theorem of Algebra

Up until now all *we've* said is about the number of zeros that a polynomial can have is the *maximum* number of zeros. Now we're going to say something more definitive about the number of zeros of a polynomial (and, by the way, it will be consistent with what has gone before). This ought to help our comprehension and our graphs.

Fundamental Theorem of Algebra

What this subsection really does is introduce us back to *complex* numbers (remember them from Chapter P?) and to consideration of complex numbers as possible zeros of higher order polynomials. We're always adding more layers of subtlety and complexity to our algebraic machinations!

❖ Fundamental Theorem of Algebra

The Fundamental Theorem of Algebra states that your polynomial is of degree greater than one (greater than linear, that is) and if it has any complex coefficients, then it will have at least one complex zero.

Number of Zeros of a Polynomial Function

If you count each zero of a polynomial according to its multiplicity (section 3.3), then every polynomial of degree n has n zeros. By extension then, the polynomial must have n linear factors.

I make no promises that these zeros and factors will be easy to find, but they do exist.

❖ existence theorems

Existence theorems are the sort of theorems that assert the existence of *something* mathematical, like the existence of a certain number of zeros for one example.

How's this for an existence theorem?

"I think, therefore 1 am."

Do you know who said that? Yep, it's our pal René Descartes *again*.

The Fundamental Theorem of Algebra: Test Prep	
How many complex zeros, counted multiplicity-wise, does $P(x) = -2x^5 + 3x^4 + x^2 - x + 12$ have?	Since the degree of $P(x)$ is 5, $P(x)$ has 5 complex zeros, counted multiplicity-wise.
Try Them	
Name all zeros, and their multiplicities, of $S(x) = x^2(x-1)^3(x+\sqrt{2})(x-\sqrt{2})(x-3-i)(x-3+i)$	• $x = 0$, multiplicity 2 • $x = 1$, multiplicity 3 • $x = -\sqrt{2}$, multiplicity 1 • $x = \sqrt{2}$, multiplicity 1 • $x = 3+i$, multiplicity 1 • $x = 3-i$, multiplicity 1
Solve $2x^4 - 3x^3 - 7x^2 - 8x + 6 = 0$	$x = 3, \dfrac{1}{2}, -1+i, -1-i$

Conjugate Pair Theorem

Remember the complex conjugate (section P.6)? The Conjugate Pair Theorem gives up two zeros for the price of one:

Conjugate Pair Theorem

If $a+bi$ (where b is not equal to zero) is a complex zero of a polynomial with real coefficients, then the conjugate, $a-bi$, is automatically a zero too.

★ For Fun

Consider that if you already know that, say, $3-2i$ is a zero of your polynomial. Then you know that its conjugate $3+2i$ is also a zero. You also know that $(x-(3-2i))$ *and* $(x-(3+2i))$ are also factors of the polynomial. If you multiply those two linear factors together, will the result be a factor of your polynomial, too? What degree will that resulting factor be? If it is a factor, can you find the reduced polynomial given this new factor? At this point, at least try simplifying and multiplying out the two factors above.

Conjugate Pair Theorem: Test Prep

Given that $x=-i$ is a zero of $P(x)=3x^4+2x^3+2x^2+2x-1$, find all the zeros.	$x=i,-i$ are zeros; therefore, $(x-i),(x+i)=(x^2+1)$ is a factor. By long division $$\frac{3x^4+2x^3+2x^2+2x-1}{x^2+1}=3x^2+2x-1$$ $$=(3x-1)(x+1)$$ So all four zeros are $x=i,-i,-1,\dfrac{1}{3}$

Try Them	
If $x = 2 - 2i$ is one zero of a polynomial with real coefficients, list another zero and the factors that correspond to these two zeros.	$x = 2 + 2i$ $(x - 2 + 2i), (x - 2 - 2i)$
Solve $x^4 - 4x^3 + 6x^2 - 16x + 8 = 0$ given that $x = 2i$ is one zero.	$x = 2i, -2i, 2 + \sqrt{2}, 2 - \sqrt{2}$

Finding a Polynomial Functions with Given Zeros

By the Factor Theorem, you know that zeros and factors go together. Also now that we know the

Conjugate Pair Theorem, we know about conjugates that are automatically included, along with their

factors, in our knowledge of the zeros of a given polynomial. Well, if you know the zeros, then you

know the factors. If you know the factors, couldn't you multiply them and "reconstitute" a polynomial

from the zeros? In order to get a more precise polynomial, you may also need information about its

degree and maybe even its leading coefficient. (In other words, specifying the zeros is not enough

information to make a polynomial that is unique to those zeros.)

Find a Polynomial Function With Given Zeros: Test Prep	
Find a polynomial function P, with real coefficients, that has zeros $x = 4, 1 + 2i$ and degree 3.	$P(x) = (x - 4)(x - 1 - 2i)(x - 1 + 2i)$ $= (x - 4)(x^2 - 2x + 5)$ $= x^3 - 6x^2 + 13x - 20$ (for example)

Try Them	
Write a polynomial P, with integer coefficients having the following zeros: $x = 2, -3, -\dfrac{1}{2}$	$P(x) = 2x^3 + 3x^2 - 11x - 6$ (for example)
Find a polynomial function P, of lowest degree with real coefficients, lead coefficient of 1, and the following zeros: $x = 3i$, multiplicity 1, and $x = 2$, multiplicity 2.	$P(x) = x^4 - 4x^3 + 13x^2 - 36x + 36$

3.5: Graphs of Rational Functions and Their Applications

Now that we have pretty thoroughly investigated finding the zeros of polynomials, we'll consider rational functions again. After all, they are nothing more than a ratio of polynomials, and knowing how to find the zeros of the numerator and denominator might come in handy!

Vertical and Horizontal Asymptotes

Finding the asymptotes in order to graph a rational function has everything to do with being able to find zeros of polynomials.

★ **For Fun**

Look up the word *asymptote*. What other English words is it related to? Make sure you understand what it means to have an asymptote on your graph. Can you make that meaning hang together with the word's meaning in English?

❖ rational function

A rational function is a ratio of polynomials, like so: $R(x) = \dfrac{P(x)}{Q(x)}$ where P and Q are polynomials.

We studied rational functions in a more general way back in Chapter P.

---

artial fractions decomposition is a method that depends on rational functions. We'll meet them in

Section 6.4.

❖ vertical asymptote

A vertical asymptote is a vertical line (of the form $x = a$) on the graph of a function that is not actually

a part of the graph of the function (hence it is a dashed line) but is a boundary that the function never

crosses. However, the function will get extremely close, horizontally, to a vertical asymptote.

Here is a graph with a vertical asymptote:

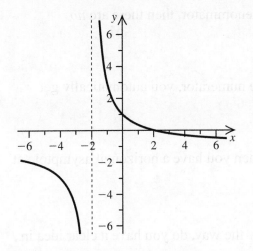

You'll notice that the vertical asymptote has to do with x values that are excluded from the domain.

What do you remember about the domain of rational functions? Remember you cannot divide by zero, and so if you find where the denominator of a rational

function goes to zero (but where the numerator does not also go to zero), then you've found that

function's vertical asymptote(s).

✤ horizontal asymptote

Horizontal asymptotes are horizontal lines $(y = b)$ also included on the graph of a function to show

horizontal lines that the actual graph of the function cannot cross. (Whereas a function never crosses its

vertical asymptote, there are some cases where a graph can cross a horizontal asymptote but only near

the origin. More about that very soon!) Again, horizontal asymptotes are also shown as dashed lines

because they are not actually part of the function.

In order to find your horizontal asymptotes, you have to be able to ascertain the degrees and sometimes

the leading coefficients of the polynomials in the numerator and denominator. There are three cases:

1. If the degree of the numerator is larger than the degree of the denominator, then there are *no*

   horizontal asymptotes.

2. If the degree of the denominator is larger than the degree of the numerator, you automatically get

   exactly one horizontal asymptote: $y = 0$ (that's the x-axis).

3. If the degree of the numerator and the denominator are equal then you have a horizontal asymptote at

   the ratio of the leading coefficients, like so:

   $R(x) = \dfrac{2x-1}{x+5}$ yields the horizontal asymptote: $y = \dfrac{2}{1} = 2$. (By the way, do you have a clear idea in

   your mind of what the line $y = 2$ looks like on a graph?)

Here's an example of a graph with a horizontal asymptote at $y = 2$.

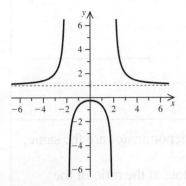

★ **For Fun**

Do you detect the presence of any vertical asymptotes in the above graph? Could you guess at what the

denominator might be like?

✓ **Questions to Ask Your Teacher**

Be sure you understand in what form you should answer questions about asymptotes. After all

asymptotes are actually lines and they ought to be expressed as equations for lines – it's the least they

deserve.

✓ **Questions to Ask Your Teacher**

Sometimes there is an "escape hatch" near the origin where a function may actually cross its horizontal

asymptote. See if you can find out why, but be sure you know how to graph these instances!

Vertical Asymptotes: Test Prep	
Identify the vertical asymptote(s) of $\dfrac{x}{x^2-1}$	$x^2-1=(x-1)(x+1)=0$ $x=1$ and $x=-1$ are the vertical asymptotes.
Try Them	
Determine the domain of $f(x) = \dfrac{x+1}{3x-5}$	$x \neq \dfrac{5}{3}$

Identify the vertical asymptote(s) of $\dfrac{3x-1}{x^3-2x}$	$x=0$, $x=\sqrt{2}$, $x=-\sqrt{2}$

Horizontal Asymptotes: Test Prep	
Find the horizontal asymptote of $R(x)=-\dfrac{6x^2-5}{2x^2+6}$ if any.	The degree of the numerator and the denominator are the same; therefore, there is a horizontal asymptote at the ratio of the leading coefficients: $$y=\frac{6}{2}$$ $y=3$ is the horizontal asymptote.
Try Them	
Find all vertical and horizontal asymptotes of $R(x)=-\dfrac{2}{x^2-7x+10}$	$v.a.: x=5$, $x=2$ $h.a.: y=0$
Find all asymptotes of $R(x)=\dfrac{2x^2}{x^2+4}$	no vertical asymptotes $h.a.: y=2$

Sign Property of Rational Functions

If you find all the zeros of the numerator and the denominator, you can set up a test number line and discover the *sign* of the rational function in each region:

(We investigated this in Section 1.4.)

For a graph of the rational function $\dfrac{x^2 - x - 12}{x+1}$, this number line lets me know that the graph will be

below the x-axis from $-\infty$ to -3, then above the x-axis from -3 to -1, and so on.

Match regions of *positive* and *negative* function value according to our test number line from the

previous page and this graph:

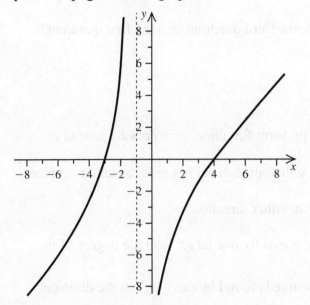

General Graphing Procedure

So when it comes time to graph a rational expression, once again we have several bits of information

available to us to help construct a more complete picture. Consider many, if not all, of these aspects

while you work:

1. asymptotes and behavior near asymptotes

2. intercepts

   (The y-intercept arises from an x-value of zero. x-intercepts arise from setting the numerator

   equal to zero.)

3. symmetry

4. sign behavior

5. plotting some extra points

    (Particularly consider each region of the domain as it might be broken up by vertical asymptotes.

    Also pay special attention near the origin where a horizontal asymptote "escape hatch" might exist.)

Slant Asymptotes

If you consider the graph of $\dfrac{x^2 - x - 12}{x + 1}$ from a page ago, do you see possible evidence of an asymptote

that is neither horizontal or vertical, but instead *slants* from the third quadrant into the first quadrant?

There is indeed a slant asymptote on that graph.

❖ **slant asymptote**

A slant asymptote is an asymptote that adheres to the general form for a line, $y = mx + b$, instead of

being strictly vertical or horizontal. Once again a function will approach but not cross a slant asymptote

as the function grows towards positive or negative infinity in either direction.

A slant asymptote arises when the degree of the numerator is exactly one larger than the degree of the

denominator, like $\dfrac{x^2 - x - 12}{x + 1}$. The equation for said asymptote is found by carrying out the division

implied by the rational function itself, which I can do by synthetic division in this case:

$$
\begin{array}{r|rrr}
-1 & 1 & -1 & -12 \\
   &   & -1 & 2 \\
\hline
   & 1 & -2 & 10
\end{array}
$$

Disregard the remainder altogether and write the *linear* result. In the case it is, from the bottom line of

my synthetic division above, $(y =)x - 2$. (I include the $y =$ to point out that this is an equation for a

line in slope-intercept form. You can graph that (dashed line again since it's an asymptote and not

actually part of the graph of the function).

Look back at our graph of the rational function $\dfrac{x^2 - x - 12}{x + 1}$ and see if this equation doesn't match the

asymptote behavior you can observe right there.

Slant Asymptotes: Test Prep	
Find the slant asymptote of $R(x) = \dfrac{x^3 - 4x}{x^2 - 1}$	By long division, $\dfrac{x^3 - 4x}{x^2 - 1} = x + \dfrac{-3x}{x^2 - 1}$ So $y = x$ is the slant asymptote (disregarding the remainder).
Try Them	
Pick the functions with slant asymptotes: • $\dfrac{x+1}{2x-1}$ • $\dfrac{3x^2 - x - 1}{x}$ • $-4x^3 - x + 8$ • $\dfrac{11 - x - 3x^3}{4 - x^2}$	$\dfrac{3x^2 - x - 1}{x}$ $\dfrac{11 - x - 3x^3}{4 - x^2}$
Determine the slant asymptote of $R(x) = \dfrac{-x^3 - 4x^2 - x + 4}{x^2 - x - 6}$	$y = -x - 5$

Can you tell when your graphing utility is/is not literally displaying a slant asymptote for a function that

does, indeed, have one?

This graph has one. Can you picture it?

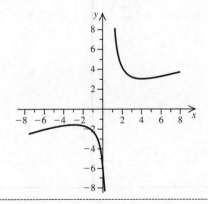

A General Procedure for Graphing Rational Functions That Have No Common Factors: Test Prep

Choose the graph of $R(x) = \dfrac{x^3 - 4x}{x^2 - 1}$:

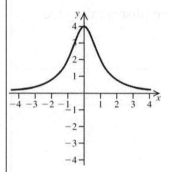

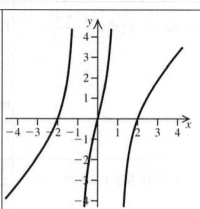

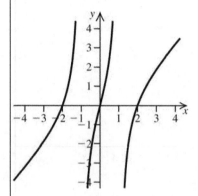

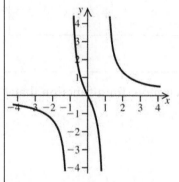

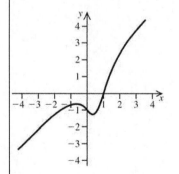

Try Them

Sketch $R(x) = \dfrac{7(x^2 - 4)}{x^3 + 3x}$

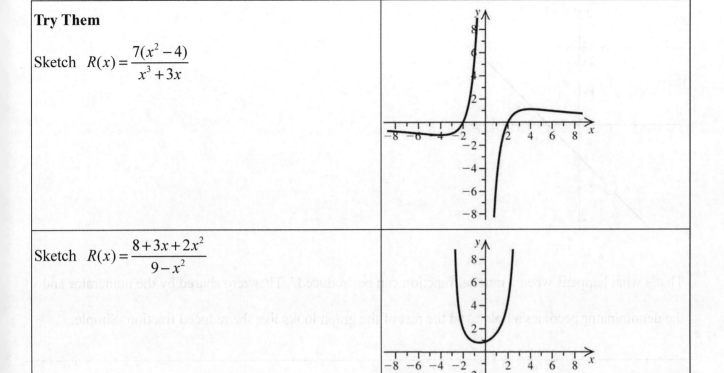

Sketch $R(x) = \dfrac{8 + 3x + 2x^2}{9 - x^2}$

Graphing Rational Functions That Have a Common Factor

There is nothing about the definition of a rational function that says it cannot be reduced like

$\dfrac{x^2 - 1}{x + 1} = \dfrac{(x - 1)(x + 1)}{x + 1}$ can be "reduced" because of the common factor of $x + 1$, right?

If you notice, both the denominator and the numerator will be equal to zero if you plug in $x = -1$. This

means there is a common factor.

This also means that although the x-value in question is still *not* in the domain of the rational function,

a vertical asymptote does *not* arise on the graph. Have you ever seen a graph with a hole in it?

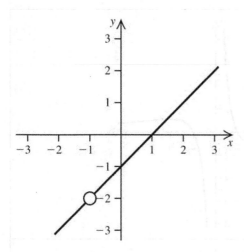

That's what happens when a rational function can be "reduced." That zero shared by the numerator and

the denominator becomes a hole! And the rest of the graph looks like the reduced fraction. Simple.

Regarding Calculators:

Graphing your rational functions by using a graphing calculator is one of the modern marvels that we

can be very thankful for.

However, be sure you thoroughly understand how your calculator displays asymptotes and the behavior

of the graph of the function near asymptotes.

1. Asymptotes are not actually part of the graph of the function itself – asymptotes are more of an aid

   to graphing. They should be dashed lines if they appear, but your calculator may show them as solid.

   Do you understand rational functions enough to tell that?

2. Except for horizontal asymptotes near the origin, graphs of rational functions do not cross their

   asymptotes. Nevertheless, your viewing window may not be able to clearly abide by that constraint.

   Just because your calculator makes it look as if a graph is touching its asymptote, will you be able to

   decipher when that is an artifact of the resolution of your graphing calculator and not actually taking

   place?

A General Procedure for Graphing Rational Functions That Have a Common Linear Factor:

Test Prep

Find the common factor(s) and sketch $R(x) = \dfrac{x^3 - 4x}{x^2 - 4}$	$R(x) = \dfrac{x^3 - 4x}{x^2 - 4} = \dfrac{x(x^2 - 4)}{(x^2 - 4)}$ $R(x) = x \quad x \neq 2, -2$ 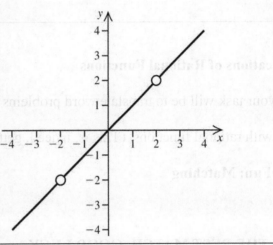
Try Them Sketch $y = \dfrac{x^2 - 3x}{x}$	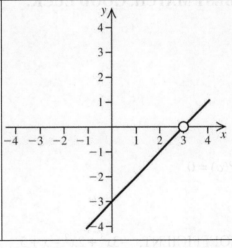

Sketch

$$f(x) = \frac{x^3 + x^2 + 11x}{3x^2 - 2x}$$

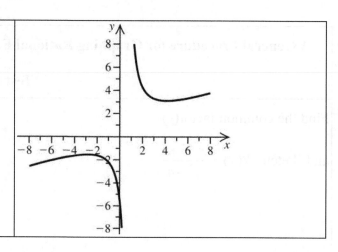

Applications of Rational Functions

Now your task will be to translate word problems into rational functions and solve them using your skills with rational functions. (This is, indeed, getting challenging!)

Final Fun: Matching

FIND THE BEST MATCH. GOOD LUCK.

CUBIC

$x = -1$

RATIO

ZERO OF $P(c) = 0$

QUARTIC

LEADING COEFFICIENT: $-3x^3 + 2x^2 - x + 5$

SMOOTH CONTINUOUS FUNCTION

$(x - 2)^k$

$$\begin{array}{r|rrr} 3 & 1 & 0 & -19 & -28 \\ & & 3 & 9 & -30 \\ \hline & 1 & 3 & -10 & -58 \end{array}$$

MULTIPLICITY

$$\frac{x^2 - 1}{x + 1}$$

VERTICAL ASYMPTOTE

SYNTHETIC DIVISION

$x^3 - x$

$x^2 - x^4 + 1$

c

$$\frac{p}{q}$$

-3

RATIONAL ZEROS   MULTIPLICITY

$$f(x) = x^3 - x + 4$$

NOW TRY THE CHAPTER 3 REVIEW EXERCISES AND

THE CHAPTER 3 PRACTICE TEST

STAY UP ON YOUR GAME – DO THE CUMULATIVE REVIEW EXERCISES

Chapter 4: Exponential and Logarithmic Functions

Word Search

```
N  I  M  A  N  C  S  T  Q  C  K  C  C  P  E
D  O  T  A  O  O  S  W  O  L  I  D  L  R  F
N  Z  I  M  G  E  I  N  A  M  S  R  A  I  I
B  A  M  T  R  N  C  T  H  D  F  I  I  N  L
Q  O  T  E  I  A  I  T  A  D  N  C  T  C  F
N  E  T  U  V  S  I  T  K  U  D  H  N  I  L
D  N  A  I  R  R  O  Y  U  J  Q  T  E  P  A
I  B  T  T  A  A  Z  P  A  D  E  E  N  A  H
P  Y  R  G  A  T  L  D  M  C  E  R  O  L  O
B  W  O  H  T  W  O  R  G  O  E  S  P  E  L
U  L  Y  T  R  E  P  O  R  P  C  D  X  S  U
U  A  W  H  I  J  I  N  V  E  R  S  E  A  Z
U  J  Q  L  Y  L  D  Z  D  P  W  T  P  B  M
S  E  I  S  M  O  G  R  A  P  H  T  M  L  W
T  E  L  C  W  S  I  B  K  B  Z  L  O  G  Q
```

BASE   EXPONENTIAL   MAGNITUDE

COMMON   GROWTH   NATURAL

COMPOSITION   HALFLIFE   PRINCIPAL

CONCAVITY   INTEREST   PROPERTY

DECAY   INVERSE   RICHTER

EQUATION   LOGARITHMIC   SEISMOGRAPH

4.1: Inverse Functions

Every time you solve even the simplest equations, you are using inverse functions to help you. The idea is that multiplication by 3 is the inverse of division by 3 and vice versa. Keep in mind that when you're solving equations, you're working towards a result that has the variable all by itself, like: $x =$ *something*. Inverse functions *do* exactly this. We'll see how a little bit more formally.

Introduction to Inverse Functions

In this section, we check into some of the implications of how inverse functions work with/against each other – in equations and in graphs. This is good stuff by itself, but it's also preparation for two new functions that we are due to meet later in the chapter – two new functions that are inverses of each other. Understanding now the implications (not just mechanics) of inverse functions will help you more fully understand what you're doing later in this very chapter.

♣ inverse function

Simply put, one function is the inverse of another if every ordered pair in one is exactly reversed in the other. The domain of one becomes the range of the other and vice versa.

If you think about it, this means that what one function does is undone by its inverse. If you put a 3 in and get a 2 out of a function – $(3,2)$ its inverse would undo that – $(2,3)$ – and you get your original 3 back. See?

★ For Fun

Think about that in terms of the generic ordered pair, $(x, f(x))$. Can you see how getting an x back is like solving an equation.

In order for a function to have an inverse, said function must be one-to-one (see Chapter Two).

Graphs of Inverse Functions

Since an inverse function swaps domain and range with its counterpart, its graph has a very distinct relationship with the original function.

That relationship is one of a reflection about the line $y = x$:

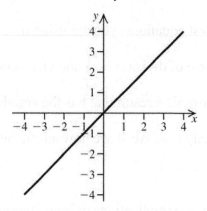

So when you graph functions that are inverses of each other on the same coordinate plane, they look something like this:

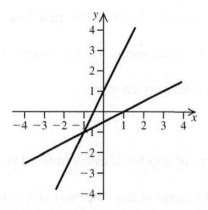

Can you see that every point on each line is a reflection of a point on the other line directly across the line $y = x$? (I've left $y = x$ off this graph to make you work a little bit harder at it.)

★ **For Fun**

If f is the inverse of g, is g necessarily the inverse of f? This is really an easy question; don't let it bog you down.

★ **For Fun**

Try out plotting point-by-point these two lines:

$$y = \frac{1}{3}x - 2$$

$$y = 3x + 6$$

Use your (old-fashioned) *T*-table to notice that the domain of one is the range of the other and that each point is a reflection across the $y = x$ line.

Graph the Inverse of a Function: Test Prep	
Graph $f(x) = 3x - 2$ and $f^{-1}(x) = \dfrac{1}{3}x + \dfrac{2}{3}$ on the same coordinate system.	Since both these are equations for lines, it's straightforward to graph them: 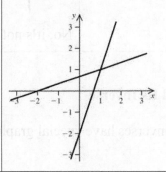
Try Them	
Given the graph of $f(x)$, plot its inverse function on the same coordinate system: 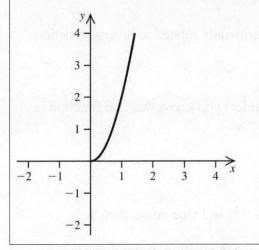	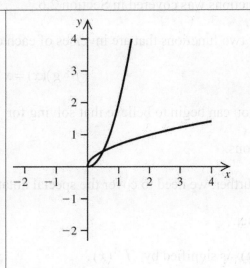

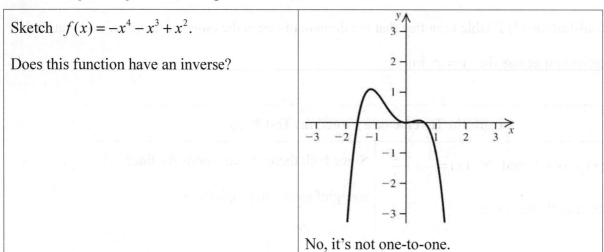

Sketch $f(x) = -x^4 - x^3 + x^2$.

Does this function have an inverse?

No, it's not one-to-one.

Composition of a Function and Its Inverse

Not only do functions and their inverses have special graphical relations, their compositions have a very special characteristic!

Composition of Functions was covered in Section 2.6.

The composition of two functions that are inverses of each other behaves like so:

$$(f \circ g)(x) = x$$

And now perhaps you can begin to believe that solving for x is intimately related to inverse functions and their compositions.

Before we go any further, we need to cover the special notation that lets you know that one function is the inverse of another.

The inverse of $f(x)$ is signified by $f^{-1}(x)$.

So, $(f \circ f^{-1})(x) = x$ is always true. Now, if f is the inverse of f^{-1} and vice versa, then $(f^{-1} \circ f)(x) = x$ is also always true. This is not a unique case where composition of functions is commutative, but it is an important instance of commutativity.

★ For Fun

Try the composition these two functions; show they are inverses of each other:

$$y = \frac{1}{3}x - 2$$

$$y = 3x + 6$$

If so, then we could name these functions so:

$$f(x) = \frac{1}{3}x - 2$$

$$f^{-1}(x) = 3x + 6$$

Do not confuse $f^{-1}(x)$ with the negative exponent. If you actually want the reciprocal of $f(x)$ then

the notation actually goes like this: $[f(x)]^{-1}$

(How's that for convoluted ... but a completely necessary distinction?)

Composition of Inverse Functions Property: Test Prep	
Use composition of functions to verify that $g(x) = \dfrac{x+1}{x}$ is the inverse function of $f(x) = \dfrac{1}{x-1}$.	$(g \circ f)(x) = g(f(x))$ $= g\left(\dfrac{1}{x-1}\right)$ $= \dfrac{\dfrac{1}{x-1} + 1}{\dfrac{1}{x-1}} \cdot \dfrac{x-1}{x-1}$ $= \dfrac{1 + x - 1}{1}$ $= x$ $(g \circ f)(x) = x$ so they are inverses of each other.

Try Them

Use composition of functions to ascertain if the following are inverse functions of each other: $f(x) = (2x + 8)^3$ $g(x) = \frac{1}{2}\sqrt[3]{x} - 8$	No.
Use composition of functions to ascertain if converting Fahrenheit temperature T to Celsius, $f(T) = \frac{5}{9}(T - 32)$, and back, $F(T) = \frac{9}{5}T + 32$, will give you the Fahrenheit temperature you started with.	Yes, $f(F(T)) = T$

Finding an Inverse Function

If you have ascertained that your function is one-to-one and that you want to know its inverse, then there is a straightforward procedure to follow (page 338 in your textbook):

1. Call $f(x)$ by y.

2. Exchange y for x everywhere in the equation.

3. Solve the resulting equation for y. That is, isolate y on one side of the equation.

4. Now name that y by $f^{-1}(x)$.

5. Check your result by performing the composition $(f \circ f^{-1})(x)$ (or $(f^{-1} \circ f)(x)$). Either way, you must find the result to be x by itself, à la $(f^{-1} \circ f)(x) = x$ for instance.

Here's an example:

$$f(x) = \frac{1}{3}x - 2$$

$$y = \frac{1}{3}x - 2$$

$$x = \frac{1}{3}y - 2$$

$$x + 2 = \frac{1}{3}y$$

$$3(x + 2) = y$$

$$f^{-1}(x) = 3(x + 2)$$

And if you check it:

$$(f \circ f^{-1})(x) = f(f^{-1}(x))$$

$$= f(3(x + 2))$$

$$= \frac{1}{3}(3(x + 2)) - 2$$

$$= x + 2 - 2$$

$$= x$$

Yes – x – perfect!

Find the Inverse of a Function: Test Prep	
Find the inverse of $f(x) = \dfrac{1}{x}$	$f(x) = \dfrac{1}{x}$ $y = \dfrac{1}{x}$ $x = \dfrac{1}{y}$ $xy = 1$ $y = \dfrac{1}{x}$

	$f^{-1}(x) = \dfrac{1}{x}$
Try Them	
Find the inverse of $\{(0,0),(1,2),(-3,-1),(5,9),(-4,1),(3,4)\}$	$f^{-1}(x) = \{(0,0),(2,1),(-1,-3),(9,5),(1.-4),(4,3)\}$
a. Does $f(x) = -3(x+1)^2 + 2$ have an inverse? *b.* Does $f(x) = -3(x+1)^2 + 2$ have an inverse if the domain is restricted to $x \le -1$?	a. No b. Yes

✓ **Questions to Ask Your Teacher**

Sometimes functions that are not one-to-one can be made one-to-one by limiting their domain. Limiting the domain of a function limits the range of its inverse, which will likely limit in turn the domain of the inverse function. This is covered in example 5 on page 340 in your textbook, and you may want to ask your teacher to do an example for you too.

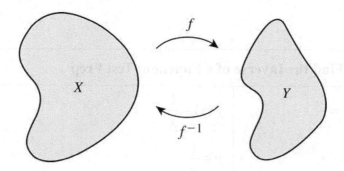

How far could we "nest" inverses?

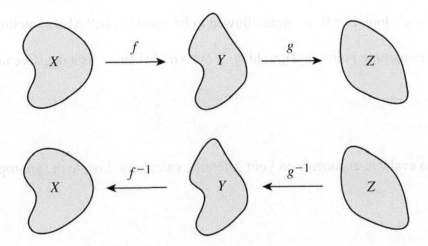

4.2: Exponential Functions and Their Applications

Believe it or not, our discussion of inverse functions has brought us to a new function. We've talked about

polynomial, radicals, and rational functions. Now we will examine something about the exponential

function.

Exponential Functions

You already know a lot about exponents, for example:

$$a^m a^n = a^{m+n}$$

$$(a^m)^n = a^{mn}$$

and so forth. All that will still apply, but now we will look at functions of this type:

$$f(x) = a^x$$

Do you notice that this is a function of x and x, the variable, is in the exponent. This is an exponential

function!

❖ **exponential function with base b**

$f(x) = b^x$ is an exponential function where b, the base, is a positive number and not equal to zero and

x, the exponent, is a real number.

★ For Fun

What would $f(x) = a^x$ look like if a were allowed to be equal to one? Also sometime when you get a

chance you might try plotting points and graphing $f(x) = a^x$ when a is a negative number – good

luck with that one!

Calculator note:

You must be able to evaluate exponents on your scientific calculator. Locate the appropriate buttons now.

Try these:

3^4

2^{-3}

7^{π}

π^{-2}

(It used to be that textbooks had tables of exponential function values in them. They no longer do. More

room for more problems. And more demand on and for calculators.)

By the way, do you know what 3^0 is? What is a^0 (as long as a is not zero)?

Properties of $f(x) = b^x$: Test Prep	
Given $f(x) = 3^x$ evaluate: a. $f(3)$ b. $f(-4)$ c. $f(x+2)$	a. $f(3) = 3^3 = 27$ b. $f(-4) = 3^{-4} = \dfrac{1}{3^4} = \dfrac{1}{81}$ c. $f(x+2) = 3^{x+2} = 3^x \cdot 3^2 = 9 \cdot 3^x$
Try Them	
Use your calculator to estimate these values to the nearest thousandth:	a. 2.297 b. 0.725

a. 2^{12}	c. 69.096
b. $5^{-0.2}$	
c. $\pi^{3.7}$	
When $f(x)=\left(\dfrac{2}{5}\right)^{x}$, find $f(3)$.	$\dfrac{8}{125}$

Graphs of Exponential Functions

Now that you know how to evaluate exponential functions on your calculator, it's a small matter to graph. At the very least you can plot points. For example:

$y = 3^x$

x	y
-2	$3^{-2}=\dfrac{1}{3^{2}}=\dfrac{1}{9}$
-1	$3^{-1}=\dfrac{1}{3}$
0	$3^{0}=1$
1	$3^{1}=3$
2	$3^{2}=9$

Plotting these gives us this picture:

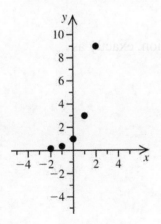

Connecting the dots, we see something like this:

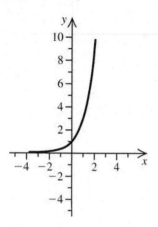

Notice the horizontal asymptote on the x-axis? Think about it: is there any x that will make 3^x become zero or negative? One prominent feature of the graph of an exponential function is that it has this asymptote.

And now you can apply all the techniques of translation, reflection, stretching, and compressing that we studied in Chapter 2. So the graph of, say, $y = 3^{-x}$ would look like this:

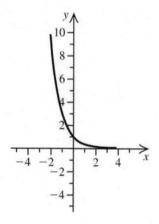

That's a reflection about the y-axis, caused by changing x into $-x$ in the function, exactly as expected.

Match these:

(being sure to notice the horizontal asymptote in each)

$y = 2 + 3^x$

$y = -2^x$

$y = 2^{-x} - 1$

$y = -3^{-x}$

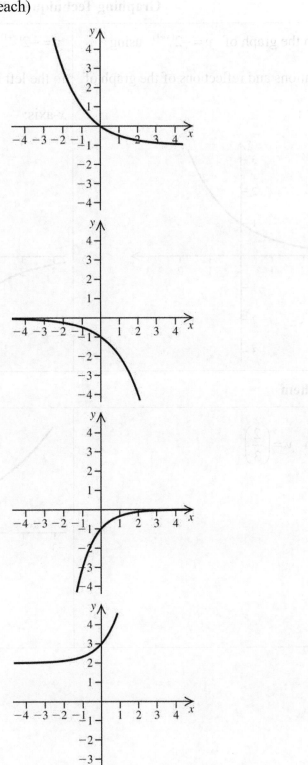

Graphing Techniques: Test Prep

Sketch the graph of $y = -2^{(x+1)}$ using translations and reflections of the graph of $y = 2^x$:	$y = -2^{(x+1)}$ is a horizontal translation of $y = 2^x$ to the left by one unit and a reflection about the x-axis:

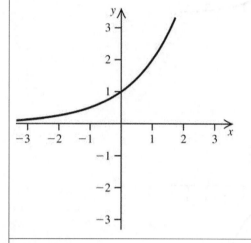

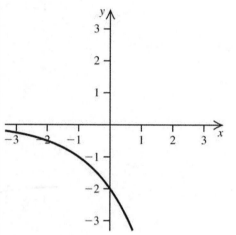

Try Them

Sketch $y = \left(\dfrac{2}{3}\right)^x$

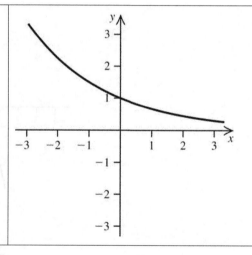

Sketch $y = 4^x - 3$

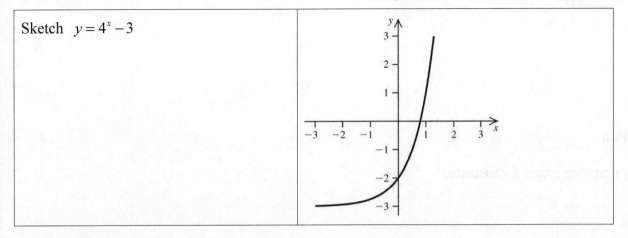

Natural Exponential Function

Every exponential function has a base that is strictly greater than zero and not equal to one. One such number is the irrational number e . e is called the natural base. Since we use a base-ten counting system, 10 is also a nice, clean number to use for an exponential base, but e has more "natural" applications.

❖ e

e is a real number between 2 and 3. It is approximately equal to 2.718281828459045. Since it is a non-repeating, non-terminating decimal, it is an irrational number (that is, you cannot express it as a ratio of integers).

★ **For Fun**

Who named it e ? Any ideas why?

❖ **natural exponential function**

$f(x) = e^x$ is the natural exponential function.

★ **For Fun**

Try these on your calculator; notice if the results are always positive. What is the range of the exponential function again?

e^2

$e^{4.289}$

e^{-5}

e^{π}

e^{1}

★ For Fun

Now try graphing some, for instance:

$y = -e^{x}$

$y = e^{2x}$

$y = 3 - e^{x}$

A <u>double exponential function</u> is a constant raised to the power of an exponential function. The general

formula is $f(x) = a^{b^{x}}$, which gets big faster than our "ordinary" exponential function.

For example, if we let both *a* and *b* *equal 10*, then:

$f(-1) \approx 1.26$

$f(0) = 10$

$f(1) = 10^{10}$

$f(2) = 10^{100}$ (a googol)

$f(3) = 10^{1000}$

$f(100) = 10^{10^{100}}$ (a googolplex.)

Natural Exponential Function: Test Prep	
Use your calculator to estimate these to four decimal places: a.· $e^{1.429}$ b. e^{π} c. $e^{-2.1}$	a. $e^{1.429} \approx 4.1745$ b. $e^{\pi} \approx 23.1407$ c.· $e^{-2.1} \approx 0.1225$

Try Them	
Sketch $y = 2e^{-x}$	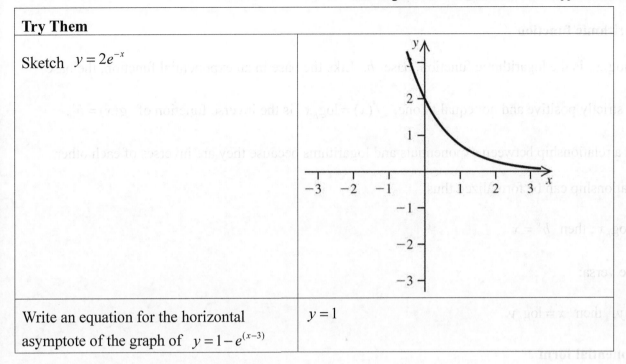
Write an equation for the horizontal asymptote of the graph of $y = 1 - e^{(x-3)}$	$y = 1$

4.3: Logarithmic Functions and Their Applications

The logarithmic function is the inverse of the exponential function, so like the exponential function, the logarithmic function has a *base*. Actual inverse exponential/logarithmic functions must have the same bases – otherwise how could they swap domains and ranges with each other? At any rate, do you already have any ideas about logarithmic functions just because you know they are the inverse of exponentials?

Logarithmic Functions

Logarithms are often called "logs" for short. Since the log is the inverse of the exponential function which you now know a bunch about, perhaps you can imagine what the graph of a logarithm might look like. Would it have an asymptote? What would be the domain and range?

❖ logarithm

log is the functional notation for a logarithm. $\log_b x$ would be a logarithm, base *b*, of *x*. *log* is an instruction to run the independent value through the *log* function and get a value out.

♣ logarithmic function

$f(x) = \log_b x$ is the logarithmic function, base b. Like the base in an exponential function, the base must be strictly positive and not equal to one. $f(x) = \log_b x$ is the inverse function of $g(x) = b^x$.

There is a relationship between exponentials and logarithms because they are inverses of each other. This relationship can be formalized thus:

If $y = \log_b x$, then $b^y = x$

And vice versa:

If $b^x = y$, then $x = \log_b y$.

♣ exponential form

A logarithmic function may be written in its equivalent exponential form, like so:

$\log_3 81 = 4$. This can be written in logarithmic form: $3^4 = 81$. (By the way, is $3^4 = 81$ true?)

♣ logarithmic form

Likewise an exponential might be rephrased in logarithmic form:

$2^{-3} = \dfrac{1}{8}$ becomes $\log_2 \dfrac{1}{8} = -3$

There is a distinct pattern to these transformations between logarithmic and exponential forms. One point to notice would be that the base of the exponential stays the base of a logarithmic equivalence. What pattern can you see involving the other 2 parameters? (I see a sort of triangle, but maybe that's just me and my visual tendencies.) Figure out for yourself a method that helps you put all the pieces in the right places.

★ For Fun

Try to rewrite this one in logarithmic form: $2^{\log_2 3} = 3$. If you can figure that one out, you might start to see the *inverse* nature of the exponential and logarithmic functions.

Exponential and Logarithmic Form: Test Prep	
Convert to logarithmic form:	The base becomes the base of the logarithm; the right-hand-side number becomes the argument of the logarithm; the exponent becomes the answer:
a. $3^{-2} = \dfrac{1}{9}$ b. $4^x = 42.5$ c. $a^m = n$	a. $3^{-2} = \dfrac{1}{9}$ $\log_3 \dfrac{1}{9} = -2$ $4^x = 42.5$ b. $\log_4 42.5 = x$ $a^m = n$ c. $\log_a n = m$
Try Them	
Convert to exponential form: a. $\log_2 16 = 4$ b. $\log 142 \approx 2.15$ c. $\log_x 3 = .004$	a. $2^4 = 16$ b. $10^{2.15} \approx 142$ c. $x^{.004} = 3$
Solve $\log_2(x-3) = 2$ by converting it into its exponential form.	$x = 7$

★ For Fun

Try these, too:

$\log_2 2 =$ _____

$\log_2 1 =$ _____

You know enough about converting between exponentials and logarithms to do these, and you'll learn some basic properties of logarithms along the way. (Or see below for help.)

Basic Properties of Logarithms:

1. $\log_b b = 1$

2. $\log_b 1 = 0$

3. $\log_b(b^x) = x$

4. $b^{\log_b x} = x$

Basic Logarithmic Properties: Test Prep	
Evaluate:	a. Since $\log_b(b^x) = x$, $\log_2 2^{-3} = -3$
a. $\log_2 2^{-3}$	b. Since $\log_b 1 = 0$, $2\log_5 1 = 2 \cdot 0 = 0$
b. $2\log_5 1$	c. Since $b^{\log_b x} = x$, $10^{\log 10} = 10$
c. $10^{\log 10}$	
Try Them	
Solve $\log_7 x = 1$	$x = 7$
Solve $\log_4 4^x = 12$	$x = 12$

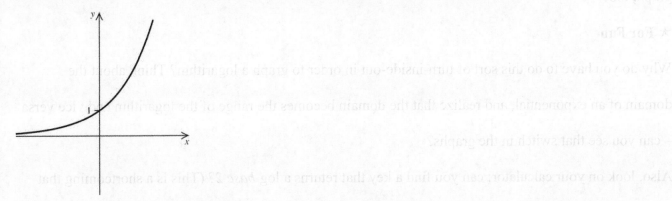

LOGARITHMS, (from λόγος *ratio*, and αριθμός *number*), the indices of the ratios of numbers to one another; being a series of numbers in arithmetical progression, corresponding to others in geometrical progression; by means of which, arithmetical calculations can be made with much more ease and expedition than otherwise.

From a 18th century Encyclopedia Britannica article.

Graphs of Logarithmic Functions

You know that $y = a^x$ has a graph that looks something like the following when $a > 1$.

And you know that the appropriate inverse function is expressed as $\log_a x = y$, so its inverse graph

would be a reflection about $y = x$:

But if you want to plot points on the graph of a function like $y = \log_2 x$, you have to use your

knowledge of its exponential form (namely $2^y = x$) in order to actually construct a T-table. This time

chose your y-values and find the values that go in the x-column:

x	y
___	-2
___	-1
___	0
___	1
___	2

Do you get points like this?

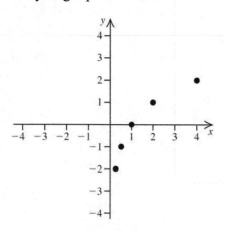

And you can see the logarithm curve now, right? By the way, can you see a *vertical* asymptote on this graph, too?

★ **For Fun**

Why do you have to do this sort of turn-inside-out in order to graph a logarithm? Think about the domain of an exponential, and realize that the domain becomes the range of the logarithm and vice versa – can you see that switch in the graphs?

Also, look on your calculator: can you find a key that returns a log *base 2*? (This is a shortcoming that we will overcome a little later in this chapter.)

Now that you know about the basic logarithmic graph, try some of these by translations, reflections, stretches, and compressions:

$y = -\log_2 x$

$y = 1 + 3\log_2 x$

$y = \log_3(-x)$

$y = \log_3(x - 2)$

Pay attention to the asymptote!

Properties of $f(x)=\log_b x$: Test Prep

Sketch $y = \log_2(x+1)$ by changing the equation into its exponential form, solving for x, and plotting points.	$y = \log_2(x+1)$ $2^y = x+1$ $2^y - 1 = x$

$$y = \log_2(x+1)$$

$$2^y = x+1$$

$$2^y - 1 = x$$

x	y
$-\dfrac{3}{4}$	-2
$-\dfrac{1}{2}$	-1
0	0
1	1
3	2

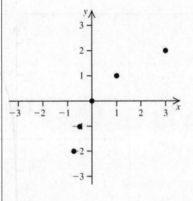

... and so...

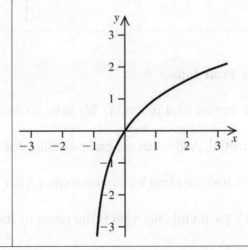

Try Them

Sketch $y = -\log(-x)$

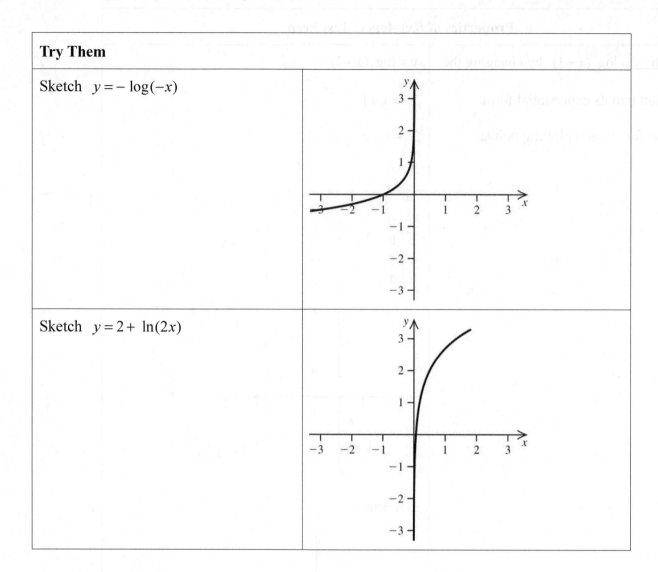

Sketch $y = 2 + \ln(2x)$

Domains of Logarithmic Functions

What is the domain of the exponential function? We have no concern when we pick out x-values for a

T-table to graph an exponential. And when we picked values for the y variable in our last T-table for

graphing the logarithm, we had the same lack of concern. (After all the domain of the exponential

becomes the *range* of the logarithm.) But what is the range of the exponential? Didn't we investigate

that horizontal asymptote?

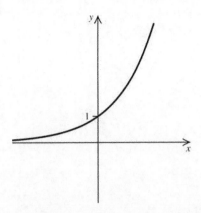

In other words, we discovered that you can never ever get zero or a negative number out of the exponential function. So if that is the range for the exponential function, what is the domain for the logarithm? Can you see that you can only put in strictly positive numbers by observing this logarithmic curve:

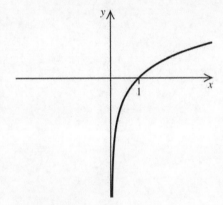

★ For Fun

Try asking for $log(-5)$ in your calculator. Or even $log(0)$. What result do you see? By the way, you *have* found the log and ln keys on your calculator, right? You know you'll need these very, very soon.

Common and Natural Logarithms

These are the logarithmic functions that actually do have keys on your scientific calculator.

❖ common logarithmic function

The common logarithm is base 10. It is usually referred to simply as "log" without denoting the base explicitly. So a missing base means "base 10."

★ For Fun

Evaluate some these in your calculator:

log3

log10

log0

$\log \pi$

❖ natural logarithmic function

The natural log, symbolized by ln, is the logarithm base e. The symbol ln means exactly the same as $\log_e$.

★ For Fun

Try these on your calculator:

ln e

ln2

ln(−1)

ln1

U.S. National Institute of Standards and Technology

NIST has proposed even newer notation for commonly used logarithms of different bases:

"ln x" means $\log_e x$

"lg x" means $\log_{10} x$

"lb x" means $\log_2 x$

However, we will stick with the much-better-known conventions of our textbook on these matters!

4.4: Properties of Logarithms and Logarithmic Scales

Since you already spent quite a bit of time in algebra classes learning all the rules of exponentials – like

$$\left(\frac{a}{b}\right)^n = \frac{a^n}{b^n}$$ – you might not realize how much more easily you can understand exponentials than

logarithms, or why. There are comparable rules for logarithms that you have to learn to use now.

Slide Rule, or how they used to look up logs

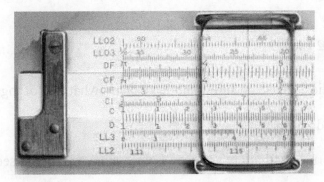

I never had to learn to read a slide rule. I'm sure you're hoping you never have to either. However, we

have another onus upon us: we must be able to get the correct results out of a scientific calculator.

Properties of logarithms

Many these properties correspond to exponential properties, but whether you can see the parallels or not,

you have to become as familiar and handy with these properties as quickly as you can. Ideally you

would know them just as well as you know that $(a^m)^n = a^{mn}$, for instance.

❖ product property of logarithms

This property says that the log of a product is equal to the sum of logs:

$$\log_b(MN) = \log_b M + \log_b N$$

For example:

$$\log_2 2x = \log_2 2 + \log_2 x = 1 + \log_2 x$$

♣ quotient property of logarithms

The quotient property says that the log of a quotient is the difference of logs:

$$\log_b\left(\frac{M}{N}\right) = \log_b M - \log_b N$$

This one you must be extra careful with because, of course, subtraction is not commutative. The log of the denominator must be subtracted from the log of the numerator (and certainly not the other way around).

For example:

$$\log_2\left(\frac{4}{3}\right) = \log_2 4 - \log_2 3 = 2 - \log_2 3$$

With these first two properties, please be sure to notice exactly what sort of logarithmic expression they work on ... and which they do not.

For instance, none of these rules will allow you to simplify $\ln(x+2)$. You see there is a rule for the log of a product but not for the log of a sum.

♣ power property of logarithms

The power property runs thus:

$$\log_b M^p = p\log_b M$$

In other words, this rule makes a log of a power into the product of logs. For example:

$$\log_3 3^4 = 4\log_3 3 = 4 \cdot 1 = 4$$

♣ logarithm-of-each-side property

This property and the next basically communicate that just like adding or multiplying both sides of an equation does not change it, taking the logarithm, same base, of both sides of an equation does not change the truth value of the equation either. Formally:

If $M = N$, then $\log_b M = \log_b N$

❖ **one-to-one property of logarithms**

Conversely, if $\log_b M = \log_b N$, then $M = N$.

❖ **expanding the logarithmic expression**

Now we can use these properties to manipulate logarithmic expressions (in preparation for solving

equations). One of the ways we can rewrite a logarithmic expression might be to expand it, for example:

$$\ln\left(\frac{x^2}{3}\right) = \ln x^2 - \ln 3$$

$$= 2 \ln x - \ln 3$$

Each remaining logarithm is the logarithm of a single variable or expression.

Properties of Logarithms: Test Prep	
Write $\dfrac{1}{2}\log x + 3 \log(x+1)$ as a single logarithm.	$\dfrac{1}{2}\log x + 3 \log(x+1)$ $= \log x^{\frac{1}{2}} + \log(x+1)^3$ $= \log(x^{\frac{1}{2}}(x+1)^3)$ $= \log(\sqrt{x}(x+1)^3)$
Try Them	
Expand the expression $\ln(5x^3 y)$	$\ln 5 + 3 \ln x + \ln y$
Condense the expression $3\log_5 x + \log_5(y-2) - 2\log_5 a$	$\log_5\left(\dfrac{x^3(y-2)}{a^2}\right)$

Change-of-Base Formula

I've already mentioned that you cannot find a key for $\log_2$, for example, on your calculator.

Only the common (base 10) and natural (base e) logarithms rate a button on your calculator. Never fear – there is a way to evaluate an expression like $\log_2 3$ using your calculator. But you must learn the change-of-base formula:

$$\log_b x = \frac{\log x}{\log b}$$

Or, if you prefer the natural log like I do:

$$\log_b x = \frac{\ln x}{\ln b}$$

This formula is not hard, but please practice. It's easy to get the formula turned upside down at which point it represents a very wrong number!

Change-of-Base Formula: Test Prep	
Use the Change-of-Base Formula and a scientific calculator to estimate the value of $\log_{50} 5120$ to three decimal places.	$\log_{50} 5120 = \dfrac{\log 5120}{\log 50} \approx 2.183$
Try Them	
Use the Change-of-Base Formula to show that the following are equivalent: $\dfrac{1}{2}\log_6 44$ $\log_6 \sqrt{44}$	Both are approximately 1.056

Use the Change-of-Base Formula to

sketch

$$y = 3 + \log_5(x-2)$$

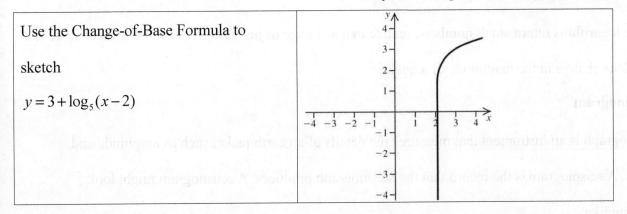

★ For Fun

Can you envision a way to use the change-of-base formula to help with plotting points in order to graph?

(So that you wouldn't have to rewrite the equation to be graphed in exponential form and chose y's

instead of x's?)

Logarithmic Scales

You might notice that $\log_3 81 = 4$ returns a much smaller number than its exponential counterpart,

$3^4 = 81$. Sometimes displaying data with the considerably smaller logarithm value is favored. Some

common uses for these logarithmic "scales" are the Richter scale, which measures earthquake intensity,

and the pH scale, which measures the acidity of a substance.

❖ **zero-level earthquake**

The smallest intensity, called I_0, that can be measured on a seismograph at the earthquake's epicenter.

This will be a constant.

❖ **Richter scale magnitude**

The Richter scale magnitude, M, of an earthquake is given by this formula:

$$M = \log\left(\frac{I}{I_0}\right)$$ where I is the measured intensity of the quake and I_0 is the known zero-level

earthquake intensity.

Because logarithms return small numbers, realize that a change of just one order on the Richter scale is a tremendous change in the magnitude of a quake.

❖ seismogram

A seismograph is an instrument that measures the details of an earthquake, such as amplitude and duration. A seismogram is the record that the seismograph produces. A seismogram might look something like

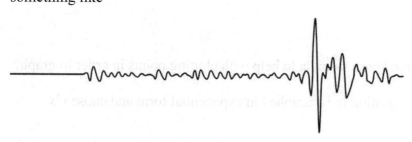

This is a record over time where earlier is to the left and later is on the right end. The individual seismograph would have its usual scales for time (the x-axis) and amplitude (the y-axis).

❖ p-waves

The p-waves are the smaller precursor waves that precede the first large s-wave. (Can you spot them in the seismogram above?)

❖ s-waves

The s-waves are the bigger waves that start after the p-waves.

Richter Scale Magnitude: Test Prep	
The 1989 San Francisco earthquake was a magnitude 7.1, but the 1906 earthquake has been estimated to have been an 8.3-magnitude earthquake based on damage done. What is the ratio of the 1906 earthquake to the 1989?	$M = \log\left(\dfrac{I}{I_0}\right) = \log I - \log I_0$ $M + \log I_0 = \log I$ $10^{M + \log I_0} = 10^{\log I}$ $10^M 10^{\log I_0} = I$

	$I = I_0 \cdot 10^M$
	so
	$\dfrac{I_{1906}}{I_{1989}} = \dfrac{I_0 \cdot 10^{83}}{I_0 \cdot 10^{7.1}}$
	$= 10^{1.2} \approx 15.85$

Try Them

If the intensity of an earthquake is determined to be 40,000 times the reference intensity I_0, what is the magnitude of the quake? Round to the nearest tenth.	4.6
The 1933 Japanese earthquake was magnitude 8.3. How does its intensity compare with the reference intensity?	$I \approx 200,000,000 I_0$

❖ **acidity**

Acidity is a property of a substance (usually a liquid) with more available hydrogen ions than pure water (hence "potential hydrogen" of the pH scale. "H" because that's the abbreviation for hydrogen on the periodic table.). Acids rank between 0 (very acid) up to 7 (neutral) on the pH scale.

❖ **alkalinity**

Alkalinity is a measure of a substances ability to neutralize an acidic solution. Alkalis measure between 14 (very alkali) to 7 (neutral) on the pH scale.

❖ **pH of a solution**

The pH of a solution is measured according to this logarithmic formula:

$pH = - \log[H^+]$ where H^+ is the measured hydrogen ion concentration in a mole of solution.

A mole is a measurement of the *amount* of stuff you have.

pH: Test Prep	
The hydrogen ion concentration, in moles per liter, of a certain hair gel is 8.9×10^{-8}. Find its pH to the nearest hundredth.	$pH = -\log(8.9 \times 10^{-8})$ $= -(\log 8.9 + \log 10^{-8})$ $\approx -(0.949 - 8)$ $= 7.05$

Try Them	
Find the pH of blood if its concentration of hydrogen ions is 4.02×10^{-8}.. Is this blood acidic or basic?	$\sim 7.40,$, basic
Determine the hydrogen ion concentration of a solution with a pH of 2.2.	$\sim 6.31 \times 10^{-3}$ moles per liter

NOW TRY THE MID-CHAPTER 4 QUIZ

4.5: Exponential and Logarithmic Equations

We made a bit of a deal out of inverse functions at the beginning of this chapter and how they are used to solve equations of all sorts. Now we're going to use these particular inverse functions, the exponential and logarithmic, to solve exponential and logarithmic equations. These procedures are just as natural as the skills you already use to solve simpler equations, but, unfortunately it's likely to take quite a bit of practice in order for these exercises to feel as natural.

For my money, I believe this section is the crux of this chapter.

✓ Questions to Ask Your Teacher

Ask your teacher what section this chapter he or she believes is the most important.

Solving Exponential Equations

We'll start off with some of the easier types of exponential and logarithmic equations to solve.

♣ **exponential equation**

An exponential equation is an equation where a variable appears in the exponent, for instance: $3^x = \dfrac{1}{27}$.

To solve such an equation, the idea is the same as it's always been – get x by itself. Now the question is: How does one get the variable out of the exponent? That's new.

One way to solve exponential equations is to equate exponents. Consider our example: $3^x = \dfrac{1}{27}$. Can you see that *both* sides of the equation can be written as a power of 3?

$$3^x = \dfrac{1}{27}$$

$$= \dfrac{1}{3^3}$$

$$= 3^{-3}$$

So $3^x = 3^{-3}$. Now we can use the Equality of Exponents Theorem to equate x and -3. That is, to express the solution: $x = -3$. (And you can check this solution in the original equation as usual.)

Equality of Exponents

If $a^x = a^y$, then $x = y$.

★ **For Fun**

Check out an example, see what you think: $3^5 = 3^{3+2}$. Does $5 = 3 + 2$?

Equality of Exponents Theorem: Test Prep	
Solve: $2^{3x} = 64$	$2^{3x} = 64$
	$2^{3x} = 2^6$

	$3x = 6$
	$x = 2$
Try Them	
Solve: $\left(\dfrac{1}{3}\right)^{x+4} = 81$	$x = -8$ no solution
Solve: $2^{-x} = -4$	

★ For Fun

Check the answers in the first two examples in the preceding Test Prep block.

Now, I promise you will not always be able to write both sides of the equation in terms of the same base.

In those cases, you will need to use the inverse function, same base *logarithm* both sides, in order to

"rescue" the variable out of the exponent. For example:

$2^{x+1} = 44$

$\ln 2^{x+1} = \ln 44$

$(x+1)\ln 2 = \ln 44$

$x + 1 = \dfrac{\ln 44}{\ln 2}$

$x = -1 + \dfrac{\ln 44}{\ln 2}$

$x \approx 4.4594$

How do we "liberate" the variable from its exponentiation onto that base 2? The inverse function

(applied on both sides of the equation) is the logarithm base 2. Remember: $\log_b b^x = x$. See how the

logarithm can rescue that x from the exponent? You can also use the power law of logarithms to rescue

the x: $\ln b^x = x \ln b$, for instance.

"log" is a function

Please do not think for one instant that you multiply both sides of your equation by some mysterious number named "log". You are performing the same function on both sides of the equation – same as if you were taking the square root of both sides. Again the log in $\log x$ is much like the f in $f(x)$. It's a function; this is functional notation.

Exponential Equations: Test Prep	
Solve: $4e^{3x+1} = 10$. Round your answer to 3 decimal places.	$4e^{3x+1} = 10$ $e^{3x+1} = \dfrac{10}{4} = \dfrac{5}{2}$ $\ln e^{3x+1} = \ln\left(\dfrac{5}{2}\right)$ $3x+1 = \ln\left(\dfrac{5}{2}\right)$ $x = \dfrac{\ln\left(\dfrac{5}{2}\right) - 1}{3}$ $x \approx -0.028$
Try Them	
Solve $e^{2x} - 4e^x + 3 = 0$ by factoring.	$x = 0, \ ln3$
Solve: $e^x - e^{-x} = 30$	$x = \ln(15 + \sqrt{226})$

Solving Logarithmic Equations

Let's try some of the same or corresponding techniques to solve logarithmic equations.

♣ logarithmic equation

Similar to an exponential equation, a logarithmic equation is an equation where a variable is inside a logarithmic function: $\log(2x - 1) = 3$.

Now, not only do you have to know how to use *exponentiation* as the inverse function to solve logarithmic equations, you have to sometimes know how best to use all the properties of logarithms in order to *condense* logarithms so that you can exponentiate meaningfully.

For example:

$$1 + \log(3x - 1) = \log(2x + 1)$$

$$1 = \log(2x + 1) - \log(3x - 1)$$

$$1 = \log\left(\frac{2x + 1}{3x - 1}\right)$$

$$10^1 = 10^{\log\left(\frac{2x+1}{3x-1}\right)}$$

$$10 = \frac{2x + 1}{3x - 1}$$

$$10(3x - 1) = 2x + 1$$

$$30x - 10 = 2x + 1$$

$$28x = 11$$

$$x = \frac{11}{28}$$

I had to use lots of steps to solve this problem. Can you follow carefully and describe what I did at each step? Do you recognize when I used exponentiation? properties of logs? plain ol' algebra?

Logarithmic Equations: Test Prep	
Solve for a : $\log a = 2 \log b$	$\log a = 2 \log b$ $10^{\log a} = 10^{\log b^2}$ $a = b^2$
Try Them	
Solve: $\log_3 124 = \log_3 4 - 2\log_3 x$	$x = \dfrac{\sqrt{31}}{31}$

Solve: $5^{2x+2} = 3^{5x-1}$. Round your answer to 3 decimal places.	$x \approx 1898$

4.6: Exponential Growth and Decay

The exponential function has some very important features. One is that it handily describes lots of systems – systems that grow and decay – from populations of bacteria to money in the bank.

Exponential Growth and Decay

The form of the exponential growth/decay functions is well known. You can learn how to put one together for yourself!

First of all, instead of x and y variables, the independent variable in an exponential growth or decay function is usually t for time. This function will model how systems change over time. Secondly, we name the function N for the *number* left in the population. Lastly, N_0 is going to be the *initial* (when $t = 0$) population, or simply "number of things."

❖ **growth rate constant**

In the exponential function, $N(t) = N_0 e^{kt}$, the k is the growth constant. Can you imagine that if it's bigger, then the effect time, t, is larger and vice versa if the growth constant is smaller?

❖ **exponential growth function**

If k is positive, then N is increasing, and that means $N(t) = N_0 e^{kt}$ describes a system that is *growing* exponentially.

❖ **exponential decay function**

Contrarily, if k is negative, then $N(t)$ is an exponential decay function because it describes a shrinking population.

❖ half-life

The half-life is the value of t (time, remember) when the population $N(t)$ is exactly half of what it started out being. This applies, obviously?, to *decay* functions. "Half-life" is usually reserved for discussions of radioactive decay in particular.

Carbon Dating

In living tissue, the ratio of the carbon isotopes-14 and-12 stay very constant. Once the tissue dies, the carbon-14, which is slightly radioactive with a half-life of 5730 years, begins to decay. That decay is modeled mathematically by the function $P(t) = 0.5^{\frac{t}{5730}}$.

★ For Fun

First of all, can you see how the half-life and the growth constant are related?

Second of all, isn't this supposed to be a *decay* function?! How come the exponent isn't negative? Well, isn't 0.5 equal to 2^{-1}? Could you rewrite $P(t)$ so that it had a negative exponent now?

Thirdly, how come this exponential decay function isn't in terms of e?

❖ carbon dating

Carbon dating is the Nobel prize-winning process that used the amount of the carbon-14 isotope present in an old bone to estimate its age.

Exponential Growth and Decay Functions: Test Prep	
At the start of an experiment, there are 100 bacteria. If the bacteria follow an exponential growth pattern with t measured in hours and rate constant $k = 0.02$, how long will it take for the population to double?	The growth equation is $N = 100e^{0.02t}$ because k is 0.02 and the initial population of bacteria is 100. When the population is doubled, we have $200 = 100e^{002t}$

	$2 = e^{0.02t}$
	$\ln 2 = 0.02t$
	$t = \dfrac{\ln 2}{0.02} \approx 35hr.$

Try Them

| Suppose that at the start of an experiment there are 8,000 bacteria. A growth inhibitor and a lethal pathogen are introduced into the colony. After two hours 1,000 bacteria are dead. If the death rates are exponential, how long will it take for the population to drop below 5,000? Round your answers to the nearest tenth. | $t \approx 7.1hr.$ |
| Say a fossil is found that has 35% carbon 14 compared to the living sample. How old is the fossil? | $\sim 8{,}680$ years old |

Compound Interest Formulas

How money makes money is a mathematical proposition. It's also exponential.

❖ interest

Interest is paid for using money. If you loan money, you earn interest. If you borrow it, unfortunately you pay interest.

❖ simple interest

Simple interest is easy to calculate. It is the amount of money earned when the percentage earned and the compounding term are constant. It's like adding 1% every 3 months (at which point the process can start all over again on the new amount). The equation for simple interest is $I = Prt$ where I is the interested earned, P is the principal invested, r is the rate of return per investment term (usually expressed as a percentage), and t is the number of terms.

♣ principal

Principal is the amount of money invested, or the amount of money to be used (or loaned) in order for it to earn interest.

♣ compound interest

Compounding interest is exactly the act of starting over again after the period of simple interest is over.

The equation for compound interest looks like this: $A = P\left(1 + \dfrac{r}{n}\right)^{nt}$ A is the total amount in the account; P is the principal invested; r is the annual rate; n is times per year that the interest will be compounded; t is the amount of time, in years, that the principal has been invested. (Do you see that this is an exponential function? The variable t is in the exponent, and all the other parameters are fixed in a given investment scenario.)

Compound Interest Formula: Test Prep	
An amount of $1,500.00 is deposited in a bank paying an annual interest rate of 4.3%, compounded quarterly. What is the balance after 6 years?	$P = 1500$ $r = 0.043$ $n = 4$ $t = 6$ $A = 1500\left(1 + \dfrac{0.043}{4}\right)^{4 \cdot 6} \approx \$1938.84$

Try Them	
The first credit card that you got charges 12.49 % interest to its customers and compounds that interest monthly. Within one day of getting your first credit card, you max out the credit limit by spending $1,200.00. If you do not buy anything else on the card and you do not make any payments, how much money would you owe the company after 6 months?	$1276.92
If 12,500 is invested at an annual interest rate of 8% for 10 years, find the balance if the interest is compounded hourly.	$27,819.16

★ compound continuously

For interest to compound continuously – that is, not stop to count up every year or month or other period then start over again – some mathematical manipulation is done to produce the Continuous Compounding Interest Formula: $A = Pe^{rt}$. Now the function is clearly the exponential growth function.

Continuous Compounding Interest Formula: Test Prep	
If I invest n dollars in a bank account that compounds 5% interest continuously, how long will it take that account 0.05 to double in value?	$2n = ne^{0.05t}$ $t = \dfrac{\ln 2}{0.05} \approx 14\,yr$

Try Them	
$ 10,000 earning 5% interest, compounded continuously, will be worth how much after one year?	$ 10.512.71
Find the balance if$12,500 is invested at an annual interest rate of 8% for 10 year and the interest is compounded continuously.	$27,819.26

Restricted Growth Models

The logistic model is a growth model, like the exponential growth model, but it takes into account limited resources.

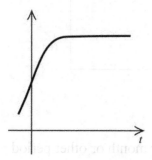

There is a portion this curve which proceeds like an exponential curve, but then the curve flattens out when the limits of the resources supporting the population are approached.

The equation representing a logistic system looks like this: $P(t) = \dfrac{c}{1 + ae^{-bt}}$. This is still an exponential function – the variable is still in the exponent some place.

❖ **logistic model**

The logistic model is also called a restricted growth model because is takes into account the limited resources available to a growing system.

❖ **carrying capacity**

The parameter c above is the carrying capacity. c is the maximum population that a system can carry.

❖ **growth rate constant**

b, a positive real number, is the growth rate constant in a logistic model. (You might notice that if b is always positive, then $-b$ is always negative in the logistic model.)

❖ **initial population**

P_0 is the initial population. It's related to the parameter a in the logistic model by the equation:

$$a = \frac{c - P_0}{P_0}.$$

Logistic Model: Test Prep	
Given the logistic population model $N = \dfrac{300}{1 + 186e^{-052t}}$, estimate, to one decimal place, the t when N will reach 170.	$140 = \dfrac{300}{1 + 186e^{-0.52t}}$ $1 + 186e^{-0.52t} = \dfrac{300}{140}$ $e^{-0.52t} = \dfrac{\frac{300}{140} - 1}{186}$ $t = \dfrac{1}{-0.52} \ln \dfrac{\frac{300}{140} - 1}{186} \approx 9.8$
Try Them	
Sketch the logistic function $P(t) = \dfrac{100}{1 + 25e^{-009t}}$	

Find the initial population for the following logistic model: $P(t) = \dfrac{320}{1+15e^{-0.12t}}$	20

4.7: Modeling Data With Exponential and Logarithmic Functions

Now, we can consider models both exponential and logistic!

Analyzing Scatter Plots

Can you imagine a scatter plot of points that might fit an exponential (or logistic) function?

Which one these do you think would be a better candidate for an exponential model?

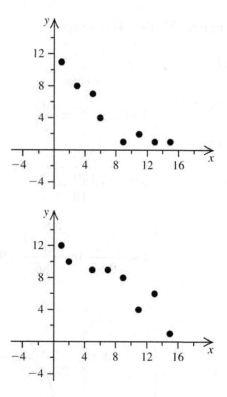

❖ **concave upward**

Technically, a graph all whose tangents lie below said graph is concave upward. You can just remember "u" for "upward" and remember that *curved* curves that open up are concave upward.

❖ **concave downwards**

Likewise, although technically concave downward means that all the tangents of a curve are above said curve, a curve that *curves* downward is concave down.

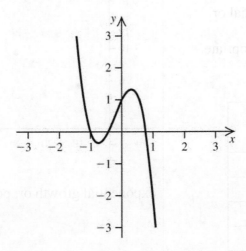

Can you pick out the regions of this graph that are concave UP or concave DOWN?

Modeling Data

Check out your spreadsheet or calculator regression options. Besides the linear, quadratic, cubic, and quartic models we've looked at, do you have exponential and logarithmic options? Now you can take those into consideration and continue to use your experience with regression parameters to make good choices of models:

1. Construct your scatter plot.

2. Chose a function to model your data.

3. Find the coefficient of determination, r^2

4. If necessary, chose a different function and find its r^2. Find the function (the one with the largest r^2) that best models your data.

Modeling Process: Test Prep

Construct a scatter plot for the following data and surmise which exponential or logarithmic model might be appropriate.

date (year AD)	population (millions)
1790	3.929
1800	5.308
1810	7.240
1820	9.638
1830	12.866
1840	17.069

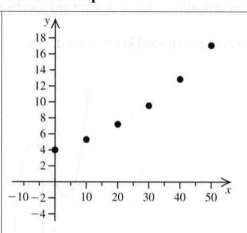

exponential growth or, possibly, logistic

Try Them

Find an exponential model for the following data:

x	y
10	6.8
12	6.9
14	15
16	16.1
18	50
19	20

$f(x) = 1.2^x$

| Find a logarithmic model for the following data: | $f(x) = 60.09 + 3.36 \ln x$ |

x	y
8	67.1
10	68.4
14	69
16	69.4

Finding a Logistic Growth Model

If you're lucky and your calculator (or spreadsheet application) includes functions for modeling data

logistically, you can include that in your repertoire also... if, of course, your data warrants a logistic model:

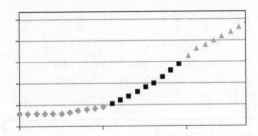

Final Fun: Matching

FIND THE BEST MATCH. GOOD LUCK.

LOGISTIC FUNCTION

$$\log_b M^p = p\log_b M$$

$$P(t) = 0.5^{\frac{t}{5730}}$$

$$P(t) = \frac{c}{1 + ae^{-bt}}$$

SEISNOGRAN

EXPONENTIAL DECAY

CHANGE OF BASE FORNULA

CONPOUNDING INTEREST FORNULA

$$A = P\left(1 + \frac{r}{n}\right)^{nt}$$

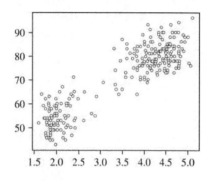

$f^{-1}(x)$

$f(x) = a^x$

LOGARITHMIC FUNCTION

$\log_4 \dfrac{1}{16}$

COMPOSITION OF INVERSES

POWER PROPERTY OF LOGARITHMS

NATURAL BASE

$- \log[H^+]$

$3^4 = 81$

EXPONENTIAL FUNCTION

pH

$f(x) = \log_b x$

-2

INVERSE FUNCTION

$\log_b x = \dfrac{\ln x}{\ln b}$

$(f^{-1} \circ f)(x)$

SCATTER PLOT

e

$\log_3 81 = 4$

NOW TRY THESE THE CHAPTER 4 REVIEW EXERCISES

AND THE

CHAPTER 4 PRACTICE TEST

STAY UP ON YOUR GAME – TRY THE CUMULATIVE REVIEW EXERCISES

Chapter 5: Trigonometric Functions

Word Search

```
E  L  G  N  A  T  U  G  T  S  C  R  A  Y  N
T  H  A  G  R  N  H  N  N  O  M  M  Y  I  O
N  R  D  C  I  E  E  G  T  I  P  D  Y  P  I
A  E  I  T  O  G  F  E  I  L  P  R  I  S  S
C  N  N  G  N  R  R  E  I  A  A  M  X  E  S
E  I  I  A  O  M  P  T  R  T  R  E  A  U  E
S  S  T  P  I  N  U  I  N  E  T  T  Y  D  R
O  O  I  N  H  D  O  E  C  R  N  U  S  O  P
C  C  A  T  E  A  M  M  E  E  Y  C  D  P  E
G  L  L  V  S  E  S  V  E  C  R  A  E  E  D
S  U  P  P  L  E  M  E  N  T  A  R  Y  R  T
I  B  Y  P  P  L  A  N  I  M  R  E  T  I  H
N  G  M  H  A  R  M  O  N  I  C  Y  J  O  G
E  O  L  A  T  N  A  R  D  A  U  Q  P  D  I
```

ACUTE	DMS	REFERENCE
AMPLITUDE	ELEVATION	RIGHT
ANGLE	HARMONIC	SECANT
ARC	INITIAL	SINE
COMPLEMENTARY	PERIOD	STRAIGHT
COSECANT	PHASE	SUPPLEMENTARY
COSINE	PSEUDOPERIOD	TANGENT

255

5.1: Angles and Arcs

This section is a lot about vocabulary and pictures and visualizing angles. There will be some "figuring" to do, but most of this material is the basic stuff and it's relatively easy. Trigonometry will be getting harder, so you might as well learn these words and basics and not have to worry about them any more.

❖ **half-line**

Any point on a line (which grows infinitely long in both directions) breaks said line into two half-lines.

★ **For Fun**

So how long is a half-line? How long is a line?

❖ **ray**

A half-line, considered by itself, is a ray. A ray that passes through the points P and A can be symbolized by $\overrightarrow{PA}$.

❖ **endpoint**

P, for instance, is the endpoint of the ray that starts at P and passes through the point A like $\overrightarrow{PA}$ above.

❖ **angle**

An angle is an expression of the "relationship" between some initial position of a ray and the position it ends at after rotating the ray while keeping its endpoint stationary.

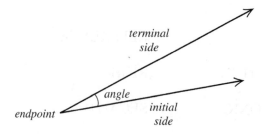

❖ **initial side**

The initial position of the ray becomes the initial side of the two sides required to specify an angle.

❖ **terminal side**

The final position of the ray is the terminal side of the angle.

★ For Fun

Do angles have a "direction" associated with them?

♣ vertex

The fixed endpoint about which the ray was rotated becomes the vertex of the angle that was formed.

♣ positive angle

Positive angles result in rotating a ray counterclockwise from initial position to terminal.

♣ negative angle

Negative angles result from a rotation in a clockwise direction.

Degree Measure

Now that we have a careful notion what an angle is, we can start measuring angles – so that we can compare them to each other. Degree measure is a popular and long-recognized measure for angles. It's interesting to ponder the origins of degrees. Have you ever wondered "Why 360 degrees?"

♣ measure

The measure of an angle is not like measuring distance from one point to another. Angles are a relationship between the two positions of a ray. The measure of an angle is an expression of that relationship – how close is it? how far apart?

♣ degree

It's agreed that in order to rotate a ray all the way around back to its initial position, so that the initial and terminal sides coincide, takes 360 degrees. So one degree is $\frac{1}{360}$ of that entire rotation.

♣ straight angle

An angle of 180 degrees is a straight angle:

❖ **right angle**

90 degrees makes a right angle, frequently illustrated by a small box inside the vertex:

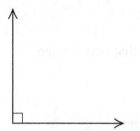

❖ **acute angle**

Any angle less than 90 degrees is an acute angle.

❖ **obtuse angle**

An angle between 90 and 180 degrees is an obtuse angle.

❖ **standard position**

If the angle is situated on a 2-dimensional Cartesian coordinate system with the vertex coincident with the origin and the initial side at the positive x-axis, then the angle is in standard position:

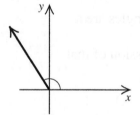

❖ **complementary angles**

Complementary is a relationship between two angles. Two angles are complements of each other if they are positive and their sum is 90 degrees.

❖ **supplementary angles**

Two angles are supplementary if they are both positive and their sum is 180 degrees.

❖ quadrantal angle

A quadrantal angle is an angle whose terminal side coincides with one of the axes on the Cartesian plane. A few of such angles would be $90°$, $180°$, $270°$, $-90°$, and $0°$, for example. Are you familiar with the position of the terminal side of all of these angles?

❖ coterminal angles

Angles in standard position all share their initial side, right? Angles that also share their terminal sides are called coterminal angles. Every angle is coterminal with itself, I suppose, but that's not very interesting. In fact, every angle has an infinite number of possible coterminal angles. Here are a couple of ways to create coterminal angles:

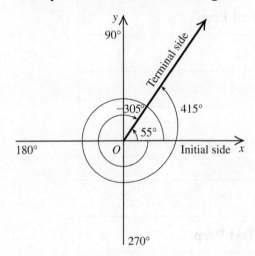

Coterminal angles all differ by $360°$ like so: when k is an integer, $x° + k(360°) = $ new angle coterminal to x.

❖ decimal degree method

Of course, angles can happen to fall between two degrees and so require a measurement that is a fraction of a degree. If that fraction is expressed as a decimal, like $97.35°$ for instance, then the decimal degree measurement method is being used.

♣ DMS (degree, minute, second) method

The DMS method breaks a degree into parts: minutes and seconds. Similar to clock measurement time, there are 60 minutes in a degree and 60 seconds in a minute. DMS angles might look something like this: $102°34'294''$

★ For Fun

Don't the hands on an analog clock make angles with each other after all?

★ For Fun

Could you convert the two previous examples into the same measurement specified in the other method?

Complementary and Supplementary Angles: Test Prep	
Find the complement of $52°$.	$90° - 52° = 38°$
Try Them	
Find the supplement of $112°$.	$68°$
Find the complement of $112°$.	undefined

Classify Angles and Find a Coterminal Angle: Test Prep	
Find one negative and one positive coterminal angle of $14°$.	$14° + 360° = 374°$ $14° - 360° = -346°$
Try Them	
Give an example of an obtuse angle.	(answers vary; any angle between $90°$ and $180°$)
Give an example of a quadrantal angle.	(answers vary; $0°, 90°, 180°, 270°, 360°$ or any angle coterminal with one of these.)

Radian Measure

Radians are also a common measurement of an angle. Be sure you know how to use your calculator in both radian and degree modes.

❖ central angle

A central angle is an angle formed by rotating a radius of a circle some portion of that circle:

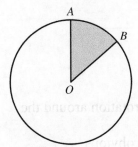

❖ arc

An arc is the portion of the circumference of a circle. A central angle naturally "goes with" a certain arc, doesn't it?

❖ radian

One radian is the measure (in radians!) of the central angle subtended by an arc length of exactly r where the circle has a radius of r. In other words the length of the arc is exactly the same as the radius of the circle.

Radian Measure and Radian-Degree Conversion: Test Prep	
Convert $\dfrac{3\pi}{8}$ to exact degree measurement.	$\dfrac{3\pi}{8} \cdot \dfrac{180^o}{\pi} = \left(\dfrac{3 \cdot 180}{8}\right)^\circ = 67.5^\circ$
Try Them	
Find the radian measure of -149° to the nearest tenth of a degree.	-2.6 radians

Sketch $-\dfrac{5\pi}{6}$. Name the quadrant where its terminal side lies.	
	quadrant III

Arcs and Arc length

It seems like the size of the circle, its radius, is obviously related to the distance and rotation around the circle, so it should be no surprise that there is a well-used formula that expresses this obvious relationship, and it comes right out of the definition of a radian in the last subsection. Bigger radius, bigger arc, same one radian, and likewise for smaller circles.

Does this sound like a proportionality to you? It is, and the proportion for any central angle looks just like this: $\theta = \dfrac{s}{r}$ where θ is the radian measure of the central angle, s is the length of the arc and r is the length of the radius.

★ For Fun

Ponder this. If $\dfrac{s}{r}$ is a length divided by a length, then what are the units of radians?

Arc Length Formula: Test Prep	
Given a circle with radius 8 cm., find the arc length subtended by a central angle of $\dfrac{2\pi}{3}$.	$\dfrac{2\pi}{3} = \dfrac{s}{8cm}$ $s = 8cm \cdot \dfrac{2\pi}{3}$ $= \dfrac{16\pi}{3} \, cm$

Try Them	
Measured on a circle of radius 5.78 inches the length of an arc is 1.74 inches. What angle does the arc subtend at the center of the circle?	approx. .3 (radians)
A round sign known to be 3 meters in diameter subtends at the eye an angle of $0.5°$. How far away is it?	about 344 meters

Linear and Angular Speed

We're used to linear speed – it's the mph your car ... or your run ... or walk ... or bike. But circular things, like tires, spin at different rates don't they? Angular speed, which measures how fast an *angle* changes, is what describes how fast something is spinning.

★ For Fun

Can you imagine a relationship between angular speed and linear speed when it comes to your car or your bike?

❖ linear speed

If a point moves on a circular path that has a radius r at some constant rate of θ radians per unit time t, the linear speed that point – its movement through space, not around and around – is going to depend on the amount of arc that has been "laid down on the road." Linear speed v is related to arc length s like so:

$$v = \frac{s}{t}$$

★ **For Fun**

Check out the units for v above. Would this formula produce expected measurement of linear velocity like mph?

♣ **angular speed**

The point moving a circular path mentioned above would have an angular speed – having to do with its 'round and 'round movement – called omega: $\omega = \dfrac{\theta}{t}$. Again θ is measured in radians. Pay attention to the units of angular speed, too.

Linear Speed – Angular Speed Relationship: Test Prep	
The radius of the magnetic disk in a 3.5-inch diskette is 1.68 inches. Find the linear speed of a point on the circumference of the disk if it is rotating at an angular speed of 360 revolutions per minute.	$\dfrac{360rev}{min} = \dfrac{2\pi \cdot 360}{min} = \dfrac{720\pi}{min}$ $\dfrac{\theta}{min} = \dfrac{s}{r \cdot min}$ $\dfrac{s}{r \cdot min} = \dfrac{(1.68in)(720\pi)}{min} = 1209.6\pi\,\dfrac{in}{min}$ $\approx 3800\,\dfrac{in}{min}$
Try Them	
A car going 65 mph has a wheel of diameter 2.5 ft. Find the number of revolutions per minute.	approx. 728 revolutions per minute
Find the linear velocity of a point on the end of a ceiling fan blade when the fan rotates 30 times per minute and has a radius of 2 feet.	approx. 4.28 mph

5.2: Right Triangle Trigonometry

Now we begin trigonometry in earnest. There are formulas and identities here that you need to memorize and practice with. The sooner these basic trigonometric functions and identities become "second nature" to you, the better you'll be at facing the more challenging material to come.

The Six Trigonometric Functions

I don't really have any idea who first came up with the idea of exploring the *ratios* of the sides of right triangles to each other, but someone did and trigonometry sprang from that notion, best I can tell. There are only six different ratios that you can form with the three sides of a triangle, and each one is now "enshrined" as a trigonometric function. Turns out they're really useful functions, too!

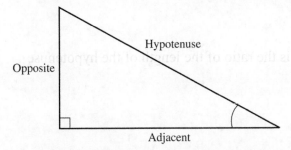

Develop the ability to pick out the three sides with respect to either acute angle of a right triangle that's in any sort of orientation. Practice on these. You'll need to find the right angle and pick an acute angle to work with for each.

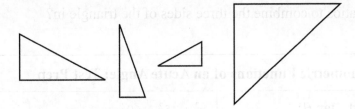

❖ **sine**

The sine of an angle, $\sin \theta$, is the ratio of the length of the side opposite the acute angle to the length of the hypotenuse.

❖ cosine

The cosine of an angle, $\cos\theta$, is the ratio of the length of the side adjacent to the acute angle to the length of the hypotenuse.

❖ tangent

The tangent of an angle, $\tan\theta$, is the ratio of the length of the side opposite the acute angle to the length of the adjacent side of the same acute angle, θ.

❖ cotangent

The reciprocal of the tangent, the cotangent of an angle, $\cot\theta$, is the ratio of the length of the side adjacent the acute angle to the length of the opposite side.

❖ secant

The reciprocal of the cosine, the secant of an angle, $\sec\theta$, is the ratio of the length of the hypotenuse to the length of the side adjacent to the acute angle, θ.

❖ cosecant

The reciprocal of the sine, the cosecant of an angle, $\csc\theta$, is the ratio of the length of the hypotenuse to the length of the side opposite θ.

★ For Fun

Can you find any other ratios to combine the three sides of the triangle in?

Trigonometric Functions of an Acute Angle: Test Prep	
Find $\sin\theta$, $\cos\theta$, and $\tan\theta$: x θ $y = 45$ m $r = 60$ m	$x^2 + 45^2 = 60^2$ $x^2 = 3600 - 2025 = 1575$ $x = 15\sqrt{7} = adjacent$ $\sin\theta = \dfrac{45}{60} = \dfrac{3}{4}$

$$\cos \theta = \frac{15\sqrt{7}}{60} = \frac{\sqrt{7}}{4}$$

$$\tan \theta = \frac{\sin \theta}{\cos \theta} = \frac{3}{\sqrt{7}}$$

Try Them

18 in

8 in

Find $\tan \gamma$ and $\cot \gamma$.

$$\tan \gamma = \frac{4}{9}$$

$$\cot \gamma = \frac{9}{4}$$

Find the values of the six trig functions of

θ:

3

9

θ

$$\sin \theta = \frac{1}{3}$$

$$\cos \theta = \frac{2\sqrt{2}}{3}$$

$$\tan \theta = \frac{\sqrt{2}}{4}$$

$$\sec \theta = \frac{3\sqrt{2}}{4}$$

$$\csc \theta = 3$$

$$\cot \theta = 2\sqrt{2}$$

Trigonometric Functions of Special Angles

Angles that are very commonly encountered, $45°$, $30°$, and $60°$, have trig function values that you

can derive (see pages 444 and 445 in your textbook) and that you can memorize for easy access. Here

are these angles, expressed in radians (can you match degree values with radian values?), and the sine,

cosine, and tangent values of these special angles:

radians	$\sin \theta$	$\cos \theta$	$\tan \theta$
0	0	1	0
$\dfrac{\pi}{6}$	$\dfrac{1}{2}$	$\dfrac{\sqrt{3}}{2}$	$\dfrac{\sqrt{3}}{3}$
$\dfrac{\pi}{4}$	$\dfrac{\sqrt{2}}{2}$	$\dfrac{\sqrt{2}}{2}$	1
$\dfrac{\pi}{3}$	$\dfrac{\sqrt{3}}{2}$	$\dfrac{1}{2}$	$\sqrt{3}$

★ For Fun

Why haven't I included cotangent, secant, and cosecant function values in my table? How quickly can you derive them? And simplify them?

♣ reciprocal functions

Cotangent, secant, and cosecant are reciprocal functions of tangent, cosine, and sine respectively. So as long as you can find the reciprocal of a number, you can derive these values quickly from the other three functions.

$$\cot \theta = \frac{1}{\tan \theta}$$

$$\sec \theta = \frac{1}{\cos \theta}$$

$$\csc \theta = \frac{1}{\sin \theta}$$

Trigonometric Functions of Special Angles: Test Prep	
Find $\sec \dfrac{\pi}{4}$.	$\sec \dfrac{\pi}{4} = \dfrac{1}{\cos \dfrac{\pi}{4}} = \dfrac{1}{\dfrac{\sqrt{2}}{2}} = \dfrac{2}{\sqrt{2}}$ $= \sqrt{2}$

Try Them	
Find the tangent and cotangent values of $30°$.	$\tan 30° = \dfrac{\sqrt{3}}{3}$ $\cot 30° = \sqrt{3}$
Find the exact value of $\sin\dfrac{\pi}{3}+\cos^2\dfrac{\pi}{3}$.	$\dfrac{2\sqrt{3}+1}{4}$

Applications Involving Right Triangles

As soon as someone realized that knowing the angle gave you information about the length of the sides of the corresponding right triangle, the application of trigonometry to measurement of distances started. That was a momentous moment for mankind, really.

❖ **angle of elevation**

The angle of elevation is the acute angle measured from a horizontal line up to a line of sight.

❖ **angle of depression**

The angle of depression is the acute angle measured from a horizontal line down to a line of sight. The angle of depression is usually not included inside the right triangle under consideration; however, you can find the complementary angle that *is* included easily.

★ **For Fun**

If you've studied geometry, then you can also use the theory of opposite interior angles to find an included acute angle of the right triangle under consideration.

Draw a picture for these trigonometry applications if one is not provided for you!

Angle of Elevation or Depression: Test Prep	
A person standing 100 feet from a building measures an angle of elevation to the top of	$\tan 75.7° = \dfrac{height}{100\,ft}$ $height = 100\tan 75.7°\,ft \approx 392\,ft$

the building of 75.7°. Find the height of the building to the nearest foot.	
Try Them	
A bike rider peddles uphill at a constant rise of 4.5° for 1.14 miles. What is the biker's increase in altitude to the nearest tenth of a mile?	0.1 mi.
Duane is anchored in ten feet of water. He sights a turtle at an angle of depression of 30 degrees. How far away is the turtle?	20 ft.

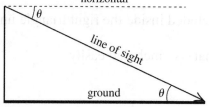

5.3: Trigonometric Functions of Any Angle

If we can rethink how we define "opposite" and "adjacent," then we can evaluate trig functions of any

angle – after all, it'd be a little dismal if we could only talk about acute angles in all of trigonometry.

Trigonometric Functions of Any Angle

Let's put our right triangle in the first quadrant, so:

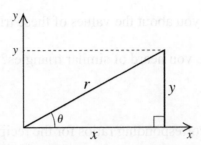

Then we can name, for instance, $\sin\theta = \dfrac{y}{r}$. Well, in fact, we can do it for angles that are not acute:

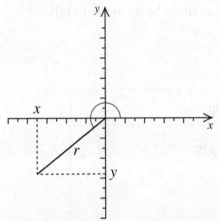

Trigonometric Functions of Any Angle:

1. $\sin\theta = \dfrac{y}{r}$

2. $\cos\theta = \dfrac{x}{r}$

3. $\tan\theta = \dfrac{y}{x}$

Where $r = \sqrt{x^2 + y^2}$ (that's just the distance formula)

Please note that the signs of x and y make all the difference to the value of the trig function now! Preserve the signs and start making conjectures about the resulting signs of the functions depending on the quadrant that the terminal side lies in.

✓ Questions to Ask Your Teacher

If the angle is denoted by a *ray* for its terminal side instead of a segment as I've used, what does the nature of a ratio, or a proportion, tell you about the values of these trig functions and the *(x, y)* point you can pick to construct them? Have you heard of similar triangles?

★ For Fun

And how quickly can you name the corresponding ratios for the reciprocal trig functions?

★ For Fun

Is it possible for r to equal zero? What does $r = 0$ mean? If $r = 0$, can there be an angle to talk about?

Trigonometric Functions of Any Angle: Test Prep	
Find the sine and cosine values of an angle whose terminal side passes through the point $(-2,5)$.	$r = \sqrt{(-2)^2 + 5^2} = \sqrt{29}$ $\sin\theta = \dfrac{y}{r} = \dfrac{5}{\sqrt{29}}$ $\cos\theta = \dfrac{x}{r} = \dfrac{-2}{\sqrt{29}}$
Try Them	
Find the values of the six trig functions of an angle 4 whose terminal side passes through the point $(3, -4)$	$\sin\theta = -\dfrac{4}{5}$ $\cos\theta = \dfrac{3}{5}$ $\tan\theta = -\dfrac{4}{3}$ $\sec\theta = \dfrac{5}{3}$ $\csc\theta = -\dfrac{5}{4}$ $\cot\theta = -\dfrac{3}{4}$

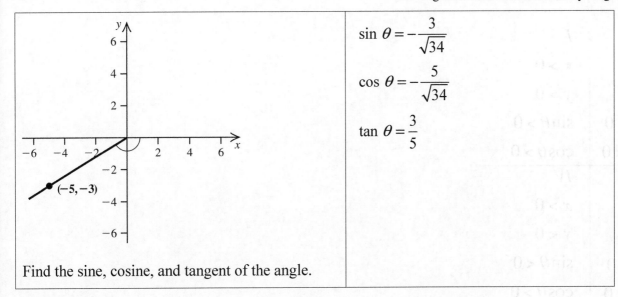

$$\sin \theta = -\frac{3}{\sqrt{34}}$$

$$\cos \theta = -\frac{5}{\sqrt{34}}$$

$$\tan \theta = \frac{3}{5}$$

Find the sine, cosine, and tangent of the angle.

Trigonometric Functions of Quadrantal Angles

The x - y - r definitions of the ratios that are the trig functions hold for quadrantal angles also. However, because the x -coordinate for an angle whose terminal side is at 90 degrees is zero, there are consequences for the resulting functions, especially if you are trying to evaluate a function like secant which is supposed to have x in the denominator. Now we wind up with some functions that are *undefined* for certain angles. Furthermore, those functions that are defined wind up have a very simple value of either *1, -1*, or *0*. These special values are summarized in Table 5.4 on page 456 of your textbook.

Signs of Trigonometric Functions

As I suggested just a page or so ago, you can definitely see a pattern in the signs of values of the trig functions according to the quadrant in which the terminal side lies. This is an important pattern to reason out and be able to use, as easily as possible. But it's not hard! You already know such things as y's are negative in the second and third quadrant and that division of a negative by a positive results in a negative, etc. Remember, too, that distance (where r comes from) is always positive.

II		*I*
$x < 0$		$x > 0$
$y > 0$		$y > 0$
$\sin\theta > 0$		$\sin\theta > 0$
$\cos\theta < 0$		$\cos\theta > 0$
III		*IV*
$x < 0$		$x > 0$
$y < 0$		$y < 0$
$\sin\theta < 0$		$\sin\theta < 0$
$\cos\theta < 0$		$\cos\theta > 0$

★ For Fun

Given the values above, can you ascertain the signs of tangents, cotangents, secants, and cosecants of θ for each quadrant?

The Reference Angle

If you know the proper signs of the trig functions per quadrant and the reference angle corresponding to the angle under consideration, then you can find the value for any angle based upon only acute angle values – and you've reduced memorization to one fourth your original job.

❖ reference angle

The reference angle θ' is the acute angle formed by the shortest and positive rotation of the terminal side to the closest portion of the x-axis. This "shortcut" idea will be revisited constantly during the rest of trigonometry!

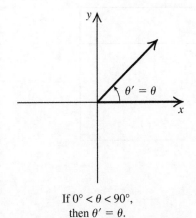

If $0° < \theta < 90°$,
then $\theta' = \theta$.

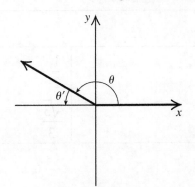

If $90° < \theta < 180°$,
then $\theta' = 180° - \theta$.

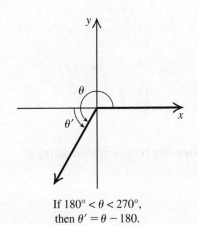

If $180° < \theta < 270°$,
then $\theta' = \theta - 180$.

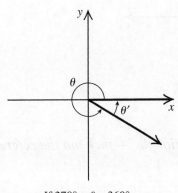

If $270° < \theta < 360°$,
then $\theta' = 360° - \theta$.

★ **For Fun**

Although the above graphics from your textbook label angles according to their *degrees*, I promise you that reference angles are easier and easier if you work in radians. Can you figure out why?

Reference Angles and the Reference Angle Evaluation Procedure: Test Prep	
Use reference angles and trig function values of special angles to evaluate $\cos\left(-\dfrac{2\pi}{3}\right)$.	The angle is a third quadrant angle, so cosines will be have negative values. The reference angle is $\dfrac{\pi}{3}$, so $$\cos\left(-\frac{2\pi}{3}\right) = -\cos\frac{\pi}{3} = -\frac{1}{2}$$

Try Them

Find $\sin\dfrac{5\pi}{6}$.	$\dfrac{1}{2}$
Find $\cot 420°$.	$\dfrac{\sqrt{3}}{3}$

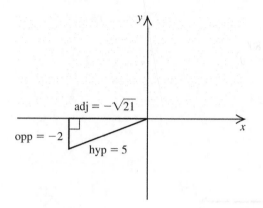

Third quadrant reference "triangle" – pick out the reference angle and explore the trig function values that go with it.

5.4: Trigonometric Functions of Real Numbers

Some applications and some reasoning about trigonometry require that we be able to put any kind of number into the trigonometric function "machines." We've already talked about angles that are larger than 360° and angles that are negative – these are exactly the kind of values that are needed if we want to use trigonometry for all the real numbers.

The Wrapping Function

The wrapping function arises from a need to move from a system of looking at right triangles to a system of using angles to "find your place" on the unit circle. The wrapping function defines the coordinates of points on the circumference of the unit circle using sines and cosines: $(\cos\theta, \sin\theta)$.

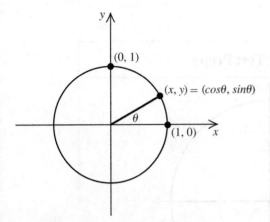

♣ unit circle

A unit circle has a radius of one. Usually *the* unit circle is the circle of radius one that is also centered at

the origin: $x^2 + y^2 = 1$.

Trigonometric Functions of Real Numbers

On the unit circle where $r = 1$, the real numbers, such as x and y, correspond to trigonometric

function values, as follows:

$$\sin \theta = y \quad \csc \theta = \frac{1}{y}$$

$$\cos \theta = x \quad \sec \theta = \frac{1}{x}$$

$$\tan \theta = \frac{y}{x} \quad \cot \theta = \frac{x}{y}$$

The moral this story is that if you know the values of the trig functions, then you can find your place on

the unit circle, and, conversely, the more you know about the unit circle and navigating around its four

quadrants, the better you'll be at reasoning out the values of trig functions. Oh, yeah, you need to

remember about reference angles, too.

✓ Questions to Ask Your Teacher

In fact, if you know enough about how the graph and its quadrants work, you can get by with only

knowing your way around the first quadrant. Everything else can be extended from there. You might

want to see all the relationships between angles and x- and y-values illustrated on the board.

Trigonometric Functions of Real Numbers: Test Prep

Find the point on the unit circle corresponding to an angle of $-\dfrac{2\pi}{3}$.	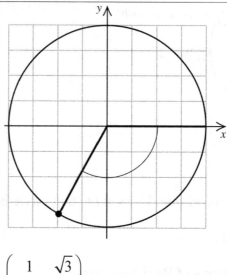 $\left(-\dfrac{1}{2}, -\dfrac{\sqrt{3}}{2}\right)$

Try Them

Find the point on the unit circle corresponding to an angle of 210°.	$\left(-\dfrac{\sqrt{3}}{2}, -\dfrac{1}{2}\right)$
Find the point on the unit circle corresponding to an angle of $-\pi$.	$(-1, 0)$

Properties of Trigonometric Functions of Real Numbers

We can observe a lot about the trig functions by their mapping onto the unit circle.

For instance, if $\sin \theta = y$, then $\sin \theta$ can only take on values that the y-coordinate takes on the

circumference of the unit circle. In other words, $-1 \leq \sin \theta \leq 1$. Similarly, $-1 \leq \cos \theta \leq 1$.

★ For Fun

If $\tan \theta = \dfrac{y}{x}$, what values do you reason the tangent function can take on? (We'll see much more of

this sort reasoning when we start graphing the functions – very soon.)

★ For Fun

Does the unit circle help you understand the domains of the trigonometric functions, too?

❖ periodic

We've already noticed the existence of coterminal angles. Well, if a coterminal angle means that you're back at the same terminal side, it means you're back at the same corresponding point on the unit circle. And that means you're back at the same values of sine and cosine and all the trig functions. This is exactly the nature of a periodic function. It comes back around and back around and back around.

★ For Fun

Does it matter to the periodic nature of a trig function if you go around clockwise or counterclockwise?

❖ period

The period is the shortest positive amount you can add to the angle that brings you back to the same value of the trig function. For sine and cosine, the period is $360°$ or 2π. That means that $\sin\theta = \sin(\theta + 2\pi)$ and likewise for cosines.

★ For Fun

Can you reason out the period for the secant and cosecant functions knowing what you know about sine and cosine periods?

★ For Fun

Challenge yourself. Given what you know about the behavior of the signs of x and y in the different quadrants, can you thoughtfully find the period of the tangent and cotangent functions? Remember, the period should be the smallest positive angle you can add and find the same function value again. By the way, trigonometric functions have even and odd properties just like we found some algebraic functions had way back in Chapter 2. The behavior here is defined the same way: what happens to the function value if you replace x with $-x$ where x here will be an angle!

✓ Questions to Ask Your Teacher

Once again the unit circle is a good way to see why these are true:

EVEN:

$$\cos(-\theta) = \cos\theta$$

ODD:

$$\sin(-\theta) = -\sin\theta$$

★ For Fun

Can you ascertain the even/odd character of the other trig functions using just these two?

Even and Odd Trigonometric Functions: Test Prep	
Is the tangent function even or odd?	$\tan\theta = \dfrac{\sin\theta}{\cos\theta}$ $\tan(-\theta) = \dfrac{\sin(-\theta)}{\cos(-\theta)} = \dfrac{-\sin\theta}{\cos\theta} = -\tan\theta$ So tangent is an odd function.
Try Them	
Is $f(x) = \sin^2 x \cdot \cos x$ an even or odd function?	Even
Is $f(x) = \tan^2 x \cdot \sin x$ an even or odd function?	Odd

Trigonometric Identities

An "identity" in math is an equation that is TRUE no matter what value you assign to the variable(s). A very simple example would be something like $x = x$. Or one you're familiar with:

$x^2 - y^2 = (x - y)(x + y)$. Of course, trigonometric identities will involve the trigonometric functions.

❖ **ratio identities**

The ratio identities, which are sometimes also called quotient identities, are relationships between

tangents, cotangents, sines, and cosines, and they follow immediately from the definitions of the

variables x and y in terms of trig functions.

Ratio Identities

$$\tan \theta = \frac{\sin \theta}{\cos \theta} \left[= \frac{y}{x} \right]$$

$$\cot \theta = \frac{\cos \theta}{\sin \theta} \left[= \frac{x}{y} \right]$$

★ **For Fun**

The relationship between tangent and cotangent could be called a "reciprocal identity," couldn't it?

❖ **Pythagorean identities**

The Pythagorean identities do come from Pythagoras' theory via the distance formula and the equation

for the unit circle. Here are these identities. A derivation of these identities can be followed in your

textbook on page 468.

Pythagorean identities

$$\cos^2 \theta + \sin^2 \theta = 1$$
$$1 + \tan^2 \theta = \sec^2 \theta$$
$$\cot^2 \theta + 1 = \csc^2 \theta$$

✓ **Questions to Ask Your Teacher**

You probably want to see and discuss the derivation of these three identities because then they'll make

more sense and you might even be able to re-derive them in a pinch, meaning you won't have to

memorize them.

Fundamental Trigonometric Identities: Test Prep	
Simplify using trig identities: $1 + \dfrac{1}{\tan^2 \gamma}$	$1 + \dfrac{1}{\tan^2 \gamma} = 1 + \left(\dfrac{1}{\tan \gamma}\right)^2$ $= 1 + \cot^2 \gamma$ $= \csc^2 \gamma$
Try Them	
Use fundamental trigonometric identities to rewrite $\cos^2 x(\sec^2 x - 1)$.	$\sin^2 x$
Simplify using trig identities: $\dfrac{1}{\tan A \csc A \sin A}$.	$\cot A$

NOW TRY THE MID-CHAPTER 5 QUIZ

5.5: Graphs of the Sine and Cosine Functions

After we went to the trouble of "wrapping" the sine and cosine functions around the unit circle, now we're going to "unwrap" them by changing the variables around so that we'll just have

$y = f(x) = \sin x$ and the similar for cosine. Notice that θ is gone; x is now the value of the angle and will change as we travel along the x-axis. We won't be going 'round and 'round any more. We're going to unwrap the angle values along the x-axis in both directions and read the function values up and down on the y-axis.

Graph of the Sine function

Starting with the simplest expression of the sine function to graph, $y = \sin x$, will first graph by using a T-table:

x	y=sin(x)
0	0
π/2	1
π	0
3π/2	-1
2π	0

Resulting in this graph:

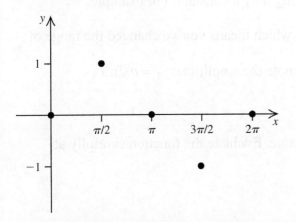

And if I told you that the sine function is a nice curvy, smooth function that makes nice rounded wave shapes, could you connect just these five dots to make a graph that looks like this:

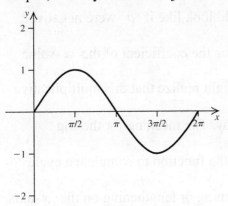

Now, can you see that in the period of 2π that the value of the function (on the *y*-axis) has gone through every value in the range of the sine function and returned to where it started, namely zero? Then, can you imagine *another* period of the sine function added onto either or even *both* ends this graph? The sine function is, after all, periodic, meaning that it just does the same thing over and over and over again – in positive and *negative* angles, which would be negative *x*-values on the other side of the *y*-axis, right?

★ **For Fun**

Try building a T-table and plotting the corresponding five points for the negative angles

$$0, -\frac{\pi}{2}, -\pi, -\frac{3\pi}{2}, -2\pi\,.$$

♣ **amplitude**

You can change the amplitude of a sine function by multiplying it by a constant. For example: $y = 2\sin x$. Amplitude changes affect the *depth* of the wave, which means you've changed the range of the basic sine function. In general, we use the letter a to denote the amplitude: $y = a\sin x$.

★ **For Fun**

Try graphing $y = 2\sin x$ for a couple of periods using a T-table. Evaluate the function carefully at $x = 0, \frac{\pi}{2}, \pi, \frac{3\pi}{2}, 2\pi$ and take note of the range.

★ **For Fun**

Since you already have studied about transformations of graphs in Chapter 2 (and more particularly about reflections of graphs), can you predict what the sine graph would look like if a were negative? The next parameter we meet for these functions is b which appears as the coefficient of the x-value in $y = a\sin bx$. Once again, remembering back to section 2.5, you might realize that this multiplicative parameter on x is going to stretch or compress the graph in some way. The meaning for the trig function is a modification of the *period*, which is "how long" it takes the function to complete a cycle and return to where it started. We should be able to see this as a shortening or lengthening on the x-axis – can you imagine it?

Here are two sine graphs with one shortened and one lengthened period:

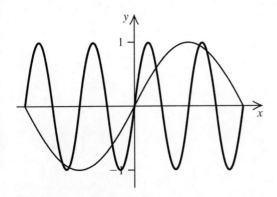

Period of the Sine Function

The period of the sine function $y = a \sin bx$ is given by $\dfrac{2\pi}{b}$.

★ For Fun

Match the given sine functions with their periods. While you're at it, try to visualize the stretching vs. compression along the x-axis.

$y = \sin 3x$ 2

$y = -2 \sin \pi x$ $\dfrac{2\pi}{3}$

$y = \dfrac{1}{3} \sin \dfrac{x}{2}$ 8

$y = -\sin \dfrac{\pi}{4} x$ 4π

In order to graph a sine function whose period has been changed from 2π, you can set up a new

T-table that goes from zero in x to exactly one of the new periods. Then break that period into exactly

4 equal intervals. For instance, here is a T-table and graph of $y = \sin \pi x$, whose period would be

$\dfrac{2\pi}{\pi} = 2$. By extension, I've included two periods of the function on the graph, too.

x	y=sinπx
0	0
½	1
1	0
¾	-1
2	0

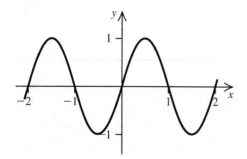

Graph of the Cosine Function

The next easiest graph is $y = \cos x$. Similarly, we can do something by plotting points

x	y=cos(x)
0	1
π/2	0
π	-1
3π/2	0
2π	1

and joining them in a gentle curve:

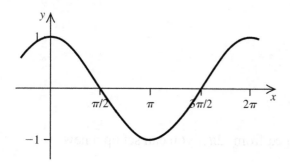

And after you've mastered graphing the cosine curve for any number of periods in any x- direction, then you can experiment with the amplitude – positive and negative and even fractional – and its effect on the range of the function.

Then add in modifications of the period which will, once again, be $\dfrac{2\pi}{b}$. Here is the graph of $y = -\cos \dfrac{3\pi}{2} x$. Can you explain all of its features from parameters in the function?

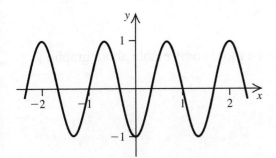

Graphs of the Sine and Cosine Functions: Test Prep

Graph $y = -2 \sin x$; include at least two periods.	This is the basic sine function reflected about the x-axis with an amplitude of 2: 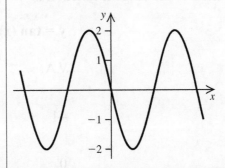

Try Them

Graph $y = \cos\left(\dfrac{\pi x}{2}\right)$	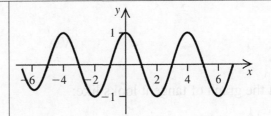
Sketch the graph of $y = \dfrac{1}{2}\sin(3\pi x)$	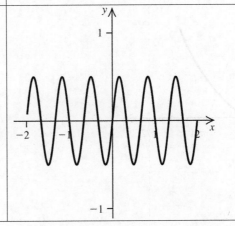

5.6: Graphs of the Other Trigonometric Functions

Now that we have the basic sine and cosine graphs down, we can be more comfortable about graphing the other trig functions. We shall start with tangent.

Graph of the Tangent Function

Since $\tan x = \dfrac{\sin x}{\cos x}$, remember that it is undefined at angles where cosine is equal to zero: …, $-\dfrac{3\pi}{2}, -\dfrac{\pi}{2}, \dfrac{\pi}{2}, \dfrac{3\pi}{2},$…. From graphs of rational functions in Chapter 3, perhaps you remember that where a function is not defined is where you get a vertical asymptote on the graph. And from the this chapter, we know the period of the tangent function is π (or 180 degrees). Let's plot some points:

x	y = tan (x)
$-\dfrac{\pi}{2}$	V.A.
$-\dfrac{\pi}{4}$	−1
0	0
$\dfrac{\pi}{4}$	1
$\dfrac{\pi}{2}$	V.A.

And this is what the graph of tangent looks like:

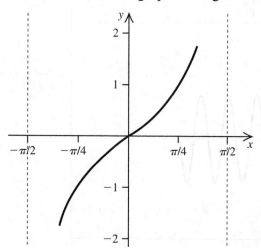

★ For Fun

Now the tangent is distinctly not wave-like, so would it have an amplitude?

Nonetheless, can you decide what effect an amplitude-like coefficient would have on the graph? And if the coefficient were negative?

★ For Fun

Don't forget that I've only graphed one period. Can you see the periodic nature (and the period) of the tangent in this graph? How about the *x*-intercepts – where do they occur? Where are the vertical asymptotes?

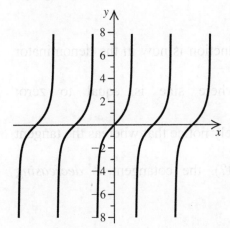

Now consider changes to the period tangent. But first you have to remember that the basic period of the tangent function is not 2π like it is for sines and cosines; the basic period for tangents and cotangents is π.

Period of the Tangent Function

The period the tangent function $y = a \tan bx$ is given by $\dfrac{\pi}{b}$.

Can you accomplish graphing a function like $y = \tan 2x$, whose period is $\dfrac{\pi}{2}$? ? Isn't $\dfrac{\pi}{2}$ simply half of the usual period for tangent? Can you visualize the graph now? Here it is:

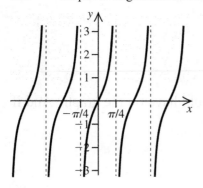

★ For Fun

Can you label on the x-axis where the rest of the vertical asymptotes occur in the graph of tangent above?

Graph of the Cotangent function

The cotangent behaves similarly to the tangent, but since the sine function is now in the denominator $\left(\cot x = \dfrac{\cos x}{\sin x} \right)$ this graph will have vertical asymptotes where sine is equal to zero: $\ldots -2\pi, -\pi, 0, \pi, 2\pi, \cdots$. And if you plot some points, you'll immediately notice that whereas the tangent function (see the graph above) is *increasing* everywhere (see it?), the cotangent is *decreasing* everywhere.

x	y=cot(x)
0	V.A.
π/4	1
π/2	0
3π/4	-1
π	V.A.

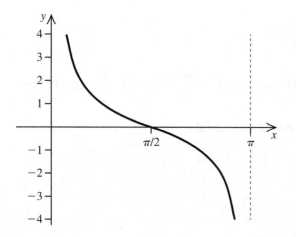

★ **For Fun**

Now can you add in the effects of a and period modifications? The more general form for cotangent is $y = a \cot bx$. Could you find the period of $y = -\cot 4\pi x$ and graph the function? Where would the asymptotes occur?

Graphs of the Tangent and Cotangent Functions: Test Prep	
Graph the cotangent function on the interval $[-2\pi, 2\pi]$.	Repeating the single period that runs from zero to pi on either side of the origin:

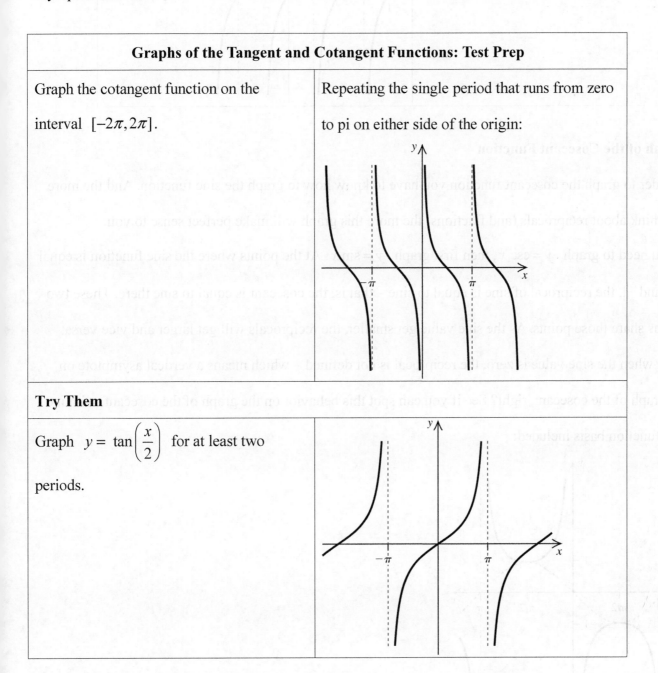

Try Them	
Graph $y = \tan\left(\dfrac{x}{2}\right)$ for at least two periods.	

Sketch the graph of $y = -2\tan(2\pi x)$ for $-1 \le x \le 1$

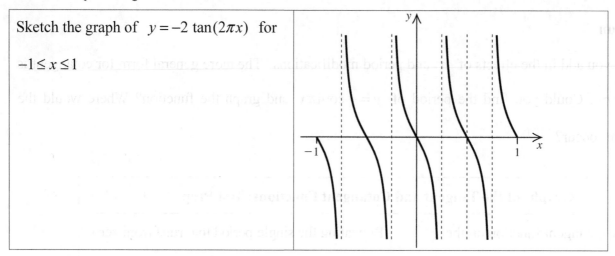

Graph of the Cosecant Function

In order to graph the cosecant function you have to know how to graph the sine function. And the more you think about reciprocals (and fractions) the more this graph will make perfect sense to you.

If you need to graph $y = \csc x$, then first graph $y = \sin x$. At the points where the sine function is equal to 1 and −1, the reciprocal of sine is equal to sine – that is, the cosecant is equal to sine there. These two graphs share those points. As the sine value get smaller, the reciprocals will get larger and vice versa. Then when the sine value is zero, the reciprocal is not defined – which means a vertical asymptote on the graph of the cosecant, right? See if you can spot this behavior on the graph of the cosecant, with its sine function basis included:

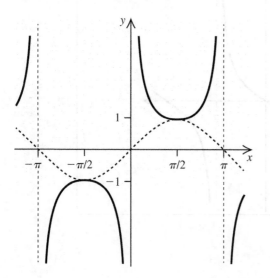

Notice that I've drawn the sine curve portion this graph with a dashed line; like an asymptote, it's not actually part of the graph of the cosecant.

★ For Fun

Now don't forget effects period and amplitude, keeping in mind the way reciprocals work. Here is a graph of $y = 2 \sin 2\pi x$ and $y = 2 \csc 2\pi x$:

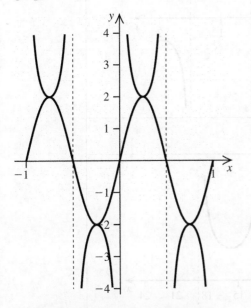

Graph of the Secant Function

The graph of the secant depends on the cosine that it "goes with" too. Here is a secant function without the cosine. Can you see the reciprocal and domain features when the cosine is not there?

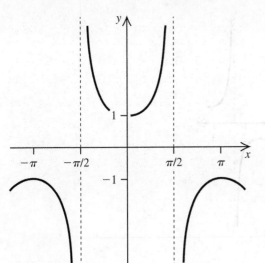

Graphs of the Cosecant and Secant Functions: Test Prep

Sketch the graph of $y = -\sec(\pi x)$ for at least two periods.	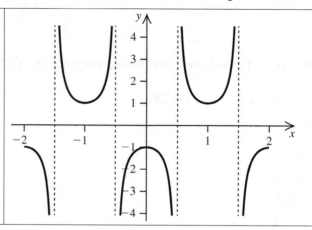

Try Them

Graph $y = 2\csc\dfrac{x}{2}$.	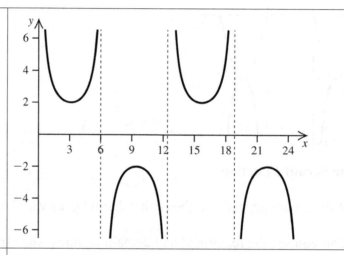
Sketch the graph of $y = \dfrac{\sec 2x}{2}$ for at least two periods.	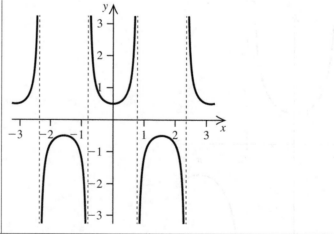

5.7: Graphing Techniques

If you recall from the sections on transformations of graphs, besides changing the shape of a graph with a constant multiplier and reflecting the graph about the x-axis, you also learned how to translate the graph horizontally and vertically. Now we need to look at this for the trigonometric graphs and assign trigonometric meaning to these parameters, too.

Translations of Trigonometric Function

Review what you know about translations, vertical and horizontal, from section 2.5. We've already handled the amplitude change for these trig graphs which stretches or compresses the graph vertically. And changing the period of these functions stretches or compresses the graph horizontally. The amplitude can also reflect the graph about the x-axis (and, frankly, a reflection about the y-axis, although it very much can happen, is not tremendously meaningful for these periodic functions). So translations it is!

In the equation $y = d + a\sin(bx + c)$, the d causes a vertical translation. Here is the graph of

$$y = 2 + \frac{1}{2}\cos \pi x:$$

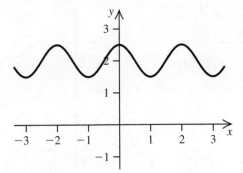

★ For Fun

Can you pick out not only the translation effect but also the amplitude and period effects in the cosine graph above?

❖ phase shift

A phase shift is a *horizontal* translation of a periodic function. It is denoted by the parameter c in

$y = d + a \sin(bx + c)$. (Does this match what you know about horizontal translations of graphs in

general?) Here is a graph of $y = \sin x$ and $y = \sin\left(x + \dfrac{\pi}{4}\right)$:

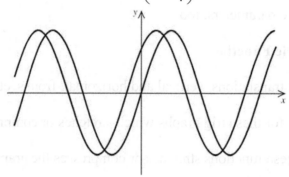

Can you pick out the phase-shifted sine function from the "normal" sine function? Can you tell the

phase shift is into the negative x direction by $\dfrac{\pi}{4}$?

★ For Fun

If you phase shift sine by $\dfrac{\pi}{2}$ in either direction, what trig function does the graph come to resemble?

Translations: Test Prep	
Use translations to graph $y = 2 + \sin\left(x - \dfrac{\pi}{3}\right)$ for at least two periods.	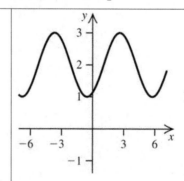
	This graph is the basic sine function moved vertically upward 2 units and horizontally to the right $\dfrac{\pi}{3}$ units.

Try Them

Graph $y = \tan\left(x + \dfrac{\pi}{6}\right)$.

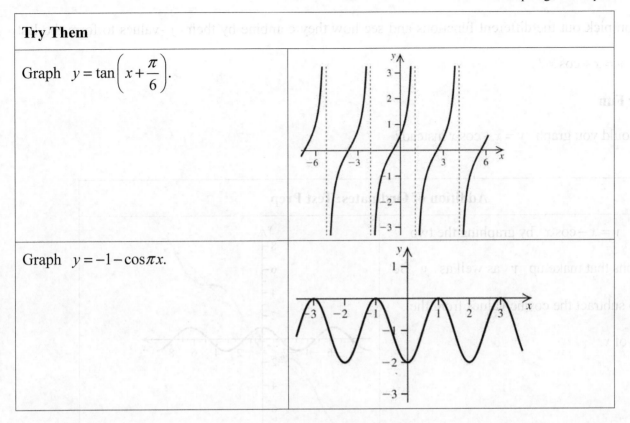

Graph $y = -1 - \cos \pi x$.

Addition of Ordinates

The ordinate was defined way, way back in Chapter P – it is the y-coordinate. If you need to graph a function that is a *combination* of other functions, for example $y = x + \cos x$, you can do so by graphing each one separately (or building a T-table for them separately) and *adding the y-values* for the new graph. Here is a graph of $y = x$, $y = \cos x$, and their combination $y = x + \cos x$:

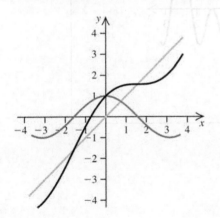

Can you pick out the different functions and see how they combine by their y-values to form the last form: $y = x + \cos x$?

★ **For Fun**

Now could you graph $y = x - \cos x$ instead?

Addition of Ordinates: Test Prep	
Graph $y = x - \cos x$ by graphing the two functions that make up y as well as y Be sure to subtract the cosine values from the values of x.	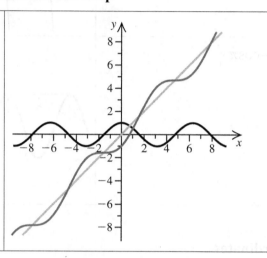
Try Them	
Sketch the graph of $y = \sin \pi x - \cos 2\pi x$.	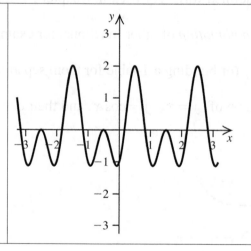

Graph $y = 1 - x + 2\cos x$.

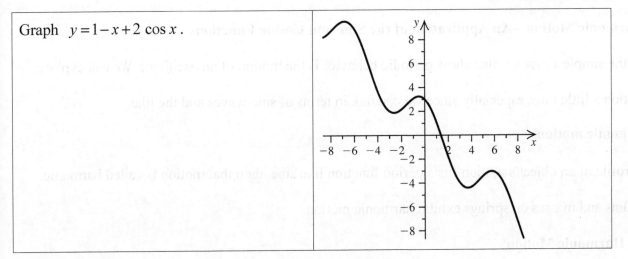

The Damping Factor

We've dealt with an amplitude change with our trig functions, but if there is a multiplicative factor in front of the function that is, itself, dependent on x, then it's a damping factor. Its effect is to change the wave of the trig function as angles get larger and larger.

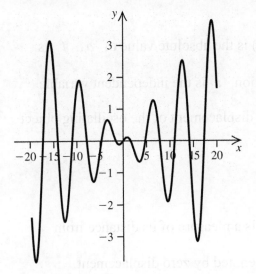

$y = .2x \cos x$

❖ **damping factor**

A factor that depends on x and multiplies an oscillatory function.

5.8: Harmonic Motion – An Application of the Sine and Cosine Functions

One of the simplest systems that show periodic behavior is the motion of an oscillator. We can explore this motion a little now, especially since we can talk in terms of sine waves and the like.

❖ harmonic motion

If the profile of an object's motion fits a period function like sine, then that motion is called harmonic. Pendulums and masses on springs exhibit harmonic motion.

Simple Harmonic Motion

Simple harmonic motion can be described using either a sine or a cosine function.

<u>Simple Harmonic Motion</u>

Simple harmonic motion can be modeled using the functions

$y = a \cos 2\pi f t$

$y = a \sin 2\pi f t$

where the amplitude of the oscillation (its maximum displacement) is the absolute value of a, f is the frequency which is the reciprocal of the period of the trig function, t is the independent variable and measures time, and y the value of the function measures the displacement of the oscillating object from equilibrium.

❖ displacement

The displacement of an object moving in simple harmonic motion is a measure of its distance from equilibrium. Equilibrium is the still, starting point, and is usually denoted by zero displacement.

❖ simple harmonic motion

An object exhibiting simple harmonic motion is under the influence of only one force and the object's motion is periodic. Simple harmonic motion does not include the possibility of a damping force, so the parameter a from the previous definitions must be a constant and not depend on the independent variable.

Simple Harmonic Motion: Test Prep	
Identify the amplitude, period, and frequency of the simple harmonic motion modeled by this equation: $y = 3.5\cos 12t$.	amplitude 3.5,, frequency $\dfrac{12}{2\pi} = \dfrac{6}{2\pi}$ period $\dfrac{\pi}{6}$

Try Them	
Write an equation for the simple harmonic motion that begins with zero displacement at $t = 0$, has an amplitude of 5 mm, and a frequency of 5 cycles per second.	$y = 5\sin 10\pi t$
Write an equation for the simple harmonic motion that begins with displacement of 2 m at $t = 0$ and period of 120 seconds.	$y = 2\cos\dfrac{\pi}{60}t$

* **spring constant**

A mass on a spring is a system that moves with simple harmonic motion (as long as its vibrations are not damped out or otherwise modified). Springs come in all different strengths or stiffnesses. The spring constant is a number that describes the strength of the spring. The frequency of the systems vibrations, or waves, depends on the spring constant like so: $f = \dfrac{1}{2\pi}\sqrt{\dfrac{k}{m}}$ where k is the spring constant and m is the mass of the object attached to the spring.

Simple Harmonic Motion of a Mass on a Spring with Spring Constant k: Test Prep	
Write the equation of simple harmonic motion for an object of 15 units of mass attached to a spring of spring constant 50	amplitude 2, $k = 50$, $m = 15$, begins with displacement:

which has been displaced from equilibrium by positive 2.	$y = 2 \cos 2\pi \cdot \dfrac{1}{2\pi}\sqrt{\dfrac{50}{15}}t$ $= 2\cos\sqrt{\dfrac{10}{3}}t$
Try Them	
Find the period of a simple harmonic oscillator with spring constant of 2 and mass of 9.	$3\pi\sqrt{2}$
If the frequency of a mass of 2.5 units is observed to be $\dfrac{3}{2\pi}$ when it oscillates attached to a spring, find the spring constant.	22.5

Damped Harmonic Motion

Damped harmonic motion is more realistic in the 3-dimensional world we consider ourselves to live in. Springs or pendulums or the like don't just go on vibrating forever in the world of our experience, do they? There are forces like friction at work that damp out the vibrations as time progresses. (Then you wind the system up again and start over!)

❖ **damped harmonic motion**

Damped harmonic motion is frequently modeled with the equation

$f(t) = ae^{-kt}\cos \varpi t$.

❖ **pseudoperiod**

Damped functions don't exhibit true periodic behavior; they don't ever return to the same values they had in the last cycle. Nonetheless, we can observe something periodic about them, and so we say they have a pseudoperiod and define it as $\dfrac{2\pi}{\varpi}$.

Damped Harmonic Motion: Test Prep

Graph the damping function $y = 2e^{-.04x}$ and the function for the damped oscillator $y = 2e^{-.04x} \sin x$ on one set of axes.	

Try Them

Find the pseudoperiod of this damped oscillator: $y = 1.14e^{-.008t} \cos \frac{\pi}{5} t$.	10 cycles per unit time
Given an oscillator modeled by $y = 5e^{-.065t} \cos 25t$ meters, find the displacement of the object at $t = 10$ seconds. Round to the nearest hundredth of a meter.	0.62 meters

Final Fun: Matching

FIND THE BEST MATCH. GOOD LUCK.

DMS ANGLE	PYTHAGOREAN ID.
SUPPLEMENTARY ANGLES	180°
$\dfrac{\cos x}{\sin x}$	125°,55°
$\sin \dfrac{\pi}{6}$	SPRING CONSTANT
	42°12′35″
UNDEFINED	sec θ
$\dfrac{hypotenuse}{adjacent}$	−90°,270°

STRAIGHT ANGLE	COT(X)
$\dfrac{y}{x}$	$\tan \theta$
$1+ \tan^2\alpha = \sec^2\alpha$	$\tan 90°$
k	$\dfrac{1}{2}$
COTERMINAL ANGLES	$45°, 45°$
CSC(X)	$\dfrac{1}{\sin x}$
COMPLEMENTARY ANGLES	

NOW TRY THESE THE CHAPTER 5 REVIEW EXERCISES AND

THE CHAPTER 5 PRACTICE TEST

STAY UP ON YOUR GAME – DO THE CUMULATIVE REVIEW EXERCISES

Chapter 6:

Trigonometric Identities, Inverse Functions, and Equations

Word Search

E	M	C	D	N	N	V	T	G	T	Q	L	R	V	O
S	U	N	O	U	O	U	E	C	Y	A	Y	E	L	K
R	N	P	M	F	F	I	U	R	D	F	G	G	O	G
E	O	S	Y	F	U	D	T	I	I	S	E	R	D	M
V	I	J	W	R	O	N	O	A	W	F	V	E	Y	T
N	T	Y	Y	R	T	S	C	M	U	S	Y	S	T	X
I	I	W	P	C	U	E	H	T	R	Q	S	S	B	U
U	S	X	E	N	Q	A	M	E	I	S	E	I	P	U
A	O	G	I	K	L	I	D	O	R	O	T	O	U	S
V	P	S	J	F	U	U	J	K	N	F	N	N	T	F
S	M	R	J	X	C	E	X	Q	M	O	S	R	D	P
O	O	W	D	T	K	C	G	K	T	Z	G	M	Q	N
L	C	D	I	F	F	E	R	E	N	C	E	I	M	W
V	D	O	U	B	L	E	A	B	D	G	Z	S	R	M
E	N	G	A	B	M	Y	T	I	T	N	E	D	I	T

COFUNCTION	HALF	REGRESSION
COMPOSITION	IDENTITY	SINUSOIDAL
DIFFERENCE	INVERSE	SOLVE
DOUBLE	PRODUCT	SUM
EQUATION	REDUCTION	TRIGONOMETRY
VERIFY		

305

6.1: Verification of Trigonometric Identities

❖ identity

An identity is an equation that is true for all of its domain values. Here is a simple trig identity:

$$\csc \theta = \frac{1}{\sin \theta}$$

Verification of Trigonometric Identities

The basic idea behind verifying a trigonometric identity is to work with one side of the equation until it looks like the other side. This is *not* the same as solving an equation where you work on both sides of the equation at the same time. And since you're not isolating a variable, you need to keep your goal in your mind – it's the other side of the equation. If you can keep that goal in mind, it might suggest the necessary steps needed to verify the identity.

Guidelines for Verifying Trigonometric Identities

- Start with the more complicated side of the equation.

- Simplify your trigonometric expression by combining like terms and expanding binomials and the like. Sometimes factoring is a good idea.

- Rewrite the trigonometric expression in terms of only sines and/ or cosines.

- Rewrite the trigonometric expression in terms of only one trigonometric function.

- Try using a basic identity to rewrite the expression, or terms within the expression

- Multiply numerator and denominator of ratios by the same trigonometric expression, especially if it appears that result will be closer to your goal.

- Keep your goal in mind – it's the other side of the equation.

★ For Fun

Do you know what *RHS* and *LHS* stand for in the following examples?

Guidelines for Verifying Trigonometric Identities: Test Prep	
Verify the identity: $\tan^2 x = \sin^2 x \tan^2 x + \sin^2 x$	$RHS = \sin^2 x \tan^2 x + \sin^2 x$ $= \sin^2 x(\tan^2 x + 1)$ $= \sin^2 x \cdot \sec^2 x$ $= \dfrac{\sin^2 x}{\cos^2 x}$ $= \tan^2 x = LHS$

Try These

| Verify the identity:

$\dfrac{\sec x + 1}{\tan x} = \dfrac{\tan x}{\sec x - 1}$ | $RHS = \dfrac{\tan x}{\sec x - 1}$

$= \dfrac{\tan x}{\sec x - 1} \cdot \dfrac{\sec x + 1}{\sec x + 1}$

$= \dfrac{\tan x(\sec x + 1)}{\sec^2 x - 1}$

$= \dfrac{\tan x(\sec x + 1)}{\tan^2 x}$

$= \dfrac{\sec x + 1}{\tan x} = LHS$ |
| Verify the identity:

$\dfrac{1}{\tan x \, \csc x \, \sin x} = \cot x$ | $LHS = \dfrac{1}{\tan x \, \csc x \, \sin x}$

$= \dfrac{1}{\tan x \dfrac{1}{\sin x} \sin x}$

$= \dfrac{1}{\tan x}$

$= \cot x = RHS$ |

Generally there is more than one way to verify an identity.

6.2: Sum, Difference, and Cofunction Identities

This section deals with trigonometric functions operating on the sums and differences of angles.

Beware

There is no "distributive property" for trig functions. An expression such as $\sin(x+y)$ *never* means you can "distribute" the "sin" on x and y – "sin" is not a number, and it is not *multiplying* $(x+y)$. "sin" is used to represent a sine function, and it must obey the appropriate rules for the sine function. The instructions in this section illustrate exactly how the sine functions (and others) work on the sums and differences of angles.

Identities That Involve $(\alpha+\beta)$

Sum and Difference Identities

$$\sin(a \pm b) = \sin a \cos b \pm \cos a \sin b$$

$$\cos(a \pm b) = \cos a \cos b \mp \sin a \sin b$$

$$\tan(a \pm b) = \frac{\tan a \pm \tan b}{1 \mp \tan a \tan b}$$

$$\cot(a \pm b) = \frac{\cot a \cot b \mp 1}{\cot b \pm \cot a}$$

Sum and Difference Identities: Test Prep

Find the exact value of:	$\sin\dfrac{\pi}{12}\cos\dfrac{5\pi}{12}+\cos\dfrac{\pi}{12}\sin\dfrac{5\pi}{12}$
$\sin\dfrac{\pi}{12}\cos\dfrac{5\pi}{12}+\cos\dfrac{\pi}{12}\sin\dfrac{5\pi}{12}$	$=\sin\left(\dfrac{\pi}{12}+\dfrac{5\pi}{12}\right)$
	$=\sin\dfrac{6\pi}{12}$
	$=\sin\dfrac{\pi}{2}=1$

Sum and Difference Identities: Test Prep	
Try These	
Identify the expression that completes an identity for $\sin\left(\dfrac{3\pi}{4} - x\right)$: a. $-\dfrac{\sqrt{2}}{2}(\cos x + \sin x)$ b. $\dfrac{\sqrt{2}}{2}(\sin x + \cos x)$ c. $\dfrac{\sqrt{2}}{2}(\cos x - \sin x)$	b.
Find the exact value of: $\sin(30° + 45°)$	$\dfrac{\sqrt{2} + \sqrt{6}}{4}$

Cofunctions

Cofunctions are trigonometric functions that are related via complementary angles. You know from work with the unit circle that $\sin 30° = \cos 60°$. Since 30° and 60° are complementary angles, this suggest that sine and cosine are cofunctions of complimentary angles.

❖ **cofunction**

Here are six trigonometric cofunction identities.

> ### Cofunction Identities
>
> $$\sin(90° - \alpha) = \cos \alpha \quad \cos(90° - \alpha) = \sin \alpha$$
>
> $$\tan(90° - \alpha) = \cot \alpha \quad \cot(90° - \alpha) = \tan \alpha$$
>
> $$\sec(90° - \alpha) = \csc \alpha \quad \csc(90° - \alpha) = \sec \alpha$$

★ **For Fun**   Rewrite the above cofunction identities using radians instead of degrees.

Cofunction Identities: Test Prep	
Using a sum/difference identity, to verify the following cofunction identity: $$\sin\left(\frac{\pi}{2} - x\right) = \cos x$$	$LHS = \sin\left(\dfrac{\pi}{2} - x\right)$ $= \sin\dfrac{\pi}{2}\cos x - \cos\dfrac{\pi}{2}\sin x$ $= 1 \cdot \cos x - 0 \cdot \sin x$ $= \cos x = RHS$

Try These

Use a cofunction identity to rewrite $\cot 23°$	$\tan 67°$
Use a cofunction identity to rewrite $\sec\dfrac{3\pi}{10}$	$\csc\dfrac{2\pi}{10}$

You can use a right triangle to see that the cofunction relationships make sense.

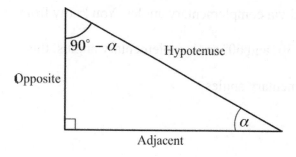

Additional Sum and Difference Identities

In this subsection we use the sum and difference formulas to evaluate expressions that involve $\alpha \pm \beta$.

Example:

Given that $\cos\alpha = -\dfrac{7}{25}$, where α is in Quadrant II, and $\sin\beta = -\dfrac{12}{13}$, with β in Quadrant IV, find $\cos(\alpha + \beta)$.

Solution: $\cos(\alpha + \beta) = \cos\alpha\cos\beta - \sin\alpha\sin\beta$, so we need to find $\cos\beta$ and $\sin\alpha$.

Use $\cos^2\alpha + \sin^2\alpha = 1$, to determine the following:

$$\cos^2\alpha + \sin^2\alpha = 1$$

$$\left(-\frac{7}{25}\right)^2 + \sin^2\alpha = 1$$

$$\sin^2\alpha = \frac{625}{625} - \frac{49}{625} = \frac{576}{625}$$

$$\sin\alpha = \pm\frac{24}{25}$$

Since α is in Quadrant II and we know $\sin\alpha = \frac{24}{25}$. Now we proceed to find $\cos\beta$.

$$\cos^2\beta + \left(-\frac{12}{13}\right)^2 = 1$$

$$\cos^2\beta = \frac{169}{169} - \frac{144}{169} = \frac{25}{169}$$

$$\cos\beta = +\frac{5}{13} \qquad \text{(Since } \beta \text{ is in Quadrant IV.)}$$

Putting all this information together produces:

$$\cos(\alpha+\beta) = \left(-\frac{7}{25}\right)\left(\frac{5}{13}\right) - \left(\frac{24}{25}\right)\left(-\frac{12}{13}\right)$$

$$= \frac{253}{325}$$

Reduction Formulas

The reduction formulas can be proven using the sum/difference identities.

❖ **reduction formula**

A trigonometric reduction formula is an identity that takes advantage of the periodic behavior of trigonometric functions to rewrite an expression with a different angle (usually smaller). For instance:

$\cos(\theta + 2k\pi) = \cos\theta$ is a reduction formula.

6.3: Double- and Half-Angle Identities

★ For Fun

The angle 6α is double the angle 3α. What angle is $\dfrac{\alpha}{12}$ half of?

Double-Angle Identities

Here are the trig double angle identities.

Double Angle Identities

$$\sin 2A = 2\sin A \cos A$$

$$\cos 2A = \cos^2 A - \sin^2 A = 1 - 2\sin^2 A = 2\cos^2 A - 1$$

$$\tan 2A = \frac{2\tan A}{1 - \tan^2 A}$$

Double Angle Identities	
Use a double-angle identity to verify: $\dfrac{\sin 2x}{1 - \sin^2 x} = 2\tan x$	$LHS = \dfrac{\sin 2x}{1 - \sin^2 x}$ $= \dfrac{2\sin x \cos x}{\cos^2 x}$ $= 2\dfrac{\sin x}{\cos x}$ $= 2\tan x = RHS$
Try These	
Given that $\tan\theta = \dfrac{15}{8}$ and that θ lies in Quadrant III, find $\sin 2\theta$.	$\dfrac{240}{289}$
Write $\cos^2 3x - \sin^2 3x$ in terms of a single trigonometric function.	$\cos 6x$

Power-Reducing Identities

The double-angle identity involving the cosine can be used to derive the following power-reducing formulas.

Power-Reducing Identities

$$\cos^2 A = \frac{1}{2}(1+\cos 2A) \qquad \sin^2 A = \frac{1}{2}(1-\cos 2A) \qquad \tan^2 A = \frac{1-\cos 2A}{1+\cos 2A}$$

Power-Reducing Identities: Test Prep

Verify $\cos^4 x = \dfrac{1}{8}(3+4\cos 2x+\cos 4x)$	$\begin{aligned} LHS &= \cos^4 x \\ &= (\cos^2 x)^2 \\ &= \left(\frac{1}{2}(1+\cos 2x)\right)^2 \\ &= \frac{1}{4}(1+2\cos 2x+\cos^2 2x) \\ &= \frac{1}{4}\left(1+2\cos 2x+\frac{1}{2}(1+\cos 4x)\right) \\ &= \frac{1}{4}\left(\frac{3}{2}+2\cos 2x+\frac{1}{2}\cos 4x\right) \\ &= \frac{1}{8}(3+4\cos 2x+\cos 4x) = RHS \end{aligned}$

Try These

Write $\sin^2\beta\cos^2\beta$ in terms of the first power of a cosine function.	$\dfrac{1}{8}(1-\cos 4\beta)$
Use the power-reducing identities to verify $\cos^4 x - \sin^4 x = \cos 2x$	$\begin{aligned} LHS &= \cos^4 x - \sin^4 x \\ &= \frac{1}{8}(3+4\cos 2x+\cos 4x)-\frac{1}{8}(3-4\cos 2x+\cos 4x) \\ &= \frac{1}{8}8\cos 2x \\ &= \cos 2x = RHS \end{aligned}$

Half-Angle Identities

The following identities are called half-angle identities.

Half-Angle Identities

$$\sin\frac{\alpha}{2} = \pm\sqrt{\frac{1-\cos\alpha}{2}} \qquad \cos\frac{\alpha}{2} = \pm\sqrt{\frac{1+\cos\alpha}{2}} \qquad \tan\frac{\alpha}{2} = \pm\sqrt{\frac{1-\cos\alpha}{1+\cos\alpha}}$$

Half-Angle Identities: Test Prep

Verify $\sin^2\frac{x}{2} = \frac{\sin^2 x}{2} + \frac{\cos x(\cos x - 1)}{2}$	$\begin{aligned} RHS &= \frac{\sin^2 x}{2} + \frac{\cos x(\cos x - 1)}{2} \\ &= \frac{\sin^2 x + \cos^2 x - \cos x}{2} \\ &= \frac{1-\cos x}{2} \\ &= \sin^2\frac{x}{2} = LHS \end{aligned}$
Try These	
Find the exact value of $\sin 112.5°$	$\dfrac{\sqrt{2+\sqrt{2}}}{2}$
If $\cos\theta = -\dfrac{12}{13}$ where θ is in Quadrant III, find $\tan\dfrac{\theta}{2}$.	5

★ For Fun

In the following graph, the black graph is the graph of $y = \sin x$. Which one of the graphs is the graph of $y = \sin 2x$?

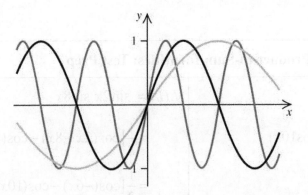

6.4: Identities Involving the Sum of Trigonometric Functions

Product-to-Sum Identities

The following product-to-sum identities are derived from the sum and difference identities.

Product-to-Sum Identities

$$\sin A \cos B = \frac{1}{2}\left[\sin(A+B) + \sin(A-B)\right]$$

$$\cos A \sin B = \frac{1}{2}\left[\sin(A+B) - \sin(A-B)\right]$$

$$\cos A \cos B = \frac{1}{2}\left[\cos(A+B) + \cos(A-B)\right]$$

$$\sin A \sin B = \frac{1}{2}\left[\cos(A-B) - \cos(A+B)\right]$$

The graph of $y = \sin 2x \sin 8x$ is shown below.

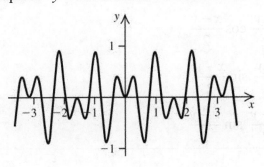

<table>
<tr><td colspan="2" align="center">Product-to-Sum Identities: Test Prep</td></tr>
<tr>
<td>

Verify

$$\sin 2x \, \sin 8x = \frac{1}{2}[\cos 6x - \cos 10x]$$

</td>
<td>

$LHS = \sin 2x \, \sin 8x$

$$= \frac{1}{2}\big[\cos(2x - 8x) - \cos(2x + 8x)\big]$$

$$= \frac{1}{2}\big[\cos(-6x) - \cos(10x)\big]$$

$$= \frac{1}{2}[\cos 6x - \cos 10x] = RHS$$

</td>
</tr>
<tr><td colspan="2">Try These</td></tr>
<tr>
<td>Find the exact value of $\sin 105°\cos 15°$</td>
<td>$\dfrac{\sqrt{3}+2}{4}$</td>
</tr>
<tr>
<td>Write $\cos 3x \, \cos 5x$ as the sum or difference of two functions.</td>
<td>$= \dfrac{1}{2}\big(\cos 8x + \cos 2x\big)$</td>
</tr>
</table>

Sum-to-Product Identities

The sum-to-product identities show how to write the sum/difference of two trigonometric functions as a product.

Sum-to-Product Identities

$$\sin x + \sin y = 2 \, \sin \frac{x+y}{2} \, \cos \frac{x-y}{2}$$

$$\cos x + \cos y = 2 \, \cos \frac{x+y}{2} \, \cos \frac{x-y}{2}$$

$$\sin x - \sin y = 2 \, \cos \frac{x+y}{2} \, \sin \frac{x-y}{2}$$

$$\cos x - \cos y = -2 \, \sin \frac{x+y}{2} \, \sin \frac{x-y}{2}$$

Sum-to-Product Identities: Test Prep	
Use a sum-to-product identity to simplify $\sin 7\theta - \sin 3\theta$.	$\sin 7\theta - \sin 3\theta = 2\sin\dfrac{4\theta}{2}\cos\dfrac{10\theta}{2}$ $= 2\sin 2\theta\cos 5\theta$
Try These	
Write $\cos 9x - \cos 7x$ as a product.	$-2\sin 8x\sin x$
Verify the identity: $\dfrac{\cos 4x - \cos 2x}{\sin 2x - \sin 4x} = \tan 3x$	$LHS = \dfrac{\cos 4x - \cos 2x}{\sin 2x - \sin 4x}$ $= \dfrac{-2\sin 3x\sin x}{2\cos 3x\sin x}$ $= -\dfrac{\sin 3x}{\cos 3x}$ $= -\tan 3x = RHS$

Functions of the form $f(x) = a\sin x + b\cos x$

The following procedure can be used to write $a\sin x + b\cos x$ as a sine function.

$$a\sin x + b\cos x = k\sin(x+\alpha)$$

where $k = \sqrt{a^2 + b^2}$ and $\sin\alpha = \dfrac{b}{\sqrt{a^2+b^2}}$ and $\cos\alpha = \dfrac{a}{\sqrt{a^2+b^2}}$

Functions of the Form $f(x) = a\sin x + b\cos x$ **: Test Prep**	
Write $y = \dfrac{\sqrt{2}}{2}\sin x + \dfrac{\sqrt{2}}{2}\cos x$ in the form $y = k\sin(x + \alpha)$ with α measured in degrees.	$y = \dfrac{\sqrt{2}}{2}\sin x + \dfrac{\sqrt{2}}{2}\cos x$ Thus $a = \dfrac{\sqrt{2}}{2}$ $b = \dfrac{\sqrt{2}}{2}$ and $k = \sqrt{\dfrac{1}{2} + \dfrac{1}{2}} = 1$ Use $\sin\alpha = \dfrac{b}{\sqrt{a^2+b^2}}$ and $\cos\alpha = \dfrac{a}{\sqrt{a^2+b^2}}$

	to show that $\sin\alpha = \dfrac{\sqrt{2}}{2} = \cos\alpha$ and
	$\alpha = 45°$.
	Thus $y = \sin(x + 45°)$.

Try These

| Write $y = \sin x - \cos x$ in the form $y = k\,\sin(x + \alpha)$ with α measured in radians. | $y = \sqrt{2}\,\sin\left(x - \dfrac{\pi}{4}\right)$ |
| Find the $y = k\sin(x + \alpha)$ form of $y = -\sqrt{2}\sin x + \sqrt{2}\cos x$. Then use a horizontal translation of the graph of the $y = k\sin(x + \alpha)$ form to sketch the graph of $y = -\sqrt{2}\sin x + \sqrt{2}\cos x$. | $y = 2\sin\left(x + \dfrac{3\pi}{4}\right)$ 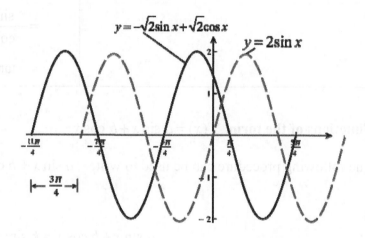 The graph of $y = -\sqrt{2}\sin x + \sqrt{2}\cos x$ is the graph of $y = 2\sin x$ shifted $3\pi/4$ units to the left. |

NOW TRY THE MID-CHAPTER 6 QUIZ.

6.5: Inverse Trigonometric Functions

In section 4.1, we studied inverse functions. Now we're going to investigate inverse trigonometric functions.

Inverse Trigonometric Functions

It's a simple matter to observe that a periodic function is not one-to-one – because, it does not pass the horizontal line test. However, if the *domain* of a given trigonometric function is restricted, then an inverse function can be defined. Here are the six inverse trigonometric functions, with their restricted domains defined.

Inverse Sine

$$y = \sin^{-1}x \text{ if and only if } x = \sin y, \text{ where } -1 \le x \le 1 \text{ and } -\frac{\pi}{2} \le y \le \frac{\pi}{2}$$

Inverse Cosine

$$y = \cos^{-1}x \text{ if and only if } x = \cos y \text{ where } -1 \le x \le 1 \text{ and } 0 \le y \le \pi$$

Inverse Tangent

$$y = \tan^{-1}x \text{ if and only if } x = \tan y, \text{ where } -\infty \le x \le \infty \text{ and } -\frac{\pi}{2} \le y \le \frac{\pi}{2}$$

Inverse Secant

$$y = \sec^{-1}x \text{ if and only if } x = \sec y, \text{ where } -1 \ge x, \; x \ge 1, \text{ and } 0 \le y \le \pi$$

Inverse Cosecant

$$y = \csc^{-1}x \text{ if and only if } x = \csc y, \text{ where } -1 \ge x, \quad x \ge 1, \text{ and } -\frac{\pi}{2} \le y \le \frac{\pi}{2}$$

Inverse Cotangent

$$y = \cot^{-1}x \text{ if and only if } x = \cot y, \text{ where } -\infty \le x \le \infty, \text{ and } 0 \le y \le \pi$$

Table 6.2 in Section 6.5 of your textbook shows the relationships between the graphs, domains, and ranges of all the inverse trigonometric functions. I recommend that you examine it closely. The statement $y = \cos^{-1}x$ if and only if $x = \cos y$ suggests a method of evaluating $\cos^{-1}\frac{1}{2}$. What *angle* (between zero and π radians) gives a cosine value of one half? Since that would be the angle $\frac{\pi}{3}$, $\cos^{-1}\frac{1}{2} = \frac{\pi}{3}$.

★ For Fun

Match each inverse trigonometric function with its angle measure.

$$\sin^{-1}\frac{\sqrt{3}}{2} \qquad \cos^{-1}(-1) \qquad \tan^{-1}\left(-\sqrt{3}\right) \qquad \sin^{-1}\left(-\frac{1}{2}\right)$$

$$-\frac{\pi}{6} \qquad -60° \qquad \pi \qquad \frac{\pi}{3}$$

★ For Fun

The inverse trig functions are also called arcsin, arccos, arctan, arccot, arcsec, and arccsc. You can use WolframAlpha to find the inverse sine of $\theta = 0.5$ by entering the text "arcsin (0.5)" into WolframAlpha's text box.

Composition of Trigonometric Functions and Their Inverses

You can often use a sketch of a triangle to find a trigonometric function of an inverse trigonometric function.

Example:

Find the exact value of $\cos\left[\sin^{-1}\left(-\frac{5}{6}\right)\right]$.

Since we're taking the inverse sine of a negative number the angle $\theta = \sin^{-1}\left(-\frac{5}{6}\right)$ must be in Quadrant IV. Draw the triangle shown below.

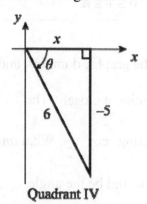

Determine the length of side x, which is the side adjacent to θ in the triangle.

$$x^2 + \left(|-5|\right)^2 = 6^2$$

$$x^2 = 36 - 25 = 11$$

$$x = \sqrt{11}$$

We chose the positive value for x because x is positive in Quadrant IV. Thus $\cos\left[\sin^{-1}\left(-\dfrac{5}{6}\right)\right] = \dfrac{\sqrt{11}}{6}$,

(which is the side adjacent over the hypotenuse of the fourth quadrant triangle we drew).

Be careful about domains and ranges when working with compositions. For instance, $\cos^{-1}\left(\tan\dfrac{2\pi}{3}\right)$ is

not defined because the value of $\tan\left(\dfrac{2\pi}{3}\right)$ is not in the domain of the inverse cosine

Composition of Trigonometric Functions and Their Inverses: Test Prep	
Find the exact value of $\tan\left(\cos^{-1}\dfrac{3}{5}\right)$.	Let $\theta = \cos^{-1}\dfrac{3}{5}$ as shown in the following reference triangle. 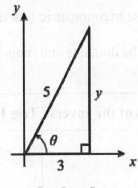 Quadrant I $$y = \sqrt{5^2 - 3^2} = \sqrt{16} = 4$$

	From the triangle we can determine that $\tan\left(\cos^{-1}\dfrac{3}{5}\right)=\tan(\theta)=\dfrac{4}{3}.$
Try These	
Find the exact value of $\cos\left(\cos^{-1}\dfrac{3}{4}\right).$	$\dfrac{3}{4}$
Find the exact value of $\sin^{-1}\left(\sin\dfrac{5\pi}{6}\right).$	$\dfrac{\pi}{6}$
Find the exact value of $\tan\left(\sin^{-1}\dfrac{\sqrt{3}}{2}\right).$	$\sqrt{3}$

★ For Fun

Since $y=\cos x$ and $y=\cos^{-1}x$ are inverse functions, you might think that $\cos^{-1}\left(\cos\dfrac{5\pi}{4}\right)=\dfrac{5\pi}{4}.$

However,. $\cos^{-1}\left(\cos\dfrac{5\pi}{4}\right)=\dfrac{3\pi}{4}.$ Explain.

Graphs of Inverse Trigonometric Functions

Let's practice graphing inverse trigonometric functions with perhaps a few transformations thrown in.

Your knowledge conserning the domains and ranges of inverse functions will help.

Graphs of the Inverse Trig Functions: Test Prep	
Graph $y=\sin^{-1}\dfrac{x}{2}$	$-1\le\dfrac{x}{2}\le1$ so $-2\le x\le2$ and $-\dfrac{\pi}{2}\le y\le\dfrac{\pi}{2}.$

Try These	
Sketch the graph of $y = \cot^{-1} x$. State its domain and range.	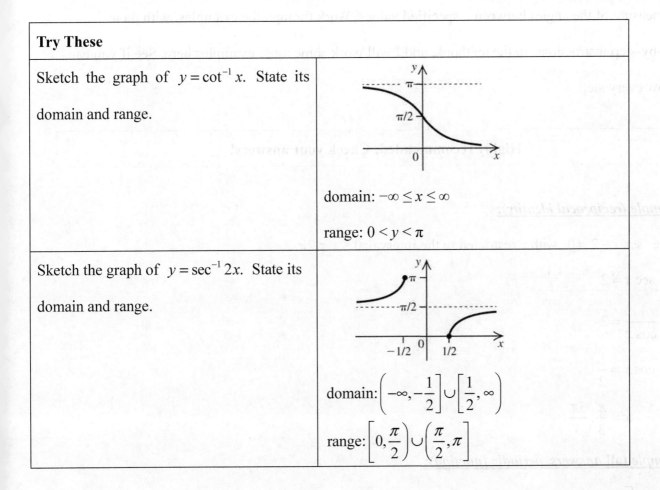 domain: $-\infty \leq x \leq \infty$ range: $0 < y < \pi$
Sketch the graph of $y = \sec^{-1} 2x$. State its domain and range.	domain: $\left(-\infty, -\dfrac{1}{2}\right] \cup \left[\dfrac{1}{2}, \infty\right)$ range: $\left[0, \dfrac{\pi}{2}\right) \cup \left(\dfrac{\pi}{2}, \pi\right]$

6.6: Trigonometric Equations

Now we have arrived at one of the most important topics in Chapter 6: Solving Trigonometric Equations.

Solving Trigonometric Equations

Remember that we're always trying to find the values of the variable that make the equation *true* when we solve equations. And now those variables represent the measure of *angles*, and so we're trying to find the measure of the angles that make our trigonometric equation true. Answers will look like $\theta =$ measure of an angle ... or the measure of several angles. In fact. with trigonometric equations you may have an infinite number of solutions. On the other hand, you may see explicit instructions to give only the measure of the angles between   specified values. Work through the examples, with their step-by-step instructions, in the textbook, and I will work some more examples here. See if you can follow every step.

Highly recommended: Check your answers!

Example (reciprocal identity):

Solve $\sec x - 2 = 0$ with x restricted to the interval: $0 \le x < 2\pi$

$$\sec x = 2$$

$$\frac{1}{\cos x} = 2$$

$$\cos x = \frac{1}{2}$$

$$x = \frac{\pi}{3}, \frac{5\pi}{3}$$

Example (all answers, periodic function):

$$\tan x + \sqrt{3} = 0 \quad -\infty < x < \infty$$

$$\tan x = -\sqrt{3}$$

$$x = \frac{2\pi}{3} + n\pi \text{ where } n \text{ is an integer.}$$

Example (square root principal and double angle):

$$2\sin^2 2x = 1 \quad 0 \le x < 2\pi$$

$$\sin^2 2x = \frac{1}{2}$$

$$\sin 2x = \pm \frac{\sqrt{2}}{2}$$

$$2x = \frac{\pi}{4}, \frac{3\pi}{4}, \frac{5\pi}{4}, \frac{7\pi}{4}$$

$$x = \frac{\pi}{8}, \frac{3\pi}{8}, \frac{5\pi}{8}, \frac{7\pi}{8}, \frac{9\pi}{8}, \frac{11\pi}{8}, \frac{13\pi}{8}, \frac{15\pi}{8}$$

Example (factoring):

$$\sec^2 x - \sec x = 2 \quad 0° \le x < 360°$$

$$\sec^2 x - \sec x - 2 = 0$$

$$(\sec x + 1)(\sec x - 2) = 0$$

$$\sec x = -1 \quad \text{or} \quad \sec x = 2$$

$$\cos x = -1 \qquad \cos x = \frac{1}{2}$$

$$x = 180° \quad \text{or} \quad x = 60°, 300°$$

$$x = 180°, 60°, 300°$$

Example (squaring both sides):

$$\sec x - \tan x = -1 \quad [0, 2\pi)$$

$$\sec^2 x = (\tan x - 1)^2$$

$$\tan^2 x + 1 = \tan^2 x - 2\tan x + 1$$

$$0 = \tan x$$

$$x = 0, \pi$$

checks

$$x = 0: \quad \sec 0 - \tan 0 \overset{?}{=} -1 \qquad\qquad x = \pi: \quad \sec \pi - \tan \pi \overset{?}{=} -1$$

$$1 - 0 \ne -1 \qquad\qquad\qquad\qquad\qquad -1 - 0 = -1 \quad \checkmark$$

Final solution: $x = \pi$

Example (no solution):

$$2\sin x + \frac{1}{\sin x} = 0 \quad [0, 2\pi)$$

$$2\sin^2 x + 1 = 0 \quad (\sin x \ne 0)$$

$$\sin^2 x = -\frac{1}{2} \quad \text{No Solution}$$

Solve Trigonometric Equations: Test Prep	
Solve the following trigonometric equation for all x in the interval $[0, 2\pi)$. $3\csc^2 x = 4$	$3\csc^2 x = 4$ $\csc^2 x = \dfrac{4}{3}$ $\csc x = \pm\dfrac{2}{\sqrt{3}}$ $\sin x = \pm\dfrac{\sqrt{3}}{2}$ $x = \dfrac{\pi}{3}, \dfrac{2\pi}{3}, \dfrac{4\pi}{3}, \dfrac{5\pi}{3}$
Try These	
Find all solutions of $\sin 2x = \dfrac{\sqrt{2}}{2}$	$x = \dfrac{\pi}{8} + n\pi, \dfrac{3\pi}{8} + n\pi$
Solve $\sin 2x = \cos x$ with $0 \le x < 360°$.	$x = 90°, 270°, 30°, 150°$

Modeling Sinusoidal Data

We've modeled data using linear and higher order polynomials, exponential, logarithmic, and even logistic regression. Now it's time to think about modeling data that looks periodic since we have an arsenal of periodic functions at hand.

❖ **sinusoidal data**

Data that can be closely modeled by a function of the form $y = d + a\,\sin(bx + c)$ is referred to as sinusoidal data.

❖ **sine regression**

Like linear, exponential, and all manner of other regression analyses, many graphing calculators can fit a sine function to sinusoidal data using a sine regression algorithm.

❖ **SinReg**

SinReg is the command you use on a TI-83/84 calculator to perform a sine regression. See Example 8 in Section 6.6 of your textbook.

Model Sinusoidal Data: Test Prep

The average monthly temperatures for a city in the southern United States were measured to be:

Month	Temp. (F)
1	48.9
2	51.8
3	59.2
4	66.0
5	73.5
6	79.1
7	81.8
8	81.0
9	76.6
10	67.3
11	59.1
12	51.7

Plot the data and use a calculator to derive a sinusoidal model.

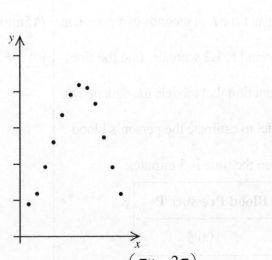

Model: $y = 16.45 \sin\left(\dfrac{\pi x}{6} - \dfrac{2\pi}{3}\right) + 65.35$

Try These

The data below represent the blood pressure P, in mm of Hg, at time t, in seconds of a person at rest. If the period is 1.2 seconds, find the sine regression function that models the data points. Use the model to estimate the person's blood pressure when the time is 5 minutes.	$P(t) \approx \sin(5.236t - 1.571) + 100$ $P(5\text{min}) = P(300) \approx 99mmHg$ $y \approx 1.53\sin(0.01x - 1.21) + 18.29$

Time t	Blood Pressure P
4	100.5
25	99.5
60	99.0
80	100.5

X	Y
1	16.9
61	18.1
122	19.2
183	19.9
245	18.8
306	17.2

Model this data using a sine regression.

Matching

FIND THE BEST MATCH.

LEFT-HAND SIDE OF EQUATION   SINUSOIDAL DATA

COFUNCTION OF COSINE $16\theta = 2(8\theta)$

$$\frac{\tan A + \tan B}{1 - \tan A \tan B}$$ SINE

DOUBLE ANGLE $\tan (A + B)$

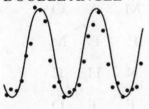   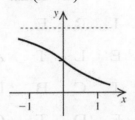

ARCCOTANGENT   LHS

NOW TRY THE

CHAPTER 6 REVIEW EXERCISES AND THE CHAPTER 6 PRACTICE TEST.

STAY UP ON YOUR GAME – TRY THE CUMULATIVE REVIEW EXERCISES.

Chapter 7: Applications of Trigonometry

Word Search

```
Y  R  A  N  I  G  A  M  I  L  W  M  A  G  E
R  T  E  R  M  I  N  A  L  A  A  A  R  N  R
D  A  D  U  M  A  N  F  Q  N  N  G  G  I  V
L  R  L  I  Q  O  R  Z  Q  O  I  N  U  D  I
X  E  L  U  R  I  R  E  I  G  N  I  M  A  O
R  H  L  E  C  E  L  T  A  O  I  T  E  E  M
S  O  H  L  A  I  C  B  D  H  T  U  N  H  E
I  W  T  L  A  E  D  T  O  T  I  D  T  T  D
N  A  H  C  J  R  T  N  I  R  A  E  O  G  T
E  L  C  O  E  D  A  I  E  O  L  D  N  H  R
H  K  R  O  W  V  E  P  N  P  N  I  E  P  A
R  P  X  E  L  P  M  O  C  U  R  O  X  N  L
A  M  B  I  G  U  O  U  S  A  R  E  X  Q  A
M  O  D  U  L  U  S  P  E  E  X  P  P  C  C
E  N  I  S  O  C  T  B  M  R  O  O  T  S  S
```

AMBIGUOUS	DOT	OBLIQUE	SINE
AREA	HEADING	ORTHOGONAL	TERMINAL
ARGUMENT	HERON	PARALLEL	THEOREM
BEARING	IMAGINARY	PERPENDICULAR	UNIT
COMPLEX	INITIAL	PROJECTION	VECTOR
COSINE	LAW	REAL	WORK
DE MOIVRE	MAGNITUDE	ROOTS	
DIRECTION	MODULUS	SCALAR	

7.1: Law of Sines

Now we will be working with *oblique* triangles, which are triangles that do not contain a right angle. In this section and the next, we'll learn how to find the measure of an angle and the length of a side of any triangle.

A Word About Notation

A side-side-side triangle, denoted SSS, is a triangle whose three side lengths are given, and so it remains for you to find the angle measurements. A side-angle-side (SAS) triangle is a triangle where two side lengths and the measure of the included angle are given. It's meaningful that the A comes *between* the two S's. A side-side-angle (SSA) triangle is a triangle where two sides lengths and the measure of an angle that is not included between the given sides is given.

Law of Sines

Some oblique triangles can be solved by using the Law of Sines. To use this law, you have to be given at least one angle and the side opposite it. (If you are given two angles, you can automatically get the third because, the sum of the angles in any triangle is 180°.)

♣ A labeling convention

Here is an illustration of the labeling convention for angles and sides of an oblique triangle. Note that the angle and the side opposite it are given the same letter; using an upper case letter to denote the measure of an angle and a lower case letter to denote the length of a side.

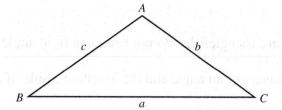

The Law of Sines

If a, b, and c are the lengths of the three sides of a triangle and A, B, and C are the

measures of the angles of that triangle, then:

$$\frac{a}{\sin A} = \frac{b}{\sin B} = \frac{c}{\sin C} \quad \text{and} \quad \frac{\sin A}{a} = \frac{\sin B}{b} = \frac{\sin C}{c}$$

The Law of Sines indicates that in any triangle the ratio of the length a side to the measure of the angle

opposite that side is equal to the ratio of the length of another side to the measure of the angle opposite

that side.

Law of Sines: Test Prep	
Given $A = 100°$, $a = 125$ inches, and $c = 10$ inches. Find B, b, and C.	$\dfrac{\sin A}{a} = \dfrac{\sin C}{c}$ $\dfrac{\sin 100°}{125} = \dfrac{\sin C}{10}$ $\sin C = \dfrac{10 \sin 100°}{125}$ $\sin C \approx 0.07878$ $C \approx \sin^{-1} 0.07878$ $C \approx 4.52°$ $B = 180° - A - C$ $B \approx 75.48°$

$$\frac{b}{\sin B} = \frac{a}{\sin A}$$

$$b = \frac{a \sin B}{\sin A} \approx \frac{125 \sin 75.48°}{\sin 100°}$$

$$b \approx 122.87 \text{ inches}$$

Law of Sines: Test Prep	
Try These	
Given $B = 28°$, $C = 104°$, and $a = 3.62$ solve the triangle. Round the lengths of the two unknown sides to two decimal places.	$A = 48°$ $b \approx 2.29$ $c \approx 4.73$
Given $a = 15\,yd$, $b = 25\,yd$ and $A = 85°$, solve the triangle.	No solution.

✓ **Question to Ask Your Instructor**

Make sure you're clear about what rounding practices your instructor wants you to follow.

Ambiguous Case (SSA)

The SSA case is known as the *ambiguous case*, because the given information may determine one

triangle, two triangles, or no triangle.

An oblique triangle can have one and only one obtuse angle if it has any.

This helps with the ambiguous case because if you recognize one angle is obtuse, then there cannot

be another obtuse angle.

The Ambiguous Case (SSA): Test Prep	
Solve the triangle with the given parameters: $a = 12$, $b = 31$, and $A = 20.5°$	$\dfrac{\sin B}{b} = \dfrac{\sin A}{a}$ $\sin B = \dfrac{31\sin 20.5°}{12} \approx 0.9047$ There are two angle B where B is between $0°$ and $180°$, such that $\sin \boldsymbol{B} = 0.9047$. One of the angles is an acute angle and the other is an obtuse angle. Use the inverse sine function to find the acute angle. $B \approx \sin^{-1}(0.9047) \approx 64.8°$ The obtuse angle is the supplement of the acute angle, so the obtuse angle is: $180° - 64.8° = 115.2°$ You can determine that $B \approx 115.2°$ is a valid result because $A = 20.5°$ and the sum of $20.5°$ and $115.2°$ is less than $180°$. Thus there are two triangles with the given dimensions. In one triangle $B \approx 64.8°$, and in the other triangle, $B \approx 115.2°$. *Case 1:* If $\boldsymbol{B} \approx 64.8°$ then: $C = 180° - A - B \approx 180° - 20.5° - 64.8° \approx 94.7°$ and $c = \dfrac{a\sin C}{\sin A} \approx \dfrac{12\sin 94.7°}{\sin 20.5°} \approx 34.2$ *Case 2:* If $\boldsymbol{B} \approx 115.2°$ then $C = 180° - A - B \approx 180° - 20.5° - 115.2° \approx 44.3°$ and $c = \dfrac{a\sin C}{\sin A} \approx \dfrac{12\sin 44.3°}{\sin 20.5°} \approx 23.9$

The Ambiguous Case (SSA): Test Prep	
Try These	
Given $A = 49.22°$ and $a = 16.92$ and $c = 24.62$, solve the triangle, if it exists.	No solution
Given $B = 41.2°$, $a = 31.5$, and $b = 21.6$, find both solution triangles.	*Case 1*: $A \approx 73.9°$, $C \approx 64.9°$, $c \approx 29.7$ *Case 1*: $A \approx 106.1°$, $C \approx 32.7°$, $c \approx 17.7$

★ **For Fun**   Check that the last two solution triangles; with

$$A \approx 73.9°, \ C \approx 64.9°, \ c \approx 29.7 \ \text{and} \ A \approx 106.1°, \ C \approx 32.7°, \ c \approx 17.7$$

are both possible by using a ruler and a protractor to sketch a scale drawing of each triangle.

Applications of the Law of Sines

❖ **heading**

The direction, specified by an angle, that a craft is pointing. Heading is expressed in terms of an angle

measured clockwise from north. The following figure shows a heading of 95°.

❖ **bearing**

The bearing is the direction towards a sighting and is measured as the acute angle to a north-south line.

The following figure shows a bearing of N52°W and a bearing of S21°E.

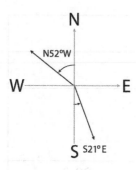

7.2: Law of Cosines and Area

Law of Cosines

If you're given a SAS triangle where the included angle is given, you won't be able to use Law of Sines to solve the triangle because you don't have a matching angle and its opposite side. The same is true for the SSS case. In these two cases, we use the following Law of Cosines to solve triangles.

Law of Cosines

If *A, B,* and *C* are the measures of the angles of a triangle and *a, b,* and *c* are the lengths of the sides opposite these angles, then

$$a^2 = b^2 + c^2 - 2bc \cos A$$

★ For Fun

The textbook gives 3 versions of the Law of Cosines. See if you can discover a pattern in the variables that allows you to use any version if you only know one version.

If you need to find angle *A* when given the lengths of the three sides, you need to solve the Law of Sines equation for *A* to produce:

$$A = \cos^{-1}\left(\frac{b^2 + c^2 - a^2}{2bc}\right)$$

★ **For Fun**

Can you isolate A in the Law of Cosines formula to produce the above result?

★ **For Fun**

Is it possible for a SSS triangle to give ambiguous results?

Law of Cosines: Test Prep	
Given a triangle where $A = 55°$, $b = 3$, and $c = 10$, solve the triangle.	$a^2 = b^2 + c^2 - 2bc\cos A$ $a^2 = 3^2 + 10^2 - 2(3)(10)\cos 55°$ $a^2 \approx 74.585$ $a \approx 8.64$ Now switch to Law of Sines to find another angle: $\sin B \approx \dfrac{3\sin 55°}{8.64} \approx 0.2846$ $B \approx \sin^{-1} 0.2846 \approx 16.53°$ $C \approx 180° - 55° - 16.53° = 108.47°$
Try these	
Given $a = 55$, $b = 25$, and $c = 72$, find the angles of this triangle. Round your answers to the nearest degree.	$A \approx 39°$ $\quad B \approx 17°$ $\quad C \approx 124°$
Solve the following triangle using the Law of Cosines. $C = 15.25°$ $\quad a = 6.25$ $\quad b = 2.15$	$A \approx 157.00°$ $\quad B \approx 7.70°$ $\quad C \approx 4.21$

Area of a Triangle

With trigonometry, we can derive an area formula that is good for all triangles.

Area of a Triangle

$$AREA = \frac{1}{2}bc\sin A = \frac{1}{2}ac\sin B = \frac{1}{2}ab\sin C$$

which is one-half the product of any two sides and the sine of their included angle. (No more

laborious work trying to discover the perpendicular height of the triangle)

Warning: Don't confuse "A" for area. We only use "A" as the measure of angle A.

Area of a Triangle: Test Prep	
Find the area of a triangle with the following measurements: $B = 130°$ $a = 62$ meters $c = 20$ meters	$AREA = \frac{1}{2}(62)(20)\sin 130° \approx 474.9\,\text{m}^2$
Try These	
Given $A = 27°$, $b = 15$ inches, and $c = 24$ inches, find the area of the triangle. Round to the nearest square inch.	82 square inches
Given $B = 46°$, $C = 53°$, and $c = 18$ inches, find the area of the triangle to the nearest tenth of a square inch.	144.1 square inches

Heron's Formula

Heron's formula is another method for finding the area of a triangle for which we know the length of

each of the sides.

❖ **semiperimeter**

One half the perimeter of a triangle is called its semiperimeter. The parameter s is used in Heron's

Formula (shown below) to denote the semiperimeter of the triangle.

Heron's Formula

If a, b, c are the lengths of the sides of a triangle, then the area of the triangle is given by

$$AREA = \sqrt{s(s-a)(s-b)(s-c)}$$

$$\text{where} \quad s = \frac{1}{2}(a+b+c)$$

Heron's Formula: Test Prep	
Given $a = 8$ yd, $b = 12$ yd, $c = 16$ yd. Find the area of triangle ABC.	$s = \dfrac{8+12+16}{2} = 18$ $AREA = \sqrt{18 \cdot 10 \cdot 6 \cdot 2} \approx 46.5 \text{ yd}^2$
Try These	
Find the area of the triangle with sides $a = 16$, $b = 12$, and $c = 14$.	≈ 81 square units
Use Heron's Formula to find the area of a triangle with side lengths of 3.6 cm, 4.2 cm, and 4.8 cm.	$\approx 7.3 \text{ cm}^2$

★ For Fun

A Heronian triangle is a triangle that has side lengths that are integers, its area is an integer, and its

inradius is an integer. Take a foray into geometry to find out what an "inradius" is?

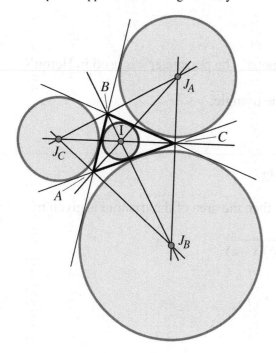

7.3: Vectors

Vectors

Vectors are useful in many fields of math and especially in physics and engineering. Vectors are line segments with the property of direction (pointing) attached. We draw vectors with arrows on one end to denote this direction.

♣ **scalar quantities, scalar**

A scalar quantity is a real number that speaks to the size of a measurement. That number is a scalar.

❖ **vector quantities, vector**

Vector quantities have both magnitude (which is described by a scalar) and direction. An object with both magnitude and direction is a vector. Vectors are denoted by using an arrow over a letter or boldface type. For instance, $\vec{v}$ and **v** each represent a vector.

❖ **magnitude of a vector**

The magnitude of a vector is its length. You can find the length of a line segment provided you know the end points and the distance formula.

❖ **initial point**

The initial point of a vector is the point that the vector begins from. It doesn't have an arrowhead in a picture.

❖ **terminal point**

The terminal point of a vector is the point where the arrowhead is affixed. The terminal point is the end of the vector – taking into account this meaning of direction for vectors.

❖ **equivalent vectors**

Any two vectors of the same magnitude and direction are equivalent vectors. They may be in different parts of the universe but they are still equivalent if they point in the same direction and have the same length.

❖ **scalar multiplication**

Scalar multiplication serves to change the magnitude of a vector. Scalar multiplication by a negative number also turns the direction of the vector around by 180 degrees.

❖ **resultant vector, resultant**

The resultant vector is the vector formed by adding two (or more) vectors together. The ultimate action of the two vectors **U** and **V** is described by the resultant vector **U** + **V**.

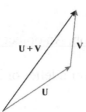

The resultant vector is **U + V**.

♣ **component**

If a vector is described by its endpoints using the component notation $\langle a, b \rangle$, then a is the component of the vector that points in the x-direction and b is the component that points in the y-direction.

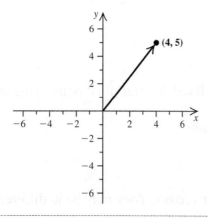

Finding the components of a vector

The component form of a vector **v** that points from an initial point (x_i, y_i) to a terminal point (x_t, y_t), is given by:

$$\mathbf{v} = \langle x_t - x_i, \ y_t - y_i \rangle$$

♣ **direction angle**

The direction angle is the angle that a vector makes with the positive x-axis.

★ For Fun

Use $\tan\theta = \dfrac{y}{x}$ to find θ in the following triangle. This procedure can be used to find the direction

angle of a vector that points at (x, y).

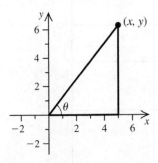

✣ zero vector, 0

The vector

$$\mathbf{0} = \langle 0,\ 0 \rangle.$$

is called the vector zero vector. Its magnitude is 0.

✣ additive inverse of a vector

The additive inverse of a vector is the vector that satisfies $\vec{v} + \underline{} = \mathbf{0}$. That vector is $-\vec{v}$ and results

from multiplying $\vec{v}$ by -1 (scalar multiplication).

Basic Vector Operations

1. Magnitude of a vector: $\left\| \vec{v} \right\| = \sqrt{\left(v_x\right)^2 + \left(v_y\right)^2}$

2. Vector addition: $\vec{v} + \vec{u} = \langle v_x,\ v_y \rangle + \langle u_x,\ u_y \rangle = \langle v_x + u_x,\ v_y + u_y \rangle$

3. Scalar multiplication: $k\vec{v} = \langle kv_x,\ kv_y \rangle$

Fundamental Vector Operations: Test Prep	
Given $\vec{u} = \langle -2, -1 \rangle$ and $\vec{v} = \langle 5, -3 \rangle$, find $\vec{u} - 3\vec{v}$.	$\vec{u} - 3\vec{v} = \langle -2, -1 \rangle - 3\langle 5, -3 \rangle$ $= \langle -2, -1 \rangle + \langle -15, 9 \rangle$ $= \langle -17, 8 \rangle$
Try These	
Find the magnitude of $\vec{v} = \langle -1, -5 \rangle$.	$\sqrt{26}$
Find the direction angle (the angle formed by a vector and the positive x-axis) of the vector $\vec{v} = \langle -2, 14 \rangle$ and sketch the vector.	Direction angle: $\theta \approx 98°$ 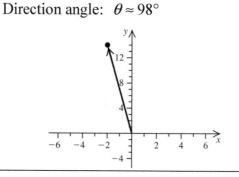

Unit Vectors

❖ **unit vector**

A unit vector is a vector that has a magnitude of one.

Standard Unit Vectors

$$\mathbf{i} = \langle 1, 0 \rangle \qquad \mathbf{j} = \langle 0, 1 \rangle$$
$$\langle a, b \rangle = a\langle 1, 0 \rangle + b\langle 0, 1 \rangle = a\mathbf{i} + b\mathbf{j}$$

❖ **horizontal component**

If θ is the direction angle of $\vec{v}$, then the horizontal component of $\vec{v}$, is $\|\vec{v}\| \cos\theta$; this also happens to be the x-component of a vector written component-wise.

❖ vertical component

The vertical component of $\vec{v}$ is $\|\vec{v}\| \sin \theta$ or the *y*-component.

Applications of Vectors

Resolving forces into one *resultant* force is an application of vectors. If you travel against a headwind, you have to work harder to get someplace than if there is no wind. Vectors give us a way to describe the effect of a headwind.

❖ **air speed**

The airspeed of a plane is the speed the plane would be traveling if there were no wind.

❖ **actual velocity**

The actual velocity is a plane's speed measured with respect to the ground.

❖ **ground speed**

The actual speed of the plane, factoring in all forces acting on it, is the plane's ground speed.

❖ **resolving the vector**

The process of finding two vectors whose sum is a given *resultant* vector is called *resolving a vector.*

❖ **vector components**

Vectors that resolve a resultant vector are the vector components of the resultant vector:

$$\vec{v} = \vec{F_1} + \vec{F_2}.$$

Dot Product

The dot product (also called the inner product and scalar product) is one way of multiplying two vectors. We have already discussed the multiplication of a vector by a scalar (scalar multiplication). However, the dot product is a different operation. Keep in mind that the result of a dot product is a scalar.

The Dot Product

$$\langle a, b \rangle \cdot \langle c, d \rangle = ac + bd$$

The *x*-components are multiplied together, the *y*-components are multiplied together and those two products are added to produce a scalar.

Dot Product: Test Prep	
Find the dot product of $\langle 6, 1 \rangle$ and $\langle -2, 3 \rangle$.	$\langle 6, 1 \rangle \cdot \langle -2, 3 \rangle = 6 \cdot (-2) + 1 \cdot 3 = -12 + 3 = -9$
Try These	
Evaluate $\langle 4\mathbf{i} - 2\mathbf{j} \rangle \cdot \langle \mathbf{i} - \mathbf{j} \rangle$	6
Evaluate $\langle 3, -1 \rangle \cdot \langle 5, 4 \rangle$	11

The Angle Between Two Vectors

If **v** and **u** are nonzero vectors and α is the smallest nonnegative angle between **v** and **u**, then

$$\cos \alpha = \frac{\mathbf{v} \cdot \mathbf{u}}{\|\mathbf{v}\|\|\mathbf{u}\|} \quad \text{and} \quad \alpha = \cos^{-1}\left(\frac{\mathbf{v} \cdot \mathbf{u}}{\|\mathbf{v}\|\|\mathbf{u}\|}\right)$$

Angle Between Two Vectors: Test Prep	
Find the angle between $\langle 3, 2 \rangle$ and $\langle 4, 0 \rangle$. Sketch the vectors and use a protractor to verify that your result is reasonable.	$\cos \alpha = \dfrac{\langle 3, 2 \rangle \cdot \langle 4, 0 \rangle}{\|\langle 3, 2 \rangle\|\|\langle 4, 0 \rangle\|}$ $= \dfrac{12}{\sqrt{13} \cdot 4} \approx 0.8321$ $\alpha \approx \cos^{-1} 0.8321 \approx 33.7^{o}$

	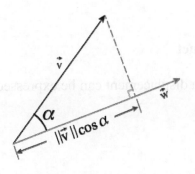

Try These

Find, to the nearest tenth of a degree, the measure of the angle between the vectors: $3i + 2j$ $-2j - j$	$172.9°$
Find the exact angle measurement between the vectors: $\vec{v} = 3\vec{i} + 6\vec{j}$ and $\vec{u} = 7\vec{i} - 3\vec{j}$	$\cos^{-1}\left(\dfrac{1}{\sqrt{290}}\right)$

Scalar Projection

Scalar Projection

If $\vec{v}$ and $\vec{w}$ are non-zero vectors and α is the angle between the vectors, then the scalar projection of $\vec{v}$ onto $\vec{w}$ is given by

$$\text{proj}_w\, \vec{v} = \left\|\vec{v}\right\| \cos\alpha.$$

Scalar Projection of $\vec{v}$ onto $\vec{w}$: Test Prep	
Given $\mathbf{v} = 3\mathbf{i} - 4\mathbf{j}$ and $\mathbf{w} = \mathbf{i} + 3\mathbf{j}$ find the scalar projection of $\mathbf{v}$ onto $\mathbf{w}$.	$\text{proj}_{\vec{w}}\vec{v} = \|\vec{v}\|\cos\alpha$ $= \sqrt{3^2 + (-4)^2} \cdot \dfrac{(3\mathbf{i}-4\mathbf{j})\cdot(\mathbf{i}+3\mathbf{j})}{\sqrt{3^2+(-4)^2}\cdot\sqrt{1^2+3^2}}$ $= \dfrac{3-12}{\sqrt{10}} = -\dfrac{9\sqrt{10}}{10}$
Try These	
Find the scalar projection of $\vec{w}$ onto $\vec{v}$ for the given the vectors above.	$-\dfrac{9}{5}$
Find, to the nearest tenth, the projection of $\vec{v}$ onto $\vec{w}$. $\vec{v} = \langle 2,4 \rangle$ $\vec{w} = \langle -1,5 \rangle$	3.5

Parallel and Perpendicular Vectors

❖ orthogonal vectors

Orthogonal vectors meet at a right angle – the angle between them is 90 degrees. If the angle between

them is 90 degrees then $\alpha = 90°$ and $\cos\alpha = 0 = \dfrac{\vec{v}\cdot\vec{u}}{\|\vec{v}\|\,\|\vec{u}\|}$ (from the angle between two vectors

formula). Since neither $\vec{v}$ nor $\vec{u}$ are zero vectors by necessity, this indicates that the dot product of

two orthogonal vectors is zero.

Work: An Application of the Dot Product

The work done by a force applied along a displacement can be expressed using the dot product:

Work

$$W = \vec{F} \cdot \vec{s} = \left\| \vec{F} \right\| \left\| \vec{s} \right\| \cos \alpha$$

where W is the work, $\vec{F}$ is the applied force, $\vec{s}$ is the displacement, and α is the angle between $\vec{F}$ and $\vec{s}$.

★ For Fun

Look up the alternative form of the dot product, in your textbook, and compare it to the above formula for work. Do you see that this alternative form arises from the formula for the angle between two vectors?

❖ work

Work is the product of force applied and distance traveled. If no distance is traveled then no work is done. Of course, if no force is applied no work gets done either!

Definition of Work: Test Prep	
A worker exerts a force on a box that is on a ramp. The ramp forms an angle of 28° with the level road, and the worker exerts a 55-pound force parallel to the road. Find the work done in sliding the box 6 feet along the ramp. Round to the nearest foot-pound.	$W = \vec{F} \cdot \vec{s} = \left\| \vec{F} \right\| \left\| \vec{s} \right\| \cos \alpha$ $= 55\,\text{lb} \cdot 6\,\text{ft} \cdot \cos 28°$ ≈ 291 ft-lb
Try This	
A force of 75 pounds pulls a sled 145 feet along a level path. The direction angle of the	≈ 8908 ft-lb

force with the level path is 35°. Find the work done. Round to the nearest foot-pound.	

NOW TRY THE MID-CHAPTER 7 QUIZ.

7.4: Trigonometric Form of Complex Numbers

Real numbers are graphed as points on a number line. Complex numbers are graphed on a complex plane.

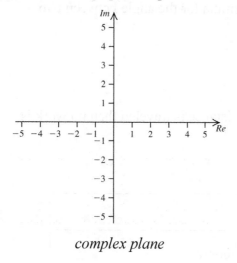

complex plane

✤ **complex plane**

The complex plane is a 2-dimensional plane with the axes labeled "real" and "imaginary" as in the previous diagram.

✤ **real axis**

The horizontal axis on a complex plane is called the real axis (*Re*) .

✤ **imaginary axis**

The vertical axis on a complex plane is called the imaginary axis (*Im*).

♣ **standard form** $a + bi$

The standard form or rectangular form of a complex number is $z = a + bi$. The complex number z $= a + bi$, can be plotted on the complex plane as the point (a, b). The complex number $z = 3 - 2i$ is plotted as a point on the complex plane in the following diagram.

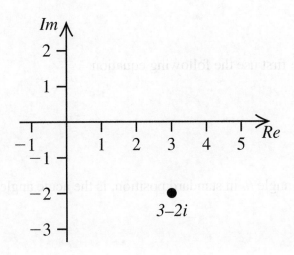

♣ **trigonometric form** $z = r(\cos\theta + i\sin\theta)$

The distance r from the origin to the point (a, b) is called the *modulus* of a complex number. The value of r is given by $r = |a + bi| = \sqrt{a^2 + b^2}$. The angle θ between the positive real axis and the line segment from the origin to (a, b) is called the *argument* of the complex number.

The figure below shows the graph of $-1 + i\sqrt{3}$. In this example the modulus r is 2 and the argument θ is 120°.

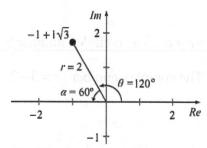

In this figure α is the
reference angle for angle θ.

To determine the argument θ of a complex number, we first use the following equation

to determine the reference angle α.

$$\alpha = \tan^{-1}\left|\frac{b}{a}\right|$$

Recall from Chapter 5 that the reference angle α of an angle θ, in standard position, is the acute angle

formed by the terminal side of θ and the x-axis.

Example: *Find the trigonometric form of* $-1 + i\sqrt{3}$

We need to determine the absolute value and the modulus of $-1 + i\sqrt{3}$.

$$\left|-1 + i\sqrt{3}\right| = \sqrt{(-1)^2 + \left(\sqrt{3}\right)^2} = \sqrt{4} = 2$$

We know $a = -1$ and $b = \sqrt{3}$. Thus the reference angle α for angle θ is

$$\alpha = \tan^{-1}\left|\frac{\sqrt{3}}{-1}\right| = 60°$$

We know from the above figure that the terminal side of θ lies in Quadrant II. Thus θ equals $180° - 60° =$

$120°$. The trigonometric form of a complex number $a + bi$ is written as $z = r\left(\cos\theta + i\sin\theta\right)$. Thus the

trig form of $z = -1 + i\sqrt{3}$ is $2(\cos 120° + i \sin 120°)$.

★ For Fun

The expression $\cos\theta + i\sin\theta$ is often abbreviated as $\text{cis}\theta$. Thus $2(\cos 120° + i \sin 120°)$ can be written

as $2 \text{ cis } 120°$.   Do you know what *cis* stands for?

Trigonometric Form of a Complex Number: Test Prep	
Write $z = 1 - i\sqrt{3}$ in trigonometric form. 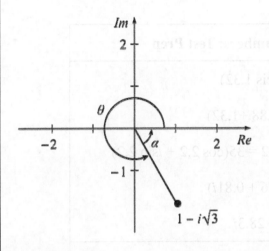	$r = \left\| -1 + i\sqrt{3} \right\| = \sqrt{(-1)^2 + \left(\sqrt{3}\right)^2} = \sqrt{4} = 2$ The reference angle for the argument is $\alpha = \tan^{-1} \left\| \dfrac{-\sqrt{3}}{1} \right\| = 60°$ The argument θ is a quadrant IV angle. Thus $\theta = 360° - 60° = 300°$. See the figure to the left. The trigonometric form of z is $z = 2(\cos 300° + i\sin 300°) = 2\operatorname{cis}300°$

Try These

Find the standard form of $z = 6\operatorname{cis}240°$.	$z = -3 - 3i\sqrt{3}$
Graph the complex number $z = 5\sqrt{2}\operatorname{cis}\dfrac{3\pi}{4}$	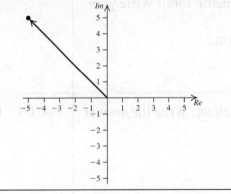

Product and Quotient of Complex Numbers Written in Trigonometric Form

There are good reasons for writing a complex number in its trigonometric form. One of those reasons is

that multiplying and dividing complex numbers in trigonometric form is easy.

❖ **product property of complex numbers**

If z_1 and z_2 are complex numbers written in trigonometric form, then their product is

$$z_1 z_2 = r_1 r_2 \left(\cos\left(\theta_1 + \theta_2\right) + i\sin\left(\theta_1 + \theta_2\right) \right) = r_1 r_2 \operatorname{cis}\left(\theta_1 + \theta_2\right).$$

❖ quotient property of complex numbers

The quotient of two complex number in trigonometric form is given by $\dfrac{z_1}{z_2} = \dfrac{r_1}{r_2} cis(\theta_1 - \theta_2)$.

Product and Quotient Properties of Complex Numbers: Test Prep	
Find the product:$(7\text{cis } 0.88) \cdot (5\text{cis } 1.32)$ Write your answer in standard form; round decimals to the nearest tenth. (Note: The angles here are measured in radians.)	$(7\text{cis } 0.88) \cdot (5\text{cis } 1.32)$ $= 35\text{cis}(0.88 + 1.32)$ $= 35\text{cis } 2.2 = 35(\cos 2.2 + i\sin 2.2)$ $\approx 35(-0.56 + 0.81i)$ $= -20.6 + 28.3i$
Try These	
Find the product $\left(-1 - i\sqrt{3}\right)\left(-\sqrt{3} - i\right)$ by converting to trigonometric form. Write your answer in standard form.	$4i$
Find the following quotient. Write the answer in standard form: $\dfrac{1 + i\sqrt{3}}{4 - 4i}$ Round constants to the nearest hundredth.	$\approx -0.09 + 0.34i$

★ For Fun

The complex plane is referred to as an Argand plane. You can learn about Jean-Robert Argand on the Internet.

7.5: De Moivre's Theorem

De Moivre's Theorem is about finding powers and roots of complex numbers. Some intense notation and careful work will pay off with some nifty results.

De Moivre's Theorem

De Moivre's Theorem can be used to find powers of complex numbers.

De Moivre's Theorem for Powers

Given $z = r\operatorname{cis}\theta$ and n is an integer, then $z^n = r^n\operatorname{cis}(n\theta)$.

De Moivre's Theorem: Test Prep	
Use De Moivre's Theorem to find $(1+i)^4$ Write the answer in the standard form.	$r = \sqrt{1^2 + 1^2} = \sqrt{2}$ $\theta = \tan^{-1}\dfrac{1}{1} = 45°$ $1+i = \sqrt{2}\operatorname{cis}45°$ $(1+i)^4 = (\sqrt{2})^4\operatorname{cis}(4\cdot45°)$ $\quad = 4\operatorname{cis}180°$ $\quad = 4(\cos180° + i\sin180°)$ $\quad = 4(-1 + i\cdot0)$ $\quad = -4$
Try These	
Use De Moivre's Theorem to find $(2+2i)^7$. Write your answer in standard form.	$1024 - 1024i$
Find the trigonometric form of $(2\operatorname{cis}330°)^4$	$-16\operatorname{cis}\dfrac{\pi}{3}$

De Moivre's Theorem for Finding Roots

De Moivre's Theorem for finding the roots (all of them!) of complex numbers is more detailed but also more astounding and clever in its results!

De Moivre's Theorem for Roots

Given $z = r cis\theta$ a complex number, there are exactly n distinct complex nth roots of z.

$$\sqrt[n]{z} = z^{\frac{1}{n}} = r^{\frac{1}{n}} cis\left(\frac{\theta + 360°k}{n}\right) \quad k = 0, 1, 2, \ldots, n-1$$

★ For Fun

Do you see that the values of k above give you n roots?

★ For Fun

Plot the answers, on a complex plane, to the next three examples and see a geometric pattern emerge!

This is the nifty part of De Moivre's Theorem for roots, that and the fact that you're actually finding *all* nth roots.

De Moivre's Theorem for Finding Roots: Test Prep	
Find all the roots of $x^3 - 2i = 0$. Write each root in the trigonometric form.	$x^3 - 2i = 0$
	$x^3 = 2i$
	$x = \sqrt[3]{2i} = (2i)^{\frac{1}{3}}$
	$z = 2i = 2 cis 90°$
	$k = 0: \quad z^{\frac{1}{3}} = 2^{\frac{1}{3}} cis 30°$

	$k = 1:$ $z^{\frac{1}{3}} = 2^{\frac{1}{3}} \operatorname{cis} \dfrac{90° + 360°}{3} = 2^{\frac{1}{3}} \operatorname{cis} 150°$
	$k = 2:$ $z^{\frac{1}{3}} = 2^{\frac{1}{3}} \operatorname{cis} \dfrac{90° + 720°}{3} = 2^{\frac{1}{3}} \operatorname{cis} 270°$

De Moivre's Theorem for Finding Roots: Test Prep	
Try These	
Find all roots of $\sqrt[6]{64}$. Write each root in trigonometric form and in standard form.	$2 \operatorname{cis} 0° = 2$ $2 \operatorname{cis} 60° = 1 + i\sqrt{3}$ $2 \operatorname{cis} 120° = -1 + i\sqrt{3}$ $2 \operatorname{cis} 180° = -2$ $2 \operatorname{cis} 240° = -1 - i\sqrt{3}$ $2 \operatorname{cis} 300° = 1 - i\sqrt{3}$
Find all cube roots of -343. Write each cube root in standard form.	$\dfrac{7}{2} + \dfrac{7\sqrt{3}}{2} i,$ $-7,$ $\dfrac{7}{2} - \dfrac{7\sqrt{3}}{2} i$

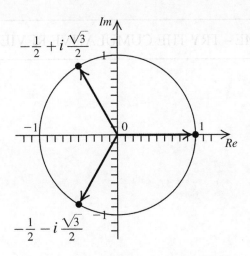

The three cube roots of 1

Matching

FIND THE BEST MATCH.

$\dfrac{a}{\sin A} = \dfrac{b}{\sin B} = \dfrac{c}{\sin C}$

DE MOIVRE'S THEOREM

$b^2 = a^2 + c^2 - 2ac \cos B$

SCALAR MULTIPLICATION

HERON'S FORMULA

$\langle 0, 1 \rangle$

DOT PRODUCT

$r = \sqrt{a^2 + b^2}$

MAGNITUDE OF A VECTOR

$z^n = r^n \operatorname{cis}(n\theta)$

$\sqrt{s(s-a)(s-b)(s-c)}$

j

$\|\vec{v}\|$

LAW OF SINES

$k\vec{v} = \langle kv_x, kv_y \rangle$

ARGUMENT OF A COMPLEX NUMBER

LAW OF COSINES

$\langle a, b \rangle \cdot \langle c, d \rangle = ac + bd$

NOW TRY THE

CHAPTER 7 REVIEW EXERCISES AND THE CHAPTER 7 PRACTICE TEST.

STAY UP ON YOUR GAME – TRY THE CUMULATIVE REVIEW EXERCISES.

Chapter 8: Topics in Analytic Geometry

Word Search

```
R  B  R  L  C  F  C  A  G  H  O  Y  E  P  E
A  Z  O  E  N  A  L  I  Y  W  T  Y  N  R  T
L  C  L  V  T  O  R  P  N  I  P  D  O  O  A
O  U  E  E  B  E  E  D  C  O  E  M  R  J  N
P  T  H  A  M  R  M  I  I  V  C  Q  H  E  I
H  E  R  X  B  N  R  A  R  O  S  E  C  C  M
O  A  R  O  W  T  I  U  R  N  I  F  O  T  I
P  Q  L  O  N  C  C  S  Q  A  U  D  T  I  L
U  A  X  E  T  R  E  V  C  L  P  T  S  L  E
R  E  C  T  A  N  G  U  L  A  R  J  I  E  L
O  C  R  O  T  A  T  I  O  N  T  M  H  W  L
E  T  O  T  P  M  Y  S  A  O  A  E  C  B  I
H  L  V  K  Y  L  X  G  S  C  X  T  A  K  P
J  M  H  Y  O  U  A  E  O  V  M  V  R  I  S
A  E  Z  F  F  Q  K  N  J  S  G  I  B  V  E
```

ASYMPTOTE	ELLIPSE	POLAR
BRACHISTOCHRONE	HYPERBOLA	PROJECTILE
CARDIOID	LEMNISCATE	RECTANGULAR
CONIC	LIMACON	ROSE
CURVE	PARABOLA	ROTATION
ECCENTRICITY	PARAMETER	VERTEX

8.1: Parabolas

We studied parabolas in chapter 2. We are going to add a little more finesse and general applicability to that knowledge now. But first, since parabolas are one of the "conic sections," we consider cones and how they can be sectioned by passing a plane through them.

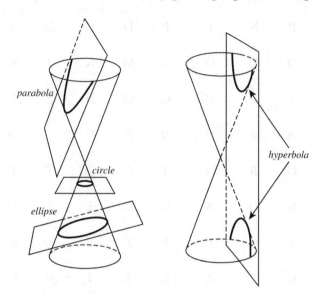

Notice how a mathematician envisions a cone? It's not just a dunce hat or an ice cream cone – the complete cone continues on in *both* directions: up and down.

Appolonious of Perga, a 3rd century B.C. Greek geometer, wrote the "Conics": the first place to show how parabolas, ellipses, hyperbolas, and circles, can be obtained by slicing a cone at various angles.

❖ degenerate conic section

There are a few special cases of passing a plane through a cone that result in something simple, such as a line, a point, or two intersecting lines. Each of these figures are referred to as degenerate conic sections.

Parabolas With Vertex at (0, 0)

We start by defining a parabola and investigating the case where the parabola's vertex is at the origin.

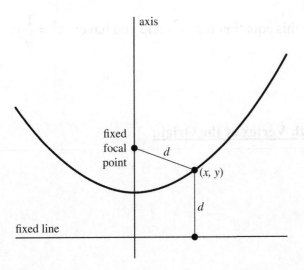

❖ parabola

You already know a parabola as a U-shaped curve. The technical definition says that a parabola is the set of all points that are the *same* distance from a fixed focal point called the focus, and a fixed line called the directrix.

❖ directrix

The fixed line mentioned in the previous paragraph is called the directrix.

❖ focus

The focus is a fixed point that, along with the directrix, defines the positions of all the points that are on the parabola. (The directrix and focus are used to define a parabola, but they are not part of the parabola.

❖ axis of symmetry

The "axis" in the previous illustration is the parabola's "axis of symmetry"..

❖ vertex

The vertex of the parabola is the (x, y) point where the axis of symmetry intersects the parabola.   In Chapter 2 we defined $f(x) = a(x-h)^2 + k$ as the standard form for the equations of a parabola. We're about to get somewhat more sophisticated. First replace $f(x)$ by y, then also let the vertex be at the origin, so both h and k are zero: $y = ax^2$. Solve this equation for x^2 and you have: $x^2 = \dfrac{1}{a}y$.

Now call the coefficient of y, $4p$.

Standard Form of the Equation of a Parabola With Vertex at the Origin

1. axis of symmetry is on the y-axis: $x^2 = 4py$

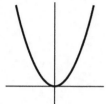

2. axis of symmetry is on the x-axis: $y^2 = 4px$

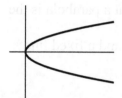

Since you know that the sign of the parameter a indicates whether the graph of the parabola opened up

or down, you know the sign of p does the same thing. In the second case, shown above, the sign of p

indicates whether the parabola opens left or right – right being reserved for positive values of p.

The distance from the vertex to the focus and the vertex to the directrix has to be the same (from the

definition of a parabola). This particular distance – the one from focus to vertex and vertex to directrix –

is the parameter p. The size of p ultimately tells you about the shape of the parabola.

Parabolas With Vertex at (h, k)

The equations for parabolas with vertex at (h, k) are as follows:

1. $(x - h)^2 = 4p(y - k)$ The graph is a parabola with a vertical axis of symmetry (it opens up or down).

2. $(y-k)^2 = 4p(x-h)$ The graph is a parabola with a horizontal axis of symmetry (it opens to the left or right).

An *interactive demonstration* that illustrates the relationship between the standard form of the equation of a parabola and its graph is available online. See the *Integrating Technology* feature that precedes Example 3 in Section 8.1 of your textbook.

Standard Forms of the Equation of a Parabola With Vertex (*h*, *k*): Test Prep

Find the standard form of the equation for the parabola with vertex at $(2,-3)$ and focus at $(0,-3)$

From the configuration of these two points, we have a parabola with horizontal axis and opening to the left. So the difference in the x-coordinates is $2-0=2=|p|$, and $p=-2$ (opening left).

Vertex at $(2,-3)$ tells us $h=2$ and $k=-3$. Therefore, we have

$$(y+3)^2 = 4(-2)(x-2)$$

$$(y+3)^2 = -8(x-2)$$

Try These

Sketch the graph of $(x-1)^2 = -2(y+2)$.

Indicate the focus and the directrix of the parabola.

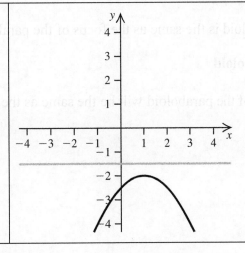

	$h = 1$ and $k = -2$. Thus the focus is $(1, -2)$. $4p = -2$, so $p = -\dfrac{1}{2}$ and the directix is a horizontal line ½ of a unit above the vertex. Thus the directrix is given by $y = -\dfrac{3}{2}$.
Find the standard form of the parabola given by $x - y^2 - 4y - 1 = 0$. Also find the x- and y-intercepts.	$(y+2)^2 = 1(x+3)$ y-intercepts: $(0, -2+\sqrt{3})$, $(0, -2-\sqrt{3})$ x-intercept: $(1,\ 0)$

Applications of Parabolas

Parabolic mirrors are one of the most well-known applications of the parabola. There might be, after all, a reason the "focus" is called the focus – light rays come together there. Or maybe television signals – a satellite dish would pick up.

❖ **paraboloid**

Revolving a 2-dimensional parabola around its axis of symmetry, produces a 3-dimensional figure called a paraboloid.

❖ **focus of a paraboloid**

The focus of a paraboloid is the same as the focus of the parabola that you created it from.

❖ **vertex of a paraboloid**

Likewise, the vertex of the paraboloid will be the same as the vertex of the parabola it is based on.

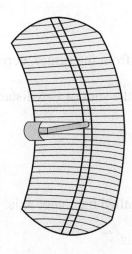

A parabolic antenna.

8.2: Ellipses

❖ ellipse

An ellipse is the set of points the sum of whose distances from two fixed foci is a constant.

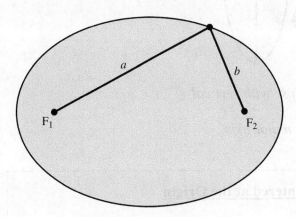

In this diagram, $a + b$ is always the same, no matter what point you pick on the ellipse.

❖ foci

The foci are the two fixed points required to describe the ellipse.

Ellipses With Center at (0,0)

Like the circle, which was based on all points being the same distance from the center, deriving the equation for an ellipse depends on the distance formula. This time it's the sum of two distances that is constant.

♣ major axis and minor axis

The graph of an ellipse has two axes of symmetry. The longer axis is called the major axis. The foci of an ellipse are on the major axis. The shorter axis is called the minor axis.

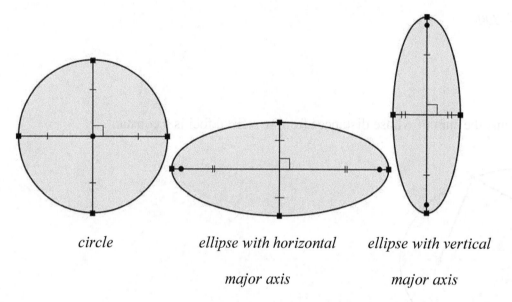

circle   *ellipse with horizontal*   *ellipse with vertical*

major axis   *major axis*

Standard Forms for an Ellipse Centered at the Origin

1. An ellipse, with center (0, 0), whose major axis is on the x-axis has an equation of the form:

$$\frac{x^2}{a^2} + \frac{y^2}{b^2} = 1 \quad \text{where} \quad a > b.$$

2. An ellipse, with center (0, 0), whose major axis is on the y-axis has an equation of the form:

$$\frac{x^2}{b^2} + \frac{y^2}{a^2} = 1 \quad \text{where} \quad a > b.$$

♣ center

The center of an ellipse is the point where the major and minor axes intersect.

❖ **vertices**

The two vertices of an ellipse are the points where the ellipse and the major axis intersect.

❖ **semimajor axis**

The length of the semimajor axis is one half of the length of the major axis that connects the center with a vertex. The length of the semimajor axis is designated as a. (What is the length the major axis?)

❖ **semiminor axis**

The length of the semiminor axis is, one half of the length of the minor axis. It connects the center of the ellipse with the points on the ellipse that are closest to the center. The length of the semiminor axis is designated as b.

Ellipse with Center (0, 0): Test Prep	
Sketch the graph of $\dfrac{x^2}{4} + \dfrac{y^2}{9} = 1$	Because $a = 3$ is greater than $b = 2$, this ellipse has a vertical major axis. The center of the ellipse is (0, 0). From (0, 0), plot the points 3 up and 3 down, 2 to the left and 2 to the right. 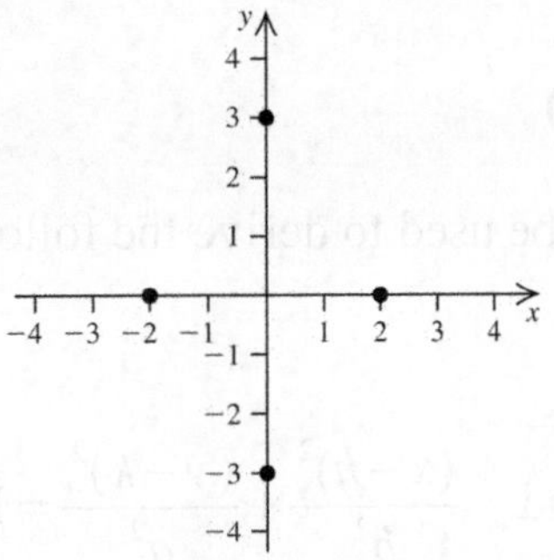

	Connect the 4 points with a smooth curve to form an ellipse. 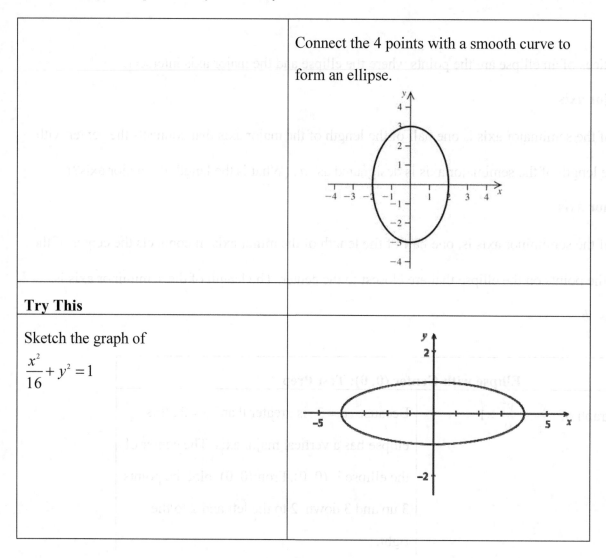
Try This	
Sketch the graph of $$\frac{x^2}{16} + y^2 = 1$$	

Ellipses With Center at (h, k)

Transformation equations can be used to derive the following standard forms for ellipses centered at (h, k).

$$\frac{(x-h)^2}{a^2} + \frac{(y-k)^2}{b^2} = 1 \quad \text{and} \quad \frac{(x-h)^2}{b^2} + \frac{(y-k)^2}{a^2} = 1 \quad \text{with} \quad a > b \quad \text{in each case.}$$

Which of the above equations is the equation of an ellipse with a horizontal major axis?

An *interactive demonstration* that illustrates the relationship between the standard form of the equation of an ellipse and its graph is available online. See the *Integrating Technology* feature that precedes Example 2 in Section 8.2 of your textbook.

Ellipse With Center at (*h*, *k*): Test Prep	
Use the method of completing the square to find the standard form of the equation for the ellipse given by $18x^2 + 12y^2 - 144x + 48y + 120 = 0$	$18x^2 + 12y^2 - 144x + 48y + 120 = 0$ $18x^2 - 144x + 12y^2 + 48y = -120$ $18(x^2 - 8x) + 12(y^2 + 4y) = -120$ $18(x^2 - 8x + 16) + 12(y^2 + 4y + 4) = -120 + 288 + 48$ $18(x-4)^2 + 12(y+2)^2 = 216$ $\dfrac{18(x-4)^2}{216} + \dfrac{12(y+2)^2}{216} = 1$ $\dfrac{(x-4)^2}{12} + \dfrac{(y+2)^2}{18} = 1$
Try These	
Use the method of completing the square to find the standard form of the equation for the ellipse given by $9x^2 + 25y^2 + 54x - 100y - 44 = 0$	$\dfrac{(x+3)^2}{25} + \dfrac{(y-2)^2}{9} = 1$
Sketch the graph of $\dfrac{(x+1)^2}{100} + \dfrac{(y-2)^2}{64} = 1$	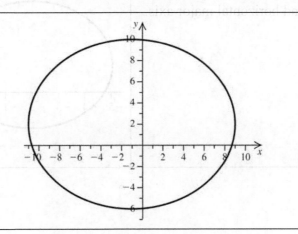

Eccentricity of an Ellipse

Some ellipses are more round than others. The measure of an ellipse's "roundness" is called its eccentricity.

♣ **eccentricity**

The eccentricity of an ellipse is given by $e = \dfrac{c}{a}$ where $c^2 = a^2 - b^2$.

The value of e is always a number between 0 and 1. (e close to 0 means an ellipse is almost circular and e close to 1 means an ellipse is elongated.)

Eccentricity of an Ellipse: Test Prep	
Find the eccentricity of the ellipse given by: $\dfrac{(y-4)^2}{289} + \dfrac{(x-3)^2}{225} = 1$	$a^2 = 289 \qquad b^2 = 225$ $e = \dfrac{\sqrt{289 - 225}}{\sqrt{289}} = \dfrac{8}{17}$
Try These	
Find the eccentricity of the ellipse given by: $4x^2 + 9y^2 + 24x + 18y + 44 = 0$	$e = \dfrac{\sqrt{5}}{3}$
Sketch the ellipse centered at $(-2,1)$ with eccentricity $\dfrac{2}{9}$ and horizontal major axis.	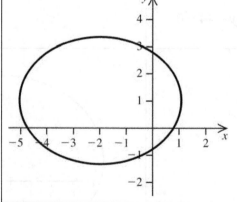

Applications of Ellipses

Planetary orbits and elliptical ceilings are two examples of applications of ellipses. There are also elliptical gears in machines and on bicycles, lithotripsy in medicine, oval tankers, and elliptical pool tables. The floor of the oval office in the White House is an ellipse. Is the President's desk is in the center of the ellipse, at one of the focuses, or is it located on the minor axis of the ellipse?

8.3: Hyperbolas

❖ **hyperbola**

This two-part curve made of flattened-out U-shapes is the conic section called the hyperbola. Defined in terms of the points that make it up, a hyperbola is all the points the *difference* of whose distances from two foci are a constant. In the following illustration $a - b$ is constant. (Whereas, it was the *sum* of these parameters that remained constant for an ellipse.)

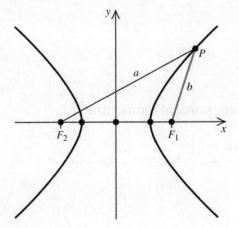

Hyperbolas With Center at (0,0)

The center of a hyperbola is the point midway between the two vertices, which are the turning-around points of the two branches of the hyperbola. So now we'll have hyperbolas that either open up-and-down or left-and-right:

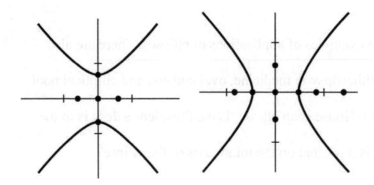

❖ **transverse axis**

With the center of the hyperbola at the origin, the transverse axis is the line segment that connects the x- or y-intercepts (depending on orientation).

Standard Forms of the Equation of a Hyperbola Centered at the Origin

1. With transverse axis on the x- axis: $\dfrac{x^2}{a^2} - \dfrac{y^2}{b^2} = 1$

2. With transverse axis on the y-axis: $\dfrac{y^2}{a^2} - \dfrac{x^2}{b^2} = 1$

Note that there is not a requirement that a be greater than b in these standard forms. Instead it is the positive term that indicates the orientation of the hyperbola.

❖ **center**

The center of the hyperbola is the midpoint of the transverse axis. The center is exactly halfway between the points where the two branches of the hyperbola are closest.

❖ **conjugate axis**

The conjugate axis is a line segment of length $2b$ perpendicular to the transverse axis and centered at the center of the hyperbola.

❖ **vertices**

The vertices of the hyperbola are the turning-around points where the two branches are nearest to each other. The vertices are each a units away from the center in the direction of the transverse axis.

❖ **asymptotes**

The asymptotes of a hyperbola are important graphing aids. They are two lines that pass through the center of the hyperbola and show the shape of the curves because each branch of the hyperbola will approach these asymptotes as you travel to the far-left or far- right.

Equations for the Asymptotes of a Hyperbola Centered at the Origin

1. Horizontal transverse axis: $y = \dfrac{b}{a}x$ and $y = -\dfrac{b}{a}x$

2. Vertical transverse axis: $y = \dfrac{a}{b}x$ and $y = -\dfrac{a}{b}x$

(The parameters a and b come from the equation of the hyperbola.)

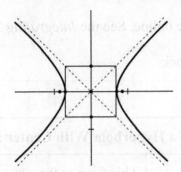

• If you make a box based on the length of the transverse axis and the conjugate axis, then your asymptotes coincide with the diagonals of that box.

Hyperbolas With Center at (h, k)

Standard Forms of the Equations of a Hyperbola With Center at (h, k)

1. Horizontal transverse axis:

$$\frac{(x-h)^2}{a^2} - \frac{(y-k)^2}{b^2} = 1 \quad \text{(with asymptotes} \quad y - k = \pm \frac{b}{a}(x-h))$$

2. Vertical transverse axis:

$$\frac{(y-k)^2}{a^2} - \frac{(x-h)^2}{b^2} = 1 \quad \text{(with asymptotes} \quad y - k = \pm \frac{a}{b}(x-h))$$

The vertices are a units away from the center in the direction of the transverse axis. The foci are c units away from the center in the direction of the transverse axis (and $c < a$). The parameters a, b, and c are related by the equation $c^2 = a^2 + b^2$.

An *interactive demonstration* that illustrates the relationship between the standard form of the equation of a hyperbola and its graph is available online. See the *Integrating Technology* feature that precedes Example 2 in Section 8.3 of your textbook.

Standard Forms of the Equation of a Hyperbola With Center at (h, k): Test Prep	
Sketch the graph of $$\frac{(x-4)^2}{9} - \frac{(y-5)^2}{16} = 1$$	This form is for a hyperbola with a horizontal transverse axis. The center is at $(4, 5)$. The asymptotes are $y = \frac{3}{4}x - \frac{1}{3}$ and $y = -\frac{3}{4}x + \frac{31}{3}$

Write the equations of the asymptotes.	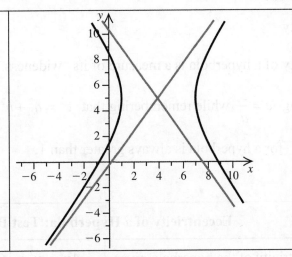

Try These

Use the method of completing the square to find the standard form of the equation for the hyperbola given by $$4x^2 - 9y^2 - 24x - 90y - 153 = 0$$	$$\dfrac{(y+5)^2}{4} - \dfrac{(x-3)^2}{9} = 1$$
Sketch the graph of the hyperbola given by $$25x^2 - 9y^2 - 100x - 72y - 269 = 0$$	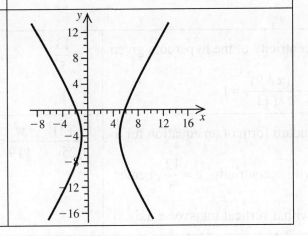

Eccentricity of a Hyperbola

Like ellipses, some hyperbolas are more eccentric than others!

❖ **eccentricity**

The eccentricity of a hyperbola is a measure of its "wideness." The eccentricity of a hyperbola is calculated using $e = \dfrac{c}{a}$ while remembering that $c^2 = a^2 + b^2$. Since c is always larger than a, the eccentricity e for a hyperbola is always greater than 1.

Eccentricity of a Hyperbola: Test Prep	
Find the eccentricity of the hyperbola given by $4x^2 - y^2 = 16$	$4x^2 - y^2 = 16$ $\dfrac{4x^2}{16} - \dfrac{y^2}{16} = 1$ $\dfrac{x^2}{4} - \dfrac{y^2}{16} = 1$ So $a^2 = 4$ and $b^2 = 16$. And $e = \dfrac{\sqrt{4+16}}{\sqrt{4}} = \dfrac{\sqrt{20}}{2} = \dfrac{2\sqrt{5}}{2} = \sqrt{5}$
Try These	
Find the eccentricity of the hyperbola given by $\dfrac{(y+5)^2}{25} - \dfrac{(x+9)^2}{144} = 1$	$e = \dfrac{13}{5}$
Find the standard form of an equation for a hyperbola with eccentricity $e = \dfrac{12}{5}$, center at $(0,-1)$, with a vertical transverse axis.	$\dfrac{(y+1)^2}{25} - \dfrac{x^2}{119} = 1$

Applications of Hyperbolas

Some comets (those that visit our sun only once) travel in hyperbolic orbits. Industry uses gears that are hyperboloids. A sonic boom is hyperbolic. Television transmission depends on the foci of hyperbolas, and some mirrors and lenses of telescopes are hyperbolic. The pattern cast on a wall by a lamp with a cylindrical shade is a hyperbola. The hyperboloid is the design standard for nuclear cooling towers.

When two stones are thrown simultaneously into a pool of still water, ripples move outward in

concentric circles, which intersect at points that form a hyperbola.

8.4: Rotation of Axes

So far we've oriented all of our conic sections so that they are "vertical" or "horizontal" – that is, they

line up with the north-south/east-west x- or y- axis. Now we'll explore how to handle conic sections

that are not constrained in their orientation. Look at the following graph of the function $y = \dfrac{1}{x}$ and see

if you recognize a hyperbola rotated 45° about the origin.

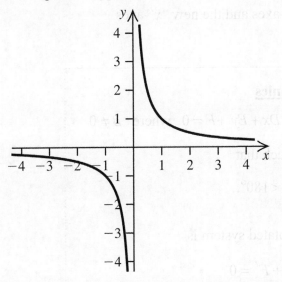

Rotation Theorem for Conics

❖ **general equation of a nondegenerate conic**

The general equation of a nondegenerate conic section is $Ax^2 + Cy^2 + Dx + Ey + F = 0$.

❖ general second-degree equation in two variables

The general second-degree equation in two variables for a conic section also includes a B term: $Ax^2 + Bxy + Cy^2 + Dx + Ey + F = 0$. This term which mixes the variables, determines the rotation of the conic into a new orientation.

❖ rotation of axes

To rotate axes means to hold the origin fixed and rotate the coordinate system in the plane and about the origin.

❖ angle of rotation

The angle of rotation is the angle between the original x-axes and the new x'-axes.

<u>Rotation Theorem for Conics</u>

Given a general conic of the form $Ax^2 + Bxy + Cy^2 + Dx + Ey + F = 0$ where $B \neq 0$

and an angle of rotation α such that

$$\cot 2\alpha = \frac{A-C}{B} \quad \text{and} \quad 0° < 2\alpha < 180°,$$

then the equation of the conic in the rotated system is

$$A'(x')^2 + C'(y')^2 + D'x' + Ey' + F' = 0$$

with the following relationships:

$$A' = A\cos^2\alpha + B\cos\alpha\sin\alpha + C\sin^2\alpha$$

$$C' = A\sin^2\alpha - B\cos\alpha\sin\alpha + C\cos^2\alpha$$

$$D' = D\cos\alpha + E\sin\alpha$$

$$E' = -D\sin\alpha + E\cos\alpha$$

$$F' = F$$

Rotation Theorem for Conics: Test Prep	
Find the angle of rotation for the conic section given by: $7x^2 - 6\sqrt{3}xy + 13y^2 - 16 = 0$	$\cot 2\alpha = \dfrac{A-C}{B} = \dfrac{7-13}{-6\sqrt{3}}$ $= \dfrac{1}{\sqrt{3}}$ $2\alpha = \cot^{-1}\dfrac{1}{\sqrt{3}} = \dfrac{\pi}{3}$ $\alpha = \dfrac{\pi}{6}$

Try These

Find the equation of the conic in the $x'y'$- system: $24x^2 + 16\sqrt{3}xy + 8y^2 - x + \sqrt{3}y - 8 = 0$	$y' = -16(x')^2 + 4$
Find the angle of rotation angle, to the nearest tenth of a degree, of the conic section given by: $32x^2 - 48xy + 18y^2 - 15x - 20y = 0$	$53.1°$

Conic Identification Theorem

Given a general second-degree equation for a conic in two dimensions, $Ax^2 + Bxy + Cy^2 + Dx + Ey + F = 0$, we can use the coefficients to determine which kind of conic section we have: parabola, ellipse, or hyperbola. In particular, it is the discriminant, $B^2 - 4AC$, that we test.

Conic Identification Theorem

- If $B^2 - 4AC < 0$, then the conic is an ellipse or circle.

- If $B^2 - 4AC = 0$, then the conic is a parabola.

- If $B^2 - 4AC > 0$, the conic is a hyperbola.

Conic Identification Theorem: Test Prep	
What conic section is described by the equation: $3xy - 8 = 0$?	$B^2 - 4AC = 9 - 4 \cdot 0 \cdot 0$ $= 9 > 0$ $\therefore$ *hyperbola*
Try These	
Identify each equation as the equation of a circle, parabola, ellipse, or hyperbola. 1. $4x^2 + 9y^2 + 3x + 9y - 6 = 0$ 2. $10x^2 + 10y^2 + 4x + 2y - 7 = 0$ 3. $8x^2 - 6y^2 + 9x + 3y - 2 = 0$ 4. $x^2 + 7x + 7y - 3 = 0$	1. ellipse 2. circle 3. hyperbola 4. parabola
Identify each equation as the equation of a circle, parabola, ellipse, or hyperbola. 1. $9x^2 + 6xy + y^2 + 6x - 3y + 5 = 0$ 2. $6xy + 2y^2 + 3x - 4y + 12 = 0$	1. parabola 2. hyperbola

Use WolframAlpha to Graph Second-Degree Equations in Two Variables

It is difficult to graph a rotated conic with a graphing calculator, but WolframAlpha (at

wolframalpha.com) can be used to graph a rotated conic just by typing in the equation of the conic into

WolframAlpha's input field. The caret symbol "^" is used to denote exponentiation.

Use WolframAlpha to Graph Conics: Test Prep	
Use WolframAlpha to graph $$8x^2 - 4xy + 3y^2 - 6x - 4y - 12 = 0$$	Enter 8x^2 - 4xy + 3y^2 - 6x - 4y – 12 = 0 in WolframAlpha's input field and click on the equal sign icon to view the graph, which is a rotated ellipse. 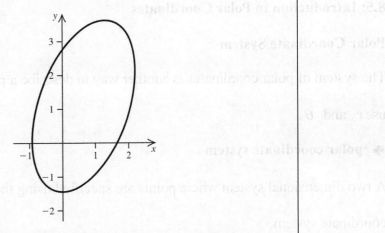 Click on the Properties icon to view the foci, vertices, center, semimajor axis length, semiminor axis length and eccentricity of the ellipse.

Try This

Use WolframAlpha to graph

$$2x^2 - 10xy + 3y^2 - x - 8y - 7 = 0$$

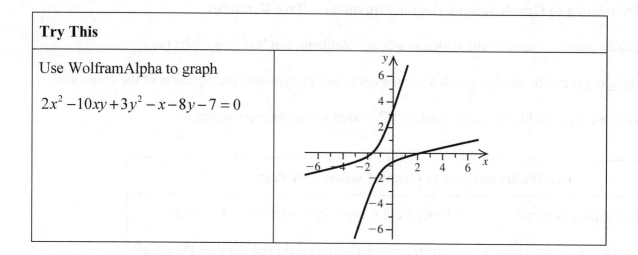

NOW TRY THE MID-CHAPTER 8 QUIZ.

8.5: Introduction to Polar Coordinates

Polar Coordinate System

The system of polar coordinates is another way to describe a point in the plane. Instead of x and y, we use r and θ.

❖ **polar coordinate system**

A two dimensional system where points are specified using the polar coordinates r and θ is a polar coordinate system.

❖ **polar axis**

The polar axis is a horizontal ray whose position specifies a polar coordinate system.

❖ **pole**

The endpoint of the polar axis (a ray) is the pole. From the pole, a line segment of length r whose direction is specified by the angle θ from the polar axis specifies a point in the polar coordinate system (just as x and y specify points in the rectangular coordinate system).

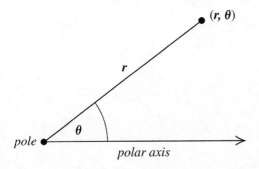

Graphs of Equations in a Polar Coordinate System

Some shapes lend themselves well to rectangular coordinates, but others can be described more gracefully with polar coordinates.

❖ **polar equation**

A polar equation is an equation in the variables r and θ. Solutions to polar equations are ordered pairs: (r, θ). Graphs of polar equations are pictures of solution sets of polar equations on the 2-dimensional polar coordinate system.

Some Familiar Shapes in Polar Coordinates

- Line through the origin: $\theta = \alpha$

- Horizontal line: $r \sin \theta = a$

- Vertical line: $r \cos \theta = a$

- Circle centered at the origin: $r = a$

❖ **limaçon**

The graph of the polar equation $r = a + b \cos \theta$ is a limaçon that is symmetric about the line $\theta = 0$:

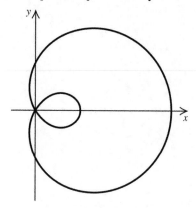

The graph of the polar equation $r = a + b \sin\theta$ is a limaçon symmetric to $\theta = 90^o$

❖ **cardioid**

The special case where $|a| = |b|$ in a polar equation describing a limaçon is a shape called a cardioid:

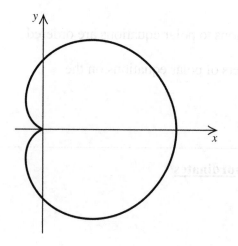

❖ **rose curves**

Rose curves are graphs of the solution sets of polar equations of the form $r = a \cos n\theta$ and

$r = a \sin n\theta$, for instance:

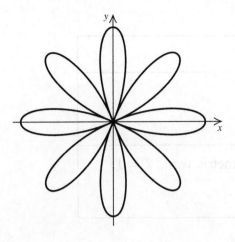

♣ **lemniscates**

The graphs of the polar equations $r^2 = a^2 \sin 2\theta$ and $r^2 = a^2 \cos 2\theta$ each have two loops that intersect at the pole. These graphs are called lemniscates.

Symmetry Tests: Test Prep	
Show that the graph of $r = 3 \sin \theta$ is symmetric with respect to (w.r.t) the line $\theta = \dfrac{\pi}{2}$.	Substitute $\pi - \theta$ for θ: $\begin{aligned} r = 3 \sin \theta &= 3 \sin(\pi - \theta) \\ &= 3(\sin \pi \cos \theta - \cos \pi \sin \theta) \\ &= 3(0 - (-\sin \theta)) \\ &= 3 \sin \theta = r \end{aligned}$ Symmetrical w.r.t. $\theta = \dfrac{\pi}{2}$

Try These	
Is the graph of $r = 2 \sin 5\theta$ symmetrical with respect to the pole?	No
Test the polar equation $r = 6 \cos \theta$ for symmetry about the line $\theta = 0$.	$r = 6 \cos \theta$ is symmetric w.r.t. $\theta = 0$.

Equations and Graphs of Lines, Circles, and Limaçons: Test Prep	
Sketch the graph of $r = 3$.	$r = 3$ is the equation for a circle of radius 3 centered at the origin: 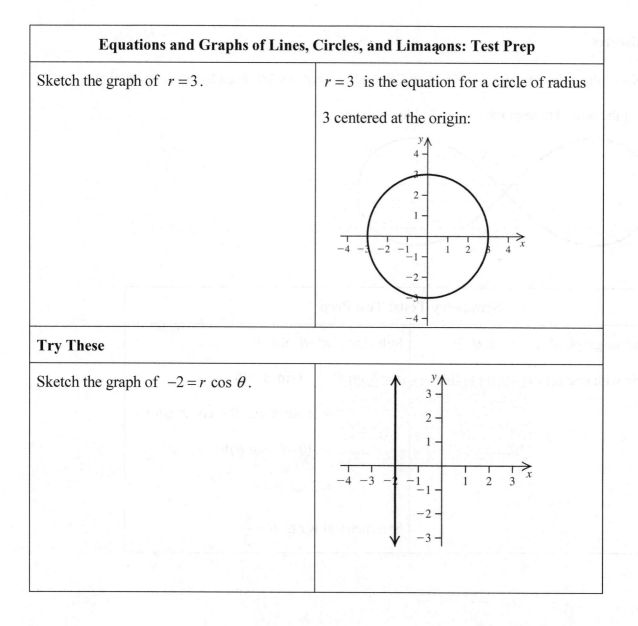
Try These	
Sketch the graph of $-2 = r \cos \theta$.	

Graph the limaçon $r = 1 + 2.6 \cos \theta$ on a polar coordinate grid.	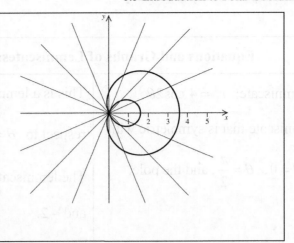

Equations and Graphs of Rose Curves: Test Prep

Graph the rose curve: $r = 5 \sin 5\theta$ This is a rose curve that is symmetric with respect to the line $\theta = \dfrac{\pi}{2}$ with 5 petals of length 5.	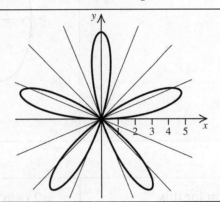

Try These

How many petals does the rose curve described by $r = 2 \sin 3\theta$ have? What line of symmetry does the graph exhibit?	3 petals; symmetry w.r.t. $\theta = \dfrac{\pi}{2}$
Find the polar equation for this rose curve: 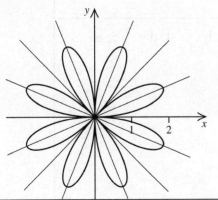	$r = 2 \sin 4\theta$

Equations and Graphs of Lemniscates: Test Prep

Graph the lemniscate: $r^2 = 4 \cos 2\theta$.

This is a lemniscate that is symmetric with respect to $\theta = 0$, $\theta = \dfrac{\pi}{2}$, and the pole.

This is a lemniscate that is symmetric with respect to $\theta = 0$, $\theta = \dfrac{\pi}{2}$, and the pole.

The lemniscate intersects the x-axis at 2, 0, and $- 2$.

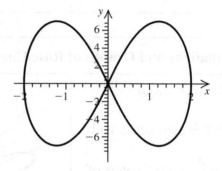

Try These

Write the polar equation for this Lemniscate.

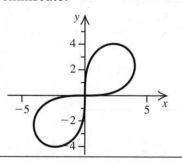

$r^2 = 25 \sin 2\theta$

Match the polar equation with its shape.

1. $r = 5 \cos 3\theta$

2. $r = 4 - 4 \sin \theta$

3. $\theta = 2$

4. $r^2 = 25 \sin 2\theta$

1. e   2. c   3. b   4. d   5. a

5. $r = 4 \sin \theta$

  a. circle   b. line   c. cardioid

  d. lemniscates   e. rose curve

Transformations Between Rectangular and Polar Coordinates

From the definitions we have already explored in trigonometry, we can transform polar and rectangular

equations into each other.

Transformations from Polar to Rectangular

$$x = r \cos \theta \qquad y = r \sin \theta$$

Transformations from Rectangular

$$r^2 = x^2 + y^2 \qquad \tan \theta = \frac{y}{x}, \; x \neq 0$$

where θ is chosen so that the point lies in the appropriate

quadrant. If $x = 0$ then $\theta = \dfrac{\pi}{2}$ or $\theta = \dfrac{3\pi}{2}$.

Transforms Between Polar and Rectangular Coordinates: Test Prep

Find the rectangular form of the equation: $r^2 \sin 2\theta = 4$.	$\begin{aligned} r^2 \sin 2\theta &= r^2 (2 \sin \theta \cos \theta) \\ &= 2(r \cos \theta)(r \sin \theta) \\ &= 2xy \\ 2xy &= 4 \\ y &= \frac{2}{x} \end{aligned}$

Transforms Between Polar and Rectangular Coordinates: Test Prep	
Try These	
Find the rectangular coordinates for the polar coordinates point given by $\left(10, \dfrac{5\pi}{3}\right)$.	$\left(5, -5\sqrt{3}\right)$
Find polar coordinates for the rectangular point given by $\left(4\sqrt{3}, -4\right)$. Let $0 \le \theta < 2\pi$ and $r > 0$.	$\left(8, \dfrac{11\pi}{6}\right)$

Writing Polar Equations as Rectangular Equations and Rectangular Equations as Polar Equations

It is possible to rewrite polar equations as rectangular equations and vice versa.

8.6: Polar Equations of the Conics

The conic sections, in rectangular, from the first three sections of this chapter (ellipses, parabolas, and hyperbolas) can be written as polar equations.

Polar Equations of the Conics

The rectangular equations for conics can be rewritten using the variables r and θ and parameters having to do with eccentricity and the directrix of the conic.

> **Standard Forms of the Polar Equations of the Conics**
>
> If the pole is a focus of a conic section, e is its eccentricity, and d is the distance its directrix is from the focus, then the equations for conics can be written in polar coordinates as follows:
>
> 1. $r = \dfrac{ed}{1 + e \cos \theta}$ (vertical directrix to the right of the pole)

2. $r = \dfrac{ed}{1 - e \cos \theta}$ (vertical directrix to the left of the pole)

3. $r = \dfrac{ed}{1 + e \sin \theta}$ (horizontal directrix above the pole)

4. $r = \dfrac{ed}{1 - e \sin \theta}$ (horizontal directrix below the pole)

- ⊙ Shapes involving cosine are symmetric with respect to $\theta = 0$

- ⊙ Shapes involving sine are symmetric with respect to $\theta = \dfrac{\pi}{2}$

Eccentricity and Conic Sections

- $e = 1$: Parabola $\bullet\ 0 < e < 1$: Ellipse $\bullet\ e > 1$: Hyperbola

Graphing a Conic Given in Polar Form

Understanding the parameters and the shapes that are made from the polar equations is, a prelude to actually sketching the graphs of the conic sections.

An *interactive demonstration* that illustrates the relationship between the polar standard form

$r = \dfrac{ed}{1 + e \cos \theta}$ and its graph is available online. See the *Exploring Concepts with Technology* feature that follows Section 8.7 in your textbook.

Standard Forms of the Polar Equations of the Conics: Test Prep	
Specify each equation as the equation of a parabola, ellipse, or hyperbola. Also specify its orientation.	1. $e = 2 > 1$, Hyperbola with focus at the pole and vertical directrix at $x = 2$; transverse axis is $\theta = 0$ and the hyperbola opens left and right.

1. $r = \dfrac{8}{2 + 4 \cos \theta}$

2. $r = \dfrac{2}{1 + \sin \theta}$

3. $r = \dfrac{15}{3 - 5 \cos \theta}$

4. $r = \dfrac{6}{3 + 2 \cos \theta}$

2. $e = 1$, Parabola with focus at the pole and horizontal directrix at $y = 2$; axis is $\theta = \dfrac{\pi}{2}$ and the parabola opens down.

3. $e = \dfrac{5}{3} > 1$, Hyperbola with focus at the pole and vertical directrix at $x = -3$, transverse axis is $\theta = 0$ and the hyperbola opens left and right.

4. $e = \dfrac{2}{3}$, $0 < \dfrac{2}{3} < 1$, Ellipse with a focus at the pole and vertical directrix at $x = 3$, major axis runs left and right.

Try These

Sketch the graph of

$$r = \dfrac{9}{3 - 3 \sin \theta}$$

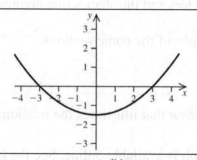

Sketch the graph of

$$r = \dfrac{8}{2 - 4 \cos \theta}$$

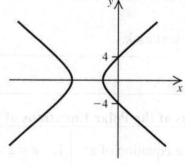

Find the Polar Equation of a Conic

Now that we've constructed graphs from the parameters e, and d, the challenge remains to come up with these parameters given pertinent information about the conic.

Example: Find the polar equation of the hyperbola with a focus at the pole, a vertex at polar point $\left(\frac{4}{3},0\right)$ and eccentricity of 2.

Solution: $e=2$ and the form of the equation will be $r=\dfrac{ed}{1+e\cos\theta}$ because the vertex is to the right of the pole (so the directrix is to the right of the pole). To find d, use the vertex $\left(\frac{4}{3},0\right)$ to substitute for r and θ into the form of the equation above, with $e=2$:

$$\frac{4}{3}=\frac{2d}{1+2\cos 0}$$

$$\frac{4}{3}=\frac{2d}{3}$$

$$d=2$$

Thus the polar equation is $r=\dfrac{4}{1+2\cos\theta}$

8.7: Parametric Equations

Parametric equations are systems of equations created by introducing a parameter (another variable that the resident variables depend on) such as t for time.

❖ **curve**

Curve and Parametric Equations

A curve is defined by a set of ordered pairs (x, y) where $x = f(t)$ and $y = g(t)$.

The t variable is a parameter whose values fall in a given interval. Choosing a value for t produces (x, y) points on the graph of the parametric equations.

Eliminating the Parameter of a Pair of Parametric Equations

Eliminating the parameter can be accomplished by substitution. Solve one parametric equation for t

and substitute that expression, into the other parametric equation.

Eliminate the Parameter: Test Prep	
Eliminate the parameter t in the following pair of parametric equations. $x = \sqrt{t}$, $y = t - 2$, $t \geq 0$	$x = \sqrt{t}$ $x^2 = t$ Thus $y = x^2 - 2$, $x \geq 0$
Try These	
Eliminate the parameter t by squaring both equations and adding them together: $x = 3 \cos t$, $y = 3 \sin t$	$x^2 = 9\cos^2 t$ and $y^2 = 9\sin^2 t$ $x^2 + y^2 = 9\cos^2 t + 9\sin^2 t$ $x^2 + y^2 = 9(\cos^2 t + 9\sin^2 t)$ $x^2 + y^2 = 9 \cdot 1$ $x^2 + y^2 = 9$ The graph of the parametric equations is a circle with radius 3 and center at (0, 0).
Eliminate the parameter θ: $x = 4 + 2\cos\theta$, $y = -1 + \sin\theta$	Let $\dfrac{x-4}{2} = \cos\theta$, and $y + 1 = \sin\theta$ Then: $\left(\dfrac{x-4}{2}\right)^2 = \cos^2\theta$, $(y+1)^2 = \sin^2\theta$ $\left(\dfrac{x-4}{2}\right)^2 = (y+1)^2 = \cos^2\theta + \sin^2\theta$ $\left(\dfrac{x-4}{2}\right)^2 + (y+1)^2 = 1$ Ellipse centered at (4, –1) with horizontal major axis.

Time as a Parameter

♣ **orientation**

The direction that a curve is traversed by increasing values of the parameter, is referred to as its orientation. We draw arrows on the curve to show the direction that a curve is traversed.

Time as a Parameter: Test Prep	
A point moves in a plane according to the equations $x = 1 - t$, $y = t^2$, and $0 \le t \le 2$ where the parameter t is time measured in seconds. Eliminate t and describe the motion of the point.	$t = \sqrt{y}$ Therefore: $x = 1 - \sqrt{y}$ $1 - x = \sqrt{y}$ $y = (1 - x)^2$ $y = (x - 1)^2$, $0 \le y \le 4; -1 \le x \le 1$ The point starts at $(1, 0)$ and moves along a parabola with vertex at $(1, 0)$ until after 2 seconds it stops at $(-1, 4)$. 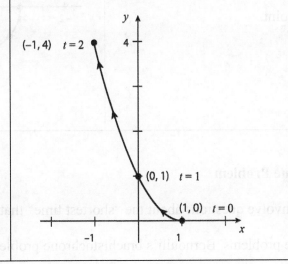

Time as a Parameter: Test Prep

Try These	
A point moves in a plane according to the equations: $x = -t,\ y = \dfrac{1}{2}t,\ \ 0 \le t \le 4$ where the parameter t is time, measured in seconds. Eliminate t and describe the motion of the point.	$y = \dfrac{1}{2}x + 2$ The point starts at $(4, 0)$ and moves along a line until after 4 seconds it stops at $(0, 2)$. 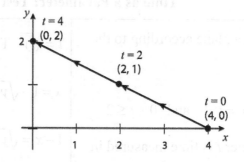
A point moves in a plane according to the equations: $x = -\cos t,\ y = \sin t,$ for $0 \le t \le \pi$ where the parameter t is time measured in minutes. Eliminate t and describe the motion of the point.	$x^2 + y^2 = 1$ The point starts at $(-1, 0)$ and moves along a semicircle until after π minutes it stops at $(1, 0)$. 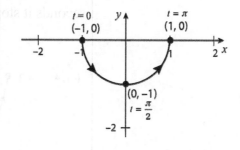

Brachistochrone Problem

Problems that involve questions about the "shortest time" that it takes something to happen are brachistochrone problems. Bernoulli's brachistochrone problem has a solution whose curve is called a cycloid. A cycloid is the shape a fixed point on a circle makes as the circle rolls, without slipping, along

a flat surface:

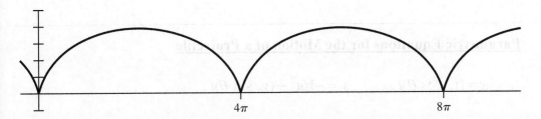

★ **For Fun**

Look up the derivation of the word "brachistochrone".

Parametric Equations of a Cycloid: Test Prep	
Graph the following cycloids on the same axis: $\begin{cases} x = \theta - \sin\theta \\ y = 1 - \cos\theta \end{cases}$ $\begin{cases} x = 0.5(\theta - \sin\theta) \\ y = 0.5(1 - \cos\theta) \end{cases}$	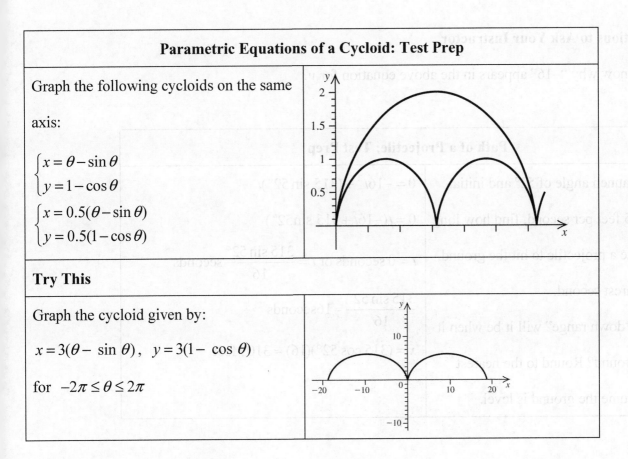
Try This	
Graph the cycloid given by: $x = 3(\theta - \sin\theta),\ \ y = 3(1 - \cos\theta)$ for $-2\pi \le \theta \le 2\pi$	

★ **For Fun**

Curves that are related to the cycloid include trochoids, epicycloids, hypocycloids, epitrochoids, and

hypotrochoids. If you've used a Spirograph, then you are familiar with these curves.

Parametric Equations and Projectile Motion

Parametric equations are often used to describe the motion of a projectile.

Parametric Equations for the Motion of a Projectile

$$x = (v_0 \cos \theta)t, \qquad y = -16t^2 + (v_0 \sin \theta)t$$

where θ is the angle of launch, v_0 is the initial velocity in feet per second, and t is time in seconds since the projectile was launched.

✓ **Questions to Ask Your Instructor**

Do you know why "–16" appears in the above equation for y?

Path of a Projectile: Test Prep	
Given a launch angle of 52° and initial speed 315 feet per second, find how long it will take a projectile to hit the ground, to the nearest second. How far "down range" will it be when it hits the ground? Round to the nearest foot? Assume the ground is level.	$0 = -16t^2 + (315 \sin 52°)t$ $0 = t(-16t + 315 \sin 52^o)$ $t = 0$ seconds or $t = \dfrac{315 \sin 52°}{16}$ seconds $\dfrac{315 \sin 52°}{16} \approx 16$ seconds $x = (315 \cos 52°)(16) \approx 3103$ feet

Try These	
Graph the trajectory of a projectile with a launch angle of 75 degrees and initial velocity of 50 feet per second.	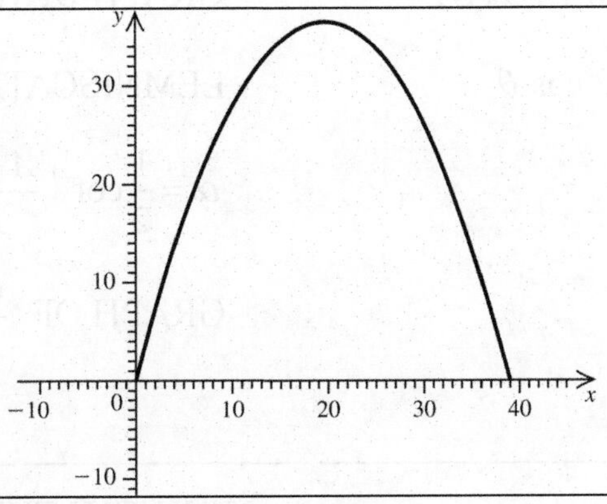
In the above example, eliminate the parameter t and find the maximum height of the projectile by algebraic means. Round to the nearest foot.	37 feet

Matching

FIND THE BEST MATCH.

POLAR COORDINATE

ECCENTRICITY OF ELLIPSE

GRAPH OF $x^2 + y^2 = r^2$

PARABOLA WITH VERTICAL AXIS

ECCENTRICITY OF HYPERBOLA

ELLIPSE WITH VERTICAL MAJOR

GRAPH OF $\dfrac{y^2}{a^2} - \dfrac{x^2}{b^2} = 1$

POLAR EQUATION

$B^2 - 4AC$

GRAPH OF $y = -\dfrac{b}{a}x$

$e = \dfrac{\sqrt{a^2 + b^2}}{a}$

LIMAÇON

HYPERBOLA

GRAPH OF $r = \dfrac{9}{3 - 3\sin\theta}$

$(3, 240°)$

DISCRIMINANT

$e = \dfrac{\sqrt{a^2 - b^2}}{a}$

GRAPH OF $16 = 9 \cos 2\theta$ CIRCLE

HYPERBOLIC ASYMPTOTE   GRAPH OF $(x - h)^2 = 4p(y - k)$

GRAPH OF $r = 3 - 2 \cos \theta$ LEMNISCATE

ROTATION ANGLE $\alpha = \dfrac{1}{2} \cot^{-1} \dfrac{A - C}{B}$

                                   GRAPH OF $\dfrac{(x - h)^2}{b^2} + \dfrac{(y - k)^2}{a^2} = 1$

NOW TRY THE

CHAPTER 8 REVIEW EXERCISES AND THE CHAPTER 8 PRACTICE TEST.

STAY UP ON YOUR GAME – TRY THE CUMULATIVE REVIEW EXERCISES.

Chapter 9: Systems of Equations and Inequalities

Word Search

```
N V E H R R O P K I R T D N I
E O I L A A A R N D N J S O N
T G I E I R E E Q E F T O I D
Q R N T T M Q N T M N J L T E
D I I I I U I S I I F E U U P
L E A A S I N A L Q G T T E
M L P L N S O R A U N R I I N
V E I E N G T P I T I O O T D
V T T O N S U L M V I O N S E
Y F C S N D I L I O H O R B N
R N E O Y B E A A S C V N U T
I B C T R S L N X R N E I S H
O P T I M I Z A T I O N D V N
N S U O E N E G O M O H M B M
Y M E Q U I V A L E N T Q S S
```

CONSTRAINTS   INCONSISTENT   SOLUTION

DECOMPOSITION   INDEPENDENT   SUBSTITUTION

DEPENDENT   INEQUALITY   SYSTEM

ELIMINATION   LINEAR   TRIANGULAR

EQUILIBRIUM   NONLINEAR   TRIVIAL

EQUIVALENT   OPTIMIZATION   HOMOGENEOUS

9.1: Systems of Linear Equations in Two Variables

❖ system of equations

A system of equations is a group of two or more equations.

❖ linear system of equations

A *linear* system of equations is a system of linear equations. A linear equation in two variables has the general form: $Ax + By = C$.

❖ solution of a system of equations

A solution to an equation, in two variables, is an ordered pair whose x- and y- values make the equation true. The solution to a system of equations, in two variables, is an ordered pair whose x- and y- values make all of the equations true.

❖ consistent system

A consistent system of equations has at least one solution.

❖ inconsistent system

An inconsistent system has no solutions.

★ For Fun

Can you picture a case where, say, two linear equations do not share an ordered pair in their solution sets?

❖ independent system

An independent system is a consistent system that has exactly *one* solution.

❖ dependent system

A dependent system of equations is a consistent system that has an infinite number of solutions.

★ For Fun

Can you picture a system of two linear equations which have an infinite number of solutions?

Substitution Method for Solving a System of Equations

One method for finding the solution(s), or lack thereof, to a system of equations is the *substitution* method.

❖ **substitution method**

The method of substitution involves the following steps:

1. If necessary, solve one of the equations for one of the variables. (It neither matters which equation nor which variable, so pick the easiest.) For example:

    $$\begin{cases} -3x+7y = 3x \ (1) \\ 2x - y = -13 \ (2) \end{cases}$$

    Solve equation (2) for y.

    $$y = 2x + 13$$

2. Rewrite equation (1) by substituting the expression that you just derived for y.

    $$-3x + 7(2x+13) = 14$$

3. Now solve that equation for the *one* variable it is now in terms of.

    $$-3x + 7(2x+13) = 14$$

    $$-3x + 14x + 91 = 14$$

    $$11x = -77$$

    $$x = -7$$

4. Substitute -7 for x in equation (2) and find the corresponding value of y.

    $$2(-7) - y = -13$$

    $$-14 - y = -13$$

    $$y = -1$$

5. Write your answer in the form of an ordered pair.

    $$(-7, -1)$$

Elimination Method for Solving Systems of Equations.

♣ **equivalent systems**

Two systems of equations are equivalent as long as they have the same solution set. (So if we can rewrite a system into an equivalent system that is easier to solve, we're still finding the same solution set for the original system.

Procedures for Producing Equivalent Systems:

1. Interchanging any two equations.

2. Multiplying both sides of any equation by the same non-zero value.

3. Replace an equation with the sum of that equation and a nonzero constant multiple of another equation in the system.

♣ **elimination method**

The elimination method uses the properties of equivalent systems to *eliminate* one of the variables and then proceed to solve the system of equations. For example, consider the system:

$$\begin{cases} 3x-8y = -6 \ (1) \\ -5x+4y = 10 \ (2) \end{cases}$$

Multiplying equation (2) by 2 yields the following system:

$$\begin{cases} 3x-8y = -6 \ (1) \\ -10x+8y = 20 \end{cases}$$

Adding the new second equation to equation (1), the left sides together and rights, respectively, eliminates the y-variable and results in the new equation:

$$-7x = 14$$

The solution of this equation is $x = -2$

Substituting -2 for x in one of our original equations gives us the value of y.

$$3(-2)-8y = -6$$

$$-6 - 8y = -6$$

$$-8y = 0$$

$$y = 0$$

The solution of our original system is (–2, 0),

You can *check* your solution in equation (2) by substituting –2 for x and 0 for y.

$$-5(-2) + 4(0) = 10 ?$$

✓ Questions to Ask Your Instructor

Both of the previous examples involved independent systems – they had exactly one ordered-pair

solution. How might you recognize an inconsistent or dependent system of equations?

★ For Fun

Besides the methods of substitution and elimination, Do you think you could find solutions by graphing?

Systems of Linear Equations in Two Variables: Test Prep	
Solve the following system of linear equations. $$\begin{cases} 3x - 4y = 8 \\ 6x - 8y = 9 \end{cases}$$	$\begin{cases} 3x - 4y = 8 & (1) \\ 6x - 8y = 9 & (2) \end{cases}$ Multiply Equation (1) by 2 and replace Equation (1) with the result. $\begin{cases} 6x - 8y = 16 & (3) \\ 6x - 8y = 9 & (2) \end{cases}$ Subtract Equation (2) from Equation (1) to produce $0 = 7$, which is a contradiction. Thus, the system has no solutions.

Try These

Solve the system of equations.	
$\begin{cases} \dfrac{3}{4}x + \dfrac{2}{5}y = 1 \\ \dfrac{1}{2}x - \dfrac{3}{5}y = -1 \end{cases}$	$\left(\dfrac{4}{13}, \dfrac{25}{13} \right)$
Solve the system of equations.	
$\begin{cases} y = 3x + 4 \\ x = 4y - 5 \end{cases}$	$(-1, 1)$

A system of linear equations can be solved using four methods:

• substitution   • elimination   • graphing   • matrices (next chapter)

Classification of Systems of Linear Equations: Test Prep

Determine whether the following system is independent, dependent, or inconsistent? $\begin{cases} 2x + y = 15 \\ x - 4y = -1 \end{cases}$	$\begin{cases} 2x + y = 15 \\ x - 4y = -1 \end{cases}$ can be rewritten as: $\begin{cases} 2x + y = 15 \\ 2x - 8y = -2 \end{cases}$ Subtracting the second equation from the first equations produces: $9y = 17$ $y = \dfrac{17}{9}$

	Substituting back in for y in the second equation of the original system yields: $$x - 4\left(\frac{17}{9}\right) = -1$$ $$x = \frac{-9 + 68}{9}$$ $$x = \frac{59}{9}$$ The lines intersect at the point, $\left(\frac{59}{9}, \frac{17}{9}\right)$. The system is an independent system.
Try These	
Is $\begin{cases} x + 3y = 4 \\ 2x - y = 1 \end{cases}$ inconsistent, dependent, or independent?	Independent system
Is $\begin{cases} \dfrac{x+3}{4} = \dfrac{2y-1}{6} \\ 3x - 4y = 2 \end{cases}$ independent, dependent, or inconsistent?	Inconsistent system

Applications of Systems of Equations

Economics, biology, chemistry, and physics are some of the places you may need to solve systems of linear equations. Any "universe" where two equations are operating simultaneously is a possible candidate.

❖ equilibrium price

The price that balances income and expenses is the equilibrium price.

★ For Fun

Equilibrium is an interesting word. You might look it up!

❖ supply-demand problems

A supply-demand problem entails two equations: a supply equation and a demand equation. Solving such a system solves the supply-demand problem by returning the correct amount to be manufactured and the best price to charge for it!

Solve a Supply-Demand Problem:

If the number of computers to be manufactured for a price p is x, and the number that will be purchased by a retailer depends on p according to the following two equations, then can you find the (x, p) data point that is the solution to this system?

$$x = \frac{9}{5}p - 1530$$

$$x = -\frac{3}{10}p + 780$$

9.2: Systems of Linear Equations in Three Variables

In the last section, we worked in two dimensions. Now we extend that practice into three dimensions.

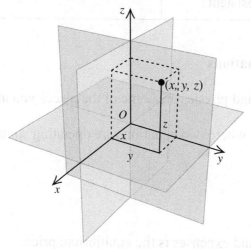

Three dimensional space is somewhat difficult to portray in two dimensions. Can you tell which way these axes are pointing? Can you pick out the three 2-dimensional planes that partition this 3-d space?

Systems of Linear Equations in Three Variables

The methods for solving linear equations in three dimensions "by hand" are exactly the same as those

we used for 2-dimensional systems ... only more complicated.

❖ **ordered triple**

An ordered triple can be the solution to a system of linear equations in three dimensions:

(x, y, z).

❖ **z-axis**

Three dimensions requires another axis, or direction, to be specified and added to the 2-dimensional

Cartesian plane. Most often that axis is called the z axis, and coordinate distances measured along the

scale on the z-axis come third in an ordered triple.

★ **For Fun**

Look carefully at the previous graphic for the 3-dimensional system. This is the usual orientation of the

three axes. Can you picture these axes in three dimensions? The x-axis would actually be poking out of

the page—straight out, orthogonal to the plane of the page. The y- and z-axes are orthogonal to each other

and in the plane of the page. Can you tell where the point (x, y, z) in the graphic is situated?

Triangular Form

Triangular form is a particular form for three or more linear equations that allows you to find solutions

by a method called "back substitution". The idea is to find a triangular system of equations *equivalent* to

the system you have to solve. This is accomplished by using the same manipulations we did earlier to

find equivalent systems.

❖ **triangular form and back substitution**

The following system of equations is written in triangular form.

$$\begin{cases} x + 2y - z = 1 \\ -5y + 3z = 4 \\ -2z = -6 \end{cases}$$

This form makes it possible to solve the bottom equation for one variable, and use that result to solve the next equation up, and use those results to solve the top equation.

★ For Fun

Use the above triangular form to show that it has the solution (2, 1, 3).

Learning how to write a system of equations in triangular form will make the material in the next chapter easer to understand. Here is an example:

$$\begin{cases} x - 2y + 3z = 5 \\ 3x - 3y + z = 9 \\ 5x + y - 3z = 3 \end{cases}$$

Multiply the top equation by 5. Subtract the new top equation from the bottom equation to produce:

$$\begin{cases} 5x - 10y + 15z = 25 \\ 3x - 3y + z = 9 \\ 11y - 18z = -22 \end{cases}$$

Divide the top equation by 5 and multiply it by 3. Subtract the new top equation from the second equation.

$$\begin{cases} 3x - 6y + 9z = 15 \\ 3y - 8z = -6 \\ 11y - 18z = -22 \end{cases}$$

Divide the first equation by 3 (gets it back to its original form). Multiply the bottom equation by 3 and the second equation by 11, subtract the second equation from the bottom equation.

$$\begin{cases} x-2y+ 3z=5 \\ 33y-88z=-66 \\ 34z=0 \end{cases}$$

Divide the new middle equation by 11 and begin the back-substitution procedure.

$$\begin{cases} x-2y+ 3z=5 \\ 3y-8z=-6 \\ 34z=0 \end{cases}$$

From the bottom equation we see that $z=0$. Substitute 0 for z in the second equation to produce $3y-8(0)=-6$. So $y=-2$. Substitute 0 for z and -2 for y in the top equation to produce $x-2(-2)+3(0)=5$. Solve the resulting equation to show that $x=1$.

Thus the solution is $(1, -2, 0)$. Lots of work, but it's fun once you get the rhythm of it.

Shall we solve a 4-dimensional system? … Yeah, I thought not! The next chapter covers methods for solving larger systems of equations.

Systems of Linear Equations in Three Variables: Test Prep	
Solve $$\begin{cases} 2x - y + z = 3 \quad (1) \\ x + 3y - 2z =11 \quad (2) \\ 3x - 2y + 4z = 1 \quad (3) \end{cases}$$	$$\begin{cases} 2x - y + z = 3 \quad (1) \\ 3x + 9y - 6z =33 \quad (4)\text{: 3 times Equation (2)} \\ 3x -2y + 4z = 1 \quad (3) \end{cases}$$ $$\begin{cases} 2x - y +z = 3 \quad (1) \\ x +3y -2z = 11 \quad (2) \\ -11y +10z = -32 \quad (5)\text{: Eq. (4) subtracted from Eq. (3)} \end{cases}$$

$$\begin{cases} 2x & -y & +z & = & 3 & \text{(1)} \\ 2x & +6y & -4z & = & 22 & \text{(6): 2 times Equation (2)} \\ & -11y & +10z & = & -32 & \text{(5)} \end{cases}$$

$$\begin{cases} 2x & -y & +z & = & 3 & \text{(1)} \\ & 7y & -5z & = & 19 & \text{(7): Equation (1) subtracted from (6)} \\ & -11y & +10z & = & -32 & \text{(5)} \end{cases}$$

$$\begin{cases} 2x & -y & +z & = & 3 & \text{(1)} \\ & 77y & -55z & = & 209 & \text{(8): 11 times Equation (7)} \\ & -77y & +70z & = & -224 & \text{(9): 7 times Equation (5)} \end{cases}$$

$$\begin{cases} 2x & -y & +z & = & 3 & \text{(1)} \\ & 7y & -5z & = & 19 & \text{(7)} \\ & & 15z & = & -15 & \text{(10): Eq. (8) added to Eq. (9)} \end{cases}$$

$z = -1$

Back substitution:

$7y + 5 = 19; \quad y = 2$

$2x - 2 - 1 = 3; \quad x = 3$

The solution of the system is:

$(3, 2, -1)$

Try These

Solve:

$$\begin{cases} 4x + 6y - 3z = 24 \\ 3x + 4y - 6z = 2 \\ 6x - 3y + 4z = 46 \end{cases}$$

$(6, 2, 4)$

Solve:

$$\begin{cases} 3x + y - z = 4 \\ x + y + 4z = 3 \\ 9x + 5y + 10z = 8 \end{cases}$$

Inconsistent system

Nonsquare Systems of Equations

A system of equations with fewer equations than variables is a nonsquare system. Nonsquare systems either have no solution or infinitely many solutions.

Nonsquare Systems of Equations: Test Prep	
Write the infinite solutions of this system as an ordered triple in terms of the parameter c. $$\begin{cases} x + 2y - 3z = 5 \\ 3x + 7y - 10z = 13 \end{cases}$$	$$\begin{cases} x + 2y - 3z = 5 \quad (1) \\ 3x + 7y - 10z = 13 \quad (2) \end{cases}$$ $$\begin{cases} 3x + 6y - 9z = 15 \quad (3)\ 3\text{ times Equation (1)} \\ 0 + y - 1z = -2 \quad (4) \end{cases}$$ $$\begin{cases} x + 2y - 3z = 5 \quad (1) \\ 0 + y - 1z = -2 \quad (4) \end{cases}$$ $$y = z - 2$$ $$x + 2z - 4 - 3z = 5; \quad x = z + 9$$ Let z be any real number c. Then the solutions are the ordered triples of the form $$(c + 9, c - 2, c).$$

Try These	
Does this nonsquare system have an infinite number of solutions or no solution? $\begin{cases} x^2 + 4x + y \leq -6 \\ x + y > -6 \end{cases}$	No solution
Does this nonsquare system have an infinite number of solutions or no solution? $\begin{cases} x - 3y + 4z = 9 \\ 3x - 8y - 2z = 4 \end{cases}$	Infinite number of solutions

Homogeneous Systems of Equations

A linear system of equations in which the constant term is 0 for all equations is called a homogeneous system of equations.

❖ **homogeneous systems of equations**

Here is an example of a homogeneous system of linear equations.

$$x - 4y + 2z = 0$$
$$-x + 7y - 4z = 0$$
$$2x - 17y + 10z = 0$$

❖ **trivial solution**

Every system of homogeneous equations in three variables has $(0, 0, 0)$ as a solution. This solution is called the trivial solution. Usually we try to discover if there are other solutions (as with the example above).

Homogeneous Systems of Equations: Test Prep

Write the infinite solutions of this system as an ordered triple in terms of the parameter c.

$$\begin{cases} 2x & -3y & +5z & = & 0 \\ 3x & +2y & +z & = & 0 \\ x & -4y & +5z & = & 0 \end{cases}$$

$$\begin{cases} 2x & -3y & +5z & = & 0 \\ 3x & +2y & +z & = & 0 \\ x & -4y & +5z & = & 0 \end{cases}$$

$$\begin{cases} x & -4y & +5z & = & 0 \\ 6x & -9y & +15z & = & 0 \\ 6x & +4y & +2z & = & 0 \end{cases}$$

$$\begin{cases} x & -4y & +5z & = & 0 \\ & -13y & +13z & = & 0 \\ & 13y & -13z & = & 0 \end{cases}$$

$$\begin{cases} x & -4y & +5z & = & 0 \\ & -y & +z & = & 0 \\ & & 0 & = & 0 \end{cases}$$

$$y = z$$

$$x - 4z + 5z = 0; \quad x = -z$$

If z is any real number c. Then the solutions of the system are the ordered triples of the form:

$$(-c \quad c \quad c).$$

Try These

Show that the trivial solution is the only solution of:

$$\begin{cases} 5x & +2y & +3z & = & 0 \\ 3x & +y & -2z & = & 0 \\ 4x & -7y & +5z & = & 0 \end{cases}$$

$$\begin{cases} 5x & +2y & +3z & = & 0 \\ 3x & +y & -2z & = & 0 \\ 4x & -7y & +5z & = & 0 \end{cases}$$

$$\begin{cases} 20x & +8y & +12z & = & 0 \\ 3x & +y & -2z & = & 0 \\ 20x & -35y & +25z & = & 0 \end{cases}$$

$$\begin{cases} 15x & +6y & +9z & = & 0 \\ 15x & +5y & -10z & = & 0 \\ & 43y & -13z & = & 0 \end{cases}$$

$$\begin{cases} 5x & +2y & +3z & = & 0 \\ & -y & -19z & = & 0 \\ & 43y & -13z & = & 0 \end{cases}$$

$$\begin{cases} 5x & +2y & +3z & = & 0 \\ & -43y & -817z & = & 0 \\ & 43y & -13z & = & 0 \end{cases}$$

$$\begin{cases} 5x & +2y & +3z & = & 0 \\ & -43y & -817z & = & 0 \\ & & -830z & = & 0 \end{cases}$$

$$z = 0$$

$$-43y - 817(0) = 0; \text{ so } y = 0$$

$$5x + 2(0) + 3(0) = 0; \text{ so } ; x = 0$$

The only solution is $(0, 0, 0)$.

Solve:

$$\begin{cases} 5x & -2y & -3z & = & 0 \\ 3x & -y & -4z & = & 0 \\ 4x & -y & -9z & = & 0 \end{cases}$$

The solutions of the system are the ordered triples of the form. $(5c, 11c, c)$

Applications of Systems of Equations

Systems of linear equations are used frequently in mathematics (curve fitting and differential equations), chemistry (simultaneous chemical processes), and business (time/cost/demand applications).

9.3: Nonlinear Systems of Equations

Solving Nonlinear Systems of Equations

You may use the method of substitution or elimination to solve nonlinear systems.

❖ **nonlinear system of equations**

A nonlinear system of equations is a system of equations, where one or more of the equations is *not* linear.

Here is an example of a nonlinear system:

$$\begin{cases} 5x + y = 3 \\ y = x^2 - 3x - 5 \end{cases}$$

A graph of the above system is shown on the next page.

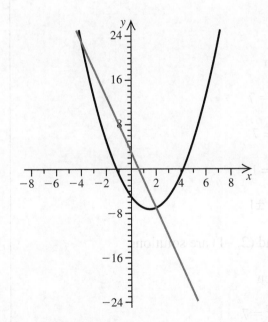

It appears from the graph that the nonlinear system has two ordered pair solutions.

★ For Fun

Is the following system of equations a nonlinear system?

$$\begin{cases} x+y=1 \\ xy\ =1 \end{cases}$$

Solutions of Nonlinear Systems of Equations: Test Prep

Solve the following nonlinear system of equations by using the elimination method:

$$\begin{cases} 2x^2\ -y^2\ =\ 7 & (1) \\ 3x^2\ +2y^2\ =\ 14 & (2) \end{cases}$$

$$\begin{cases} 4x^2-2y^2=14 & (3):\ \text{Equation (1) multiplied by 2} \\ 3x^2+2y^2=14 & (2) \end{cases}$$

Adding equation (3) to equation (2) equations yields:

$$7x^2 = 28$$

$$x^2 = 4$$

$$x = \pm 2$$

If $x = 2$, then

$$2(2)^2 - y^2 = 7$$

$$8 - y^2 = 7$$

$$y^2 = 1$$

$$y = \pm 1$$

Thus (2, 1) and (2, −1) are solutions.

If $x = -2$, then

$$2(-2)^2 - y^2 = 7$$

$$8 - y^2 = 7$$

$$y^2 = 1$$

	$y = \pm 1$ Thus $(-2, 1)$ and $(-2, -1)$ are also solutions. Therefore, this system of equations has the following four solutions. $(2, 1), (2, -1), (-2, 1), (-2, -1)$

Try These

Graph and solve: $\begin{cases} x^2 + y^2 = 25 \\ x + 2y = 10 \end{cases}$	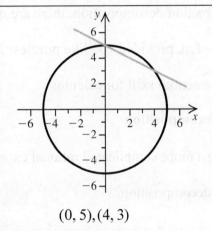 $(0, 5), (4, 3)$
Solve by using the substitution method. $\begin{cases} x^2 + y^2 = 40 \\ xy = 12 \end{cases}$	$(6, 2), (-6, -2), (2, 6), (-2, -6)$

NOW TRY THE MID-CHAPTER 9 QUIZ.

9.4: Partial Fractions

Partial fractions are about taking the right-hand side of an equation, like those below, and *decomposing*

it into the corresponding left-hand side.

$$\frac{3}{5} - \frac{2}{7} = \frac{11}{35} \qquad\qquad \frac{5}{x-1} + \frac{1}{x+2} = \frac{6x+9}{(x-1)(x+2)}$$

Are you "up" on your factoring skills? They are essential here, just like they were with rational expressions and functions. You need to start by factoring the denominators of every ratio you need to "decompose." Can you see that the factors, in the denominators, on the right-hand side of the equations above appear in the left- hand side of the equations?

Partial Fraction Decomposition

In order to do partial fraction decomposition, there are quite a few cases and procedures to learn. Practicing these can be fun, provided you like puzzles! If you need more motivation than fun, then be assured that this is an essential skill for calculus.

❖ **partial fraction decomposition**

The process of writing a more complicated rational expression in terms of sums of simpler ratios is called partial fraction decomposition.

Case 1: Nonrepeated Linear Factors

Each non-repeated linear factor, $ax+b$, in a denominator will be included in the decomposition in the form $\dfrac{A}{ax+b}$.

In the following example we need to determine the values of A and B so that

$$\frac{x-5}{x^2-1} = \frac{A}{x+1} + \frac{B}{x-1}$$

Start by factoring x^2-1.

$$\frac{x-5}{x^2-1} = \frac{x-5}{(x+1)(x-1)} = \frac{A}{x+1} + \frac{B}{x-1}$$

$$(x+1)(x-1)\left(\frac{x-5}{(x+1)(x-1)}\right) = (x+1)(x-1)\left(\frac{A}{x+1} + \frac{B}{x-1}\right) \qquad \text{Multiply each side by } (x+1)(x-1).$$

$$x - 5 = A(x-1) + B(x+1)$$

Let $x = 1$ in the equation above.

$$-4 = A \cdot 0 + B \cdot 2$$

$$B = -2$$

Now let $x = -1$ in the same equation: $x - 5 = A(x-1) + B(x+1)$

$$-6 = A \cdot (-2) + B \cdot 0$$

$$A = 3$$

Thus, $\dfrac{x-5}{x^2-1} = \dfrac{A}{x+1} + \dfrac{B}{x-1} = \dfrac{3}{x+1} + \dfrac{-2}{x-1}$

Case 2: Repeated Linear Factors

Each repeated linear factor in the denominator is included in the decomposition in the form

$$\frac{A_1}{ax+b} + \frac{A_2}{(ax+b)^2} + \ldots + \frac{A_m}{(ax+b)^m}$$ where m is the multiplicity of the factor in the

denominator.

Example

$$\frac{-7x+27}{x(x-3)^2} = \frac{A}{x} + \frac{B}{x-3} + \frac{C}{(x-3)^2}$$

$$x(x-3)^2 \left(\frac{-7x+27}{x(x-3)^2} \right) = x(x-3)^2 \left(\frac{A}{x} + \frac{B}{x-3} + \frac{C}{(x-3)^2} \right)$$ ¥Multiply each side by $x(x-3)^2$.

$$-7x+27 = A(x-3)^2 + Bx(x-3) + Cx$$

Now we'll use the definition of equality of polynomials from your textbook (above Example

1 in Section 9.4) which states that:

If two polynomials are equal then the coefficients of equal-degree terms are equal.

$$-7x+27 = A(x-3)^2 + Bx(x-3)+Cx$$

$$-7x+27 = A(x^2-6x+9)+B(x^2-3x)+Cx \quad \text{Expand the right side of the equation.}$$

$$-7x+27 = Ax^2-6Ax+9A+Bx^2-3Bx+Cx \quad \text{Simplify.}$$

$$-7x+27 = (A+B)x^2+(-6A-3B+C)x+9A \quad (1)$$

Now we can write a system of equations that arises out of equating coefficients. Equation (2) on the next page, was produced by equating the quadratic coefficients on each side of equation (1) above.

Equation (3) on the next page was produced by equating the linear coefficients of equation (1), and

equation (3) on the next page was produced by equating constant terms of equation (1):

$$\begin{array}{rcl} A +B & = & 0 \quad (2) \\ -6A -3B +C & = & -7 \quad (3) \\ 9A & = & 27 \quad (4) \end{array}$$

From equation (4) we determine that $A = 3$. Substitute 3 for A in $A + B = 0$ and you have:

$$3+B=0$$

$$B=-3$$

Substitute for 3 for A and -3 for B in $-6A - 3B + C = -7$, to find C.

$$-6(3)-3(-3)+C=-7$$

$$C=2$$

Now that we know the values of A, B, and C, we can write the partial fraction decomposition as:

$$\frac{-7x+27}{x(x-3)^2} = \frac{3}{x}+\frac{-3}{x-3}+\frac{2}{(x-3)^2}$$

★ For Fun

Add the rational expressions on the right-hand side of the above equation to produce the left-hand side.

Case 3: Nonrepeated quadratic factors

Each nonrepeated quadratic factor in the denominator will be included in the decomposition

in the form $\dfrac{Ax + B}{ax^2 + bx + c}$.

Example

$$\frac{3x^2 + 6x - 21}{(x-5)(x^2+3)} = \frac{A}{x-5} + \frac{Bx+C}{x^2+3}$$

$$(x-5)(x^2+3)\left(\frac{3x^2+6x-21}{(x-5)(x^2+3)}\right) = (x-5)(x^2+3)\left(\frac{A}{x-5} + \frac{Bx+C}{x^2+3}\right)$$

$$3x^2 + 6x - 21 = A(x^2+3) + (Bx+C)(x-5)$$

$$3x^2 + 6x - 21 = Ax^2 + 3A + Bx^2 - 5Bx + Cx - 5C$$

$$3x^2 + 6x - 21 = (A+B)x^2 + (-5B+C)x + (3A-5C)$$

Equate the coefficients of like terms to produce the following system of equations:

$$\begin{cases} A + B && = & 3 \\ & -5B + C & = & 6 \\ 3A & -5C & = & -21 \end{cases}$$

The above system can be written in triangular form as:

$$\begin{array}{rrcr} A & +B & = & 3 \\ & -5B +C & = & 6 \\ & C & = & 6 \end{array}$$

Use back substitution to determine that, $C = 6$, $B = 0$, and $A = 3$. Thus the partial fraction decomposition

is:

$$\frac{3x^2 + 6x - 21}{(x-5)(x^2+3)} = \frac{3}{x-5} + \frac{6}{x^2+3}$$

Case 4: Repeated Quadratic Factors

Each repeated quadratic factor in the denominator will be included in the decomposition in

the form $\dfrac{A_1 x + B_1}{ax^2 + bx + c} + \dfrac{A_2 x + B_2}{(ax^2 + bx + c)^2} + \ldots + \dfrac{A_m x + B_m}{(ax^2 + bx + c)^m}$ where, m is the multiplicity

of the quadratic factor in the denominator.

Example

$$\frac{2x^3 - x^2 - 6x - 7}{(x^2 - x - 3)^2} = \frac{Ax + B}{x^2 - x - 3} + \frac{Cx + D}{(x^2 - x - 3)^2}$$

$$(x^2 - x - 3)^2 \left(\frac{2x^3 - x^2 - 6x - 7}{(x^2 - x - 3)^2} \right) = (x^2 - x - 3)^2 \left(\frac{Ax + B}{x^2 - x - 3} + \frac{Cx + D}{(x^2 - x - 3)^2} \right)$$

$$2x^3 - x^2 - 6x - 7 = (Ax + B)(x^2 - x - 3) + Cx + D$$

$$2x^3 - x^2 - 6x - 7 = Ax^3 - Ax^2 - 3Ax + Bx^2 - Bx - 3B + Cx + D$$

$$2x^3 - x^2 - 6x - 7 = Ax^3 + (-A + B)x^2 + (-3A - B + C)x + (-3B + D)$$

Equating coefficients of like terms produces the following system of equations:

$$
\begin{array}{rcl}
A & = & 2 \\
-A + B & = & -1 \\
-3A - B + C & = & -6 \\
-3B + D & = & -7
\end{array}
$$

Solve the above system to show that $A = 2$, $B = 1$, $C = 1$, and $D = -4$.

Thus the partial fraction decomposition is:

$$\frac{2x^3 - x^2 - 6x - 7}{(x^2 - x - 3)^2} = \frac{2x + 1}{x^2 - x - 3} + \frac{x - 4}{(x^2 - x - 3)^2}$$

Partial Fraction Decomposition: Test Prep	
Give the *form* of the partial fraction decomposition of: $\dfrac{x - 1}{x^3 - 3x^2 + 7x - 21}$.	Factoring the denominator by grouping and choosing the appropriate numerators for the factors gives us:

Do not solve for the unknown constants.	$\dfrac{x-1}{x^3-3x^2+7x-21} = \dfrac{Ax+B}{x^2+7} + \dfrac{C}{x-3}$
Try These	
Find the partial fraction decomposition of: $\dfrac{-2x+15}{x^2-x-12}$	$\dfrac{-2x+15}{x^2-x-12} = \dfrac{1}{x-4} + \dfrac{-3}{x+3}$
Find the partial fraction decomposition of: $\dfrac{3x^3+13x^2+18x+14}{(x^2+3x+4)(x-1)(x+2)}$	$\dfrac{3x^3+13x^2+18x+14}{(x^2+3x+4)(x-1)(x+2)}$ $= \dfrac{2x+3}{x^2+3x+4} + \dfrac{2}{x-1} + \dfrac{-1}{x+2}$

The Degree of Numerator Larger than the Degree of the Denominator Case

If the degree of the numerator is *larger* than the degree of the denominator, then use long division to write the rational expression as a polynomial plus a remainder over the denominator. Then find the partial fraction decomposition of the remainder over the denominator.

Example

$$\frac{x^3-13x-9}{x^2-x-12} = x+1+\frac{3}{x^2-x-12}$$

The partial fraction decomposition of $\dfrac{3}{x^2-x-12}$ is $\dfrac{3}{7(x-4)} + \dfrac{-3}{(x+3)}$.

Thus the partial fraction decomposition of

$\dfrac{x^3-13x-9}{x^2-x-12}$ is $x+1+\dfrac{3}{7(x-4)} + \dfrac{-3}{(x+3)}$

9.5: Inequalities in Two Variables and Systems of Inequalities

❖ solution set of an inequality

The solution set of an inequality is all the ordered pairs that make the inequality true.

<u>To Graph an Inequality in Two Variables</u>

1. Graph the corresponding equality.

    a. If the inequality *includes* the equality (for example, $\leq$ or $\geq$), then graph the corresponding equation with a solid line.

    b. If the inequality does not include the equality (for example, $<$ or $>$), then graph the corresponding equality using a dashed line.

2. For each region defined by the graph, pick a test point (ordered pair) that does not fall on the curve you just graphed. Test the truth-value of the inequality using the ordered pair you chose.

    a. If the inequality is true, shade in the region that contains the test point.

    b. If the inequality is false at the test point, do not shade in the region that contains the test point.

For example, the graph of $y < |x-1|$ looks like this:

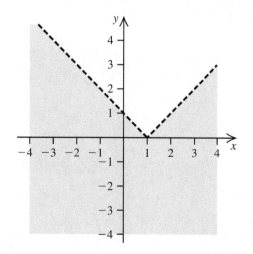

★ For Fun

Check the test point (0, 0) in the inequality $y < |x-1|$. Does it render the inequality true or false? Is the region which includes the origin in the shaded region of the graph?

Systems of Inequalities in Two Variables

You can illustrate and solve a system of inequalities graphically. A region on the graph that is shaded in for *every* inequality in the system will be the solution set. If you recall the meaning of an intersection of sets, then you'll realize that the region of overlapping shading is the *intersection* of the solution sets of all the inequalities in the system. (If there is no overlapping region, then there is no solution.)

❖ solution set of a system of inequalities

The graph of the solution set of a system of inequalities is the region or regions that makes all the inequalities in the system true.

Inequalities in Two Variables: Test Prep	
Graph $xy \geq 4$	1. Graph $y = \dfrac{4}{x}$ using a solid curve.
	2. Test the point (0, 0) in the inequality $xy > 4 : 0$ is not greater than 4.
	3. Do not shade in the region that includes the point (0, 0).

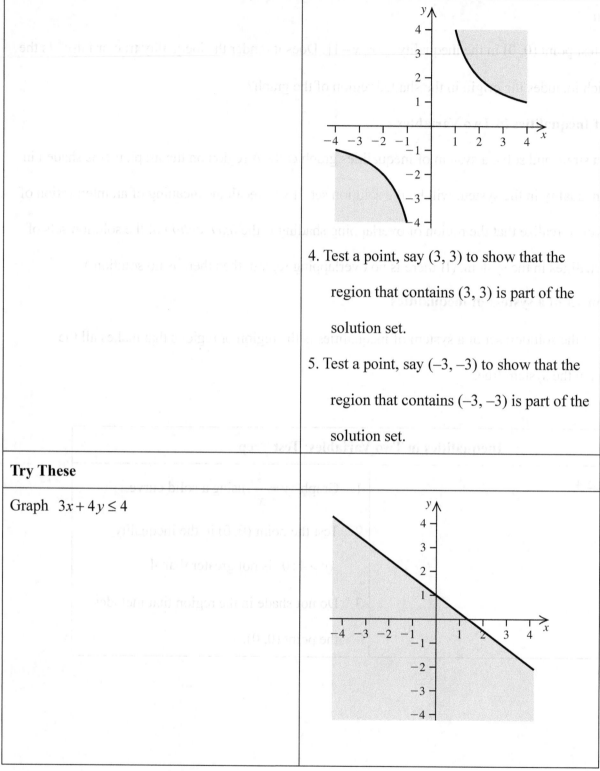

4. Test a point, say (3, 3) to show that the region that contains (3, 3) is part of the solution set.

5. Test a point, say (–3, –3) to show that the region that contains (–3, –3) is part of the solution set.

Try These

Graph $3x + 4y \leq 4$

Graph $y \geq 2^x + 1$

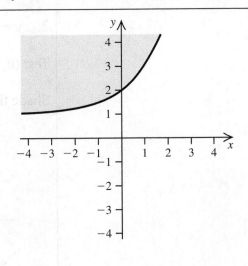

Nonlinear Systems of Inequalities

Systems of Inequalities: Test Prep

Graph

$$\begin{cases} x^2 + 4x + y \leq -6 \\ x + y > -6 \end{cases}$$

1. $x^2 + 4x + y = -6$

$y = -x^2 - 4x - 6$

$y = -(x+2)^2 - 2$

Test $(-2, -3)$: $-7 \leq -6$

Shade the interior region of the parabola.

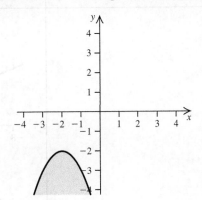

2. $y = -x - 6$

Test (0, 0):   0 is greater than −6:

Shade the region above the line.

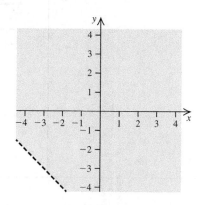

3.   The graph of the system is the region

where the above two graphs intersect.

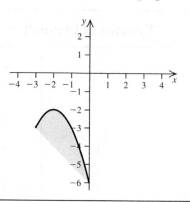

Try These

Graph $\begin{cases} 3x+y & \geq 7 \\ 2x+5y & \leq 9 \end{cases}$

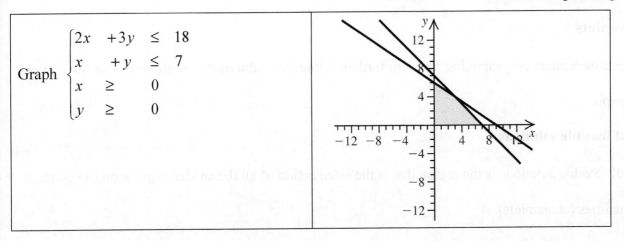

Graph $\begin{cases} 2x & +3y & \leq & 18 \\ x & +y & \leq & 7 \\ x & & \geq & 0 \\ y & & \geq & 0 \end{cases}$

9.6: Linear Programming

Graphing systems of linear inequalities is a skill you need to solve "linear programming" problems. The

task of linear programming is to find the optimal solution of an objective function that is subject to a set

of inequalities that are *constraints* on the objective function. It turns out, by the Fundamental Theorem

of Linear Programming, that the solution will be at one of the *vertices* where the graphs of the system of

linear inequalities intersect. So graph and test all the ordered pairs that are vertices of the graph of the

inequalities, and you will be able to determine the solution.

❖ **optimization problems**

Optimization problems are require you to find the minimum or maximum of a situation.

❖ **linear programming**

Linear programming is a method for finding the optimal solution (minimum or maximum) of an

objective function subject to a system of constraints in the form of inequalities.

❖ **linear objective form**

The objective form is the equation in the system that needs to be optimized. The linear inequalities of the

system are constraints on the system.

❖ constraints

Constraints on a linear programming problem further define the realm of the solution set – or the region

on the graph.

❖ set of feasible solutions

The set of feasible solutions is the region that is the intersection of all the shaded regions on the graph of

the inequalities (constraints).

Solving Optimization Problems

Example The objective function to maximize is $p = 3x + 5y$. The system of constraints is given by:

$$\begin{cases} x+y \leq 5 \\ 2x+y \leq 6 \\ x \geq 0 \ y \geq 0 \end{cases}$$

This is the graph of the above system of inequalities. The shaded region is the set of feasible solutions.

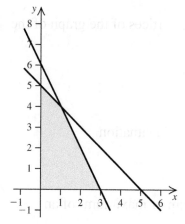

The vertices of the shaded region are (0, 0), (3, 0), (0, 5) and (1, 4). The last point can be found by

solving the system: $\begin{cases} x+y=5 \\ 2x+y=6 \end{cases}$

Now test vertices in the object function $p = 3x + 5y$:

(0, 0): $p = 0$

(3, 0): $p = 9$

(0, 5): $p = 25$ • maximum

$(1, 4): p = 23$

The maximum value of the objective function occurs at the point $(0, 5)$ and results in a value of 25 for p.

Optimization: Test Prep

Maximize $C = 2x + 6y$ with the following constraints: $$\begin{cases} x & +y & \leq 12 \\ 3x & +4y & \leq 40 \\ x & +2y & \leq 18 \end{cases}$$	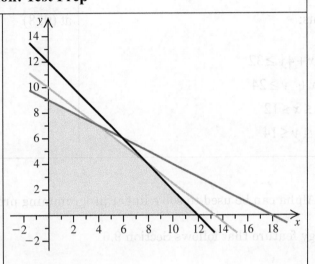 Testing the vertices: $(0, 9): C = 54$ • maximum $(12, 0): C = 24$ $(8, 4): C = 40$ $(4, 7): C = 50$ The maximum value of C is 54 and it occurs at the point $(0, 9)$.

Try These

Maximize $R = 40x + 30y$ subject to the constraints:	The maximum value of R is 10,000 and it occurs at $(100, 200)$.

$\begin{cases} 4x+2y\leq800 \\ x+y\leq300 \\ x\geq0 \\ y\geq0 \end{cases}$	
Minimize $C=5x+4y$ subject to the constraints: THE $\begin{cases} 3x+4y\geq32 \\ x+y\geq24 \\ 0\leq x\leq12 \\ 0\leq y\leq14 \end{cases}$	The minimum value of C is 32 and it occurs at $(0, 8)$.

WolframAlpha can be used to solve linear programming problems. See the Exploring Concepts with Technology feature that follows Section 9.6.

Matching

FIND THE BEST MATCH.

$\begin{cases} 4x-2y=9 \\ 2x-y=3 \end{cases}$

QUADRATIC FACTOR MULTIPLICITY 2

$\begin{cases} 2x-4y+z=-3 \\ 3y-2z=9 \\ 3z=-9 \end{cases}$

NONSQUARE SYSTEM

FEASIBLE SOLUTIONS

$(0, 0, 0)$

PARTIAL FRACTIONS

$\begin{cases} x-4y=0 \\ -x+7y=0 \end{cases}$

TRIVIAL SOLUTION

$\begin{cases} 3x-2y>6 \\ 2x-5y\leq10 \end{cases}$

INCONSISTENT SYSTEM

TRIANGULAR FORM

$\begin{cases} x^2+y^2-4y=4 \\ 5x-2y=2 \end{cases}$

HOMOGENEOUS SYSTEM

$Ax + B$

NONLINEAR SYSTEM

$$\frac{x^3 + 2x}{(x^2 + 1)^2} = \frac{Ax + B}{x^2 + 1} + \frac{Cx + D}{(x^2 + 1)^2}$$

SYSTEM OF INEQUALITIES

LINEAR FACTOR

$(x^2 + 3)^2$

$$\begin{cases} x + 2y - 3z = 5 \\ 3x + 7y - 10z = 13 \end{cases}$$

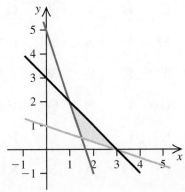

NOW TRY

CHAPTER 9 REVIEW EXERCISES AND THE CHAPTER 9 PRACTICE TEST.

STAY UP ON YOUR GAME – TRY THE CUMULATIVE REVIEW EXERCISES.

Chapter 10: Matrices

Word Search

```
W  S  D  E  S  R  E  V  N  I  N  M  G  M  Y
N  K  I  H  Z  C  O  X  Z  O  O  N  C  J  C
E  O  M  N  H  M  E  X  N  I  I  H  O  T  N
D  E  I  E  G  T  I  S  W  T  T  N  F  R  E
G  X  L  T  R  U  I  N  A  V  A  Y  A  A  C
E  O  I  E  C  N  L  L  O  I  M  R  C  N  A
N  D  V  R  G  E  O  A  S  R  R  A  T  S  J
M  R  S  U  T  P  L  S  R  O  O  L  O  L  D
Z  Q  L  H  R  A  U  F  W  L  F  A  R  A  A
B  A  U  E  A  A  M  K  E  L  S  C  H  T  T
R  O  T  V  G  Q  L  K  V  R  N  S  O  I  G
L  N  L  M  E  L  I  M  I  N  A  T  I  O  N
I  C  R  A  M  E  R  G  P  A  R  E  H  N  G
Y  W  T  N  A  N  I  M  R  E  T  E  D  E  C
E  F  R  O  T  A  T  I  O  N  C  R  R  Z  D
```

ADJACENCY	GAUSSIAN	ROTATION
COFACTOR	INTERPOLATING	ROW
CRAMER	INVERSE	SCALAR
DETERMINANT	MATRIX	SINGULAR
ECHELON	MINOR	TRANSFORMATION
EDGE	NONSINGULAR	TRANSLATION
ELIMINATION	REFLECTION	VERTEX

10.1: Gaussian Elimination Method

The Gaussian Elimination Method for matrices is a convenient and pretty powerful tool for solving systems of equations. We're going to build up to it slowly in this section.

✓ **Questions to Ask Your Instructor**

Have a look-see in your calculator manual and find out what it can do with matrices. Then check with your instructor to make certain sure you know how much and when you have to show your own work ... and when it might be okay to get your calculator to manipulate a matrix for you. **Introduction to Matrices**

Here we will study some of the basic configurations of matrices and their elements and learn the vocabulary that goes with these ideas.

♣ **matrix**

A matrix is an array, a table, or a rectangular "holder" of numbers made up of rows (they go horizontally) and columns (they go vertically). By the way, working with matrices rather naturally suggests the use of graph paper since you probably need to keep your "holder" well organized.

♣ **element**

What's in each "compartment" of a matrix is called an element. You can specify an element by which row and column it's in – where the row/column intersects – it takes two pieces of information to locate an element. For example, the "first" element in a matrix, the element in the top left-hand position, is indicated by specifying its column and row in a subscript like so: a_{11}. Moving one element to the right of a_{11} means we have remained on the same row, but added one to the column tally, and we specify this element like so: a_{12}. The element below a_{11} is element a_{21}. Do you see that the first subscript denotes the row and the second denotes the column?

♣ order *m x n*, dimension *m x n*

The "size" of a matrix is called its order or dimension. Like the area of a rectangle, the dimension of a matrix is like multiplying length times width, only it's *m x n* ("m by n") where *m* is the number of rows and *n* is the number of columns ... and you don't actually carry out the multiplication either. This is, for instance, a 2 by 4 matrix:

$$\begin{bmatrix} -1 & 0 & 3 & 5 \\ -7 & 2 & -2 & 0 \end{bmatrix}$$

♣ square matrix of order *n*

If a matrix is square, then the number of rows is the same as the number of columns, and one need only specify one dimension *n* and that the matrix is square. Here is a square matrix of dimension 2:

$$\begin{bmatrix} 2 & -1 \\ 3 & 0 \end{bmatrix}$$

♣ main diagonal

The main diagonal of a matrix is the diagonal that runs from the upper left diagonally towards the lower right. (It will not end in the lower right unless the matrix is square.)

♣ augmented matrix

You can translate a system of equations (Chapter 6) into an array using the coefficients and constants. This is how it's done:

$$\begin{aligned} 2x - y &= 9 \\ x + 3y &= -7 \end{aligned} \quad \text{becomes} \quad \begin{bmatrix} 2 & -1 & 9 \\ 1 & 3 & -7 \end{bmatrix}$$

This matrix is an augmented matrix because it includes the column of "answers" from the system – the constants on the left-hand side of the equations.

★ For Fun

Make an augmented matrix of this system. Be careful! All the x's and y's and constants have to line up correctly – a sort of "place value" for matrices.

$x = -2y + 1$

$y - x = 4$

♣ coefficient matrix

If you leave off the answer/constant column from your augmented matrix, you'll have a matrix made up of only the coefficients of the equations in your system.	$\begin{bmatrix} 2 & -1 \\ 1 & 3 \end{bmatrix}$

★ For Fun

Can you name the a_{12} element of this coefficient matrix? ... the a_{21}? ... the elements of the main diagonal?

♣ constant matrix

The (left-over) constant column that you took off your augmented matrix makes up a *column* matrix (of one column only) that looks like this.	$\begin{bmatrix} 9 \\ -7 \end{bmatrix}$

★ For Fun

What are the dimensions of this constant matrix? Get the order of m and n correct.

Matrix: Test Prep	
What order is this matrix? $$\begin{bmatrix} 0 & 1 & 3 & 9 & 0 \\ 2 & -2 & 5 & 0 & -1 \\ 3 & 11 & -7 & 2 & -3 \end{bmatrix}$$	Since the matrix has 3 rows and 5 columns it is an order *3* *x* *5* matrix
Try these	
Matching, matrix to its dimension: 1. $\begin{bmatrix} 4 & 0 & -1 \\ 2 & 3 & -7 \\ 5 & 0 & 9 \end{bmatrix}$ 2. $\begin{bmatrix} 4 \\ 1 \\ -1 \\ 0 \\ -2 \end{bmatrix}$ 3. $\begin{bmatrix} 0 & 9 \\ -2 & 3 \\ 5 & -7 \\ 10 & 1 \end{bmatrix}$ 4. $\begin{bmatrix} \frac{1}{2} & 0 & -1 & \frac{2}{3} \\ -4 & 3 & \frac{2}{3} & 0 \\ 9 & -5 & 1 & \frac{1}{6} \end{bmatrix}$ A. *4 x 2* B. *5 x 1* C. *3 x 3* D. *3 x 4*	1. *C* 2. *B* 3. *A* 4. *D*
Rewrite this system of equations as an augmented matrix: $$2x - 6y + 5z = 1$$ $$8x + 4y + z = 13$$ $$-2x + 3y + 4z = 5$$	$$\begin{bmatrix} 2 & -6 & 5 & | & 1 \\ 8 & 4 & 1 & | & 13 \\ -2 & 3 & 4 & | & 5 \end{bmatrix}$$

♣ row echelon form

A matrix in a form that corresponds to the *triangular* form for a set of equations is in row echelon form. Row echelon form is desirable for the same reason triangular form was – you can "back substitute" to solve the system of equations (that corresponds to your matrix). A matrix is in row echelon form if all the elements "under" the main diagonal are zero.

Elementary Row Operations

Just like we learned how to use equivalence of equations to manipulate systems of equations, there is an analogous set of operations that will help us to get a matrix into, say, row echelon form.

♣ elementary row operations

Elementary row operations are those analogous manipulations allowed to be performed on a matrix without changing the matrix's underlying validity.

Elementary Row Operations

Here are the operations you can perform on a matrix that correspond to a (still) equivalent system of equations:

1. You may interchange any two rows.

2. You may multiply all the elements in a row by the same (non-zero) constant.

3. You can replace any row with the sum of itself and a non-zero multiple of any other row.

★ For Fun

Can you see the correspondences between these three allowed row operations and the operations that were allowed for equivalent systems (page 735 of your textbook)?

★ For Fun

See if you can name the operation I perform on each step of reducing this matrix to row echelon form.

$$\begin{bmatrix} 5 & 1 & -2 & 4 \\ -2 & 4 & 4 & 6 \\ 3 & 5 & 2 & 10 \end{bmatrix} \rightarrow \begin{bmatrix} 5 & 1 & -2 & 4 \\ -6 & 12 & 12 & 18 \\ 6 & 10 & 4 & 20 \end{bmatrix} \rightarrow \begin{bmatrix} 5 & 1 & -2 & 4 \\ -2 & 4 & 4 & 6 \\ 0 & 22 & 16 & 38 \end{bmatrix} \rightarrow$$

$$\begin{bmatrix} 10 & 2 & -4 & 8 \\ -10 & 20 & 20 & 30 \\ 0 & 11 & 8 & 19 \end{bmatrix} \rightarrow \begin{bmatrix} 5 & 1 & -2 & 4 \\ 0 & 22 & 16 & 38 \\ 0 & 11 & 8 & 19 \end{bmatrix} \rightarrow \begin{bmatrix} 5 & 1 & -2 & 4 \\ 0 & 11 & 8 & 19 \\ 0 & 0 & 0 & 0 \end{bmatrix}$$

★ For Fun

What does that last row of all-zeros in the example above tell you about the system of equations that might correspond to this matrix? (consistent, dependent, independent?)

<table>
<tr><td colspan="2" align="center">**Row Echelon Form: Test Prep**</td></tr>
<tr>
<td>Use elementary row operations to rewrite this matrix in row echelon form:

$$\begin{bmatrix} 2 & 6 & -1 & 1 \\ 1 & 3 & -1 & 1 \\ 3 & 10 & -2 & 1 \end{bmatrix}$$</td>
<td>$$\begin{bmatrix} 2 & 6 & -1 & 1 \\ 1 & 3 & -1 & 1 \\ 3 & 10 & -2 & 1 \end{bmatrix} \rightarrow \begin{bmatrix} 2 & 6 & -2 & 2 \\ 3 & 10 & -2 & 1 \\ 2 & 6 & -1 & 1 \end{bmatrix}$$

$$\rightarrow \begin{bmatrix} 3 & 9 & -3 & 3 \\ 3 & 10 & -2 & 1 \\ 0 & 0 & 1 & -1 \end{bmatrix} \rightarrow \begin{bmatrix} 1 & 3 & -1 & 1 \\ 0 & 1 & 1 & -2 \\ 0 & 0 & 1 & -1 \end{bmatrix}$$</td>
</tr>
<tr>
<td colspan="2">**Try these**</td>
</tr>
<tr>
<td>Write in row echelon form:

$$\begin{bmatrix} -3 & 6 & -2 \\ 6 & 6 & -8 \end{bmatrix}$$</td>
<td>possibly:

$$\begin{bmatrix} 1 & -2 & \dfrac{2}{3} \\ 0 & 1 & -\dfrac{2}{3} \end{bmatrix}$$</td>
</tr>
<tr>
<td>Write in row echelon form:

$$\begin{bmatrix} 2 & 5 & 2 & -1 \\ 1 & 2 & -3 & 5 \\ 5 & 12 & 1 & 10 \end{bmatrix}$$</td>
<td>possibly:

$$\begin{bmatrix} 1 & 2 & -3 & 5 \\ 0 & 1 & 8 & -11 \\ 0 & 0 & 0 & 1 \end{bmatrix}$$</td>
</tr>
</table>

Gaussian Elimination Method

The goal of Gaussian Elimination Method is to rewrite an augmented matrix in row echelon form so as to solve the corresponding system of equations.

★ For Fun

You might pay careful attention to just how this "elimination" method is an awful lot like the "Method of Elimination" we used to solve systems of equations in Chapter 6.

♣ Gaussian elimination method

The Gaussian elimination method is (simply) an algorithm using matrices to solve a system of equations by getting the corresponding matrix into a row echelon form.

★ For Fun

What is the difference between the words "algebra" and "algorithm?"

Gaussian Elimination Method: Test Prep	
Solve using Gaussian Elimination method: $x - 3y + 2z = 6$ $4x - y + 3z = 10$ $7x + y + 4z = -2$	$\begin{bmatrix} 1 & -3 & 2 & 6 \\ 4 & -1 & 3 & 10 \\ 7 & 1 & 4 & -2 \end{bmatrix} \rightarrow \begin{bmatrix} 7 & -21 & 14 & 42 \\ 4 & -1 & 3 & 10 \\ 0 & 22 & -10 & -44 \end{bmatrix}$ $\rightarrow \begin{bmatrix} 4 & -12 & 8 & 24 \\ 0 & 11 & -5 & -14 \\ 0 & 11 & -5 & -22 \end{bmatrix} \rightarrow \begin{bmatrix} 1 & -3 & 2 & 6 \\ 0 & 11 & -5 & -14 \\ 0 & 0 & 0 & -8 \end{bmatrix}$ $0 \neq -8$; therefore, no solution.
Try these	
Solve using Gaussian Elimination method: $2x - 5y = 10$ $5x + 2y = 4$	$\left(\dfrac{40}{29}, -\dfrac{42}{29} \right)$
Solve using Gaussian Elimination method:	$(3c - 5, -7c + 14, 4 - 3c, c)$

$$w + 2x - 3y + 2z = 11$$
$$2w + 5x - 8y + 5z = 28$$
$$-2w - 4x + 7y - z = -18$$

Application: Interpolating Polynomials

Here's a neat little application of matrices that help us find a polynomial model ... without a calculator!

♣ **interpolating polynomial**

An interpolating polynomial is a sort of best-guess polynomial that models an application based on knowledge of a finite number of data points that have been observed to hold for the situation. After all, each ordered pair assumes a function. We then assume the function is a polynomial of an arbitrary degree (depending on the number of ordered pairs we have). Then we use a matrix to find the unknown coefficients of said polynomial like we used systems of equations to find the unknown constants in partial fraction decomposition (section 6.4).

Interpolating Polynomial: Test Prep	
Write a system of equations to describe a second degree polynomial passing through the points (-3, 28), (-1, 6), and (2, 3).	$a_2(-3)^2 + a_1(-3) + a_0 = 28$ $a_2(-1)^2 + a_1(-1) + a_0 = 6$ $a_2(2)^2 + a_1(2) + a_0 = 3$ becomes $9a_2 - 3a_1 + a_0 = 28$ $a_2 - a_1 + a_0 = 6$ $4a_2 + 2a_1 + a_0 = 3$
Try these	
Write the above polynomial by solving the system.	$p(x) = 2x^2 - 3x + 1$

Find a polynomial whose graph passes through the points $(-2,-3), (0,-1)$, and $(3,17)$.	$p(x) = x^2 + 3x - 1$

10.2: Algebra of Matrices

Matrices are, after all, just arrays of numbers. In the last section we considered them mostly as stand-ins for systems of equations. However, in many cases, they are nothing more or less than a table of values, and as such we may wish to do any number of arithmetic or algebraic operations with tables-cum-matrices.

Addition and Subtraction of Matrices

If you use a table to represent data, such as sales figures or rainfall, you might want to add two matrices together to get a grand total. You might want to subtract matrices of data if you wanted to know how much you saved from one month to the next or something similar. Adding and subtracting matrices is easy ... just be careful to apply each operation to each element correctly.

First we have to set up the whole arena matrices for arithmetic operations by talking about the same sorts of underlying ideas we had to address for arithmetic with real numbers – additive identities and so forth.

♣ **zero matrix, _0_**

The zero matrix is a matrix completely filled with zeros. It's the additive identity for matrix addition:

$$\begin{bmatrix} a & b \\ c & d \end{bmatrix} + \begin{bmatrix} 0 & 0 \\ 0 & 0 \end{bmatrix} = \begin{bmatrix} 0 & 0 \\ 0 & 0 \end{bmatrix} + \begin{bmatrix} a & b \\ c & d \end{bmatrix} = \begin{bmatrix} a & b \\ c & d \end{bmatrix}$$

If you consider that matrix addition is just element-wise addition of the numbers in the matrices, this property really follows from the fact that zero is the additive identity for real numbers. Matrix addition, after all, works just like this:

$$\begin{bmatrix} a & b \\ c & d \end{bmatrix} + \begin{bmatrix} 0 & 0 \\ 0 & 0 \end{bmatrix} = \begin{bmatrix} a+0 & b+0 \\ c+0 & d+0 \end{bmatrix} = \begin{bmatrix} a & b \\ c & d \end{bmatrix}$$

♣ commutative

Likewise, since addition of real numbers is commutative, matrix addition would be commutative, too right? So if A is an $m \times n$ matrix as is B, then $A + B = B + A$.

♣ associative

Again, since adding matrices involves adding the real numbers element-wise in the matrices, then matrix addition is associative: $A + (B + C) = (A + B) + C.$

★ For Fun

Do you understand why the matrices being added must be the same order/dimensions?

♣ additive inverse

In order to talk about subtraction of matrices, we can first define the additive inverse (which is formally what we have to do to define subtraction of real numbers, isn't it?). So in order to construct the additive inverse of matrix A, we'll simply take the additive inverse of every element in A, and we can call that $-A$:

$$A = \begin{bmatrix} a & b \\ c & d \end{bmatrix} \quad \text{and} \quad -A = \begin{bmatrix} -a & -b \\ -c & -d \end{bmatrix}$$

So, to subtract matrices, you wind up simply adding the additive inverse of each element:

$$\begin{bmatrix} 3 & -1 \\ 0 & 2 \end{bmatrix} - \begin{bmatrix} 0 & 3 \\ 9 & -3 \end{bmatrix} = \begin{bmatrix} 3 & -1 \\ 0 & 2 \end{bmatrix} + \begin{bmatrix} 0 & -3 \\ -9 & 3 \end{bmatrix} = \begin{bmatrix} 3+0 & -1+(-3) \\ 0+(-9) & 2+3 \end{bmatrix} = \begin{bmatrix} 3 & -4 \\ -9 & 5 \end{bmatrix}$$

(This is not mysterious stuff!)

♣ additive identity

The additive identity matrix is the zero matrix. The dimensions of the additive identity matrix will match exactly the dimensions of the matrix you are adding it to – that is, add *nothing* to.

Scalar Multiplication

If we need to be able to change every element in a matrix (or table) by some constant factor, we need to have a way of wholesale multiplying that number by every element in that matrix. This is scalar multiplication.

★ **For Fun**

What does the word "scalar" mean in this context?

♣ **scalar multiplication**

Scalar multiplication is basically the operation of multiplying each element in a given matrix by a constant (or scalar). For example:

$$-2\begin{bmatrix} 3 & -1 \\ 0 & 2 \end{bmatrix} = \begin{bmatrix} -6 & 2 \\ 0 & -4 \end{bmatrix}$$

★ **For Fun**

Could we describe the additive inverse matrix as a product of scalar multiplication?

Operations on Matrices: Test Prep	
Given $\mathbf{A} = \begin{bmatrix} 0 & -1 & 2 \\ 5 & 7 & 9 \\ 1 & 3 & 1 \end{bmatrix}$ and $\mathbf{B} = \begin{bmatrix} 4 & 1 & 1 \\ 2 & 3 & 1 \\ 0 & 1 & 0 \end{bmatrix}$, find $-3A + 2B$.	$-3\mathbf{A} + 2\mathbf{B} = \begin{bmatrix} 0 & 3 & -6 \\ -15 & -21 & -27 \\ -3 & -9 & -3 \end{bmatrix} + \begin{bmatrix} 8 & 2 & 2 \\ 4 & 6 & 2 \\ 0 & 2 & 0 \end{bmatrix}$ $= \begin{bmatrix} 8 & 5 & -4 \\ -11 & -15 & -25 \\ -3 & -7 & -3 \end{bmatrix}$
Try these	
Given $\mathbf{A} = \begin{bmatrix} 2 & 0 \\ -1 & 3 \end{bmatrix}$ and $\mathbf{B} = [3 \ 4 \ 0]$, find $A + B$.	Cannot add matrices of different order.

Simplify: $2\begin{bmatrix} 4 & 0 \\ -1 & 1 \end{bmatrix} - \begin{bmatrix} 5 & 7 \\ 1 & 3 \end{bmatrix} + 5\begin{bmatrix} 1 & -2 \\ 0 & 4 \end{bmatrix}$ | $\begin{bmatrix} 8 & -17 \\ -3 & 19 \end{bmatrix}$

Matrix Multiplication

Scalar multiplication is *one* kind of multiplication that is carried out on matrices. In real numbers (scalars), we really have only one kind of multiplication. Matrices have another kind: matrix multiplication where two or more matrices are multiplied by each other. This is the beginning of a strangeness in matrices that is unlike anything you've done with real numbers. This process will take some careful place-keeping and practice.

♣ **column matrix**

A column matrix is a matrix with only one column (but any number of rows), for instance:

$$\begin{bmatrix} 2 \\ -5 \\ 0 \end{bmatrix}$$

♣ **row matrix**

Similarly, a row matrix is a matrix with only one row:

$[1 \quad -1 \quad 0 \quad 2.5]$

★ **For Fun**

Can you give the dimensions of these last two matrices?

♣ **associative property**

Without defining matrix multiplication (yet), we can go ahead and affirm that if the matrices can be multiplied together in this order, then the multiplication follows the associative rule: $A(BC) = (AB)C$.

In general, matrix multiplication is not commutative.

❖ **distributive property**

Given that the matrices can be multiplied together in the order given, then matrix multiplication does follow a distributive property of matrix multiplication over matrix addition: $A(B+C) = AB + AC$ or $(A+B)C = AC + BC$.

❖ **identity matrix**

The identity matrix is always a square matrix. The identity matrix of order or dimension n is the $n \times n$ square matrix with all 1's on the main diagonal and all 0's everywhere else.

Order/Dimension Considerations in Matrix Multiplication

- In order for two matrices to be multiplied together, the number of columns in the first matrix must match the number of rows in the second matrix. For example, a 3×2 matrix can be multiplied by a 2×4 matrix, but a 3×2 matrix cannot be multiplied by a 3×4 matrix.

- The matrix that results from matrix multiplication will have the number of rows of the first matrix in the multiplication and the number of columns in the second matrix in the multiplication. So multiplying a 3×2 matrix by a 2×4 matrix will result in a new matrix of order 3×4.

Method of Matrix Multiplication

- The elements of the first row of the first matrix get multiplied in a one-by-one matching with the elements in the first column of the second matrix. Those products are added together, and that sum becomes the new a_{11} of the product matrix.

- The elements of the first row of the first matrix are multiplied one-by-one by the elements of the second column of the second matrix, and the sum of those products becomes the a_{12} element of the product matrix.

- … and so on.

For example:

$$\begin{bmatrix} 1 & -2 \\ 0 & 3 \\ -1 & 5 \end{bmatrix} \begin{bmatrix} -3 & 2 \\ -2 & 0 \end{bmatrix} = \begin{bmatrix} (1)(-3)+(-2)(-2) & (1)(2)+(-2)(0) \\ (0)(-3)+(3)(-2) & (0)(2)+(3)(0) \\ (-1)(-3)+(5)(-2) & (-1)(2)+(5)(0) \end{bmatrix}$$

$$= \begin{bmatrix} -3+4 & 2+0 \\ 0-6 & 0+0 \\ 3-10 & -2+0 \end{bmatrix}$$

$$= \begin{bmatrix} 1 & 2 \\ -6 & 0 \\ -7 & -2 \end{bmatrix}$$

★ **For Fun**

Can you do the dimensional analysis of these matrices to verify that they are multipliable and that my result matrix has the correct order?

Matrix multiplication is a pattern-heavy, almost kinesthetic, surely visual operation. Matrix multiplication involves lots of small steps, so it is error- prone. Practice is my advice.

Think About It:

Do you think that matrix multiplication could be used to *encrypt* messages? How about a series of matrix multiplications?

Multiplication of Two Matrices: Test Prep

Multiply, if possible: $\begin{bmatrix} 3 & 6 & 1 \\ 0 & -1 & -2 \end{bmatrix} \begin{bmatrix} 5 \\ -4 \\ 0 \end{bmatrix}$	$\begin{bmatrix} 3 & 6 & 1 \\ 0 & -1 & -2 \end{bmatrix} \begin{bmatrix} 5 \\ -4 \\ 0 \end{bmatrix} = \begin{bmatrix} 15-24+0 \\ 0+4+0 \end{bmatrix}$ $= \begin{bmatrix} -9 \\ 4 \end{bmatrix}$

Try these

Name the dimension of these two matrices and the dimension of their product, assuming they are multiplied in the order given: $\begin{bmatrix} 3 & 0 & 1 \\ 5 & -1 & 4 \end{bmatrix},$ $\begin{bmatrix} 4 & 5 & 0 & -1 \\ 5 & 1 & -1 & 0 \\ 2 & 3 & 7 & -4 \end{bmatrix}$	$2 \times 3,\ 3 \times 4,\ 2 \times 4$
Carry out the matrix multiplication, if possible: $\begin{bmatrix} 1 & -2 & 3 \\ 2 & -1 & 8 \\ -1 & 3 & -2 \end{bmatrix} \begin{bmatrix} -1 & 3 & 2 \\ 1 & 4 & -1 \end{bmatrix}$	not possible

Matrix Products and Systems of Equations

Systems of equations can be expressed as multiplication of matrices in the following manner.

$$\begin{cases} 3x - y + 2z = 5 \\ -x + y = -2 \\ 2x - 3y + z = 1 \end{cases} \rightarrow \begin{bmatrix} 3 & -1 & 2 \\ -1 & 1 & 0 \\ 2 & -3 & 1 \end{bmatrix} \begin{bmatrix} x \\ y \\ z \end{bmatrix} = \begin{bmatrix} 5 \\ -2 \\ 1 \end{bmatrix}$$

Check the orders of the matrices. Can they be multiplied together? Can you go ahead and multiply the matrices together? Can you rewrite the system as an augmented matrix? Can you recover the system of equations from the augmented matrix?

Matrix Form of a System of Equations: Test Prep	
Rewrite this matrix equation as a system of equations. Use matrix multiplication. $$\begin{bmatrix} 2 & 0 & -3 & 5 \\ 0 & 1 & -1 & 3 \\ -4 & 3 & 2 & -1 \\ 0 & 5 & 4 & 0 \end{bmatrix} \begin{bmatrix} w \\ x \\ y \\ z \end{bmatrix} = \begin{bmatrix} 15 \\ -3 \\ 5 \\ 11 \end{bmatrix}$$	$$\begin{bmatrix} 2 & 0 & -3 & 5 \\ 0 & 1 & -1 & 3 \\ -4 & 3 & 2 & -1 \\ 0 & 5 & 4 & 0 \end{bmatrix} \begin{bmatrix} w \\ x \\ y \\ z \end{bmatrix} = \begin{bmatrix} 15 \\ -3 \\ 5 \\ 11 \end{bmatrix}$$ $$\begin{bmatrix} 2w - 3y + 5z \\ x - y + 3z \\ -4w + 3x + 2y - z \\ 5x + 4y \end{bmatrix} = \begin{bmatrix} 15 \\ -3 \\ 5 \\ 11 \end{bmatrix}$$ $$\begin{aligned} 2w - 3y + 5z &= 15 \\ x - y + 3z &= -3 \\ -4w + 3x + 2y - z &= 5 \\ 5x + 4y &= 11 \end{aligned}$$ by equality of matrices
Try these	
Write a system of equations from the matrix equation:	$$\begin{aligned} 2x - y &= 3 \\ 5y &= -2 \end{aligned}$$

$$\begin{bmatrix} 2 & -1 \\ 0 & 5 \end{bmatrix} \begin{bmatrix} x \\ y \end{bmatrix} = \begin{bmatrix} 3 \\ -2 \end{bmatrix}$$

| Write a system of equations from the matrix equation: $$\begin{bmatrix} 3 & 0 & \frac{1}{2} \\ -1 & 5 & -6 \\ \frac{2}{3} & 4 & 0 \end{bmatrix} \begin{bmatrix} x \\ y \\ z \end{bmatrix} = \begin{bmatrix} \frac{1}{3} \\ -2 \\ -7 \end{bmatrix}$$ | $$\begin{bmatrix} 3 & 0 & \frac{1}{2} \\ -1 & 5 & -6 \\ \frac{2}{3} & 4 & 0 \end{bmatrix} \begin{bmatrix} x \\ y \\ z \end{bmatrix} = \begin{bmatrix} \frac{1}{3} \\ -2 \\ -7 \end{bmatrix}$$ $$3x + \frac{1}{2}z = \frac{1}{3}$$ $$-x + 5y - 6z = -2$$ $$\frac{2}{3}x + 4y = -7$$ |

Transformation Matrices

Using matrix multiplication and some special matrices you can transform points (and thereby objects that are defined by their vertices or similarly) by moving (translating) them through space, reflecting them, and rotating them. There are a set of well defined, well used matrices available to accomplish these feats!

♣ **transformation matrices**

A matrix that, when it is multiplied by another (frequently column) matrix, transforms that (column) matrix into a different matrix in a regular way.

♣ **translation matrix** $\mathbf{T}_{a,b}$

| The translation matrix in three dimensions usually looks like this: | $$\begin{bmatrix} 1 & 0 & a \\ 0 & 1 & b \\ 0 & 0 & 1 \end{bmatrix}$$ |

In this form, the translation matrix will move a 3×1 three dimensional point a units in the x-direction and b units in the y-direction. For example:

$$\begin{bmatrix} 1 & 0 & 3 \\ 0 & 1 & -2 \\ 0 & 0 & 1 \end{bmatrix}\begin{bmatrix} 1 \\ 1 \\ 1 \end{bmatrix} = \begin{bmatrix} 1+3 \\ 1-2 \\ 1 \end{bmatrix} = \begin{bmatrix} 4 \\ -1 \\ 1 \end{bmatrix}$$

... and we've moved the ordered triple $(1, 1, 1)$ $+3$ in the x-direction and -2 in the y-direction so that the triple is now $(4, -1, 1)$.

♣ reflection matrix R_x

The reflection matrix that reflects a point across the x-axis looks like this:

$$\begin{bmatrix} 1 & 0 & 0 \\ 0 & -1 & 0 \\ 0 & 0 & 1 \end{bmatrix}$$

♣ reflection matrix R_y

A matrix the accomplishes the reflection of a point about the y-axis is

$$\begin{bmatrix} -1 & 0 & 0 \\ 0 & 1 & 0 \\ 0 & 0 & 1 \end{bmatrix}$$

♣ reflection matrix R_{xy}

And the matrix that will reflect a point about the line $y = x$ look like this:

$$\begin{bmatrix} 0 & 1 & 0 \\ 1 & 0 & 0 \\ 0 & 0 & 1 \end{bmatrix}$$

♣ rotation matrices

There are several versions of rotational matrices that depend on which way and how much you wish to rotate your point. For example, this matrix rotates a point ninety degrees about the origin:

$$\mathbf{R}_{90} = \begin{bmatrix} 0 & -1 & 0 \\ 1 & 0 & 0 \\ 0 & 0 & 1 \end{bmatrix}$$

Some other popular rotation matrices are given on page 819 in your textbook.

★ **For Fun**

Try some these matrices out on a point in 3 dimensions written in the form of a 3×1 column matrix.

Combinations of Translations, Reflections, and Rotations

A new matrix that does many transformations can be built by matrix multiplication of the individual

transformation matrices in a sort of reverse order (remember that matrix multiplication is not

commutative so the order can matter). The first transformation you wish to use will be the right-most,

and each successive transformation desired will be one more place to the left. For example, translating in

x and y and then rotating by 90 degrees might look like this:

$$\begin{bmatrix} 0 & -1 & 0 \\ 1 & 0 & 0 \\ 0 & 0 & 1 \end{bmatrix}\begin{bmatrix} 1 & 0 & 3 \\ 0 & 1 & -2 \\ 0 & 0 & 1 \end{bmatrix} = \begin{bmatrix} 0 & -1 & 2 \\ 1 & 0 & 3 \\ 0 & 0 & 1 \end{bmatrix}$$

★ **For Fun**

Try the above transformation matrix on a column matrix. See if it does the expected translation and

rotation.

✓ **Questions to Ask Your Instructor**

These transformations can be used on whole shapes, too, not just points in the form of a 3×1 column

matrix. This is discussed further in section 10.2 in your textbook. Ask you instructor to go over these

expanded examples. It may be required that you know how to move an entire shape in one

transformation matrix.

Transformation Matrices: Test Prep	
Rotate the point $\begin{bmatrix} 5 \\ -3 \\ 2 \end{bmatrix}$ about the origin of the 2-dimensional cartesian plane 180 degrees counterclockwise.	$\begin{bmatrix} -1 & 0 & 0 \\ 0 & -1 & 0 \\ 0 & 0 & 1 \end{bmatrix}\begin{bmatrix} 5 \\ -3 \\ 2 \end{bmatrix} = \begin{bmatrix} -5+0+0 \\ 0+3+0 \\ 0+0+2 \end{bmatrix} = \begin{bmatrix} -5 \\ 3 \\ 2 \end{bmatrix}$
Try these	
Define a matrix that will rotate a 3-dimensional point 90 degrees counterclockwise about the point $(2, -3, 0)$.	$\begin{bmatrix} 0 & -1 & 3 \\ 1 & 0 & 2 \\ 0 & 0 & 1 \end{bmatrix}$
Define a matrix that will reflect a point about the x-axis and about the y-axis.	$\begin{bmatrix} -1 & 0 & 0 \\ 0 & -1 & 0 \\ 0 & 0 & 1 \end{bmatrix}$

Adjacency Matrices

Adjacency matrices refer to ways that of points can be connected to each other. This is part of what is called "Graph Theory."

♣ **graph**

A graph in this context is a set of points, vertices, and the line segments, edges, that connect those points.

Here is an example of a graph with vertices labeled:

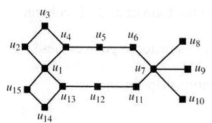

♣ vertices

The vertices of a graph are the points.

♣ edges

The edges of the graph are the line segments that connect the vertices.

♣ step

A step on a graph is a movement from one vertex to another along the edge that connects them. Only two vertices are involved in a step.

♣ walk

A walk is a string of several steps.

♣ length

The number of steps in a walk is the length of the walk. Frequently, graphs are used as a way to explore finding walks that are of the shortest possible length – naturally!

♣ adjacency matrix

The elements of an adjacency matrix signify how many vertices connect one vertex to another. In other words, the element $a_{13} = 2$ means that there are two vertices between vertex #1 and vertex #3. $a_{13} = 0$ would mean that there are no edges between these two vertices (though one may still find a *walk* from vertex #1 and #3).

★ For Fun

What does the word "adjacent" mean?

★ For Fun

Think about it: would an adjacency matrix have to have some *symmetries*?

★ For Fun

Here is an example of an adjacency matrix for a graph with 4 vertices:

$$\begin{bmatrix} 0 & 1 & 1 & 0 \\ 1 & 1 & 2 & 0 \\ 1 & 2 & 0 & 1 \\ 0 & 0 & 1 & 0 \end{bmatrix}$$

Could you sketch a graph that would correspond to this adjacency matrix?

Adjacency Matrix: Test Prep	
Define the adjacency matrix for this graph:	by inspection
	$$\begin{bmatrix} 0 & 1 & 0 & 1 & 0 & 0 \\ 1 & 0 & 1 & 1 & 0 & 0 \\ 0 & 1 & 0 & 0 & 1 & 0 \\ 1 & 1 & 0 & 0 & 1 & 0 \\ 0 & 0 & 1 & 1 & 0 & 1 \\ 0 & 0 & 0 & 0 & 1 & 0 \end{bmatrix}$$
Try these	
What would be the dimension of the adjacency matrix of this graph:	15×15
Define the adjacency matrix for this *complete* graph:	$$\begin{bmatrix} 0 & 1 & 1 & 1 & 1 \\ 1 & 0 & 1 & 1 & 1 \\ 1 & 1 & 0 & 1 & 1 \\ 1 & 1 & 1 & 0 & 1 \\ 1 & 1 & 1 & 1 & 0 \end{bmatrix}$$

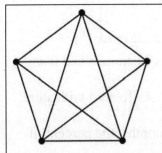

You may label the vertices any way you wish.

Applications of Matrices

Matrices are used frequently in the sense of repeated transformations – when the same action is required over and over again. This lends itself to cyclical phenomena like those in the natural world. Physics makes use of matrix algebra to do such things as describe what a series of several lenses might do overall to a beam of light. Graph theory is used to explore quickest-route problems for emergency services. Computer graphics of all sorts use transformational matrix algebra to change images in regular and uniform ways. Matrices have even been used in the arts as generators of patterns to be exploited in musical or visual pieces.

Applications: Test Prep	
If each of these matrices give sales figures for the summer months at a retail store, what would the result of adding these two matrices together and what would the meaning of that result be? $\begin{bmatrix} 12 \\ 4 \\ 7 \\ 16 \end{bmatrix} \begin{bmatrix} 2 \\ 10 \\ 8 \\ 20 \end{bmatrix}$	$\begin{bmatrix} 14 \\ 14 \\ 15 \\ 36 \end{bmatrix}$ The resulting matrix shows total sales for two years.

Try these

A town has two grocery stores, A and B. Each month store A retains 98% of is customers and loses 2% to store B, and store B retains 95% of its customers and loses 5% to store A. To start with store B has 75% of the town's customers and store A has the other 25%. After n months, the percent of the customers who shop at store A, denoted by a, and the percent of customers who shop at store B, denoted by b, are given by $[0.25 \quad 0.75]\begin{bmatrix} 0.98 & 0.02 \\ 0.05 & 0.95 \end{bmatrix}^n = [a \quad b]$. Find the percent, to the nearest 0.1% of the customers who shop at store A after 5 months.   **Ans: 39.10%**

Experiment with different values of n in the above problem to find out how long, to the nearest month, it will be before store A has 50% of the customers.	11 months

10.3: Inverse of a Matrix

Much like inverse functions are used to solve equations, inverse matrices have a major use in the solutions of systems of equations. Be sure your elementary row operation skills are in good working order as we embark on this exercise.

Finding the Inverse of a Matrix

One of the main reasons for finding the inverse of a matrix is to use it to solve systems of equations. This is a really neat trick, but first you have to find the inverse... which is labor-intensive (when done by hand).

♣ inverse

Some square matrices have inverse matrices. The inverse of the matrix **A** is signified by $\mathbf{A}^{-1}$. Much as with multiplicative inverses of real numbers result in 1 when multiplied together, matrix multiplication of a matrix with its inverse results in the identity matrix.

The method for finding the inverse of a matrix (if it has one) involves identity matrices and elementary row operations. Here is an example of finding the inverse of a 3×3 square matrix:

First augment the matrix with the identity matrix of the same order:

$$\begin{bmatrix} 1 & 1 & 4 \\ 2 & 3 & 6 \\ -1 & -1 & 2 \end{bmatrix}\begin{bmatrix} 1 & 0 & 0 \\ 0 & 1 & 0 \\ 0 & 0 & 1 \end{bmatrix}$$

Now, go about row operations with an eye to turning the left-hand matrix into the 3×3 identity matrix. Meanwhile, let the right-hand matrix evolve thereby into whatever it must become according to those row operations.

$$\left[\begin{array}{ccc|ccc} 1 & 1 & 4 & 1 & 0 & 0 \\ 2 & 3 & 6 & 0 & 1 & 0 \\ -1 & -1 & 2 & 0 & 0 & 1 \end{array}\right] \rightarrow \left[\begin{array}{ccc|ccc} 2 & 2 & 8 & 2 & 0 & 0 \\ 2 & 3 & 6 & 0 & 1 & 0 \\ 0 & 0 & 6 & 1 & 0 & 1 \end{array}\right] \rightarrow$$

$$\left[\begin{array}{ccc|ccc} 1 & 1 & 4 & 1 & 0 & 0 \\ 0 & 1 & -2 & -2 & 1 & 0 \\ 0 & 0 & 2 & \dfrac{1}{3} & 0 & \dfrac{1}{3} \end{array}\right] \rightarrow \left[\begin{array}{ccc|ccc} 1 & 0 & 6 & 3 & -1 & 0 \\ 0 & 1 & 0 & -\dfrac{5}{3} & 1 & \dfrac{1}{3} \\ 0 & 0 & 6 & 1 & 0 & 1 \end{array}\right] \rightarrow$$

$$\left[\begin{array}{ccc|ccc} 1 & 0 & 0 & 2 & -1 & -1 \\ 0 & 1 & 0 & -\dfrac{5}{3} & 1 & \dfrac{1}{3} \\ 0 & 0 & 1 & \dfrac{1}{6} & 0 & \dfrac{1}{6} \end{array}\right]$$

The resulting right-hand matrix *is* the inverse matrix. You can check it by performing the matrix multiplication $\mathbf{A} \cdot \mathbf{A}^{-1}$ or $\mathbf{A}^{-1} \cdot \mathbf{A}$. They're inverses if the result is the 3×3 identity matrix.

✓ **Questions to Ask Your Instructor**

If you cannot decipher all the steps I took in the above row operations, you might want to step through it in class with your instructor and for the benefit of all your classmates. Probably several examples will help everybody in class!

Finding the Inverse of a Matrix

We are using Gaussian elimination to find the inverse of a matrix. There are other methods that you are welcome to research!

1. Analytically, using Cramer's Rule (section 10.5)

2. Neumann Series

3. Blockwise Inversion

Inverse of a Matrix: Test Prep

Verify that these matrices are inverses of each other by using matrix multiplication:

$$\begin{bmatrix} 1 & 1 & 4 \\ 2 & 3 & 6 \\ -1 & -1 & 2 \end{bmatrix}$$

$$\begin{bmatrix} 2 & -1 & -1 \\ -\dfrac{5}{3} & 1 & \dfrac{1}{3} \\ \dfrac{1}{6} & 0 & \dfrac{1}{6} \end{bmatrix}$$

$$\begin{bmatrix} 1 & 1 & 4 \\ 2 & 3 & 6 \\ -1 & -1 & 2 \end{bmatrix} \begin{bmatrix} 2 & -1 & -1 \\ -\dfrac{5}{3} & 1 & \dfrac{1}{3} \\ \dfrac{1}{6} & 0 & \dfrac{1}{6} \end{bmatrix}$$

$$= \begin{bmatrix} 2-\dfrac{5}{3}+\dfrac{4}{6} & -1+1+0 & -1+\dfrac{1}{3}+\dfrac{4}{6} \\ 4-\dfrac{15}{3}+\dfrac{6}{6} & -2+3+0 & -2+\dfrac{3}{3}+\dfrac{6}{6} \\ -2+\dfrac{5}{3}+\dfrac{2}{6} & 1-1+0 & 1-\dfrac{1}{3}+\dfrac{2}{6} \end{bmatrix}$$

	$$= \begin{bmatrix} 1 & 0 & 0 \\ 0 & 1 & 0 \\ 0 & 0 & 1 \end{bmatrix} = \mathbf{I}$$
Try these	
Find the inverse of the matrix, if it exists: $\begin{bmatrix} 3 & 4 \\ 2 & 3 \end{bmatrix}$ Verify the result using matrix multiplication.	$\begin{bmatrix} 3 & -4 \\ -2 & 3 \end{bmatrix}$
Find the inverse of the matrix, if it exists: $\begin{bmatrix} -3 & -9 & 0 \\ -1 & -5 & -2 \\ 5 & 9 & -6 \end{bmatrix}$ Verify the result using matrix multiplication.	The matrix has no inverse.

♣ singular matrix

A singular matrix does not have an inverse. You'll discover that your matrix is singular if, during the row manipulations to find the inverse, one of the rows of the original left-hand matrix becomes all zeros.

★ For Fun

If one row becomes all zeros, wouldn't the system be dependent? Would the system have a unique solution?

♣ nonsingular matrix

A nonsingular matrix is a square matrix that does have an inverse.

★ For Fun

Do you imagine that nonsquare matrices can have inverses?

Singular Matrix: Test Prep	
Show that this matrix is singular: $$\begin{bmatrix} 1 & 3 & 4 \\ 2 & 5 & 3 \\ 1 & 4 & 9 \end{bmatrix}$$	$$\left[\begin{array}{ccc|ccc} 1 & 3 & 4 & 1 & 0 & 0 \\ 2 & 5 & 3 & 0 & 1 & 0 \\ 1 & 4 & 9 & 0 & 0 & 1 \end{array}\right] \rightarrow$$ $$\left[\begin{array}{ccc|ccc} 1 & 3 & 4 & 1 & 0 & 0 \\ 0 & 1 & 5 & 2 & -1 & 0 \\ 0 & 1 & 5 & -1 & 0 & 1 \end{array}\right] \rightarrow$$ $$\left[\begin{array}{ccc|ccc} 1 & 0 & -11 & -5 & -3 & 0 \\ 0 & 1 & 5 & 2 & 1 & 0 \\ 0 & 0 & 0 & -3 & 1 & 1 \end{array}\right]$$ Zero row on left-hand matrix means a singular matrix.
Try these	
Find the inverse of the matrix, if it exists: $$\begin{bmatrix} 1 & -6 & 4 \\ 3 & 4 & 2 \\ 5 & 3 & 5 \end{bmatrix}$$	The matrix is singular.
Given $\mathbf{A} = \begin{bmatrix} 2 & 5 & 4 \\ 1 & 4 & 3 \\ 1 & -3 & -2 \end{bmatrix}$ find $\mathbf{A}^{-1}$. Is A singular or nonsingular?	$\mathbf{A}^{-1} = \begin{bmatrix} -1 & 2 & 1 \\ -5 & 8 & 2 \\ 7 & -11 & -3 \end{bmatrix}$ A is nonsingular.

Solving Systems of Equations Using Inverse Matrices

If you rewrite a 2-dimensional system of equations symbolically like this $\mathbf{A} \cdot \begin{bmatrix} x \\ y \end{bmatrix} = \begin{bmatrix} a \\ b \end{bmatrix}$, notice what can

be done with the inverse of the *coefficient* matrix $\mathbf{A}$.

$$\mathbf{A}^{-1} \cdot \mathbf{A} \cdot \begin{bmatrix} x \\ y \end{bmatrix} = \mathbf{A}^{-1} \begin{bmatrix} a \\ b \end{bmatrix}$$

$$\begin{bmatrix} 1 & 0 \\ 0 & 1 \end{bmatrix} \begin{bmatrix} x \\ y \end{bmatrix} = \mathbf{A}^{-1} \cdot \begin{bmatrix} a \\ b \end{bmatrix}$$

$$\begin{bmatrix} x \\ y \end{bmatrix} = \mathbf{A}^{-1} \cdot \begin{bmatrix} a \\ b \end{bmatrix}$$

And you might notice that you have solved the system for the "coordinate" column matrix. That boils

down to having solved for the variables x and y. (Of course, finding the inverse of the coefficient

matrix can still be a big job.)

This can clearly be extended to arbitrarily large systems of equations with many variables to solve for.

Solving Systems of Equations Using Inverse Matrices: Test Prep	
Solve the matrix equation: $\begin{bmatrix} 7 & -5 \\ 2 & -3 \end{bmatrix} \begin{bmatrix} x \\ y \end{bmatrix} = \begin{bmatrix} 12 \\ 6 \end{bmatrix}$	If $\mathbf{A} = \begin{bmatrix} 7 & -5 \\ 2 & -3 \end{bmatrix}$, then $\mathbf{A}^{-1} = \begin{bmatrix} \dfrac{3}{11} & \dfrac{-5}{11} \\ \dfrac{2}{11} & \dfrac{-7}{11} \end{bmatrix} = -\dfrac{1}{11} \begin{bmatrix} -3 & 5 \\ -2 & 7 \end{bmatrix}$ and

$$\begin{bmatrix} x \\ y \end{bmatrix} = -\frac{1}{11}\begin{bmatrix} -3 & 5 \\ -2 & 7 \end{bmatrix}\begin{bmatrix} 12 \\ 6 \end{bmatrix}$$

$$= -\frac{1}{11}\begin{bmatrix} -36+30 \\ -24+42 \end{bmatrix} = \begin{bmatrix} \dfrac{6}{11} \\ -\dfrac{18}{11} \end{bmatrix}$$

Try these

Use inverse matrices to solve the system of equations: $\begin{cases} 4x+7y=2 \\ 3x+5y=1 \end{cases}$	$(-3, 2)$
Use inverse matrices to solve the system of equations: $\begin{cases} x+y-z=-3 \\ x+2y-z=6 \\ -x-y+2z=4 \end{cases}$	$(-11, 9, 1)$

Input-Output Analysis

The labor required to find the inverse of a coefficient matrix is similar to the effort needed to simply put the matrix form of a system of equations into row-echelon form and back-substitute in order to solve for the variables under consideration. But if the result matrix changes while the coefficient matrix doesn't, then it's well worth your time to find the one key that will solve every variable with just some matrix multiplication and algebraic interpretation. Input-output analysis makes use of this convenience.

♣ input-output analysis

Input-output analysis is a discipline that seeks to understand quantitatively the relationships between the demands of industries and customers and their outputs.

❖ **input-output matrix**

A particular matrix that gives the monetary relationship between input and output costs. Each column represents the cost of an input that will generate $1 of output.

❖ **final demand**

Final demand is represented by a column matrix and signifies the amount of output from the industries that consumers (and other industries) will purchase.

Applications: Test Prep	
Each edge of a metal plate is kept at a constant temperature as shown below. Find the temperature at x_1 and x_2. Round to the nearest tenth of a degree. 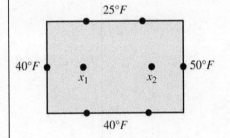	Write a system of equations: $$\begin{cases} x_1 = \dfrac{40+40+25+x_2}{4} = \dfrac{105+x_2}{4} \\ x_2 = \dfrac{25+60+40+x_1}{4} = \dfrac{125+x_1}{4} \end{cases}$$ Rewrite the system of equations: $$\begin{cases} 4x_1 - x_2 = 105 \\ -x_1 + 4x_2 = 125 \end{cases}$$ Solve the equations using the inverse matrix: $$\begin{bmatrix} 4 & -1 \\ -1 & 4 \end{bmatrix}\begin{bmatrix} x_1 \\ x_2 \end{bmatrix} = \begin{bmatrix} 105 \\ 125 \end{bmatrix}$$ $$\begin{bmatrix} x_1 \\ x_2 \end{bmatrix} = \frac{1}{15}\begin{bmatrix} 4 & 1 \\ 1 & 4 \end{bmatrix}\begin{bmatrix} 105 \\ 125 \end{bmatrix}$$ $$\begin{bmatrix} x_1 \\ x_2 \end{bmatrix} = \begin{bmatrix} \dfrac{109}{3} \\ \dfrac{121}{3} \end{bmatrix}$$ And the temperatures, to the nearest tenth of a degree, are 36.3° and 40.3°.

Try these

The following table shows the carbohydrate, fat, and protein content of three food types in grams:

	carbo-hydrate	fat	protein
type I	13	10	13
type II	4	4	3
type III	1	0	10

A nutritionist must prepare two diets from these three food groups. The first diet must contain 23 grams of carbohydrate, 18 grams of fat, and 39 grams of protein. The second diet must contain 35 grams of carbohydrate, 28 grams of fat, and 42 grams of protein. How many grams of each food type are required for the first diet, and how many grams of each food type are required for the second diet?

For the first diet, 100g of type I, 200g of type II, and 200g of type III; for the second diet, 200g of type I, 200g of type II, and 100g of type III.

A four-sector economy consists of manufacturing, agriculture, service, and transportation. The input-output matrix for this economy is

$$\begin{bmatrix} 0.10 & 0.05 & 0.200 & 0.15 \\ 0.20 & 0.10 & 0.30 & 0.10 \\ 0.05 & 0.30 & 0.20 & 0.40 \\ 0.10 & 0.20 & 0.15 & 0.20 \end{bmatrix}$$

$219.0 million of manufacturing, $294.3 million agriculture, $316.7 million service, and $260.3 million transportation

Find the gross output needed to satisfy the consumer

demand of $80 million worth of transportation.

TRY THE MID-CHAPTER 10 QUIZ

10.4: Determinants

A determinant is some sort of "characteristic value" associated with a matrix. First we learn how to

calculate them, then what they are useful for.

Determinant of a 2×2 Matrix

At the root of most methods of calculating the value of a determinant is the value of a 2×2 matrix's

determinant. The determinant for a 2×2 is easy, so it's a good place to start.

Ultimately it will generalize to larger matrices, too.

♣ determinant

The determinant of a matrix A is denoted by $|A|$ or sometimes by $\det(A)$. The absolute value-like

symbol can help to remind you that the determinant of a matrix is a scalar– that is, a number. What this

number tells you remains to be seen; however, in the meantime, here is how one calculates the

determinant of a 2×2 matrix:

$$\begin{vmatrix} a_{11} & a_{12} \\ a_{21} & a_{22} \end{vmatrix} = a_{11} \cdot a_{22} - a_{12} \cdot a_{21}$$

For example,

$$\begin{vmatrix} 1 & -2 \\ 3 & 0 \end{vmatrix} = 1 \cdot 0 - (-2) \cdot 3 = 6$$

Maybe you can see from this example that this is a place to be very mindful of the positive/negative

natures of the numbers involved. Be careful with the signs of the numbers and subtracting!

★ For Fun

Can you figure out how to get your calculator to calculate a 2×2 determinant?

Determinant of a 2×2 Matrix: Test Prep	
Find the determinant of $\begin{bmatrix} 9 & 1 \\ 8 & 2 \end{bmatrix}$	$\begin{vmatrix} 9 & 1 \\ 8 & 2 \end{vmatrix} 9 \cdot 2 - 1 \cdot 8 = 18 - 8 = 10$
Try these	
Evaluate $\begin{vmatrix} -1 & 5 \\ 3 & -7 \end{vmatrix}$	-8
Evaluate $\begin{vmatrix} 3 & 7 \\ 0 & 0 \end{vmatrix}$	0

Minors and Cofactors

Minors and cofactors are going to help us work up to figuring the determinant of matrices of order larger than 2×2.

♣ minor

The minor of a matrix is the determinant of a smaller portion of the matrix. Consider this 3×3 matrix:

$$\begin{bmatrix} 3 & 0 & -1 \\ 1 & 2 & -3 \\ 5 & 7 & -4 \end{bmatrix}$$

The M_{13} minor is, as follows: $\begin{vmatrix} 1 & 2 \\ 5 & 7 \end{vmatrix}$

Do you see those elements in the above matrix? It is the determinant of what's leftover when you remove the 1st row and 1st column of the matrix, and the subscripts of M_{11} tell you which row/column to "remove".

★ **For Fun**

Could you calculate M_{11} for the above matrix? How about a different minor like, maybe, M_{32}?

♣ **cofactor**

The cofactor C_{ij} of a matrix is the minor determinant with a ± 1 factor in front of it. The ± 1 depends on which minor like this:

$$C_{ij} = (-1)^{i+j} M_{ij}$$

So the cofactor C_{22}, for instance, would have entail a factor of positive 1; whereas, the C_{32} cofactor would get a negative one.

★ **For Fun**

Figure that ± 1 factor for every element in a 3×3 matrix, for instance. You could use my 3×3 matrix above. Look for a pattern.

★ **For Fun**

Now look at the formula for calculating the determinant of a 2×2. Do you see an alternating of signs even in it?

	Minor and Cofactor: Test Prep	
Given the matrix $\begin{bmatrix} 5 & -2 & 1 \\ 0 & -1 & 3 \\ -4 & 2 & 0 \end{bmatrix}$ find M_{23} and C_{23}	$M_{23=} \begin{vmatrix} 5 & -2 \\ -4 & 2 \end{vmatrix} = 10 - 8 = 2$ $C_{23} = (-1)^5 \begin{vmatrix} 5 & -2 \\ -4 & 2 \end{vmatrix} = -2$	
Try these		
Find M_{11} and C_{11} for the matrix above.	$M_{11} = -6 \quad C_{11} = -6$	
Find M_{21} and C_{21} for the matrix above.	$M_{21} = -2 \quad C_{21} = 2$	

Evaluating a Determinant Using Expanding by Cofactors

Now we'll see how we can carefully combine cofactors, which include their corresponding minors, and the elements of a 3×3 or larger square matrix to calculate a determinant. Again, to me this is a very visual exercise, so watch plenty of examples and then do plenty of examples of your own.

First of all, you have to pick one particular row *or* column to expand your determinant. Personally, I like to pick a row with as many small and positive numbers as I can with special preference given to zeros.

For instance, considering the matrix

$$\begin{bmatrix} 2 & 0 & -1 \\ -2 & 0 & -3 \\ 5 & 7 & -4 \end{bmatrix}$$

I would definitely choose to expand using the 2nd column. Try a few, you'll develop a sense for this quickly.

Having chosen the 2nd column above, this is how the determinant would be calculated:

$$\begin{vmatrix} 2 & 0 & -1 \\ -2 & 0 & -3 \\ 5 & 7 & -4 \end{vmatrix} = (0) \cdot (-1)^{1+2} \cdot M_{12} + (0) \cdot (-1)^{2+2} \cdot M_{22} + (7) \cdot (-1)^{3+2} \cdot M_{32}$$

Do you see the elements of the column multiplied by their corresponding cofactors? Here's the next step:

$$(0) \cdot (-1)^{1+2} \cdot \begin{vmatrix} -2 & -3 \\ 5 & -4 \end{vmatrix} + (0) \cdot (-1)^{2+2} \cdot \begin{vmatrix} 2 & -1 \\ 5 & -4 \end{vmatrix} + (7) \cdot (-1)^{3+2} \cdot \begin{vmatrix} 2 & -1 \\ -2 & -3 \end{vmatrix}$$

See the minors, which are now determinants of 2×2's? Do you see that the elements of those minors are the elements of the matrix once you've "removed" the corresponding row/column specified for the minor? Next step:

$-7 \cdot \begin{vmatrix} 2 & -1 \\ -2 & -3 \end{vmatrix} = -7[(2) \cdot (-3) - (-1) \cdot (-2)] = -7(-6 - 2) = 56$

So $\begin{vmatrix} 2 & 0 & -1 \\ -2 & 0 & -3 \\ 5 & 7 & -4 \end{vmatrix} = 56$.

This will work for any square matrix.

★ For Fun

Use cofactors to derive the general rule for a 3×3 determinant of $\begin{bmatrix} a & b & c \\ d & e & f \\ g & h & i \end{bmatrix}$.

Evaluate a Determinant by Expanding by Cofactors: Test Prep	
Evaluate the determinant by expanding the cofactors: $\begin{bmatrix} 0 & 1 & -2 \\ -1 & 2 & 1 \\ 4 & 0 & 3 \end{bmatrix}$	$\begin{vmatrix} 0 & 1 & -2 \\ -1 & 2 & 1 \\ 4 & 0 & 3 \end{vmatrix} = 1 \cdot 0 \cdot \begin{vmatrix} 2 & 1 \\ 0 & 3 \end{vmatrix} - 1 \cdot 1 \cdot \begin{vmatrix} -1 & 1 \\ 4 & 3 \end{vmatrix} + 1 \cdot (-2) \cdot \begin{vmatrix} -1 & 2 \\ 4 & 0 \end{vmatrix}$ $= -1(-3 - 4) - 2(0 - 8) = 7 + 16 = 23$
Try these	
Find the determinant of this matrix by expanding the cofactors: $\begin{bmatrix} 6 & 0 & 0 \\ 2 & -3 & 0 \\ 7 & -8 & 2 \end{bmatrix}$	-36

Evaluate: $\begin{vmatrix} -2 & 3 & 9 \\ 4 & -2 & -6 \\ 0 & -8 & -24 \end{vmatrix}$	0

Evaluating a Determinant Using Elementary Row Operations

We can use elementary row operations to help us simplify finding the determinant of larger matrices. First some rules about row operations and how they effect the determinant itself.

Elementary Row Operations on Determinants:

1. Interchanging any two rows of the matrix A changes the sign of the determinant of A.

   $|\mathbf{A}| \quad \rightarrow \quad -|\mathbf{A}|$

2. Multiplying a row of A by a constant k changes the determinant of A by the same factor.

   $|\mathbf{A}| \quad \rightarrow \quad k|\mathbf{A}|$

3. Adding a multiple of a row of A to another row of A doesn't change the value of the determinant of A.

   $|\mathbf{A}| \quad \rightarrow \quad |\mathbf{A}|$

♣ **triangular form**

We've met the triangular form of a matrix before. For instance:

$$\begin{bmatrix} 2 & -3 & 0 & 1 \\ 0 & 1 & -1 & 5 \\ 0 & 0 & 3 & -2 \\ 0 & 0 & 0 & 4 \end{bmatrix}$$

We'll expand what we mean as a matrix in triangular form to include matrices like this now:

$$\begin{bmatrix} -1 & 0 & 0 & 0 \\ 5 & 2 & 0 & 0 \\ 1 & -3 & 4 & 0 \\ 9 & 7 & 8 & 1 \end{bmatrix}$$

In other words, any square matrix whose elements either below *or* above the main diagonal are all zeros qualify for the triangular form label.

If your matrix is in triangular form, then the determinant of that matrix is simply the product of

the elements on the main diagonal.

(Think about it.)

Evaluate a Determinant by Using Elementary Row Operations: Test Prep	
Find the determinant of $\begin{bmatrix} 2 & 1 & 3 \\ 1 & 0 & -4 \\ -2 & 0 & 10 \end{bmatrix}$ by using row operations and their rules for determinants to turn this into a determinant of a triangular form.	$\begin{vmatrix} 2 & 1 & 3 \\ 1 & 0 & -4 \\ -2 & 0 & 10 \end{vmatrix} = (-1)\begin{vmatrix} 1 & 0 & -4 \\ 2 & 1 & 3 \\ -2 & 0 & 10 \end{vmatrix} \quad R_1 \leftrightarrow R_2$ $= (-1)\begin{vmatrix} 1 & 0 & -4 \\ 0 & 1 & 11 \\ 0 & 0 & 2 \end{vmatrix} \quad \begin{matrix} -2R_1 + R_2 \\ 2R_1 + R_3 \end{matrix}$ $= (-1)(1)(1)(2) = -2$
Try these	
Evaluate: $\begin{vmatrix} 3 & 0 & 0 \\ 2 & -1 & 0 \\ 3 & 4 & 5 \end{vmatrix}$	-15

Evaluate the determinant by first rewriting it in triangular form: $\begin{vmatrix} 1 & 2 & 0 & -2 \\ -1 & 1 & 3 & 5 \\ 2 & 1 & 4 & 0 \\ -2 & 5 & 2 & 6 \end{vmatrix}$	0

Condition for a Square Matrix to Have a Multiplicative Inverse

Determinants are handy for discovering if a square matrix has an inverse. And we've seen that an inverse means there exists a unique solution to the system back in section 7.3. A singular matrix does not have an inverse. This is exactly analogous to a matrix whose determinant is zero. Here are the conditions when the determinant will be equal to zero.

Given that A is a square matrix, $|A| = 0$ when ...

1. Any row or column is all-zeros.

2. Any two rows or columns are the same.

3. Any one row is a constant multiple of another row. Or any one column is a constant multiple of another column.

These conditions need not hold for the original matrix but do guarantee a zero determinant if they appear after row reductions!

Now that we know how to tell if a determinant is zero, we can say the following:

If A is a square matrix of order n, then A has a multiplicative inverse if and only if $|A|$ is not equal to zero, which also implies $|A^{-1}| = \dfrac{1}{|A|}$.

★ For Fun

So do you imagine that a determinant that is equal to zero implies anything about the solution set of a

system of equations?

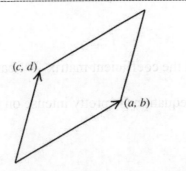

(c, d)

(a, b)

The area of a parallelogram is related to the determinant of the matrix formed from the points of the

vertices of the shape.

10.5: Cramer's Rule

This is Cramer's Rule, from page 846 of your textbook:

$$
\text{Let}
\begin{cases}
a_{11}x_1 + a_{12}x_2 + a_{13}x_3 + \ldots + a_{1n}x_n = b_1 \\
a_{21}x_1 + a_{22}x_2 + a_{23}x_3 + \ldots + a_{2n}x_n = b_2 \\
a_{31}x_1 + a_{32}x_2 + a_{33}x_3 + \ldots + a_{3n}x_n = b_3 \\
\quad\vdots \qquad\quad \vdots \qquad\quad \vdots \qquad\quad \vdots \qquad \vdots \quad = \vdots \\
a_{n1}x_1 + a_{n2}x_2 + a_{n3}x_3 + \ldots + a_{nn}x_n = b_n
\end{cases}
$$

be a system of n equations in n variables. The solution of the system is given by $(x_1, x_2, x_3, \ldots, x_n)$,

where

$$
x_1 = \frac{D_1}{D} \quad x_2 = \frac{D_2}{D} \qquad \ldots \qquad x_j = \frac{D_j}{D} \quad \ldots \quad x_n = \frac{D_n}{D}
$$

and D is the determinant of the coefficient matrix, $D \neq 0$. D_j is the determinant formed by

replacing the ith column of the coefficient matrix with the column of constants $b_1, b_2, b_3, \ldots, b_n$.

✓ Questions to Ask Your Instructor

Read carefully. Cramer's Rule tells you how to solve a system of equations using determinants, provided the system can be solved uniquely. This rule includes some pretty intense notation and lots of subscripts. Ask your instructor to help you understand every little bit if you are unclear about any of the pieces.

★ For Fun

Can you see why the determinant of the coefficient matrix D cannot be equal to zero?

This method of solving a system of equations is pretty intense on the determinant front. Have you practiced your determinants?

Cramer's Rule: Test Prep	
Solve the following system of equations using Cramer's Rule: $$\begin{cases} 4x_1 - 5x_2 = 12 \\ 3x_1 - 4x_2 = 10 \end{cases}$$	$$x_1 = \frac{\begin{vmatrix} 12 & -5 \\ 10 & 4 \end{vmatrix}}{\begin{vmatrix} 4 & -5 \\ 3 & 4 \end{vmatrix}} = \frac{2}{-1} = -2$$ $$x_2 = \frac{\begin{vmatrix} 4 & 12 \\ 3 & 10 \end{vmatrix}}{\begin{vmatrix} 4 & -5 \\ 3 & -4 \end{vmatrix}} = \frac{4}{-1} = -4$$ $(-2, -4)$
Try these	
Consider the system $$\begin{cases} 4x - 6y = 8 \\ 2x - 3y = 4 \end{cases}$$	1. no 2. yes
1. Can this system be solved using Cramer's Rule? 2. Can this system be solved using Gaussian elimination method?	

Solve the following system of equations using Cramer's Rule: $\begin{cases} x & +3y & & = -2 \\ 2x & -3y & +z & = 1 \\ 4x & +5y & -2z & = 0 \end{cases}$	$\left(\dfrac{4}{25}, -\dfrac{18}{25}, -\dfrac{37}{25} \right)$

Final Fun: Matching

FIND THE BEST MATCH. GOOD LUCK.

AUGMENTED MATRIX

$\begin{vmatrix} 2 & -1 & 0.5 \\ 3 & 5 & -4 \\ 0 & 7 & -0 \end{vmatrix}$

TRIANGULAR MATRIX

ROW MATRIX

DETERMINANT OF 2×2

$(-1)^{i+j} \mathbf{M}_{ij}$

IDENTITY

$\begin{bmatrix} 2 & -3 & 1 & 0 & 5 \\ -5 & 2 & -2 & 1 & 3 \end{bmatrix}$

$\mathbf{B}^{-1}$

$\begin{bmatrix} 0 & 0 \\ 0 & 0 \end{bmatrix}$

COFACTOR

$\begin{bmatrix} 1 & 0 \\ -2 & 3 \end{bmatrix}$

DETERMINANT

GRAPH

$[-3\ 0\ 1\ -2]$

ADDITIVE IDENTITY

INVERSE MATRIX

2 X 5 MATRIX

$\begin{bmatrix} 2 & -1 & 9 \\ 1 & 3 & -7 \end{bmatrix}$

$\begin{bmatrix} 1 & 0 & 0 \\ 0 & 1 & 0 \\ 0 & 0 & 1 \end{bmatrix}$

$a_{11} \cdot a_{22} - a_{12} \cdot a_{21}$

NOW TRY THESE CHAPTER 10 REVIEW EXERCISES AND

THE CHAPTER 10 PRACTICE TEST

STAY UP ON YOUR GAME – TRY THE CUMULATIVE REVIEW EXERCISES

Chapter 11: Sequences, Series, and Probability

Word Search

```
L  A  I  M  O  N  I  B  H  Z  D  J  X  F  I
S  C  I  N  D  E  P  E  N  D  E  N  T  O  N
S  X  S  W  B  K  V  C  G  N  S  A  E  N  F
N  E  I  P  H  G  O  E  U  W  T  H  X  O  I
A  O  Q  V  N  U  O  Q  N  C  E  Y  P  I  N
S  R  I  U  N  M  U  N  A  T  C  T  E  T  I
Z  E  I  T  E  E  H  S  E  V  G  I  R  C  T
N  O  I  T  A  N  I  B  M  O  C  L  I  U  E
K  N  R  R  H  T  C  M  W  I  N  I  M  D  T
G  I  F  N  E  M  U  E  Z  H  A  B  E  N  I
C  B  X  C  B  S  E  M  X  R  W  A  N  I  N
P  A  S  C  A  L  U  T  R  S  K  B  T  V  I
Z  K  T  J  O  E  J  K  I  E  G  O  X  C  F
L  A  I  R  O  T  C  A  F  C  P  R  S  C  N
E  X  C  L  U  S  I  V  E  B  P  P  J  F  M
```

ARITHMETIC	EXPERIMENT	INFINITE
BINOMIAL	FACTORIAL	PASCAL
COMBINATION	FINITE	PERMUTATION
COUNTING	GEOMETRIC	PROBABILITY
EVENT	INDEPENDENT	SEQUENCE
EXCLUSIVE	INDUCTION	SERIES

11.1: Infinite Sequences and Summation Notation

This section is really just about learning a new notation: the summation notation. Just like x_2 means something entirely different from x^2, you need to be able to assign the correct *meaning* to a subscript, or an *index*, in an expression. By the way, you are clear about the difference between x_2 and x^2, right?

Infinite Sequences

A sequence is a set of elements that come in a particular order. Remember from the discussion of sets way back in Chapter P that there is more than one way to describe a set. Now we get the way one can describe an infinite *ordered* set.

♣ infinite sequence

An infinite sequence can be described in terms of functions: An infinite sequence is a function whose domain is the positive integers and whose range is a set of real numbers (page 862 in your textbook). The thing with sequences is that we're usually very interested in the order of the elements of the range of the function that defines the sequence, and we are usually very particular about starting off our domain at one (or sometimes zero). This way, the correspondence between the input and the output can become meaningful.

♣ terms

A term is an element of the range of the sequence. If the function defining the sequence is something like $f(n) = (n+1)^2$, then the first *term* of the sequence is $f(1) = 4$.

♣ first term

Unless otherwise noted, the "positive integers" of the domain of a sequence starts at one. So the first term of a sequence is $f(1)$.

♣ second term

The second term will be $f(2)$ … and so on.

❖ *nth* term, general term

The *nth* term definition of an element of a sequence is just exactly the definition of the sequence that depends on *n*. The *nth* term, or general term, of $f(n) = (n+1)^2$ is just $(n+1)^2$.

★ For Fun

So finding a particular term of a sequence is pretty much the same proposition as evaluating a function, right?

❖ alternating sequence

An alternating sequence is a sequence whose terms alternate between negative and positive. (Frequently such a sequence will include an expression like $(-1)^n$ or some other evident dependence on -1.

❖ recursively-defined sequence

A recursive definition is one that depends on itself. If you are programming and your function calls itself, that is a recursive relationship. A recursively-defined term in a sequence will depend on itself something like this:

$$a_n = \frac{a_{n-1} + a_{n+1}}{2}$$

❖ Fibonacci sequence

The famous Fibonacci sequence is 1, 1, 2, 3, 5, 8, 13, 21, ...

★ For Fun

Find out more about this Fibonacci character and his tremendous contributions to mathematics. You might found out how the sequence of Fibonacci numbers is defined also!

Infinite Sequence: Test Prep	
Find the eleventh term of this sequence: $a_n = \dfrac{(-1)^{n+1}}{n^2}$	$a_{11} = \dfrac{(-1)^{11+1}}{11^2} = \dfrac{1}{121}$

Try these	
List the first five terms of this infinite sequence: $a_n = \left(\dfrac{3}{5}\right)^n$	$\dfrac{3}{5}, \dfrac{9}{25}, \dfrac{27}{125}, \dfrac{81}{625}, \dfrac{243}{3125}$
List the first three terms of the recursively defined sequence: $a_1 = 2, \quad a_n = (-3)na_{n-1}$	$2, -12, 108$

Factorials

Although you probably have a factorial function on your calculator (it looks like an exclamation mark), it's necessary to understand how factorials work ... because they get really big really quickly. Your calculator is limited in how well it can handle big numbers.

♣ *n* factorial

n-factorial is defined (as a function of *n*) like this:

$$n! = n \cdot (n-1) \cdot (n-2) \cdot \ldots \cdot 2 \cdot 1.$$

So, for instance, $5! = 5 \cdot 4 \cdot 3 \cdot 2 \cdot 1 = 120$.

★ For Fun

Does the definition of the factorial seem recursive to you?

n Factorial: Test Prep	
Evaluate the factorial expression without using a calculator: $\dfrac{26!}{24!}$	$\dfrac{26!}{24!} = \dfrac{26 \cdot 25 \cdot 24 \cdot \ldots \cdot 3 \cdot 2 \cdot 1}{24 \cdot \ldots \cdot 3 \cdot 2 \cdot 1} = 26 \cdot 25 = 650$
Try these	
Evaluate $5! + 4!$	144
Find the fifth term in the infinite sequence:	-30

$$a_n = \frac{(-1)^n n!}{n-1}$$

Partial Sums and Summation Notation

So far every time we've mentioned sequence, we've been referring to an infinite sequence, like the Fibonacci sequence whose ellipses say "This goes on forever." Partial sums happen when you take a finite part of a sequence *and add the terms up*.

❖ *n*th partial sum

S_n is the nth partial sum of the a sequence. For instance, if you need to find S_4, you would add the first four terms of your sequence together to find the fourth partial sum.

❖ sequence of partial sums

Okay, now you can make a new infinite sequence out of the successive partial sums of a sequence. That is $S_1, S_2, S_3, \cdots, S_n, \cdots$

★ For Fun

You wanna try to write the first few terms of the sequence of partial sums of

$$f(n) = (n+1)^2 ?$$

❖ summation notation

And now it's time to dissect and reconstitute summation notation which might look like this: $\displaystyle\sum_{i=1}^{\infty} a_i$.

★ For Fun

Do you know which Greek letter $\sum$ is? Does it make sense that this stands for "summation"?

❖ index of the summation

In the example $\displaystyle\sum_{i=1}^{\infty} a_j$, i is the index of the summation. The index starts where the statement at the bottom of the summation notation informs you (1 in this case) and continues in integer steps to the value at the top of the summation symbol. This example is an *infinite* sum because the index goes until ∞ (not that ∞ is a number, mind you!).

♣ upper limit

The number at the top of the summation symbol is the upper limit. That's the value of the index that arrests

the sum (if it is not an infinite sum).

♣ lower limit

The lower limit is the value where the index starts. This is given at the bottom of the summation symbol.

nth Partial Sum: Test Prep	
Find the 3rd partial sum of $a_n = (-1)^n \dfrac{n}{n!}$	$a_1 = -1; \quad a_2 = \dfrac{2}{2!} = 1; \quad a_3 = -\dfrac{3}{3!} = -\dfrac{1}{2}$
	$a_1 + a_2 + a_3 = -1 + 1 - \dfrac{1}{2} = -\dfrac{1}{2}$
Try these	
Evaluate the series: $\displaystyle\sum_{i=1}^{3} \dfrac{i-1}{i+1}$	$\dfrac{5}{6}$
Evaluate the series: $\displaystyle\sum_{k=3}^{6} (-1)^k k!$	618

11.2: Arithmetic Sequences and Series

The next couple of sections deal with some standard examples of sequences and series (sums). They're

good practice ... and you'll see them in calculus again.

Arithmetic Sequences

A sequence whose terms differ by a constant quantity is an arithmetic sequence – I guess because you have

to do arithmetic to figure out the next term!

♣ arithmetic sequence

If you calculate the first few terms of a sequence defined like, say, $5n - 2$, you'll find that each term

differs from the next by five. That makes it an arithmetic sequence: a sequence where $a_{i+1} - a_j = d$, where

d is a real number.

★ For Fun

Try a few: $-2n+3$, $n-1$, $3(2n-1)$.

❖ common difference

d in the definition above is the common difference between the terms of an arithmetic sequence.

★ For Fun

What is the numeric value of the common differences in the examples given above?

Arithmetic Sequence: Test Prep	
Consider the arithmetic sequence $7, 9, 11, \ldots, 2n+5, \ldots$. What is the common difference?	Since all the terms of the series increase by 2, it is 2; also the coefficient of the nth term is 2.
Try these	
Find the 21st term of the above sequence.	47
Find the 13th element of the sequence: $-3, 4, 11, \ldots, 7n-10$	81

Arithmetic Series

Now we're adding terms together that have a common difference. Don't forget how to do partial sums.

❖ arithmetic series

The third partial sum of an arithmetic series might look something like this:

$$\sum_{n=1}^{3}(-4n+1)$$

★ For Fun

Compute the above partial arithmetic sum. What is the common difference of the terms?

Formulae for the *n*th Partial Sum of an Arithmetic Series:

1. $S_n = \dfrac{n}{2}(a_1 + a_n)$

2. $S_n = \dfrac{n}{2}\left[2a_1 + (n-1)d\right]$

[Practice these formulae.

They will help clarify variables and indices in your mind.]

Sum of an Arithmetic Series: Test Prep	
Find the 14th partial sum of the arithmetic sequence $a_n = 7n$	$S_{14} = \dfrac{14}{2}(7+98) = 7 \cdot 105 = 735$
Try these	
Find the sum of the first 75 terms of the arithmetic sequence whose first 3 terms are $\dfrac{1}{2}, \dfrac{9}{4}, 4$.	$\dfrac{19979}{4}$
Find the 25th partial sum of the arithmetic sequence $a_n = n - 4$	225

Arithmetic Means

Can you build an arithmetic sequence from a series of halfway points?

♣ arithmetic mean

The arithmetic mean is what we usually refer to simply as the mean: the arithmetic mean of two numbers a and b is $\dfrac{a+b}{2}$. If you think about it, since the mean is situated exactly between a and b, there is a common difference between a and b and their mean. So those three numbers form an arithmetic sequence. Since you can splint any interval exactly in half, then one can create any arithmetic sequence using successive means. Do you see what I mean?

★ **For Fun**

This reminds me of Zeno's paradox, "Arguments Against Motion." Are you familiar with it?

Arithmetic Mean: Test Prep	
Insert five arithmetic means between 4 and 20.	$a = 4, c_1, c_2, c_3, c_4, c_5, 20 = b$ $n = 7$ $20 = 4 + (7-1)d = 4 + 6d$ $d = \dfrac{16}{6} = \dfrac{8}{3}$ $c_1 = 4 + \dfrac{8}{3} = \dfrac{20}{3}$ $c_2 = 4 + 2\left(\dfrac{8}{3}\right) = \dfrac{12}{3} + \dfrac{16}{3} = \dfrac{28}{3}$ $c_3 = 4 + 3\left(\dfrac{8}{3}\right) = 12$ $c_4 = \dfrac{44}{3}$ $c_5 = \dfrac{52}{3}$ $4, \dfrac{20}{3}, \dfrac{28}{3}, 12, \dfrac{44}{3}, \dfrac{52}{3}, 20$
Try these	
Insert 5 arithmetic means between 7 and 19.	$7, 9, 11, 13, 15, 17, 19$
Insert 4 arithmetic means between $\dfrac{11}{3}$ and 6.	$\dfrac{11}{3}, \dfrac{62}{15}, \dfrac{69}{15}, \dfrac{76}{15}, \dfrac{83}{15}, 6$

★ **For Fun**

There's a famous comparison of arithmetic means with geometric means. You could look it up!

11.3: Geometric Sequences and Series

More on common sequences and series: the geometric sequence and series shows up again in calculus.

Meanwhile, it's great practice sorting out notation.

Geometric Sequences

The separation between the terms in a geometric series is not a common difference like what holds sway in the land of arithmetic sequences; the relation between geometric terms is more of a common *ratio*.

♣ **geometric sequence**

Here is the relation between terms in a geometric sequence:

$$\frac{a_{j+1}}{a_j} = r.$$

★ **For Fun**

Do you see a common ratio between terms?

♣ **common ratio**

The parameter r in the above definition is the common ratio of the geometric sequence (or series).

The *n*th Term of a Geometric Sequence:

$$a_n = a_1 r^{n-1}$$

Geometric Sequence: Test Prep	
Find the *nth* term of the geometric sequence whose first three terms are 36, 30, 25.	$r = \dfrac{30}{36} = \dfrac{5}{6}$ $a_1 = 36$ $a_n = 36\left(\dfrac{5}{6}\right)^{n-1}$
Try these	
Determine if this sequence is geometric:	No (it is arithmetic, however)

$16, 9, 2, -5, \ldots, -7n+23, \ldots$	
Determine if the sequence is geometric, arithmetic, or neither: a. e^n b. $\dfrac{(-1)^n}{n}$ c. $3n+3$	a. Geometric b. Neither c. Arithmetic

Finite Geometric Series

We can begin by practicing with the partial sums of geometric series, but next we will progress to using and understanding the infinite geometric series. It has some interesting usefulness.

❖ **finite geometric series**

The nth partial sum of a geometric series is a finite geometric series. An example might appear like this:

$$\sum_{n=1}^{7} 2^n$$

★ **For Fun**

Which partial sum is the above? Could you evaluate it term-by-term? Can you discover that the common ratio is 2? Could you compare the term-by-term result with the result obtained by trying the following formula from page 878 in your textbook?

Formula for the nth Partial Sum of a Geometric Series:
$$S_n = \frac{a_1(1-r^n)}{1-r} \text{ where } r \neq 1$$

Sum of a Finite Geometric Series: Test Prep	
Find the sum of the first 5 terms of the geometric sequence	$a_1 = -4$ $r = -5$

$-4, 20, -100, \ldots, -4(-5)^{n-1}, \ldots$	$n = 5$ $$S_5 = \frac{-4\left[1-(-5)^5\right]}{1-(-5)} = \frac{-4(1+3125)}{6}$$ $$= \frac{-2(3126)}{3} = -2(1042)$$ $$= -2084$$
Try these	
Evaluate the finite geometric series: $$\sum_{n=0}^{6} 5\left(\frac{2}{3}\right)^n$$	$$\frac{10295}{729}$$
Find the sum of the finite geometric series: $$\sum_{n=1}^{7} 2^n$$	254

Infinite Geometric Series

Under certain circumstances the *infinite* sum of a geometric series is absolutely defined and easy to evaluate.

★ For Fun

Look up a proof for fun.

♣ infinite series

The upper limit of an infinite series is infinity: ∞.

♣ infinite geometric series

A geometric series that includes *all* its terms (whose upper limit is ∞) is an infinite geometric series.

> ### The Sum of an Infinite Geometric Series
>
> Given an infinite geometric series with terms a_n and where $|r| < 1$, then the
>
> $$\text{sum } S \text{ is } \frac{a_1}{1-r}.$$

Sum of an Infinite Geometric Series: Test Prep	
Evaluate $\displaystyle\sum_{n=1}^{\infty}\left(\frac{3}{8}\right)^{n}$	$a_1 = \dfrac{3}{8}$ $r = \dfrac{3}{8}$ $\displaystyle\sum_{n=1}^{\infty}\left(\frac{3}{8}\right)^{n} = \dfrac{\frac{3}{8}}{1-\frac{3}{8}} \cdot \dfrac{8}{8} = \dfrac{3}{8-3} = \dfrac{3}{5}$
Try these	
Evaluate $\displaystyle\sum_{n=1}^{\infty}(0.5)^{n}$	1
Write $0.\overline{63}$ as a ratio of two integers in simplest form.	$\dfrac{7}{11}$

Applications of Geometric Sequences and Series

There are some financial applications of these geometric sequences and series that we will explore briefly

now.

❖ future value

The future value of an investment, or the value after some *last* deposit, is the sum of the values of all the

deposits.

❖ annuities

Equal amounts deposited at equal intervals of time are called annuities.

❖ ordinary annuity

Ordinary annuities entail deposits made at the end of investment periods – for example, the 31st of

December.

Future Value of an Ordinary Annuity

Let $r = \dfrac{i}{n}$ and $m = nt$, where i is the annual interest rate, n is the number of compounding periods per year, and t is the number of years. Then the future value A of an ordinary annuity after m compounding periods is given by

$$A = \dfrac{P\left[(1+r)^m - 1\right]}{r} \text{ where } P \text{ is the amount of each deposit}$$

(page 882 of your textbook).

♣ **Gordon model of stock valuation**

This model of stock valuation is used to determine the value of a stock whose dividend is expected to increase by the same percentage each year.

♣ **multiplier effect**

The multiplier effect has to do with an idea that an initial amount of spending leads to an increase in consumer spending that ideally results in an increase in overall income that outweighs the initial expense. In other words, the increase is a multiple of the initial expense.

TRY THE MID-CHAPTER 11 QUIZ

11.4: Mathematical Induction

Mathematical induction – not the same as deduction – is based on this Induction Axiom from page 888 of your textbook. It is deceptively simple.

Induction Axiom

Suppose S is a set of positive integers with the following two properties:

1. 1 is an element of S.
2. If the positive integer k is in S, then $k + 1$ is in S.

Then S contains all positive integers.

Principle of Mathematical Induction

Expanding the Induction Axiom, so that it can be used to prove the general case from specific cases in mathematics, forms the Principle of Mathematical Induction.

Principle of Mathematical Induction

Let P_n be a statement about a positive integer n. If **1.** P_1 is true and **2.** the truth of P_k implies the truth of P_{k+1}, then P_n is true for all positive integers.

It has been suggested that a mathematical proof by induction is like knocking over dominoes – if knocking over any domino will cause the next to fall and you knock over the first one, then all the dominoes will fall.

❖ induction hypothesis

So there are basically three steps to an induction proof. First show the premise holds for the number 1.

Then assume it holds true for k – this step is the induction hypothesis.

Third, show that given truth for k that the premise is still true for $k+1$.

Principle of Mathematical Induction: Test Prep	
Prove that $1^2 + 3^2 + 5^2 + \cdots + (2n-1)^2$ $= \dfrac{n(2n+1)(2n-1)}{3}$ by mathematical induction.	a. Let $n=1$: $\quad 1^2 = 1$ $\quad \dfrac{1(2+1)(2-1)}{3} = 1$ $\quad$ (satisfied) b. Assume $\quad 1^2 + 3^2 + 5^2 + \ldots + (2k-1)^2$ $\quad = \dfrac{k(2k+1)(2k-1)}{3}$ Show: $\quad 1^2 + 3^2 + 5^2 + \ldots + (2k-1)^2 + (2k+1)^2$

$$= \frac{(k+1)(2(k+1)+1)(2(k+1)-1)}{3}$$

$$= \frac{4k^3 + 12k^2 + 11k + 3}{3}$$

Adding the new term onto both sides of the assumption from step two:

$$1^2 + 3^2 + 5^2 + \ldots + (2k-1)^2 + (2k+1)^2$$

$$= \frac{k(2k+1)(2k-1)}{3} + (2k+1)^2$$

$$= \frac{4k^3 - k}{3} + \frac{3(4k^2 + 4k + 1)}{3}$$

$$= \frac{4k^3 + 12k^2 + 11k + 3}{3}$$

(satisfied)

Try these

Prove that	a. Let $n = 1$:
$\left(1 - \dfrac{1}{2}\right)\left(1 - \dfrac{2}{3}\right)\left(1 - \dfrac{3}{4}\right) \times \ldots \times \left(1 - \dfrac{n}{n+1}\right)$ $= \dfrac{1}{(n+1)!}$ for all positive integers n.	$\left(1 - \dfrac{1}{2}\right) = \dfrac{1}{(1+1)!} = \dfrac{1}{2}$ (satisfied)

b. Assume:

$$\left(1 - \frac{1}{2}\right)\left(1 - \frac{2}{3}\right)\left(1 - \frac{3}{4}\right) \times \ldots \times \left(1 - \frac{k}{k+1}\right)$$

$$= \frac{1}{(k+1)!}$$

c. Show

$$\left(1 - \frac{1}{2}\right)\left(1 - \frac{2}{3}\right)\left(1 - \frac{3}{4}\right) \times \ldots \times \left(1 - \frac{k}{k+1}\right)\left(1 - \frac{k+1}{k+2}\right)$$

$$= \frac{1}{(k+2)!}$$

Substituting from step two:

	$$\left(\frac{1}{(k+1)!}\right)\left(1-\frac{k+1}{k+2}\right)$$ $$=\left(\frac{1}{(k+1)!}\right)\left(\frac{k+2-k-1}{k+2}\right)$$ $$=\left(\frac{1}{(k+1)!}\right)\left(\frac{1}{k+2}\right)=\frac{1}{(k+2)!}$$ (satisfied)
Prove that $2n-1<2^n$ for all positive integers n.	a. Let $n=1$: $2-1<2$ (satisfied)
	b. Assume: $2k-1<2^k$
	c. Show $2(k+1)-1<2^{k+1}$. But since $2(2k-1)=4k-2>2k+1$, then $2^{k+1}>2k+1$ (satisfied)

Extended Principle of Mathematical Induction

In the case where the beginning index for the desired induction is greater than 1, we have the Extended Principle of Mathematical Induction (page 891 in your textbook). Look for similarities between the structure of this argument and the structure of the previous induction principle.

The Extended Principle of Mathematical Induction

Let P_n be statement about a positive integer n. If **1.** P_j is true for some positive integer j, and **2.** for $k \geq j$ the truth of P_k implies the truth of P_{k+1}, then P_n is true for all positive integers $n \geq j$

Extended Principle of Mathematical Induction: Test Prep	
Prove that for each natural number $n \geq 6$, $(n+1)^2 \leq 2^n$	Let m be a natural number such that $(m+1)^2 \leq 2^m$. The number 6 is a possible value of m because $(6+1)^2 = 49 \leq 64 = 2^6$.

Suppose that $n \geq 6$ and $(n+1)^2 \leq 2^n$. Then

$\dfrac{(n+1)^2}{2^n} \leq 1$. Compute

$$\dfrac{(n+2)^2}{2^{n+1}} = \dfrac{1}{2}\left(\dfrac{n+2}{n+1}\right)^2 \dfrac{(n+1)^2}{2^n} \leq \dfrac{1}{2}\left(1+\dfrac{1}{n+1}\right)^2.$$

Since $n \geq 6$, $1 + \dfrac{1}{n+1} \leq 1.4 \leq \sqrt{2}$. Therefore,

$\dfrac{(n+2)^2}{2^{n+1}} \leq 1$, and we have proved $m = n+1$ is valid. By the Extended Principle of Mathematical Induction we have shown $(n+1)^2 \leq 2^n$ for $n \geq 6$.

Try these

Prove that for natural numbers $n \geq 2$ we have $2^{n+1} \leq 3^n$.	For $n=2$ we have $2^{2+1} = 8 \leq 9 = 3^2$. Now assume for $k \geq 2$ that $2^{k+1} \leq 3^k$. For $k+1$, we have $2^{k+2} = 2 \cdot 2^{k+1} \leq 2 \cdot 3^k \leq 3 \cdot 3^k = 3^{k+1}$. So the assumption holds for $k+1$, and by the Extended Principle of Mathematical Induction we've shown that $2^{n+1} \leq 3^n$ for $n \geq 2$.
Prove that $n^2 \geq 3n$ for $n \geq 3$.	For $n=3$ we have $3^2 = 9 = 3 \cdot 3$. Assume that for $n \geq 3$, $n^2 \geq 3n$. For $n+1$: $(n+1)^2 = n^2 + 2n + 1 \geq 3n + 2n + 1 \geq 3n + 3$ for $n \geq 3$. And so $(n+1)^2 \geq 3n + 3 = 3(n+1)$ or $(n+1)^2 \geq 3(n+1)$ for $n \geq 3$ by the Extended Principle of Mathematical Induction.

★ For Fun

The following is a statement of the Induction Axiom in logical symbols. Do you know any of these symbols? Look them up. At the very least, you have to admit that it's a compact way to state the Axiom (even if incomprehensible):

$$\forall X(1 \in X \land \forall x(x \in X \Rightarrow x+1 \in X) \Rightarrow) \forall x(x \in X))$$

11.5: Binomial Theorem

Not only will this section be about learning a pattern to simplify a big multiplication problem, but the notation carries over into calculus, statistics, and probability. I actually like to think of this section as "fun with numbers"!

Binomial Theorem

Actually carrying out the multiplication of a binomial like $(x+3)^7$ can be a pretty daunting proposition. Do you want to perform this operation:

$$(x+3)(x+3)(x+3)(x+3)(x+3)(x+3)(x+3)$$

And be assured that you haven't made an error?

There are some very regular patterns in this enterprise that we can learn and use.

♣ expanding the binomial

Expanding the binomial refers to any process of carrying out the multiplication implied by the exponent applied to a binomial: $(a+b)^n$

♣ binomial coefficient

The binomial coefficient is symbolized like this $\binom{n}{k}$

This is not a fraction – although it looks somewhat like a fraction. What this symbol means is $\dfrac{n!}{k!(n-k)!}$

The Binomial Coefficient

$$\binom{n}{k} = \frac{n!}{k!(n-k)!}$$

★ For Fun

Are factorials defined for negative integers? Can you tell which *has* to be bigger, *n* or *k*, from the denominator of the definition of the binomial coefficient?

Binomial Coefficient: Test Prep	
Evaluate $\binom{10}{3}$	$\binom{10}{3} = \dfrac{10!}{3!(10-3)!} = \dfrac{10!}{3!7!}$ $= \dfrac{10 \cdot 9 \cdot 8 \cdot 7 \cdot 6 \cdot 5 \cdot 4 \cdot 3 \cdot 2 \cdot 1}{(3 \cdot 2 \cdot 1) \cdot (7 \cdot 6 \cdot 5 \cdot 4 \cdot 3 \cdot 2 \cdot 1)}$ $= \dfrac{10 \cdot 9 \cdot 8}{3 \cdot 2 \cdot 1} = 120$
Try these	
Evaluate $\binom{7}{0}$	1
Evaluate $\binom{15}{13}$	105

The following formula is one place the binomial coefficient is used. This is how you can expand a binomial.

The Binomial Theorem for Positive Integers:

If *n* is a positive integer, then

$$(a+b)^n = \sum_{i=0}^{n} \binom{n}{i} a^{n-i} b^i$$

$$= \binom{n}{0} a^n + \binom{n}{1} a^{n-1}b + \binom{n}{2} a^{n-2}b^2 + \ldots + \binom{n}{n} b^n$$

The above formula has it all for expanding binomials. Every element of the expansion is described in the formula, but the formula is a little complicated to most newcomers. I would suggest some serious practice.

It gets easier pretty quickly. And then you can amaze your friends with just how fast you can multiply

something like $\left(x-\sqrt{2}\right)^6$.

Binomial Theorem for Positive Integers: Test Prep	
Expand $(x-\sqrt{2})^6$ using the binomial theorem.	$\left(x-\sqrt{2}\right)^6 = \sum_{i=0}^{6} \binom{6}{i} x^{6-i} \left(-\sqrt{2}\right)^i$ $= \binom{6}{0} x^6 + \binom{6}{1} x^5 \left(-\sqrt{2}\right) + \binom{6}{2} x^4 \left(-\sqrt{2}\right)^2$ $+ \binom{6}{3} x^3 \left(-\sqrt{2}\right)^3 + \binom{6}{4} x^2 \left(-\sqrt{2}\right)^4$ $+ \binom{6}{5} x \left(-\sqrt{2}\right)^5 + \binom{6}{6} \left(-\sqrt{2}\right)^6$ $= x^6 - 6x^5\sqrt{2} + 30x^4 - 40x^3\sqrt{2} + 60x^2 - 24x\sqrt{2} + 8$
Try these	
Expand $(a-2)^4$ using the binomial theorem.	$a^4 - 8a^3 + 24a^2 - 32a + 16$
Expand $(2x-1)^5$ using the binomial theorem.	$32x^5 - 80x^4 + 80x^3 - 40x^2 + 10x - 1$

*i*th Term of a Binomial Expansion

You can find a *particular* term in a binomial expansion using the following formula. All this takes is

careful attention to indices and exponential evaluation.

The *i*th term of the binomial expansion of $(a+b)^n$ is given by $\binom{n}{i-1} a^{n-i+1} b^{i-1}$

*i*th Term of a Binomial Expansion: Test Prep	
Find the 3rd term of the expansion of $(x+8)^7$	$a=x$, $b=8$, $n=7$, $i=3$ $\binom{7}{2}x^5 \cdot 8^2 = \dfrac{7!}{2!5!}64x^5 = 21\cdot 64x^5 = 1344x^5$
Try these	
Find the sixth term of $(5x^2 - 2y^3)^8$	$-224000x^6 y^{15}$
Find the 8th term of $\left(\dfrac{3}{x} - \dfrac{x}{3}\right)^{13}$	$\dfrac{5148}{x}$

Pascal's Triangle

Pascal's Triangle was known well before Blaise Pascal documented it. This Triangle has a lot of neat features and patterns. We'll use Pascal's Triangle as it pertains to the binomial expansion.

♣ Pascal's Triangle

Named for Blaise Pascal, Pascal's Triangle is a way to arrange the binomial coefficients. The Triangle also makes it possible to derive the smaller binomial coefficients quickly because each number in the triangle is the sum of the two diagonally above it.

If you consider the coefficient $\binom{n}{k}$ and label the rows as n's and the diagonals as k's – in both cases you must start labeling with zero – then you can discover the correspondence between these numbers and the binomial coefficients.

```
                  1
               1     1
            1     2     1
         1     3     3     1
      1     4     6     4     1
   1     5    10    10     5     1
 1     6    15    20    15     6     1
1    7    21    35    35    21     7     1
```

★ For Fun

Can you find all the values of the next row or two of the triangle above?

Here are a few of things you can find in Pascal's Triangle:

- Hockey stick pattern
- Prime numbers
- Fibonacci sequence
- Magic 11's
- Powers of 2
- Triangular numbers
- Square numbers

It's amazing!

11.6: Permutations and Combinations

Counting, permutations, and combinations are the beginning of probability theory. But to a large extent you may find that it is still "fun with numbers."

Fundamental Counting Principle

The Fundamental Counting Principle is a way to find out how many ways events – each with many ways of happening – can all combine to a larger list of choices. For instance, if you have 5 shirts, 3 pants, and 2 pair of shoes, you have $5 \times 3 \times 2$ possible outfits:

Fundamental Counting Principle

Given a sequence of n events $E_1, E_2, E_3, E_4, E_5, \ldots, E_n$ where each event can occur in several ways w_n with n showing the correspondence to the event in the sequence, then the number of ways that all events can occur is given by $w_1 \cdot w_2 \cdot w_3 \cdot \ldots \cdot w_n$.

Fundamental Counting Principle: Test Prep	
If ID numbers are generated by choosing 4 capital letters and 2 digits, how many different ID numbers are possible?	$26 \cdot 26 \cdot 26 \cdot 26 \cdot 10 \cdot 10$ $= 45697600$
Try these	
A home improvement store is offering a package on Christmas trees. The trees come in three colors, decoration packs in 5 colors, skirts in 2 patterns, and stands in 3 models. How many possible combinations are available in a package deal that includes one of each of the above?	90
A computer monitor produces color by blending colors on a palette. If a monitor has four palettes and each palette has four colors, how many blended colors can be formed? (Every palette has to be used in each blend.)	256

Permutations

The Fundamental Counting Principle can help us discover how many ways a distinct set of elements can be arranged in a definite order.

♣ permutation

A permutation is an arrangement of distinct objects in a definite order.

The number of permutations of n distinct objects taken r at a time is given by

$$P(n,r) = \frac{n!}{(n-r)!}$$

★ For Fun

If the objects were *not* distinct, do you think there would be more ways to combine them or fewer?

<table>
<tr><th colspan="2" align="center">Permutations: Test Prep</th></tr>
<tr>
<td>Evaluate $P(10,3)$</td>
<td>$$P(10,3) = \frac{10!}{(10-3)!} = \frac{10!}{7!} = 10 \cdot 9 \cdot 8$$
$$= 720$$</td>
</tr>
<tr><td colspan="2">Try these</td></tr>
<tr>
<td>How many ways can you choose 4 people to sit on a committee from a group of 12 people?</td>
<td>11880</td>
</tr>
<tr>
<td>How many ways can first, second, and third prizes be awarded to 100 entrants in a raffle?</td>
<td>970200</td>
</tr>
</table>

Combinations

Combinations have to do with selecting objects that are not distinct. There is a difference between choosing a president, a vice president, and a secretary and simply choosing three people.

❖ combination

A combination is an arrangement of objects where the order of selection is not meaningful.

> The number of combinations of n objects taken r at a time is given by
> $$C(n,r) = \frac{n!}{r!(n-r)!}$$

★ For Fun

Does the combinations formula remind you of another formula we've seen recently?

Combinations: Test Prep	
Evaluate $C(27,10)$.	$C(27,10) = \dfrac{27!}{10!(27-10)!} = \dfrac{27!}{10!17!}$ $= \dfrac{27 \cdot 26 \cdot 25 \cdot 24 \cdot 23 \cdot 22 \cdot 21 \cdot 20 \cdot 19 \cdot 18}{10 \cdot 9 \cdot 8 \cdot 7 \cdot 6 \cdot 5 \cdot 4 \cdot 3 \cdot 2 \cdot 1}$ $= 8,436,285$
Try these	
Evaluate $C(20,17)$.	1140
How many 13-card hands can be chosen from a standard 52-card deck?	635,013,559,600

✓ Questions to Ask Your Instructor

It's difficult to get the hang of when to use the counting principle, permutations, or combinations. There are guidelines on page 902 of your textbook, and your instructor may have more examples to help.

> Please make sure you never cancel factorials incorrectly:
>
> $$\frac{7!}{5!} \neq 2!$$
>
> Where in the order of operations does the factorial operation fall?

11.7: Introduction to Probability

The study of probabilities came into being the first time someone decided to *insure* something – a ship, a life, a cargo. The pastime of gambling has also informed a lot of the interest and study of the science of predicting outcomes: Probability.

♣ probability

Officially, probability is the mathematical study of random patterns. The word is derived from a Latin word that means "credible". "Probability" has the feel of "likelihood". I think we have a different sense of what we mean when we use "probability" in an English sentence. In math its meaning tends to surprise. Try to have an open mind about what probability means and signifies in math.

Sample Spaces and Events

The basic ideas probability are a beginning to our study. What are we trying to predict with our *probabilities*? What are we looking at? What sort of systems does probability study pertain to? How is the subject organized?

♣ experiment

An occurrence with an observable outcome is an experiment.

♣ sample space

A sample space is the set of *all* possible outcomes of an experiment. For instance, you might write the sample space for flipping a coin like so: $\{T, H\}$

Sample Space: Test Prep	
List the sample space elements, in set notation, for flipping one coin and rolling one die.	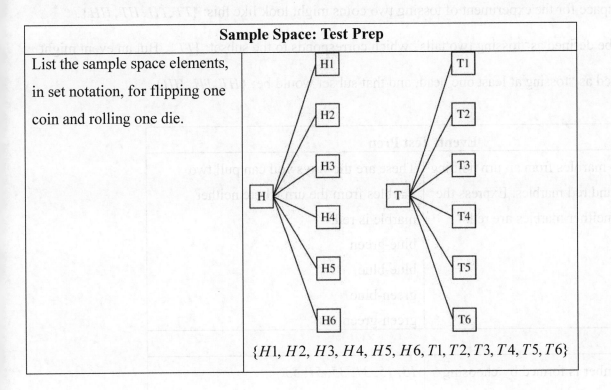 $\{H1, H2, H3, H4, H5, H6, T1, T2, T3, T4, T5, T6\}$

Try these	
List the sample space element for tossing a fair coin four times.	$\left\{\begin{array}{l} HHHH, HHHT, HHTH, HHTT, HTHH, \\ HTHT, HTTH, HTTT, THHH, THHT, \\ THTH, THTT, TTHH, TTHT, TTTH, TTTT \end{array}\right\}$
List the element in the sample space of finding the color of a card drawn from a standard 52-card deck.	$\left\{\text{black, red}\right\}$

♣ **event**

An event, frequently symbolized by E, is a specific subset of the sample space. One might ask what the probability of an event is.

The sample space for the experiment of tossing two coins might look like this: $\{TT, TH, HT, HH\}$.

event might be defined as "tossing two tails" which corresponds to the subset: $\{TT\}$. But an event might also be defined as "tossing at least one head, and that subset would be: $\{HT, TH, HH\}$.

Event: Test Prep	
You pull two marbles from an urn holding blue, green, and red marbles. Express the event where neither marbles are red.	These are the ways you can pull two marbles from the urn where neither marble is red: blue-green blue-blue green-blue green-green
Try these	
A 2-digit number is formed by choosing from the digits *1, 2, 3, 4, 5,* and *6* without replacing the first choice and where both digits are odd. Describe the event space.	*13, 15, 31, 35, 51, 53*

Express the event space of three or more tails from flipping four fair coins.	$\{HTTT, TTTH, THTT, TTHT, TTTT\}$

Probability of an Event

Calculating the probability of an event can be just as simple as counting up the elements in an event subset and dividing it by the number of elements in the sample space.

Probability of an Event

Let $n(S)$ and $n(E)$ represent the number of elements in a sample space S and in an event E, respectively. The probability of an event E is given by

$$P(E) = \frac{n(E)}{n(S)}$$

Because of the relationship between the sizes of a set and its possible subsets, possible values for probabilities all fall between 0 and 1.

Choose the valid probabilities:

0.03   1.3%

2             $\frac{2}{5}$

21%   .123

$\frac{5}{3}$ 1

★ For Fun

Can you name an event whose probability is 1, or 100%? Can you name an event whose probability is 0 ?

Probability of an Event: Test Prep	
What is the probability of getting exactly two heads when flipping four fair coins?	Sample space S is

	$\left\{ \begin{array}{l} HHHH, HHHT, HHTH, HHTT, HTHH, \\ HTHT, HTTH, HTTT, THHH, THHT, \\ THTH, THTT, TTHH, TTHT, TTTH, TTTT \end{array} \right\}.$
	Event space E is
	$\{HHTT, HTHT, HTTH, THHT, THTH, TTHH\}.$
	$P(2H) = \dfrac{n(E)}{n(S)} = \dfrac{6}{16} = \dfrac{3}{8}$
Try these	
What is the probability of tossing two fair dice whose sum is greater than three?	$\dfrac{11}{12}$
What is the probability of drawing any king from a standard 52-card deck?	$\dfrac{1}{13}$

❖ **mutually exclusive**

Two (or more) events that cannot occur at the same time are mutually exclusive. (Does that give you an idea of how to describe an event of zero probability?) For instance, if you have red marbles and yellow marbles, and you pull one out of a bag, it is impossible for the marble you have drawn to be both red *and* yellow – "red" and "yellow" are mutually exclusive events – there is no overlap in their Venn diagrams (remember sets?).

> **Addition Rules for Probabilities:**
>
> If E_1 and E_2 are two events, then the probability of both happening is given
>
> by
>
> $$P(E_1 \cup E_2) = P(E_1) + P(E_2) - P(E_1 \cap E_2)$$

If E_1 and E_2 are mutually exclusive then their intersection $(E_1 \cap E_2)$ is empty and the probability of two mutually exclusive events happening is given by

$$P(E_1 \cup E_2) = P(E_1) + P(E_2)$$

Addition Rules for Probabilities: Test Prep	
Toss a coin three times. Let E_1 be the event that the coin lands on heads two times and E_2 be the event that the third toss is a head. Find the probability of E_1 or E_2.	The events are not mutually exclusive. $S = \{HHH, HHT, HTH, THH, HTT, THT, TTH, TTT\}$ $P(E_1) = \dfrac{3}{8}$ $P(E_2) = \dfrac{4}{8} = \dfrac{1}{4}$ $P(E_1 \cap E_2) = \dfrac{1}{8}$ $P(E_1 \cup E_2) = \dfrac{3}{8} + \dfrac{1}{2} - \dfrac{1}{8} = \dfrac{3}{4}$
Try these	
Toss a coin three times. Let E_1 be the event that the coin is a head all 3 times; let E_2 be the event that the coin is a tail all 3 times. Find the probability of E_1 or E_2.	$\dfrac{1}{4}$
Toss a pair of dice. Find the probability that one of the die is a two or the sum is six.	$\dfrac{7}{18}$

Independent Events

The question of whether events are independent or not is very important (and subtle) in the calculation of probabilities. If the outcome of one experiment effects the next experiment, wouldn't that possibly greatly influence the probability of an event in the second experiment?

♣ independent

Independent events are events whose outcomes do *not* have effects on each other. Rolling a fair dice twice is an example of independent events – the result from the first roll has no effect on the second roll. Pulling a marble out of a box then pulling another marble out without replacing the first one is an example of events that are not independent – the outcome of the first draw affects the probabilities of the second.

Probability Rule for Independent Events:

If E_1 and E_2 are independent events, then the probability that both E_1 and E_2 happen is give by the product of their individual probabilities:

$$P(E_1) \cdot P(E_2)$$

Independent Events: Test Prep	
The probability Joe hits a target at the shooting range is $\frac{1}{4}$; the probability Ann hits is $\frac{2}{3}$. Find the probability that Joe and Ann hit the target.	Assuming the events are independent, the probability is $$\frac{1}{4} \cdot \frac{2}{3} = \frac{1}{6}$$

Try these	
A card is drawn from a standard 52-card deck then a coin is tossed. What is the probability the card was an ace and the coin was a head?	$\dfrac{1}{26}$
Let E_1 and E_2 be independent events. Let $P(E_1)=\dfrac{1}{4}$ and $P(E_2)=\dfrac{1}{3}$. Find the probability either one happens.	$\dfrac{1}{2}$

Binomial Probabilities

Binomial experiments are experiments with two and only two possible outcomes. (Most experiments can be re-phrased as binomial experiments, but then there is no partial credit!) This binomial scenario is well studied, modeled by a formula, and involves many calculations similar to those we learned in the previous section on The Binomial Theorem.

Binomial Probability Formula:

Given an experiment that consists of n independent trials where the probability of success on a single trial is p and the probability of failure is $q = 1 - p$, the probability of k successes out of n trials is

$$P(k) = \binom{n}{k} p^k q^{n-k}$$

There are quite a few "little pieces" in this formula, but once you get the hang of defining your experiment, it becomes pretty easy to use. Oh, yes, you need to be able to enter all these parameters into your calculator correctly and interpret the result the calculator gives you, too!

Binomial Probability Formula: Test Prep	
A true/false test consists of four questions. Given random guessing, what is the probability of getting exactly 2 questions correct?	$n = 4$ $p = q = \dfrac{1}{2}$ $k = 2$ $P(k = 2) = \dbinom{4}{2}\left(\dfrac{1}{2}\right)^2\left(\dfrac{1}{2}\right)^2 = 4 \cdot 3 \cdot \dfrac{1}{4} \cdot \dfrac{1}{4}$ $= \dfrac{3}{4}$
Try these	
Calculate a binomial probability given $n = 20$, $p = 0.7$, and $k = 15$	approximately 0.1787
The probability Duane scores a 3-point basketball shot is 0.45. He shoots from the 3-point zone 8 times. What is the probability he scores exactly one time?	approximately 0.0055

★ For Fun

How do you think that last problem would change if you wanted to know the probability that Duane would score three or fewer 3-point goals?

Final Fun: Matching

FIND THE BEST MATCH. GOOD LUCK.

FIBONACCI SEQUENCE $\dbinom{7}{3}$

$\displaystyle\sum_{i=1}^{\infty}(i-1)^2$ BINOMIAL EXPANSION

                                ARITHMETIC MEAN

PASCAL'S TRIANGLE   q

$$\frac{X+Y}{2}$$

$$\frac{10!}{(10-7)!}$$

$$1, 1, 2, 3, 5, 8, \dots$$

BINOMIAL COEFFICIENT

INFINITE SERIES

$$\binom{6}{0}x^6 + \binom{6}{1}x^5 y + \binom{6}{2}x^4 y^2 + \dots + \binom{6}{6}b^6$$

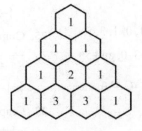

PERMUTATION

PROBABILITY OF FAILURE

NOW TRY THE CHAPTER 11 REVIEW EXERCISES AND

THE CHAPTER 11 PRACTICE TEST

STAY UP ON YOUR GAME – TRY THE CUMULATIVE REVIEW EXERCISES

Chapter P Preliminary Concepts

Section P.1 Exercises

1. Identify the prime numbers.

   ii 53 is prime; and iv 97 is prime.

3. List numbers common to both A and B.

   Since $A = \{-7, -3, 0, 2, 5, 8\}$ and $B = \{-3, -1, 0, 1, 3, 5, 7\}$, then the numbers common to both are -3, 0, and 5.

5. If $a < 0$, then a is a negative number. Thus, a^2 is a positive number.

7. Classify each number.

   $-\dfrac{1}{5}$: rational, real;   0: integer, rational, real;

   -44: integer, rational, real; π: irrational, real;

   3.14: rational, real;

   5.05005000500005…: irrational, real;

   $\sqrt{81} = 9$: integer, rational, real;

   53: integer, rational, prime, real

9. List the four smallest elements of the set.

   Let $x = 1, 2, 3, 4$.

   Then $\{2x \mid x$ is a positive integer$\} = \{2, 4, 6, 8\}$

11. List the four smallest elements of the set.

    Let $x = 1, 2, 3, 4$. (Recall 0 is not a natural number.)

    Then $\{y \mid y = 2x + 1, x$ is a natural number$\} = \{3, 5, 7, 9\}$

13. List the four smallest elements of the set.

    Let $x = 0, 1, 2, 3$.

    (We could have used $x = -3, -2, -1, 0$.)

    Then $\{z \mid z = |x|, x$ is an integer$\} = \{0, 1, 2, 3\}$

15. Perform the operation.

    $A \cup B = \{-3, -2, -1, 0, 1, 2, 3, 4, 6\}$

17. Perform the operation.

    $A \cap C = \{0, 1, 2, 3\}$

19. Perform the operation.

    $B \cap D = \varnothing$

21. Perform the operation.

    $(B \cup C) = \{-2, 0, 1, 2, 3, 4, 5, 6\}$

    $D \cap (B \cup C) = \{1, 3\}$

23. Perform the operation.

    $(B \cup C) \cap (B \cup D)$

    $= \{-2, 0, 1, 2, 3, 4, 5, 6\} \cap \{-3, -2, -1, 0, 1, 2, 3, 4, 6\}$

    $= \{-2, 0, 1, 2, 3, 4, 6\}$

25. Graph the set and write in set builder notation.

    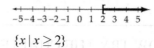

    $\{x \mid -2 < x < 3\}$

27. Graph the set and write in set builder notation.

    $\{x \mid -5 \le x \le -1\}$

29. Graph the set and write in set builder notation.

    $\{x \mid x \ge 2\}$

31. Graph the set and write in interval notation.

    $(3, 5)$

33. Graph the set and write in interval notation.

    $[-2, \infty)$

35. Graph the set and write in interval notation.

    $[0, 1]$

37. Graph the set.

    $(-\infty, 0) \cup [2, 4]$

    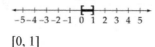

39. Graph the set.

    $(-4, 0) \cap [-2, 5] = [-2, 0)$

41. Graph the set.

    $(1, \infty) \cup (-2, \infty) = (-2, \infty)$

43. Graph the set.

$$(1, \infty) \cap (-2, \infty) = (1, \infty)$$

45. Graph the set.

$$[-2, 4] \cap [4, 5] = \{4\}$$

47. Graph the set.

$$(-2, 4) \cap (4, 5) = \varnothing$$

49. Graph the set.

$$\{x \mid x < -3\} \cup \{x \mid 1 < x < 2\}$$

51. Graph the set.

$$\{x \mid x < -3\} \cup \{x \mid x < 2\} = \{x \mid x < 2\}$$

53. Write the expression without absolute value symbols.

$$-|-5| = -5$$

55. Write the expression without absolute value symbols.

$$|3| \cdot |-4| = 3(4) = 12$$

57. Write the expression without absolute value symbols.

$$|\pi^2 + 10| = \pi^2 + 10$$

59. Write the expression without absolute value symbols.

$$|x - 4| + |x + 5| = 4 - x + x + 5 = 9$$

61. Write the expression without absolute value symbols.

$$|2x| - |x - 1| = 2x - (1 - x)$$
$$= 2x - 1 + x$$
$$= 3x - 1$$

63. Write in absolute value notation.

$$|m - n|$$

65. Write in absolute value notation.

$$|x - 3|$$

67. Write in absolute value notation.

$$|x - (-2)| = 4$$
$$|x + 2| = 4$$

69. Write in absolute value notation.

$$|a - 4| < 5$$

71. Write in absolute value notation.

$$|x + 2| > 4$$

73. Write in absolute value notation.

$$0 < |x - 4| < 1$$

75. Evaluate the expression.

$$-5^3(-4)^2 = -5 \cdot 5 \cdot 5(-4)(-4) = -125(16) = -2000$$

77. Evaluate the expression.

$$4 + (3 - 8)^2 = 4 + (-5)^2$$
$$= 4 + (-5)(-5)$$
$$= 4 + 25$$
$$= 29$$

79. Evaluate the expression.

$$28 \div (-7 + 5)^2 = 28 \div (-2)^2$$
$$= 28 \div (-2)(-2)$$
$$= 28 \div 4$$
$$= 7$$

81. Evaluate the expression.

$$7 + 2[3(-2)^3 - 4^2 \div 8]$$
$$= 7 + 2[3(-2)(-2)(-2) - (4 \cdot 4) \div 8]$$
$$= 7 + 2[-24 - 16 \div 8]$$
$$= 7 + 2[-24 - 2]$$
$$= 7 + 2[-26]$$
$$= 7 + -52$$
$$= -45$$

83. Evaluate.

$$-(-2)^3 = -(-8) = 8$$

85. Evaluate.

$$2(3)(-2)(-1) = 12$$

87. Evaluate.

$$-2(3)^2(-2)^2 = -2(9)(4) = -72$$

89. Evaluate.

$$3(-2)-(-1)[3-(-2)]^2 = 3(-2)-(-1)[3+2]^2$$
$$= (3)(-2)-(-1)[5]^2$$
$$= (3)(-2)-(-1)(25)$$
$$= -6+25 = 19$$

91. Evaluate.

$$\frac{3^2+(-2)^2}{3+(-2)} = \frac{9+4}{1} = \frac{13}{1} = 13$$

93. Evaluate.

$$\frac{3(-2)}{3} - \frac{2(-1)}{-2} = \frac{-6}{3} - \frac{-2}{-2} = -2-1 = -3$$

95. State the property used.

$$(ab^2)c = a(b^2c)$$

Associative property of multiplication

97. State the property used.

$$4(2a-b) = 8a-4b$$

Distributive property

99. State the property used.

$$(3x)y = y(3x)$$

Commutative property of multiplication

101. State the property used.

$$1 \cdot (4x) = 4x$$

Identity property of multiplication

103. **State** the property used.

$$x^2 + 1 = x^2 + 1$$

Reflexive property of equality

105. State the property used.

If $2x + 1 = y$ and $y = 3x - 2$, then $2x + 1 = 3x - 2$

Transitive property of equality

107. State the property used.

$$4 \cdot \frac{1}{4} = 1$$

Inverse property of multiplication

109. No.

$$(8 \div 4) \div 2 = 2 \div 2 = 1$$

$$8 \div (4 \div 2) = 8 \div 2 = 4$$

111. All but the multiplicative inverse property

113. Simplify the variable expression.

$$2 + 3(2x - 5)$$
$$= 2 + 6x - 15$$
$$= 6x - 13$$

115. Simplify the variable expression.

$$5 - 3(4x - 2y)$$
$$= 5 - 12x + 6y$$
$$= -12x + 6y + 5$$

117. Simplify the variable expression.

$$3(2a - 4b) - 4(a - 3b)$$
$$= 6a - 12b - 4a + 12b$$
$$= 6a - 4a - 12b + 12b$$
$$= 2a$$

119. Simplify the variable expression.

$$5a - 2[3 - 2(4a + 3)]$$
$$= 5a - 2(3 - 8a - 6)$$
$$= 5a - 2(-8a - 3)$$
$$= 5a + 16a + 6$$
$$= 21a + 6$$

121. Simplify the variable expression.

$$\frac{3}{4}(5a + 2) - \frac{1}{2}(3a - 5)$$
$$= \frac{15a}{4} + \frac{6}{4} - \frac{3a}{2} + \frac{5}{2}$$
$$= \frac{15a}{4} - \frac{6a}{4} + \frac{3}{2} + \frac{5}{2}$$
$$= \frac{9a}{4} + 4$$

123. Find the area.

$$\text{Area} = \frac{1}{2}bh = \frac{1}{2}(3 \text{ in})(4 \text{ in}) = 6 \text{ in}^2$$

125. Find the heart rate.

$$\text{Heart rate} = 65 + \frac{53}{4t + 1}$$
$$= 65 + \frac{53}{4(10) + 1}$$
$$= 65 + \frac{53}{41}$$
$$\approx 66$$

Heart rate is about 66 beats per minute.

127. Find the height.

$$\text{Height} = -16t^2 + 80t + 4$$
$$= -16(2)^2 + 80(2) + 4$$
$$= -16(4) + 80(2) + 4$$
$$= -64 + 160 + 4$$
$$= 100$$

After 2 seconds, the ball will have a height of 100 feet.

129. Find the SLG.

$$\text{SLG} = \frac{\text{singles} + 2 \cdot 2B + 3 \cdot 3B + 4 \cdot 4B}{AB}$$
$$= \frac{197 + 2 \cdot 48 + 3 \cdot 0 + 4 \cdot 30}{572}$$
$$= \frac{413}{572}$$
$$= 0.722$$

131. Perform the operation.

If A and B are two sets and $A \cap B = B$, then all

elements of B are contained in A. So B is a subset of A.

133. Perform the operation.

For any set A, $A \cup A = A$.

135. Perform the operation.

For any set A, $A \cup \varnothing = A$.

137. Give the example.

Answers may vary. One example is $\dfrac{a+b}{2}$.

139. Write the deleted delta neighborhood.

$$0 < |x - a| < \delta$$

Prepare for Section P.2

P1. Simplify.

$$2^2 \cdot 2^3 = 4 \cdot 8 = 32$$

Alternate method: $2^2 \cdot 2^3 = 2^{2+3} = 2^5 = 32$

P3. Simplify.

$$(2^3)^2 = 8^2 = 64$$

Alternate method: $(2^3)^2 = 2^{3(2)} = 2^6 = 64$

P5. False

$$3^4 \cdot 3^2 = 3^6, \text{ not } 9^6.$$

Section P.2 Exercises

1. Evaluate the expression.

$$-5^3 = -(5^3) = -125$$

3. Evaluate the expression.

$$\left(\frac{2}{3}\right)^0 = 1$$

5. Evaluate the expression.

$$4^{-2} = \frac{1}{4^2} = \frac{1}{16}$$

7. Evaluate the expression.

$$\frac{1}{2^{-5}} = 2^5 = 32$$

9. Write in scientific notation.

$$2,011,000,000,000 = 2.011 \times 10^{12}$$

11. Write in scientific notation.

$$0.000000000562 = 5.62 \times 10^{-10}$$

13. Write in decimal notation.

$$3.14 \times 10^7 = 31,400,000$$

15. Write in decimal notation.

$$-2.3 \times 10^{-6} = -0.0000023$$

17. Evaluate the expression.

$$4^{3/2} = \sqrt{4}^3 = 2^3 = 8$$

19. Evaluate the expression.

$$-64^{2/3} = -\sqrt[3]{64}^2 = -4^2 = -16$$

21. Evaluate the expression.

$$9^{-3/2} = \frac{1}{9^{3/2}} = \frac{1}{\left(\sqrt{9}\right)^3} = \frac{1}{3^3} = \frac{1}{27}$$

23. Evaluate the expression.

$$\frac{2^{-3}}{6^{-3}} = \left(\frac{2}{6}\right)^{-3} = \left(\frac{1}{3}\right)^{-3} = \left(\frac{3}{1}\right)^3 = 3^3 = 27$$

25. Evaluate the expression.

$$-2x^0 = -2$$

27. Write in simplest form.

$$2x^{-4} = 2\left(x^{-4}\right) = \frac{2}{x^4}$$

29. Write in simplest form.

$$\frac{5}{z^{-6}} = 5z^6$$

31. Write in simplest form.

$$(x^3y^2)(xy^5) = x^{3+1}y^{2+5} = x^4y^7$$

33. Write in simplest form.

$$(-2ab^4)(-3a^2b^5) = (-2)(-3)a^{1+2}b^{4+5} = 6a^3b^9$$

35. Write in simplest form.

$$\left(-4x^{-3}y\right)\left(7x^5y^{-2}\right) = (-4)(7)x^{-3+5}y^{1-2}$$
$$= -28x^2y^{-1}$$
$$= -\frac{28x^2}{y}$$

37. Write in simplest form.

$$\frac{6a^4}{8a^8} = \frac{6}{8}a^{4-8} = \frac{3}{4}a^{-4} = \frac{3}{4a^4}$$

39. Write in simplest form.

$$\frac{12x^3y^4}{18x^5y^2} = \frac{12}{18}x^{3-5}y^{4-2} = \frac{2}{3}x^{-2}y^2 = \frac{2y^2}{3x^2}$$

41. Write in simplest form.

$$\frac{36a^{-2}b^3}{3ab^4} = \frac{36}{3}a^{-2-1}b^{3-4} = 12a^{-3}b^{-1} = \frac{12}{a^3b}$$

43. Write in simplest form.

$$(-2m^3n^2)(-3mn^2)^2 = (-2m^3n^2)(9m^2n^4)$$
$$= (-2)(9)m^{3+2}n^{2+4}$$
$$= -18m^5n^6$$

45. Write in simplest form.

$$(x^{-2}y)^2(xy)^{-2} = (x^{-4}y^2)(x^{-2}y^{-2})$$
$$= x^{-4-2}y^{2-2}$$
$$= x^{-6}y^0$$
$$= \frac{1}{x^6}$$

47. Write in simplest form.

$$\left(\frac{3a^2b^3}{6a^4b^4}\right)^2 = \frac{9a^4b^6}{36a^8b^8} = \frac{9}{36}a^{4-8}b^{6-8}$$
$$= \frac{1}{4}a^{-4}b^{-2}$$
$$= \frac{1}{4a^4b^2}$$

49. Write in simplest form.

$$\frac{(-4x^2y^3)^2}{(2xy^2)^3} = \frac{16x^4y^6}{8x^3y^6}$$
$$= 2x^{4-3}y^{6-6}$$
$$= 2x$$

51. Write in simplest form.

$$\left(\frac{a^{-2}b}{a^3b^{-4}}\right)^2 = \frac{a^{-4}b^2}{a^6b^{-8}}$$
$$= a^{-4-6}b^{2-(-8)}$$
$$= a^{-4-6}b^{2+8}$$
$$= a^{-10}b^{10}$$
$$= \frac{b^{10}}{a^{10}}$$

53. Perform the operation and write in scientific notation.

$$(3\times10^{12})(9\times10^{-5}) = (3)(9)\times10^{12-5}$$
$$= 27\times10^7$$
$$= 2.7\times10^8$$

55. Perform the operation and write in scientific notation.

$$\frac{9\times10^{-3}}{6\times10^8} = \frac{9}{6}\times10^{-3-8}$$
$$= 1.5\times10^{-11}$$

57. Perform the operation and write in scientific notation.

$$\frac{(3.2\times10^{-11})(2.7\times10^{18})}{1.2\times10^{-5}} = \frac{(3.2)(2.7)}{1.2}\times10^{-11+18-(-5)}$$
$$= 7.2\times10^{-11+18+5}$$
$$= 7.2\times10^{12}$$

59. Perform the operation and write in scientific notation.

$$\frac{(4.0\times10^{-9})(8.4\times10^5)}{(3.0\times10^{-6})(1.4\times10^{18})} = \frac{(4.0)(8.4)}{(3.0)(1.4)}\times10^{-9+5+6-18}$$
$$= 8\times10^{-16}$$

61. Evaluate the expression.

$$\left(\frac{4}{9}\right)^{1/2} = \sqrt{\frac{4}{9}} = \frac{\sqrt{4}}{\sqrt{9}} = \frac{2}{3}$$

63. Evaluate the expression.

$$\left(\frac{1}{8}\right)^{-4/3} = 8^{4/3} = \sqrt[3]{8^4} = 2^4 = 16$$

65. Evaluate the expression.

$$(4a^{2/3}b^{1/2})(2a^{1/3}b^{3/2}) = (4)(2)a^{2/3+1/3}b^{1/2+3/2}$$
$$= 8a^{3/3}b^{4/2} = 8ab^2$$

67. Evaluate the expression.

$$(-3x^{2/3})(4x^{1/4}) = (-3)(4)x^{2/3+1/4}$$
$$= -12x^{8/12+3/12}$$
$$= -12x^{11/12}$$

69. Evaluate the expression.

$$(81x^8y^{12})^{1/4} = 81^{1/4}x^{8/4}y^{12/4} = \sqrt[4]{81}x^2y^3 = 3x^2y^3$$

71. Evaluate the expression.

$$\frac{16z^{3/5}}{12z^{1/5}} = \frac{16z^{3/5-1/5}}{12} = \frac{4z^{2/5}}{3}$$

73. Evaluate the expression.

$$(2x^{2/3}y^{1/2})(3x^{1/6}y^{1/3}) = (2)(3)x^{2/3+1/6}y^{1/2+1/3}$$
$$= 6x^{5/6}y^{5/6}$$

75. Evaluate the expression.

$$\frac{9a^{3/4}b}{3a^{2/3}b^2} = \frac{9a^{3/4-2/3}b^{1-2}}{3} = 3a^{9/12-8/12}b^{-1} = \frac{3a^{1/12}}{b}$$

77. Simplify the radical expression.

$$\sqrt{45} = \sqrt{3^2 \cdot 5} = 3\sqrt{5}$$

79. Simplify the radical expression.

$$\sqrt[3]{24} = \sqrt[3]{2^3 \cdot 3} = 2\sqrt[3]{3}$$

81. Simplify the radical expression.

$$\sqrt[3]{-135} = \sqrt[3]{(-3)^3 \cdot 5} = -3\sqrt[3]{5}$$

83. Simplify the radical expression.

$$\sqrt{24x^2y^3} = \sqrt{2^2x^2y^2} \cdot \sqrt{6y} = 2|xy|\sqrt{6y}$$

85. Simplify the radical expression.

$$\sqrt[3]{16a^3y^7} = \sqrt[3]{2^3a^3y^6} \cdot \sqrt[3]{2y} = 2ay^2\sqrt[3]{2y}$$

87. Simplify and combine like radicals.

$$2\sqrt{32} - 3\sqrt{98} = 2\sqrt{16} \cdot \sqrt{2} - 3\sqrt{49} \cdot \sqrt{2}$$
$$= 2(4)\sqrt{2} - 3(7)\sqrt{2}$$
$$= 8\sqrt{2} - 21\sqrt{2}$$
$$= -13\sqrt{2}$$

89. Simplify and combine like radicals.

$$-8\sqrt[4]{48} + 2\sqrt[4]{243} = -8\sqrt[4]{16 \cdot 3} + 2\sqrt[4]{81 \cdot 3}$$
$$= -8\sqrt[4]{16} \cdot \sqrt[4]{3} + 2\sqrt[4]{81} \cdot \sqrt[4]{3}$$
$$= -8\sqrt[4]{2^4} \cdot \sqrt[4]{3} + 2\sqrt[4]{3^4} \cdot \sqrt[4]{3}$$
$$= -8(2)\sqrt[4]{3} + 2(3)\sqrt[4]{3}$$
$$= -16\sqrt[4]{3} + 6\sqrt[4]{3}$$
$$= -10\sqrt[4]{3}$$

91. Simplify and combine like radicals.

$$4\sqrt[3]{32y^4} + 3y\sqrt[3]{108y} = 4\sqrt[3]{8y^3 \cdot 4y} + 3y\sqrt[3]{27 \cdot 4y}$$
$$= 4\sqrt[3]{8y^3} \cdot \sqrt[3]{4y} + 3y\sqrt[3]{27} \cdot \sqrt[3]{4y}$$
$$= 4\sqrt[3]{2^3y^3} \cdot \sqrt[3]{4y} + 3y\sqrt[3]{3^3} \cdot \sqrt[3]{4y}$$
$$= 4(2y)\sqrt[3]{4y} + 3y(3)\sqrt[3]{4y}$$
$$= 8y\sqrt[3]{4y} + 9y\sqrt[3]{4y}$$
$$= 17y\sqrt[3]{4y}$$

93. Simplify and combine like radicals.

$$x\sqrt[3]{8x^3y^4} - 4y\sqrt[3]{64x^6y}$$
$$= x\sqrt[3]{8x^3y^3 \cdot y} - 4y\sqrt[3]{64x^6 \cdot y}$$
$$= x\sqrt[3]{8x^3y^3} \cdot \sqrt[3]{y} - 4y\sqrt[3]{64x^6} \cdot \sqrt[3]{y}$$
$$= x\sqrt[3]{2^3x^3y^3} \cdot \sqrt[3]{y} - 4y\sqrt[3]{4^3x^6} \cdot \sqrt[3]{y}$$
$$= x(2xy)\sqrt[3]{y} - 4y(4x^2)\sqrt[3]{y}$$
$$= 2x^2y\sqrt[3]{y} - 16x^2y\sqrt[3]{y}$$
$$= -14x^2y\sqrt[3]{y}$$

95. Find the product and write in simplest form.

$$(\sqrt{5} + 3)(\sqrt{5} + 4) = \sqrt{5}^2 + 4\sqrt{5} + 3\sqrt{5} + (3)(4)$$
$$= 5 + 7\sqrt{5} + 12$$
$$= 17 + 7\sqrt{5}$$

97. Find the product and write in simplest form.

$$(\sqrt{2} - 3)(\sqrt{2} + 3) = \sqrt{2}^2 + 3\sqrt{2} - 3\sqrt{2} + (-3)(3)$$
$$= 2 - 9$$
$$= -7$$

99. Find the product and write in simplest form.

$$(3\sqrt{z} - 2)(4\sqrt{z} + 3)$$
$$= (3)(4)\sqrt{z}^2 + 3(3\sqrt{z}) - 2(4\sqrt{z}) + (-2)(3)$$
$$= 12z + 9\sqrt{z} - 8\sqrt{z} - 6$$
$$= 12z + \sqrt{z} - 6$$

101. Find the product and write in simplest form.

$$(\sqrt{x}+2)^2 = (\sqrt{x}+2)(\sqrt{x}+2)$$
$$= x+4\sqrt{x}+4$$

103. Find the product and write in simplest form.

$$(\sqrt{x-3}+2)^2 = (\sqrt{x-3}+2)(\sqrt{x-3}+2)$$
$$= x-3+4\sqrt{x-3}+4$$
$$= x+4\sqrt{x-3}+1$$

105. Rationalize the denominator.

$$\frac{2}{\sqrt{2}} = \frac{2}{\sqrt{2}}\cdot\frac{\sqrt{2}}{\sqrt{2}} = \frac{2\sqrt{2}}{\sqrt{2}^2} = \frac{2\sqrt{2}}{2} = \frac{\cancel{2}\sqrt{2}}{\cancel{2}} = \sqrt{2}$$

107. Rationalize the denominator.

$$\sqrt{\frac{5}{18}} = \sqrt{\frac{5}{2\cdot 3^2}} = \sqrt{\frac{5}{2\cdot 3^2}}\cdot\sqrt{\frac{2}{2}}$$
$$= \sqrt{\frac{5\cdot 2}{2^2\cdot 3^2}} = \frac{\sqrt{5\cdot 2}}{\sqrt{2^2\cdot 3^2}} = \frac{\sqrt{10}}{2\cdot 3}$$
$$= \frac{\sqrt{10}}{6}$$

109. Rationalize the denominator.

$$\frac{3}{\sqrt[3]{2}} = \frac{3}{\sqrt[3]{2}}\cdot\frac{\sqrt[3]{2^2}}{\sqrt[3]{2^2}} = \frac{3\sqrt[3]{2^2}}{\sqrt[3]{2^3}} = \frac{3\sqrt[3]{4}}{2}$$

111. Rationalize the denominator.

$$\frac{4}{\sqrt[3]{8x^2}} = \frac{4}{\sqrt[3]{2^3 x^2}} = \frac{4}{2\sqrt[3]{x^2}} = \frac{\cancel{4}^2}{\cancel{2}\sqrt[3]{x^2}} = \frac{2}{\sqrt[3]{x^2}}$$
$$= \frac{2}{\sqrt[3]{x^2}}\cdot\frac{\sqrt[3]{x}}{\sqrt[3]{x}} = \frac{2\sqrt[3]{x}}{\sqrt[3]{x^3}}$$
$$= \frac{2\sqrt[3]{x}}{x}$$

113. Rationalize the denominator.

$$\frac{3}{\sqrt{3}+4} = \frac{3}{\sqrt{3}+4}\cdot\frac{\sqrt{3}-4}{\sqrt{3}-4} = \frac{3(\sqrt{3}-4)}{(\sqrt{3}+4)(\sqrt{3}-4)}$$
$$= \frac{3(\sqrt{3}-4)}{\sqrt{3}^2-4^2} = \frac{3(\sqrt{3}-4)}{3-16}$$
$$= \frac{3\sqrt{3}-12}{-13}$$
$$= -\frac{3\sqrt{3}-12}{13}$$

115. Rationalize the denominator.

$$\frac{6}{2\sqrt{5}+2} = \frac{6}{2(\sqrt{5}+1)} = \frac{\cancel{6}^3}{\cancel{2}(\sqrt{5}+1)} = \frac{3}{\sqrt{5}+1}$$
$$= \frac{3}{\sqrt{5}+1}\cdot\frac{\sqrt{5}-1}{\sqrt{5}-1}$$
$$= \frac{3(\sqrt{5}-1)}{(\sqrt{5}+1)(\sqrt{5}-1)} = \frac{3(\sqrt{5}-1)}{\sqrt{5}^2-1}$$
$$= \frac{3(\sqrt{5}-1)}{5-1}$$
$$= \frac{3\sqrt{5}-3}{4}$$

117. Rationalize the denominator.

$$\frac{3+2\sqrt{5}}{5-3\sqrt{5}} = \frac{3+2\sqrt{5}}{5-3\sqrt{5}}\cdot\frac{5+3\sqrt{5}}{5+3\sqrt{5}}$$
$$= \frac{(3+2\sqrt{5})(5+3\sqrt{5})}{(5-3\sqrt{5})(5+3\sqrt{5})}$$
$$= \frac{3(5+3\sqrt{5})+2\sqrt{5}(5+3\sqrt{5})}{(5-3\sqrt{5})(5+3\sqrt{5})}$$
$$= \frac{15+9\sqrt{5}+10\sqrt{5}+30}{5^2-(3\sqrt{5})^2}$$
$$= \frac{45+19\sqrt{5}}{25-45}$$
$$= \frac{45+19\sqrt{5}}{-20} \text{ or } -\frac{45+19\sqrt{5}}{20}$$

119. Rationalize the denominator.

$$\frac{6\sqrt{3}-11}{4\sqrt{3}-7} = \frac{6\sqrt{3}-11}{4\sqrt{3}-7}\cdot\frac{4\sqrt{3}+7}{4\sqrt{3}+7}$$
$$= \frac{(6\sqrt{3}-11)(4\sqrt{3}+7)}{(4\sqrt{3}-7)(4\sqrt{3}+7)}$$
$$= \frac{6\sqrt{3}(4\sqrt{3}+7)-11(4\sqrt{3}+7)}{(4\sqrt{3}-7)(4\sqrt{3}+7)}$$
$$= \frac{72+42\sqrt{3}-44\sqrt{3}-77}{(4\sqrt{3})^2-7^2}$$
$$= \frac{-5-2\sqrt{3}}{48-49} = \frac{-5-2\sqrt{3}}{-1}$$
$$= 5+2\sqrt{3}$$

121. Rationalize the denominator.

$$\frac{2+\sqrt{x}}{3-2\sqrt{x}} = \frac{2+\sqrt{x}}{3-2\sqrt{x}} \cdot \frac{3+2\sqrt{x}}{3+2\sqrt{x}}$$

$$= \frac{(2+\sqrt{x})(3+2\sqrt{x})}{(3-2\sqrt{x})(3+2\sqrt{x})}$$

$$= \frac{2(3+2\sqrt{x})+\sqrt{x}(3+2\sqrt{x})}{(3-2\sqrt{x})(3+2\sqrt{x})}$$

$$= \frac{6+4\sqrt{x}+3\sqrt{x}+2x}{3^2-(2\sqrt{x})^2}$$

$$= \frac{6+7\sqrt{x}+2x}{9-4x}$$

123. Rationalize the denominator.

$$\frac{x-\sqrt{5}}{x+2\sqrt{5}} = \frac{x-\sqrt{5}}{x+2\sqrt{5}} \cdot \frac{x-2\sqrt{5}}{x-2\sqrt{5}}$$

$$= \frac{(x-\sqrt{5})(x-2\sqrt{5})}{(x+2\sqrt{5})(x-2\sqrt{5})}$$

$$= \frac{x(x-2\sqrt{5})-\sqrt{5}(x-2\sqrt{5})}{(x+2\sqrt{5})(x-2\sqrt{5})}$$

$$= \frac{x^2-2x\sqrt{5}-x\sqrt{5}+10}{x^2-(2\sqrt{5})^2}$$

$$= \frac{x^2-3x\sqrt{5}+10}{x^2-20}$$

125. Rationalize the denominator.

$$\frac{3}{\sqrt{5}+\sqrt{x}} = \frac{3}{\sqrt{5}+\sqrt{x}} \cdot \frac{\sqrt{5}-\sqrt{x}}{\sqrt{5}-\sqrt{x}}$$

$$= \frac{3(\sqrt{5}-\sqrt{x})}{(\sqrt{5}+\sqrt{x})(\sqrt{5}-\sqrt{x})}$$

$$= \frac{3\sqrt{5}-3\sqrt{x}}{(\sqrt{5})^2-(\sqrt{x})^2}$$

$$= \frac{3\sqrt{5}-3\sqrt{x}}{5-x}$$

127. Find the number of seeds.

$$\frac{1 \text{ seed}}{3.2\times10^{-8} \text{ ounce}} \cdot \frac{1 \text{ ounce}}{\text{package}}$$

$$\frac{1}{3.2\times10^{-8}} \cdot 1 = \frac{1}{3.2}\times10^8$$

$$= 0.3125\times10^8$$

$$= 3.13 \times10^7 \text{ seeds per package}$$

129. Find the red shift.

$$\text{Red shift} = \frac{\lambda_r - \lambda_s}{\lambda_s}$$

$$= \frac{5.13\times10^{-7} - 5.06\times10^{-7}}{5.06\times10^{-7}}$$

$$= \frac{(5.13 - 5.06)\times10^{-7}}{5.06\times10^{-7}}$$

$$= \frac{0.07}{5.06} \cdot \frac{10^{-7}}{10^{-7}}$$

$$\approx 1.38\times10^{-2}$$

131. Find the number of minutes.

$$\frac{1 \text{ sec}}{3\times10^8 \text{ m}} \cdot 1.44\times10^{11} \text{ m} \cdot \frac{1 \text{ min}}{60 \text{ sec}}$$

$$\frac{1}{3\times10^8} \cdot 1.44\times10^{11} \cdot \frac{1}{60} = \frac{(1)(1.44)(1)\times10^{11}}{3(60)\times10^8}$$

$$= \frac{1.44}{180}\times10^{11-8}$$

$$= 0.008\times10^3$$

$$= 8 \text{ minutes}$$

133. Find the BSA.

$$\frac{1 \text{ gram}}{6.023\times10^{23} \text{ atoms}} \cdot 1 \text{ atom}$$

$$\text{BSA} = 0.0235h^{0.3964} \cdot w^{0.51456}$$

$$= 0.0235(178)^{0.3964} \cdot (73)^{0.51456}$$

$$\approx 1.67$$

The BSA is approximately 1.67 m^2.

135. a. Evaluate P when $d = 10$.

$$P = 10^{2-d/40} = 10^{2-10/40} = 10^{2-0.25} = 10^{1.75} \approx 56$$

The amount of light that will pass to a depth of 10 feet below the ocean's surface is about 56%.

b. Evaluate P when $d = 25$.

$$P = 10^{2-d/40} = 10^{2-25/40} = 10^{2-0.625} = 10^{1.375} \approx 24$$

The amount of light that will pass to a depth of 25 feet below the ocean's surface is about 24%.

137. Rationalize the numerator.

$$\sqrt{n^2+4}-n=\frac{\sqrt{n^2+4}-n}{1}\cdot\frac{\sqrt{n^2+4}+n}{\sqrt{n^2+4}+n}$$

$$=\frac{(\sqrt{n^2+4})^2-n^2}{\sqrt{n^2+4}+n}=\frac{n^2+4-n^2}{\sqrt{n^2+4}+n}$$

$$=\frac{4}{\sqrt{n^2+4}+n}$$

139. Rationalize the numerator.

$$\frac{\sqrt{4+h}-2}{h}=\frac{\sqrt{4+h}-2}{h}\cdot\frac{\sqrt{4+h}+2}{\sqrt{4+h}+2}$$

$$=\frac{(\sqrt{4+h})^2-2^2}{h(\sqrt{4+h}+2)}$$

$$=\frac{4+h-4}{h(\sqrt{4+h}+2)}=\frac{h}{h(\sqrt{4+h}+2)}$$

$$=\frac{1}{\sqrt{4+h}+2}$$

Prepare for Section P.3

P1. Simplify.

$$-3(2a-4b)=-6a+12b$$

P3. Simplify.

$$2x^2+3x-5+x^2-6x-1$$
$$=2x^2+x^2+3x-6x-5-1$$
$$=(2+1)x^2+(3-6)x-(5+1)$$
$$=3x^2-3x-6$$

P5. $4-3x-2x^2\overset{?}{=}-2x^2-4x+4$

$-2x^2-3x+4\overset{?}{=}-2x^2-4x+4$

False.

Section P.3 Exercises

1. D

3. H

5. G

7. B

9. J

11. Perform the indicated operation using special product formula.

$$(x-4)(x+4)=x^2-4^2=x^2-16$$

13. Perform the indicated operation using special product formula.

$$(z-4)^2=z^2-2(4z)+4^2$$
$$=z^2-8z+16$$

15. a. x^2+2x-7

b. 2

c. 1, 2, –7

d. 1

e. $x^2,\ 2x,\ -7$

17. a. x^3-1

b. 3

c. 1, –1

d. 1

e. $x^3,-1$

19. a. $2x^4+3x^3+4x^2+5$

b. 4

c. 2, 3, 4, 5

d. 2

e. $2x^4,3x^3,4x^2,5$

21. Determine the degree of the polynomial.

3

23. Determine the degree of the polynomial.

5

25. Determine the degree of the polynomial.

2

27. Perform operations, simplify, write in standard form.

$(3x^2+4x+5)+(2x^2+7x-2)=5x^2+11x+3$

29. Perform operations, simplify, write in standard form.

$(4w^3-2w+7)+(5w^3+8w^2-1)=9w^3+8w^2-2w+6$

31. Perform operations, simplify, write in standard form.

$$(r^2-2r-5)-(3r^2-5r+7)$$
$$=r^2-2r-5-3r^2+5r-7$$
$$=-2r^2+3r-12$$

33. Perform operations, simplify, write in standard form.

$$(u^3-3u^2-4u+8)-(u^3-2u+4)$$
$$=u^3-3u^2-4u+8-u^3+2u-4$$
$$=-3u^2-2u+4$$

35. Perform operations, simplify, write in standard form.

$$(2x^2 + 7x - 8)(4x - 5)$$
$$= 2x^2(4x - 5) + 7x(4x - 5) - 8(4x - 5)$$
$$= 8x^3 - 10x^2 + 28x^2 - 35x - 32x + 40$$
$$= 8x^3 + 18x^2 - 67x + 40$$

37. Perform operations, simplify, write in standard form.

$$(3x^2 - 5x + 6)(3x - 1)$$
$$= (3x^2 - 5x + 6)(3x) + (3x^2 - 5x + 6)(-1)$$
$$= 9x^3 - 15x^2 + 18x - 3x^2 + 5x - 6$$
$$= 9x^3 - 18x^2 + 23x - 6$$

39. Perform operations, simplify, write in standard form.

$$(2x + 6)(5x^3 - 6x^2 + 4)$$
$$= 2x(5x^3 - 6x^2 + 4) + 6(5x^3 - 6x^2 + 4)$$
$$= 10x^4 - 12x^3 + 8x + 30x^3 - 36x^2 + 24$$
$$= 10x^4 + 18x^3 - 36x^2 + 8x + 24$$

41. Perform operations, simplify, write in standard form.

$$(x^3 - 4x^2 + 9x - 6)(2x + 5)$$
$$= (x^3 - 4x^2 + 9x - 6)(2x) + (x^3 - 4x^2 + 9x - 6)(5)$$
$$= 2x^4 - 8x^3 + 18x^2 - 12x + 5x^3 - 20x^2 + 45x - 30$$
$$= 2x^4 - 3x^3 - 2x^2 + 33x - 30$$

43. Perform operations, simplify, write in standard form.

$$\begin{array}{r} 3x^2 - 2x + 5 \\ 2x^2 - 5x + 2 \\ \hline + 6x^2 - 4x + 10 \\ -15x^3 + 10x^2 - 25x \\ + 6x^4 - 4x^3 + 10x^2 \\ \hline 6x^4 - 19x^3 + 26x^2 - 29x + 10 \end{array}$$

45. Find the product using the FOIL method.

$$(y + 2)(y + 1) = y^2 + y + 2y + 2$$
$$= y^2 + 3y + 2$$

47. Find the product using the FOIL method.

$$(2x + 4)(5x + 1) = 10x^2 + 2x + 20x + 4$$
$$= 10x^2 + 22x + 4$$

49. Find the product using the FOIL method.

$$(4z - 3)(z - 4) = 4z^2 - 16z - 3z + 12$$
$$= 4z^2 - 19z + 12$$

51. Find the product using the FOIL method.

$$(a + 6)(a - 3) = a^2 - 3a + 6a - 18$$
$$= a^2 + 3a - 18$$

53. Find the product using the FOIL method.

$$(5x - 11y)(2x - 7y) = 10x^2 - 35xy - 22xy + 77y^2$$
$$= 10x^2 - 57xy + 77y^2$$

55. Find the product using the FOIL method.

$$(9x + 5y)(2x + 5y) = 18x^2 + 45xy + 10xy + 25y^2$$
$$= 18x^2 + 55xy + 25y^2$$

57. Find the product using the FOIL method.

$$(3p + 5q)(2p - 7q) = 6p^2 - 21pq + 10pq - 35q^2$$
$$= 6p^2 - 11pq - 35q^2$$

59. Use the special product formulas.

$$(3x + 5)(3x - 5) = (3x)^2 - 5^2 = 9x^2 - 25$$

61. Use the special product formulas.

$$(3x^2 - y)^2 = (3x^2)^2 - 2(3x^2y) - y^2$$
$$= 9x^4 - 6x^2y + y^2$$

63. Use the special product formulas.

$$(4w + z)^2 = (4w)^2 + 2(4wz) + z^2$$
$$= 16w^2 + 8wz + z^2$$

65. Use the special product formulas.

$$[(x + 5) + y][(x + 5) - y] = (x + 5)^2 - y^2$$
$$= x^2 + 10x + 25 - y^2$$

67. Evaluate $x^2 + 7x - 1$ for $x = 3$.

$$(3)^2 + 7(3) - 1 = 9 + 21 - 1 = 29$$

69. Evaluate $-x^2 + 5x - 3$ for $x = -2$.

$$-(-2)^2 + 5(-2) - 3 = -4 - 10 - 3 = -17$$

71. Evaluate $3x^3 - 2x^2 - x + 3$ for $x = -1$.

$3(-1)^3 - 2(-1)^2 - (-1) + 3$
$= 3(-1) - 2(1) + 1 + 3$
$= -3 - 2 + 1 + 3$
$= -1$

73. Evaluate $1 - x^5$ for $x = -2$.

$1 - (-2)^5 = 1 - (-32) = 1 + 32 = 33$

75. Substitute the given value of v into $0.016v^2$. Then simplify.

a. $0.016v^2$

$0.016(10)^2 = 1.6$

The air resistance is 1.6 pounds.

b. $0.016v^2$

$0.016(15)^2 = 3.6$

The air resistance is 3.6 pounds.

77. Substitute the given value of h and r into $\pi r^2 h$. Then simplify

a. $\pi r^2 h$

$\pi(3)^2(8) = 72\pi$

The volume is 72π in^3.

b. $\pi r^2 h$

$\pi(5)^2(12) = 300\pi$

The volume 300π cm^3.

79. Substitute the given value of v into $0.005x^2 - 0.32x + 12$.

a. $0.005x^2 - 0.32x + 12$

$0.005(20)^2 - 0.32(20) + 12 = 7.6$

The reaction time is 7.6 hundredths of a second or 0.076 second.

b. $0.005x^2 - 0.32x + 12$

$0.005(50)^2 - 0.32(50) + 12 = 8.5$

The reaction time is 8.5 hundredths of a second or 0.085 second.

81. Find the number of chess matches.

$\frac{1}{2}n^2 - \frac{1}{2}n = \frac{1}{2}(150)^2 - \frac{1}{2}(150) = 11{,}175$ matches

83. Estimate the times.

a. $1.9 \times 10^{-6}(4000)^2 - 3.9 \times 10^{-3}(4000) = 14.8$ s

b. $1.9 \times 10^{-6}(8000)^2 - 3.9 \times 10^{-3}(8000) = 90.4$ s

85. Evaluate $-16t^2 + 4.7881t + 6$ when $t = 0.5$

$\text{height} = -16t^2 + 4.7881t + 6$
$= -16(0.5)^2 + 4.7881(0.5)t + 6$
$= 4.39$

Yes. The ball is approximately 4.4 feet high when it crosses home plate.

87. Perform the operations and simplify.

$(4d - 1)^2 - (2d - 3)^2$
$= (16d^2 - 8d + 1) - (4d^2 - 12d + 9)$
$= 16d^2 - 8d + 1 - 4d^2 + 12d - 9$
$= 12d^2 + 4d - 8$

89. Perform the operations and simplify.

$(3c - 2)(4c + 1)(5c - 2) = (12c^2 - 5c - 2)(5c - 2)$

$$
\begin{array}{r}
12c^2 - 5c - 2 \\
5c - 2 \\
\hline
-24c^2 + 10c + 4 \\
60c^3 - 25c^2 - 10c \\
\hline
60c^3 - 49c^2 \qquad + 4
\end{array}
$$

91. The product of a polynomial with degree 3 and a polynomial with degree 5 results in a polynomial of degree 8.

Mid-Chapter P Quiz

1. Evaluate $2x^3 - 4(3xy - z^2)$ for the given values.

$2(-2)^3 - 4(3(-2)(3) - (-4)^2)$
$= 2(-8) - 4(-18 - 16)$
$= -16 - 4(-34)$
$= -16 + 136$
$= 120$

3. Simplify.

$\dfrac{24x^{-3}y^4}{6x^2y^{-3}} = 4x^{-3-2}y^{4+3} = 4x^{-5}y^7 = \dfrac{4y^7}{x^5}$

5. Simplify.

$$\sqrt[3]{16a^4b^9c^8} = \sqrt[3]{8a^3b^9c^6} \cdot \sqrt[3]{2ac^2} = 2ab^3c^2\sqrt[3]{2ac^2}$$

7. Multiply.

$$(3x-4y)(2x+5y) = 6x^2 + 15xy - 8xy - 20y^2$$
$$= 6x^2 + 7xy - 20y^2$$

9. Multiply.

$$(2x-3)(4x^2+5x-7)$$
$$= 2x(4x^2+5x-7) - 3(4x^2+5x-7)$$
$$= 8x^3 + 10x^2 - 14x - 12x^2 - 15x + 21$$
$$= 8x^3 - 2x^2 - 29x + 21$$

Prepare for Section P.4

P1. Simplify.

$$\frac{6x^3}{2x} = 3x^{3-1} = 3x^2$$

P3. a. $x^6 = (x^2)^3$

b. $x^6 = (x^3)^2$

P5. Replace "?" to make a true statement.

$$-3(5a-?) = -15a + 21 = -3(5a-7)$$

Thus, $? = 7$

Section P.4 Exercises

1. Find the GCF.

$$9x^2y^3 = 3^2x^2y^3$$
$$12xy^4 = 2^2 \cdot 3^1 xy^4$$
$$\text{GCF} = 3xy^3$$

3. Find the GCF.

$$12x^3 = 2^2 \cdot 3^1 x^3$$
$$35y^3 = 5 \cdot 7y^3$$
$$\text{GCF} = 1$$

5. Find the GCF.

$$2v^3(v-4) = 2v^3(w-4)$$
$$12v(w-4)^2 = 2^2 \cdot 3v(w-4)^2$$
$$\text{GCF} = 2v(w-4)$$

7. State whether the polynomial is factored completely.

No, $x^2 + 4x$ can be factored as $x(x+4)$.

9. State whether the polynomial is factored completely.

No, $x^3 + 8$ is the sum of cubes and can be factored as

$(x+2)(x^2 - 2x + 4)$.

11. Factor out the GCF.

$$5x + 20 = 5(x+4)$$

13. Factor out the GCF.

$$-15x^2 - 12x = -3x(5x+4)$$

15. Factor out the GCF.

$$10x^2y + 6xy - 14xy^2 = 2xy(5x+3-7y)$$

17. Factor out the GCF.

$$(x-3)(a+b) + (x-3)(a+2b)$$
$$= (x-3)(a+b+a+2b)$$
$$= (x-3)(2a+3b)$$

19. Factor.

$$x^2 + 7x + 12 = (x+3)(x+4)$$

21. Factor.

$$a^2 - 10a - 24 = (a-12)(a+2)$$

23. Factor.

$$x^2 + 6x + 5 = (x+5)(x+1)$$

25. Factor.

$$6x^2 + 25x + 4 = (6x+1)(x+4)$$

27. Factor.

$$51x^2 - 5x - 4 = (17x+4)(3x-1)$$

29. Factor.

$$6x^2 + xy - 40y^2 = (3x+8y)(2x-5y)$$

31. Factor.

$$6x^2 + 23x + 15 = (6x+5)(x+3)$$

33. Determine whether the trinomial is factorable.

$$b^2 - 4ac = 26^2 - 4(8)(15) = 196 = 14^2$$

The trinomial is factorable over the integers.

35. Determine whether the trinomial is factorable.

$$b^2 - 4ac = (-5)^2 - 4(4)(6) = -71$$

The trinomial is not factorable over the integers.

37. Determine whether the trinomial is factorable.

$$b^2 - 4ac = (-14)^2 - 4(6)(5) = 76$$

The trinomial is not factorable over the integers.

39. Factor the difference of squares.

$$x^2 - 9 = (x+3)(x-3)$$

41. Factor the difference of squares.

$$4a^2 - 49 = (2a+7)(2a-7)$$

43. Factor the difference of squares.

$$1 - 100x^2 = (1+10x)(1-10x)$$

45. Factor the difference of squares.

$$(x+1)^2 - 4 = [(x+1)+2][(x+1)-2]$$
$$= (x+3)(x-1)$$

47. Factor the difference of squares.

$$6x^2 - 216 = 6(x^2 - 36)$$
$$= 6(x+6)(x-6)$$

49. Factor the difference of squares.

$$x^4 - 625 = (x^2 + 25)(x^2 - 25)$$
$$= (x^2 + 25)(x+5)(x-5)$$

51. Factor the difference of squares.

$$x^5 - 81x = x(x^4 - 81)$$
$$= x(x^2 + 9)(x^2 - 9)$$
$$= x(x^2 + 9)(x+3)(x-3)$$

53. Factor the perfect-square trinomial.

$$x^2 + 10x + 25 = (x+5)^2$$

55. Factor the perfect-square trinomial.

$$a^2 - 14a + 49 = (a-7)^2$$

57. Factor the perfect-square trinomial.

$$4x^2 + 12x + 9 = (2x+3)^2$$

59. Factor the perfect-square trinomial.

$$z^4 + 4z^2w^2 + 4w^4 = (z^2)^2 + 2(2z^2w^2) + (2w^2)^2$$
$$= (z^2 + 2w^2)^2$$

61. Factor the sum or difference of cubes.

$$x^3 - 8 = (x-2)(x^2 + 2x + 4)$$

63. Factor the sum or difference of cubes.

$$8x^3 - 27y^3 = (2x)^3 - (3y)^3$$
$$= (2x - 3y)(4x^2 + 6xy + 9y^2)$$

65. Factor the sum or difference of cubes.

$$8 - x^6 = 2^3 - (x^2)^3$$
$$= (2 - x^2)(4 + 2x + x^4)$$

67. Factor the sum or difference of cubes.

$$(x-2)^3 - 1 = [(x-2)-1][(x-2)^2 + (x-2) + 1]$$
$$= (x-3)[x^2 - 4x + 4 + x - 2 + 1]$$
$$= (x-3)(x^2 - 3x + 3)$$

69. Factor polynomials that are quadratic in form.

$$x^4 - x^2 - 6 = (x^2)^2 - x^2 - 6 \quad \text{Let } u = x^2.$$
$$= u^2 - u - 6$$
$$= (u-3)(u+2) \quad \text{Replace } u \text{ with } x^2.$$
$$= (x^2 - 3)(x^2 + 2)$$

71. Factor polynomials that are quadratic in form.

$$x^2y^2 + 4xy - 5 = (xy)^2 + 4xy - 5 \quad \text{Let } u = xy.$$
$$= u^2 + 4u - 5$$
$$= (u+5)(u-1) \quad \text{Replace } u \text{ with } xy.$$
$$= (xy + 5)(xy - 1)$$

73. Factor polynomials that are quadratic in form.

$$4x^5 - 4x^3 - 8x$$
$$= 4x(x^4 - x^2 - 2)$$
$$= 4x[(x^2)^2 - x^2 - 2] \quad \text{Let } u = x^2.$$
$$= 4x[u^2 - u - 2]$$
$$= 4x(u-2)(u+1) \quad \text{Replace } u \text{ with } x^2.$$
$$= 4x(x^2 - 2)(x^2 + 1)$$

75. Factor polynomials that are quadratic in form.

$$z^4 + z^2 - 20 = (z^2)^2 + z^2 - 20 \quad \text{Let } u = z^2.$$
$$= u^2 + u - 20$$
$$= (u-4)(u+5) \quad \text{Replace } u \text{ with } z^2.$$
$$= (z^2 - 4)(z^2 + 5)$$
$$= (z+2)(z-2)(z^2 + 5)$$

77. Factor by grouping.

$$3x^3 + x^2 + 6x + 2 = x^2(3x+1) + 2(3x+1)$$
$$= (3x+1)(x^2+2)$$

79. Factor by grouping.

$$ax^2 - ax + bx - b = ax(x-1) + b(x-1)$$
$$= (x-1)(ax+b)$$

81. Factor by grouping.

$$6w^3 + 4w^2 - 15w - 10 = 2w^2(3w+2) - 5(3w+2)$$
$$= (3w+2)(2w^2-5)$$

83. Factor the polynomial, if not state nonfactorable.

$$18x^2 - 2 = 2(9x^2-1) \quad \text{difference of squares}$$
$$= 2(3x+1)(3x-1)$$

85. Factor the polynomial, if not state nonfactorable.

$$16x^4 - 1 = (4x^2)^2 - 1 \quad \text{difference of squares}$$
$$= (4x^2+1)(4x^2-1) \quad \text{difference of squares}$$
$$= (4x^2+1)(2x+1)(2x-1)$$

87. Factor the polynomial, if not state nonfactorable.

$$12ax^2 - 23axy + 10ay^2 = a(12x^2 - 23xy + 10y^2)$$
$$= a(3x-2y)(4x-5y)$$

89. Factor the polynomial, if not state nonfactorable.

$$3bx^3 + 4bx^2 - 3bx - 4b$$
$$= b(3x^3 + 4x^2 - 3x - 4) \quad \text{factor by grouping}$$
$$= b[x^2(3x+4) - (3x+4)]$$
$$= b(3x+4)(x^2-1)$$
$$= b(3x+4)(x+1)(x-1)$$

91. Factor the polynomial, if not state nonfactorable.

$$72bx^2 + 24bxy + 2by^2$$
$$= 2b(36x^2 + 12xy + y^2) \quad \text{perfect-square trinomial}$$
$$= 2b(6x+y)^2$$

93. Factor the polynomial, if not state nonfactorable.

$$(w-5)^3 + 8 \quad \text{sum of cubes}$$
$$= [(w-5)+2][(w-5)^2 - 2(w-5)+4]$$
$$= (w-3)(w^2 - 10w + 25 - 2w + 10 + 4)$$
$$= (w-3)(w^2 - 12w + 39)$$

95. Factor the polynomial, if not state nonfactorable.

$$x^2 + 6xy + 9y^2 - 1 \quad \text{perfect-square trinomial}$$
$$= (x+3y)^2 - 1 \quad \text{difference of squares}$$
$$= (x+3y+1)(x+3y-1)$$

97. Factor the polynomial, if not state nonfactorable.

$$8x^2 + 3x - 4 \text{ is nonfactorable over the integers.}$$

99. Factor the polynomial, if not state nonfactorable.

$$5x(2x-5)^2 - (2x-5)^3 \quad \text{factor by grouping}$$
$$= (2x-5)^2[5x - (2x-5)]$$
$$= (2x-5)^2(3x+5)$$

101. Factor the polynomial, if not state nonfactorable.

$$4x^2 + 2x - y - y^2$$
$$= 4x^2 - y^2 + 2x - y \quad \text{factor by grouping}$$
$$= (2x+y)(2x-y) + (2x-y)$$
$$= (2x-y)(2x+y+1)$$

103. Find k.

$$x^2 + kx + 16$$
$$(x+4)^2 = x^2 + 8x + 16$$
$$(x-4)^2 = x^2 - 8x + 16$$

Thus, $k = -8, 8$.

105. Find k.

$$x^2 + 16x + k$$
$$(x+8)^2 = x^2 + 16x + 64$$

Thus, $k = 64$.

107. Find k.

$$x^2 + kx - 6$$
$$(x+1)(x-6) = x^2 - 5x - 6$$
$$(x-1)(x+6) = x^2 + 5x - 6$$
$$(x+2)(x-3) = x^2 - 1x - 6$$
$$(x-2)(x+3) = x^2 + 1x - 6$$

Thus, $k = -5, -1, 1, 5$.

109. Find k.

$$2x^2 + kx - 10$$

$$(2x+1)(x-10) = 2x^2 - 19x - 10$$

$$(2x+10)(x-1) = 2x^2 + 8x - 10$$

$$(2x-1)(x+10) = 2x^2 + 19x - 10$$

$$(2x-10)(x+1) = 2x^2 - 8x - 10$$

$$(2x+2)(x-5) = 2x^2 - 8x - 10$$

$$(2x-2)(x+5) = 2x^2 + 8x - 10$$

$$(2x-5)(x+2) = 2x^2 - 1x - 10$$

$$(2x+5)(x-2) = 2x^2 + 1x - 10$$

Thus, $k = -19, -8, -1, 1, 8, 19$.

Prepare for Section P.5

P1. Simplify.

$$1 + \frac{1}{2 - \frac{1}{3}} = 1 + \frac{1}{2 - \frac{1}{3}} \cdot \left(\frac{3}{3}\right)$$

$$= 1 + \frac{1 \cdot 3}{2 \cdot 3 - \frac{1}{3} \cdot 3}$$

$$= 1 + \frac{3}{6 - 1} = 1 + \frac{3}{5}$$

$$= 1\frac{3}{5} \text{ or } \frac{8}{5}$$

P3. Find the common binomial factor.

$$x^2 + 2x - 3 = (x+3)(x-1)$$

$$x^2 + 7x + 12 = (x+4)(x+3)$$

The common factor is $x + 3$.

P5. Factor completely.

$$x^2 - 5x - 6 = (x-6)(x+1)$$

Section P.5 Exercises

1. Determine whether the value is in the domain.

For $x = -3$, $\dfrac{x+5}{x^2 - 9} = \dfrac{-3+5}{(-3)^2 - 9} = \dfrac{2}{0}$ undefined

No, the value is not in the domain.

3. Determine whether the value is in the domain.

For $x = 1$, $\dfrac{3x-7}{x^2 + x - 2} = \dfrac{3(1)-7}{(1)^2 + 1 - 2} = \dfrac{-4}{0}$ undefined

No, the value is not in the domain.

5. Find the LCD.

$\dfrac{3}{xy}$ and $\dfrac{4}{xz^2}$ has LCD of xyz^2.

7. Find the LCD.

$\dfrac{2x-1}{x+5}$ and $\dfrac{8}{x-4}$ has LCD of $(x+5)(x-4)$.

9. Simplify.

$$\frac{x^2 - x - 20}{3x - 15} = \frac{(x+4)(x-5)}{3(x-5)} = \frac{x+4}{3}$$

11. Simplify.

$$\frac{x^3 - 9x}{x^3 + x^2 - 6x} = \frac{x(x^2 - 9)}{x(x^2 + x - 6)} = \frac{x(x-3)(x+3)}{x(x+3)(x-2)} = \frac{x-3}{x-2}$$

13. Simplify.

$$\frac{a^3 + 8}{a^2 - 4} = \frac{(a+2)(a^2 - 2a + 4)}{(a-2)(a+2)} = \frac{a^2 - 2a + 4}{a-2}$$

15. Simplify.

$$\frac{x^2 + 3x - 40}{-(x^2 - 3x - 10)} = \frac{(x-5)(x+8)}{-(x-5)(x+2)} = -\frac{x+8}{x+2}$$

17. Simplify.

$$\frac{4y^3 - 8y^2 + 7y - 14}{-y^2 - 5y + 14} = \frac{4y^2(y-2) + 7(y-2)}{-(y^2 + 5y - 14)}$$

$$= \frac{(y-2)(4y^2 + 7)}{-(y+7)(y-2)}$$

$$= -\frac{4y^2 + 7}{y+7}$$

19. Simplify.

$$\left(-\frac{4a}{3b^2}\right)\left(\frac{6b}{a^4}\right) = -\frac{24ab}{3a^4b^2} = -\frac{8}{a^3b}$$

21. Simplify.

$$\left(\frac{6p^2}{5q^2}\right)^{-1}\left(\frac{2p}{3q^2}\right)^2 = \frac{5q^2}{6p^2} \cdot \frac{4p^2}{9q^4} = \frac{10}{27q^2}$$

23. Simplify.

$$\frac{x^2 + x}{2x+3} \cdot \frac{3x^2 + 19x + 28}{x^2 + 5x + 4} = \frac{x(x+1)}{2x+3} \cdot \frac{(3x+7)(x+4)}{(x+4)(x+1)}$$

$$= \frac{x(3x+7)}{2x+3}$$

25. Simplify.

$$\frac{3x-15}{2x^2-50}\cdot\frac{2x^2+16x+30}{6x+9}$$

$$=\frac{3(x-5)}{2(x^2-25)}\cdot\frac{2(x^2+8x+15)}{3(2x+3)}$$

$$=\frac{3(x-5)}{2(x-5)(x+5)}\cdot\frac{2(x+3)(x+5)}{3(2x+3)}$$

$$=\frac{x+3}{2x+3}$$

27. Simplify.

$$\frac{12y^2+28y+15}{6y^2+35y+25}\div\frac{2y^2-y-3}{3y^2+11y-20}$$

$$=\frac{(6y+5)(2y+3)}{(6y+5)(y+5)}\cdot\frac{(3y-4)(y+5)}{(2y-3)(y+1)}$$

$$=\frac{(2y+3)(3y-4)}{(2y-3)(y+1)}$$

29. Simplify.

$$\frac{a^2+9}{a^2-64}\div\frac{a^3-3a^2+9a-27}{a^2+5a-24}$$

$$=\frac{a^2+9}{(a-8)(a+8)}\cdot\frac{(a-3)(a+8)}{a^2(a-3)+9(a-3)}$$

$$=\frac{a^2+9}{(a-8)(a+8)}\cdot\frac{(a-3)(a+8)}{(a-3)(a^2+9)}$$

$$=\frac{1}{a-8}$$

31. Simplify.

$$\frac{p+5}{r}+\frac{2p-7}{r}=\frac{p+5+2p-7}{r}=\frac{3p-2}{r}$$

33. Simplify.

$$\frac{5y-7}{y+4}-\frac{2y-3}{y+4}=\frac{(5y-7)-(2y-3)}{y+4}$$

$$=\frac{5y-7-2y+3}{y+4}$$

$$=\frac{3y-4}{y+4}$$

35. Simplify.

$$\frac{x}{x-5}+\frac{7x}{x+3}=\frac{x(x+3)+7x(x-5)}{(x-5)(x+3)}$$

$$=\frac{x^2+3x+7x^2-35x}{(x-5)(x+3)}$$

$$=\frac{8x^2-32x}{(x-5)(x+3)}$$

$$=\frac{8x(x-4)}{(x-5)(x+3)}$$

37. Simplify.

$$\frac{4z}{2z-3}+\frac{5z}{z-5}=\frac{4z(z-5)+5z(2z-3)}{(2z-3)(z-5)}$$

$$=\frac{4z^2-20z+10z^2-15z}{(2z-3)(z-5)}$$

$$=\frac{14z^2-35z}{(2z-3)(z-5)}$$

$$=\frac{7z(2z-5)}{(2z-3)(z-5)}$$

39. Simplify.

$$\frac{x}{x^2-9}-\frac{3x-1}{x^2+7x+12}=\frac{x}{(x-3)(x+3)}-\frac{3x-1}{(x+3)(x+4)}$$

$$=\frac{x(x+4)-(3x-1)(x-3)}{(x-3)(x+3)(x+4)}$$

$$=\frac{(x^2+4x)-(3x^2-10x+3)}{(x-3)(x+3)(x+4)}$$

$$=\frac{x^2+4x-3x^2+10x-3}{(x-3)(x+3)(x+4)}$$

$$=\frac{-2x^2+14x-3}{(x-3)(x+3)(x+4)}$$

41. Simplify.

$$\frac{1}{x}+\frac{2}{3x-1}\cdot\frac{3x^2+11x-4}{x-5}=\frac{1}{x}+\frac{2}{3x-1}\cdot\frac{(3x-1)(x+4)}{(x-5)}$$

$$=\frac{1}{x}+\frac{2(x+4)}{x-5}$$

$$=\frac{1(x-5)+x[2(x+4)]}{x(x-5)}$$

$$=\frac{x-5+2x^2+8x}{x(x-5)}$$

$$=\frac{2x^2+9x-5}{x(x-5)}$$

$$=\frac{(2x-1)(x+5)}{x(x-5)}$$

43. Simplify.

$$\frac{q+1}{q-3}-\frac{2q}{q-3}\div\frac{q+5}{q-3}=\frac{q+1}{q-3}-\frac{2q}{q-3}\cdot\frac{q-3}{q+5}$$

$$=\frac{q+1}{q-3}-\frac{2q}{q+5}$$

$$=\frac{(q+1)(q+5)-2q(q-3)}{(q-3)(q+5)}$$

$$=\frac{q^2+6q+5-2q^2+6q}{(q-3)(q+5)}$$

$$=\frac{-q^2+12q+5}{(q-3)(q+5)}$$

45. Simplify.

$$\frac{1}{x^2+7x+12}+\frac{1}{x^2-9}+\frac{1}{x^2-16}$$

$$=\frac{1}{(x+3)(x+4)}+\frac{1}{(x-3)(x+3)}+\frac{1}{(x-4)(x+4)}$$

$$=\frac{1(x-3)(x-4)+1(x-4)(x+4)+1(x-3)(x+3)}{(x+3)(x+4)(x-3)(x-4)}$$

$$=\frac{x^2-7x+12+x^2-16+x^2-9}{(x+3)(x+4)(x-3)(x-4)}$$

$$=\frac{3x^2-7x-13}{(x+3)(x+4)(x-3)(x-4)}$$

47. Simplify.

$$\left(1+\frac{2}{x}\right)\left(3-\frac{1}{x}\right)=\left(\frac{x}{x}+\frac{2}{x}\right)\left(\frac{3x}{x}-\frac{1}{x}\right)$$

$$=\left(\frac{x+2}{x}\right)\left(\frac{3x-1}{x}\right)$$

$$=\frac{(x+2)(3x-1)}{x^2}$$

49. Simplify the complex fraction.

$$\frac{4+\dfrac{1}{x}}{1-\dfrac{1}{x}}=\frac{\left(4+\dfrac{1}{x}\right)x}{\left(1-\dfrac{1}{x}\right)x}=\frac{4x+1}{x-1}$$

51. Simplify the complex fraction.

$$\frac{\dfrac{x}{y}-2}{y-x}=\frac{\left(\dfrac{x}{y}-2\right)y}{(y-x)y}=\frac{x-2y}{y(y-x)}$$

53. Simplify the complex fraction.

$$\frac{5-\dfrac{1}{x+2}}{1+\dfrac{3}{1+\dfrac{3}{x}}}=\frac{5-\dfrac{1}{x+2}}{1+\dfrac{3}{\dfrac{1(x)}{x}+\dfrac{3}{x}}}=\frac{5-\dfrac{1}{x+2}}{1+\dfrac{3}{\dfrac{x+3}{x}}}$$

$$=\frac{5-\dfrac{1}{x+2}}{1+3\div\dfrac{x+3}{x}}=\frac{5-\dfrac{1}{x+2}}{1+3\cdot\dfrac{x}{x+3}}$$

$$=\frac{5-\dfrac{1}{x+2}}{1+\dfrac{3x}{x+3}}=\frac{\dfrac{5(x+2)}{x+2}-\dfrac{1}{x+2}}{\dfrac{1(x+3)}{x+3}+\dfrac{3x}{x+3}}$$

$$=\frac{\dfrac{5(x+2)-1}{x+2}}{\dfrac{1(x+3)+3x}{x+3}}=\frac{\dfrac{5x+10-1}{x+2}}{\dfrac{x+3+3x}{x+3}}=\frac{\dfrac{5x+9}{x+2}}{\dfrac{4x+3}{x+3}}$$

$$=\frac{5x+9}{x+2}\div\frac{4x+3}{x+3}=\frac{5x+9}{x+2}\cdot\frac{x+3}{4x+3}$$

$$=\frac{(5x+9)(x+3)}{(x+2)(4x+3)}$$

55. Simplify the complex fraction.

$$\frac{1+\dfrac{1}{b-2}}{1-\dfrac{1}{b+3}}=\frac{\left(1+\dfrac{1}{b-2}\right)}{\left(1-\dfrac{1}{b+3}\right)}\cdot\frac{(b-2)(b+3)}{(b-2)(b+3)}$$

$$=\frac{1(b-2)(b+3)+1(b+3)}{1(b-2)(b+3)-1(b-2)}$$

$$=\frac{b^2+b-6+b+3}{b^2+b-6-b+2}=\frac{b^2+2b-3}{b^2-4}$$

$$=\frac{(b+3)(b-1)}{(b-2)(b+2)}$$

57. Simplify the complex fraction.

$$\frac{1-\dfrac{1}{x^2}}{1+\dfrac{1}{x}}=\frac{\left(1-\dfrac{1}{x^2}\right)}{\left(1+\dfrac{1}{x}\right)}\cdot\frac{x^2}{x^2}=\frac{x^2-1}{x^2+x}=\frac{(x-1)(x+1)}{x(x+1)}=\frac{x-1}{x}$$

59. Simplify the complex fraction.

$$2 - \cfrac{m}{1 - \cfrac{1-m}{-m}} = 2 - \cfrac{m}{1 + \cfrac{1-m}{m}}$$

$$= 2 - \cfrac{m}{\cfrac{1(m)}{m} + \cfrac{1-m}{m}} = 2 - \cfrac{m}{\cfrac{m+1-m}{m}}$$

$$= 2 - \cfrac{m}{\cfrac{1}{m}} = 2 - \cfrac{(m)}{\left(\cfrac{1}{m}\right)} \cdot \cfrac{m}{m}$$

$$= 2 - m^2$$

61. Simplify the complex fraction.

$$\cfrac{\left(\cfrac{1}{x} - \cfrac{x-4}{x+1}\right)}{\cfrac{x}{x+1}} \cdot \cfrac{x(x+1)}{x(x+1)} = \cfrac{x+1-x(x-4)}{x(x)}$$

$$= \cfrac{x+1-x^2+4x}{x^2}$$

$$= \cfrac{-x^2+5x+1}{x^2}$$

63. Simplify the complex fraction.

$$\cfrac{\left(\cfrac{1}{x+3} - \cfrac{2}{x-1}\right)}{\left(\cfrac{x}{x-1} + \cfrac{3}{x+3}\right)} \cdot \cfrac{(x+3)(x-1)}{(x+3)(x-1)} = \cfrac{1(x-1)-2(x+3)}{x(x+3)+3(x-1)}$$

$$= \cfrac{x-1-2x-6}{x^2+3x+3x-3}$$

$$= \cfrac{-x-7}{x^2+6x-3}$$

65. Simplify the complex fraction.

$$\cfrac{\cfrac{x^2+3x-10}{x^2+x-6}}{\cfrac{x^2-x-30}{2x^2-15x+18}} = \cfrac{\cfrac{(x-2)(x+5)}{(x-2)(x+3)}}{\cfrac{(x+5)(x-6)}{(2x-3)(x-6)}} = \cfrac{\cfrac{x+5}{x+3}}{\cfrac{x+5}{2x-3}}$$

$$= \cfrac{x+5}{x+3} \div \cfrac{x+5}{2x-3} = \cfrac{x+5}{x+3} \cdot \cfrac{2x-3}{x+5}$$

$$= \cfrac{2x-3}{x+3}$$

67. Simplify and write answer with positive exponents.

$$\cfrac{a^{-1}+b^{-1}}{a-b} = \cfrac{\cfrac{1}{a}+\cfrac{1}{b}}{a-b} = \cfrac{\cfrac{1b}{ab}+\cfrac{1a}{ab}}{a-b} = \cfrac{\cfrac{b+a}{ab}}{a-b}$$

$$= \cfrac{b+a}{ab} \div (a-b) = \cfrac{b+a}{ab} \cdot \cfrac{1}{a-b}$$

$$= \cfrac{a+b}{ab(a-b)}$$

69. Simplify and write answer with positive exponents.

$$\cfrac{a^{-1}b-ab^{-1}}{a^2+b^2} = \cfrac{\cfrac{b}{a}-\cfrac{a}{b}}{a^2+b^2} = \cfrac{\cfrac{(b)b}{(b)a}-\cfrac{a(a)}{b(a)}}{a^2+b^2}$$

$$= \cfrac{\cfrac{b^2}{ab}-\cfrac{a^2}{ab}}{a^2+b^2} = \cfrac{\cfrac{b^2-a^2}{ab}}{a^2+b^2}$$

$$= \cfrac{b^2-a^2}{ab} \div a^2+b^2$$

$$= \cfrac{b^2-a^2}{ab} \cdot \cfrac{1}{a^2+b^2}$$

$$= \cfrac{b^2-a^2}{ab(a^2+b^2)}$$

$$= \cfrac{(b-a)(b+a)}{ab(a^2+b^2)}$$

71. a.
$$\cfrac{2}{\cfrac{1}{180}+\cfrac{1}{110}} = \cfrac{2}{\cfrac{110+180}{180(110)}}$$

$$= 2 \div \cfrac{290}{180(110)} = 2 \cdot \cfrac{(180)(110)}{290}$$

$$\approx 136.55 \text{ mph (to the nearest hundredth)}$$

b.
$$\cfrac{2}{\cfrac{1}{v_1}+\cfrac{1}{v_2}} = \cfrac{2}{\cfrac{v_2+v_1}{v_1v_2}} = \cfrac{2v_1v_2}{v_2+v_1} = \cfrac{2v_1v_2}{v_1+v_2}$$

73.a. $3 + \cfrac{1}{6+\cfrac{3^2}{6+\cfrac{5^2}{6+\cfrac{7^2}{6}}}}$

b. $3 + \cfrac{1}{6+\cfrac{3^2}{6+\cfrac{5^2}{6+\cfrac{7^2}{6}}}} \approx 3.1397$

Prepare for Section P.6

P1. Simplify.

$$(2-3x)(4-5x) = 8-10x-12x+15x^2$$
$$= 15x^2-22x+8$$

P3. Simplify.

$$\sqrt{96} = \sqrt{16 \cdot 6} = 4\sqrt{6}$$

P5. Simplify.

$$\frac{5+\sqrt{2}}{3-\sqrt{2}} = \frac{5+\sqrt{2}}{3-\sqrt{2}} \cdot \frac{3+\sqrt{2}}{3+\sqrt{2}} = \frac{15+8\sqrt{2}+2}{9-2} = \frac{17+8\sqrt{2}}{7}$$

Section P.6 Exercises

1. Write the real part and imaginary part.

   For $12-17i$, 12 is real and -17 is imaginary.

3. Write the real part and imaginary part.

   For $\sqrt{7}$, $\sqrt{7}$ is real and 0 is imaginary.

5. Write the conjugate.

   The conjugate of $4-9i$ is $4+9i$.

7. Write the conjugate.

   The conjugate of $-11i$ is $11i$.

9. Write the complex number in standard form.

   $\sqrt{-81} = i\sqrt{81} = 9i$

11. Write the complex number in standard form.

    $\sqrt{-98} = i\sqrt{98} = 7i\sqrt{2}$

13. Write the complex number in standard form.

    $\sqrt{16} + \sqrt{-81} = 4 + i\sqrt{81} = 4 + 9i$

15. Write the complex number in standard form.

    $5 + \sqrt{-49} = 5 + i\sqrt{49} = 5 + 7i$

17. Write the complex number in standard form.

    $8 - \sqrt{-18} = 8 - i\sqrt{18} = 8 - 3i\sqrt{2}$

19. Simplify and write in standard form.

    $(5+2i)+(6-7i) = 5+2i+6-7i$
    $= (5+6)+(2i-7i)$
    $= 11-5i$

21. Simplify and write in standard form.

    $(-2-4i)-(5-8i) = -2-4i-5+8i$
    $= (-2-5)+(-4i+8i)$
    $= -7+4i$

23. Simplify and write in standard form.

    $(1-3i)+(7-2i) = 1-3i+7-2i$
    $= (1+7)+(-3i-2i)$
    $= 8-5i$

25. Simplify and write in standard form.

    $(-3-5i)-(7-5i) = -3-5i-7+5i$
    $= (-3-7)+(-5i+5i)$
    $= -10$

27. Simplify and write in standard form.

    $8i-(2-8i) = 8i-2+8i$
    $= -2+(8i+8i)$
    $= -2+16i$

29. Simplify and write in standard form.

    $5i \cdot 8i = 40i^2 = 40(-1) = -40$

31. Simplify and write in standard form.

    $\sqrt{-50} \cdot \sqrt{-2} = i\sqrt{50} \cdot i\sqrt{2} = 5i\sqrt{2} \cdot i\sqrt{2}$
    $= 5i^2(\sqrt{2})^2 = 5(-1)(2)$
    $= -10$

33. Simplify and write in standard form.

    $3(2+5i)-2(3-2i) = 6+15i-6+4i$
    $= (6-6)+(15i+4i)$
    $= 19i$

35. Simplify and write in standard form.

    $(4+2i)(3-4i) = 4(3-4i)+(2i)(3-4i)$
    $= 12-16i+6i-8i^2$
    $= 12-16i+6i-8(-1)$
    $= 12-16i+6i+8$
    $= (12+8)+(-16i+6i)$
    $= 20-10i$

37. Simplify and write in standard form.

    $(-3-4i)(2+7i) = -3(2+7i)-4i(2+7i)$
    $= -6-21i-8i-28i^2$
    $= -6-21i-8i-28(-1)$
    $= -6-21i-8i+28$
    $= (-6+28)+(-21i-8i)$
    $= 22-29i$

39. Simplify and write in standard form.

    $(4-5i)(4+5i) = 4(4+5i)-5i(4+5i)$
    $= 16+20i-20i-25i^2$
    $= 16+20i-20i-25(-1)$
    $= 16+20i-20i+25$
    $= (16+25)+(20i-20i)$
    $= 41$

41. Write as a complex number in standard form.

$$\frac{6}{i} = \frac{6}{i} \cdot \frac{i}{i} = \frac{6i}{i^2} = \frac{6i}{-1} = -6i$$

43. Write as a complex number in standard form.

$$\frac{6+3i}{i} = \frac{6+3i}{i} \cdot \frac{i}{i} = \frac{6i+3i^2}{i^2}$$

$$= \frac{6i+3(-1)}{-1} = \frac{6i-3}{-1}$$

$$= 3-6i$$

45. Write as a complex number in standard form.

$$\frac{1}{7+2i} = \frac{1}{7+2i} \cdot \frac{7-2i}{7-2i}$$

$$= \frac{1(7-2i)}{(7+2i)(7-2i)}$$

$$= \frac{7-2i}{49-4i^2}$$

$$= \frac{7-2i}{49-4(-1)} = \frac{7-2i}{49+4}$$

$$= \frac{7-2i}{53} = \frac{7}{53} - \frac{2}{53}i$$

47. Write as a complex number in standard form.

$$\frac{2i}{1+i} = \frac{2i}{1+i} \cdot \frac{1-i}{1-i} = \frac{2i(1-i)}{(1+i)(1-i)}$$

$$= \frac{2i-2i^2}{1-i^2} = \frac{2i-2(-1)}{1-(-1)}$$

$$= \frac{2i+2}{1+1} = \frac{2+2i}{2} = \frac{2}{2} + \frac{2}{2}i$$

$$= 1+i$$

49. Write as a complex number in standard form.

$$\frac{5-i}{4+5i} = \frac{5-i}{4+5i} \cdot \frac{4-5i}{4-5i} = \frac{(5-i)(4-5i)}{(4+5i)(4-5i)}$$

$$= \frac{5(4-5i) - i(4-5i)}{4(4-5i) + 5i(4-5i)}$$

$$= \frac{20-25i-4i+5i^2}{16-20i+20i-25i^2}$$

$$= \frac{20-25i-4i+5(-1)}{16-25(-1)}$$

$$= \frac{20-25i-4i-5}{16+25}$$

$$= \frac{(20-5)+(-25i-4i)}{16+25}$$

$$= \frac{15-29i}{41} = \frac{15}{41} - \frac{29}{41}i$$

51. Write as a complex number in standard form.

$$\frac{3+2i}{3-2i} = \frac{3+2i}{3-2i} \cdot \frac{3+2i}{3+2i} = \frac{(3+2i)^2}{(3-2i)(3+2i)}$$

$$= \frac{3^2+2(3)(2i)+(2i)^2}{3^2-(2i)^2} = \frac{9+12i+4i^2}{9-4i^2}$$

$$= \frac{9+12i+4(-1)}{9-4(-1)}$$

$$= \frac{9+12i-4}{9+4} = \frac{5+12i}{13}$$

$$= \frac{5}{13} + \frac{12}{13}i$$

53. Write as a complex number in standard form.

$$\frac{-7+26i}{4+3i} = \frac{-7+26i}{4+3i} \cdot \frac{4-3i}{4-3i} = \frac{(-7+26i)(4-3i)}{(4+3i)(4-3i)}$$

$$= \frac{-7(4-3i)+26i(4-3i)}{4^2-(3i)^2}$$

$$= \frac{-28+21i+104i-78i^2}{16-9i^2}$$

$$= \frac{-28+21i+104i-78(-1)}{16-9(-1)}$$

$$= \frac{-28+21i+104i+78}{16+9} = \frac{50+125i}{25}$$

$$= \frac{50}{25} + \frac{125}{25}i$$

$$= 2+5i$$

55. Write as a complex number in standard form.

$$(3-5i)^2 = 3^3 + 2(3)(-5i) + (-5i)^2$$

$$= 9-30i+25i^2$$

$$= 9-30i+25(-1)$$

$$= 9-30i-25$$

$$= -16-30i$$

57. Write as a complex number in standard form.

$$(1+2i)^3 = (1+2i)(1+2i)^2$$

$$= (1+2i)[1^2 + 2(1)(2i) + (2i)^2]$$

$$= (1+2i)[1+4i+4i^2]$$

$$= (1+2i)[1+4i-4]$$

$$= (1+2i)(-3+4i)$$

$$= 1(-3+4i) + 2i(-3+4i)$$

$$= -3+4i-6i+8i^2$$

$$= -3+4i-6i+8(-1)$$

$$= -3+4i-6i-8$$

$$= -11-2i$$

59. Evaluate the power of i.

Use the Powers of i Theorem.

The remainder of $15 \div 4$ is 3.

$i^{15} = i^3 = -i$

61. Evaluate the power of i.

Use the Powers of i Theorem.

The remainder of $40 \div 4$ is 0.

$-i^{40} = -(i^0) = -1$

63. Evaluate the power of i.

Use the Powers of i Theorem.

The remainder of $25 \div 4$ is 1.

$\dfrac{1}{i^{25}} = \dfrac{1}{i} = \dfrac{1}{i} \cdot \dfrac{i}{i} = \dfrac{i}{i^2} = \dfrac{i}{-1} = -i$

65. Evaluate the power of i.

Use the Powers of i Theorem.

The remainder of $34 \div 4$ is 2.

$i^{-34} = \dfrac{1}{i^{34}} = \dfrac{1}{i^2} = \dfrac{1}{-1} = -1$

67. Evaluate.

Use $a = 2$, $b = 4$, $c = 4$.

$\dfrac{-b + \sqrt{b^2 - 4ac}}{2a} = \dfrac{-(4) + \sqrt{(4)^2 - 4(2)(4)}}{2(2)}$

$= \dfrac{-4 + \sqrt{16 - 32}}{4} = \dfrac{-4 + \sqrt{-16}}{4}$

$= \dfrac{-4 + i\sqrt{16}}{4} = \dfrac{-4 + 4i}{4}$

$= \dfrac{-4}{4} + \dfrac{4i}{4} = -1 + i$

69. Evaluate.

Use $a = 3$, $b = -3$, $c = 3$.

$\dfrac{-b + \sqrt{b^2 - 4ac}}{2a} = \dfrac{-(-3) + \sqrt{(-3)^2 - 4(3)(3)}}{2(3)}$

$= \dfrac{3 + \sqrt{9 - 36}}{6} = \dfrac{3 + \sqrt{-27}}{6}$

$= \dfrac{3 + i\sqrt{27}}{6} = \dfrac{3 + 3i\sqrt{3}}{6}$

$= \dfrac{3}{6} + \dfrac{3\sqrt{3}}{6}i = \dfrac{1}{2} + \dfrac{\sqrt{3}}{2}i$

71. Expand.

$(1 - 3i)^3 = [(1 - 3i)(1 - 3i)](1 - 3i)$

$= (1 - 6i + 9i^2)(1 - 3i)$

$= (1 - 6i - 9)(1 - 3i)$

$= (-8 - 6i)(1 - 3i)$

$= -8 + 24i - 6i + 18i^2$

$= -8 + 18i - 18$

$= -26 + 18i$

73. Show the statement $\sqrt{i} = \dfrac{1}{\sqrt{2}} + \dfrac{i}{\sqrt{2}}$ is true.

Square the left-hand side.

$\left(\sqrt{i}\right)^2 = i$

Square the right-hand side.

$\left(\dfrac{1}{\sqrt{2}} + \dfrac{i}{\sqrt{2}}\right)^2 = \left(\dfrac{1}{\sqrt{2}} + \dfrac{i}{\sqrt{2}}\right)\left(\dfrac{1}{\sqrt{2}} + \dfrac{i}{\sqrt{2}}\right)$

$= \dfrac{1}{2} + \dfrac{i}{2} + \dfrac{i}{2} + \dfrac{i^2}{2}$

$= \dfrac{1}{2} + i - \dfrac{1}{2}$

$= i$

Since the left-hand side and the right-hand side are equal, the statement $\sqrt{i} = \dfrac{1}{\sqrt{2}} + \dfrac{i}{\sqrt{2}}$ is true.

Exploring Concepts with Technology

1. Use a calculator to find the first 20 iterations.

Below is a chart of a selection of those iterations.

Iteration	$p + 3p(1 - p)$	$4p - 3p^2$
1	0.5	0.5
3	0.95703125	0.95703125
5	0.8198110957	0.8198110957
10	0.3846309658	0.3846309658
15	0.5610061236	0.5610061236
19	1.218765181	1.218765181
20	0.4188950251	0.4188950245

Chapter P Review Exercises

1. Classify the number. [P.1]

3: integer, rational number, real number, prime number

3. Classify the number. [P.1]

$-\dfrac{1}{2}$: rational number, real number

5. List the four smallest elements of the set. [P.1]

Let $x = 0, 1, 2, 3$. (We could have used $x = -3, -2, -1,$

0.) Then $\left\{ y \mid y = x^2, \ x \in \text{integers} \right\} = \{0, 1, 4, 9\}$

7. Perform the operation. [P.1]

$A \cup B = \{1, 2, 3, 5, 7, 11\}$

9. Graph the interval, write in set-builder notation. [P.1]

$[-3, 2)$

$\left\{ x \mid -3 \le x < 2 \right\}$

11. Graph the set, write in interval notation. [P.1]

$\left\{ x \mid -4 < x \le 2 \right\}$

$(-4, 2]$

13. Write without absolute value symbols. [P.1]

$\mid 7 \mid = 7$

15. Write without absolute value symbols. [P.1]

$\mid 4 - \pi \mid = 4 - \pi$, because $4 > \pi$

17. Write without absolute value symbols. [P.1]

$\mid x - 2 \mid + \mid x + 1 \mid, \ -1 < x < 2$

$= 2 - x + x + 1$

$= 3$

19. Find the distance between -3 and 7. [P.1]

$\mid -3 - 7 \mid = \mid -10 \mid = 10$

21. Evaluate. [P.1]

$-4^4 = -4 \cdot 4 \cdot 4 \cdot 4 = -256$

23. Evaluate. [P.1]

$-5 \cdot 3^2 + 4\left\{ 5 - 2[-6 - (-4)] \right\}$

$= -5 \cdot 3^2 + 4\left\{ 5 - 2[-6 + 4] \right\}$

$= -5 \cdot 3^2 + 4\left\{ 5 - 2[-2] \right\}$

$= -5 \cdot 3^2 + 4\left\{ 5 + 4 \right\}$

$= -5 \cdot 3 \cdot 3 + 4\left\{ 9 \right\}$

$= -45 + 36$

$= -9$

25. Evaluate. [P.1]

$-3x^3 - 4xy - z^2, \quad x = -2, \ y = 3, \ z = -5$

$-3(-2)^3 - 4(-2)(3) - (-5)^2$

$= -3(-2)(-2)(-2) - 4(-2)(3) - (-5)(-5)$

$= 24 + 24 - 25$

$= 23$

27. Identify the property illustrated. [P.1]

Distributive property

29. Identify the property illustrated. [P.1]

Associative property of multiplication

31. Identify the property illustrated. [P.1]

Identity property of addition

33. Identify the property illustrated. [P.1]

Symmetric property of equality

35. Simplify the expression. [P.1]

$8 - 3(2x - 5) = 8 - 6x + 15 = -6x + 23$

37. Simplify the exponential expression. [P.2]

$-2^{-5} = -\dfrac{1}{2^5} = -\dfrac{1}{2 \cdot 2 \cdot 2 \cdot 2 \cdot 2} = -\dfrac{1}{32}$

39. Simplify the exponential expression. [P.2]

$\dfrac{2}{z^{-4}} = 2z^4$

41. Write the number in scientific notation. [P.2]

$620,000 = 6.2 \times 10^5$

43. Write the number in decimal form. [P.2]

$3.5 \times 10^4 = 35,000$

45. Evaluate the exponential expression. [P.2]

$25^{1/2} = \sqrt{25} = 5$

47. Evaluate the exponential expression. [P.2]

$36^{-1/2} = \dfrac{1}{\sqrt{36}} = \dfrac{1}{6}$

49. Simplify the expression. [P.2]

$\left(-4x^3 y^2 \right)\left(6x^4 y^3 \right) = -4(6)x^{3+4} y^{2+3} = -24x^7 y^5$

51. Simplify the expression. [P.2]

$$\left(-3x^{-2}y^3\right)^{-3} = (-3)^{-3}\left(x^{-2}\right)^{-3}\left(y^3\right)^{-3}$$

$$= -\frac{1}{27}x^6 y^{-9}$$

$$= -\frac{x^6}{27y^9}$$

53. Simplify the expression. [P.2]

$$\left(-4x^{-3}y^2\right)^{-2}\left(8x^{-2}y^{-3}\right)^2$$

$$= \left(\frac{1}{-4x^{-3}y^2}\right)^2\left(8x^{-2}y^{-3}\right)^2$$

$$= \frac{1}{16x^{-6}y^4}\cdot 64x^{-4}y^{-6}$$

$$= 4x^{6-4}y^{-4-6} = 4x^2 y^{-10}$$

$$= \frac{4x^2}{y^{10}}$$

55. Simplify the expression. [P.2]

$$\left(x^{-1/2}\right)\left(x^{3/4}\right) = x^{-1/2+3/4} = x^{1/4}$$

57. Simplify the expression. [P.2]

$$\left(\frac{8x^{5/4}}{x^{1/2}}\right)^{2/3} = \left(8x^{5/4-1/2}\right)^{2/3} = \left(8x^{5/4-2/4}\right)^{2/3}$$

$$= \left(8x^{3/4}\right)^{2/3} = 8^{2/3}x^{(3/4)(2/3)}$$

$$= \left(2^3\right)^{2/3}x^{(3/4)(2/3)} = 2^2 x^{1/2}$$

$$= 4x^{1/2}$$

59. Simplify the radical expression. [P.2]

$$\sqrt{48a^2 b^7} = \sqrt{16a^2 b^6 \cdot 3b} = 4ab^3 \sqrt{3b}$$

61. Simplify the radical expression. [P.2]

$$\sqrt[3]{-135x^2 y^7} = \sqrt[3]{-27y^6 \cdot 5x^2 y} = -3y^2 \sqrt[3]{5x^2 y}$$

63. Simplify the radical expression. [P.2]

$$b\sqrt{8a^4 b^3} + 2a\sqrt{18a^2 b^5}$$

$$= b\sqrt{4a^4 b^2}\sqrt{2b} + 2a\sqrt{9a^2 b^4}\sqrt{2b}$$

$$= 2a^2 b^2 \sqrt{2b} + 6a^2 b^2 \sqrt{2b}$$

$$= 8a^2 b^2 \sqrt{2b}$$

65. Simplify the radical expression. [P.2]

$$\left(3+2\sqrt5\right)\left(7-3\sqrt5\right) = 21-9\sqrt5+14\sqrt5-30$$

$$= -9+5\sqrt5$$

67. Simplify the radical expression. [P.2]

$$\left(4-2\sqrt7\right)^2 = 16-2\left(4\cdot2\sqrt7\right)+28$$

$$= 44-16\sqrt7$$

69. Simplify the radical expression. [P.2]

$$\frac{6}{\sqrt8} = \frac{6}{2\sqrt2} = \frac{3}{\sqrt2} = \frac{3}{\sqrt2}\cdot\frac{\sqrt2}{\sqrt2} = \frac{3\sqrt2}{2}$$

71. Simplify the radical expression. [P.2]

$$\frac{3+2\sqrt7}{9-3\sqrt7} = \frac{3+2\sqrt7}{9-3\sqrt7}\cdot\frac{9+3\sqrt7}{9+3\sqrt7}$$

$$= \frac{27+9\sqrt7+18\sqrt7+42}{81-63}$$

$$= \frac{69+27\sqrt7}{18} = \frac{3\left(23+9\sqrt7\right)}{3\cdot6}$$

$$= \frac{23+9\sqrt7}{6}$$

73. Write the polynomial in standard for, identify the degree, leading coefficient and constant term. [P.3]

$$-x^3 - 7x^2 + 4x + 5$$

degree: 3; leading coefficient: –1; constant: 5

75. Perform the operation. [P.3]

$$\left(2a^2 + 3a - 7\right) + \left(-3a^2 - 5a + 6\right)$$

$$= \left[2a^2 + \left(-3a^2\right)\right] + \left[3a + \left(-5a\right)\right] + \left[-7+6\right]$$

$$= -a^2 - 2a - 1$$

77. Perform the operation. [P.3]

$$(3x-2)\left(2x^2 + 4x - 9\right)$$

$$= 3x\left(2x^2 + 4x - 9\right) - 2\left(2x^2 + 4x - 9\right)$$

$$= 6x^3 + 12x^2 - 27x - 4x^2 - 8x + 18$$

$$= 6x^3 + 8x^2 - 35x + 18$$

79. Perform the operation. [P.3]

$$(3x-4)(x+2) = 3x^2 + 6x - 4x - 8$$

$$= 3x^2 + 2x - 8$$

81. Perform the operation. [P.3]

$$(2x+5)^2 = 4x^2 + 2(2x\cdot5) + 25$$

$$= 4x^2 + 20x + 25$$

83. Factor out the GCF. [P.4]

$$12x^3 y^4 + 10x^2 y^3 - 34xy^2 = 2xy^2\left(6x^2 y^2 + 5xy - 17\right)$$

85. Factor out the GCF. [P.4]

$$(2x+7)(3x-y)-(3x+2)(3x-y)$$
$$=(3x-y)[2x+7-(3x+2)]$$
$$=(3x-7)(2x+7-3x-2)$$
$$=(3x-7)(-x+5)$$

87. Factor. [P.4]

$$x^2+7x-18=(x-2)(x+9)$$

89. Factor. [P.4]

$$2x^2+11x+12=(2x+3)(x+4)$$

91. Factor. [P.4]

$$6x^3y^2-12x^2y^2-144xy^2=6xy^2(x^2-2x-24)$$
$$=6xy^2(x-6)(x+4)$$

93. Factor. [P.4]

$$9x^2-100=(3x-10)(3x+10)$$

95. Factor. [P.4]

$$x^4-5x^2-6=(x^2)^2-5x^2-6=(x^2-6)(x^2+1)$$

99. Factor. [P.4]

$$4x^4-x^2-4x^2y^2+y^2$$
$$=x^2(4x^2-1)-y^2(4x^2-1)$$
$$=(4x^2-1)(x^2-y^2)$$
$$=(2x+1)(2x-1)(x+y)(x-y)$$

101. Factor. [P.4]

$$24a^2b^2-14ab^3-90b^4$$
$$=2b^2(12a^2-7ab-45b^2)$$
$$=2b^2(3a+5b)(4a-9b)$$

103. Simplify the rational expression. [P.5]

$$\frac{6x^2-19x+10}{2x^2+3x-20}=\frac{(3x-2)(2x-5)}{(2x-5)(x+4)}=\frac{3x-2}{x+4}$$

105. Perform the operation and simplify. [P.5]

$$\frac{10x^2+13x-3}{6x^2-13x-5}\cdot\frac{6x^2+5x+1}{10x^2+3x-1}$$
$$=\frac{(2x+3)(5x-1)}{(2x-5)(3x+1)}\cdot\frac{(2x+1)(3x+1)}{(2x+1)(5x-1)}$$
$$=\frac{2x+3}{2x-5}$$

107. Perform the operation and simplify. [P.5]

$$\frac{x}{x^2-9}+\frac{2x}{x^2+x-12}=\frac{x}{(x-3)(x+3)}+\frac{2x}{(x+4)(x-3)}$$
$$=\frac{x(x+4)+2x(x+3)}{(x-3)(x+3)(x+4)}$$
$$=\frac{x^2+4x+2x^2+6x}{(x-3)(x+3)(x+4)}$$
$$=\frac{3x^2+10x}{(x-3)(x+3)(x+4)}$$
$$=\frac{x(3x+10)}{(x-3)(x+3)(x+4)}$$

109. Simplify the complex fraction. [P.5]

$$\frac{2+\dfrac{1}{x-5}}{3-\dfrac{2}{x-5}}=\frac{\left(2+\dfrac{1}{x-5}\right)}{\left(3-\dfrac{2}{x-5}\right)}\cdot\frac{x-5}{x-5}$$
$$=\frac{2(x-5)+1}{3(x-5)-2}=\frac{2x-10+1}{3x-15-2}$$
$$=\frac{2x-9}{3x-17}$$

111. Write the complex number in standard form. [P.6]

$$5+\sqrt{-64}=5+8i$$

113. Perform the operation and simplify. [P.6]

$$(2-3i)+(4+2i)=2-3i+4+2i$$
$$=(2+4)+(-3i+2i)$$
$$=6-i$$

115. Perform the operation and simplify. [P.6]

$$2i(3-4i)=6i-8i^2$$
$$=6i-8(-1)$$
$$=6i+8$$
$$=8+6i$$

117. Perform the operation and simplify. [P.6]

$$(3+i)^2=3^2+2(3)(i)+i^2$$
$$=9+6i+(-1)$$
$$=8+6i$$

119. Perform the operation and simplify. [P.6]

$$\frac{4-6i}{2i}=\frac{2(2-3i)}{2i}=\frac{\cancel{2}(2-3i)}{\cancel{2}i}=\frac{2-3i}{i}$$
$$=\frac{2-3i}{i}\cdot\frac{i}{i}=\frac{2i-3i^2}{i^2}=\frac{2i-3(-1)}{-1}=\frac{2i+3}{-1}$$
$$=-3-2i$$

Chapter P Test

1. Distributive property [P.1]

3. Simplify. [P.1]

$$|x+1|-|x-5|, \quad -1 < x < 4$$
$$= x+1+x-5$$
$$= 2x-4$$

5. Simplify. [P.2]

$$\frac{\left(2a^{-1}bc^{-2}\right)^2}{\left(3^{-1}b\right)\left(2^{-1}ac^{-2}\right)^3} = \frac{2^2\,a^{-2}b^2c^{-4}}{\left(3^{-1}b\right)\left(2^{-3}a^3c^{-6}\right)}$$
$$= \frac{2^2 \cdot 2^3 \cdot 3^1 \cdot b^2 c^6}{ba^3 a^2 c^4}$$
$$= \frac{2^5 \cdot 3 \cdot bc^2}{a^5}$$
$$= \frac{96bc^2}{a^5}$$

7. Simplify. [P.2]

$$\frac{x^{1/3}y^{-3/4}}{x^{-1/2}y^{3/2}} = x^{1/3-(-1/2)}\,y^{-3/4-3/2}$$
$$= x^{1/3+1/2}\,y^{-3/4-3/2}$$
$$= x^{2/6+3/6}\,y^{-3/4-6/4}$$
$$= x^{5/6}\,y^{-9/4}$$
$$= \frac{x^{5/6}}{y^{9/4}}$$

9. Simplify. [P.2]

$$\left(2\sqrt{3}-4\right)\left(5\sqrt{3}+2\right) = 30+4\sqrt{3}-20\sqrt{3}-8$$
$$= 22-16\sqrt{3}$$

11. Simplify. [P.2]

$$\frac{x}{\sqrt[4]{2x^3}} = \frac{x}{\sqrt[4]{2x^3}} \cdot \frac{\sqrt[4]{8x}}{\sqrt[4]{8x}} = \frac{x\sqrt[4]{8x}}{2x} = \frac{\sqrt[4]{8x}}{2}$$

13. Simplify. [P.2]

$$\frac{2+\sqrt{5}}{4-2\sqrt{5}} = \frac{2+\sqrt{5}}{4-2\sqrt{5}} \cdot \frac{4+2\sqrt{5}}{4+2\sqrt{5}}$$
$$= \frac{8+4\sqrt{5}+4\sqrt{5}+10}{16-20}$$
$$= \frac{18+8\sqrt{5}}{-4} = -\frac{2\left(9+4\sqrt{5}\right)}{2\cdot2}$$
$$= -\frac{9+4\sqrt{5}}{2}$$

15. Multiply. [P.3]

$$(3a+7b)(2a-9b) = 6a^2-27ab+14ab-63b^2$$
$$= 6a^2-13ab-63b^2$$

17. Factor. [P.4]

$$7x^2+34x-5 = (7x-1)(x+5)$$

19. Factor. [P.4]

$$16x^4-2xy^3 = 2x(8x^3-y^3)$$
$$= 2x(2x-y)(4x^2+2xy+y^2)$$

21. Simplify. [P.5]

$$\frac{x^2-2x-15}{25-x^2} = \frac{(x-5)(x+3)}{(5-x)(5+x)} = \left(\frac{x-5}{5-x}\right)\left(\frac{x+3}{x+5}\right)$$
$$= -1\cdot\left(\frac{x+3}{x+5}\right) = -\frac{x+3}{x+5}$$

23. Multiply. [P.5]

$$\frac{x^2-3x-4}{x^2+x-20}\cdot\frac{x^2+3x-10}{x^2+2x-8} = \frac{(x+1)(x-4)}{(x+5)(x-4)}\cdot\frac{(x+5)(x-2)}{(x+4)(x-2)}$$
$$= \frac{x+1}{x+4}$$

25. Simplify. [P.5]

$$x-\frac{x}{x+\dfrac{1}{2}} = x-\frac{x}{\dfrac{2x}{2}+\dfrac{1}{2}} = x-\frac{x}{\dfrac{2x+1}{2}}$$
$$= x-x\div\frac{2x+1}{2} = x-x\cdot\frac{2}{2x+1}$$
$$= x-\frac{2x}{2x+1} = \frac{x(2x+1)}{2x+1}-\frac{2x}{2x+1}$$
$$= \frac{2x^2+x}{2x+1}-\frac{2x}{2x+1} = \frac{2x^2+x-2x}{2x+1}$$
$$= \frac{2x^2-x}{2x+1} = \frac{x(2x-1)}{2x+1}$$

27. Write in simplest form. [P.6]

$$(4-3i)-(2-5i) = 4-3i-2+5i$$
$$= (4-2)+(-3i+5i)$$
$$= 2+2i$$

29. Write in simplest form. [P.6]

$$\frac{3+4i}{5-i} = \frac{3+4i}{5-i}\cdot\frac{5+i}{5+i} = \frac{(3+4i)(5+i)}{(5-i)(5+i)}$$
$$= \frac{3(5+i)+4i(5+i)}{5^2-i^2} = \frac{15+3i+20i+4i^2}{25-i^2}$$
$$= \frac{15+3i+20i+4(-1)}{25-(-1)}$$
$$= \frac{15+3i+20i-4}{25+1} = \frac{(15-4)+(3i+20i)}{26}$$
$$= \frac{11+23i}{26} = \frac{11}{26}+\frac{23}{26}i$$

Chapter 1 Equations and Inequalities

Section 1.1 Exercises

1. Determine whether 2 is a solution.

$$3(x-4)+5 = 2x-5$$

$$3(2-4)+5 \overset{?}{=} 2(2)-5$$

$$3(-2)+5 \overset{?}{=} 4-5$$

$$-6+5 \overset{?}{=} -1$$

$$-1 = -1$$

Yes, 2 is a solution.

3. State the difference.

A conditional equation is true for some values of the variable, but not true for other values of the variable. An identity has an infinite number of solutions.

5. Solve the equation.

$$2x+10 = 40$$

$$2x = 40-10$$

$$2x = 30$$

$$x = 15$$

7. Solve the equation.

$$5x+2 = 2x-10$$

$$5x-2x = -10-2$$

$$3x = -12$$

$$x = -4$$

9. Solve the equation.

$$2(x-3)-5 = 4(x-5)$$

$$2x-6-5 = 4x-20$$

$$2x-11 = 4x-20$$

$$2x-4x = -20+11$$

$$-2x = -9$$

$$x = \frac{9}{2}$$

11. Solve the equation.

$$3x+5(1-2x) = 4-3(x+1)$$

$$3x+5-10x = 4-3x-3$$

$$-7x+5 = 1-3x$$

$$-4x = -4$$

$$x = 1$$

13. Solve the equation.

$$4(2r-17)+5(3r-8) = 0$$

$$8r-68+15r-40 = 0$$

$$23r-108 = 0$$

$$23r = 108$$

$$r = \frac{108}{23}$$

15. Solve the equation.

$$\frac{3}{4}x+\frac{1}{2} = \frac{2}{3}$$

$$12 \cdot \left(\frac{3}{4}x+\frac{1}{2}\right) = 12 \cdot \left(\frac{2}{3}\right)$$

$$9x+6 = 8$$

$$9x = 8-6$$

$$9x = 2$$

$$x = \frac{2}{9}$$

17. Solve the equation.

$$\frac{2}{3}x-5 = \frac{1}{2}x-3$$

$$6 \cdot \left(\frac{2}{3}x-5\right) = 6 \cdot \left(\frac{1}{2}x-3\right)$$

$$4x-30 = 3x-18$$

$$4x-3x = -18+30$$

$$x = 12$$

19. Solve the equation.

$$0.2x+0.4 = 3.6$$

$$0.2x = 3.6-0.4$$

$$0.2x = 3.2$$

$$x = 16$$

21. Solve the equation.

$$x+0.08(60) = 0.20(60+x)$$

$$x+4.8 = 12+0.20x$$

$$x-0.20x = 12-4.8$$

$$0.80x = 7.2$$

$$x = 9$$

23. Solve the equation.

$$5[x-(4x-5)]=3-2x$$
$$5(x-4x+5)=3-2x$$
$$5(-3x+5)=3-2x$$
$$-15x+25=3-2x$$
$$-15x+2x=3-25$$
$$-13x=-22$$
$$x=\frac{22}{13}$$

25. Solve the equation.

$$\frac{40-3x}{5}=\frac{6x+7}{8}$$
$$40\cdot\left(\frac{40-3x}{5}\right)=40\cdot\left(\frac{6x+7}{8}\right)$$
$$8(40-3x)=5(6x+7)$$
$$320-24x=30x+35$$
$$-24x-30x=35-320$$
$$-54x=-285$$
$$x=\frac{95}{18}$$

27. Classify: contradiction, conditional equation, identity.

$$-3(x-5)=-3x+15$$
$$-3x+15=-3x+15$$

Identity

29. Classify: contradiction, conditional equation, identity.

$$2x+7=3(x-1)$$
$$2x+7=3x-3$$
$$2x-3x=-3-7$$
$$-x=-10$$
$$x=10$$

Conditional equation

31. Classify: contradiction, conditional equation, identity.

$$\frac{4x+8}{4}=x+8$$
$$4x+8=4(x+8)$$
$$4x+8=4x+32$$
$$8=32$$

Contradiction

33. Classify: contradiction, conditional equation, identity.

$$3[x-2(x-5)]-1=-3x+29$$
$$3[x-2x+10]-1=-3x+29$$
$$3[-x+10]-1=-3x+29$$
$$-3x+30-1=-3x+29$$
$$-3x+29=-3x+29$$

Identity

35. Classify: contradiction, conditional equation, identity.

$$2x-8=-x+9$$
$$3x=17$$
$$x=\frac{17}{3}$$

Conditional equation

37. Solve for x.

$$|x|=4$$
$$x=4\quad\text{or}\quad x=-4$$

39. Solve for x.

$$|x-5|=2$$
$$x-5=2\quad\text{or}\quad x-5=-2$$
$$x=7\qquad\qquad x=3$$

41. Solve for x.

$$|2x-5|=11$$
$$2x-5=11\quad\text{or}\quad 2x-5=-11$$
$$2x=16\qquad\qquad 2x=-6$$
$$x=8\qquad\qquad x=-3$$

43. Solve for x.

$$|2x+6|=10$$
$$2x+6=10\quad\text{or}\quad 2x+6=-10$$
$$2x=4\qquad\qquad 2x=-16$$
$$x=2\qquad\qquad x=-8$$

45. Solve for x.

$$\left|\frac{x-4}{2}\right|=8$$
$$\frac{x-4}{2}=8\qquad\text{or}\qquad\frac{x-4}{2}=-8$$
$$x-4=8(2)\qquad\qquad x-4=-8(2)$$
$$x-4=16\qquad\qquad x-4=-16$$
$$x=20\qquad\qquad x=-12$$

47. Solve for x.

$$|2x+5| = -8$$

$$|2x+5| \geq 0$$

$$-8 \geq 0$$

Contradiction. There is no solution.

49. Solve for x.

$$2|x+3| + 4 = 34$$

$$2|x+3| = 30$$

$$|x+3| = 15$$

$$x+3 = 15 \qquad \text{or} \qquad x+3 = -15$$

$$x = 12 \qquad\qquad\qquad x = -18$$

51. Solve for x.

$$3|2x-5| + 2 = 11$$

$$3|2x-5| = 9$$

$$|2x-5| = 3$$

$$2x-5 = 3 \quad \text{or} \quad 2x-5 = -3$$

$$2x = 8 \quad \text{or} \qquad 2x = 2$$

$$x = 4 \quad \text{or} \qquad\; x = 1$$

53. Estimate the volume of the pouch.

$$F = -5.5V + 5400$$

$$550 = -5.5V + 5400$$

$$-4850 = -5.5V$$

$$881.81\overline{81} = V$$

$$V \approx 882 \text{ cm}^3$$

55. Find the time.

$$d = |210 - 50t|$$

$$60 = 210 - 50t \quad \text{or} \quad -60 = 210 - 50t$$

$$-150 = -50t \qquad\qquad -270 = -50t$$

$$t = 3 \qquad\qquad\qquad\quad t = 5.4$$

5.4 hours = 5 hours 24 minutes

Ruben will be exactly 60 miles from Barstow

after 3 hours and after 5 hours and 24 minutes.

57. Find the number of minutes.

$$g = 18 - \frac{1}{36}t$$

$$0 = 18 - \frac{1}{36}t$$

$$\frac{1}{36}t = 18$$

$$t = 648$$

The engine will run out of gas in 648 min.

59. Find the number of minutes. Round to the nearest tenth

of a minute.

$$p = 100 - \frac{30}{N}t$$

$$25 = 100 - \frac{30}{110}t$$

$$\frac{30}{110}t = 75$$

$$t = 275 \text{ seconds}$$

$$275 \text{ sec} \cdot \left(\frac{1 \text{ min}}{60 \text{ sec}}\right) \approx 4.6 \text{ min}$$

61. Find the MHR for a male and female. Round to the

nearest beat per minute.

$$\text{male} = 202 - 0.55a \qquad \text{female} = 216 - 1.09a$$

$$= 202 - 0.55(25) \qquad\quad = 216 - 1.09(25)$$

$$= 202 - 13.75 \qquad\qquad = 216 - 27.25$$

$$= 188.25 \qquad\qquad\qquad = 188.75$$

The MHR for a male who is 25 years of age is about

188 beats per minute.

The MHR for a female who is 25 years of age is about

189 beats per minute.

63. Solve.

$$\frac{\frac{1}{2}}{1+x} + \frac{1}{x} = \frac{x+3}{2}$$

$$\frac{1+x}{2} + \frac{1}{x} = \frac{x+3}{2} \qquad \text{LCD} = 2x$$

$$x(1+x) + 2 = x(x+3)$$

$$x + x^2 + 2 = x^2 + 3x$$

$$2 = 2x$$

$$1 = x$$

65. Solve.

$$|x| + |x-1| = 3$$

For $x < 0$,

$$x + x - 1 = -3$$
$$2x - 1 = -3$$
$$2x = -2$$
$$x = -1$$

For $0 \le x \le 1$, (one is positive and one is negative)

$$x - (x-1) = 3$$
$$\qquad 1 \ne 3 \quad \text{contradiction (no solution)}$$

For $x > 1$,

$$x + x - 1 = 3$$
$$2x - 1 = 3$$
$$2x = 4$$
$$x = 2$$

Prepare for Section 1.2

P1. Evaluate.

$$32 - x$$

$$32 - 8\frac{1}{2} = 23\frac{1}{2}$$

P3. Determine the property.

$$2l + 2w = 2(l + w)$$

Distributive property

P5. Add.

$$\frac{2}{5}x + \frac{1}{3}x = \frac{6}{15}x + \frac{5}{15}x = \frac{11}{15}x$$

Section 1.2 Exercises

1. Write the first step to solve $A = P + Prt$ for r.

Subtract P from each side.

3. Write the ratio.

For similar triangles, the ratio of two sides is equal to the ratio of the same two sides of the similar triangle.

$$\frac{x}{6} = \frac{2}{3}$$

5. Write the expression representing the total value.

Value of almonds: $7x$

Value of walnuts: $9(10)$, or 90

Total value of nut mixture: $7x + 90$

7. Solve the formula.

$$V = \frac{1}{3}\pi r^2 h$$

$$3V = \pi r^2 h$$

$$\frac{3V}{\pi r^2} = h$$

9. Solve the formula.

$$I = Prt$$

$$\frac{I}{Pr} = t$$

11. Solve the formula.

$$F = \frac{Gm_1 m_2}{d^2}$$

$$Fd^2 = Gm_1 m_2$$

$$\frac{Fd^2}{Gm_2} = m_1$$

13. Solve the formula.

$$a_n = a_1 + (n-1)d$$

$$a_n - a_1 = (n-1)d$$

$$\frac{a_n - a_1}{n-1} = d$$

15. Solve the formula.

$$S = \frac{a_1}{1-r}$$

$$S(1-r) = a_1$$

$$S - Sr = a_1$$

$$S - a_1 = Sr$$

$$\frac{S - a_1}{S} = r$$

17. Find Schaub's quarterback rating.

$$\text{QB rating} = \frac{100}{6}[0.05(C-30) + 0.25(Y-3)$$
$$+ 0.2T + (2.375 - 0.25I)]$$
$$= \frac{100}{6}[0.05(61.0-30) + 0.25(8.49-3)$$
$$+ 0.2(5.1) + (2.375 - 0.25 \cdot 2.05)]$$
$$= 96.75$$
$$\approx 96.8$$

19. Estimate the reading level. Round to the nearest tenth.

$$\text{SMOG} = \sqrt{42} + 3 \approx 6.5 + 3 = 9.5$$

21. Estimate the reading level. Round to the nearest tenth.

GFI $= 0.4(14.8+15.1) \approx 12.0$

23. Find the width and length.

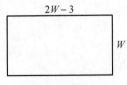

$$P = 2L+2W$$
$$174 = 2(2W-3)+2W$$
$$174 = 4W-6+2W$$
$$180 = 6W$$
$$W = 30 \text{ ft}$$
$$L = 2W-3 = 2(30)-3 = 60-3 = 57 \text{ ft}$$

25. Find the length of each side.

$$3x+3x+x = 84$$
$$7x = 84$$
$$x = 12$$
$$3x = 3(12) = 36$$

The shortest side is 12 cm.

The longer sides are each 36 cm.

27. Use similar triangles to find the height of the tree.

Let h = the height of the tree.

$$\frac{h}{10} = \frac{6}{4}$$
$$h = 15 \text{ ft}$$

29. Find the distance from the building.

Let x = the distance from the building.

$$\frac{20}{50} = \frac{20-x}{6}$$
$$300\left(\frac{20}{50}\right) = 300\left(\frac{20-x}{6}\right)$$
$$120 = 1000-50x$$
$$50x = 880$$
$$x = 17.6 \text{ ft}$$

31. Find the number of pairs of sunglasses.

Let x = the number of sunglasses.

Profit = Revenue − Cost

$$17,884 = 29.99x-8.95x$$
$$17,884 = 21.04x$$
$$x = 850$$

The manufacturer must sell 850 pairs of sunglasses to make a profit of $17,884.

33. Find the price of the book and the bookmark.

Let x = price of book

$10.10 - x$ = price of bookmark.

$$x = 10+(10.10-x)$$
$$2x = 20.10$$
$$x = 10.05$$
$$10.10-10.05 = 0.05$$

The price of the book is $10.05.

The price of the bookmark is $0.05.

35. Find the cost of the computer last year.

Let x = cost last year.

$$x-0.20x = 750$$
$$0.80x = 750$$
$$x = 937.50$$

The cost of a computer last year was $937.50.

37. Find how much was invested in each account.

Let x = amount invested at 8%.

$(14,000-x)$ = amount invested at 6.5%.

$$0.08x+0.065(14,000-x) = 1024$$
$$0.08x+910-0.065x = 1024$$
$$0.015x = 114$$
$$x = 7600$$
$$14,000-x = 6400$$

$7600 was invested at 8%.

$6400 was invested at 6.5%.

39. Find the amount invested at 8%.

5.5%	2500
8%	x
7%	$2500+x$

$$0.055(2500)+0.08x = 0.07(2500+x)$$
$$137.5+0.08x = 175+0.07x$$
$$0.01x = 37.5$$
$$x = 3750$$

$3750 additional investment at 8%.

41. Find the length of the track.

$$d = 6t \longrightarrow$$

$$\longleftarrow d = 2(160 - t)$$

Let t = the time to run to the end of the track.

Let $160 - t$ = the time in seconds to jog back.

$$6t = 2(160 - t)$$

$$6t = 320 - 2t$$

$$8t = 320$$

$$t = 40$$

$$d = 6(40) = 240 \text{ meters}$$

43. Find the time.

$$d = 240(t + 3) \longrightarrow$$

$$\longleftarrow d = 600t$$

Let t = the time (in hours) of the second plane.

Let $t + 3$ = the time (in hours) of the first plane.

$$d = 240(t + 3)$$

$$d = 600t$$

$$240(t + 3) = 600t$$

$$240t + 720 = 600t$$

$$720 = 360t$$

$$2 = t$$

$$t = 2 \text{ hours}$$

45. Find the distance.

$$d = 1865t \longrightarrow$$

$$\longleftarrow d = 1100(2 - t)$$

$$1865t = 1100(2 - t)$$

$$1865t = 2200 - 1100t$$

$$2965t = 2200$$

$$t \approx 0.741989$$

$$d = 1865t$$

$$d \approx 1383.81$$

The distance to the target is 1384 feet (to the nearest foot).

47. Find the speed of the second car.

30 seconds = 0.5 minutes = $\frac{0.5}{60}$ hour = $\frac{1}{120}$ hour.

500 meters = $\frac{1}{2}$ km.

	rate	time
Faster car	x	$\frac{1}{120}$
Slower car	80	$\frac{1}{120}$

$$x\left(\frac{1}{120}\right) - 80\left(\frac{1}{120}\right) = \frac{1}{2}$$

$$120\left[x\left(\frac{1}{120}\right) - 80\left(\frac{1}{120}\right)\right] = 120\left(\frac{1}{2}\right)$$

$$x - 80 = 60$$

$$x = 140$$

140 km/h

49. Find the number of grams of pure silver.

1.00	x
0.45	$200 - x$
0.50	200

$$x + 0.45(200 - x) = 0.50(200)$$

$$x + 90 - 0.45x = 100$$

$$0.55x = 10$$

$$x = 18\frac{2}{11} \text{ g pure silver}$$

51. Find the number of liters of water.

0	x
0.12	160
0.20	$160 - x$

$$0.12(160) - 0 = 0.20(160 - x)$$

$$19.2 = 32 - 0.20x$$

$$0.20x = 12.8$$

$$x = 64$$

64 liters of water

53. Find the number of grams of pure gold.

1.00	x
$\frac{14}{24}$ or $\frac{7}{12}$	15
$\frac{18}{24}$ or $\frac{3}{4}$	$15 + x$

$$x + \frac{7}{12}(15) = \frac{3}{4}(15 + x)$$

$$12 \cdot \left[x + \frac{7}{12}(15)\right] = 12 \cdot \left[\frac{3}{4}(15 + x)\right]$$

$$12x + 7(15) = 9(15 + x)$$

$$12x + 105 = 135 + 9x$$

$$3x = 30$$

$$x = 10$$

10 g of pure gold

55. Find the amount of each grade of tea.

\$6.50	x
\$4.25	$20-x$
\$5.60	20

$$6.50x + 4.25(20 - x) = 5.60(20)$$
$$6.50x + 85 - 4.25x = 112$$
$$2.25x + 85 = 112$$
$$2.25x = 27$$
$$x = 12 \text{ lbs of \$6.50 grade}$$
$$20 - x = 8 \text{ lbs of \$4.25 grade}$$

57. Find the number of pounds of each.

\$6	x
\$3	$25-x$
\$3.84	25

$$6x + 3(25 - x) = 3.84(25)$$
$$6x + 75 - 3x = 96$$
$$3x = 21$$
$$x = 7 \text{ lbs of cranberries}$$
$$25 - x = 18 \text{ lb of granola}$$

59. Find the number of pounds of each kind of coffee.

\$8	x
\$4	$50-x$
\$5.50	50

$$8x + 4(50 - x) = 5.50(50)$$
$$8x + 200 - 4x = 275$$
$$4x = 75$$
$$x = 18.75 \text{ lb of \$8 coffee}$$
$$50 - x = 31.25 \text{ lb of \$4 coffee}$$

61. Find the time working together.

Let t = the time it takes both electricians working together to wire the house.

The first electrician does $\frac{1}{14}$ of the job every hour.

The second electrician does $\frac{1}{18}$ of the job every hour.

$$\frac{1}{14}t + \frac{1}{18}t = 1$$
$$126\left[\frac{1}{14}t + \frac{1}{18}t\right] = 126 \cdot 1$$
$$9t + 7t = 126$$
$$16t = 126$$
$$t = 7\frac{7}{8} \text{ hours}$$

63. Find the time working together.

Let t = the time it takes both painters working together to paint the kitchen.

The painter can paint $\frac{1}{10}$ of the kitchen every hour.

The apprentice can paint $\frac{1}{15}$ of the kitchen every hour.

$$\frac{1}{10}t + \frac{1}{15}t = 1$$
$$30\left[\frac{1}{10}t + \frac{1}{15}t\right] = 30 \cdot 1$$
$$3t + 2t = 30$$
$$5t = 30$$
$$t = 6 \text{ hours}$$

65. Find the time for the older machine.

Let t = time it takes the older machine to finish the job.

The new machine does $\frac{1}{12}$ of the job every hour.

The old machine does $\frac{1}{16}$ of the job every hour.

The new machine works for 4 hours: $4\left(\frac{1}{12}\right) = \frac{1}{3}$.

The old machine completes the job.

$$\frac{1}{3} + \frac{1}{16}t = 1$$
$$\frac{1}{16}t = \frac{2}{3}$$
$$t = 10\frac{2}{3} \text{ hours}$$

67. Find the percent salt concentration.

8 kg of 30% salt has a weight of $8(0.30) = 2.4$ kg

$$8(0.30) = 2.4 \text{ kg salt}$$

$$8(0.70) = 5.6 \text{ kg water}$$

After 2 kg of water evaporates, the solution has

2.4 kg salt and 3.6 kg water, for a total weight of 6 kg.

New solution: $\dfrac{2.4}{6} = 40\%$ salt and $\dfrac{3.6}{6} = 60\%$ water.

Finally, 2 kg of 30% salt solution is added to 6 kg.

$2(0.30) = 0.6$ kg salt

The solution has a total weight of 8 kg. The weight of salt is 2.4 kg + 0.6 kg = 3 kg. The percent salt concentration is $\dfrac{3}{8} = 37.5\%$.

69. Find the distance.

To Jon's house	rate	time	distance
Up	4	$\dfrac{d_2}{4}$	d_2
Level	6	$\dfrac{d_1}{6}$	d_1
Down	12	$\dfrac{d_2}{12}$	d_2

$$\dfrac{d_2}{4} + \dfrac{d_1}{6} + \dfrac{d_1}{6} + \dfrac{d_2}{12} = 1$$
$$12\left(\dfrac{d_2}{4} + \dfrac{d_1}{6} + \dfrac{d_1}{6} + \dfrac{d_2}{12}\right) = 12(1)$$
$$3d_2 + 2d_1 + 2d_1 + d_2 = 12$$
$$4d_1 + 4d_2 = 12$$
$$4(d_1 + d_2) = 12$$
$$d_1 + d_2 = 3$$

The distance is 3 mi.

Prepare for Section 1.3

P1. Factor.

$$x^2 - x - 42 = (x+6)(x-7)$$

P3. Write in $a + bi$ form.

$$3 + \sqrt{-16} = 3 + 4i$$

P5. Evaluate.

$$\dfrac{-(-3) + \sqrt{(-3)^2 - 4(2)(1)}}{2(2)} = \dfrac{3 + \sqrt{1}}{4} = 1$$

Section 1.3 Exercises

1. From the equation $(x-8)(x+5) = 0$, the zero product property says either $x - 8 = 0$ or $x + 5 = 0$.

3. Complete the square.

a. $\left(\dfrac{1}{2} \cdot 12\right)^2 = 36$

b. $\left(\dfrac{1}{2} \cdot 9\right)^2 = \dfrac{81}{4}$

5. State the quadratic formula. Identify the discriminant.

$$x = \dfrac{-b \pm \sqrt{b^2 - 4ac}}{2a} \; ; \; b^2 - 4ac \text{ is the discriminant.}$$

7. Solve: factor and apply the zero product principle.

$$x^2 - 2x - 15 = 0$$
$$(x+3)(x-5) = 0$$
$$x + 3 = 0 \quad \text{or} \quad x - 5 = 0$$
$$x = -3 \qquad\qquad x = 5$$

9. Solve: factor and apply the zero product principle.

$$2x^2 - x = 1$$
$$2x^2 - x - 1 = 0$$
$$(2x+1)(x-1) = 0$$
$$2x + 1 = 0 \quad \text{or} \quad x - 1 = 0$$
$$2x = -1 \qquad\qquad x = 1$$
$$x = -\dfrac{1}{2}$$

11. Solve: factor and apply the zero product principle.

$$8x^2 + 189x - 72 = 0$$
$$(8x-3)(x+24) = 0$$
$$8x - 3 = 0 \quad \text{or} \quad x + 2 = 0$$
$$8x = 3 \qquad\qquad x = -24$$
$$x = \dfrac{3}{8}$$

13. Solve: factor and apply the zero product principle.

$$(x-3)(x+4) = 8$$
$$x^2 + x - 12 = 8$$
$$x^2 + x - 20 = 0$$
$$(x+5)(x-4) = 0$$
$$x + 5 = 0 \quad \text{or} \quad x - 4 = 0$$
$$x = -5 \qquad\qquad x = 4$$

15. Solve: factor and apply the zero product principle.

$$3x^2 + x - 1 = (2x+9)(x-1)$$
$$3x^2 + x - 1 = 2x^2 + 7x - 9$$
$$x^2 - 6x + 8 = 0$$
$$(x-2)(x-4) = 0$$

$$x - 2 = 0 \quad \text{or} \quad x - 4 = 0$$
$$x = 2 \qquad\qquad x = 4$$

17. Solve by the square root procedure.

$$y^2 = 24$$
$$y = \pm\sqrt{24}$$
$$y = \pm 2\sqrt{6}$$

19. Solve by the square root procedure.

$$z^2 = -16$$
$$z = \pm\sqrt{-16}$$
$$z = \pm 4i$$

21. Solve by the square root procedure.

$$(x-5)^2 = 36$$
$$x - 5 = \pm\sqrt{36}$$
$$x - 5 = \pm 6$$
$$x = 5 \pm 6$$

$$x = 5 + 6 \quad \text{or} \quad x = 5 - 6$$
$$x = 11 \qquad\qquad x = -1$$

23. Solve by the square root procedure.

$$(x+2)^2 = 27$$
$$x + 2 = \pm\sqrt{27}$$
$$x + 2 = \pm 3\sqrt{3}$$
$$x = -2 \pm 3\sqrt{3}$$

$$x = -2 + 3\sqrt{3} \quad \text{or} \quad x = -2 - 3\sqrt{3}$$

25. Solve by the square root procedure.

$$(z-4)^2 + 25 = 0$$
$$(z-4)^2 = -25$$
$$z - 4 = \pm\sqrt{-25}$$
$$z - 4 = \pm 5i$$
$$z = 4 \pm 5i$$

$$z = 4 + 5i \quad \text{or} \quad z = 4 - 5i$$

27. Solve by the square root procedure.

$$(y-6)^2 - 4 = 14$$
$$(y-6)^2 = 18$$
$$y - 6 = \pm\sqrt{18}$$
$$y - 6 = \pm 3\sqrt{2}$$
$$y = 6 \pm 3\sqrt{2}$$

$$y = 6 + 3\sqrt{2} \quad \text{or} \quad y = 6 - 3\sqrt{2}$$

29. Solve by the square root procedure.

$$5(x+6)^2 + 60 = 0$$
$$5(x+6)^2 = -60$$
$$(x+6)^2 = -12$$
$$x + 6 = \pm\sqrt{-12}$$
$$x + 6 = \pm 2i\sqrt{3}$$
$$x = -6 \pm 2i\sqrt{3}$$

$$x = -6 + 2i\sqrt{3} \quad \text{or} \quad x = -6 - 2i\sqrt{3}$$

31. Solve by the square root procedure.

$$2(x+4)^2 = 9$$
$$(x+4)^2 = \frac{9}{2}$$
$$x + 4 = \pm\sqrt{\frac{9}{2}}$$
$$x + 4 = \pm\frac{\sqrt{9}}{\sqrt{2}} \cdot \frac{\sqrt{2}}{\sqrt{2}}$$
$$x + 4 = \pm\frac{3\sqrt{2}}{2}$$
$$x = -4 \pm \frac{3\sqrt{2}}{2} = \frac{-8 \pm 3\sqrt{2}}{2}$$

$$x = \frac{-8 + 3\sqrt{2}}{2} \quad \text{or} \quad x = \frac{-8 - 3\sqrt{2}}{2}$$

33. Solve by the square root procedure.

$$4(x-2)^2 + 15 = 0$$
$$4(x-2)^2 = -15$$
$$(x-2)^2 = -\frac{15}{4}$$
$$x - 2 = \pm\sqrt{-\frac{15}{4}}$$
$$x - 2 = \pm\frac{i\sqrt{15}}{2}$$

$$x = 2 \pm \frac{i\sqrt{15}}{2} = \frac{4 \pm i\sqrt{15}}{2}$$

$$x = \frac{4 + i\sqrt{15}}{2} \quad \text{or} \quad x = \frac{4 - i\sqrt{15}}{2}$$

35. Solve by completing the square.

$$x^2 - 2x - 15 = 0$$

$$x^2 - 2x + 1 = 15 + 1$$

$$(x - 1)^2 = 16$$

$$x - 1 = \pm\sqrt{16}$$

$$x = 1 \pm 4$$

$$x = 1 + 4 \quad \text{or} \quad x = 1 - 4$$

$$x = 5 \qquad\qquad x = -3$$

37. Solve by completing the square.

$$2x^2 - 5x - 12 = 0$$

$$2x^2 - 5x = 12$$

$$x^2 - \frac{5}{2}x = 6$$

$$x^2 - \frac{5}{2}x + \frac{25}{16} = 6 + \frac{25}{16}$$

$$\left(x - \frac{5}{4}\right)^2 = \frac{121}{16}$$

$$x - \frac{5}{4} = \pm\sqrt{\frac{121}{16}}$$

$$x = \frac{5}{4} \pm \frac{11}{4}$$

$$x = \frac{5}{4} + \frac{11}{4} \quad \text{or} \quad x = \frac{5}{4} - \frac{11}{4}$$

$$x = 4 \qquad\qquad x = -\frac{3}{2}$$

39. Solve by completing the square.

$$x^2 + 6x + 1 = 0$$

$$x^2 + 6x + 9 = -1 + 9$$

$$(x + 3)^2 = 8$$

$$x + 3 = \pm\sqrt{8}$$

$$x = -3 \pm 2\sqrt{2}$$

$$x = -3 + 2\sqrt{2} \quad \text{or} \quad x = -3 - 2\sqrt{2}$$

41. Solve by completing the square.

$$x^2 + 3x - 1 = 0$$

$$x^2 + 3x + \frac{9}{4} = 1 + \frac{9}{4}$$

$$\left(x + \frac{3}{2}\right)^2 = \frac{13}{4}$$

$$x + \frac{3}{2} = \pm\sqrt{\frac{13}{4}}$$

$$x = -\frac{3}{2} \pm \frac{\sqrt{13}}{2} = \frac{-3 \pm \sqrt{13}}{2}$$

$$x = \frac{-3 + \sqrt{13}}{2} \quad \text{or} \quad x = \frac{-3 - \sqrt{13}}{2}$$

43. Solve by completing the square.

$$3x^2 - 8x = -1$$

$$x^2 - \frac{8}{3}x = -\frac{1}{3}$$

$$x^2 - \frac{8}{3}x + \frac{16}{9} = -\frac{1}{3} + \frac{16}{9}$$

$$\left(x - \frac{4}{3}\right)^2 = \frac{13}{9}$$

$$x - \frac{4}{3} = \pm\sqrt{\frac{13}{9}}$$

$$x = \frac{4}{3} \pm \frac{\sqrt{13}}{3} = \frac{4 \pm \sqrt{13}}{3}$$

$$x = \frac{4 + \sqrt{13}}{3} \quad \text{or} \quad x = \frac{4 - \sqrt{13}}{3}$$

45. Solve by completing the square.

$$x^2 + 4x + 5 = 0$$

$$x^2 + 4x + 4 = -5 + 4$$

$$(x + 2)^2 = -1$$

$$x + 2 = \pm\sqrt{-1}$$

$$x + 2 = \pm i$$

$$x = -2 \pm i$$

$$x = -2 - i \quad \text{or} \quad x = -2 + i$$

47. Solve by completing the square.

$$4x^2 + 4x + 2 = 0$$

$$4x^2 + 4x = -2$$

$$x^2 + x = -\frac{1}{2}$$

$$x^2 + x + \frac{1}{4} = -\frac{1}{2} + \frac{1}{4}$$

$$\left(x + \frac{1}{2}\right)^2 = -\frac{1}{4}$$

$$x + \frac{1}{2} = \pm\sqrt{-\frac{1}{4}}$$

$$x = -\frac{1}{2} \pm \frac{1}{2}i \text{ or } \frac{-1 \pm i}{2}$$

$$x = \frac{-1+i}{2} \text{ or } x = \frac{-1-i}{2}$$

49. Solve by completing the square.

$$3x^2 + 2x + 1 = 0$$

$$3x^2 + 2x = -1$$

$$x^2 + \frac{2}{3}x = -\frac{1}{3}$$

$$x^2 + \frac{2}{3}x + \frac{1}{9} = -\frac{1}{3} + \frac{1}{9}$$

$$\left(x + \frac{1}{3}\right)^2 = -\frac{2}{9}$$

$$x + \frac{1}{3} = \pm\sqrt{-\frac{2}{9}}$$

$$x = -\frac{1}{3} \pm \frac{\sqrt{2}}{3}i \text{ or } \frac{-1 \pm i\sqrt{2}}{3}$$

$$x = \frac{-1+i\sqrt{2}}{3} \text{ or } x = \frac{-1-i\sqrt{2}}{3}$$

51. Solve by using the quadratic formula.

$$x^2 - 2x - 15 = 0, \ a = 1, \ b = -2, \ c = -15$$

$$x = \frac{-b \pm \sqrt{b^2 - 4ac}}{2a}$$

$$x = \frac{-(-2) \pm \sqrt{(-2)^2 - 4(1)(-15)}}{2(1)}$$

$$x = \frac{2 \pm \sqrt{4+60}}{2} = \frac{2 \pm \sqrt{64}}{2}$$

$$x = \frac{2 \pm 8}{2}$$

$$x = \frac{2+8}{2} = \frac{10}{2} = 5 \quad \text{or} \quad x = \frac{2-8}{2} = \frac{-6}{2} = -3$$

$$x = 5 \text{ or } x = -3$$

53. Solve by using the quadratic formula.

$$12x^2 - 11x - 15 = 0, \ a = 12, \ b = -11, \ c = -15$$

$$x = \frac{-b \pm \sqrt{b^2 - 4ac}}{2a}$$

$$x = \frac{-(-11) \pm \sqrt{(-11)^2 - 4(12)(-15)}}{2(12)}$$

$$x = \frac{11 \pm \sqrt{121+720}}{24} = \frac{11 \pm \sqrt{841}}{24}$$

$$x = \frac{11 \pm 29}{24}$$

$$x = \frac{11+29}{24} = \frac{40}{24} = \frac{5}{3} \quad \text{or} \quad x = \frac{11-29}{24} = \frac{-18}{24} = -\frac{3}{4}$$

$$x = \frac{5}{3} \text{ or } x = -\frac{3}{4}$$

55. Solve by using the quadratic formula.

$$x^2 - 2x = 2$$

$$x^2 - 2x - 2 = 0, \ a = 1, \ b = -2, \ c = -2$$

$$x = \frac{-b \pm \sqrt{b^2 - 4ac}}{2a}$$

$$x = \frac{-(-2) \pm \sqrt{(-2)^2 - 4(1)(-2)}}{2(1)}$$

$$x = \frac{2 \pm \sqrt{4+8}}{2} = \frac{2 \pm \sqrt{12}}{2}$$

$$x = \frac{2 \pm 2\sqrt{3}}{2} = 1 \pm \sqrt{3}$$

$$x = 1 + \sqrt{3} \quad \text{or} \quad x = 1 - \sqrt{3}$$

57. Solve by using the quadratic formula.

$$x^2 = -x + 1$$

$$x^2 + x - 1 = 0, \ a = 1, \ b = 1, \ c = -1$$

$$x = \frac{-1 \pm \sqrt{1^2 - 4(1)(-1)}}{2(1)}$$

$$x = \frac{-1 \pm \sqrt{1+4}}{2} = \frac{-1 \pm \sqrt{5}}{2}$$

$$x = \frac{-1+\sqrt{5}}{2} \quad \text{or} \quad x = \frac{-1-\sqrt{5}}{2}$$

59. Solve by using the quadratic formula.

$$4x^2 = 41 - 8x$$

$$4x^2 + 8x - 41 = 0, \ a = 4, \ b = 8, \ c = -41$$

$$x = \frac{-8 \pm \sqrt{8^2 - 4(4)(-41)}}{2(4)}$$

$$x = \frac{-8 \pm \sqrt{64+656}}{8} = \frac{-8 \pm \sqrt{720}}{8}$$

$$x = \frac{-8 \pm 12\sqrt{5}}{8} = \frac{-2 \pm 3\sqrt{5}}{2}$$

$$x = \frac{-2+3\sqrt{5}}{2} \quad \text{or} \quad x = \frac{-2-3\sqrt{5}}{2}$$

61. Solve by using the quadratic formula.

$$\frac{1}{2}x^2 + \frac{3}{4}x - 1 = 0$$

$$4\left(\frac{1}{2}x^2 + \frac{3}{4}x - 1\right) = 4(0)$$

$$2x^2 + 3x - 4 = 0$$

$$x = \frac{-3 \pm \sqrt{3^2 - 4(2)(-4)}}{2(2)}$$

$$x = \frac{-3 \pm \sqrt{9 + 32}}{4}$$

$$x = \frac{-3 \pm \sqrt{41}}{4}$$

$$x = \frac{-3 + \sqrt{41}}{4} \quad \text{or} \quad x = \frac{-3 - \sqrt{41}}{4}$$

63. Solve by using the quadratic formula.

$$x^2 + 6x + 13 = 0, \ a = 1, \ b = 6, \ c = 13$$

$$x = \frac{-6 \pm \sqrt{6^2 - 4(1)(13)}}{2(1)}$$

$$x = \frac{-6 \pm \sqrt{36 - 52}}{2} = \frac{-6 \pm \sqrt{-16}}{2}$$

$$x = \frac{-6 \pm 4i}{2} = -3 \pm 2i$$

$$x = -3 - 2i \quad \text{or} \quad x = -3 + 2i$$

65. Solve by using the quadratic formula.

$$2x^2 = 2x - 13$$

$$2x^2 - 2x + 13 = 0, \ a = 1, \ b = -2, \ c = 13$$

$$x = \frac{-(-2) \pm \sqrt{(-2)^2 - 4(2)(13)}}{2(2)}$$

$$x = \frac{2 \pm \sqrt{4 - 104}}{4} = \frac{2 \pm \sqrt{-100}}{4}$$

$$x = \frac{2 \pm 10i}{4} = \frac{1 \pm 5i}{2}$$

$$x = \frac{1 - 5i}{2} \quad \text{or} \quad x = \frac{1 + 5i}{2}$$

67. Solve by using the quadratic formula.

$$x^2 + 2x + 29 = 0, \ a = 1, \ b = 2, \ c = 29$$

$$x = \frac{-2 \pm \sqrt{2^2 - 4(1)(29)}}{2(1)}$$

$$x = \frac{-2 \pm \sqrt{4 - 116}}{2} = \frac{-2 \pm \sqrt{-112}}{2}$$

$$x = \frac{-2 \pm 4i\sqrt{7}}{2} = -1 \pm 2i\sqrt{7}$$

$$x = -1 - 2i\sqrt{7} \quad \text{or} \quad x = -1 + 2i\sqrt{7}$$

69. Solve by using the quadratic formula.

$$4x^2 + 4x + 13 = 0, \ a = 4, \ b = 4, \ c = 13$$

$$x = \frac{-4 \pm \sqrt{4^2 - 4(4)(13)}}{2(4)}$$

$$x = \frac{-4 \pm \sqrt{16 - 208}}{8} = \frac{-4 \pm \sqrt{-192}}{8}$$

$$x = \frac{-4 \pm 8i\sqrt{3}}{8} = \frac{-1 \pm 2i\sqrt{3}}{2}$$

$$x = \frac{-1 - 2i\sqrt{3}}{2} \quad \text{or} \quad x = \frac{-1 + 2i\sqrt{3}}{2}$$

71. Determine the discriminant, state the number of real solutions. $2x^2 - 5x - 7 = 0$

$$b^2 - 4ac = (-5)^2 - 4(2)(-7)$$

$$= 25 + 56 = 81 > 0$$

Two real solutions

73. Determine the discriminant, state the number of real solutions. $3x^2 - 2x + 10 = 0$

$$b^2 - 4ac = (-2)^2 - 4(3)(10)$$

$$= 4 - 120 = -116 < 0$$

No real solutions

75. Determine the discriminant, state the number of real solutions. $x^2 - 20x + 100 = 0$

$$b^2 - 4ac = (-20)^2 - 4(1)(100)$$

$$= 400 - 400 = 0$$

One real solution

77. Determine the discriminant, state the number of real solutions.

$$24x^2 + 10x - 21 = 0$$

$$b^2 - 4ac = (10)^2 - 4(24)(-21)$$

$$= 100 + 2016 = 2116 > 0$$

Two real solutions

79. Determine the discriminant, state the number of real

solutions. $12x^2 + 15x + 7 = 0$

$$b^2 - 4ac = (15)^2 - 4(12)(7)$$
$$= 225 - 336 = -111 < 0$$

No real solutions

81. Find the distance from the left side. Round to the nearest

tenth of a foot.

$$h = 0.045x^2 - 1.33x + 20$$
$$11 = 0.045x^2 - 1.33x + 20$$
$$0 = 0.045x^2 - 1.33x + 9$$
$$a = 0.045, \ b = -1.33, \ c = 9$$

$$x = \frac{1.33 \pm \sqrt{(-1.33)^2 - 4(0.045)(9)}}{2(0.045)}$$

$$= \frac{1.33 \pm \sqrt{0.1489}}{0.09} \approx \frac{1.33 \pm 0.4}{0.09}$$

$$x = \frac{1.33 + 0.4}{0.09} \qquad \text{or} \qquad x = \frac{1.33 - 0.4}{0.09}$$

$$x \approx 19.1 \text{ ft} \qquad\qquad x \approx 10.5 \text{ ft}$$

The suspension cable hangs 11 ft above the footbridge

10.5 ft and 19.1 ft from the left side of the bridge.

83. Find the depth of the water. Round to the nearest tenth

of a foot.

$$V = 32d^2 + 32d$$
$$850 = 32d^2 + 32d$$
$$0 = 32d^2 + 32d - 850$$
$$0 = 2(16d^2 + 16d - 425)$$
$$a = 16, \ b = 16 \ c = -425$$

$$x = \frac{-16 \pm \sqrt{16^2 - 4(16)(-425)}}{2(16)}$$

$$= \frac{-16 \pm \sqrt{27,456}}{32} \approx \frac{-16 \pm 165.7}{32}$$

$$x = \frac{-16 + 165.7}{32} \qquad \text{or} \qquad x = \frac{-16 - 165.7}{32}$$

$$= \frac{149.7}{32} \approx 4.7 \qquad\qquad = \frac{-181.7}{32} \approx -5.7$$

The depth is 4.7 ft.

(Reject negative x-value; x must be positive.)

85. Find the distance.

$$h = -0.0114x^2 + 1.732x$$
$$10 = -0.0114x^2 + 1.732x$$
$$0 = -0.0114x^2 + 1.732x - 10$$
$$a = -0.0114, \ b = 1.732, \ c = -10$$

$$x = \frac{-1.732 \pm \sqrt{1.732^2 - 4(-0.0114)(-10)}}{2(-0.0114)}$$

$$x = \frac{-1.732 \pm \sqrt{2.999824 - 0.456}}{-0.0228}$$

$$x = \frac{-1.732 \pm \sqrt{2.543824}}{-0.0228}$$

$$x = \frac{-1.732 + \sqrt{2.543824}}{-0.0228} \approx 6.0$$

$$x = \frac{-1.732 - \sqrt{2.543824}}{-0.0228} \approx 145.9$$

Since $x = 6.0$ ft is not realistic, it is not a solution.

Convert from feet to yards: $145.9 \text{ ft} \cdot \dfrac{1 \text{ yd}}{3 \text{ ft}} \approx 48.6$ yd

The kicker can be up to 48.6 yd from the goalpost.

87. Find the number of items sold.

$$R = xp$$
$$16,500 = x(26 - 0.01x)$$
$$16,500 = 26x - 0.01x^2$$
$$0 = -0.01x^2 + 26x - 16,500$$
$$a = -0.01, \ b = 26, \ c = -16,500$$

$$x = \frac{-26 \pm \sqrt{26^2 - 4(-0.01)(-16,500)}}{2(-0.01)}$$

$$= \frac{-26 \pm \sqrt{16}}{-0.02} = \frac{-26 \pm 4}{-0.02}$$

$$x = \frac{-26 + 4}{-0.02} = 1100 \qquad \text{or} \qquad x = \frac{-26 - 4}{-0.02} = 1500$$

1100 or 1500 items must be sold.

89. Find the dimensions.

Let w = width of region

Then $\dfrac{132 - 3w}{2}$ = length.

$$\text{Area} = \text{length(width)}$$

$$576 = \frac{132 - 3w}{2} \cdot w$$

$$1152 = 132w - 3w^2$$

$$3w^2 - 132w + 1152 = 0$$
$$3(w^2 - 44w + 384) = 0$$
$$w^2 - 44w + 384 = 0$$
$$(w - 32)(w - 12) = 0$$

$$w - 32 = 0 \quad \text{or} \quad w - 12 = 0$$
$$w = 32 \qquad\qquad w = 12$$
$$\frac{132 - 3w}{2} = \frac{132 - 3(32)}{2} \quad \frac{132 - 3w}{2} = \frac{132 - 3(12)}{2}$$
$$= 18 \qquad\qquad\qquad = 48$$

The region is either 32 feet wide and 18 feet long, or 12 feet wide and 48 feet long.

91. Find the distances.

Solve $D = -45x^2 + 190x + 200$ for x with $D = 250$.

$$250 = -45x^2 + 190x + 200$$
$$0 = -45x^2 + 190x - 50$$
$$a = -45, \ b = 190, \ c = -50$$
$$x = \frac{-190 \pm \sqrt{190^2 - 4(-45)(-50)}}{2(-45)}$$
$$= \frac{-190 \pm \sqrt{27100}}{-90} \approx \frac{-190 \pm 164.2}{-90}$$
$$x \approx \frac{-190 + 164.2}{-90} \quad \text{or} \quad x \approx \frac{-190 - 164.2}{-90}$$
$$\approx 0.3 \text{ mile} \qquad\qquad \approx 3.9 \text{ miles}$$

93. Find the time in the air.

Solve $h = -16t^2 + 25.3t + 20$ for t where $h = 17$.

$$17 = -16t^2 + 25.3t + 20$$
$$0 = -16t^2 + 25.3t + 3$$
$$t = \frac{-25.3 \pm \sqrt{(25.3)^2 - 4(-16)(3)}}{2(-16)} = \frac{-25.3 \pm \sqrt{832.09}}{-32}$$
$$t = 1.7 \quad \text{or} \quad t = -0.11$$

He was in the air for 1.7 s.

95. Find the times.

Solve $h = -16t^2 + 220t$ for t where $h = 350$.

$$350 = -16t^2 + 220t$$
$$0 = -16t^2 + 220t - 350$$
$$a = -16, \ b = 220, \ c = -350$$

$$t = \frac{-220 \pm \sqrt{220^2 - 4(-16)(-350)}}{2(-16)}$$
$$= \frac{-220 \pm \sqrt{26000}}{-32} \approx \frac{-220 \pm 161.245}{-32}$$
$$t \approx \frac{-220 + 161.245}{-32} \quad \text{or} \quad t \approx \frac{-220 - 161.245}{-32}$$
$$\approx 1.8 \text{ seconds} \qquad\qquad \approx 11.9 \text{ seconds}$$

97. Determine whether the baseball will clear the fence.

Solve $s = 103.9t$ for t where $s = 360$ to find the time it takes the ball to reach the fence.

$$360 = 103.9t$$
$$t \approx 3.465 \text{ seconds}$$

Next, evaluate $h = -16t^2 + 50t + 4.5$ where $t = 3.465$ to determine if the ball is at least 10 feet in the air when it reaches the fence.

$$h = -16(3.465)^2 + 50(3.465) + 4.5$$
$$h \approx -14.3$$

No, the ball will not clear the fence.

99. Find the year.

Solve $M = 0.0001t^2 + 0.16t + 4.34$ for t where $M = 10$.

$$10 = 0.0001t^2 + 0.16t + 4.34$$
$$0 = 0.0001t^2 + 0.16t - 5.66$$
$$a = 0.0001, \ b = 0.16, \ c = -5.66$$

$$t = \frac{-0.16 \pm \sqrt{0.16^2 - 4(0.0001)(-5.66)}}{2(0.0001)}$$
$$= \frac{-0.16 \pm \sqrt{0.027864}}{0.0002}$$
$$= \frac{-0.16 \pm 0.166925}{0.0002}$$
$$= \frac{-0.16 + 0.166925}{0.0002} \quad \text{reject the negative value}$$
$$\approx 34.5 \text{ years from 2000, or 2034}$$

101. Find the year.

Solve $N = 0.3453x^2 - 9.417x + 164.1$ for x,

where $N = 200$.

$$200 = 0.3453x^2 - 9.417x + 164.1$$

$$0 = 0.3453x^2 - 9.417x - 35.9$$

$$a = 0.3453, \ b = -9.417, \ c = -35.9$$

$$x = \frac{-(-9.417) \pm \sqrt{(-9.417)^2 - 4(0.3453)(-35.9)}}{2(0.3453)}$$

$$= \frac{9.417 \pm \sqrt{138.265}}{0.6906} \approx \frac{9.417 \pm 11.759}{0.6906}$$

$$x \approx \frac{9.417 + 11.759}{0.6906} \quad \text{or} \quad x \approx \frac{9.417 - 11.759}{0.6906}$$

$$\approx 30.663 \qquad\qquad \approx -3.39 \text{ (not in the future)}$$

There will be more than 200,000 centenarians living in the US in about 31 years from 2000, or 2031.

103. Verify the discriminant is the same for both definitions.

a. coefficient definition:

$$x^2 - x - 6 = 0$$

$$a = 1, \ b = -1, \ c = -6$$

$$b^2 - 4ac = (-1)^2 - 4(1)(-6) = 25$$

roots definition:

$$x^2 - x - 6 = 0$$

$$(x + 2)(x - 3) = 0$$

$$x + 2 = 0 \quad \text{or} \quad x - 3 = 0$$

$$x = -2 \qquad\qquad x = 3$$

$$a = 1, \ r_1 = -2, \ r_2 = 3$$

$$a^2(r_1 - r_2)^2 = (1)^2(-2 - 3)^2 = 25$$

b. coefficient definition:

$$9x^2 - 6x - 1 = 0$$

$$a = 9, \ b = -6, \ c = -1$$

$$b^2 - 4ac = (-6)^2 - 4(9)(-1) = 72$$

roots definition:

$$x = \frac{-(-6) \pm \sqrt{(-6)^2 - 4(9)(-1)}}{2(9)}$$

$$= \frac{6 \pm \sqrt{72}}{18} = \frac{6 \pm 6\sqrt{2}}{18}$$

$$= \frac{1 \pm \sqrt{2}}{3}$$

$$a = 9, \ r_1 = \frac{1 + \sqrt{2}}{3}, \ r_2 = \frac{1 - \sqrt{2}}{3}$$

$$a^2(r_1 - r_2)^2 = (9)^2 \left[\frac{1 + \sqrt{2}}{3} - \frac{1 - \sqrt{2}}{3} \right]^2$$

$$= 81 \left(\frac{1 + \sqrt{2} - 1 + \sqrt{2}}{3} \right)^2$$

$$= 81 \left(\frac{2\sqrt{2}}{3} \right)^2 = 81 \cdot \frac{8}{9}$$

$$= 72$$

c. coefficient definition:

$$x^2 + 4x + 4 = 0$$

$$a = 1, \ b = 4, \ c = 4$$

$$b^2 - 4ac = (4)^2 - 4(1)(4) = 0$$

roots definition:

$$x^2 + 4x + 4 = 0$$

$$(x + 2)(x + 2) = 0$$

$$x + 2 = 0 \quad \text{or} \quad x + 2 = 0$$

$$x = -2 \qquad\qquad x = -2$$

$$a = 1, \ r_1 = -2, \ r_2 = -2$$

$$a^2(r_1 - r_2)^2 = (1)^2[-2 - (-2)]^2 = 0$$

105. Find the discriminant.

$$x^3 - 4x^2 - 4x + 16 = 0$$

$$x^2(x - 4) - 4(x - 4) = 0$$

$$(x^2 - 4)(x - 4) = 0$$

$$(x + 2)(x - 2)(x - 4) = 0$$

$$x + 2 = 0 \quad \text{or} \quad x - 2 = 0 \quad \text{or} \quad x - 4 = 0$$

$$x = -2 \qquad\qquad x = 2 \qquad\qquad x = 4$$

$$a = 1, \ r_1 = -2, \ r_2 = 2, \ r_3 = 4$$

$$a^4(r_1 - r_2)^2(r_1 - r_3)^2(r_2 - r_3)^2$$

$$= 1^4(-2 - 2)^2(-2 - 4)^2(2 - 4)^2$$

$$= (16)(36)(4)$$

$$= 2304$$

Mid-Chapter 1 Quiz

1. Solve.

$$6 - 4(2x+1) = 5(3-2x)$$
$$6 - 8x - 4 = 15 - 10x$$
$$2 - 8x = 15 - 10x$$
$$2x = 13$$
$$x = \frac{13}{2}$$

3. Solve by factoring.

$$x^2 - 5x = 6$$
$$x^2 - 5x - 6 = 0$$
$$(x+1)(x-6) = 0$$

$x = -1$ or $x = 6$

5. Solve by quadratic formula.

$$x^2 - 6x + 12 = 0$$
$$a = 1, \ b = -6, \ c = 12$$

$$x = \frac{-(-6) \pm \sqrt{(-6)^2 - 4(1)(12)}}{2(1)}$$

$$x = \frac{6 \pm \sqrt{-12}}{2} = \frac{6 \pm 2i\sqrt{3}}{2} = 3 \pm i\sqrt{3}$$

$$x = 3 - i\sqrt{3} \ \text{or} \ x = 3 + i\sqrt{3}$$

7. Find the amount of each solution.

9%	x
4%	$500 - x$
6%	500

$$0.09x + 0.04(500 - x) = 0.06(500)$$
$$0.09x + 20 - 0.04x = 30$$
$$0.05x = 10$$
$$x = 200$$
$$500 - x = 300$$

200 ml of 9% solution and 300 ml of 4% solution.

Prepare for Section 1.4

P1. Factor.

$$x^3 - 16x = x(x^2 - 16)$$
$$= x(x+4)(x-4)$$

P3. Evaluate.

$$8^{2/3} = \left(\sqrt[3]{8}\right)^2 = 2^2 = 4$$

P5. Multiply.

$$\left(1 + \sqrt{x-5}\right)^2 = 1^2 + 2\sqrt{x-5} + \left(\sqrt{x-5}\right)^2$$
$$= 1 + 2\sqrt{x-5} + x - 5$$
$$= x + 2\sqrt{x-5} - 4$$

Section 1.4 Exercises

1. Write the solutions.

Using the principle of zero products,

$$x(x-3)(x+2) = 0$$

$x = 0, \ x = 3, \ \text{or} \ x = -2$

3. Find the expression to clear fractions.

For $\dfrac{1}{x-5} + 3 = \dfrac{x+1}{x-5}$, multiply each side by $x - 5$.

5. To solve $x^{2/3} = 9$ for x, raise each side to the $\dfrac{3}{2}$

power.

7. Solve by factoring and using principle of zero products.

$$x^3 - 25x = 0$$
$$x(x^2 - 25) = 0$$
$$x(x-5)(x+5) = 0$$

$x = 0, \ x = 5, \ \text{or} \ x = -5$

9. Solve by factoring and using principle of zero products.

$$x^3 - 2x^2 - x + 2 = 0$$
$$x^2(x-2) - (x-2) = 0$$
$$(x-2)(x^2 - 1) = 0$$
$$(x-2)(x-1)(x+1) = 0$$

$x = 2, \ x = 1, \ \text{or} \ x = -1$

11. Solve by factoring and using principle of zero products.

$$x^3 - 3x^2 - 5x + 15 = 0$$
$$x^2(x-3) - 5(x-3) = 0$$
$$(x-3)(x^2 - 5) = 0$$

$x - 3 = 0 \quad \text{or} \quad x^2 - 5 = 0$
$\qquad x = 3 \qquad\qquad x^2 = 5$
$\qquad\qquad\qquad\qquad x = \pm\sqrt{5}$

$x = 3, \ x = -\sqrt{5}, \ x = \sqrt{5}$

13. Solve by factoring and using principle of zero products.

$$3x^3 + 2x^2 - 27x - 18 = 0$$
$$x^2(3x+2) - 9(3x+2) = 0$$
$$(3x+2)(x^2-9) = 0$$
$$(3x+2)(x+3)(x-3) = 0$$
$$x = -\frac{2}{3}, \ x = -3, \ x = 3$$

15. Solve by factoring and using principle of zero products.

$$x^3 - 8 = 0$$
$$(x-2)(x^2+2x+4) = 0$$

$$x = 2, \text{ or } x^2 + 2x + 4 = 0$$

$$x^2 + 2x = -4 \quad \text{Use completing the square.}$$
$$x^2 + 2x + 1 = -4 + 1$$
$$(x+1)^2 = -3$$
$$x + 1 = \pm\sqrt{-3}$$
$$x = -1 \pm i\sqrt{3}$$

Thus the solutions are $2, \ -1 + i\sqrt{3}, \ -1 - i\sqrt{3}$.

17. Solve the rational equation.

$$\frac{5}{x+4} - 2 = \frac{7x+18}{x+4}$$
$$(x+4)\left(\frac{5}{x+4} - 2\right) = (x+4)\left(\frac{7x+18}{x+4}\right)$$
$$5 - 2(x+4) = 7x + 18$$
$$5 - 2x - 8 = 7x + 18$$
$$-2x - 3 = 7x + 18$$
$$-9x = 21$$
$$x = -\frac{7}{3}$$

19. Solve the rational equation.

$$2 + \frac{9}{r-3} = \frac{3r}{r-3}$$
$$(r-3)\left(2 + \frac{9}{r-3}\right) = (r-3)\left(\frac{3r}{r-3}\right)$$
$$2(r-3) + 9 = 3r$$
$$2r - 6 + 9 = 3r$$
$$2r + 3 = 3r$$
$$2r - 3r = -3$$
$$-r = -3$$
$$r = 3$$

No solution because each side is undefined when $r = 3$.

21. Solve the rational equation.

$$\frac{3}{x+2} = \frac{5}{2x-7}$$
$$3(2x-7) = 5(x+2)$$
$$6x - 21 = 5x + 10$$
$$6x - 5x = 10 + 21$$
$$x = 31$$

23. Solve the rational equation.

$$x - \frac{2x+3}{x+3} = \frac{2x+9}{x+3}$$
$$(x+3)\left(x - \frac{2x+3}{x+3}\right) = (x+3)\left(\frac{2x+9}{x+3}\right)$$
$$x(x+3) - (2x+3) = 2x + 9$$
$$x^2 + 3x - 2x - 3 = 2x + 9$$
$$x^2 - x - 12 = 0$$
$$(x+3)(x-4) = 0$$

$$x = -3 \text{ or } x = 4$$

4 checks as a solution. -3 is not in the domain and does not check as a solution.

25. Solve the rational equation.

$$\frac{5}{x-3} - \frac{3}{x-2} = \frac{4}{x-3}$$
$$(x-3)(x-2)\left(\frac{5}{x-3} - \frac{3}{x-2}\right) = (x-3)(x-2)\left(\frac{4}{x-3}\right)$$
$$5(x-2) - 3(x-3) = 4(x-2)$$
$$5x - 10 - 3x + 9 = 4x - 8$$
$$2x - 1 = 4x - 8$$
$$2x - 4x = -8 + 1$$
$$-2x = -7$$
$$x = \frac{7}{2}$$

27. Solve the rational equation.

$$\frac{x}{x+1} - \frac{x+2}{x-1} = \frac{x-12}{x+1}$$
$$(x+1)(x-1)\left(\frac{x}{x+1} - \frac{x+2}{x-1}\right) = (x+1)(x-1)\left(\frac{x-12}{x+1}\right)$$
$$x(x-1) - (x+1)(x+2) = (x-1)(x-12)$$
$$x^2 - x - x^2 - 3x - 2 = x^2 - 13x + 12$$
$$-4x - 2 = x^2 - 13x + 12$$
$$0 = x^2 - 9x + 14$$
$$0 = (x-2)(x-7)$$
$$x = 2 \text{ or } x = 7$$

29. Solve the rational equation.

$$\frac{3-2x}{x+3} - \frac{2x+1}{x-4} = \frac{5x-29}{x-4}$$

$$(x+3)(x-4)\left(\frac{3-2x}{x+3} - \frac{2x+1}{x-4}\right) = (x+3)(x-4)\left(\frac{5x-29}{x-4}\right)$$

$$(x-4)(3-2x)-(x+3)(2x+1) = (x+3)(5x-29)$$

$$-2x^2+11x-12-2x^2-7x-3 = 5x^2-14x-87$$

$$-4x^2+4x-15 = 5x^2-14x-87$$

$$0 = 9x^2-18x-72$$

$$0 = 9(x^2-2x-8)$$

$$0 = 9(x-4)(x+2)$$

$$x=4 \text{ or } x=-2$$

–2 checks as a solution. 4 is not in the domain and does not check as a solution.

31. Solve the radical equation.

$$\sqrt{x-4}-6 = 0$$

$$\sqrt{x-4} = 6$$

$$x-4 = 36$$

$$x = 40$$

Check $\sqrt{40-4}-6 = 0$

$$\sqrt{36}-6 = 0$$

$$6-6 = 0$$

$$0 = 0$$

The solution is 40.

33. Solve the radical equation.

$$\sqrt{9x-20} = x$$

$$\left(\sqrt{9x-20}\right)^2 = x^2$$

$$9x-20 = x^2$$

$$0 = x^2-9x+20$$

$$0 = (x-4)(x-5)$$

$$x=4 \text{ or } x=5$$

Check $\sqrt{9(4)-20} = 4$ $\sqrt{9(5)-20} = 5$

$$\sqrt{16} = 4 \qquad\qquad \sqrt{25} = 5$$

$$4 = 4 \qquad\qquad\quad 5 = 5$$

4 and 5 check as solutions.

35. Solve the radical equation.

$$\sqrt{-7x+2}+x = 2$$

$$\sqrt{-7x+2} = 2-x$$

$$\left(\sqrt{-7x+2}\right)^2 = (2-x)^2$$

$$-7x+2 = 4-4x+x^2$$

$$0 = x^2+3x+2$$

$$0 = (x+2)(x+1)$$

$$x=-2 \text{ or } x=-1$$

Check

$$\sqrt{-7(-2)+2}+(-2) = 2 \qquad \sqrt{-7(-1)+2}+(-1) = 2$$

$$\sqrt{16}-2 = 2 \qquad\qquad\qquad \sqrt{9}-1 = 2$$

$$4-2 = 2 \qquad\qquad\qquad\qquad 3-1 = 2$$

$$2 = 2 \qquad\qquad\qquad\qquad\quad 2 = 2$$

–2 and –1 check as solutions.

37. Solve the radical equation.

$$\sqrt{3x-5}-\sqrt{x+2} = 1$$

$$(\sqrt{3x-5})^2 = (1+\sqrt{x+2})^2$$

$$3x-5 = 1+2\sqrt{x+2}+x+2$$

$$2x-8 = 2\sqrt{x+2}$$

$$(x-4)^2 = (\sqrt{x+2})^2$$

$$x^2-8x+16 = x+2$$

$$x^2-9x+14 = 0$$

$$(x-7)(x-2) = 0$$

$$x=7 \text{ or } x=2$$

Check $\sqrt{3(7)-5}-\sqrt{7+2} = 1$

$$\sqrt{16}-\sqrt{9} = 1$$

$$4-3 = 1$$

$$1 = 1$$

$$\sqrt{3(2)-5}-\sqrt{2+2} = 1$$

$$\sqrt{1}-\sqrt{4} = 1$$

$$1-2 = 1$$

$$-1 = 1 \quad \text{(No)}$$

The solution is 7.

39. Solve the radical equation.

$$\sqrt{2x+11} - \sqrt{2x-5} = 2$$
$$(\sqrt{2x+11})^2 = (2+\sqrt{2x-5})^2$$
$$2x+11 = 4+4\sqrt{2x-5}+2x-5$$
$$12 = 4\sqrt{2x-5}$$
$$(3)^2 = (\sqrt{2x-5})^2$$
$$9 = 2x-5$$
$$14 = 2x$$
$$7 = x$$

Check $\sqrt{2(7)+11} - \sqrt{2(7)-5} = 2$
$$\sqrt{25} - \sqrt{9} = 2$$
$$5-3 = 2$$
$$2 = 2$$

7 checks as the solution.

41. Solve the radical equation.

$$\sqrt{x-4} + \sqrt{x+1} = 1$$
$$\sqrt{x-4} = 1-\sqrt{x+1}$$
$$(\sqrt{x-4})^2 = (1-\sqrt{x+1})^2$$
$$x-4 = 1-2\sqrt{x+1}+x+1$$
$$2\sqrt{x+1} = 6$$
$$(2\sqrt{x+1})^2 = 6^2$$
$$4(x+1) = 36$$
$$4x+4 = 36$$
$$4x = 32$$
$$x = 8$$

Check $\sqrt{8-4} + \sqrt{8+1} = 1$
$$\sqrt{4} + \sqrt{9} = 1$$
$$2+3 = 1$$
$$5 = 1 \text{ (No)}$$

There is no solution.

43. Solve the radical equation.

$$\sqrt{2x-1} - \sqrt{x-1} = 1$$
$$\sqrt{2x-1} = 1+\sqrt{x-1}$$
$$(\sqrt{2x-1})^2 = (1+\sqrt{x-1})^2$$
$$2x-1 = 1+2\sqrt{x-1}+x-1$$
$$x-1 = 2\sqrt{x-1}$$
$$(x-1)^2 = (2\sqrt{x-1})^2$$
$$x^2-2x+1 = 4(x-1)$$
$$x^2-2x+1 = 4x-4$$
$$x^2-6x+5 = 0$$
$$(x-1)(x-5) = 0$$
$$x = 1 \text{ or } x = 5$$

Check $\sqrt{2(1)-1} - \sqrt{(1)-1} = 1 \qquad \sqrt{2(5)-1} - \sqrt{5-1} = 1$
$$\sqrt{1} - \sqrt{0} = 1 \qquad\qquad \sqrt{9} - \sqrt{4} = 1$$
$$1 = 1 \qquad\qquad\qquad 3-2 = 1$$
$$1 = 1$$

1 and 5 check as solutions.

45. Solve.

$$x^{1/3} = 2$$
$$(x^{1/3})^3 = (2)^3$$
$$x = 8$$

47. Solve.

$$x^{2/5} = 9$$
$$(x^{2/5})^{5/2} = (9)^{5/2}$$
$$|x| = 243$$
$$x = -243, \ 243$$

49. Solve.

$$x^{3/2} = 27$$
$$(x^{3/2})^{2/3} = (27)^{2/3}$$
$$x = 9$$

51. Solve.

$$3x^{2/3} - 16 = 59$$
$$3x^{2/3} = 75$$
$$x^{2/3} = 25$$
$$(x^{2/3})^{3/2} = (25)^{3/2}$$
$$|x| = 125$$
$$x = -125, \ 125$$

53. Solve.

$$4x^{3/4} - 31 = 77$$
$$4x^{3/4} = 108$$
$$x^{3/4} = 27$$
$$\left(x^{3/4}\right)^{4/3} = (27)^{4/3}$$
$$x = 81$$

55. Find all real solutions.

$$x^4 - 9x^2 + 14 = 0$$

Let $u = x^2$.

$$u^2 - 9u + 14 = 0$$
$$(u - 7)(u - 2) = 0$$

$$u = 7 \qquad \text{or} \qquad u = 2$$
$$x^2 = 7 \qquad\qquad x^2 = 2$$
$$x = \pm\sqrt{7} \qquad\qquad x = \pm\sqrt{2}$$

The solutions are $\sqrt{7}, \ -\sqrt{7}, \ \sqrt{2}, \ -\sqrt{2}$.

57. Find all real solutions.

$$2x^4 - 11x^2 + 12 = 0$$

Let $u = x^2$.

$$2u^2 - 11u + 12 = 0$$
$$(2u - 3)(u - 4) = 0$$

$$u = \frac{3}{2} \qquad\qquad \text{or} \quad u = 4$$
$$x^2 = \frac{3}{2} \qquad\qquad\qquad x^2 = 4$$
$$x = \pm\sqrt{\frac{3}{2}} = \pm\frac{\sqrt{6}}{2} \qquad x = \pm 2$$

The solutions are $\dfrac{\sqrt{6}}{2}, \ -\dfrac{\sqrt{6}}{2}, \ 2, \ -2$.

59. Find all real solutions.

$$x^6 + x^3 - 6 = 0$$

Let $u = x^3$.

$$u^2 + u - 6 = 0$$
$$(u - 2)(u + 3) = 0$$

$$u = 2 \quad \text{or} \quad u = -3$$
$$x^3 = 2 \qquad\quad x^3 = -3$$
$$x = \sqrt[3]{2} \qquad x = \sqrt[3]{-3} = -\sqrt[3]{3}$$

The solutions are $\sqrt[3]{2}$ and $-\sqrt[3]{3}$.

61. Find all real solutions.

$$x^{1/2} - 3x^{1/4} + 2 = 0$$

Let $u = x^{1/4}$.

$$u^2 - 3u + 2 = 0$$
$$(u - 1)(u - 2) = 0$$

$$u = 1 \quad \text{or} \quad u = 2$$
$$x^{1/4} = 1 \qquad x^{1/4} = 2$$
$$x = 1 \qquad\quad x = 16$$

The solutions are 1 and 16.

63. Find all real solutions.

$$3x^{2/3} - 11x^{1/3} - 4 = 0$$

Let $u = x^{1/3}$.

$$3u^2 - 11u - 4 = 0$$
$$(3u + 1)(u - 4) = 0$$

$$u = -\frac{1}{3} \quad \text{or} \quad u = 4$$
$$x^{1/3} = -\frac{1}{3} \qquad x^{1/3} = 4$$
$$x = -\frac{1}{27} \qquad\quad x = 64$$

The solutions are $-\dfrac{1}{27}$ and 64.

65. Find all real solutions.

$$x^4 + 8x^2 - 9 = 0$$

Let $u = x^2$.

$$u^2 + 8u - 9 = 0$$
$$(u + 9)(u - 1) = 0$$

$$u = -9 \qquad \text{or} \quad u = 1$$
$$x^2 = -9 \qquad\qquad x^2 = 1$$
$$x = \pm\sqrt{-9} \qquad\quad x = \pm 1$$
$$x = \pm 3i$$

The solutions are $-1, 1, -3i, 3i$.

67. Find all real solutions.

$$x^{2/5} - x^{1/5} - 2 = 0$$

Let $u = x^{1/5}$.

$$u^2 - u - 2 = 0$$
$$(u+1)(u-2) = 0$$

$$u = -1 \quad \text{or} \quad u = 2$$
$$x^{1/5} = -1 \qquad x^{1/5} = 2$$
$$x = -1 \qquad x = 32$$

The solutions are -1 and 32.

69. Find all real solutions.

$$9x - 52\sqrt{x} + 64 = 0$$

Let $\sqrt{x} = u$.

$$9u^2 - 52u + 64 = 0$$
$$(9u - 16)(u - 4) = 0$$

$$u = \frac{16}{9} \quad \text{or} \quad u = 4$$
$$\sqrt{x} = \frac{16}{9} \qquad \sqrt{x} = 4$$
$$x = \frac{256}{81} \qquad x = 16$$

The solutions are $\dfrac{256}{81}$ and 16.

71. Find the speed of the boat.

Using the formula $t = \dfrac{d}{r}$, we get the equation.

$$\frac{24}{v+3} + \frac{24}{v-3} = 6$$
$$(v+3)(v-3)\left(\frac{24}{v+3} + \frac{24}{v-3}\right) = (v+3)(v-3)6$$
$$24(v-3) + 24(v+3) = (v^2 - 9)(6)$$
$$24v - 72 + 24v + 72 = 6v^2 - 54$$
$$48v = 6v^2 - 54$$
$$0 = 6v^2 - 48v - 54$$
$$0 = 6(v^2 - 8v - 9)$$
$$0 = 6(v+1)(v-9)$$
$$v = -1 \ (\text{No}) \quad \text{or} \quad v = 9$$

The speed is 9 mph.

73. Find the time it takes the assistant, working alone.

Let $x =$ the number of hours the assistant would take to build the fence working alone.

The worker does $\frac{1}{8}$ of the job per hour; the assistant does $\frac{1}{x}$ of the job per hour.

worker	$\frac{1}{8}$	5
assistant	$\frac{1}{x}$	5

$$\left(\frac{1}{8}\right)(5) + \left(\frac{1}{x}\right)(5) = 1$$
$$\frac{5}{8} + \frac{5}{x} = 1$$
$$8x\left(\frac{5}{8} + \frac{5}{x}\right) = 1(8x)$$
$$5x + 40 = 8x$$
$$40 = 3x$$
$$\frac{40}{3} = x$$
$$x = 13\frac{1}{3} \text{ hours}$$

75. Find the time for the experienced painter.

Let $t =$ the time for the experienced painter.

$2t - 5$ is the time for the apprentice painter.

$\dfrac{1}{t}$ is the rate for the experienced painter.

$\dfrac{1}{2t-5}$ is the rate for the apprentice painter.

$$\frac{1}{t}\cdot 6 + \frac{1}{2t-5}\cdot 6 = 1$$
$$t(2t-5)\left(\frac{6}{t} + \frac{6}{2t-5}\right) = t(2t-5)\cdot 1$$
$$6(2t-5) + 6t = 2t^2 - 5t$$
$$12t - 30 + 6t = 2t^2 - 5t$$
$$0 = 2t^2 - 23t + 30$$
$$0 = (2t-3)(t-10)$$
$$t = \frac{3}{2} \ (\text{No}) \quad \text{or} \quad t = 10$$

It takes the experience painter 10 hours.

77. Find the time.

$$T = \frac{\sqrt{s}}{4} + \frac{s}{1100}$$

$$T = \frac{\sqrt{7100}}{4} + \frac{7100}{1100}$$

$$T \approx 27.5 \text{ seconds}$$

79. Find the radius.

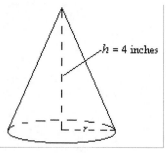

$$L = \pi r \sqrt{r^2 + h^2}$$

$$15\pi = \pi r \sqrt{r^2 + 4^2}$$

$$15 = r \sqrt{r^2 + 16}$$

$$225 = r^2(r^2 + 16)$$

$$0 = r^4 + 16r^2 - 225$$

Let $u = r^2$.

$$u^2 + 16u - 225 = 0$$

$$(u + 25)(u - 9) = 0$$

$$u = 9 \quad \text{or} \quad u = -25 \text{ (No)}$$

$$r^2 = 9$$

$$r = 3$$

The radius is 3 in.

81. Find the length of the edge.

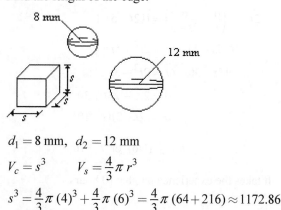

$$d_1 = 8 \text{ mm}, \quad d_2 = 12 \text{ mm}$$

$$V_c = s^3 \qquad V_s = \frac{4}{3}\pi r^3$$

$$s^3 = \frac{4}{3}\pi (4)^3 + \frac{4}{3}\pi (6)^3 = \frac{4}{3}\pi (64 + 216) \approx 1172.86$$

$$s \approx 10.5 \text{ mm}$$

The side is approximately 10.5 mm.

83. Find the height.

$$d = \sqrt{1.5h}$$

$$14 = \sqrt{1.5h}$$

$$14^2 = 1.5h$$

$$196 = 1.5h$$

$$131 \approx h$$

The height is approximately 131 ft.

85. Find the distance.

$$\frac{16 - x}{22} + \frac{\sqrt{16 + x^2}}{7} = 2$$

$$154\left(\frac{16 - x}{22} + \frac{\sqrt{16 + x^2}}{7}\right) = 154(2)$$

$$7(16 - x) + 22\sqrt{16 + x^2} = 308$$

$$112 - 7x + 22\sqrt{16 + x^2} = 308$$

$$22\sqrt{16 + x^2} = 7x + 196$$

$$\left(22\sqrt{16 + x^2}\right)^2 = (7x + 196)^2$$

$$484(16 + x^2) = 49x^2 + 2744x + 38,416$$

$$7744 + 484x^2 = 49x^2 + 2744x + 38,416$$

$$435x^2 - 2744x - 30,672 = 0$$

Use quadratic formula to solve for x.

$$x = \frac{-(-2744) \pm \sqrt{(-2744)^2 - 4(435)(-30,672)}}{2(435)}$$

$$x = \frac{2744 \pm \sqrt{60,898,816}}{870}$$

$$x = \frac{2744 - \sqrt{60,898,816}}{870} \approx -5.8 \text{ km (No)}$$

$$x = \frac{2744 + \sqrt{60,898,816}}{870} \approx 12.1 \text{ km}$$

The distance is about 12.1 km.

87. Solve.

$$x^4 - 2x^3 + 27x - 54 = 0$$

$$x^3(x - 2) + 27(x - 2) = 0$$

$$(x - 2)(x^3 + 27) = 0$$

$$(x - 2)(x + 3)(x^2 - 3x + 9) = 0$$

$$x = \frac{-(-3) \pm \sqrt{(-3)^2 - 4(1)(9)}}{2(1)}$$

$$x = \frac{3 \pm \sqrt{-27}}{2} = \frac{3 \pm 3i\sqrt{3}}{2}$$

The solutions are $2, -3, \dfrac{3 - 3i\sqrt{3}}{2}, \dfrac{3 + 3i\sqrt{3}}{2}$.

89. Solve.

$$\sqrt[3]{x^3 - 2x - 13} = x - 1$$

$$\left(\sqrt[3]{x^3 - 2x - 13}\right)^3 = (x-1)^3$$

$$x^3 - 2x - 13 = x^3 - 3x^2 + 3x - 1$$

$$3x^2 - 5x - 12 = 0$$

$$(3x + 4)(x - 3) = 0$$

$$3x + 4 = 0 \quad \text{or} \quad x - 3 = 0$$

$$x = -\frac{4}{3} \qquad\qquad x = 3$$

Check $\sqrt[3]{\left(-\frac{4}{3}\right)^3 - 2\left(-\frac{4}{3}\right) - 13} = -\frac{4}{3} - 1$

$$\sqrt[3]{-\frac{343}{27}} = -\frac{7}{3}$$

$$-\frac{7}{3} = -\frac{7}{3}$$

$$\sqrt[3]{(3)^3 - 2(3) - 13} = 3 - 1$$

$$\sqrt[3]{8} = 2$$

$$2 = 2$$

Both $-\frac{4}{3}$ and 3 check as solutions.

91. Show that $\dfrac{AD}{AB} = \phi$.

Label the point G that is between B and C.

Note that $EG = ED$, by construction.

Let $x = FD$.

Then $EG = 1 + x$.

$$EG = \sqrt{EF^2 + FG^2} = \sqrt{1^2 + 2^2} = \sqrt{5}$$

$$\sqrt{5} = 1 + x$$

$$x = \sqrt{5} - 1$$

$$\frac{AD}{AB} = \frac{2 + x}{2} = \frac{2 + \sqrt{5} - 1}{2} = \frac{1 + \sqrt{5}}{2} = \phi$$

Prepare for Section 1.5

P1. Find $\{x | x > 2\} \cap \{x | x > 5\}$.

$$\{x \mid x > 5\}$$

P3. Evaluate.

$$\frac{7+3}{7-2} = 2$$

P5. Find the value of x where the expression is undefined.

$$\frac{x-3}{2x-7}, \quad 2x - 7 \neq 0$$

It is undefined for $x = \dfrac{7}{2}$.

Section 1.5 Exercises

1. State whether the inequalities are equivalent.

a. No, $x > -3$ is not equivalent to $x > 0$.

b. $3x \leq -6$ can be simplified as $x \leq -2$ which is not equivalent to $x \geq -2$.

c. $-2x < 0$ can be simplified as $x > 0$ which is equivalent to $x > 0$.

d. $\dfrac{2}{3}x \geq -6$ can be simplified as $x \geq -9$ which is equivalent to $x \geq -9$.

3. Rewrite $|x - 5| \leq 8$.

$$x - 5 \geq -8 \text{ and } x - 5 \leq 8$$

5. Find the solution set in set builder notation and graph.

$$2x + 3 < 11$$

$$2x < 11 - 3$$

$$2x < 8$$

$$x < 4$$

$$\{x | x < 4\}$$

7. Find the solution set in set builder notation and graph.

$$x + 4 > 3x + 16$$

$$-2x > 12$$

$$x < -6$$

$$\{x | x < -6\}$$

9. Find the solution set in set builder notation and graph.

$$-3(x + 2) \leq 5x + 7$$

$$-3x - 6 \leq 5x + 7$$

$$-8x \leq 13$$

$$x \geq -\frac{13}{8}$$

$$\left\{x \middle| x \geq -\frac{13}{8}\right\}$$

11. Find the solution set in set builder notation and graph.

$$-4(3x-5)>2(x-4)$$
$$-12x+20>2x-8$$
$$-14x>-28$$
$$x<2$$

$$\{x|x<2\}$$

13. Find the solution set in set builder notation and graph.

$$4x+1>-2 \quad \text{and} \quad 4x+1\leq17$$
$$4x>-3 \quad \text{and} \quad 4x\leq16$$
$$x>-\frac{3}{4} \quad \text{and} \quad x\leq4$$

$$\left\{x\middle|x>-\frac{3}{4}\right\}\cap\left\{x\middle|x\leq4\right\}=\left\{x\middle|-\frac{3}{4}<x\leq4\right\}$$

15. Find the solution set in set builder notation and graph.

$$10\geq3x-1\geq0$$
$$11\geq \quad 3x \quad \geq1$$
$$\frac{11}{3}\geq \quad x \quad \geq\frac{1}{3}$$

$$\left\{x\middle|\frac{1}{3}\leq x\leq\frac{11}{3}\right\}$$

17. Find the solution set in set builder notation and graph.

$$x+2<-1 \quad \text{or} \quad x+3\geq2$$
$$x<-3 \quad \text{or} \quad x\geq-1$$

$$\{x|x<-3\}\cup\{x|x\geq-1\}=\{x|x<-3 \quad \text{or} \quad x\geq-1\}$$

19. Find the solution set in set builder notation and graph.

$$-4x+5>9 \quad \text{or} \quad 4x+1<5$$
$$-4x>4 \quad \text{or} \quad 4x<4$$
$$x<-1 \quad \text{or} \quad x<1$$

$$\{x|x<-1\}\cup\{x|x<1\}=\{x|x<1\}$$

21. Find the solution set in interval notation

$$|2x-1|>4$$

$$2x-1<-4 \quad \text{or} \quad 2x-1>4$$
$$2x<-3 \quad \text{or} \quad 2x>5$$
$$x<-\frac{3}{2} \quad \text{or} \quad x>\frac{5}{2}$$

$$\left(-\infty, -\frac{3}{2}\right)\cup\left(\frac{5}{2}, \infty\right)$$

23. Find the solution set in interval notation

$$|x+3|\geq5$$

$$x+3\leq-5 \quad \text{or} \quad x+3\geq5$$
$$x\leq-8 \quad \text{or} \quad x\geq2$$

$$(-\infty, -8]\cup[2, \infty)$$

25. Find the solution set in interval notation

$$|3x-10|\leq14$$

$$-14\leq3x-10\leq14$$
$$-4\leq \quad 3x \quad \leq24$$
$$-\frac{4}{3}\leq \quad x \quad \leq8$$

$$\left[-\frac{4}{3}, 8\right]$$

27. Find the solution set in interval notation

$$|4-5x|\geq24$$

$$4-5x\leq-24 \quad \text{or} \quad 4-5x\geq24$$
$$-5x\leq-28 \quad \text{or} \quad -5x\geq20$$
$$x\geq\frac{28}{5} \quad \text{or} \quad x\leq-4$$

$$(-\infty, -4]\cup\left[\frac{28}{5}, \infty\right)$$

29. Find the solution set in interval notation

$$|x-5|\geq0 \qquad \text{(Note: The absolute value of } any$$

real number is greater than or equal to 0.)

$$(-\infty, \infty)$$

31. Find the solution set in interval notation

$$|x-4|\leq0$$

(Note: No absolute value is less than 0.)

$$x-4=0$$
$$x=4$$

$$\{4\}$$

33. Use the critical value method to solve. Write solution in interval notation.

$$x^2 + 7x > 0$$
$$x(x+7) > 0$$

The product $x(x+7)$ is positive.

$x = 0$ is a critical value.
$x + 7 = 0 \Rightarrow x = -7$ is a critical value.

$x(x+7)$ $\quad + + + + | - - - - - - - | + + + +$
$\qquad\qquad\qquad \overset{-7}{} \qquad \overset{0}{}$

$(-\infty,\ -7) \cup (0,\ \infty)$

35. Use the critical value method to solve. Write solution in interval notation.

$$x^2 - 16 \le 0$$
$$(x-4)(x+4) \le 0$$

The product $(x-4)(x+4)$ is negative or zero.

$x - 4 = 0 \Rightarrow x = 4$ is a critical value.
$x + 4 = 0 \Rightarrow x = -4$ is a critical value.

$(x-4)(x+4)$ $\quad + + + + | - - - - - - - | + + +$
$\qquad\qquad\qquad \overset{-4}{} \quad \overset{0}{} \quad \overset{4}{}$

$[-4,\ 4]$

37. Use the critical value method to solve. Write solution in interval notation.

$$x^2 + 7x + 10 < 0$$
$$(x+5)(x+2) < 0$$

The product $(x+5)(x+2)$ is negative.

$x + 5 = 0 \Rightarrow x = -5$ is a critical value.
$x + 2 = 0 \Rightarrow x = -2$ is a critical value.

$(x+5)(x+2)$ $\quad + + + + | - - - | + + + + + +$
$\qquad\qquad\qquad \overset{-5}{} \quad \overset{-2}{} \ \overset{0}{}$

$(-5, -2)$

39. Use the critical value method to solve. Write solution in interval notation.

$$x^2 - 3x \ge 28$$
$$x^2 - 3x - 28 \ge 0$$
$$(x-7)(x+4) \ge 0$$

The product $(x-7)(x+4)$ is positive or zero.

$x - 7 = 0 \Rightarrow x = 7$ is a critical value.
$x + 4 = 0 \Rightarrow x = -4$ is a critical value.

$(x-7)(x+4)$ $\quad + + | - - - - - - - - - - - | + +$
$\qquad\qquad\qquad \overset{-4}{} \quad \overset{0}{} \qquad \overset{7}{}$

$(-\infty,\ -4] \cup [7,\ \infty)$

41. Use the critical value method to solve. Write solution in interval notation.

$$x^3 - x^2 - 16x + 16 < 0$$
$$x^2(x-1) - 16(x-1) < 0$$
$$(x-1)(x^2 - 16) < 0$$
$$(x-1)(x+4)(x-4) < 0$$

The product $(x-1)(x+4)(x-4)$ is negative.

$x - 1 = 0 \Rightarrow x = 1$ is a critical value.
$x - 4 = 0 \Rightarrow x = 4$ is a critical value.
$x + 4 = 0 \Rightarrow x = -4$ is a critical value.

The critical values are –4, 1, and 4.

$(x-1)(x+4)(x-4)$ $\quad - - | + + + + | - - - - | + +$
$\qquad\qquad\qquad\qquad \overset{-4}{} \quad \overset{1}{} \quad \overset{4}{}$

$(-\infty,\ -4) \cup (1,\ 4)$

43. Use the critical value method to solve. Write solution in interval notation.

$$x^4 - 20x^2 + 64 \ge 0$$
$$(x^2 - 16)(x^2 - 4) \ge 0$$
$$(x+4)(x-4)(x+2)(x-2) \ge 0$$

The product is positive or zero.

$x + 4 = 0 \Rightarrow x = -4$ is a critical value.
$x - 4 = 0 \Rightarrow x = 4$ is a critical value.
$x + 2 = 0 \Rightarrow x = -2$ is a critical value.
$x - 2 = 0 \Rightarrow x = 2$ is a critical value.

The critical values are –4, –2, 2, and 4.

$(x+4)(x-4)(x+2)(x-2)$ $\quad + + | - - - - | + + + | - - - - | + +$
$\qquad\qquad\qquad\qquad\qquad \overset{-4}{} \quad \overset{-2}{} \quad \overset{2}{} \quad \overset{4}{}$

$(-\infty,\ -4] \cup [-2,\ 2] \cup [4,\ \infty)$

45. Use the critical value method to solve. Write solution in interval notation.

$$\frac{x+4}{x-1} < 0$$

The quotient $\dfrac{x+4}{x-1}$ is negative.

$$x+4=0 \Rightarrow x=-4$$
$$x-1=0 \Rightarrow x=1$$

The critical values are –4 and 1.

$\dfrac{x+4}{x-1}$ $\quad$ + + + + + + |– – – – –|+ + +

$\qquad\qquad$ –4 $\quad$ 1

(–4, 1)

47. Use the critical value method to solve. Write solution in interval notation.

$$\frac{x-5}{x+8} \geq 3$$
$$\frac{x-5}{x+8} - 3 \geq 0$$
$$\frac{x-5-3(x+8)}{x+8} \geq 0$$
$$\frac{x-5-3x-24}{x+8} \geq 0$$
$$\frac{-2x-29}{x+8} \geq 0$$

The quotient $\dfrac{-2x-29}{x+8}$ is positive or zero.

$$-2x-29=0 \Rightarrow x=-\frac{29}{2}$$
$$x+8=0 \Rightarrow x=-8$$

The critical values are $-\dfrac{29}{2}$ and -8.

$\dfrac{-2x-29}{x+8}$ $\quad$ – – –|+ + + + +|– – – – – –

$\qquad\qquad -\frac{29}{2} \quad -8 \quad 0$

The denominator cannot equal zero $\Rightarrow x \neq -8$.

$$\left[-\frac{29}{2},\ -8\right)$$

49. Use the critical value method to solve. Write solution in interval notation.

$$\frac{x}{2x+7} \geq 4$$
$$\frac{x}{2x+7} - 4 \geq 0$$
$$\frac{x-4(2x+7)}{2x+7} \geq 0$$
$$\frac{x-8x-28}{2x+7} \geq 0$$
$$\frac{-7x-28}{2x+7} \geq 0$$

The quotient $\dfrac{-7x-28}{2x+7}$ is positive or zero.

$$-7x-28=0 \Rightarrow x=-4$$
$$2x+7=0 \Rightarrow x=-\frac{7}{2}$$

The critical values are -4 and $-\dfrac{7}{2}$.

$\dfrac{-7x-28}{2x+7}$ $\quad$ – – –|+|– – – – – – – – – – –

$\qquad\qquad -4\,-\frac{7}{2} \qquad 0$

The denominator cannot equal zero $\Rightarrow x \neq -\dfrac{7}{2}$.

$$\left[-4, -\frac{7}{2}\right)$$

51. Use the critical value method to solve. Write solution in interval notation.

$$\frac{(x+1)(x-4)}{x-2} < 0$$

The quotient $\dfrac{(x+1)(x-4)}{x-2}$ is negative.

$$x+1=0 \Rightarrow x=-1$$
$$x-4=0 \Rightarrow x=4$$
$$x-2=0 \Rightarrow x=2$$

The critical values are –1, 4, and 2.

$\dfrac{(x+1)(x-4)}{x-2}$ $\quad$ – – – –|+ + +|– –|+ + + +

$\qquad\qquad$ –1 0 2 $\quad$ 4

$(-\infty,\ -1) \cup (2,\ 4)$

53. Use the critical value method to solve. Write solution in interval notation.

$$\frac{x+2}{x-5} \leq 2$$
$$\frac{x+2}{x-5} - 2 \leq 0$$
$$\frac{x+2-2(x-5)}{x-5} \leq 0$$
$$\frac{x+2-2x+10}{x-5} \leq 0$$
$$\frac{-x+12}{x-5} \leq 0$$

The quotient $\dfrac{-x+12}{x-5}$ is negative or zero.

$$-x+12=0 \Rightarrow x=12$$
$$x-5=0 \Rightarrow x=5$$

The critical values are 12 and 5.

$\dfrac{-x+12}{x-5}$

The denominator cannot equal zero $\Rightarrow x \neq 5$.

$(-\infty, 5) \cup [12, \infty)$

55. Use the critical value method to solve. Write solution in interval notation.

$$\dfrac{6x^2 - 11x - 10}{x} > 0$$

$$\dfrac{(3x+2)(2x-5)}{x} > 0$$

The quotient $\dfrac{(3x+2)(2x-5)}{x}$ is positive.

$3x + 2 = 0 \Rightarrow x = -\dfrac{2}{3}$

$2x - 5 = 0 \Rightarrow x = \dfrac{5}{2}$

$x = 0$

The critical values are $-\dfrac{2}{3}$, $\dfrac{5}{2}$, and 0.

$\dfrac{(3x+2)(2x-5)}{x}$

$\left(-\dfrac{2}{3}, 0\right) \cup \left(\dfrac{5}{2}, \infty\right)$

57. Use the critical value method to solve. Write solution in interval notation.

$$\dfrac{x^2 - 6x + 9}{x-5} \leq 0$$

$$\dfrac{(x-3)(x-3)}{x-5} \leq 0$$

The quotient $\dfrac{(x-3)(x-3)}{x-5}$ is negative or zero.

$x - 3 = 0 \Rightarrow x = 3$

$x - 5 = 0 \Rightarrow x = 5$

The critical values are 3 and 5.

$\dfrac{(x-3)(x-3)}{x-5}$

The denominator cannot equal zero $\Rightarrow x \neq 5$.

$(-\infty, 5)$

59. Find the conditions to use the LowCharge plan.

LowCharge: $5 + 0.01x$

FeeSaver: $1 + 0.08x$

$5 + 0.01x < 1 + 0.08x$

$\qquad 4 < 0.07x$

$\quad 57.1 < x$

LowCharge is less expensive if you use more than 57 checks.

61. Find the range of heights, h.

Let h = the height of the package.

$$\text{length} + \text{girth} \leq 165$$

$$\text{length} + 2(\text{width}) + 2(\text{height}) \leq 165$$

$$47 + 2(22) + 2h \leq 165$$

$$47 + 44 + 2h \leq 165$$

$$91 + 2h \leq 165$$

$$2h \leq 74$$

$$h \leq 37$$

The height must be more than 0 but less than or equal to 37 inches.

63. Find the mileage if the value is $41,000.

$$41,000 = -176.05m + 50,520$$

$$-9520 = -176.05m$$

$$m \approx 54$$

The mileage is about 54,000 miles.

65. Estimate the range of heights.

Solve $|h - (2.47f + 54.10)| \leq 3.72$ for h

where $f = 32.24$.

$$|h - (2.47f + 54.10)| \leq 3.72$$

$$|h - [2.47(32.24) + 54.10]| \leq 3.72$$

$$|h - (79.6328 + 54.10)| \leq 3.72$$

$$|h - 133.7328| \leq 3.72$$

$h - 133.7328 \leq 3.72 \qquad$ or $\qquad h - 133.7328 \geq -3.72$

$\qquad h \leq 137.4528 \qquad\qquad\qquad h \geq 130.0128$

The height, to the nearest 0.1 cm, is from 130.0 cm to 137.5 cm.

67. Find the interval where monthly revenue is greater than zero.

$$R = 420x - 2x^2$$

$$420x - 2x^2 > 0$$

$$2x(210 - x) > 0$$

The product is positive.

$$2x = 0 \Rightarrow x = 0$$

$$210 - x = 0 \Rightarrow x = 210$$

Critical values are 0 and 210.

$2x(210 - x)$ `----|+++++++++|---`

($0, $210)

69. Find the number of books.

$$\frac{14.25x + 350,000}{x} < 50$$

$$14.25x + 350,000 < 50x$$

$$-35.75x < -350,000$$

$$x > 9790.2$$

At least 9791 books must be published.

71. Find the time interval.

$$s = -16t^2 + v_0t + s_0, \quad s > 96, \ t > 0, \ v_0 = 80, \ s_0 = 32$$

$$-16t^2 + 80t + 32 > 96$$

$$-16t^2 + 80t - 64 > 0$$

$$-16(t^2 - 5t + 4) > 0$$

$$-16(t - 1)(t - 4) > 0$$

The product is positive. The critical values are 1 and 4.

$(t - 1)(t - 4)$ `|-|+++|----`

1 second $< t <$ 4 seconds

The ball is higher than 96 ft between 1 and 4 seconds.

73. Find the range of mean weights of men.

$$-2.575 < \frac{190 - \mu}{2.45} < 2.575$$

$$-6.30875 < 190 - \mu < 6.30875$$

$$-196.30875 < -\mu < -183.69125$$

$$196.30875 > \mu > 183.69125$$

$$183.7 \text{ lb} < \mu < 196.3 \text{ lb}$$

Prepare for Section 1.6

P1. Solve.

$$1820 = k(28)$$

$$65 = k$$

P3. Evaluate.

$$k \cdot \frac{3}{5^2}$$

$$225 \cdot \frac{3}{5^2} = 27$$

P5. The area becomes 4 times as large.

Section 1.6 Exercises

1. Write the equation of variation.

$$d = kt$$

3. Write the equation of variation.

$$y = \frac{k}{x}$$

5. Write the equation of variation.

$$m = knp$$

7. Write the equation of variation.

$$V = klwh$$

9. Write the equation of variation.

$$A = ks^2$$

11. Write the equation of variation.

$$F = \frac{km_1m_2}{d^2}$$

13. Write the equation and solve for k.

$$y = kx$$

$$64 = k \cdot 48$$

$$\frac{64}{48} = k$$

$$\frac{4}{3} = k$$

15. Write the equation and solve for k.

$$r = kt^2$$

$$144 = k \cdot 108^2$$

$$\frac{144}{108^2} = k$$

$$\frac{2^4 \cdot 3^2}{2^4 \cdot 3^6} = k$$

$$\frac{1}{81} = k$$

17. Write the equation and solve for k.

$$T = krs^2$$
$$210 = k \cdot 30 \cdot 5^2$$
$$\frac{210}{30 \cdot 5^2} = k$$
$$\frac{7}{25} = k$$
$$0.28 = k$$

19. Write the equation and solve for k.

$$V = klwh$$
$$240 = k \cdot 8 \cdot 6 \cdot 5$$
$$\frac{240}{8 \cdot 6 \cdot 5} = k$$
$$1 = k$$

21. Find the volume of the balloon.

$$V = kT$$
$$0.85 = k \cdot 270$$
$$\frac{0.85}{270} = k$$
$$\frac{0.17}{54} = k$$

Thus $V = \dfrac{0.17}{54}T = \dfrac{0.17}{54} \cdot 324 = (0.17)6 = 1.02$ liters

23. Find the number of semester hours.

$$s = k \cdot q$$
$$34 = k \cdot 51$$
$$\frac{2}{3} = k$$
$$p = \frac{2}{3} \cdot 93$$
$$p = 62 \text{ semester hours}$$

25. Find the amount of juice. Round to the nearest tenth of a fluid ounce.

$$j = k \cdot d^3$$
$$6 = k \cdot (4)^3$$
$$\frac{3}{32} = k$$
$$p = \frac{3}{32} \cdot (5)^3$$
$$p \approx 11.7 \text{ fl oz}$$

27.
$$T = k\sqrt{l}$$
$$1.8 = k\sqrt{3}$$
$$k = \frac{1.8}{\sqrt{3}} \approx 1.03923$$

a. $T = \dfrac{1.8}{\sqrt{3}}\sqrt{10} = \dfrac{1.8\sqrt{30}}{3} = 0.6\sqrt{30} \approx 3.3$ seconds

b.
$$T = k\sqrt{l}$$
$$\frac{T}{k} = \sqrt{l}$$
$$\frac{2}{1.03923} = \sqrt{l}$$
$$l = \frac{4}{1.03923^2} \approx 3.7 \text{ ft}$$

29. Find the speed of the gear with 48 teeth.

$$r = \frac{k}{t}$$
$$30 = \frac{k}{64}$$
$$1920 = k$$
$$r = \frac{1920}{48}$$
$$r = 40 \text{ revolutions per minute}$$

31. Find the sound intensity.

$$I = \frac{k}{d^2}$$
$$0.5 = \frac{k}{7^2}$$
$$24.5 = k$$
$$I = \frac{24.5}{d^2}$$
$$I = \frac{24.5}{10^2}$$
$$I = 0.245 \text{ W/m}^2$$

33. a. $V = kr^2h$
$$V_1 = k(3r)^2 h$$
$$= 9(kr^2h)$$
$$= 9V$$

Thus the new volume is 9 times the original volume.

b. $V_2 = kr^2(3h)$
$$= 3(kr^2h)$$
$$= 3V$$

Thus the new volume is 3 times the original volume.

c. $V_3 = k(3r)^2(3h)$
$$= k9r^2 \cdot 3 \cdot h$$
$$= 27(kr^2h)$$
$$= 27V$$

Thus the new volume is 27 times the original volume.

35. Find what happens to the volume.

$$V = \frac{knT}{P}$$

$$V_1 = \frac{k(3n)T}{\left(\frac{1}{2}p\right)}$$

$$= 6\left(\frac{knT}{p}\right)$$

$$= 6V$$

Thus the new volume is 6 times larger than the original volume.

37. a. Find the variation constant.

For Kershaw,

$$\text{ERA} = \frac{kr}{i}$$

$$2.28 = \frac{k(59)}{(233.1)}$$

$$9.0 = k$$

b. Find Wilson's ERA.

$$\text{ERA} = \frac{9(73)}{(223.1)} = 2.94$$

39. Find the force. Round to the nearest 10 lb.

$$F = \frac{kws^2}{r}$$

$$2800 = \frac{k \cdot 1800 \cdot 45^2}{425}$$

$$\frac{2800 \cdot 425}{1800 \cdot 45^2} = k$$

$$\frac{14 \cdot 425}{9 \cdot 45^2} = k$$

$$0.3264746 \approx k$$

Thus $F = \dfrac{(0.3264746) \cdot 1800 \cdot 55^2}{450} \approx 3950$ pounds

41. Find the distance. Round to the nearest million miles.

$$t = kd^{3/2}$$

$$365 = k \cdot 93^{3/2}$$

$$\frac{365}{93^{3/2}} = k$$

Thus $\quad 686 = \dfrac{365}{93^{3/2}} \cdot d^{3/2}$

$$\frac{686 \cdot 93^{3/2}}{365} = d^{3/2}$$

$$\left(\frac{686 \cdot 93^{3/2}}{365}\right)^{2/3} = d$$

$$93\left(\frac{686}{365}\right)^{2/3} = d$$

$$142 \text{ million miles} \approx d$$

Chapter 1 Review Exercises

1. Solve. [1.1]

$$4 - 5x = 3x + 14$$

$$-8x = 10$$

$$x = -\frac{5}{4}$$

3. Solve. [1.1]

$$\frac{4x}{3} - \frac{4x-1}{6} = \frac{1}{2}$$

$$6\left(\frac{4x}{3} - \frac{4x-1}{6}\right) = 6\left(\frac{1}{2}\right)$$

$$2(4x) - (4x-1) = 3$$

$$8x - 4x + 1 = 3$$

$$4x + 1 = 3$$

$$4x = 2$$

$$x = \frac{1}{2}$$

5. Solve. [1.1]

$$|x - 3| = 2$$

$$x - 3 = 2 \quad \text{or} \quad x - 3 = -2$$

$$x = 5 \qquad\qquad x = 1$$

7. Solve. [1.1]

$$|2x + 1| = 5$$

$$2x + 1 = 5 \quad \text{or} \quad 2x + 1 = -5$$

$$2x = 4 \qquad\qquad 2x = -6$$

$$x = 2 \qquad\qquad x = -3$$

9. Solve. [1.2]

$$V = \pi r^2 h$$

$$\frac{V}{\pi r^2} = h$$

11. Solve. [1.2]

$$A = \frac{h}{2}(b_1 + b_2)$$
$$2A = h(b_1 + b_2)$$
$$2A = hb_1 + hb_2$$
$$2A - hb_2 = hb_1$$
$$\frac{2A - hb_2}{h} = b_1$$

13. Solve. [1.3]

$$x^2 - 5x + 6 = 0$$
$$(x - 2)(x - 3) = 0$$

$$x - 2 = 0 \quad \text{or} \quad x - 3 = 0$$
$$x = 2 \qquad\qquad x = 3$$

15. Solve. [1.3]

$$(x - 2)^2 = 50$$
$$x - 2 = \pm\sqrt{50}$$
$$x = 2 \pm 5\sqrt{2}$$

$$x = 2 + 5\sqrt{2} \quad \text{or} \quad x = 2 - 5\sqrt{2}$$

17. Solve. [1.3]

$$x^2 - 6x - 1 = 0$$
$$x^2 - 6x = 1$$
$$x^2 - 6x + 9 = 1 + 9$$
$$(x - 3)^2 = 10$$
$$x - 3 = \pm\sqrt{10}$$
$$x = 3 \pm \sqrt{10}$$

$$x = 3 + \sqrt{10} \quad \text{or} \quad x = 3 - \sqrt{10}$$

19. Solve. [1.3]

$$3x^2 - x - 1 = 0 \quad \text{Use the quadratic formula.}$$

$$x = \frac{-(-1) \pm \sqrt{(-1)^2 - 4(3)(-1)}}{2(3)}$$

$$x = \frac{1 \pm \sqrt{13}}{6}$$

$$x = \frac{1 + \sqrt{13}}{6} \quad \text{or} \quad x = \frac{1 - \sqrt{13}}{6}$$

21. Determine whether the equations has real or complex solutions. [1.3]

$$2x^2 + 4x = 5$$
$$2x^2 + 4x - 5 = 0$$

$$b^2 - 4ac = (4)^2 - 4(2)(-5)$$
$$= 16 + 40 = 56 > 0$$

There are two real number solutions.

23. Solve the equation. [1.4]

$$3x^3 - 5x^2 = 0$$
$$x^2(3x - 5) = 0$$

$$x^2 = 0 \Rightarrow x = 0$$
$$3x - 5 = 0 \Rightarrow x = \frac{5}{3}$$
$$x = 0 \quad \text{or} \quad x = \frac{5}{3}$$

25. Solve the equation. [1.4]

$$2x^3 + 3x^2 - 8x - 12 = 0$$
$$x^2(2x + 3) - 4(2x + 3) = 0$$
$$(2x + 3)(x^2 - 4) = 0$$
$$(2x + 3)(x + 2)(x - 2) = 0$$

$$x = -\frac{3}{2}, \ x = -2, \ \text{or} \ x = 2$$

27. Solve the equation. [1.4]

$$\frac{x}{x + 2} + \frac{1}{4} = 5$$

$$4(x + 2)\left(\frac{x}{x + 2} + \frac{1}{4}\right) = 5(4)(x + 2)$$

$$4x + x + 2 = 20(x + 2)$$
$$5x + 2 = 20x + 40$$
$$-15x = 38$$

$$x = -\frac{38}{15}$$

29. Solve the equation. [1.4]

$$3x + \frac{2}{x - 2} = \frac{4x - 1}{x - 2}$$

$$(x - 2)\left(3x + \frac{2}{x - 2}\right) = (x - 2)\left(\frac{4x - 1}{x - 2}\right)$$

$$3x(x - 2) + 2 = 4x - 1$$
$$3x^2 - 6x + 2 = 4x - 1$$
$$3x^2 - 10x + 3 = 0$$
$$(3x - 1)(x - 3) = 0$$

$$3x - 1 = 0 \quad \text{or} \quad x - 3 = 0$$
$$x = \frac{1}{3} \qquad \qquad x = 3$$

31. Solve the equation. [1.4]

$$\sqrt{2x + 6} - 1 = 3$$
$$\sqrt{2x + 6} = 4$$
$$\left(\sqrt{2x + 6}\right)^2 = 4^2$$
$$2x + 6 = 16$$
$$2x = 10$$
$$x = 5$$

Check $\sqrt{2(5) + 6} - 1 = 3$
$$\sqrt{16} - 1 = 3$$
$$4 - 1 = 3$$
$$3 = 3$$

5 checks as a solution.

33. Solve the equation. [1.4]

$$\sqrt{-2x - 7} + 2x = -7$$
$$\sqrt{-2x - 7} = -2x - 7$$
$$\left(\sqrt{-2x - 7}\right)^2 = (-2x - 7)^2$$
$$-2x - 7 = 4x^2 + 28x + 49$$
$$0 = 4x^2 + 30x + 56$$
$$0 = 2\left(2x^2 + 15x + 28\right)$$
$$0 = 2(2x + 7)(x + 4)$$

$$x = -\frac{7}{2} \text{ or } x = -4$$

Check $\sqrt{-2\left(-\frac{7}{2}\right) - 7} + 2\left(-\frac{7}{2}\right) = -7$
$$\sqrt{7 - 7} - 7 = -7$$
$$-7 = -7$$
$$\sqrt{-2(-4) - 7} + 2(-4) = -7$$
$$\sqrt{8 - 7} - 8 = -7$$
$$1 - 8 = -7$$
$$-7 = -7$$

$-\frac{7}{2}$ and -4 check as solutions.

35. Solve the equation. [1.4]

$$\sqrt{3x + 4} + \sqrt{x - 3} = 5$$
$$\sqrt{3x + 4} = 5 - \sqrt{x - 3}$$
$$\left[\sqrt{3x + 4}\right]^2 = \left[5 - \sqrt{x - 3}\right]^2$$
$$3x + 4 = 25 - 10\sqrt{x - 3} + x - 3$$
$$2x - 18 = -10\sqrt{x - 3}$$
$$x - 9 = -5\sqrt{x - 3}$$
$$(x - 9)^2 = \left[-5\sqrt{x - 3}\right]^2$$
$$x^2 - 18x + 81 = 25(x - 3)$$
$$x^2 - 18x + 81 = 25x - 75$$
$$x^2 - 43x + 156 = 0$$
$$(x - 4)(x - 39) = 0$$

$$x = 4 \quad \text{or} \quad x = 39$$

Check $\sqrt{3(4) + 4} + \sqrt{4 - 3} = 5$
$$\sqrt{16} + \sqrt{1} = 5$$
$$4 + 1 = 5$$
$$5 = 5$$
$$\sqrt{3(39) + 4} + \sqrt{39 - 3} = 5$$
$$\sqrt{121} + \sqrt{36} = 5$$
$$11 + 6 = 5$$
$$17 = 5 \ (\text{No})$$

The solution is 4.

37. Solve the equation. [1.4]

$$x^{5/4} - 32 = 0$$
$$x^{5/4} = 32$$
$$\left(x^{5/4}\right)^{4/5} = (32)^{4/5}$$
$$x = 16$$

39. Solve the equation. [1.4]

$$6x^4 - 23x^2 + 20 = 0$$

Let $u = x^2$.

$$6u^2 - 23u + 20 = 0$$
$$(3u - 4)(2u - 5) = 0$$

$$a = \frac{4}{3} \qquad \text{or} \qquad u = \frac{5}{2}$$

$$x^2 = \frac{4}{3} \qquad\qquad x^2 = \frac{5}{2}$$

$$x = \pm\sqrt{\frac{4}{3}} \qquad\qquad x = \pm\sqrt{\frac{5}{2}}$$

$$x = \pm\frac{2}{\sqrt{3}}\left(\frac{\sqrt{3}}{\sqrt{3}}\right) \qquad x = \pm\frac{\sqrt{5}}{\sqrt{2}}\left(\frac{\sqrt{2}}{\sqrt{2}}\right)$$

$$x = \pm\frac{2\sqrt{3}}{3} \qquad\qquad x = \pm\frac{\sqrt{10}}{2}$$

41. Solve and write the solution in interval notation. [1.5]

$$-3x + 4 \geq -2$$
$$-3x \geq -2 - 4$$
$$-3x \geq -6$$
$$x \leq 2$$

$$(-\infty,\ 2]$$

43. Solve and write the solution in interval notation. [1.5]

$$3x + 1 > 7 \quad \text{or} \quad 3x + 2 < -7$$
$$3x > 6 \qquad\qquad 3x < -9$$
$$x > 2 \qquad\qquad x < -3$$
$$(-\infty, -3) \cup (2, \infty)$$

45. Solve and write the solution in interval notation. [1.5]

$$61 \leq \frac{9}{5}C + 32 \leq 95$$
$$29 \leq \quad \frac{9}{5}C \quad \leq 63$$
$$\frac{145}{9} \leq \quad C \quad \leq 35$$
$$\left[\frac{145}{9},\ 35\right]$$

47. Solve and write the solution in interval notation. [1.5]

$$|3x - 4| < 2$$

$$-2 < 3x - 4 < 2$$
$$2 < \quad 3x \quad < 6$$
$$\frac{2}{3} < \quad x \quad < 2$$
$$\left(\frac{2}{3},\ 2\right)$$

49. Solve and write the solution in interval notation. [1.5]

$$0 < |x - 2| < 1$$

If $x - 2 \geq 0$, then $2 < x < 3$.

If $x - 2 < 0$, then $0 < x - 2 < -1$

$$2 > \quad x \quad > 1.$$

$$(1, 2) \cup (2, 3)$$

51. Solve and write the solution in interval notation. [1.5]

$$x^2 + x - 6 \geq 0$$
$$(x + 3)(x - 2) \geq 0$$

The product is positive or zero.

$$x + 3 = 0 \Rightarrow x = -3$$
$$x - 2 = 0 \Rightarrow x = 2$$

Critical values are –3 and 2.

$$(x+3)(x-2) \quad \overset{++\,|\text{---------}|++}{\underset{-3 \qquad\quad 2}{\rule{3.5cm}{0.4pt}}}$$

$$(-\infty, -3] \cup [2, \infty)$$

53. Solve and write the solution in interval notation. [1.5]

$$\frac{x + 3}{x - 4} > 0$$

The quotient is positive.

$$x + 3 = 0 \Rightarrow x = -3$$
$$x - 4 = 0 \Rightarrow x = 4$$

The critical values are –3 and 4.

$$\frac{x+3}{x-4} \quad \overset{++++|\text{-------}|+++++}{\underset{-3 \quad 0 \qquad 4}{\rule{3.5cm}{0.4pt}}}$$

$$(-\infty,\ -3) \cup (4, \infty)$$

55. Solve and write the solution in interval notation. [1.5]

$$\frac{2x}{3 - x} \leq 10$$

$$\frac{2x}{3 - x} - 10 \leq 0$$

$$\frac{2x - 10(3 - x)}{3 - x} \leq 0$$

$$\frac{2x - 30 + 10x}{3 - x} \leq 0$$

$$\frac{12x - 30}{3 - x} \leq 0$$

The quotient is negative or zero.

$$12x - 30 = 0 \Rightarrow x = \frac{5}{2}$$
$$3 - x = 0 \Rightarrow x = 3$$

The critical values are $\frac{5}{2}$ and 3.

$$\frac{12x - 30}{3 - x} \quad \overset{\text{-------}|+|\text{------}}{\underset{0 \qquad\; \frac{5}{2}\,3}{\rule{3.5cm}{0.4pt}}}$$

Denominator $\neq 0 \Rightarrow x \neq 3$.

$$\left(-\infty,\ \frac{5}{2}\right] \cup (3, \infty)$$

57. Find the width and length. [1.2]

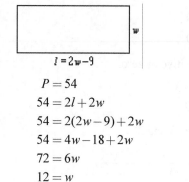

$$P = 54$$
$$54 = 2l + 2w$$
$$54 = 2(2w - 9) + 2w$$
$$54 = 4w - 18 + 2w$$
$$72 = 6w$$
$$12 = w$$
$$2w - 9 = 2(12) - 9 = 24 - 9 = 15$$

width = 12 ft, length = 15 ft

59. Find the height of the tree. [1.2]

$$\frac{h}{15} = \frac{9}{6}$$
$$h = 22.5 \text{ ft}$$

61. Find the diameter. Round to the nearest foot. [1.2]

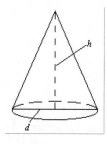

$$h = \frac{1}{4}d$$
$$V = 144$$
$$d = 2r$$
$$r = \frac{d}{2}$$

$$V = \frac{1}{3}\pi r^2 h$$
$$144 = \frac{1}{3}\pi \left(\frac{d}{2}\right)^2 \left(\frac{1}{4}d\right)$$
$$144 = \frac{\pi d^3}{48}$$
$$\frac{144(48)}{\pi} = d^3$$
$$d \approx 13 \text{ ft}$$

63. Find the monthly maintenance cost. [1.2]

Let x = monthly maintenance cost per owner

$$18x = 24(x - 12)$$
$$18x = 24x - 288$$
$$-6x = -288$$
$$x = 48$$
$$18x = 864$$

The total monthly maintenance cost is $864.

65. Find the distance. [1.2]

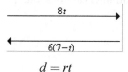

$$d = rt$$
$$d = 8t \qquad d = 6(7 - t)$$
$$6(7 - t) = 8t$$
$$42 - 6t = 8t$$
$$42 = 14t$$
$$3 = t$$
$$d = 8(3) = 24 \text{ nautical miles}$$

67. Find the amount of each solution. [1.2]

5%	x
11%	$600 - x$
7%	600

$$0.05x + 0.11(600 - x) = 0.07(600)$$
$$0.05x + 66 - 0.11x = 42$$
$$-0.06x = -24$$
$$x = 400 \text{ ml of } 5\%$$
$$600 - x = 200 \text{ ml of } 11\%$$

69. Find the amount of gold alloy. [1.2]

$460	x
$220	25
$310	$x + 25$

$$460x + 220(25) = 310(x + 25)$$
$$460x + 5500 = 310x + 7750$$
$$150x = 2250$$
$$x = 15 \text{ oz}$$

71. Find the time it would take the apprentice, working alone. [1.4]

Let x = time for apprentice

$x - 9$ = time for mason

$\dfrac{1}{x}$ = rate for apprentice

$\dfrac{1}{x-9}$ = rate for mason

$$6\left(\frac{1}{x}+\frac{1}{x-9}\right)=1$$

$$6x(x-9)\left(\frac{1}{x}+\frac{1}{x-9}\right)=1x(x-9)$$

$$6(x-9)+6x=x^2-9x$$

$$6x-54+6x=x^2-9x$$

$$0=x^2-21x+54$$

$$0=(x-18)(x-3)$$

$$x=18 \quad \text{or} \quad x=3$$

Note: $x = 3 \Rightarrow$ mason's time $= -6$ hours. Thus $x \neq 3$.

Apprentice takes 18 hours to build the wall.

73. Find the distance of AX. [1.3]

Let x = the distance of AX.

$$\sqrt{60^2+x^2}=\sqrt{40^2+(100-x)^2}$$

$$\left(\sqrt{60^2+x^2}\right)^2=\left(\sqrt{40^2+(100-x)^2}\right)^2$$

$$3600+x^2=1600+10{,}000-200x+x^2$$

$$200x=8000$$

$$x=40$$

The distance AX is 40 yards.

75. Find the time. [1.3]

$$5=-4.9t^2+7.5t+10$$

$$0=-4.9t^2+7.5t+5$$

$$t=\frac{-7.5\pm\sqrt{(7.5)^2-4(-4.9)(5)}}{2(-4.9)}$$

$$t=\frac{-7.5\pm\sqrt{154.25}}{-9.8}$$

$$t\approx 2.0 \quad \text{or} \quad t\approx -0.5 \ (\text{No})$$

In about 2.0 seconds.

77. Find the range of mean heights. [1.5]

$$-1.645<\frac{63.8-\mu}{0.45}<1.645$$

$$-0.74025<63.8-\mu<0.74025$$

$$-64.54025<\ -\mu\ <-63.05975$$

$$64.5>\quad \mu\quad >63.1$$

$$63.1<\quad \mu\quad <64.5 \text{ lb}$$

79. Find the range of diameters. [1.5]

Let C = the circumference, r = the radius, and d = the diameter.

$$C=2\pi r=\pi d$$

$$29.5\leq C\leq 30.0$$

$$29.5\leq \pi d\leq 30.0$$

$$\frac{29.5}{\pi}\leq d\leq \frac{30.0}{\pi}$$

$$9.39\leq d\leq 9.55$$

The diameter of the basketball is from 9.39 to 9.55 inches.

81. Find the acceleration. [1.6]

$$F=ka$$

$$10=k(2)$$

$$5=k$$

$$F=5a$$

$$15=5a$$

$$3=a$$

$$a=3 \text{ ft/s}^2$$

83. Find the number of players. [1.6]

$$N=\frac{k}{p}$$

$$5000=\frac{k}{150}$$

$$750{,}000=k$$

$$N=\frac{750{,}000}{p}$$

$$N=\frac{750{,}000}{125}$$

$$N=6000 \text{ players}$$

85. Find the acceleration. Round to the nearest hundredth of a meter per second squared. [1.6]

$$a = \frac{km}{r^2}$$

$$9.8 = \frac{k(5.98 \times 10^{26})}{(6,370,000)^2}$$

$$9.8(6,370,000)^2 = k(5.98 \times 10^{26})$$

$$\frac{9.8(6,370,000)^2}{5.98 \times 10^{26}} = k$$

$$k \approx 6.6497 \times 10^{-13}$$

$$a = \frac{6.6497 \times 10^{-13} \, m}{r^2}$$

$$a = \frac{(6.6497 \times 10^{-13})(7.46 \times 10^{24})}{(1,740,000)^2}$$

$$a \approx 1.64 \text{ meters/sec}^2$$

Chapter 1 Test

1. Solve. [1.1]

$$\frac{2x}{3} + \frac{1}{2} = \frac{x}{2} - \frac{3}{4}$$

$$12\left(\frac{2x}{3} + \frac{1}{2}\right) = 12\left(\frac{x}{2} - \frac{3}{4}\right)$$

$$8x + 6 = 6x - 9$$

$$2x = -15$$

$$x = -\frac{15}{2}$$

3. Solve for x. [1.2]

$$ax - c = c(x - d)$$

$$ax - c = cx - cd$$

$$ax - cx = c - cd$$

$$x(a - c) = c - cd$$

$$x = \frac{c - cd}{a - c}, \ a \neq c$$

5. Solve by completing the square. [1.3]

$$2x^2 - 8x + 1 = 0$$

$$x^2 - 4x = -\frac{1}{2}$$

$$x^2 - 4x + 4 = -\frac{1}{2} + 4$$

$$(x - 2)^2 = \frac{7}{2}$$

$$x - 2 = \pm\sqrt{\frac{7}{2}} = \pm\frac{\sqrt{14}}{2}$$

$$x = 2 \pm \frac{\sqrt{14}}{2} = \frac{4 \pm \sqrt{14}}{2}$$

The solutions are $\dfrac{4 - \sqrt{14}}{2}$ and $\dfrac{4 + \sqrt{14}}{2}$.

7. Find the discriminant and number of solutions. [1.3]

$$2x^2 + 3x + 1 = 0$$

$$a = 2, \ b = 3, \ c = 1$$

$$b^2 - 4ac = (3)^2 - 4(2)(1) = 9 - 8 = 1$$

The discriminant, 1, is a positive number. Therefore, there are two real solutions

9. Solve. [1.4]

$$\sqrt{3x + 1} - \sqrt{x - 1} = 2$$

$$\sqrt{3x + 1} = \sqrt{x - 1} + 2$$

$$\left(\sqrt{3x + 1}\right)^2 = \left(\sqrt{x - 1} + 2\right)^2$$

$$3x + 1 = x - 1 + 4\sqrt{x - 1} + 4$$

$$2x - 2 = 4\sqrt{x - 1}$$

$$x - 1 = 2\sqrt{x - 1} \quad \text{Remove factor of 2.}$$

$$(x - 1)^2 = \left(2\sqrt{x - 1}\right)^2$$

$$x^2 - 2x + 1 = 4(x - 1)$$

$$x^2 - 2x + 1 = 4x - 4$$

$$x^2 - 6x + 5 = 0$$

$$(x - 1)(x - 5) = 0$$

$$x = 1 \ \text{ or } \ x = 5$$

$$\text{Check } \sqrt{3(1) + 1} - \sqrt{1 - 1} = 2$$

$$\sqrt{4} - \sqrt{0} = 2$$

$$2 = 2$$

$$\sqrt{3(5) + 1} - \sqrt{5 - 1} = 2$$

$$\sqrt{16} - \sqrt{4} = 2$$

$$4 - 2 = 2$$

$$2 = 2$$

1 and 5 check as solutions.

11. Solve. [1.4]

$$\frac{3}{x + 2} - \frac{3}{4} = \frac{5}{x + 2}$$

$$4(x + 2)\left(\frac{3}{x + 2} - \frac{3}{4}\right) = 4(x + 2)\left(\frac{5}{x + 2}\right)$$

$$4(3) - 3(x + 2) = 4(5)$$

$$12 - 3x - 6 = 20$$

$$-3x = 14$$

$$x = -\frac{14}{3}$$

13. Solve. [1.4]

$$x^3 - 64 = 0$$

$$(x-4)(x^2 + 4x + 16) = 0$$

$$x - 4 = 0 \quad \text{or} \quad x^2 + 4x + 16 = 0$$

$$x = 4$$

$$x = \frac{-4 \pm \sqrt{4^2 - 4(1)(16)}}{2(1)}$$

$$x = \frac{-4 \pm \sqrt{-48}}{2} = \frac{-4 \pm 4i\sqrt{3}}{2} = -2 \pm 2i\sqrt{3}$$

The solutions are $4, \; -2 - 2i\sqrt{3}, \; -2 + 2i\sqrt{3}$.

15. Solve and write in interval notation. [1.5]

$$|3x - 4| > 5$$

$$3x - 4 > 5 \quad \text{or} \quad 3x - 4 < -5$$

$$3x > 9 \qquad\qquad 3x < -1$$

$$x > 3 \qquad\qquad x < -\frac{1}{3}$$

$$\left(-\infty, -\frac{1}{3}\right) \cup (3, \, \infty)$$

17. Solve and write in interval notation. [1.5]

$$\frac{x^2 + x - 12}{x + 1} \geq 0$$

$$\frac{(x+4)(x-3)}{x+1} \geq 0$$

The quotient is positive or zero.

$$x + 4 = 0 \Rightarrow x = -4$$

$$x - 3 = 0 \Rightarrow x = 3$$

$$x + 1 = 0 \Rightarrow x = -1$$

Critical values are –4, 3, and –1.

$$\frac{(x+4)(x-3)}{x+1} \qquad \begin{array}{c} - - - |+ + +|- - - -|+ + + \\ \xleftrightarrow{\;+\!+\!+\!+\!+\!+\!+\!+\!+\!+\!+\!+\!+\!+\;} \\ \quad -4 \;\; -1\; 0 \qquad 3 \end{array}$$

Denominator $\neq 0 \Rightarrow x \neq -1$.

$$[-4, \; -1) \cup [3, \, \infty)$$

19. Find the time for the assistant working alone. [1.2]

Let x = number of hours the assistant needs to cover the parking lot.

$$6\left[\frac{1}{10} + \frac{1}{x}\right] = 1$$

$$10x(6)\left[\frac{1}{10} + \frac{1}{x}\right] = 10x(1)$$

$$6x + 60 = 10x$$

$$-4x = -60$$

$$x = 15$$

The assistant takes 15 hours to cover the parking lot.

21. Find the amount of each. [1.2]

$3.45	x
$2.70	$50 - x$
$3.15	50

$$3.45x + 2.7(50 - x) = 3.15(50)$$

$$3.45x + 135 - 2.70x = 157.50$$

$$0.75x = 22.5$$

$$x = 30$$

$$50 - x = 20$$

30 lb of ground beef and 20 lb of sausage.

23. Find Zoey's speed. [1.4]

Let $x + 4$ = Zoey's speed.

$$\frac{15}{x} - \frac{15}{x+4} = 1$$

$$x(x+4)\left(\frac{15}{x} - \frac{15}{x+4}\right) = x(x+4)(1)$$

$$15(x+4) - 15x = x(x+4)$$

$$15x + 60 - 15x = x^2 + 4x$$

$$0 = x^2 + 4x - 60$$

$$0 = (x+10)(x-6)$$

$$x = -10 \;(\text{No}) \quad \text{or} \quad x = 6$$

$$x + 4 = 10$$

Zoey's speed is 10 mph.

25. Find the velocity of the meteorite. [1.6]

$$v = \frac{k}{\sqrt{d}}$$

$$4 = \frac{k}{\sqrt{3000}}$$

$$k = 4\sqrt{3000} = 40\sqrt{30}$$

$$v = \frac{40\sqrt{30}}{\sqrt{2500}} = \frac{40\sqrt{30}}{50}$$

$$v = \frac{4\sqrt{30}}{5} \approx 4.4 \text{ miles/second}$$

Cumulative Review Exercises

1. Evaluate. [P.1]

$$4 + 3(-5) = 4 - 15 = -11$$

3. Perform the operations and simplify. [P.3]

$$(3x - 5)^2 - (x + 4)(x - 4)$$
$$= (9x^2 - 30x + 25) - (x^2 - 16)$$
$$= 9x^2 - 30x + 25 - x^2 + 16$$
$$= 8x^2 - 30x + 41$$

5. Simplify. [P.5]

$$\frac{7x - 3}{x - 4} - 5 = \frac{7x - 3 - 5x + 20}{x - 4} = \frac{2x + 17}{x - 4}$$

7. Simplify. [P.6]

$$(2 + 5i)(2 - 5i) = 4 - 25i^2 = 4 + 25 = 29$$

9. Solve using the quadratic formula. [1.3]

$$2x^2 - 4x = 3$$
$$2x^2 - 4x - 3 = 0$$
$$x = \frac{-(-4) \pm \sqrt{(-4)^2 - 4(2)(-3)}}{2(2)} = \frac{4 \pm 2\sqrt{10}}{4}$$
$$= \frac{2 \pm \sqrt{10}}{2}$$

11. Solve. [1.4]

$$x = 3 + \sqrt{9 - x}$$
$$x - 3 = \sqrt{9 - x}$$
$$(x - 3)^2 = \left(\sqrt{9 - x}\right)^2$$
$$x^2 - 6x + 9 = 9 - x$$
$$x^2 - 5x = 0$$
$$x(x - 5) = 0$$
$$x = 0 \quad \text{or} \quad x = 5$$

Check 0:   Check 5:

$$0 = 3 + \sqrt{9 - 0} \qquad 5 = 3 + \sqrt{9 - 5}$$
$$0 = 3 + \sqrt{9} \qquad\quad 5 = 3 + \sqrt{4}$$
$$0 = 3 + 3 \qquad\qquad 5 = 3 + 2$$
$$0 = 6 \text{ No} \qquad\qquad 5 = 5$$

The solution is 5.

13. Solve. [1.4]

$$2x^4 - 11x^2 + 15 = 0 \quad \text{Let } u = x^2.$$
$$2u^2 - 11u + 15 = 0$$
$$(2u - 5)(u - 3) = 0$$

$$2u - 5 = 0 \qquad\qquad \text{or} \qquad u - 3 = 0$$
$$u = x^2 = \frac{5}{2} \qquad\qquad\qquad\qquad u = 3$$
$$x = \pm\sqrt{\frac{5}{2}} = \pm\frac{\sqrt{10}}{2} \qquad\qquad x^2 = 3$$
$$x = \pm\sqrt{3}$$

The solutions are $-\dfrac{\sqrt{10}}{2}, \dfrac{\sqrt{10}}{2}, -\sqrt{3}, \sqrt{3}$.

15. Solve. Write the solution in interval notation. [1.5]

$$|x - 6| \geq 2$$

$$x - 6 \geq 2 \qquad \text{or} \qquad x - 6 \leq -2$$
$$x \geq 8 \qquad\qquad\qquad\qquad x \leq 4$$

The solution is $(-\infty, \, 4] \cup [8, \, \infty)$.

17. Find the dimensions of the fence. [1.2]

$$\text{Perimeter} = 2(\text{Length}) + 2(\text{Width})$$
$$200 = 2(w + 16) + 2w$$
$$200 = 2w + 32 + 2w$$
$$168 = 4w$$
$$42 = w$$
$$w = 42$$
$$w + 16 = 58$$

The width is 42 feet; the length is 58 feet.

19. Find the range of scores. [1.5]

Let x = the score on the fourth test.

$$80 \leq \frac{86 + 72 + 94 + x}{4} < 90 \quad \text{and} \quad 0 \leq x \leq 100$$
$$80 \leq \frac{252 + x}{4} < 90$$
$$320 \leq 252 + x < 360$$
$$68 \leq x < 108$$

$$[68, \, 108) \cap [0, \, 100] = [68, \, 100]$$

The fourth test score must be from 68 to 100.

Chapter 2 Functions and Graphs

Section 2.1 Exercises

1. Plot the points:

3. a. Find the decrease: The average debt decreased between 2006 and 2007, and 2008 and 2009.

b. Find the average debt in 2011:

Increase between 2009 to 2010: $22.0 - 20.1 = 1.9$

Then the increase from 2010 to 2011:

$22.0 + 1.9 = 23.9$, or $23,900.

5. Determine whether the ordered pair is a solution

$$2x + 5y = 16$$
$$2(-2) + 5(4) \overset{?}{=} 16$$
$$-4 + 20 \overset{?}{=} 16$$
$$16 = 16 \quad \text{True}$$

$(-2, 4)$ is a solution.

7. Determine whether the ordered pair is a solution

$$y = 3x^2 - 4x + 2$$
$$17 \overset{?}{=} 3(-3)^2 - 4(-3) + 2$$
$$17 \overset{?}{=} 27 + 12 + 2$$
$$17 = 41 \quad \text{False}$$

$(-3, 17)$ is not a solution.

9. Find the distance: $(6, 4)$, $(-8, 11)$

$$d = \sqrt{(-8-6)^2 + (11-4)^2}$$
$$= \sqrt{(-14)^2 + (7)^2}$$
$$= \sqrt{196 + 49}$$
$$= \sqrt{245}$$
$$= 7\sqrt{5}$$

11. Find the distance: $(-4, -20)$, $(-10, 15)$

$$d = \sqrt{(-10-(-4))^2 + (15-(-20))^2}$$
$$= \sqrt{(-6)^2 + (35)^2}$$
$$= \sqrt{36 + 1225}$$
$$= \sqrt{1261}$$

13. Find the distance: $(5, -8)$, $(0, 0)$

$$d = \sqrt{(0-5)^2 + (0-(-8))^2}$$
$$= \sqrt{(-5)^2 + (8)^2}$$
$$= \sqrt{25 + 64}$$
$$= \sqrt{89}$$

15. Find the distance: $\left(\sqrt{3}, \sqrt{8}\right)$, $\left(\sqrt{12}, \sqrt{27}\right)$

$$d = \sqrt{(\sqrt{12} - \sqrt{3})^2 + (\sqrt{27} - \sqrt{8})^2}$$
$$= \sqrt{(2\sqrt{3} - \sqrt{3})^2 + (3\sqrt{3} - 2\sqrt{2})^2}$$
$$= \sqrt{(\sqrt{3})^2 + (3\sqrt{3} - 2\sqrt{2})^2}$$
$$= \sqrt{3 + (27 - 12\sqrt{6} + 8)}$$
$$= \sqrt{3 + 27 - 12\sqrt{6} + 8}$$
$$= \sqrt{38 - 12\sqrt{6}}$$

17. Find the distance: (a, b), $(-a, -b)$

$$d = \sqrt{(-a-a)^2 + (-b-b)^2}$$
$$= \sqrt{(-2a)^2 + (-2b)^2}$$
$$= \sqrt{4a^2 + 4b^2}$$
$$= \sqrt{4(a^2 + b^2)}$$
$$= 2\sqrt{a^2 + b^2}$$

19. Find the distance: $(x, 4x)$, $(-2x, 3x)$

$$d = \sqrt{(-2x-x)^2 + (3x-4x)^2} \quad \text{with } x < 0$$
$$= \sqrt{(-3x)^2 + (-x)^2}$$
$$= \sqrt{9x^2 + x^2}$$
$$= \sqrt{10x^2}$$
$$= -x\sqrt{10} \quad (\text{Note: since } x < 0, \sqrt{x^2} = -x)$$

579

21. Find the midpoint: $(1, -1), (5, 5)$

$$M = \left(\frac{x_1 + x_2}{2}, \ \frac{y_1 + y_2}{2} \right)$$

$$= \left(\frac{1+5}{2}, \ \frac{-1+5}{2} \right)$$

$$= \left(\frac{6}{2}, \ \frac{4}{2} \right)$$

$$= (3, \ 2)$$

23. Find the midpoint: $(6, -3), (6, 11)$

$$M = \left(\frac{6+6}{2}, \ \frac{-3+11}{2} \right)$$

$$= \left(\frac{12}{2}, \ \frac{8}{2} \right)$$

$$= (6, \ 4)$$

25. Find the midpoint: $(1.75, 2.25), (-3.5, 5.57)$

$$M = \left(\frac{1.75 + (-3.5)}{2}, \ \frac{2.25 + 5.57}{2} \right)$$

$$= \left(-\frac{1.75}{2}, \ \frac{7.82}{2} \right)$$

$$= (-0.875, \ 3.91)$$

27. Find other endpoint: endpoint $(5, 1)$, midpoint $(9, 3)$

$$\left(\frac{x+5}{2}, \ \frac{y+1}{2} \right) = (9, \ 3)$$

therefore $\dfrac{x+5}{2} = 9$ and $\dfrac{y+1}{2} = 3$

$$x + 5 = 18 \qquad y + 1 = 6$$

$$x = 13 \qquad\quad y = 5$$

Thus $(13, 5)$ is the other endpoint.

29. Find other endpoint: endpoint $(-3, -8)$,

midpoint $(2, -7)$

$$\left(\frac{x + (-3)}{2}, \ \frac{y + (-8)}{2} \right) = (2, \ -7)$$

therefore $\dfrac{x-3}{2} = 2$ and $\dfrac{y-8}{2} = -7$

$$x - 3 = 4 \qquad y - 8 = -14$$

$$x = 7 \qquad\quad y = -6$$

Thus $(7, -6)$ is the other endpoint.

31. Graph the equation: $x - y = 4$

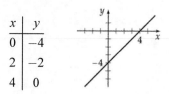

x	y
0	−4
2	−2
4	0

33. Graph the equation: $y = 0.25x^2$

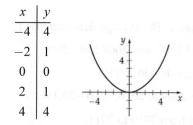

x	y
−4	4
−2	1
0	0
2	1
4	4

35. Graph the equation: $y = -2|x - 3|$

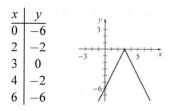

x	y
0	−6
2	−2
3	0
4	−2
6	−6

37. Graph the equation: $y = x^2 - 3$

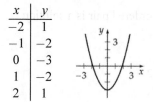

x	y
−2	1
−1	−2
0	−3
1	−2
2	1

39. Graph the equation: $y = \dfrac{1}{2}(x-1)^2$

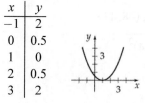

x	y
−1	2
0	0.5
1	0
2	0.5
3	2

41. Graph the equation: $y = x^2 + 2x - 8$

x	y
−4	0
−2	−8
−1	−9
0	−8
2	0

43. Graph the equation: $y = -x^2 + 2$

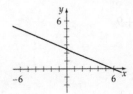

x	y
-2	-2
-1	1
0	2
1	1
2	-2

45. Find the x- and y-intercepts and graph: $2x + 5y = 12$

For the y-intercept, let $x = 0$ and solve for y.

$$2(0) + 5y = 12$$
$$y = \frac{12}{5}, \quad y\text{-intercept: } \left(0, \frac{12}{5}\right)$$

For the x-intercept, let $y = 0$ and solve for x.

$$2x + 5(0) = 12$$
$$x = 6, \quad x\text{-intercept: } (6, 0)$$

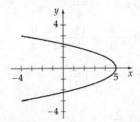

47. Find the x- and y-intercepts and graph: $x = -y^2 + 5$

For the y-intercept, let $x = 0$ and solve for y.

$$0 = -y^2 + 5$$
$$y = \pm\sqrt{5}, \quad y\text{-intercepts: } \left(0, -\sqrt{5}\right), \left(0, \sqrt{5}\right)$$

For the x-intercept, let $y = 0$ and solve for x.

$$x = -(0)^2 + 5$$
$$x = 5, \quad x\text{-intercept: } (5, 0)$$

49. Find the x- and y-intercepts and graph: $x = |y| - 4$

For the y-intercept, let $x = 0$ and solve for y.

$$0 = |y| - 4$$
$$y = \pm 4, \quad y\text{-intercepts: } (0, -4), (0, 4)$$

For the x-intercept, let $y = 0$ and solve for x.

$$x = |0| - 4$$
$$x = -4, \quad x\text{-intercept: } (-4, 0)$$

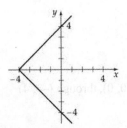

51. Find the x- and y-intercepts and graph: $x^2 + y^2 = 4$

For the y-intercept, let $x = 0$ and solve for y.

$$(0)^2 + y^2 = 4$$
$$y = \pm 2, \quad y\text{-intercepts: } (0, -2), (0, 2)$$

For the x-intercept, let $y = 0$ and solve for x.

$$x^2 + (0)^2 = 4$$
$$x = \pm 2, \quad x\text{-intercepts: } (-2, 0), (2, 0)$$

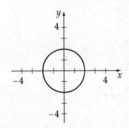

53. Find center and radius: $x^2 + y^2 = 36$

center $(0, 0)$, radius 6

55. Find center and radius: $(x-1)^2 + (y-3)^2 = 49$

center $(1, 3)$, radius 7

57. Find center and radius: $(x+2)^2 + (y+5)^2 = 25$

center $(-2, -5)$, radius 5

59. Find center and radius: $(x-8)^2 + y^2 = \frac{1}{4}$

center $(8, 0)$, radius $\frac{1}{2}$

61. Find circle equation: center $(4, 1)$, radius 2

$$(x-4)^2 + (y-1)^2 = 2^2$$
$$(x-4)^2 + (y-1)^2 = 4$$

63. Find circle equation: center $\left(\frac{1}{2}, \frac{1}{4}\right)$, radius $\sqrt{5}$

$$\left(x - \frac{1}{2}\right)^2 + \left(y - \frac{1}{4}\right)^2 = \left(\sqrt{5}\right)^2$$
$$\left(x - \frac{1}{2}\right)^2 + \left(y - \frac{1}{4}\right)^2 = 5$$

65. Find circle equation: center $(0, 0)$, through $(-3, 4)$

$$(x - 0)^2 + (y - 0)^2 = r^2$$
$$(-3 - 0)^2 + (4 - 0)^2 = r^2$$
$$(-3)^2 + 4^2 = r^2$$
$$9 + 16 = r^2$$
$$25 = 5^2 = r^2$$
$$(x - 0)^2 + (y - 0)^2 = 25$$

67. Find circle equation: center $(1, 3)$, through $(4, -1)$

$$(x + 2)^2 + (y - 5)^2 = r^2$$
$$(x - 1)^2 + (y - 3)^2 = r^2$$
$$(4 - 1)^2 + (-1 - 3)^2 = r^2$$
$$3^2 + (-4)^2 = r^2$$
$$9 + 16 = r^2$$
$$25 = 5^2 = r^2$$
$$(x - 1)^2 + (y - 3)^2 = 25$$

69. Find circle equation: center $(-2, 5)$, diameter 10

diameter 10 means the radius is $5 \Rightarrow r^2 = 25$.

$$(x + 2)^2 + (y - 5)^2 = 25$$

71. Find circle equation: endpoints $(2, 3)$ and $(-4, 11)$

$$d = \sqrt{(-4 - 2)^2 + (11 - 3)^2}$$
$$= \sqrt{36 + 64} = \sqrt{100}$$
$$= 10$$

Since the diameter is 10, the radius is 5.

The center is the midpoint of the line segment from $(2, 3)$ to $(-4, 11)$.

$$\left(\frac{2 + (-4)}{2}, \frac{3 + 11}{2}\right) = (-1, 7) \text{ center}$$

$$(x + 1)^2 + (y - 7)^2 = 25$$

73. Find circle equation: endpoints $(5, -3)$ and $(-1, -5)$

$$d = \sqrt{(-5 - (-3))^2 + (-1 - 5)^2} = \sqrt{4 + 36} = \sqrt{40}$$

Since the diameter is $\sqrt{40}$, the radius is $\frac{\sqrt{40}}{2} = \sqrt{10}$.

Center is $\left(\frac{5 + (-1)}{2}, \frac{(-3) + (-5)}{2}\right) = (2, -4)$

$$(x - 2)^2 + (y + 4)^2 = \left(\sqrt{10}\right)^2$$
$$(x - 2)^2 + (y + 4)^2 = 10$$

75. Find circle equation: center $(7, 11)$, tangent to x-axis

Since it is tangent to the x-axis, its radius is 11.

$$(x - 7)^2 + (y - 11)^2 = 11^2$$

77. Find center and radius: $x^2 + y^2 - 6x + 5 = 0$

$$x^2 - 6x \qquad + y^2 = -5$$
$$x^2 - 6x + 9 + y^2 = -5 + 9$$
$$(x - 3)^2 + y^2 = 2^2$$

center $(3, 0)$, radius 2

79. Find center and radius: $x^2 + y^2 - 14x + 8y + 53 = 0$

$$x^2 - 14x \qquad + y^2 + 8y \qquad = -53$$
$$x^2 - 14x + 49 + y^2 + 8y + 16 = -53 + 49 + 16$$
$$(x - 7)^2 + (y + 4)^2 = 12$$

center $(7, -4)$, radius $\sqrt{12} = 2\sqrt{3}$

81. Find center and radius: $x^2 + y^2 - x + 3y - \frac{15}{4} = 0$

$$x^2 - x \qquad + y^2 + 3y \qquad = \frac{15}{4}$$
$$x^2 - x + \frac{1}{4} + y^2 + 3y + \frac{9}{4} = \frac{15}{4} + \frac{1}{4} + \frac{9}{4}$$
$$\left(x - \frac{1}{2}\right)^2 + \left(y + \frac{3}{2}\right)^2 = \left(\frac{5}{2}\right)^2$$

center $\left(\frac{1}{2}, -\frac{3}{2}\right)$, radius $\frac{5}{2}$

83. Find center and radius: $x^2 + y^2 + 3x - 6y + 2 = 0$

$$x^2 + 3x \qquad + y^2 - 6y \qquad = -2$$

$$x^2 + 3x + \frac{9}{4} + y^2 - 6y + 9 = -2 + \frac{9}{4} + 9$$

$$\left(x + \frac{3}{2}\right)^2 + (y-3)^2 = \left(\frac{\sqrt{37}}{2}\right)^2$$

center $\left(-\frac{3}{2},\ 3\right)$, radius $\dfrac{\sqrt{37}}{2}$

85. Find the points:

$$\sqrt{(4-x)^2 + (6-0)^2} = 10$$

$$\left(\sqrt{(4-x)^2 + (6-0)^2}\right)^2 = 10^2$$

$$16 - 8x + x^2 + 36 = 100$$

$$x^2 - 8x - 48 = 0$$

$$(x - 12)(x + 4) = 0$$

$$x = 12 \quad \text{or} \quad x = -4$$

The points are $(12, 0),\ (-4, 0)$.

87. Find the x- and y-intercepts and graph: $|x| + |y| = 4$

Intercepts: $(0,\ \pm 4),\ (\pm 4,\ 0)$

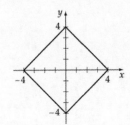

89. Find the formula:

$$\sqrt{(3-x)^2 + (4-y)^2} = 5$$

$$(3-x)^2 + (4-y)^2 = 5^2$$

$$9 - 6x + x^2 + 16 - 8y + y^2 = 25$$

$$x^2 - 6x + y^2 - 8y = 0$$

Prepare for Section 2.2

P1. $x^2 + 3x - 4$

$$(-3)^2 + 3(-3) - 4 = 9 - 9 - 4 = -4$$

P3. $d = \sqrt{(3 - (-4))^2 + (-2 - 1)^2} = \sqrt{49 + 9} = \sqrt{58}$

P5. $\quad x^2 - x - 6 = 0$

$$(x + 2)(x - 3) = 0$$

$$x + 2 = 0 \qquad x - 3 = 0$$

$$x = -2 \qquad x = 3$$

$-2, 3$

Section 2.2 Exercises

1. Write the domain and range. State whether a relation.

Domain: $\{-4, 2, 5, 7\}$; range: $\{1, 3, 11\}$

Yes. The set of ordered pairs defines y as a function of x since each x is paired with exactly one y.

3. Write the domain and range. State whether a relation.

Domain: $\{4, 5, 6\}$; range: $\{-3, 1, 4, 5\}$

No. The set of ordered pair does not define y as a function of x since 4 is paired with 4 and 5.

5. Determine if the value is in the domain.

$$f(0) = \frac{3(0)}{0 + 4} = 0$$

Yes, 0 is in the domain of the function.

7. Determine if the value is in the domain.

$$F(0) = \frac{-1 - 1}{-1 + 1} = \frac{-2}{0} \text{ undefined}$$

No, -1 is not in the domain of the function.

9. Determine if the value is in the domain.

$$g(-1) = \frac{5(-1) - 1}{(-1)^2 + 1} = \frac{-6}{2} = -3$$

Yes, -1 is in the domain of the function.

11. Is y a function of x?

$$2x + 3y = 7$$

$$3y = -2x + 7$$

$$y = -\frac{2}{3}x + \frac{7}{3},\ y \text{ is a function of } x.$$

13. Is y a function of x?

$$-x + y^2 = 2$$

$$y^2 = x + 2$$

$$y = \pm\sqrt{x + 2},\ y \text{ is a not function of } x.$$

15. Is y a function of x?

$$x^2 + y^2 = 9$$
$$y^2 = 9 - x^2$$
$$y = \pm\sqrt{9 - x^2}, \quad y \text{ is a not function of } x.$$

17. Is y a function of x?

$$y = |x| + 5, \quad y \text{ is a function of } x.$$

19. Determine if the value is a zero.

$$f(-2) = 3(-2) + 6 = 0$$

Yes, -2 is a zero.

21. Determine if the value is a zero.

$$G\left(-\frac{1}{3}\right) = 3\left(-\frac{1}{3}\right)^2 + 2\left(-\frac{1}{3}\right) - 1 = -\frac{4}{3}$$

No, $-\dfrac{1}{3}$ is not a zero.

23. Determine if the value is a zero.

$$y(1) = 5(1)^2 - 2(1) - 2 = 1$$

No, 1 is not a zero.

25. Evaluate the function $f(x) = 3x - 1$,

a. $f(2) = 3(2) - 1$
$$= 6 - 1$$
$$= 5$$

b. $f(-1) = 3(-1) - 1$
$$= -3 - 1$$
$$= -4$$

c. $f(0) = 3(0) - 1$
$$= 0 - 1$$
$$= -1$$

d. $f\left(\dfrac{2}{3}\right) = 3\left(\dfrac{2}{3}\right) - 1$
$$= 2 - 1$$
$$= 1$$

e. $f(k) = 3(k) - 1$
$$= 3k - 1$$

f. $f(k + 2) = 3(k + 2) - 1$
$$= 3k + 6 - 1$$
$$= 3k + 5$$

27. Evaluate the function $A(w) = \sqrt{w^2 + 5}$,

a. $A(0) = \sqrt{(0)^2 + 5} = \sqrt{5}$

b. $A(2) = \sqrt{(2)^2 + 5} = \sqrt{9} = 3$

c. $A(-2) = \sqrt{(-2)^2 + 5} = \sqrt{9} = 3$

d. $A(4) = \sqrt{4^2 + 5} = \sqrt{21}$

e. $A(r + 1) = \sqrt{(r+1)^2 + 5}$
$$= \sqrt{r^2 + 2r + 1 + 5}$$
$$= \sqrt{r^2 + 2r + 6}$$

f. $A(-c) = \sqrt{(-c)^2 + 5} = \sqrt{c^2 + 5}$

29. Evaluate the function $f(x) = \dfrac{1}{|x|}$,

a. $f(2) = \dfrac{1}{|2|} = \dfrac{1}{2}$

b. $f(-2) = \dfrac{1}{|-2|} = \dfrac{1}{2}$

c. $f\left(-\dfrac{3}{5}\right) = \dfrac{1}{\left|-\dfrac{3}{5}\right|}$
$$= \dfrac{1}{\frac{3}{5}}$$
$$= 1 \div \dfrac{3}{5} = 1 \cdot \dfrac{5}{3}$$
$$= \dfrac{5}{3}$$

d. $f(2) + f(-2) = \dfrac{1}{2} + \dfrac{1}{2} = 1$

e. $f(c^2 + 4) = \dfrac{1}{|c^2 + 4|} = \dfrac{1}{c^2 + 4}$

f. $f(2 + h) = \dfrac{1}{|2 + h|}$

31. Evaluate the function $s(x) = \dfrac{x}{|x|}$,

a. $s(4) = \dfrac{4}{|4|} = \dfrac{4}{4} = 1$

b. $s(5) = \dfrac{5}{|5|} = \dfrac{5}{5} = 1$

c. $s(-2) = \dfrac{-2}{|-2|} = \dfrac{-2}{2} = -1$

d. $s(-3) = \dfrac{-3}{|-3|} = \dfrac{-3}{3} = -1$

e. Since $t > 0$, $|t| = t$.

$s(t) = \dfrac{t}{|t|} = \dfrac{t}{t} = 1$

f. Since $t < 0$, $|t| = -t$.

$s(t) = \dfrac{t}{|t|} = \dfrac{t}{-t} = -1$

33. a. Since $x = -4 < 2$, use $P(x) = 3x + 1$.

$P(-4) = 3(-4) + 1 = -12 + 1 = -11$

b. Since $x = \sqrt{5} \geq 2$, use $P(x) = -x^2 + 11$.

$P(\sqrt{5}) = -(\sqrt{5})^2 + 11 = -5 + 11 = 6$

c. Since $x = c < 2$, use $P(x) = 3x + 1$.

$P(c) = 3c + 1$

d. Since $k \geq 1$, then $x = k + 1 \geq 2$,

so use $P(x) = -x^2 + 11$.

$$P(k+1) = -(k+1)^2 + 11 = -(k^2 + 2k + 1) + 11$$
$$= -k^2 - 2k - 1 + 11$$
$$= -k^2 - 2k + 10$$

35. For $f(x) = 3x - 4$, the domain is the set of all real numbers.

37. For $f(x) = x^2 + 2$, the domain is the set of all real numbers.

39. For $f(x) = \dfrac{4}{x+2}$, the domain is $\{x \mid x \neq -2\}$.

41. For $f(x) = \sqrt{7+x}$, the domain is $\{x \mid x \geq -7\}$.

43. For $f(x) = \sqrt{4 - x^2}$, the domain is $\{x \mid -2 \leq x \leq 2\}$.

45. For $f(x) = \dfrac{1}{\sqrt{x+4}}$, the domain is $\{x \mid x > -4\}$.

47. To graph $f(x) = 3x - 4$, plot points and draw a smooth graph.

x	-1	0	1	2
$y = f(x) = 3x - 4$	-7	-4	-1	2

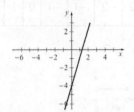

49. To graph $g(x) = x^2 - 1$, plot points and draw a smooth graph.

x	-2	-1	0	1	2
$y = g(x) = x^2 - 1$	3	0	-1	0	3

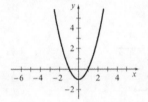

51. To graph $f(x) = \sqrt{x+4}$, plot points and draw a smooth graph.

x	-4	-2	0	2	5
$y = f(x) = \sqrt{x+4}$	0	$\sqrt{2}$	2	$\sqrt{6}$	3

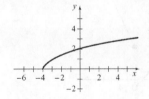

53. To graph $f(x) = |x - 2|$, plot points and draw a smooth graph.

x	-3	0	2	4	6
$y = f(x) = |x-2|$	5	2	0	2	4

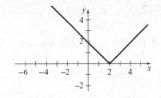

55. To graph $L(x) = \left[\!\left[\dfrac{1}{3}x\right]\!\right]$ for $-6 \le x \le 6$, plot points

and draw a smooth graph.

x	-6	-4	-3	-1	0	4	6
$y = L(x) = \left[\!\left[\dfrac{1}{3}x\right]\!\right]$	-2	-2	-1	-1	0	1	2

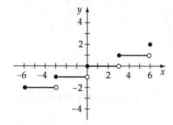

57. To graph $N(x) = \text{int}(-x)$ for $-3 \le x \le 3$, plot points and draw a smooth graph.

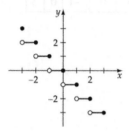

59. Graph $f(x) = \begin{cases} 1 - x, & x < 2 \\ 2x, & x \ge 2 \end{cases}$.

Graph $y = 1 - x$ for $x < 2$ and graph $y = 2x$ for $x \ge 2$.

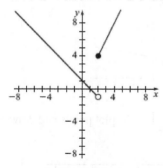

61. Graph $r(x) = \begin{cases} -x^2 + 4, & x < -1 \\ -x + 2, & -1 \le x \le 1 \\ 3x - 2, & x > 1 \end{cases}$.

Graph $y = -x^2 + 4$ for $x < -1$, graph $y = -x + 2$ for

$-1 \le x \le 1$, and graph $y = 3x - 2$ for $x > 1$.

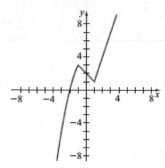

63. Find the value of a in the domain of $f(x) = 3x - 2$ for which $f(a) = 10$.

$3a - 2 = 10$ Replace $f(a)$ with $3a - 2$
$3a = 12$
$a = 4$

65. Find the values of a in the domain of

$f(x) = x^2 + 2x - 2$ for which $f(a) = 1$.

$a^2 + 2a - 2 = 1$ Replace $f(a)$ with $a^2 + 2a - 2$
$a^2 + 2a - 3 = 0$
$(a + 3)(a - 1) = 0$

$a + 3 = 0 \quad a - 1 = 0$
$a = -3 \quad a = 1$

67. Find the values of a in the domain of $f(x) = |x|$ for which $f(a) = 4$.

$|a| = 4$ Replace $f(a)$ with $|a|$
$a = -4 \quad a = 4$

69. Find the values of a in the domain of $f(x) = x^2 + 2$ for which $f(a) = 1$.

$a^2 + 2 = 1$ Replace $f(a)$ with $a^2 + 2$
$a^2 = -1$

There are no real values of a.

71. Find the zeros of f for $f(x) = 3x - 6$.

$f(x) = 0$
$3x - 6 = 0$
$3x = 6$
$x = 2$

73. Find the zeros of f for $f(x) = 5x + 2$.

$$f(x) = 0$$
$$5x + 2 = 0$$
$$5x = -2$$
$$x = -\frac{2}{5}$$

75. Find the zeros of f for $f(x) = x^2 - 4$.

$$f(x) = 0$$
$$x^2 - 4 = 0$$
$$(x + 2)(x - 2) = 0$$
$$x + 2 = 0 \qquad x - 2 = 0$$
$$x = -2 \qquad x = 2$$

77. Find the zeros of f for $f(x) = x^2 - 5x - 24$.

$$f(x) = 0$$
$$x^2 - 5x - 24 = 0$$
$$(x + 3)(x - 8) = 0$$
$$x + 3 = 0 \qquad x - 8 = 0$$
$$x = -3 \qquad x = 8$$

79. Determine which graphs are functions.

a. Yes; every vertical line intersects the graph in one point.

b. Yes; every vertical line intersects the graph in one point.

c. No; some vertical lines intersect the graph at more than one point.

d. Yes; every vertical line intersects the graph in one point.

81. Determine where the graph is increasing, constant, or decreasing. Decreasing on $(-\infty, 0]$; increasing on $[0, \infty)$

83. Determine where the graph is increasing, constant, or decreasing. Increasing on $(-\infty, \infty)$

85. Determine where the graph is increasing, constant, or decreasing. Decreasing on $(-\infty, -3]$; increasing on $[-3, 0]$; decreasing on $[0, 3]$; increasing on $[3, \infty)$

87. Determine where the graph is increasing, constant, or decreasing. Constant on $(-\infty, 0]$; increasing on $[0, \infty)$

89. Determine where the graph is increasing, constant, or decreasing. Decreasing on $(-\infty, 0]$; constant on $[0, 1]$; increasing on $[1, \infty)$

91. Determine which functions from 77-81 are one-to-one. g and F are one-to-one since every horizontal line intersects the graph at one point.

f, V, and p are not one-to-one since some horizontal lines intersect the graph at more than one point.

93. a. $C(2.8) = 0.90 - 0.20\text{int}(1 - 2.8)$
$$= 0.90 - 0.20\text{int}(-1.8)$$
$$= 0.90 - 0.20(-2)$$
$$= 0.90 + 0.4$$
$$= \$1.30$$

b. Graph $C(w)$.

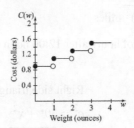

95. a. Write the width.
$$2l + 2w = 50$$
$$2w = 50 - 2l$$
$$w = 25 - l$$

b. Write the area.
$$A = lw$$
$$A = l(25 - l)$$
$$A = 25l - l^2$$

97. Write the function.
$$v(t) = 80,000 - 6500t, \qquad 0 \le t \le 10$$

99. a. Write the total cost function.
$$C(x) = 5(400) + 22.80x$$
$$= 2000 + 22.80x$$

b. Write the revenue function. $R(x) = 37.00x$

c. Write the profit function.

$$P(x) = 37.00x - C(x)$$
$$= 37.00 - [2000 + 22.80x]$$
$$= 37.00x - 2000 - 22.80x$$
$$= 14.20x - 2000$$

Note x is a natural number.

101. Write the function.

$$\frac{15}{3} = \frac{15 - h}{r}$$
$$5 = \frac{15 - h}{r}$$
$$5r = 15 - h$$
$$h = 15 - 5r$$
$$h(r) = 15 - 5r$$

103. Write the function.

$$d = \sqrt{(3t)^2 + (50)^2}$$
$$d = \sqrt{9t^2 + 2500} \text{ meters, } 0 \le t \le 60$$

105. Write the function.

$$d = \sqrt{(45 - 8t)^2 + (6t)^2} \text{ miles}$$

where t is the number of hours after 12:00 noon

107. a. Write the function.

Left side triangle
$$c^2 = 20^2 + (40 - x)^2$$
$$c = \sqrt{400 + (40 - x)^2}$$

Right side triangle
$$c^2 = 30^2 + x^2$$
$$c = \sqrt{900 + x^2}$$

Total length $= \sqrt{900 + x^2} + \sqrt{400 + (40 - x)^2}$

b. Complete the table.

x	0	10	20	30	40
Total Length	74.72	67.68	64.34	64.79	70

c. Find the domain. Domain: [0, 40].

109. Complete the table.

x	5	10	12.5	15	20
$Y(x)$	275	375	385	390	394

Answers accurate to the nearest apple.

111. Find c.

$$f(c) = c^2 - c - 5 = 1$$
$$c^2 - c - 6 = 0$$
$$(c - 3)(c + 2) = 0$$
$$c - 3 = 0 \quad \text{or} \quad c + 2 = 0$$
$$c = 3 \qquad\qquad c = -2$$

113. Determine if 1 is in the range.

1 is not in the range of $f(x)$, since

$$1 = \frac{x - 1}{x + 1} \text{ only if } x + 1 = x - 1 \text{ or } 1 = -1.$$

115. Graph functions. Explain how the graphs are related.

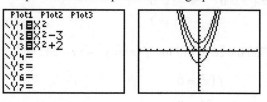

The graph of $g(x) = x^2 - 3$ is the graph of $f(x) = x^2$

shifted down 3 units. The graph of $h(x) = x^2 + 2$ is

the graph of $f(x) = x^2$ shifted up 2 units.

117. Find all fixed points.

$$a^2 + 3a - 3 = a$$
$$a^2 + 2a - 3 = 0$$
$$(a - 1)(a + 3) = 0$$
$$a = 1 \quad \text{or} \quad a = -3$$

119. a. Write the function.

$$A = xy$$
$$A(x) = x\left(-\frac{1}{2}x + 4\right)$$
$$A(x) = -\frac{1}{2}x^2 + 4x$$

b. Complete the table.

x	1	2	4	6	7
Area	3.5	6	8	6	3.5

c. Find the domain. Domain: [0, 8].

121. a. Write the function.

Circle	Square
$C = 2\pi r$	$C = 4s$
$x = 2\pi r$	$20 - x = 4s$
$r = \dfrac{x}{2\pi}$	$s = 5 - \dfrac{x}{4}$
$\text{Area} = \pi r^2 = \pi\left(\dfrac{x}{2\pi}\right)^2$	$\text{Area} = s^2 = \left(5 - \dfrac{x}{4}\right)^2$
$= \dfrac{x^2}{4\pi}$	$= 25 - \dfrac{5}{2}x + \dfrac{x^2}{16}$

$$\text{Total Area} = \dfrac{x^2}{4\pi} + 25 - \dfrac{5}{2}x + \dfrac{x^2}{16}$$

$$= \left(\dfrac{1}{4\pi} + \dfrac{1}{16}\right)x^2 - \dfrac{5}{2}x + 25$$

b. Complete the table.

x	0	4	8	12	16	20
Total Area	25	17.27	14.09	15.46	21.37	31.83

c. Find the domain. Domain: [0, 20].

Prepare for Section 2.3

P1. $d = 5 - (-2) = 7$

P3. $\dfrac{-4 - 4}{2 - (-3)} = \dfrac{-8}{5}$

P5. $3x - 5y = 15$

$-5y = -3x + 15$

$y = \dfrac{3}{5}x - 3$

Section 2.3 Exercises

1. If a line has a negative slope, then as the value of y increases, the value of x <u>decreases</u>.

3. The graph of a line with zero slope is <u>horizontal</u>.

5. Determine the slope and y-intercept.

$y = 4x - 5$: $m = 4$, y-intercept: $(0, -5)$

7. Determine the slope and y-intercept.

$f(x) = \dfrac{2x}{3}$: $m = \dfrac{2}{3}$, y-intercept: $(0, 0)$

9. Determine whether the graphs are parallel, perpendicular, or neither.

$y = 3x - 4$: $m = 3$,
$y = -3x + 2$: $m = -3$

The graphs are neither parallel nor perpendicular

11. Determine whether the graphs are parallel, perpendicular, or neither.

$f(x) = 3x - 1$: $m = 3$,

$y = -\dfrac{x}{3} - 1$: $m = -\dfrac{1}{3}$

$3\left(-\dfrac{1}{3}\right) = -1$

The graphs are perpendicular.

13. Find the slope.

$m = \dfrac{y_2 - y_1}{x_2 - x_1} = \dfrac{7 - 4}{1 - 3} = \dfrac{3}{-2} = -\dfrac{3}{2}$

15. Find the slope.

$m = \dfrac{2 - 0}{0 - 4} = -\dfrac{1}{2}$

17. Find the slope.

$m = \dfrac{-7 - 2}{3 - 3} = \dfrac{-9}{0}$ undefined

19. Find the slope.

$m = \dfrac{-2 - 4}{-4 - (-3)} = \dfrac{-6}{-1} = 6$

21. Find the slope.

$m = \dfrac{\dfrac{7}{2} - \dfrac{1}{2}}{\dfrac{7}{3} - (-4)} = \dfrac{\dfrac{6}{2}}{\dfrac{19}{3}} = 3 \cdot \dfrac{3}{19} = \dfrac{9}{19}$

23. Find the slope, y-intercept, and graph. $y = 2x - 4$

$m = 2$, y-intercept $(0, -4)$

25. Find the slope, y-intercept, and graph. $y = \dfrac{3}{4}x + 1$

$m = \dfrac{3}{4}$, y-intercept $(0, 1)$

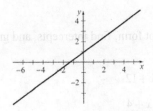

27. Find the slope, y-intercept, and graph. $y = -2x + 3$

$m = -2$, y-intercept $(0, 3)$

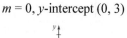

29. Find the slope, y-intercept, and graph. $y = 3$

$m = 0$, y-intercept $(0, 3)$

31. Find the slope, y-intercept, and graph. $y = 2x$

$m = 2$, y-intercept $(0, 0)$

33. Find the slope, y-intercept, and graph. $y = x$

$m = 1$, y-intercept $(0, 0)$

35. Write slope-intercept form, find intercepts, and graph.

$2x + y = 5$

$y = -2x + 5$

x-intercept $\left(\dfrac{5}{2}, 0 \right)$, y-intercept $(0, 5)$

37. Write slope-intercept form, find intercepts, and graph.

$4x + 3y - 12 = 0$

$3y = -4x + 12$

$y = -\dfrac{4}{3}x + 4$

x-intercept $(3, 0)$, y-intercept $(0, 4)$

39. Write slope-intercept form, find intercepts, and graph.

$2x - 5y = -15$

$-5y = -2x - 15$

$y = \dfrac{2}{5}x + 3$

x-intercept $\left(-\dfrac{15}{2}, 0 \right)$, y-intercept $(0, 3)$

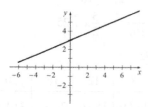

41. Write slope-intercept form, find intercepts, and graph.

$x + 2y = 6$

$y = -\dfrac{1}{2}x + 3$

x-intercept $(6, 0)$, y-intercept $(0, 3)$

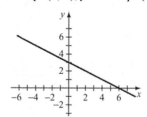

43. Find the equation.

Use $y = mx + b$ with $m = 1$, $b = 3$.

$y = x + 3$

45. Find the equation.

Use $y = mx + b$ with $m = \dfrac{3}{4}$, $b = \dfrac{1}{2}$.

$y = \dfrac{3}{4}x + \dfrac{1}{2}$

47. Find the equation.

Use $y = mx + b$ with $m = 0$, $b = 4$.

$y = 4$

49. Find the equation.

$$y - 2 = -4(x - (-3))$$
$$y - 2 = -4x - 12$$
$$y = -4x - 10$$

51. Find the equation.

$$m = \frac{4-1}{-1-3} = \frac{3}{-4} = -\frac{3}{4}$$

$$y - 1 = -\frac{3}{4}(x - 3)$$

$$y = -\frac{3}{4}x + \frac{9}{4} + \frac{4}{4}$$

$$y = -\frac{3}{4}x + \frac{13}{4}$$

53. Find the equation.

$$m = \frac{-1-11}{2-7} = \frac{-12}{-5} = \frac{12}{5}$$

$$y - 11 = \frac{12}{5}(x - 7)$$

$$y - 11 = \frac{12}{5}x - \frac{84}{5}$$

$$y = \frac{12}{5}x - \frac{84}{5} + \frac{55}{5}$$

$$= \frac{12}{5}x - \frac{29}{5}$$

55. Find the equation.

$$y = 2x + 3 \text{ has slope } m = 2.$$

$$y - y_1 = 2(x - x_1)$$

$$y + 4 = 2(x - 2)$$

$$y + 4 = 2x - 4$$

$$y = 2x - 8$$

57. Find the equation.

$$y = -\frac{3}{4}x + 3 \text{ has slope } m = -\frac{3}{4}.$$

$$y - y_1 = -\frac{3}{4}(x - x_1)$$

$$y - 2 = -\frac{3}{4}(x + 4)$$

$$y - 2 = -\frac{3}{4}x - 3$$

$$y = -\frac{3}{4}x - 1$$

59. Find the equation.

$$2x - 5y = 2$$

$$-5y = -2x + 2$$

$$y = \frac{2}{5}x - \frac{2}{5} \text{ has slope } m = \frac{2}{5}.$$

$$y - y_1 = \frac{2}{5}(x - x_1)$$

$$y - 2 = \frac{2}{5}(x - 5)$$

$$y - 2 = \frac{2}{5}x - 2$$

$$y = \frac{2}{5}x$$

61. Find the equation.

$$y = 2x - 5 \text{ has perpendicular slope } m = -\frac{1}{2}.$$

$$y - y_1 = -\frac{1}{2}(x - x_1)$$

$$y + 4 = -\frac{1}{2}(x - 3)$$

$$y + 4 = -\frac{1}{2}x + \frac{3}{2}$$

$$y = -\frac{1}{2}x - \frac{5}{2}$$

63. Find the equation.

$$y = -\frac{3}{4}x + 1 \text{ has perpendicular slope } m = \frac{4}{3}.$$

$$y - y_1 = \frac{4}{3}(x - x_1)$$

$$y - 0 = \frac{4}{3}(x + 6)$$

$$y = \frac{4}{3}x + 8$$

65. Find the equation.

$$-x - 4y = 6$$

$$-4y = x + 6$$

$$y = -\frac{1}{4}x - \frac{3}{2} \text{ has perpendicular slope } m = 4.$$

$$y - y_1 = 4(x - x_1)$$

$$y - 2 = 4(x - 5)$$

$$y - 2 = 4x - 20$$

$$y = 4x - 18$$

67. Find the zero of f.

$$f(x) = 3x - 12$$
$$3x - 12 = 0$$
$$3x = 12$$
$$x = 4$$

The x-intercept of the graph of $f(x)$ is $(4, 0)$.

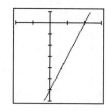

Xmin = -4, Xmax = 6, Xscl=2,

Ymin = -12.2, Ymax = 2, Yscl = 2

69. Find the zero of f.

$$f(x) = \frac{1}{4}x + 5$$
$$\frac{1}{4}x + 5 = 0$$
$$\frac{1}{4}x = -5$$
$$x = -20$$

The x-intercept of the graph of $f(x)$ is $(-20, 0)$.

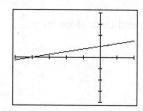

Xmin = -30, Xmax = 30, Xscl = 10,

Ymin = -10, Ymax = 10, Yscl = 1

71. Find the slope and explain the meaning.

$$m = \frac{1505 - 1482}{28 - 20} = 2.875$$

The value of the slope indicates that the speed of sound in water increases 2.875 m/s for a one-degree Celsius increase in temperature.

73. a. $m = \dfrac{31 - 20}{23 - 12} = 1$

$$H(c) - 20 = 1(c - 12)$$
$$H(c) = c + 8$$

b. $H(19) = (19) + 8 = 27$ mpg

75. a. $m = \dfrac{316,500 - 279,200}{2020 - 2010} = 3730$

$$N(t) - 279,200 = 3730(t - 2010)$$
$$N(t) = 3730t - 7,218,100$$

b. $300,000 = 3730t - 7,218,100$

$$7,518,100 = 3730t$$
$$2015.6 \approx t$$

The number of jobs will exceed 300,000 in 2015.

77. a. $m = \dfrac{240 - 180}{18 - 16} = 30$

$$B(d) - 180 = 30(d - 16)$$
$$B(d) = 30d - 300$$

b. The value of the slope means that a 1-inch increase in the diameter of a log 32 ft long results in an increase of 30 board-feet of lumber that can be obtained from the log.

c. $B(19) = 30(19) - 300 = 270$ board feet

79. Line A represents Michelle.

Line B represents Amanda.

Line C represents the distance between Michelle and Amanda.

81. a. $m = \dfrac{80.5 - 19.9}{0 - 65} \approx -0.9323$

$$y - 80.5 = -0.9323(x - 0)$$
$$y = -0.9323x + 80.5$$

b. $y = -0.9323(25) + 80.5 = 57.19 \approx 57$ years

83. Determine the profit function and break-even point.

$$P(x) = 92.50x - (52x + 1782)$$
$$P(x) = 92.50x - 52x - 1782$$
$$P(x) = 40.50x - 1782$$

$$40.50x - 1782 = 0$$
$$40.50x = 1782$$
$$x = \frac{1782}{40.50}$$
$$x = 44, \text{ the break-even point}$$

85. Determine the profit function and break-even point.

$$P(x) = 259x - (180x + 10,270)$$
$$P(x) = 259x - 180x - 10,270$$
$$P(x) = 79x - 10,270$$

$$79x - 10,270 = 0$$
$$79x = 10,270$$
$$x = \frac{10,270}{79}$$
$$x = 130, \text{ the break-even point}$$

87. a. $C(0) = 8(0) + 275 = 0 + 275 = \$275$

b. $C(1) = 8(1) + 275 = 8 + 275 = \$283$

c. $C(10) = 8(10) + 275 = 80 + 275 = \$355$

d. The marginal cost is the slope of $C(x) = 8x + 275$,

which is $\$8$ per unit.

89. a. $C(t) = 19,500.00 + 6.75t$

b. $R(t) = 55.00t$

c. $P(t) = R(t) - C(t)$
$$P(t) = 55.00t - (19,500.00 + 6.75t)$$
$$P(t) = 55.00t - 19,500.00 - 6.75t$$
$$P(t) = 48.25t - 19,500.00$$

d. $48.25t = 19,500.00$
$$t = \frac{19,500.00}{48.25}$$
$$t = 404.1451 \text{ days} \approx 405 \text{ days}$$

91. The equation of the line through $(0,0)$ and $P(3,4)$ has

slope $\frac{4}{3}$.

The path of the rock is on the line through $P(3,4)$ with

slope $-\frac{3}{4}$, so $y - 4 = -\frac{3}{4}(x - 3)$.

$$y - 4 = -\frac{3}{4}x + \frac{9}{4}$$
$$y = -\frac{3}{4}x + \frac{9}{4} + 4$$
$$y = -\frac{3}{4}x + \frac{25}{4}$$

The point where the rock hits the wall at $y = 10$ is the

point of intersection of $y = -\frac{3}{4}x + \frac{25}{4}$ and $y = 10$.

$$-\frac{3}{4}x + \frac{25}{4} = 10$$
$$-3x + 25 = 40$$
$$-3x = 15$$
$$x = -5 \text{ feet}$$

Therefore the rock hits the wall at $(-5, 10)$.

The x-coordinate is -5.

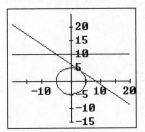

93. Find a.
$$f(a) = 2a + 3 = -1$$
$$2a = -4$$
$$a = -2$$

95. Find a.
$$f(a) = 1 - 4a = 3$$
$$-4a = 2$$
$$a = -\frac{1}{2}$$

97. a. $h = 1$ so
$$Q(1 + h, [1 + h]^2 + 1) = Q(2, 2^2 + 1) = Q(2, 5)$$
$$m = \frac{5 - 2}{2 - 1} = \frac{3}{1} = 3$$

b. $h = 0.1$ so
$$Q(1 + h, [1 + h]^2 + 1) = Q(1.1, 1.1^2 + 1) = Q(1.1, 2.21)$$
$$m = \frac{2.21 - 2}{1.1 - 1} = \frac{0.21}{0.1} = 2.1$$

c. $h = 0.01$ so
$$Q(1 + h, [1 + h]^2 + 1) = Q(1.01, 1.01^2 + 1)$$
$$= Q(1.01, 2.0201)$$
$$m = \frac{2.0201 - 2}{1.01 - 1} = \frac{0.0201}{0.01} = 2.01$$

d. As h approaches 0, the slope of PQ seems to be

approaching 2.

e. $x_1 = 1, y_1 = 2, x_2 = 1+h, y_2 = [1+h]^2 + 1$

$$m = \frac{y_2 - y_1}{x_2 - x_1} = \frac{[1+h]^2 + 1 - 2}{(1+h) - 1} = \frac{(1+2h+h^2) + 1 - 2}{h}$$

$$= \frac{2h + h^2}{h} = 2 + h$$

99. The slope of the line through $(3, 9)$ and (x, y)

is $\frac{15}{2}$, so $\frac{y-9}{x-3} = \frac{15}{2}$.

Therefore

$$2(y-9) = 15(x-3)$$
$$2y - 18 = 15x - 45$$
$$2y - 15x + 27 = 0$$
$$2x^2 - 15x + 27 = 0 \quad \text{Substituting } y = x^2$$
$$(2x-9)(x-3) = 0$$
$$x = \frac{9}{2} \quad \text{or} \quad x = 3$$

If $x = \frac{9}{2}$, $y = x^2 = \left(\frac{9}{2}\right)^2 = \frac{81}{4} \Rightarrow \left(\frac{9}{2}, \frac{81}{4}\right)$.

If $x = 3$, $y = x^2 = (3)^2 = 9 \Rightarrow (3, 9)$, but this is the

point itself. The point $\left(\frac{9}{2}, \frac{81}{4}\right)$ is on the graph of

$y = x^2$, and the slope of the line containing $(3, 9)$ and

$\left(\frac{9}{2}, \frac{81}{4}\right)$ is $\frac{15}{2}$.

Mid-Chapter 2 Quiz

1. Find the midpoint and length.

$$M = \left(\frac{-3+1}{2}, \frac{4-2}{2}\right) = \left(\frac{-2}{2}, \frac{2}{2}\right) = (-1, 1)$$

$$d = \sqrt{(1-(-3))^2 + (-2-4)^2} = \sqrt{(4)^2 + (-6)^2}$$
$$= \sqrt{16 + 36} = \sqrt{52}$$
$$= 2\sqrt{13}$$

3. Evaluate.

$$f(x) = x^2 - 6x + 1$$
$$f(-3) = (-3)^2 - 6(-3) + 1 = 9 + 18 + 1 = 28$$

5. Find the zeros of f for $f(x) = x^2 - x - 12$.

$$f(x) = 0$$
$$x^2 - x - 12 = 0$$
$$(x+3)(x-4) = 0$$
$$x+3 = 0 \quad x-4 = 0$$
$$x = -3 \quad\quad x = 4$$

7. Find the equation.

$$2x + 3y = 5$$
$$3y = -2x + 5$$
$$y = -\frac{2}{3}x + \frac{5}{3} \quad \text{has slope } m = -\frac{2}{3}.$$

$$y - y_1 = -\frac{2}{3}(x - x_1)$$
$$y - (-1) = -\frac{2}{3}(x - 3)$$
$$y + 1 = -\frac{2}{3}x + 2$$
$$y = -\frac{2}{3}x + 1$$

Prepare for Section 2.4

P1. $3x^2 + 10x - 8 = (3x - 2)(x + 4)$

P3. $f(-3) = 2(-3)^2 - 5(-3) - 7$
$$= 18 + 15 - 7$$
$$= 26$$

P5. $x^2 + 3x - 2 = 0$

$$x = \frac{-3 \pm \sqrt{(3)^2 - 4(1)(-2)}}{2(1)}$$
$$= \frac{-3 \pm \sqrt{17}}{2}$$

Section 2.4 Exercises

1. d

3. b

5. g

7. c

9. Write in standard form, find the vertex, the axis of symmetry and graph.

$$f(x) = (x^2 + 4x) + 1$$
$$= (x^2 + 4x + 4) + 1 - 4$$
$$= (x + 2)^2 - 3 \quad \text{standard form,}$$

vertex $(-2, -3)$, axis of symmetry $x = -2$

11. Write in standard form, find the vertex, the axis of symmetry and graph.

$$f(x) = (x^2 - 8x) + 5$$
$$= (x^2 - 8x + 16) + 5 - 16$$
$$= (x - 4)^2 - 11 \quad \text{standard form,}$$

vertex $(4, -11)$, axis of symmetry $x = 4$

13. Write in standard form, find the vertex, the axis of symmetry and graph.

$$f(x) = (x^2 + 3x) + 1$$
$$= \left(x^2 + 3x + \frac{9}{4}\right) + 1 - \frac{9}{4}$$
$$= \left(x + \frac{3}{2}\right)^2 + \frac{4}{4} - \frac{9}{4}$$
$$= \left(x + \frac{3}{2}\right)^2 - \frac{5}{4} \quad \text{standard form,}$$

vertex $\left(-\frac{3}{2}, -\frac{5}{4}\right)$, axis of symmetry $x = -\frac{3}{2}$

15. Write in standard form, find the vertex, the axis of symmetry and graph.

$$f(x) = -x^2 + 4x + 2$$
$$= -(x^2 - 4x) + 2$$
$$= -(x^2 - 4x + 4) + 2 + 4$$
$$= -(x - 2)^2 + 6 \quad \text{standard form,}$$

vertex $(2, 6)$, axis of symmetry $x = 2$

17. Write in standard form, find the vertex, the axis of symmetry and graph.

$$f(x) = -3x^2 + 3x + 7$$
$$= -3(x^2 - 1x) + 7$$
$$= -3\left(x^2 - 1x + \frac{1}{4}\right) + 7 + \frac{3}{4}$$
$$= -3\left(x - \frac{1}{2}\right)^2 + \frac{28}{4} + \frac{3}{4}$$
$$= -3\left(x - \frac{1}{2}\right)^2 + \frac{31}{4} \quad \text{standard form,}$$

vertex $\left(\frac{1}{2}, \frac{31}{4}\right)$, axis of symmetry $x = \frac{1}{2}$

19. Find the vertex, write the function in standard form.

$$x = \frac{-b}{2a} = \frac{10}{2(1)} = 5$$

$$y = f(5) = (5)^2 - 10(5)$$
$$= 25 - 50 = -25$$

vertex $(5, -25)$
$$f(x) = (x - 5)^2 - 25$$

21. Find the vertex, write the function in standard form.

$$x = \frac{-b}{2a} = \frac{0}{2(1)} = 0$$

$$y = f(0) = (0)^2 - 10 = -10$$

vertex $(0, -10)$
$$f(x) = x^2 - 10$$

23. Find the vertex, write the function in standard form.

$$x = \frac{-b}{2a} = \frac{-6}{2(-1)} = \frac{-6}{-2} = 3$$

$$y = f(3) = -(3)^2 + 6(3) + 1$$
$$= -9 + 18 + 1$$
$$= 10$$

vertex (3, 10)

$$f(x) = -(x-3)^2 + 10$$

25. Find the vertex, write the function in standard form.

$$x = \frac{-b}{2a} = \frac{3}{2(2)} = \frac{3}{4}$$

$$y = f\left(\frac{3}{4}\right) = 2\left(\frac{3}{4}\right)^2 - 3\left(\frac{3}{4}\right) + 7$$

$$= 2\left(\frac{9}{16}\right) - \frac{9}{4} + 7$$

$$= \frac{9}{8} - \frac{9}{4} + 7 = \frac{9}{8} - \frac{18}{8} + \frac{56}{8}$$

$$= \frac{47}{8}$$

vertex $\left(\frac{3}{4}, \frac{47}{8}\right)$

$$f(x) = 2\left(x - \frac{3}{4}\right)^2 + \frac{47}{8}$$

27. Find the vertex, write the function in standard form.

$$x = \frac{-b}{2a} = \frac{-1}{2(-4)} = \frac{1}{8}$$

$$y = f\left(\frac{1}{8}\right) = -4\left(\frac{1}{8}\right)^2 + \left(\frac{1}{8}\right) + 1$$

$$= -4\left(\frac{1}{64}\right) + \frac{1}{8} + 1$$

$$= -\frac{1}{16} + \frac{1}{8} + 1 = -\frac{1}{16} + \frac{2}{16} + \frac{16}{16}$$

$$= \frac{17}{16}$$

vertex $\left(\frac{1}{8}, \frac{17}{16}\right)$

$$f(x) = -4\left(x - \frac{1}{8}\right)^2 + \frac{17}{16}$$

29. Find the range, find x.

$$f(x) = x^2 - 2x - 1$$
$$= (x^2 - 2x) - 1$$
$$= (x^2 - 2x + 1) - 1 - 1$$
$$= (x-1)^2 - 2$$

vertex (1, −2)

The y-value of the vertex is −2.

The parabola opens up since $a = 1 > 0$.

Thus the range is $\{y | y \geq -2\}$.

31. Find the range, find x.

$$f(x) = -2x^2 + 5x - 1$$

$$= -2\left(x^2 - \frac{5}{2}x\right) - 1$$

$$= -2\left(x^2 - \frac{5}{2}x + \frac{25}{16}\right) - 1 + 2\left(\frac{25}{16}\right)$$

$$= -2\left(x - \frac{5}{4}\right)^2 - \frac{8}{8} + \frac{25}{8}$$

$$= -2\left(x - \frac{5}{4}\right)^2 + \frac{17}{8}$$

vertex $\left(\frac{5}{4}, \frac{17}{8}\right)$

The y-value of the vertex is $\frac{17}{8}$.

The parabola opens down since $a = -2 < 0$.

Thus the range is $\left\{y \middle| y \leq \frac{17}{8}\right\}$.

33. Find the real zeros and x-intercepts.

$$f(x) = x^2 + 2x - 24$$
$$= (x+6)(x-4)$$

$$x + 6 = 0 \qquad x - 4 = 0$$
$$x = -6 \qquad x = 4$$
$$(-6, 0) \qquad (4, 0)$$

35. Find the real zeros and x-intercepts.

$$f(x) = 2x^2 + 11x + 12$$
$$= (x+4)(2x+3)$$

$$x + 4 = 0 \qquad 2x + 3 = 0$$
$$x = -4 \qquad x = -\frac{3}{2}$$
$$(-4, 0) \qquad \left(-\frac{3}{2}, 0\right)$$

37. Find the minimum or maximum.

$$f(x) = x^2 + 8x$$
$$= (x^2 + 8x + 16) - 16$$
$$= (x+4)^2 - 16$$

minimum value of −16 when $x = -4$

39. Find the minimum or maximum.

$$f(x) = -x^2 + 6x + 2$$
$$= -(x^2 - 6x) + 2$$
$$= -(x^2 - 6x + 9) + 2 + 9$$
$$= -(x-3)^2 + 11$$

maximum value of 11 when $x = 3$

41. Find the minimum or maximum.

$$f(x) = 2x^2 + 3x + 1$$
$$= 2\left(x^2 + \frac{3}{2}x\right) + 1$$
$$= 2\left(x^2 + \frac{3}{2}x + \frac{9}{16}\right) + 1 - 2\left(\frac{9}{16}\right)$$
$$= 2\left(x + \frac{3}{4}\right)^2 + \frac{8}{8} - \frac{9}{8}$$
$$= 2\left(x + \frac{3}{4}\right)^2 - \frac{1}{8}$$

minimum value of $-\frac{1}{8}$ when $x = -\frac{3}{4}$

43. Find the minimum or maximum.

$$f(x) = 5x^2 - 11$$
$$= 5(x^2) - 11$$
$$= 5(x-0)^2 - 11$$

minimum value of -11 when $x = 0$

45. Find the minimum or maximum.

$$f(x) = -\frac{1}{2}x^2 + 6x + 17$$
$$= -\frac{1}{2}(x^2 - 12x) + 17$$
$$= -\frac{1}{2}(x^2 - 12x + 36) + 17 + 18$$
$$= -\frac{1}{2}(x-6)^2 + 35$$

maximum value of 35 when $x = 6$

47. $A(t) = -4.9t^2 + 90t + 9000$

Microgravity begins and ends at a height of 9000 m.

$$9000 = -4.9t^2 + 90t + 9000$$
$$4.9t^2 - 90t = 0$$
$$t(4.9t - 90) = 0$$

$$t = 0 \quad 4.9t - 90 = 0$$
$$t = 0 \qquad t = \frac{90}{4.9} \approx 18.4$$

The time of microgravity is 18.4 seconds.

49. $h(x) = -\frac{3}{64}x^2 + 27 = -\frac{3}{64}(x-0)^2 + 27$

a. The maximum height of the arch is 27 feet.

b. $h(10) = -\frac{3}{64}(10)^2 + 27$
$$= -\frac{3}{64}(100) + 27$$
$$= -\frac{75}{16} + 27$$
$$= -\frac{75}{16} + \frac{432}{16}$$
$$= \frac{357}{16} = 22\frac{5}{16} \quad \text{feet}$$

c. $h(x) = 8 = -\frac{3}{64}x^2 + 27$

$$8 - 27 = -\frac{3}{64}x^2$$
$$-19 = -\frac{3}{64}x^2$$
$$64(-19) = -3x^2$$
$$\frac{64(-19)}{-3} = x^2$$
$$\sqrt{\frac{64(-19)}{-3}} = x$$
$$8\sqrt{\frac{19}{3}} = x$$
$$\frac{8\sqrt{19}\sqrt{3}}{3} = x$$
$$\frac{8\sqrt{57}}{3} = x$$
$$20.1 \approx x$$

$h(x) = 8$ when $x \approx 20.1$ feet

51. a. $3w + 2l = 600$
$$3w = 600 - 2l$$
$$w = \frac{600 - 2l}{3}$$

b. $A = w \cdot l$
$$A = \left(\frac{600 - 2l}{3}\right) l$$
$$= 200l - \frac{2}{3}l^2$$

c. $A = -\dfrac{2}{3}(l^2 - 300l)$

$A = -\dfrac{2}{3}(l^2 - 300l + 150^2) + 15,000$

In standard form,

$A = -\dfrac{2}{3}(l - 150)^2 + 15,000$

The maximum area of $15,000 \text{ ft}^2$ is produced when

$l = 150$ ft and the width $w = \dfrac{600 - 2(150)}{3} = 100$ ft .

53. a. $T(t) = -0.7t^2 + 9.4t + 59.3$

$= -0.7\left(t^2 - \dfrac{9.4}{0.7}t\right) + 59.3$

$= -0.7\left(t^2 - \dfrac{94}{7}t\right) + 59.3$

$= -0.7\left(t^2 - \dfrac{94}{7}t + \left[\dfrac{47}{7}\right]^2\right) + 59.3 + 0.7\left[\dfrac{47}{7}\right]^2$

$\approx -0.7\left(t - \dfrac{47}{7}\right)^2 + 90.857$

$\approx -0.7\left(t - 6\dfrac{5}{7}\right)^2 + 91$

The temperature is a maximum when $t = \dfrac{47}{7} = 6\dfrac{5}{7}$

hours after 6:00 A.M. Note $\dfrac{5}{7}(60 \text{ min}) \approx 43$ min.

Thus the temperature is a maximum at 12:43 P.M.

b. The maximum temperature is approximately $91°F$.

55. $t = -\dfrac{b}{2a} = -\dfrac{82.86}{2(-279.67)} = 0.14814$

$E(0.14814) = -279.67(0.14814)^2 + 82.86(0.14814)$
≈ 6.1

The maximum energy is 6.1 joules.

57. a. $E(v) = -0.018v^2 + 1.476v + 3.4$

$= -0.018\left(v^2 - \dfrac{1.476}{0.018}v\right) + 3.4$

$= -0.018(v^2 - 82v) + 3.4$

$= -0.018\left(v^2 - 82v + 41^2\right) + 3.4 + 0.018(41)^2$

$= -0.018(v - 41)^2 + 33.658$

The maximum fuel efficiency is obtained at a speed of
41 mph.

b. The maximum fuel efficiency for this car, to the nearest mile per gallon, is 34 mpg.

59. $-\dfrac{b}{2a} = -\dfrac{296}{2(-0.2)} = 740$

$R(740) = 296(740) - 0.2(740)^2 = 109,520$

Thus, 740 units yield a maximum revenue of $109,520.

61. $-\dfrac{b}{2a} = -\dfrac{1.7}{2(-0.01)} = 85$

$P(85) = -0.01(85)^2 + 1.7(85) - 48 = 24.25$

Thus, 85 units yield a maximum profit of $24.25.

63. $P(x) = R(x) - C(x)$

$= x(102.50 - 0.1x) - (52.50x + 1840)$

$= -0.1x^2 + 50x - 1840$

The break-even points occur when $R(x) = C(x)$

or $P(x) = 0$.

Thus, $0 = -0.1x^2 + 50x - 1840$

$x = \dfrac{-50 \pm \sqrt{50^2 - 4(-0.1)(-1840)}}{2(-0.1)}$

$= \dfrac{-50 \pm \sqrt{1764}}{-0.2}$

$= \dfrac{-50 \pm 42}{-0.2}$

$x = 40 \quad \text{or} \quad x = 460$

The break-even points occur when $x = 40$ or $x = 460$.

65. Let x = the number of people that take the tour.

a. $R(x) = x(15.00 + 0.25(60 - x))$

$= x(15.00 + 15 - 0.25x)$

$= -0.25x^2 + 30.00x$

b. $P(x) = R(x) - C(x)$

$= (-0.25x^2 + 30.00x) - (180 + 2.50x)$

$= -0.25x^2 + 27.50x - 180$

c. $-\dfrac{b}{2a} = -\dfrac{27.50}{2(-0.25)} = 55$

$P(55) = -0.25(55)^2 + 27.50(55) - 180 = \$576.25$

d. The maximum profit occurs when $x = 55$ tickets.

67. Let x = the number of $0.05 price reductions.

Then the price per gallon is

$$p(x) = 3.95 - 0.05x$$

The number of gallons sold each day is

$$q(x) = 10,000 + 500x$$

Then the revenue is

$$R(x) = (3.95 - 0.05x)(10,000 + 500x)$$
$$= -25x^2 + 1475x + 39,500$$

The cost is

$$C(x) = 2.75(10,000 + 500x) = 27,500 + 1375x$$

The profit equals revenue minus cost.

$$P(x) = R(x) - C(x)$$
$$= -25x^2 + 1475x + 39,500 - (27,500 + 1375x)$$
$$= -25x^2 + 100x + 12,000$$

The maximum profit occurs when

$$x = -\frac{b}{2a} = -\frac{100}{2(-25)} = 2$$

The price per gallon that maximizes profit is

$$p(2) = 3.95 - 0.05(2) = \$3.85$$

69. $h(t) = -16t^2 + 128t$

a. $-\dfrac{b}{2a} = -\dfrac{128}{2(-16)} = 4$ seconds

b. $h(4) = -16(4)^2 + 128(4) = 256$ feet

c. $0 = -16t^2 + 128t$

$$0 = -16t(t - 8)$$
$$-16t = 0 \quad \text{or} \quad t - 8 = 0$$
$$t = 0 \qquad\qquad t = 8$$

The projectile hits the ground at t = 8 seconds.

71. $y(x) = -0.014x^2 + 1.19x + 5$

$$-\frac{b}{2a} = -\frac{1.19}{2(-0.014)}$$
$$= 42.5$$

$$y(42.5) = -0.014(42.5)^2 + 1.19(42.5) + 5$$
$$= 30.2875$$
$$\approx 30 \text{ feet}$$

73. $h(x) = -0.002x^2 - 0.03x + 8$

$$h(39) = -0.002(39)^2 - 0.03(39) + 8 = 3.788 > 3$$

Solve for x using quadratic formula.

$$-0.002x^2 - 0.03x + 8 = 0$$
$$x^2 + 15x - 4000 = 0$$

$$x = \frac{-15 \pm \sqrt{(15)^2 - 4(1)(-4000)}}{2(1)}$$

$$= \frac{-15 \pm \sqrt{16,225}}{2}, \text{ use positive value of } x$$

$$x \approx 56.2$$

Yes, the conditions are satisfied.

75. Find height and radius.

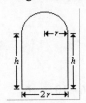

The perimeter is $48 = \pi r + h + 2r + h$.

Solve for h.

$$48 - \pi r - 2r = 2h$$
$$\frac{1}{2}(48 - \pi r - 2r) = h$$

Area = semicircle + rectangle

$$A = \frac{1}{2}\pi r^2 + 2rh$$
$$= \frac{1}{2}\pi r^2 + 2r\left(\frac{1}{2}\right)(48 - \pi r - 2r)$$
$$= \frac{1}{2}\pi r^2 + r(48 - \pi r - 2r)$$
$$= \frac{1}{2}\pi r^2 + 48r - \pi r^2 - 2r^2$$
$$= \left(\frac{1}{2}\pi - \pi - 2\right)r^2 + 48r$$
$$= \left(-\frac{1}{2}\pi - 2\right)r^2 + 48r$$

Graph the function A to find that its maximum occurs when $r \approx 6.72$ feet.

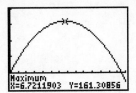

Xmin = 0, Xmax = 14, Xscl = 1

Ymin = −50, Ymax = 200, Yscl = 50

$$h = \frac{1}{2}(48 - \pi r - 2r)$$

$$\approx \frac{1}{2}(48 - \pi(6.72) - 2(6.72))$$

$$\approx 6.72 \text{ feet}$$

Hence the optimal window has its semicircular radius equal to its height.

Note: Using calculus it can be shown that the exact value of $r = h = \dfrac{48}{\pi + 4}$.

Prepare for Section 2.5

P1. $f(x) = x^2 + 4x - 6$

$$-\frac{b}{2a} = -\frac{4}{2(1)} = -2$$

$$x = -2$$

P3. $f(-2) = 2(-2)^3 - 5(-2) = -16 + 10 = -6$

$$-f(2) = -[2(2)^3 - 5(2)] = -[16 - 10] = -6$$

$$f(-2) = -f(2)$$

P5. $\dfrac{-a + a}{2} = 0, \quad \dfrac{b + b}{2} = b$

midpoint is $(0, b)$

Section 2.5 Exercises

1. Plot the points.

3. Plot the points.

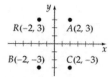

5. Plot the points.

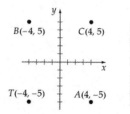

7. Determine how the graph is symmetric.

The graph is symmetric with respect to the origin.

9. Determine how the graph is symmetric.

The graph is symmetric with respect to the x-axis, the y-axis, and the origin.

11. Sketch the graph symmetric to the x-axis.

13. Sketch the graph symmetric to the y-axis.

15. Sketch the graph symmetric to the origin.

17. Determine if the graph is symmetric.

a. No

b. Yes

19. Determine if the graph is symmetric.

a. No

b. No

21. Determine if the graph is symmetric.

a. Yes

b. Yes

23. Determine if the graph is symmetric.

a. Yes

b. Yes

25. Determine if the graph is symmetric.

a. Yes

b. Yes

27. Determine if the graph is symmetric to the origin.

No, since $(-y) = 3(-x) - 2$ simplifies to

$(-y) = -3x - 2$, which is not equivalent to the original

equation $y = 3x - 2$.

29. Determine if the graph is symmetric to the origin.

Yes, since $(-y) = -(-x)^3$ implies

$-y = x^3$ or $y = -x^3$, which is the original equation.

31. Determine if the graph is symmetric to the origin.

Yes, since $(-x)^2 + (-y)^2 = 10$ simplifies to the original equation.

33. Determine if the graph is symmetric to the origin.

Yes, since $-y = \dfrac{-x}{\left|-x\right|}$ simplifies to the original

equation.

35. Determine if the function is odd, even or neither.

Even since $g(-x) = (-x)^2 - 7 = x^2 - 7 = g(x)$.

37. Determine if the function is odd, even or neither.

Odd, since $F(-x) = (-x)^5 + (-x)^3$
$$= -x^5 - x^3$$
$$= -F(x).$$

39. Determine if the function is odd, even or neither.

Even

41. Determine if the function is odd, even or neither.

Even

43. Determine if the function is odd, even or neither.

Even

45. Determine if the function is odd, even or neither.

Even

47. Determine if the function is odd, even or neither.

Neither

49. Sketch the graphs.

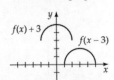

51. Sketch the graphs.

a. $f(x+2)$

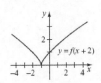

b. $f(x) + 2$

53. Sketch the graphs.

a. $y = f(x-2) + 1$

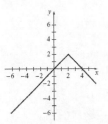

b. $y = f(x+3) - 2$

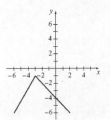

55. a. Give three points on the graph.

$f(x+3)$
$(-2-3, 5) = (-5, 5)$
$(0-3, -2) = (-3, -2)$
$(1-3, 0) = (-2, 0)$

b. Give three points on the graph.

$f(x) + 1$
$(-2, 5+1) = (-2, 6)$
$(0, -2+1) = (0, -1)$
$(1, 0+1) = (1, 1)$

57. a. Give two points on the graph.

$f(-x)$
$(--1, 3) = (1, 3)$
$(-2, -4)$

b. Give two points on the graph.

$-f(x)$
$(-1, -3)$
$(2, --4) = (2, 4)$

59. Sketch the graphs.

a. $f(-x)$

b. $-f(x)$

61. Sketch the graphs.

63. Sketch the graphs.

a. $2g(x)$

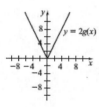

b. $\frac{1}{2}g(x)$

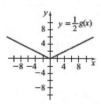

65. Sketch the graphs.

a. $3h(x)$

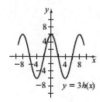

b. $\frac{1}{2}h(x)$

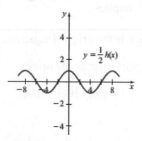

67. Sketch the graphs.

a. $f(2x)$

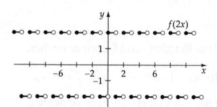

b. $f\left(\frac{1}{3}x\right)$

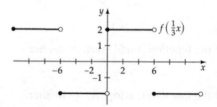

69. Sketch the graphs.

a. $h(2x)$

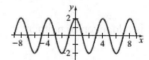

b. $h\left(\frac{1}{2}x\right)$

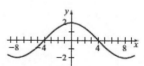

71. Sketch the graph.

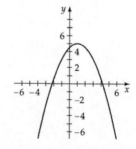

73. Sketch the graphs.

a. $y = -\dfrac{1}{2}j(x)+1$

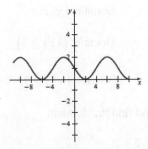

b. $y = 2j(x)-1$

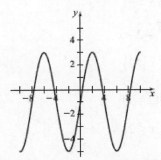

75. Graph using a graphing utility.

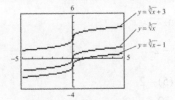

77. Graph using a graphing utility.

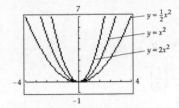

79. Reflect the graph about the *y*-axis and then about the origin.

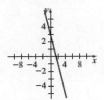

81. Reflect the graph about the *y*-axis and then about the *x*-axis.

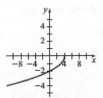

Prepare for Section 2.6

P1. $(2x^2+3x-4)-(x^2+3x-5) = x^2+1$

P3. $f(3a) = 2(3a)^2 - 5(3a)+2$

$\qquad = 18a^2 - 15a+2$

P5. Domain: all real numbers except $x = 1$

Section 2.6 Exercises

1. Evaluate.

$(f+g)(-2) = f(-2)+g(-2)$

$\qquad\qquad = 3+(-6)$

$\qquad\qquad = -3$

3. Evaluate.

$(f \cdot g)(-2) = f(-2) \cdot g(-2)$

$\qquad\qquad = 3(-6)$

$\qquad\qquad = -18$

5. Evaluate.

$g\big[f(-5)\big] = g[7] = -2$

7. Simplify.

$f(2+h) = 3(2+h)-4$

$\qquad\qquad = 6+3h-4$

$\qquad\qquad = 2+3h$

9. Perform the operations and find the domain.

$f(x)+g(x) = (x^2-2x-15)+(x+3)$

$\qquad\qquad = x^2-x-12$ Domain all real numbers

$f(x)-g(x) = (x^2-2x-15)-(x+3)$

$\qquad\qquad = x^2-3x-18$ Domain all real numbers

$f(x)g(x) = (x^2-2x-15)(x+3)$

$\qquad\qquad = x^3+x^2-21x-45$

$\qquad$ Domain all real numbers

$f(x)/g(x) = (x^2-2x-15)/(x+3)$

$\qquad\qquad = x-5$ Domain $\{x \mid x \neq -3\}$

11. Perform the operations and find the domain.

$$f(x)+g(x)=(2x+8)+(x+4)$$
$$=3x+12 \text{ Domain all real numbers}$$

$$f(x)-g(x)=(2x+8)-(x+4)$$
$$=x+4 \text{ Domain all real numbers}$$

$$f(x)g(x)=(2x+8)(x+4)$$
$$=2x^2+16x+32 \text{ Domain all real numbers}$$

$$f(x)/g(x)=(2x+8)/(x+4)$$
$$=[2(x+4)]/(x+4)$$
$$=2 \text{ Domain } \{x\,|\,x\neq-4\}$$

13. Perform the operations and find the domain.

$$f(x)+g(x)=(x^3-2x^2+7x)+x$$
$$=x^3-2x^2+8x \text{ Domain all real numbers}$$

$$f(x)-g(x)=(x^3-2x^2+7x)-x$$
$$=x^3-2x^2+6x \text{ Domain all real numbers}$$

$$f(x)g(x)=(x^3-2x^2+7x)x$$
$$=x^4-2x^3+7x^2 \text{ Domain all real numbers}$$

$$f(x)/g(x)=(x^3-2x^2+7x)/x$$
$$=x^2-2x+7 \text{ Domain } \{x\,|\,x\neq0\}$$

15. Perform the operations and find the domain.

$$f(x)+g(x)=(4x-7)+(2x^2+3x-5)$$
$$=2x^2+7x-12 \text{ Domain all real numbers}$$

$$f(x)-g(x)=(4x-7)-(2x^2+3x-5)$$
$$=-2x^2+x-2 \text{ Domain all real numbers}$$

$$f(x)g(x)=(4x-7)(2x^2+3x-5)$$
$$=8x^3-14x^2+12x^2-20x-21x+35$$
$$=8x^3-2x^2-41x+35$$
$$\text{Domain all real numbers}$$

$$f(x)/g(x)=(4x-7)/(2x^2+3x-5)$$
$$=\frac{4x-7}{2x^2+3x-5}$$
$$\text{Domain } \left\{x\,|\,x\neq1, x\neq-\frac{5}{2}\right\}$$

17. Perform the operations and find the domain.

$$f(x)+g(x)=\sqrt{x-3}+x \qquad \text{Domain } \{x\,|\,x\geq3\}$$

$$f(x)-g(x)=\sqrt{x-3}-x \qquad \text{Domain } \{x\,|\,x\geq3\}$$

$$f(x)g(x)=x\sqrt{x-3} \qquad \text{Domain } \{x\,|\,x\geq3\}$$

$$f(x)/g(x)=\frac{\sqrt{x-3}}{x} \qquad \text{Domain } \{x\,|\,x\geq3\}$$

19. Perform the operations and find the domain.

$$f(x)+g(x)=\sqrt{4-x^2}+2+x$$
$$\text{Domain } \{x\,|-2\leq x\leq2\}$$

$$f(x)-g(x)=\sqrt{4-x^2}-2-x$$
$$\text{Domain } \{x\,|-2\leq x\leq2\}$$

$$f(x)g(x)=\left(\sqrt{4-x^2}\right)(2+x)$$
$$\text{Domain } \{x\,|-2\leq x\leq2\}$$

$$f(x)/g(x)=\frac{\sqrt{4-x^2}}{2+x} \text{ Domain } \{x\,|-2< x\leq2\}$$

21. Evaluate the function.

$$(f+g)(x)=x^2-x-2$$

$$(f+g)(5)=(5)^2-(5)-2$$
$$=25-5-2$$
$$=18$$

23. Evaluate the function.

$$(f+g)(x)=x^2-x-2$$

$$(f+g)\left(\frac{1}{2}\right)=\left(\frac{1}{2}\right)^2-\left(\frac{1}{2}\right)-2$$
$$=\frac{1}{4}-\frac{1}{2}-2$$
$$=-\frac{9}{4}$$

25. Evaluate the function.

$$(f-g)(x)=x^2-5x+6$$
$$(f-g)(-3)=(-3)^2-5(-3)+6$$
$$=9+15+6$$
$$=30$$

27. Evaluate the function.

$$(f-g)(x) = x^2 - 5x + 6$$
$$(f-g)(-1) = (-1)^2 - 5(-1) + 6$$
$$= 1 + 5 + 6$$
$$= 12$$

29. Evaluate the function.

$$(fg)(x) = \left(x^2 - 3x + 2\right)(2x - 4)$$
$$= 2x^3 - 6x^2 + 4x - 4x^2 + 12x - 8$$
$$= 2x^3 - 10x^2 + 16x - 8$$
$$(fg)(7) = 2(7)^3 - 10(7)^2 + 16(7) - 8$$
$$= 686 - 490 + 112 - 8$$
$$= 300$$

31. Evaluate the function.

$$(fg)(x) = 2x^3 - 10x^2 + 16x - 8$$
$$(fg)\left(\frac{2}{5}\right) = 2\left(\frac{2}{5}\right)^3 - 10\left(\frac{2}{5}\right)^2 + 16\left(\frac{2}{5}\right) - 8$$
$$= \frac{16}{125} - \frac{40}{25} + \frac{32}{5} - 8$$
$$= \frac{-384}{125} = -3.072$$

33. Evaluate the function.

$$\left(\frac{f}{g}\right)(x) = \frac{x^2 - 3x + 2}{2x - 4}$$
$$\left(\frac{f}{g}\right)(x) = \frac{1}{2}x - \frac{1}{2}$$
$$\left(\frac{f}{g}\right)(-4) = \frac{1}{2}(-4) - \frac{1}{2}$$
$$= -2 - \frac{1}{2}$$
$$= -2\frac{1}{2} \text{ or } -\frac{5}{2}$$

35. Evaluate the function.

$$\left(\frac{f}{g}\right)(x) = \frac{1}{2}x - \frac{1}{2}$$
$$\left(\frac{f}{g}\right)\left(\frac{1}{2}\right) = \frac{1}{2}\left(\frac{1}{2}\right) - \frac{1}{2}$$
$$= \frac{1}{4} - \frac{1}{2}$$
$$= -\frac{1}{4}$$

37. Find the difference quotient.

$$\frac{f(x+h) - f(x)}{h} = \frac{[2(x+h) + 4] - (2x + 4)}{h}$$
$$= \frac{2x + 2(h) + 4 - 2x - 4}{h}$$
$$= \frac{2h}{h}$$
$$= 2$$

39. Find the difference quotient.

$$\frac{f(x+h) - f(x)}{h} = \frac{\left[(x+h)^2 - 6\right] - (x^2 - 6)}{h}$$
$$= \frac{x^2 + 2x(h) + (h)^2 - 6 - x^2 + 6}{h}$$
$$= \frac{2x(h) + h^2}{h}$$
$$= 2x + h$$

41. Find the difference quotient.

$$\frac{f(x+h) - f(x)}{h}$$
$$= \frac{2(x+h)^2 + 4(x+h) - 3 - (2x^2 + 4x - 3)}{h}$$
$$= \frac{2x^2 + 4xh + 2h^2 + 4x + 4h - 3 - 2x^2 - 4x + 3}{h}$$
$$= \frac{4xh + 2h^2 + 4h}{h}$$
$$= 4x + 2h + 4$$

43. Find the difference quotient.

$$\frac{f(x+h) - f(x)}{h} = \frac{-4(x+h)^2 + 6 - (-4x^2 + 6)}{h}$$
$$= \frac{-4x^2 - 8xh - 4h^2 + 6 + 4x^2 - 6}{h}$$
$$= \frac{-8xh - 4h^2}{h}$$
$$= -8x - 4h$$

45. a. On $[0, 1]$, $a = 0$

$$\Delta t = 1 - 0 = 1$$
$$C(a + \Delta t) = C(1) = 99.8 \text{ (mg/L)/h}$$
$$C(a) = C(0) = 0$$

Average rate of change $= \dfrac{C(1) - C(0)}{1} = 99.8 - 0 = 99.8$

This is identical to the slope of the line through $(0, C(0))$ and $(1, C(1))$ since

$$m = \frac{C(1) - C(0)}{1 - 0} = C(1) - C(0)$$

b. On $[0, 0.5]$, $a = 0$, $\Delta t = 0.5$

Average rate of change

$$= \frac{C(0.5) - C(0)}{0.5} = \frac{78.1 - 0}{0.5} = 156.2 \text{ (mg/L)/h}$$

c. On $[1, 2]$, $a = 1$, $\Delta t = 2 - 1 = 1$

Average rate of change

$$= \frac{C(2) - C(1)}{1} = \frac{50.1 - 99.8}{1} = -49.7 \text{ (mg/L)/h}$$

d. On $[1, 1.5]$, $a = 1$, $\Delta t = 1.5 - 1 = 0.5$

Average rate of change

$$= \frac{C(1.5) - C(1)}{0.5} = \frac{84.4 - 99.8}{0.5} = \frac{-15.4}{0.5} = -30.8 \text{ (mg/L)/h}$$

e. On $[1, 1.25]$, $a = 1$, $\Delta t = 1.25 - 1 = 0.25$

Average rate of change

$$= \frac{C(1.25) - C(1)}{0.25} = \frac{95.7 - 99.8}{0.25} = \frac{-4.1}{0.25} = -16.4 \text{ (mg/L)/h}$$

f. On $[1, 1 + \Delta t]$,

$Con(1 + \Delta t)$

$$= 25(1 + \Delta t)^3 - 150(1 + \Delta t)^2 + 225(1 + \Delta t)$$

$$= 25(1 + 3\Delta t + 3(\Delta t)^2 + 1(\Delta t)^3) - 150(1 + 2(\Delta t) + (\Delta t)^2)$$
$$+ 225(1 + \Delta t)$$

$$= 25 + 75(\Delta t) + 75(\Delta t)^2 + 25(\Delta t)^3$$

$$\quad - 150 - 300(\Delta t) - 150(\Delta t)^2 + 225 + 225(\Delta t)$$

$$= 100 - 75(\Delta t)^2 + 25(\Delta t)^3$$

$Con(1) = 100$

Average rate of change

$$= \frac{Con(1 + \Delta t) - Con(1)}{\Delta t}$$

$$= \frac{100 - 75(\Delta t)^2 + 25(\Delta t)^3 - 100}{\Delta t}$$

$$= \frac{-75(\Delta t)^2 + 25(\Delta t)^3}{\Delta t}$$

$$= -75(\Delta t) + 25(\Delta t)^2$$

As Δt approaches 0, the average rate of change over

$[1, 1 + \Delta t]$ seems to approach 0 (mg/L)/h.

47. Find the composite functions.

$$(g \circ f)(x) = g[f(x)] \qquad (f \circ g)(x) = f[g(x)]$$
$$\qquad\qquad = g[3x + 5] \qquad\qquad\qquad = f[2x - 7]$$
$$\qquad\qquad = 2[3x + 5] - 7 \qquad\qquad = 3[2x - 7] + 5$$
$$\qquad\qquad = 6x + 10 - 7 \qquad\qquad\quad = 6x - 21 + 5$$
$$\qquad\qquad = 6x + 3 \qquad\qquad\qquad\quad = 6x - 16$$

49. Find the composite functions.

$$(g \circ f)(x) = g[x^2 + 4x - 1]$$
$$= [x^2 + 4x - 1] + 2$$
$$= x^2 + 4x + 1$$

$$(f \circ g)(x) = f[x + 2]$$
$$= [x + 2]^2 + 4[x + 2] - 1$$
$$= x^2 + 4x + 4 + 4x + 8 - 1$$
$$= x^2 + 8x + 11$$

51. Find the composite functions.

$$(g \circ f)(x) = g[f(x)]$$
$$= g[x^3 + 2x]$$
$$= -5[x^3 + 2x]$$
$$= -5x^3 - 10x$$

$$(f \circ g)(x) = f[g(x)]$$
$$= f[-5x]$$
$$= [-5x]^3 + 2[-5x]$$
$$= -125x^3 - 10x$$

53. Find the composite functions.

$$(g \circ f)(x) = g[f(x)]$$
$$= g\left[\frac{2}{x + 1}\right]$$
$$= 3\left[\frac{2}{x + 1}\right] - 5$$
$$= \frac{6}{x + 1} - \frac{5(x + 1)}{x + 1}$$
$$= \frac{6 - 5x - 5}{x + 1}$$
$$= \frac{1 - 5x}{x + 1}$$

$$(f \circ g)(x) = f[g(x)]$$
$$= f[3x - 5]$$
$$= \frac{2}{[3x - 5] + 1}$$
$$= \frac{2}{3x - 4}$$

55. Find the composite functions.

$$(g \circ f)(x) = g[f(x)] = g\left[\frac{1}{x^2}\right] \qquad (f \circ g)(x) = f[g(x)]$$

$$= f\left[\sqrt{x-1}\right]$$

$$= \sqrt{\left[\frac{1}{x^2}\right] - 1}$$

$$= \frac{1}{\left[\sqrt{x-1}\right]^2}$$

$$= \sqrt{\frac{1-x^2}{x^2}}$$

$$= \frac{1}{x-1}$$

$$= \frac{\sqrt{1-x^2}}{|x|}$$

57. Find the composite functions.

$$(g \circ f)(x) = g\left[\frac{3}{|5-x|}\right]$$

$$= -\frac{2}{\left[\dfrac{3}{|5-x|}\right]}$$

$$= \frac{-2|5-x|}{3}$$

$$(f \circ g)(x) = f\left[-\frac{2}{x}\right]$$

$$= \frac{3}{\left|5 - \left[-\dfrac{2}{x}\right]\right|}$$

$$= \frac{3}{\left|5 + \dfrac{2}{x}\right|}$$

$$= \frac{3}{\dfrac{|5x+2|}{|x|}}$$

$$= \frac{3|x|}{|5x+2|}$$

59. Evaluate the composite function.

$$(g \circ f)(x) = 4x^2 + 2x - 6$$

$$(g \circ f)(4) = 4(4)^2 + 2(4) - 6$$

$$= 64 + 8 - 6$$

$$= 66$$

61. Evaluate the composite function.

$$(f \circ g)(x) = 2x^2 - 10x + 3$$

$$(f \circ g)(-3) = 2(-3)^2 - 10(-3) + 3$$

$$= 18 + 30 + 3$$

$$= 51$$

63. Evaluate the composite function.

$$(g \circ h)(x) = 9x^4 - 9x^2 - 4$$

$$(g \circ h)(0) = 9(0)^4 - 9(0)^2 - 4$$

$$= -4$$

65. Evaluate the composite function.

$$(f \circ f)(x) = 4x + 9$$

$$(f \circ f)(8) = 4(8) + 9$$

$$= 41$$

67. Evaluate the composite function.

$$(h \circ g)(x) = -3x^4 + 30x^3 - 75x^2 + 4$$

$$(h \circ g)\left(\frac{2}{5}\right) = -3\left(\frac{2}{5}\right)^4 + 30\left(\frac{2}{5}\right)^3 - 75\left(\frac{2}{5}\right)^2 + 4$$

$$= -\frac{48}{625} + \frac{240}{125} - \frac{300}{25} + 4$$

$$= \frac{-48 + 1200 - 7500 + 2500}{625}$$

$$= -\frac{3848}{625}$$

69. Evaluate the composite function.

$$(g \circ f)(x) = 4x^2 + 2x - 6$$

$$(g \circ f)(\sqrt{3}) = 4(\sqrt{3})^2 + 2(\sqrt{3}) - 6$$

$$= 12 + 2\sqrt{3} - 6$$

$$= 6 + 2\sqrt{3}$$

71. Evaluate the composite function.

$$(g \circ f)(x) = 4x^2 + 2x - 6$$

$$(g \circ f)(2c) = 4(2c)^2 + 2(2c) - 6$$

$$= 16c^2 + 4c - 6$$

73. Evaluate the composite function.

$$(g \circ h)(x) = 9x^4 - 9x^2 - 4$$

$$(g \circ h)(k+1)$$

$$= 9(k+1)^4 - 9(k+1)^2 - 4$$

$$= 9(k^4 + 4k^3 + 6k^2 + 4k + 1) - 9k^2 - 18k - 9 - 4$$

$$= 9k^4 + 36k^3 + 54k^2 + 36k + 9 - 9k^2 - 18k - 13$$

$$= 9k^4 + 36k^3 + 45k^2 + 18k - 4$$

75. Show $(g \circ f)(x) = (f \circ g)(x)$.

$$
\begin{aligned}
(g \circ f)(x) && (f \circ g)(x) \\
= g[f(x)] && = f[g(x)] \\
= g[2x+3] && = f[5x+12] \\
= 5(2x+3)+12 && = 2(5x+12)+3 \\
= 10x+15+12 && = 10x+24+3 \\
= 10x+27 && = 10x+27
\end{aligned}
$$

$$(g \circ f)(x) = (f \circ g)(x)$$

77. Show $(g \circ f)(x) = (f \circ g)(x)$.

$$
\begin{aligned}
(g \circ f)(x) && (f \circ g)(x) \\
= g[f(x)] && = f[g(x)] \\
= g\left[\dfrac{6x}{x-1}\right] && = f\left[\dfrac{5x}{x-2}\right] \\
= \dfrac{5\left(\frac{6x}{x-1}\right)}{\frac{6x}{x-1}-2} && = \dfrac{6\left(\frac{5x}{x-2}\right)}{\frac{5x}{x-2}-1} \\
= \dfrac{\frac{30x}{x-1}}{\frac{6x-2x+2}{x-1}} = \dfrac{\frac{30x}{x-1}}{\frac{4x+2}{x-1}} && = \dfrac{\frac{30x}{x-2}}{\frac{5x-x+2}{x-2}} = \dfrac{\frac{30x}{x-2}}{\frac{4x+2}{x-2}} \\
= \dfrac{30x}{x-1} \cdot \dfrac{x-1}{2(2x+1)} && = \dfrac{30x}{x-2} \cdot \dfrac{x-2}{2(2x+1)} \\
= \dfrac{15x}{2x+1} && = \dfrac{15x}{2x+1}
\end{aligned}
$$

$$(g \circ f)(x) = (f \circ g)(x)$$

79. Show $(g \circ f)(x) = x$ and $(f \circ g)(x) = x$.

$$
\begin{aligned}
(g \circ f)(x) = g[f(x)] && (f \circ g)(x) = f[g(x)] \\
= g[2x+3] && = f\left[\dfrac{x-3}{2}\right] \\
= \dfrac{[2x+3]-3}{2} && = 2\left[\dfrac{x-3}{2}\right]+3 \\
= x && = x
\end{aligned}
$$

81. Show $(g \circ f)(x) = x$ and $(f \circ g)(x) = x$.

$$
\begin{aligned}
(g \circ f)(x) = g[f(x)] && (f \circ g)(x) = f[g(x)] \\
= g\left[\dfrac{4}{x+1}\right] && = f\left[\dfrac{4-x}{x}\right] \\
= \dfrac{4 - \left[\frac{4}{x+1}\right]}{\left[\frac{4}{x+1}\right]} && = \dfrac{4}{\left[\frac{4-x}{x}\right]+1} \\
= \dfrac{\frac{4x+4-4}{x+1}}{\frac{4}{x+1}} && = \dfrac{4}{\frac{4-x+x}{x}} \\
= \dfrac{4x}{x+1} \cdot \dfrac{x+1}{4} && = \dfrac{4}{\frac{4}{x}} \\
= x && = 4 \cdot \dfrac{x}{4} \\
&& = x
\end{aligned}
$$

83. $(Y \circ F)(x) = Y(F(x))$ converts x inches to yards. F takes x inches to feet, and then Y takes feet to yards.

85. a. $r = 1.5t$ and $A = \pi r^2$

so $A(t) = \pi[r(t)]^2$

$$
\begin{aligned}
&= \pi(1.5t)^2 \\
A(2) &= 2.25\pi(2)^2 \\
&= 9\pi \text{ square feet} \\
&\approx 28.27 \text{ square feet}
\end{aligned}
$$

b. $r = 1.5t$

$h = 2r = 2(1.5t) = 3t$ and

$V = \dfrac{1}{3}\pi r^2 h$ so

$$
\begin{aligned}
V(t) &= \dfrac{1}{3}\pi(1.5t)^2 [3t] \\
&= 2.25\pi t^3
\end{aligned}
$$

Note: $V = \dfrac{1}{3}\pi r^2 h = \dfrac{1}{3}(\pi r^2 h) = \dfrac{1}{3}hA$

$$
\begin{aligned}
&= \dfrac{1}{3}(3t)(2.25\pi t^2) = 2.25\pi t^3 \\
V(3) &= 2.25\pi(3)^3 \\
&= 60.75\pi \text{ cubic feet} \\
&\approx 190.85 \text{ cubic feet}
\end{aligned}
$$

87. a. Since

$$
\begin{aligned}
d^2 + 4^2 &= s^2, \\
d^2 &= s^2 - 16 \\
d &= \sqrt{s^2 - 16} \\
d &= \sqrt{(48-t)^2 - 16} \quad \text{Substitute } 48-t \text{ for } s \\
&= \sqrt{2304 - 96t + t^2 - 16} \\
&= \sqrt{t^2 - 96t + 2288}
\end{aligned}
$$

b. $s(35) = 48 - 35 = 13$ ft

$$
\begin{aligned}
d(35) &= \sqrt{35^2 - 96(35) + 2288} \\
&= \sqrt{153} \approx 12.37 \text{ ft}
\end{aligned}
$$

Prepare for Section 2.7

P1. Slope: $-\dfrac{1}{3}$; y-intercept: $(0, 4)$

P3. $y = -0.45x + 2.3$

P5. $f(2) = 3(2)^2 + 4(2) - 1 = 12 + 8 - 1 = 19$

Section 2.7 Exercises

1. The scatter diagram suggests no relationship between x and y.

3. The scatter diagram suggests a linear relationship between x and y.

5. Figure A better approximates a graph that can be modeled by an equation than does Figure B. Thus Figure A has a coefficient of determination closer to 1.

7. Enter the data on your calculator. The technique for a TI-83 calculator is illustrated here. Press STAT.

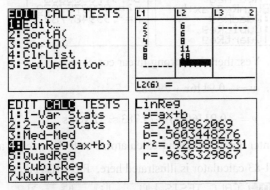

$$y = 2.00862069x + 0.5603448276$$

9. Enter the data on your calculator. The technique for a TI-83 calculator is illustrated here. Press STAT.

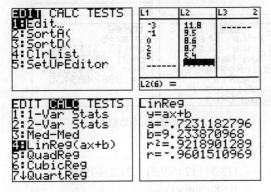

$$y = -0.7231182796x + 9.233870968$$

11. Enter the data on your calculator. The technique for a TI-83 calculator is illustrated here. Press STAT.

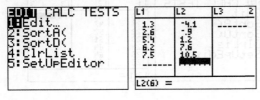

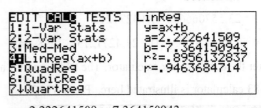

$$y = 2.222641509x - 7.364150943$$

13. Enter the data on your calculator. The technique for a TI-83 calculator is illustrated here. Press STAT.

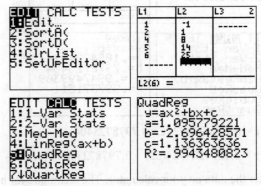

$$y = 1.095779221x^2 - 2.696428571x + 1.136363636$$

15. Enter the data on your calculator. The technique for a TI-83 calculator is illustrated here. Press STAT.

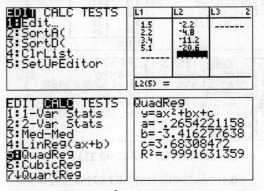

$$y = -0.2987274717x^2 - 3.20998141x + 3.416463667$$

17. Enter the data on your calculator. The technique for a TI-83 calculator is illustrated here. Press STAT.

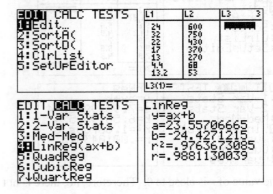

a. $y = 23.55706665x - 24.4271215$

b. $y = 23.55706665(54) - 24.4271215 \approx 1248$ cm

19. Enter the data on your calculator. The technique for a TI-83 calculator is illustrated here. Press STAT.

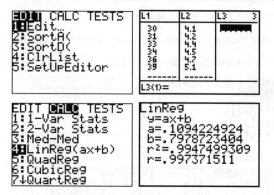

a. $y = 0.1094224924x + 0.7978723404$

b. $y = 0.1094224924(32) + 0.7978723404 \approx 4.3$ m/s

21. Enter the data on your calculator. The technique for a TI-83 calculator is illustrated here. Press STAT.

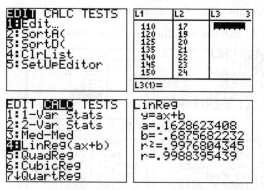

a. $y = 0.1628623408x - 0.6875682232$

b. $y = 0.1628623408(158) - 0.6875682232 \approx 25$

23. Enter the data on your calculator. The technique for a TI-83 calculator is illustrated here. Press STAT.

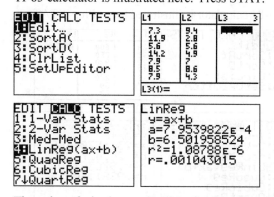

The value of r is close to 0. Therefore, no, there is not

a strong linear relationship between the current and the torque.

25. Enter the data on your calculator. The technique for a TI-83 calculator is illustrated here. Press STAT.

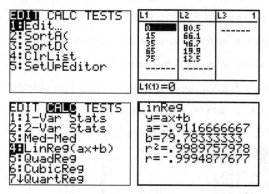

a. Yes, there is a strong linear correlation.

b. $y = -0.91\overline{16}x + 79.78\overline{3}$

c. $y = -0.91\overline{16}(25) + 79.78\overline{3} \approx 57$ years

27. Enter the data on your calculator. The technique for a TI-83 calculator is illustrated here. Press STAT.

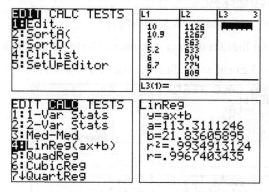

$y = 113.3111246x + 21.83605895$

a. Positively

b. $y = 113.3111246(9.5) + 21.83605895$
≈ 1098 calories

29. Enter the data on your calculator. The technique for a TI-83 calculator is illustrated here. Press STAT.

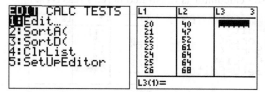

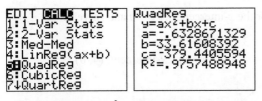

$$y = -0.6328671329x^2 + 33.61608329x - 379.4405594$$

31. Enter the data on your calculator. The technique for a TI-83 calculator is illustrated here. Press STAT.

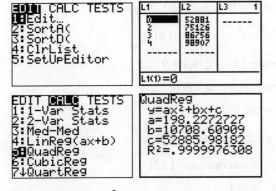

$$y = 198.2272727x^2 + 10,708.60909x + 52,885.98182$$

For 2006, $x = 1$,

$$y = 198.2272727(1)^2 + 10,708.60909(1) + 52,885.98182$$

$$\approx 63,793 \text{ thousand gallons}$$

33. Enter the data on your calculator. The technique for a TI-83 calculator is illustrated here. Press STAT.

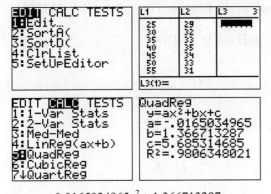

a. $y = -0.0165034965x^2 + 1.366713287x$

    $+ 5.685314685$

b. $y = -0.0165034965(50)^2 + 1.366713287(50)$

    $+ 5.685314685$

    ≈ 32.8 mpg

35. a. Enter the data on your calculator. The technique for a TI-83 calculator is illustrated here. Press STAT.

5-lb ball

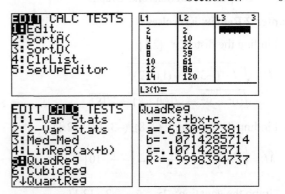

$$y = 0.6130952381t^2 - 0.0714285714t + 0.1071428571$$

10-lb ball

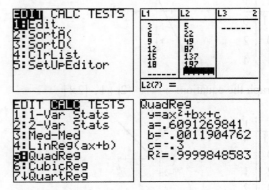

$$y = 0.6091269841t^2 - 0.0011904762t - 0.3$$

15-lb ball

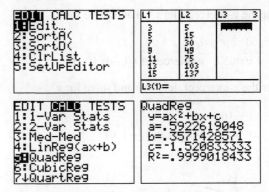

$$y = 0.5922619048t^2 + 0.3571428571t - 1.520833333$$

b. All the regression equations are approximately the same. Therefore, there is one equation of motion.

Chapter 2 Review Exercises

1. Finding the distance. [2.1]

$$d = \sqrt{(7-(-3))^2 + (11-2)^2}$$
$$= \sqrt{10^2 + 9^2} = \sqrt{100+81} = \sqrt{181}$$

3. Finding the midpoint: (2, 8), (–3, 12). [2.1]

$$M = \left(\frac{2+(-3)}{2}, \frac{8+12}{2}\right) = \left(-\frac{1}{2}, 10\right)$$

5. Graph the equation: $2x - y = -2$. [2.1]

x	y
−1	0
0	2
1	4

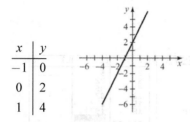

7. Graph the equation: $y = |x-2| + 1$. [2.1]

x	y
−1	4
0	3
2	1
3	2
4	3

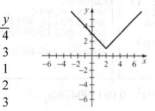

9. Finding x- and y-intercepts and graph: $x = y^2 - 1$ [2.1]

For the y-intercept, let $x = 0$ and solve for y.

$$0 = y^2 - 1$$
$$y = \pm 1, \ \ y\text{-intercepts: } (0, -1), (0, 1)$$

For the x-intercept, let $y = 0$ and solve for x.

$$x = (0)^2 - 1$$
$$x = -1, \ \ x\text{-intercept: } (-1, 0)$$

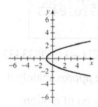

11. Finding x- and y-intercepts and graph: [2.1]

$$3x + 4y = 12$$

For the y-intercept, let $x = 0$ and solve for y.

$$3(0) + 4y = 12$$
$$y = 3, \ \ y\text{-intercept: } (0, 3)$$

For the x-intercept, let $y = 0$ and solve for x.

$$3x + 4(0) = 12$$
$$x = 4, \ \ x\text{-intercept: } (4, 0)$$

13. Finding the center and radius. [2.1]

$$(x-3)^2 + (y+4)^2 = 81$$

center (3, −4), radius 9

15. Finding the equation. [2.1]

Center: (2, −3), radius 5

$$(x-2)^2 + (y+3)^2 = 5^2$$

17. Is y a function of x? [2.2]

$$x - y = 4$$
$$y = x - 4, \ \ y \text{ is a function of } x.$$

19. Is y a function of x? [2.2]

$$|x| + |y| = 4$$
$$|y| = -|x| + 4$$
$$y = \pm(-|x| + 4), \ \ y \text{ is a not function of } x.$$

21. Evaluate the function $f(x) = 3x^2 + 4x - 5$, [2.2]

a. $f(1) = 3(1)^2 + 4(1) - 5$
$$= 3(1) + 4 - 5$$
$$= 3 + 4 - 5$$
$$= 2$$

b. $f(-3) = 3(-3)^2 + 4(-3) - 5$
$$= 3(9) - 12 - 5$$
$$= 27 - 12 - 5$$
$$= 10$$

c. $f(t) = 3t^2 + 4t - 5$

d. $f(x+h) = 3(x+h)^2 + 4(x+h) - 5$
$$= 3(x^2 + 2xh + h^2) + 4x + 4h - 5$$
$$= 3x^2 + 6xh + 3h^2 + 4x + 4h - 5$$

e. $3f(t) = 3(3t^2 + 4t - 5)$
$$= 9t^2 + 12t - 15$$

f. $f(3t) = 3(3t)^2 + 4(3t) - 5$

$= 3(9t^2) + 12t - 5$

$= 27t^2 + 12t - 5$

23. Evaluate the function. [2.2]

a. Since $x = 3 \geq 0$, use $f(x) = x^2 - 3$.

$f(3) = (3)^2 - 3 = 9 - 3 = 6$

b. Since $x = -2 < 0$, use $f(x) = 3x + 2$.

$f(-2) = 3(-2) + 2 = -6 + 2 = -4$

c. Since $x = 0 \geq 0$, use $f(x) = x^2 - 3$.

$f(0) = (0)^2 - 3 = 0 - 3 = -3$

25. Find the domain of $f(x) = -2x^2 + 3$. [2.2]

Domain $\{x | x \text{ is a real number}\}$

27. Find the domain of $f(x) = \sqrt{25 - x^2}$. [2.2]

Domain $\{x | -5 \leq x \leq 5\}$

29. Find the values of a in the domain of

$f(x) = x^2 + 2x - 4$ for which $f(a) = -1$. [2.2]

$a^2 + 2a - 4 = -1$ Replace $f(a)$ with $a^2 + 2a - 4$

$a^2 + 2a - 3 = 0$

$(a + 3)(a - 1) = 0$

$a + 3 = 0 \qquad a - 1 = 0$

$a = -3 \qquad a = 1$

31. Graph $f(x) = |x - 1| - 1$ [2.2]

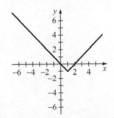

33. Find the zeros of f for $f(x) = 2x + 6$. [2.2]

$f(x) = 0$

$2x + 6 = 0$

$2x = -6$

$x = -3$

35. Evaluate the function $g(x) = [\![2x]\!]$. [2.2]

a. $g(\pi) = [\![2\pi]\!] \approx [\![6.283185307]\!] = 6$

b. $g\left(-\dfrac{2}{3}\right) = \left[\!\!\left[2\left(-\dfrac{2}{3}\right)\right]\!\!\right] = \left[\!\!\left[-\dfrac{4}{3}\right]\!\!\right] \approx [\![-1.333333]\!] = -2$

c. $g(-2) = [\![2(-2)]\!] = -4$

37. Find the slope. [2.3]

$m = \dfrac{-1 - 6}{4 + 3} = \dfrac{-7}{7} = -1$

39. Find the slope. [2.3]

$m = \dfrac{-2 + 2}{-3 - 4} = \dfrac{0}{-7} = 0$

41. Graph $f(x) = -\dfrac{3}{4}x + 2$. [2.3]

$m = -\dfrac{3}{4}$, y-intercept $(0, 2)$

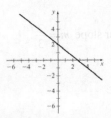

43. Graph $3x - 4y = 8$. [2.3]

$-4y = -3x + 8$

$y = \dfrac{3}{4}x - 2$

x-intercept $\left(\dfrac{8}{3}, 0\right)$, y-intercept $(0, -2)$

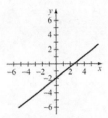

45. Find the equation. [2.3]

$y - 2 = -\dfrac{2}{3}(x + 3)$

$y - 2 = -\dfrac{2}{3}x - 2$

$y = -\dfrac{2}{3}x$

47. Find the equation. [2.3]

$$m = \frac{6-3}{1+2} = \frac{3}{3} = 1$$

$$y - 6 = 1(x-1)$$
$$y - 6 = x - 1$$
$$y = x + 5$$

49. Find the equation. [2.3]

$y = \frac{2}{3}x - 1$ has slope $m = \frac{2}{3}$.

$$y - y_1 = \frac{2}{3}(x - x_1)$$
$$y - (-5) = \frac{2}{3}(x - 3)$$
$$y + 5 = \frac{2}{3}x - 2$$
$$y = \frac{2}{3}x - 7$$

51. Find the equation. [2.3]

$y = -\frac{3}{2}x - 2$ has perpendicular slope $m = \frac{2}{3}$.

$$y - y_1 = \frac{2}{3}(x - x_1)$$
$$y - (-1) = \frac{2}{3}(x - 3)$$
$$y + 1 = \frac{2}{3}x - 2$$
$$y = \frac{2}{3}x - 3$$

53. Find the function. [2.3]

$$m = \frac{175 - 155}{118 - 106} = \frac{20}{12} = \frac{5}{3}$$

$$f(x) - 175 = \frac{5}{3}(x - 118)$$
$$f(x) - 175 = \frac{5}{3}x - \frac{590}{3}$$
$$f(x) = \frac{5}{3}x - \frac{65}{3}$$

55. Write the quadratic equation in standard form. [2.4]

$$f(x) = (x^2 + 6x) + 10$$
$$f(x) = (x^2 + 6x + 9) + 10 - 9$$
$$f(x) = (x + 3)^2 + 1$$

57. Write the quadratic equation in standard form. [2.4]

$$f(x) = -x^2 - 8x + 3$$
$$f(x) = -(x^2 + 8x) + 3$$
$$f(x) = -(x^2 + 8x + 16) + 3 + 16$$
$$f(x) = -(x + 4)^2 + 19$$

59. Write the quadratic equation in standard form. [2.4]

$$f(x) = -3x^2 + 4x - 5$$
$$f(x) = -3\left(x^2 - \frac{4}{3}x\right) - 5$$
$$f(x) = -3\left(x^2 - \frac{4}{3}x + \frac{4}{9}\right) - 5 + \frac{4}{3}$$
$$f(x) = -3\left(x - \frac{2}{3}\right)^2 - \frac{11}{3}$$

61. Find the vertex. [2.4]

$$\frac{-b}{2a} = \frac{-(-6)}{2(3)} = \frac{6}{6} = 1$$

$$f(1) = 3(1)^2 - 6(1) + 11$$
$$= 3(1) - 6 + 11$$
$$= 3 - 6 + 11$$
$$= 8$$

Thus the vertex is (1, 8).

63. Find the vertex. [2.4]

$$\frac{-b}{2a} = \frac{-(60)}{2(-6)} = \frac{-60}{-12} = 5$$

$$f(5) = -6(5)^2 + 60(5) + 11$$
$$= -6(25) + 300 + 11$$
$$= -150 + 300 + 11$$
$$= 161$$

Thus the vertex is (5, 161).

65. Find the value. [2.4]

$$f(x) = -x^2 + 6x - 3$$
$$= -(x^2 - 6x) - 3$$
$$= -(x^2 - 6x + 9) - 3 + 9$$
$$= -(x - 3)^2 + 6$$

maximum value of 6

67. Find the maximum height. [2.4]

$$h(t) = -16t^2 + 50t + 4$$

$$-\frac{b}{2a} = -\frac{50}{2(-16)} = \frac{25}{16}$$

$$h\left(\frac{25}{16}\right) = -16\left(\frac{25}{16}\right)^2 + 50\left(\frac{25}{16}\right) + 4 = 43.0625$$

The ball reaches a maximum height of 43.0625 ft.

69. Find the maximum area. [2.4]

Let x be the width. Using the formula for perimeter for three sides, $P = 2w + l \Rightarrow 700 = 2x + l$

$$l = 700 - 2x$$

Using the formula for area, $A = lw$. Then

$$A(x) = x(700 - 2x)$$

$$A(x) = -2x^2 + 700x$$

$$-\frac{b}{2a} = -\frac{700}{2(-2)} = 175$$

$$A(175) = -2(175)^2 + 700(175) = 61{,}250 \text{ ft}^2$$

71. Sketch a graph with different kinds of symmetry. [2.5]

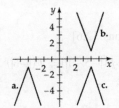

73. The graph of $x = y^2 + 3$ is symmetric with respect to the x-axis. [2.5]

75. The graph of $y^2 = x^2 + 4$ is symmetric with respect to the x-axis, y-axis, and the origin. [2.5]

77. The graph of $xy = 8$ is symmetric with respect to the origin. [2.5]

79. The graph of $|x + y| = 4$ is symmetric with respect to the origin. [2.5]

81. Sketch the graph $g(x) = -2x - 4$. [2.5]

a. Domain all real numbers

Range all real numbers

b. g is neither even nor odd

83. Sketch the graph $g(x) = \sqrt{16 - x^2}$. [2.5]

a. Domain $\{x \mid -4 \le x \le 4\}$

Range $\{y \mid 0 \le y \le 4\}$

b. g is an even function

85. Sketch the graph $g(x) = 2[\![x]\!]$. [2.5]

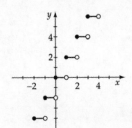

a. Domain all real numbers

Range $\{y \mid y \text{ is an even integer}\}$

b. g is neither even nor odd

87. $g(x) = f(x + 3)$ [2.5]

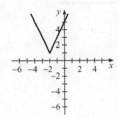

89. $g(x) = f(x + 2) - 1$ [2.5]

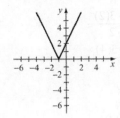

91. $g(x) = -f(x)$ [2.5]

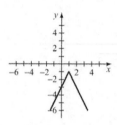

93. $g(x) = \dfrac{1}{2}f(x)$ [2.5]

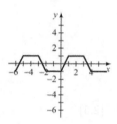

95. $g(x) = f\left(\dfrac{1}{2}x\right)$ [2.5]

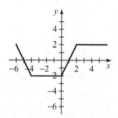

97. Find the difference quotient. [2.6]

$$\frac{f(x+h) - f(x)}{h}$$

$$= \frac{4(x+h)^2 - 3(x+h) - 1 - (4x^2 - 3x - 1)}{h}$$

$$= \frac{4x^2 + 8xh + 4h^2 - 3x - 3h - 1 - 4x^2 + 3x + 1}{h}$$

$$= \frac{8xh + 4h^2 - 3h}{h}$$

$$= 8x + 4h - 3$$

99. $s(t) = 3t^2$ [2.4]

a. Average velocity $= \dfrac{3(4)^2 - 3(2)^2}{4 - 2}$

$$= \frac{3(16) - 3(4)}{2}$$

$$= \frac{48 - 12}{2}$$

$$= \frac{36}{2} = 18 \text{ ft/sec}$$

b. Average velocity $= \dfrac{3(3)^2 - 3(2)^2}{3 - 2}$

$$= \frac{3(9) - 3(4)}{1}$$

$$= \frac{27 - 12}{1} = 15 \text{ ft/sec}$$

c. Average velocity $= \dfrac{3(2.5)^2 - 3(2)^2}{2.5 - 2}$

$$= \frac{3(6.25) - 3(4)}{0.5}$$

$$= \frac{18.75 - 12}{0.5}$$

$$= \frac{6.75}{0.5} = 13.5 \text{ ft/sec}$$

d. Average velocity $= \dfrac{3(2.01)^2 - 3(2)^2}{2.01 - 2}$

$$= \frac{3(4.0401) - 3(4)}{0.01}$$

$$= \frac{12.1203 - 12}{0.01}$$

$$= \frac{0.1203}{0.01} = 12.03 \text{ ft/sec}$$

e. It appears that the average velocity of the ball approaches 12 ft/sec.

101. Evaluate the composite functions. [2.6]

a. $(f \circ g)(-5) = f\big(g(-5)\big) = f(|-5-1|) = f(|-6|)$

$$= f(6) = 2(6)^2 + 7$$

$$= 72 + 7 = 79$$

b. $(g \circ f)(-5) = g\big(f(-5)\big) = g\big(2(-5)^2 + 7\big)$

$$= g(57) = |57 - 1|$$

$$= 56$$

c. $(f \circ g)(x) = f\big(g(x)\big)$

$$= 2|x - 1|^2 + 7$$

$$= 2x^2 - 4x + 2 + 7$$

$$= 2x^2 - 4x + 9$$

d. $(g \circ f)(x) = g\big(f(x)\big)$

$$= |2x^2 + 7 - 1|$$

$$= |2x^2 + 6|$$

$$= 2x^2 + 6$$

103. a. Enter the data on your calculator. The technique for a TI-83 calculator is illustrated here. Press STAT. [2.7]

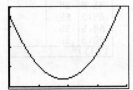

$$h = 0.0047952048t^2 - 1.756843157t + 180.4065934$$

b. Empty $\Rightarrow y = 0 \Rightarrow$ the graph intersects the x-axis.

Graph the equation, and notice that it never intersects the x-axis.

Xmin = 0, Xmax = 400, Xscl = 100

Ymin = 0, Ymax = 200, Yscl = 50

Thus, no, on the basis of this model, the can never empties.

c. The regression line is a model of the data and is not based on physical principles.

Chapter 2 Test

1. Finding the midpoint and length. [2.1]

$$\text{midpoint} = \left(\frac{x_1 + x_2}{2}, \frac{y_1 + y_2}{2} \right)$$

$$= \left(\frac{-2 + 4}{2}, \frac{3 + (-1)}{2} \right)$$

$$= \left(\frac{2}{2}, \frac{2}{2} \right) = (1,\ 1)$$

$$\text{length} = d = \sqrt{(x_1 - x_2)^2 + (y_1 - y_2)^2}$$

$$= \sqrt{(-2 - 4)^2 + (3 - (-1))^2}$$

$$= \sqrt{(-6)^2 + 4^2} = \sqrt{36 + 16} = \sqrt{52}$$

$$= 2\sqrt{13}$$

3. Graphing $y = |x + 2| + 1$. [2.1]

x	y
-4	3
-3	2
-2	1
-1	2
0	3

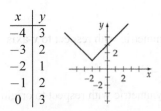

5. Determining the domain of the function. [2.2]

$$x^2 - 16 \geq 0$$

$$(x - 4)(x + 4) \geq 0$$

The product is positive or zero.

The critical values are 4 and -4.

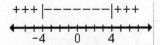

The domain is $\{x \mid x \geq 4 \text{ or } x \leq -4\}$.

7. Find the slope. [2.3]

$$m = \frac{3 - (-2)}{-1 - 5} = \frac{5}{-6} = -\frac{5}{6}$$

9. Finding the equation in slope-intercept form. [2.3]

$$3x - 2y = 4$$

$$-2y = -3x + 4$$

$$y = \frac{3}{2}x - 2$$

Slope of perpendicular line is $-\frac{2}{3}$.

$$y - y_1 = m(x - x_1)$$

$$y + 2 = -\frac{2}{3}(x - 4)$$

$$y + 2 = -\frac{2}{3}x + \frac{8}{3}$$

$$y = -\frac{2}{3}x + \frac{8}{3} - \frac{6}{3}$$

$$y = -\frac{2}{3}x + \frac{2}{3}$$

11. Finding the maximum or minimum value. [2.4]

$$-\frac{b}{2a} = -\frac{-4}{2(1)} = 2$$

$$f(2) = 2^2 - 4(2) - 8$$

$$= 4 - 8 - 8$$

$$= -12$$

The minimum value of the function is -12.

618 Chapter 2 Functions and Graphs

13. Identify the type of symmetry. [2.5]

a. $(-y)^2 = x + 1$

$y^2 = x + 1$ symmetric with respect to x-axis

b. $-y = 2(-x)^3 + 3(-x)$

$y = 2x^3 + 3x$ symmetric with respect to origin

c. $y = 3(-x)^2 - 2$

$y = 3x^2 - 2$ symmetric with respect to y-axis

15. $g(x) = f\left(\frac{1}{2}x\right)$ [2.5]

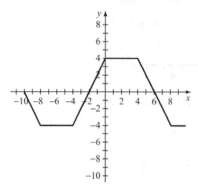

17. $g(x) = f(x-1) + 3$ [2.5]

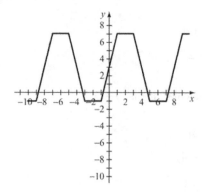

19. Perform the operations. [2.6]

a. $(f-g)(x) = (x^2 - x + 2) - (2x - 1) = x^2 - 3x + 3$

b. $(f \cdot g)(-2) = ((-2)^2 - (-2) + 2)(2(-2) - 1)$

$= (8)(-5) = -40$

c. $(f \circ g)(3) = f(g(3)) = f(2(3) - 1)$

$= f(5) = 5^2 - 5 + 2$

$= 22$

d. $(g \circ f)(x) = g(f(x)) = 2(x^2 - x + 2) - 1$

$= 2x^2 - 2x + 3$

21. Find the maximum area. [2.4]

Using the formula for perimeter for three sides,

$P = 2w + l \Rightarrow 80 = 2x + y$

$y = 80 - 2x$

Using the formula for area, $A = xy$. Then

$A(x) = x(80 - 2x)$

$A(x) = -2x^2 + 80x$

$-\dfrac{b}{2a} = -\dfrac{80}{2(-2)} = 20$

$y = 80 - 2(20) = 40$

$x = 20$ ft and $y = 40$ ft

23. a. Enter the data on your calculator. The technique for a TI-83 calculator is illustrated here. Press STAT. [2.7]

L1	L2	L3	3
93.2	28	▬▬▬	
92.3	26		
91.9	39		
89.5	56		
89.6	56		
90.5	36		
L3(1)=			

```
EDIT CALC TESTS
1:Edit
2:SortA(
3:SortD(
4:ClrList
5:SetUpEditor
```

```
EDIT CALC TEST
1:1-Var Stats
2:2-Var Stats
3:Med-Med
4:LinReg (ax+b)
5:QuadReg
6:CubicReg
7↓QuartReg
```

```
LinReg
y=ax+b
a=-7.98245614
b=767.122807
r2=.805969575
r=-.8977580826
```

$y = -7.98245614x + 767.122807$

b. Evaluating the equation from part (a) at 89.

$y = -7.98245614(89) + 767.122807$

≈ 57 calories

Cumulative Review Exercises

1. Determine the property for $3(a+b) = 3(b+a)$. [P.1]

Commutative Property of Addition

3. Simplifying. [P.1]

$3 + 4(2x - 9)$

$= 3 + 8x - 36$

$= 8x - 33$

5. Simplifying. [P.2]

$\dfrac{24a^4b^3}{18a^4b^5} = \dfrac{4a^{4-4}b^{3-5}}{3} = \dfrac{4b^{-2}}{3} = \dfrac{4}{3b^2}$

7. Simplifying. [P.5]

$$\frac{x^2+6x-27}{x^2-9}=\frac{(x+9)(x-3)}{(x+3)(x-3)}=\frac{x+9}{x+3}$$

9. Solving for x. [1.1]

$$6-2(2x-4)=14$$
$$6-4x+8=14$$
$$-4x=0$$
$$x=0$$

11. Solving for x. [1.3]

$$(2x-1)(x+3)=4$$
$$2x^2+5x-3=4$$
$$2x^2+5x-7=0$$
$$(2x+7)(x-1)=0$$

$$x=-\frac{7}{2}\ \text{or}\ x=1$$

13. Solving for x. [1.4]

$$x^4-x^2-2=0$$

Let $u=x^2$.

$$u^2-u-2=0$$
$$(u-2)(u+1)=0$$

$$u-2=0\quad\text{or}\quad u+1=0$$
$$u=2\qquad\qquad u=-1$$
$$x^2=2\qquad\qquad x^2=-1$$
$$x=\pm\sqrt{2}\qquad\quad x=\pm i$$

15. Finding the distance. [2.1]

$$\text{distance}=\sqrt{[-2-2]^2+[-4-(-3)]^2}$$
$$=\sqrt{(-4)^2+(-1)^2}=\sqrt{16+1}$$
$$=\sqrt{17}$$

17. Finding the equation of the line. [2.3]

The slope is $m=\dfrac{-1-(-3)}{-2-2}=\dfrac{-1+3}{-2-2}=\dfrac{2}{-4}=-\dfrac{1}{2}$

The equation is $y-(-3)=-\dfrac{1}{2}(x-2)$

$$y=-\frac{1}{2}x-2$$

19. Evaluating a quadratic function. [2.4]

$$h(x)=-0.002x^2-0.03x+8$$
$$h(39)=-0.002(39)^2-0.03(39)+8=3.788\ \text{ft}$$

Yes.

Chapter 3 Polynomial and Rational Functions

Section 3.1 Exercises

1. Determine whether synthetic division can be used.
Yes, since the divisor $x-2$ is in the form $x-c$.

3. Determine whether synthetic division can be used.
No, since the divisor x^2+1 is not in the form $x-c$.

5. By the remainder theorem, if $P(x)$ is divided by $(x-4)$, then $P(4)=7$.

7. Use long division to divide.

$$
\begin{array}{r}
5x^2 - 9x + 10 \\
x+3\overline{\smash{)}5x^3 + 6x^2 - 17x + 20} \\
\underline{5x^3 + 15x^2} \\
-9x^2 - 17x \\
\underline{-9x^2 - 27x} \\
10x + 20 \\
\underline{10x + 30} \\
-10
\end{array}
$$

Answer: $5x^2 - 9x + 10 - \dfrac{10}{x+3}$

9. Use long division to divide.

$$
\begin{array}{r}
x^3 + 2x^2 - x + 1 \\
x-2\overline{\smash{)}x^4 + 0x^3 - 5x^2 + 3x - 1} \\
\underline{x^4 - 2x^3} \\
2x^3 - 5x^2 \\
\underline{2x^3 - 4x^2} \\
-x^2 + 3x \\
\underline{-x^2 + 2x} \\
x - 1 \\
\underline{x - 2} \\
1
\end{array}
$$

Answer: $x^3 + 2x^2 - x + 1 + \dfrac{1}{x-2}$

11. Use long division to divide.

$$
\begin{array}{r}
x^2 + 3x - 2 \\
2x^2 - x + 1\overline{\smash{)}2x^4 + 5x^3 - 6x^2 + 4x + 3} \\
\underline{2x^4 - x^3 + x^2} \\
6x^3 - 7x^2 + 4x \\
\underline{6x^3 - 3x^2 + 3x} \\
-4x^2 + x + 3 \\
\underline{-4x^2 + 2x - 2} \\
-x + 5
\end{array}
$$

Answer: $x^2 + 3x - 2 + \dfrac{-x+5}{2x^2 - x + 1}$

13. Use long division to divide.

$$
\begin{array}{r}
x^3 - x^2 + 2x - 1 \\
2x^2 + 2x - 3\overline{\smash{)}2x^5 \quad - x^3 + 5x^2 - 9x + 6} \\
\underline{2x^5 + 2x^4 - 3x^3} \\
-2x^4 + 2x^3 + 5x^2 \\
\underline{-2x^4 - 2x^3 + 3x^2} \\
4x^3 + 2x^2 - 9x \\
\underline{4x^3 + 4x^2 - 6x} \\
-2x^2 - 3x + 6 \\
\underline{-2x^2 - 2x + 3} \\
-x + 3
\end{array}
$$

Answer: $x^3 - x^2 + 2x - 1 + \dfrac{-x+3}{2x^2 + x - 3}$

15. Use synthetic division to divide.

$$
\begin{array}{r|rrrr}
2 & 4 & -5 & 6 & -7 \\
  &   & 8 & 6 & 24 \\
\hline
  & 4 & 3 & 12 & 17
\end{array}
$$

Answer: $4x^2 + 3x + 12 + \dfrac{17}{x-2}$

17. Use synthetic division to divide.

$$
\begin{array}{r|rrrr}
-1 & 4 & 0 & -2 & 3 \\
   &   & -4 & 4 & -2 \\
\hline
   & 4 & -4 & 2 & 1
\end{array}
$$

Answer: $4x^2 - 4x + 2 + \dfrac{1}{x+1}$

19. Use synthetic division to divide.

$$
\begin{array}{r|rrrrrr}
4 & 1 & 0 & -10 & 0 & 5 & -1 \\
  &   & 4 & 16 & 24 & 96 & 404 \\
\hline
  & 1 & 4 & 6 & 24 & 101 & 403
\end{array}
$$

Answer: $x^4 + 4x^3 + 6x^2 + 24x + 101 + \dfrac{403}{x-4}$

21. Use synthetic division to divide.

$$
\begin{array}{r|rrrrrr}
1 & 1 & 0 & 0 & 0 & 0 & -1 \\
  &   & 1 & 1 & 1 & 1 & 1 \\
\hline
  & 1 & 1 & 1 & 1 & 1 & 0
\end{array}
$$

Answer: $x^4 + x^3 + x^2 + x + 1$

23. Use synthetic division to divide.

$$
\begin{array}{r|rrrr}
\frac{1}{2} & 8 & -4 & 6 & -3 \\
            &   & 4 & 0 & 3 \\
\hline
            & 8 & 0 & 6 & 0
\end{array}
$$

Answer: $8x^2 + 6$

25. Use synthetic division to divide.

$$\begin{array}{r|rrrrrrrrr} 2 & 1 & 0 & 1 & 0 & 1 & 0 & 1 & 0 & 4 \\ & & 2 & 4 & 10 & 20 & 42 & 84 & 170 & 340 \\ \hline & 1 & 2 & 5 & 10 & 21 & 42 & 85 & 170 & 344 \end{array}$$

Answer: $x^7 + 2x^6 + 5x^5 + 10x^4 + 21x^3 + 42x^2$
$$+\, 85x + 170 + \dfrac{344}{x-2}$$

27. Use synthetic division to divide.

$$\begin{array}{r|rrrrrrr} -3 & 1 & 0 & 0 & 0 & 0 & 1 & -10 \\ & & -3 & 9 & -27 & 81 & -243 & 726 \\ \hline & 1 & -3 & 9 & -27 & 81 & -242 & 716 \end{array}$$

Answer: $x^5 - 3x^4 + 9x^3 - 27x^2 + 81x - 242 + \dfrac{716}{x+3}$

29. Find $P(c)$ using synthetic division and Remainder Theorem.

$$\begin{array}{r|rrrr} 3 & 5 & 2 & -1 & -7 \\ & & 15 & 51 & 150 \\ \hline & 5 & 17 & 50 & 143 \end{array}$$

$P(c) = P(3) = 143$

31. Find $P(c)$ using synthetic division and Remainder Theorem.

$$\begin{array}{r|rrrrr} -3 & 3 & 0 & -5 & 0 & 7 \\ & & -9 & 27 & -66 & 198 \\ \hline & 3 & -9 & 22 & -66 & 205 \end{array}$$

$P(c) = P(-3) = 205$

33. Find $P(c)$ using synthetic division and Remainder Theorem.

$$\begin{array}{r|rrrr} 8 & -4 & -1 & 3 & -11 \\ & & -32 & -264 & -2088 \\ \hline & -4 & -33 & -261 & -2099 \end{array}$$

$P(c) = P(8) = -2099$

35. Find $P(c)$ using synthetic division and Remainder Theorem.

$$\begin{array}{r|rrrrr} 6 & -1 & 0 & 0 & 1 & -2 \\ & & -6 & -36 & -216 & -1290 \\ \hline & -1 & -6 & -36 & -215 & -1292 \end{array}$$

$P(c) = P(6) = -1292$

37. Find $P(c)$ using synthetic division and Remainder Theorem.

$$\begin{array}{r|rrrrrrr} 2 & -1 & 0 & -8 & 0 & 0 & -3 & 5 \\ & & -2 & -4 & -24 & -48 & -96 & -198 \\ \hline & -1 & -2 & -12 & -24 & -48 & -99 & -193 \end{array}$$

$P(c) = P(2) = -193$

39. Determine whether the binomial is a factor using synthetic division and the Factor Theorem.

$$\begin{array}{r|rrrr} 2 & 1 & 2 & -5 & -6 \\ & & 2 & 8 & 6 \\ \hline & 1 & 4 & 3 & 0 \end{array}$$

A remainder of 0 implies that $x - 2$ is a factor of $P(x)$.

41. Determine whether the binomial is a factor using synthetic division and the Factor Theorem.

$$\begin{array}{r|rrrr} -1 & 2 & 1 & -3 & -1 \\ & & -2 & 1 & 2 \\ \hline & 2 & -1 & -2 & 1 \end{array}$$

A remainder of 1 implies that $x + 1$ is not a factor of $P(x)$.

43. Determine whether the binomial is a factor using synthetic division and the Factor Theorem.

$$\begin{array}{r|rrrrr} -4 & 1 & 1 & -2 & 5 & -140 \\ & & -4 & 12 & -40 & 140 \\ \hline & 1 & -3 & 10 & -35 & 0 \end{array}$$

A remainder of 0 implies that $x + 4$ is a factor of $P(x)$.

45. Determine whether the binomial is a factor using synthetic division and the Factor Theorem.

$$\begin{array}{r|rrrrrr} 5 & 1 & 2 & -22 & -50 & -75 & 0 \\ & & 5 & 35 & 65 & 75 & 0 \\ \hline & 1 & 7 & 13 & 15 & 0 & 0 \end{array}$$

A remainder of 0 implies that $x - 5$ is a factor of $P(x)$.

47. Determine whether the binomial is a factor using synthetic division and the Factor Theorem.

$$\begin{array}{r|rrrrr} \frac{1}{4} & 16 & -8 & 9 & 14 & 4 \\ & & 4 & -1 & 2 & 4 \\ \hline & 16 & -4 & 8 & 16 & 8 \end{array}$$

A remainder of 8 implies that $x - 1/4$ is not a factor of $P(x)$.

49. Use synthetic division to show c is a zero.

$$\begin{array}{r|rrrr} 2 & 3 & -8 & -10 & 28 \\ & & 6 & -4 & -28 \\ \hline & 3 & -2 & -14 & 0 \end{array}$$

51. Use synthetic division to show c is a zero.

$$
\begin{array}{r|rrrrr}
1 & 1 & 0 & 0 & 0 & -1 \\
 & & 1 & 1 & 1 & 1 \\
\hline
 & 1 & 1 & 1 & 1 & 0
\end{array}
$$

53. Use synthetic division to show c is a zero.

$$
\begin{array}{r|rrrrr}
-2 & 3 & 8 & 10 & 2 & -20 \\
 & & -6 & -4 & -12 & 20 \\
\hline
 & 3 & 2 & 6 & -10 & 0
\end{array}
$$

55. Use synthetic division to show c is a zero.

$$
\begin{array}{r|rrrr}
11 & 2 & -18 & -50 & 66 \\
 & & 22 & 44 & -66 \\
\hline
 & 2 & 4 & -6 & 0
\end{array}
$$

56. Use synthetic division to show c is a zero.

$$
\begin{array}{r|rrrrr}
15 & 2 & -34 & 70 & -153 & 45 \\
 & & 30 & -60 & 150 & -45 \\
\hline
 & 2 & -4 & 10 & -3 & 0
\end{array}
$$

57. Verify the given binomial is a factor.

$$
\begin{array}{r|rrrr}
2 & 1 & 1 & 1 & -14 \\
 & & 2 & 6 & 14 \\
\hline
 & 1 & 3 & 7 & 0
\end{array}
$$

A remainder of 0 implies that $x - 2$ is a factor of $P(x)$.

$$P(x) = (x - 2)(x^2 + 3x + 7)$$

59. Verify the given binomial is a factor.

$$
\begin{array}{r|rrrrr}
4 & 1 & -1 & -9 & -11 & -4 \\
 & & 4 & 12 & 12 & 4 \\
\hline
 & 1 & 3 & 3 & 1 & 0
\end{array}
$$

A remainder of 0 implies that $x - 4$ is a factor of $P(x)$.

$$P(x) = (x - 4)(x^3 + 3x^2 + 3x + 1)$$

61. a. Find how many different ways to select 3 cards.

$$
\begin{array}{r|rrrr}
8 & 1 & -3 & 2 & 0 \\
 & & 8 & 40 & 336 \\
\hline
 & 1 & 5 & 42 & 336
\end{array}
$$

336 ways

b. Evaluate for $n = 8$ and compare to part a.

$$
\begin{aligned}
P(8) &= 8^3 - 3(8)^2 + 2(8) \\
&= 512 - 3(64) + 2(8) \\
&= 512 - 192 + 16 \\
&= 336 \text{ ways}
\end{aligned}
$$

They are the same.

63. a. Find the number of cards needed for 7 rows.

$$
\begin{array}{r|rrr}
7 & 1.5 & 0.5 & 0 \\
 & & 10.5 & 77 \\
\hline
 & 1.5 & 11 & 77
\end{array}
$$

77 cards

b. Find the number of cards needed for 12 rows.

$$
\begin{array}{r|rrr}
12 & 1.5 & 0.5 & 0 \\
 & & 18 & 222 \\
\hline
 & 1.5 & 18.5 & 222
\end{array}
$$

222 cards

65. a. Find the number of ways officers can be selected from 8 students.

$$
\begin{array}{r|rrrrr}
8 & 1 & -6 & 11 & -6 & 0 \\
 & & 8 & 16 & 216 & 1680 \\
\hline
 & 1 & 2 & 27 & 210 & 1680
\end{array}
$$

1680 ways

b. Find the number of ways officers can be selected from 18 students.

$$
\begin{array}{r|rrrrr}
18 & 1 & -6 & 11 & -6 & 0 \\
 & & 18 & 216 & 4086 & 73440 \\
\hline
 & 1 & 12 & 227 & 4080 & 73440
\end{array}
$$

73,440 ways

67. a. Find the volume for $x = 6$ in.

$$
\begin{array}{r|rrrr}
6 & 1 & 1 & 10 & -8 \\
 & & 6 & 42 & 312 \\
\hline
 & 1 & 7 & 52 & 304
\end{array}
$$

304 cubic inches

b. Find the volume for $x = 9$ in.

$$
\begin{array}{r|rrrr}
9 & 1 & 1 & 10 & -8 \\
 & & 9 & 90 & 900 \\
\hline
 & 1 & 10 & 100 & 892
\end{array}
$$

892 cubic inches

69. Divide each by $x - 1$ and write as the quotient.

$$
\begin{array}{r|rrrr}
1 & 1 & 0 & 0 & -1 \\
 & & 1 & 1 & 1 \\
\hline
 & 1 & 1 & 1 & 0
\end{array}
$$

So, $(x^3 - 1) \div (x - 1) = x^2 + x + 1$

$$
\begin{array}{r|rrrrrr}
1 & 1 & 0 & 0 & 0 & 0 & -1 \\
 & & 1 & 1 & 1 & 1 & 1 \\
\hline
 & 1 & 1 & 1 & 1 & 1 & 0
\end{array}
$$

So, $(x^5 - 1) \div (x - 1) = x^4 + x^3 + x^2 + x + 1$

$$\begin{array}{r|rrrrrrr} 1 & 1 & 0 & 0 & 0 & 0 & 0 & -1 \\ & & 1 & 1 & 1 & 1 & 1 & 1 & 1 \\ \hline & 1 & 1 & 1 & 1 & 1 & 1 & 1 & 0 \end{array}$$

So, $(x^7 - 1) \div (x - 1) = x^6 + x^5 + x^4 + x^3 + x^2 + x + 1$

$(x^9 - 1) \div (x - 1)$
$= x^8 + x^7 + x^6 + x^5 + x^4 + x^3 + x^2 + x + 1$

71. Find k.

$$\begin{array}{r|rrrr} 3 & 2 & 1 & -25 & k \\ & & 6 & 21 & -12 \\ \hline & 2 & 7 & -4 & 0 \end{array}$$

$k = 12$

73. Find k.

$$\begin{array}{r|rrrrr} -4 & 1 & 3 & -8 & k & 16 \\ & & -4 & 4 & 16 & -4k - 64 \\ \hline & 1 & -1 & -4 & k+16 & 0 \end{array}$$

$16 - 4k - 64 = 0$
$-4k = 48$
$k = -12$

75. Find the remainder by using the Remainder Theorem.

$5(1)^{48} + 6(1)^{10} - 5(1) + 7 = 5 + 6 - 5 + 7 = 13$

77. Determine whether i is a zero of the polynomial.

$$\begin{array}{r|rrrr} i & 1 & -3 & 1 & -3 \\ & & i & -1-3i & 3 \\ \hline & 1 & -3+i & -3i & 0 \end{array}$$

A remainder of 0 implies that $x - i$ is a factor of

$x^3 - 3x^2 + x - 3$.

Prepare for Section 3.2

P1. Find the minimum value.

$P(x) = x^2 - 4x + 6$

$-\dfrac{b}{2a} = -\dfrac{-4}{2(1)} = -\dfrac{-4}{2} = 2$

$P(2) = (2)^2 - 4(2) + 6 = 4 - 8 + 6 = 2$

The minimum value is 2.

P3. Find the interval where the function is increasing.

$P(x) = x^2 + 2x + 7$ is a parabola that opens up.

The x-value of the vertex is

$-\dfrac{b}{2a} = -\dfrac{2}{2(1)} = -\dfrac{2}{2} = -1$

The graph decreases from the left until it reaches the vertex, and then it increases.

$P(x)$ increases on the interval $[-1, \infty)$.

P5. Factor.

$x^4 - 5x^2 + 4$
$(x^2 - 4)(x^2 - 1)$
$(x + 2)(x - 2)(x + 1)(x - 1)$

Section 3.2 Exercises

1. The degree of $P(x) = 4x^3 - x^2 + 3x - 5$ is 3.

3. The largest number of turning points for a polynomial of degree 4 is 3.

5. Determine the far-left and far-right behavior.
Since $a_n = 3$ is positive and $n = 4$ is even, the graph of P goes up to its far left and up to its far right.

7. Determine the far-left and far-right behavior.
Since $a_n = 5$ is positive and $n = 5$ is odd, the graph of P goes down to its far left and up to its far right.

9. Determine the far-left and far-right behavior.

$P(x) = x^6 - 450$

Since $a_n = 1$ is positive and $n = 6$ is even, the graph of P goes up to its far left and up to its far right.

11. Determine the far-left and far-right behavior.

$P(x) = -x^5 + 4x^2 + 9$

Since $a_n = -1$ is negative and $n = 5$ is odd, the graph of P goes up to its far left and down to its far right.

13. Determine the leading coefficient from the far-left and far-right behavior of the graph.
Up to the far left and down to the far right. $a < 0$.

15. Use a graphing utility to graph, estimate any relative maxima or minima.

```
Plot1 Plot2 Plot3      WINDOW
\Y1=X^3+X^2-9X-9       Xmin=-10
                       Xmax=10
\Y2=                   Xscl=2
\Y3=                   Ymin=-20
\Y4=                   Ymax=10
\Y5=                   Yscl=2
\Y6=                   Xres=1
```

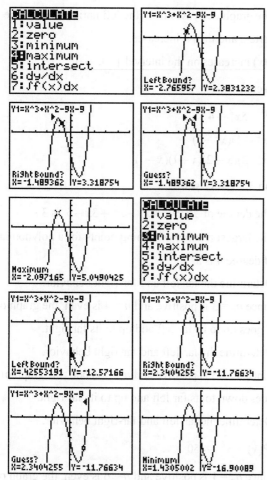

On a TI-83 calculator, the CALC feature is located above the TRACE key.

Relative maximum of $y \approx 5.0$ at $x \approx -2.1$.

Relative minimum of $y \approx -16.9$ at $x \approx 1.4$.

17. Use a graphing utility to graph, estimate any relative maxima or minima.

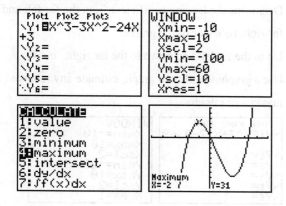

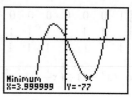

The step-by-step technique for a TI-83 calculator is illustrated in the solution to Exercise **11.**

The CALC feature is located above the TRACE key.

Relative maximum of $y \approx 31.0$ at $x \approx -2.0$.

Relative minimum of $y \approx -77.0$ at $x \approx 4$.

19. Use a graphing utility to graph, estimate any relative maxima or minima.

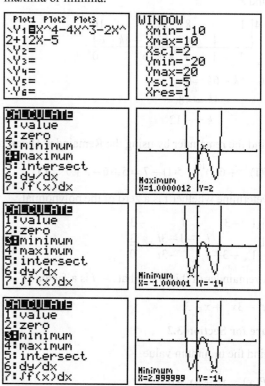

The step-by-step technique for a TI-83 calculator is illustrated in the solution to Exercise **11.**

The CALC feature is located above the TRACE key.

Relative maximum of $y \approx 2.0$ at $x \approx 1.0$.

Relative minimum of $y \approx -14.0$ at $x \approx -1.0$, and

another relative minimum of $y \approx -14.0$ at $x \approx 3.0$.

21. Find the real zeros by factoring.

$$P(x) = x^3 - 2x^2 - 15x$$
$$0 = x(x^2 - 2x - 15)$$
$$0 = x(x-5)(x+3)$$

The zeros are 0, 5, –3.

23. Find the real zeros by factoring.

$$P(x) = x^4 - 13x^2 + 36$$
$$0 = (x^2 - 9)(x^2 - 4)$$
$$0 = (x+3)(x-3)(x+2)(x-2)$$

The zeros are –3, 3, –2, 2.

25. Find the real zeros by factoring.

$$P(x) = x^5 - 5x^3 + 4x$$
$$0 = x(x^4 - 5x^2 + 4)$$
$$0 = x(x^2 - 4)(x^2 - 1)$$
$$0 = x(x+2)(x-2)(x+1)(x-1)$$

The zeros are 0, –2, 2, –1, 1.

27. Verify the zero between a and b using Intermediate Value Theorem.

$$P(x) = 2x^3 + 3x^2 - 23x - 42$$

```
3| 2   3   -23   -42
  |     6    27    12
   2   9    4    -30

4| 2   3   -23   -42
  |     8    44    84
   2  11   21    42
```

$P(3)$ is negative; $P(4)$ is positive.

Therefore $P(x)$ has a zero between 3 and 4.

29. Verify the zero between a and b using Intermediate Value Theorem.

$$P(x) = 3x^3 + 7x^2 + 3x + 7$$

```
-3| 3    7    3     7
   |     -9    6   -27
    3   -2    9   -20

-2| 3    7    3     7
   |     -6   -2    -2
    3    1    1     5
```

$P(-3)$ is negative; $P(-2)$ is positive.

Therefore $P(x)$ has a zero between –3 and –2.

31. Verify the zero between a and b using Intermediate Value Theorem.

$$P(x) = 4x^4 + 7x^3 - 11x^2 + 7x - 15$$

```
1| 4    7   -11    7   -15
 |      4    11    0     7
  4   11     0    7    -8

1½ = 1.5| 4    7   -11     7      -15
        |      6   19.5  12.75  29.625
         4   13    8.5  19.75  14.625
```

$P(1)$ is negative; $P(1.5)$ is positive.

Therefore $P(x)$ has a zero between 1 and 1.5.

33. Verify the zero between a and b using Intermediate Value Theorem.

$$P(x) = x^4 - x^2 - x - 4$$

```
1.7| 1    0    -1      -1       -4
   |      1.7  2.89   3.213    3.7621
    1   1.7   1.89   2.213   -0.2379

1.8| 1    0    -1      -1       -4
   |      1.8  3.24   4.032    5.4576
    1   1.8   2.24   3.032    1.4526
```

$P(1.7)$ is negative; $P(1.8)$ is positive.

Therefore $P(x)$ has a zero between 1.7 and 1.8.

35. Verify the zero between a and b using Intermediate Value Theorem.

$$P(x) = -x^4 + x^3 + 5x - 1$$

```
0.1| -1     1      0       5        -1
   |       -0.1   0.09    0.009    0.5009
    -1    0.9    0.09    5.009   -0.4991

0.2| -1     1      0       5        -1
   |       -0.2   0.16    0.032    1.0064
    -1    0.8    0.16    5.032    0.0064
```

$P(0.1)$ is negative; $P(0.2)$ is positive.

Therefore $P(x)$ has a zero between 0.1 and 0.2.

37. Determine x-intercepts. State which cross x-axis.

$$P(x) = (x-1)(x+1)(x-3)$$
$$0 = (x-1)(x+1)(x-3)$$

$$x - 1 = 0 \quad \text{or} \quad x + 1 = 0 \quad \text{or} \quad x - 3 = 0$$
$$x = 1 \qquad\qquad x = -1 \qquad\qquad x = 3$$

The exponents on $(x + 1)$, $(x - 1)$, and $(x - 3)$ are odd integers. Therefore the graph of $P(x)$ will cross the x-axis at $(-1, 0)$, $(1, 0)$, and $(3, 0)$.

39. Determine x-intercepts. State which cross x-axis.

$$P(x) = x(x-5)^2(x-3)$$
$$0 = x(x-5)^2(x-3)$$

$x = 0$ or $x - 5 = 0$ or $x - 3 = 0$
$\qquad\qquad\quad x = 5 \qquad\qquad x = 3$

The exponent on x is an odd integer. Therefore the graph of $P(x)$ will cross the x-axis at $(0, 0)$.

The exponent on $(x - 5)$ is an even integer. Therefore the graph of $P(x)$ will intersect but not cross the x-axis at $(5, 0)$.

The exponent on $(x - 3)$ is an odd integer. Therefore the graph of $P(x)$ will cross the x-axis at $(3, 0)$.

41. Determine x-intercepts. State which cross x-axis.

$$P(x) = x^2(x-15)(2x-7)^2$$
$$0 = x^2(x-15)(2x-7)^2$$

$x = 0$ or $x - 15 = 0$ or $2x - 7 = 0$
$\qquad\qquad\quad x = 15 \qquad\qquad x = \dfrac{7}{2}$

The exponent on x is an even integer. Therefore the graph of $P(x)$ will intersect but not cross the x-axis at $(0, 0)$.

The exponent on $(x - 15)$ is an odd integer. Therefore the graph of $P(x)$ will cross the x-axis at $(15, 0)$.

The exponent on $(2x - 7)$ is an even integer. Therefore the graph of $P(x)$ will intersect but not cross the x-axis $\left(\dfrac{7}{2}, 0\right)$.

43. Determine x-intercepts. State which cross x-axis.

$$P(x) = x^3 - 6x^2 + 9x$$
$$0 = x(x^2 - 6x + 9)$$
$$0 = x(x-3)^2$$

$x = 0$ or $x - 3 = 0$
$\qquad\qquad\quad x = 3$

The exponent on x is an odd integer. Therefore the graph of $P(x)$ will cross the x-axis at $(0, 0)$.

The exponent on $(x - 3)$ is an even integer. Therefore the graph of $P(x)$ will intersect but not cross the x-axis at $(3, 0)$.

45. Sketch the graph of the function.

Let $P(x) = 0$.

$$x^3 - x^2 - 2x = 0$$
$$x(x^2 - x - 2) = 0$$
$$x(x-2)(x+1) = 0$$

$x = 0, x = 2, x = -1$

The graph crosses the x-axis at $(0, 0)$, $(2, 0)$, and $(-1, 0)$.

Let $x = 0$. $P(0) = 0^3 - 0^2 - 2(0) = 0$

The y-intercept is $(0, 0)$.

x^3 has a positive coefficient and an odd exponent. Therefore, the graph goes down to the far left and up to the far right.

47. Sketch the graph of the function.

Let $P(x) = 0$.

$$(x-2)(x+3)(x+1) = 0$$

$x = 2, x = -3, x = -1$

The graph crosses the x-axis at $(2, 0)$, $(-3, 0)$, and $(-1, 0)$.

Let $x = 0$. $P(0) = -(0)^3 - 2(0)^2 + 5(0) + 6 = 6$

The y-intercept is $(0, 6)$.

$-x^3$ has a negative coefficient and an odd exponent. Therefore, the graph goes up to the far left and down to the far right.

49. Sketch the graph of the function.

Let $P(x) = 0$.

$$(x-1)^2(x-3)(x+1) = 0$$

$x = 1, x = 3, x = -1$

The graph intersects the x-axis but does not cross it at $(1, 0)$.

The graph crosses the x-axis at $(3, 0)$ and $(-1, 0)$.

Let $x = 0$.

$P(0) = (0)^4 - 4(0)^3 + 2(0)^2 + 4(0) - 3 = -3$

The y-intercept is $(0, -3)$.

x^4 has a positive coefficient and an even exponent.

Therefore, the graph goes up to the far left and up to the far right.

51. Sketch the graph of the function.

Let $P(x) = 0$.

$x^3 + 6x^2 + 5x - 12 = 0$
$(x - 1)(x + 3)(x + 4) = 0$

$x = 1, x = -3, x = -4$

The graph crosses the x-axis at $(1, 0)$, $(-3, 0)$, and $(-4, 0)$.

Let $x = 0$. $P(0) = (0)^3 + 6(0)^2 + 5(0) - 12 = -12$

The y-intercept is $(0, -12)$.

x^3 has a positive coefficient and an odd exponent.

Therefore, the graph goes down to the far left and up to the far right.

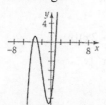

53. Sketch the graph of the function.

Let $P(x) = 0$.

$$
\begin{array}{r|rrrr}
1 & -1 & 0 & 7 & -6 \\
  &   & -1 & -1 & 6 \\
\hline
  & -1 & -1 & 6 & 0
\end{array}
$$

$-x^3 + 7x - 6 = 0$
$(x - 1)(-x^2 - x + 6) = 0$
$(x - 1)(-x - 3)(x - 2) = 0$

$x = 1, x = -3, x = 2$

The graph crosses the x-axis at $(1, 0)$, $(-3, 0)$, and $(2, 0)$.

Let $x = 0$. $P(0) = -(0)^3 + 7(0) - 6 = -6$

The y-intercept is $(0, -6)$.

$-x^3$ has a negative coefficient and an odd exponent.

Therefore, the graph goes up to the far left and down to the far right.

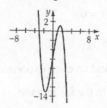

55. Sketch the graph of the function.

Let $P(x) = 0$.

$-x^3 + 4x^2 - 4x = 0$
$-x(x - 2)^2 = 0$

$x = 0, x = 2$

The graph crosses the x-axis at $(0, 0)$, and $(2, 0)$.

Let $x = 0$. $P(0) = -(0)^3 + 4(0)^2 - 4(0) = 0$

The y-intercept is $(0, 0)$.

$-x^3$ has a negative coefficient and an odd exponent.

Therefore, the graph goes up to the far left and down to the far right.

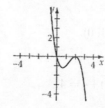

57. Sketch the graph of the function.

Let $P(x) = 0$.

$$-x^4 + 3x^3 + x^2 - 3x = 0$$
$$-x[x^3 - 3x^2 - x + 3] = 0$$
$$-x[x^2(x-3) - 1(x-3)] = 0$$
$$-x[(x-3)(x^2-1)] = 0$$
$$-x(x-3)(x+1)(x-1) = 0$$

$x = 0, x = 3, x = -1, x = 1$

The graph crosses the x-axis at $(0, 0)$, $(3, 0)$, $(-1, 0)$ and $(1, 0)$.

Let $x = 0$.

$$P(0) = -(0)^4 + 3(0)^3 + 0^2 - 3(0) = 0$$

The y-intercept is $(0, 0)$.

$-x^4$ has a negative coefficient and an even exponent. Therefore, the graph goes down to the far left and down to the far right.

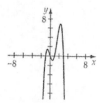

59. Sketch the graph of the function.

Let $P(x) = 0$.

$$x^5 - x^4 - 5x^3 + x^2 + 8x + 4 = 0$$
$$(x+1)^3(x-2)^2 = 0$$

$x = -1, x = 2$

The graph intersects the x-axis but does not cross it at $(2, 0)$.

The graph crosses the x-axis at $(-1, 0)$.

Let $x = 0$.

$$P(0) = 0^5 - 0^4 - 5(0)^3 + 0^2 + 8(0) + 4 = 4$$

The y-intercept is $(0, 4)$.

x^5 has a positive coefficient and an odd exponent. Therefore, the graph goes down to the far left and up to

the far right.

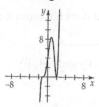

61. Explain how to produce graph P from graph Q.

Shift the graph of P vertically upward 2 units.

63. Explain how to produce graph P from graph Q.

Shift the graph of P horizontally 1 unit to the right.

65. Explain how to produce graph P from graph Q.

Shift the graph of P horizontally 2 units to the right and reflect this graph about the x-axis. Then shift the resulting graph vertically upward 3 units.

67. Use a graphing utility to find the maximum.

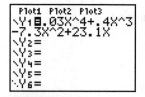 

 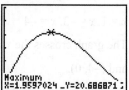

a. 20.69 milligrams

b. 1.968 hours $\times 60$ minutes per hour
≈ 118 minutes after taking the medication

69. a. Express the volume V as a function of x.

Volume = length × width × height
$$V(x) = (15 - 2x)(10 - 2x)x$$
$$= [15(10 - 2x) - 2x(10 - 2x)]x$$
$$= [150 - 30x - 20x + 4x^2]x$$
$$= [4x^2 - 50x + 150]x$$
$$= 4x^3 - 50x^2 + 150x$$

b. Find the value of x that maximizes the volume.

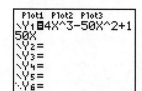

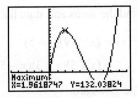

$x = 1.96$ inches (to the nearest 0.01 inch) maximizes the volume of the box.

71. a. Find the power.

$$P(v) = 4.95v^3$$

$$P(8) = 4.95(8)^3 = 2534.4 \approx 2530$$

The power is about 2530 W.

b. Find the wind speed.

$$10{,}000 = 4.95v^3$$

$$2020.20202 = v^3$$

$$12.6 \approx v$$

The speed is about 12.6 m/s.

c. Describe the effect if the wind speed is doubled.

The power increases by a factor of 8.

d. Describe the effect if the wind speed is tripled.

The power increases by a factor of 27.

73. a. Find the cubic regression function.

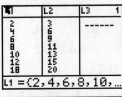

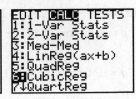

$$P(x) = 0.00072333177x^3 - 0.0481273792x^2$$
$$+ 1.749672272x - 0.2386484211$$

b. Write the coefficient of determination.

$R^2 \approx 0.9995341654$

c. Write the indication of the coefficient of determination.

The cubic regression function provides a very good model of the data.

d. Estimate the human age of a 5-month-old cat. Round to the nearest tenth of a year.

$$P(5) = 0.00072333177(5)^3 - 0.0481273792(5)^2$$
$$+ 1.749672272(5) - 0.2386484211$$
$$\approx 7.4 \text{ years}$$

75. a. Find the cubic and quartic model for the data.

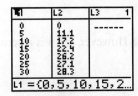

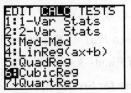

Cubic: $f(x) = 0.00015385409x^3 - 0.0297742717x^2$
$$+ 1.674968246x + 2.136506708$$

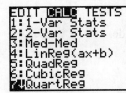

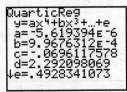

Quartic: $f(x) = -0.000005619394x^4$
$$+ 0.00099676312x^3 - 0.0696117578x^2$$
$$+ 2.292098069x + 0.4928341073$$

b. Use the cubic and quartic models to predict the fuel efficiency for a car traveling at 80 mph.

Cubic:

$$f(80) = 0.00015385409(80)^3 - 0.0297742717(80)^2$$
$$+ 1.674968246(80) + 2.136506708$$
$$\approx 24.4 \text{ mpg}$$

Quartic:

$$f(80) = -0.000005619394(80)^4$$
$$+ 0.00099676312(80)^3 - 0.0696117578(80)^2$$
$$+ 2.292098069(80) + 0.4928341073$$
$$\approx 18.5 \text{ mpg}$$

c. Determine which value from part b is more realistic.

Answers will vary; however, the downward trend for speeds greater than 50 mph suggests that 18.5 mpg is the more realistic value.

77. $P(x-3)$ shifts the graph horizontally three points to the right.

$(2+3, 0) = (5, 0)$

79. Shift the graph of $y = x^3$ horizontally two units to the right and vertically upward 1 unit.

81. False. Let $P(x) = x^2 - 2x - 8$,

$a = -3$, and $b = -5$.

Then $P(a) = 7$ and $P(b) = 7$. However, $x = 4$ is a zero of P and $-3 < 4 < 5$.

Prepare for Section 3.3

P1. Find the zeros.

$P(x) = 6x^2 - 25x + 14$

$0 = 6x^2 - 25x + 14$

$0 = (3x - 2)(2x - 7)$

$3x - 2 = 0 \quad \text{or} \quad 2x - 7 = 0$

$x = \dfrac{2}{3} \qquad\qquad x = \dfrac{7}{2}$

P3. Use synthetic division to divide.

$$
\begin{array}{r|rrrrr}
3 & 3 & 0 & -21 & -3 & -5 \\
  &   & 9 & 27 & 18 & 45 \\
\hline
  & 3 & 9 & 6   & 15 & 40
\end{array}
$$

$3x^3 + 9x^2 + 6x + 15 + \dfrac{40}{x-3}$

P5. List all integer factors of 27.

$\pm 1, \ \pm 3, \ \pm 9, \ \pm 27$

Section 3.3 Exercises

1. Yes. All of the zeros could be irrational numbers, complex nonreal numbers, or a combination of the two.

3. Find the zeros and state the multiplicity.

$P(x) = (x-4)(x+2)^2$

The zeros are:

4 (multiplicity 1), –2 (multiplicity 2).

5. Find the zeros and state the multiplicity.

$P(x) = x^2(3x+5)^2$

The zeros are:

$-\dfrac{5}{3}$ (multiplicity 2), 0 (multiplicity 2).

7. Find the zeros and state the multiplicity.

$P(x) = (x^2 - 9)(x+5)^2$

$\qquad = (x+3)(x-3)(x+5)^2$

The zeros are:

–3 (multiplicity 1), 3 (multiplicity 1),

–5 (multiplicity 2).

9. Find the zeros and state the multiplicity.

$P(x) = (3x-5)^2(2x-7)$

The zeros are:

$\dfrac{5}{3}$ (multiplicity 2), $\dfrac{7}{2}$ (multiplicity 1).

11. List possible rational zeros using Rational Zero Theorem.

$P(x) = x^3 + 3x^2 - 6x - 8$

$p = \pm \text{ factors of } 8 = \pm 1, \ \pm 2, \ \pm 4, \ \pm 8$

$q = \pm \text{ factors of } 1 = \pm 1$

$\dfrac{p}{q} = \text{possible rational zeros} = \pm 1, \ \pm 2, \ \pm 4, \ \pm 8$

13. List possible rational zeros using Rational Zero Theorem.

$P(x) = 2x^3 + x^2 - 25x + 12$

$p = \pm \text{ factors of } 12 = \pm 1, \ \pm 2, \ \pm 3, \ \pm 4,$

$\qquad\qquad\qquad\qquad\qquad \pm 6, \ \pm 12$

$q = \pm \text{ factors of } 2 = \pm 1, \ \pm 2$

$\dfrac{p}{q} = \text{possible rational zeros} = \pm 1, \ \pm 2, \ \pm 3,$

$\qquad\qquad \pm 4, \ \pm 6, \ \pm 12, \ \pm\dfrac{1}{2}, \ \pm\dfrac{3}{2}$

15. List possible rational zeros using Rational Zero Theorem.

$P(x) = 6x^4 + 23x^3 + 19x^2 - 8x - 4$

$p = \pm \text{ factors of } 4 = \pm 1, \ \pm 2, \ \pm 4$

$q = \pm \text{ factors of } 6 = \pm 1, \ \pm 2, \ \pm 3, \ \pm 6$

$\dfrac{p}{q} = \text{possible rational zeros} = \pm 1, \ \pm 2,$

$\qquad \pm 4, \ \pm\dfrac{1}{2}, \ \pm\dfrac{1}{3}, \ \pm\dfrac{1}{6}, \ \pm\dfrac{2}{3}, \ \pm\dfrac{4}{3}$

17. List possible rational zeros using Rational Zero Theorem.

$$P(x) = 4x^4 - 12x^3 - 3x^2 + 12x - 7$$

$p = \pm \text{ factors of } 7 = \pm 1, \ \pm 7$

$q = \pm \text{ factors of } 4 = \pm 1, \ \pm 2, \ \pm 4$

$\dfrac{p}{q} = \text{ possible rational zeros} = \pm 1, \ \pm 7,$

$$\pm \frac{1}{2}, \ \pm \frac{7}{2}, \ \pm \frac{1}{4}, \ \pm \frac{7}{4}$$

19. List possible rational zeros using Rational Zero Theorem.

$$P(x) = x^5 - 10x^4 + 27x^3 + 14x^2 - 128x + 96$$

$p = \pm \text{ factors of } 96 = \pm 1, \ \pm 2, \ \pm 3, \ \pm 4, \ \pm 6,$

$\quad \pm 8, \ \pm 12, \ \pm 16, \ \pm 24, \ \pm 32, \ \pm 48, \ \pm 96$

$q = \pm \text{ factors of } 1 = \pm 1$

$\dfrac{p}{q} = \text{ possible rational zeros} = \pm 1, \ \pm 2, \ \pm 3, \ \pm 4,$

$\quad \pm 6, \ \pm 8, \ \pm 12, \ \pm 16, \ \pm 32, \ \pm 24, \ \pm 48, \ \pm 96$

21. Find the upper and lower bounds.

```
1 | 1   3   -6   -6
  |     1    4
  --------------------
    1   4   -2
```

Don't finish dividing. 1 is not an upper bound.

```
2 | 1   3   -6   -6
  |     2   10    8
  --------------------
    1   5    4    2
```

The smallest integer that is an upper bound is 2.

```
-1 | 1   3   -6   -6
   |    -1
   --------------------
     1   2
```

Don't finish dividing. −1 is not a lower bound.

```
-2 | 1   3   -6   -6
   |    -2
   --------------------
     1   1
```

Don't finish dividing. −2 is not a lower bound.

```
-3 | 1   3   -6   -6
   |    -3    0
   --------------------
     1   0   -6
```

Don't finish dividing. −3 is not a lower bound.

```
-4 | 1   3   -6   -6
   |    -4    4
   --------------------
     1  -1   -2
```

Don't finish dividing. −4 is not a lower bound.

```
-5 | 1   3   -6    -6
   |    -5   10   -20
   --------------------
     1  -2    4   -26
```

The largest integer that is a lower bound is −5.

23. Find the upper and lower bounds.

```
3 | 2   1   -25   10
  |     6    21
  --------------------
    2   7    -4
```

Don't finish dividing. 3 is not an upper bound.

```
4 | 2   1   -25   10
  |     8    36   44
  --------------------
    2   9    11   54
```

The smallest integer that is an upper bound is 4.

```
-3 | 2    1   -25   10
   |     -6    15
   --------------------
     2   -5   -10
```

Don't finish dividing. −3 is not a lower bound.

```
-4 | 2    1   -25    10
   |     -8    28   -12
   --------------------
     2   -7     3    -2
```

The largest integer that is a lower bound is −4.

25. Find the upper and lower bounds.

```
3 | 3   -9   2   -5   -8
  |      9   0    6    3
  ----------------------
    3    0   2    1   -5
```

3 is not an upper bound.

Find the upper and lower bounds.

```
4 | 3   -9    2   -5    -8
  |     12   12   56   204
  -----------------------
    3    3   14   51   196
```

The smallest integer that is an upper bound is 4.

```
-1 | 3   -9    2    -5    -8
   |     -3   12   -14    19
   -----------------------
     3  -12   14   -19    11
```

The largest integer that is a lower bound is −1.

27. Find the upper and lower bounds.

$$\begin{array}{r|rrrrr}
3 & -5 & 14 & 2 & -12 & 325 \\
 & & -15 & -3 & -3 & -45 \\
\hline
 & -5 & -1 & -1 & -15 & 280
\end{array}$$

3 is not an upper bound.

$$\begin{array}{r|rrrrr}
4 & -5 & 14 & 2 & -12 & 325 \\
 & & -20 & -24 & -88 & -400 \\
\hline
 & -5 & -6 & -22 & -100 & -75
\end{array}$$

The smallest integer that is an upper bound is 4.

$$\begin{array}{r|rrrrr}
-2 & -5 & 14 & 2 & -12 & 325 \\
 & & 10 & -48 & 92 & -160 \\
\hline
 & -5 & 24 & -46 & 80 & 165
\end{array}$$

–2 is not a lower bound.

$$\begin{array}{r|rrrrr}
-3 & -5 & 14 & 2 & -12 & 325 \\
 & & 15 & -87 & 255 & -729 \\
\hline
 & -5 & 29 & -85 & 243 & -404
\end{array}$$

The largest integer that is a lower bound is –3.

29. Find the upper and lower bounds.

$$\begin{array}{r|rrrrrr}
1 & 1 & 0 & 0 & 0 & 0 & -32 \\
 & & 1 & 1 & 1 & 1 & 1 \\
\hline
 & 1 & 1 & 1 & 1 & 1 & -31
\end{array}$$

1 is not an upper bound.

$$\begin{array}{r|rrrrrr}
2 & 1 & 0 & 0 & 0 & 0 & -32 \\
 & & 2 & 4 & 8 & 16 & 32 \\
\hline
 & 1 & 2 & 4 & 8 & 16 & 0
\end{array}$$

The smallest integer that is an upper bound is 2.

$$\begin{array}{r|rrrrrr}
-1 & 1 & 0 & 0 & 0 & 0 & -32 \\
 & & -1 & 1 & -1 & 1 & -1 \\
\hline
 & 1 & -1 & 1 & -1 & 1 & -33
\end{array}$$

The largest integer that is a lower bound is –1.

31. State the number of positive and negative real zeros using Descartes' Rule of Signs.

$P(x) = x^3 + 3x^2 - 6x - 8$ has 1 change in sign.

Therefore there is one positive zero.

$P(-x) = -x^3 + 3x^2 + 6x - 8$ has 2 changes in sign.

Therefore there are two or no negative zeros.

33. State the number of positive and negative real zeros using Descartes' Rule of Signs.

$P(x) = 2x^3 + x^2 - 25x + 12$ has 2 changes in sign.

Therefore there are two or no positive zeros.

$P(-x) = -2x^3 + x^2 + 25x + 12$ has 1 change in sign.

Therefore there is one negative zero.

35. State the number of positive and negative real zeros using Descartes' Rule of Signs.

$P(x) = 2x^5 - 37x^4 + 258x^3 - 820x^2 + 1100x - 375$

has 5 changes in sign. Therefore there are five, three or one positive zeros.

$P(-x) = -2x^5 - 37x^4 - 258x^3 - 820x^2 - 1100x - 375$

has no changes in sign. Therefore there are no negative zeros.

37. State the number of positive and negative real zeros using Descartes' Rule of Signs.

$P(x) = 2x^5 + 23x^4 + 90x^3 + 152x^2 + 116x + 33$ has no changes in sign. . Therefore there are no positive zeros.

$P(-x) = -2x^5 + 23x^4 - 90x^3 + 152x^2 - 116x + 33$

has 5 changes in sign. Therefore there are five, three or one negative zeros.

39. State the number of positive and negative real zeros using Descartes' Rule of Signs.

$P(x) = x^5 - 2x^4 - 35x^3 + 158x^2 - 242x + 168$ has 4 changes in sign. Therefore there are four, two or no positive zeros.

$P(-x) = -x^5 - 2x^4 + 35x^3 + 158x^2 + 242x + 168$ has one change in sign. Therefore there is one negative zero.

41. State the number of positive and negative real zeros using Descartes' Rule of Signs.

$P(x) = 2x^6 - 19x^5 + 49x^4 - 88x^3 - 158x^2 + 656x - 280$

has 5 changes in sign. Therefore there are five, three or one positive zeros.

$P(-x) = 2x^6 + 19x^5 + 49x^4 + 88x^3 - 158x^2 - 656x - 280$

has 1 change in sign. Therefore there is one negative zero.

43. State the number of positive and negative real zeros using Descartes' Rule of Signs.

$$P(x) = 12x^7 - 112x^6 + 421x^5 - 840x^4$$
$$+ 1038x^3 - 938x^2 + 629x - 210$$

has 7 changes in sign. Therefore there are seven, five, three or one positive zeros.

$$P(-x) = -12x^7 - 112x^6 - 421x^5 - 840x^4$$
$$- 1038x^3 - 938x^2 - 629x - 210$$

has no changes in sign. Therefore there are no negative zeros.

45. Find the zeros and state multiplicity if more than one.

$$P(x) = x^3 + 3x^2 - 6x - 8$$

one positive and two or no negative real zeros

$$\frac{p}{q} = \pm 1, \ \pm 2, \ \pm 4, \ \pm 8$$

$$\begin{array}{r|rrrr} 2 & 1 & 3 & -6 & -8 \\ & & 2 & 10 & 8 \\ \hline & 1 & 5 & 4 & 0 \end{array}$$

$$x^2 + 5x + 4 = 0$$
$$(x+4)(x+1) = 0$$
$$x = -4, -1$$

The zeros of $P(x)$ are 2, –4, and –1.

47. Find the zeros and state multiplicity if more than one.

$P(x) = 2x^3 + x^2 - 25x + 12$ has two or no positive and one negative real zero.

$$\frac{p}{q} = \pm 1, \ \pm 2, \ \pm 3, \ \pm 4, \ \pm 6, \ \pm 12, \ \pm \frac{1}{2}, \ \pm \frac{3}{2}$$

$$\begin{array}{r|rrrr} 3 & 2 & 1 & -25 & 12 \\ & & 6 & 21 & -12 \\ \hline & 2 & 7 & -4 & 0 \end{array}$$

$$2x^2 + 7x - 4 = 0$$
$$(2x-1)(x+4) = 0$$
$$x = \frac{1}{2}, \ -4$$

The zeros of $P(x)$ are 3, $\frac{1}{2}$, –4.

49. Find the zeros and state multiplicity if more than one.

$$P(x) = 6x^4 + 23x^3 + 19x^2 - 8x - 4$$

one positive and three or one negative real zero

$$\begin{array}{r|rrrrr} -2 & 6 & 23 & 19 & -8 & -4 \\ & & -12 & -22 & 6 & 4 \\ \hline & 6 & 11 & -3 & -2 & 0 \end{array}$$

$$\begin{array}{r|rrrr} -2 & 6 & 11 & -3 & -2 \\ & & -12 & 2 & 2 \\ \hline & 6 & -1 & -1 & 0 \end{array}$$

$$6x^2 - x - 1 = 0$$
$$(3x+1)(2x-1) = 0$$
$$x = -\frac{1}{3}, \ \frac{1}{2}$$

The zeros of $P(x)$ are –2 (multiplicity 2), $-\frac{1}{3}$, $\frac{1}{2}$.

51. Find the zeros and state multiplicity if more than one.

$$P(x) = 2x^4 - 9x^3 - 2x^2 + 27x - 12$$

three or one positive and one negative real zero

$$\frac{p}{q} = \pm 1, \ \pm 2, \ \pm 3, \ \pm 4, \ \pm 6, \ \pm 12, \ \pm \frac{1}{2}, \ \pm \frac{3}{2}$$

$$\begin{array}{r|rrrrr} 4 & 2 & -9 & -2 & 27 & -12 \\ & & 8 & -4 & -24 & 12 \\ \hline & 2 & -1 & -6 & 3 & 0 \end{array}$$

$$\begin{array}{r|rrrr} \frac{1}{2} & 2 & -1 & -6 & 3 \\ & & 1 & 0 & -3 \\ \hline & 2 & 0 & -6 & 0 \end{array}$$

$$2x^2 - 6 = 0$$
$$2(x^2 - 3) = 0$$
$$x = \pm\sqrt{3}$$

The zeros of $P(x)$ are 4, $\frac{1}{2}$, $\sqrt{3}$, $-\sqrt{3}$.

53. Find the zeros and state multiplicity if more than one.

$$P(x) = x^3 - 8x^2 + 8x + 24$$

two or no positive and one negative real zero

$$\frac{p}{q} = \pm 1, \ \pm 2, \ \pm 3, \ \pm 4, \ \pm 6, \ \pm 8, \ \pm 12, \ \pm 24$$

$$\begin{array}{r|rrrr} 6 & 1 & -8 & 8 & 24 \\ & & 6 & -12 & -24 \\ \hline & 1 & -2 & -4 & 0 \end{array}$$

$$x^2 - 2x - 4 = 0$$

$$x = \frac{-(-2) \pm \sqrt{(-2)^2 - 4(1)(-4)}}{2(1)}$$

$$= \frac{2 \pm \sqrt{20}}{2} = \frac{2 \pm 2\sqrt{5}}{2} = 1 \pm \sqrt{5}$$

The zeros of $P(x)$ are 6, $1+\sqrt{5}$, $1-\sqrt{5}$.

55. Find the zeros and state multiplicity if more than one.

$$P(x) = 2x^4 - 19x^3 + 51x^2 - 31x + 5$$

four, two or no positive and no negative real zeros

$$\frac{p}{q} = \pm 1, \ \pm 5, \ \pm\frac{1}{2}, \ \pm\frac{5}{2}$$

$$
\begin{array}{r|rrrrr}
5 & 2 & -19 & 51 & -31 & 5 \\
  &   & 10 & -45 & 30 & -5 \\
\hline
  & 2 & -9 & 6 & -1 & 0
\end{array}
$$

$$
\begin{array}{r|rrrr}
\frac{1}{2} & 2 & -9 & 6 & -1 \\
  &   & 1 & -4 & 1 \\
\hline
  & 2 & -8 & 2 & 0
\end{array}
$$

$$2x^2 - 8x + 2 = 2(x^2 - 4x + 1) = 0$$

$$x = \frac{-(-4) \pm \sqrt{(-4)^2 - 4(1)(1)}}{2(1)} = \frac{4 \pm \sqrt{12}}{2} = \frac{4 \pm 2\sqrt{3}}{2}$$

$$= 2 \pm \sqrt{3}$$

The zeros of $P(x)$ are 5, $\frac{1}{2}$, $2 + \sqrt{3}$, $2 - \sqrt{3}$.

57. Find the zeros and state multiplicity if more than one.

$$P(x) = 3x^6 - 10x^5 - 29x^4 + 34x^3 + 50x^2 - 24x - 24$$

three or one positive and three or one negative real zeros

$$\frac{p}{q} = \pm 1, \ \pm 2, \ \pm 3, \ \pm 4, \ \pm 6, \ \pm 8, \ \pm 12, \ \pm 24,$$
$$\pm\frac{1}{3}, \ \pm\frac{2}{3}, \ \pm\frac{4}{3}, \ \pm\frac{3}{2}, \ \pm\frac{8}{3}$$

$$
\begin{array}{r|rrrrrrr}
1 & 3 & -10 & -29 & 34 & 50 & -24 & -24 \\
  &   & 3 & -7 & -36 & -2 & 48 & 24 \\
\hline
  & 3 & -7 & -36 & -2 & 48 & 24 & 0
\end{array}
$$

$$
\begin{array}{r|rrrrrr}
-1 & 3 & -7 & -36 & -2 & 48 & 24 \\
   &   & -3 & 10 & 26 & -24 & -24 \\
\hline
   & 3 & -10 & -26 & 24 & 24 & 0
\end{array}
$$

$$
\begin{array}{r|rrrrr}
-2 & 3 & -10 & -26 & 24 & 24 \\
   &   & -6 & 32 & -12 & -24 \\
\hline
   & 3 & -16 & 6 & 12 & 0
\end{array}
$$

$$
\begin{array}{r|rrrr}
-\frac{2}{3} & 3 & -16 & 6 & 12 \\
   &   & -2 & 12 & -12 \\
\hline
   & 3 & -18 & 18 & 0
\end{array}
$$

$$3x^2 - 18x + 18 = 0$$

$$3(x^2 - 6x + 6) = 0$$

$$x^2 - 6x + 6 = 0$$

$$x = \frac{-(-6) \pm \sqrt{(-6)^2 - 4(1)(6)}}{2(1)} = \frac{6 \pm \sqrt{12}}{2} = \frac{6 \pm 2\sqrt{3}}{2}$$

$$= 3 \pm \sqrt{3}$$

The zeros of $P(x)$ are 1, -1, -2, $-\frac{2}{3}$, $3 + \sqrt{3}$, $3 - \sqrt{3}$.

59. Find the zeros and state multiplicity if more than one.

$$P(x) = x^3 - 3x - 2$$

one positive and two or no negative real zeros

$$\frac{p}{q} = \pm 1, \ \pm 2$$

$$
\begin{array}{r|rrrr}
2 & 1 & 0 & -3 & -2 \\
  &   & 2 & 4 & 2 \\
\hline
  & 1 & 2 & 1 & 0
\end{array}
$$

$$x^2 + 2x + 1 = 0$$

$$(x + 1)^2 = 0$$

$$x = -1$$

The zeros of $P(x)$ are 2, -1 (multiplicity 2).

61. Find the zeros and state multiplicity if more than one.

$$P(x) = x^4 - 5x^2 - 2x = x(x^3 - 5x - 2)$$

one positive and two or no negative real zeros

$$\frac{p}{q} = \pm 1, \ \pm 2$$

$$
\begin{array}{r|rrrr}
-2 & 1 & 0 & -5 & -2 \\
   &   & -2 & 4 & 2 \\
\hline
   & 1 & -2 & -1 & 0
\end{array}
$$

$$x^2 - 2x - 1 = 0$$

$$x = \frac{-(-2) \pm \sqrt{(-2)^2 - 4(1)(-1)}}{2(1)}$$

$$= \frac{2 \pm \sqrt{8}}{2} = \frac{2 \pm 2\sqrt{2}}{2} = 1 \pm \sqrt{2}$$

The zeros of $P(x)$ are 0, -2, $1 + \sqrt{2}$, $1 - \sqrt{2}$.

63. Find the zeros and state multiplicity if more than one.

$$P(x) = x^4 + x^3 - 3x^2 - 5x - 2$$

one positive and three or one negative real zeros

$$\frac{p}{q} = \pm 1, \ \pm 2$$

$$
\begin{array}{r|rrrrr}
-1 & 1 & 1 & -3 & -5 & -2 \\
   &   & -1 & 0 & 3 & 2 \\
\hline
   & 1 & 0 & -3 & -2 & 0
\end{array}
$$

$$-1 \begin{array}{|rrrr} 1 & 0 & -3 & -2 \\ & -1 & 1 & 2 \\ \hline 1 & -1 & -2 & 0 \end{array}$$

$$x^2 - x - 2 = 0$$
$$(x-2)(x+1) = 0$$
$$x = 2, -1$$

The zeros of $P(x)$ are 2, -1 (multiplicity 3).

65. Find the zeros and state multiplicity if more than one.

$$P(x) = 2x^4 - 17x^3 + 4x^2 + 35x - 24$$

three or one positive and one negative real zeros

$$\frac{p}{q} = \pm 1, \ \pm 2, \ \pm 3, \ \pm 4, \ \pm 6, \ \pm 8, \ \pm 12, \ \pm 24,$$

$$\pm \frac{1}{2}, \ \pm \frac{3}{2}$$

$$1 \begin{array}{|rrrrr} 2 & -17 & 4 & 35 & -24 \\ & 2 & -15 & -11 & 24 \\ \hline 2 & -15 & -11 & 24 & 0 \end{array}$$

$$1 \begin{array}{|rrrr} 2 & -15 & -11 & 24 \\ & 2 & -13 & -24 \\ \hline 2 & -13 & -24 & 0 \end{array}$$

$$2x^2 - 13x - 24 = 0$$
$$(2x+3)(x-8) = 0$$
$$x = -\frac{3}{2}, \ 8$$

The zeros of $P(x)$ are 1(multiplicity 2), $-\frac{3}{2}$, 8.

67. Find the zeros and state multiplicity if more than one.

$$P(x) = x^3 - 16x = x(x^2 - 16)$$

one positive and one negative real zeros

$$\frac{p}{q} = \pm 1, \ \pm 2, \ \pm 4, \ \pm 8, \ \pm 16$$

$$x(x^2 - 16) = 0$$
$$x(x+4)(x-4) = 0$$
$$x = -4, \ 0, \ 4$$

The zeros of $P(x)$ are -4, 0, and 4.

69. Find n.

The original cube's dimensions are $n \times n \times n$.

The resulting solid measures $n \cdot n \cdot (n-2)$.

$$n^2(n-2) = 567$$
$$n^3 - 2n^2 - 567 = 0$$

$$9 \begin{array}{|rrrr} 1 & -2 & 0 & -567 \\ & 9 & 63 & 567 \\ \hline 1 & 7 & 63 & 0 \end{array}$$

$n = 9$ inches.

71. Find the value of x.

$$[(x)(x+1)(x+2)] - [(2)(1)(x)] = 112$$
$$x(x^2 + 3x + 2) - 2x = 112$$
$$x^3 + 3x^2 + 2x - 2x = 112$$
$$x^3 + 3x^2 - 112 = 0$$

$$4 \begin{array}{|rrrr} 1 & 3 & 0 & -112 \\ & 4 & 28 & 112 \\ \hline 1 & 7 & 28 & 0 \end{array}$$

$x = 4$ inches

73. a. Find the maximum number of pieces.

$$P(5) = \frac{5^3 + 5(5) + 6}{6} = \frac{125 + 25 + 6}{6}$$

$$= \frac{156}{6} = 26 \text{ pieces}$$

b. Find the fewest number of straight cuts.

$$\frac{n^3 + 5n + 6}{6} = 64$$

$$n^3 + 5n + 6 = 384$$
$$n^3 + 5n - 378 = 0$$

$$7 \begin{array}{|rrrr} 1 & 0 & 5 & -378 \\ & 7 & 49 & 378 \\ \hline 1 & 7 & 54 & 0 \end{array}$$

At least 7 cuts are needed to produce 64 pieces.

75. Find the number of rows of the pyramid.

```
Plot1 Plot2 Plot3        WINDOW
\Y1◼(1/6)(2X^3+3         Xmin=-20
X^2+X)                   Xmax=20
\Y2◼140                  Xscl=5
\Y3=                     Ymin=-590
\Y4=                     Ymax=330
\Y5=                     Yscl=100
\Y6=                     Xres=1
```

```
CALCULATE
1:value
2:zero
3:minimum
4:maximum
5:intersect             Intersection
6:dy/dx                 X=7   Y=140
7:∫f(x)dx
```

If 140 cannonballs are used, there are 7 rows in the pyramid.

77. Find x.

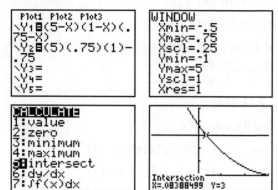

The company should decrease each dimension by 0.084 inch.

79. Find the possible lengths of l.

$$4w + l = 81$$
$$l = 81 - 4w$$

Volume = $w^2 l = 4900$

$$w^2(81 - 4w) = 4900$$

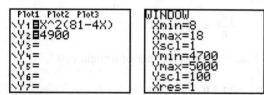

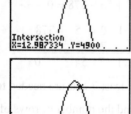

When $w = 12.9875$, then

$l = 81 - 4w = 81 - 4(12.9875) = 29.05$ in.

When $w = 14$, then $l = 81 - 4w = 81 - 4(14) = 25$ in.

Thus, the lengths can be 25 in. or 29.05 in.

81. Find the giraffe's height, rounding to the nearest tenth of a foot.

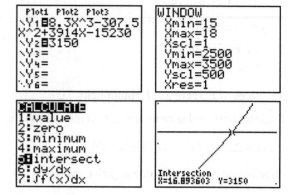

The giraffe is 16.9 feet tall.

83. Find B and verify the absolute value is less than B.

$$B = \left(\frac{\text{max of} (|-5|, |-28|, |15|)}{|2|} + 1 \right)$$

$$B = \left(\frac{28}{2} + 1 \right)$$

$$B = 15$$

$$|-3| = 3 < 15$$

$$\left| \frac{1}{2} \right| = \frac{1}{2} < 15$$

$$|5| = 5 < 15$$

The absolute value of each zero is less than B.

85. Find B and verify the absolute value is less than B.

$$B = \left(\frac{\text{max of} (|-2|, |9|, |2|, |-10|)}{|1|} + 1 \right)$$

$$B = \left(\frac{10}{1} + 1 \right)$$

$$B = 11$$

$$|1 + 3i| = \sqrt{10} < 11$$

$$|1 - 3i| = \sqrt{10} < 11$$

$$|1| = 1 < 11$$

$$|-1| = 1 < 11$$

The absolute value of each zero is less than B.

Mid-Chapter 3 Quiz

1. Use the Remainder Theorem to find $P(5)$.

$$\begin{array}{r|rrrr}
5 & 3 & 7 & -2 & -5 \\
  &   & 15 & 110 & 540 \\
\hline
  & 3 & 22 & 108 & 535 \\
\end{array}$$

$$P(5) = 535$$

3. Determine the far-right and far-left behavior.

Since $a_n = 4$ is positive and $n = 3$ is odd, the graph of

P goes down to the far left and up to the far right.

5. Verify the zero between 3 and 4 using Intermediate

Value Theorem.

$$P(x) = 2x^4 - 5x^3 + x^2 - 20x - 28$$

$$
\begin{array}{r|rrrrr}
3 & 2 & -5 & 1 & -20 & -28 \\
  &   & 6 & 3 & 12 & -24 \\
\hline
  & 2 & 1 & 4 & -8 & -52 \\
\end{array}
$$

$$
\begin{array}{r|rrrrr}
4 & 2 & -5 & 1 & -20 & -28 \\
  &   & 8 & 12 & 52 & 128 \\
\hline
  & 2 & 3 & 13 & 32 & 100 \\
\end{array}
$$

$P(3)$ is negative; $P(4)$ is positive.

Therefore $P(x)$ has a zero between 3 and 4.

7. Find the relative maximum. [3.2]

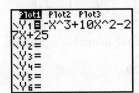 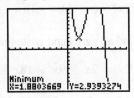

The relative minimum is $y \approx 2.94$ at $x \approx 1.88$.

Prepare for Section 3.4

P1. Find the conjugate of $3 - 2i$.

$3 + 2i$

P3. Multiply.

$$(x-1)(x-3)(x-4)$$

$$= (x-1)(x^2 - 7x + 12)$$

$$= x(x^2 - 7x + 12) - 1(x^2 - 7x + 12)$$

$$= x^3 - 7x^2 + 12x - x^2 + 7x - 12$$

$$= x^3 - 8x^2 + 19x - 12$$

P5. Solve.

$$x^2 + 9 = 0$$

$$x^2 = -9$$

$$x = \pm\sqrt{-9}$$

$$x = \pm 3i$$

The solutions are $3i$ and $-3i$.

Section 3.4 Exercises

1. Since $P(x) = x^3 - 5x^2 + 6x - 3$ has degree 3, then by

the by the number of zeros of a polynomial function,

the polynomial has 3 complex zeros.

3. Since $P(x) = -3x^3 + 5x^2 + 6x + 2$ has degree 3, then

by the by the number of zeros of a polynomial

function, the polynomial has 3 complex zeros.

5. Find all zeros and write the polynomial as a product.

$$P(x) = x^4 + x^3 - 2x^2 + 4x - 24$$

$$
\begin{array}{r|rrrrr}
2 & 1 & 1 & -2 & 4 & -24 \\
  &   & 2 & 6 & 8 & 24 \\
\hline
  & 1 & 3 & 4 & 12 & 0 \\
\end{array}
$$

$$
\begin{array}{r|rrrr}
-3 & 1 & 3 & 4 & 12 \\
   &   & -3 & 0 & -12 \\
\hline
   & 1 & 0 & 4 & 0 \\
\end{array}
$$

$$x^2 + 4 = 0$$

$$x^2 = -4$$

$$x = \pm\sqrt{-4} = \pm 2i$$

The zeros are $2, -3, 2i, -2i$.

$$P(x) = (x-2)(x+3)(x-2i)(x+2i)$$

7. Find all zeros and write the polynomial as a product.

$$P(x) = 2x^4 - x^3 - 4x^2 + 10x - 4$$

$$
\begin{array}{r|rrrrr}
\frac{1}{2} & 2 & -1 & -4 & 10 & -4 \\
  &   & 1 & 0 & -2 & 4 \\
\hline
  & 2 & 0 & -4 & 8 & 0 \\
\end{array}
$$

$$
\begin{array}{r|rrrr}
-2 & 2 & 0 & -4 & 8 \\
   &   & -4 & 8 & -8 \\
\hline
   & 2 & -4 & 4 & 0 \\
\end{array}
$$

$$2x^2 - 4x + 4 = 2(x^2 - 2x + 2) = 0$$

$$x = \frac{-(-2) \pm \sqrt{(-2)^2 - 4(1)(2)}}{2(1)}$$

$$= \frac{2 \pm \sqrt{-4}}{2} = \frac{2 \pm 2i}{2} = 1 \pm i$$

The zeros are $\frac{1}{2}, -2, 1+i, 1-i$.

$$P(x) = 2\left(x - \frac{1}{2}\right)(x+2)(x-1-i)(x-1+i)$$

9. Find all zeros and write the polynomial as a product.

$$P(x) = x^5 - 9x^4 + 34x^3 - 58x^2 + 45x - 13$$

$$
\begin{array}{r|rrrrrr}
1 & 1 & -9 & 34 & -58 & 45 & -13 \\
  &   & 1 & -8 & 26 & -32 & 13 \\
\hline
  & 1 & -8 & 26 & -32 & 13 & 0 \\
\end{array}
$$

$$
\begin{array}{r|rrrrr}
1 & 1 & -8 & 26 & -32 & 13 \\
  &   & 1 & -7 & 19 & -13 \\
\hline
  & 1 & -7 & 19 & -13 & 0 \\
\end{array}
$$

$$
\begin{array}{r|rrrr}
1 & 1 & -7 & 19 & -13 \\
  &   & 1 & -6 & 13 \\
\hline
  & 1 & -6 & 13 & 0 \\
\end{array}
$$

$$x^2 - 6x + 13 = 0$$

$$x = \frac{-(-6) \pm \sqrt{(-6)^2 - 4(1)(13)}}{2(1)} = \frac{6 \pm \sqrt{36 - 52}}{2}$$

$$= \frac{6 \pm \sqrt{-16}}{2} = \frac{6 \pm 4i}{2} = 3 \pm 2i$$

The zeros are 1 (multiplicity 3), $3 + 2i$, $3 - 2i$.

$$P(x) = (x-1)^3(x-3-2i)(x-3+2i)$$

11. Find all zeros and write the polynomial as a product.

$$P(x) = 2x^4 - x^3 - 15x^2 + 23x + 15$$

$$
\begin{array}{r|rrrrr}
-3 & 2 & -1 & -15 & 23 & 15 \\
   &   & -6 & 21 & -18 & -15 \\
\hline
   & 2 & -7 & 6 & 5 & 0 \\
\end{array}
$$

$$
\begin{array}{r|rrrr}
-\frac{1}{2} & 2 & -7 & 6 & 5 \\
             &   & -1 & 4 & -5 \\
\hline
             & 2 & -8 & 10 & 0 \\
\end{array}
$$

$$2x^2 - 8x + 10 = 2(x^2 - 4x + 5) = 0$$

$$x = \frac{-(-4) \pm \sqrt{(-4)^2 - 4(1)(5)}}{2(1)}$$

$$= \frac{4 \pm \sqrt{-4}}{2} = \frac{4 \pm 2i}{2} = 2 \pm i$$

The zeros are -3, $-\frac{1}{2}$, $2+i$, $2-i$.

$$P(x) = 2(x+3)\left(x+\frac{1}{2}\right)(x-2-i)(x-2+i)$$

13. Find all zeros and write the polynomial as a product.

$$P(x) = 2x^4 - 14x^3 + 33x^2 - 46x + 40$$

$$
\begin{array}{r|rrrrr}
4 & 2 & -14 & 33 & -46 & 40 \\
  &   & 8 & -24 & 36 & -40 \\
\hline
  & 2 & -6 & 9 & -10 & 0 \\
\end{array}
$$

$$
\begin{array}{r|rrrr}
2 & 2 & -6 & 9 & -10 \\
  &   & 4 & -4 & 10 \\
\hline
  & 2 & -2 & 5 & 0 \\
\end{array}
$$

$$2x^2 - 2x + 5 = 0$$

$$x = \frac{-(-2) \pm \sqrt{(-2)^2 - 4(2)(5)}}{2(2)}$$

$$= \frac{2 \pm \sqrt{-36}}{4} = \frac{2 \pm 6i}{4} = \frac{1}{2} \pm \frac{3}{2}i$$

The zeros are 4, 2, $\frac{1}{2}+\frac{3}{2}i$, $\frac{1}{2}-\frac{3}{2}i$.

$$P(x) = (x-4)(x-2)\left(x-\frac{1}{2}-\frac{3}{2}i\right)\left(x-\frac{1}{2}+\frac{3}{2}i\right)$$

15. Find all zeros and write the polynomial as a product.

$$P(x) = 2x^3 - 9x^2 + 18x - 20$$

$$
\begin{array}{r|rrrr}
\frac{5}{2} & 2 & -9 & 18 & -20 \\
            &   & 5 & -10 & 20 \\
\hline
            & 2 & -4 & 8 & 0 \\
\end{array}
$$

$$2x^2 - 4x + 8 = 2(x^2 - 2x + 4) = 0$$

$$x = \frac{-(-2) \pm \sqrt{(-2)^2 - 4(1)(4)}}{2(1)}$$

$$= \frac{2 \pm \sqrt{-12}}{2} = \frac{2 \pm 2i\sqrt{3}}{2} = 1 \pm i\sqrt{3}$$

The zeros are $\frac{5}{2}$, $1 + i\sqrt{3}$, $1 - i\sqrt{3}$.

$$P(x) = 2\left(x-\frac{5}{2}\right)(x-1-i\sqrt{3})(x-1+i\sqrt{3})$$

17. Find all zeros and write the polynomial as a product.

$$P(x) = 2x^4 - x^3 - 2x^2 + 13x - 6$$

$$
\begin{array}{r|rrrrr}
-2 & 2 & -1 & -2 & 13 & -6 \\
   &   & -4 & 10 & -16 & 6 \\
\hline
   & 2 & -5 & 8 & -3 & 0 \\
\end{array}
$$

$$
\begin{array}{r|rrrr}
\frac{1}{2} & 2 & -5 & 8 & -3 \\
            &   & 1 & -2 & 3 \\
\hline
            & 2 & -4 & 6 & 0 \\
\end{array}
$$

$2x^2 - 4x + 6 = 2(x^2 - 2x + 3) = 0$

$x = \dfrac{-(-2) \pm \sqrt{(-2)^2 - 4(1)(3)}}{2(1)}$

$= \dfrac{2 \pm \sqrt{-8}}{2} = \dfrac{2 \pm 2i\sqrt{2}}{2} = 1 \pm i\sqrt{2}$

The zeros are -2, $\frac{1}{2}$, $1 + i\sqrt{2}$, $1 - i\sqrt{2}$.

$P(x) = 2(x+2)\left(x - \dfrac{1}{2}\right)(x - 1 - i\sqrt{2})(x - 1 + i\sqrt{2})$

19. Find all zeros and write the polynomial as a product.

$P(x) = 3x^5 + 2x^4 + 10x^3 + 6x^2 - 25x - 20$

$$\begin{array}{r|rrrrrr} -1 & 3 & 2 & 10 & 6 & -25 & -20 \\ & & -3 & 1 & -11 & 5 & 20 \\ \hline & 3 & -1 & 11 & -5 & -20 & 0 \end{array}$$

$$\begin{array}{r|rrrrr} -1 & 3 & -1 & 11 & -5 & -20 \\ & & -3 & 4 & -15 & 20 \\ \hline & 3 & -4 & 15 & -20 & 0 \end{array}$$

$$\begin{array}{r|rrrr} \frac{4}{3} & 3 & -4 & 15 & -20 \\ & & 4 & 0 & 20 \\ \hline & 3 & 0 & 15 & 0 \end{array}$$

$3x^2 + 15 = 0$

$3(x^2 + 5) = 0$

$x^2 + 5 = 0$

$x^2 = -5$

$x = \pm\sqrt{-5} = \pm\, i\sqrt{5}$

The zeros are -1, (multiplicity 2), $\frac{4}{3}$, $i\sqrt{5}$, $-i\sqrt{5}$.

$P(x) = 3(x+1)^2 \left(x - \dfrac{4}{3}\right)(x - i\sqrt{5})(x + i\sqrt{5})$

21. Find the remaining zeros.

$$\begin{array}{r|rrrr} 1+i & 2 & -5 & 6 & -2 \\ & & 2+2i & -5-i & 2 \\ \hline & 2 & -3+2i & 1-i & 0 \end{array}$$

$$\begin{array}{r|rrr} 1-i & 2 & -3+2i & 1-i \\ & & 2-2i & -1+i \\ \hline & 2 & -1 & 0 \end{array}$$

$2x - 1 = 0$

$x = \dfrac{1}{2}$

The remaining zeros are $1 - i$, $\dfrac{1}{2}$.

23. Find the remaining zeros.

$$\begin{array}{r|rrrr} -i & 1 & 3 & 1 & 3 \\ & & -i & -1-3i & -3 \\ \hline & 1 & 3-i & -3i & 0 \end{array}$$

$$\begin{array}{r|rrr} i & 1 & 3-i & -3i \\ & & i & 3i \\ \hline & 1 & 3 & 0 \end{array}$$

$x + 3 = 0$

$x = -3$

The remaining zeros are i, -3.

25. Find the remaining zeros.

$$\begin{array}{r|rrrrr} 2-3i & 1 & -4 & 14 & -4 & 13 \\ & & 2-3i & -13 & 2-3i & -13 \\ \hline & 1 & -2-3i & 1 & -2-3i & 0 \end{array}$$

$$\begin{array}{r|rrrr} 2+3i & 1 & -2-3i & 1 & -2-3i \\ & & 2+3i & 0 & 2+3i \\ \hline & 1 & -2-3i & 1 & 0 \end{array}$$

$x^2 + 1 = 0$

$x^2 = -1$

$x = \pm i$

The remaining zeros are $2 + 3i$, i, $-i$.

27. Find the remaining zeros.

$$\begin{array}{r|rrrrr} 1+3i & 1 & -4 & 19 & -30 & 50 \\ & & 1+3i & -12-6i & 25+15i & -50 \\ \hline & 1 & -3+3i & 7-6i & -5+15i & 0 \end{array}$$

$$\begin{array}{r|rrrr} 1-3i & 1 & -3+3i & 7-6i & -5+15i \\ & & 1-3i & -2+6i & 5-15i \\ \hline & 1 & -2 & 5 & 0 \end{array}$$

$x^2 - 2x + 5 = 0$

$x = \dfrac{-(-2) \pm \sqrt{(-2)^2 - 4(1)(5)}}{2(1)} = \dfrac{2 \pm \sqrt{-16}}{2} = \dfrac{2 \pm 4i}{2}$

$= 1 \pm 2i$

The remaining zeros are $1 - 3i$, $1 + 2i$, $1 - 2i$.

29. Find the remaining zeros.

$$\begin{array}{r|rrrrrr} -2i & 1 & -3 & 7 & -13 & 12 & -4 \\ & & -2i & -4+6i & 12-6i & -12+2i & 4 \\ \hline & 1 & -3-2i & 3+6i & -1-6i & 2i & 0 \end{array}$$

$$\begin{array}{r|rrrrr} 2i & 1 & -3-2i & 3+6i & -1-6i & 2i \\ & & 2i & -6i & 6i & -2i \\ \hline & 1 & -3 & 3 & -1 & 0 \end{array}$$

$$\begin{array}{r|rrrr} 1 & 1 & -3 & 3 & -1 \\ & & 1 & -2 & 1 \\ \hline & 1 & -2 & 1 & 0 \end{array}$$

$$x^2 - 2x + 1 = 0$$

$$(x-1)^2 = 0$$

$$x = 1$$

The remaining zeros are $2i$, 1 (multiplicity 3).

31. Find the remaining zeros.

$$\begin{array}{r|rrrrr} 5+2i & 1 & -17 & 112 & -333 & 377 \\ & & 5+2i & -64-14i & 268+26i & -377 \\ \hline & 1 & -12+2i & 48-14i & -65+26i & 0 \end{array}$$

$$\begin{array}{r|rrrr} 5-2i & 1 & -12+2i & 48-14i & -65+26i \\ & & 5-2i & -35+14i & 65-26i \\ \hline & 1 & -7 & 13 & 0 \end{array}$$

$$x^2 - 7x + 13 = 0$$

$$x = \frac{-(-7) \pm \sqrt{(-7)^2 - 4(1)(13)}}{2(1)} = \frac{7 \pm \sqrt{-3}}{2} = \frac{7 \pm i\sqrt{3}}{2}$$

$$= \frac{7}{2} \pm \frac{\sqrt{3}}{2} i$$

The remaining zeros are $5-2i$, $\frac{7}{2}+\frac{\sqrt{3}}{2}i$, $\frac{7}{2}-\frac{\sqrt{3}}{2}i$.

33. Determine the exact values of the solutions.

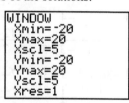

 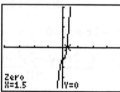

$$\begin{array}{r|rrrr} 1.5 & 2 & -1 & 1 & -6 \\ & & 3 & 3 & 6 \\ \hline & 2 & 2 & 4 & 0 \end{array}$$

$$2x^2 + 2x + 4 = 0$$

$$2(x^2 + x + 2) = 0$$

$$x = \frac{-1 \pm \sqrt{(1)^2 - 4(1)(2)}}{2(1)} = \frac{-1 \pm \sqrt{-7}}{2} = \frac{-1 \pm i\sqrt{7}}{2}$$

$$= -\frac{1}{2} \pm \frac{\sqrt{7}}{2} i$$

The solutions are 1.5, $-\frac{1}{2}+\frac{\sqrt{7}}{2}i$, $-\frac{1}{2}-\frac{\sqrt{7}}{2}i$.

35. Determine the exact values of the solutions.

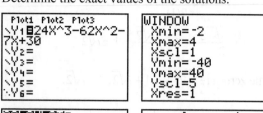

 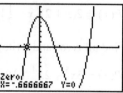

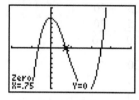 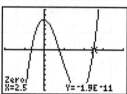

The solutions are $-0.\overline{6}$, 0.75, 2.5 or $-\frac{2}{3}$, $\frac{3}{4}$, $\frac{5}{2}$.

37. Determine the exact values of the solutions.

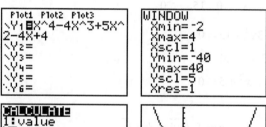

 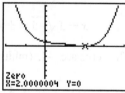

$$\begin{array}{r|rrrrr} 2 & 1 & -4 & 5 & -4 & 4 \\ & & 2 & -4 & 2 & -4 \\ \hline & 1 & -2 & 1 & -2 & 0 \end{array}$$

$$\begin{array}{r|rrrr} 2 & 1 & -2 & 1 & -2 \\ & & 2 & 0 & 2 \\ \hline & 1 & 0 & 1 & 0 \end{array}$$

$$x^2 + 1 = 0$$

$$x^2 = -1$$

$$x = \pm\sqrt{-1} = \pm i$$

The solutions are 2 (multiplicity 2), i, $-i$.

39. Determine the exact values of the solutions.

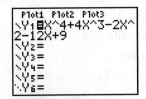 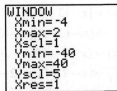

 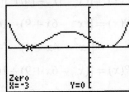

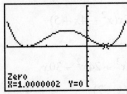

The solutions are -3 (multiplicity 2), 1 (multiplicity 2).

41. Find the polynomial function from the given zeros.

$$P(x) = (x-4)(x+3)(x-2)$$
$$P(x) = (x-4)(x^2+x-6)$$
$$P(x) = x(x^2+x-6) - 4(x^2+x-6)$$
$$P(x) = x^3 + x^2 - 6x - 4x^2 - 4x + 24$$
$$P(x) = x^3 - 3x^2 - 10x + 24$$

43. Find the polynomial function from the given zeros.

$$P(x) = (x-3)(x-2i)(x+2i)$$
$$P(x) = (x-3)(x^2 - [2i]^2)$$
$$P(x) = (x-3)(x^2 - 4i^2)$$
$$P(x) = (x-3)(x^2 - 4[-1])$$
$$P(x) = (x-3)(x^2 + 4)$$
$$P(x) = x^3 - 3x^2 + 4x - 12$$

45. Find the polynomial function from the given zeros.

$$P(x) = [x-(3+i)][x-(3-i)][x-(2+5i)][x-(2-5i)]$$
$$P(x) = (x-3-i)(x-3+i)(x-2-5i)(x-2+5i)$$
$$P(x) = [(x-3)^2 - i^2][(x-2)^2 - 25i^2]$$
$$P(x) = [(x^2-6x+9)-(-1)][(x^2-4x+4)-(25[-1])]$$
$$P(x) = [(x^2-6x+9)+1][(x^2-4x+4)+25]$$
$$P(x) = (x^2-6x+9+1)(x^2-4x+4+25)$$
$$P(x) = (x^2-6x+10)(x^2-4x+29)$$

$$P(x) = x^2(x^2-4x+29) - 6x(x^2-4x+29)$$
$$\qquad + 10(x^2-4x+29)$$
$$P(x) = x^4 - 4x^3 + 29x^2 - 6x^3 + 24x^2 - 174x$$
$$\qquad + 10x^2 - 40x + 290$$
$$P(x) = x^4 - 10x^3 + 63x^2 - 214x + 290$$

47. Find the polynomial function from the given zeros.

$$P(x) = [x-(6+5i)][x-(6-5i)](x-2)(x-3)(x-5)$$
$$P(x) = [x-6-5i][x-6+5i](x-2)(x^2-8x+15)$$
$$P(x) = [(x-6)^2 - (5i)^2]$$
$$\qquad \times [x(x^2-8x+15) - 2(x^2-8x+15)]$$
$$P(x) = [(x^2-12x+36) - (25i^2)]$$
$$\qquad \times (x^3 - 8x^2 + 15x - 2x^2 + 16x - 30)$$
$$P(x) = [(x^2-12x+36) - (25[-1])]$$
$$\qquad \times (x^3 - 10x^2 + 31x - 30)$$
$$P(x) = (x^2-12x+36+25)(x^3-10x^2+31x-30)$$
$$P(x) = (x^2-12x+61)(x^3-10x^2+31x-30)$$
$$P(x) = x^2(x^3-10x^2+31x-30)$$
$$\qquad - 12x(x^3-10x^2+31x-30)$$
$$\qquad + 61(x^3-10x^2+31x-30)$$
$$P(x) = x^5 - 10x^4 + 31x^3 - 30x^2$$
$$\qquad - 12x^4 + 120x^3 - 372x^2 + 360x$$
$$\qquad + 61x^3 - 610x^2 + 1891x - 1830$$
$$P(x) = x^5 - 22x^4 + 212x^3 - 1012x^2 + 2251x - 1830$$

49. Find the polynomial function from the given zeros.

Note: $4x-3=0$ if and only if $x = \frac{3}{4}$. Therefore, if

$x = \frac{3}{4}$ (that is, $\frac{3}{4}$ is a zero), then $4x-3=0$.

$$P(x) = \left(x - \frac{3}{4}\right)[x-(2+7i)][x-(2-7i)]$$
$$P(x) = (4x-3)[x-2-7i][x-2+7i]$$
$$P(x) = (4x-3)[(x-2)^2 - (7i)^2]$$
$$P(x) = (4x-3)[(x^2-4x+4) - 49i^2]$$
$$P(x) = (4x-3)[(x^2-4x+4) - 49(-1)]$$
$$P(x) = (4x-3)(x^2-4x+4+49)$$
$$P(x) = (4x-3)(x^2-4x+53)$$
$$P(x) = 4x(x^2-4x+53) - 3(x^2-4x+53)$$
$$P(x) = 4x^3 - 16x^2 + 212x - 3x^2 + 12x - 159$$
$$P(x) = 4x^3 - 19x^2 + 224x - 159$$

51. Find the polynomial function from the given zeros.

$$P(x) = [x-(2-i\sqrt{2})][x-(2+i\sqrt{2})](x-3)$$
$$P(x) = [x-2+i\sqrt{2}][x-2-i\sqrt{2}](x-3)$$
$$P(x) = [(x-2)^2 - (i\sqrt{2})^2](x-3)$$
$$P(x) = [(x^2-4x+4)-(2i^2)](x-3)$$
$$P(x) = [(x^2-4x+4)-(2[-1])](x-3)$$
$$P(x) = (x^2-4x+4+2)(x-3)$$
$$P(x) = (x^2-4x+6)(x-3)$$
$$P(x) = x(x^2-4x+6)-3(x^2-4x+6)$$
$$P(x) = x^3-4x^2+6x-3x^2+12x-18$$
$$P(x) = x^3-7x^2+18x-18$$

53. Find the polynomial function from the given zeros.

Note: $2x-11=0$ if and only if $x = \dfrac{11}{2}$, or 5.5.

Therefore, if $x = 5.5 = \dfrac{11}{2}$ (that is, 5.5 is a zero), then

$2x-11=0$. Similarly, if $x=-3.5$, then $2x+7=0$.

$$P(x) = (2x-11)(2x+7)[x-(1-i)][(x-(1+i)]$$
$$P(x) = (4x^2-8x-77)[x^2-1+i][x-1-i]$$
$$P(x) = (4x^2-8x-77)[(x-1)^2 - i^2]$$
$$P(x) = (4x^2-8x-77)[x^2-2x+1-(-1)]$$
$$P(x) = (4x^2-8x-77)(x^2-2x+2)$$
$$P(x) = 4x^2(x^2-2x+2)-8x(x^2-2x+2)$$
$$\qquad -77(x^2-2x+2)$$
$$P(x) = 4x^4-8x^3+8x^2-8x^3+16x^2-16x$$
$$\qquad -77x^2+154x-154$$
$$P(x) = 4x^4-16x^3-53x^2+138x-154$$

55. Find the polynomial function from the given zeros.

$$P(x) = [x-(3+4i)][x-(3-4i)]$$
$$\qquad \times[x-(2-i)][x-(2+i)](x-7)$$
$$P(x) = (x-3-4i)(x-3+4i)$$
$$\qquad \times(x-2+i)(x-2-i)(x-7)$$
$$P(x) = [(x-3)^2-(4i)^2][(x-2)^2-i^2](x-7)$$
$$P(x) = [(x^2-6x+9)-16(-1)]$$
$$\qquad \times[(x^2-4x+4)-(-1)](x-7)$$
$$P(x) = [(x^2-6x+9)+16][(x^2-4x+4)+1](x-7)$$
$$P(x) = (x^2-6x+25)(x^2-4x+5)(x-7)$$
$$P(x) = [x^2(x^2-4x+5)-6x(x^2-4x+5)$$
$$\qquad +25(x^2-4x+5)](x-7)$$
$$P(x) = [x^4-4x^3+5x^2-6x^3+24x^2-30x$$
$$\qquad +25x^2-100x+125](x-7)$$
$$P(x) = (x^4-10x^3+54x^2-130x+125)(x-7)$$
$$P(x) = x(x^4-10x^3+54x^2-130x+125)$$
$$\qquad -7(x^4-10x^3+54x^2-130x+125)$$
$$P(x) = x^5-10x^4+54x^3-130x^2+125x$$
$$\qquad -7x^4+70x^3-378x^2+910x-875$$
$$P(x) = x^5-17x^4+124x^3-508x^2+1035x-875$$

57. Find the polynomial function from the given zeros.

If $x = 1.5-i = \dfrac{3}{2}-i$ (that is, $1.5-i$ is a zero), then

$2x-3-2i=0$.

$$P(x) = [x-(1.5-i)][(x-(1.5+i)](x-4)(x+1)$$
$$P(x) = [2x-3+2i][2x-3-2i](x^2-3x-4)$$
$$P(x) = [(2x-3)^2-(2i)^2](x^2-3x-4)$$
$$P(x) = [4x^2-12x+9-4(-1)](x^2-3x-4)$$
$$P(x) = (4x^2-12x+13)(x^2-3x-4)$$
$$P(x) = 4x^2(x^2-3x-4)-12x(x^2-3x-4)$$
$$\qquad +13(x^2-3x-4)$$
$$P(x) = 4x^4-12x^3-16x^2-12x^3+36x^2+48x$$
$$\qquad +13x^2-39x-52$$
$$P(x) = 4x^4-24x^3+33x^2+9x-52$$

59. Find the polynomial function.

If $2 - 5i$ is a zero, then $2 + 5i$ is also a zero.

$P(x) = [x - (2 - 5i)][x - (2 + 5i)](x + 4)$

$P(x) = [x - 2 + 5i][x - 2 - 5i](x + 4)$

$P(x) = [(x - 2)^2 - (5i)^2](x + 4)$

$P(x) = [x^2 - 4x + 4 - 25i^2](x + 4)$

$P(x) = [x^2 - 4x + 4 - 25(-1)](x + 4)$

$P(x) = [x^2 - 4x + 4 + 25](x + 4)$

$P(x) = (x^2 - 4x + 29)(x + 4)$

$P(x) = x^2(x + 4) - 4x(x + 4) + 29(x + 4)$

$P(x) = x^3 + 4x^2 - 4x^2 - 16x + 29x + 116$

$P(x) = x^3 + 13x + 116$

61. Find the polynomial function.

If $4 + 3i$ and $5 - i$ are zeros, then $4 - 3i$ and $5 + i$ are

also zeros.

$P(x) = [x - (4 + 3i)][x - (4 - 3i)]$
$\qquad \times [x - (5 - i)][x - (5 + i)]$

$P(x) = (x^2 - 8x + 25)(x^2 - 10x + 26)$

$P(x) = x^4 - 10x^3 + 26x^2 - 8x^3 + 80x^2 - 208x$
$\qquad + 25x^2 - 250x + 650$

$P(x) = x^4 - 18x^3 + 131x^2 - 458x + 650$

63. Find the polynomial function.

$P(x) = (x + 2)(x - 1)(x - 3)[x - (1 + 4i)][x - (1 - 4i)]$

$P(x) = (x^2 + x - 2)(x - 3)[x - 1 - 4i][x - 1 + 4i]$

$P(x) = [x^3 - 2x^2 - 5x + 6][(x - 1)^2 - (4i)^2]$

$P(x) = (x^3 - 2x^2 - 5x + 6)[x^2 - 2x + 1 - 16i^2]$

$P(x) = (x^3 - 2x^2 - 5x + 6)[x^2 - 2x + 1 + 16]$

$P(x) = (x^3 - 2x^2 - 5x + 6)[x^2 - 2x + 17]$

$P(x) = x^5 - 4x^4 + 16x^3 - 18x^2 - 97x + 102$

65. Find the polynomial function.

$P(x) = a(x + 1)(x - 2)(x - 3)$

$P(1) = a(1 + 1)(1 - 2)(1 - 3)$

$\quad 12 = a(2)(-1)(-2)$

$\quad 12 = 4a$

$\quad \ 3 = a$

**

$P(x) = 3(x + 1)(x - 2)(x - 3)$

$P(x) = (3x + 3)(x^2 - 5x + 6)$

$P(x) = 3x(x^2 - 5x + 6) + 3(x^2 - 5x + 6)$

$P(x) = 3x^3 - 15x^2 + 18x + 3x^2 - 15x + 18$

$P(x) = 3x^3 - 12x^2 + 3x + 18$

67. Find the polynomial function.

$P(x) = a(x - 3)(x + 5)[x - (2 + i)][x - (2 - i)]$

$P(x) = a(x^2 + 2x - 15)[x - 2 - i][x - 2 + i]$

$P(x) = a(x^2 + 2x - 15)[(x - 2)^2 - i^2]$

$P(x) = a(x^2 + 2x - 15)[x^2 - 4x + 4 - (-1)]$

$P(x) = a(x^2 + 2x - 15)[x^2 - 4x + 4 + 1]$

$P(x) = a(x^2 + 2x - 15)(x^2 - 4x + 5)$

$P(1) = a(1^2 + 2[1] - 15)(1^2 - 4[1] + 5)$

$\quad 48 = a(1 + 2 - 15)(1 - 4 + 5)$

$\quad 48 = a(-12)(2)$

$\quad 48 = -24a$

$\quad -2 = a$

$P(x) = -2(x^2 + 2x - 15)(x^2 - 4x + 5)$

$P(x) = -2[x^2(x^2 - 4x + 5) + 2x(x^2 - 4x + 5)$
$\qquad - 15(x^2 - 4x + 5)]$

$P(x) = -2[x^4 - 4x^3 + 5x^2 + 2x^3 - 8x^2 + 10x$
$\qquad - 15x^2 + 60x - 75)]$

$P(x) = -2(x^4 - 2x^3 - 18x^2 + 70x - 75)$

$P(x) = -2x^4 + 4x^3 + 36x^2 - 140x + 150$

69. Verify and explain.

$P(x) = x^3 - x^2 - ix^2 - 9x + 9 + 9i$
$\qquad = x^3 + (-1 - i)x^2 - 9x + (9 + 9i)$

$$
\begin{array}{r|rrrr}
1 + i & 1 & -1 - i & -9 & 9 + 9i \\
      &   & 1 + i & 0 & -9 - 9i \\
\hline
      & 1 & 0   & -9 & 0
\end{array}
$$

Zero remainder implies $1 + i$ is a zero.

$$
\begin{array}{r|rrr}
1 - i & 1 & 0   & -9 \\
      &   & 1 - i & 0 \\
\hline
      & 1 & 1 - i & -9
\end{array}
$$

Non-zero remainder implies $1 - i$ is not a zero.

The Conjugate Pair Theorem does not apply because

some of the coefficients of the polynomial function are

not real numbers.

71. a. Answers will vary.

b. Answers will vary.

c. No such polynomial function exists because the nonreal complex zeros must occur in pairs.

d. Answers will vary.

Prepare for Section 3.5

P1. Simplify.

$$\frac{x^2-9}{x^2-2x-15}=\frac{(x+3)(x-3)}{(x+3)(x-5)}=\frac{x-3}{x-5}, \; x\neq-3$$

P3. Evaluate.

$$\frac{2(-3)^2+4(-3)-5}{-3+6}=\frac{2(9)-12-5}{3}=\frac{18-12-5}{3}=\frac{1}{3}$$

P5. Find the degree of the numerator and denominator.

The degree of the numerator, x^3+3x^2-5, is 3;

the degree of the denominator, x^2-4, is 2.

Section 3.5 Exercises

1. State the rational functions.

$F(x)=\dfrac{3x-5}{x^2+3}$ and $H(x)=\dfrac{1}{2x-3}$ are rational functions.

3. Find the domain.

$F(x)=\dfrac{1}{x}$ Then $x\neq0$.

The domain is all real numbers except 0.

5. Find the domain.

$F(x)=\dfrac{x^2-3}{x^2+1}$ Then no restrictions on denominator.

The domain is all real numbers.

7. Find the domain.

$F(x)=\dfrac{2x-1}{2x^2-15x+18}=\dfrac{2x-1}{(2x-3)(x-6)}$

Then $x\neq\dfrac{3}{2}$, 6.

The domain is all real numbers except $\dfrac{3}{2}$ and 6.

9. Find the domain.

$$F(x)=\frac{2x^2}{x^3-4x^2-12x}=\frac{2x^2}{x(x-6)(x+2)}$$

Then $x\neq0,\;6,-2$.

The domain is all real numbers except 0, 6 and –2.

11. Find the domain.

$$F(x)=\frac{x^2+1}{x^3-8}=\frac{x^2+1}{(x-2)(x^2+4x+2)}$$

Then $x\neq2$.

The domain is all real numbers except 2.

13. Find all vertical asymptotes.

$x^2+3x=0$

$x(x+3)=0$

$x=0$ or $x+3=0$

$\qquad\qquad\qquad x=-3$

Vertical asymptotes: $x=0,\;x=-3$

15. Find all vertical asymptotes.

$6x^2-5x-4=0$

$(3x-4)(2x+1)=0$

$3x-4=0$ or $2x+1=0$

$x=\dfrac{4}{3}$ $x=-\dfrac{1}{2}$

Vertical asymptotes: $x=-\dfrac{1}{2},\;x=\dfrac{4}{3}$

17. Find all vertical asymptotes.

$4x^3-25x^2+6x=0$

$x(4x^2-25x+6)=0$

$x(4x-1)(x-6)=0$

$x=0$ or $4x-1=0$ or $x-6=0$

$\qquad\qquad x=\dfrac{1}{4}\qquad\qquad x=6$

Vertical asymptotes: $x=0,\;x=\dfrac{1}{4},\;x=6$

19. Find all vertical asymptotes.

$x^2-2=0$

$x^2=2$

$x=\pm\sqrt{2}$

Vertical asymptotes: $x=\sqrt{2},\;x=-\sqrt{2}$

21. Find all vertical asymptotes.

$$3x^2 + 5 = 0$$
$$3x^2 = -5$$
$$x^2 = -\frac{5}{3}$$

There are no real values of x which make this statement true.

Vertical asymptote: none

23. Find the horizontal asymptote.

Horizontal asymptote: $y = \frac{4}{1} = 4$

25. Find the horizontal asymptote.

Horizontal asymptote: $y = \frac{15,000}{500} = 30$

27. Find the horizontal asymptote.

Horizontal asymptote: $y = \frac{4}{\frac{1}{3}} = 12$

29. Find the horizontal asymptote.

$$F(x) = \frac{2x^2 + x + 7}{\frac{1}{2}x^2 + 5x - 4}$$

Horizontal asymptote: $y = \frac{2}{\frac{1}{2}} = 2 \cdot \frac{2}{1} = 4$

31. Find the horizontal asymptote.

$$F(x) = \frac{-3x^2 + 8}{x^3 + x^2 - 5x + 2}$$

The degree of the numerator is less than the degree of the denominator. By the Theorem on Horizontal Asymptotes, the x-axis is the horizontal asymptote.

Horizontal asymptote: $y = 0$

33. Find all vertical and horizontal asymptotes. Sketch the graph and label intercepts and asymptotes.

$$F(x) = \frac{1}{x + 4}$$

Vertical asymptote: $x = -4$

Horizontal asymptote: $y = 0$

x-intercepts: none

y-intercept: $\left(0, \frac{1}{4}\right)$

35. Find all vertical and horizontal asymptotes. Sketch the graph and label intercepts and asymptotes.

$$F(x) = \frac{-4}{x - 3}$$

Vertical asymptote: $x = 3$

Horizontal asymptote: $y = 0$

x-intercepts: none

y-intercept: $\left(0, \frac{4}{3}\right)$

37. Find all vertical and horizontal asymptotes. Sketch the graph and label intercepts and asymptotes.

$$F(x) = \frac{4}{x}$$

Vertical asymptotes: $x = 0$

Horizontal asymptote: $y = 0$

x-intercepts: none

y-intercept: none

39. Find all vertical and horizontal asymptotes. Sketch the graph and label intercepts and asymptotes.

$$F(x) = \frac{x}{x + 4}$$

Vertical asymptote: $x = -4$

Horizontal asymptote: $y = 1$

x-intercept: $(0, 0)$

y-intercept: $(0, 0)$

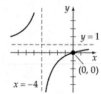

41. Find all vertical and horizontal asymptotes. Sketch the graph and label intercepts and asymptotes.

$$F(x) = \frac{x+4}{2-x}$$

Vertical asymptote: $x = 2$

Horizontal asymptote: $y = -1$

x-intercept: $(-4, 0)$

y-intercept: $(0, 2)$

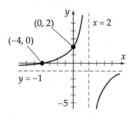

43. Find all vertical and horizontal asymptotes. Sketch the graph and label intercepts and asymptotes.

$$F(x) = \frac{1}{x^2 - 9}$$

Vertical asymptotes: $x = 3, x = -3$

Horizontal asymptote: $y = 0$

x-intercept: none

y-intercept: $\left(0, -\frac{1}{9}\right)$

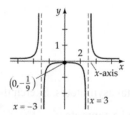

45. Find all vertical and horizontal asymptotes. Sketch the graph and label intercepts and asymptotes.

$$F(x) = \frac{1}{x^2 + 2x - 3}$$

Vertical asymptotes: $x = -3, x = 1$

Horizontal asymptote: $y = 0$

x-intercept: none

y-intercept: $\left(0, -\frac{1}{3}\right)$

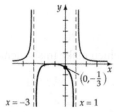

47. Find all vertical and horizontal asymptotes. Sketch the graph and label intercepts and asymptotes.

$$F(x) = \frac{x^2}{x^2 + 4x + 4}$$

Vertical asymptote: $x = -2$

Horizontal asymptote: $y = 1$

x-intercept: $(0, 0)$

y-intercept: $(0, 0)$

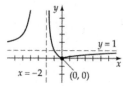

49. Find all vertical and horizontal asymptotes. Sketch the graph and label intercepts and asymptotes.

$$F(x) = \frac{10}{x^2 + 2}$$

Vertical asymptote: none

Horizontal asymptote: $y = 0$

x-intercept: none

y-intercept: none

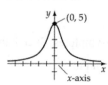

51. Find all vertical and horizontal asymptotes. Sketch the graph and label intercepts and asymptotes.

$$F(x) = \frac{2x^2 - 2}{x^2 - 9}$$

Vertical asymptotes: $x = 3, x = -3$

Horizontal asymptote: $y = 2$

x-intercepts: $(-1, 0)$ and $(1, 0)$

y-intercept: $\left(0, \dfrac{2}{9}\right)$

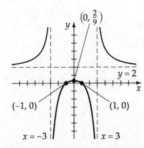

53. Find all vertical and horizontal asymptotes. Sketch the graph and label intercepts and asymptotes.

$$F(x) = \dfrac{x^2 + x + 4}{x^2 + 2x - 1}$$

Vertical asymptotes: $x = -1 + \sqrt{2}, x = -1 - \sqrt{2}$

Horizontal asymptote: $y = 1$

x-intercept: none

y-intercept: $(0, -4)$

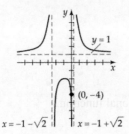

55. Find the slant asymptote.

$$F(x) = \dfrac{3x^2 + 5x - 1}{x + 4}$$

$$
\begin{array}{r|rrr}
-4 & 3 & 5 & -1 \\
   &   & -12 & 28 \\
\hline
   & 3 & -7 & 27
\end{array}
$$

$$F(x) = 3x - 7 + \dfrac{27}{x + 4}$$

Slant asymptote: $y = 3x - 7$

57. Find the slant asymptote.

$$F(x) = \dfrac{x^3 - 1}{x^2} = \dfrac{x^3}{x^2} - \dfrac{1}{x^2} = x - \dfrac{1}{x^2}$$

Slant asymptote: $y = x$

59. Find the slant asymptote.

$$
\begin{array}{r|rrr}
5 & -4 & 15 & 8 \\
  &   & -20 & -25 \\
\hline
  & -4 & -5 & -17
\end{array}
$$

$$F(x) = -4x - 5 + \dfrac{-17}{x - 5}$$

Slant asymptote: $y = -4x - 5$

61. Find the slant asymptote.

$$F(x) = \dfrac{x^3 - 2}{x^4 + 1}$$

$$
\begin{array}{r}
x \\
x^4 + 1 \overline{) x^5 + 0x + 1} \\
\underline{x^5 \quad + x} \\
-x + 1
\end{array}
$$

$$F(x) = x + \dfrac{-x + 1}{x^4 + 1}$$

Slant asymptote: $y = x$

63. Find the slant asymptote.

$$F(x) = \dfrac{\frac{1}{3}x^2 - 1}{2x + 3}$$

$$
\begin{array}{r}
\frac{1}{6}x - \frac{1}{4} \\
2x + 3 \overline{) \frac{1}{3}x^2 + 0x - 1} \\
\underline{\frac{1}{3}x^2 + \frac{1}{2}x} \\
-\frac{1}{2}x - 1 \\
\underline{-\frac{1}{2}x - \frac{3}{4}} \\
-\frac{1}{4}
\end{array}
$$

$$F(x) = \dfrac{1}{6}x - \dfrac{1}{4} + \dfrac{-\frac{1}{4}}{2x + 3}$$

Slant asymptote: $y = \dfrac{1}{6}x - \dfrac{1}{4}$

65. Find the vertical and slant asymptotes and graph.

$$F(x) = \dfrac{x^2 - 4}{x} = x - \dfrac{4}{x}$$

Slant asymptote: $y = x$

Vertical asymptote: $x = 0$

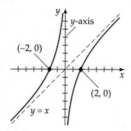

67. Find the vertical and slant asymptotes and graph.

$$\begin{array}{r|rrr} -3 & 1 & -3 & -4 \\ & & -3 & 18 \\ \hline & 1 & -6 & 14 \end{array}$$

$$F(x) = \frac{x^2 - 3x - 4}{x + 3} = x - 6 + \frac{14}{x + 3}$$

Slant asymptote: $y = x - 6$

Vertical asymptote: $x = -3$

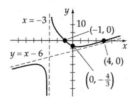

69. Find the vertical and slant asymptotes and graph.

$$\begin{array}{r|rrr} 4 & 2 & 5 & 3 \\ & & 8 & 52 \\ \hline & 2 & 13 & 55 \end{array}$$

$$F(x) = \frac{2x^2 + 5x + 3}{x - 4} = 2x + 13 + \frac{55}{x - 4}$$

Slant asymptote: $y = 2x + 13$

Vertical asymptote: $x = 4$

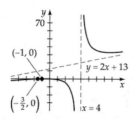

71. Find the vertical and slant asymptotes and graph.

$$\begin{array}{r|rrr} -2 & 1 & -1 & 0 \\ & & -2 & 6 \\ \hline & 1 & -3 & 6 \end{array}$$

$$F(x) = \frac{x^2 - x}{x + 2} = x - 3 + \frac{6}{x + 2}$$

Slant asymptote: $y = x - 3$

Vertical asymptote: $x = -2$

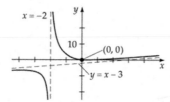

73. Find the vertical and slant asymptotes and graph.

$$x^2 - 4 \overline{)\begin{array}{l} x \\ x^3 + 1 \\ \underline{x^3 - 4x} \\ 4x + 1 \end{array}}$$

$$F(x) = \frac{x^3 + 1}{x^2 - 4} = x + \frac{4x + 1}{x^2 - 4}$$

Slant asymptote: $y = x$

Vertical asymptotes: $x = 2, x = -2$

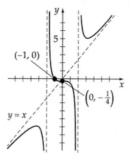

75. Sketch the graph of the rational function.

$$F(x) = \frac{x^2 + x}{x + 1}$$

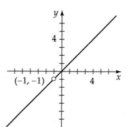

77. Sketch the graph of the rational function.

$$F(x) = \frac{2x^3 + 4x^2}{2x + 4}$$

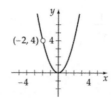

79. Sketch the graph of the rational function.

$$F(x) = \frac{-2x^3 + 6x}{2x^2 - 6x}$$

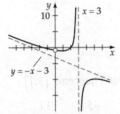

81. Sketch the graph of the rational function.

$$F(x) = \frac{x^2 - 3x - 10}{x^2 + 4x + 4}$$

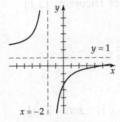

83. a. Find the current.

$$I(3) = \frac{9}{3 + 4.5} = 1.2 \text{ amps}$$

b. Find the resistance.

$$0.24 = \frac{9}{x + 4.5}$$
$$0.24x + 1.08 = 9$$
$$0.24x = 7.92$$
$$x = 33 \text{ ohms}$$

c. Find the horizontal asymptote and explain the meaning.

Horizontal asymptote: $y = 0$.

The current approaches 0 amps as the resistance of the variable resistor increases without bound.

85. a. Find the average costs per golf ball.

$$\overline{C}(1000) = \frac{0.43(1000) + 76,000}{1000} = \frac{430 + 76,000}{1000}$$
$$= \frac{76,430}{1000} = \$76.43$$

$$\overline{C}(10,000) = \frac{0.43(10,000) + 76,000}{10,000}$$
$$= \frac{4,300 + 76,000}{10,000} = \frac{80,300}{10,000} = \$8.03$$

$$\overline{C}(100,00) = \frac{0.43(100,000) + 76,000}{100,000}$$
$$= \frac{43,000 + 76,000}{100,000} = \frac{119,000}{100,000} = \$1.19$$

b. Find the horizontal asymptote and explain the meaning.

Horizontal asymptote: $y = 0.43$.

As the number of golf balls that are produced increases, the average cost per golf ball approaches $0.43.

87. a. Find the cost of removing 40% salt.

$$C(40) = \frac{2000(40)}{100 - 40} = \frac{80,000}{60} = \$1,333.33$$

b. Find the cost of removing 80% salt.

$$C(80) = \frac{2000(80)}{100 - 80} = \frac{160,000}{20} = \$8,000$$

c. Sketch the graph.

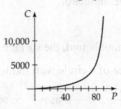

89. a. Find the average wait time, in minutes.

$$W(25) = \frac{7.5}{25(25 - 15)} = \frac{7.5}{250} = 0.03 \text{ hour}$$

$$0.03 \text{ hr} = 0.03 \text{ hr} \cdot \frac{60 \text{ min}}{1 \text{ hr}} = 1.8 \text{ min}$$

b. Find the month where there is a maximum sales.

$$W(x) = \frac{7.5}{x(x - 15)}$$
$$0.0125 = \frac{7.5}{x(x - 15)}$$
$$0.0125x(x - 15) = 7.5$$
$$0.0125x^2 - 0.1875x = 7.5$$
$$0.0125x^2 - 0.1875x - 7.5 = 0$$
$$x^2 - 15x - 600 = 0$$

$$x = \frac{-(-15) \pm \sqrt{(-15)^2 - 4(1)(-600)}}{2(1)}$$

$$x = \frac{15 \pm \sqrt{225 + 2400}}{2} = \frac{15 \pm \sqrt{2625}}{2}$$

$$x = \frac{15 + \sqrt{2625}}{2} \quad \text{reject negative solution}$$

$$x \approx 33.1 \text{ customers per hour}$$

c. Explain why W decreases as x increases.

If more customers can be served per hour, then the average wait time will decrease.

91. a. Graph, estimate r that makes a minimum.

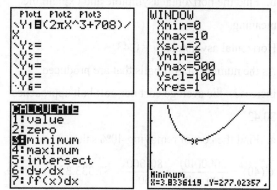

$r \approx 3.8$ centimeters

b. Determine whether there is a slant asymptote.

No. The degree of the numerator is not exactly one more than the degree of the denominator.

c. Explain the statement.

As the radius r increases without bound, the surface area approaches twice the area of a circle with radius r.

Since $V = \pi r^2 h$, then $h = \dfrac{V}{\pi r^2}$. $h \to 0$ as $r \to \infty$ so as the radius increases without bound, the surface area of the can approaches the area of the top and bottom of the can, two circles with radius r.

93. Determine where the graph intersects the horizontal asymptote.

Horizontal asymptote: $y = 2$

$$2 = \frac{2x^2 + 3x + 4}{x^2 + 4x + 7}$$

$$2x^2 + 8x + 14 = 2x^2 + 3x + 4$$

$$5x + 10 = 0$$

$$x = -2$$

The graph of F intersects its horizontal asymptote at $(-2, 2)$.

95. Create a rational function with given characteristics.

One possible function: $f(x) = \dfrac{2x^2}{x^2 - 9}$

Chapter 3 Review Exercises

1. Use synthetic division to divide. [3.1]

$$\begin{array}{r|rrrr} 3 & 4 & -11 & 5 & -2 \\ & & 12 & 3 & 24 \\ \hline & 4 & 1 & 8 & 22 \end{array}$$

Answer: $4x^2 + x + 8 + \dfrac{22}{x - 3}$

3. Find $P(c)$ using the Remainder Theorem. [3.1]

$$\begin{array}{r|rrrrrr} 5 & -2 & 0 & 3 & -1 & -5 & 11 \\ & & -10 & -50 & -235 & -1180 & -5925 \\ \hline & -2 & -10 & -47 & -236 & -1185 & -5914 \end{array}$$

$P(5) = -5914$

5. Find $P(c)$ using the Remainder Theorem. [3.1]

$$\begin{array}{r|rrrrr} -2 & 6 & 0 & -12 & 8 & 1 \\ & & -12 & 24 & -24 & 32 \\ \hline & 6 & -12 & 12 & -16 & 33 \end{array}$$

$P(-2) = 33$

7. Use synthetic division to show c is a zero. [3.1]

$$\begin{array}{r|rrrr} 3 & 1 & 2 & -26 & 33 \\ & & 3 & 15 & -33 \\ \hline & 1 & 5 & -11 & 0 \end{array}$$

9. Use synthetic division to show c is a zero. [3.1]

$$\begin{array}{r|rrrrr} 1 & 1 & -1 & 0 & -2 & 1 & 1 \\ & & 1 & 0 & 0 & -2 & -1 \\ \hline & 1 & 0 & 0 & -2 & -1 & 0 \end{array}$$

11. Determine whether the binomial is a factor using the Factor Theorem. [3.1]

$$\begin{array}{r|rrrr} 5 & 1 & -11 & 39 & -45 \\ & & 5 & -30 & 45 \\ \hline & 1 & -6 & 9 & 0 \end{array}$$

A remainder of 0 implies that $x - 5$ is a factor of $P(x)$.

13. Determine the far-left and far-right behavior. [3.2]

$$P(x) = -2x^3 - 5x^2 + 6x - 3$$

Since $a_n = -2$ is negative and $n = 3$ is odd, the graph of P goes up to its far left and down to its far right.

15. Use a graphing utility to graph, estimate any relative maxima or minima. [3.2]

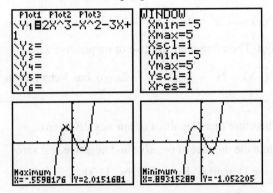

The step-by-step technique for a TI-83 calculator is illustrated in the solution to Exercise **15** from [3.2]. The CALC feature is located above the TRACE key.

Relative maximum of $y \approx 2.015$ at $x \approx -0.560$.

Relative minimum of $y \approx -1.052$ at $x \approx 0.893$.

17. Verify the zero between a and b using Intermediate Value Theorem. [3.2]

$$P(x) = 3x^3 - 7x^2 - 3x + 7$$

```
2 | 3   -7   -3    7
  |      6   -2  -10
  ‾‾‾‾‾‾‾‾‾‾‾‾‾‾‾‾‾‾‾
    3   -1   -5   -3
```

```
3 | 3   -7   -3    7
  |      9    6    9
  ‾‾‾‾‾‾‾‾‾‾‾‾‾‾‾‾‾‾‾
    3    2    3   16
```

$P(2)$ is negative; $P(3)$ is positive.

Therefore $P(x)$ has a zero between 2 and 3.

19. Determine x-intercepts. State which cross x-axis. [3.2]

$$P(x) = (x+3)(x-5)^2$$
$$0 = (x+3)(x-5)^2$$

$x + 3 = 0$ or $x - 5 = 0$
$x = -3$ $x = 5$

The exponent on $(x + 3)$ is an odd integer. Therefore the graph of $P(x)$ will cross the x-axis at $(-3, 0)$.

The exponent on $(x - 5)$ is an even integer. Therefore the graph of $P(x)$ will intersect but not cross the x-axis at $(5, 0)$.

21. Sketch the graph of the function. [3.2]

Let $P(x) = 0$.

$$x^3 - x = 0$$
$$x(x^2 - 1) = 0$$
$$x(x+1)(x-1) = 0$$
$$x = 0, x = -1, x = 1$$

The graph crosses the x-axis at $(0, 0)$, $(-1, 0)$, and $(1, 0)$.

Let $x = 0$. $P(0) = 0^3 - (0) = 0$

The y-intercept is $(0, 0)$.

x^3 has a positive coefficient and an odd exponent. Therefore, the graph goes down to the far left and up to the far right.

23. Sketch the graph of the function. [3.2]

Let $P(x) = 0$.

$$x^4 - 6 = 0$$
$$\left(x^2 + \sqrt{6}\right)\left(x^2 - \sqrt{6}\right) = 0$$
$$\left(x^2 + \sqrt{6}\right)\left(x + \sqrt{\sqrt{6}}\right)\left(x - \sqrt{\sqrt{6}}\right) = 0$$
$$\sqrt{\sqrt{6}} \approx 1.565$$

$x = -1.565, x = 1.565$

The graph crosses the x-axis at $(-1.565, 0)$, $(1.565, 0)$.

Let $x = 0$.

$$P(0) = (0)^4 - 6 = -6$$

The y-intercept is $(0, -6)$.

x^4 has a positive coefficient and an even exponent. Therefore, the graph goes up to the far left and up to the far right.

25. Sketch the graph of the function. [3.2]

Let $P(x) = 0$.

$$x^4 - 10x^2 + 9 = 0$$
$$(x^2 - 1)(x^2 - 9) = 0$$
$$(x+1)(x-1)(x+3)(x-3) = 0$$

$x = -1, x = 1, x = -3, x = 3$

The graph crosses the x-axis at $(-3, 0), (-1, 0), (1, 0)$ and $(3, 0)$.

Let $x = 0$.

$$P(0) = (0)^4 - 10(0)^2 + 9 = 9$$

The y-intercept is $(0, 9)$.

x^4 has a positive coefficient and an even exponent. Therefore, the graph goes up to the far left and up to the far right.

27. List the possible rational zeros. [3.3]

$P(x) = x^3 - 7x - 6$

$p = \pm 1, \ \pm 2, \ \pm 3, \ \pm 6$

$q = \pm 1$

$\dfrac{p}{q} = \pm 1, \ \pm 2, \ \pm 3, \ \pm 6$

29. List the possible rational zeros. [3.3]

$P(x) = 3x^3 - 20x^2 + 23x + 10$

$p = \pm 1, \ \pm 2, \ \pm 5, \ \pm 10$

$q = \pm 1, \ \pm 3$

$\dfrac{p}{q} = \pm 1, \ \pm 2, \ \pm 5, \ \pm 10, \ \pm \dfrac{1}{3}, \ \pm \dfrac{2}{3}, \ \pm \dfrac{5}{3}, \ \pm \dfrac{10}{3}$

31. List the possible rational zeros. [3.3]

$P(x) = x^3 + x^2 - x - 1$

$p = \pm 1$

$q = \pm 1$

$\dfrac{p}{q} = \pm 1$

33. State the number of positive and negative real zeros using Descartes' Rule of Signs. [3.3]

$P(x) = -2x^5 + 4x^3 + 5x^2 - 2x - 6$ has 2 changes in sign. Therefore there are two or no positive zeros.

$P(-x) = 2x^5 - 4x^3 + 5x^2 + 2x - 6$ has 3 changes in sign.

Therefore there are three or one negative zeros.

35. State the number of positive and negative real zeros using Descartes' Rule of Signs. [3.3]

$P(x) = x^4 - x - 1$ has 1 change in sign. Therefore there is one positive zero.

$P(-x) = x^4 + x - 1$ has 1 change in sign. Therefore there is one negative zero.

37. Find the zeros. [3.3]

$$\begin{array}{r|rrrr} 6 & 2 & -7 & -33 & 18 \\ & & 12 & 30 & -18 \\ \hline & 2 & 5 & -3 & 0 \end{array}$$

$$2x^2 + 5x - 3 = 0$$
$$(x+3)(2x-1) = 0$$

$$x = -3 \quad \text{or} \quad x = \dfrac{1}{2}$$

The zeros of $2x^3 - 7x^2 - 33x + 18$ are -3, $\dfrac{1}{2}$, and 6.

39. Find the zeros. [3.3]

$$\begin{array}{r|rrrrr} -2 & 6 & 35 & 72 & 60 & 16 \\ & & -12 & -46 & -52 & -16 \\ \hline & 6 & 23 & 26 & 8 & 0 \end{array}$$

$$\begin{array}{r|rrrr} -2 & 6 & 23 & 26 & 8 \\ & & -12 & -22 & -8 \\ \hline & 6 & 11 & 4 & 0 \end{array}$$

$$6x^2 + 11x + 4 = 0$$
$$(3x+4)(2x+1) = 0$$

$$x = -\dfrac{4}{3} \quad \text{or} \quad x = -\dfrac{1}{2}$$

The zeros of $6x^4 + 35x^3 + 72x^2 + 60x + 16$ are -2 (multiplicity 2), $-\dfrac{4}{3}$, and $-\dfrac{1}{2}$.

41. Find the zeros. [3.3]

$$
\begin{array}{r|rrrrr}
1 & 1 & -4 & 6 & -4 & 1 \\
  &   & 1 & -3 & 3 & -1 \\
\hline
  & 1 & -3 & 3 & -1 & 0
\end{array}
$$

$$
\begin{array}{r|rrrr}
1 & 1 & -3 & 3 & -1 \\
  &   & 1 & -2 & 1 \\
\hline
  & 1 & -2 & 1 & 0
\end{array}
$$

$$x^2 - 2x + 1 = 0$$
$$(x-1)(x-1) = 0$$
$$x = 1 \quad \text{or} \quad x = 1$$

The zero of $x^4 - 4x^3 + 6x^2 - 4x + 1$ is 1 (multiplicity 4).

43. Find all zeros and write as a product. [3.4]

$$P(x) = 2x^4 - 9x^3 + 22x^2 - 29x + 10$$

$$
\begin{array}{r|rrrrr}
\frac{1}{2} & 2 & -9 & 22 & -29 & 10 \\
  &   & 1 & -4 & 9 & -10 \\
\hline
  & 2 & -8 & 18 & -20 & 0
\end{array}
$$

$$
\begin{array}{r|rrrr}
2 & 2 & -8 & 18 & -20 \\
  &   & 4 & -8 & 20 \\
\hline
  & 2 & -4 & 10 & 0
\end{array}
$$

$$2x^2 - 4x + 10 = 0$$
$$2(x^2 - 2x + 5) = 0$$

$$x = \frac{-(-2) \pm \sqrt{(-2)^2 - 4(1)(5)}}{2(1)} = \frac{2 \pm \sqrt{4 - 20}}{2}$$

$$= \frac{2 \pm \sqrt{-16}}{2} = \frac{2 \pm 4i}{2} = 1 \pm 2i$$

The zeros are $\frac{1}{2}$, 2, $1 + 2i$, $1 - 2i$.

$$P(x) = 2\left(x - \frac{1}{2}\right)(x - 2)(x - 1 - 2i)(x - 1 + 2i)$$

45. Find the remaining zeros. [3.4]

$$
\begin{array}{r|rrrrr}
1-2i & 1 & -4 & 6 & -4 & -15 \\
  &   & 1-2i & -7+4i & 7+6i & 15 \\
\hline
  & 1 & -3-2i & -1+4i & 3+6i & 0
\end{array}
$$

$$
\begin{array}{r|rrrr}
1+2i & 1 & -3-2i & -1+4i & 3+6i \\
  &   & 1+2i & -2-4i & -3-6i \\
\hline
  & 1 & -2 & -3 & 0
\end{array}
$$

$$x^2 - 2x - 3 = 0$$
$$(x-3)(x+1) = 0$$
$$x = 3, \; x = -1$$

The remaining zeros are $1 + 2i$, 3, and -1.

47. Find the polynomial function. [3.4]

$$
\begin{aligned}
P(x) &= (x-4)(x+3)(2x-1) \\
&= (x^2 - x - 12)(2x - 1) \\
&= 2x^3 - x^2 - 2x^2 + x - 24x + 12 \\
&= 2x^3 - 3x^2 - 23x + 12
\end{aligned}
$$

49. Find the polynomial function. [3.4]

$$
\begin{aligned}
P(x) &= (x-1)(x-2)(x-5i)(x+5i) \\
&= (x^2 - 3x + 2)(x^2 + 25) \\
&= x^4 + 25x^2 - 3x^3 - 75x + 2x^2 + 50 \\
&= x^4 - 3x^3 + 27x^2 - 75x + 50
\end{aligned}
$$

51. Find the domain. [3.5]

$$F(x) = \frac{2x-3}{x^2 - 25} = \frac{2x-3}{(x+5)(x-5)}$$

Then $x \neq -5, 5$.

The domain is all real numbers except -5 and 5.

53. Find all vertical asymptotes. [3.5]

$$6x^2 - x - 35 = 0$$
$$(3x + 7)(2x - 5) = 0$$

The vertical asymptotes are $x = -\frac{7}{3}$, $x = \frac{5}{2}$.

55. Find the horizontal asymptote. [3.5]

The degree of the numerator is less than the degree of the denominator. By the Theorem on Horizontal Asymptotes, the x-axis is the horizontal asymptote.

Horizontal asymptote: $y = 0$

57. Find the slant asymptote. [3.5]

$$f(x) = \frac{x^3 - 5x^2 + x + 12}{x^2 - 2x - 5}$$

$$
\begin{array}{r}
x - 3 \\
x^2 - 2x - 5 \overline{) x^3 - 5x^2 + x + 12 } \\
\underline{x^3 - 2x^2 - 5x } \\
-3x^2 + 6x + 12 \\
\underline{-3x^2 + 6x + 15} \\
-3
\end{array}
$$

$$f(x) = x - 3 + \frac{-3}{x^2 - 2x - 5}$$

Slant asymptote: $y = x - 3$

59. Graph the rational function. [3.5]

$$f(x) = \frac{3x - 2}{x}$$

Vertical asymptote: $x = 0$

Horizontal asymptote: $y = 3$

61. Graph the rational function. [3.5]

$$f(x) = \frac{12x - 24}{x^2 - 4}$$

Vertical asymptotes: $x = -2, x = 2$

Horizontal asymptote: $y = 0$

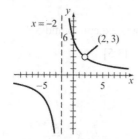

63. Graph the rational function. [3.5]

$$f(x) = \frac{2x^3 - 4x + 6}{x^2 - 4}$$

Vertical asymptote: $x = -2, x = 2$

Horizontal asymptote: none

Slant asymptote: $y = 2x$

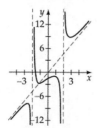

65. Graph the rational function. [3.5]

$$f(x) = \frac{3x^2 - 6}{x^2 - 9}$$

Vertical asymptotes: $x = -3, x = 3$

Horizontal asymptote: $y = 3$

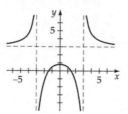

67. a. Find the average cost per skateboard. [3.5]

$$C(5000) = \frac{5.75(5000) + 34,200}{5000} = \frac{62,950}{5000}$$

$$= \$12.59$$

$$C(50,000) = \frac{5.75(50,000) + 34,200}{50,000} = \frac{321,700}{50,000}$$

$$\approx \$6.43$$

b. Find the horizontal asymptote and explain.

$y = 5.75$. As the number of skateboards produced increases, the average cost per skateboard approaches $5.75.

69. a. Find the quartic regression function. [3.2]

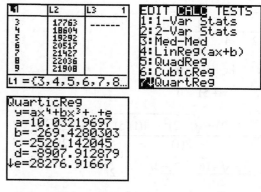

$$f(x) = 10.03219697x^4 - 269.4280303x^3$$
$$+ 2526.142045x^2 - 8907.912879x$$
$$+ 28,276.91667$$

b. Estimate the cost for 2013–2014. Round to the nearest hundred dollars.

$$f(13) = 10.03219697(13)^4 - 269.4280303(13)^3$$
$$+ 2526.142045(13)^2 - 8907.912879(13)$$
$$+ 28,276.91667$$
$$\approx \$34,000$$

Chapter 3 Test

1. Use synthetic division to divide. [3.1]

$$\begin{array}{r|rrrr} -2 & 3 & 5 & 4 & -1 \\ & & -6 & 2 & -12 \\ \hline & 3 & -1 & 6 & -13 \end{array}$$

Answer: $3x^2 - x + 6 + \dfrac{-13}{x+2}$

3. Determine whether the binomial is a factor using the Factor Theorem. [3.1]

$$\begin{array}{r|rrrrr} 1 & 1 & -4 & 7 & -6 & 2 \\ & & 1 & -3 & 4 & -2 \\ \hline & 1 & -3 & 4 & -2 & 0 \end{array}$$

A remainder of 0 implies that $x - 1$ is a factor of

$x^4 - 4x^3 + 7x^2 - 6x + 2$.

5. Find the real zeros. [3.2]

$3x^3 + 7x^2 - 6x = 0$

$x(3x^2 + 7x - 6) = 0$

$x(3x - 2)(x + 3) = 0$

$x = 0, \quad 3x - 2 = 0, \text{ or } x + 3 = 0$

$$x = \frac{2}{3} \qquad x = -3$$

The zeros of $3x^3 + 7x^2 - 6x = 0$ are 0, $\dfrac{2}{3}$, and -3.

7. Find the zeros and state multiplicity. [3.3]

$P(x) = (x^2 - 4)^2(2x - 3)(x + 1)^3$

$P(x) = (x - 2)^2(x + 2)^2(2x - 3)(x + 1)^3$

The zeros of P are 2 (multiplicity 2),

-2 (multiplicity 2), $\dfrac{3}{2}$ (multiplicity 1), and

-1 (multiplicity 3).

9. State the number of positive and negative real zeros using Descartes' Rule of Signs. [3.3]

$P(x) = x^4 - 5x^3 + 12x^2 - 23x - 15$ has 3 changes in

sign. Therefore there are three or one positive zeros.

$P(-x) = x^4 + 5x^3 + 12x^2 + 23x - 15$ has one change

in sign. Therefore there is one negative zero.

11. Find the remaining zeros. [3.4]

$$\begin{array}{r|rrrrr} 2+3i & 6 & -5 & 12 & 207 & 130 \\ & & 12+18i & -40+57i & -227+30i & -130 \\ \hline & 6 & 7+18i & -28+57i & -20+30i & 0 \end{array}$$

$$\begin{array}{r|rrrr} 2-3i & 6 & 7+18i & -28+57i & -20+30i \\ & & 12-18i & 38-57i & 20-30i \\ \hline & 6 & 19 & 10 & 0 \end{array}$$

$6x^2 + 19x + 10 = 0$

$(3x + 2)(2x + 5) = 0$

$3x + 2 = 0 \qquad 2x + 5 = 0$

$$x = -\frac{2}{3} \qquad\quad x = -\frac{5}{2}$$

The remaining zeros are $2 - 3i$, $-\dfrac{2}{3}$ and $-\dfrac{5}{2}$.

13. Find the polynomial function. [3.4]

$$\begin{aligned} P(x) &= [x - (3 + 2i)][x - (3 - 2i)](x - 2)(2x + 1) \\ &= [(x - 3)^2 - (2i)^2](2x^2 - 3x - 2) \\ &= (x^2 - 6x + 9 + 4)(2x^2 - 3x - 2) \\ &= (x^2 - 6x + 13)(2x^2 - 3x - 2) \\ &= x^2(2x^2 - 3x - 2) - 6x(2x^2 - 3x - 2) \\ &\quad + 13(2x^2 - 3x - 2) \\ &= 2x^4 - 3x^3 - 2x^2 - 12x^3 + 18x^2 + 12x \\ &\quad + 26x^2 - 39x - 26 \\ &= 2x^4 - 15x^3 + 42x^2 - 27x - 26 \end{aligned}$$

15. Graph. [3.2]

$P(x) = x^3 - 6x^2 + 9x + 1 = 0$

Let $x = 0$. $P(0) = 0^3 - 6(0)^2 + 9(0) + 1 = 1$

The y-intercept is $(0, 1)$.

x^3 has a positive coefficient and an odd exponent.

Therefore, the graph goes down to the far left and up to the far right.

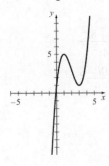

17. Graph. [3.5]

$$f(x) = \frac{2x^2 + 2x + 1}{x + 1}$$

Vertical asymptote: $x = -1$

Horizontal asymptote: none

Slant asymptote: $y = 2x$

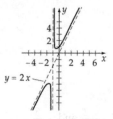

19. a. Evaluate the function at the values. [3.5]

$$w(t) = \frac{70t + 120}{t + 40}$$

$$w(1) = \frac{70(1) + 120}{1 + 40} = \frac{70 + 120}{41} = \frac{190}{41}$$

$$\approx 5 \text{ words per minute}$$

$$w(10) = \frac{70(10) + 120}{10 + 40} = \frac{700 + 120}{50} = \frac{820}{50}$$

$$\approx 16 \text{ words per minute}$$

$$w(20) = \frac{70(20) + 120}{20 + 40} = \frac{1400 + 120}{60} = \frac{1520}{60}$$

$$\approx 25 \text{ words per minute}$$

b. Find the number of hours.

$$60 = \frac{70t + 120}{t + 40}$$

$$60t + 2400 = 70t + 120$$

$$2280 = 10t$$

$$t = 228 \text{ hours}$$

c. Explain the far-right behavior.

As $t \to \infty$, $w(t) \to \dfrac{70}{1} = 70$ words per minute

Cumulative Review Exercises

1. Write in $a + bi$ form. [P.6]

$$\frac{3 + 4i}{1 - 2i} = \frac{3 + 4i}{1 - 2i} \cdot \frac{1 + 2i}{1 + 2i} = \frac{3 + 10i + 8i^2}{1^2 - 4i^2}$$

$$= \frac{3 + 10i + 8(-1)}{1 - 4(-1)} = \frac{3 + 10i - 8}{1 + 4}$$

$$= \frac{-5 + 10i}{5} = -1 + 2i$$

3. Solve. [1.4]

$$\sqrt{2x + 5} - \sqrt{x - 1} = 2$$

$$\sqrt{2x + 5} = 2 + \sqrt{x - 1}$$

$$\left(\sqrt{2x + 5}\right)^2 = \left(2 + \sqrt{x - 1}\right)^2$$

$$2x + 5 = 4 + 4\sqrt{x - 1} + x - 1$$

$$2x + 5 = 3 + 4\sqrt{x - 1} + x$$

$$x + 2 = 4\sqrt{x - 1}$$

$$(x + 2)^2 = \left(4\sqrt{x - 1}\right)^2$$

$$x^2 + 4x + 4 = 16(x - 1)$$

$$x^2 + 4x + 4 = 16x - 16$$

$$x^2 - 12x + 20 = 0$$

$$(x - 2)(x - 10) = 0$$

$$x = 2, x = 10$$

Check 2:

$$\sqrt{2(2) + 5} - \sqrt{(2) - 1} = 2$$

$$\sqrt{4 + 5} - \sqrt{2 - 1} = 2$$

$$\sqrt{9} - \sqrt{1} = 2$$

$$3 - 1 = 2$$

$$2 = 2$$

Yes

Check 10:

$$\sqrt{2(10) + 5} - \sqrt{(10) - 1} = 2$$

$$\sqrt{20 + 5} - \sqrt{10 - 1} = 2$$

$$\sqrt{25} - \sqrt{9} = 2$$

$$5 - 3 = 2$$

$$2 = 2$$

Yes

The solutions are $x = 2$, $x = 10$.

5. Find the distance between the points. [2.1]

$$d = \sqrt{(2 - 7)^2 + [5 - (-11)]^2}$$

$$= \sqrt{(2 - 7)^2 + (5 + 11)^2}$$

$$= \sqrt{(-5)^2 + (16)^2}$$

$$= \sqrt{25 + 256}$$

$$= \sqrt{281}$$

7. Find the difference quotient. [2.6]

$$P(x) = x^2 - 2x - 3$$

$$\frac{P(x+h) - P(x)}{h}$$

$$= \frac{[(x+h)^2 - 2(x+h) - 3] - (x^2 - 2x - 3)}{h}$$

$$= \frac{x^2 + 2xh + h^2 - 2x - 2h - 3 - x^2 + 2x + 3}{h}$$

$$= \frac{2xh + h^2 - 2h}{h}$$

$$= 2x + h - 2$$

9. Find $(f - g)(x)$. [2.6]

$$(f - g)(x) = f(x) - g(x)$$

$$= x^3 - 2x + 7 - (x^2 - 3x - 4)$$

$$= x^3 - 2x + 7 - x^2 + 3x + 4$$

$$= x^2 - x^2 + x + 11$$

11. Find $P(3)$ using Remainder Theorem. [3.1]

$$P(x) = 2x^4 - 3x^2 + 4x - 6$$

$$\begin{array}{r|rrrrr} 3 & 2 & 0 & -3 & 4 & -6 \\ & & 6 & 18 & 45 & 147 \\ \hline & 2 & 6 & 15 & 49 & 141 \end{array}$$

$$P(3) = 141$$

13. Find the relative maximum. [3.2]

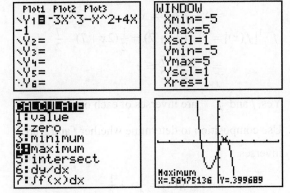

The relative maximum (to the nearest 0.0001) is 0.3997.

15. State the number of positive and negative real zeros using Descartes' Rule of Signs. [3.3]

$P(x) = x^3 + x^2 + 2x + 4$ has no changes of sign.

There are no positive real zeros.

$P(-x) = -x^3 + x^2 - 2x + 4$ has three changes of sign.

There are three or one negative real zeros.

17. Find the polynomial with given zeros. [3.4]

If $3 + i$ is a zero of $P(x)$, then $3 - i$ is also a zero.

$$P(x) = [x - (3 + i)][x - (3 - i)](x + 2)$$

$$= [x - 3 - i][x - 3 + i](x + 2)$$

$$= [(x - 3)^2 - i^2](x + 2)$$

$$= [x^2 - 6x + 9 - (-1)](x + 2)$$

$$= [x^2 - 6x + 9 + 1](x + 2)$$

$$= (x^2 - 6x + 10)(x + 2)$$

$$= x^2(x + 2) - 6x(x + 2) + 10(x + 2)$$

$$= x^3 + 2x^2 - 6x^2 - 12x + 10x + 20$$

$$= x^3 - 4x^2 - 2x + 20$$

19. Find the vertical and horizontal asymptotes. [3.5]

$$F(x) = \frac{4x^2}{x^2 + x - 6}$$

Vertical asymptotes:

$$x^2 + x - 6 = 0$$

$$(x + 3)(x - 2) = 0$$

$$x = -3, x = 2$$

Horizontal asymptote:

$$y = \frac{4}{1} \text{ Then } y = 4.$$

Chapter 4 Exponential and Logarithmic Functions

Section 4.1 Exercises

1. Write the definition of a one-to-one function.

A function f is one-to-one if and only if $f(a) = f(b)$ implies $a = b$.

3. Describe the relationship between the inverse of a function and the graph of the function.

The graph of the inverse is a reflection of the graph of the function about the line $x = y$.

5. If $f(3) = 7$, then $f^{-1}(7) = 3$.

7. If $h^{-1}(-3) = -4$, then $h(-4) = -3$.

9. If 3 is in the domain of f^{-1}, then $f[f^{-1}(3)] = 3$.

11. The domain of the inverse function f^{-1} is the range of f.

13. Draw the graph of the inverse relation, determine if it is a function.

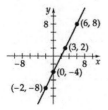

Yes, the inverse is a function.

15. Draw the graph of the inverse relation, determine if it is a function.

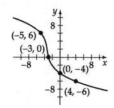

Yes, the inverse is a function.

17. Draw the graph of the inverse relation, determine if it is a function.

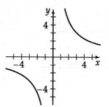

Yes, the inverse is a function.

19. Draw the graph of the inverse relation, determine if it is a function.

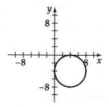

No, the inverse relation is not a function.

21. Use composition to determine whether f and f^{-1} are inverses.

$$f(x) = 2x + 7; \; f^{-1}(x) = \frac{1}{2}x - \frac{7}{2}$$
$$f[f^{-1}(x)] = f\left(\frac{1}{2}x - \frac{7}{2}\right) = 2\left(\frac{1}{2}x - \frac{7}{2}\right) + 7$$
$$= x - 7 + 7 = x$$
$$f^{-1}[f(x)] = f^{-1}(2x + 7) = \frac{1}{2}(2x + 7) - \frac{7}{2}$$
$$= 2 + \frac{7}{2} - \frac{7}{2} = x$$

Yes, f and f^{-1} are inverses of each other.

23. Use composition to determine whether f and g are inverses.

$$f(x) = 4x - 1; \; g(x) = \frac{1}{4}x + \frac{1}{4}$$
$$f[g(x)] = f\left(\frac{1}{4}x + \frac{1}{4}\right)$$
$$= 4\left(\frac{1}{4}x + \frac{1}{4}\right) - 1 = x + 1 - 1$$
$$= x$$
$$g[f(x)] = g(4x - 1)$$
$$= \frac{1}{4}(4x - 1) + \frac{1}{4} = x - \frac{1}{4} + \frac{1}{4}$$
$$= x$$

Yes, f and g are inverses of each other.

658

25. Use composition to determine whether f and g are inverses.

$$f(x) = -\frac{1}{2}x - \frac{1}{2};\ g(x) = -2x + 1$$

$$f[g(x)] = f(-2x + 1)$$

$$= -\frac{1}{2}(-2x + 1) - \frac{1}{2}$$

$$= x - \frac{1}{2} - \frac{1}{2}$$

$$= x - 1$$

$$\neq x$$

No, f and g are not inverses of each other.

27. Use composition to determine whether f and g are inverses.

$$f(x) = \frac{5}{x - 3};\ g(x) = \frac{5}{x} + 3$$

$$f[g(x)] = f\left(\frac{5}{x} + 3\right)$$

$$= \frac{5}{\frac{5}{x} + 3 - 3} = \frac{5}{\frac{5}{x}} = 5 \cdot \frac{x}{5}$$

$$= x$$

$$g[f(x)] = g\left(\frac{5}{x - 3}\right)$$

$$= \frac{5}{\frac{5}{x - 3}} + 3 = x - 3 + 3$$

$$= x$$

Yes, f and g are inverses of each other.

29. Use composition to determine whether f and g are inverses.

$$f(x) = x^3 + 2;\ g(x) = \sqrt[3]{x - 2}$$

$$f[g(x)] = f\left(\sqrt[3]{x - 2}\right)$$

$$= \left(\sqrt[3]{x - 2}\right)^3 + 2 = x - 2 + 2$$

$$= x$$

$$g[f(x)] = g\left(x^3 + 2\right)$$

$$= \sqrt[3]{x^3 + 2 - 2} = \sqrt[3]{x^3}$$

$$= x$$

Yes, f and g are inverses of each other.

31. Find the inverse.

The inverse of $\{(-3,\ 1),\ (-2,\ 2),\ (1,\ 5),\ (4,\ -7)\}$ is

$\{(1,\ -3),\ (2,\ -2),\ (5,\ 1),\ (-7,\ 4)\}$.

33. Find the inverse.

The inverse of $\{(0,\ 1),\ (1,\ 2),\ (2,\ 4),\ (3,\ 8),\ (4,\ 16)\}$

is $\{(1,\ 0),\ (2,\ 1),\ (4,\ 2),\ (8,\ 3),\ (16,\ 4)\}$.

35. Find $f^{-1}(x)$.

$$f(x) = 2x + 4$$

$$x = 2y + 4$$

$$x - 4 = 2y$$

$$\frac{1}{2}x - 2 = y$$

$$f^{-1}(x) = \frac{1}{2}x - 2$$

37. Find $f^{-1}(x)$.

$$f(x) = 3x - 7$$

$$x = 3y - 7$$

$$x + 7 = 3y$$

$$\frac{1}{3}x + \frac{7}{3} = y$$

$$f^{-1}(x) = \frac{1}{3}x + \frac{7}{3}$$

39. Find $f^{-1}(x)$.

$$f(x) = -3x + 11$$

$$x = -3y + 11$$

$$x - 11 = -3y$$

$$-\frac{1}{3}x + \frac{11}{3} = y$$

$$f^{-1}(x) = -\frac{1}{3}x + \frac{11}{3}$$

41. Find $f^{-1}(x)$.

$$f(x) = \frac{2x}{x - 1},\ x \neq 1$$

$$x = \frac{2y}{y - 1}$$

$$x(y - 1) = xy - x = 2y$$

$$xy - 2y = y(x - 2) = x$$

$$y = \frac{x}{x - 2}$$

$$f^{-1}(x) = \frac{x}{x - 2},\ x \neq 2$$

43. Find $f^{-1}(x)$.

$$f(x) = \frac{x-1}{x+1}, \ x \neq -1$$

$$x = \frac{y-1}{y+1}$$

$$x(y+1) = xy + x = y - 1$$

$$xy - y = -x - 1$$

$$y - xy = y(1-x) = x + 1$$

$$y = \frac{x+1}{1-x}$$

$$f^{-1}(x) = \frac{x+1}{1-x}, \ x \neq 1$$

45. Find $f^{-1}(x)$.

$$f(x) = x^2 + 1, \ x \geq 0$$

$$x = y^2 + 1$$

$$x - 1 = y^2$$

$$\sqrt{x-1} = y$$

$$f^{-1}(x) = \sqrt{x-1}, \ x \geq 1$$

Note: Do not use $\pm$ with the radical because the domain of f, and thus the range of f^{-1}, is nonnegative.

47. Find $f^{-1}(x)$.

$$f(x) = \sqrt{x-2}, \ x \geq 2$$

$$x = \sqrt{y-2}$$

$$x^2 = y - 2$$

$$x^2 + 2 = y$$

$$f^{-1}(x) = x^2 + 2, \ x \geq 0$$

Note: The range of f, is nonnegative, therefore the domain of f^{-1} is also nonnegative.

49. Find $f^{-1}(x)$.

$$f(x) = x^2 + 4x, \ x \geq -2$$

$$x = y^2 + 4y$$

$$x + 4 = y^2 + 4y + 4$$

$$x + 4 = (y+2)^2$$

$$\sqrt{x+4} = y + 2$$

$$y = \sqrt{x+4} - 2$$

$$f^{-1}(x) = \sqrt{x+4} - 2, \ x \geq -4$$

51. Find $f^{-1}(x)$.

$$f(x) = x^2 + 4x - 1, \ x \leq -2$$

$$x = y^2 + 4y - 1$$

$$x + 1 = y^2 + 4y$$

$$x + 1 + 4 = y^2 + 4y + 4$$

$$x + 5 = (y+2)^2$$

$$-\sqrt{x+5} = y + 2$$

$$-\sqrt{x+5} - 2 = y$$

$$f^{-1}(x) = -\sqrt{x+5} - 2, \ x \geq -5$$

53. Find $V^{-1}(x)$ and explain what it represents.

$$V(x) = x^3$$

$$x = y^3$$

$$\sqrt[3]{x} = y$$

$$V^{-1}(x) = \sqrt[3]{x}$$

$V^{-1}(x)$ is the length of a side of a cube that has a volume of x cubic units.

55. Find f^{-1} and explain how it is used.

$$f(x) = \frac{5}{9}(x - 32)$$

$$x = \frac{5}{9}(y - 32)$$

$$\frac{9}{5}x = y - 32$$

$$\frac{9}{5}x + 32 = y$$

$$f^{-1}(x) = \frac{9}{5}x + 32$$

$f^{-1}(x)$ is used to convert x degrees Celsius to an equivalent Fahrenheit temperature.

57. Find $s^{-1}(x)$.

$$s(x) = 2x + 24$$

$$x = 2y + 24$$

$$x - 24 = 2y$$

$$\frac{1}{2}x - 12 = y$$

$$s^{-1}(x) = \frac{1}{2}x - 12$$

59. a. Find $c(30)$.

$$c(x) = \frac{300 + 12x}{x}$$

$$c(30) = \frac{300 + 12(30)}{30} = \$22$$

The company charges \$22 per person to cater a dinner for 30 people.

b. Find $c^{-1}(x)$.

$$c(x) = \frac{300 + 12x}{x}$$

$$x = \frac{300 + 12y}{y}$$

$$xy = 300 + 12y$$

$$xy - 12y = 300$$

$$y(x - 12) = 300$$

$$y = \frac{300}{x - 12}$$

$$c^{-1}(x) = \frac{300}{x - 12}$$

c. $c^{-1}(\$15.00) = \dfrac{300}{15.00 - 12} = 100$ people

61. Find and explain how it could be used.

$$E(s) = 0.05s + 2500$$

$$s = 0.05y + 2500$$

$$s - 2500 = 0.05y$$

$$\frac{1}{0.05}s - \frac{2500}{0.05} = y$$

$$20s - 50,000 = y$$

$$E^{-1}(s) = 20s - 50,000$$

The executive can use the inverse function to determine the value of the software that must be sold in order to achieve a given monthly income.

63. a. $p(10) \approx 0.12 = 12\%$; $p(30) \approx 0.71 = 71\%$

b. The graph of p, for $1 \le n \le 60$, is an increasing function. Thus p has an inverse that is a function.

c. $p^{-1}(0.223)$ represents the number of people required to be in the group for a 22.3% probability that at least two of the people will share a birthday.

65. a. Encode the message.

D $f(13) = 2(13) - 1 = 25$
O $f(24) = 2(24) - 1 = 47$
(space) $f(36) = 2(36) - 1 = 71$
Y $f(34) = 2(34) - 1 = 67$
O $f(24) = 47$
U $f(30) = 2(30) - 1 = 59$
R $f(27) = 2(27) - 1 = 53$
(space) $f(36) = 71$
H $f(17) = 2(17) - 1 = 33$
O $f(24) = 47$
M $f(22) = 2(22) - 1 = 43$
E $f(14) = 2(14) - 1 = 27$
W $f(32) = 2(32) - 1 = 63$
O $f(24) = 47$
R $f(27) = 53$
K $f(20) = 2(20) - 1 = 39$

The code is

25 47 71 67 47 59 53 71 33 47 43 27 63 47 53 39.

b. Decode the message.

$$f^{-1}(49) = \frac{49 + 1}{2} = 25 \quad P$$

$$f^{-1}(33) = \frac{33 + 1}{2} = 17 \quad H$$

$$f^{-1}(47) = \frac{47 + 1}{2} = 24 \quad O$$

$$f^{-1}(45) = \frac{45 + 1}{2} = 23 \quad N$$

$$f^{-1}(27) = \frac{27 + 1}{2} = 14 \quad E$$

$$f^{-1}(71) = \frac{71 + 1}{2} = 36 \quad \text{(space)}$$

$$f^{-1}(33) = 17 \quad H$$

$$f^{-1}(47) = 24 \quad O$$

$$f^{-1}(43) = \frac{43 + 1}{2} = 22 \quad M$$

$$f^{-1}(27) = 14 \quad E$$

The message is PHONE HOME.

c. Answers will vary.

67. $f(2) = 7$, $f(5) = 12$, and $f(4) = c$. Because f is an increasing linear function, and 4 is between 2 and 5, then $f(4)$ is between $f(2)$ and $f(5)$. Thus, c is between 7 and 12.

69. f is a linear function, therefore f^{-1} is a linear function.

$$f(2) = 3 \Rightarrow f^{-1}(3) = 2$$
$$f(5) = 9 \Rightarrow f^{-1}(9) = 5$$

Since 6 is between 3 and 9, $f^{-1}(6)$ is between 2 and 5.

71. A horizontal line intersects the graph of the function at more than one point. Thus, the function is not a one-to-one function.

73. A horizontal line intersects the graph of the function at more than one point. Thus, the function is not a one-to-one function.

75. A horizontal line intersects the graph of the function at more than one point. Thus, the function is not a one-to-one function.

77. There is at most one point where each horizontal line intersects the graph of the function. The function is a one-to-one function.

79. The reflection of f across the line given by $y=x$ yields f. Thus f is its own inverse.

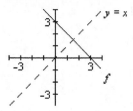

81. Find the slope and y-intercept of the inverse.

$$f(x) = mx + b, \quad m \ne 0$$
$$y = mx + b$$
$$x = my + b$$
$$x - b = my$$
$$\frac{x - b}{m} = y$$
$$f^{-1}(x) = \frac{1}{m}x - \frac{b}{m}$$

The slope is $\dfrac{1}{m}$ and the y-intercept is $\left(0, -\dfrac{b}{m}\right)$.

Prepare for Section 4.2

P1. Evaluate.

$$2^3 = 2 \cdot 2 \cdot 2 = 8$$

P3. Evaluate.

$$\frac{2^2 + 2^{-2}}{2} = \frac{4 + \frac{1}{4}}{2} = \frac{16 + 1}{8} = \frac{17}{8}$$

P5. Evaluate.

$$f(x) = 10^x$$
$$f(-1) = 10^{-1} = \frac{1}{10}$$
$$f(0) = 10^0 = 1$$
$$f(1) = 10^1 = 10$$
$$f(2) = 10^2 = 100$$

Section 4.2 Exercises

1. Name two characteristics.

Answers will vary.

3. Explain how to get the second function from the first.

Shift the graph of $f(x) = b^x$ horizontally to the right 2 units to produce the graph of $g(x) = b^{x-2}$.

5. Evaluate.

$$f(0) = 3^0 = 1$$
$$f(4) = 3^4 = 81$$

7. Evaluate.

$$g(-3) = 10^{-3} = \frac{1}{1000}$$
$$g(2) = 10^2 = 100$$

9. Evaluate.

$$h(3) = \left(\frac{5}{3}\right)^3 = \frac{125}{27}$$
$$h(-2) = \left(\frac{5}{3}\right)^{-2} = \frac{9}{25}$$

11. Evaluate.

$$j(-2) = \left(\frac{1}{2}\right)^{-2} = 4$$
$$j(4) = \left(\frac{1}{2}\right)^4 = \frac{1}{16}$$

13. Evaluate using a calculator. Round to the hundredth.

$$f(3.2) = 2^{3.2} \approx 9.19$$

15. Evaluate using a calculator. Round to the hundredth.

$$g(-3) = e^{-3} \approx 0.05$$

17. Evaluate using a calculator. Round to the hundredth.

$$h(\sqrt{3}) = 3.5^{\sqrt{3}} \approx 8.76$$

19. $f(x) = 5^x$ is a basic exponential graph.

$g(x) = 1 + 5^{-x}$ is the graph of $f(x)$ reflected across the y-axis and moved up 1 unit.

$h(x) = 5^{x+3}$ is the graph of $f(x)$ moved to the left 3 units.

$k(x) = 5^x + 3$ is the graph of $f(x)$ moved up 3 units.

a. $k(x)$

b. $g(x)$

c. $h(x)$

d. $f(x)$

21. Sketch the graph.

$$f(x) = 3^x$$

23. Sketch the graph.

$$f(x) = 10^x$$

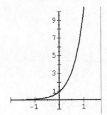

25. Sketch the graph.

$$f(x) = \left(\frac{3}{2}\right)^x$$

27. Sketch the graph.

$$f(x) = \left(\frac{1}{3}\right)^x$$

29. Explain how to get the second function from the first. Shift the graph of f vertically upward 3 units.

31. Explain how to get the second function from the first. Shift the graph of f horizontally to the right 3 units.

33. Explain how to get the second function from the first. Reflect the graph of f across the y-axis.

35. Explain how to get the second function from the first. Stretch the graph of f vertically away from the x-axis by a factor of 2.

37. Explain how to get the second function from the first. Shift the graph of f horizontally 3 units to the left, and then shift this graph vertically downward 2 units.

39. Explain how to get the second function from the first. Shift the graph of f horizontally 4 units to the right, and then reflect this graph across the x-axis.

41. Explain how to get the second function from the first. Reflect the graph of f across the y-axis, and then shift this graph vertically upward 3 units.

43. Graph the function using a graphing utility and state the equation of existing horizontal asymptotes.

$$f(x) = \frac{3^x + 3^{-x}}{2}$$

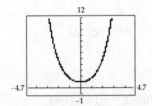

No horizontal asymptote

45. Graph the function using a graphing utility and state the equation of existing horizontal asymptotes.

$$f(x) = \frac{e^x - e^{-x}}{2}$$

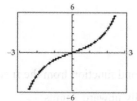

No horizontal asymptote

47. Graph the function using a graphing utility and state the equation of existing horizontal asymptotes.

$$f(x) = -e^{(x-4)}$$

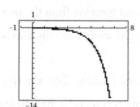

Horizontal asymptote: $y = 0$

49. Graph the function using a graphing utility and state the equation of existing horizontal asymptotes.

$$f(x) = \frac{10}{1 + 0.4e^{-0.5x}}, \text{ with } x \geq 0$$

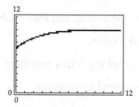

Horizontal asymptote: $y = 10$

51. a. $P(t) = 100 \cdot 2^{2t}$

$$P(3) = 100 \cdot 2^{2(3)} = 100 \cdot 2^6 = 100 \cdot 64$$
$$= 6400 \text{ bacteria}$$

$$P(t) = 100 \cdot 2^{2t}$$
$$P(6) = 100 \cdot 2^{2(6)} = 100 \cdot 2^{12} = 100 \cdot 4096$$
$$= 409,600 \text{ bacteria}$$

b. Use a graphing utility to solve.

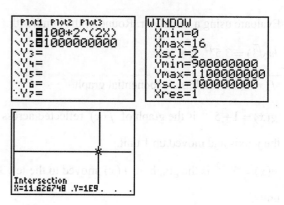

11.6 hours

53. a. $d(p) = 880e^{-0.18p}$

$$d(10) = 880e^{-0.18(10)} \approx 145 \text{ items per month}$$
$$d(p) = 880e^{-0.18p}$$
$$d(18) = 880e^{-0.18(18)} \approx 34 \text{ items per month}$$

b. As $p \to \infty$, $d(p) \to 0$. The demand will approach 0 items per month.

55. a. $P(x) = (0.9)^x$

$$P(3.5) = (0.9)^{3.5} \approx 69.2\%$$

b. Use a graphing utility to solve.

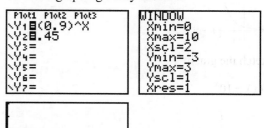

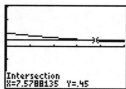

For a transparency of 45%, the UV index is 7.6.

57. a. $B(n) = \frac{3^{n+1} - 3}{2}$

$$B(5) = \frac{3^{5+1} - 3}{2} = \frac{3^6 - 3}{2} = \frac{729 - 3}{2} = \frac{726}{2}$$
$$= 363 \text{ beneficiaries}$$

$$B(n) = \frac{3^{n+1} - 3}{2}$$
$$B(10) = \frac{3^{10+1} - 3}{2} = \frac{3^{11} - 3}{2} = \frac{177147 - 3}{2} = \frac{177144}{2}$$
$$= 88,572 \text{ beneficiaries}$$

b. Use a graphing utility to solve.

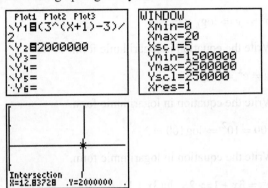

13 rounds

59. a. $T(t) = 65 + 115e^{-0.042t}$

$T(10) = 65 + 115e^{-0.042(10)} = 65 + 115e^{-0.42}$

$\approx 141° \text{ F}$

b. Use a graphing utility to solve.

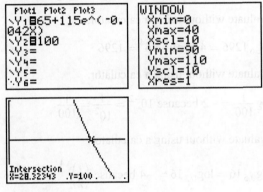

28.3 minutes

61. a. $f(n) = (27.5)2^{(n-1)/12}$

$f(40) = (27.5)2^{(40-1)/12} = (27.5)2^{39/12}$

$= (27.5)2^{3.25}$

$\approx 261.63 \text{ vibrations per second}$

b. No. The function $f(n)$ is not a linear function.

Therefore, the graph of $f(n)$ does not increase at a

constant rate.

63. $f(x) = \dfrac{e^x - e^{-x}}{2}$ is an odd function. That is, prove

$f(-x) = -f(x)$.

Proof: $f(x) = \dfrac{e^x - e^{-x}}{2}$

$f(-x) = \dfrac{e^{-x} - e^x}{2}$

$= \dfrac{-e^{-x} + e^x}{2}$

$= \dfrac{(e^x - e^{-x})}{2}$

$= -f(x)$

65. Draw the graph as described.

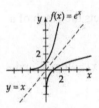

67. Determine the domain in interval notation.

domain: $(-\infty, \infty)$

69. Determine the domain in interval notation.

domain: $[0, \infty)$

71. By definition the average of two numbers is their sum

divided by 2. The expression $\dfrac{e^x + e^{-x}}{2}$ shows that f is

the average of $g(x) = e^x$ and $h(x) = e^{-x}$.

Prepare for Section 4.3

P1. Determine the value of x.

$2^x = 16$

$2^x = 2^4$

$x = 4$

P3. Determine the value of x.

$$x^4 = 625$$
$$x^4 = 5^4$$
$$x = 5$$

P5. State the domain.

$$g(x) = \sqrt{x-2}$$
$$x - 2 \geq 0$$
$$x \geq 2$$

The domain is $\{x \mid x \geq 2\}$.

Section 4.3 Exercises

1. The function $f(x) = \log_b x$ is a one-to-one function since every horizontal line intersects the graph of a function at most once.

3. State the relationship between $f(x) = \log_b x$ and $g(x) = b^x$.

   The graph of f is a reflection of the graph of g across the line given by $y = x$.

5. Write the equation in exponential form.

   $$1 = \log 10 \Rightarrow 10^1 = 10$$

7. Write the equation in exponential form.

   $$2 = \log_8 64 \Rightarrow 8^2 = 64$$

9. Write the equation in exponential form.

   $$3 = \log_5 125 \Rightarrow 5^3 = 125$$

11. Write the equation in exponential form.

    $$\ln x = 7 \Rightarrow e^7 = x$$

13. Write the equation in exponential form.

    $$\ln 1 = 0 \Rightarrow e^0 = 1$$

15. Write the equation in exponential form.

    $$3 = \log(5x+2) \Rightarrow 10^3 = 5x+2$$

17. Write the equation in logarithmic form.

    $$3^2 = 9 \Rightarrow \log_3 9 = 2$$

19. Write the equation in logarithmic form.

    $$7^2 = 49 \Rightarrow \log_7 49 = 2$$

21. Write the equation in logarithmic form.

    $$b^x = y \Rightarrow \log_b y = x$$

23. Write the equation in logarithmic form.

    $$y = e^x \Rightarrow \ln y = x$$

25. Write the equation in logarithmic form.

    $$100 = 10^2 \Rightarrow \log 100 = 2$$

27. Write the equation in logarithmic form.

    $$e^2 = 3x+1 \Rightarrow 2 = \ln(3x+1)$$

29. Evaluate without using a calculator.

    $$\log_4 16 = 2 \text{ because } 4^2 = 16$$

31. Evaluate without using a calculator.

    $$\log_3 \frac{1}{243} = -5 \text{ because } 3^{-5} = \left(\frac{1}{3}\right)^5 = \frac{1}{243}$$

33. Evaluate without using a calculator.

    $$\log_6 1296 = 4 \text{ because } 6^4 = 1296$$

35. Evaluate without using a calculator.

    $$\log \frac{1}{100} = -2 \text{ because } 10^{-2} = \frac{1}{10^2} = \frac{1}{100}$$

37. Evaluate without using a calculator.

    $$\log_{0.5} 16 = \log_{1/2} 16 = -4 \text{ because } \left(\frac{1}{2}\right)^{-4} = 2^4 = 16$$

39. Evaluate without using a calculator.

    $$4\log 1000 = 12 \Rightarrow \log 1000^4 = 12$$
    $$\text{because } 10^{12} = \left(10^3\right)^4 = (1000)^4$$

41. Evaluate without using a calculator.

    $$3\log_7 49 = 6 \Rightarrow \log_7 49^3 = 6$$
    $$\text{because } 7^9 = \left(7^2\right)^3 = (49)^2$$

43. Evaluate without using a calculator.

    $$\log \sqrt{1000} = \frac{3}{2} \Rightarrow \log_{10} 1000^{1/2} = \frac{3}{2}$$
    $$\text{because } 10^{3/2} = \left(10^3\right)^{1/2} = (1000)^{1/2}$$

45. Evaluate without using a calculator.

    $$\ln e^4 = 4 \text{ because } e^4 = e^4$$

47. Graph the function using its exponential form.

$y = \log_4 x$

$x = 4^y$

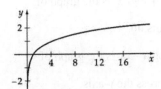

49. Graph the function using its exponential form.

$y = \log_{12} x$

$x = 12^y$

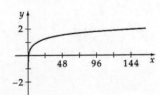

51. Graph the function using its exponential form.

$y = \log_{1/2} x$

$x = (1/2)^y$

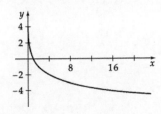

53. Graph the function using its exponential form.

$y = \log_{5/2} x$

$x = (5/2)^y$

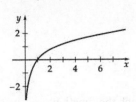

55. Find the domain, write in interval notation.

$f(x) = \log_5(x-3)$

$x - 3 > 0$

$x > 3$

The domain is $(3, \infty)$.

57. Find the domain, write in interval notation.

$k(x) = \log_{2/3}(11 - x)$

$11 - x > 0$

$-x > -11$

$x < 11$

The domain is $(-\infty, 11)$.

59. Find the domain, write in interval notation.

$P(x) = \ln(x^2 - 4)$

$x^2 - 4 > 0$

$(x+2)(x-2) > 0$

The critical values are -2 and 2.

The product is positive.

The domain is $(-\infty, -2) \cup (2, \infty)$.

61. Find the domain, write in interval notation.

$h(x) = \ln\left(\dfrac{x^2}{x-4}\right)$

$\dfrac{x^2}{x-4} > 0$

The critical values are 0 and 4.

The quotient is positive.

$x > 4$

The domain is $(4, \infty)$.

63. Find the domain, write in interval notation.

$x^3 - x > 0$

$x(x^2 - 1) > 0$

$x(x+1)(x-1) > 0$

Critical values are 0, -1 and 1.

Product is positive.

$-1 < x < 0$ or $x > 1$

$(-1, 0) \cup (1, \infty)$

65. Find the domain, write in interval notation.

$2x - 11 > 0$

$2x > 11$

$x > \dfrac{11}{2}$

The domain is $\left(\dfrac{11}{2}, \infty\right)$.

67. Find the domain, write in interval notation.

$$3x - 7 > 0$$
$$3x > 7$$
$$x > \frac{7}{3}$$

The domain is $\left(\frac{7}{3}, \infty\right)$.

69. Explain how to get the second function from the first.

Shift the graph of $f(x) = \log_4 x$ horizontally 3 units to the right.

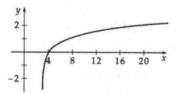

71. Explain how to get the second function from the first.

Shift the graph of $f(x) = \log_{12} x$ vertically 2 units upward.

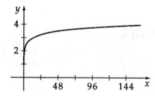

73. Explain how to get the second function from the first.

Shift the graph of $f(x) = \log_{1/2} x$ vertically 3 units upward.

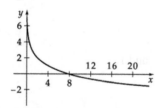

75. Explain how to get the second function from the first.

Shift the graph of $f(x) = \log_{5/2} x$ horizontally 4 units to the right and vertically 1 unit upward.

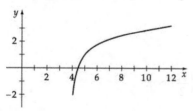

77. The graph of $f(x) = \log_5(x - 2)$ is the graph of

$y = \log_5 x$ shifted 2 units to the right.

The graph of $g(x) = 2 + \log_5 x$ is the graph of

$y = \log_5 x$ shifted 2 units up.

The graph of $h(x) = \log_5(-x)$ is the graph of

$y = \log_5 x$ reflected across the y-axis.

The graph of $k(x) = -\log_5(x + 3)$ is the graph of

$y = \log_5 x$ reflected across the x-axis and shifted

left 3 units.

a. $k(x)$

b. $f(x)$

c. $g(x)$

d. $h(x)$

79. Use a graphing utility to graph the function.

$$f(x) = -2 \ln x$$

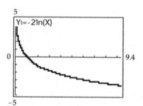

81. Use a graphing utility to graph the function.

$$f(x) = |\ln x|$$

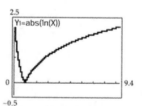

83. Use a graphing utility to graph the function.

$$f(x) = \log \sqrt[3]{x}$$

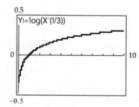

85. Use a graphing utility to graph the function.

$$f(x) = \log(x + 10)$$

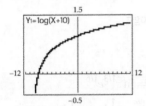

87. Use a graphing utility to graph the function.

$$f(x) = 3\log|2x + 10|$$

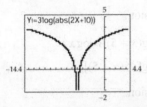

89. a. $r(t) = 0.69607 + 0.60781\ln t$
  $r(9) = 0.69607 + 0.60781\ln 9 \approx 2.0\%$

b. Use a graphing calculator to solve.

 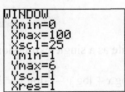

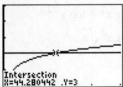

45 months

91. $N(x) = 2750 + 180\ln\left(\dfrac{x}{1000} + 1\right)$

a. $N(20,000) = 2750 + 180\ln\left(\dfrac{20,000}{1000} + 1\right)$
  $= 2750 + 180\ln(21) \approx 3298$ units

  $N(40,000) = 2750 + 180\ln\left(\dfrac{40,000}{1000} + 1\right)$
  $= 2750 + 180\ln(41) \approx 3418$ units

  $N(60,000) = 2750 + 180\ln\left(\dfrac{60,000}{1000} + 1\right)$
  $= 2750 + 180\ln(61) \approx 3490$ units

b. $N(0) = 2750 + 180\ln\left(\dfrac{0}{1000} + 1\right)$
  $= 2750 + 180\ln(1) = 2750 + 180(0)$
  $= 2750 + 0 = 2750$ units

93. $f(x) = 76.42 - 6.25\ln x$

a. Find the world record time in 2000.

  $f(90) = 76.42 - 6.25\ln 90$
  ≈ 48.296

The time is about 46.30 seconds.

b. Predict the world record time in 2030.

  $f(120) = 76.42 - 6.25\ln 120$
  ≈ 46.498

The time is about 46.50 seconds.

95. $N = \text{int}(x \log b) + 1$

a. $N = \text{int}(10\log 2) + 1 = 3 + 1 = 4$ digits

b. $N = \text{int}(200\log 3) + 1 = 95 + 1 = 96$ digits

c. $N = \text{int}(4005\log 7) + 1 = 3384 + 1 = 3385$ digits

d. $N = \text{int}(57,885,161\log 2) + 1 = 17,425,169 + 1$
  $= 17,425,170$ digits

97. $f(x)$ and $g(x)$ are inverse functions

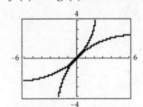

99. Domain: all real numbers; range $\{y \mid 0 < y \le 1\}$.

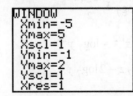 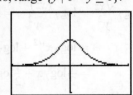

Prepare for Section 4.4

P1. Use a calculator to compare the values.

  $\log 3 + \log 2 \approx 0.77815$
  $\log 6 \approx 0.77815$

P3. Use a calculator to compare the values.

  $3\log 4 \approx 1.80618$
  $\log(4^3) \approx 1.80618$

P5. Use a calculator to compare the values.

$\ln 5 \approx 1.60944$

$\dfrac{\log 5}{\log e} \approx 1.60944$

Section 4.4 Exercises

1. The product property of logarithms states that

$\log_b (MN)$ equals $\log_b M + \log_b N$.

3. The power property of logarithms states that

$\log_b (M^p)$ equals $p \log_b M$.

5. Expand the expression.

$\log_b (xyz) = \log_b x + \log_b y + \log_b z$

7. Expand the expression.

$$\log_4 \left(\dfrac{y^2}{16x} \right) = \log_4 y^2 - \log_4 16x$$

$$= \log_4 y^2 - \log_4 16 - \log_4 x$$

$$= \log_4 y^2 - \log_4 4^2 - \log_4 x$$

$$= \log_4 y^2 - 2\log_4 4 - \log_4 x$$

$$= \log_4 y^2 - 2 - \log_4 x$$

9. Expand the expression.

$$\log_2 \dfrac{\sqrt{x}}{y^3} = \log_2 \sqrt{x} - \log_2 y^3$$

$$= \log_2 x^{1/2} - \log_2 y^3$$

$$= \dfrac{1}{2} \log_2 x - 3\log_2 y$$

11. Expand the expression.

$$\log_7 \dfrac{\sqrt{xz}}{y^2} = \log_7 \dfrac{(xz)^{1/2}}{y^2} = \log_7 \dfrac{x^{1/2} z^{1/2}}{y^2}$$

$$= \log_7 x^{1/2} + \log_7 z^{1/2} - \log_7 y^2$$

$$= \dfrac{1}{2} \log_7 x + \dfrac{1}{2} \log_7 z - 2\log_7 y$$

13. Expand the expression.

$\ln(e^2 z) = \ln e^2 + \ln z = 2\ln e + \ln z = 2 + \ln z$

15. Expand the expression.

$$\log_6 \left(\dfrac{x^3 \sqrt[3]{y}}{216 z^4} \right) = \log_6 x^3 + \log_6 y^{1/3} - \log_6 6^3 - \log_6 z^4$$

$$= 3\log_6 x + \dfrac{1}{3} \log_6 y - 3\log_6 6 - 4\log_6 z$$

$$= 3\log_6 x + \dfrac{1}{3} \log_6 y - 3 - 4\log_6 z$$

17. Expand the expression.

$$\log \sqrt{x\sqrt{z}} = \log \left(xz^{1/2} \right)^{1/2} = \log x^{1/2} z^{1/4}$$

$$= \dfrac{1}{2} \log x + \dfrac{1}{4} \log z$$

19. Expand the expression.

$$\ln \left(\sqrt[3]{z\sqrt{e}} \right) = \ln \left(ze^{1/2} \right)^{1/3} = \ln z^{1/3} e^{1/6}$$

$$= \ln z^{1/3} + \ln e^{1/6} = \dfrac{1}{3} \ln z + \dfrac{1}{6} \ln e$$

$$= \dfrac{1}{3} \ln z + \dfrac{1}{6}$$

21. Write as a single expression.

$$\log(x+5) + 2\log x = \log(x+5) + \log x^2$$

$$= \log[x^2 (x+5)]$$

23. Write as a single expression.

$$\ln(x^2 - y^2) - \ln(x-y) = \ln \dfrac{x^2 - y^2}{x-y}$$

$$= \ln \dfrac{(x+y)(x-y)}{x-y}$$

$$= \ln(x+y)$$

25. Write as a single expression.

$$2\log x + \log y - \dfrac{1}{2} \log y = \log x^2 + \dfrac{1}{2} \log y$$

$$= \log x^2 + \log y^{1/2}$$

$$= \log x^2 \sqrt{y}$$

27. Write as a single expression.

$$\log(xy^2) - \log z = \log \left(\dfrac{xy^2}{z} \right)$$

29. Write as a single expression.

$$2\left(\log_6 x + \log_6 y^2 \right) - \log_6 (x+2)$$

$$= \log_6 x^2 + \log_6 y^4 - \log_6 (x+2)$$

$$= \log_6 \left(\dfrac{x^2 y^4}{x+2} \right)$$

31. Write as a single expression.

$$2\ln(x+4) - \ln x - \ln(x^2 - 3)$$

$$= \ln(x+4)^4 - \ln x - \ln(x^2 - 3)$$

$$= \ln \left[\dfrac{(x+4)^4}{x(x^2 - 3)} \right]$$

33. Write as a single expression.

$\ln(2x+5) - \ln y - 2\ln z + \dfrac{1}{2}\ln w$

$= \ln(2x+5) - \ln y - \ln z^2 + \ln w^{1/2}$

$= \ln\left[\dfrac{(2x+5)\sqrt{w}}{yz^2}\right]$

35. Write as a single expression.

$\ln(x^2 - 9) - 2\ln(x-3) + 3\ln y$

$= \ln(x+3)(x-3) - \ln(x-3)^2 + \ln y^3$

$= \ln\left[\dfrac{(x+3)(x-3)y^3}{(x-3)^2}\right]$

$= \ln\left[\dfrac{(x+3)y^3}{x-3}\right]$

37. Approximate to the nearest ten-thousandth.

$\log_7 20 = \dfrac{\log 20}{\log 7} \approx 1.5395$

39. Approximate to the nearest ten-thousandth.

$\log_2 14 = \dfrac{\log 14}{\log 2} \approx 3.8074$

41. Approximate to the nearest ten-thousandth.

$\log_8\left(\dfrac{3}{5}\right) = \dfrac{\log\left(\dfrac{3}{5}\right)}{\log 8} \approx -0.2457$

43. Approximate to the nearest ten-thousandth.

$\log_9 \sqrt{17} = \dfrac{\log \sqrt{17}}{\log 9} \approx 0.6447$

45. Approximate to the nearest ten-thousandth.

$\log_{\sqrt{2}} 17 = \dfrac{\log 17}{\log \sqrt{2}} \approx 8.1749$

47. Approximate to the nearest ten-thousandth.

$\log_\pi e = \dfrac{\log e}{\log \pi} \approx 0.8736$

49. Graph the function using a graphing utility.

$f(x) = \log_4 x = \dfrac{\log x}{\log 4}$

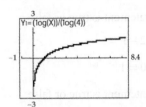

51. Graph the function using a graphing utility.

$g(x) = \log_8(x-3) = \dfrac{\log(x-3)}{\log 8}$

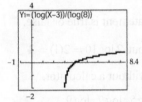

53. Graph the function using a graphing utility.

$h(x) = \log_3(x-3)^2$

$= \dfrac{\log(x-3)^2}{\log 3}$

$= \dfrac{2\log(x-3)}{\log 3}$

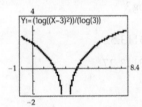

55. Graph the function using a graphing utility.

$F(x) = -\log_5|x-2| = -\dfrac{\log|x-2|}{\log 5}$

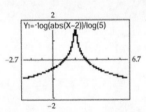

57. Determine whether the statement is true or false.

False. $\log 10 + \log 10 = 1 + 1 = 2$

but $\log(10+10) = \log 20 \neq 2$.

59. Determine whether the statement is true or false.

True.

61. Determine whether the statement is true or false.

False. $\log 100 - \log 10 = 2 - 1 = 1$

but $\log(100 - 10) = \log 90 \neq 1$

63. Determine whether the statement is true or false.

False. $\dfrac{\log 100}{\log 10} = \dfrac{2}{1} = 2$

but $\log 100 - \log 10 = 2 - 1$
$$= 1$$

65. Determine whether the statement is true or false.

False. $(\log 10)^2 = 1^2 = 1$ but $2 \log 10 = 2(1) = 2$

67. Evaluate the expression without a calculator.

$$\log_3 5 \cdot \log_5 7 \cdot \log_7 9 = \frac{\log 5}{\log 3} \cdot \frac{\log 7}{\log 5} \cdot \frac{\log 9}{\log 7}$$

$$= \frac{\log 5}{\log 3} \cdot \frac{\log 7}{\log 5} \cdot \frac{\log 9}{\log 7}$$

$$= \frac{\log 9}{\log 3} = \frac{\log 3^2}{\log 3} = \frac{2\log 3}{\log 3}$$

$$= \frac{2\log 3}{\log 3} = 2$$

69. $\ln 500^{501} = 501 \ln 500 \approx 3113.52$

$\ln 506^{500} = 500 \ln 506 \approx 3113.27$

$\ln 500^{501}$ is larger.

71. $M = \log\left(\dfrac{I}{I_0}\right)$

$M = \log\left(\dfrac{7,943,300 I_0}{I_0}\right)$

$= \log 7,943,300$

≈ 6.9

73. $\log\left(\dfrac{I}{I_0}\right) = M$

$\log\left(\dfrac{I}{I_0}\right) = 8.6$

$\dfrac{I}{I_0} = 10^{8.6}$

$I = 10^{8.6} I_0$

$I \approx 398,107,170.6 I_0$

75. $M = \log\left(\dfrac{I}{I_0}\right)$

$M_5 = \log\left(\dfrac{I_5}{I_0}\right) \qquad M_3 = \log\left(\dfrac{I_3}{I_0}\right)$

$5 = \log\left(\dfrac{I_5}{I_0}\right) \qquad 3 = \log\left(\dfrac{I_3}{I_0}\right)$

$10^5 = \dfrac{I_5}{I_0} \qquad\qquad 10^3 = \dfrac{I_3}{I_0}$

$10^5 I_0 = I_5 \qquad\qquad 10^3 I_0 = I_3$

$\dfrac{I_5}{I_3} = \dfrac{10^5 I_0}{10^3 I_0} = \dfrac{10^5}{10^3} = 10^{5-3} = 10^2 = 100$ to 1

77. $\dfrac{10^{9.0}}{10^{7.1}} = \dfrac{10^{9.0-7.1}}{1} = 10^{1.9}$ to $1 \approx 79.4$ to 1

79. $M = \log A + 3\log 8t - 2.92$
$= \log 18 + 3\log[8(31)] - 2.92 \approx 5.5$

81. $\text{pH} = -\log[\text{H}^+]$

$\text{pH} = -\log[3.97 \times 10^{-11}]$

$\text{pH} = 10.4$

$10.4 > 7$; milk of magnesia is a base

83. $\text{pH} = -\log[\text{H}^+]$

$9.5 = -\log[\text{H}^+]$

$-9.5 = \log[\text{H}^+]$

$10^{-9.5} = 10^{\log[\text{H}^+]}$

$[\text{H}^+] = 3.16 \times 10^{-10}$ mole per liter

85. $dB(I) = 10\log\left(\dfrac{I}{I_0}\right)$

a. $dB(1.58 \times 10^8 \cdot I_0) = 10\log\left(\dfrac{1.58 \times 10^8 \cdot I_0}{I_0}\right)$

$= 10\log(1.58 \times 10^8)$

≈ 82.0 decibels

b. $dB(10,800 \cdot I_0) = 10\log\left(\dfrac{10,800 \cdot I_0}{I_0}\right)$

$= 10\log(10,800)$

≈ 40.3 decibels

c. $dB(3.16 \times 10^{11} \cdot I_0) = 10\log\left(\dfrac{3.16 \times 10^{11} \cdot I_0}{I_0}\right)$

$= 10\log(3.16 \times 10^{11})$

≈ 115.0 decibels

d. $dB(1.58 \times 10^{15} \cdot I_0) = 10 \log\left(\dfrac{1.58 \times 10^{15} \cdot I_0}{I_0}\right)$

$\qquad\qquad\qquad\quad = 10 \log(1.58 \times 10^{15})$

$\qquad\qquad\qquad\quad \approx 152.0 \text{ decibels}$

87. $dB(I) = 10 \log\left(\dfrac{I}{I_0}\right)$

$\qquad 120 = 10 \log\left(\dfrac{I_{120}}{I_0}\right) \qquad\qquad 110 = 10 \log\left(\dfrac{I_{110}}{I_0}\right)$

$\qquad 12 = \log\left(\dfrac{I_{120}}{I_0}\right) \qquad\qquad 11 = \log\left(\dfrac{I_{110}}{I_0}\right)$

$\qquad 10^{12} = \dfrac{I_{120}}{I_0} \qquad\qquad\qquad 10^{11} = \dfrac{I_{110}}{I_0}$

$\qquad 10^{12} \cdot I_0 = I_{120} \qquad\qquad 10^{11} \cdot I_0 = I_{110}$

$\qquad \dfrac{I_{120}}{I_{110}} = \dfrac{10^{12} \cdot I_0}{10^{11} \cdot I_0}$

$\qquad\quad\;\; = \dfrac{10^{12}}{10^{11}} = 10^{12-11} = 10^1$

$\qquad\quad\;\; = 10 \text{ times more intense}$

89. Determine the number of scales for each stage.

$S_n = S_0 \cdot 10^{\frac{n}{N}(\log S_f - \log S_0)}$

$S_1 = 1{,}000{,}000 \cdot 10^{\frac{1}{5}(\log 500{,}000 - \log 1{,}000{,}000)}$

$\quad = 870{,}551$

$S_2 = 1{,}000{,}000 \cdot 10^{\frac{2}{5}(\log 500{,}000 - \log 1{,}000{,}000)}$

$\quad = 757{,}858$

$S_3 = 1{,}000{,}000 \cdot 10^{\frac{3}{5}(\log 500{,}000 - \log 1{,}000{,}000)}$

$\quad = 659{,}754$

$S_4 = 1{,}000{,}000 \cdot 10^{\frac{4}{5}(\log 500{,}000 - \log 1{,}000{,}000)}$

$\quad = 574{,}349$

$S_5 = 1{,}000{,}000 \cdot 10^{\frac{5}{5}(\log 500{,}000 - \log 1{,}000{,}000)}$

$\quad = 500{,}000$

The scales are 1:870,551; 1:757,858; 1:659,754; 1:574,349; 1:500,000.

91. Let $r = \log_b M$ and $s = \log_b N$.

Then $M = b^r$ and $N = b^s$.

Consider the quotient of M and N

$\qquad \dfrac{M}{N} = \dfrac{b^r}{b^s}$

$\qquad \dfrac{M}{N} = b^{r-s}$

$\qquad \log_b \dfrac{M}{N} = r - s$

$\qquad \log_b \dfrac{M}{N} = \log_b M - \log_b N$

Mid-Chapter 4 Quiz

1. Use composition to determine whether f and g are inverses.

$$f(x) = \dfrac{500 + 120x}{x}; \; g(x) = \dfrac{500}{x-120}$$

$f[g(x)] = f\left(\dfrac{500}{x-120}\right)$

$\qquad\quad = \dfrac{500 + 120\left(\dfrac{500}{x-120}\right)}{\dfrac{500}{x-120}}$

$\qquad\quad = \dfrac{500 + 120\left(\dfrac{500}{x-120}\right)}{\dfrac{500}{x-120}} \cdot \dfrac{x-120}{x-120}$

$\qquad\quad = \dfrac{500(x-120) + 120(500)}{500} = x - 120 + 120$

$\qquad\quad = x$

$g[f(x)] = g\left(\dfrac{500 + 120x}{x}\right)$

$\qquad\quad = \dfrac{500}{\dfrac{500 + 120x}{x} - 120} = \dfrac{500}{\dfrac{500 + 120x}{x} - \dfrac{120x}{x}}$

$\qquad\quad = \dfrac{500}{\dfrac{500}{x}} = 500 \cdot \dfrac{x}{500}$

$\qquad\quad = x$

Yes, f and g are inverses of each other.

3. Evaluate.

$\qquad f(x) = e^x$

$\qquad f(-2.4) = e^{-2.4} \approx 0.0907$

5. Graph $f(x) = \log_3(x+3)$.

Shift 3 units to the left.

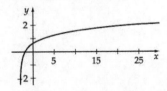

7. Write as a single logarithm.

$$\log_3 x^4 - 2\log_3 z + \log_3\left(xy^2\right)$$
$$= \log_3 x^4 - \log_3 z^2 + \log_3\left(xy^2\right)$$
$$= \log_3\left(\frac{x^5 y^2}{z^2}\right)$$

9. Find the magnitude.

$$M = \log\left(\frac{I}{I_0}\right)$$
$$M = \log\left(\frac{789,251 I_0}{I_0}\right)$$
$$= \log 789,251$$
$$\approx 5.9$$

Prepare for Section 4.5

P1. Write the expression in logarithmic form.

$$3^6 = 729 \Rightarrow \log_3 729 = 6$$

P3. Write the expression in logarithmic form.

$$a^{x+2} = b \Rightarrow \log_a b = x + 2$$

P5. Solve for x.

$$165 = \frac{300}{1 + 12x}$$
$$165(1 + 12x) = 300$$
$$165 + 1980x = 300$$
$$1980x = 135$$
$$x = \frac{135}{1980} = \frac{3}{44}$$

Section 4.5 Exercises

1. State the Equality of Exponents Theorem.

If $b^x = b^y$, then $x = y$, provided $b > 0$, and $b \neq 1$.

3. State the way to solve $3^{2x+7} = 8^{x+6}$ without a graphing utility.

Logarithm-of-each-side property

5. Find the exact solution.

$$2^x = 64$$
$$2^x = 2^6$$
$$x = 6$$

7. Find the exact solution.

$$49^x = \frac{1}{343}$$
$$7^{2x} = 7^{-3}$$
$$2x = -3$$
$$x = -\frac{3}{2}$$

9. Find the exact solution.

$$3^{2x-1} = 81$$
$$3^{2x-1} = 3^4$$
$$2x - 1 = 4$$
$$2x = 5$$
$$x = \frac{5}{2}$$

11. Find the exact solution.

$$\left(\frac{2}{5}\right)^x = \frac{8}{125}$$
$$\left(\frac{2}{5}\right)^x = \left(\frac{2}{5}\right)^3$$
$$x = 3$$

13. Find the exact solution.

$$5^x = 70$$
$$\log(5^x) = \log 70$$
$$x \log 5 = \log 70$$
$$x = \frac{\log 70}{\log 5}$$

15. Find the exact solution.

$$3^{x+1} = 251$$
$$\ln(3^{x+1}) = \ln 251$$
$$(x+1)\ln 3 = \ln 251$$
$$x + 1 = \frac{\ln 251}{\ln 3}$$
$$x = \frac{\ln 251}{\ln 3} - 1$$

17. Find the exact solution.

$$10^{2x+3} = 315$$
$$\log 10^{2x+3} = \log 315$$
$$(2x+3)\log 10 = \log 315$$
$$2x + 3 = \log 315$$
$$x = \frac{\log 315 - 3}{2}$$

19. Find the exact solution.

$$e^x = 10$$

$$\ln e^x = \ln 10$$

$$x = \ln 10$$

21. Find the exact solution.

$$2^{1-x} = 3^{x+1}$$

$$\ln 2^{1-x} = \ln 3^{x+1}$$

$$(1-x)\ln 2 = (x+1)\ln 3$$

$$\ln 2 - x\ln 2 = x\ln 3 + \ln 3$$

$$\ln 2 - x\ln 2 - x\ln 3 = \ln 3$$

$$-x\ln 2 - x\ln 3 = \ln 3 - \ln 2$$

$$-x(\ln 2 + \ln 3) = \ln 3 - \ln 2$$

$$x = -\frac{(\ln 3 - \ln 2)}{(\ln 2 + \ln 3)}$$

$$x = \frac{\ln 2 - \ln 3}{\ln 2 + \ln 3} \text{ or } \frac{\ln 2 - \ln 3}{\ln 6}$$

23. Find the exact solution.

$$2^{2x-3} = 5^{-x-1}$$

$$\log 2^{2x-3} = \log 5^{-x-1}$$

$$(2x-3)\log 2 = (-x-1)\log 5$$

$$2x\log 2 - 3\log 2 = -x\log 5 - \log 5$$

$$2x\log 2 + x\log 5 - 3\log 2 = -\log 5$$

$$2x\log 2 + x\log 5 = 3\log 2 - \log 5$$

$$x(2\log 2 + \log 5) = 3\log 2 - \log 5$$

$$x = \frac{3\log 2 - \log 5}{2\log 2 + \log 5}$$

25. Find the exact solution.

$$\log(4x - 18) = 1$$

$$4x - 18 = 10^1$$

$$4x - 18 = 10$$

$$4x = 28$$

$$x = 7$$

27. Find the exact solution.

$$\ln(x^2 - 9) = \ln(x + 11)$$

$$x^2 - 9 = x + 11$$

$$x^2 - x - 20 = 0$$

$$(x+4)(x-5) = 0$$

$$x = -4 \text{ or } x = 5$$

29. Find the exact solution.

$$\log_2 x + \log_2 (x - 4) = 2$$

$$\log_2 x(x - 4) = 2$$

$$\log_2 (x^2 - 4x) = 2$$

$$2^2 = x^2 - 4x$$

$$0 = x^2 - 4x - 4$$

$$x = \frac{4 \pm \sqrt{16 - 4(1)(-4)}}{2}$$

$$x = \frac{4 \pm 4\sqrt{2}}{2}$$

$$x = 2 \pm 2\sqrt{2}$$

$2 - 2\sqrt{2}$ is not a solution because the logarithm of a negative number is not defined. The solution is

$$x = 2 + 2\sqrt{2}.$$

31. Find the exact solution.

$$\log(5x - 1) = 2 + \log(x - 2)$$

$$\log(5x - 1) - \log(x - 2) = 2$$

$$\log \frac{(5x - 1)}{(x - 2)} = 2$$

$$10^2 = \frac{(5x - 1)}{(x - 2)}$$

$$100(x - 2) = 5x - 1$$

$$100x - 200 = 5x - 1$$

$$95x = 199$$

$$x = \frac{199}{95}$$

33. Find the exact solution.

$$\ln(1 - x) + \ln(3 - x) = \ln 8$$

$$\ln[(1 - x)(3 - x)] = \ln 8$$

$$(1 - x)(3 - x) = 8$$

$$3 - 4x + x^2 = 8$$

$$x^2 - 4x - 5 = 0$$

$$(x + 1)(x - 5) = 0$$

$$x = -1 \text{ or } x = 5 \text{ (No; not in domain.)}$$

The solution is $x = -1$.

35. Find the exact solution.

$$\log\sqrt{x^3 - 17} = \frac{1}{2}$$

$$\frac{1}{2}\log(x^3 - 17) = \frac{1}{2}$$

$$10^1 = x^3 - 17$$

$$27 = x^3$$

$$\sqrt[3]{27} = \sqrt[3]{x^3}$$

$$3 = x$$

The solution is $x = 3$.

37. Find the exact solution.

$$\ln(\ln x) = -1$$

$$e^{-1} = \ln x$$

$$\frac{1}{e} = \ln x$$

$$e^{1/e} = x$$

39. Find the exact solution.

$$\ln(e^{3x}) = 6$$

$$3x \ln e = 6$$

$$3x(1) = 6$$

$$3x = 6$$

$$x = 2$$

41. Find the exact solution.

$$\ln(2x + 5) = \ln(x + 3) + \ln(x - 1)$$

$$\ln(2x + 5) = \ln(x + 3)(x - 1)$$

$$2x + 5 = x^2 + 2x - 3$$

$$x^2 = 8$$

$$x = \pm 2\sqrt{2}$$

$x = -2\sqrt{2}$ is not a solution because $\ln(-2\sqrt{2} - 1)$ is undefined.

The solution is $2\sqrt{2}$.

43. Find the exact solution.

$$e^{\ln(x-1)} = 4$$

$$\ln e^{\ln(x-1)} = \ln 4$$

$$\ln(x - 1)\ln e = \ln 4$$

$$\ln(x - 1)(1) = \ln 4$$

$$(x - 1) = 4$$

$$x = 5$$

45. Find the exact solution.

$$\frac{10^x - 10^{-x}}{2} = 20$$

$$10^x(10^x - 10^{-x}) = 40(10^x)$$

$$10^{2x} - 1 = 40(10^x)$$

$$10^{2x} - 40(10)^x - 1 = 0$$

Let $u = 10^x$.

$$u^2 - 40u - 1 = 0$$

$$u = \frac{40 \pm \sqrt{40^2 - 4(1)(-1)}}{2}$$

$$= \frac{40 \pm \sqrt{1600 + 4}}{2}$$

$$= \frac{40 \pm \sqrt{1604}}{2}$$

$$= \frac{40 \pm 2\sqrt{401}}{2}$$

$$= 20 \pm \sqrt{401} \qquad \text{Reject negative}$$

$$10^x = 20 + \sqrt{401}$$

$$\log 10^x = \log(20 + \sqrt{401})$$

$$x = \log(20 + \sqrt{401})$$

47. Find the exact solution.

$$\frac{10^x + 10^{-x}}{10^x - 10^{-x}} = 5$$

$$10^x + 10^{-x} = 5(10^x - 10^{-x})$$

$$10^x(10^x + 10^{-x}) = 5(10^x - 10^{-x})10^x$$

$$10^{2x} + 1 = 5(10^{2x} - 1)$$

$$4(10^{2x}) = 6$$

$$2(10^{2x}) = 3$$

$$(10^x)^2 = \frac{3}{2}$$

$$10^x = \sqrt{\frac{3}{2}}$$

$$x \log 10 = \log\sqrt{\frac{3}{2}}$$

$$x = \log\sqrt{\frac{3}{2}}$$

$$x = \frac{1}{2}\log\left(\frac{3}{2}\right)$$

49. Find the exact solution.

$$\frac{e^x + e^{-x}}{2} = 15$$

$$e^x\left(e^x + e^{-x}\right) = (30)e^x$$

$$e^{2x} + 1 = e^x(30)$$

$$e^{2x} - 30e^x + 1 = 0$$

Let $u = e^x$.

$$u^2 - 30u + 1 = 0$$

$$u = \frac{30 \pm \sqrt{900 - 4}}{2}$$

$$u = \frac{30 \pm \sqrt{896}}{2}$$

$$u = \frac{30 \pm 8\sqrt{14}}{2}$$

$$u = 15 \pm 4\sqrt{14}$$

$$e^x = 15 \pm 4\sqrt{14}$$

$$x \ln e = \ln(15 \pm 4\sqrt{14})$$

$$x = \ln(15 \pm 4\sqrt{14})$$

51. Find the exact solution.

$$\frac{1}{e^x - e^{-x}} = 4$$

$$1 = 4(e^x - e^{-x})$$

$$1(e^x) = 4(e^x)(e^x - e^{-x})$$

$$e^x = 4(e^{2x} - 1)$$

$$e^x = 4e^{2x} - 4$$

$$0 = 4e^{2x} - e^x - 4$$

Let $u = e^x$.

$$0 = 4u^2 - u - 4$$

$$u = \frac{1 \pm \sqrt{1 - 4(4)(-4)}}{8}$$

$$u = \frac{1 \pm \sqrt{65}}{8} \quad \text{Reject negative}$$

$$e^x = \frac{1 + \sqrt{65}}{8}$$

$$x \ln e = \ln\left(\frac{1 + \sqrt{65}}{8}\right)$$

$$x = \ln(1 + \sqrt{65}) - \ln 8$$

53. Use a graphing utility to approximate the solution.

$$2^{-x+3} = x + 1$$

Graph $f = 2^{-x+3} - (x + 1)$.

Its x-intercept is the solution.

$$x \approx 1.61$$

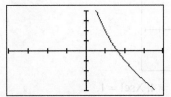

Xmin $= -4$, Xmax $= 4$, Xscl $= 1$,

Ymin $= -4$, Ymax $= 4$, Yscl $= 1$

55. Use a graphing utility to approximate the solution.

$$e^{3-2x} - 2x = 1$$

Graph $f = e^{3-2x} - 2x - 1$.

Its x-intercept is the solution.

$$x \approx 0.96$$

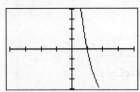

Xmin $= -4$, Xmax $= 4$, Xscl $= 1$,

Ymin $= -4$, Ymax $= 4$, Yscl $= 1$

57. Use a graphing utility to approximate the solution.

$$3 \log_2(x - 1) = -x + 3$$

Graph $f = \frac{3 \log(x - 1)}{\log 2} + x - 3$.

Its x-intercept is the solution.

$$x \approx 2.20$$

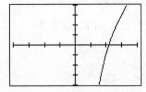

Xmin $= -4$, Xmax $= 4$, Xscl $= 1$,

Ymin $= -4$, Ymax $= 4$, Yscl $= 1$

59. Use a graphing utility to approximate the solution.

$$\ln(2x+4)+\frac{1}{2}x=-3$$

Graph $f = \ln(2x+4)+\frac{1}{2}x+3$.

Its x-intercept is the solution. $x \approx -1.93$

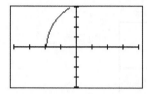

Xmin $= -4$, Xmax $= 4$, Xscl $= 1$,
Ymin $= -4$, Ymax $= 4$, Yscl $= 1$

61. Use a graphing utility to approximate the solution.

$$2^{x+1}=x^2-1$$

Graph $f = 2^{x+1}-x^2+1$.

Its x-intercept is the solution. $x \approx -1.34$

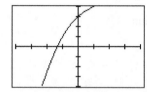

Xmin $= -4$, Xmax $= 4$, Xscl $= 1$,
Ymin $= -4$, Ymax $= 4$, Yscl $= 1$

63. a. $P(0) = 8500(1.1)^0 = 8500(1) = 8500$

$P(2) = 8500(1.1)^2 = 10{,}285$

b.
$$15{,}000 = 8500(1.1)^t$$
$$\ln 15{,}000 = 8500(1.1)^t$$
$$\ln 51{,}000 = \ln 8500 + t\ln(1.1)$$
$$\frac{\ln 15{,}000 - \ln 8500}{\ln(1.1)} = t$$
$$6 \approx t$$

The population will reach 15,000 in 6 years.

65. a. $T(10) = 36+43e^{-0.058(10)} = 36+43e^{-0.58}$

$T \approx 60°\,F$

b.
$$45 = 36+43e^{-0.058t}$$
$$\ln(45-36) = \ln 43 - 0.058t \ln e$$
$$\frac{\ln(45-36)-\ln 43}{-0.058} = t$$
$$t \approx 27 \text{ minutes}$$

67.
$$114 = 198-(198-0.9)e^{-0.23x}$$
$$-84 = -197.1e^{-0.23x}$$
$$\frac{84}{197.1} = e^{-0.23x}$$
$$\ln\left(\frac{84}{197.1}\right) = -0.23x$$
$$x = \frac{\ln\left(\frac{84}{197.1}\right)}{-0.23}$$
$$x \approx 3.7 \text{ years}$$

69. Solve $5+29\ln(t+1) = 65$ for t.

$$5+29\ln(t+1) = 65$$
$$29\ln(t+1) = 60$$
$$\ln(t+1) = \frac{60}{29}$$
$$t+1 = e^{60/29}$$
$$t = -1+e^{60/29}$$
$$t \approx 6.9 \text{ months}$$

71. Consider the first function for time less than 10 seconds.

$$275 = -2.25x^2+56.26x-0.28$$
$$0 = -2.25x^2+56.26x-275.28$$
$$x = \frac{-56.26 \pm \sqrt{(56.26)^2-4(-2.25)(-275.28)}}{2(-2.25)}$$
$$x = 6.67 \text{ or } 18.33$$

18.33 s > 10 s, so it is not a solution. The solution is 6.67 s.

Consider the second function for time greater than 10 seconds.

$$275 = 8320(0.73)^x$$
$$\frac{275}{8320} = (0.73)^x$$
$$\ln\left(\frac{275}{8320}\right) = \ln 0.73^x$$
$$\ln\left(\frac{275}{8320}\right) = x\ln 0.73$$
$$x = \frac{\ln\left(\frac{275}{8320}\right)}{\ln 0.73} \approx 10.83$$

The solutions are 6.67 s and 10.83 s.

73. a. Use a graphing utility to graph.

Hours of training

b. 48 hours

c. $P = 100$

d. As the number of hours of training increases, the test scores approach 100%.

75. a. Use a graphing utility to graph.

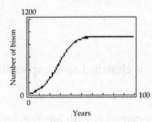

Years

b. In 27 years or 2026

c. $B = 1000$

d. As the number of years increases, the bison population approaches but never exceeds 1000.

77. a. Use a graphing utility to graph.

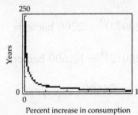

Percent increase in consumption

b. When $r = 3\%$, or 0.03, $T \approx 78$ years

c. When $T = 100$, $r \approx 0.019$, or 1.9%

79. a.
$$v = 100\left(\frac{e^{0.64t} - 1}{e^{0.64t} + 1}\right)$$

$$50 = 100\left(\frac{e^{0.64t} - 1}{e^{0.64t} + 1}\right)$$

$$\frac{50}{100} = \frac{e^{0.64t} - 1}{e^{0.64t} + 1}$$

$$0.5 = \frac{e^{0.64t} - 1}{e^{0.64t} + 1}$$

$$0.5(e^{0.64t} + 1) = e^{0.64t} - 1$$

$$0.5e0^{0.64t} + 0.5 = e^{0.64t} - 1$$

$$0.5e^{0.64t} - e^{0.64t} = -1.5$$

$$-0.5e^{0.64t} = -1.5$$

$$e^{0.64t} = 3$$

$$0.64t = \ln 3$$

$$t = \frac{\ln 3}{0.64}$$

$$t \approx 1.72$$

In approximately 1.72 seconds, the velocity will be 50 feet per second.

b. The horizontal asymptote is the value of

$$100\left[\frac{e^{0.64t} - 1}{e^{0.64t} + 1}\right] \text{ as } t \to \infty. \text{ Therefore, the horizontal}$$

asymptote is $v = 100$ feet per second.

c. The object cannot fall faster than 100 feet per second.

81. Graph $V = 400,000 - 150,000(1.005)^x$
and $V = 100,000$.

They intersect when $x \approx 138.97$.

After 138 withdrawals, the account has $101,456.39.

After 139 withdrawals, the account has $99,963.67.

The designer can make at most 138 withdrawals and still have $100,000.

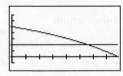

$Xmin = 0, Xmax = 200, Xscl = 25$
$Ymin = -50000, Ymax = 350000, Yscl = 50000$

83. The second step because $\log 0.5 < 0$. Thus the inequality sign must be reversed.

85. $\log(x + y) = \log x + \log y$

$\log(x + y) = \log xy$

Therefore $x + y = xy$

$$x - xy = -y$$

$$x(1 - y) = -y$$

$$x = \frac{-y}{1 - y}$$

$$x = \frac{y}{y - 1}$$

87. Since $e^{0.336} \approx 1.4,$

$$F(x) = (1.4)^x \approx (e^{0.336})^x = e^{0.336x} = G(x)$$

89. a. No. If $w = v$, the equation becomes $y = 1 - (x^2 / 4)$ and when $x = 0$, $y = 1$. So the boat will arrive at $(0, 1)$.

b. No. If $w > v$, then $1 - (w / v)$ is negative and the expression $(x / 2)^{1 - (w/v)}$ becomes larger and larger as x approaches 0. Thus, the path of the boat curves north and never reaches O.

c. Yes. If $w < v$, then $y = 0$ when $x = 0$, and the boat reaches O.

Prepare for Section 4.6

P1. Evaluate. Round to the nearest hundredth.

$$A = 1000\left(1 + \frac{0.1}{12}\right)^{12(2)} = 1220.39$$

P3. Solve. Round to the nearest ten-thousandth.

$$0.5 = e^{14k}$$

$$\ln 0.5 = \ln e^{14k}$$

$$\ln 0.5 = 14k$$

$$\frac{\ln 0.5}{14} = k$$

$$-0.0495 \approx k$$

P5. Solve. Round to the nearest thousandth.

$$6 = \frac{70}{5 + 9e^{-12k}}$$

$$6(5 + 9e^{-12k}) = 70$$

$$30 + 54e^{-12k} = 70$$

$$54e^{-12k} = 40$$

$$e^{-12k} = \frac{20}{27}$$

$$\ln e^{-12k} = \ln \frac{20}{27}$$

$$-12k = \ln \frac{20}{27}$$

$$k = -\frac{1}{12}\ln \frac{20}{27}$$

$$k \approx 0.025$$

Section 4.6 Exercises

1. State how $N(t) = N_0 e^{kt}$ is classified as an exponential growth or decay function.

If $k > 0$, then N is an exponential growth function and if $k < 0$, then N is an exponential decay function.

3. Define compound continuously.

To compound continuously means to increase the number of compounding periods per year, without bound.

5. a. $t = 0$ hours, $N(0) = 2200(2)^0 = 2200$ bacteria

b. $t = 3$ hours, $N(3) = 2200(2)^3 = 17,600$ bacteria

7. a. $N(t) = N_0 e^{kt}$ where $N_0 = 24600$

$$N(5) = 22,600e^{k(5)}$$

$$24,200 = 22,600e^{5k}$$

$$\frac{24,200}{22,600} = e^{5k}$$

$$\ln\left(\frac{24,200}{22,600}\right) = \ln\left(e^{5k}\right)$$

$$\ln\left(\frac{24,200}{22,600}\right) = 5k$$

$$\frac{1}{5}\left[\ln\frac{24,200}{22,600}\right] = k$$

$$0.01368 \approx k$$

$$N(t) = 22,600e^{0.01368t}$$

b. $t = 15$

$$N(15) = 22,600e^{0.01368(15)}$$

$$= 22,600e^{0.2052}$$

$$\approx 27,700$$

9. $N(t) = N_0 e^{kt}$ where $N_0 = 143,110$

$$N(10) = 143,110e^{k(10)}$$

$$212,375 = 143,110e^{10k}$$

$$\frac{212,375}{143,110} = e^{10k}$$

$$\ln\left(\frac{212,375}{143,110}\right) = \ln\left(e^{10k}\right)$$

$$\ln\left(\frac{212,375}{143,110}\right) = 10k$$

$$\frac{1}{10}\left[\ln\frac{212,375}{143,110}\right] = k$$

$$0.0394740015 \approx k$$

$$N(t) = 143,110e^{0.0394740015\,t}$$

$$t = 16$$

$$N(16) = 143,110e^{0.0394740015(16)}$$

$$= 143,110e^{0.63158}$$

$$\approx 269,000$$

11. a. Graph the equation.

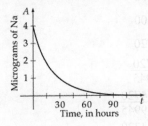

b. $A(5) = 4e^{-0.23} \approx 3.18$ micrograms

c. Since $A = 4$ micrograms are present when $t = 0$, find the time t at which half remains—that is when $A = 2$.

$$2 = 4e^{-0.046t}$$

$$\frac{1}{2} = e^{-0.046t}$$

$$\ln\left(\frac{1}{2}\right) = -0.046t$$

$$\frac{\ln\left(\frac{1}{2}\right)}{-0.046} = t$$

$$15.07 \approx t$$

The half-life of sodium-24 is about 15.07 hours.

d. $1 = 4e^{-0.046t}$

$$\frac{1}{4} = e^{-0.046t}$$

$$\ln\left(\frac{1}{4}\right) = -0.046t$$

$$\frac{\ln\left(\frac{1}{4}\right)}{-0.046t} = t$$

$$30.14 \approx t$$

The amount of sodium-24 will be 1 microgram after 30.14 hours.

13. $$N(t) = N_0(0.5)^{t/5730}$$

$$N(t) = 0.45N_0$$

$$0.45N_0 = N_0(0.5)^{t/5730}$$

$$\ln(0.45) = \frac{t}{5730}\ln 0.5$$

$$5730\frac{\ln 0.45}{\ln 0.5} = t$$

$$6601 \approx t$$

The bone is about 6601 years old.

15. $$N(t) = N_0(0.5)^{t/5730}$$

$$N(t) = 0.75N_0$$

$$0.75N_0 = N_0(0.5)^{t/5730}$$

$$\ln 0.75 = \frac{t}{5730}\ln 0.5$$

$$5730\frac{\ln 0.75}{\ln 0.5} = t$$

$$2378 \approx t$$

The Rhind papyrus is about 2378 years old.

17. Find the balance.

a. $P = 4500$, $r = 0.025$, $t = 5$, $n = 1$

$$B = 4500\left(1 + \frac{0.025}{1}\right)^5 \approx \$5091.34$$

b. $P = 4500$, $r = 0.025$, $t = 12$, $n = 1$

$$B = 4500\left(1 + \frac{0.025}{1}\right)^{12} \approx \$6052.00$$

19. Find the balance.

a. $P = 22,000$, $r = 0.0275$, $t = 5$, $n = 12$

$$B = 22,000\left(1 + \frac{0.0275}{12}\right)^{5(12)} \approx \$25,238.87$$

b. $P = 22,000$, $r = 0.0275$, $t = 5$, $n = 365$

$$B = 22,000\left(1 + \frac{0.0275}{365}\right)^{5(365)} \approx \$25,242.71$$

21. Find the balance.

$P = 3200, \ r = 0.04, \ t = 10$

$B = 3200e^{10(0.04)} \approx \$4773.84$

23. Find the doubling time. Round to the nearest tenth of a year.

$r = 0.020, \ n = 365$

$$2P = P\left(1 + \frac{0.020}{365}\right)^{t(365)}$$

$$2 = \left(1 + \frac{0.020}{365}\right)^{365t}$$

$$\ln 2 = \ln\left(1 + \frac{0.020}{365}\right)^{365t}$$

$$\ln 2 = 365t \ln\left(1 + \frac{0.020}{365}\right)$$

$$365t = \frac{\ln 2}{\ln\left(1 + \frac{0.020}{365}\right)}$$

$$t = \frac{1}{365} \cdot \frac{\ln 2}{\ln\left(1 + \frac{0.020}{365}\right)}$$

$$t \approx 34.658$$

It will take about 34.7 years to double your money.

25. $B = Pe^{rt}$ Let $B = 3P$

$3P = Pe^{rt}$

$3 = e^{rt}$

$\ln 3 = rt \ln e$

$t = \dfrac{\ln 3}{r}$

27. $t = \dfrac{\ln 3}{r}$ $r = 0.076$

$t = \dfrac{\ln 3}{0.076}$

$t \approx 14$ years

29. a. Find the carrying capacity.

1900

b. Find the growth rate constant.

0.16

c. Find the initial population.

$$P(0) = \frac{1900}{1 + 8.5e^{-0.16(0)}} = 200$$

31. a. Find the carrying capacity.

157,500

b. Find the growth rate constant.

0.04

c. Find the initial population.

$$P(0) = \frac{157,500}{1 + 2.5e^{-0.04(0)}} = 45,000$$

33. a. Find the carrying capacity.

2400

b. Find the growth rate constant.

0.12

c. Find the initial population.

$$P(0) = \frac{2400}{1 + 7e^{-0.12(0)}} = 300$$

35. $a = \dfrac{c - P_0}{P_0} = \dfrac{5500 - 400}{400} = 12.75$

$$P(t) = \frac{c}{1 + ae^{-bt}}$$

$$P(2) = \frac{5500}{1 + 12.75e^{-b(2)}}$$

$$780 = \frac{5500}{1 + 12.75e^{-2b}}$$

$$780\left(1 + 12.75e^{-2b}\right) = 5500$$

$$780 + 9945e^{-2b} = 5500$$

$$9945e^{-2b} = 4720$$

$$e^{-2b} = \frac{4720}{9945}$$

$$\ln e^{-2b} = \ln \frac{4720}{9945}$$

$$-2b = \ln \frac{4720}{9945}$$

$$b = -\frac{1}{2}\ln \frac{4720}{9945}$$

$$b \approx 0.37263$$

$$P(t) = \frac{5500}{1 + 12.75e^{-0.37263t}}$$

37. $a = \dfrac{c - P_0}{P_0} = \dfrac{100 - 18}{18} = 4.55556$

$$P(t) = \dfrac{c}{1 + ae^{-bt}}$$

$$P(3) = \dfrac{100}{1 + 4.55556e^{-b(3)}}$$

$$30 = \dfrac{100}{1 + 4.55556e^{-3b}}$$

$$30\left(1 + 4.55556e^{-3b}\right) = 100$$

$$30 + 136.67e^{-3b} = 100$$

$$136.67e^{-3b} = 70$$

$$e^{-3b} = \dfrac{70}{136.67}$$

$$\ln e^{-3b} = \ln \dfrac{70}{136.67}$$

$$-3b = \ln \dfrac{70}{136.67}$$

$$b = -\dfrac{1}{3}\ln \dfrac{70}{136.67}$$

$$b \approx 0.22302$$

$$P(t) = \dfrac{100}{1 + 4.55556e^{-0.22302t}}$$

39. a. $R(t) = \dfrac{625{,}000}{1 + 3.1e^{-0.045t}}$

$$R(1) = \dfrac{625{,}000}{1 + 3.1e^{-0.045(1)}} \approx \$158{,}000$$

$$R(2) = \dfrac{625{,}000}{1 + 3.1e^{-0.045(2)}} \approx \$163{,}000$$

b. $R(t) = \dfrac{625{,}000}{1 + 3.1e^{-0.045t}}$, as

$t \to \infty,\ R(t) \to \$625{,}000$

41. a. $a = \dfrac{c - P_0}{P_0} = \dfrac{1600 - 312}{312} = 4.12821$

$$P(t) = \dfrac{c}{1 + ae^{-bt}}$$

$$P(6) = \dfrac{1600}{1 + 4.12821e^{-b(6)}}$$

$$416 = \dfrac{1600}{1 + 4.12821e^{-6b}}$$

$$416\left(1 + 4.12821e^{-6b}\right) = 1600$$

$$416 + 1717.34e^{-6b} = 1600$$

$$1717.34e^{-6b} = 1184$$

$$e^{-6b} = \dfrac{1184}{1717.34}$$

$$\ln e^{-6b} = \ln \dfrac{1184}{1717.34}$$

$$-6b = \ln \dfrac{1184}{1717.34}$$

$$b = -\dfrac{1}{6}\ln \dfrac{1184}{1717.34}$$

$$b \approx 0.06198$$

$$P(t) = \dfrac{1600}{1 + 4.12821e^{-0.06198t}}$$

b. $P(10) = \dfrac{1600}{1 + 4.12821e^{-0.06198(10)}} \approx 497$ wolves

43. a. $a = \dfrac{c - P_0}{P_0} = \dfrac{8500 - 1500}{1500} = 4.66667$

$$P(t) = \dfrac{c}{1 + ae^{-bt}}$$

$$P(2) = \dfrac{8500}{1 + 4.66667e^{-b(2)}}$$

$$1900 = \dfrac{8500}{1 + 4.66667e^{-2b}}$$

$$1900\left(1 + 4.66667e^{-2b}\right) = 8500$$

$$1900 + 8866.673e^{-2b} = 8500$$

$$8866.673e^{-2b} = 6600$$

$$e^{-2b} = \dfrac{6600}{8866.673}$$

$$\ln e^{-2b} = \ln \dfrac{6600}{8866.673}$$

$$-2b = \ln \dfrac{6600}{8866.673}$$

$$b = -\dfrac{1}{2}\ln \dfrac{6600}{8866.673}$$

$$b \approx 0.14761$$

$$P(t) = \dfrac{8500}{1 + 4.66667e^{-0.14761t}}$$

b.
$$4000 = \dfrac{8500}{1 + 4.66667e^{-0.14761t}}$$

$$4000\left(1 + 4.66667e^{-0.14761t}\right) = 8500$$

$$1 + 4.66667e^{-0.14761t} = 2.125$$

$$4.66667e^{-0.14761t} = 1.125$$

$$e^{-0.14761t} = \dfrac{1.125}{4.66667}$$

$$4000\left(1 + 4.66667e^{-0.14761t}\right) = 8500$$

$$1 + 4.66667e^{-0.14761t} = 2.125$$

$$4.66667e^{-0.14761t} = 1.125$$

$$e^{-0.14761t} = \frac{1.125}{4.66667}$$

$$\ln e^{-0.14761t} = \ln \frac{1.125}{4.66667}$$

$$-0.14761t = \ln \frac{1.125}{4.66667}$$

$$t = -\frac{1}{0.14761} \ln \frac{1.125}{4.66667}$$

$$t \approx 9.6$$

The population will exceed 4000 in $2010 + 9 = 2019$.

45. a. $A = 34°F$, $T_0 = 75°F$, $T_t = 65°F$, $t = 5$. Find k.

$$65 = 34 + (75 - 34)e^{-5k}$$

$$31 = 41e^{-5k}$$

$$\frac{31}{41} = e^{-5k}$$

$$\ln\left(\frac{31}{41}\right) = -5k$$

$$k = -\frac{1}{5}\ln\left(\frac{31}{41}\right)$$

$$k \approx 0.056$$

b. $A = 34°F$, $k = 0.056$, $T_0 = 75°F$, $t = 30$

$$T_t = 34 + (75 - 34)e^{-30(0.056)}$$

$$T_t = 34 + (41)e^{-1.68}$$

$$T_t \approx 42°F$$

c. $T_t = 36°F$, $k = 0.056$, $T_t = 75°F$, $A = 34°F$

$$36 = 34 + (75 - 34)e^{-0.056t}$$

$$2 = 41e^{-0.056t}$$

$$t \approx 54 \text{ minutes}$$

47. a. 10% of 80,000 is 8000.

$$8000 = 80,000\left(1 - e^{-0.0005t}\right)$$

$$0.1 = 1 - e^{-0.0005t}$$

$$-0.9 = -e^{-0.0005t}$$

$$0.9 = e^{-0.0005t}$$

$$\ln 0.9 = -0.0005t \ln e$$

$$\ln 0.9 = -0.0005t$$

$$\frac{\ln 0.9}{-0.0005} = t$$

$$211 \text{ h} \approx t$$

b. 50% of 80,000 is 40,000.

$$40,000 = 80,000\left(1 - e^{-0.0005t}\right)$$

$$0.5 = 1 - e^{-0.0005t}$$

$$-0.5 = -e^{-0.0005t}$$

$$0.5 = e^{-0.0005t}$$

$$\ln 0.5 = \ln\left(e^{-0.0005t}\right)$$

$$\ln 0.5 = -0.0005t$$

$$\frac{\ln 0.5}{-0.0005} = t$$

$$1386 \text{ h} \approx t$$

49. $V(t) = V_0(1 - r)^t$

$$0.5V_0 = V_0(1 - 0.20)^t$$

$$0.5 = (1 - 0.20)^t$$

$$0.5 = 0.8^t$$

$$\ln 0.5 = \ln 0.8^t$$

$$\ln 0.5 = t \ln 0.8$$

$$\frac{\ln 0.5}{\ln 0.8} = t$$

$$3.1 \text{ years} \approx t$$

51. a. Graph the equation.

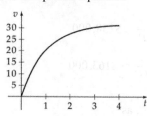

b. $$20 = 32(1 - e^{-t})$$

$$0.625 = 1 - e^{-t}$$

$$e^{-t} = 0.375$$

$$-t = \ln 0.375$$

$$t \approx 0.98 \text{ seconds}$$

c. The horizontal asymptote is $v = 32$.

d. As time increases, the object's velocity approaches but never exceeds 32 ft/sec.

53. a. Graph the equation.

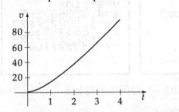

b. The graphs of $s = 32t + 32(e^{-t} - 1)$ and $s = 50$ intersect when $t \approx 2.5$ seconds.

c. The slope m of the secant line containing $(1, s(1))$ and $(2, s(2))$ is $m = \dfrac{s(2) - s(1)}{2 - 1} \approx 24.56$ ft/sec

d. The average speed of the object was 24.56 feet per second between $t = 1$ and $t = 2$.

55. Use the graph to solve.

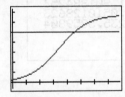

Xmin $= 0$, Xmax $= 80$, Xscl $= 10$,
Ymin $= -10$, Ymax $= 110$, Yscl $= 15$
When $P = 75\%$, $t \approx 45$ hours.

57. a. $A(1) = 0.5^{1/2}$

≈ 0.71 gram

b. $A(4) = 0.5^{4/2} + 0.5^{(4-3)/2}$

$= 0.5^2 + 0.5^{1/2}$

≈ 0.96 gram

c. $A(9) = 0.5^{9/2} + 0.5^{(9-3)/2} + 0.5^{(9-6)/2}$

$= 0.5^{4.5} + 0.5^3 + 0.5^{1.5}$

≈ 0.52 gram

59. $N(t) = 22{,}755e^{0.0287\,t}$

$= 22{,}755\left(e^{0.0287}\right)^t$

$\approx 22{,}755(1.0291)^t$

The annual growth rate is 2.91%.

61. a. $P_0 = \dfrac{1000}{1 + (-0.3333)} = \dfrac{1000}{0.6667} = 1500$ fish

b. As $t \to \infty$, $P(t) \to 1000$ fish.

63. a. $WR(107) = \dfrac{199.13}{1 + (-0.21726)(0.42536)}$

$= 219.41$ sec $= 3$ min, 39.41 sec

$WR(137) = \dfrac{199.13}{1 + (-0.21726)(0.33471)}$

$= 214.75$ sec $= 3$ min, 34.75 sec

b. As $t \to \infty$, $WR(t) \to 199.13$ sec $= 3$ min, 19.13 sec.

Prepare for Section 4.7

P1. This is a decreasing function.

P3. Evaluate.

$P(0) = \dfrac{108}{1 + 2e^{-0.1(0)}} = \dfrac{108}{1 + 2} = 36$

P5. Solve. Round to the nearest tenth.

$$10 = \frac{20}{1 + 2.2e^{-0.05t}}$$

$$10(1 + 2.2e^{-0.05t}) = 20$$

$$10 + 22e^{-0.05t} = 20$$

$$e^{-0.05t} = \frac{10}{22}$$

$$\ln e^{-0.05t} = \ln \frac{10}{22}$$

$$-0.05t = \ln \frac{10}{22}$$

$$t = -20 \ln \frac{10}{22}$$

$$t \approx 15.8$$

Section 4.7 Exercises

1. The curve is concave up.

3. a. The domain is all positive real numbers.

b. The function is decreasing.

c. The function is concave downward.

5. Use a scatter plot to find the model function.

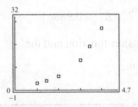

increasing exponential function

7. Use a scatter plot to find the model function.

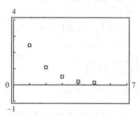

decreasing exponential function;

decreasing logarithmic function

9. Use a scatter plot to find the model function.

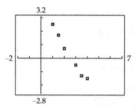

decreasing logarithmic function

11. Find the exponential regression function and the coefficient of determination.

TI:

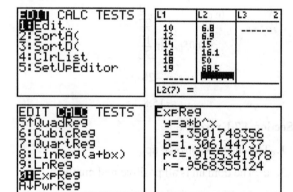

$y \approx 0.3501748356(1.306144737)^x$;

$r^2 \approx 0.9155341978$

WA: $y \approx 0.0681768e^{0.36394x}$; $R^2 \approx 0.988617$

13. Find the logarithmic regression function and the coefficient of determination.

TI:

 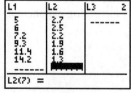

 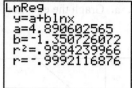

$y \approx 4.890602565 - 1.350726072 \ln x$

$r^2 \approx 0.9984239966$

WA: $y \approx -1.35073\ln(0.0267634x)$; $R^2 \approx 0.999914$

15. Find the logistic regression function.

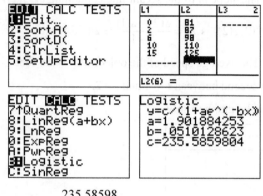

$y \approx \dfrac{235.58598}{1 + 1.90188e^{-0.05101x}}$

17. a. Find the exponential regression function.

TI:

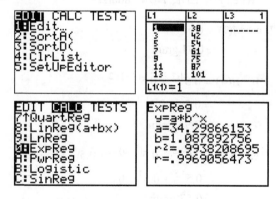

$y \approx 34.29866153(1.087892756)^x$

WA: $y \approx 34.6224e^{0.0831662x}$

b. Find the price for 2016. Round to the nearest dollar.

TI: $y \approx 34.29866153(1.087892756)^{16} \approx \$132$

WA: $y \approx 34.6224e^{0.0831662(16)} \approx \$131$

19. a. Find the exponential regression model.

TI:

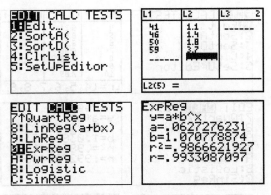

$$T(F) \approx 0.0627276231(1.079805006)^F$$

WA: $T(F) \approx 0.048971e^{0.0731739F}$

b. Estimate the time. Round to the nearest tenth of an hour.

TI: $T(34) \approx 0.0627276231(1.079805006)^{34}$
≈ 5.3 hours

WA: $T(65) \approx 0.048971e^{0.0731739(65)} \approx 5.7$ hours

21. a. Find the exponential regression model.

TI:

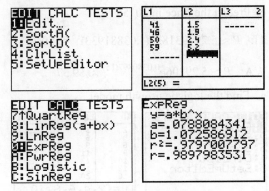

$$T \approx 0.0788084341(1.072586912)^F$$

WA: $T \approx 0.0568099e^{0.076395F}$

b. Estimate the time. Round to the nearest tenth of an hour. Compare to Exercise 19b.

TI: $T \approx 0.0788084341(1.072586912)^{65} \approx 7.5$ hours
$7.5 - 5.3 = 2.2$ hours

WA: $T \approx 0.0568099e^{0.076395(65)} \approx 8.1$ hours
$8.1 - 5.7 = 2.4$ hours

23. a. Find exponential and logarithmic regression models.

TI:

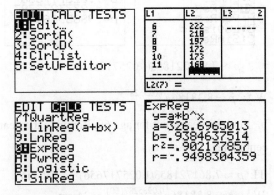

Exponential: $P \approx 326.6965013(0.9384637514)^t$

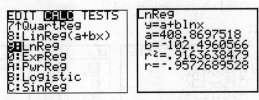

Logarithmic: $P \approx 408.8697518 - 102.4960566 \ln t$

WA:

Exponential: $P \approx 330.965e^{-0.0649969t}$

Logarithmic: $P \approx -102.496 \ln(0.0185159t)$

b. Estimate the price in 2014 using exponential model. Round to the nearest thousand dollars.

TI: $P \approx 326.6965013(0.9384637514)^{14} \approx \$134,000$

WA: $P \approx 330.965e^{-0.0649969(14)} \approx \$133,000$

c. Estimate the price in 2014 using exponential model. Round to the nearest thousand dollars.

TI: $P \approx 408.8697518 - 102.4960566 \ln 14 \approx \$138,000$

WA: $P \approx -102.496 \ln(0.0185159(14)) \approx \$138,000$

25. a. Find the exponential regression model.

TI:

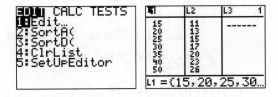

$$p \approx 7.861721853(1.025717656)^y$$

WA: $p \approx 8.38186e^{0.0235287y}$

b. Find the near point for age 60. Round to the nearest centimeter.

TI: $p \approx 7.861721853(1.025717656)^{60} \approx 36$ cm
WA: $p \approx 8.38186e^{0.0235287(60)} \approx 34$ cm

27. a. Find a logarithmic model.

TI:

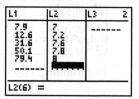
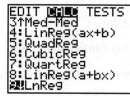
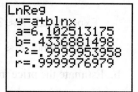

Logarithmic: $\text{pH} \approx 6.102513175 + 0.4336881498\ln q$

WA: $\text{pH} \approx 0.433688\ln(1.29135 \times 10^6 q)$

b. Find q when pH is 8.2.

TI:

$$8.2 \approx 6.102513175 + 0.4336881498\ln q$$
$$2.097486825 \approx 0.4336881498\ln q$$
$$\frac{2.097486825}{0.4336881498} \approx \ln q$$
$$q \approx e^{\frac{2.097486825}{0.4336881498}} \approx 126.0$$

WA:

$$8.2 \approx 0.433688\ln(1.29135 \times 10^6 q)$$
$$\frac{8.2}{0.433688} \approx \ln(1.29135 \times 10^6 q)$$
$$1.29135 \times 10^6 q \approx e^{\frac{8.2}{0.433688}}$$
$$q \approx \frac{e^{\frac{8.2}{0.433688}}}{1.29135 \times 10^6} \approx 126.0$$

29. a. Find the exponential and logarithmic models.

TI:

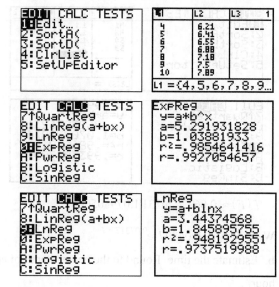

Exponential: $P \approx 5.291931828(1.03881933)^t$;

Logarithmic: $P \approx 3.44374568 + 1.845895755\ln t$

WA: Exponential: $P \approx 5.29206e^{0.0380897t}$;

Logarithmic: $P \approx 1.8459\ln(6.45996t)$

b. The exponential model provides the better fit.

c. Predict the price in 2015.

TI: $P \approx 5.291931828(1.03881933)^{15} \approx \$9.37$

WA: $P \approx 5.29206e^{0.0380897(15)} \approx \$9.37$

31. a. Find the logistic growth model.

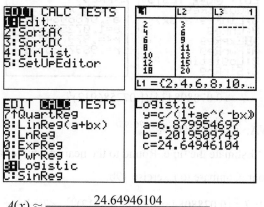

$$A(x) \approx \frac{24.64946104}{1 + 6.879954697e^{-0.2019509749x}}$$

b. Estimate the human age of 5-month-old cat. Round to the nearest tenth of a year. Is this realistic?

$$A(5) \approx \frac{24.64946104}{1 + 6.879954697e^{-0.2019509749(5)}} \approx 7.0 \text{ years}$$

Answers may vary.

c. Estimate the human age of 72-month-old cat. Round to the nearest tenth of a year. Is this realistic?

$$A(72) \approx \frac{24.64946104}{1 + 6.879954697e^{-0.2019509749(72)}} \approx 24.6 \text{ years}$$

Answers may vary.

d. Answers may vary.

33. Determine whether A and B have the same function.

For A:

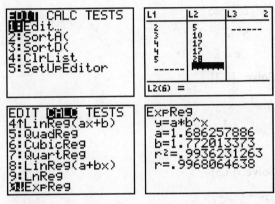

For B:

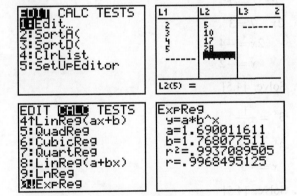

A and B have different exponential regression functions.

35. a. Find the exponential and power regression functions.

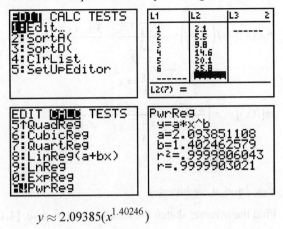

$$y \approx 2.09385(x^{1.40246})$$

37. Determine the x-coordinate of the inflection point of f.

$$f(x) = \frac{c}{1 + ae^{-bx}}$$

$$\frac{c}{2} = \frac{c}{1 + ae^{-bx}}$$

$$2 = 1 + ae^{-bx}$$

$$1 = ae^{-bx}$$

$$\frac{1}{a} = e^{-bx}$$

$$\ln\left(\frac{1}{a}\right) = \ln e^{-bx}$$

$$\ln 1 - \ln a = -bx$$

$$-\ln a = -bx$$

$$\frac{\ln a}{b} = x$$

Chapter 4 Review Exercises

1. Draw the graph of the inverse. [4.1]

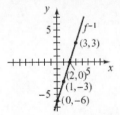

3. Determine whether the functions are inverses. [4.1]

$$F[G(x)] = F\left(\frac{x+5}{2}\right) = 2\left(\frac{x+5}{2}\right) - 5 = x + 5 - 5 = x$$

$$G[F(x)] = G(2x - 5) = \frac{2x - 5 + 5}{2} = \frac{2x}{2} = x$$

Yes, F and G are inverses.

5. Determine whether the functions are inverses. [4.1]

$$l[m(x)] = l\left(\frac{3}{x-1}\right) = \frac{\dfrac{3}{x-1}+3}{\dfrac{3}{x-1}} = \frac{3+3(x-1)}{3}$$

$$= \frac{3+3x-3}{3} = \frac{3x}{3} = x$$

$$m[l(x)] = m\left(\frac{x+3}{x}\right) = \frac{3}{\dfrac{x+3}{x}-1} = \frac{3x}{x+3-x}$$

$$= \frac{3x}{3} = x$$

Yes, l and m are inverses.

7. Find the inverse; sketch the function and inverse. [4.1]

$$y = 3x - 4$$
$$x = 3y - 4$$
$$x + 4 = 3y$$
$$\frac{x+4}{3} = y$$
$$f^{-1}(x) = \frac{1}{3}x + \frac{4}{3}$$

9. Find the inverse; sketch the function and inverse. [4.1]

$$y = -\frac{1}{2}x - 2$$
$$x = -\frac{1}{2}y - 2$$
$$x + 2 = -\frac{1}{2}y$$
$$-2(x+2) = y$$
$$h^{-1}(x) = -2x - 4$$

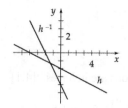

11. Find the inverse. [4.1]

$$f(x) = \frac{2x}{x-1}, \; x > 1$$
$$x = \frac{2y}{y-1}$$
$$x(y-1) = 2y$$
$$xy - x = 2y$$
$$xy - 2y = x$$
$$y(x-2) = x$$
$$y = \frac{x}{x-2}$$
$$f^{-1}(x) = \frac{x}{x-2}, \; x > 2$$

13. Solve. [4.3]

$$\log_5 25 = x$$
$$5^x = 25$$
$$5^x = 5^2$$
$$x = 2$$

15. Solve. [4.3]

$$\ln e^3 = x$$
$$e^x = e^3$$
$$x = 3$$

17. Solve. [4.5]

$$3^{2x+7} = 27$$
$$3^{2x+7} = 3^3$$
$$2x + 7 = 3$$
$$2x = -4$$
$$x = -2$$

19. Solve. [4.5]

$$3^x = \frac{1}{243}$$
$$3^x = 3^{-5}$$
$$x = -5$$

21. Solve. [4.5]

$$\log x^2 = 2$$
$$10^2 = x^2$$
$$100 = x^2$$
$$\pm\sqrt{100} = x$$
$$\pm 10 = x$$

23. Solve. [4.5]

$$10^{\log 2x} = 14$$
$$2x = 14$$
$$x = 7$$

25. Sketch the graph of the function.

$$f(x) = (2.5)^x$$

27. Sketch the graph of the function.

$$f(x) = 3^{|x|}$$

29. Sketch the graph of the function.

$$f(x) = 2^x - 3$$

31. Sketch the graph of the function.

$$f(x) = \log_5 x$$

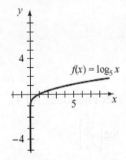

33. Sketch the graph of the function.

$$f(x) = \frac{1}{3}\log x$$

35. Sketch the graph of the function.

$$f(x) = -\frac{1}{2}\ln x$$

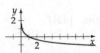

37. Use a graphing utility to graph the function.

$$f(x) = \frac{4^x + 4^{-x}}{2}$$

39. Change to exponential form. [4.3]

$$\log 1000 = 3$$
$$10^3 = 1000$$

41. Change to exponential form. [4.3]

$$\log_{\sqrt{2}} 4 = 4$$
$$\left(\sqrt{2}\right)^4 = 4$$

43. Change to logarithmic form. [4.3]

$$5^3 = 125$$
$$\log_5 125 = 3$$

45. Change to logarithmic form. [4.3]

$$10^0 = 1$$
$$\log_{10} 1 = 0$$

47. Expand the expression. [4.4]

$$\log_b\left(\frac{x\sqrt{y}}{z^3}\right) = \log_b x + \log_b y^{1/2} - \log_b z^3$$
$$= \log_b x + \frac{1}{2}\log_b y - 3\log_b z$$

49. Expand the expression. [4.4]

$$\ln xy^3 = \ln x + 3\ln y$$

51. Write as a single logarithm with coefficient 1. [4.4]

$$2\log x + \frac{1}{3}\log(x+1) = \log\left(x^2\sqrt[3]{x+1}\right)$$

53. Write as a single logarithm with coefficient 1. [4.4]

$$\frac{1}{2}\ln 2xy - 3\ln z = \ln\frac{\sqrt{2xy}}{z^3}$$

55. Approximate to six significant digits. [4.4]

$$\log_2 551 = \frac{\log 551}{\log 2} \approx 9.10591$$

57. Approximate to six significant digits. [4.4]

$$\log_4 0.85 = \frac{\log 0.85}{\log 4} \approx -0.117233$$

59. Solve for x. [4.5]

$$4^x = 30$$
$$\log 4^x = \log 30$$
$$x \log 4 = \log 30$$
$$x = \frac{\log 30}{\log 4}$$

61. Solve for x. [4.5]

$$\ln(3x) - \ln(x-1) = \ln 4$$
$$\ln \frac{3x}{x-1} = \ln 4$$
$$\frac{3x}{x-1} = 4$$
$$3x = 4(x-1)$$
$$3x = 4x - 4$$
$$4 = x$$

63. Solve for x. [4.5]

$$e^{\ln(x+2)} = 6$$
$$(x+2) = 6$$
$$x + 2 = 6$$
$$x = 4$$

65. Solve for x. [4.5]

$$\frac{4^x + 4^{-x}}{4^x - 4^{-x}} = 2$$
$$4^x \left(4^x + 4^{-x}\right) = 2\left(4^x - 4^{-x}\right)4^x$$
$$4^{2x} + 1 = 2\left(4^{2x} - 1\right)$$
$$4^{2x} + 1 = 2\left(4^{2x} - 1\right)$$
$$4^{2x} - 2 \cdot 4^{2x} + 3 = 0$$
$$4^{2x} = 3$$
$$2^x \ln 4 = \ln 3$$
$$x = \frac{\ln 3}{2 \ln 4}$$

67. Solve for x. [4.5]

$$\log(\log x) = 3$$
$$10^3 = \log x$$
$$10^{(10^3)} = x$$
$$10^{1000} = x$$

69. Solve for x. [4.5]

$$\log \sqrt{x-5} = 3$$
$$10^3 = \sqrt{x-5}$$
$$10^6 = x - 5$$
$$10^6 + 5 = x$$
$$x = 1,000,005$$

71. Solve for x. [4.5]

$$\log_4(\log_3 x) = 1$$
$$4 = \log_3 x$$
$$3^4 = x$$
$$81 = x$$

73. Solve for x. [4.5]

$$\log_5 x^3 = \log_5 16x$$
$$x^3 = 16x$$
$$x^2 = 16$$
$$x = 4 \quad \text{Reject negative}$$

75. Find the magnitude. [4.4]

$$m = \log\left(\frac{I}{I_0}\right)$$
$$= \log\left(\frac{63,280,000 I_0}{I_0}\right)$$
$$= \log 63,280,000$$
$$\approx 7.8$$

77. Find the ratio of larger to smaller intensities. [4.4]

$$\log\left(\frac{I_1}{I_0}\right) = 7.2 \quad \text{and} \quad \log\left(\frac{I_2}{I_0}\right) = 3.7$$
$$\frac{I_1}{I_0} = 10^{7.2} \qquad \qquad \frac{I_2}{I_0} = 10^{3.7}$$
$$I_1 = 10^{7.2} I_0 \qquad \qquad I_2 = 10^{3.7} I_0$$

$$\frac{I_1}{I_2} = \frac{10^{7.2} I_0}{10^{3.7} I_0} = \frac{10^{3.5}}{1} \approx \frac{3162}{1}$$

3162 to 1

79. Find the pH. [4.4]

$$\text{pH} = -\log\left[\text{H}_3\text{O}^+\right]$$
$$= -\log\left[3.16 \times 10^{-11}\right]$$
$$\approx 10.5$$

81. Find the balance. [4.6]

$P = 3750$, $r = 0.025$, $t = 5$

a. $B = 3750\left(1 + \dfrac{0.025}{12}\right)^{5(12)} \approx \$4248.75$

b. $B = 3750\left(1 + \dfrac{0.025}{365}\right)^{5(365)} \approx \$4249.29$

c. $B = 3750e^{0.025(5)} \approx \$4249.31$

83. Find the value. [4.6]

$$S(n) = P(1-r)^n, \; P = 12,400, \; r = 0.29, \; t = 3$$
$$S(n) = 12,400(1 - 0.29)^3 \approx \$4438.10$$

85. Find the exponential growth or decay function. [4.6]

$N(0) = 1$ $\qquad$ $N(2) = 5$

$1 = N_0 e^{k(0)}$ $\qquad$ $5 = e^{2k}$

$1 = N_0$ $\qquad\qquad$ $\ln 5 = 2k$

$\qquad\qquad\qquad$ $k = \dfrac{\ln 5}{2} \approx 0.8047$

Thus $N(t) = e^{0.8047t}$

87. Find the exponential growth or decay function. [4.6]

$4 = N(1) = N_0 e^k$ and thus $\dfrac{4}{N_0} = e^k$. Now, we also

have $N(5) = 5 = N_0 e^{5k} = N_0\left(\dfrac{4}{N_0}\right)^5 = \dfrac{1024}{N_0^4}$.

$N_0 = \sqrt[4]{\dfrac{1024}{5}} \approx 3.783$

Then $4 = 3.783e^k$

$\qquad k = \ln\left(\dfrac{4}{3.783}\right) \approx 0.0558$

Thus $N_0 = 3.783e^{0.0558t}$.

89. a. Find the exponential growth function. [4.6]

$$N(1) = 25,200e^{k(1)} = 26,800$$
$$e^k = \frac{26,800}{25,200}$$
$$\ln e^k = \ln\left(\frac{26,800}{25,200}\right)$$
$$k \approx 0.061557893$$

$N(t) = 25,200e^{0.061557893\,t}$

b. $N(7) = 25,200e^{0.061557893(7)}$

$\qquad = 25,200e^{0.430905251}$

$\qquad \approx 38,800$

91. a. Find the exponential and logarithmic models. [4.7]

TI:

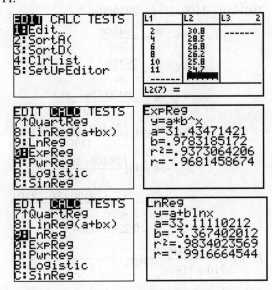

Exponential: $N \approx 31.43471421(0.9783185172)^t$;

$\qquad\qquad r^2 \approx 0.9373064206$

Logarithmic: $N \approx 33.11110212 - 3.367402012\ln t$

$\qquad\qquad r^2 \approx 0.9834023567$

WA: Exponential: $N \approx 31.546e^{-0.0224299t}$

$\qquad\qquad R^2 \approx 0.999662$

$\qquad$ Logarithmic: $N \approx -3.3674\ln(0.0000536605t)$

$\qquad\qquad R^2 \approx 0.99991$

b. Since the coefficient of determination is closer to 1 for the logarithmic model, it provides a better fit to the data.

c. Predict the admissions for 2015 using model from part b.

TI: $N \approx 33.11110212 - 3.367402012 \ln(15)$
≈ 24.0 million admissions per week

WA: $N \approx -3.3674 \ln(0.0000536605(15))$
≈ 24.0 million admissions per week

93. a. Find the logistic model. [4.6]

$$P(t) = \frac{mP_0}{P_0 + (m - P_0)e^{-kt}}$$

$$P(3) = 360 = \frac{1400(210)}{210 + (1400 - 210)e^{-k(3)}}$$

$$360 = \frac{294000}{210 + 1190e^{-3k}}$$

$$360(210 + 1190e^{-3k}) = 294000$$

$$210 + 1190e^{-3k} = \frac{294000}{360}$$

$$1190e^{-3k} = \frac{29400}{36} - 210$$

$$e^{-3k} = \frac{29400/36 - 210}{1190}$$

$$\ln e^{-3k} = \ln\left(\frac{29400/36 - 210}{1190}\right)$$

$$-3k = \ln\left(\frac{29400/36 - 210}{1190}\right)$$

$$k = -\frac{1}{3}\ln\left(\frac{29400/36 - 210}{1190}\right)$$

$$k \approx 0.2245763649$$

$$P(t) = \frac{294000}{210 + 1190e^{-0.22458t}} = \frac{1400}{1 + \frac{17}{3}e^{-0.22458t}}$$

b. $P(13) = \frac{294000}{210 + 1190e^{-0.22458(13)}}$
≈ 1070 coyotes

Chapter 4 Test

1. Find the inverse; sketch the function and inverse. [4.1]

$$y = 2x - 3$$
$$x = 2y - 3$$
$$x + 3 = 2y$$
$$\frac{1}{2}x + \frac{3}{2} = y$$
$$f^{-1}(x) = \frac{1}{2}x + \frac{3}{2}$$

3. a. Write in exponential form. [4.3]

$$\log_b (5x - 3) = c$$

$$b^c = 5x - 3$$

b. Write in logarithmic form.

$$3^{x/2} = y$$

$$\log_3 y = \frac{x}{2}$$

5. Write as a single logarithm with coefficient of 1. [4.4]

$$\log_{10}(2x + 3) - 3\log_{10}(x - 2)$$
$$= \log_{10}(2x + 3) - \log_{10}(x - 2)^3$$
$$= \log_{10}\frac{2x + 3}{(x - 2)^3}$$

7. Sketch the graph of the function.

$$f(x) = 3^{-x/2}$$

9. Solve. [4.5]

$$5^x = 22$$
$$x \log 5 = \log 22$$
$$x = \frac{\log 22}{\log 5}$$
$$x \approx 1.9206$$

11. Solve. [4.5]

$$\log(x + 99) - \log(3x - 2) = 2$$
$$\log\frac{x + 99}{3x - 2} = 2$$
$$\frac{x + 99}{3x - 2} = 10^2$$
$$x + 99 = 100(3x - 2)$$
$$x + 99 = 300x - 200$$
$$-299x = -299$$
$$x = 1$$

13. Find the balance. [4.6]

$P = 2800, r = 0.0225, t = 10$

a. $B = 2800\left(1 + \dfrac{0.0225}{12}\right)^{10(12)} \approx \$3505.76$

b. $B = 2800\left(1 + \dfrac{0.0225}{365}\right)^{10(365)} \approx \$3506.48$

c. $B = 2800e^{0.0225(10)} \approx \$3506.50$

15. a. Find the magnitude. [4.4]

$$M = \log\left(\dfrac{I}{I_0}\right)$$
$$= \log\left(\dfrac{42,304,000 I_0}{I_0}\right)$$
$$= \log 42,304,000$$
$$\approx 7.6$$

b. Compare the intensities.

$\log\left(\dfrac{I_1}{I_0}\right) = 6.3 \qquad$ and $\qquad \log\left(\dfrac{I_2}{I_0}\right) = 4.5$

$\dfrac{I_1}{I_0} = 10^{6.3} \qquad\qquad\qquad \dfrac{I_2}{I_0} = 10^{4.5}$

$I_1 = 10^{6.3} I_0 \qquad\qquad\qquad I_2 = 10^{4.5} I_0$

$\dfrac{I_1}{I_2} = \dfrac{10^{6.3} I_0}{10^{4.5} I_0} = \dfrac{10^{1.8}}{1} \approx \dfrac{63}{1}$

Therefore the ratio is 63 to 1.

17. Find the age of the bone. [4.6]

$$P(t) = 0.5^{\,t/5730} = 0.92$$
$$\log 0.5^{\,t/5730} = \log 0.92$$
$$\dfrac{t}{5730}\log 0.5 = \log 0.92$$
$$\dfrac{t}{5730} = \dfrac{\log 0.92}{\log 0.5}$$
$$t = 5730\left(\dfrac{\log 0.92}{\log 0.5}\right)$$
$$t \approx 690 \text{ years}$$

19. a. Find the logarithmic and logistic models. [4.7]

TI:

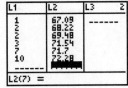
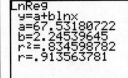

$d \approx 67.53180722 + 2.24539645\ln t$

WA: $d \approx 2.2454\ln(1.15265\times10^{13}\,t)$

b. TI: $d \approx 67.53180722 + 2.24539645\ln(17)$
≈ 73.89 meters

WA: $d \approx 2.2454\ln(1.15265\times10^{13}\times17)$
≈ 73.89 meters

Cumulative Review Exercises

1. Solve. [1.5]

$|x-4| \le 2 \Rightarrow -2 \le x-4 \le 2 \Rightarrow 2 \le x \le 6$.

The solution is [2, 6].

3. Find the distance. [2.1]

$$d = \sqrt{(11-5)^2 + (7-2)^2}$$
$$= \sqrt{6^2 + 5^2} = \sqrt{36+25}$$
$$= \sqrt{61} \approx 7.8$$

5. Find $(g \circ f)(x)$ for $f(x) = 2x+1, g(x) = x^2-5$. [2.6]

$$(g \circ f)(x) = g[f(x)]$$
$$= g(2x+1)$$
$$= (2x+1)^2 - 5$$
$$= 4x^2 + 4x + 1 - 5$$
$$= 4x^2 + 4x - 4$$

7. Find the weight. [1.6]

$$L = kwd^2$$
$$1500 = k(4)(8)^2$$
$$1500 = 256k$$
$$\dfrac{1500}{256} = \dfrac{375}{64} = k$$
$$L = \dfrac{375}{64}wd^2$$
$$L = \dfrac{375}{64}(6)(10)^2$$
$$L \approx 3500 \text{ pounds}$$

9. Find the zeros. [3.3]

$P(x) = x^4 - 5x^3 + x^2 + 15x - 12$ has three or one

positive and one negative real zeros.

$\dfrac{p}{q} = \pm 1, \pm 2, \pm 3, \pm 4, \pm 6, \pm 12$ are the possible

rational zeros.

$$\begin{array}{r|rrrr} 1 & 1 & -5 & 1 & 15 & -12 \\ & & 1 & -4 & -3 & 12 \\ \hline & 1 & -4 & -3 & 12 & 0 \end{array}$$

$$\begin{array}{r|rrrr} 4 & 1 & -4 & -3 & 12 \\ & & 4 & 0 & -12 \\ \hline & 1 & 0 & -3 & 0 \end{array}$$

$$x^2 - 3 = 0$$
$$x^2 = 3$$
$$x = \pm\sqrt{3}$$

The zeros are 1, 4, $-\sqrt{3}$, $\sqrt{3}$.

11. Find the asymptotes. [3.5]

$$r(x) = \frac{3x - 5}{x - 4}$$

Vertical asymptote: $x - 4 = 0$
$$x = 4$$

Horizontal asymptote: $(n = m)$
$$y = \frac{3}{1}$$
$$y = 3$$

13. State if the function is increasing or decreasing. [4.2]

$f(x) = 0.4^x$ is a decreasing function since $0.4 < 1$.

15. Write in logarithmic form. [4.3]

$$5^3 = 125 \Rightarrow \log_5 125 = 3$$

17. Solve. Round to the nearest ten-thousandth. [4.5]

$$2e^x = 15$$
$$e^x = 7.5$$
$$\ln e^x = \ln 7.5$$
$$x = \ln 7.5$$
$$x \approx 2.0149$$

19. Solve. Round to the nearest ten-thousandth. [4.5]

$$\frac{e^x - e^{-x}}{2} = 12$$
$$e^x\left(e^x - e^{-x}\right) = (24)e^x$$
$$e^{2x} - 1 = e^x(24)$$
$$e^{2x} - 24e^x - 1 = 0$$

Let $u = e^x$.

$$u^2 - 24u - 1 = 0$$

$$u = \frac{24 \pm \sqrt{576 - 4(-1)}}{2}$$

$$u = \frac{24 \pm \sqrt{580}}{2}$$

$$u = \frac{24 \pm 2\sqrt{145}}{2}$$

$$u = 12 \pm \sqrt{145}$$

$$e^x = 12 \pm \sqrt{145}$$

$$x \ln e = \ln\left(12 + \sqrt{145}\right) \quad \text{cannot take ln of a negative}$$

$$x = \ln\left(12 + \sqrt{145}\right)$$

$$x \approx 3.1798$$

Chapter 5 Trigonometric Functions

Section 5.1 Exercises

1. Find the complement and supplement.

    $90° - 15° = 75°$

    $180° - 15° = 165°$

3. Find the complement and supplement.

    $\begin{array}{cc} 90° & 89°60' \\ -70°15' = -70°15' \\ \hline 19°45' \end{array} \quad \begin{array}{cc} 180° & 179°60' \\ -70°15' = -70°15' \\ \hline 109°45' \end{array}$

5. Find the complement and supplement.

    $\begin{array}{cc} 90° & 89°59'60'' \\ -56°33'15'' = -56°33'15'' \\ \hline 33°26'45'' \end{array}$

    $\begin{array}{cc} 180° & 179°59'60'' \\ -56°33'15'' = -56°33'15'' \\ \hline 123°26'45'' \end{array}$

7. Find the complement and supplement.

    $\frac{\pi}{2} - 1$

    $\pi - 1$

9. Find the complement and supplement.

    $\frac{\pi}{2} - \frac{\pi}{4} = \frac{\pi}{4}$

    $\pi - \frac{\pi}{4} = \frac{3\pi}{4}$

11. Find the complement and supplement.

    $\frac{\pi}{2} - \frac{2\pi}{5} = \frac{\pi}{10}$

    $\pi - \frac{2\pi}{5} = \frac{3\pi}{5}$

13. Find the coterminal angle and state the quadrant.

    $610° = 250° + 360°$

    α is a quadrant III angle coterminal with an angle of measure $250°$.

15. Find the coterminal angle and state the quadrant.

    $-975° = 105° - 3 \cdot 360°$

α is a quadrant II angle coterminal with an angle of measure $105°$.

17. Find the coterminal angle and state the quadrant.

    $2456° = 296° + 6 \cdot 360°$

    α is a quadrant IV angle coterminal with an angle of measure $296°$.

19. Convert from degrees to exact radians.

    $30° = 30°\left(\frac{\pi}{180°}\right) = \frac{\pi}{6}$

21. Convert from degrees to exact radians.

    $90° = 90°\left(\frac{\pi}{180°}\right) = \frac{\pi}{2}$

23. Convert from degrees to exact radians.

    $165° = 165°\left(\frac{\pi}{180°}\right) = \frac{11\pi}{12}$

25. Convert from degrees to exact radians.

    $420° = 420°\left(\frac{\pi}{180°}\right) = \frac{7\pi}{3}$

27. Convert from degrees to exact radians.

    $585° = 585°\left(\frac{\pi}{180°}\right) = \frac{13\pi}{4}$

29. Convert from degrees to exact radians.

    $-9° = -9°\left(\frac{\pi}{180°}\right) = -\frac{\pi}{20}$

31. Convert from radians to exact degrees.

    $\frac{7\pi}{3} = \frac{7\pi}{3}\left(\frac{180°}{\pi}\right) = 420°$

33. Convert from radians to exact degrees.

    $\frac{\pi}{5} = \frac{\pi}{5}\left(\frac{180°}{\pi}\right) = 36°$

35. Convert from radians to exact degrees.

    $\frac{\pi}{6} = \frac{\pi}{6}\left(\frac{180°}{\pi}\right) = 30°$

37. Convert from radians to exact degrees.

    $\frac{3\pi}{8} = \frac{3\pi}{8}\left(\frac{180°}{\pi}\right) = 67.5°$

39. Convert from radians to exact degrees.

$$\frac{11\pi}{3} = \frac{11\pi}{3}\left(\frac{180°}{\pi}\right) = 660°$$

41. Convert from radians to exact degrees.

$$-\frac{5\pi}{12} = -\frac{5\pi}{12}\left(\frac{180°}{\pi}\right) = -75°$$

43. Convert from radians to degrees. Round to the nearest tenth.

$$1.5 = 1.5\left(\frac{180°}{\pi}\right) \approx 85.94°$$

45. Convert from degrees to radians. Round to the nearest tenth.

$$133° = 133°\left(\frac{\pi}{180°}\right) \approx 2.32$$

47. Convert from radians to degrees. Round to the nearest tenth.

$$8.25 = 8.25\left(\frac{180°}{\pi}\right) \approx 472.69°$$

49. Find the central angle in degrees and radians. Round to the nearest hundredth.

$$\theta = \frac{s}{r}$$
$$= \frac{8}{2} = 4 \text{ radians}$$
$$= 4\left(\frac{180°}{\pi}\right) \approx 229.18°$$

51. Find the central angle in degrees and radians. Round to the nearest hundredth.

$$\theta = \frac{s}{r}$$
$$= \frac{12.4}{5.2} \approx 2.38 \text{ radians}$$
$$= 2.38\left(\frac{180°}{\pi}\right) \approx 136.63°$$

53. Find the length of the arc. Round to the nearest hundredth.

$$s = r\theta$$
$$= (8)\frac{\pi}{4}$$
$$\approx 6.28 \text{ in.}$$

55. Find the length of the arc. Round to the nearest hundredth.

$$s = r\theta$$
$$= 25\cdot(42°)\left(\frac{\pi}{180°}\right)$$
$$\approx 18.33 \text{ cm}$$

57. $s = r\theta$

$$= 93,000,000(31')\left(\frac{1°}{60'}\right)\left(\frac{\pi}{180°}\right)$$
$$\approx 840,000 \text{ mi}$$

59. $s = r\theta$

$$= 100(7°)\left(\frac{\pi}{180°}\right)$$
$$\approx 12 \text{ ft}$$

61. $\theta_2 = \frac{r_1}{r_2}\theta_1$

$$= \frac{14}{28}(150°)\left(\frac{\pi}{180°}\right)$$
$$= \frac{5\pi}{12} \text{ radians or } 75°$$

63. $\omega = \frac{\theta}{t}$

$$= \frac{2\pi}{60}$$
$$= \frac{\pi}{30} \text{ radian/sec}$$

65. $\omega = \frac{\theta}{t}$

$$= \frac{50(2\pi)}{60}$$
$$= \frac{5\pi}{3} \text{ radians/sec}$$

67. $\omega = \frac{v}{r}$

$$= \frac{\frac{10}{12}}{\frac{394}{2}} = \frac{10}{12}\cdot\frac{2}{394}$$
$$\approx 0.004 \text{ radians/sec}$$

69. $v = \omega r$

$$= \frac{450\cdot2\pi\cdot60\cdot15}{12\cdot5280}$$
$$\approx 40 \text{ mph}$$

71. Find the linear speed.

$$v = \frac{s}{t}$$

$$= \frac{2\pi r}{24 \text{ hours}}$$

$$= \frac{2\pi(3960 \text{ mi})\left(\dfrac{5280 \text{ ft}}{1 \text{ mi}}\right)}{(24 \text{ hr})\left(\dfrac{60 \text{ min}}{1 \text{ hr}}\right)\left(\dfrac{60 \text{ s}}{1 \text{ min}}\right)}$$

$$\approx 1520 \text{ ft/s}$$

73. Find the linear speed.

$$v = \frac{s}{t}$$

$$= \frac{2\pi r}{24 \text{ hours}}$$

$$= \frac{2\pi(35,790 \text{ km} + 6400 \text{ km})}{24 \text{ hours}}$$

$$\approx 11,045 \text{ km/h}$$

75.

$$r_1\theta_1 = r_2\theta_2$$

$$(3.5)(150 \cdot 2\pi) = (1.75)\theta_2$$

$$600\pi = \theta_2$$

$$300(2\pi) = \theta_2$$

The rear gear is making 300 revolutions.

The rear gear and tire are making same number of revolutions.

Tire is 12 inches = 1 ft

$$s = 1 \text{ ft}(300)(2\pi) = 1885 \text{ ft}$$

77. a. When the rear tire makes one revolution, the bicycle travels $s = r\theta = 2\pi r = 2\pi(30 \text{ inches}) = 60\pi$ inches.

The angular velocity of point A is

$$\omega = \frac{\theta}{t} = \frac{2\pi}{t} = 2\left(\frac{\pi}{t}\right).$$

When the bicycle travels 60π inches, point B on the front tire travels though an angle of

$$\theta = \frac{s}{r} = \frac{60\pi \text{ inches}}{20 \text{ inches}} = 3\pi.$$

The angular velocity of point B is

$$\omega = \frac{\theta}{t} = \frac{3\pi}{t} = 3\left(\frac{\pi}{t}\right).$$

Thus, point B has the greater angular velocity.

b. Point A and point B travel a linear distance of 60π inches in the same amount of time. Therefore, both points have the same linear velocity.

79. a. $\omega = \dfrac{\theta}{t}$

$$= \frac{2\pi}{1.61 \text{ hours}}$$

$$\approx 3.9 \text{ radians per hour}$$

b. $v = \dfrac{s}{t}$

$$= \frac{2\pi r}{1.61 \text{ hours}}$$

$$= \frac{2\pi(625 \text{ km} + 6370 \text{ km})}{1.61 \text{ hours}}$$

$$= \frac{2\pi(6995 \text{ km})}{1.61 \text{ hours}}$$

$$\approx 27,300 \text{ km per hour}$$

81. Find the area.

$$A = \frac{1}{2}r^2\theta$$

$$= \frac{1}{2}(5^2)\left(\frac{\pi}{3}\right)$$

$$\approx 13 \text{ in}^2$$

83. Find the area.

$$A = \frac{1}{2}r^2\theta$$

$$= \frac{1}{2}(120)^2\, 0.65$$

$$= 4680 \text{ cm}^2$$

85. Find the distance from the city to the equator.

$$25°47' = 25° + 47'\left(\frac{1°}{60'}\right)$$

$$= 25\frac{47}{60}°$$

Convert to radians.

$$25\frac{47}{60}° \cdot \frac{\pi}{180°} = \frac{1547\pi}{10,800} \text{ radians}$$

$$s = r\theta$$

$$s = 3960\left(\frac{1547\pi}{10,800}\right)$$

$$\approx 1780$$

To the nearest 10 miles, Miami is 1780 miles north of the equator.

87. $7.5° = 7.5° \left(\dfrac{\pi}{180°} \right) = \dfrac{\pi}{24}$

Solving the formula $\theta = \dfrac{s}{r}$ for r,

we have $r = \dfrac{s}{\theta}$

$= \dfrac{520 \text{ miles}}{\pi/24}$

$\approx 3970 \text{ miles}$

Prepare for Section 5.2

P1. Rationalize the denominator.

$\dfrac{1}{\sqrt{3}} \cdot \dfrac{\sqrt{3}}{\sqrt{3}} = \dfrac{\sqrt{3}}{3}$

P3. Simplify.

$a \div \left(\dfrac{a}{2} \right) = a \cdot \left(\dfrac{2}{a} \right) = 2$

P5. Solve. Round to the nearest hundredth.

$\dfrac{\sqrt{2}}{2} = \dfrac{x}{5}$

$\dfrac{5\sqrt{2}}{2} = x$

$x \approx 3.54$

Section 5.2 Exercises

1. Find the value of the 6 trigonometric functions.

$r = \sqrt{5^2 + 12^2}$

$r = \sqrt{25 + 144} = \sqrt{169}$

$r = 13$

$\sin \theta = \dfrac{y}{r} = \dfrac{12}{13}$ $\csc \theta = \dfrac{r}{y} = \dfrac{13}{12}$

$\cos \theta = \dfrac{x}{r} = \dfrac{5}{13}$ $\sec \theta = \dfrac{r}{x} = \dfrac{13}{5}$

$\tan \theta = \dfrac{y}{x} = \dfrac{12}{5}$ $\cot \theta = \dfrac{x}{y} = \dfrac{5}{12}$

3. Find the value of the 6 trigonometric functions.

$x = \sqrt{7^2 - 4^2}$

$x = \sqrt{49 - 16} = \sqrt{33}$

$\sin \theta = \dfrac{y}{r} = \dfrac{4}{7}$ $\csc \theta = \dfrac{r}{y} = \dfrac{7}{4}$

$\cos \theta = \dfrac{x}{r} = \dfrac{\sqrt{33}}{7}$ $\sec \theta = \dfrac{r}{x} = \dfrac{7}{\sqrt{33}} = \dfrac{7\sqrt{33}}{33}$

$\tan \theta = \dfrac{y}{x} = \dfrac{4}{\sqrt{33}} = \dfrac{4\sqrt{33}}{33}$ $\cot \theta = \dfrac{x}{y} = \dfrac{\sqrt{33}}{4}$

5. Find the value of the 6 trigonometric functions.

$r = \sqrt{2^2 + 5^2}$

$r = \sqrt{4 + 25} = \sqrt{29}$

$\sin \theta = \dfrac{y}{r} = \dfrac{5}{\sqrt{29}} = \dfrac{5\sqrt{29}}{29}$ $\csc \theta = \dfrac{r}{y} = \dfrac{\sqrt{29}}{5}$

$\cos \theta = \dfrac{x}{r} = \dfrac{2}{\sqrt{29}} = \dfrac{2\sqrt{29}}{29}$ $\sec \theta = \dfrac{r}{x} = \dfrac{\sqrt{29}}{2}$

$\tan \theta = \dfrac{y}{x} = \dfrac{5}{2}$ $\cot \theta = \dfrac{x}{y} = \dfrac{2}{5}$

7. Find the value of the 6 trigonometric functions.

$r = \sqrt{2^2 + \left(\sqrt{3} \right)^2}$

$r = \sqrt{4 + 3} = \sqrt{7}$

$\sin \theta = \dfrac{y}{r} = \dfrac{\sqrt{3}}{\sqrt{7}} = \dfrac{\sqrt{21}}{7}$ $\csc \theta = \dfrac{r}{y} = \dfrac{\sqrt{7}}{\sqrt{3}} = \dfrac{\sqrt{21}}{3}$

$\cos \theta = \dfrac{x}{r} = \dfrac{2}{\sqrt{7}} = \dfrac{2\sqrt{7}}{7}$ $\sec \theta = \dfrac{r}{x} = \dfrac{\sqrt{7}}{2}$

$\tan \theta = \dfrac{y}{x} = \dfrac{\sqrt{3}}{2}$ $\cot \theta = \dfrac{x}{y} = \dfrac{2}{\sqrt{3}} = \dfrac{2\sqrt{3}}{3}$

9. Find the value of the 6 trigonometric functions.

$\text{opposite side} = \sqrt{6^2 - 3^2}$

$= \sqrt{36 - 9}$

$= \sqrt{27} = 3\sqrt{3}$

$\sin \theta = \dfrac{\text{opp}}{\text{hyp}} = \dfrac{3\sqrt{3}}{6} = \dfrac{\sqrt{3}}{2}$

$\cos \theta = \dfrac{\text{adj}}{\text{hyp}} = \dfrac{3}{6} = \dfrac{1}{2}$

$\tan \theta = \dfrac{\text{opp}}{\text{adj}} = \dfrac{3\sqrt{3}}{3} = \sqrt{3}$

$\csc \theta = \dfrac{\text{hyp}}{\text{opp}} = \dfrac{6}{3\sqrt{3}} = \dfrac{2}{\sqrt{3}} = \dfrac{2\sqrt{3}}{3}$

$\sec \theta = \dfrac{\text{hyp}}{\text{adj}} = \dfrac{6}{3} = 2$

$\cot \theta = \dfrac{\text{adj}}{\text{opp}} = \dfrac{3}{3\sqrt{3}} = \dfrac{1}{\sqrt{3}} = \dfrac{\sqrt{3}}{3}$

11. Find the value of the 6 trigonometric functions.

$\text{hypotenuse} = \sqrt{5^2 + 6^2}$

$= \sqrt{25 + 36}$

$= \sqrt{61}$

$$\sin\theta = \frac{opp}{hyp} = \frac{6}{\sqrt{61}} = \frac{6\sqrt{61}}{61} \qquad \csc\theta = \frac{hyp}{opp} = \frac{\sqrt{61}}{6}$$

$$\cos\theta = \frac{adj}{hyp} = \frac{5}{\sqrt{61}} = \frac{5\sqrt{61}}{61} \qquad \sec\theta = \frac{hyp}{adj} = \frac{\sqrt{61}}{5}$$

$$\tan\theta = \frac{opp}{adj} = \frac{6}{5} \qquad \cot\theta = \frac{adj}{opp} = \frac{5}{6}$$

For exercises 13 and 15, since

$$\sin\theta = \frac{y}{r} = \frac{3}{5}, y = 3, r = 5, \text{ and } x = \sqrt{5^2 - 3^2} = 4 \ .$$

13. $\tan\theta = \frac{y}{x} = \frac{3}{4}$

15. $\cos\theta = \frac{x}{r} = \frac{4}{5}$

For exercise 17, since

$$\tan\theta = \frac{y}{x} = \frac{4}{3}, y = 4, x = 3, \text{ and } r = \sqrt{3^2 + 4^2} = 5 \ .$$

17. $\cot\theta = \frac{x}{y} = \frac{3}{4}$

For exercises 19 and 21, since $\sec\beta = \frac{r}{x} = \frac{13}{12}, r = 13,$

$$x = 12, \text{ and } y = \sqrt{13^2 - 12^2} = \sqrt{25} = 5 \ .$$

19. $\cos\beta = \frac{x}{r} = \frac{12}{13}$

21. $\csc\beta = \frac{r}{y} = \frac{13}{5}$

For exercise 23, since $\cos\theta = \frac{x}{r} = \frac{2}{3}, x = 2, r = 3,$

$$\text{and } y = \sqrt{3^2 - 2^2} = \sqrt{9-4} = \sqrt{5} \ .$$

23. $\sec\theta = \frac{r}{x} = \frac{3}{2}$

25. Find the exact value of the expression.

$$\sin 45° + \cos 45° = \frac{\sqrt{2}}{2} + \frac{\sqrt{2}}{2} = \sqrt{2}$$

27. Find the exact value of the expression.

$$\sin 30° \cos 60° - \tan 45° = \frac{1}{2} \cdot \frac{1}{2} - 1$$
$$= \frac{1}{4} - 1$$
$$= -\frac{3}{4}$$

29. Find the exact value of the expression.

$$\sin 30° \cos 60° + \tan 45° = \frac{1}{2} \cdot \frac{1}{2} + 1 = \frac{1}{4} + 1 = \frac{5}{4}$$

31. Find the exact value of the expression.

$$\sin\frac{\pi}{3} + \cos\ \frac{\pi}{6} = \frac{\sqrt{3}}{2} + \frac{\sqrt{3}}{2} = 2 \cdot \frac{\sqrt{3}}{2} = \sqrt{3}$$

33. Find the exact value of the expression.

$$\sin\frac{\pi}{4} + \tan\ \frac{\pi}{6} = \frac{\sqrt{2}}{2} + \frac{\sqrt{3}}{3} = \frac{3\sqrt{2} + 2\sqrt{3}}{6}$$

35. Find the exact value of the expression.

$$\sec\frac{\pi}{3} \cos\ \frac{\pi}{3} - \tan\frac{\pi}{6} = 2\ \cdot\frac{1}{2} - \frac{\sqrt{3}}{3} = 1 - \frac{\sqrt{3}}{3} = \frac{3-\sqrt{3}}{3}$$

37. Find the exact value of the expression.

$$2\csc\ \frac{\pi}{4} - \sec\frac{\pi}{3}\cos\frac{\pi}{6} = 2 \cdot \sqrt{2} - 2 \cdot \frac{\sqrt{3}}{2} = 2\sqrt{2} - \sqrt{3}$$

39. Use a calculator to find the value to 4 decimal places.

$$\tan 32° \approx 0.6249$$

41. Use a calculator to find the value to 4 decimal places.

$$\cos 63°20' \approx 0.4488$$

43. Use a calculator to find the value to 4 decimal places.

$$\cos 34.7° \approx 0.8221$$

45. Use a calculator to find the value to 4 decimal places.

$$\sec 5.9° \approx 1.0053$$

47. Use a calculator to find the value to 4 decimal places.

$$\tan\frac{\pi}{7} \approx 0.4816$$

49. Use a calculator to find the value to 4 decimal places.

$$\csc 1.2 \approx 1.0729$$

51. We make a drawing and solve.

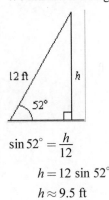

$$\sin 52° = \frac{h}{12}$$
$$h = 12 \sin 52°$$
$$h \approx 9.5 \text{ ft}$$

53. We make a drawing and solve.

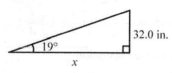

$$\tan 19^\circ = \frac{32.0}{x}$$

$$x = \frac{32.0}{\tan 19^\circ} = 92.9 \text{ in.}$$

55. We make a drawing and solve.

$$\cos 38^\circ = \frac{d - 0.33}{6}$$

$$d = 6 \cos 38^\circ + 0.33$$

$$d \approx 5.1 \text{ ft}$$

57. We make a drawing and solve.

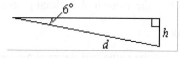

$$d = \frac{240 \text{ mi}}{\text{hr}} \left(\frac{1 \text{ hr}}{60 \text{ min}}\right) 4 \text{ min}$$

$$d = 16 \text{ mi}$$

$$\sin 6^\circ = \frac{h}{d}$$

$$h = 16 \sin 6^\circ$$

$$h \approx 1.7 \text{ mi}$$

59. We make a drawing and solve.

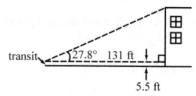

$$\tan 27.8^\circ = \frac{x}{131}$$

$$x = 131 \tan 27.8^\circ = 69.1 \text{ ft}$$

$$h = 69.1 + 5.5 = 74.6 \text{ ft}$$

61. $\sin 0.056^\circ = \dfrac{670,900}{d}$

$$d = \frac{670,900}{\sin 0.056^\circ} \text{ km}$$

$$\approx 686,000,000 \text{ km}$$

63. Find the values of x and y.

From the second figure,

$$\tan(90^\circ - 33^\circ) = \frac{x}{y}, \text{ then } x = y \tan 57^\circ$$

From the first figure,

$$\tan(90^\circ - 77^\circ) = \frac{x}{6 + y}$$

$$\tan 13^\circ = \frac{y \tan 57^\circ}{6 + y} \quad \begin{array}{l}\text{Substitute } y \tan 57^\circ \\ \text{for } x.\end{array}$$

$$\tan 13^\circ (6 + y) = y \tan 57^\circ$$

$$6 \tan 13^\circ + y \tan 13^\circ = y \tan 57^\circ$$

$$6 \tan 13^\circ = y \tan 57^\circ - y \tan 13^\circ$$

$$6 \tan 13^\circ = y(\tan 57^\circ - \tan 13^\circ)$$

$$y = \frac{6 \tan 13^\circ}{\tan 57^\circ - \tan 13^\circ}$$

$$y \approx 1.068 \approx 1.1 \text{ ft}$$

$$x = y \tan 57^\circ = 1.068 \tan 57^\circ \approx 1.6 \text{ ft}$$

65. We make a drawing and solve.

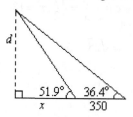

$$\tan 36.4 = \frac{d}{350 + x} \qquad \tan 51.9 = \frac{d}{x}$$

$$x = \frac{d}{\tan 51.9}$$

$$\tan 36.4^\circ = \frac{d}{350 + \frac{d}{\tan 51.9^\circ}}$$

$$d = \frac{350 \tan 36.4^\circ}{1 - \frac{\tan 36.4^\circ}{\tan 51.9^\circ}}$$

$$d \approx 612 \text{ ft}$$

67. Let h represent the height of the Washington Monument.

$$\tan 42^\circ = \frac{h}{x} \qquad \tan 37.77^\circ = \frac{h}{100 + x}$$

$$x = \frac{h}{\tan 42^\circ} \qquad \tan 37.77^\circ = \frac{h}{100 + \frac{h}{\tan 42^\circ}}$$

$$h = \frac{100 \tan 37.77^\circ}{1 - \frac{\tan 37.77^\circ}{\tan 42.0^\circ}}$$

$$h \approx 555.6 \text{ ft}$$

69. Let d represent the horizontal distance of the drive.

$$\tan 32.0° = \frac{211}{d}$$

$$d = \frac{211}{\tan 32.0°}$$

$$d \approx 338 \text{ m}$$

71. We make a drawing and solve.

a.

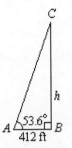

$$\tan 53.6° = \frac{h}{412}$$

$$h = 412 \tan 53.6°$$

$$h \approx 559 \text{ feet}$$

b. $(AC)^2 = 412^2 + 559^2$

$$AC = \sqrt{412^2 + 559^2}$$

$$AC = \sqrt{482,225}$$

$$AC \approx 694.4 \text{ feet}$$

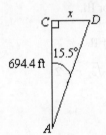

$$\tan 15.5° = \frac{x}{694.4}$$

$$x = 694.4 \tan 15.5°$$

$$x \approx 193 \text{ feet}$$

73. We make a drawing and solve.

if $\theta = 45°$ $d_1 = d_2$

$$\sin\theta = \frac{3}{d_1}$$

$$d_1 = \frac{3}{\sin 45°}$$

$$d = 2d_1 = \frac{6}{\sin 45°}$$

$$d \approx 8.5 \text{ ft}$$

Prepare for Section 5.3

P1. Find the reciprocal.

$$-\frac{4}{3}$$

P3. Evaluate.

$$|120 - 180| = |-60| = 60$$

P5. Simplify.

$$\frac{3}{2}\pi - \frac{1}{2}\pi = \frac{2}{2}\pi = \pi$$

Section 5.3 Exercises

1. Find the value of the 6 trigonometric functions.

$$x = 2, y = 3, r = \sqrt{2^2 + 3^2} = \sqrt{13}$$

$$\sin\theta = \frac{y}{r} = \frac{3}{\sqrt{13}} = \frac{3\sqrt{13}}{13} \quad \csc\theta = \frac{\sqrt{13}}{3}$$

$$\cos\theta = \frac{x}{r} = \frac{2}{\sqrt{13}} = \frac{2\sqrt{13}}{13} \quad \sec\theta = \frac{\sqrt{13}}{2}$$

$$\tan\theta = \frac{y}{x} = \frac{3}{2} \quad\quad\quad \cot\theta = \frac{2}{3}$$

3. Find the value of the 6 trigonometric functions.

$$x = -2, y = 3, r = \sqrt{(-2)^2 + (3)^2} = \sqrt{13}$$

$$\sin\theta = \frac{y}{r} = \frac{3}{\sqrt{13}} = \frac{3\sqrt{13}}{13} \quad \csc\theta = \frac{\sqrt{13}}{3}$$

$$\cos\theta = \frac{x}{r} = \frac{-2}{\sqrt{13}} = -\frac{2\sqrt{13}}{13} \quad \sec\theta = -\frac{\sqrt{13}}{2}$$

$$\tan\theta = \frac{y}{x} = \frac{3}{-2} = -\frac{3}{2} \quad \cot\theta = -\frac{2}{3}$$

5. Find the value of the 6 trigonometric functions.

$$x = -8, y = -5, r = \sqrt{(-8)^2 + (-5)^2} = \sqrt{89}$$

$$\sin\theta = \frac{y}{r} = \frac{-5}{\sqrt{89}} = -\frac{5\sqrt{89}}{89} \quad \csc\theta = -\frac{\sqrt{89}}{5}$$

$$\cos\theta = \frac{x}{r} = \frac{-8}{\sqrt{89}} = -\frac{8\sqrt{89}}{89} \quad \sec\theta = -\frac{\sqrt{89}}{8}$$

$$\tan\theta = \frac{y}{x} = \frac{-5}{-8} = \frac{5}{8} \quad\quad \cot\theta = \frac{8}{5}$$

7. Find the value of the 6 trigonometric functions.

$$x = -5, y = 0, r = \sqrt{(-5)^2 + (0)^2} = 5$$

$\sin\theta = \dfrac{y}{r} = \dfrac{0}{5} = 0 \qquad \csc\theta$ is undefined

$\cos\theta = \dfrac{x}{r} = \dfrac{-5}{5} = -1 \quad \sec\theta = -1$

$\tan\theta = \dfrac{y}{x} = \dfrac{0}{-5} = 0 \qquad \cot\theta$ is undefined

9. Evaluate the function.

$\sin 180° = 0$

11. Evaluate the function.

$\tan 180° = 0$

13. Evaluate the function.

$\csc 90° = 1$

15. Evaluate the function.

$\cos\dfrac{\pi}{2} = 0$

17. Evaluate the function.

$\tan\dfrac{\pi}{2} =$ undefined

19. Evaluate the function.

$\sin\dfrac{\pi}{2} = 1$

21. State the quadrant in which θ lies.

$\sin\theta > 0$ in quadrants I and II.
$\cos\ \theta\ >\ 0$ in quadrants I and IV.
quadrant I

23. State the quadrant in which θ lies.

$\cos\theta > 0$ in quadrants I and IV.
$\tan\ \theta\ <\ 0$ in quadrants II and IV.
quadrant IV

25. State the quadrant in which θ lies.

$\sin\theta < 0$ in quadrants III and IV.
$\cos\ \theta\ <\ 0$ in quadrants II and III.
quadrant III

27. Find the exact value of the expression.

$\sin\theta = -\dfrac{1}{2} = \dfrac{y}{r},\ y = -1,\ r = 2,$

$x = \pm\sqrt{2^2 - (-1)^2} = \pm\sqrt{3},$

$x = -\sqrt{3}$ in quadrant III,

$\tan\theta = \dfrac{y}{x} = \dfrac{-1}{-\sqrt{3}} = \dfrac{\sqrt{3}}{3}$

29. Find the exact value of the expression.

$\csc\theta = \sqrt{2} = \dfrac{r}{y},\ r = \sqrt{2},\ y = 1,$

$x = \pm\sqrt{\left(\sqrt{2}\right)^2 - 1^2} = \pm 1$

$x = -1$ in quadrant II

$\cot\theta = \dfrac{-1}{1} = -1$

31. Find the exact value of the expression.

$\cos\theta = \dfrac{1}{2},\ \theta$ is in quadrant I or IV.

$\tan\theta = \sqrt{3},\ \theta$ is in quadrant I or III.

θ is in quadrant I, $x = 1,\ y = \sqrt{3},\ r = 2$

$\csc\theta = \dfrac{r}{y} = \dfrac{2}{\sqrt{3}} = \dfrac{2\sqrt{3}}{3}$

33. Find the exact value of the expression.

$\cos\theta = -\dfrac{1}{2}, \theta$ is in quadrant II or III.

$\sin\theta = \dfrac{\sqrt{3}}{2},\ \theta$ is in quadrant I or II.

θ is in quadrant II, $x = -1,\ y = \sqrt{3},\ r = 2$

$\cot\theta = \dfrac{x}{y} = \dfrac{-1}{\sqrt{3}} = -\dfrac{\sqrt{3}}{3}$

35. Find $\tan\theta$.

θ is in quadrant I, $\sin\theta = \dfrac{1}{2} = \dfrac{y}{r},\ y = 1,\ r = 2,$

$x = \sqrt{2^2 - 1^2} = \sqrt{3}$

$\tan\theta = \dfrac{1}{\sqrt{3}} = \dfrac{\sqrt{3}}{3}$

37. Find $\csc\theta$.

θ is in quadrant IV, $\cos\theta = \dfrac{3}{5} = \dfrac{x}{r},\ x = 3,\ r = 5,$

$y = \sqrt{5^2 - 3^2} = -4$

$\csc\theta = -\dfrac{5}{4}$

39. Find $\sec\theta$.

θ is in quadrant IV, $\cot\theta = -\dfrac{1}{5} = \dfrac{x}{y}$, $x = 1$, $y = -5$,

$r = \sqrt{1^2 + (-5)^2} = \sqrt{26}$

$\sec\theta = \dfrac{\sqrt{26}}{1} = \sqrt{26}$

41. Find the reference angle.

$\theta = 160°$

Since $90° < \theta < 180°$,

$\theta + \theta' = 180°$

$\theta' = 20°$

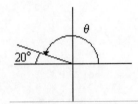

43. Find the reference angle.

$\theta = 351°$

Since $270° < \theta < 360°$,

$\theta + \theta' = 360°$

$\theta' = 9°$

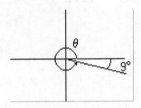

45. Find the reference angle.

$\theta = \dfrac{11\pi}{5}$

$\theta > 2\pi = \dfrac{10\pi}{5}$,

θ is coterminal with $\alpha = \dfrac{11\pi}{5} - \dfrac{10\pi}{5} = \dfrac{\pi}{5}$.

Since $0 < \alpha < \dfrac{\pi}{2}$,

$\alpha' = \alpha = \theta'$

$\theta' = \dfrac{\pi}{5}$

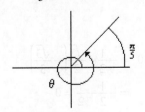

47. Find the reference angle.

$\theta = \dfrac{8}{3}$

Since $\dfrac{\pi}{2} < \theta < \pi$,

$\theta + \theta' = \pi$

$\theta' = \pi - \dfrac{8}{3}$

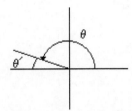

49. Find the reference angle.

$\theta = 1406° = 326° + 3 \cdot 360°$

θ is coterminal with $\alpha = 326°$.

Since $270° < \alpha < 360°$,

$\alpha + \alpha' = 360°$

$\alpha' = 34°$

$\theta' = 34°$

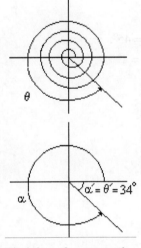

51. Find the reference angle.

$\theta = -475° = 245° - 2 \cdot 360°$

θ is coterminal with $\alpha = 245°$

Since $180° < \alpha < 270°$,

$\alpha' + 180° = \alpha$

$\alpha' = 245° - 180°$

$\alpha' = 65°$

$\theta' = 65°$

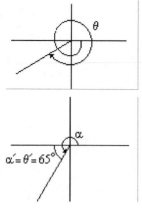

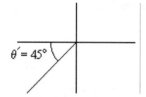

53. Find the exact value.

$\theta = 225°$ is in quadrant III.

$225° - 180° = 45°$ so $\theta' = 45°$.

Thus, $\sin 225° = -\sin 45° = -\dfrac{\sqrt{2}}{2}$.

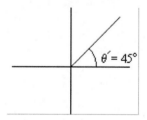

55. Find the exact value.

$\theta = 405°$ is in quadrant I.

$405° - 360° = 45°$ so $\theta' = 45°$.

Thus, $\tan 405° = \tan 45° = 1$.

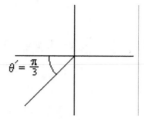

57. Find the exact value.

$\theta = \dfrac{4}{3}\pi$ is in quadrant III.

$\dfrac{4}{3}\pi - \pi = \dfrac{\pi}{3}$ so $\theta' = \dfrac{\pi}{3}$.

Thus, $\csc \dfrac{4\pi}{3} = \dfrac{1}{\sin 4\pi/3} = \dfrac{1}{-\sin \pi/3}$

$= \dfrac{1}{-\sqrt{3}/2} = -\dfrac{2}{\sqrt{3}} = -\dfrac{2\sqrt{3}}{3}$.

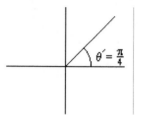

59. Find the exact value.

$\theta = \dfrac{17\pi}{4} = \dfrac{16\pi}{4} + \dfrac{\pi}{4}$ is coterminal

with $\dfrac{\pi}{4}$ in quadrant I and $\theta' = \dfrac{\pi}{4}$,

so $\cos \dfrac{17\pi}{4} = \cos \dfrac{\pi}{4} = \dfrac{\sqrt{2}}{2}$.

61. Find the exact value.

$\theta = 765° = 720° + 45°$ is coterminal

with $45°$ in quadrant I and $\theta' = 45°$,

so $\sec 765° = \sec 45° = \dfrac{1}{\cos 45°} = \dfrac{1}{\sqrt{2}/2} = \sqrt{2}$.

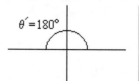

63. Find the exact value.

$\theta = 540° = 360° + 180°$ is coterminal

with $180°$, so $\cot 540° = \cot 180° = \dfrac{\cos 180°}{\sin 180°} = \dfrac{-1}{0}$,

which is undefined.

65. Find the exact value.

$\sin 210° - \cos 330° \tan 330° = -\dfrac{1}{2} - \dfrac{\sqrt{3}}{2}\left(-\dfrac{\sqrt{3}}{3}\right)$

$= -\dfrac{1}{2} + \dfrac{1}{2} = 0$

67. Find the exact value.

$$\sin^2 30° + \cos^2 30° = \left(\frac{1}{2}\right)^2 + \left(\frac{\sqrt{3}}{2}\right)^2 = \frac{1}{4} + \frac{3}{4} = 1$$

69. Find the exact value.

$$\sin\frac{3\pi}{2}\tan\frac{\pi}{4} - \cos\frac{\pi}{3} = (-1)(1) - \frac{1}{2} = -1 - \frac{1}{2} = -\frac{3}{2}$$

71. Find the exact value.

$$\sin^2\frac{5\pi}{4} + \cos^2\frac{5\pi}{4} = \left(-\frac{\sqrt{2}}{2}\right)^2 + \left(-\frac{\sqrt{2}}{2}\right)^2 = \frac{1}{2} + \frac{1}{2} = 1$$

73. Find two values of θ.

$\sin\theta = \frac{1}{2}$, θ is in quadrant I or quadrant II

$\theta = 30°,\ 150°$

75. Find two values of θ.

$\cos\theta = \frac{-\sqrt{3}}{2}$, θ is in quadrant II or quadrant III

$\theta = 150°,\ 210°$

77. Find two values of θ.

$\csc\theta = -\sqrt{2}$

θ is in quadrant III or IV

$\theta = 225°,\ 315°$

79. Find two values of θ.

$\tan\theta = -1$

θ is in quadrant II or IV

$\theta = \frac{3\pi}{4}, \frac{7\pi}{4}$

81. Find two values of θ.

$\tan\theta = \frac{-\sqrt{3}}{3}$

θ is in quadrant II or IV

$\theta = \frac{5\pi}{6}, \frac{11\pi}{6}$

83. Find two values of θ.

$\sin\theta = \frac{\sqrt{3}}{2}$

θ is in quadrant I or II

$\theta = \frac{\pi}{3}, \frac{2\pi}{3}$

85. $1 + \tan^2\theta = \sec^2\theta$

$$1 + \frac{y^2}{x^2} = \frac{x^2 + y^2}{x^2}$$

$$= \frac{r^2}{x^2}$$

$$= \sec^2\theta$$

87.

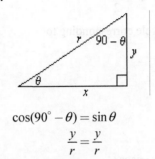

$\cos(90° - \theta) = \sin\theta$

$$\frac{y}{r} = \frac{y}{r}$$

Prepare for Section 5.4

P1. Is the point on the circle?

$$x^2 + y^2 = 1$$
$$(0)^2 + (1)^2 = 1$$

Yes

P3. Is the point on the circle?

$$x^2 + y^2 = 1$$

$$\left(\frac{\sqrt{2}}{2}\right)^2 + \left(\frac{\sqrt{3}}{2}\right)^2 = 1$$

$$\frac{2}{4} + \frac{3}{4} \neq 1$$

No

P5. Is the function even, odd, or neither.

even

Section 5.4 Exercises

1. Evaluate.

$$t = \frac{\pi}{6}$$

$$y = \sin t \qquad\qquad x = \cos t$$

$$= \sin\frac{\pi}{6} \qquad\qquad = \cos\frac{\pi}{6}$$

$$= \frac{1}{2} \qquad\qquad = \frac{\sqrt{3}}{2}$$

The point on the unit circle corresponding to

$$t = \frac{\pi}{6} \text{ is } \left(\frac{\sqrt{3}}{2},\ \frac{1}{2}\right).$$

3. Evaluate.

$t = \dfrac{7\pi}{6}$

$y = \sin t$ $\qquad$ $x = \cos t$

$\quad = \sin\dfrac{7\pi}{6}$ $\qquad = \cos\dfrac{7\pi}{6}$

$\quad = -\dfrac{1}{2}$ $\qquad\quad = -\dfrac{\sqrt{3}}{2}$

The point on the unit circle corresponding to

$t = \dfrac{7\pi}{6}$ is $\left(-\dfrac{\sqrt{3}}{2}, -\dfrac{1}{2}\right)$.

5. Evaluate.

$t = \dfrac{11\pi}{6}$

$y = \sin t$ $\qquad$ $x = \cos t$

$\quad = \sin\dfrac{11\pi}{6}$ $\qquad = \cos\dfrac{11\pi}{6}$

$\quad = -\dfrac{1}{2}$ $\qquad\quad = \dfrac{\sqrt{3}}{2}$

The point on the unit circle corresponding to

$t = \dfrac{11\pi}{6}$ is $\left(\dfrac{\sqrt{3}}{2}, -\dfrac{1}{2}\right)$.

7. Evaluate.

$t = \pi$

$y = \sin t$ $\qquad x = \cos t$

$\quad = \sin\pi$ $\qquad = \cos\pi$

$\quad = 0$ $\qquad\quad = -1$

The point on the unit circle corresponding to $t = \pi$ is $(-1, 0)$.

9. a. Estimate $W(t)$ to the nearest tenth.

On the unit circle, start at $(1, 0)$, move in the positive direction to reach $t = 0.8$, then estimate the coordinates.

$W(0.8) \approx (0.7,\ 0.7)$

b. Estimate $\cos t$ and $\sin t$ to the nearest tenth.

$\cos 0.8 \approx 0.7$ $\quad$ and $\quad \sin 0.8 \approx 0.7$

c. Use a calculator to find $\cos t$ and $\sin t$ to the nearest ten-thousandths.

$\cos 0.8 \approx 0.7648$ $\quad$ and $\quad \sin 0.8 \approx 0.7174$

11. a. Estimate $W(t)$ to the nearest tenth.

On the unit circle, start at $(1, 0)$, move in the positive direction to reach $t = 2.1$, then estimate the coordinates.

$W(2.1) \approx (-0.5,\ 0.9)$

b. Estimate $\cos t$ and $\sin t$ to the nearest tenth.

$\cos 2.1 \approx -0.5$ $\quad$ and $\quad \sin 2.1 \approx 0.9$

c. Use a calculator to find $\cos t$ and $\sin t$ to the nearest ten-thousandths.

$\cos 2.1 \approx -0.5048$ $\quad$ and $\quad \sin 2.1 \approx 0.8632$

13. a. Estimate $W(t)$ to the nearest tenth.

On the unit circle, start at $(1, 0)$, move in the negative direction to reach $t = -5.6$, then estimate the coordinates.

$W(-5.6) \approx (0.8,\ 0.6)$

b. Estimate $\cos t$ and $\sin t$ to the nearest tenth.

$\cos(-5.6) \approx 0.8$ $\quad$ and $\quad \sin(-5.6) \approx 0.6$

c. Use a calculator to find $\cos t$ and $\sin t$ to the nearest ten-thousandths.

$\cos(-5.6) \approx 0.7756$ $\quad$ and $\quad \sin(-5.6) \approx 0.6313$

15. State the period of the function.

2π

17. State the period of the function.

π

19. State the period of the function.

2π

21. Find the exact value.

$\tan\dfrac{11\pi}{6} = -\tan\dfrac{\pi}{6} = -\dfrac{\sqrt{3}}{3}$

23. Find the exact value.

$\cos\left(-\dfrac{2\pi}{3}\right) = -\cos\dfrac{\pi}{3} = -\dfrac{1}{2}$

25. Find the exact value.

$\csc\left(-\dfrac{\pi}{3}\right) = -\csc\dfrac{\pi}{3} = -\dfrac{2\sqrt{3}}{3}$

27. Find the exact value.

$\sin\dfrac{3\pi}{2} = -\sin\dfrac{\pi}{2} = -1$

29. Find the exact value.

$$\sec\left(-\frac{7\pi}{6}\right) = -\sec\frac{\pi}{6} = -\frac{2\sqrt{3}}{3}$$

31. Determine whether the function is odd, even, or neither.

$$f(-x) = -4\sin(-x)$$
$$= 4\sin x$$
$$= -f(x)$$

The function defined by $f(x) = -4\sin x$ is an odd function.

33. Determine whether the function is odd, even, or neither.

$$G(-x) = \sin(-x) + \cos(-x)$$
$$= -\sin x + \cos x$$

The function defined by $G(x) = \sin x + \cos x$ is neither an even nor an odd function.

35. Determine whether the function is odd, even, or neither.

$$S(-x) = \frac{\sin(-x)}{-x}$$
$$= -\frac{\sin x}{-x} = \frac{\sin x}{x}$$
$$= S(x)$$

The function defined by $S(x) = \frac{\sin(x)}{x}$ is an even function.

37. Determine whether the function is odd, even, or neither.

$$v(-x) = 2\sin(-x)\cos(-x)$$
$$= -2\sin x \cos x$$
$$= -v(x)$$

The function defined by $v(x) = 2\sin x \cos x$ is an odd function.

39. Verify the identity.

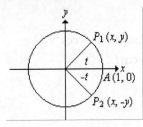

$$\cos t = x$$
$$\cos(-t) = x$$
$$\cos(-t) = \cos t$$

41. Verify the identity.

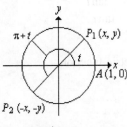

$$\cos t = x$$
$$\cos(\pi + t) = -x$$
$$\cos t = -\cos(\pi + t)$$

43. Verify the identity.

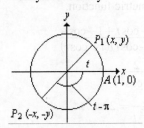

$$\sin(t - \pi) = -y$$
$$\sin t = y$$
$$\sin(t - \pi) = -\sin t$$

45. Write as a single trigonometric function.

$$\tan t \cos t = \frac{\sin t}{\cos t} \cdot \cos t$$
$$= \sin t$$

47. Write as a single trigonometric function.

$$\frac{\csc t}{\cot t} = \frac{\frac{1}{\sin t}}{\frac{\cos t}{\sin t}}$$
$$= \frac{1}{\sin t} \cdot \frac{\sin t}{\cos t}$$
$$= \frac{1}{\cos t} = \sec t$$

49. Write as a single trigonometric function.

$$1 - \sec^2 t = 1 - \frac{1}{\cos^2 t}$$
$$= \frac{\cos^2 t - 1}{\cos^2 t}$$
$$= \frac{-\sin^2 t}{\cos^2 t} = -\tan^2 t$$

51. Write as a single trigonometric function.

$$\tan t - \frac{\sec^2 t}{\tan t} = \tan t - \frac{1 + \tan^2 t}{\tan t}$$

$$= \tan t - \frac{1}{\tan t} - \frac{\tan^2 t}{\tan t}$$

$$= \tan t - \cot t - \tan t$$

$$= -\cot t$$

53. Write as a single trigonometric function.

$$\frac{1 - \cos^2 t}{\tan^2 t} = \frac{\sin^2 t}{\dfrac{\sin^2 t}{\cos^2 t}}$$

$$= \sin^2 t \cdot \frac{\cos^2 t}{\sin^2 t}$$

$$= \cos^2 t$$

55. Write as a single trigonometric function.

$$\frac{1}{1 - \cos t} + \frac{1}{1 + \cos t} = \frac{1 + \cos t + 1 - \cos t}{(1 - \cos t)(1 + \cos t)}$$

$$= \frac{2}{1 - \cos^2 t}$$

$$= \frac{2}{\sin^2 t} = 2\csc^2 t$$

57. Write as a single trigonometric function.

$$\frac{\tan t + \cot t}{\tan t} = \frac{\dfrac{\sin t}{\cos t} + \dfrac{\cos t}{\sin t}}{\dfrac{\sin t}{\cos t}}$$

$$= \left(\frac{\sin^2 t + \cos^2 t}{\sin t \cdot \cos t} \right) \frac{\cos t}{\sin t}$$

$$= \frac{\sin^2 t + \cos^2 t}{\sin^2 t}$$

$$= \frac{1}{\sin^2 t}$$

$$= \csc^2 t$$

59. Write as a single trigonometric function.

$$\sin^2 t (1 + \cot^2 t) = \sin^2 t (\csc^2 t)$$

$$= \sin^2 t \cdot \frac{1}{\sin^2 t}$$

$$= 1$$

61. $\sin^2 t + \cos^2 t = 1$

$$\sin^2 t = 1 - \cos^2 t$$

$$\sin t = \pm\sqrt{1 - \cos^2 t}$$

Because $0 < t < \dfrac{\pi}{2}$, $\sin t$ is positive.

Thus, $\sin t = \sqrt{1 - \cos^2 t}$.

63. $\csc^2 t = 1 + \cot^2 t$

$$\csc t = \pm\sqrt{1 + \cot^2 t}$$

Because $\dfrac{\pi}{2} < t < \pi$, $\csc t$ is positive.

Thus, $\csc t = \sqrt{1 + \cot^2 t}$.

65. $d(t) = 1970 \cos\left(\dfrac{\pi}{64} t\right)$

$$d(24) = 1970 \cos\left(\frac{\pi}{64} \cdot 24\right)$$

$$= 1970 \cos\left(\frac{3\pi}{8}\right)$$

$$\approx 750 \text{ miles}$$

67. Find the angle after 2.5 seconds. Round to the nearest hundredth.

$$\theta(2.5) = \frac{1}{5}\cos(4 \cdot 2.5) \approx -0.1678$$

The angle of the rod from the vertical after 2.5 seconds is −0.17 radian. The rod is to the left of the equilibrium position.

69. Simplify.

$$\cos t - \frac{1}{\cos t} = \frac{\cos^2 t - 1}{\cos t} = -\frac{\sin^2 t}{\cos t}$$

71. Simplify.

$$\cot t + \frac{1}{\cot t} = \frac{\cot^2 t + 1}{\cot t} = \frac{\csc^2 t}{\cot t} = \frac{\dfrac{1}{\sin^2 t}}{\dfrac{\cos t}{\sin t}}$$

$$= \frac{1}{\sin^2 t} \cdot \frac{\sin t}{\cos t} = \frac{1}{\sin t \cos t}$$

$$= \frac{1}{\sin t} \cdot \frac{1}{\cos t} = \csc t \sec t$$

73. Simplify.

$$(1 - \sin t)^2 = 1 - 2\sin t + \sin^2 t$$

75. Simplify.

$$(\sin t - \cos t)^2 = \sin^2 t - 2\sin t \cos t + \cos^2 t$$

$$= 1 - 2\sin t \cos t$$

77. Simplify.

$$(1 - \sin t)(1 + \sin t) = 1 - \sin^2 t$$
$$= \cos^2 t$$

79. Find the value of the function.

$$\csc t = \sqrt{2}, \quad 0 < t < \frac{\pi}{2}$$

$$\sin t = \frac{1}{\csc t} = \frac{\sqrt{2}}{2}$$

$$\cos^2 t + \sin^2 t = 1$$

$$\cos t = \pm\sqrt{1 - \sin^2 t}$$

$\cos t$ is positive in quadrant I.

$$\cos t = \sqrt{1 - \left(\frac{\sqrt{2}}{2}\right)^2} = \sqrt{\frac{1}{2}}$$

$$\cos t = \frac{\sqrt{2}}{2}$$

81. Find the value of the function.

$$\sin t = \frac{1}{2}, \quad \frac{\pi}{2} < t < \pi$$

$$\tan t = \frac{\sin t}{\cos t}$$

$$\tan t = \frac{\sin t}{\pm\sqrt{1 - \sin^2 t}}$$

Because $\frac{\pi}{2} < t < \pi$, $\tan t$ is negative.

$$\tan t = -\frac{\frac{1}{2}}{\sqrt{1 - \left(\frac{1}{2}\right)^2}}$$

$$\tan t = -\frac{\frac{1}{2}}{\frac{\sqrt{3}}{2}}$$

$$\tan t = -\frac{\sqrt{3}}{3}$$

83. Identify the length of the given line segment.

$$\cos \theta = \frac{OA}{1}$$

85. Identify the length of the given line segment.

$$\sin \theta = \frac{1}{OR}$$

$$OR = \frac{1}{\sin \theta} = \csc \theta$$

87. Identify the length of the given line segment.

$$\tan \theta = \frac{PB}{1}$$

Mid-Chapter 5 Quiz

1. a. Convert to radians.

$$105° = 105°\left(\frac{\pi}{180°}\right) = \frac{7\pi}{12}$$

b. Convert to degrees.

$$\frac{3\pi}{5} = \frac{3\pi}{5}\left(\frac{180°}{\pi}\right) = 108°$$

3. $v = \omega r$

$$= \frac{3 \cdot 2\pi \cdot 60 \cdot 60 \cdot 12.5}{12 \cdot 5280}$$

$$\approx 13 \text{ mph}$$

5. Find the exact value.

$$\sin^2 \frac{\pi}{6} + \tan^2 \frac{\pi}{6} = \left(\frac{1}{2}\right)^2 + \left(\frac{\sqrt{3}}{3}\right)^2 = \frac{1}{4} + \frac{1}{3} = \frac{7}{12}$$

7. Find the exact value.

a. $\theta = 480°$ is in quadrant II.

$480° - 360° = 120°$ so $\theta' = 120°$.

Thus, $\cos 120° = -\frac{1}{2}$.

b. $\theta = \frac{7\pi}{6}$ is in quadrant III.

$$\frac{7\pi}{6} - \pi = \frac{\pi}{6} \text{ so } \theta' = \frac{\pi}{6}.$$

Thus, $\tan \frac{7\pi}{6} = \tan \frac{\pi}{6} = \frac{\sin \pi/6}{\cos \pi/6} = \frac{1/2}{\sqrt{3}/2} = \frac{\sqrt{3}}{3}$.

9. Evaluate $W\left(\frac{11\pi}{6}\right)$.

$$t = \frac{11\pi}{6}$$

$$\begin{array}{ll} y = \sin t & x = \cos t \\ \quad = \sin \frac{11\pi}{6} & \quad = \cos \frac{11\pi}{6} \\ \quad = -\frac{1}{2} & \quad = \frac{\sqrt{3}}{2} \end{array}$$

The point on the unit circle corresponding to

$t = \frac{11\pi}{6}$ is $\left(\frac{\sqrt{3}}{2}, -\frac{1}{2}\right)$.

Prepare for Section 5.5

P1. Estimate to the nearest tenth.

$$\sin \frac{3\pi}{4} \approx 0.7$$

P3. Reflect the graph of $y = f(x)$ across the x-axis to produce $y = -f(x)$.

P5. Simplify.

$$\frac{2\pi}{\frac{1}{3}} = \frac{2\pi}{1} \cdot \frac{3}{1} = 6\pi$$

Section 5.5 Exercises

1. State the amplitude and period.

$$y = 2\sin x$$
$$a = 2,\ p = 2\pi$$

3. State the amplitude and period.

$$y = \sin 2x$$
$$a = 1,\ p = \frac{2\pi}{2} = \pi$$

5. State the amplitude and period.

$$y = \frac{1}{2}\sin 2\pi x$$
$$a = \frac{1}{2},\ p = \frac{2\pi}{2\pi} = 1$$

7. State the amplitude and period.

$$y = -2\sin\frac{x}{2}$$
$$a = |-2| = 2,\ p = \frac{2\pi}{1/2} = 4\pi$$

9. State the amplitude and period.

$$y = \cos\frac{x}{4}$$
$$a = 1,\ p = \frac{2\pi}{1/4} = 8\pi$$

11. State the amplitude and period.

$$y = 2\cos\frac{\pi x}{3}$$
$$a = 2,\ p = \frac{2\pi}{\pi/3} = 6$$

13. State the amplitude and period.

$$y = -3\cos\frac{2x}{3}$$
$$a = |-3| = 3,\ p = \frac{2\pi}{2/3} = 3\pi$$

15. Write an equation for a sine function.

$$a = 2$$
$$p = \frac{2\pi}{b} = 3\pi$$
$$b = \frac{2}{3}$$
$$y = 2\sin\frac{2}{3}x$$

17. Write an equation for a sine function.

$$a = 2.5$$
$$p = \frac{2\pi}{b} = 3.2$$
$$b = \frac{5\pi}{8}$$
$$y = 2.5\sin\frac{5\pi}{8}x$$

19. Write an equation for a cosine function.

$$a = 3$$
$$p = \frac{2\pi}{b} = 2.5$$
$$b = \frac{4\pi}{5}$$
$$y = 3\cos\frac{4\pi}{5}x$$

21. Graph one full period.

$$y = \frac{1}{2}\sin x,\ a = \frac{1}{2},\ p = 2\pi$$

23. Graph one full period.

$$y = 3\cos x,\ a = 3,\ p = 2\pi$$

25. Graph one full period.

$$y = -\frac{7}{2}\cos x,\ a = \left|-\frac{7}{2}\right| = \frac{7}{2},\ p = 2\pi$$

27. Graph one full period.

$$y = -4\sin x, \ a = |-4| = 4, \ p = 2\pi$$

29. Graph one full period.

$$y = \cos 3x, \ a = 1, \ p = \frac{2\pi}{3}$$

31. Graph one full period.

$$y = \sin \frac{3x}{2}, \ a = 1, \ p = \frac{4\pi}{3}$$

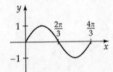

33. Graph one full period.

$$y = \cos \frac{\pi}{2}x, \ a = 1, \ p = 4$$

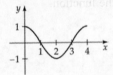

35. Graph one full period.

$$y = \sin 2\pi x, \ a = 1, \ p = 1$$

37. Graph one full period.

$$y = 4\cos \frac{x}{2}, \ a = 4, \ p = \frac{2\pi}{1/2} = 4\pi$$

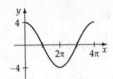

39. Graph one full period.

$$y = -2\cos \frac{x}{3}, \ a = |-2| = 2, \ p = \frac{2\pi}{1/3} = 6\pi$$

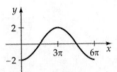

41. Graph one full period.

$$y = 2\sin \pi x, \ a = 2, \ p = \frac{2\pi}{\pi} = 2$$

43. Graph one full period.

$$y = \frac{3}{2}\cos \frac{\pi x}{2}, \ a = \frac{3}{2}, \ p = \frac{2\pi}{\pi/2} = 4$$

45. Graph one full period.

$$y = 4\sin \frac{2\pi}{3}x, \ a = |4| = 4, \ p = \frac{2\pi}{2\pi/3} = 3$$

47. Graph one full period.

$$y = 2\cos 2x, \ a = 2, \ p = \frac{2\pi}{2} = \pi$$

49. Graph one full period.

$$y = -2\sin 1.5x, \ a = |-2| = 2, \ p = \frac{2\pi}{1.5} = \frac{4\pi}{3}$$

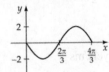

51. Graph one full period.

$$y = \left| 2\sin\frac{x}{2} \right|$$

53. Graph one full period.

$$y = \left| -2\cos 3x \right|$$

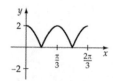

55. Graph one full period.

$$y = -\left| 2\sin\frac{x}{3} \right|$$

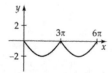

57. Graph one full period.

$$y = -\left| 3\cos \pi x \right|$$

59. Find an equation for the graph.

$$\frac{2\pi}{b} = \pi, \ b = 2, \ a = 1$$

$$y = \cos 2x$$

61. Find an equation for the graph.

$$\frac{2\pi}{b} = 3\pi, \ b = \frac{2}{3}, \ a = 2$$

$$y = 2\sin\frac{2}{3}x$$

63. a. Amplitude, $a = 4$, period, $b = \frac{2\pi}{2} = \pi$

$$V = 4\sin \pi t, \ 0 \le t \le 8 \text{ ms}$$

b. Frequency $= \dfrac{4 \text{ cycles}}{8 \text{ ms}} = \dfrac{1}{2}$ cycle/ms

65. Sketch the graph.

$$f(x) = 2\sin\frac{2x}{3}, \ a = 2, \ p = \frac{2\pi}{2/3} = 3\pi$$

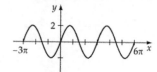

67. Sketch the graph.

$$y_1 = 2\cos\frac{x}{2}, \ a = 2, \ p = \frac{2\pi}{1/2} = 4\pi$$
$$y_2 = 2\cos x, \ a = 2, \ p = 2\pi$$

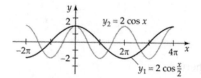

69. Use a graphing utility to graph the function.

$$y = \frac{1}{2}x\sin x$$

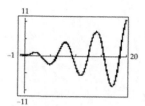

71. Use a graphing utility to graph the function.

$$y = -x\cos x$$

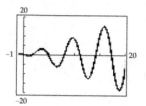

73. $y = e^{\sin x}$

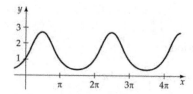

The maximum value of $e^{\sin x}$ is e.

The minimum value of $e^{\sin x}$ is $\dfrac{1}{e} \approx 0.3679$.

The function $y = e^{\sin x}$ is periodic with a period of 2π.

Prepare for Section 5.6

P1. Estimate to the nearest tenth.

$$\tan\frac{\pi}{3} \approx 1.7$$

P3. Stretch the position of each point on the graph of $y = f(x)$ away from the x-axis by a factor of 2 to produce $y = 2f(x)$.

P5. Simplify.

$$\frac{\pi}{\frac{1}{2}} = \frac{\pi}{1} \cdot \frac{2}{1} = 2\pi$$

Section 5.6 Exercises

1. Find where x is undefined.

$y = \tan x$ is undefined for $\frac{\pi}{2} + k\pi$, k an integer.

3. Find where x is undefined.

$y = \sec x$ is undefined for $\frac{\pi}{2} + k\pi$, k an integer.

5. State the period.

$$p = 2\pi$$

7. State the period.

$$p = \pi$$

9. State the period.

$$p = \frac{\pi}{1/2} = 2\pi$$

11. State the period.

$$p = \frac{2\pi}{3}$$

13. State the period.

$$p = \frac{2\pi}{1/4} = 8\pi$$

15. State the period.

$$p = \frac{\pi}{\pi/5} = 5$$

17. Write the trigonometric form.

$$\frac{\pi}{b} = \frac{\pi}{3}, \; b = 3$$

$$y = \tan 3x$$

19. Write the trigonometric form.

$$\frac{2\pi}{b} = \frac{3\pi}{4}, \; b = \frac{8}{3}$$

$$y = \sec\frac{8}{3}x$$

21. Write the trigonometric form.

$$\frac{\pi}{b} = 2, \; b = \frac{\pi}{2}$$

$$y = \cot\frac{\pi}{2}x$$

23. Sketch one full period of the graph.

$$y = 3\tan x, \; p = \pi$$

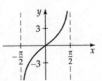

25. Sketch one full period of the graph.

$$y = \frac{3}{2}\cot x, \; p = \pi$$

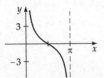

27. Sketch one full period of the graph.

$$y(x) = 2\sec x, \; p = 2\pi$$

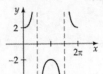

29. Sketch one full period of the graph.

$$y = \frac{1}{2}\csc x, \; p = 2\pi$$

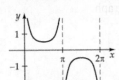

31. Sketch one full period of the graph.

$$y = 2\tan\frac{x}{2}, \; p = \frac{\pi}{1/2} = 2\pi$$

33. Sketch one full period of the graph.

$$y = -3\cot\frac{x}{2}, \ p = \frac{\pi}{1/2} = 2\pi$$

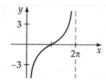

35. Sketch one full period of the graph.

$$y = -2\csc\frac{x}{3}, \ p = \frac{2\pi}{1/3} = 6\pi$$

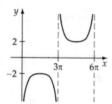

37. Sketch one full period of the graph.

$$y = \frac{1}{2}\sec 2x, \ p = \frac{2\pi}{2} = \pi$$

39. Sketch one full period of the graph.

$$y = -2\sec\pi x, \ p = \frac{2\pi}{\pi} = 2$$

41. Sketch one full period of the graph.

$$y = 3\tan 2\pi x, \ p = \frac{\pi}{2\pi} = \frac{1}{2}$$

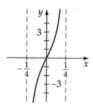

43. Graph each equation.

$$y = 2\csc 3x, \ p = \frac{2\pi}{3}$$

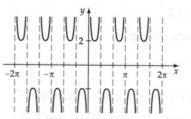

45. Graph each equation.

$$y = 3\sec\pi x, \ p = \frac{2\pi}{\pi} = 2$$

47. Graph each equation.

$$y = 2\cot 2x, \ p = \frac{\pi}{2}$$

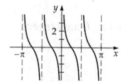

49. Graph each equation.

$$y = 3\tan\pi x, \ p = \frac{\pi}{\pi} = 1$$

51. Find the equation for the graph.

$$\frac{\pi}{b} = \frac{2\pi}{3}, \ b = \frac{3}{2}$$

$$y = \cot\frac{3}{2}x$$

53. Find the equation for the graph.

$$\frac{2\pi}{b} = 3\pi, \ b = \frac{2}{3}$$

$$y = \csc\frac{2}{3}x$$

55. Find the equation for the graph.

$$\frac{2\pi}{b} = \frac{8\pi}{3}, \ b = \frac{3}{4}$$

$$y = \sec\frac{3}{4}x$$

57. Graph the equation.

$y = \tan|x|$

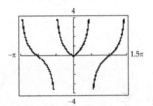

59. Graph the equation.

$y = |\csc x|$

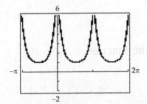

61. a. Find the function.

Let h = the underwater distance. Then, $h = \sec\theta$.

The cost of the underwater distance is $0.4\sec\theta$.

The over land distance is $6 - \tan\theta$. The cost of the

over land distance is $0.2(6 - \tan\theta) = 1.2 - 0.2\tan\theta$.

The cost function is $C(\theta) = 0.4\sec\theta - 0.2\tan\theta + 1.2$

b. Find the angle. Round to the nearest tenth of a

degree.

Using a graphing utility,

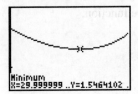

Xmin=0,   Xmax=50,   Xscl=1

Ymin=1.45, Ymax=1.65, Yscl=0.5

The angle is about 30.0°.

c. Find the minimum cost. Round to the nearest

hundredth of a million dollars.

$C(30.0) = 0.4\sec 30.0° - 0.2\tan 30.0° + 1.2$

$C(30.0) \approx 1.55$

The minimum cost is about $1.55 million.

63. Find the length of the shortest ladder. Round to the

nearest tenth of a foot.

$\cos\theta = \dfrac{4}{L_1}$, so $L_1 = \dfrac{4}{\cos\theta}$

$\sin\theta = \dfrac{6}{L_2}$, so $L_2 = \dfrac{6}{\sin\theta}$

Then, $L_1 + L_2 = \dfrac{4}{\cos\theta} + \dfrac{6}{\sin\theta}$

Use a graphing utility to graph $y = \dfrac{4}{\cos\theta} + \dfrac{6}{\sin\theta}$ and

find the minimum of the function.

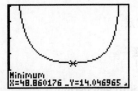

Xmin=0, Xmax=90, Xscl=10

Ymin=0, Ymax=50, Yscl=10

The shortest ladder should be 14.0 feet.

65. a. $\tan x = \dfrac{h}{1.4}$

$h = 1.4\tan x$

b. $\cos x = \dfrac{1.4}{d}$

$d = \dfrac{1.4}{\cos x} = 1.4\sec x$

c.

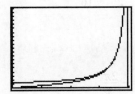

d. The graph of d is above the graph of h, however, the

distance between the graphs approaches 0 as x

approaches $\dfrac{\pi}{2}$.

Prepare for Section 5.7

P1. Find the amplitude and period.

$y = 2\sin 2x$

amplitude = 2, period = π

P3. Find the amplitude and period.

$$y = -4\sin 2\pi x$$

amplitude = 4, period = 1

P5. Find the maximum value.

minimum at –3

Section 5.7 Exercises

1. Find the amplitude, phase shift and period.

$$a = 1, \ p = \frac{2\pi}{2} = \pi, \ \text{phase shift} = \frac{\pi/4}{2} = \frac{\pi}{8}$$

3. Find the amplitude, phase shift and period.

$$a = |-4| = 4, \ p = \frac{2\pi}{2/3} = 3\pi,$$

$$\text{phase shift} = \frac{-\pi/6}{2/3} = -\frac{\pi}{4}$$

5. Find the amplitude, phase shift and period.

$$a = \frac{5}{4}, \ p = \frac{2\pi}{3}, \ \text{phase shift} = \frac{2\pi}{3}$$

7. Find the phase shift and period.

$$p = \frac{\pi}{2}, \ \text{phase shift} = \frac{\pi/4}{2} = \frac{\pi}{8}$$

9. Find the phase shift and period.

$$p = \frac{2\pi}{1/3} = 6\pi, \ \text{phase shift} = \frac{-\pi}{1/3} = -3\pi$$

11. Find the phase shift and period.

$$p = \frac{2\pi}{2} = \pi, \ \text{phase shift} = \frac{\pi/8}{2} = \frac{\pi}{16}$$

13. sine function, $a = 2$, $p = \pi$

$$\frac{2\pi}{b} = \pi, \ b = 2$$

$$\text{phase shift} = -\frac{c}{b} = \frac{\pi}{3}, \ c = -\frac{2\pi}{3}$$

$$y = 2\sin\left(2x - \frac{2\pi}{3}\right)$$

15. tangent function, $p = 2\pi$

$$\frac{\pi}{b} = 2\pi \Rightarrow b = \frac{1}{2}$$

$$\text{phase shift} = -\frac{c}{b} = \frac{\pi}{2} \Rightarrow c = -\frac{\pi}{4}$$

$$y = \tan\left(\frac{x}{2} - \frac{\pi}{4}\right)$$

17. Graph one full period of the function.

$$y = \sin\left(x - \frac{\pi}{2}\right)$$

$$0 \le x - \frac{\pi}{2} \le 2\pi$$

$$\frac{\pi}{2} \le x \le \frac{5\pi}{2}$$

$$\text{period} = 2\pi, \ \text{phase shift} = \frac{\pi}{2}$$

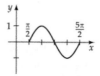

19. Graph one full period of the function.

$$y = \cos\left(\frac{x}{2} + \frac{\pi}{3}\right)$$

$$0 \le \frac{x}{2} + \frac{\pi}{3} \le 2\pi$$

$$-\frac{\pi}{3} \le \frac{x}{2} \le \frac{5\pi}{3}$$

$$-\frac{2\pi}{3} \le x \le \frac{10\pi}{3}$$

$$\text{period} = 4\pi, \ \text{phase shift} = -\frac{2\pi}{3}$$

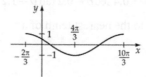

21. Graph one full period of the function.

$$y = \tan\left(x + \frac{\pi}{4}\right)$$

$$-\frac{\pi}{2} < x + \frac{\pi}{4} < \frac{\pi}{2}$$

$$-\frac{3\pi}{4} < x < \frac{\pi}{4}$$

$$\text{period} = \pi, \ \text{phase shift} = -\frac{\pi}{4}$$

23. Graph one full period of the function.

$$y = 2\cot\left(\frac{x}{2} - \frac{\pi}{8}\right)$$

$$0 < \frac{x}{2} - \frac{\pi}{8} < \pi$$

$$\frac{\pi}{8} < \frac{x}{2} < \frac{9\pi}{8}$$

$$\frac{\pi}{4} < x < \frac{9\pi}{4}$$

period $= 2\pi$, phase shift $= \frac{\pi}{4}$

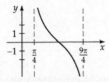

25. Graph one full period of the function.

$$y = \sec\left(x + \frac{\pi}{4}\right)$$

$$0 \le x + \frac{\pi}{4} \le 2\pi$$

$$-\frac{\pi}{4} \le x \le \frac{7\pi}{4}$$

period $= 2\pi$, phase shift $= -\frac{\pi}{4}$

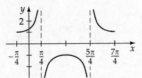

27. Graph one full period of the function.

$$y = \csc\left(\frac{x}{3} - \frac{\pi}{2}\right)$$

$$0 \le \frac{x}{3} - \frac{\pi}{2} \le 2\pi$$

$$\frac{\pi}{2} \le \frac{x}{3} \le \frac{5\pi}{2}$$

$$\frac{3\pi}{2} \le x \le \frac{15\pi}{2}$$

period $= 6\pi$, phase shift $= \frac{3\pi}{2}$

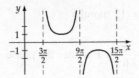

29. Graph one full period of the function.

$$y = -2\sin\left(\frac{x}{3} - \frac{2\pi}{3}\right)$$

$$0 \le \frac{x}{3} - \frac{2\pi}{3} \le 2\pi$$

$$\frac{2\pi}{3} \le \frac{x}{3} \le \frac{8\pi}{3}$$

$$2\pi \le x \le 8\pi$$

period $= 6\pi$, phase shift $= 2\pi$

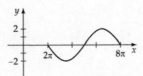

31. Graph one full period of the function.

$$y = -3\cos\left(3x + \frac{\pi}{4}\right)$$

$$0 \le 3x + \frac{\pi}{4} \le 2\pi$$

$$-\frac{\pi}{4} \le 3x \le \frac{7\pi}{4}$$

$$-\frac{\pi}{12} \le x \le \frac{7\pi}{12}$$

period $= \frac{2\pi}{3}$, phase shift $= -\frac{\pi}{12}$

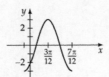

33. Graph the function.

$$y = \sin x + 1, \; p = 2\pi$$

35. Graph the function.

$$y = -\cos x - 2, \; p = 2\pi$$

37. Graph the function.

$$y = \sin 2x - 2, \; p = \pi$$

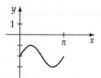

39. Graph the function.

$$y = 4\cos(\pi x - 2) + 1, \; p = 2$$

phase shift $= \dfrac{2}{\pi}$

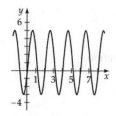

41. Graph the function.

$$y = -\sin(\pi x + 1) - 2, \; p = 2$$

phase shift $= -\dfrac{1}{\pi}$

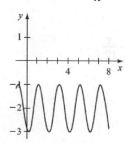

43. Graph the function.

$$y = \sin\left(x - \dfrac{\pi}{2}\right) - \dfrac{1}{2}, \; p = 2\pi$$

phase shift $= \dfrac{\pi}{2}$

45. Graph the function.

$$y = \tan\dfrac{x}{2} - 4, \; p = 2\pi$$

47. Graph the function.

$$y = \sec 2x - 2, \; p = \pi$$

49. Graph the function.

$$y = \csc\dfrac{x}{2} - 1, \; p = 4\pi$$

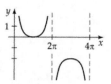

51. Graph the function using addition-of-ordinates method.

$$y = x - \sin x$$

53. Graph the function using addition-of-ordinates method.

$$y = x + \sin 2x$$

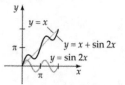

55. Graph the function using addition-of-ordinates method.

$$y = \sin x + \cos x$$

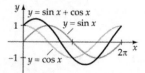

57. a. phase shift:

$$-\dfrac{c}{b} = -\dfrac{-1.25\pi}{\pi/6} = 1.25(6) = 7.5 \text{ months}$$

period: $\dfrac{2\pi}{b} = \dfrac{2\pi}{\pi/6} = 2(6) = 12$ months

b. First graph $y_1 = 4.1\cos\left(\dfrac{\pi}{6}t\right)$. Because the phase shift is 7.5 months, shift the graph of y_1 7.5 units to the right to produce the graph of y_2. Now shift the graph of y_2 upward 7 units to product the graph of S.

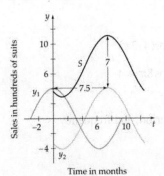

Time in months

c. 7.5 months after January 1 is the middle of August.

59. $y = 2.3\sin 2\pi t + 1.25t + 315$

$t = 16$ between 1994 and 2010

$y = 2.3\sin 2\pi(16) + 1.25(16) + 315$

$y \approx 335$ ppm

difference $\approx 335 - 315$

$\phantom{\text{difference}} \approx 20$ ppm

61. 5 rpm $a = 7$

$p = 0.2$ min $p = \dfrac{2\pi}{b}$

$\phantom{p = 0.2 \text{ min}}\quad 0.2 = \dfrac{2\pi}{b}$

$\phantom{p = 0.2 \text{ min}}\quad b = 10\pi$

$\phantom{p = 0.2 \text{ min}}\quad s = 7\cos 10\pi t + 5$

63. Change 6 rpm to radians/second.

$6 \text{ rpm} = \dfrac{6 \cdot 2\pi \text{ radians}}{1 \text{ minute}} \cdot \dfrac{1 \text{ minute}}{60 \text{ sec}}$

$\phantom{6 \text{ rpm}} = \dfrac{\pi}{5} \text{ radians/sec}$

in t seconds, θ increased by $\dfrac{\pi}{5}t$ radians.

$\tan \dfrac{\pi}{5}t = \dfrac{s}{400}$

$400 \tan \dfrac{\pi}{5}t = s$

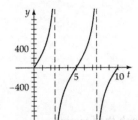

65. amplitude $A = 3$

$k = 9$

$24 = 2$ cycles

$12 = 1$ cycle

$\dfrac{2\pi}{B} = 12 \Rightarrow B = \dfrac{\pi}{6}$

$y = 3\cos\dfrac{\pi}{6}t + 9$

At 6:00 P.M., $t = 12$.

$y = 3\cos\dfrac{\pi}{6} \cdot 12 + 9$

$y = 3\cos 2\pi + 9$

$y = 3 + 9$

$y = 12$ ft

67. Use a graphing utility to graph the function.

$y = \sin x - \cos\dfrac{x}{2}$

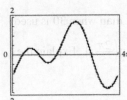

69. Use a graphing utility to graph the function.

$y = 2\cos x + \sin\dfrac{x}{2}$

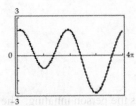

71. Use a graphing utility to graph the function.

$y = \dfrac{x}{2}\sin x$

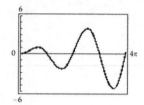

73. Use a graphing utility to graph the function.

$$y = \frac{\sin x}{x}$$

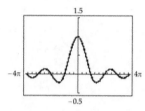

75. State how the frequency of the beats changes.

Using WolframAlpha, for 30,

Using WolframAlpha, for 60,

The beats are more frequent than when 30 is used.

77. Graph $V(t) = 2500 + 250\sin\left(\frac{2\pi}{5}t\right)$. Explain the

significance.

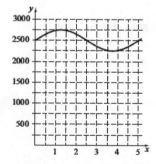

The interval from 0 to 2.5 is the person inhaling. The
interval from 2.5 to 5 is the person exhaling.

Prepare for Section 5.8

P1. Find the reciprocal.

$$\frac{3}{2\pi}$$

P3. Evaluate.

$$4\cos 2\pi(0) = 4\cos 0 = 4$$

P5. Evaluate.

$$4\cos\left(\sqrt{\frac{16}{4}} \cdot 2\pi\right) = 4\cos(4 \cdot 2\pi)$$
$$= 4\cos 8\pi = 4$$

Section 5.8 Exercises

1. Find the amplitude, period, and frequency.

$y = 2\sin 2t$

amplitude $= 2$

$p = \dfrac{2\pi}{2} = \pi$

frequency $= \dfrac{1}{p} = \dfrac{1}{\pi}$

3. Find the amplitude, period, and frequency.

$y = 3\cos\dfrac{2t}{3}$

amplitude $= 3$

$p = \dfrac{2\pi}{2/3} = 3\pi$

frequency $= \dfrac{1}{p} = \dfrac{1}{3\pi}$

5. Find the amplitude, period, and frequency.

$y = 4\cos \pi t$

amplitude $= 4$

$p = \dfrac{2\pi}{\pi} = 2$

frequency $= \dfrac{1}{p} = \dfrac{1}{2}$

7. Find the amplitude, period, and frequency.

$y = \dfrac{3}{4}\sin\dfrac{\pi t}{2}$

amplitude $= \dfrac{3}{4}$

$p = \dfrac{2\pi}{\pi/2} = 4$

frequency $= \dfrac{1}{p} = \dfrac{1}{4}$

9. Write the equation and graph.

$a = 4$

$p = \dfrac{1}{\text{frequency}} = \dfrac{1}{1.5} = \dfrac{2}{3}$

$\dfrac{2\pi}{b} = \dfrac{2}{3} \Rightarrow b = 3\pi$

$y = 4\cos 3\pi t$

11. Write the equation and graph.

$p = 1.5$

$\text{amplitude} = \dfrac{3}{2}$

$\dfrac{2\pi}{b} = 1.5 \Rightarrow b = \dfrac{4\pi}{3}$

$y = \dfrac{3}{2}\cos\dfrac{4\pi}{3}t$

13. Write the equation and graph.

$\text{amplitude} = 2$

$p = \pi$

$\dfrac{2\pi}{b} = \pi \Rightarrow b = 2$

$y = 2\sin 2t$

15. Write the equation and graph.

$\text{amplitude} = 1$

$p = 2$

$\dfrac{2\pi}{b} = 2 \Rightarrow b = \pi$

$y = \sin \pi t$

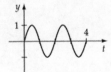

17. Write the equation and graph.

$\text{amplitude} = 2$

$\text{frequency} = 1$

$p = \dfrac{1}{\text{frequency}} = \dfrac{1}{1} = 1$

$\dfrac{2\pi}{b} = 1 \Rightarrow b = 2\pi$

$y = 2\sin 2\pi t$

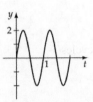

19. Write the equation.

$\text{amplitude} = \dfrac{1}{2}$

$\text{frequency} = \dfrac{2}{\pi} \Rightarrow p = \dfrac{\pi}{2}$

$\dfrac{2\pi}{b} = \dfrac{\pi}{2} \Rightarrow b = 4$

$y = \dfrac{1}{2}\cos 4t$

21. Write the equation.

$\text{amplitude} = 2.5$

$\text{frequency} = 0.5 \Rightarrow p = 2$

$\dfrac{2\pi}{b} = 2 \Rightarrow b = \pi$

$y = 2.5\cos \pi t$

23. Write the equation.

$\text{amplitude} = \dfrac{1}{2}$

$p = 3$

$\dfrac{2\pi}{b} = 3 \Rightarrow b = \dfrac{2}{3}\pi$

$y = \dfrac{1}{2}\cos\dfrac{2\pi}{3}t$

25. Write the equation.

$\text{amplitude} = 4$

$p = \dfrac{\pi}{2}$

$\dfrac{2\pi}{b} = \dfrac{\pi}{2} \Rightarrow b = 4$

$y = 4\cos 4t$

27. Write an equation of motion.

amplitude = 2 feet, $a = -2$

frequency $= f = \dfrac{1}{2\pi}\sqrt{\dfrac{k}{m}} = \dfrac{1}{2\pi}\sqrt{\dfrac{8}{32}} = \dfrac{1}{2\pi}\cdot\dfrac{1}{2} = \dfrac{1}{4\pi}$

period $= p = \dfrac{1}{f} = 4\pi$

$$y = a\cos 2\pi ft = -2\cos 2\pi\left(\dfrac{1}{4\pi}\right)t$$
$$= -2\cos\dfrac{1}{2}t$$

29. a. frequency $= f = \dfrac{392\pi}{2\pi} = 196$ cycles/s

period $= p = \dfrac{1}{196}$ s

b. The amplitude needs to increase.

31. amplitude $= a = \dfrac{78-4}{2} = 37$

period $= p = \dfrac{2\pi}{45} = \dfrac{\pi}{22.5}$

$$h = -37\cos\left(\dfrac{\pi}{22.5}t\right) + 41$$

33. a. The pseudoperiod is $\dfrac{2\pi}{2} = \pi$. There are $\dfrac{10}{\pi} \approx 3$ complete oscillations of length π in $0 \le t \le 10$.

b. $\left|f(t)\right| < 0.01$ for all $t > 59.8$ (nearest tenth).

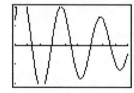

Xmin $= 56$, Xmax $= 65$, Xscl $= 1$,
Ymin $= -0.01$, Ymax $= 0.01$, Yscl $= 0.005$

35. a. The pseudoperiod is $\dfrac{2\pi}{2\pi} = 1$. There are 10 complete oscillations of length 1 in $0 \le t \le 10$.

b. $\left|f(t)\right| < 0.01$ for all $t > 71.0$ (nearest tenth).

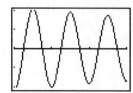

Xmin $= 70$, Xmax $= 73$, Xscl $= 1$
Ymin $= -0.01$, Ymax $= 0.01$, Yscl $= 0.005$

37. a. The pseudoperiod is $\dfrac{2\pi}{2\pi} = 1$. There are 10 complete oscillations of length 1 in $0 \le t \le 10$.

b. $\left|f(t)\right| < 0.01$ for all $t > 9.1$ (nearest tenth).

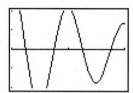

Xmin $= 8$, Xmax $= 10$, Xscl $= 1$,
Ymin $= -0.01$, Ymax $= 0.01$, Yscl $= 0.005$

39. a. The pseudoperiod is $\dfrac{2\pi}{2\pi} = 1$. There are 10 complete oscillations of length 1 in $0 \le t \le 10$.

b. $\left|f(t)\right| < 0.01$ for all $t > 6.1$ (nearest tenth).

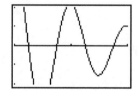

Xmin $= 5$, Xmax $= 7$, Xscl $= 1$,
Ymin $= -0.01$, Ymax $= .01$, Yscl $= 0.005$

41. $p_1 = 2\pi\sqrt{\dfrac{m}{k}}$, $p_2 = 2\pi\sqrt{\dfrac{9m}{k}} = 3p_1$

Increasing the main mass to $9m$ will triple the period.

43. a. Graph $y_1(t)$.

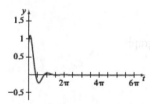

b. Graph $y_2(t)$.

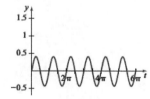

c. Graph $y_3(t)$.

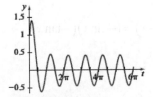

d. Explain.

After a few seconds, $y_1(t)$ is approximately zero, so the equation of motion of the system is basically given by $y_2(t)$.

Exploring Concepts with Technology

1. Using a graphing calculator,

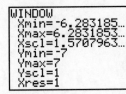

All three sine graphs have, a period of 2π, x-intercepts at $n\pi$, and no phase shift, but their amplitudes are 2, 4, and 6 respectively.

3. Using a graphing calculator,

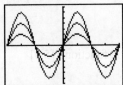

All three sine graphs have a period of 2π and an amplitude of 1, but their phase shifts are

$-\pi/4$, $-\pi/6$, and $-\pi/12$, respectively.

Chapter 5 Review Exercises

1. Find the complement and supplement. [5.1]

complement: $90° - 62° = 28°$

supplement: $180° - 62° = 118°$

3. Find the coterminal angle. [5.1]

$505° = 145° + 360°$

α is a quadrant II angle coterminal with an angle of measure $145°$.

5. Convert to radians. [5.1]

$$315° = 315°\left(\frac{\pi}{180°}\right) = \frac{7\pi}{4}$$

7. Find the arc length. [5.1]

$$s = r\theta = 3(75°)\left(\frac{\pi}{180°}\right) = 3.93 \text{ m}$$

9. Find the angular speed. [5.1]

$$\omega = \frac{\theta}{t} = \frac{4(2\pi)}{1} = 8\pi \text{ radians/sec}$$

11. Find the angular speed. [5.1]

$$\frac{V}{r} = \frac{50}{16} \cdot \frac{63360}{3600} \approx 55 \text{ rad/sec}$$

For exercise 13, $\csc\theta = \frac{3}{2} = \frac{r}{y}$, $r = 3$, $y = 2$,

and $x = \sqrt{3^2 - 2^2} = \sqrt{5}$.

13. Evaluate. [5.2]

$$\cot\theta = \frac{x}{y} = \frac{\sqrt{5}}{2}$$

15. Find the exact value. [5.2]

$$\sin 45° \cos 30° + \tan 60° = \frac{\sqrt{2}}{2} \cdot \frac{\sqrt{3}}{2} + \sqrt{3}$$

$$= \frac{\sqrt{6} + 4\sqrt{3}}{4}$$

17. Find the exact value. [5.2]

$$\sin^2 30° + \sec 60° = \left(\frac{1}{2}\right)^2 + 2 = 2\frac{1}{4}$$

19. Find the six trigonometric functions.

$$x = 1, \ y = -3, \ r = \sqrt{1^2 + (-3)^2} = \sqrt{10}$$

$$\sin\theta = -\frac{3}{\sqrt{10}} = -\frac{3\sqrt{10}}{10}$$

21. Find the reference angle. [5.3]

$\theta = 290°$

Since $270° < \theta < 360°$,

$\theta + \theta' = 360°$

$\theta' = 70°$

23. Find the reference angle. [5.3]

$\theta = \dfrac{3\pi}{4}$

Since $\dfrac{\pi}{2} < \theta < \pi$,

$\theta + \theta' = \pi$

$\theta' = \dfrac{\pi}{4}$

25. Find the exact value. [5.3]

  a. $\sec 150° = \dfrac{2}{-\sqrt{3}} = -\dfrac{2\sqrt{3}}{3}$

  b. $\tan\left(-\dfrac{3\pi}{4}\right) = 1$

  c. $\cot(-225°) = -1$

  d. $\cos\dfrac{2\pi}{3} = -\dfrac{1}{2}$

27. Find the exact value. [5.3]

$\cos\phi = -\dfrac{\sqrt{3}}{2} = \dfrac{x}{r},\ x = -\sqrt{3},\ r = 2,$

$y = -\sqrt{2^2 - \left(-\sqrt{3}\right)^2} = -1$

  a. $\sin\phi = \dfrac{y}{r} = -\dfrac{1}{2}$

  b. $\tan\phi = \dfrac{y}{x} = \dfrac{-1}{-\sqrt{3}} = \dfrac{\sqrt{3}}{3}$

29. Find the exact value. [5.3]

$\sin\phi = -\dfrac{\sqrt{2}}{2},\ y = -\sqrt{2},\ r = 2,$

$x = -\sqrt{2^2 - \left(-\sqrt{2}\right)^2} = \sqrt{2}$

  a. $\cos\phi = \dfrac{x}{r} = \dfrac{\sqrt{2}}{2}$

  b. $\cot\phi = \dfrac{x}{y} = \dfrac{\sqrt{2}}{-\sqrt{2}} = -1$

31. Find if the function is odd, even or neither. [5.4]

$f(x) = \sin(x)\tan(x)$

$f(-x) = \sin(-x)\tan(-x) = (-\sin x)(-\tan x)$

$\quad = \sin x \tan x$

$\quad = f(x)$

The function defined by $f(x) = \sin(x)\tan(x)$ is an even function.

33. Use the unit circle to show the identity.

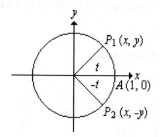

$\tan(-t) = -\dfrac{y}{x}$

$\tan t = \dfrac{y}{x}$

$\tan(-t) = -\tan t$

35. Write as a single trigonometric function. [5.4]

$\dfrac{\tan\phi + 1}{\cot\phi + 1} = \dfrac{\dfrac{\sin\phi}{\cos\phi} + 1}{\dfrac{\cos\phi}{\sin\phi} + 1}$

$\quad = \dfrac{\dfrac{\sin\phi + \cos\phi}{\cos\phi}}{\dfrac{\cos\phi + \sin\phi}{\sin\phi}}$

$\quad = \dfrac{\sin\phi(\sin\phi + \cos\phi)}{\cos\phi(\cos\phi + \sin\phi)}$

$\quad = \tan\phi$

37. Write as a single trigonometric function. [5.4]

$\sin^2\phi(\tan^2\phi + 1) = \sin^2\phi\sec^2\phi$

$\quad = \dfrac{\sin^2\phi}{\cos^2\phi}$

$\quad = \tan^2\phi$

39. Write as a single trigonometric function. [5.4]

$\dfrac{\cos^2\phi}{1 - \sin^2\phi} - 1 = \dfrac{1 - \sin^2\phi}{1 - \sin^2\phi} - 1$

$\quad = 1 - 1$

$\quad = 0$

41. State the amplitude, period, and phase shift. [5.6]

$$y = 2\tan 3x$$

no amplitude; period $= \dfrac{\pi}{b} = \dfrac{\pi}{3}$

phase shift $= 0$

43. State the amplitude, period, and phase shift. [5.5]

$$y = \cos\left(2x - \dfrac{2\pi}{3}\right) + 2$$

$a = |1| = 1$; period $= \dfrac{2\pi}{b} = \dfrac{2\pi}{2} = \pi$

phase shift $= -\dfrac{c}{b} = -\dfrac{-2\pi/3}{2} = \dfrac{\pi}{3}$

45. State the amplitude, period, and phase shift. [5.6]

$$y = 2\csc\left(x - \dfrac{\pi}{4}\right) - 3$$

no amplitude; period $= \dfrac{2\pi}{b} = \dfrac{2\pi}{1} = 2\pi$

phase shift $= -\dfrac{c}{b} = -\dfrac{-\pi/4}{1} = \dfrac{\pi}{4}$

47. Graph the function. [5.5]

$$y = -\sin\dfrac{2x}{3}, \; p = \dfrac{2\pi}{2/3} = 3\pi$$

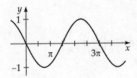

49. Graph the function. [5.7]

$$y = \cos\left(x - \dfrac{\pi}{2}\right), \; p = 2\pi$$

phase shift $= \dfrac{\pi}{2}$

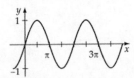

51. Graph the function. [5.7]

$$y = 3\cos 3(x - \pi), \; p = \dfrac{2\pi}{3}$$

phase shift $= \pi$

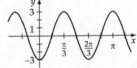

53. Graph the function. [5.6]

$$y = 2\cot 2x, \; p = \dfrac{\pi}{2}$$

55. Graph the function. [5.6]

$$y = 3\sec\dfrac{1}{2}x, \; p = 4\pi$$

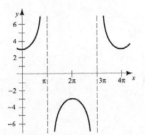

57. Graph the function. [5.7]

$$y = -\cot\left(2x + \dfrac{\pi}{4}\right), \; p = \dfrac{\pi}{2}$$

phase shift $= -\dfrac{\pi}{8}$

59. Graph the function. [5.7]

$$y = 3\sec\left(x + \dfrac{\pi}{4}\right), \; p = 2\pi$$

phase shift $= -\dfrac{\pi}{4}$

61. Graph the function. [5.7]

$$y = 2\cos 3x + 3$$

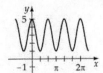

63. Graph the function. [5.7]

$$y = 3\sin\left(4x - \frac{2\pi}{3}\right) - 3$$

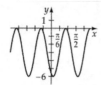

65. Graph the function. [5.7]

$$y = \sin x - \sqrt{3}\cos x$$

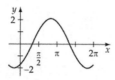

67. Graph the function. [5.7]

$$y = 3\cos x + \sin 2x$$

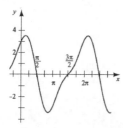

69. Find the height of the tree. [5.2]

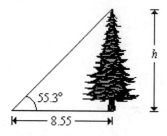

$$\tan 55.3° = \frac{h}{8.55}$$

$$h = 8.55\tan 55.3 \approx 12.3 \text{ feet}$$

71. Find the height of the building. [5.2]

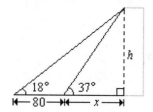

(1) $\cot 18° = \dfrac{80 + x}{h} = \dfrac{80}{h} + \dfrac{x}{h}$

(2) $\cot 37° = \dfrac{x}{h}$

Substitute for $\dfrac{x}{h}$ in equation (1).

$$\cot 18° = \frac{80}{h} + \cot 37°$$

Solve for h. $\dfrac{80}{h} = \cot 18° - \cot 37°$

$$\frac{h}{80} = \frac{1}{\cot 18° - \cot 37°}$$

$$h = \frac{80}{\cot 18° - \cot 37°} \approx 46 \text{ ft}$$

73. Write the simple harmonic motion equation. [5.8]

amplitude $= 5$, zero displacement

$p = 1$

$$\frac{2\pi}{b} = 1 \Rightarrow b = 2\pi$$

$$y = 5\sin 2\pi t$$

75. Write an equation of motion. [5.8]

amplitude $= 0.5$

$$f = \frac{1}{2\pi}\sqrt{\frac{k}{m}} = \frac{1}{2\pi}\sqrt{\frac{20}{5}} = \frac{1}{\pi}$$

$p = \pi$

$$y = -0.5\cos 2\pi ft = -0.5\cos 2\pi\left(\frac{1}{\pi}\right)t$$

$$y = -0.5\cos 2t$$

Chapter 5 Test

1. Convert. [5.1]

  a. $150° = 150°\left(\dfrac{\pi}{180°}\right) = \dfrac{5\pi}{6}$

  b. $\dfrac{2\pi}{5} = \dfrac{2\pi}{5}\left(\dfrac{180°}{\pi}\right) = 72°$

3. Find the arc length. [5.1]

$s = r\theta$

$$s = 10(75°)\left(\frac{\pi}{180°}\right)$$

$$s \approx 13.1 \text{ cm}$$

5. Find the linear speed. [5.1]

$v = rw$

$= 8 \cdot 10$

$= 80 \text{ cm/sec}$

7. Find the six trigonometric functions.

$$x = -2, \; y = -4, \; r = \sqrt{(-2)^2 + (-4)^2} = 2\sqrt{5}$$

$$\sin\theta = -\frac{2\sqrt{5}}{5} \quad \csc\theta = -\frac{\sqrt{5}}{2}$$

$$\cos\theta = -\frac{\sqrt{5}}{5} \quad \sec\theta = -\sqrt{5}$$

$$\tan\theta = 2 \quad\quad \cot\theta = \frac{1}{2}$$

9. Find the exact coordinates. [5.4]

$$t = \frac{11\pi}{6}$$

$$x = \cos t = \frac{\sqrt{3}}{2} \quad\quad y = \sin t = -\frac{1}{2}$$

$$W(x, y) = W\left(\frac{\sqrt{3}}{2}, -\frac{1}{2}\right)$$

11. State the period. [5.6]

$$\text{period} = \frac{\pi}{b} = \frac{\pi}{3}$$

13. State the period and phase shift. [5.7]

$$\text{period} = \frac{\pi}{\pi/3} = 3$$

$$\text{phase shift} = -\frac{c}{b} = -\frac{\pi/6}{\pi/3} = -\frac{1}{2}$$

15. Graph one full period.

$$y = -2\sec\frac{1}{2}x, \quad p = 4\pi$$

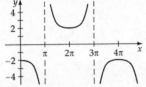

17. Graph one full period.

$$y = 2 - \sin\frac{x}{2}$$

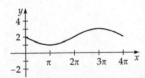

19. Find the height of the tree. [5.2]

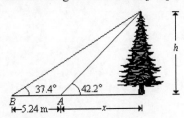

$$\tan 42.2° = \frac{h}{x}$$

$$x = \frac{h}{\tan 42.2°}$$

$$= h\cot 42.2°$$

$$\tan 37.4° = \frac{h}{5.24 + x}$$

$$= \frac{h}{5.24 + h\cot 42.2°}$$

Solve for h.

$$\tan 37.4° = \frac{h}{5.24 + h\cot 42.2°}$$

$$\tan 37.4°(5.24 + h\cot 42.2°) = h$$

$$5.24\tan 37.4° + h\tan 37.4°\cot 42.2° = h$$

$$5.24\tan 37.4° = h - h\tan 37.4°\cot 42.2°$$

$$5.24\tan 37.4° = h(1 - \tan 37.4°\cot 42.2°)$$

$$\frac{5.24\tan 37.4°}{1 - \tan 37.4°\cot 42.2°} = h$$

$$h \approx 25.5 \text{ meters}$$

The height of the tree is approximately 25.5 meters.

Cumulative Review Exercises

1. Factor. [P.4]

$$x^2 - y^2 = (x + y)(x - y)$$

3. Find the area. [1.2]

$$A = \frac{1}{2}bh$$

$$= \frac{1}{2}(4)(6)$$

$$= 12 \text{ in}^2$$

5. Find the inverse. [4.1]

$$f(x) = \frac{x}{2x - 3}$$

$$x = \frac{y}{2y - 3}$$

$$x(2y - 3) = 2xy - 3x = y$$

$$2xy - y = y(2x - 1) = 3x$$

$$y = \frac{3x}{2x - 1}$$

$$f^{-1}(x) = \frac{3x}{2x - 1}$$

7. State the range. [2.2]

Range: [0, 2]

9. Explain. [2.5]

Reflect the graph of $y = f(x)$ across the y-axis.

11. Convert to degrees. [5.1]

$$\frac{5\pi}{4} = \frac{5\pi}{4}\left(\frac{180°}{\pi}\right) = 225°$$

13. Evaluate. [5.2]

$$f\left(\frac{\pi}{3}\right) = \sin\left(\frac{\pi}{3}\right) + \sin\left(\frac{\pi}{6}\right) = \frac{\sqrt{3}}{2} + \frac{1}{2} = \frac{\sqrt{3}+1}{2}$$

15. Determine the sign. [5.3]

negative

17. Find the reference angle. [5.4]

$$\theta = \frac{2\pi}{3}$$

Since $\frac{\pi}{2} < \theta < \pi$,

$$\theta + \theta' = \pi$$

$$\theta' = \frac{\pi}{3}$$

19. Find the range. [5.4]

Range: $[-1, 1]$

Chapter 6 Trigonometric Identities and Equations

Section 6.1 Exercises

1. Determine if $\cos|x| = |\cos x|$ is an identity. Explain.

No. The left side of the equation does not equal the right side for $x = \pi$.

3. State the other two Pythagorean identities.

$\tan^2 x + 1 = \sec^2 x$ and $1 + \cot^2 x = \csc^2 x$

5. Verify the identity.

$$\tan x \csc x \cos x = \frac{\sin x}{\cos x} \cdot \frac{1}{\sin x} \cdot \cos x = 1$$

7. Verify the identity.

$$\sin(-x) - \cos(-x) = -\sin x - \cos x$$
$$= -(\sin x + \cos x)$$

9. Verify the identity.

$$(\sin x - \cos x)(\sin x + \cos x) = \sin^2 x - \cos^2 x$$
$$= 1 - \cos^2 x - \cos^2 x$$
$$= 1 - 2\cos^2 x$$

11. Verify the identity.

$$\frac{1}{\sin x} - \frac{1}{\cos x} = \frac{\cos x}{\sin x \cos x} - \frac{\sin x}{\sin x \cos x}$$
$$= \frac{\cos x - \sin x}{\sin x \cos x}$$

13. Verify the identity.

$$\frac{\cos x}{1 - \sin x} = \frac{\cos x(1 + \sin x)}{(1 - \sin x)(1 + \sin x)}$$
$$= \frac{\cos x(1 + \sin x)}{1 - \sin^2 x}$$
$$= \frac{\cos x(1 + \sin x)}{\cos^2 x}$$
$$= \frac{(1 + \sin x)}{\cos x} = \frac{1}{\cos x} + \frac{\sin x}{\cos x}$$
$$= \sec x + \tan x$$

15. Verify the identity.

$$\frac{1 - \tan^4 x}{\sec^2 x} = \frac{(1 + \tan^2 x)(1 - \tan^2 x)}{\sec^2 x}$$
$$= \frac{\sec^2 x(1 - \tan^2 x)}{\sec^2 x}$$
$$= 1 - \tan^2 x$$

17. Verify the identity.

$$\frac{1 + \tan^3 x}{1 + \tan x} = \frac{(1 + \tan x)(1 - \tan x + \tan^2 x)}{1 + \tan x}$$
$$= 1 - \tan x + \tan^2 x$$

19. Verify the identity.

$$\frac{\sin x - 2 + \frac{1}{\sin x}}{\sin x - \frac{1}{\sin x}} = \frac{\sin x - 2 + \frac{1}{\sin x}}{\sin x - \frac{1}{\sin x}} \cdot \frac{\sin x}{\sin x}$$
$$= \frac{\sin^2 x - 2\sin x + 1}{\sin^2 x - 1}$$
$$= \frac{(\sin x - 1)(\sin x - 1)}{(\sin x - 1)(\sin x + 1)}$$
$$= \frac{\sin x - 1}{\sin x + 1}$$

21. Verify the identity.

$$(\sin x + \cos x)^2 = \sin^2 x + 2\sin x \cos x + \cos^2 x$$
$$= \sin^2 x + \cos^2 x + 2\sin x \cos x$$
$$= 1 + 2\sin x \cos x$$

23. Verify the identity.

$$\frac{1}{\tan x + \sec x} = \frac{1}{\frac{\sin x}{\cos x} + \frac{1}{\cos x}}$$
$$= \frac{1}{\frac{\sin x + 1}{\cos x}}$$
$$= \frac{1}{1} \cdot \frac{\cos x}{\sin x + 1}$$
$$= \frac{\cos x}{1 + \sin x}$$

25. Verify the identity.

$$\frac{\cot x + \tan x}{\sec x} = \frac{\frac{\cos x}{\sin x} + \frac{\sin x}{\cos x}}{\frac{1}{\cos x}}$$
$$= \frac{\frac{\cos x}{\sin x} + \frac{\sin x}{\cos x}}{\frac{1}{\cos x}} \cdot \frac{\sin x \cos x}{\sin x \cos x}$$
$$= \frac{\cos^2 x + \sin^2 x}{\sin x}$$
$$= \frac{1}{\sin x}$$
$$= \csc x$$

27. Verify the identity.

$$\frac{\cos x \tan x + 2\cos x - \tan x - 2}{\tan x + 2}$$

$$= \frac{\cos x(\tan x + 2) - (\tan x + 2)}{\tan x + 2}$$

$$= \frac{(\tan x + 2)(\cos x - 1)}{\tan x + 2}$$

$$= \cos x - 1$$

29. Verify the identity.

$$\frac{1 - \sin x}{\cos x} = \frac{1}{\cos x} - \frac{\sin x}{\cos x} = \sec x - \tan x$$

31. Verify the identity.

$$1 - \sin^2 x \cot^2 x = 1 - \sin^2 x \frac{\cos^2 x}{\sin^2 x}$$

$$= 1 - \cos^2 x$$

$$= \sin^2 x$$

$$= \sin^2 x \cdot \frac{\cos^2 x}{\cos^2 x}$$

$$= \cos^2 x \cdot \frac{\sin^2 x}{\cos^2 x}$$

$$= \cos^2 x \tan^2 x$$

33. Verify the identity.

$$\frac{1}{\sin^2 x} + \frac{1}{\cos^2 x} = \frac{\cos^2 x + \sin^2 x}{\sin^2 x \cos^2 x}$$

$$= \frac{1}{\sin^2 x \cos^2 x}$$

$$= \csc^2 x \sec^2 x$$

35. Verify the identity.

$$\sec x - \cos x = \frac{1}{\cos x} - \cos x$$

$$= \frac{1 - \cos^2 x}{\cos x}$$

$$= \frac{\sin^2 x}{\cos x}$$

$$= \sin x \tan x$$

37. Verify the identity.

$$\frac{\dfrac{1}{\sin x} + 1}{\dfrac{1}{\sin x} - 1} = \frac{\dfrac{1}{\sin x} + 1}{\dfrac{1}{\sin x} - 1} \cdot \frac{\sin x}{\sin x}$$

$$= \frac{1 + \sin x}{1 - \sin x}$$

$$= \frac{(1 + \sin x)}{1 - \sin x} \cdot \frac{(1 + \sin x)}{1 + \sin x}$$

$$= \frac{1 + 2\sin x + \sin^2 x}{1 - \sin^2 x}$$

$$= \frac{1 + 2\sin x + \sin^2 x}{\cos^2 x}$$

$$= \frac{1}{\cos^2 x} + \frac{2\sin x}{\cos^2 x} + \frac{\sin^2 x}{\cos^2 x}$$

$$= \sec^2 x + 2\tan x \sec x + \tan^2 x$$

39. Verify the identity.

$$\sin^4 x - \cos^4 x = (\sin^2 x + \cos^2 x)(\sin^2 x - \cos^2 x)$$

$$= 1(\sin^2 x - \cos^2 x)$$

$$= \sin^2 x - (1 - \sin^2 x)$$

$$= \sin^2 x - 1 + \sin^2 x$$

$$= 2\sin^2 x - 1$$

41. Verify the identity.

$$\frac{1}{1 - \cos x} = \frac{1}{1 - \cos x} \cdot \frac{1 + \cos x}{1 + \cos x}$$

$$= \frac{1 + \cos x}{1 - \cos^2 x}$$

$$= \frac{1 + \cos x}{\sin^2 x}$$

43. Verify the identity.

$$\frac{\sin x}{1 - \sin x} - \frac{\cos x}{1 - \sin x} = \frac{\sin x - \cos x}{1 - \sin x}$$

$$= \frac{\dfrac{\sin x}{\sin x} - \dfrac{\cos x}{\sin x}}{\dfrac{1}{\sin x} - \dfrac{\sin x}{\sin x}}$$

$$= \frac{1 - \cot x}{\csc x - 1}$$

45. Verify the identity.

$$\frac{1}{1 + \sin \beta} + \frac{1}{1 - \sin \beta} = \frac{(1 - \sin \beta) + (1 + \sin \beta)}{(1 + \sin \beta)(1 - \sin \beta)}$$

$$= \frac{1 - \sin \beta + 1 + \sin \beta}{1 - \sin^2 \beta}$$

$$= \frac{2}{\cos^2 \beta}$$

$$= 2\sec^2 \beta$$

47. Verify the identity.

$$\frac{\dfrac{1}{\sin x} + \csc x}{\dfrac{1}{\sin x} - \sin x} = \frac{\dfrac{1}{\sin x} + \csc x}{\dfrac{1}{\sin x} - \sin x} \cdot \frac{\sin x}{\sin x}$$

$$= \frac{1 + 1}{1 - \sin^2 x}$$

$$= \frac{2}{\cos^2 x}$$

49. Verify the identity.

$$\frac{\cot x}{1+\csc x}+\frac{1+\csc x}{\cot x}=\frac{\dfrac{\cos x}{\sin x}}{1+\dfrac{1}{\sin x}}+\frac{1+\dfrac{1}{\sin x}}{\dfrac{\cos x}{\sin x}}$$

$$=\frac{\cos x}{\sin x+1}+\frac{\sin x+1}{\cos x}$$

$$=\frac{\cos^2 x+(\sin x+1)^2}{\cos x(\sin x+1)}$$

$$=\frac{\cos^2 x+\sin^2 x+2\sin x+1}{\cos x(\sin x+1)}$$

$$=\frac{1+2\sin x+1}{\cos x(\sin x+1)}$$

$$=\frac{2(1+\sin x)}{\cos x(\sin x+1)}$$

$$=2\sec x$$

51. Verify the identity.

$$\sqrt{\frac{1+\sin x}{1-\sin x}}=\sqrt{\frac{1+\sin x}{1-\sin x}\cdot\frac{1+\sin x}{1+\sin x}}$$

$$=\sqrt{\frac{(1+\sin x)^2}{1-\sin^2 x}}$$

$$=\sqrt{\frac{(1+\sin x)^2}{\cos^2 x}}$$

$$=\frac{1+\sin x}{\cos x},\ \cos x>0$$

53. Verify the identity.

$$\frac{\sin^3 x+\cos^3 x}{\sin x+\cos x}$$

$$=\frac{(\sin x+\cos x)(\sin^2 x-\sin x\cos x+\cos^2 x)}{\sin x+\cos x}$$

$$=\sin^2 x-\sin x\cos x+\cos^2 x$$

$$=1-\sin x\cos x$$

55. Verify the identity.

$$\frac{\sec x-1}{\sec x+1}-\frac{\sec x+1}{\sec x-1}$$

$$=\frac{(\sec x-1)^2-(\sec x+1)^2}{(\sec x-1)(\sec x+1)}$$

$$=\frac{\sec^2 x-2\sec x+1-\sec^2 x-2\sec x-1}{\sec^2 x-1}$$

$$=\frac{-4\sec x}{\tan^2 x}$$

$$=\frac{-4}{\cos x}\cdot\frac{\cos^2 x}{\sin^2 x}$$

$$=-4\csc x\cot x$$

57. Verify the identity.

$$\frac{1+\sin x}{\cos x}-\frac{\cos x}{1-\sin x}$$

$$=\frac{(1+\sin x)(1-\sin x)-\cos x(\cos x)}{\cos x(1-\sin x)}$$

$$=\frac{1-\sin^2 x-\cos^2 x}{\cos x(1-\sin x)}$$

$$=\frac{\cos^2 x-\cos^2 x}{\cos x(1-\sin x)}$$

$$=0$$

59. Verify the identity.

$$\frac{\sec x+\tan x}{\sec x-\tan x}=\frac{\dfrac{1}{\cos x}+\dfrac{\sin x}{\cos x}}{\dfrac{1}{\cos x}-\dfrac{\sin x}{\cos x}}\cdot\frac{\cos x}{\cos x}$$

$$=\frac{1+\sin x}{1-\sin x}$$

$$=\frac{1+\sin x}{1-\sin x}\cdot\frac{1+\sin x}{1+\sin x}$$

$$=\frac{(1+\sin x)^2}{1-\sin^2 x}$$

$$=\frac{(1+\sin x)^2}{\cos^2 x}$$

61. Compare the graphs; state whether identity.

Identity

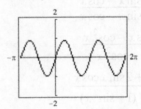

63. Compare the graphs; state whether identity.

Identity

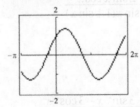

65. Compare the graphs; state whether identity.

Identity

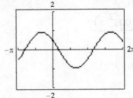

67. Compare the graphs; state whether identity.

Not an identity

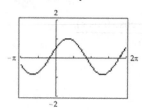

69. Find an x value to verify the equation is *not* an identity.

Not an identity. If $x = \pi/4$, the left side is 2 and the right side is 1.

71. Find an x value to verify the equation is *not* an identity.

Not an identity. If $x = 0°$, the left side is $\sqrt{3}/2$ and the right side is $(2 + \sqrt{3})/2$.

73. Find an x value to verify the equation is *not* an identity.

Not an identity. If $x = 0$, the left side is -1 and the right side is 1.

75. Verify the identity.

$$\frac{1 - \sin x + \cos x}{1 + \sin x + \cos x}$$

$$= \frac{1 - \sin x + \cos x}{1 + \sin x + \cos x} \cdot \frac{1 + \sin x - \cos x}{1 + \sin x - \cos x}$$

$$= \frac{1 - \sin^2 x + 2\sin x \cos x - \cos^2 x}{1 + 2\sin x + \sin^2 x - \cos^2 x}$$

$$= \frac{1 - (\sin^2 x + \cos^2 x) + 2\sin x \cos x}{1 + 2\sin x + \sin^2 x - (1 - \sin^2 x)}$$

$$= \frac{1 - 1 + 2\sin x \cos x}{1 + 2\sin x + \sin^2 x - 1 + \sin^2 x}$$

$$= \frac{2\sin x \cos x}{2\sin x + 2\sin^2 x} = \frac{2\sin x \cos x}{2\sin x(1 + \sin x)}$$

$$= \frac{\cos x}{1 + \sin x}$$

77. Verify the identity.

$$\frac{2\sin^4 x + 2\sin^2 x \cos^2 x - 3\sin^2 x - 3\cos^2 x}{2\sin^2 x}$$

$$= \frac{2\sin^2 x(\sin^2 x + \cos^2 x) - 3(\sin^2 x + \cos^2 x)}{2\sin^2 x}$$

$$= \frac{(2\sin^2 x - 3)(\sin^2 x + \cos^2 x)}{2\sin^2 x}$$

$$= \frac{2\sin^2 x - 3}{2\sin^2 x}$$

$$= \frac{2\sin^2 x}{2\sin^2 x} - \frac{3}{2\sin^2 x}$$

$$= 1 - \frac{3}{2}\csc^2 x$$

79. Verify the identity.

$$\frac{\sin x(\tan x + 1) - 2\tan x \cos x}{\sin x - \cos x}$$

$$= \frac{\sin x \tan x + \sin x - 2\dfrac{\sin x}{\cos x}\cos x}{\sin x - \cos x}$$

$$= \frac{\sin x \tan x + \sin x - 2\sin x}{\sin x - \cos x}$$

$$= \frac{\sin x(\tan x - 1)}{\sin x - \cos x}$$

$$= \frac{\sin x(\tan x - 1)}{\cos x}$$
$$= \frac{\dfrac{\sin x(\tan x - 1)}{\cos x}}{\dfrac{\sin x}{\cos x} - \dfrac{\cos x}{\cos x}}$$

$$= \frac{\tan x(\tan x - 1)}{\tan x - 1}$$

$$= \tan x$$

81. Verify the identity.

$$\sin^4 x + \cos^4 x$$

$$= \sin^4 x + 2\sin^2 x \cos^2 x + \cos^4 x - 2\sin^2 x \cos^2 x$$

$$= (\sin^2 x + \cos^2 x)^2 - 2\sin^2 x \cos^2 x$$

$$= 1 - 2\sin^2 x \cos^2 x$$

Prepare for Section 6.2

P1. Compare.

$$\cos\left(\frac{\pi}{2} - \frac{\pi}{6}\right) = \cos\left(\frac{\pi}{3}\right) = \frac{1}{2}$$

$$\cos\left(\frac{\pi}{2}\right)\cos\left(\frac{\pi}{6}\right) + \sin\left(\frac{\pi}{2}\right)\sin\left(\frac{\pi}{6}\right) = 0 \cdot \frac{\sqrt{3}}{2} + 1 \cdot \frac{1}{2} = \frac{1}{2}$$

Both functional values equal $\dfrac{1}{2}$.

P3. Compare.

$$\sin(90° - 30°) = \sin(60°) = \frac{\sqrt{3}}{2} = \cos(30°)$$

$$\sin(90° - 45°) = \sin(45°) = \frac{\sqrt{2}}{2} = \cos(45°)$$

$$\sin(90° - 120°) = \sin(-30°) = -\frac{1}{2} = \cos(120°)$$

For each of the given values of θ, the functional values are equal.

P5. Compare.

$$\tan\left(\frac{\pi}{3}-\frac{\pi}{6}\right)=\tan\left(\frac{\pi}{6}\right)=\frac{\sqrt{3}}{3}$$

$$\frac{\tan\left(\frac{\pi}{3}\right)-\tan\left(\frac{\pi}{6}\right)}{1+\tan\left(\frac{\pi}{3}\right)\tan\left(\frac{\pi}{6}\right)}=\frac{\sqrt{3}-\frac{\sqrt{3}}{3}}{1+\sqrt{3}\cdot\frac{\sqrt{3}}{3}}$$

$$=\frac{\frac{3\sqrt{3}}{3}-\frac{\sqrt{3}}{3}}{1+1}=\frac{\frac{2\sqrt{3}}{3}}{2}=\frac{\sqrt{3}}{3}$$

Both functional values equal $\frac{\sqrt{3}}{3}$.

Section 6.2 Exercises

1. Determine if $\sin(\alpha+\beta)=\sin\alpha+\sin\beta$ is an identity.

No. For a counterexample, let $\alpha=60°$ and $\beta=30°$.

Then $\sin(60°+30°)=\sin 90°=1$, whereas

$$\sin 60°+\sin 30°=\frac{\sqrt{3}+1}{2}.$$

3. For k is any integer, the value of $(2k+1)$ will result in odd integers.

5. Find the exact value of the expression.

$$\sin(45°+30°)=\sin 45°\cos 30°+\cos 45°\sin 30°$$

$$=\frac{\sqrt{2}}{2}\cdot\frac{\sqrt{3}}{2}+\frac{\sqrt{2}}{2}\cdot\frac{1}{2}$$

$$=\frac{\sqrt{6}}{4}+\frac{\sqrt{2}}{4}=\frac{\sqrt{6}+\sqrt{2}}{4}$$

7. Find the exact value of the expression.

$$\cos(45°-30°)=\cos 45°\cos 30°+\sin 45°\sin 30°$$

$$=\frac{\sqrt{2}}{2}\cdot\frac{\sqrt{3}}{2}+\frac{\sqrt{2}}{2}\cdot\frac{1}{2}$$

$$=\frac{\sqrt{6}}{4}+\frac{\sqrt{2}}{4}=\frac{\sqrt{6}+\sqrt{2}}{4}$$

9. Find the exact value of the expression.

$$\tan(135°-60°)=\frac{\tan 135°-\tan 60°}{1+\tan 135°\tan 60°}$$

$$=\frac{-1-\sqrt{3}}{1-1(\sqrt{3})}$$

$$=\frac{(-1-\sqrt{3})(1+\sqrt{3})}{(1-\sqrt{3})(1+\sqrt{3})}$$

$$=\frac{-1-2\sqrt{3}-3}{-2}=\frac{-4-2\sqrt{3}}{-2}$$

$$=2+\sqrt{3}$$

11. Find the exact value of the expression.

$$\sin\left(\frac{5\pi}{4}-\frac{\pi}{6}\right)=\sin\frac{5\pi}{4}\cos\frac{\pi}{6}-\cos\frac{5\pi}{4}\sin\frac{\pi}{6}$$

$$=-\frac{\sqrt{2}}{2}\cdot\frac{\sqrt{3}}{2}-\left(-\frac{\sqrt{2}}{2}\right)\cdot\frac{1}{2}$$

$$=-\frac{\sqrt{6}}{4}+\frac{\sqrt{2}}{4}=\frac{-\sqrt{6}+\sqrt{2}}{4}$$

13. Find the exact value of the expression.

$$\cos\left(\frac{3\pi}{4}+\frac{\pi}{6}\right)=\cos\frac{3\pi}{4}\cos\frac{\pi}{6}-\sin\frac{3\pi}{4}\sin\frac{\pi}{6}$$

$$=-\frac{\sqrt{2}}{2}\cdot\frac{\sqrt{3}}{2}-\frac{\sqrt{2}}{2}\cdot\frac{1}{2}$$

$$=-\frac{\sqrt{6}}{4}-\frac{\sqrt{2}}{4}=-\frac{\sqrt{6}+\sqrt{2}}{4}$$

15. Find the exact value of the expression.

$$\tan\left(\frac{\pi}{6}+\frac{\pi}{4}\right)=\frac{\tan\frac{\pi}{6}+\tan\frac{\pi}{4}}{1-\tan\frac{\pi}{6}\tan\frac{\pi}{4}}$$

$$=\frac{\frac{\sqrt{3}}{3}+1}{1-\frac{\sqrt{3}}{3}\cdot 1}=\frac{\frac{\sqrt{3}+3}{3}}{\frac{3-\sqrt{3}}{3}}$$

$$=\frac{\sqrt{3}+3}{3-\sqrt{3}}\cdot\frac{3+\sqrt{3}}{3+\sqrt{3}}=\frac{9+6\sqrt{3}+3}{9-3}$$

$$=\frac{12+6\sqrt{3}}{6}=2+\sqrt{3}$$

17. Find the exact value of the expression.

$$\cos 60°\cos 15°+\sin 60°\sin 15°$$

$$=\cos(60°-15°)=\cos 45°=\frac{\sqrt{2}}{2}$$

19. Find the exact value of the expression.

$$\sin\frac{5\pi}{12}\cos\frac{\pi}{4}-\cos\frac{5\pi}{12}\sin\frac{\pi}{4}=\sin\left(\frac{5\pi}{12}-\frac{\pi}{4}\right)=\sin\frac{\pi}{6}=\frac{1}{2}$$

21. Find the exact value of the expression.

$$\frac{\tan\frac{7\pi}{12}-\tan\frac{\pi}{4}}{1+\tan\frac{7\pi}{12}\tan\frac{\pi}{4}}=\tan\left(\frac{7\pi}{12}-\frac{\pi}{4}\right)=\tan\frac{\pi}{3}=\sqrt{3}$$

23. Write an equivalent expression.

$$\sin 21°=\cos(90°-21°)=\cos 69°$$

25. Write an equivalent expression.

$$\tan 55°=\cot(90°-55°)=\cot 35°$$

27. Write an equivalent expression.

$$\sec 8° = \csc(90° - 8°) = \csc 82°$$

29. Write the expression as a single trigonometric function.

$$\sin 7x \cos 2x - \cos 7x \sin 2x = \sin(7x - 2x) = \sin 5x$$

31. Write the expression as a single trigonometric function.

$$\cos x \cos 2x + \sin x \sin 2x = \cos(x - 2x)$$
$$= \cos(-x) = \cos x$$

33. Write the expression as a single trigonometric function.

$$\sin 5x \cos 4x + \cos 5x \sin 4x = \sin(5x + 4x) = \sin 9x$$

35. Write the expression as a single trigonometric function.

$$\cos 2x \cos(-x) - \sin 2x \sin(-x)$$
$$= \cos 2x \cos x + \sin 2x \sin x$$
$$= \cos(2x - x)$$
$$= \cos x$$

37. Write the expression as a single trigonometric function.

$$\sin \frac{x}{3} \cos \frac{2x}{3} + \cos \frac{x}{3} \sin \frac{2x}{3} = \sin\left(\frac{x}{3} + \frac{2x}{3}\right) = \sin x$$

39. Write the expression as a single trigonometric function.

$$\frac{\tan 3x + \tan 4x}{1 - \tan 3x \tan 4x} = \tan(3x + 4x) = \tan 7x$$

41. Find the exact value.

$$\tan \alpha = -\frac{4}{3}, \; \sin \alpha = \frac{4}{5}, \; \cos \alpha = -\frac{3}{5},$$

$$\tan \beta = \frac{15}{8}, \; \sin \beta = -\frac{15}{17}, \; \cos \beta = -\frac{8}{17}$$

a. $\sin(\alpha - \beta) = \sin \alpha \cos \beta - \cos \alpha \sin \beta$

$$= \left(\frac{4}{5}\right)\left(-\frac{8}{17}\right) - \left(-\frac{3}{5}\right)\left(-\frac{15}{17}\right)$$

$$= -\frac{32}{85} - \frac{45}{85} = -\frac{77}{85}$$

b. $\cos(\alpha + \beta) = \cos \alpha \cos \beta - \sin \alpha \sin \beta$

$$= \left(-\frac{3}{5}\right)\left(-\frac{8}{17}\right) - \left(\frac{4}{5}\right)\left(-\frac{15}{17}\right)$$

$$= \frac{24}{85} + \frac{60}{85} = \frac{84}{85}$$

c. $\tan(\alpha - \beta) = \dfrac{\tan \alpha - \tan \beta}{1 + \tan \alpha \tan \beta}$

$$= \frac{-\dfrac{4}{3} - \dfrac{15}{8}}{1 + \left(-\dfrac{4}{3}\right)\left(\dfrac{15}{8}\right)} = \frac{-\dfrac{4}{3} - \dfrac{15}{8}}{1 - \dfrac{60}{24}} \cdot \frac{24}{24}$$

$$= \frac{-32 - 45}{24 - 60} = \frac{77}{36}$$

43. Find the exact value.

$$\sin \alpha = \frac{3}{5}, \; \cos \alpha = \frac{4}{5}, \; \tan \alpha = \frac{3}{4},$$

$$\cos \beta = -\frac{5}{13}, \; \sin \beta = \frac{12}{13}, \; \tan \beta = -\frac{12}{5}$$

a. $\sin(\alpha - \beta) = \sin \alpha \cos \beta - \cos \alpha \sin \beta$

$$= \frac{3}{5}\left(-\frac{5}{13}\right) - \frac{4}{5}\left(\frac{12}{13}\right)$$

$$= -\frac{15}{65} - \frac{48}{65} = -\frac{63}{65}$$

b. $\cos(\alpha + \beta) = \cos \alpha \cos \beta - \sin \alpha \sin \beta$

$$= \frac{4}{5}\left(-\frac{5}{13}\right) - \frac{3}{5}\left(\frac{12}{13}\right)$$

$$= -\frac{20}{65} - \frac{36}{65} = -\frac{56}{65}$$

c. $\tan(\alpha - \beta) = \dfrac{\tan \alpha - \tan \beta}{1 + \tan \alpha \tan \beta}$

$$= \frac{\dfrac{3}{4} - \left(-\dfrac{12}{5}\right)}{1 + \left(\dfrac{3}{4}\right)\left(-\dfrac{12}{5}\right)} = \frac{\dfrac{3}{4} + \dfrac{12}{5}}{1 - \dfrac{36}{20}} \cdot \frac{20}{20}$$

$$= \frac{15 + 48}{20 - 36} = -\frac{63}{16}$$

45. Find the exact value.

$$\sin \alpha = -\frac{4}{5}, \; \cos \alpha = -\frac{3}{5}, \; \tan \alpha = \frac{4}{3},$$

$$\cos \beta = -\frac{12}{13}, \; \sin \beta = \frac{5}{13}, \; \tan \beta = -\frac{5}{12}$$

a. $\sin(\alpha - \beta) = \sin \alpha \cos \beta - \cos \alpha \sin \beta$

$$= \left(-\frac{4}{5}\right)\left(-\frac{12}{13}\right) - \left(-\frac{3}{5}\right)\frac{5}{13}$$

$$= \frac{48}{65} + \frac{15}{65} = \frac{63}{65}$$

b. $\cos(\alpha + \beta) = \cos \alpha \cos \beta - \sin \alpha \sin \beta$

$$= \left(-\frac{3}{5}\right)\left(-\frac{12}{13}\right) - \left(-\frac{4}{5}\right)\frac{5}{13}$$

$$= \frac{36}{65} + \frac{20}{65} = \frac{56}{65}$$

c. $\tan(\alpha + \beta) = \dfrac{\tan \alpha + \tan \beta}{1 - \tan \alpha \tan \beta}$

$$= \frac{\dfrac{4}{3} + \left(-\dfrac{5}{12}\right)}{1 - \left(\dfrac{4}{3}\right)\left(-\dfrac{5}{12}\right)} = \frac{\dfrac{4}{3} - \dfrac{5}{12}}{1 + \dfrac{20}{36}} = \frac{\dfrac{4}{3} - \dfrac{5}{12}}{1 + \dfrac{20}{36}} \cdot \frac{36}{36}$$

$$= \frac{48 - 15}{36 + 20} = \frac{33}{56}$$

47. Find the exact value.

$$\cos\alpha=\frac{15}{17},\ \sin\alpha=\frac{8}{17},\ \tan\alpha=\frac{8}{15},$$

$$\sin\beta=-\frac{3}{5},\ \cos\beta=-\frac{4}{5},\ \tan\beta=\frac{3}{4}$$

a. $\sin(\alpha+\beta)=\sin\alpha\cos\beta+\cos\alpha\sin\beta$

$$=\frac{8}{17}\left(-\frac{4}{5}\right)+\frac{15}{17}\left(-\frac{3}{5}\right)$$

$$=-\frac{32}{85}-\frac{45}{85}=-\frac{77}{85}$$

b. $\cos(\alpha-\beta)=\cos\alpha\cos\beta+\sin\alpha\sin\beta$

$$=\frac{15}{17}\left(-\frac{4}{5}\right)+\frac{8}{17}\left(-\frac{3}{5}\right)$$

$$=-\frac{60}{85}-\frac{24}{85}=-\frac{84}{85}$$

c. $\tan(\alpha-\beta)=\dfrac{\tan\alpha-\tan\beta}{1+\tan\alpha\tan\beta}$

$$=\frac{\frac{8}{15}-\frac{3}{4}}{1+\frac{8}{15}\left(\frac{3}{4}\right)}=\frac{\frac{8}{15}-\frac{3}{4}}{1+\frac{24}{60}}=\frac{\frac{8}{15}-\frac{3}{4}}{1+\frac{24}{60}}\cdot\frac{60}{60}$$

$$=\frac{32-45}{60+24}=-\frac{13}{84}$$

49. Find the exact value.

$$\cos\alpha=-\frac{3}{5},\ \sin\alpha=-\frac{4}{5},\ \tan\alpha=\frac{4}{3},$$

$$\sin\beta=\frac{5}{13},\ \cos\beta=\frac{12}{13},\ \tan\beta=\frac{5}{12}$$

a. $\sin(\alpha-\beta)=\sin\alpha\cos\beta-\cos\alpha\sin\beta$

$$=\left(-\frac{4}{5}\right)\frac{12}{13}-\left(-\frac{3}{5}\right)\left(\frac{5}{13}\right)$$

$$=-\frac{48}{65}+\frac{15}{65}=-\frac{33}{65}$$

b. $\cos(\alpha+\beta)=\cos\alpha\cos\beta-\sin\alpha\sin\beta$

$$=\left(-\frac{3}{5}\right)\frac{12}{13}-\left(-\frac{4}{5}\right)\frac{5}{13}$$

$$=-\frac{36}{65}+\frac{20}{65}=-\frac{16}{65}$$

c. $\tan(\alpha+\beta)=\dfrac{\tan\alpha+\tan\beta}{1-\tan\alpha\tan\beta}$

$$=\frac{\frac{4}{3}+\frac{5}{12}}{1-\frac{4}{3}\left(\frac{5}{12}\right)}=\frac{\frac{4}{3}+\frac{5}{12}}{1-\frac{20}{36}}$$

$$=\frac{\frac{4}{3}+\frac{5}{12}}{1-\frac{20}{36}}\cdot\frac{36}{36}=\frac{48+15}{36-20}=\frac{63}{16}$$

51. Find the exact value.

$$\sin\alpha=\frac{3}{5},\ \cos\alpha=\frac{4}{5},\ \tan\alpha=\frac{3}{4},$$

$$\tan\beta=\frac{5}{12},\ \sin\beta=-\frac{5}{13},\ \cos\beta=-\frac{12}{13}$$

a. $\sin(\alpha+\beta)=\sin\alpha\cos\beta+\cos\alpha\sin\beta$

$$=\left(\frac{3}{5}\right)\left(-\frac{12}{13}\right)+\left(\frac{4}{5}\right)\left(-\frac{5}{13}\right)$$

$$=-\frac{36}{65}-\frac{20}{65}=-\frac{56}{65}$$

b. $\cos(\alpha-\beta)=\cos\alpha\cos\beta+\sin\alpha\sin\beta$

$$=\left(\frac{4}{5}\right)\left(-\frac{12}{13}\right)+\left(\frac{3}{5}\right)\left(-\frac{5}{13}\right)$$

$$=-\frac{48}{65}-\frac{15}{65}=-\frac{63}{65}$$

c. $\tan(\alpha-\beta)=\dfrac{\tan\alpha-\tan\beta}{1+\tan\alpha\tan\beta}$

$$=\frac{\frac{3}{4}-\frac{5}{12}}{1+\left(\frac{3}{4}\right)\left(\frac{5}{12}\right)}=\frac{\frac{3}{4}-\frac{5}{12}}{1+\frac{15}{48}}\cdot\frac{48}{48}$$

$$=\frac{36-20}{48+15}=\frac{16}{63}$$

53. Verify the identity.

$$\cos\left(\frac{\pi}{2}-\theta\right)=\cos\frac{\pi}{2}\cos\theta+\sin\frac{\pi}{2}\sin\theta$$

$$=0\cdot\cos\theta+1\cdot\sin\theta$$

$$=\sin\theta$$

55. Verify the identity.

$$\sin\left(\theta+\frac{\pi}{2}\right)=\sin\theta\cos\frac{\pi}{2}+\cos\theta\sin\frac{\pi}{2}$$

$$=\sin\theta(0)+\cos\theta(1)$$

$$=\cos\theta$$

57. Verify the identity.

$$\tan\left(\theta+\frac{\pi}{4}\right)=\frac{\tan\theta+\tan\frac{\pi}{4}}{1-\tan\theta\tan\frac{\pi}{4}}$$

$$=\frac{\tan\theta+1}{1-\tan\theta}$$

59. Verify the identity.

$$\cos\left(\frac{3\pi}{2}-\theta\right)=\cos\frac{3\pi}{2}\cos\theta+\sin\frac{3\pi}{2}\sin\theta$$

$$=0(\cos\theta)+(-1)\sin\theta$$

$$=-\sin\theta$$

61. Verify the identity.

$$\cot\left(\frac{\pi}{2}-\theta\right) = \frac{\cos(\pi/2-\theta)}{\sin(\pi/2-\theta)}$$

$$= \frac{\left(\cos\frac{\pi}{2}\right)\cos\theta + \left(\sin\frac{\pi}{2}\right)\sin\theta}{\left(\sin\frac{\pi}{2}\right)\cos\theta - \left(\cos\frac{\pi}{2}\right)\sin\theta}$$

$$= \frac{(0)\cos\theta + (1)\sin\theta}{(1)\cos\theta - (0)\sin\theta}$$

$$= \frac{\sin\theta}{\cos\theta}$$

$$= \tan\theta$$

63. Verify the identity.

$$\csc(\pi-\theta) = \frac{1}{\sin(\pi-\theta)}$$

$$= \frac{1}{\sin\pi\cos\theta - \cos\pi\sin\theta}$$

$$= \frac{1}{(0)\cos\theta - (-1)\sin\theta}$$

$$= \frac{1}{\sin\theta}$$

$$= \csc\theta$$

65. Verify the identity.

$$\sin 6x\cos 2x - \cos 6x\sin 2x$$

$$= \sin(6x-2x)$$

$$= \sin 4x$$

$$= \sin(2x+2x)$$

$$= \sin 2x\cos 2x + \cos 2x\sin 2x$$

$$= 2\sin 2x\cos 2x$$

67. Verify the identity.

$$\cos(\alpha+\beta)+\cos(\alpha-\beta)$$

$$= \cos\alpha\cos\beta - \sin\alpha\sin\beta + \cos\alpha\cos\beta + \sin\alpha\sin\beta$$

$$= 2\cos\alpha\cos\beta$$

69. Verify the identity.

$$\sin(\alpha+\beta)+\sin(\alpha-\beta)$$

$$= \sin\alpha\cos\beta + \cos\alpha\sin\beta + \sin\alpha\cos\beta - \cos\alpha\sin\beta$$

$$= 2\sin\alpha\cos\beta$$

71. Verify the identity.

$$\frac{\cos(\alpha-\beta)}{\sin(\alpha+\beta)} = \frac{\cos\alpha\cos\beta + \sin\alpha\sin\beta}{\sin\alpha\cos\beta + \cos\alpha\sin\beta}$$

$$= \frac{\dfrac{\cos\alpha\cos\beta}{\sin\alpha\cos\beta} + \dfrac{\sin\alpha\sin\beta}{\sin\alpha\cos\beta}}{\dfrac{\sin\alpha\cos\beta}{\sin\alpha\cos\beta} + \dfrac{\cos\alpha\sin\beta}{\sin\alpha\cos\beta}}$$

$$= \frac{\cot\alpha + \tan\beta}{1 + \cot\alpha\tan\beta}$$

73. Verify the identity.

$$\frac{\sin(x+h)-\sin x}{h} = \frac{\sin x\cos h + \cos x\sin h - \sin x}{h}$$

$$= \frac{\sin x(\cos h - 1)}{h} + \frac{\cos x\sin h}{h}$$

$$= \sin x\frac{\cos h - 1}{h} + \cos x\frac{\sin h}{h}$$

75. Verify the identity.

$$\sin\left(\frac{\pi}{2}+\alpha-\beta\right) = \sin\left[\frac{\pi}{2}+(\alpha-\beta)\right]$$

$$= \sin\frac{\pi}{2}\cos(\alpha-\beta) + \cos\frac{\pi}{2}\sin(\alpha-\beta)$$

$$= (1)\cos(\alpha-\beta) + (0)\sin(\alpha-\beta)$$

$$= \cos(\alpha-\beta)$$

$$= \cos\alpha\cos\beta + \sin\alpha\sin\beta$$

77. Verify the identity.

$$\sin 3x = \sin(2x+x)$$

$$= \sin 2x\cos x + \cos 2x\sin x$$

$$= \sin(x+x)\cos x + \cos(x+x)\sin x$$

$$= (\sin x\cos x + \cos x\sin x)\cos x$$

$$\quad + (\cos x\cos x - \sin x\sin x)\sin x$$

$$= 2\sin x\cos^2 x + \sin x\cos^2 x - \sin^3 x$$

$$= 3\sin x\cos^2 x - \sin^3 x$$

$$= 3\sin x(1-\sin^2 x) - \sin^3 x$$

$$= 3\sin x - 3\sin^3 x - \sin^3 x$$

$$= 3\sin x - 4\sin^3 x$$

79. Simplify the expression.

$$\cos(\theta+3\pi) = \cos\theta\cos 3\pi - \sin\theta\sin 3\pi$$

$$= (\cos\theta)(-1) - (\sin\theta)(0)$$

$$= -\cos\theta$$

81. Simplify the expression.

$$\tan(\theta+\pi) = \frac{\tan\theta + \tan\pi}{1 - \tan\theta\tan\pi}$$

$$= \frac{\tan\theta + 0}{1 - (\tan\theta)(0)}$$

$$= \tan\theta$$

83. Simplify the expression.

$$\sin(\theta+2k\pi) = \sin\theta\cos(2k\pi) + \cos\theta\sin 2k\pi$$

$$= (\sin\theta)(1) + (\cos\theta)(0)$$

$$= \sin\theta$$

85. Compare the graphs.

$y = \sin\left(\dfrac{\pi}{2} - x\right)$ and $y = \cos x$ both have the following graph.

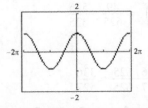

87. Compare the graphs.

$y = \sin 7x \ \cos 2x - \cos 7x \ \sin 2x$ and $y = \sin 5x$ both have the following graph.

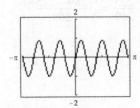

89. Verify the identity.

$\sin(x - y) \cdot \sin(x + y)$
$= (\sin x \cos y - \cos x \sin y)(\sin x \cos y + \cos x \sin y)$
$= \sin^2 x \cos^2 y - \cos^2 x \sin^2 y$

91. Verify the identity.

$\cos(x + y + z)$
$= \cos[x + (y + z)]$
$= \cos x \cos(y + z) - \sin x \sin(y + z)$
$= \cos x[\cos y \cos z - \sin y \sin z]$
$\quad - \sin x[\sin y \cos z + \cos y \sin z]$
$= \cos x \cos y \cos z - \cos x \sin y \sin z$
$\quad - \sin x \sin y \cos z - \sin x \cos y \sin z$

93. Verify the identity.

$\dfrac{\cos(x - y)}{\cos x \sin y} = \dfrac{\cos x \cos y + \sin x \sin y}{\cos x \sin y}$

$\quad = \dfrac{\cos x \cos y}{\cos x \sin y} + \dfrac{\sin x \sin y}{\cos x \sin y}$

$\quad = \cot y + \tan x$

Prepare for Section 6.3

P1. Use the identity to rewrite the expression.

$\sin 2\alpha = \sin(\alpha + \alpha)$
$\quad = \sin \alpha \cos \alpha + \cos \alpha \sin \alpha$
$\quad = 2 \sin \alpha \cos \alpha$

P3. Use the identity to rewrite the expression.

$\tan 2\alpha = \tan(\alpha + \alpha)$
$\quad = \dfrac{\tan \alpha + \tan \alpha}{1 - \tan \alpha \tan \alpha}$
$\quad = \dfrac{2 \tan \alpha}{1 - \tan^2 \alpha}$

P5. Verify equation is not an identity.

Let $\alpha = 45°$; then the left side of the equation is 1, and the right side of the equation is $\sqrt{2}$.

Section 6.3 Exercises

1. Determine if $\tan 2\alpha = \dfrac{2 \tan \alpha}{1 - \tan^2 \alpha}$ is a double-angle identity.

Yes.

3. Explain.

Both $1 - 2 \sin^2 x$ and $2 \cos^2 x - 1$ are equal to $\cos 2x$.

5. Explain.

The choice on whether to use the plus sign or minus sign depends on the quadrant in which $\dfrac{\alpha}{2}$ lies.

7. Write as a single expression.

$2 \sin 2\alpha \cos 2\alpha = \sin 2(2\alpha) = \sin 4\alpha$

9. Write as a single expression.

$\cos^2 \dfrac{\beta}{2} - \sin^2 \dfrac{\beta}{2} = \cos 2\left(\dfrac{\beta}{2}\right) = \cos \beta$

11. Write as a single expression.

$\cos^2 3\alpha - \sin^2 3\alpha = \cos 2(3\alpha) = \cos 6\alpha$

13. Write as a single expression.

$\dfrac{2 \tan 3\alpha}{1 - \tan^2 3\alpha} = \tan 2(3\alpha) = \tan 6\alpha$

15. Find the exact value of $\sin 2\alpha, \ \cos 2\alpha, \ \tan 2\alpha$.

$\text{hypotenuse} = \sqrt{1^2 + 3^2} = \sqrt{10}$

$\sin \alpha = \dfrac{1}{\sqrt{10}}, \quad \cos \alpha = \dfrac{3}{\sqrt{10}},$

$\tan \alpha = \dfrac{\sin \alpha}{\cos \alpha} = \dfrac{1/\sqrt{10}}{3/\sqrt{10}} = \dfrac{1}{3}$

$\sin 2\alpha = 2 \sin \alpha \cos \alpha$

$\quad = 2\left(\dfrac{1}{\sqrt{10}}\right)\left(\dfrac{3}{\sqrt{10}}\right) = \dfrac{6}{10} = \dfrac{3}{5}$

$$\cos 2\alpha = \cos^2\alpha - \sin^2\alpha$$
$$= \left(\frac{3}{\sqrt{10}}\right)^2 - \left(\frac{1}{\sqrt{10}}\right)^2$$
$$= \frac{9-1}{10} = \frac{4}{5}$$

$$\tan 2\alpha = \frac{\sin 2\alpha}{\cos 2\alpha} = \frac{3/5}{4/5} = \frac{3}{4}$$

17. Find the exact values of $\sin 2\alpha,\ \cos 2\alpha,\ \tan 2\alpha$.

$$\cos\alpha = -\frac{4}{5},\ \sin\alpha = \sqrt{1-\left(\frac{4}{5}\right)^2} = \frac{3}{5},$$

$$\tan\alpha = \frac{\sin\alpha}{\cos\alpha} = \frac{3/5}{-4/5} = -\frac{3}{4}$$

$$\sin 2\alpha = 2\sin\alpha\cos\alpha$$
$$= 2\left(\frac{3}{5}\right)\left(-\frac{4}{5}\right) = -\frac{24}{25}$$

$$\cos 2\alpha = \cos^2\alpha - \sin^2\alpha$$
$$= \left(-\frac{4}{5}\right)^2 - \left(\frac{3}{5}\right)^2 = \frac{16}{25} - \frac{9}{25} = \frac{7}{25}$$

$$\tan 2\alpha = \frac{\sin 2\alpha}{\cos 2\alpha} = \frac{-24/25}{7/25} = -\frac{24}{7}$$

19. Find the exact values of $\sin 2\alpha,\ \cos 2\alpha,\ \tan 2\alpha$.

$$\sin\alpha = \frac{3}{4},\ \cos\alpha = -\sqrt{1-\left(\frac{3}{4}\right)^2} = -\frac{\sqrt{7}}{4}$$

$$\sin 2\alpha = 2\sin\alpha\cos\alpha$$
$$= 2\left(\frac{3}{4}\right)\left(-\frac{\sqrt{7}}{4}\right) = -\frac{3\sqrt{7}}{8}$$

$$\cos 2\alpha = \cos^2\alpha - \sin^2\alpha$$
$$= \left(-\frac{\sqrt{7}}{4}\right)^2 - \left(\frac{3}{4}\right)^2 = \frac{7}{16} - \frac{9}{16} = -\frac{1}{8}$$

$$\tan 2\alpha = \frac{\sin 2\alpha}{\cos 2\alpha} = \frac{-\dfrac{3\sqrt{7}}{8}}{-\dfrac{1}{8}} = 3\sqrt{7}$$

21. Find the exact values of $\sin 2\alpha,\ \cos 2\alpha,\ \tan 2\alpha$.

$$\tan\alpha = -\frac{24}{7},$$

$$r = \sqrt{24^2 + 7^2} = \sqrt{576+49} = \sqrt{625} = 25$$

$$\sin\alpha = -\frac{24}{25},\ \cos\alpha = \frac{7}{25}$$

$$\sin 2\alpha = 2\sin\alpha\cos\alpha$$
$$= 2\left(-\frac{24}{25}\right)\left(\frac{7}{25}\right) = -\frac{336}{625}$$

$$\cos 2\alpha = \cos^2\alpha - \sin^2\alpha$$
$$= \left(\frac{7}{25}\right)^2 - \left(-\frac{24}{25}\right)^2 = \frac{49}{625} - \frac{576}{625}$$
$$= -\frac{527}{625}$$

$$\tan 2\alpha = \frac{\sin 2\alpha}{\cos 2\alpha} = \frac{-336/625}{-527/625} = \frac{336}{527}$$

23. Find the exact values of $\sin 2\alpha,\ \cos 2\alpha,\ \tan 2\alpha$.

$$\sin\alpha = \frac{15}{17},$$

$$\cos\alpha = \sqrt{1-\left(\frac{15}{17}\right)^2} = \sqrt{1-\frac{225}{289}} = \sqrt{\frac{289-225}{289}}$$
$$= \sqrt{\frac{64}{289}} = \frac{8}{17},$$

$$\tan\alpha = \frac{15/17}{8/17} = \frac{15}{8}$$

$$\sin 2\alpha = 2\sin\alpha\cos\alpha$$
$$= 2\left(\frac{15}{17}\right)\left(\frac{8}{17}\right) = \frac{240}{289}$$

$$\cos 2\alpha = \cos^2\alpha - \sin^2\alpha$$
$$= \left(\frac{8}{17}\right)^2 - \left(\frac{15}{17}\right)^2 = \frac{64}{289} - \frac{225}{289}$$
$$= -\frac{161}{289}$$

$$\tan 2\alpha = \frac{\sin 2\alpha}{\cos 2\alpha} = \frac{240/289}{-161/289} = -\frac{240}{161}$$

25. Find the exact values of $\sin 2\alpha,\ \cos 2\alpha,\ \tan 2\alpha$.

$$\cos\alpha = -\frac{2}{3},\ \sin\alpha = \sqrt{1-\left(-\frac{2}{3}\right)^2} = \frac{\sqrt{5}}{3},$$

$$\tan\alpha = \frac{-2/3}{\sqrt{5}/3} = -\frac{2}{\sqrt{5}}$$

$$\sin 2\alpha = 2\sin\alpha\cos\alpha$$
$$= 2\left(-\frac{2}{3}\right)\left(\frac{\sqrt{5}}{3}\right) = -\frac{4\sqrt{5}}{9}$$

$$\cos 2\alpha = \cos^2\alpha - \sin^2\alpha$$
$$= \left(-\frac{2}{3}\right)^2 - \left(\frac{\sqrt{5}}{3}\right)^2 = \frac{4}{9} - \frac{5}{9} = -\frac{1}{9}$$

$$\tan 2\alpha = \frac{\sin 2\alpha}{\cos 2\alpha} = \frac{-\dfrac{4\sqrt{5}}{9}}{-\dfrac{1}{9}} = 4\sqrt{5}$$

27. Rewrite the function as one or more cosine functions in the first power.

$$\cos^2 x \sin^2 x = \left(\frac{1+\cos 2x}{2}\right)\left(\frac{1-\cos 2x}{2}\right)$$

$$= \frac{1-\cos^2 2x}{4}$$

$$= \frac{1-\left(\frac{1+\cos 4x}{2}\right)}{4} = \frac{\frac{2-1-\cos 4x}{2}}{\frac{4}{1}}$$

$$= \frac{1-\cos 4x}{2} \cdot \frac{1}{4}$$

$$= \frac{1}{8}(1-\cos 4x)$$

29. Rewrite the function as one or more cosine functions in the first power.

$$\cos^4 x = \left(\frac{1+\cos 2x}{2}\right)^2$$

$$= \frac{1}{4}(1+2\cos 2x + \cos^2 2x)$$

$$= \frac{1}{4}\left(1+2\cos 2x + \frac{1+\cos 4x}{2}\right)$$

$$= \frac{1}{8}(3+4\cos 2x + \cos 4x)$$

31. Rewrite the function as one or more cosine functions in the first power.

$$\sin^4 x \ \cos^2 x = \left(\frac{1-\cos 2x}{2}\right)^2\left(\frac{1+\cos 2x}{2}\right)$$

$$= \frac{1}{8}(1-\cos^2 2x)(1-\cos 2x)$$

$$= \frac{1}{8}\left(1-\frac{1+\cos 4x}{2}\right)(1-\cos 2x)$$

$$= \frac{1}{8}\left(\frac{1}{2}-\frac{1}{2}\cos 4x\right)(1-\cos 2x)$$

$$= \frac{1}{16}(1-\cos 2x - \cos 4x + \cos 4x \cos 2x)$$

33. Find the exact value using half-angle identity.

$$\sin 75° = \sin\frac{150°}{2}$$

$$= +\sqrt{\frac{1-\cos 150°}{2}}$$

$$= \sqrt{\frac{1-(-\sqrt{3}/2)}{2}}$$

$$= \sqrt{\frac{2+\sqrt{3}}{4}}$$

$$= \frac{\sqrt{2+\sqrt{3}}}{2}$$

35. Find the exact value using half-angle identity.

$$\tan 67.5° = \tan\frac{135°}{2}$$

$$= \frac{1-\cos 135°}{\sin 135°}$$

$$= \frac{1-(-\sqrt{2}/2)}{\sqrt{2}/2}$$

$$= \frac{2+\sqrt{2}}{\sqrt{2}} \cdot \frac{\sqrt{2}}{\sqrt{2}}$$

$$= \frac{2\sqrt{2}+2}{2}$$

$$= \sqrt{2}+1$$

37. Find the exact value using half-angle identity.

$$\cos 157.5° = \cos\frac{315°}{2}$$

$$= -\sqrt{\frac{1+\cos 315°}{2}}$$

$$= -\sqrt{\frac{1+(\sqrt{2}/2)}{2}}$$

$$= -\sqrt{\frac{2+\sqrt{2}}{4}}$$

$$= -\frac{\sqrt{2+\sqrt{2}}}{2}$$

39. Find the exact value using half-angle identity.

$$\tan 105° = \frac{1-\cos 210°}{\sin 210°}$$

$$= \frac{1+\frac{\sqrt{3}}{2}}{-\frac{1}{2}} = \frac{2+\sqrt{3}}{2}\left(-\frac{2}{1}\right)$$

$$= -2-\sqrt{3}$$

41. Find the exact value using half-angle identity.

$$\sin\frac{7\pi}{8} = \sin\frac{7\pi/4}{2}$$

$$= +\sqrt{\frac{1-\cos 7\pi/4}{2}}$$

$$= \sqrt{\frac{1-\sqrt{2}/2}{2}}$$

$$= \sqrt{\frac{2-\sqrt{2}}{4}}$$

$$= \frac{\sqrt{2-\sqrt{2}}}{2}$$

43. Find the exact value using half-angle identity.

$$\cos\frac{5\pi}{12} = \cos\frac{5\pi/6}{2}$$

$$= +\sqrt{\frac{1+\cos 5\pi/6}{2}}$$

$$= \sqrt{\frac{1-\sqrt{3}/2}{2}}$$

$$= \sqrt{\frac{2-\sqrt{3}}{4}}$$

$$= \frac{\sqrt{2-\sqrt{3}}}{2}$$

45. Find the exact value of $\sin\frac{\alpha}{2}$, $\cos\frac{\alpha}{2}$, $\tan\frac{\alpha}{2}$.

$$\text{hypotenuse} = \sqrt{24^2+7^2} = \sqrt{625} = 25$$

$$\sin\alpha = \frac{7}{25}, \ \cos\alpha = \frac{24}{25}, \ \tan\alpha = \frac{7}{25}.$$

$$\sin\frac{\alpha}{2} = \sqrt{\frac{1-\cos\alpha}{2}}$$

$$= \sqrt{\frac{1-24/25}{2}}$$

$$= \sqrt{\frac{1}{50}} = \sqrt{\frac{1}{50}\left(\frac{2}{2}\right)} = \sqrt{\frac{2}{100}}$$

$$= \frac{\sqrt{2}}{10}$$

$$\cos\frac{\alpha}{2} = \sqrt{\frac{1+\cos\alpha}{2}}$$

$$= \sqrt{\frac{1+24/25}{2}}$$

$$= \sqrt{\frac{49}{50}} = \sqrt{\frac{49}{50}\left(\frac{2}{2}\right)} = \sqrt{\frac{7^2(2)}{100}}$$

$$= \frac{7\sqrt{2}}{10}$$

$$\tan\frac{\alpha}{2} = \frac{1-\cos\alpha}{\sin\alpha}$$

$$= \frac{1-\dfrac{24}{25}}{\dfrac{7}{25}} = \frac{\dfrac{25}{25}-\dfrac{24}{25}}{\dfrac{7}{25}}$$

$$= \frac{25-24}{7}$$

$$= \frac{1}{7}$$

47. Find the exact value of $\sin\frac{\alpha}{2}$, $\cos\frac{\alpha}{2}$, $\tan\frac{\alpha}{2}$.

$$\sin\alpha = \frac{5}{13}, \ \cos\alpha = -\sqrt{1-\left(\frac{5}{13}\right)^2} = -\sqrt{1-\frac{25}{169}} = -\frac{12}{13}.$$

$$\sin\frac{\alpha}{2} = \sqrt{\frac{1-\cos\alpha}{2}}$$

$$= \sqrt{\frac{1-(-12/13)}{2}}$$

$$= \sqrt{\frac{13+12}{26}}$$

$$= \sqrt{\frac{25}{26}} = \frac{5}{\sqrt{26}}$$

$$= \frac{5\sqrt{26}}{26}$$

$$\cos\frac{\alpha}{2} = \sqrt{\frac{1+\cos\alpha}{2}}$$

$$= \sqrt{\frac{1-12/13}{2}}$$

$$= \sqrt{\frac{13-12}{26}} = \sqrt{\frac{1}{26}} = \frac{1}{\sqrt{26}}$$

$$= \frac{\sqrt{26}}{26}$$

$$\tan\frac{\alpha}{2} = \frac{1-\cos\alpha}{\sin\alpha}$$

$$= \frac{1+\dfrac{12}{13}}{\dfrac{5}{13}}$$

$$= \frac{13+12}{5}$$

$$= 5$$

49. Find the exact value of $\sin\frac{\alpha}{2}$, $\cos\frac{\alpha}{2}$, $\tan\frac{\alpha}{2}$.

$$\cos\alpha = -\frac{3}{5},$$

$$\sin\alpha = -\sqrt{1-\left(-\frac{3}{5}\right)^2} = -\sqrt{1-\frac{9}{25}} = -\frac{4}{5}$$

$$\sin\frac{\alpha}{2} = \sqrt{\frac{1-\cos\alpha}{2}}$$

$$= \sqrt{\frac{1-(-3/5)}{2}}$$

$$= \sqrt{\frac{5+3}{10}} = \sqrt{\frac{4}{5}}$$

$$= \frac{2\sqrt{5}}{5}$$

$$\cos\frac{\alpha}{2} = -\sqrt{\frac{1+\cos\alpha}{2}}$$

$$= -\sqrt{\frac{1-3/5}{2}}$$

$$= -\sqrt{\frac{5-3}{10}} = -\sqrt{\frac{1}{5}}$$

$$= -\frac{\sqrt{5}}{5}$$

$$\tan\frac{\alpha}{2} = \frac{1-\cos\alpha}{\sin\alpha}$$

$$= \frac{1+\frac{3}{5}}{-\frac{4}{5}} = \frac{\frac{5}{5}+\frac{3}{5}}{-\frac{4}{5}}$$

$$= \frac{5+3}{-4}$$

$$= -2$$

51. Find the exact value of $\sin\frac{\alpha}{2}$, $\cos\frac{\alpha}{2}$, $\tan\frac{\alpha}{2}$.

$$\tan\alpha = \frac{4}{3}, \ r = \sqrt{3^2+4^2} = \sqrt{25} = 5,$$

$$\sin\alpha = \frac{4}{5}, \ \cos = \frac{3}{5}$$

$$\sin\frac{\alpha}{2} = \sqrt{\frac{1-\cos\alpha}{2}}$$

$$= \sqrt{\frac{1-3/5}{2}}$$

$$= \sqrt{\frac{5-3}{10}} = \sqrt{\frac{1}{5}}$$

$$= \frac{\sqrt{5}}{5}$$

$$\cos\frac{\alpha}{2} = \sqrt{\frac{1+\cos\alpha}{2}}$$

$$= \sqrt{\frac{1+3/5}{2}}$$

$$= \sqrt{\frac{5+3}{10}} = \sqrt{\frac{4}{5}}$$

$$= \frac{2\sqrt{5}}{5}$$

$$\tan\frac{\alpha}{2} = \frac{1-\cos\alpha}{\sin\alpha}$$

$$= \frac{1-\frac{3}{5}}{\frac{4}{5}}$$

$$= \frac{5-3}{4}$$

$$= \frac{1}{2}$$

53. Find the exact value of $\sin\frac{\alpha}{2}$, $\cos\frac{\alpha}{2}$, $\tan\frac{\alpha}{2}$.

$$\cos\alpha = \frac{24}{25}, \ \sin\alpha = -\sqrt{1-\left(\frac{24}{25}\right)^2} = -\sqrt{1-\frac{576}{625}} = -\frac{7}{25}$$

$$\sin\frac{\alpha}{2} = \sqrt{\frac{1-\cos\alpha}{2}}$$

$$= \sqrt{\frac{1-24/25}{2}}$$

$$= \sqrt{\frac{25-24}{50}} = \sqrt{\frac{1}{50}}$$

$$= \frac{\sqrt{2}}{10}$$

$$\cos\frac{\alpha}{2} = -\sqrt{\frac{1+\cos\alpha}{2}}$$

$$= -\sqrt{\frac{1+24/25}{2}}$$

$$= -\sqrt{\frac{25+24}{50}} = -\sqrt{\frac{49}{50}}$$

$$= -\frac{7\sqrt{2}}{10}$$

$$\tan\frac{\alpha}{2} = \frac{1-\cos\alpha}{\sin\alpha}$$

$$= \frac{1-\frac{24}{25}}{-\frac{7}{25}} = \frac{25-24}{-7}$$

$$= -\frac{1}{7}$$

55. Verify the identity.

$$\sin 3x\cos 3x = \frac{1}{2}(2\sin 3x\cos 3x)$$

$$= \frac{1}{2}\sin 2(3x)$$

$$= \frac{1}{2}\sin 6x$$

57. Verify the identity.

$$\sin^2 x + \cos 2x = \sin^2 x + \cos^2 x - \sin^2 x$$

$$= \cos^2 x$$

59. Verify the identity.

$$\frac{1+\cos 2x}{\sin 2x} = \frac{1+2\cos^2 x - 1}{2\sin x\cos x}$$

$$= \frac{2\cos^2 x}{2\sin x\cos x}$$

$$= \cot x$$

61. Verify the identity.

$$\frac{\sin 2x}{1-\sin^2 x} = \frac{2\sin x\cos x}{\cos^2 x} = 2\tan x$$

63. Verify the identity.

$$\frac{\cos 2x}{\cos^2 x} = \frac{\cos^2 x - \sin^2 x}{\cos^2 x}$$
$$= \frac{\cos^2 x}{\cos^2 x} - \frac{\sin^2 x}{\cos^2 x}$$
$$= 1 - \tan^2 x$$

65. Verify the identity.

$$\sin 2x - \tan x = 2\sin x \cos x - \frac{\sin x}{\cos x}$$
$$= \frac{2\sin x \cos^2 x - \sin x}{\cos x}$$
$$= \frac{\sin x(2\cos^2 x - 1)}{\cos x}$$
$$= \tan x \cos 2x$$

67. Verify the identity.

$$\cos^4 x - \sin^4 x = (\cos^2 x + \sin^2 x)(\cos^2 x - \sin^2 x)$$
$$= \cos^2 x - \sin^2 x$$
$$= \cos 2x$$

69. Verify the identity.

$$\cos^2 x - 2\sin^2 x \cos^2 x - \sin^2 x + 2\sin^4 x$$
$$= \cos^2 x(1 - 2\sin^2 x) - \sin^2 x(1 - 2\sin^2 x)$$
$$= (1 - 2\sin^2 x)(\cos^2 x - \sin^2 x)$$
$$= \cos 2x \cos 2x$$
$$= \cos^2 2x$$

71. Verify the identity.

$$\cos 4x = \cos 2(2x)$$
$$= 2\cos^2 2x - 1$$
$$= 2(2\cos^2 x - 1)^2 - 1$$
$$= 2(4\cos^4 x - 4\cos^2 x + 1) - 1$$
$$= 8\cos^4 x - 8\cos^2 x + 1$$

73. Verify the identity.

$$\cos 3x - \cos x$$
$$= \cos(2x + x) - \cos x$$
$$= \cos 2x \cos x - \sin 2x \sin x - \cos x$$
$$= (2\cos^2 x - 1)\cos x - 2\sin x \cos x \cdot \sin x - \cos x$$
$$= 2\cos^3 x - \cos x - 2\sin^2 x \cos x - \cos x$$
$$= 2\cos^3 x - 2\cos x - 2\sin^2 x \cos x$$
$$= 2\cos^3 x - 2\cos x - 2(1 - \cos^2 x)\cos x$$
$$= 2\cos^3 x - 2\cos x - 2\cos x + 2\cos^3 x$$
$$= 4\cos^3 x - 4\cos x$$

75. Verify the identity.

$$\sin^3 x + \cos^3 x$$
$$= (\sin x + \cos x)(\sin^2 x - \sin x \cos x + \cos^2 x)$$
$$= (\sin x + \cos x)\left(\sin^2 x + \cos^2 x - \frac{2\sin x \cos x}{2}\right)$$
$$= (\sin x + \cos x)\left(1 - \frac{1}{2}\sin 2x\right)$$

77. Verify the identity.

$$\sin^2 \frac{x}{2} = \left[\pm\sqrt{\frac{1 - \cos x}{2}}\right]^2$$
$$= \frac{1 - \cos x}{2}$$
$$= \frac{1 - \cos x}{2} \cdot \frac{\sec x}{\sec x}$$
$$= \frac{\sec x - 1}{2\sec x}$$

79. Verify the identity.

$$\tan \frac{x}{2} = \frac{1 - \cos x}{\sin x} = \frac{1}{\sin x} - \frac{\cos x}{\sin x} = \csc x - \cot x$$

81. Verify the identity.

$$2\sin \frac{x}{2} \cos \frac{x}{2} = \sin 2\left(\frac{x}{2}\right) = \sin x$$

83. Verify the identity.

$$\left(\cos \frac{x}{2} + \sin \frac{x}{2}\right)^2 = \cos^2 \frac{x}{2} + 2\sin \frac{x}{2}\cos \frac{x}{2} + \sin^2 \frac{x}{2}$$
$$= \cos^2 \frac{x}{2} + \sin^2 \frac{x}{2} + \sin 2\left(\frac{x}{2}\right)$$
$$= 1 + \sin x$$

85. Verify the identity.

$$\sin^2 \frac{x}{2} \sec x = \left(\pm\sqrt{\frac{1 - \cos x}{2}}\right)^2 \sec x$$
$$= \frac{1 - \cos x}{2} \cdot \sec x$$
$$= \frac{1}{2}(\sec x - 1)$$

87. Verify the identity.

$$\cos^2 \frac{x}{2} - \cos x = \left(\pm\sqrt{\frac{1 + \cos x}{2}}\right)^2 - \cos x$$
$$= \frac{1 + \cos x}{2} - \cos x$$
$$= \frac{1 + \cos x - 2\cos x}{2}$$
$$= \frac{1 - \cos x}{2}$$
$$= \sin^2 \frac{x}{2}$$

89. Verify the identity.

$$\sin^2 \frac{x}{2} - \cos^2 \frac{x}{2} = -\left(\cos^2 \frac{x}{2} - \sin^2 \frac{x}{2}\right)$$

$$= -\cos 2\left(\frac{x}{2}\right)$$

$$= -\cos x$$

91. Verify the identity.

$$\sin 2x - \cos x = 2\sin x \cos x - \cos x$$

$$= (\cos x)(2\sin x - 1)$$

93. Verify the identity.

$$\tan 2x = \frac{2\tan x}{1 - \tan^2 x}$$

$$= \frac{\dfrac{2\tan x}{\tan x}}{\dfrac{1}{\tan x} - \dfrac{\tan^2 x}{\tan x}}$$

$$= \frac{2}{\cot x - \tan x}$$

95. Verify the identity.

$$\frac{\sin^2 x + 1 - \cos^2 x}{\sin x(1 + \cos x)} = \frac{1 - \cos^2 x + 1 - \cos^2 x}{\sin x(1 + \cos x)}$$

$$= \frac{2(1 - \cos^2 x)}{\sin x(1 + \cos x)}$$

$$= \frac{2(1 - \cos x)(1 + \cos x)}{\sin x(1 + \cos x)}$$

$$= \frac{2(1 - \cos x)}{\sin x}$$

$$= 2\tan \frac{x}{2}$$

97. Verify the identity.

$$\csc 2x = \frac{1}{\sin 2x} = \frac{1}{2\sin x \cos x} = \frac{1}{2}\csc x \sec x$$

99. Verify the identity.

$$\cos \frac{x}{5} = \cos 2\left(\frac{x}{10}\right) = 1 - 2\sin^2 \frac{x}{10}$$

101.a. $$M = \frac{1}{\sin \dfrac{(\pi/4)}{2}} = 1 \div \sqrt{\frac{1 - \cos \frac{\pi}{4}}{2}}$$

$$= 1 \div \sqrt{\frac{1 - \frac{\sqrt{2}}{2}}{2}} = 1 \div \sqrt{\frac{2 - \sqrt{2}}{4}}$$

$$= \frac{2}{\sqrt{2 - \sqrt{2}}}$$

$$\approx 2.61$$

b. $$M \sin \frac{\alpha}{2} = 1$$

$$\sin \frac{\alpha}{2} = \frac{1}{M}$$

$$\frac{\alpha}{2} = \sin^{-1}\left(\frac{1}{M}\right)$$

$$\alpha = 2\sin^{-1}\left(\frac{1}{M}\right)$$

c. As M increases $\dfrac{1}{M}$ decreases. So if $\sin^{-1}\left(\dfrac{1}{M}\right)$

decreases, then α decreases.

103. Compare the graphs and predict whether the equation is an identity.

$$y = \frac{\sin 2x}{1 - \sin^2 x} \text{ and } y = 2\tan x \text{ both have the following}$$

graph. Identity.

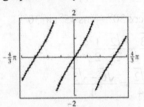

105. Compare the graphs and predict whether the equation is an identity.

$$y = \left(\cos \frac{x}{2} + \sin \frac{x}{2}\right)^2 \text{ and } y = 1 + \sin x \text{ both have the}$$

following graph. Identity.

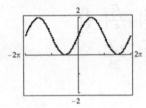

107. Verify the identity.

$$\cos^4 x = \cos^2 x \cdot \cos^2 x$$

$$= \frac{\cos 2x + 1}{2} \cdot \frac{\cos 2x + 1}{2}$$

$$= \frac{1}{4}\left(\cos^2 2x + 2\cos 2x + 1\right)$$

$$= \frac{1}{4}\left(\frac{\cos 4x + 1}{2} + 2\cos 2x + 1\right)$$

$$= \frac{1}{8}\cos 4x + \frac{1}{8} + \frac{1}{2}\cos 2x + \frac{1}{4}$$

$$= \frac{1}{8}\cos 4x + \frac{1}{2}\cos 2x + \frac{3}{8}$$

109.a. $\cos\theta = \dfrac{x}{1}$, so $x = \cos\theta$

b. $\sin\theta = \dfrac{y}{1}$, so $y = \sin\theta$

c. $\tan\dfrac{\theta}{2} = \dfrac{y}{1+x}$, so $\tan\dfrac{\theta}{2} = \dfrac{\sin\theta}{1+\cos\theta}$

Mid-Chapter 6 Quiz

1. Verify the identity.

$$\sec^2 x - \sin^2 x \sec^2 x = \sec^2 x(1-\sin^2 x)$$
$$= \frac{1-\sin^2 x}{\cos^2 x}$$
$$= \frac{\cos^2 x}{\cos^2 x}$$
$$= 1$$

3. Find the exact value.

$$\sin(45° - 30°) = \sin 45° \cos 30° - \cos 45° \sin 30°$$
$$= \frac{\sqrt{2}}{2}\cdot\frac{\sqrt{3}}{2} - \frac{\sqrt{2}}{2}\cdot\frac{1}{2}$$
$$= \frac{\sqrt{6}}{4} - \frac{\sqrt{2}}{4} = \frac{\sqrt{6}-\sqrt{2}}{4}$$

5. Find the exact value.

$$\sin\alpha = \frac{\sqrt{5}}{5},\quad \cos\alpha = -\frac{2\sqrt{5}}{5},\quad \tan\alpha = -\frac{1}{2},$$
$$\tan\beta = \frac{4}{3},\quad \sin\beta = -\frac{4}{5},\quad \cos\beta = -\frac{3}{5}$$

$$\cos(\alpha - \beta) = \cos\alpha \cos\beta + \sin\alpha \sin\beta$$
$$= -\frac{2\sqrt{5}}{5}\left(-\frac{3}{5}\right) + \frac{\sqrt{5}}{5}\left(-\frac{4}{5}\right)$$
$$= \frac{6\sqrt{5}}{25} - \frac{4\sqrt{5}}{25} = \frac{2\sqrt{5}}{25}$$

7. Find the exact value.

$$\cos 75° = \cos\frac{150°}{2}$$
$$= +\sqrt{\frac{1+\cos 150°}{2}}$$
$$= \sqrt{\frac{1+(-\sqrt{3}/2)}{2}}$$
$$= \sqrt{\frac{2-\sqrt{3}}{4}}$$
$$= \frac{\sqrt{2-\sqrt{3}}}{2}$$

Prepare for Section 6.4

P1. Use the identities to rewrite the expression.

$$\frac{1}{2}\big[\sin(\alpha+\beta) + \sin(\alpha-\beta)\big]$$
$$= \frac{1}{2}\big[\sin\alpha\cos\beta + \cos\alpha\sin\beta + \sin\alpha\cos\beta - \cos\alpha\sin\beta\big]$$
$$= \sin\alpha\cos\beta$$

P3. Compare the expressions at the values.

$$\sin\pi - \sin\frac{\pi}{6} = 0 - \frac{1}{2} = -\frac{1}{2}$$

$$2\cos\left(\frac{\pi+\frac{\pi}{6}}{2}\right)\sin\left(\frac{\pi-\frac{\pi}{6}}{2}\right)$$
$$= 2\cos\left(\frac{7\pi}{12}\right)\sin\left(\frac{5\pi}{12}\right)$$
$$= 2\left(-\sqrt{\frac{1+\cos\left(\frac{7\pi}{6}\right)}{2}}\right)\left(\sqrt{\frac{1-\cos\left(\frac{5\pi}{6}\right)}{2}}\right)$$
$$= -2\sqrt{\frac{1-\frac{\sqrt{3}}{2}}{2}}\sqrt{\frac{1+\frac{\sqrt{3}}{2}}{2}}$$
$$= -\sqrt{1-\frac{3}{4}} = -\sqrt{\frac{1}{4}}$$
$$= -\frac{1}{2}$$

Both functional values equal $-\dfrac{1}{2}$.

P5. Verify that the equation is not an identity

Answers will vary.

Section 6.4 Exercises

1. Determine which equation is an identity.

iii. $\cos x + \cos y = 2\cos\dfrac{x+y}{2}\cos\dfrac{x-y}{2}$

3. Determine the amplitude, phase shift, and period.

Amplitude: 2; phase shift: $-\dfrac{\pi}{3}$; period: 2π

5. Write the expression as a sum or difference.

$$2\sin x \cos 2x = 2\cdot\frac{1}{2}\big[\sin(x+2x) + \sin(x-2x)\big]$$
$$= \sin 3x + \sin(-x)$$
$$= \sin 3x - \sin x$$

7. Write the expression as a sum or difference.

$$\cos 2x \sin x = \frac{1}{2}\left[\sin(2x+x) - \sin(2x-x)\right]$$
$$= \frac{1}{2}\left[\sin 3x - \sin x\right]$$

9. Write the expression as a sum or difference.

$$\frac{1}{2}\sin 3x \cos x = \frac{1}{4}\left[\sin(3x+x) + \sin(3x-x)\right]$$
$$= \frac{1}{4}\left[\sin 4x + \sin 2x\right]$$

11. Write the expression as a sum or difference.

$$\sin x \sin 5x = \frac{1}{2}\left[\cos(x-5x) - \cos(x+5x)\right]$$
$$= \frac{1}{2}\left[\cos(-4x) - \cos 6x\right]$$
$$= \frac{1}{2}(\cos 4x - \cos 6x)$$

13. Find the exact value of the expression.

$$\cos 75° \cos 15° = \frac{1}{2}\left[\cos(75°+15°) + \cos(75°-15°)\right]$$
$$= \frac{1}{2}(\cos 90° + \cos 60°)$$
$$= \frac{1}{2}\left(0 + \frac{1}{2}\right)$$
$$= \frac{1}{4}$$

15. Find the exact value of the expression.

$$\cos 165° \sin 195°$$
$$= \frac{1}{2}\left[\sin(165°+195°) - \sin(165°-195°)\right]$$
$$= \frac{1}{2}(\sin 360° - \sin(-30°))$$
$$= \frac{1}{2}\left(0 - \left(-\frac{1}{2}\right)\right)$$
$$= \frac{1}{4}$$

17. Find the exact value of the expression.

$$\sin\frac{13\pi}{12}\cos\frac{\pi}{12} = \frac{1}{2}\left[\sin\left(\frac{13\pi}{12}+\frac{\pi}{12}\right) + \sin\left(\frac{13\pi}{12}-\frac{\pi}{12}\right)\right]$$
$$= \frac{1}{2}\left(\sin\frac{7\pi}{6} + \sin\pi\right)$$
$$= \frac{1}{2}\left(-\frac{1}{2}+0\right)$$
$$= -\frac{1}{4}$$

19. Find the exact value of the expression.

$$\cos\frac{7\pi}{24}\cos\frac{11\pi}{24} = \frac{1}{2}\left[\cos\left(\frac{7\pi}{24}+\frac{11\pi}{24}\right) + \cos\left(\frac{7\pi}{24}-\frac{11\pi}{24}\right)\right]$$
$$= \frac{1}{2}\left(\cos\frac{3\pi}{4} + \cos\left(-\frac{\pi}{6}\right)\right)$$
$$= \frac{1}{2}\left(-\frac{\sqrt{2}}{2} + \frac{\sqrt{3}}{2}\right)$$
$$= \frac{\sqrt{3}-\sqrt{2}}{4}$$

21. Find the exact value of the expression.

$$\sin 165° + \sin 105°$$
$$= 2\sin\frac{(165°+105°)}{2}\cos\frac{(165°-105°)}{2}$$
$$= 2\sin 135° \cos 30°$$
$$= 2\cdot\frac{\sqrt{2}}{2}\cdot\frac{\sqrt{3}}{2}$$
$$= \frac{\sqrt{6}}{2}$$

23. Find the exact value of the expression.

$$\sin\frac{\pi}{12} - \sin\frac{5\pi}{12}$$
$$= 2\cos\frac{\left(\frac{\pi}{12}+\frac{5\pi}{12}\right)}{2}\sin\frac{\left(\frac{\pi}{12}-\frac{5\pi}{12}\right)}{2}$$
$$= 2\cos\frac{\pi}{4}\sin\left(-\frac{\pi}{6}\right)$$
$$= 2\cdot\frac{\sqrt{2}}{2}\cdot\left(-\frac{1}{2}\right)$$
$$= -\frac{\sqrt{2}}{2}$$

25. Write the expression as a product of two functions.

$$\sin 4\theta + \sin 2\theta = 2\sin\frac{4\theta+2\theta}{2}\cos\frac{4\theta-2\theta}{2}$$
$$= 2\sin 3\theta \cos\theta$$

27. Write the expression as a product of two functions.

$$\cos 3\theta + \cos\theta = 2\cos\frac{3\theta+\theta}{2}\cos\frac{3\theta-\theta}{2}$$
$$= 2\cos 2\theta \cos\theta$$

29. Write the expression as a product of two functions.

$$\cos 6\theta - \cos 2\theta = -2\sin\frac{6\theta+2\theta}{2}\sin\frac{6\theta-2\theta}{2}$$
$$= -2\sin 4\theta \sin 2\theta$$

31. Write the expression as a product of two functions.

$$\cos\frac{\theta}{2}+\cos\frac{5\theta}{2}=2\cos\frac{\frac{\theta}{2}+\frac{5\theta}{2}}{2}\cos\frac{\frac{\theta}{2}-\frac{5\theta}{2}}{2}$$

$$=2\cos\frac{3\theta}{2}\cos(-\theta)$$

$$=2\cos\frac{3\theta}{2}\cos\theta$$

33. Write the expression as a product of two functions.

$$\sin 5\theta+\sin 9\theta=2\sin\frac{5\theta+9\theta}{2}\cos\frac{5\theta-9\theta}{2}$$

$$=2\sin 7\theta\cos(-2\theta)$$

$$=2\sin 7\theta\cos 2\theta$$

35. Write the expression as a product of two functions.

$$\sin\frac{\theta}{5}+\sin\frac{\theta}{2}=2\sin\frac{\frac{\theta}{5}+\frac{\theta}{2}}{2}\cos\frac{\frac{\theta}{5}-\frac{\theta}{2}}{2}$$

$$=2\sin\frac{7\theta}{20}\cos\left(-\frac{3\theta}{20}\right)$$

$$=2\sin\frac{7\theta}{20}\cos\frac{3\theta}{20}$$

37. Write the expression as a product of two functions.

$$\cos\frac{\theta}{2}-\cos\theta=-2\sin\frac{\frac{\theta}{2}+\theta}{2}\sin\frac{\frac{\theta}{2}-\theta}{2}$$

$$=-2\sin\frac{3}{4}\theta\sin\left(-\frac{1}{4}\theta\right)$$

$$=2\sin\frac{3}{4}\theta\sin\frac{1}{4}\theta$$

39. Write the expression as a product of two functions.

$$\sin\frac{\theta}{2}-\sin\frac{\theta}{3}=2\cos\frac{\frac{\theta}{2}+\frac{\theta}{3}}{2}\sin\frac{\frac{\theta}{2}-\frac{\theta}{3}}{2}$$

$$=2\cos\frac{5}{12}\theta\sin\frac{1}{12}\theta$$

41. Verify the identity.

$$\cos(\alpha+\beta)+\cos(\alpha-\beta)$$

$$=\cos\alpha\cos\beta-\sin\alpha\sin\beta+\cos\alpha\cos\beta+\sin\alpha\sin\beta$$

$$=2\cos\alpha\cos\beta$$

43. Verify the identity.

$$2\cos 3x\sin x$$

$$=2\cdot\frac{1}{2}\left[\sin(3x+x)-\sin(3x-x)\right]$$

$$=\sin 4x-\sin 2x$$

$$=2\sin 2x\cos 2x-\sin 2x$$

$$=\sin 2x(2\cos 2x-1)$$

$$=2\sin x\cos x\left[2(1-2\sin^2 x)-1\right]$$

$$=4\sin x\cos x-8\sin^3 x\cos x-2\sin x\cos x$$

$$=2\sin x\cos x-8\cos x\sin^3 x$$

45. Verify the identity.

$$2\cos 5x\cos 7x=2\cdot\frac{1}{2}\left[\cos(5x+7x)+\cos(5x-7x)\right]$$

$$=\cos 12x+\cos(-2x)$$

$$=\cos 12x+\cos 2x$$

$$=\cos^2 6x-\sin^2 6x+2\cos^2 x-1$$

47. Verify the identity.

$$\sin 3x-\sin x=2\cos\frac{3x+x}{2}\sin\frac{3x-x}{2}$$

$$=2\cos 2x\sin x$$

$$=2(1-2\sin^2 x)\sin x$$

$$=2\sin x-4\sin^3 x$$

49. Verify the identity.

$$\sin 2x+\sin 4x$$

$$=2\sin\frac{2x+4x}{2}\cos\frac{2x-4x}{2}$$

$$=2\sin 3x\cos(-x)$$

$$=2\sin 3x\cos x=2\cos x\sin 3x$$

$$=2\cos x\sin(2x+x)$$

$$=2\cos x(\sin 2x\cos x+\cos 2x\sin x)$$

$$=2\cos x[(2\sin x\cos x)\cos x+(2\cos^2 x-1)\sin x]$$

$$=2\cos x\sin x(4\cos^2 x-1)$$

51. Verify the identity.

$$\frac{\sin 3x-\sin x}{\cos 3x-\cos x}=\frac{2\cos\frac{3x+x}{2}\sin\frac{3x-x}{2}}{-2\sin\frac{3x+x}{2}\sin\frac{3x-x}{2}}$$

$$=-\frac{\cos 2x}{\sin 2x}$$

$$=-\cot 2x$$

53. Verify the identity.

$$\frac{\sin 5x+\sin 3x}{4\sin x\cos^3 x-4\sin^3 x\cos x}$$

$$=\frac{2\sin\frac{5x+3x}{2}\cos\frac{5x-3x}{2}}{4\sin x\cos x(\cos^2 x-\sin^2 x)}$$

$$=\frac{\sin 4x\cos x}{2\sin x\cos x\cos 2x}$$

$$=\frac{2\sin 2x\cos 2x\cos x}{\sin 2x\cos 2x}$$

$$=2\cos x$$

55. Verify the identity.

$$\sin(x+y)\cos(x-y)$$
$$=\frac{1}{2}[\sin(x+y+x-y)+\sin(x+y-x+y)]$$
$$=\frac{1}{2}[\sin 2x+\sin 2y]$$
$$=\frac{1}{2}[2\sin x\cos x+2\sin y\cos y]$$
$$=\sin x\cos x+\sin y\cos y$$

57. Write the equation in the form $y=k\sin(x+\alpha)$.

$$a=-1,\ b=-1,\ k=\sqrt{(-1)^2+(-1)^2}=\sqrt{2},$$

α is a third quadrant angle.

$$\sin\beta=\left|\frac{-1}{\sqrt{2}}\right|=\frac{1}{\sqrt{2}}$$
$$\beta=45°$$
$$\alpha=-180°+45°=-135°$$
$$y=\sqrt{2}\sin(x-135°)$$

59. Write the equation in the form $y=k\sin(x+\alpha)$.

$$a=\frac{1}{2},\ b=-\frac{\sqrt{3}}{2},\ k=\sqrt{\left(\frac{1}{2}\right)^2+\left(-\frac{\sqrt{3}}{2}\right)^2}=1,$$

α is a fourth quadrant angle.

$$\sin\beta=\left|\frac{-\frac{\sqrt{3}}{2}}{1}\right|=\frac{\sqrt{3}}{2}$$
$$\beta=60°$$
$$\alpha=-60°$$
$$y=\sin(x-60°)$$

61. Write the equation in the form $y=k\sin(x+\alpha)$.

$$a=\frac{1}{2},\ b=-\frac{1}{2},\ k=\sqrt{\left(\frac{1}{2}\right)^2+\left(\frac{1}{2}\right)^2}=\frac{\sqrt{2}}{2},$$

α is a fourth quadrant angle.

$$\sin\beta=\left|\frac{-1/2}{\sqrt{2}/2}\right|=\frac{\sqrt{2}}{2}$$
$$\beta=45°$$
$$\alpha=-45°$$
$$y=\frac{\sqrt{2}}{2}\sin(x-45°)$$

63. Write the equation in the form $y=k\sin(x+\alpha)$.

$$a=-3,\ b=3,\ k=\sqrt{(-3)^2+3^2}=3\sqrt{2},$$

α is a second quadrant angle.

$$\sin\beta=\left|\frac{3}{3\sqrt{2}}\right|=\frac{1}{\sqrt{2}}=\frac{\sqrt{2}}{2}$$
$$\beta=45°$$
$$\alpha=180°-45°=135°$$
$$y=3\sqrt{2}\sin(x+135°)$$

65. Write the equation in the form $y=k\sin(x+\alpha)$.

$$a=\pi,\ b=-\pi,\ k=\sqrt{\pi^2+(-\pi)^2}=\pi\sqrt{2},$$

α is a fourth quadrant angle.

$$\sin\beta=\left|\frac{-\pi}{\pi\sqrt{2}}\right|=\frac{1}{\sqrt{2}}=\frac{\sqrt{2}}{2}$$
$$\beta=45°$$
$$\alpha=-45°$$
$$y=\pi\sqrt{2}\sin(x-45°)$$

67. Write the equation in the form $y=k\sin(x+\alpha)$.

$$a=-1,\ b=1,\ k=\sqrt{(-1)^2+1^2}=\sqrt{2},$$

α is a second quadrant angle.

$$\sin\beta=\left|\frac{1}{\sqrt{2}}\right|=\frac{1}{\sqrt{2}}=\frac{\sqrt{2}}{2}$$
$$\beta=\frac{\pi}{4}$$
$$\alpha=\pi-\frac{\pi}{4}=\frac{3\pi}{4}$$
$$y=\sqrt{2}\sin\left(x+\frac{3\pi}{4}\right)$$

69. Write the equation in the form $y=k\sin(x+\alpha)$.

$$a=\frac{\sqrt{3}}{2},\ b=\frac{1}{2},\ k=\sqrt{\left(\frac{\sqrt{3}}{2}\right)^2+\left(\frac{1}{2}\right)^2}=1,$$

α is a first quadrant angle.

$$\sin\beta=\left|\frac{1/2}{1}\right|=\frac{1}{2}$$
$$\beta=\frac{\pi}{6}$$
$$\alpha=\frac{\pi}{6}$$
$$y=\sin\left(x+\frac{\pi}{6}\right)$$

71. Write the equation in the form $y = k\sin(x + \alpha)$.

$a = -10,\ b = 10\sqrt{3},\ k = \sqrt{(-10)^2 + (10\sqrt{3})^2} = 20,$
α is a second quadrant angle.

$\sin\beta = \dfrac{10\sqrt{3}}{20} = \dfrac{\sqrt{3}}{2}$

$\beta = \dfrac{\pi}{3}$

$\alpha = \pi - \dfrac{\pi}{3} = \dfrac{2\pi}{3}$

$y = 20\sin\left(x + \dfrac{2\pi}{3}\right)$

73. Write the equation in the form $y = k\sin(x + \alpha)$.

$a = -5,\ b = 5,\ k = \sqrt{(-5)^2 + 5^2} = 5\sqrt{2},$
α is a second quadrant angle.

$\sin\beta = \left|\dfrac{5}{5\sqrt{2}}\right| = \dfrac{\sqrt{2}}{2}$

$\beta = \dfrac{\pi}{4}$

$\alpha = \pi - \dfrac{\pi}{4} = \dfrac{3\pi}{4}$

$y = 5\sqrt{2}\sin\left(x + \dfrac{3\pi}{4}\right)$

75. Graph one cycle of the function.

$y = -\sin x - \sqrt{3}\cos x$

$y = 2\sin\left(x - \dfrac{2\pi}{3}\right)$

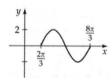

77. Graph one cycle of the function.

$y = 2\sin x + 2\cos x$

$y = 2\sqrt{2}\sin\left(x + \dfrac{\pi}{4}\right)$

79. Graph one cycle of the function.

$y = -\sqrt{3}\sin x - \cos x$

$y = 2\sin\left(x - \dfrac{5\pi}{6}\right)$

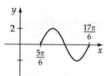

81. Graph one cycle of the function.

$y = -5\sin x + 5\sqrt{3}\cos x$

$y = 10\sin\left(x + \dfrac{2\pi}{3}\right)$

83. Graph one cycle of the function.

$y = 6\sqrt{3}\sin x - 6\cos x$

$y = 12\sin\left(x - \dfrac{\pi}{6}\right)$

85. a. $s(t) = \cos(2\pi 1336t) + \cos(2\pi 770t)$

b. $s(t)$

$= 2\cos\left(\dfrac{2\pi\cdot 1336t + 2\pi\cdot 770t}{2}\right)\cos\left(\dfrac{2\pi\cdot 1336t - 2\pi\cdot 770t}{2}\right)$

$= 2\cos(2106\pi t)\cos(566\pi t)$

c. $\dfrac{1336 + 770}{2} = \dfrac{2106}{2} = 1053$ cycles per second

87. Compare the graphs and predict whether the equation is an identity.

Identity

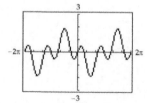

89. Compare the graphs and predict whether the equation is an identity.

Identity

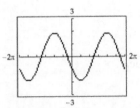

91. Compare the graphs and predict whether the equation is an identity.

Identity

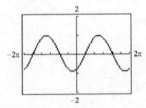

93. Derive the identity.

Let $x = \alpha + \beta$ and $y = \alpha - \beta$.

$x + y = \alpha + \beta + \alpha - \beta$ and $x - y = \alpha + \beta - (\alpha - \beta)$

$x + y = 2\alpha$ $x - y = 2\beta$

$$\alpha = \frac{x+y}{2} \qquad\qquad \beta = \frac{x-y}{2}$$

$\cos(\alpha - \beta) + \cos(\alpha + \beta) = 2\cos\alpha\cos\beta$

$\cos\left[\dfrac{x+y}{2} - \dfrac{x-y}{2}\right] + \cos\left[\dfrac{x+y}{2} + \dfrac{x-y}{2}\right]$

$$= 2\cos\frac{x+y}{2}\cos\frac{x-y}{2}$$

$$\cos y + \cos x = 2\cos\frac{x+y}{2}\cos\frac{x-y}{2}$$

95. Verify the identity.

$\sin 2x + \sin 4x + \sin 6x$

$= 2\sin\dfrac{2x+4x}{2}\cos\dfrac{2x-4x}{2} + 2\sin 3x\cos 3x$

$= 2\sin 3x\cos x + 2\sin 3x\cos 3x$

$= 2\sin 3x(\cos x + \cos 3x)$

$= 2\sin 3x\left(2\cos\dfrac{x+3x}{2}\cos\dfrac{x-3x}{2}\right)$

$= 4\sin 3x\cos 2x\cos x$

97. Verify the identity.

$\dfrac{\cos 10x + \cos 8x}{\sin 10x - \sin 8x} = \dfrac{2\cos\dfrac{10x+8x}{2}\cos\dfrac{10x-8x}{2}}{2\cos\dfrac{10x+8x}{2}\sin\dfrac{10x-8x}{2}}$

$$= \frac{2\cos 9x\cos x}{2\cos 9x\sin x}$$

$$= \cot x$$

99. Verify the identity.

$\dfrac{\sin 2x + \sin 4x + \sin 6x}{\cos 2x + \cos 4x + \cos 6x}$

$= \dfrac{\sin 2x + \sin 6x + \sin 4x}{\cos 2x + \cos 6x + \cos 4x}$

$= \dfrac{2\sin\dfrac{2x+6x}{2}\cos\dfrac{2x-6x}{2} + \sin 4x}{2\cos\dfrac{2x+6x}{2}\cos\dfrac{2x-6x}{2} + \cos 4x}$

$= \dfrac{2\sin 4x\cos 2x + \sin 4x}{2\cos 4x\cos 2x + \cos 4x}$

$= \dfrac{\sin 4x(2\cos 2x + 1)}{\cos 4x(2\cos 2x + 1)}$

$= \dfrac{\sin 4x}{\cos 4x}$

$= \tan 4x$

101. Verify the identity.

$\cos^2 x - \sin^2 x$

$= \cos x \cdot \cos x - \sin x \cdot \sin x$

$= \dfrac{1}{2}[\cos(x+x) + \cos(x-x)]$

$\qquad - \dfrac{1}{2}[\cos(x-x) - \cos(x+x)]$

$= \dfrac{1}{2}\cos 2x + \dfrac{1}{2}\cos 0 - \dfrac{1}{2}\cos 0 + \dfrac{1}{2}\cos 2x$

$= \cos 2x$

103. Show the result.

$x + y = 180°$

$\quad y = 180° - x$

$\sin x + \sin y = \sin x + \sin(180° - x)$

$\qquad\qquad = \sin x + \sin 180°\cos x - \cos 180°\sin x$

$\qquad\qquad = \sin x + 0(\cos x) - (-1)\sin x$

$\qquad\qquad = 2\sin x$

105. Verify the identity.

Let $k = \sqrt{a^2 + b^2}$, $\tan \alpha = \dfrac{a}{b}$

$a \sin x + b \cos x$

$= \dfrac{\sqrt{a^2 + b^2}}{\sqrt{a^2 + b^2}} (a \sin x + b \cos x)$

$= \sqrt{a^2 + b^2} \left(\dfrac{a}{\sqrt{a^2 + b^2}} \sin x + \dfrac{b}{\sqrt{a^2 + b^2}} \cos x \right)$

$= k(\sin \alpha \sin x + \cos \alpha \cos x)$

because $\sin \alpha = \dfrac{a}{\sqrt{a^2 + b^2}}$ and $\cos \alpha = \dfrac{b}{\sqrt{a^2 + b^2}}$

$= k(\cos x \cos \alpha + \sin x \sin \alpha) = k \cos(x - \alpha)$

Prepare for Section 6.5

P1. A one-to-one function is a function for which each range value (y-value) is paired with one and only one domain value (x-value).

P3. $f[g(x)] = f \left[\dfrac{1}{2} x - 2 \right]$

$\qquad = 2 \left(\dfrac{1}{2} x - 2 \right) + 4$

$\qquad = x - 4 + 4$

$\qquad = x$

P5. The graph of f^{-1} is the reflection of the graph of f across the line given by $y = x$.

Section 6.5 Exercises

1. State the domain and range.

a. Domain: $\{x | -1 \le x \le 1\}$

range: $\left\{ y \middle| -\dfrac{\pi}{2} \le y \le \dfrac{\pi}{2} \right\}$

b. Domain: $\{x | -1 \le x \le 1\}$

range: $\{y | 0 \le y \le \pi\}$

c. Domain: $\{x | -\infty \le x \le \infty\}$

range: $\left\{ y \middle| -\dfrac{\pi}{2} < y < \dfrac{\pi}{2} \right\}$

3. Determine which statement is true.

iii. $-\dfrac{\pi}{2} \le x \le \dfrac{\pi}{2}$, then $\sin^{-1}(\sin x) = x$.

5. Find the exact radian value.

$y = \sin^{-1} 1$

$\sin y = 1$ with $-\dfrac{\pi}{2} \le y \le \dfrac{\pi}{2}$

$y = \dfrac{\pi}{2}$

7. Find the exact radian value.

$y = \cos^{-1} \left(\dfrac{\sqrt{2}}{2} \right)$

$\cos y = \dfrac{\sqrt{2}}{2} \qquad 0 \le y \le \pi$

$y = \dfrac{\pi}{4}$

9. Find the exact radian value.

$y = \tan^{-1} -\dfrac{\sqrt{3}}{3}$

$\tan y = -\dfrac{\sqrt{3}}{3} \qquad -\dfrac{\pi}{2} < y < \dfrac{\pi}{2}$

$y = -\dfrac{\pi}{6}$

11. Find the exact radian value.

$y = \cot^{-1} \dfrac{\sqrt{3}}{3}$

$\cot y = \dfrac{\sqrt{3}}{3} \qquad 0 < y < \pi$

$y = \dfrac{\pi}{3}$

13. Find the exact radian value.

$y = \sec^{-1} \dfrac{2\sqrt{3}}{3}$

$\sec y = \dfrac{2\sqrt{3}}{3} \qquad 0 \le y \le \pi$

$y = \dfrac{\pi}{6}$

15. Find the exact radian value.

$y = \sec^{-1} \sqrt{2}$

$\sec y = \sqrt{2} \qquad 0 \le y \le \pi$

$y = \dfrac{\pi}{4}$

17. Find the exact radian value.

$y = \sin^{-1} \left(-\dfrac{\sqrt{3}}{2} \right)$

$\sin y = -\dfrac{\sqrt{3}}{2} \qquad -\dfrac{\pi}{2} \le y \le \dfrac{\pi}{2}$

$y = -\dfrac{\pi}{3}$

19. Find the exact radian value.

$$y = \sec^{-1}\left(-\sqrt{2}\right)$$
$$\sec y = -\sqrt{2} \qquad \pi \le y \le 2\pi$$
$$y = \frac{3\pi}{4}$$

21. Find the exact radian value.

$$y = \tan^{-1}(-1)$$
$$\tan y = -1 \qquad -\frac{\pi}{2} < y < \frac{\pi}{2}$$
$$y = -\frac{\pi}{4}$$

23. Use a calculator to approximate to 4 decimal places.

a. $\sin^{-1}(0.4255) \approx 0.4395$

b. $\tan^{-1}(51.7505) \approx 1.5515$

25. Use a calculator to approximate to 4 decimal places.

a. $\cot^{-1}(15.2525) = \tan^{-1}\left(\dfrac{1}{15.2525}\right) \approx 0.0655$

b. $\sec^{-1}(-5.8572) = \cos^{-1}\left(\dfrac{1}{-5.8572}\right) \approx 1.7424$

27. Express θ as a function of x.

$$\cos\theta = \frac{x}{7} \text{ or } \theta = \cos^{-1}\left(\frac{x}{7}\right)$$

29. Find the exact value, or round to 4 decimal places.

$$y = \cos\left(\cos^{-1}\frac{1}{2}\right)$$
$$y = \cos\frac{\pi}{3}$$
$$y = \frac{1}{2}$$

31. Find the exact value, or round to 4 decimal places.

$$y = \tan(\tan^{-1}2)$$
$$y = 2$$

33. Find the exact value, or round to 4 decimal places.

$$y = \sin\left(\tan^{-1}\frac{3}{4}\right)$$
$$y = \frac{3}{5}$$

35. Find the exact value, or round to 4 decimal places.

$$y = \tan\left(\sin^{-1}\frac{\sqrt{2}}{2}\right)$$
$$y = 1$$

37. Find the exact value, or round to 4 decimal places.

$$y = \cos(\sec^{-1}2)$$
$$y = \frac{1}{2}$$

39. Find the exact value, or round to 4 decimal places.

$$y = \sin^{-1}\left(\sin\frac{\pi}{6}\right) = \sin^{-1}\frac{1}{2}$$
$$y = \frac{\pi}{6}$$

41. Find the exact value, or round to 4 decimal places.

$$y = \cos^{-1}\left(\sin\frac{\pi}{4}\right) = \cos^{-1}\frac{\sqrt{2}}{2}$$
$$y = \frac{\pi}{4}$$

43. Find the exact value, or round to 4 decimal places.

$$y = \sin^{-1}\left(\tan\frac{\pi}{3}\right) = \sin^{-1}\sqrt{3}$$

y is not defined.

45. Find the exact value, or round to 4 decimal places.

$$y = \sec^{-1}\left(\sin\frac{\pi}{6}\right) = \sec^{-1}\frac{1}{2}$$

y is not defined.

47. Find the exact value, or round to 4 decimal places.

$$y = \sin^{-1}\left(\cos\left[-\frac{2\pi}{3}\right]\right) = \sin^{-1}\left(-\frac{1}{2}\right)$$
$$y = -\frac{\pi}{6}$$

49. Find the exact value, or round to 4 decimal places.

Let $\theta = \sin^{-1}\dfrac{1}{2}$ and find $y = \tan\theta$.

Then $\sin\theta = \dfrac{1}{2}$ and $-\dfrac{\pi}{2} \le \theta \le \dfrac{\pi}{2}$.

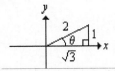

Thus $\tan\theta = \dfrac{1}{\sqrt{3}} = \dfrac{\sqrt{3}}{3}$ and $y = \dfrac{\sqrt{3}}{3}$.

51. Find the exact value, or round to 4 decimal places.

Let $\theta = \sin^{-1}\dfrac{1}{4}$ and find $y = \sec\theta$.

Then $\sin\theta = \dfrac{1}{4}$, and $-\dfrac{\pi}{2} \le \theta \le \dfrac{\pi}{2}$.

Thus $\sec\theta = \dfrac{4}{\sqrt{15}} = \dfrac{4\sqrt{15}}{15}$ and $y = \dfrac{4\sqrt{15}}{15}$.

53. Find the exact value, or round to 4 decimal places.

Let $\theta = \sin^{-1}\dfrac{7}{25}$ and find $y = \cos\theta$.

Then $\sin\theta = \dfrac{7}{25}$, and $-\dfrac{\pi}{2} \le \theta \le \dfrac{\pi}{2}$.

Thus $\cos\theta = \dfrac{24}{25}$ and $y = \dfrac{24}{25}$.

55. Find the exact value, or round to 4 decimal places.

Let $\alpha = \sin^{-1}\dfrac{\sqrt{2}}{2}$, $\alpha = \dfrac{\pi}{4}$, $\sin\alpha = \dfrac{\sqrt{2}}{2}$, $\cos\alpha = \dfrac{\sqrt{2}}{2}$.

$$y = \cos\left(2\sin^{-1}\dfrac{\sqrt{2}}{2}\right)$$
$$= \cos 2\alpha$$
$$= \cos^2\alpha - \sin^2\alpha$$
$$= \left(\dfrac{\sqrt{2}}{2}\right)^2 - \left(\dfrac{\sqrt{2}}{2}\right)^2$$
$$= 0$$

57. Find the exact value, or round to 4 decimal places.

Let $\alpha = \sin^{-1}\dfrac{4}{5}$, $\sin\alpha = \dfrac{4}{5}$, $\cos\alpha = \sqrt{1-\left(\dfrac{4}{5}\right)^2} = \dfrac{3}{5}$.

$$y = \sin\left(2\sin^{-1}\dfrac{4}{5}\right)$$
$$= \sin 2\alpha = 2\sin\alpha\cos\alpha$$
$$= 2\left(\dfrac{4}{5}\right)\left(\dfrac{3}{5}\right) = \dfrac{24}{25}$$

59. Find the exact value, or round to 4 decimal places.

$$y = \sin\left(\sin^{-1}\dfrac{2}{3} + \cos^{-1}\dfrac{1}{2}\right)$$

Let $\alpha = \sin^{-1}\dfrac{2}{3}$, $\sin\alpha = \dfrac{2}{3}$, $\cos\alpha = \sqrt{1-\left(\dfrac{2}{3}\right)^2} = \dfrac{\sqrt{5}}{3}$.

$\beta = \cos^{-1}\dfrac{1}{2}$, $\cos\beta = \dfrac{1}{2}$, $\sin\beta = \sqrt{1-\left(\dfrac{1}{2}\right)^2} = \dfrac{\sqrt{3}}{2}$.

$$y = \sin(\alpha+\beta)$$
$$= \sin\alpha\cos\beta + \cos\alpha\sin\beta$$
$$= \dfrac{2}{3}\left(\dfrac{1}{2}\right) + \dfrac{\sqrt{5}}{3}\left(\dfrac{\sqrt{3}}{2}\right)$$
$$= \dfrac{1}{3} + \dfrac{\sqrt{15}}{6} = \dfrac{2+\sqrt{15}}{6}$$

61. Find the exact value, or round to 4 decimal places.

$$y = \tan\left(\cos^{-1}\dfrac{1}{2} - \sin^{-1}\dfrac{3}{4}\right)$$

Let $\alpha = \cos^{-1}\dfrac{1}{2}$, $\cos\alpha = \dfrac{1}{2}$,

$$\sin\alpha = \sqrt{1-\left(\dfrac{1}{2}\right)^2} = \dfrac{\sqrt{3}}{2}, \quad \tan\alpha = \dfrac{\frac{\sqrt{3}}{2}}{\frac{1}{2}} = \sqrt{3}.$$

$\beta = \sin^{-1}\dfrac{3}{4}$, $\sin\beta = \dfrac{3}{4}$,

$$\cos\beta = \sqrt{1-\left(\dfrac{3}{4}\right)^2} = \dfrac{\sqrt{7}}{4}, \quad \tan\beta = \dfrac{\frac{3}{4}}{\frac{\sqrt{7}}{4}} = \dfrac{3}{\sqrt{7}} = \dfrac{3\sqrt{7}}{7}.$$

$$y = \tan(\alpha-\beta)$$
$$= \dfrac{\tan\alpha - \tan\beta}{1 + \tan\alpha\tan\beta}$$
$$= \dfrac{\sqrt{3} - \frac{3\sqrt{7}}{7}}{1 + \sqrt{3}\cdot\frac{3\sqrt{7}}{7}} = \dfrac{\sqrt{3} - \frac{3\sqrt{7}}{7}}{1 + \sqrt{3}\cdot\frac{3\sqrt{7}}{7}}\cdot\dfrac{7}{7}$$
$$= \dfrac{7\sqrt{3} - 3\sqrt{7}}{7 + 3\sqrt{21}} = \dfrac{7\sqrt{3} - 3\sqrt{7}}{7 + 3\sqrt{21}}\cdot\dfrac{7 - 3\sqrt{21}}{7 - 3\sqrt{21}}$$
$$= \dfrac{112\sqrt{3} - 84\sqrt{7}}{-140} = \dfrac{3\sqrt{7} - 4\sqrt{3}}{5}$$
$$= \dfrac{1}{5}(3\sqrt{7} - 4\sqrt{3})$$

63. Solve for x algebraically.

$$\sin^{-1}x = \cos^{-1}\dfrac{5}{13}$$
$$\sin(\sin^{-1}x) = \sin\left(\cos^{-1}\dfrac{5}{13}\right)$$
$$x = \dfrac{12}{13}$$

65. Solve for x algebraically.

$$\sin^{-1}(x-1) = \dfrac{\pi}{2}$$
$$(x-1) = \sin\dfrac{\pi}{2}$$
$$(x-1) = 1$$
$$x = 2$$

67. Solve for x algebraically.

$$\tan^{-1}\left(x+\frac{\sqrt{2}}{2}\right)=\frac{\pi}{4}$$

$$\left(x+\frac{\sqrt{2}}{2}\right)=\tan\frac{\pi}{4}$$

$$x=1-\frac{\sqrt{2}}{2}$$

$$x=\frac{2-\sqrt{2}}{2}$$

69. Solve for x algebraically.

$$\sin^{-1}\frac{3}{5}+\cos^{-1}x=\frac{\pi}{4}$$

$$\cos^{-1}x=\frac{\pi}{4}-\sin^{-1}\frac{3}{5}$$

$$x=\cos\left(\frac{\pi}{4}-\sin^{-1}\frac{3}{5}\right)$$

Let $\alpha=\sin^{-1}\frac{3}{5}, \sin\alpha=\frac{3}{5}, \cos\alpha=\frac{4}{5}$.

$$x=\cos\left(\frac{\pi}{4}-\alpha\right)$$

$$x=\cos\frac{\pi}{4}\cos\alpha+\sin\frac{\pi}{4}\sin\alpha$$

$$x=\frac{\sqrt{2}}{2}\cdot\frac{4}{5}+\frac{\sqrt{2}}{2}\cdot\frac{3}{5}$$

$$x=\frac{4\sqrt{2}}{10}+\frac{3\sqrt{2}}{10}=\frac{7\sqrt{2}}{10}$$

71. Solve for x algebraically.

$$\sin^{-1}\frac{\sqrt{2}}{2}+\cos^{-1}x=\frac{2\pi}{3}$$

$$\cos^{-1}x=\frac{2\pi}{3}-\sin^{-1}\frac{\sqrt{2}}{2}$$

$$x=\cos\left(\frac{2\pi}{3}-\sin^{-1}\frac{\sqrt{2}}{2}\right)$$

Let $\alpha=\sin^{-1}\frac{\sqrt{2}}{2}, \sin\alpha=\frac{\sqrt{2}}{2}, \cos\alpha=\frac{\sqrt{2}}{2}$.

$$x=\cos\left(\frac{2\pi}{3}-\alpha\right)$$

$$x=\cos\frac{2\pi}{3}\cos\alpha+\sin\frac{2\pi}{3}\sin\alpha$$

$$x=-\frac{1}{2}\cdot\frac{\sqrt{2}}{2}+\frac{\sqrt{3}}{2}\cdot\frac{\sqrt{2}}{2}=-\frac{\sqrt{2}}{4}+\frac{\sqrt{6}}{4}$$

$$x=\frac{-\sqrt{2}+\sqrt{6}}{4}\approx0.2588$$

Note: Since $\sin\alpha=\frac{\sqrt{2}}{2}$ and $\cos\alpha=\frac{\sqrt{2}}{2}$,

then $\alpha=\frac{\pi}{4}$.

Thus, $\cos\left(\frac{2\pi}{3}-\alpha\right)=\cos\left(\frac{2\pi}{3}-\frac{\pi}{4}\right)=\cos\left(\frac{5\pi}{12}\right)\approx0.2588$.

73. Write in terms of x.

$$\tan(\cos^{-1}x)=\frac{\sqrt{1-x^2}}{x}$$

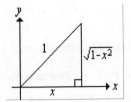

75. Verify the identity.

Let $\alpha=\sin^{-1}x, \sin\alpha=x, \cos\alpha=\sqrt{1-x^2}$.

Let $\beta=\sin^{-1}(-x), \sin\beta=-x, \cos\beta=\sqrt{1-x^2}$.

$$\sin^{-1}x+\sin^{-1}(-x)=\alpha+\beta$$

$$=\sin^{-1}\left[\sin(\alpha+\beta)\right]$$

$$=\sin^{-1}(\sin\alpha\cos\beta+\cos\alpha\sin\beta)$$

$$=\sin^{-1}\left[x\sqrt{1-x^2}+\sqrt{1-x^2}(-x)\right]$$

$$=\sin^{-1}0$$

$$=0$$

77. Verify the identity.

Let $\alpha=\tan^{-1}x, \tan\alpha=x, \beta=\tan^{-1}\frac{1}{x}, \tan\beta=\frac{1}{x}$.

$$\tan^{-1}x+\tan^{-1}\frac{1}{x}=\alpha+\beta$$

$$=\tan^{-1}\left[\tan(\alpha+\beta)\right]$$

$$=\tan^{-1}\left[\frac{\tan\alpha+\tan\beta}{1-\tan\alpha\tan\beta}\right]$$

$$=\tan^{-1}\left[\frac{x+\frac{1}{x}}{1-x\cdot\frac{1}{x}}\right]$$

$$=\tan^{-1}\frac{\frac{x^2+1}{x}}{1-1}, \text{ which is undefined}$$

Thus $x=\frac{\pi}{2}$.

79. Graph using stretching, shrinking and translation.

The graph of $y=\sin^{-1}(x)+2$ (shown as a black graph) is the graph of $y=\sin^{-1}x$ (shown as a gray graph) moved two units up.

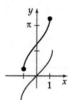

81. Graph using stretching, shrinking and translation.

The graph of $y = \sin^{-1}(x+1) - 2$ (shown as a black graph) is the graph of $y = \sin^{-1} x$ (shown as a gray graph) moved one unit to the left and two units down.

83. Graph using stretching, shrinking and translation.

The graph of $y = 2\cos^{-1} x$ (shown as a black graph) is the graph of $y = \cos^{-1} x$ (shown as a gray graph) stretched.

85. Graph using stretching, shrinking and translation.

The graph of $y = \tan^{-1}(x+1) - 2$ (shown as a black graph) is the graph of $y = \tan^{-1} x$ (shown as a gray graph) moved one unit to the left and two units down.

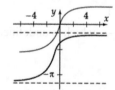

87. a. $s = 3960\theta$

$$\cos\theta = \frac{3960}{a+3960}$$

$$\theta = \cos^{-1}\left(\frac{3960}{a+3960}\right)$$

$$s = 3960\cos^{-1}\left(\frac{3960}{a+3960}\right)$$

b. Solve $5500 = 3960\cos^{-1}\left(\dfrac{3960}{a+3960}\right)$ for a.

$$5500 = 3960\cos^{-1}\left(\frac{3960}{a+3960}\right)$$

$$\frac{5500}{3960} = \cos^{-1}\left(\frac{3960}{a+3960}\right)$$

$$\cos\left(\frac{5500}{3960}\right) = \frac{3960}{a+3960}$$

$$a+3960 = \frac{3960}{\cos\left(\dfrac{5500}{3960}\right)}$$

$$a = \frac{3960}{\cos\left(\dfrac{5500}{3960}\right)} - 3960$$

$$a \approx 17{,}930 \text{ mi}$$

89. [Note: $f(x) = \cos^{-1} x$ is neither odd nor even.

$g(x) = \sin^{-1}\sqrt{1-x^2}$ is an even function.]

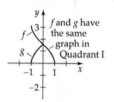

f and g have the same graph in Quadrant I

No, $f(x) \neq g(x)$ on the interval $[-1, 1]$.

91. Use a graphing utility to graph each equation.

$y = \csc^{-1} 2x$

$\csc y = 2x$

$-\dfrac{\pi}{2} \le y \le \dfrac{\pi}{2}, \ y \neq 0$

$2x \le -1$ or $2x \ge 1$

$x \le -\dfrac{1}{2}$ or $x \ge \dfrac{1}{2}$

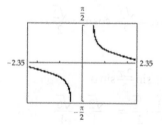

93. Use a graphing utility to graph each equation.

$y = \sec^{-1}(x-1)$

$\sec y = x - 1$

$0 < y < \pi$

$y \neq \dfrac{\pi}{2}$

$x - 1 \le -1 \qquad x - 1 \ge 1$

$x \le 0 \qquad\quad x \ge 2$

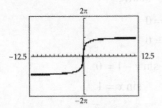

95. Use a graphing utility to graph each equation.

$$y = 2\tan^{-1} 2x$$

$$\frac{y}{2} = \tan^{-1} 2x$$

$$\tan y = 2x$$

$$-\frac{\pi}{2} < \frac{y}{2} < \frac{\pi}{2}$$

$$-\pi < y < \pi$$

$$-\infty < 2x < \infty$$

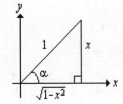

97. Solve for y in terms of x.

$$5x = \tan^{-1} 3y$$

$$\tan 5x = 3y$$

$$y = \frac{1}{3}\tan 5x$$

99. Solve for y in terms of x.

$$x - \frac{\pi}{3} = \cos^{-1}(y - 3)$$

$$\cos\left(x - \frac{\pi}{3}\right) = y - 3$$

$$y = 3 + \cos\left(x - \frac{\pi}{3}\right)$$

101. Verify the identity.

Let $\alpha = \sin^{-1} x$

$$\sin\alpha = x$$

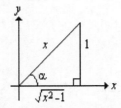

$$\cos\left(\sin^{-1} x\right) = \cos\alpha$$

$$= \frac{\sqrt{1 - x^2}}{1}$$

$$= \sqrt{1 - x^2}$$

103. Verify the identity.

Let $\alpha = \csc^{-1} x$

$$\csc\alpha = x$$

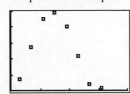

$$\tan(\csc^{-1} x) = \tan\alpha$$

$$= \frac{1}{\sqrt{x^2 - 1}}$$

$$= \frac{\sqrt{x^2 - 1}}{x^2 - 1}, \ x > 1$$

Prepare for Section 6.6

P1. Solve using the quadratic formula.

$$x = \frac{5 \pm \sqrt{(-5)^2 - 4(3)(-4)}}{2(3)} = \frac{5 \pm \sqrt{73}}{6}$$

P3. Evaluate for $k = 1, 2,$ and 3.

$$\frac{\pi}{2} + 2(1)\pi = \frac{5}{2}\pi$$

$$\frac{\pi}{2} + 2(2)\pi = \frac{9}{2}\pi$$

$$\frac{\pi}{2} + 2(3)\pi = \frac{13}{2}\pi$$

P5. Graph the scatter plot.

Section 6.6 Exercises

1. Determine the number of solutions. State the procedure used.

There are 3 solutions to $\sin x = \frac{1}{2}x$. One method is to graph $y = \sin x$ and $y = \dfrac{1}{2}x$ on the same coordinate system to determine the number of intersections.

3. Write the solutions of $\tan x = \sqrt{3}$.

The solutions are $x = \dfrac{\pi}{3} + k\pi$, where k is an integer.

5. Solve for solutions in $0 \leq x < 2\pi$.

$$\sec x - \sqrt{2} = 0$$
$$\sec x = \sqrt{2}$$
$$x = \dfrac{\pi}{4}, \dfrac{7\pi}{4}$$

7. Solve for solutions in $0 \leq x < 2\pi$.

$$\tan x - \sqrt{3} = 0$$
$$\tan x = \sqrt{3}$$
$$x = \dfrac{\pi}{3}, \dfrac{4\pi}{3}$$

9. Solve for solutions in $0 \leq x < 2\pi$.

$$2\sin x \cos x = \sqrt{2}\cos x$$
$$2\sin x \cos x - \sqrt{2}\cos x = 0$$
$$\cos x(2\sin x - \sqrt{2}) = 0$$

$$\cos x = 0 \qquad 2\sin x - \sqrt{2} = 0$$
$$\sin x = \dfrac{\sqrt{2}}{2}$$
$$x = \dfrac{\pi}{2}, \dfrac{3\pi}{2} \qquad x = \dfrac{\pi}{4}, \dfrac{3\pi}{4}$$

The solutions are $\dfrac{\pi}{4}, \dfrac{\pi}{2}, \dfrac{3\pi}{4}, \dfrac{3\pi}{2}$.

11. Solve for solutions in $0 \leq x < 2\pi$.

$$\sec x - \dfrac{2\sqrt{3}}{3} = 0$$
$$\sec x = \dfrac{2\sqrt{3}}{3}$$
$$x = \dfrac{\pi}{6}, \dfrac{11\pi}{6}$$

13. Solve for solutions in $0 \leq x < 2\pi$.

$$4\sin x \cos x - 2\sqrt{3}\sin x - 2\sqrt{2}\cos x + \sqrt{6} = 0$$
$$2\sin x(2\cos x - \sqrt{3}) - \sqrt{2}(2\cos x - \sqrt{3}) = 0$$
$$(2\cos x - \sqrt{3})(2\sin x - \sqrt{2}) = 0$$

$$2\cos x - \sqrt{3} = 0 \qquad 2\sin x - \sqrt{2} = 0$$
$$\cos x = \dfrac{\sqrt{3}}{2} \qquad \sin x = \dfrac{\sqrt{2}}{2}$$
$$x = \dfrac{\pi}{6}, \dfrac{11\pi}{6} \qquad x = \dfrac{\pi}{4}, \dfrac{3\pi}{4}$$

The solutions are $\dfrac{\pi}{6}, \dfrac{\pi}{4}, \dfrac{3\pi}{4}, \dfrac{11\pi}{6}$.

15. Solve for solutions in $0 \leq x < 2\pi$.

$$\csc x - \sqrt{2} = 0$$
$$\csc x = \sqrt{2}$$
$$x = \dfrac{\pi}{4}, \dfrac{3\pi}{4}$$

17. Solve for solutions in $0 \leq x < 2\pi$.

$$2\sin^2 x + 1 = 3\sin x$$
$$2\sin^2 x - 3\sin x + 1 = 0$$
$$(2\sin x - 1)(\sin x - 1) = 0$$

$$2\sin x - 1 = 0 \qquad \sin x - 1 = 0$$
$$\sin x = \dfrac{1}{2} \qquad \sin x = 1$$
$$x = \dfrac{\pi}{6}, \dfrac{5\pi}{6} \qquad x = \dfrac{\pi}{2}$$

The solutions are $\dfrac{\pi}{6}, \dfrac{\pi}{2}, \dfrac{5\pi}{6}$.

19. Solve for solutions in $0 \leq x < 2\pi$.

$$4\cos^2 x - 3 = 0$$
$$\cos^2 x = \dfrac{3}{4}$$
$$\cos x = \pm\dfrac{\sqrt{3}}{2}$$
$$x = \dfrac{\pi}{6}, \dfrac{5\pi}{6}, \dfrac{7\pi}{6}, \dfrac{11\pi}{6}$$

21. Solve for solutions in $0 \leq x < 2\pi$.

$$2\sin^3 x = \sin x$$
$$2\sin^3 x - \sin x = 0$$
$$\sin x(2\sin^2 x - 1) = 0$$

$$\sin x = 0 \qquad 2\sin^2 x = 1$$
$$x = 0, \pi \qquad \sin x = \pm\dfrac{\sqrt{2}}{2}$$
$$\qquad x = \dfrac{\pi}{4}, \dfrac{3\pi}{4}, \dfrac{5\pi}{4}, \dfrac{7\pi}{4}$$

The solutions are $0, \dfrac{\pi}{4}, \dfrac{3\pi}{4}, \pi, \dfrac{5\pi}{4}, \dfrac{7\pi}{4}$.

23. Solve for solutions in $0 \le x < 2\pi$.

$$4\sin^2 x + 2\sqrt{3}\sin x - \sqrt{3} = 2\sin x$$
$$4\sin^2 x + 2\sqrt{3}\sin x - 2\sin x - \sqrt{3} = 0$$
$$2\sin x(2\sin x + \sqrt{3}) - (2\sin x + \sqrt{3}) = 0$$
$$(2\sin x + \sqrt{3})(2\sin x - 1) = 0$$

$$\begin{array}{ll} 2\sin x + \sqrt{3} = 0 & 2\sin x - 1 = 0 \\ \sin x = -\dfrac{\sqrt{3}}{2} & \sin x = \dfrac{1}{2} \\ x = \dfrac{4\pi}{3}, \dfrac{5\pi}{3} & x = \dfrac{\pi}{6}, \dfrac{5\pi}{6} \end{array}$$

The solutions are $\dfrac{\pi}{6}, \dfrac{5\pi}{6}, \dfrac{4\pi}{3}, \dfrac{5\pi}{3}$.

25. Solve for solutions in $0 \le x < 2\pi$.

$$\sin^4 x = \sin^2 x$$
$$\sin^4 x - \sin^2 x = 0$$
$$\sin^2 x(\sin^2 x - 1) = 0$$

$$\begin{array}{ll} \sin^2 x = 0 & \sin^2 x - 1 = 0 \\ \sin x = 0 & \sin x = \pm 1 \\ x = 0, \pi & x = \dfrac{\pi}{2}, \dfrac{3\pi}{2} \end{array}$$

The solutions are $0, \dfrac{\pi}{2}, \pi, \dfrac{3\pi}{2}$.

27. Solve for solutions in $0° \le x < 360°$. Round approximate solutions to the nearest tenth.

$$\tan x - 25.5 = 0$$
$$\tan x = 25.5$$
$$x \approx 87.8°, 267.8°$$

29. Solve for solutions in $0° \le x < 360°$. Round approximate solutions to the nearest tenth.

$$3\sin x - 5 = 0$$
$$3\sin x = 5$$
$$\sin x = \dfrac{5}{3}$$

no solution

31. Solve for solutions in $0° \le x < 360°$. Round approximate solutions to the nearest tenth.

$$2\cot x - 11 = 0$$
$$2\cot x = 11$$
$$\cot x = \dfrac{11}{2}$$
$$\dfrac{1}{\tan x} = \dfrac{11}{2}$$
$$\tan x = \dfrac{2}{11}$$
$$x \approx 10.3°, 190.3°$$

33. Solve for solutions in $0° \le x < 360°$. Round approximate solutions to the nearest tenth.

$$3 - 5\sin x = 4\sin x + 1$$
$$-9\sin x = -2$$
$$\sin x = \dfrac{2}{9}$$
$$x \approx 12.8°, 167.2°$$

35. Solve for solutions in $0° \le x < 360°$. Round approximate solutions to the nearest tenth.

$$3\tan^2 x - 2\tan x = 0$$
$$\tan x(3\tan x - 2) = 0$$

$$\begin{array}{ll} \tan x = 0 & 3\tan x - 2 = 0 \\ x = 0, 180° & \tan x = \dfrac{2}{3} \\ & x \approx 33.7°, 213.7° \end{array}$$

The solutions are $0°, 33.7°, 180°, 213.7°$.

37. Solve for solutions in $0° \le x < 360°$. Round approximate solutions to the nearest tenth.

$$2\sin^2 x = 1 - \cos x$$
$$2(1 - \cos^2 x) = 1 - \cos x$$
$$2 - 2\cos^2 x = 1 - \cos x$$
$$0 = 2\cos^2 x - \cos x - 1$$
$$0 = (2\cos x + 1)(\cos x - 1)$$

$$\begin{array}{ll} 2\cos x + 1 = 0 & \cos x - 1 = 0 \\ \cos x = -\dfrac{1}{2} & \cos x = 1 \\ x = 120°, 240° & x = 0° \end{array}$$

The solutions are $0°, 120°, 240°$.

39. Solve for solutions in $0° \leq x < 360°$. Round approximate solutions to the nearest tenth.

$3\cos^2 x + 5\cos x - 2 = 0$

$$\cos x = \frac{-5 \pm \sqrt{5^2 - 4(3)(-2)}}{2 \cdot 3}$$

$$= \frac{-5 \pm \sqrt{49}}{6} = \frac{-5 \pm 7}{6}$$

$\cos x = \frac{1}{3}$ $\qquad$ $\cos x = -2$

$x \approx 70.5°, 289.5°$ $\qquad$ no solution

The solutions are $70.5°, 289.5°$.

41. Solve for solutions in $0° \leq x < 360°$. Round approximate solutions to the nearest tenth.

$2\tan^2 x - \tan x - 10 = 0$

$(\tan x + 2)(2\tan x - 5) = 0$

$\tan x + 2 = 0$ $\qquad$ $2\tan x - 5 = 0$

$\tan x = -2$ $\qquad$ $\tan x = \frac{5}{2}$

$x \approx 116.6°, 296.6°$ $\qquad$ $x \approx 68.2°, 248.2°$

The solutions are $68.2°, 116.6°, 248.2°, 296.6°$.

43. Solve for solutions in $0° \leq x < 360°$. Round approximate solutions to the nearest tenth.

$3\sin x \cos x - \cos x = 0$

$\cos x(3\sin x - 1) = 0$

$\cos x = 0$ $\qquad$ $3\sin x - 1 = 0$

$x = 90°, 270°$ $\qquad$ $\sin x = \frac{1}{3}$

$\qquad\qquad$ $x \approx 19.5°, 160.5°$

The solutions are $19.5°, 90°, 160.5°, 270°$.

45. Solve for solutions in $0° \leq x < 360°$. Round approximate solutions to the nearest tenth.

$2\sin x \cos x - \sin x - 2\cos x + 1 = 0$

$\sin x(2\cos x - 1) - (2\cos x - 1) = 0$

$(2\cos x - 1)(\sin x - 1) = 0$

$2\cos x - 1 = 0$ $\qquad$ $\sin x - 1 = 0$

$\cos x = \frac{1}{2}$ $\qquad$ $\sin x = 1$

$x = 60°, 300°$ $\qquad$ $x = 90°$

The solutions are $60°, 90°, 300°$.

47. Solve for solutions in $0° \leq x < 360°$. Round approximate solutions to the nearest tenth.

$2\sin x - \cos x = 1$

$2\sin x - 1 = \cos x$

$(2\sin x - 1)^2 = (\cos x)^2$

$4\sin^2 x - 4\sin x + 1 = \cos^2 x$

$4\sin^2 x - 4\sin x + 1 = 1 - \sin^2 x$

$5\sin^2 x - 4\sin x = 0$

$\sin x(5\sin x - 4) = 0$

$\sin x = 0$ $\qquad$ $5\sin x - 4 = 0$

$x = 180°$ $\qquad$ $\sin x = \frac{4}{5}$

$\qquad\qquad$ $x \approx 53.1°$ or $126.9°$

$126.9°$ does not check.

The solutions are $53.1°, 180°$.

49. Solve for solutions in $0° \leq x < 360°$. Round approximate solutions to the nearest tenth.

$2\sin x - 3\cos x = 1$

$2\sin x = 3\cos x + 1$

$(2\sin x)^2 = (3\cos x + 1)^2$

$4\sin^2 x = 9\cos^2 x + 6\cos x + 1$

$4(1 - \cos^2 x) = 9\cos^2 x + 6\cos x + 1$

$0 = 13\cos^2 x + 6\cos x - 3$

$$\cos x = \frac{-6 \pm \sqrt{6^2 - 4(13)(-3)}}{2(13)}$$

$$= \frac{-6 \pm \sqrt{192}}{26}$$

$\cos x \approx 0.3022$ $\qquad$ $\cos x \approx -0.7637$

$x \approx 72.4°$ or $287.6°$ $\qquad$ $x \approx 139.8°$ or $220.2°$

$287.6°$ and $139.8°$ do not check.

The solutions are $72.4°, 220.2°$.

51. Solve for solutions in $0° \leq x < 360°$. Round approximate solutions to the nearest tenth.

$$\cos^2 x - 3\sin x + 2\sin^2 x = 0$$
$$1 - \sin^2 x - 3\sin x + 2\sin^2 x = 0$$
$$\sin^2 x - 3\sin x + 1 = 0$$

$$\sin x = \frac{3 \pm \sqrt{(-3)^2 - 4(1)(1)}}{2(1)} = \frac{3 \pm \sqrt{5}}{2}$$

$\sin x = 2.6180$ $\sin x = 0.3820$

no solution $x = 22.5°, \ 157.5°$

The solutions are $22.5°, \ 157.5°$.

53. Find the exact solutions, in radians.

$$\tan 2x - 1 = 0$$
$$\tan 2x = 1$$
$$2x = \frac{\pi}{4} + k\pi$$
$$x = \frac{\pi}{8} + \frac{k\pi}{2}, \text{ where } k \text{ is an integer}$$

55. Find the exact solutions, in radians.

$$\sin 5x = 1$$
$$5x = \frac{\pi}{2} + 2k\pi$$
$$x = \frac{\pi}{10} + \frac{2}{5}k\pi, \text{ where } k \text{ is an integer}$$

57. Find the exact solutions, in radians.

$$\sin 2x - \sin x = 0$$
$$2\sin x \cos x - \sin x = 0$$
$$\sin x(2\cos x - 1) = 0$$

$\sin x = 0$ $2\cos x - 1 = 0$

$x = 0 + 2k\pi$ $\cos x = \dfrac{1}{2}$

or $x = \dfrac{\pi}{3} + 2k\pi$

$x = \pi + 2k\pi$

or

$x = \dfrac{5\pi}{3} + 2k\pi$

The solutions are $0 + 2k\pi, \ \dfrac{\pi}{3} + 2k\pi, \ \pi + 2k\pi, \ \dfrac{5\pi}{3} + 2k\pi$ where k is an integer.

59. Find the exact solutions, in radians.

$$\sin\left(2x + \frac{\pi}{6}\right) = -\frac{1}{2}$$

$2x + \dfrac{\pi}{6} = \dfrac{7\pi}{6} + 2k\pi$ or $2x + \dfrac{\pi}{6} = \dfrac{11\pi}{6} + 2k\pi$

$2x = \pi + 2k\pi$ $2x = \dfrac{5\pi}{3} + 2k\pi$

$x = \dfrac{\pi}{2} + k\pi$ $x = \dfrac{5\pi}{6} + k\pi$

The solutions are $\dfrac{\pi}{2} + k\pi, \ \dfrac{5\pi}{6} + k\pi$ where k is an integer.

61. Find the exact solutions, in radians.

$$\sin^2 \frac{x}{2} + \cos x = 1$$
$$\left(\pm\sqrt{\frac{1 - \cos x}{2}}\right)^2 + \cos x = 1$$
$$\frac{1 - \cos x}{2} + \cos x = 1$$
$$1 - \cos x + 2\cos x = 2$$
$$\cos x = 1$$

$$x = 0 + 2k\pi \text{ where } k \text{ is an integer}$$

63. Solve for solutions in $0 \leq x < 2\pi$.

$$\cos 2x = 1 - 3\sin x$$
$$1 - 2\sin^2 x = 1 - 3\sin x$$
$$0 = 2\sin^2 x - 3\sin x$$
$$0 = \sin x(2\sin x - 3)$$

$\sin x = 0$ $2\sin x - 3 = 0$

$x = 0, \ \pi$ $\sin x = \dfrac{3}{2}$

no solution.

The solutions are $0, \ \pi$.

65. Solve for solutions in $0 \leq x < 2\pi$.

$$\sin 4x - \sin 2x = 0$$
$$2\sin 2x \cos 2x - \sin 2x = 0$$
$$\sin 2x(2\cos 2x - 1) = 0$$

$\sin 2x = 0$ $2\cos 2x - 1 = 0$

$2x = 0 + 2k\pi$ $\cos 2x = \dfrac{1}{2}$

or $2x = \dfrac{\pi}{3} + 2k\pi$

$2x = \pi + 2k\pi$

or

$2x = \dfrac{5\pi}{3} + 2k\pi$

$x = 0 + k\pi, \ \dfrac{\pi}{2} + k\pi, \ \dfrac{\pi}{6} + k\pi, \ \dfrac{5\pi}{6} + k\pi$

The solutions are $0, \ \dfrac{\pi}{6}, \ \dfrac{\pi}{2}, \ \dfrac{5\pi}{6}, \ \pi, \ \dfrac{7\pi}{6}, \ \dfrac{3\pi}{2}, \ \dfrac{11\pi}{6}.$

67. Solve for solutions in $0 \le x < 2\pi$.

$$\tan\dfrac{x}{2} = \sin x$$

$$\dfrac{1 - \cos x}{\sin x} = \sin x$$

$$1 - \cos x = \sin^2 x$$

$$1 - \cos x = 1 - \cos^2 x$$

$$\cos^2 x - \cos x = 0$$

$$\cos x(\cos x - 1) = 0$$

$\cos x = 0 \qquad \cos x = 1$

$x = \dfrac{\pi}{2}, \dfrac{3\pi}{2} \qquad x = 0$

The solutions are $0, \ \dfrac{\pi}{2}, \ \dfrac{3\pi}{2}.$

69. Solve for solutions in $0 \le x < 2\pi$.

$$\sin 2x \cos x + \cos 2x \sin x = 0$$

$$\sin(2x + x) = 0$$

$$\sin 3x = 0$$

$3x = 0 + 2k\pi \quad \text{or} \quad 3x = \pi + 2k\pi$

$x = 0 + \dfrac{2}{3}k\pi \quad \text{or} \quad x = \dfrac{\pi}{3} + \dfrac{2}{3}k\pi$

The solutions are $0, \ \dfrac{\pi}{3}, \ \dfrac{2}{3}\pi, \ \pi, \ \dfrac{4}{3}\pi \ \dfrac{5}{3}\pi.$

71. Solve for solutions in $0 \le x < 2\pi$.

$$\sin x \cos 2x - \cos x \sin 2x = \dfrac{\sqrt{3}}{2}$$

$$\sin(x - 2x) = \dfrac{\sqrt{3}}{2}$$

$$\sin(-x) = \dfrac{\sqrt{3}}{2}$$

$$\sin x = -\dfrac{\sqrt{3}}{2}$$

$$x = \dfrac{4\pi}{3}, \dfrac{5\pi}{3}$$

73. Solve for solutions in $0 \le x < 2\pi$.

$$\sin 3x - \sin x = 0$$

$$2\cos\dfrac{3x + x}{2}\sin\dfrac{3x - x}{2} = 0$$

$$2\cos 2x \sin x = 0$$

$$2(1 - 2\sin^2 x)\sin x = 0$$

$\sin x = 0 \qquad 1 - 2\sin^2 x = 0$

$x = 0, \ \pi \qquad \sin^2 x = \dfrac{1}{2}$

$$\sin x = \pm\dfrac{\sqrt{2}}{2}$$

$$x = \dfrac{\pi}{4}, \ \dfrac{3\pi}{4}, \ \pi, \ \dfrac{5\pi}{4}, \ \dfrac{7\pi}{4}$$

The solutions are $0, \ \dfrac{\pi}{4}, \ \dfrac{3\pi}{4}, \ \pi, \ \dfrac{5\pi}{4}, \ \dfrac{7\pi}{4}.$

75. Solve for solutions in $0 \le x < 2\pi$.

$$2\sin x \cos x + 2\sin x - \cos x - 1 = 0$$

$$2\sin x(\cos x + 1) - (\cos x + 1) = 0$$

$$(\cos x + 1)(2\sin x - 1) = 0$$

$\cos x + 1 = 0 \qquad 2\sin x - 1 = 0$

$\cos x = -1 \qquad \sin x = \dfrac{1}{2}$

$x = \pi \qquad x = \dfrac{\pi}{6}, \ \dfrac{5\pi}{6}$

The solutions are $\dfrac{\pi}{6}, \ \dfrac{5\pi}{6}, \ \pi.$

77. Use a graphing utility to solve each equation and state approximate solutions to 4 decimal places.

$-1.8955, 0, 1.8955$

79. Use a graphing utility to solve each equation and state approximate solutions to 4 decimal places.

$-3.2957, 3.2957$

81. Use a graphing utility to solve and approximate to the nearest hundredth.

1.16

83. Set your graphing utility to "degree" mode and graph

$$d = \dfrac{(288)^2}{16}\sin\theta\cos\theta \ \text{and} \ d = 1295 \ \text{for}$$

$0° \le \theta \le 90°, \ -500 \le d \le 3000.$

Use the TRACE or INTERSECT feature of your graphing utility to determine the intersection of the two graphs.

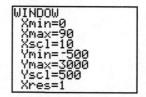

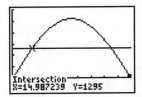 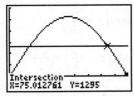

Thus, $d = 1295$ for $\theta \approx 14.99°$ and $\theta \approx 75.01°$.

85. a. Find the regression function.

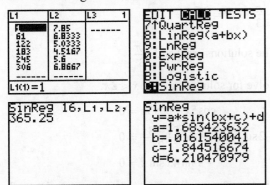

$$y \approx 1.683424\sin(0.016154x + 1.844517) + 6.210471$$

b. Set your graphing utility to "radian" mode.

$$f(38) \approx 1.683424\sin(0.016154(38) + 1.844517)$$
$$+ 6.210471$$
$$\approx 7.2732 \text{ hours}$$
$$\approx 7 + 0.2732(60) \rightarrow 7:16 \text{ (h: min)}$$

87. a. Find the regression function.

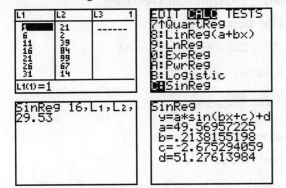

$$y \approx 49.569572\sin(0.213816x - 2.675294) + 51.27614$$

b. Set your graphing utility to "radian" mode.

$$f(45) \approx 49.569572\sin(0.213816(45) - 2.675294)$$
$$+ 51.27614$$
$$\approx 81.794$$
$$\approx 82\%$$

89. a. Find the regression function.

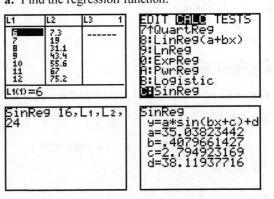

$$y \approx 35.038234\sin(0.407966x + 2.794923) + 38.119377$$

b. Set your graphing utility to "radian" mode.

$$f(10.5) \approx 35.038234\sin(0.407966(10.5) + 2.794923)$$
$$+ 38.119377$$
$$\approx 63.1°$$

91. a. $\sin\theta = \dfrac{y}{3}$, so $y = 3\sin\theta$

$\cos\theta = \dfrac{x}{3}$, so $x = 3\cos\theta$

$$A = 2\left(\frac{1}{2}bh\right) + 3y$$
$$= (3\sin\theta)(3\cos\theta) + 3(3\sin\theta)$$
$$= 9\sin\theta\cos\theta + 9\sin\theta$$
$$= 9\sin\theta(\cos\theta + 1)$$

b. Make sure your calculator is in degree mode.

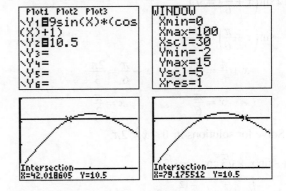

The solutions are $42°$ and $79°$.

c. Make sure your calculator is in degree mode.

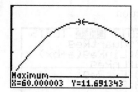

The value of θ that would maximize the area is $60°$.

93. $\theta = 20°$

$$x = \sqrt{(4 + 18\cot 20°)^2 + 100} - (4 + 18\cot 20°)$$

$$x \approx 0.93 \text{ ft.}$$

$\theta = 30°$

$$x = \sqrt{(4 + 18\cot 30°)^2 + 100} - (4 + 18\cot 30°)$$

$$x \approx 1.39 \text{ ft.}$$

95. Solve for solutions in $0 \le x < 2\pi$.

$$\sqrt{3} \sin x + \cos x = \sqrt{3}$$

$a = \sqrt{3}, \ b = 1, \ k = \sqrt{(\sqrt{3})^2 + 1} = 2,$

$\qquad\qquad \alpha$ is in first quadrant

$$\tan\beta = \frac{1}{\sqrt{3}}$$

$$\beta = \frac{\pi}{6}$$

$$\alpha = \frac{\pi}{6}$$

$$2\sin\left(x + \frac{\pi}{6}\right) = \sqrt{3}$$

$$\sin\left(x + \frac{\pi}{6}\right) = \frac{\sqrt{3}}{2}$$

$$x + \frac{\pi}{6} = \frac{\pi}{3} \qquad x + \frac{\pi}{6} = \frac{2\pi}{3}$$

$$x = \frac{\pi}{6} \qquad\qquad x = \frac{\pi}{2}$$

97. Solve for solutions in $0 \le x < 2\pi$.

$$-\sin x + \sqrt{3}\cos x = \sqrt{3}$$

$$k = \sqrt{(-1)^2 + (\sqrt{3})^2} = 2$$

$$\tan\beta = \left|\frac{\sqrt{3}}{-1}\right| = \sqrt{3}$$

$$\beta = \frac{\pi}{3}$$

$$\alpha = \frac{2\pi}{3} \quad \text{second quadrant}$$

$$2\sin\left(x + \frac{2\pi}{3}\right) = \sqrt{3}$$

$$\sin\left(x + \frac{2\pi}{3}\right) = \frac{\sqrt{3}}{2}$$

$$x + \frac{2\pi}{3} = \frac{\pi}{3} \qquad x + \frac{2\pi}{3} = \frac{2\pi}{3}$$

$$x = -\frac{\pi}{3} \qquad\qquad x = 0$$

$$-\frac{\pi}{3} + 2\pi = \frac{5\pi}{3}$$

The solutions are 0 and $\dfrac{5\pi}{3}$.

99. Solve for solutions in $0 \le x < 2\pi$.

$$\cos 5x - \cos 3x = 0$$

$$-2\sin\frac{5x + 3x}{2}\sin\frac{5x - 3x}{2} = 0$$

$$-2\sin 4x \sin x = 0$$

$\sin 4x = 0 \qquad\qquad\qquad \sin x = 0$

$4x = 0 + 2k\pi \qquad 4x = \pi + 2k\pi \qquad x = 0, \ \pi$

$$x = 0 + \frac{1}{2}k\pi \qquad x = \frac{\pi}{4} + \frac{1}{2}k\pi$$

$$x = 0, \ \frac{\pi}{2}, \ \pi, \ \frac{3\pi}{2} \qquad x = \frac{\pi}{4}, \ \frac{3\pi}{4}, \ \frac{5\pi}{4}, \ \frac{7\pi}{4}$$

The solutions are $0, \ \dfrac{\pi}{4}, \ \dfrac{\pi}{2}, \ \dfrac{3\pi}{4}, \ \pi, \ \dfrac{5\pi}{4}, \ \dfrac{3\pi}{2}, \ \dfrac{7\pi}{4}$.

Exploring Concepts with Technology

1. Find the number of beats.

$223 - 220 = 3$ beats per second

3. Find the frequency.

$$\frac{600 + 800}{2} = 700 \text{ Hz}$$

Chapter 6 Review Exercises

1. Find the exact value. [6.2]

$$\cos(45° + 30°) = \cos 45° \cos 30° - \sin 45° \sin 30°$$

$$= \frac{\sqrt{2}}{2} \cdot \frac{\sqrt{3}}{2} - \frac{\sqrt{2}}{2} \cdot \frac{1}{2}$$

$$= \frac{\sqrt{6}}{4} - \frac{\sqrt{2}}{4}$$

$$= \frac{\sqrt{6} - \sqrt{2}}{4}$$

3. Find the exact value. [6.2]

$$\cos\left(\frac{5\pi}{6}+\frac{\pi}{3}\right)=\cos\frac{5\pi}{6}\cos\frac{\pi}{3}-\sin\frac{5\pi}{6}\sin\frac{\pi}{3}$$

$$=-\frac{\sqrt{3}}{2}\cdot\frac{1}{2}-\frac{1}{2}\cdot\frac{\sqrt{3}}{2}$$

$$=-\frac{\sqrt{3}}{4}-\frac{\sqrt{3}}{4}$$

$$=-\frac{\sqrt{3}}{2}$$

5. Find the exact value. [6.2]

$$\sin(60°-135°)=\sin60°\cos135°-\cos60°\sin135°$$

$$=\frac{\sqrt{3}}{2}\left(-\frac{\sqrt{2}}{2}\right)-\frac{1}{2}\cdot\frac{\sqrt{2}}{2}$$

$$=-\frac{\sqrt{6}}{4}-\frac{\sqrt{2}}{4}$$

$$=-\frac{\sqrt{6}+\sqrt{2}}{4}$$

7. Find the exact value. [6.3]

$$\sin22.5°=\sin\frac{45°}{2}$$

$$=\sqrt{\frac{1-\cos45°}{2}}$$

$$=\sqrt{\frac{1-\frac{\sqrt{2}}{2}}{2}}$$

$$=\sqrt{\frac{2-\sqrt{2}}{4}}$$

$$=\frac{\sqrt{2-\sqrt{2}}}{2}$$

9. Find the exact value. [6.3]

$$\tan\frac{7\pi}{12}=\frac{1-\cos\frac{7\pi}{6}}{\sin\frac{7\pi}{6}}$$

$$=\frac{1+\frac{\sqrt{3}}{2}}{-\frac{1}{2}}=\frac{2+\sqrt{3}}{2}\left(-\frac{2}{1}\right)$$

$$=-2-\sqrt{3}$$

11. Find the exact value. [6.2/6.3]

$$\sin\alpha=\frac{1}{2}, \cos\alpha=\frac{\sqrt{3}}{2}, \text{quadrant I}, \tan\alpha=\frac{\sqrt{3}}{3}$$

$$\cos\beta=\frac{1}{2}, \sin\beta=-\frac{\sqrt{3}}{2}, \text{quadrant IV}, \tan\beta=-\sqrt{3}$$

a. $\cos(\alpha-\beta)=\cos\alpha\cos\beta+\sin\alpha\sin\beta$

$$=\frac{\sqrt{3}}{2}\cdot\frac{1}{2}+\frac{1}{2}\left(-\frac{\sqrt{3}}{2}\right)$$

$$=\frac{\sqrt{3}}{4}-\frac{\sqrt{3}}{4}=0$$

b. $\tan2\alpha=\dfrac{2\tan\alpha}{1-\tan^2\alpha}$

$$=\frac{2\left(\frac{\sqrt{3}}{3}\right)}{1-\left(\frac{\sqrt{3}}{3}\right)^2}=\frac{\frac{2\sqrt{3}}{3}}{1-\frac{1}{3}}\cdot\frac{3}{3}$$

$$=\frac{2\sqrt{3}}{3-1}=\sqrt{3}$$

c. $\sin\dfrac{\beta}{2}=\sqrt{\dfrac{1-\cos\beta}{2}}$

$$=\sqrt{\frac{1-\frac{1}{2}}{2}}=\sqrt{\frac{1}{4}}$$

$$=\frac{1}{2}$$

13. Find the exact value. [6.2/6.3]

$$\sin\alpha=-\frac{1}{2}, \quad\cos\alpha=\frac{\sqrt{3}}{2}, \text{quadrant IV}, \quad\tan\alpha=-\frac{\sqrt{3}}{3}$$

$$\cos\beta=-\frac{\sqrt{3}}{2}, \quad\sin\beta=-\frac{1}{2}, \text{quadrant III}$$

a. $\sin(\alpha-\beta)=\sin\alpha\cos\beta-\cos\alpha\sin\beta$

$$=-\frac{1}{2}\left(-\frac{\sqrt{3}}{2}\right)-\frac{\sqrt{3}}{2}\left(-\frac{1}{2}\right)$$

$$=\frac{\sqrt{3}}{4}+\frac{\sqrt{3}}{4}=\frac{2\sqrt{3}}{4}$$

$$=\frac{\sqrt{3}}{2}$$

b. $\tan2\alpha=\dfrac{2\tan\alpha}{1-\tan^2\alpha}=\dfrac{2\left(-\frac{\sqrt{3}}{3}\right)}{1-\left(-\frac{\sqrt{3}}{3}\right)^2}$

$$=\frac{-\frac{2\sqrt{3}}{3}}{1-\frac{1}{3}}\cdot\frac{3}{3}$$

$$=\frac{-2\sqrt{3}}{3-1}$$

$$=-\sqrt{3}$$

c. $\cos\dfrac{\beta}{2} = -\sqrt{\dfrac{1+\cos\beta}{2}}$

$\qquad = -\sqrt{\dfrac{1+\left(-\dfrac{\sqrt{3}}{2}\right)}{2}}$

$\qquad = -\sqrt{\dfrac{2-\sqrt{3}}{4}}$

$\qquad = -\dfrac{\sqrt{2-\sqrt{3}}}{2}$

15. Write an equivalent expression. [6.2]

$\cos 41° = \sin(90° - 41°) = \sin 49°$

17. Write an equivalent expression. [6.2]

$\tan\dfrac{\pi}{8} = \cot\left(\dfrac{\pi}{2} - \dfrac{\pi}{8}\right) = \cot\dfrac{3\pi}{8}$

19. Write as a single trigonometric expression. [6.3]

$2\sin 3x\cos 3x = \sin 2(3x)$

$\qquad = \sin 6x$

21. Write as a single trigonometric expression. [6.2]

$\sin 4x\cos x - \cos 4x\sin x = \sin(4x - x)$

$\qquad = \sin 3x$

23. Write as a single trigonometric expression. [6.1]

$\dfrac{\sin 2\theta}{\cos 2\theta} = \tan 2\theta$

25. Verify the identity. [6.3]

$\sin^2 x\cos^2 x = \dfrac{1-\cos 2x}{2}\cdot\dfrac{1+\cos 2x}{2}$

$\qquad = \dfrac{1-\cos^2 2x}{4}$

$\qquad = \dfrac{1-\dfrac{1+\cos 4x}{2}}{4}$

$\qquad = \dfrac{2-1-\cos 4x}{8}$

$\qquad = \dfrac{1-\cos 4x}{8}$

27. Verify the identity. [6.4]

$\sin 2x\cos 3x = \dfrac{1}{2}\big[\sin(2x+3x)+\sin(2x-3x)\big]$

$\qquad = \dfrac{1}{2}(\sin 5x - \sin x)$

29. Write as a product. [6.4]

$\sin 3x + \sin x = 2\sin\dfrac{3x+x}{2}\cos\dfrac{3x-x}{2}$

$\qquad = 2\sin 2x\cos x$

31. Write as a product. [6.4]

$\cos 8\theta + \cos 2\theta = 2\cos\dfrac{8\theta+2\theta}{2}\cos\dfrac{8\theta-2\theta}{2}$

$\qquad = 2\cos 5\theta\cos 3\theta$

33. Write as a product. [6.4]

$\cos(2\pi 442t) + \cos(2\pi 440t)$

$\qquad = 2\cos\dfrac{2\pi 442t+2\pi 440t}{2}\cos\dfrac{2\pi 442t-2\pi 440t}{2}$

$\qquad = 2\cos(2\pi 441t)\cos(2\pi t)$, or $2\cos(882\pi t)\cos(2\pi t)$

35. Verify the identity.

$\dfrac{1+\sin x}{\cos^2 x} = \dfrac{1}{\cos^2 x} + \dfrac{\sin x}{\cos^2 x}$

$\qquad = \sec^2 x + \tan x\sec x$

$\qquad = \tan^2 x + 1 + \tan x\sec x$

37. Verify the identity.

$\dfrac{1}{\cos x} - \cos x = \dfrac{1-\cos^2 x}{\cos x}$

$\qquad = \dfrac{\sin^2 x}{\cos x}$

$\qquad = \tan x\sin x$

39. Verify the identity.

$\sin\left(\dfrac{\pi}{4} - \alpha\right) = \sin\dfrac{\pi}{4}\cos\alpha - \cos\dfrac{\pi}{4}\sin\alpha$

$\qquad = \dfrac{\sqrt{2}}{2}\cos\alpha - \dfrac{\sqrt{2}}{2}\sin\alpha$

$\qquad = \dfrac{\sqrt{2}}{2}(\cos\alpha - \sin\alpha)$

41. Verify the identity.

$\dfrac{\sin 4x - \sin 2x}{\cos 4x - \cos 2x} = \dfrac{2\cos\dfrac{4x+2x}{2}\sin\dfrac{4x-2x}{2}}{-2\sin\dfrac{4x+2x}{2}\sin\dfrac{4x-2x}{2}}$

$\qquad = -\dfrac{\cos 3x\sin x}{\sin 3x\sin x}$

$\qquad = -\cot 3x$

43. Verify the identity.

$\sin x - \cos 2x = \sin x - (1 - 2\sin^2 x)$

$\qquad = 2\sin^2 x + \sin x - 1$

$\qquad = (2\sin x - 1)(\sin x + 1)$

45. Verify the identity.

$$\tan 4x = \frac{2\tan 2x}{1-\tan^2 2x}$$

$$= \frac{2\left(\dfrac{2\tan x}{1-\tan^2 x}\right)}{1-\left(\dfrac{2\tan x}{1-\tan^2 x}\right)^2}$$

$$= \frac{\dfrac{4\tan x}{1-\tan^2 x}}{\dfrac{(1-\tan^2 x)^2-(2\tan x)^2}{(1-\tan^2 x)^2}} \cdot \frac{(1-\tan^2 x)^2}{(1-\tan^2 x)^2}$$

$$= \frac{4\tan x(1-\tan^2 x)}{(1-\tan^2 x)^2-4\tan^2 x}$$

$$= \frac{4\tan x-4\tan^3 x}{1-2\tan^2 x+\tan^4 x-4\tan^2 x}$$

$$= \frac{4\tan x-4\tan^3 x}{1-6\tan^2 x+\tan^4 x}$$

47. Verify the identity.

$$2\sin 3x\cos 3x-2\sin x\cos x = \sin 6x-\sin 2x$$

$$= 2\cos\frac{6x+2x}{2}\sin\frac{6x-2x}{2}$$

$$= 2\cos 4x\sin 2x$$

49. Verify the identity.

$$\cos(x+y)\cos(x-y)$$

$$= \frac{1}{2}[\cos(x+y+x-y)+\cos(x+y-x+y)]$$

$$= \frac{1}{2}(\cos 2x+\cos 2y)$$

$$= \frac{1}{2}[2\cos^2 x-1+2\cos^2 y-1]$$

$$= \cos^2 x+\cos^2 y-1$$

51. Evaluate the expression. [6.5]

$$y=\sec\left(\sin^{-1}\frac{12}{13}\right),\quad \alpha=\sin^{-1}\frac{12}{13},\quad \sin\alpha=\frac{12}{13},$$

$$\cos\alpha=\frac{5}{13},\quad \sec\alpha=\frac{13}{5}$$

$$y=\sec\alpha=\frac{13}{5}$$

53. Evaluate the expression. [6.5]

Let $x=\cos^{-1}\dfrac{8}{17}$

$$\cos x=\frac{8}{17}\ \text{ and }\ \sin x=\sqrt{1-\left(\frac{8}{17}\right)^2}=\frac{15}{17}$$

$$y=\tan\left(\cos^{-1}\frac{8}{17}\right)=\tan x=\frac{\sin x}{\cos x}=\frac{15/17}{8/17}=\frac{15}{8}$$

55. Solve the equation. [6.6]

$$2\sin^{-1}(x-1)=\frac{\pi}{3}$$

$$\sin^{-1}(x-1)=\frac{\pi}{6}$$

$$x-1=\sin\frac{\pi}{6}$$

$$x-1=\frac{1}{2}$$

$$x=\frac{3}{2}$$

57. Find the solutions for $0°\le x<360°$. [6.6]

$$4\sin^2 x+2\sqrt{3}\sin x-2\sin x-\sqrt{3}=0$$
$$2\sin x(2\sin x+\sqrt{3})-(2\sin x+\sqrt{3})=0$$
$$(2\sin x+\sqrt{3})(2\sin x-1)=0$$

$$2\sin x+\sqrt{3}=0 \qquad\qquad 2\sin x-1=0$$

$$\sin x=-\frac{\sqrt{3}}{2} \qquad\qquad \sin x=\frac{1}{2}$$

$$x=240°,\ 300° \qquad\qquad x=30°,\ 150°$$

The solutions are $30°$, $150°$, $240°$, $300°$.

59. Solve the equation. Round approximate solutions to 4 decimal places. [6.6]

$$3\cos^2 x+\sin x=1$$
$$3(1-\sin^2 x)+\sin x=1$$
$$0=3\sin^2 x-\sin x-2$$
$$0=(3\sin x+2)(\sin x-1)$$

$$3\sin x+2=0 \qquad\qquad \sin x-1=0$$

$$\sin x=-\frac{2}{3} \qquad\qquad \sin x=1$$

$$x=3.8713\ \text{or}\ 5.553 \qquad\qquad x=\frac{\pi}{2}$$

The solutions are $\dfrac{\pi}{2}+2k\pi$, $3.8713+2k\pi$, $5.553+2k\pi$ where k is an integer.

61. Find the solutions for $0\le x<2\pi$. [6.6]

$$\sin 3x\cos x-\cos x 3x\sin x=\frac{1}{2}$$

$$\sin(3x-x)=\frac{1}{2}$$

$$\sin 2x=\frac{1}{2}$$

$$2x=\frac{\pi}{6}+2k\pi \qquad\qquad 2x=\frac{5\pi}{6}+2k\pi$$

$$x=\frac{\pi}{12}+k\pi \qquad\qquad x=\frac{5\pi}{12}+k\pi$$

The solutions are $\dfrac{\pi}{12}$, $\dfrac{5\pi}{12}$, $\dfrac{13\pi}{12}$, $\dfrac{17\pi}{12}$.

63. Rewrite the equation and graph. [6.4]

$$f(x) = \sqrt{3}\sin x + \cos x$$
$$y = 2\sin\left(x + \frac{\pi}{6}\right)$$

amplitude $= 2$

phase shift $= -\dfrac{\pi}{6}$

65. Rewrite the equation and graph. [6.4]

$$f(x) = -\sin x - \sqrt{3}\cos x$$
$$y = 2\sin\left(x + \frac{4\pi}{3}\right)$$

amplitude $= 2$

phase shift $= -\dfrac{4\pi}{3}$

67. Find the exact value. [6.5]

$$\sin\left(\sin^{-1} 0.4\right) = 0.4$$

69. Graph the function.

$$f(x) = 2\cos^{-1} x$$

71. Graph the function.

$$f(x) = \sin^{-1}\frac{x}{2}$$

73. a. The sine regression function was obtained on a TI-83 Plus calculator by using an iteration value of 16. The use of a different iteration factor may produce a sine regression function that varies from the regression function listed below.

L1	L2	L3	1
1	7.25	------	
60	6.8		
121	5.7		
182	5.4833		
244	6.0167		
305	6.5667		
-----	-----		

L1(1)=1

```
EDIT CALC TESTS
7↑QuartReg
8:LinReg(a+bx)
9:LnReg
0:ExpReg
A:PwrReg
B:Logistic
C:SinReg
```

```
SinReg 16,L1,L2,
365.25
```

```
SinReg
y=a*sin(bx+c)+d
a=.9114076795
b=.0155058916
c=1.895514074
d=6.389234854
```

$$y \approx 0.911408\sin(0.015506x + 1.895514) + 6.389235$$

b. $f(162) \approx 0.911408\sin(0.015506(162) + 1.895514)$
$$+ 6.389235$$
$$\approx 5.5198696 \text{ hours}$$
$$\approx 5 + 0.5198696(60) = 5 : 31 \text{ (h: min)}$$

Chapter 6 Test

1. Verify the identity.

$$1 + \sin^2 x \sec^2 x = 1 + \sin^2 x \frac{1}{\cos^2 x}$$
$$= 1 + \tan^2 x$$
$$= \sec^2 x$$

3. Verify the identity.

$$\cos^3 x + \cos x \sin^2 x = \cos x(\cos^2 x + \sin^2 x)$$
$$= \cos x$$

5. Find the exact value. [6.2]

$$\cos 165° = \cos(120° + 45°)$$
$$= \cos 120° \cos 45° - \sin 120° \sin 45°$$
$$= -\frac{1}{2}\left(\frac{\sqrt{2}}{2}\right) - \left(\frac{\sqrt{3}}{2}\right)\left(\frac{\sqrt{2}}{2}\right)$$
$$= \frac{-\sqrt{2} - \sqrt{6}}{4} = -\frac{\sqrt{2} + \sqrt{6}}{4}$$

7. Verify the identity.

$$\sin\left(\theta - \frac{3\pi}{2}\right) = \sin\theta\cos\frac{3\pi}{2} - \cos\theta\sin\frac{3\pi}{2}$$
$$= \sin\theta(0) - \cos\theta(-1)$$
$$= \cos\theta$$

9. Find the exact value of $\cos 2\theta$. [6.3]

$$\sin\theta = \frac{4}{5}, \quad \cos\theta = -\frac{3}{5}$$

$$\cos 2\theta = 2\cos^2\theta - 1$$
$$= 2\left(-\frac{3}{5}\right)^2 - 1$$
$$= \frac{18}{25} - 1 = -\frac{7}{25}$$

11. Verify the identity.

$$\sin^2 2x + 4\cos^4 x = (2\sin x \cos x)^2 + 4\cos^4 x$$
$$= 4\sin^2 x \cos^2 x + 4\cos^4 x$$
$$= 4\cos^2 x(\sin^2 x + \cos^2 x)$$
$$= 4\cos^2 x$$

13. Rewrite in the form $y = k\sin(x + \alpha)$. [6.4]

$$y = -\frac{\sqrt{3}}{2}\sin x + \frac{1}{2}\cos x$$

$$a = -\frac{\sqrt{3}}{2}, \ b = \frac{1}{2}$$

$$k = \sqrt{\left(-\frac{\sqrt{3}}{2}\right)^2 + \left(\frac{1}{2}\right)^2} = 1$$

$$\sin\beta = \left|\frac{\frac{1}{2}}{1}\right| = \frac{1}{2}$$

$$\beta = \frac{\pi}{6}$$

$$\alpha = \pi - \frac{\pi}{6} = \frac{5\pi}{6}$$

$$y = \sin\left(x + \frac{5\pi}{6}\right)$$

14. Approximate to the nearest thousandth. [6.5]

$$\theta = \cos^{-1}(0.7644)$$
$$\theta = 0.701$$

15. Find the exact value of $\tan\left(\cos^{-1}\left(-\frac{12}{37}\right)\right)$. [6.5]

Let $x = \cos^{-1}\left(-\frac{12}{37}\right)$

$$\cos x = -\frac{12}{37} \ \text{ and } \ \sin x = \sqrt{1 - \left(-\frac{12}{37}\right)^2} = \frac{35}{37}$$

$$y = \tan\left(\cos^{-1}\frac{3}{5}\right) = \tan x = \frac{\sin x}{\cos x} = \frac{35/37}{-12/37} = -\frac{35}{12}$$

17. Solve and round to the nearest tenth. [6.6]

$$3\sin x - 2 = 0$$
$$\sin x = \frac{2}{3}$$
$$x = 41.8°, \ 138.2°$$

19. Find the exact solutions for $0 \le x < 2\pi$. [6.6]

$$\sin 2x + \sin x - 2\cos x - 1 = 0$$
$$2\sin x \cos x + \sin x - 2\cos x - 1 = 0$$
$$\sin x(2\cos x + 1) - (2\cos x + 1) = 0$$
$$(2\cos x + 1)(\sin x - 1) = 0$$

$$\cos x = -\frac{1}{2} \qquad\qquad \sin x = 1$$

$$x = \frac{2\pi}{3}, \frac{4\pi}{3} \qquad\qquad x = \frac{\pi}{2}$$

The solutions are $\frac{\pi}{2}, \frac{2\pi}{3}, \frac{4\pi}{3}$.

Cumulative Review Exercises

1. Factor. [P.4]

$$x^3 - y^3 = (x - y)(x^2 + xy + y^2)$$

3. Shift the graph of $y = f(x)$ horizontally 1 unit to the left and up 2 units. [2.5]

5. Find the vertical asymptote. [3.5]

$$x - 2 = 0$$
$$x = 2$$

7. Find the inverse. [4.1]

$$f(x) = \frac{5x}{x-1}$$
$$y = \frac{5x}{x-1}$$
$$x = \frac{5y}{y-1}$$
$$x(y-1) = 5y$$
$$xy - 5y = x$$
$$y(x - 5) = x$$
$$y = \frac{x}{x-5}$$
$$f^{-1}(x) = \frac{x}{x-5}$$

9. Evaluate. [4.3]

$$\log_{10} 1000 = \log_{10} 10^3$$
$$= 3\log_{10} 10$$
$$= 3$$

11. Convert to degrees. [5.1]

$$\frac{5\pi}{3} = \frac{5\pi}{3}\left(\frac{180°}{\pi}\right) = 300°$$

13. Determine the sign of $\cot\theta$. [5.3]

$\cot\theta > 0$ in quadrant III

Positive

15. Find the reference angle. [5.4]

$\theta = \dfrac{5\pi}{3}$

Since $\dfrac{3\pi}{2} < \theta < 2\pi$,

$\theta = \theta' = 2\pi$

$\theta' = \dfrac{\pi}{3}$

17. Find the amplitude, period, and phase shift. [5.7]

$y = 0.43\cos\left(2x - \dfrac{\pi}{6}\right)$

amplitude: 0.43

$0 \le 2x - \dfrac{\pi}{6} \le 2\pi$

$\dfrac{\pi}{6} \le 2x \le \dfrac{13\pi}{6}$

$\dfrac{\pi}{12} \le x \le \dfrac{13\pi}{12}$

period $= \pi$, phase shift $= \dfrac{\pi}{12}$

19. State the domain. [6.5]

Domain: $[-1, 1]$.

Chapter 7 Applications of Trigonometry

Section 7.1 Exercises

1. No. There is an infinite number of triangles with the given angles.

3. The SSA case may yield no solution, one solution or two solutions.

5. Solve the triangles.

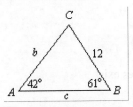

$C = 180° - 42° - 61° = 77°$

$$\frac{b}{\sin B} = \frac{a}{\sin A}$$
$$\frac{b}{\sin 61°} = \frac{12}{\sin 42°}$$
$$b = \frac{12\sin 61°}{\sin 42°}$$
$$b \approx 16$$

$$\frac{c}{\sin C} = \frac{a}{\sin A}$$
$$\frac{c}{\sin 77°} = \frac{12}{\sin 42°}$$
$$c = \frac{12\sin 77°}{\sin 42°}$$
$$c \approx 17$$

7. Solve the triangles.

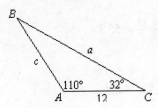

$B = 180° - 110° - 32°$
$B = 38°$

$$\frac{b}{\sin B} = \frac{a}{\sin A}$$
$$\frac{12}{\sin 38°} = \frac{a}{\sin 110°}$$
$$a = \frac{12\sin 110°}{\sin 38°}$$
$$a \approx 18$$

$$\frac{c}{\sin C} = \frac{b}{\sin B}$$
$$\frac{c}{\sin 32°} = \frac{12}{\sin 38°}$$
$$c = \frac{12\sin 32°}{\sin 38°}$$
$$c \approx 10$$

9. Solve the triangles.

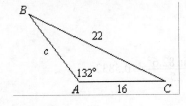

$$\frac{a}{\sin A} = \frac{b}{\sin B}$$
$$\frac{22}{\sin 132°} = \frac{16}{\sin B}$$

$$\sin B = \frac{16\sin 132°}{22}$$
$$\sin B \approx 0.5405$$
$$B \approx 33°$$

$$C \approx 180° - 33° - 132° \approx 15°$$

$$\frac{a}{\sin A} = \frac{c}{\sin C}$$
$$\frac{22}{\sin 132°} \approx \frac{c}{\sin 15°}$$
$$c \approx \frac{22\sin 15°}{\sin 132°}$$
$$c \approx 7.7$$

11. Solve the triangles.

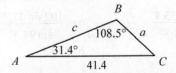

$C = 180° - 31.4° - 108.5° = 40.1°$

$$\frac{b}{\sin B} = \frac{a}{\sin A}$$
$$\frac{41.4}{\sin 108.5°} = \frac{a}{\sin 31.4°}$$
$$a = \frac{41.4\sin 31.4°}{\sin 108.5°}$$
$$a \approx 22.7$$

$$\frac{c}{\sin C} = \frac{b}{\sin B}$$
$$\frac{c}{\sin 40.1°} = \frac{41.4}{\sin 108.5°}$$
$$c = \frac{41.4\sin 40.1°}{\sin 108.5°}$$
$$c \approx 28.1$$

13. Solve the triangles.

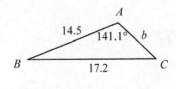

$$\frac{c}{\sin C} = \frac{a}{\sin A}$$
$$\frac{14.5}{\sin C} = \frac{17.2}{\sin 141.1}$$
$$\sin C = \frac{14.5\sin 141.1°}{17.2} \approx 0.5294$$
$$C \approx 32.0°$$

$B \approx 180° - 141.1° - 32.0°$

$B \approx 6.9°$

$$\frac{b}{\sin B} = \frac{a}{\sin A}$$

$$\frac{b}{\sin 6.9°} \approx \frac{17.2}{\sin 141.1°}$$

$$b = \frac{17.2 \sin 6.9°}{\sin 141.1°} \approx 3.3$$

15. Solve the triangles.

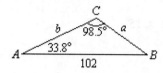

$B = 180° - 98.5° - 33.8°$

$B = 47.7°$

$$\frac{a}{\sin A} = \frac{c}{\sin C} \qquad \frac{b}{\sin B} = \frac{c}{\sin C}$$

$$\frac{a}{\sin 33.8°} = \frac{102}{\sin 98.5°} \qquad \frac{b}{\sin 47.7°} = \frac{102}{\sin 98.5°}$$

$$a = \frac{102 \sin 33.8°}{\sin 98.5°} \qquad b = \frac{102 \sin 47.7°}{\sin 98.5°}$$

$$a \approx 57.4 \qquad b \approx 76.3$$

17. Solve the triangles.

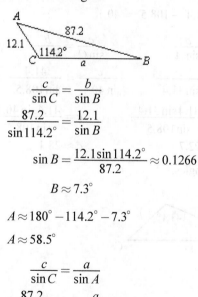

$$\frac{c}{\sin C} = \frac{b}{\sin B}$$

$$\frac{87.2}{\sin 114.2°} = \frac{12.1}{\sin B}$$

$$\sin B = \frac{12.1 \sin 114.2°}{87.2} \approx 0.1266$$

$$B \approx 7.3°$$

$A \approx 180° - 114.2° - 7.3°$

$A \approx 58.5°$

$$\frac{c}{\sin C} = \frac{a}{\sin A}$$

$$\frac{87.2}{\sin 114.2°} \approx \frac{a}{\sin 58.5°}$$

$$a = \frac{87.2 \sin 58.5°}{\sin 114.2°} \approx 81.5$$

19. Solve the triangles that exist.

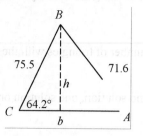

$$\sin 64.2° = \frac{h}{75.5}$$

$$h = 75.5 \sin 64.2°$$

$$h \approx 68$$

Since $h < 75.5$, two triangles exist.

$$\frac{c}{\sin C} = \frac{a}{\sin A}$$

$$\frac{71.6}{\sin 64.2°} = \frac{75.5}{\sin A}$$

$$\sin A = \frac{75.5 \sin 64.2°}{71.6} = 0.9493$$

$$A \approx 71.7° \text{ or } 108.3°$$

For $A = 71.7°$

$$B = 180° - 71.7° - 64.2° = 44.1°$$

$$\frac{b}{\sin 44.1°} = \frac{71.6}{\sin 64.2°}$$

$$b = \frac{71.6 \sin 44.1°}{\sin 64.2°} = 55.3$$

For $A = 108.3°$

$$B = 180° - 108.3° - 64.2° = 7.5°$$

$$\frac{b}{\sin 7.5°} = \frac{71.6}{\sin 64.2°}$$

$$b = \frac{71.6 \sin 7.5°}{\sin 64.2°} \approx 10.4$$

21. Solve the triangles that exist.

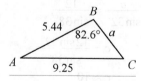

Since $b > c$, one triangle exists.

$$\frac{b}{\sin B} = \frac{c}{\sin C}$$

$$\frac{9.25}{\sin 82.6°} = \frac{5.44}{\sin C}$$

$$\sin C = \frac{5.44 \sin 82.6°}{9.25} \approx 0.5832$$

$$C = 35.7°$$

$A = 180° - 82.6° - 35.7° = 61.7°$

$$\frac{a}{\sin A} = \frac{b}{\sin B}$$

$$\frac{a}{\sin 61.7°} = \frac{9.25}{\sin 82.6°}$$

$$a = \frac{9.25 \sin 61.7°}{\sin 82.6°}$$

$$a \approx 8.21$$

23. Solve the triangles that exist.

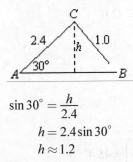

$$\sin 30° = \frac{h}{2.4}$$

$$h = 2.4 \sin 30°$$

$$h \approx 1.2$$

Since $h > 1$, no triangle is formed.

25. Solve the triangles that exist.

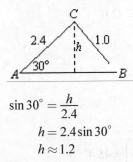

$$\sin 14.8° = \frac{h}{6.35}$$

$$h = 6.35 \sin 14.8°$$

$$h \approx 1.62$$

Since $h < 4.80$, two solutions exist.

$$\frac{c}{\sin C} = \frac{a}{\sin A}$$

$$\frac{6.35}{\sin C} = \frac{4.80}{\sin 14.8°}$$

$$\sin C = \frac{6.35 \sin 14.8°}{4.80}$$

$$\sin C = 0.3379$$

$$C \approx 19.8° \text{ or } 160.2°$$

For $C = 19.8°$

$$B = 180° - 19.8° - 14.8°$$

$$= 145.4°$$

$$\frac{b}{\sin 145.4°} = \frac{4.80}{\sin 14.8°}$$

$$b = \frac{4.80 \sin 145.4°}{\sin 14.8°}$$

$$b \approx 10.7$$

For $C = 160.2°$

$$B = 180° - 160.2° - 14.8°$$

$$= 5.0°$$

$$\frac{b}{\sin 5.0°} = \frac{4.80}{\sin 14.8°}$$

$$b = \frac{4.80 \sin 5.0°}{\sin 14.8°}$$

$$b \approx 1.64$$

27. Solve the triangles that exist.

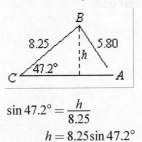

$$\sin 47.2° = \frac{h}{8.25}$$

$$h = 8.25 \sin 47.2°$$

$$h \approx 6.05$$

Since $h > 5.80$, no triangle is formed.

29. Solve the triangles that exist.

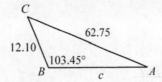

Since $b > a$, one triangle exists.

$$\frac{a}{\sin A} = \frac{b}{\sin B}$$

$$\frac{12.10}{\sin A} = \frac{62.75}{\sin 103.45°}$$

$$\sin A = \frac{12.10 \sin 103.45°}{62.75} \approx 0.1875$$

$$A = 10.81°$$

$$C = 180° - 10.81° - 103.45°$$

$$C = 65.74°$$

$$\frac{c}{\sin C} = \frac{b}{\sin B}$$

$$\frac{c}{\sin 65.74°} = \frac{62.75}{\sin 103.45°}$$

$$c = \frac{62.75 \sin 65.74°}{\sin 103.45°}$$

$$c \approx 58.82$$

31. Solve the triangles that exist.

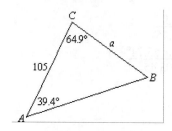

$$\sin 20.5° = \frac{h}{14.1}$$
$$h = 14.1\sin 20.5°$$
$$h \approx 4.9$$

Since $h < 10.3$, two solutions exist.

$$\frac{a}{\sin A} = \frac{c}{\sin C}$$
$$\frac{10.3}{\sin 20.5°} = \frac{14.1}{\sin C}$$
$$\sin C = \frac{14.1\sin 20.5°}{10.3}$$
$$\sin C = 0.4794$$
$$C \approx 28.6° \text{ or } 151.4°$$

For $C = 28.6°$
$$B = 180° - 28.6° - 20.5°$$
$$B = 130.9°$$
$$\frac{b}{\sin 130.9°} = \frac{10.3}{\sin 20.5°}$$
$$b = \frac{10.3\sin 130.9°}{\sin 20.5°}$$
$$b \approx 22.2$$

For $C = 151.4°$
$$B = 180° - 151.4° - 20.5°$$
$$B = 8.1°$$
$$\frac{b}{\sin 8.1°} = \frac{10.3}{\sin 20.5°}$$
$$b = \frac{10.3\sin 8.1°}{\sin 20.5°}$$
$$b \approx 4.1$$

33. Find the distance.

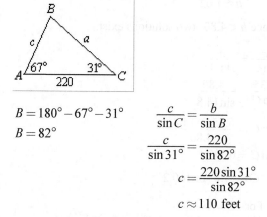

$$B = 180° - (39.4° + 64.9°)$$
$$B = 75.7°$$

$$\frac{a}{\sin A} = \frac{b}{\sin B}$$
$$\frac{a}{\sin 39.4°} = \frac{105}{\sin 75.7°}$$
$$a = \frac{105\sin 39.4°}{\sin 75.7°}$$
$$a \approx 68.8 \text{ miles}$$

35. $a = 155$ yd, $c = 165$ yd, $A = 42.0°$

$$\frac{c}{\sin C} = \frac{a}{\sin A}$$
$$\frac{165}{\sin C} = \frac{155}{\sin 42.0°}$$
$$\sin C = \frac{165\sin 42.0°}{155}$$
$$C = \sin^{-1}\left(\frac{165\sin 42.0°}{155}\right) = 45.4°$$

$$B = 180° - 42.0° - 45.4° = 92.6°$$

$$\frac{b}{\sin B} = \frac{a}{\sin A}$$
$$\frac{b}{\sin 92.6°} = \frac{155}{\sin 42.0°}$$
$$b = \frac{155\sin 92.6°}{\sin 42.0°} = 231 \text{ yd}$$

37. Find the distance.

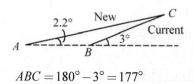

$$B = 180° - 67° - 31°$$
$$B = 82°$$

$$\frac{c}{\sin C} = \frac{b}{\sin B}$$
$$\frac{c}{\sin 31°} = \frac{220}{\sin 82°}$$
$$c = \frac{220\sin 31°}{\sin 82°}$$
$$c \approx 110 \text{ feet}$$

39. Find the length of the new runway.

$$ABC = 180° - 3° = 177°$$

$$\frac{AC}{\sin ABC} = \frac{BC}{\sin CAB}$$

$$\frac{AC}{\sin 177°} = \frac{3550}{\sin 2.2°}$$

$$AC = \frac{3550\sin 177°}{\sin 2.2°} \approx 4840 \text{ ft}$$

41. Find the length of the guy wire.

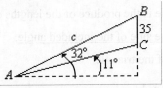

$A = 21°$

$B = 180° - 90° - 32°$

$B = 58°$

$C = 180° - 58° - 21°$

$C = 101°$

$$\frac{c}{\sin C} = \frac{a}{\sin A}$$

$$\frac{c}{\sin 101°} = \frac{35}{\sin 21°}$$

$$c = \frac{35\sin 101°}{\sin 21°}$$

$$c \approx 96 \text{ feet}$$

43. Find the height of the hill.

$A = 5°$

$B = 180° - 90° - 75°$

$B = 15°$

$C = 180° - 15° - 5°$

$C = 160°$

$$\frac{b}{\sin B} = \frac{a}{\sin A} \qquad \sin 70° = \frac{h}{b}$$

$$\frac{b}{\sin 15°} = \frac{12}{\sin 5°} \qquad h = b\sin 70°$$

$$b = \frac{12\sin 15°}{\sin 5°} \qquad h = \left(\frac{12\sin 15°}{\sin 5°}\right)\sin 70°$$

$$h \approx 33 \text{ feet}$$

45. Find the distance.

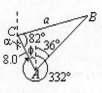

$\phi = 360° - 332°$

$\phi = 28°$

$\alpha = 28°$

$C = 82° - 28° = 54°$

$A = 28° + 36° = 64°$

$B = 180° - 64° - 54° = 62°$

$$\frac{a}{\sin A} = \frac{b}{\sin B}$$

$$\frac{a}{\sin 64°} = \frac{8.0}{\sin 62°}$$

$$a = \frac{8.0\sin 64°}{\sin 62°}$$

$$a \approx 8.1 \text{ miles}$$

47. Find the distance.

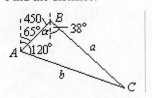

$A = 120° - 65° = 55°$

$\alpha = 65°$

$B = 38° + 65° = 103°$

$C = 180° - 103° - 55° = 22°$

$$\frac{b}{\sin B} = \frac{c}{\sin C}$$

$$\frac{b}{\sin 103°} = \frac{450}{\sin 22°}$$

$$b = \frac{450\sin 103°}{\sin 22°}$$

$$b \approx 1200 \text{ miles}$$

49. Find the distance.

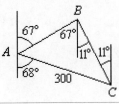

$A = 180° - 67° - 68° = 45°$

$B = 67° + 11° = 78°$

$C = 180° - 45° - 78° = 57°$

$$\frac{c}{\sin C} = \frac{b}{\sin B}$$

$$\frac{c}{\sin 57°} = \frac{300}{\sin 78°}$$

$$c = \frac{300\sin 57°}{\sin 78°}$$

$$c \approx 260 \text{ meters}$$

51. $\dfrac{a}{\sin A} = \dfrac{b}{\sin B}$

$\dfrac{a}{b} = \dfrac{\sin A}{\sin B}$

$\dfrac{a}{b} + 1 = \dfrac{\sin A}{\sin B} + 1$

$\dfrac{a+b}{b} = \dfrac{\sin A + \sin B}{\sin B}$

53. Using the figure below,

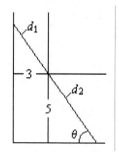

$\dfrac{3}{d_1} = \cos\theta, \; \dfrac{5}{d_2} = \sin\theta$

$d_1 = \dfrac{3}{\cos\theta}, \; d_2 = \dfrac{5}{\sin\theta}$

$L = d_1 + d_2$

$L(\theta) = \dfrac{3}{\cos\theta} + \dfrac{5}{\sin\theta}$

The graph of L is shown.

The minimum value of L is approximately 11.19 m.

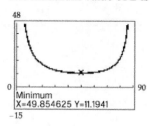

Prepare for Section 7.2

P1. Evaluate.

$\sqrt{(10.0)^2 + (15.0)^2 - 2(10.0)(15.0)\cos 110.0^\circ} \approx 20.7$

P3. Solve for C.

$c^2 = a^2 + b^2 - 2ab\cos C$

$c^2 - a^2 - b^2 = -2ab\cos C$

$\cos C = \dfrac{c^2 - a^2 - b^2}{-2ab}$

$C = \cos^{-1}\left(\dfrac{c^2 - a^2 - b^2}{-2ab}\right)$

$C = \cos^{-1}\left(\dfrac{a^2 + b^2 - c^2}{2ab}\right)$

P5. Evaluate.

$s = \dfrac{a+b+c}{2} = \dfrac{3+4+5}{2} = 6$

$\sqrt{6(6-3)(6-4)(6-5)} = \sqrt{6(3)(2)(1)} = 6$

Section 7.2 Exercises

1. It is an example of SAS case.

3. The area of a triangle is the produce of the lengths of any two sides and <u>the sine of the included angle</u>.

5. Find the third side of the triangle.

$a^2 = b^2 + c^2 - bc\cos A$

$a^2 = 53.4^2 + 71.4^2 - 2(53.4)(71.4)\cos 25.9^\circ$

$a^2 = 7949.52 - 7625.52\cos 25.9^\circ$

$a = \sqrt{7949.52 - 7625.52\cos 25.9^\circ}$

$a \approx 33.0$

7. Find the third side of the triangle.

$c^2 = a^2 + b^2 - 2ab\cos C$

$c^2 = 9.0^2 + 7.0^2 - 2(9.0)(7.0)\cos 72^\circ$

$c^2 = 130 - 126\cos 72^\circ$

$c = \sqrt{130 - 126\cos 72^\circ}$

$c \approx 9.5$

9. Find the third side of the triangle.

$c^2 = a^2 + b^2 - 2ab\cos C$

$c^2 = 1378^2 + 2411^2 - 2(1378)(2411)\cos 118.25^\circ$

$c^2 = 7,711,805 - 6,644,716\cos 118.25^\circ$

$c = \sqrt{7,711,805 - 6,644,716\cos 118.25^\circ}$

$c \approx 3295$

11. Find the third side of the triangle.

$b^2 = a^2 + c^2 - 2ac\cos B$

$b^2 = 25.9^2 + 33.4^2 - 2(25.9)(33.4)\cos 84.0^\circ$

$b^2 = 1786.37 - 1730.12\cos 84.0^\circ$

$b = \sqrt{1786.37 - 1730.12\cos 84.0^\circ}$

$b \approx 40.1$

13. Find the third side of the triangle.

$b^2 = a^2 + c^2 - 2ac\cos B$

$b^2 = 122^2 + 55.9^2 - 2(122)(55.9)\cos 44.2^\circ$

$b^2 = 18,008.81 - 13,639.6\cos 44.2^\circ$

$b = \sqrt{18,008.81 - 13,639.6\cos 44.2^\circ}$

$b \approx 90.7$

15. Find the specified angle.

$$\cos C = \frac{a^2 + b^2 - c^2}{2ab}$$

$$\cos C = \frac{23^2 + 29^2 - 42^2}{2(23)(29)}$$

$$\cos C = \frac{-394}{1334}$$

$$C = \cos^{-1}\left(\frac{-394}{1334}\right) \approx 107°$$

17. Find the specified angle.

$$\cos C = \frac{a^2 + b^2 - c^2}{2ab}$$

$$\cos C = \frac{82.15^2 + 61.45^2 - 72.84^2}{2(82.15)(61.45)}$$

$$\cos C = \frac{5219.0594}{10,096.235}$$

$$C = \cos^{-1}\left(\frac{5219.0594}{10,096.235}\right) \approx 58.87°$$

19. Find the specified angle.

$$\cos B = \frac{a^2 + c^2 - b^2}{2ac}$$

$$\cos B = \frac{80^2 + 124^2 - 92^2}{2(80)(124)}$$

$$\cos B = \frac{13,312}{19,840}$$

$$B = \cos^{-1}\left(\frac{13312}{19840}\right) \approx 47.9°$$

21. Find the specified angle.

$$\cos B = \frac{a^2 + c^2 - b^2}{2ac}$$

$$\cos B = \frac{19.2^2 + 29.1^2 - 14.3^2}{2(19.2)(29.1)}$$

$$\cos B = \frac{1010.96}{1117.44}$$

$$B = \cos^{-1}\left(\frac{1010.96}{1117.44}\right) \approx 25.2°$$

23. Find the specified angle.

$$\cos B = \frac{a^2 + c^2 - b^2}{2ac}$$

$$\cos B = \frac{32.5^2 + 29.6^2 - 40.1^2}{2(32.5)(29.6)}$$

$$\cos B = \frac{324.4}{1924}$$

$$B = \cos^{-1}\left(\frac{324.4}{1924}\right) \approx 80.3°$$

25. Solve the triangle.

$$a^2 = b^2 + c^2 - 2bc \cos A$$

$$a = \sqrt{27.4^2 + 41.4^2 - 2(27.4)(41.4)\cos 27.6°}$$

$$a \approx 21.3$$

$$\frac{b}{\sin B} = \frac{a}{\sin A}$$

$$\frac{27.4}{\sin B} = \frac{21.3}{\sin 27.6°}$$

$$B = \sin^{-1}\left(\frac{27.4 \sin 27.6°}{21.3}\right)$$

$$B \approx 36.6°$$

$$C = 180° - 27.6° - 36.6°$$

$$C = 115.8°$$

27. Solve the triangle.

$$b^2 = a^2 + c^2 - 2ac \cos B$$

$$b = \sqrt{9^2 + 5^2 - 2(9)(5)\cos 39°}$$

$$b \approx 6$$

$$\frac{b}{\sin B} = \frac{c}{\sin C}$$

$$\frac{6}{\sin 39°} = \frac{5}{\sin C}$$

$$C = \sin^{-1}\left(\frac{5 \sin 39°}{6}\right)$$

$$C \approx 32°$$

$$A = 180° - 39° - 32°$$

$$A = 109°$$

29. Solve the triangle.

$$\cos B = \frac{a^2 + c^2 - b^2}{2ac}$$

$$\cos B = \frac{205^2 + 157^2 - 321^2}{2(205)(157)}$$

$$\cos B = \frac{-36,367}{67,370}$$

$$B = \cos^{-1}\left(\frac{-36,367}{67,370}\right) \approx 124.4°$$

$$\frac{b}{\sin B} = \frac{a}{\sin A}$$

$$\frac{321}{\sin 124.4°} = \frac{205}{\sin A}$$

$$A = \sin^{-1}\left(\frac{205 \sin 124.4°}{321}\right) \approx 31.8°$$

$$C = 180° - 31.8° - 124.4° = 23.8°$$

31. Solve the triangle.

$$\cos B = \frac{a^2 + c^2 - b^2}{2ac}$$

$$\cos B = \frac{1135^2 + 1462^2 - 1725^2}{2(1135)(1462)}$$

$$\cos B = \frac{450{,}044}{3{,}318{,}740}$$

$$B = \cos^{-1}\left(\frac{450{,}044}{3{,}318{,}740}\right) \approx 82.21°$$

$$\frac{b}{\sin B} = \frac{a}{\sin A}$$

$$\frac{1725}{\sin 82.21°} = \frac{1135}{\sin A}$$

$$A = \sin^{-1}\left(\frac{1135\sin 82.21°}{1725}\right)$$

$$A \approx 40.68°$$

$$C = 180° - 40.68° - 82.21°$$

$$C = 57.11°$$

33. Find the area of the triangle. Round using the rules of significant digits.

$$K = \frac{1}{2}bc\sin A$$

$$K = \frac{1}{2}(12)(24)\sin 105°$$

$$K \approx 140 \text{ square units}$$

35. Find the area of the triangle. Round using the rules of significant digits.

$$A = 180° - 47.2° - 62.4°$$

$$A = 70.4°$$

$$K = \frac{a^2 \sin B \sin C}{2\sin A}$$

$$K = \frac{22.4^2 \sin 47.2° \sin 62.4°}{2\sin 70.4°}$$

$$K \approx 173 \text{ square units}$$

37. Find the area of the triangle. Round using the rules of significant digits.

$$s = \frac{1}{2}(a+b+c)$$

$$s = \frac{1}{2}(16+12+14)$$

$$s = 21$$

$$K = \sqrt{s(s-a)(s-b)(s-c)}$$

$$K = \sqrt{21(21-16)(21-12)(21-14)}$$

$$K = \sqrt{21(5)(9)(7)}$$

$$K \approx 81 \text{ square units}$$

39. Find the area of the triangle. Round using the rules of significant digits.

$$\frac{a}{\sin A} = \frac{b}{\sin B}$$

$$\frac{22.4}{\sin A} = \frac{26.9}{\sin 54.3°}$$

$$\sin A = \frac{22.4\sin 54.3°}{26.9}$$

$$\sin A \approx 0.6762$$

$$A \approx 42.5°$$

$$C = 180° - 42.5° - 54.3°$$

$$C = 83.2°$$

$$K = \frac{1}{2}ab\sin C$$

$$K = \frac{1}{2}(22.4)(26.9)\sin 83.2°$$

$$K \approx 299 \text{ square units}$$

41. Find the area of the triangle. Round using the rules of significant digits.

$$C = 180° - 116° - 34°$$

$$C = 30°$$

$$K = \frac{c^2 \sin A \sin B}{2\sin C}$$

$$K = \frac{8.5^2 \sin 116° \sin 34°}{2\sin 30°}$$

$$K \approx 36 \text{ square units}$$

43. Find the area of the triangle. Round using the rules of significant digits.

$$s = \frac{1}{2}(a+b+c)$$

$$s = \frac{1}{2}(3.6+4.2+4.8)$$

$$s = 6.3$$

$$K = \sqrt{s(s-a)(s-b)(s-c)}$$

$$K = \sqrt{6.3(6.3-3.6)(6.3-4.2)(6.3-4.8)}$$

$$K = \sqrt{6.3(2.7)(2.1)(1.5)}$$

$$K \approx 7.3 \text{ square units}$$

45. Find the distance.

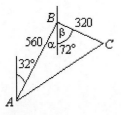

$\alpha = 32°$

$\beta = 72°$

$B = 72° + 32°$

$B = 104°$

$b^2 = a^2 + c^2 - 2ac\cos 104°$

$b^2 = 320^2 + 560^2 - 2(320)(560)\cos 104°$

$b^2 = 416{,}000 - 358{,}400\cos 104°$

$b = \sqrt{416{,}000 - 358{,}400\cos 104°}$

$b \approx 710$ miles

47. Find the distance from the ball to first base.

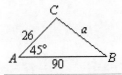

$a^2 = b^2 + c^2 - 2bc\cos A$

$a^2 = 26^2 + 90^2 - 2(26)(90)\cos 45°$

$a^2 = 8776 - 4680\cos 45°$

$a = \sqrt{8776 - 4680\cos 45°}$

$a \approx 74$ feet

49. Find the angle.

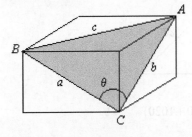

Let a = the length of the diagonal on the front of the box.

Let b = the length of the diagonal on the right side of the box.

Let c = the length of the diagonal on the top of the box.

$a^2 = (4.75)^2 + (6.50)^2 = 64.8125$

$a = \sqrt{64.8125}$

$b^2 = (3.25)^2 + (4.75)^2 = 33.125$

$b = \sqrt{33.125}$

$c^2 = (6.50)^2 + (3.25)^2 = 52.8125$

$\theta = C$

$\cos C = \dfrac{a^2 + b^2 - c^2}{2ab}$

$\cos\theta = \dfrac{64.8125 + 33.125 - 52.8125}{2\sqrt{64.8125}\sqrt{33.125}}$

$\cos\theta = \dfrac{45.125}{2\sqrt{64.8125}\sqrt{33.125}}$

$\theta = \cos^{-1}\left(\dfrac{45.125}{2\sqrt{64.8125}\sqrt{33.125}}\right)$

$\theta \approx 60.9°$

51. Find the distance between ships after 10 hrs.

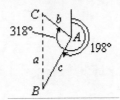

$b = (18\text{ mph})(10\text{ hours}) = 180$ miles

$c = (22\text{ mph})(10\text{ hours}) = 220$ miles

$A = 318° - 198°$

$A = 120°$

$a^2 = b^2 + c^2 - 2bc\cos A$

$a^2 = 180^2 + 220^2 - 2(180)(220)\cos 120°$

$a^2 = 120{,}400$

$a \approx 350$ miles

53. Find the length of one side.

$A = \dfrac{360°}{6}$

$A = 60°$

$a^2 = 40^2 + 40^2 - 2(40)(40)\cos 60°$

$a^2 = 1600$

$a = 40$ cm

55. Find the distance.

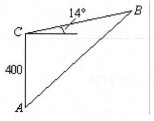

$$C = 90° + 14° = 104°$$

$$a = \frac{180(5280)}{3600} \cdot 10 = 2640 \text{ feet}$$

$$c^2 = 2640^2 + 400^2 - 2(2640)(400)(\cos 104°)$$

$$c^2 \approx 7{,}640{,}539$$

$$c \approx 2800 \text{ feet}$$

57. Find the distance and bearing.

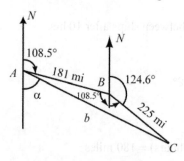

$$B = 108.5 + (180 - 124.6)$$

$$B = 163.9°$$

$$b^2 = a^2 + c^2 - 2ac \cos B$$

$$b^2 = 225^2 + 181^2 - 2(225)(181) \cos 163.9°$$

$$b^2 = 83{,}386 - 81{,}450 \cos 163.9°$$

$$b = \sqrt{83{,}386 - 81{,}450 \cos 163.9°}$$

$$b \approx 402.046592$$

$$b \approx 402 \text{ mi}$$

$$\cos A = \frac{b^2 + c^2 - a^2}{2bc}$$

$$\cos A = \frac{402.046592^2 + 181^2 - 225^2}{2(402.046592)(181)}$$

$$A = \cos^{-1}\left(\frac{143{,}777.4621}{145{,}540.8663}\right)$$

$$A \approx 8.9°$$

$$\alpha = 180 - (108.5 + 8.9)$$

$$\alpha = 62.6°$$

The distance is 402 mi and the bearing is S62.6°E.

59. Find the area.

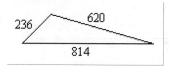

$$s = \frac{1}{2}(a + b + c)$$

$$s = \frac{1}{2}(236 + 620 + 814)$$

$$s = 835$$

$$K = \sqrt{s(s-a)(s-b)(s-c)}$$

$$K = \sqrt{835(835 - 236)(835 - 620)(835 - 814)}$$

$$K = \sqrt{835(599)(215)(21)}$$

$$K = \sqrt{2{,}258{,}244{,}975}$$

$$K \approx 47{,}500 \text{ square meters}$$

61. Find the cost.

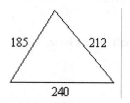

$$s = \frac{1}{2}(185 + 212 + 240)$$

$$s = 318.5$$

$$K = \sqrt{318.5(318.5 - 185)(318.5 - 212)(318.5 - 240)}$$

$$K \approx 18{,}854 \text{ ft}^2$$

$$\text{cost} = 2.20(18{,}854)$$

$$\text{cost} \approx \$41{,}000$$

63. Find the cost.

$$s = \frac{1}{2}(680 + 800 + 1020)$$

$$s = 1250$$

$$K = \sqrt{1250(1250 - 680)(1250 - 800)(1250 - 1020)}$$

$$K \approx 271{,}558 \text{ ft}^2$$

$$\text{Acres} = \frac{271{,}558}{43{,}560}$$

$$\text{Acres} \approx 6.23$$

65. Determine whether triangles ABC or DEF have correction dimensions.

For ABC,

$$\frac{13.0 - 16.1}{10.4} \approx -0.2981$$

$$\frac{\sin\left(\frac{53.5 - 86.5}{2}\right)}{\cos\left(\frac{40.0}{2}\right)} \approx -0.3022$$

Triangle ABC has correct dimensions.

For DEF,

$$\frac{17.2 - 21.3}{22.8} \approx -0.1798$$

$$\frac{\sin\left(\frac{52.1 - 59.9}{2}\right)}{\cos\left(\frac{68.0}{2}\right)} \approx -0.0820$$

Triangle DEF has an incorrect dimension.

67. Find the volume.

$V = 18K$, where K = area of triangular base

$$V = \frac{1}{2}(4)(4)(\sin 72°)(18)$$

$$V \approx 140 \text{ in}^3$$

Prepare for Section 7.3

P1. Evaluate.

$$\sqrt{\left(\frac{3}{5}\right)^2 + \left(-\frac{4}{5}\right)^2} = \sqrt{\frac{9}{25} + \frac{16}{25}} = \sqrt{\frac{25}{25}} = 1$$

P3. Solve.

$$\tan \alpha = \left|\frac{-\sqrt{3}}{3}\right|$$

$$\tan \alpha = \frac{\sqrt{3}}{3}$$

$$\alpha = \tan^{-1}\left(\frac{\sqrt{3}}{3}\right) = 30°$$

P5. Rationalize the denominator.

$$\frac{1}{\sqrt{5}} \cdot \frac{\sqrt{5}}{\sqrt{5}} = \frac{\sqrt{5}}{5}$$

Section 7.3 Exercises

1. The dot product of two vectors is a scalar.

3. Yes, the vector $\langle a, b \rangle$ is perpendicular to the vector $\langle -b, a \rangle$.

5. Find the amount of work.

$$W = F \cdot s = \|\mathbf{F}\|\|\mathbf{s}\|\cos \alpha$$
$$= \|16\|\|3\|\cos 0°$$
$$= 48 \text{ ft-lb}$$

7. Find the components, then write the vector.

$a = 5 - 1 = 4$
$b = 4 - 2 = 2$

A vector equivalent to $\mathbf{P_1P_2}$ is $\mathbf{v} = \langle 4, 2 \rangle$.

9. Find the components, then write the vector.

$a = -3 - 2 = -5$
$b = 5 - 1 = 4$

A vector equivalent to $\mathbf{P_1P_2}$ is $\mathbf{v} = \langle -5, 4 \rangle$.

11. Find the components, then write the vector.

$a = 5 - (-7) = 12$
$b = 8 - 11 = -3$

A vector equivalent to $\mathbf{P_1P_2}$ is $\mathbf{v} = \langle 12, -3 \rangle$.

13. Find the components, then write the vector.

$a = 7.9 - 2.5 = 5.4$
$b = -6.5 - 3.1 = -9.6$

A vector equivalent to $\mathbf{P_1P_2}$ is $\mathbf{v} = \langle 5.4, -9.6 \rangle$.

15. Find the components, then write the vector.

$a = 2 - 2 = 0$
$b = 3 - (-5) = 8$

A vector equivalent to $\mathbf{P_1P_2}$ is $\mathbf{v} = \langle 0, 8 \rangle$.

17. Find the magnitude, direction, and the unit vector.

$$\|\mathbf{v}\| = \sqrt{(-3)^2 + 4^2} \qquad \alpha = \tan^{-1}\left|\frac{4}{-3}\right| = \tan^{-1}\frac{4}{3}$$
$$\|\mathbf{v}\| = \sqrt{9 + 16} \qquad \alpha \approx 53.1°$$
$$\|\mathbf{v}\| = 5 \qquad \theta = 180° - \alpha$$
$$\theta \approx 180° - 53.1°$$
$$\theta \approx 126.9°$$

$$\mathbf{u} = \left\langle \frac{-3}{5}, \frac{4}{5} \right\rangle$$

A unit vector in the direction of $\mathbf{v}$ is $\mathbf{u} = \left\langle -\frac{3}{5}, \frac{4}{5} \right\rangle$.

19. Find the magnitude, direction, and the unit vector.

$$\|\mathbf{v}\| = \sqrt{8^2 + (-15)^2} \qquad \alpha = \tan^{-1}\left|\frac{-15}{8}\right| = \tan^{-1}1.875$$
$$\|\mathbf{v}\| = \sqrt{64 + 225} \qquad\qquad \alpha \approx 61.9°$$
$$\|\mathbf{v}\| = \sqrt{289} = 17 \qquad\qquad \theta = 360° - \alpha$$
$$\qquad\qquad\qquad\qquad\qquad \theta \approx 360° - 61.9°$$
$$\qquad\qquad\qquad\qquad\qquad \theta \approx 298.1°$$

$$\mathbf{u} = \left\langle \frac{8}{17}, \frac{-15}{17} \right\rangle$$

A unit vector in the direction of $\mathbf{v}$ is $\mathbf{u} = \left\langle \frac{8}{17}, \frac{-15}{17} \right\rangle$.

21. Find the magnitude, direction, and the unit vector.

$$\|\mathbf{v}\| = \sqrt{36^2 + 77^2} \qquad \alpha = \tan^{-1}\left|\frac{77}{36}\right| = \tan^{-1}\frac{77}{36}$$
$$\|\mathbf{v}\| = \sqrt{1296 + 5929} \qquad \alpha \approx 64.9°$$
$$\|\mathbf{v}\| = \sqrt{7225} = 85 \qquad\quad \theta = 64.9°$$

$$\mathbf{u} = \left\langle \frac{36}{85}, \frac{77}{85} \right\rangle$$

A unit vector in the direction of $\mathbf{v}$ is $\mathbf{u} = \left\langle \frac{36}{85}, \frac{77}{85} \right\rangle$.

23. Find the magnitude, direction, and the unit vector.

$$\|\mathbf{v}\| = \sqrt{5^2 + (-2)^2} \qquad \alpha = \tan^{-1}\left|\frac{-2}{5}\right| = \tan^{-1}\frac{2}{5}$$
$$\|\mathbf{v}\| = \sqrt{25 + 4} \qquad\qquad \alpha \approx 21.8°$$
$$\|\mathbf{v}\| = \sqrt{29} \qquad\qquad\quad \theta = 360° - \alpha$$
$$\qquad\quad \approx 5.4 \qquad\qquad\qquad \theta \approx 360° - 21.8°$$
$$\qquad\qquad\qquad\qquad\qquad \theta \approx 338.2°$$

$$\mathbf{u} = \left\langle \frac{5}{\sqrt{29}}, \frac{-2}{\sqrt{29}} \right\rangle = \left\langle \frac{5\sqrt{29}}{29}, -\frac{2\sqrt{29}}{29} \right\rangle$$

A unit vector in the direction of $\mathbf{v}$ is

$$\mathbf{u} = \left\langle \frac{5\sqrt{29}}{29}, -\frac{2\sqrt{29}}{29} \right\rangle.$$

25. Perform the indicated operations.

$$3\mathbf{u} = 3\langle -2, 4 \rangle = \langle -6, 12 \rangle$$

27. Perform the indicated operations.

$$2\mathbf{u} - \mathbf{v} = 2\langle -2, 4 \rangle - \langle -3, -2 \rangle$$
$$= \langle -4, 8 \rangle - \langle -3, -2 \rangle$$
$$= \langle -1, 10 \rangle$$

29. Perform the indicated operations.

$$\frac{2}{3}\mathbf{u} + \frac{1}{6}\mathbf{v} = \frac{2}{3}\langle -2, 4 \rangle + \frac{1}{6}\langle -3, -2 \rangle$$
$$= \left\langle -\frac{4}{3}, \frac{8}{3} \right\rangle + \left\langle -\frac{1}{2}, -\frac{1}{3} \right\rangle$$
$$= \left\langle -\frac{11}{6}, \frac{7}{3} \right\rangle$$

31. Perform the indicated operations.

$$\|\mathbf{u}\| = \sqrt{(-2)^2 + 4^2} = \sqrt{20} = 2\sqrt{5}$$

33. Perform the indicated operations.

$$\|\mathbf{u}\|\|\mathbf{v}\|\cos 60° = \sqrt{(-2)^2 + 4^2}\,\sqrt{(-3)^2 + (-2)^2}\,\cos 60°$$
$$= \sqrt{20}\sqrt{13}\,\frac{1}{2} = 2\sqrt{5}\sqrt{13}\,\frac{1}{2}$$
$$= \sqrt{65} \approx 8.06$$

35. Perform the indicated operations.

$$4\mathbf{v} = 4(-2\mathbf{i} + 3\mathbf{j})$$
$$= -8\mathbf{i} + 12\mathbf{j}$$

37. Perform the indicated operations.

$$6\mathbf{u} + 2\mathbf{v} = 6(3\mathbf{i} - 2\mathbf{j}) + 2(-2\mathbf{i} + 3\mathbf{j})$$
$$= (18\mathbf{i} - 12\mathbf{j}) + (-4\mathbf{i} + 6\mathbf{j})$$
$$= (18 - 4)\mathbf{i} + (-12 + 6)\mathbf{j}$$
$$= 14\mathbf{i} - 6\mathbf{j}$$

39. Perform the indicated operations.

$$\frac{2}{3}\mathbf{v} + \frac{3}{4}\mathbf{u} = \frac{2}{3}(-2\mathbf{i} + 3\mathbf{j}) + \frac{3}{4}(3\mathbf{i} - 2\mathbf{j})$$
$$= \left(-\frac{4}{3}\mathbf{i} + 2\mathbf{j}\right) + \left(\frac{9}{4}\mathbf{i} - \frac{3}{2}\mathbf{j}\right)$$
$$= \left(-\frac{4}{3} + \frac{9}{4}\right)\mathbf{i} + \left(2 - \frac{3}{2}\right)\mathbf{j}$$
$$= \frac{11}{12}\mathbf{i} + \frac{1}{2}\mathbf{j}$$

41. Perform the indicated operations.

$$\mathbf{u} \cdot \mathbf{v} = (3i - 2j) \cdot (-2i + 3j)$$
$$= (3)(-2) + (-2)(3)$$
$$= -12$$

43. Find the horizontal and vertical components and write an equivalent vector in the form $\mathbf{v} = a_1\mathbf{i} + a_2\mathbf{j}$.

$$a_1 = 5\cos 27° \approx 4.5$$
$$a_2 = 5\sin 27° \approx 2.3$$
$$\mathbf{v} = a_1\mathbf{i} + a_2\mathbf{j} \approx 4.5\mathbf{i} + 2.3\mathbf{j}$$

45. Find the horizontal and vertical components and write an equivalent vector in the form $\mathbf{v} = a_1\mathbf{i} + a_2\mathbf{j}$.

$$a_1 = 4\cos\frac{\pi}{4} \approx 2.8$$
$$a_2 = 4\sin\frac{\pi}{4} \approx 2.8$$
$$\mathbf{v} = a_1\mathbf{i} + a_2\mathbf{j} \approx 2.8\mathbf{i} + 2.8\mathbf{j}$$

47. Find the ground speed.

heading $= 124° \Rightarrow$ direction angle $= -34°$

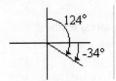

wind from the west $\Rightarrow$ direction angle $= 0°$

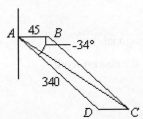

$\mathbf{AB} = 45\mathbf{i}$
$\mathbf{AD} = 340\cos(-34°)\mathbf{i} + 340\sin(-34°)\mathbf{j}$
$\mathbf{AD} \approx 281.9\mathbf{i} - 190.1\mathbf{j}$
$\mathbf{AC} = \mathbf{AB} + \mathbf{AD}$
$\mathbf{AC} = 45\mathbf{i} + 281.9\mathbf{i} - 190.1\mathbf{j}$
$\mathbf{AC} \approx 327\mathbf{i} - 190\mathbf{j}$
$\|\mathbf{AC}\| = \sqrt{327^2 + (-190)^2}$
$\|\mathbf{AC}\| \approx 380$ mph

The ground speed of the plane is approximately 380 mph.

49. Find the ground speed and course.

heading $= 96° \Rightarrow$ direction angle $= -6°$

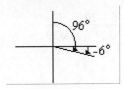

heading $= 37° \Rightarrow$ direction angle $= 53°$

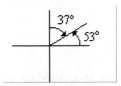

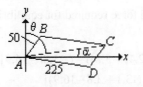

$\mathbf{AB} = 50\cos53°\mathbf{i} + 50\sin53°\mathbf{j}$
$\quad \approx 30.1\mathbf{i} + 39.9\mathbf{j}$
$\mathbf{AD} = 225\cos(-6°)\mathbf{i} + 225\sin(-6°)\mathbf{j}$
$\quad \approx 223.8\mathbf{i} - 23.5\mathbf{j}$
$\mathbf{AC} = \mathbf{AB} + \mathbf{AD}$
$\quad \approx 30.1\mathbf{i} + 39.9\mathbf{j} + 223.8\mathbf{i} - 23.5\mathbf{j}$
$\quad \approx 253.9\mathbf{i} + 16.4\mathbf{j}$
$\|\mathbf{AC}\| = \sqrt{(253.9)^2 + (16.4)^2} \approx 250$

$$\alpha = \tan^{-1}\left|\frac{16.4}{253.9}\right|$$
$$\quad = \tan^{-1}\frac{16.4}{253.9}$$
$$\alpha \approx 4°$$
$$\theta = 90° - \alpha$$
$$\theta \approx 90° - 4°$$
$$\theta \approx 86°$$

The ground speed of the plane is about 250 mph at a heading of approximately 86°.

51. Find the magnitude of the force.

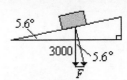

$$\sin 5.6° = \frac{F}{3000}$$
$$F = 3000\sin 5.6°$$
$$F \approx 293 \text{ lb}$$

53. a. $\sin 22.4° = \dfrac{\|\mathbf{F_1}\|}{345}$

$\quad\quad \|\mathbf{F_1}\| = 345\sin 22.4°$
$\quad\quad \|\mathbf{F_1}\| \approx 131 \text{ lb}$

b. $\cos 22.4° = \dfrac{\|\mathbf{F_2}\|}{345}$

$\|\mathbf{F_2}\| = 345\cos 22.4°$

$\|\mathbf{F_2}\| \approx 319$ lb

55. Determine whether the forces are in equilibrium, if not determine the additional force required for equilibrium.

$\mathbf{F_1} + \mathbf{F_2} + \mathbf{F_3}$
$= (18.2\mathbf{i} + 13.1\mathbf{j}) + (-12.4\mathbf{i} + 3.8\mathbf{j}) + (-5.8\mathbf{i} - 16.9\mathbf{j})$
$= (18.2 - 12.4 - 5.8)\mathbf{i} + (13.1 + 3.8 - 16.9)\mathbf{j}$
$= 0$

The forces are in equilibrium.

57. Determine whether the forces are in equilibrium, if not determine the additional force required for equilibrium.

$\mathbf{F_1} + \mathbf{F_2} + \mathbf{F_3}$
$= (155\mathbf{i} - 257\mathbf{j}) + (-124\mathbf{i} + 149\mathbf{j}) + (-31\mathbf{i} + 98\mathbf{j})$
$= (155 - 124 - 31)\mathbf{i} + (-257 + 149 + 98)\mathbf{j}$
$= 0\mathbf{i} - 10\mathbf{j}$

The forces are not in equilibrium. $\mathbf{F_4} = 0\mathbf{i} + 10\mathbf{j}$

59. Determine whether the forces are in equilibrium, if not determine the additional force required for equilibrium.

$\mathbf{F_1} + \mathbf{F_2} + \mathbf{F_3} = (189.3\mathbf{i} + 235.7\mathbf{j}) + (45.8\mathbf{i} - 205.6\mathbf{j})$
$\qquad + (-175.2\mathbf{i} - 37.7\mathbf{j}) + (-59.9\mathbf{i} + 7.6\mathbf{j})$
$= (189.3 + 45.8 - 175.2 - 59.9)\mathbf{i}$
$\qquad + (235.7 - 205.6 - 37.7 + 7.6)\mathbf{j}$
$= 0$

The forces are in equilibrium.

61. Find the dot product of the vectors.

$\mathbf{v} \cdot \mathbf{w} = \langle 3, -2 \rangle \cdot \langle 1, 3 \rangle$
$= 3(1) + (-2)3$
$= 3 - 6$
$= -3$

63. Find the dot product of the vectors.

$\mathbf{v} \cdot \mathbf{w} = (\mathbf{i} + 2\mathbf{j}) \cdot (-\mathbf{i} + \mathbf{j})$
$= 1(-1) + 2(1)$
$= -1 + 2$
$= 1$

65. Find the angle. Round to the nearest tenth.

$\cos\theta = \dfrac{\mathbf{v} \cdot \mathbf{w}}{\|\mathbf{v}\|\,\|\mathbf{w}\|}$

$\cos\theta = \dfrac{\langle 2, -1 \rangle \cdot \langle 3, 4 \rangle}{\sqrt{2^2 + (-1)^2}\,\sqrt{3^2 + 4^2}}$

$\cos\theta = \dfrac{2(3) + (-1)4}{\sqrt{5}\,\sqrt{25}}$

$\cos\theta = \dfrac{2}{5\sqrt{5}} \approx 0.1789$

$\theta \approx 79.7°$

67. Find the angle. Round to the nearest tenth.

$\cos\theta = \dfrac{\mathbf{v} \cdot \mathbf{w}}{\|\mathbf{v}\|\,\|\mathbf{w}\|}$

$\cos\theta = \dfrac{(5\mathbf{i} - 2\mathbf{j}) \cdot (2\mathbf{i} + 5\mathbf{j})}{\sqrt{5^2 + (-2)^2}\,\sqrt{2^2 + 5^2}}$

$\cos\theta = \dfrac{5(2) + (-2)(5)}{\sqrt{29}\,\sqrt{29}}$

$\cos\theta = \dfrac{0}{\sqrt{29}\,\sqrt{29}} = 0$

$\theta = 90°$

Thus, the vectors are orthogonal.

69. Find the angle. Round to the nearest tenth.

$\cos\theta = \dfrac{\mathbf{v} \cdot \mathbf{w}}{\|\mathbf{v}\|\,\|\mathbf{w}\|}$

$\cos\theta = \dfrac{(5\mathbf{i} + 2\mathbf{j}) \cdot (-5\mathbf{i} - 2\mathbf{j})}{\sqrt{5^2 + 2^2}\,\sqrt{(-5)^2 + (-2)^2}}$

$\cos\theta = \dfrac{5(-5) + 2(-2)}{\sqrt{29}\,\sqrt{29}}$

$\cos\theta = \dfrac{-29}{\sqrt{29}\,\sqrt{29}} = -1$

$\theta = 180°$

71. Find $\text{proj}_{\mathbf{w}}\mathbf{v}$.

$\text{proj}_{\mathbf{w}}\mathbf{v} = \dfrac{\mathbf{v} \cdot \mathbf{w}}{\|\mathbf{w}\|} = \dfrac{\langle 6, 7 \rangle \cdot \langle 3, 4 \rangle}{\sqrt{3^2 + 4^2}} = \dfrac{18 + 28}{\sqrt{25}} = \dfrac{46}{5}$

73. Find $\text{proj}_{\mathbf{w}}\mathbf{v}$.

$\text{proj}_{\mathbf{w}}\mathbf{v} = \dfrac{\mathbf{v} \cdot \mathbf{w}}{\|\mathbf{w}\|}$

$\text{proj}_{\mathbf{w}}\mathbf{v} = \dfrac{\langle -3, 4 \rangle \cdot \langle 2, 5 \rangle}{\sqrt{2^2 + 5^2}} = \dfrac{-6 + 20}{\sqrt{29}} = \dfrac{14}{\sqrt{29}}$

$= \dfrac{14\sqrt{29}}{29} \approx 2.6$

75. Find $\text{proj}_{\mathbf{w}}\mathbf{v}$.

$$\text{proj}_{\mathbf{w}}\mathbf{v} = \frac{\mathbf{v} \cdot \mathbf{w}}{\|\mathbf{w}\|}$$

$$\text{proj}_{\mathbf{w}}\mathbf{v} = \frac{(2\mathbf{i}+\mathbf{j})\cdot(6\mathbf{i}+3\mathbf{j})}{\sqrt{6^2+3^2}} = \frac{12+3}{\sqrt{45}} = \frac{5}{\sqrt{5}}$$

$$= \sqrt{5} \approx 2.2$$

77. Find $\text{proj}_{\mathbf{w}}\mathbf{v}$.

$$\text{proj}_{\mathbf{w}}\mathbf{v} = \frac{\mathbf{v} \cdot \mathbf{w}}{\|\mathbf{w}\|}$$

$$\text{proj}_{\mathbf{w}}\mathbf{v} = \frac{(3\mathbf{i}-4\mathbf{j})\cdot(-6\mathbf{i}+12\mathbf{j})}{\sqrt{(-6)^2+12^2}} = \frac{-18-48}{\sqrt{180}} = -\frac{11}{\sqrt{5}}$$

$$= -\frac{11\sqrt{5}}{5} \approx -4.9$$

79. Find the work done.

$$W = \mathbf{F} \cdot \mathbf{s}$$
$$W = \|\mathbf{F}\|\,\|\mathbf{s}\|\cos\alpha$$
$$W = (75)(15)(\cos 32°)$$
$$W \approx 954 \text{ foot-pounds}$$

81. Find the work done.

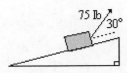

$$W = \mathbf{F} \cdot \mathbf{s}$$
$$W = \|\mathbf{F}\|\,\|\mathbf{s}\|\cos\alpha$$
$$W = (75)(12)(\cos 30°)$$
$$W \approx 779 \text{ foot-pounds}$$

83. Find the sum geometrically.

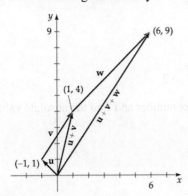

Thus, the sum is $\langle 6,\ 9 \rangle$.

85. Find the vector.

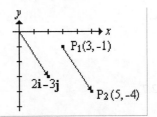

The vector from $P_1(3, -1)$ to $P_2(5, -4)$ is equivalent to $2\mathbf{i} - 3\mathbf{j}$.

87. Is $\mathbf{v}$ perpendicular to $\mathbf{w}$? Why or why not?

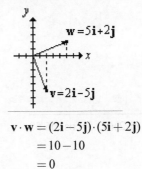

$$\mathbf{v} \cdot \mathbf{w} = (2\mathbf{i}-5\mathbf{j})\cdot(5\mathbf{i}+2\mathbf{j})$$
$$= 10-10$$
$$= 0$$

The two vectors are perpendicular.

89. $\mathbf{v} = \langle -2,\ 7 \rangle$

$$\langle -2,\ 7 \rangle \cdot \langle a,\ b \rangle = 0$$
$$-2a + 7b = 0$$
$$a = \frac{7}{2}b$$
Let $b = 2$
$$a = 7$$

Thus, $\mathbf{u} = \langle 7,\ 2 \rangle$ is one example.

91. Let $\mathbf{u} = c\mathbf{i}+b\mathbf{j}$, $\mathbf{v} = c\mathbf{i}+d\mathbf{j}$, and $\mathbf{w} = e\mathbf{i}+f\mathbf{j}$.

$$(\mathbf{u}\cdot\mathbf{v})\cdot\mathbf{w} = \left[(a\mathbf{i}+b\mathbf{j})\cdot(c\mathbf{i}+d\mathbf{j})\right]\cdot(e\mathbf{i}+f\mathbf{j})$$
$$= (ac+bd)\cdot(e\mathbf{i}+f\mathbf{j})$$

$ac + bd$ is a scalar quantity. The product of a scalar and a vector is not defined. Therefore, no,

$(\mathbf{u}\cdot\mathbf{v})\cdot\mathbf{w}$ does not equal $\mathbf{u}\cdot(\mathbf{v}\cdot\mathbf{w})$.

93. Let $\mathbf{v} = \langle a,b \rangle$ and $\mathbf{w} = \langle d,e \rangle$

$$c\mathbf{v} = \langle ca,cb \rangle$$

$$c(\mathbf{v}\cdot\mathbf{w}) = c\langle a,b \rangle \cdot \langle d,e \rangle = c(ad+be) = cad+cbe$$
$$(c\mathbf{v}\cdot\mathbf{w}) = \langle ca,cb \rangle \cdot \langle d,e \rangle = cad+cbe$$

Therefore, $c(\mathbf{v}\cdot\mathbf{w}) = (c\mathbf{v})\cdot\mathbf{w}$.

95. Neither. If the force and the distance are the same, the work will be the same.

Mid-Chapter 7 Quiz

1. Find b.

$$C = 180° - 71.2° - 62.7°$$
$$C = 46.1°$$

$$\frac{b}{\sin B} = \frac{c}{\sin C}$$
$$\frac{b}{\sin 62.7°} = \frac{18.5}{\sin 46.1°}$$
$$b = \frac{18.5 \sin 62.7°}{\sin 46.1°}$$
$$b \approx 22.8 \text{ in.}$$

3. Find A.

$$b^2 = a^2 + c^2 - 2ac \cos B$$
$$b = \sqrt{21.2^2 + 32.5^2 - 2(21.2)(32.5)\cos 105.5°}$$
$$b \approx 43.2890804$$
$$b \approx 43.3$$

$$\cos A = \frac{b^2 + c^2 - a^2}{2bc}$$
$$\cos A = \frac{43.2890804^2 + 32.5^2 - 21.2^2}{2(43.2890804)(32.5)}$$
$$A = \cos^{-1}\left(\frac{2480.754482}{2813.790226}\right)$$
$$A \approx 28.2°$$

5. Find the area of the triangle.

$$K = \frac{1}{2}bc \sin A$$
$$K = \frac{1}{2}(15.4)(16.5)\sin 34.5°$$
$$K \approx 72 \text{ m}^2$$

7. Find the magnitude of the force.

$$\sin 9.5° = \frac{F}{725}$$
$$F = 725 \sin 9.5°$$
$$F \approx 120 \text{ lb}$$

Prepare for Section 7.4

P1. Simplify.

$$(1+i)(2+i) = 2 + 3i + i^2 = 1 + 3i$$

P3. Find the conjugate.

$$2 - 3i$$

P5. Find the solutions using the quadratic formula.

$$x = \frac{-1 \pm \sqrt{1^2 - 4(1)(1)}}{2(1)} = \frac{-1 \pm \sqrt{-3}}{2} = -\frac{1}{2} \pm \frac{\sqrt{3}}{2}i$$

Section 7.4 Exercises

1. The absolute value of the product is 1.

3. Write the conjugate of $4(\cos 60° + i \sin 60°)$.

$$4(\cos 60° - i \sin 60°)$$

5. Explain the relationship between the trigonometric form of a complex number and the polar form.

They are exactly the same.

7. Graph the complex number and find the absolute value.

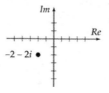

$$|z| = \sqrt{(-2)^2 + (-2)^2} = \sqrt{8} = 2\sqrt{2}$$

9. Graph the complex number and find the absolute value.

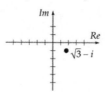

$$|z| = \sqrt{(\sqrt{3})^2 + (-1)^2} = \sqrt{3+1} = \sqrt{4} = 2$$

11. Graph the complex number and find the absolute value.

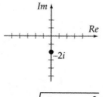

$$|z| = \sqrt{0^2 + (-2)^2} = 2$$

13. Graph the complex number and find the absolute value.

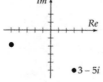

$$|z| = \sqrt{3^2 + (-5)^2} = \sqrt{34}$$

15. Write the complex number in trigonometric form.

$$r = \sqrt{1^2 + (-1)^2}$$
$$r = \sqrt{2}$$

$$\alpha = \tan^{-1}\left|\frac{-1}{1}\right|$$
$$= \tan^{-1} 1$$
$$= 45°$$

$$\theta = 360° - 45° = 315°$$

$$z = \sqrt{2} \text{ cis } 315°$$

17. Write the complex number in trigonometric form.

$$r = \sqrt{(\sqrt{3})^2 + (-1)^2}$$
$$r = 2$$

$$\alpha = \tan^{-1}\left|\frac{-1}{\sqrt{3}}\right|$$
$$= \tan^{-1}\frac{1}{\sqrt{3}} = 30°$$

$$\alpha = 360° - 30° = 330°$$

$$z = 2 \text{ cis } 330°$$

19. Write the complex number in trigonometric form.

$$r = \sqrt{0^2 + 5^2}$$
$$r = 5$$

$$\theta = 90°$$

$$z = 5 \text{ cis } 90°$$

21. Write the complex number in trigonometric form.

$$r = \sqrt{(-17)^2 + 0^2} = 17$$

$$\theta = 180°$$

$$z = 17 \text{ cis } 180°$$

23. Write the complex number in trigonometric form.

$$r = \sqrt{(-3\sqrt{3})^2 + (3)^2} = 6$$

$$\alpha = \tan^{-1}\left|\frac{3}{-3\sqrt{3}}\right|$$
$$= \tan^{-1}\frac{\sqrt{3}}{3} = 30°$$

$$\theta = 180° - 30° = 150°$$

$$z = 6 \text{ cis } 150°$$

25. Write the complex number in trigonometric form.

$$r = \sqrt{(-2)^2 + (-2\sqrt{3})^2}$$
$$r = 4$$

$$\alpha = \tan^{-1}\left|\frac{-2\sqrt{3}}{-2}\right|$$
$$= \tan^{-1}\sqrt{3} = 60°$$

$$\theta = 180° + 60° = 240°$$

$$z = 4 \text{ cis } 240°$$

27. Write the complex number in standard form.

$$z = 2(\cos 45° + i \sin 45°)$$
$$z = 2\left(\frac{\sqrt{2}}{2} + \frac{\sqrt{2}}{2}i\right)$$
$$z = \sqrt{2} + i\sqrt{2}$$

29. Write the complex number in standard form.

$$z = \cos 315° + i \sin 315°$$
$$z = \frac{\sqrt{2}}{2} - \frac{\sqrt{2}}{2}i$$

31. Write the complex number in standard form.

$$z = 6 \text{ cis } 135°$$
$$z = 6(\cos 135° + i \sin 135°)$$
$$z = 6\left(-\frac{\sqrt{2}}{2} + \frac{\sqrt{2}}{2}i\right)$$
$$z = -3\sqrt{2} + 3i\sqrt{2}$$

33. Write the complex number in standard form.

$$z = 8 \text{ cis } 0°$$
$$z = 8(\cos 0° + i \sin 0°)$$
$$z = 8(1 + 0i)$$
$$z = 8$$

35. Write the complex number in standard form.

$$z = 2\left(\cos\frac{5\pi}{6} + i \sin\frac{5\pi}{6}\right)$$
$$z = 2\left(-\frac{\sqrt{3}}{2} + \frac{1}{2}i\right)$$
$$z = -\sqrt{3} + i$$

37. Write the complex number in standard form.

$$z = 3\left(\cos\frac{3\pi}{2} + i \sin\frac{3\pi}{2}\right)$$
$$z = 3(0 - i)$$
$$z = -3i$$

39. Write the complex number in standard form.

$$z = 8 \text{ cis } \frac{3\pi}{4}$$

$$= 8 \left(\cos \frac{3\pi}{4} + i \sin \frac{3\pi}{4} \right)$$

$$z = 8 \left(-\frac{\sqrt{2}}{2} + \frac{i\sqrt{2}}{2} \right)$$

$$z = -4\sqrt{2} + 4i\sqrt{2}$$

41. Write the complex number in standard form.

$$z = 9 \text{ cis } \frac{11\pi}{6}$$

$$z = 9 \left(\cos \frac{11\pi}{6} + i \sin \frac{11\pi}{6} \right)$$

$$z = 9 \left(\frac{\sqrt{3}}{2} - \frac{1}{2} i \right)$$

$$z = \frac{9\sqrt{3}}{2} - \frac{9}{2} i$$

43. Multiply, write answer in trigonometric form.

$$z_1 z_2 = 2 \text{ cis } 30° \cdot 3 \text{ cis } 225°$$

$$z_1 z_2 = 6 \text{ cis}(30° + 225°)$$

$$z_1 z_2 = 6 \text{ cis } 255°$$

45. Multiply, write answer in trigonometric form.

$$z_1 z_2 = 3(\cos 122° + i \sin 122°) \cdot 4(\cos 213° + i \sin 213°)$$

$$z_1 z_2 = 12[\cos(122° + 213°) + i \sin(122° + 213°)]$$

$$z_1 z_2 = 12(\cos 335° + i \sin 335°)$$

$$z_1 z_2 = 12 \text{ cis } 335°$$

47. Multiply, write answer in trigonometric form.

$$z_1 z_2 = 5 \left(\cos \frac{2\pi}{3} + i \sin \frac{2\pi}{3} \right) \cdot 2 \left(\cos \frac{2\pi}{5} + i \sin \frac{2\pi}{5} \right)$$

$$z_1 z_2 = 10 \left[\cos\left(\frac{2\pi}{3} + \frac{2\pi}{5} \right) + i \sin\left(\frac{2\pi}{3} + \frac{2\pi}{5} \right) \right]$$

$$z_1 z_2 = 10 \left(\cos \frac{16\pi}{15} + i \sin \frac{16\pi}{15} \right)$$

$$z_1 z_2 = 10 \text{ cis } \frac{16\pi}{15}$$

49. Multiply, write answer in trigonometric form.

$$z_1 z_2 = 4 \text{ cis } 2.4 \cdot 6 \text{ cis } 4.1$$

$$z_1 z_2 = 24 \text{ cis } (2.4 + 4.1)$$

$$z_1 z_2 = 24 \text{ cis } 6.5$$

51. Divide, write answer in standard form, round answer to 3 decimal places.

$$\frac{z_1}{z_2} = \frac{32 \text{ cis } 30°}{4 \text{ cis } 150°}$$

$$\frac{z_1}{z_2} = 8 \text{ cis}(30° - 150°)$$

$$\frac{z_1}{z_2} = 8 \text{ cis}(-120°)$$

$$\frac{z_1}{z_2} = 8 (\cos 120° - i \sin 120°)$$

$$\frac{z_1}{z_2} = 8 \left(-\frac{1}{2} - \frac{i\sqrt{3}}{2} \right) = -4 - 4i\sqrt{3}$$

53. Divide, write answer in standard form, round answer to 3 decimal places.

$$\frac{z_1}{z_2} = \frac{27(\cos 315° + i \sin 315°)}{9(\cos 225° + i \sin 225°)}$$

$$\frac{z_1}{z_2} = 3 [\cos(315° - 225°) + i \sin(315° - 225°)]$$

$$\frac{z_1}{z_2} = 3(\cos 90° + i \sin 90°) = 3(0 + i) = 3i$$

55. Divide, write answer in standard form, round answer to 3 decimal places.

$$\frac{z_1}{z_2} = \frac{12\left(\cos \frac{2\pi}{3} + i \sin \frac{2\pi}{3} \right)}{4\left(\cos \frac{11\pi}{6} + i \sin \frac{11\pi}{6} \right)}$$

$$\frac{z_1}{z_2} = 3 \left[\cos\left(\frac{2\pi}{3} - \frac{11\pi}{6} \right) + i \sin\left(\frac{2\pi}{3} - \frac{11\pi}{6} \right) \right]$$

$$\frac{z_1}{z_2} = 3\left(\cos \frac{7\pi}{6} - i \sin \frac{7\pi}{6} \right)$$

$$\frac{z_1}{z_2} = 3 \left[-\frac{\sqrt{3}}{2} - \left(-\frac{1}{2} i \right) \right]$$

$$\frac{z_1}{z_2} = -\frac{3\sqrt{3}}{2} + \frac{3}{2} i$$

57. Divide, write answer in standard form, round answer to 3 decimal places.

$$\frac{z_1}{z_2} = \frac{25 \text{ cis } 3.5}{5 \text{ cis } 1.5}$$

$$\frac{z_1}{z_2} = 5 \text{ cis } (3.5 - 1.5)$$

$$\frac{z_1}{z_2} = 5 \text{ cis } 2$$

$$\frac{z_1}{z_2} = 5 (\cos 2 + i \sin 2)$$

$$\frac{z_1}{z_2} \approx 5 (-0.4161 + 0.9093i)$$

$$\frac{z_1}{z_2} \approx -2.081 + 4.546i$$

59. Perform indicated operation, write answer in standard form, round constants to 4 decimal places.

For $z_1 = 1 - i\sqrt{3}$

$r_1 = \sqrt{1^2 + (\sqrt{3})^2}$ $\alpha = \tan^{-1}\left|\dfrac{-\sqrt{3}}{1}\right| = 60°$

$r_1 = 2$ $\theta_1 = 300°$

$z_1 = 2(\cos 300° + i\,\sin 300°)$

For $z_2 = 1 + i$

$r_2 = \sqrt{1^2 + 1^2}$ $\alpha = \tan^{-1}\left|\dfrac{1}{1}\right| = 45°$

$r_2 = \sqrt{2}$ $\theta_2 = 45°$

$z_2 = \sqrt{2}(\cos 45° + i\,\sin 45°)$

$z_1 z_2 = 2(\cos 300° + i\,\sin 300°) \cdot \sqrt{2}(\cos 45° + i\,\sin 45°)$

$z_1 z_2 = 2\sqrt{2}[\cos(300° + 45°) + i\,\sin(300° + 45°)]$

$z_1 z_2 = 2\sqrt{2}(\cos 345° + i\,\sin 345°)$

$z_1 z_2 \approx 2.732 - 0.732i$

61. Perform indicated operation, write answer in standard form, round constants to 4 decimal places.

For $z_1 = 3 - 3i$

$r_1 = \sqrt{3^2 + (-3)^2}$ $\alpha = \tan^{-1}\left|\dfrac{-3}{3}\right| = 45°$

$r_1 = 3\sqrt{2}$ $\theta_1 = 315°$

$z_1 = 3\sqrt{2}(\cos 315° + i\,\sin 315°)$

For $z_2 = 1 + i$

$r_2 = \sqrt{1^2 + 1^2}$ $\alpha = \tan^{-1}\left|\dfrac{1}{1}\right| = 45°$

$r_1 = \sqrt{2}$ $\theta_1 = 45°$

$z_2 = \sqrt{2}(\cos 45° + i\,\sin 45°)$

$z_1 z_2 = 3\sqrt{2}(\cos 315° + i\,\sin 315°) \cdot \sqrt{2}(\cos 45° + i\sin 45°)$

$z_1 z_2 = 6[\cos(315° + 45°) + i\,\sin(315° + 45°)]$

$z_1 z_2 = 6(\cos 360° + i\,\sin 360°)$

$z_1 z_2 = 6 + 0i$

$z_1 z_2 = 6$

63. Perform indicated operation, write answer in standard form, round constants to 4 decimal places.

For $z_1 = 1 + i\sqrt{3}$

$r_1 = \sqrt{1^2 + (\sqrt{3})^2}$ $\alpha_1 = \tan^{-1}\left|\dfrac{\sqrt{3}}{1}\right| = 60°$

$r_1 = 2$ $\theta_1 = 60°$

$z_1 = 2(\cos 60° + i\,\sin 60°)$

For $z_2 = 1 - i\sqrt{3}$

$r_2 = \sqrt{1^2 + (\sqrt{3})^2}$ $\alpha = \tan^{-1}\left|\dfrac{-\sqrt{3}}{1}\right| = 60°$

$r_2 = 2$ $\theta_2 = 300°$

$z_2 = 2(\cos 300° + i\,\sin 300°)$

$\dfrac{z_1}{z_2} = \dfrac{2(\cos 60° + i\,\sin 60°)}{2(\cos 300° + i\,\sin 300°)}$

$\dfrac{z_1}{z_2} = \cos(60° - 300°) + i\,\sin(60° - 300°)$

$\dfrac{z_1}{z_2} = \cos 240° - i\,\sin 240° = -\dfrac{1}{2} + \dfrac{\sqrt{3}}{2}i$

65. Perform indicated operation, write answer in standard form, round constants to 4 decimal places.

For $z_1 = \sqrt{2} - i\sqrt{2}$

$r_1 = \sqrt{(\sqrt{2})^2 + (\sqrt{2})^2}$ $\alpha_1 = \tan^{-1}\left|\dfrac{-\sqrt{2}}{2}\right| = 45°$

$r_1 = 2$ $\theta_1 = 315°$

$z_1 = 2(\cos 315° + i\,\sin 315°)$

For $z_2 = 1 + i$

$r_2 = \sqrt{1^2 + 1^2}$ $\alpha_2 = \tan^{-1}\left|\dfrac{1}{1}\right| = 45°$

$r_2 = \sqrt{2}$ $\theta_2 = 45°$

$z_2 = \sqrt{2}(\cos 45° + i\,\sin 45°)$

$\dfrac{z_1}{z_2} = \dfrac{2(\cos 315° + i\,\sin 315°)}{\sqrt{2}(\cos 45° + i\,\sin 45°)}$

$\dfrac{z_1}{z_2} = \sqrt{2}[\cos(315° - 45°) + i\,\sin(315° - 45°)]$

$\dfrac{z_1}{z_2} = \sqrt{2}(\cos 270° + i\,\sin 270°)$

$\dfrac{z_1}{z_2} = \sqrt{2}[0 + i(-1)] = \sqrt{2}(0 - 1i) = 0 - \sqrt{2}i = -\sqrt{2}i$

or $-i\sqrt{2}$

67. Perform indicated operation in trigonometric form, write answer in standard form.

For $z_1 = \sqrt{3} - 1$

$$r_1 = \sqrt{(\sqrt{3})^2 + (-1)^2} \qquad \alpha_1 = \tan^{-1}\left|\frac{-1}{\sqrt{3}}\right| = 30°$$
$$r_1 = 2 \qquad\qquad\qquad \theta_1 = 330°$$

$$z_1 = 2(\cos 330° + i \sin 330°)$$

For $z_2 = 2 + 2i$

$$r_2 = \sqrt{2^2 + 2^2} \qquad \alpha_2 = \tan^{-1}\left|\frac{2}{2}\right| = 45°$$
$$r_2 = 2\sqrt{2} \qquad\qquad \theta_2 = 45°$$

$$z_2 = 2\sqrt{2}(\cos 45° + i \sin 45°)$$

For $z_3 = 2 - 2i\sqrt{3}$

$$r_3 = \sqrt{2^2 + (-2\sqrt{3})^2} \qquad \alpha_3 = \tan^{-1}\left|\frac{-2\sqrt{3}}{2}\right| = 60°$$
$$r_3 = 4 \qquad\qquad\qquad \theta_3 = 300°$$

$$z_3 = 4(\cos 300° + i \sin 300°)$$

$$z_1 z_2 z_3 = 2(\cos 330° + i \sin 330°)$$
$$\cdot 2\sqrt{2}(\cos 45° + i \sin 45°)$$
$$\cdot 4(\cos 300° + i \sin 300°)$$
$$z_1 z_2 z_3 = 16\sqrt{2}[\cos(330° + 45° + 300°)$$
$$\qquad\qquad + i \sin(330° + 45° + 300°)]$$
$$z_1 z_2 z_3 = 16\sqrt{2}(\cos 675° + i \sin 675°)$$
$$z_1 z_2 z_3 = 16\sqrt{2}(\cos 315° + i \sin 315°)$$
$$z_1 z_2 z_3 = 16\sqrt{2}\left(\frac{1}{\sqrt{2}} - \frac{1}{\sqrt{2}}i\right) = 16 - 16i$$

69. Perform indicated operation in trigonometric form, write answer in standard form.

For $z_1 = \sqrt{3} + i\sqrt{3}$

$$r_1 = \sqrt{(\sqrt{3})^2 + (\sqrt{3})^2} \qquad \alpha_1 = \tan^{-1}\left|\frac{\sqrt{3}}{\sqrt{3}}\right| = 45°$$
$$r_1 = \sqrt{6} \qquad\qquad\qquad \theta_1 = 45°$$

$$z_1 = \sqrt{6}(\cos 45° + i \sin 45°)$$

For $z_2 = 1 - i\sqrt{3}$

$$r_2 = \sqrt{1^2 + (-\sqrt{3})^2} \qquad \alpha_2 = \tan^{-1}\left|\frac{-\sqrt{3}}{1}\right| = 60°$$
$$r_2 = 2 \qquad\qquad\qquad \theta_2 = 300°$$

$$z_2 = 2(\cos 300° + i \sin 300°)$$

For $z_3 = 2 - 2i$

$$r_3 = \sqrt{2^2 + (-2)^2} \qquad \alpha_3 = \tan^{-1}\left|\frac{-2}{2}\right| = 45°$$
$$r_3 = 2\sqrt{2} \qquad\qquad \theta_3 = 315°$$

$$z_3 = 2\sqrt{2}(\cos 315° + i \sin 315°)$$

$$\frac{z_1}{z_2 z_3}$$

$$= \frac{\sqrt{6}(\cos 45° + i \sin 45°)}{2(\cos 300° + i \sin 300°)\cdot 2\sqrt{2}(\cos 315° + i \sin 315°)}$$

$$\frac{z_1}{z_2 z_3} = \frac{\sqrt{6}(\cos 45° + i \sin 45°)}{4\sqrt{2}[\cos(300° + 315°) + i \sin(300° + 315°)]}$$

$$\frac{z_1}{z_2 z_3} = \frac{\sqrt{6}(\cos 45° + i \sin 45°)}{4\sqrt{2}(\cos 255° + i \sin 255°)}$$

$$\frac{z_1}{z_2 z_3} = \frac{\sqrt{3}}{4}[\cos(45° - 255°) + i \sin(45° - 255°)]$$

$$\frac{z_1}{z_2 z_3} = \frac{\sqrt{3}}{4}(\cos 210° - i \sin 210°) = \frac{\sqrt{3}}{4}\left(-\frac{\sqrt{3}}{2} + \frac{i}{2}\right)$$

$$= -\frac{3}{8} + \frac{\sqrt{3}}{8}i$$

71. Perform indicated operation in trigonometric form, write answer in standard form.

For $z_1 = 1 - 3i$

$$r_1 = \sqrt{1^2 + (-3)^2} \qquad \alpha_1 = \tan^{-1}\left|\frac{-3}{1}\right| \approx 71.57°$$
$$r_1 = \sqrt{10} \qquad\qquad \theta_1 = 288.43°$$

$$z_1 = \sqrt{10}(\cos 288.4° + i \sin 288.4°)$$

For $z_2 = 2 + 3i$

$$r_2 = \sqrt{2^2 + 3^2} \qquad \alpha_2 = \tan^{-1}\left|\frac{3}{2}\right| \approx 56.31°$$
$$r_2 = \sqrt{13} \qquad\qquad \theta_2 = 56.31°$$

$$z_2 = \sqrt{13}(\cos 56.3° + i \sin 56.3°)$$

For $z_3 = 4 + 5i$

$$r_3 = \sqrt{4^2 + 5^2} \qquad \alpha_3 = \tan^1\left|\frac{5}{4}\right| \approx 51.34°$$
$$r_3 = \sqrt{41} \qquad\qquad \theta_3 = 51.34°$$

$$z_3 = \sqrt{41}(\cos 51.3° + i \sin 51.3°)$$

$$z_1 z_2 z_3 = \sqrt{10}(\cos 288.4° + i\sin 288.4°)$$
$$\cdot \sqrt{13}(\cos 56.3° + i\sin 56.3°)$$
$$\cdot \sqrt{41}(\cos 51.3° + i\sin 51.3°)$$

$$z_1 z_2 z_3 = \sqrt{10}\cdot\sqrt{13}\cdot\sqrt{41}[\cos(288.43° + 56.31°$$
$$+ 51.34°) + i\sin(288.43° + 56.31° + 51.34°)]$$

$$z_1 z_2 z_3 \approx 73.0(\cos 396.08° + i\sin 396.08°)$$
$$z_1 z_2 z_3 = 73.0(\cos 36.08° + i\sin 36.08°)$$
$$z_1 z_2 z_3 \approx 59.0 + 43.0i$$

73. $z = r(\cos\theta + i\sin\theta)$ $\overline{z} = r(\cos\theta - i\sin\theta)$

$$z\cdot\overline{z} = r(\cos\theta + i\sin\theta)\cdot r(\cos\theta - i\sin\theta)$$
$$z\cdot\overline{z} = r(\cos\theta + i\sin\theta)\cdot r[\cos(-\theta) + i\sin(-\theta)]$$
$$z\cdot\overline{z} = r^2[\cos(\theta-\theta) + i\sin(\theta-\theta)]$$
$$z\cdot\overline{z} = r^2(\cos 0 + i\sin 0)$$
$$z\cdot\overline{z} = r^2 \text{ or } a^2 + b^2$$

75. $(2 \text{ cis } 60°)^6 = 2^6 (\text{cis } 60°)^6$

$$2^6(\text{cis } 60° + 60° + 60° + 60° + 60° + 60°)$$
$$= 2^6[\text{cis } 6(60°)]$$
$$= 2^6(\text{cis } 360°)$$
$$= 64(\cos 360° + i\sin 360°)$$
$$= 64 + 0i$$

Prepare for Section 7.5

P1. Simplify.

$$\left(\frac{\sqrt{2}}{2} + \frac{\sqrt{2}}{2}i\right)^2 = \frac{2}{4} + 2\frac{2}{4}i + \frac{2}{4}i^2 = i$$

P3. Find the real root.

$$x^5 - 243 = (x-3)(x^4 + 3x^3 + 9x^2 + 27x + 81)$$
$$(x^4 + 3x^3 + 9x^2 + 27x + 81) \text{ yields } 4$$
$$\text{complex solutions}$$
$$(x-3) \text{ yields 1 real solution}$$

The real root is 3.

P5. Write in standard form.

$$2(\cos 150° + i\sin 150°) = 2\left(-\frac{\sqrt{3}}{2} + \frac{1}{2}i\right) = -\sqrt{3} + i$$

Section 7.5 Exercises

1. Determine the number of solutions.

There is just 1 solution, since z is to the power of 1.

3. Yes, the student is correct.

5. Find the indicated power.

$$[2(\cos 30° + i\sin 30°)]^8 = 2^8[\cos(8\cdot 30°) + i\sin(8\cdot 30°)]$$
$$= 256(\cos 240° + i\sin 240°)$$
$$= -128 - 128i\sqrt{3}$$

7. Find the indicated power.

$$[5(\cos 10° + i\sin 10°)]^3$$
$$= 5^3[\cos(3\cdot 10°) + i\sin(3\cdot 10°)]$$
$$= 125[\cos 30° + i\sin 30°]$$
$$= 125\left(\frac{\sqrt{3}}{2} + i\frac{1}{2}\right)$$
$$= \frac{125\sqrt{3}}{2} + \frac{125}{2}i$$

9. Find the indicated power.

$$[\text{cis}(150°)]^4 = \text{cis}(4\cdot 150°)$$
$$= \cos 600° + i\sin 600°$$
$$= \cos 240° + i\sin 240°$$
$$= -\frac{1}{2} - \frac{\sqrt{3}}{2}i$$

11. Find the indicated power.

$$[2\text{cis}(120°)]^6 = 2^6 \text{cis}(6\cdot 2\pi/3)$$
$$= 64(\cos 720° + i\sin 720°)$$
$$= 64(\cos 0° + i\sin 0°)$$
$$= 64$$

13. Find the indicated power.

$$z = \sqrt{3} - i$$
$$r = \sqrt{(\sqrt{3})^2 + (-1)^2} \quad \alpha = \tan^{-1}\left|\frac{-1}{\sqrt{3}}\right| = 30°$$
$$r = 2$$
$$\theta = 360 - 30° = 330°$$
$$z = 2(\cos 330° + i\sin 330°)$$
$$(\sqrt{3} - i)^4 = [2(\cos 330° + i\sin 330°)]^4$$
$$= 2^4[\cos(4\cdot 330°) + i\sin(4\cdot 330°)]$$
$$= 16(\cos 1320° + i\sin 1320°)$$
$$= 16(\cos 240° + i\sin 240°)$$
$$= 16\left(-\frac{1}{2} - \frac{\sqrt{3}}{2}\right)$$
$$= -8 - 8i\sqrt{3}$$

15. Find the indicated power.

$z = 1 + i$

$r = \sqrt{1^2 + 1^2}$ $\qquad$ $\alpha = \tan^{-1}\left|\dfrac{1}{1}\right| = 45°$

$r = \sqrt{2}$

$\qquad\qquad\qquad\qquad \theta = 45°$

$z = \sqrt{2}(\cos 45° + i \sin 45°)$

$(1 + i)^4 = [\sqrt{2}(\cos 45° + i \sin 45°)]^4$

$\qquad\quad = (\sqrt{2})^4 [\cos(4 \cdot 45°) + i \sin(4 \cdot 45°)]$

$\qquad\quad = 4(\cos 180° + i \sin 180°)$

$\qquad\quad = -4 + 0i = -4$

17. Find the indicated power.

$z = 2 + 2i$

$r = \sqrt{2^2 + 2^2}$ $\qquad\qquad$ $\alpha = \tan^{-1}\left|\dfrac{2}{2}\right| = 45°$

$r = 2\sqrt{2}$

$\qquad\qquad\qquad\qquad\quad \theta = 45°$

$z = 2\sqrt{2}(\cos 45° + i \sin 45°)$

$(2 + 2i)^7 = [2\sqrt{2}(\cos 45° + i \sin 45°)]^7$

$\qquad\qquad = 1024\sqrt{2}[\cos(7 \cdot 45°) + i \sin(7 \cdot 45°)]$

$\qquad\qquad = 1024\sqrt{2}(\cos 315° + i \sin 315°)$

$\qquad\qquad = 1024 - 1024i$

19. Find the indicated power.

$z = \dfrac{\sqrt{2}}{2} + i\dfrac{\sqrt{2}}{2}$

$r = \sqrt{(\sqrt{2}/2)^2 + (\sqrt{2}/2)^2}$ $\quad$ $\alpha = \tan^{-1}\left|\dfrac{\sqrt{2}/2}{\sqrt{2}/2}\right| = 45°$

$r = 1$

$\qquad\qquad\qquad\qquad\qquad\quad \theta = 45°$

$z = \cos 45° + i \sin 45°$

$\left(\dfrac{\sqrt{2}}{2} - i\dfrac{\sqrt{2}}{2}\right)^6 = (\cos 45° + i \sin 45°)^6$

$\qquad\qquad\quad = \cos(6 \cdot 45°) + i \sin(6 \cdot 45°)$

$\qquad\qquad\quad = \cos 270° + i \sin 270°$

$\qquad\qquad\quad = 0 - 1i = -i$

21. Find all the indicated roots.

$9 = 9(\cos 0° + i \sin 0°)$

$w_k = 9^{1/2}\left(\cos\dfrac{0° + 360°k}{2} + i \sin\dfrac{0° + 360°k}{2}\right)$ $\quad k = 0, 1$

$w_0 = 3(\cos 0° + i \sin 0°)$

$\qquad = 3 + 0i = 3$

$w_1 = 3\left(\cos\dfrac{0° + 360°}{2} + i \sin\dfrac{0° + 360°}{2}\right)$

$\qquad = 3(\cos 180° + i \sin 180°)$

$\qquad = -3 + 0i = -3$

23. Find all the indicated roots.

$64 = 64(\cos 0° + i \sin 0°)$

$w_k = 64^{1/6}\left(\cos\dfrac{0° + 360°k}{6} + i \sin\dfrac{0° + 360°k}{6}\right)$

$\qquad\qquad\qquad k = 0, 1, 2, 3, 4, 5$

$w_0 = 2(\cos 0° + i \sin 0°)$

$\qquad = 2 + 0i = 2$

$w_1 = 2\left(\cos\dfrac{0 + 360°}{6} + i \sin\dfrac{0 + 360°}{6}\right)$

$\qquad = 2(\cos 60° + i \sin 60°)$

$\qquad = 2\left(\dfrac{1}{2} + i\dfrac{\sqrt{3}}{2}\right)$

$\qquad = 1 + i\sqrt{3}$

$w_2 = 2\left(\cos\dfrac{0° + 360° \cdot 2}{6} + i \sin\dfrac{0° + 360° \cdot 2}{6}\right)$

$\qquad = 2(\cos 120° + i \sin 120°)$

$\qquad = 2\left(-\dfrac{1}{2} + i\dfrac{\sqrt{3}}{2}\right)$

$\qquad = -1 + i\sqrt{3}$

$w_3 = 2\left(\cos\dfrac{0° + 360° \cdot 3}{6} + i \sin\dfrac{0° + 360° \cdot 3}{6}\right)$

$\qquad = 2(\cos 180° + i \sin 180°)$

$\qquad = 2(-1 + 0i)$

$\qquad = -2 + 0i = -2$

$w_4 = 2\left(\cos\dfrac{0° + 360° \cdot 4}{6} + i \sin\dfrac{0° + 360° \cdot 4}{6}\right)$

$\qquad = 2(\cos 240° + i \sin 240°)$

$\qquad = 2\left(-\dfrac{1}{2} - i\dfrac{\sqrt{3}}{2}\right)$

$\qquad = -1 - i\sqrt{3}$

$w_5 = 2\left(\cos\dfrac{0° + 360° \cdot 5}{6} + i \sin\dfrac{0° + 360° \cdot 5}{6}\right)$

$\qquad = 2(\cos 300° + i \sin 300°)$

$\qquad = 2\left(\dfrac{1}{2} - i\dfrac{\sqrt{3}}{2}\right)$

$\qquad = 1 - i\sqrt{3}$

25. Find all the indicated roots.

$$-1 = 1(\cos 180° + i\sin 180°)$$

$$w_k = 1^{1/5}\left(\cos\frac{180° + 360°k}{5} + i\sin\frac{180° + 360°k}{5}\right)$$

$$k = 0, 1, 2, 3, 4$$

$$w_0 = 1(\cos 36° + i\sin 36°)$$
$$\approx 0.809 + 0.588i$$

$$w_1 = \cos\frac{180° + 360°}{5} + i\sin\frac{180° + 360°}{5}$$
$$= \cos 108° + i\sin 108°$$
$$\approx -0.309 + 0.951i$$

$$w_2 = \cos\frac{180° + 360°\cdot 2}{5} + i\sin\frac{180° + 360°\cdot 2}{5}$$
$$= \cos 180° + i\sin 180°$$
$$= -1 + 0i = -1$$

$$w_3 = \cos\frac{180° + 360°\cdot 3}{5} + i\sin\frac{180° + 360°\cdot 3}{5}$$
$$= \cos 252° + i\sin 252°$$
$$\approx -0.309 - 0.951i$$

$$w_4 = \cos\frac{180° + 360°\cdot 4}{5} + i\sin\frac{180° + 360°\cdot 4}{5}$$
$$= \cos 324° + i\sin 324°$$
$$\approx 0.809 - 0.588i$$

27. Find all the indicated roots.

$$1 = \cos 0° + i\sin 0°$$

$$w_k = \cos\frac{0° + 360°k}{3} + i\sin\frac{0° + 360°k}{3} \qquad k = 0, 1, 2$$

$$w_0 = \cos\frac{0°}{3} + i\sin\frac{0°}{3}$$
$$= \cos 0° + i\sin 0°$$
$$= 1 + 0i = 1$$

$$w_1 = \cos\frac{0° + 360°}{3} + i\sin\frac{0° + 360°}{3}$$
$$= \cos 120° + i\sin 120°$$
$$= -\frac{1}{2} + \frac{\sqrt{3}}{2}i$$

$$w_2 = \cos\frac{0° + 360°\cdot 2}{3} + i\sin\frac{0° + 360°\cdot 2}{3}$$
$$= \cos 240° + i\sin 240°$$
$$= -\frac{1}{2} - \frac{\sqrt{3}}{2}i$$

29. Find all the indicated roots.

$$1 + i = \sqrt{2}(\cos 45° + i\sin 45°)$$

$$w_k = \left(\sqrt{2}\right)^{1/4}\left(\cos\frac{45° + 360°k}{4} + i\sin\frac{45° + 360°k}{4}\right)$$

$$k = 0, 1, 2, 3$$

$$w_0 = 2^{1/8}\left(\cos\frac{45°}{4} + i\sin\frac{45°}{4}\right)$$
$$= 2^{1/8}(\cos 11.25° + i\sin 11.25°)$$
$$\approx 1.070 + 0.213i$$

$$w_1 = 2^{1/8}\left(\cos\frac{45° + 360°}{4} + i\sin\frac{45° + 360°}{4}\right)$$
$$= 2^{1/8}(\cos 101.25° + i\sin 101.25°)$$
$$\approx -0.213 - 1.070i$$

$$w_2 = 2^{1/8}\left(\cos\frac{45° + 360°\cdot 2}{4} + i\sin\frac{45° + 360°\cdot 2}{4}\right)$$
$$= 2^{1/8}(\cos 191.25° + i\sin 191.25°)$$
$$\approx -1.070 - 0.213i$$

$$w_3 = 2^{1/8}\left(\cos\frac{45° + 360°\cdot 3}{4} + i\sin\frac{45° + 360°\cdot 3}{4}\right)$$
$$= 2^{1/8}(\cos 281.25° + i\sin 281.25°)$$
$$\approx 0.213 - 1.070i$$

31. Find all the indicated roots.

$$2 - 2i\sqrt{3} = 4(\cos 300° + i\sin 300°) \quad k = 0, 1, 2$$

$$w_k = 4^{1/3}\left(\cos\frac{300° + 360°k}{3} + i\sin\frac{300° + 360°k}{3}\right)$$

$$w_0 = 4^{1/3}\left(\cos\frac{300°}{3} + i\sin\frac{300°}{3}\right)$$
$$= 4^{1/3}(\cos 100° + i\sin 100°)$$
$$\approx -0.276 + 1.563i$$

$$w_1 = 4^{1/3}\left(\cos\frac{300° + 360°}{3} + i\sin\frac{300° + 360°}{3}\right)$$
$$= 4^{1/3}(\cos 220° + i\sin 220°)$$
$$\approx -1.216 - 1.020i$$

$$w_2 = 4^{1/3}\left(\cos\frac{300° + 360°\cdot 2}{3} + i\sin\frac{300° + 360°\cdot 2}{3}\right)$$
$$= 4^{1/3}(\cos 340° + i\sin 340°)$$
$$\approx 1.492 - 0.543i$$

33. Find all the indicated roots.

$$-16+16i\sqrt{3}=32(\cos 120°+i\sin 120°)$$

$$w_k=32^{1/2}\left(\cos\frac{120°+360°k}{2}+i\sin\frac{120°+360°k}{2}\right)$$

$$k=0,\,1$$

$$w_0=4\sqrt{2}\left(\cos\frac{120°}{2}+i\sin\frac{120°}{2}\right)$$

$$=4\sqrt{2}(\cos 60°+i\sin 60°)$$

$$\approx 2\sqrt{2}+2i\sqrt{6}$$

$$w_1=4\sqrt{2}\left(\cos\frac{120°+360°}{2}+i\sin\frac{120°+360°}{2}\right)$$

$$=4\sqrt{2}(\cos 240°+i\sin 240°)$$

$$\approx -2\sqrt{2}-2i\sqrt{6}$$

35. Find all the roots.

$$x^3+8=0$$
$$x^3=-8$$

Find the three cube roots of –8.

$$x_k=8^{1/3}\left(\cos\frac{180°+360°k}{3}+i\sin\frac{180°+360°k}{3}\right)$$
$$k=0,\,1,\,2$$

$$w_0=2\left(\cos\frac{180°}{3}+i\sin\frac{180°}{3}\right)$$

$$=2(\cos 60°+i\sin 60°)$$

$$=2\text{ cis }60°$$

$$w_1=2\left(\cos\frac{180°+360°}{3}+i\sin\frac{180°+360°}{3}\right)$$

$$=2(\cos 180°+i\sin 180°)$$

$$=2\text{ cis }180°$$

$$w_2=2\left(\cos\frac{180°+360°·2}{3}+i\sin\frac{180°+360°·2}{3}\right)$$

$$=2(\cos 300°+i\sin 300°)$$

$$=2\text{ cis }300°$$

37. Find all the roots.

$$x^4+i=0$$
$$x^4=-i$$

Find the four fourth roots of –i.

$$-i=(\cos 270°+i\sin 270°)$$

$$w_k=\cos\frac{270°+360°k}{4}+i\sin\frac{270°+360°k}{4}$$
$$k=0,\,1,\,2,\,3$$

$$w_0=\text{cis }\frac{270°}{4}=\text{cis }67.5°$$

$$w_1=\text{cis }\frac{270°+360°}{4}=\text{cis }157.5°$$

$$w_2=\text{cis }\frac{270°+360°·2}{4}=\text{cis }247.5°$$

$$w_3=\text{cis }\frac{270°+360°·3}{4}=\text{cis }337.5°$$

39. Find all the roots.

$$x^3-27=0$$
$$x^3=27$$

Find the three cube roots of 27.

$$27=27(\cos 0°+i\sin 0°)$$

$$w_k=3\left(\cos\frac{0°+360°k}{3}+i\sin\frac{0°+360°k}{3}\right)\quad k=0,\,1,\,2$$

$$w_0=3\text{ cis }\frac{0°}{3}=3\text{ cis }0°$$

$$w_1=3\text{ cis }\frac{0°+360°}{3}=3\text{ cis }120°$$

$$w_2=3\text{ cis }\frac{0°+360°·2}{3}=3\text{ cis }240°$$

41. Find all the roots.

$$x^4+81=0$$
$$x^4=-81$$

Find the four fourth roots of –81.

$$-81=81(\cos 180°+i\sin 180°)$$

$$w_k=81^{1/4}\left(\cos\frac{180°+360°k}{4}+i\sin\frac{180°+360°k}{4}\right)$$
$$k=0,\,1,\,2,\,3$$

$$w_0=3\text{ cis }\frac{180°}{4}=3\text{ cis }45°$$

$$w_1=3\text{ cis }\frac{180°+360°}{4}=3\text{ cis }135°$$

$$w_2=3\text{ cis }\frac{180°+360°·2}{4}=3\text{ cis }225°$$

$$w_3=3\text{ cis }\frac{180°+360°·3}{4}=3\text{ cis }315°$$

43. Find all the roots.

$$x^4 - (1 - i\sqrt{3}) = 0$$
$$x^4 = 1 - i\sqrt{3}$$

Find the four fourth roots of $1 - i\sqrt{3}$.

$$1 - i\sqrt{3} = 2(\cos 300° + i\sin 300°)$$

$$w_k = 2^{1/4}\left(\cos\frac{300° + 360°k}{4} + i\sin\frac{300° + 360°k}{4}\right)$$
$$k = 0, 1, 2, 3$$

$$w_0 = \sqrt[4]{2}\ \text{cis}\ \frac{300°}{4}$$
$$= \sqrt[4]{2}\ \text{cis}\ 75°$$

$$w_1 = \sqrt[4]{2}\ \text{cis}\ \frac{300° + 360°}{4}$$
$$= \sqrt[4]{2}\ \text{cis}\ 165°$$

$$w_2 = \sqrt[4]{2}\ \text{cis}\ \frac{300° + 360° \cdot 2}{4}$$
$$= \sqrt[4]{2}\ \text{cis}\ 255°$$

$$w_3 = \sqrt[4]{2}\ \text{cis}\ \frac{300° + 360° \cdot 3}{4}$$
$$= \sqrt[4]{2}\ \text{cis}\ 345°$$

45. Find all the roots.

$$x^3 + (1 + i\sqrt{3}) = 0$$
$$x^3 = -1 - i\sqrt{3}$$

Find the three cube roots of $-1 - i\sqrt{3}$.

$$-1 - i\sqrt{3} = 2(\cos 240° + i\sin 240°)$$

$$w_k = 2^{1/3}\left(\cos\frac{240° + 360°k}{3} + i\sin\frac{240° + 360°k}{3}\right)$$
$$k = 0, 1, 2$$

$$w_0 = \sqrt[3]{2}\ \text{cis}\ \frac{240°}{3}$$
$$= \sqrt[3]{2}\ \text{cis}\ 80°$$

$$w_1 = \sqrt[3]{2}\ \text{cis}\ \frac{240° + 360°}{3}$$
$$= \sqrt[3]{2}\ \text{cis}\ 200°$$

$$w_2 = \sqrt[3]{2}\ \text{cis}\ \frac{240° + 360° \cdot 2}{3}$$
$$= \sqrt[3]{2}\ \text{cis}\ 320°$$

47. Let $z = a + bi$. Then $\bar{z} = a - bi$ by definition.

Substitute $a = r\cos\theta$ and $b = r\sin\theta$.

Thus $\bar{z} = r\cos\theta - ri\sin\theta = r(\cos\theta - i\sin\theta)$

49.
$$z = r(\cos\theta + i\sin\theta)$$
$$z^2 = r^2(\cos 2\theta + i\sin 2\theta)$$
$$\frac{1}{z^2} = \frac{1}{r^2(\cos 2\theta + i\sin 2\theta)}$$
$$= \frac{\cos 2\theta - i\sin 2\theta}{r^2(\cos 2\theta + i\sin 2\theta)(\cos 2\theta - i\sin 2\theta)}$$
$$= \frac{\cos 2\theta - i\sin 2\theta}{r^2(\cos^2 2\theta - i^2\sin^2 2\theta)} = \frac{\cos 2\theta - i\sin 2\theta}{r^2(\cos^2 2\theta + \sin^2 2\theta)}$$
$$z^{-2} = r^{-2}(\cos 2\theta - i\sin 2\theta)$$

51. For $n = 2$, the two square roots of 1 are

1 and -1.

The sum of these roots is

$$1 + (-1) = 0.$$

For $n = 3$, the three cube roots of 1 are

(from exercise 23)

$$1, \ -\frac{1}{2} + \frac{\sqrt{3}}{2}i \text{ and } -\frac{1}{2} - \frac{\sqrt{3}}{2}i.$$

The sum of these roots is

$$1 - \frac{1}{2} + \frac{\sqrt{3}}{2}i - \frac{1}{2} - \frac{\sqrt{3}}{2}i = 0.$$

For $n = 4$, the four fourth roots of 1 are

$1, -1, i,$ and $-i$.

The sum of these roots is

$$1 - 1 + i - i = 0$$

For $n = 5$, the five fifth roots of 1 are

1, cis 72°, cis 144°, cis 216°, cis 288°

The sum of these roots is

$$1 + \text{cis}\ 72° + \text{cis}\ 144° + \text{cis}\ 216° + \text{cis}\ 288° = 0$$

For $n = 6$, the six sixth roots of 1 are

$$1, -1, -\frac{1}{2} + \frac{\sqrt{3}}{2}i, \ -\frac{1}{2} - \frac{\sqrt{3}}{2}i, \ \frac{1}{2} + \frac{\sqrt{3}}{2}i, \ \frac{1}{2} - \frac{\sqrt{3}}{2}i$$

The sum of these roots is

$$1 - 1 - \frac{1}{2} + \frac{\sqrt{3}}{2}i - \frac{1}{2} - \frac{\sqrt{3}}{2}i + \frac{1}{2} + \frac{\sqrt{3}}{2}i + \frac{1}{2} - \frac{\sqrt{3}}{2}i = 0$$

For $n \geq 2$, the sum of the nth roots of 1 is 0.

Exploring Concepts with Technology

1. Find the force using WolframAlpha.

$\mathbf{a} = A\langle \cos 97.5°, \ \sin 97.5°\rangle$

$\mathbf{b} = B\langle \cos 10.1°, \ \sin 10.1°\rangle$

$\mathbf{a} + \mathbf{b} + \mathbf{c} = \langle 0, \ 0\rangle$

$\langle 0, \ 0\rangle = A\langle \cos 97.5°, \ \sin 97.5°\rangle$
$\qquad + B\langle \cos 10.1°, \ \sin 10.1°\rangle + \langle 0, -58\rangle$

Using WolframAlpha, $A \approx 57.16$ and $B \approx 7.57832$.

The force exerted by cable A is about 57.2 lb.

The force exerted by cable B is about 7.58 lb.

3. Find the force using WolframAlpha.

$\mathbf{F}_1 = 1280\langle \cos 119°, \ \sin 119°\rangle$

$\mathbf{F}_2 = 945\langle \cos 153°, \ \sin 153°\rangle$

Using WolframAlpha, $\mathbf{F}_1 + \mathbf{F}_2$, the magnitude is

approximately 2130 lb, and the direction angle is

approximately 133°.

Chapter 7 Review Exercises

1. Solve the triangle.

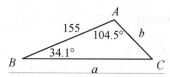

$C = 180° - 34.1° - 104.5° = 41.4°$

$\dfrac{a}{\sin A} = \dfrac{c}{\sin C}$ $\qquad$ $\dfrac{b}{\sin B} = \dfrac{c}{\sin C}$

$\dfrac{a}{\sin 104.5°} = \dfrac{155}{\sin 41.4°}$ $\qquad$ $\dfrac{b}{\sin 34.1°} = \dfrac{155}{\sin 41.4°}$

$\qquad a = \dfrac{155\sin 104.5°}{\sin 41.4°}$ $\qquad$ $b = \dfrac{155\sin 34.1°}{\sin 41.4°}$

$\qquad a \approx 227$ $\qquad\qquad$ $b \approx 131$

3. Solve the triangle. [7.2]

$\cos B = \dfrac{12^2 + 20^2 - 15^2}{2(12)(20)}$

$\cos B \approx 0.6646$

$\quad B \approx 48°$

$\cos C = \dfrac{12^2 + 15^2 - 20^2}{2(12)(15)}$

$\cos C \approx -0.0861$

$\quad C \approx 95°$

$A = 180° - 48° - 95° = 37°$

5. Solve the triangle. [7.2]

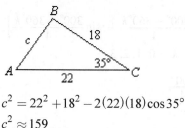

$c^2 = 22^2 + 18^2 - 2(22)(18)\cos 35°$

$c^2 \approx 159$

$\quad c = \sqrt{159} \approx 13$

$\dfrac{18}{\sin A} = \dfrac{\sqrt{159}}{\sin 35°}$

$\sin A = \dfrac{18\sin 35°}{\sqrt{159}}$

$\sin A \approx 0.8188 \approx 55°$

$B \approx 180 - 35° - 55° \approx 90°$

7. Solve the triangle. [7.1]

$\dfrac{10}{\sin C} = \dfrac{8}{\sin 105°}$

$\sin C = \dfrac{10\sin 105°}{8}$

$\sin C \approx 1.207$

No triangle is formed.

9. Solve the triangle.

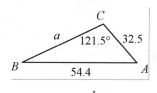

$\dfrac{c}{\sin C} = \dfrac{b}{\sin B}$

$\dfrac{54.4}{\sin 121.5°} = \dfrac{32.5}{\sin B}$

$\qquad B = \sin^{-1}\left(\dfrac{32.5\sin 121.5°}{54.4}\right)$

$\qquad B \approx 30.6°$

$A = 180° - 30.6° - 121.5° = 27.9°$

$$\frac{c}{\sin C} = \frac{a}{\sin A}$$

$$\frac{54.4}{\sin 121.5°} = \frac{a}{\sin 27.9°}$$

$$a = \frac{54.4 \sin 27.9°}{\sin 121.5°}$$

$$a \approx 29.9$$

11. Find the area of the triangle. [7.2]

$$s = \frac{1}{2}(a + b + c)$$

$$s = \frac{1}{2}(24 + 30 + 36)$$

$$s = 45$$

$$K = \sqrt{s(s-a)(s-b)(s-c)}$$

$$K = \sqrt{45(45-24)(45-30)(45-36)}$$

$$K = \sqrt{127,575}$$

$$K \approx 360 \text{ square units}$$

13. Find the area of the triangle. [7.2]

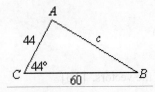

$$K = \frac{1}{2}ab\sin C$$

$$K = \frac{1}{2}(60)(44)\sin 44°$$

$$K \approx 920 \text{ square units}$$

15. Find the area of the triangle. [7.2]

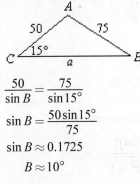

$$\frac{50}{\sin B} = \frac{75}{\sin 15°}$$

$$\sin B = \frac{50 \sin 15°}{75}$$

$$\sin B \approx 0.1725$$

$$B \approx 10°$$

$$A \approx 180° - 10° - 15° \approx 155°$$

$$K \approx \frac{1}{2}(50)(75)\sin 155°$$

$$K \approx 790 \text{ square units}$$

17. Find the area of the triangle. [7.2]

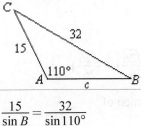

$$\frac{15}{\sin B} = \frac{32}{\sin 110°}$$

$$\sin B = \frac{15 \sin 110°}{32}$$

$$\sin B \approx 0.4405$$

$$B \approx 26°$$

$$C \approx 180° - 110° - 26° \approx 44°$$

$$K \approx \frac{1}{2}(15)(32)\sin 44°$$

$$K \approx 170 \text{ square units}$$

19. Find the components of each vector. Write an equivalent vector. [7.3]

Let $\mathbf{P_1P_2} = a_1\mathbf{i} + a_2\mathbf{j}$.

$$a_1 = 3 - (-2) = 5$$

$$a_2 = 7 - 4 = 3$$

A vector equivalent to $\mathbf{P_1P_2}$ is $\mathbf{v} = \langle 5, 3 \rangle$.

21. Find the magnitude and direction. [7.3]

$\|\mathbf{v}\| = \sqrt{(-4)^2 + 2^2}$ $\alpha \approx \tan^{-1}\left|\frac{2}{-4}\right| = \tan^{-1}\frac{1}{2}$

$\|\mathbf{v}\| = \sqrt{16 + 4}$ $\alpha \approx 26.6°$

$\|\mathbf{v}\| \approx 4.5$ $\theta \approx 180° - 26.6° \approx 153.4°$

23. Find the magnitude and direction. [7.3]

$\|\mathbf{u}\| = \sqrt{(-2)^2 + 3^2}$ $\alpha = \tan^{-1}\left|\frac{3}{-2}\right| = \tan^{-1}\frac{3}{2}$

$\|\mathbf{u}\| = \sqrt{4 + 9}$ $\alpha \approx 56.3°$

$\|\mathbf{u}\| \approx 3.6$ $\theta \approx 180° - 56.3° \approx 123.7°$

25. Find the unit vector. [7.3]

$$\|\mathbf{w}\| = \sqrt{(-8)^2 + 5^2}$$

$$\|\mathbf{w}\| = \sqrt{89}$$

$$\|\mathbf{u}\| = \left\langle \frac{-8}{\sqrt{89}}, \frac{5}{\sqrt{89}} \right\rangle = \left\langle -\frac{8\sqrt{89}}{89}, \frac{5\sqrt{89}}{89} \right\rangle$$

A unit vector in the direction of

$\|\mathbf{w}\|$ is $\|\mathbf{u}\| = \left\langle -\frac{8\sqrt{89}}{89}, \frac{5\sqrt{89}}{89} \right\rangle$.

27. Find the unit vector. [7.3]

$$\|\mathbf{v}\| = \sqrt{5^2 + 1^2}$$
$$\|\mathbf{v}\| = \sqrt{26}$$

$$\mathbf{u} = \frac{5}{\sqrt{26}}\mathbf{i} + \frac{1}{\sqrt{26}}\mathbf{j} = \frac{5\sqrt{26}}{26}\mathbf{i} + \frac{\sqrt{26}}{26}\mathbf{j}$$

A unit vector in the direction of

$\mathbf{v}$ is $\mathbf{u} = \dfrac{5\sqrt{26}}{26}\mathbf{i} + \dfrac{\sqrt{26}}{26}\mathbf{j}$.

29. Perform the indicated operation. [7.3]

$$\mathbf{v} - \mathbf{u} = \langle -4, -1 \rangle - \langle 3,\ 2 \rangle = \langle -7, -3 \rangle$$

31. Perform the indicated operation. [7.3]

$$-\mathbf{u} + \frac{1}{2}\mathbf{v} = -(10\mathbf{i} + 6\mathbf{j}) + \frac{1}{2}(8\mathbf{i} - 5\mathbf{j})$$

$$= (-10\mathbf{i} - 6\mathbf{j}) + \left(4\mathbf{i} - \frac{5}{2}\mathbf{j}\right)$$

$$= (-10 + 4)\mathbf{i} + \left(-6 - \frac{5}{2}\right)\mathbf{j}$$

$$= -6\mathbf{i} - \frac{17}{2}\mathbf{j}$$

33. Find the distance from A to C. [7.1]

From the drawing, we find A.

$$A = 360° - 84.5° - 222.1° = 53.4°$$

From the drawing, the angle formed from the north line, vertex B and point C is

$$NBC = 360° - 332.4° = 27.6°$$

We make a drawing by extending the north lines, which are parallel, at both A and B. We find the angle, 42.1°, by subtracting 180° from 222.1°.

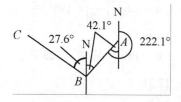

Next we see that alternate interior angles of 42.1° are formed. Therefore, we can find angle B and then C.

$$B = 27.6° + 42.1° = 69.7°$$
$$C = 180° - A - B = 180° - 53.4° - 69.7° = 56.9°$$

$$\frac{b}{\sin B} = \frac{c}{\sin C}$$
$$\frac{b}{\sin 69.7°} = \frac{345}{\sin 56.9°}$$
$$b = \frac{345 \sin 69.7°}{\sin 56.9°}$$
$$b \approx 386$$

The distance between A and C is about 386 miles.

35. $\mathbf{v} = 400\sin 204°\mathbf{i} + 400\cos 204°\mathbf{j}$ [7.3]

$$\mathbf{v} \approx -162.7\mathbf{i} - 365.4\mathbf{j}$$
$$\mathbf{w} = -45\mathbf{i}$$
$$\mathbf{R} = \mathbf{v} + \mathbf{w}$$
$$\mathbf{R} \approx -162.7\mathbf{i} - 365.4\mathbf{j} - 45\mathbf{i}$$
$$\mathbf{R} \approx -207.7\mathbf{i} - 365.4\mathbf{j}$$

$$\|\mathbf{R}\| \approx \sqrt{(-207.7)^2 + (-365.4)^2}$$
$$\|\mathbf{R}\| \approx 420 \text{ mph}$$

The ground speed is approximately 420 mph.

37. Find the dot product of the vectors. [7.3]

$$\mathbf{u} \cdot \mathbf{v} = \langle 3,\ 7 \rangle \cdot \langle -1,\ 3 \rangle$$
$$= (3)(-1) + (7)(3)$$
$$= 18$$

39. Find the dot product of the vectors. [7.3]

$$\mathbf{v} \cdot \mathbf{u} = (-4\mathbf{i} - \mathbf{j}) \cdot (2\mathbf{i} + \mathbf{j})$$
$$= (-4)(2) + (-1)(1)$$
$$= -9$$

41. Find the angle between the vectors. [7.3]

$$\cos \alpha = \frac{\langle 7, -4 \rangle \cdot \langle 2, 3 \rangle}{\sqrt{7^2 + (-4)^2}\sqrt{2^2 + 3^2}}$$

$$\cos \alpha = \frac{14 + (-12)}{\sqrt{65}\sqrt{13}}$$

$$\cos \alpha \approx 0.0688$$

$$\alpha \approx 86°$$

43. Find the angle between the vectors. [7.3]

$$\cos \alpha = \frac{(6\mathbf{i} - 11\mathbf{j}) \cdot (2\mathbf{i} + 4\mathbf{j})}{\sqrt{6^2 + (-11)^2}\sqrt{2^2 + 4^2}}$$

$$\cos \alpha = \frac{12 - 44}{\sqrt{157}\sqrt{20}}$$

$$\cos \alpha \approx -0.5711$$

$$\cos \alpha \approx 125°$$

45. Find $\text{proj}_{\mathbf{w}}\mathbf{v}$. [7.3]

$$\text{proj}_{\mathbf{w}}\mathbf{v} = \frac{\mathbf{v}\cdot\mathbf{w}}{\|\mathbf{w}\|}$$

$$\text{proj}_{\mathbf{w}}\mathbf{v} = \frac{\langle -2,\ 5\rangle\cdot\langle 5,\ 4\rangle}{\sqrt{5^2+4^2}}$$

$$= \frac{-10+20}{\sqrt{41}} = \frac{10}{\sqrt{41}}$$

$$= \frac{10\sqrt{41}}{41}$$

47. $\mathbf{w} = \|\mathbf{F}\|\ \|\mathbf{S}\|\cos\theta$ [7.3]

$\mathbf{w} = 60\cdot 14\cos 38°$

$\mathbf{w} \approx 662$ foot-pounds

49. Find the modulus and the argument, and graph. [7.4]

$$r = \sqrt{(-5)^2 + \left(\sqrt{3}\right)^2}$$

$$r = \sqrt{28} \approx 5.29$$

$$\alpha = \tan^{-1}\left|\frac{\sqrt{3}}{-5}\right| = \tan^{-1}\frac{\sqrt{3}}{5}$$

$$\alpha \approx 19°$$

$$\theta \approx 180° - 19° \approx 161°$$

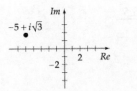

51. Write the number in trigonometric form. [7.4]

$$z = -\sqrt{3} + 3i$$

$$r = \sqrt{\left(-\sqrt{3}\right)^2 + 3^2} = \sqrt{12} = 2\sqrt{3}$$

$$\alpha = \tan^{-1}\left|\frac{3}{-\sqrt{3}}\right| = \tan^{-1}\frac{3}{\sqrt{3}}$$

$$\alpha = 60°$$

$$\theta = 180° - 60° = 120°$$

$$z = 2\sqrt{3}\ \text{cis}\ 120°$$

53. Write the number in standard form. [7.4]

$$z = 6\left(\cos\frac{4\pi}{3} + i\sin\frac{4\pi}{3}\right)$$

$$z = 6\left(-\frac{1}{2} - \frac{i\sqrt{3}}{2}\right)$$

$$z = -3 - 3i\sqrt{3}$$

55. Multiply the numbers, write in standard form. [7.4]

$$z_1 z_2 = 3\ \text{cis}\ 12°\cdot 4\ \text{cis}\ 126°$$

$$z_1 z_2 = 12\ \text{cis}(12° + 126°)$$

$$z_1 z_2 = 12\ \text{cis}\ 138°$$

$$z_1 z_2 = 12(\cos 138° + i\sin 138°)$$

$$z_1 z_2 \approx -8.918 + 8.030i$$

57. Multiply the numbers, write in standard form. [7.4]

$$z_1 z_2 = (3\ \text{cis}\ 1.8)\cdot(5\ \text{cis}\ 2.5)$$

$$z_1 z_2 = 15\ \text{cis}(1.8 + 2.5)$$

$$z_1 z_2 = 15\ \text{cis}\ 4.3$$

$$z_1 z_2 = 15(\cos 4.3 + i\sin 4.3)$$

$$z_1 z_2 \approx -6.012 - 13.742i$$

59. Divide the numbers, write in trigonometric form. [7.4]

$$\frac{z_1}{z_2} = \frac{30\ \text{cis}\ 165°}{10\ \text{cis}\ 55°}$$

$$\frac{z_1}{z_2} = 3\ \text{cis}(165° - 55°)$$

$$\frac{z_1}{z_2} = 3\ \text{cis}\ 110°$$

61. Divide the numbers, write in trigonometric form. [7.4]

$$\frac{z_1}{z_2} = \frac{\sqrt{3} - i}{1 + i}$$

$$= \frac{2\ \text{cis}\ 330°}{\sqrt{2}\ \text{cis}\ 45°}$$

$$= \sqrt{2}\ \text{cis}(330° - 45°)$$

$$= \sqrt{2}\ \text{cis}\ 285°$$

$$\sqrt{3} - i = 2\ \text{cis}\ 330° \qquad 1 + i = \sqrt{2}\ \text{cis}\ 45°$$

63. Find the indicated power, write in standard form. [7.5]

$$\left(\text{cis}\ \frac{11\pi}{6}\right)^8 = \text{cis}\ 8\cdot\frac{11\pi}{6} = \text{cis}\ \frac{44\pi}{3} = \text{cis}\ \frac{2\pi}{3}$$

$$= \cos\frac{2\pi}{3} + i\sin\frac{2\pi}{3}$$

$$= -\frac{1}{2} + \frac{\sqrt{3}}{2}i \ \text{ or } \ -\frac{1}{2} + \frac{i\sqrt{3}}{2}$$

65. Find the indicated power, write in standard form. [7.5]

$$(-2 - 2i)^{10}$$

$$= \left(2\sqrt{2}\ \text{cis}\ 225°\right)^{10} = 32,768\ \text{cis}\ 10\cdot 225°$$

$$= 32,768\ \text{cis}\ 2250°$$

$$= 32,768(\cos 2250° + i\sin 2250°)$$

$$= 0 + 32,768i$$

$$= 32,768i$$

67. Find all the roots, write in trigonometric form. [7.5]

$$8i = 8 \text{ cis } 90°$$

$$w_k = 8^{1/4} \text{ cis } \frac{90° + 360°k}{4} = \sqrt[4]{8} \text{ cis } \frac{90° + 360°k}{4}$$

$$k = 0, 1, 2, 3$$

$$w_0 = \sqrt[4]{8} \text{ cis } \frac{90°}{4}$$
$$= \sqrt[4]{8} \text{ cis } 22.5°$$

$$w_1 = \sqrt[4]{8} \text{ cis } \frac{90° + 360°}{4}$$
$$= \sqrt[4]{8} \text{ cis } 112.5°$$

$$w_2 = \sqrt[4]{8} \text{ cis } \frac{90° + 360° \cdot 2}{4}$$
$$= \sqrt[4]{8} \text{ cis } 202.5°$$

$$w_3 = \sqrt[4]{8} \text{ cis } \frac{90° + 360° \cdot 3}{4}$$
$$= \sqrt[4]{8} \text{ cis } 292.5°$$

69. Find all the roots, write in trigonometric form. [7.5]

$$\sqrt{\left(\frac{1}{2}\right)^2 + \left(\frac{\sqrt{3}}{2}\right)^2} = \sqrt{\frac{1}{4} + \frac{3}{4}} = 1$$

$$z = 1 \text{ cis } 60°$$

$$w_k = (1)^{1/5} \text{ cis } \frac{60° + 360°k}{5} = 1 \text{ cis } \frac{60° + 360°k}{5}$$

$$k = 0, 1, 2, 3, 4$$

$$w_0 = \text{cis } \frac{60°}{5}$$
$$= \text{cis } 12°$$

$$w_1 = \text{cis } \frac{60° + 360°}{5}$$
$$= \text{cis } 84°$$

$$w_2 = \text{cis } \frac{60° + 360° \cdot 2}{5}$$
$$= \text{cis } 156°$$

$$w_3 = \text{cis } \frac{60° + 360° \cdot 3}{5}$$
$$= \text{cis } 228°$$

$$w_4 = \text{cis } \frac{60° + 360° \cdot 4}{5}$$
$$= \text{cis } 300°$$

Chapter 7 Test

1. Solve the triangle. [7.1]

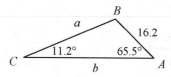

$$B = 180° - 11.2° - 65.5°$$
$$B = 103.3°$$

$$\frac{a}{\sin A} = \frac{c}{\sin C} \qquad \frac{b}{\sin B} = \frac{c}{\sin C}$$
$$\frac{a}{\sin 65.5°} = \frac{16.2}{\sin 11.2°} \qquad \frac{b}{\sin 103.3°} = \frac{16.2}{\sin 11.2°}$$
$$a = \frac{16.2 \sin 65.5°}{\sin 11.2°} \qquad b = \frac{16.2 \sin 103.3°}{\sin 11.2°}$$
$$a \approx 75.9 \qquad\qquad b \approx 81.2$$

3. Solve the triangle. [7.2]

$$b^2 = a^2 + c^2 - 2ac \cos B$$
$$b = \sqrt{36.5^2 + 42.4^2 - 2(36.5)(42.4)\cos 51.5°}$$
$$b \approx 34.7$$

$$\cos A = \frac{b^2 + c^2 - a^2}{2bc}$$
$$\cos A = \frac{34.7^2 + 42.4^2 - 36.5^2}{2(34.7)(42.4)}$$
$$A = \cos^{-1}\left(\frac{1669.6}{2942.56}\right)$$
$$A \approx 55.4°$$

$$C = 180° - 55.4° - 51.5° = 73.1°$$

5. Find the area. [7.2]

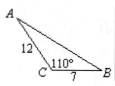

$$K = \frac{1}{2} ab \sin C$$
$$K = \frac{1}{2}(7)(12)(\sin 110°)$$
$$K \approx 39 \text{ square units}$$

7. Write an equivalent vector. [7.3]

$$a_1 = 12 \cos 220° \approx -9.2$$
$$a_2 = 12 \sin 220° \approx -7.7$$
$$\mathbf{v} = a_1 \mathbf{i} + a_2 \mathbf{j}$$
$$\mathbf{v} = -9.2 \mathbf{i} - 7.7 \mathbf{j}$$

9. Find the dot product. [7.3]

$$\mathbf{u} \cdot \mathbf{v} = (-2\mathbf{i} + 3\mathbf{j}) \cdot (5\mathbf{i} + 3\mathbf{j})$$
$$= (-2 \cdot 5) + (3 \cdot 3) = -10 + 9$$
$$= -1$$

11. Write in trigonometric form. [7.4]

$$z = -3\sqrt{2} + 3i$$
$$|z| = \sqrt{(-3\sqrt{2})^2 + 3^2} = 3\sqrt{3}$$

$$\alpha = \tan^{-1}\left|\frac{3}{-3\sqrt{2}}\right| = \tan^{-1}\frac{\sqrt{2}}{2}$$

$$\alpha \approx 35°$$
$$\theta \approx 180° - 35°$$
$$\theta \approx 145°$$

$$z \approx 3\sqrt{3} \text{ cis } 145°$$

13. Simplify, write in standard form. [7.5]

$$z = \frac{1}{2} + \frac{\sqrt{3}}{2}i$$

$$r = \sqrt{(1/2)^2 + (\sqrt{3}/2)^2} = 1$$

$$\alpha = \tan^{-1}\left|\frac{\sqrt{3}/2}{1/2}\right| = 60°$$

$$z = \cos 60° + i \sin 60°$$

$$\left(\frac{1}{2} + i\frac{\sqrt{3}}{2}\right)^3 = (\cos 60° + i \sin 60°)^3$$
$$= \cos(3 \cdot 60°) + i \sin(3 \cdot 60°)$$
$$= \cos 180° + i \sin 180°$$
$$= -1 + 0i = -1$$

15. Simplify, write in standard form. [7.5]

$$z = \sqrt{2} - i$$

$$|z| = \sqrt{\sqrt{2}^2 + (-1)^2} = \sqrt{3}$$

$$\alpha = \tan^{-1}\left|\frac{-1}{\sqrt{2}}\right| = \tan^{-1}\frac{\sqrt{2}}{2}$$

$$\alpha \approx 35.2644°$$
$$\theta \approx 360° - 35.2644°$$
$$\theta \approx 324.7356°$$

$$z \approx \sqrt{3} \text{ cis } 324.7356°$$
$$z^5 \approx (\sqrt{3})^5 \text{ cis } (5 \cdot 324.7356°)$$
$$z^5 \approx 9\sqrt{3}(\cos 1623.678° + i \sin 1623.678°)$$
$$z^5 \approx -15.556 - 1.000i$$

17. Find the distance. [7.3]

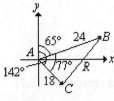

$$A = 142° - 65° = 77°$$
$$R^2 = 24^2 + 18^2 - 2(24)(18)\cos 77°$$
$$R \approx 27 \text{ miles}$$

19. Find the roots, write in trigonometric form. [7.5]

$$\frac{\sqrt{2}}{2} + \frac{\sqrt{2}}{2}i = 1(\cos 45° + i \sin 45°)$$

$$w_k = \cos\frac{45° + 360°k}{5} + i \sin\frac{45° + 360°k}{5}$$
$$k = 0, 1, 2, 3, 4$$

$$w_0 = \cos\frac{45°}{5} + i \sin\frac{45°}{5}$$
$$= (\cos 9° + i \sin 9°)$$
$$= \text{cis } 9°$$

$$w_1 = \cos\frac{45° + 360°}{5} + i \sin\frac{45° + 360°}{5}$$
$$= (\cos 81° + i \sin 81°)$$
$$= \text{cis } 81°$$

$$w_2 = \cos\frac{45° + 360° \cdot 2}{5} + i \sin\frac{45° + 360° \cdot 2}{5}$$
$$= \cos 153° + i \sin 153°$$
$$= \text{cis } 153°$$

$$w_3 = \cos\frac{45° + 360° \cdot 3}{5} + i \sin\frac{45° + 360° \cdot 3}{5}$$
$$= \cos 225° + i \sin 225°$$
$$= \text{cis } 225°$$

$$w_4 = \cos\frac{45° + 360° \cdot 4}{5} + i \sin\frac{45° + 360° \cdot 4}{5}$$
$$= \cos 297° + i \sin 297°$$
$$= \text{cis } 297°$$

Cumulative Review Exercises

1. Find $(f \circ g)(x)$. [2.6]

$$(f \circ g)(x) = f[g(x)] = f[x^2 + 1] = \cos(x^2 + 1)$$

3. Convert to degrees. [5.1]

$$\frac{3\pi}{2}\left(\frac{180°}{\pi}\right) = 270°$$

5. Find c. [5.2]

$$\cos 26.0° = \frac{15.0}{c}$$

$$c = \frac{15.0}{\cos 26.0°} \approx 16.7 \text{ cm}$$

7. Find the amplitude, period and phase shift. [5.7]

$$y = 4\cos\left(2x - \frac{\pi}{2}\right)$$

$$0 \le 2x - \frac{\pi}{2} \le 2\pi$$

$$\frac{\pi}{2} \le 2x \le \frac{5\pi}{2}$$

$$\frac{\pi}{4} \le x \le \frac{5\pi}{4}$$

amplitude = 4, period = π, phase shift = $\frac{\pi}{4}$

9. Determine whether the function is odd, even, or neither. [5.4]

$y = \sin x$ is an odd function.

11. Evaluate. [6.5]

$$\tan\left(\sin^{-1}\left(\frac{12}{13}\right)\right) = \tan(67.38°) = \frac{12}{5}$$

13. Solve. [6.6]

$$2\cos^2 x + \sin x - 1 = 0$$

$$2(1 - \sin^2 x) + \sin x - 1 = 0$$

$$2\sin^2 - \sin x + 1 = 0$$

$$(2\sin x + 1)(\sin x - 1) = 0$$

$2\sin x + 1 = 0$	$\sin x - 1 = 0$
$\sin x = -\dfrac{1}{2}$	$\sin x = 1$
$x = \dfrac{7\pi}{6}, \dfrac{11\pi}{6}$	$x = \dfrac{\pi}{2}$

The solutions are $\frac{\pi}{2}, \frac{7\pi}{6}, \frac{11\pi}{6}$.

15. Find the magnitude and direction angle. [7.3].

$$\|\mathbf{v}\| = \sqrt{(-3)^2 + 4^2}$$
$$\|\mathbf{v}\| = \sqrt{9 + 16}$$
$$\|\mathbf{v}\| = 5$$

$$\alpha = \tan^{-1}\left|\frac{4}{-3}\right| = \tan^{-1}\frac{4}{3}$$
$$\alpha \approx 53.1°$$
$$\theta = 180° - \alpha$$
$$\theta \approx 180° - 53.1°$$
$$\theta \approx 126.9°$$

magnitude: 5, angle: 126.9°

17. Find the ground speed and course. [7.2]

$$\mathbf{AB} = 415(\cos 42\mathbf{i} + \sin 42\mathbf{j}) \approx 308.4\mathbf{i} + 277.6\mathbf{j}$$
$$\mathbf{AD} = 55[\cos(-25°)\mathbf{i} + \sin(-25°)\mathbf{j}] \approx 49.8\mathbf{i} - 23.2\mathbf{j}$$
$$\mathbf{AC} = \mathbf{AB} + \mathbf{AD}$$
$$\mathbf{AC} = 308.4\mathbf{i} + 277.6\mathbf{j} + 49.8\mathbf{i} - 23.2\mathbf{j}$$
$$\mathbf{AC} \approx 358.2\mathbf{i} + 254.4\mathbf{j}$$
$$\|\mathbf{AC}\| = \sqrt{358.2^2 + (254.4)^2}$$
$$\|\mathbf{AC}\| \approx 439 \text{ mph}$$

$$\alpha = 90° - \theta = 90° - \tan^{-1}\left(\frac{254.4}{358.2}\right) \approx 54.6°$$

19. Simplify. [7.5]

$$z = 1 - i$$

$$r = \sqrt{1^2 + (-1)^2} \qquad \alpha = \tan^{-1}\left|\frac{-1}{1}\right| = 45°$$
$$r = \sqrt{2} \qquad\qquad \theta = 315°$$

$$z = \sqrt{2}(\cos 315° + i\sin 315°)$$
$$(1 - i)^8 = [\sqrt{2}(\cos 315° + i\sin 315°)]^8$$
$$= (\sqrt{2})^8[\cos(8 \cdot 315°) + i\sin(8 \cdot 315°)]$$
$$= 16(\cos 2520° + i\sin 2520°)$$
$$= 16(\cos 0° + i\sin 0°)$$
$$= 16 - 0i = 16$$

Chapter 8 Topics in Analytic Geometry

Section 8.1 Exercises

1. Find the focus of $(x-h)^2 = 4p(y-k)$.

$(h,\ k+p)$

3. Explain when $x^2 = 4py$ opens up or down.

If $p > 0$, the parabola opens up. If $p < 0$, the parabola opens down.

5. a. iii

   b. i

   c. iv

   d. ii

7. Find the vertex, focus, and directrix; sketch the graph.

$x^2 = -4y$

$4p = -4$

$p = -1$

vertex $= (0,\ 0)$

focus $= (0,\ -1)$

directrix: $y = 1$

9. Find the vertex, focus, and directrix; sketch the graph.

$y^2 = \dfrac{1}{3}x$

$4p = \dfrac{1}{3}$

$p = \dfrac{1}{12}$

vertex $= (0,\ 0)$

focus $= \left(\dfrac{1}{12},\ 0\right)$

directrix: $x = -\dfrac{1}{12}$

11. Find the vertex, focus, and directrix; sketch the graph.

$(x-2)^2 = 8(y+3)$

vertex $= (2,-3)$

$4p = 8,\quad p = 2$

$(h, k+p) = (2,-3+2) = (2,-1)$

focus $= (2,-1)$

$k - p = -3 - 2 = -5$

directrix: $y = -5$

13. Find the vertex, focus, and directrix; sketch the graph.

$(y+4)^2 = -4(x-2)$

vertex $= (2,\ -4)$

$4p = -4\quad p = -1$

$(h+p,\ k) = (2-1,\ -4) = (1,\ -4)$

focus $= (1,\ -4)$

$h - p = 2 + 1 = 3$

directrix: $x = 3$

15. Find the vertex, focus, and directrix; sketch the graph.

$(y-1)^2 = 2x+8$

vertex $= (-4,\ 1)$

$4p = 2,\quad p = \dfrac{1}{2}$

$(h+p,\ k) = \left(-4+\dfrac{1}{2},\ 1\right) = \left(-\dfrac{7}{2},\ 1\right)$

focus $= \left(-\dfrac{7}{2},\ 1\right)$

$h - p = -4 - \dfrac{1}{2} = -\dfrac{9}{2}$

directrix: $x = -\dfrac{9}{2}$

17. Find the vertex, focus, and directrix; sketch the graph.

$$(2x-4)^2 = 8y-16 \implies (x-2)^2 = 2(y-2)$$

vertex $= (2, 2)$

$$4p = 2, \quad p = \frac{1}{2}$$

$$(h, k+p) = \left(2, 2+\frac{1}{2}\right) = \left(2, \frac{5}{2}\right)$$

focus $= \left(2, \frac{5}{2}\right)$

$$k - p = 2 - \frac{1}{2} = \frac{3}{2}$$

directrix: $y = \frac{3}{2}$

19. Find the vertex, focus, and directrix; sketch the graph.

$$x^2 + 8x - y + 6 = 0$$

$$x^2 + 8x = y - 6$$

$$x^2 + 8x + 16 = y - 6 + 16$$

$$(x+4)^2 = y + 10$$

vertex $= (-4, -10)$

$$4p = 1, \quad p = \frac{1}{4}$$

focus $= \left(-4, -\frac{39}{4}\right)$

directrix: $y = -\frac{41}{4}$

21. Find the vertex, focus, and directrix; sketch the graph.

$$x + y^2 - 3y + 4 = 0$$

$$y^2 - 3y = -x - 4$$

$$y^2 - 3y + \frac{9}{4} = -x - 4 + \frac{9}{4}$$

$$\left(y - \frac{3}{2}\right)^2 = -\left(x + \frac{7}{4}\right)$$

vertex $= \left(-\frac{7}{4}, \frac{3}{2}\right)$

$$4p = -1, \quad p = -\frac{1}{4}$$

focus $= \left(-2, \frac{3}{2}\right)$

directrix: $x = -\frac{3}{2}$

23. Find the vertex, focus, and directrix; sketch the graph.

$$2x - y^2 - 6y + 1 = 0$$

$$-y^2 - 6y = -2x - 1$$

$$y^2 + 6y = 2x + 1$$

$$y^2 + 6y + 9 = 2x + 1 + 9$$

$$(y+3)^2 = 2(x+5)$$

vertex $= (-5, -3)$

$$4p = 2, \quad p = \frac{1}{2}$$

focus $= \left(-\frac{9}{2}, -3\right)$

directrix: $x = -\frac{11}{2}$

25. Find the vertex, focus, and directrix; sketch the graph.

$$x^2 + 3x + 3y - 1 = 0$$

$$x^2 + 3x = -3y + 1$$

$$x^2 + 3x + \frac{9}{4} = -3y + 1 + \frac{9}{4}$$

$$\left(x + \frac{3}{2}\right)^2 = -3\left(y - \frac{13}{12}\right)$$

vertex $\left(-\frac{3}{2}, \frac{13}{12}\right)$

$$4p = -3, \quad p = -\frac{3}{4}$$

focus $= \left(-\frac{3}{2}, \frac{1}{3}\right)$

directrix: $y = \frac{11}{6}$

27. Find the vertex, focus, and directrix; sketch the graph.

$$2x^2 - 8x - 4y + 3 = 0$$
$$2(x^2 - 4x) = 4y - 3$$
$$2(x^2 - 4x + 4) = 4y - 3 + 8$$
$$2(x - 2)^2 = 4y + 5$$
$$(x - 2)^2 = 2y + \frac{5}{2}$$
$$(x - 2)^2 = 2\left(y + \frac{5}{4}\right)$$

vertex $= \left(2, \ -\frac{5}{4}\right)$

$4p = 2, \ p = \frac{1}{2}$

focus $= \left(2, \ -\frac{3}{4}\right)$

directrix $y = -\frac{7}{4}$

29. Find the vertex, focus, and directrix; sketch the graph.

$$2x + 4y^2 + 8y - 5 = 0$$
$$4y^2 + 8y = -2x + 5$$
$$4(y^2 + 2y) = -2x + 5$$
$$4(y^2 + 2y + 1) = -2x + 5 + 4$$
$$4(y + 1)^2 = -2x + 9$$
$$(y + 1)^2 = -\frac{1}{2}x + \frac{9}{4}$$
$$(y + 1)^2 = -\frac{1}{2}\left(x - \frac{9}{2}\right)$$

vertex $= \left(\frac{9}{2}, \ -1\right)$

$4p = -\frac{1}{2}, \ p = -\frac{1}{8}$

focus $= \left(\frac{35}{8}, \ -1\right)$

directrix $x = \frac{37}{8}$

31. Find the vertex, focus, and directrix; sketch the graph.

$$3x^2 - 6x - 9y + 4 = 0$$
$$x^2 - 2x = 3y - \frac{4}{3}$$
$$x^2 - 2x + 1 = 3y - \frac{1}{3}$$
$$(x - 1)^2 = 3\left(y - \frac{1}{9}\right)$$

vertex $= \left(1, \ \frac{1}{9}\right)$

$4p = 3, \ p = \frac{3}{4}$

focus $= \left(1, \ \frac{31}{36}\right)$

directrix $y = -\frac{23}{36}$

33. Find the equation of the parabola in standard form.

vertex $(0, 0), \ $ focus $(0, 2)$

$$x^2 = 4py$$
$$p = -4 \text{ since focus is } (0, p)$$
$$x^2 = 4(2)y$$
$$x^2 = 8y$$

35. Find the equation of the parabola in standard form.

vertex $(1, 3), \ $ focus $(1, 0)$

$$(x - h)^2 = 4p(y - k)$$
$$h = 1, \ k = 3.$$

The distance p from the vertex to the focus is -3.

$$(x - 1)^2 = 4(-3)(y - 3)$$
$$(x - 1)^2 = -12(y - 3)$$

37. Find the equation of the parabola in standard form.

focus $(2, 5), \ $ directrix $y = 3$

The vertex is the midpoint of the line segment joining $(2, 5)$ and the point $(2, 3)$ on the directrix.

$$(h, \ k) = \left(\frac{2 + 2}{2}, \ \frac{5 + 3}{2}\right) = (2, \ 4)$$

The distance p from the vertex to the focus is 1.

$$4p = 4(1) = 4$$
$$(x-h)^2 = 4p(y-k)$$
$$(x-2)^2 = 4(y-4)$$

39. Find the equation of the parabola in standard form.

vertex = $(-4, 1)$, point: $(-2, 2)$ on the parabola.

Axis of symmetry $x = -4$.

If $P_1 = (-2, 2)$, $(x+4)^2 = 4p(y-1)$. Since $(-2, 2)$

is on the curve, we get

$$(-2+4)^2 = 4p(2-1)$$
$$4 = 4p \Rightarrow p = 1$$

Thus the equation in standard form is $(x+4)^2 = 4(y-1)$.

41. The equation of the mirrored parabolic trough is

$$x^2 = 4py \quad -5 \le x \le 5$$

Because $(5, 1.5)$ is a point on the parabola,

$(5, 1.5)$ must be a solution of the equation. Thus

$$5^2 = 4p(1.5)$$
$$25 = 6p$$
$$4\frac{1}{6} = p$$

The focus is $4\frac{1}{6}$ ft , or 4 ft 2 inches above the vertex.

43. Place the satellite dish on an xy-coordinate system with

its vertex at $(0, -1)$ as shown.

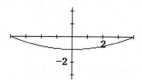

The equation of the parabola is

$$x^2 = 4p(y+1) \quad -1 \le y \le 0$$

Because $(4, 0)$ is a point on this graph, $(4, 0)$ must be a

solution of the equation of the parabola. Thus,

$$16 = 4p(0+1)$$
$$16 = 4p$$
$$4 = p$$

Because p is the distance from the vertex to the focus,

the focus is on the x-axis 4 feet above the vertex.

45. The focus of the parabola is $(p, 0)$ where $y^2 = 4px$.

Half of 18.75 inches is 9.375 inches.

Therefore, the point $(3.66, 9.375)$ is on the parabola.

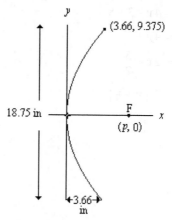

$$(9.375)^2 = 4p(3.66)$$
$$87.890625 = 14.64p$$
$$\frac{87.890625}{14.64} = p$$
$$p \approx 6.0 \text{ inches}$$

47. $S = \frac{\pi r}{6d^2}\left[\left(r^2 + 4d^2\right)^{3/2} - r^3\right]$

a. $r = 40.5$ feet

$d = 16$ feet

$$S = \frac{\pi(40.5)}{6(16)^2}\left[\left([40.5]^2 + 4[16]^2\right)^{3/2} - (40.5)^3\right]$$
$$= \frac{40.5\pi}{1536}\left[(2664.25)^{3/2} - 66430.125\right]$$
$$= \frac{40.5\pi}{1536}\left[137518.9228 - 66430.125\right]$$
$$= \frac{40.5\pi}{1536}\left[71088.79775\right]$$
$$\approx 5900 \text{ square feet}$$

b. $r = 125$ feet

$d = 52$ feet

$$S = \frac{\pi(125)}{6(52)^2}\left[\left([125]^2 + 4[52]^2\right)^{3/2} - (125)^3\right]$$
$$= \frac{125\pi}{16224}\left[(26441)^{3/2} - 1953125\right]$$
$$= \frac{125\pi}{16224}\left[4299488.724 - 1953125\right]$$
$$= \frac{125\pi}{16224}\left[2346363.724\right]$$
$$\approx 56,800 \text{ square feet}$$

49. The equation of the mirror is given by

$$x^2 = 4py \quad -60 \le x \le 60$$

Because p is the distance from the vertex to the focus and the coordinates of the focus are $(0, 600)$, $p = 600$. Therefore,

$$x^2 = 4(600)y$$
$$x^2 = 2400y$$

To determine a, substitute $(60, a)$ into the equation $x^2 = 2400y$ and solve for a.

$$x^2 = 2400y$$
$$60^2 = 2400a$$
$$3600 = 2400a$$
$$1.5 = a$$

The concave depth of the mirror is 1.5 inches.

51. a. Find the height.

$$h(6) = -\frac{1}{9}(6)^2 + \frac{8}{3}(6) = 12 \text{ ft}$$

The height is 12 ft.

b. Find the height.

$$h(15) = -\frac{1}{9}(15)^2 + \frac{8}{3}(15) = 15 \text{ ft}$$

The height is 15 ft.

c. Find the maximum height. Find the vertex.

$$\frac{-b}{2a} = \frac{-\frac{8}{3}}{2\left(-\frac{1}{9}\right)} = \frac{8}{3}\left(\frac{9}{2}\right) = 12$$

$$h(12) = -\frac{1}{9}(12)^2 + \frac{8}{3}(12) = 16 \text{ ft}$$

The maximum height is 16 ft.

53. $x^2 = 10y$

$$4p = 10, \quad p = \frac{5}{2}$$

$$\text{focus} = \left(\frac{5}{2}, 0\right)$$

Substituting the vertical coordinate of the focus for y to obtain x-coordinates of endpoints (x_1, y_1), (x_2, y_2), we have

$$x^2 = 10\left(\frac{5}{2}\right), \quad \text{or } x^2 = 25$$
$$x = \pm\sqrt{25}$$
$$x_1 = -5 \quad x_2 = 5$$

Length of latus rectum $= |x_2 - x_1| = 5 - (-5) = 10$.

55. $(x - h)^2 = 4p(y - k)$

$$\text{focus} = (h, \ k + p)$$

Substituting the vertical coordinate of the focus for y to obtain x-coordinates of endpoints (x_1, y_1), (x_2, y_2), we have

$$(x - h)^2 = 4p(k + p - k)$$
$$(x - h)^2 = 4p^2$$
$$x - h = \pm 2p$$

$$x_1 = h - 2p \quad x_2 = h + 2p$$

Solving for $|x_2 - x_1|$, we obtain

$$\Delta x = |x_2 - x_1| = |h + 2p - h + 2p| = 4|p|$$
$$\text{or } (y - k)^2 = 4p(x - h)$$

$$\text{focus} = (h + p, \ k)$$

Substituting the horizontal coordinate of the focus for x to obtain the y-coordinates of the endpoints (x_1, y_1), (x_2, y_2), we have

$$(y - k)^2 = 4p(h + p - h)$$
$$(y - k)^2 = 4p^2$$
$$y - k = \pm 2p$$

Solving for $|y_2 - y_1|$, we obtain

$$\Delta y = |y_2 - y_1| = |k + 2p - k + 2p| = 4|p|$$

Thus, the length of the latus rectum is $4|p|$.

57. Graph $(y + 4)^2 = -(x - 1)$.

$$4p = -1, \quad p = -\frac{1}{4}$$

focus $\left(\frac{3}{4}, -4\right)$

one point: $\left(\frac{3}{4}, k+2p\right) = \left(\frac{3}{4}, -\frac{9}{2}\right)$

one point: $\left(\frac{3}{4}, k-2p\right) = \left(\frac{3}{4}, -\frac{7}{2}\right)$

59. Graph $y = \frac{7}{4} + \frac{1}{4}x|x|$.

61. By definition, any point on the curve (x, y) will be equidistant from both the focus $(1, 1)$ and the directrix, $(y_2 = -x_2 - 2)$.

If we let d_1 equal the distance from the focus to the point (x, y), we get $d_1 = \sqrt{(x-1)^2 + (y-1)^2}$

To determine the distance d_2 from the point (x, y) to the line $y = -x - 2$, draw a line segment from (x, y) to the directrix so as to meet the directrix at a $90°$ angle. Now drop a line segment parallel to the y-axis from (x, y) to the directrix. This segment will meet the directrix at a $45°$ angle, thus forming a right isosceles triangle with the directrix and the line segment perpendicular to the directrix from (x, y). The length of this segment, which is the hypotenuse of the triangle, is the difference between y and the y-value of the directrix at x, or $-x - 2$. Thus, the hypotenuse has a length of $y + x + 2$, and since the right triangle is also isosceles, each leg has a length of $\frac{y+x+2}{\sqrt{2}}$.

But since d_2 is the length of the leg drawn from (x, y) to the directrix, $d_2 = \frac{y+x+2}{\sqrt{2}}$.

Thus, $d_1 = \sqrt{(x-1)^2 + (y-1)^2}$ and $d_2 = \frac{x+y+2}{\sqrt{2}}$.

By definition, $d_1 = d_2$. So by substitution,

$$\sqrt{(x-1)^2 + (y-1)^2} = \frac{x+y+2}{\sqrt{2}}$$

$$\sqrt{2}\sqrt{(x-1)^2 + (y-1)^2} = x+y+2$$

$$2\left[(x-1)^2 + (y-1)^2\right] = x^2 + y^2 + 4x + 4y + 2xy + 4$$

$$2\left[x^2 - 2x + 1 + y^2 - 2y + 1\right] = x^2 + y^2 + 4x + 4y + 2xy + 4$$

$$2x^2 - 4x + 2y^2 - 4y + 4 = x^2 + y^2 + 4x + 4y + 2xy + 4$$

$$x^2 + y^2 - 8x - 8y - 2xy = 0$$

63. a. Find the equation. Find the focal length. Round to the nearest tenth.

$$x^2 = 4p_1 y$$

$$(11.0)^2 = 4p_1(3.9)$$

$$p_1 \approx 7.8$$

The equation is $x^2 = 4(7.8)y$.

The focal length is approximately 7.8 cm.

b. Find the equation. Find the focal length. Round to the nearest tenth.

$$x^2 = 4p_2(y - 7.8)$$

$$(11.0)^2 = 4p_2(3.9 - 7.8)$$

$$p_2 \approx -7.8$$

The equation is $x^2 = 4(-7.8)(y - 7.8)$.

The focal length is approximately 7.8 cm.

c. The approximate vertical distance between a given point on the raspberry and the corresponding point in its image is about 7.8 cm.

This distance and the focal length of the mirrors are the same.

Prepare for Section 8.2

P1. Find the midpoint and length.

midpoint: $\left(\frac{x_1 + x_2}{2}, \frac{y_1 + y_2}{2}\right) = \left(\frac{5 + (-1)}{2}, \frac{1 + 5}{2}\right)$

The midpoint is $(2, 3)$.

length: $\sqrt{(x_2 - x_1)^2 + (y_2 - y_1)^2}$

$$\sqrt{(-1 - 5)^2 + (5 - 1)^2} = \sqrt{36 + 16} = \sqrt{52}$$

The length is $2\sqrt{13}$.

P3. Solve $x^2 - 2x = 2$.

$$x^2 - 2x = 2$$
$$x^2 - 2x + 1 = 2 + 1$$
$$(x-1)^2 = 3$$
$$x - 1 = \pm\sqrt{3}$$
$$x = 1 \pm \sqrt{3}$$

P5. Solve $(x-2)^2 + y^2 = 4$ for y.

$$(x-2)^2 + y^2 = 4$$
$$y^2 = 4 - (x-2)^2$$
$$y = \pm\sqrt{4 - (x-2)^2}$$

Section 8.2 Exercises

1. Find the vertices of the ellipse.

$(h,\ k+a)$ and $(h,\ k-a)$

3. Find the eccentricity of the ellipse.

$$e = \frac{c}{a} = \frac{\sqrt{a^2 - b^2}}{a}$$

5. a. iv

 b. i

 c. ii

 d. iii

7. Find the center, vertices, and foci; sketch the graph.

$$\frac{x^2}{16} + \frac{y^2}{25} = 1$$

$$a^2 = 25 \rightarrow a = 5$$
$$b^2 = 16 \rightarrow b = 4$$
$$c = \sqrt{a^2 - b^2} = \sqrt{25 - 16} = \sqrt{9} = 3$$

Center $(0, 0)$

Vertices $(0, \pm 5)$

Foci $(0, \pm 3)$

9. Find the center, vertices, and foci; sketch the graph.

$$\frac{x^2}{9} + \frac{y^2}{4} = 1$$

$$a^2 = 9 \rightarrow a = 3$$
$$b^2 = 4 \rightarrow b = 2$$
$$c = \sqrt{a^2 - b^2} = \sqrt{9 - 4} = \sqrt{5}$$

Center $(0, 0)$

Vertices $(\pm 3, 0)$

Foci $(\pm\sqrt{5},\ 0)$

11. Find the center, vertices, and foci; sketch the graph.

$$\frac{x^2}{9} + \frac{y^2}{7} = 1$$

$$a^2 = 9 \rightarrow a = 3$$
$$b^2 = 7 \rightarrow b = \sqrt{7}$$
$$c = \sqrt{a^2 - b^2} = \sqrt{9 - 7} = \sqrt{2}$$

Center $(0, 0)$

Vertices $(\pm 3,\ 0)$

Foci $(\pm\sqrt{2},\ 0)$

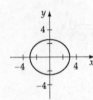

13. Find the center, vertices, and foci; sketch the graph.

$$\frac{(x-3)^2}{25} + \frac{(y+2)^2}{16} = 1$$

Center $(3,\ -2)$

Vertices $(3 \pm 5,\ -2) = (8,\ -2),\ (-2,\ -2)$

Foci $(3 \pm 3,\ -2) = (6,\ -2),\ (0,\ -2)$

15. Find the center, vertices, and foci; sketch the graph.

$$\frac{(x+2)^2}{9}+\frac{y^2}{25}=1$$

Center $(-2,\,0)$

Vertices $(-2,\,5),\ (-2,-5)$

Foci $(-2,\,4),\ (-2,-4)$

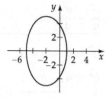

17. Find the center, vertices, and foci; sketch the graph.

$$\frac{(x-1)^2}{21}+\frac{(y-3)^2}{4}=1$$

Center $(1,\,3)$

Vertices $\left(1\pm\sqrt{21},\,3\right)$

Foci $\left(1\pm\sqrt{17},\,3\right)$

19. Find the center, vertices, and foci; sketch the graph.

$$\frac{9(x-1)^2}{16}+\frac{(y+1)^2}{9}=1$$

Center $(1,\,-1)$

Vertices $(1,\ -1\pm3)=(1,\,2),\ (1,\,-4)$

Foci $\left(1,\ -1\pm\dfrac{\sqrt{65}}{3}\right)$

21. Find the center, vertices, and foci; sketch the graph.

$$3x^2+4y^2=12$$

$$\frac{x^2}{4}+\frac{y^2}{3}=1$$

Center $(0,\,0)$

Vertices $(\pm2,\,0)$

Foci $(\pm1,\,0)$

23. Find the center, vertices, and foci; sketch the graph.

$$25x^2+16y^2=400$$

$$\frac{x^2}{16}+\frac{y^2}{25}=1$$

Center $(0,0)$

Vertices $(0,\,\pm5)$

Foci $(0,\,\pm3)$

25. Find the center, vertices, and foci; sketch the graph.

$$64x^2+25y^2=400$$

$$\frac{x^2}{25/4}+\frac{y^2}{16}=1$$

Center $(0,0)$

Vertices $(0,\,\pm4)$

Foci $\left(0,\ \pm\dfrac{\sqrt{39}}{2}\right)$

27. Find the center, vertices, and foci; sketch the graph.

$$4x^2+y^2-24x-8y+48=0$$
$$4\left(x^2-6x\right)+\left(y^2-8y\right)=-48$$
$$4\left(x^2-6x+9\right)+\left(y^2-8y+16\right)=-48+36+16$$
$$4(x-3)^3+(y-4)^2=4$$
$$\frac{(x-3)^2}{1}+\frac{(y-4)^2}{4}=1$$

Center $(3,\,4)$

Vertices $(3,\,4\pm2)=(3,\,6),\ (3,\,2)$

Foci $\left(3,\,4\pm\sqrt{3}\right)$

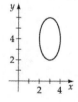

29. Find the center, vertices, and foci; sketch the graph.

$$5x^2 + 9y^2 - 20x + 54y + 56 = 0$$

$$5(x^2 - 4x) + 9(y^2 + 6y) = -56$$

$$5(x^2 - 4x + 4) + 9(y^2 + 6y + 9) = -56 + 20 + 81$$

$$5(x - 2)^2 + 9(y + 3)^2 = 45$$

$$\frac{(x-2)^2}{9} + \frac{(y+3)^2}{5} = 1$$

Center $(2, -3)$

Vertices $(2 \pm 3, -3) = (-1, -3), (5, -3)$

Foci $(2 \pm 2, 3) = (0, -3), (4, -3)$

31. Find the center, vertices, and foci; sketch the graph.

$$16x^2 + 9y^2 - 64x - 80 = 0$$

$$16(x^2 - 4x) + 9y^2 = 80$$

$$16(x^2 - 4x + 4) + 9y^2 = 80 + 64$$

$$16(x - 2)^2 + 9y^2 = 144$$

$$\frac{(x-2)^2}{9} + \frac{y^2}{16} = 1$$

Center $(2, 0)$

Vertices $(2, \pm 4) = (2, 4), (2, -4)$

Foci $(2 \pm \sqrt{7})$

33. Find the center, vertices, and foci; sketch the graph.

$$25x^2 + 16y^2 + 50x - 32y - 359 = 0$$

$$25(x^2 + 2x) + 16(y^2 - 2y) = 359$$

$$25(x^2 + 2x + 1) + 16(y^2 - 2y + 1) = 359 + 25 + 16$$

$$25(x + 1)^2 + 16(y - 1)^2 = 400$$

$$\frac{(x+1)^2}{16} + \frac{(y-1)^2}{25} = 1$$

Center $(-1, 1)$

Vertices $(-1, 1 \pm 5) = (-1, 6), (-1, -4)$

Foci $(-1, 1 \pm 3) = (-1, 4), (-1, -2)$

35. Find the equation of the ellipse in standard form.

$$2a = 10, \quad a = 5, \quad a^2 = 25$$

$$c = 4$$

$$c^2 = a^2 - b^2$$

$$16 = 25 - b^2$$

$$b^2 = 9$$

$$\frac{x^2}{25} + \frac{y^2}{9} = 1$$

37. Find the equation of the ellipse in standard form.

$$a = 6, \quad a^2 = 36$$

$$b = 4, \quad b^2 = 16$$

$$\frac{x^2}{36} + \frac{y^2}{16} = 1$$

39. Find the equation of the ellipse in standard form.

$$2a = 8$$

$$a = 4$$

$$a^2 = 16$$

$$\frac{x^2}{16} + \frac{y^2}{b^2} = 1$$

$$\frac{(2)^2}{16} + \frac{(\sqrt{3})^2}{b^2} = 1$$

$$\frac{4}{16} + \frac{3}{b^2} = 1$$

$$\frac{3}{b^2} = \frac{3}{4}$$

$$b^2 = 4$$

$$\frac{x^2}{16} + \frac{y^2}{4} = 1$$

41. Find the equation of the ellipse in standard form.

$$c = 3$$

$$2a = 8$$

$$a = 4$$

$$a^2 = 16$$

$$c^2 = a^2 - b^2$$
$$9 = 16 - b^2$$
$$b^2 = 7$$
$$\frac{(x+2)^2}{16} + \frac{(y-4)^2}{7} = 1$$

43. Find the equation of the ellipse in standard form.

$$2a = 12$$
$$a = 6$$
$$a^2 = 36$$

Since the center of the ellipse is (5, 1) and the point (7, –2) is on the ellipse, we have

$$\frac{(x-5)^2}{b^2} + \frac{(y-1)^2}{a^2} = 1$$
$$\frac{(7-5)^2}{b^2} + \frac{(-2-1)^2}{36} = 1$$
$$\frac{4}{b^2} = 1 - \frac{9}{36}$$
$$\frac{4}{b^2} = \frac{3}{4}$$
$$b^2 = \frac{16}{3}$$
$$\frac{(x-5)^2}{16/3} + \frac{(y-1)^2}{36} = 1$$

45. Find the equation of the ellipse in standard form.

center (3, 3)

$$c = 1$$
$$2a = 8$$
$$a = 4$$
$$a^2 = 16$$
$$c^2 = a^2 - b^2$$
$$1 = 16 - b^2$$
$$b^2 = 15$$
$$\frac{(x-3)^2}{15} + \frac{(y-3)^2}{16} = 1$$

47. Find the eccentricity of the ellipse.

$$a^2 = 25,\ b^2 = 4$$
$$c^2 = a^2 - b^2$$
$$c^2 = 25 - 4$$
$$c = \sqrt{21}$$

$$e = \frac{c}{a} = \frac{\sqrt{21}}{5} \approx 0.92$$

49. Find the eccentricity of the ellipse.

$$a^2 = 85^2,\ b^2 = 60^2$$
$$c^2 = a^2 - b^2$$
$$c^2 = 7225 - 3600$$
$$c = \sqrt{3625} = 5\sqrt{145}$$
$$e = \frac{c}{a} = \frac{5\sqrt{145}}{85} = \frac{\sqrt{145}}{17} \approx 0.71$$

51. Find the equation of the ellipse in standard form.

center (1, 3)

$$c = 2$$
$$\frac{c}{a} = \frac{2}{5}$$
$$\frac{2}{a} = \frac{2}{5}$$
$$a = 5$$
$$c^2 = a^2 - b^2$$
$$4 = 25 - b^2$$
$$b^2 = 21$$
$$\frac{(x-1)^2}{25} + \frac{(y-3)^2}{21} = 1$$

53. Find the area.

$$\frac{x^2}{9} + \frac{y^2}{4} = 1$$
$$a^2 = 9,\ a = 3$$
$$b^2 = 4,\ b = 2$$
$$A = \pi ab = \pi(3)(2) = 6\pi \text{ square units}$$

55. a. Find the area of the Ellipse. Round to the nearest ten thousand square feet.

$$2a = 1058,\ a = 529$$
$$2b = 903,\ b = 451.5$$
$$A = \pi ab = \pi(529)(451.5) \approx 750{,}000 \text{ ft}^2$$

b. Find the number of times larger than a football field.

$$A = lw = (360)(160) = 57{,}600 \text{ ft}^2$$
$$\frac{750{,}000 \text{ ft}^2}{57{,}600 \text{ ft}^2} \approx 13.0$$

The area of the Ellipse is 13.0 times larger than the area of a football field.

57. Estimate the perimeter of the ellipse. Round to the nearest tenth of a unit.

$$\frac{x^2}{49} + \frac{y^2}{9} = 1$$

$a^2 = 49, \; b^2 = 9$

$$P = \pi\sqrt{2(a^2 + b^2)}$$
$$= \pi\sqrt{2(49 + 9)}$$
$$= \pi\sqrt{116}$$
$$\approx 33.8 \text{ units}$$

59. $484 = 64 + c^2$

$\quad\; c^2 = 420$

$\quad\;\; c = 20.494$

$\quad 2c = 40.9878 \approx 41$

The emitter should be placed 41 cm away.

61. a. Find the lengths of the major and minor axes of the roof of the turret. Round to the nearest tenth of a foot.

The minor axis is 60.0 ft.

$$a = \sqrt{60^2 + 60^2} \approx 84.9 \text{ ft}$$

The major axis is approximately 84.9 ft.

b. Find the lengths of the major and minor axes of the skylight. Round to the nearest tenth of a foot.

The minor axis is 40.0 ft.

$$a = \sqrt{40^2 + 40^2} \approx 56.6 \text{ ft}$$

c. Find each eccentricity. Write a conclusion.

$$e = \frac{c}{a} = \frac{60}{84.9} \approx 0.71$$

$$e = \frac{c}{a} = \frac{40}{56.6} \approx 0.71$$

The roof and the skylight each have eccentricity of approximately 0.71. They are different in size, but they are similar in shape.

d. Find the area of the skylight. Round to the nearest ten feet.

$$A = \pi ab = \pi\left(\frac{56.6}{2}\right)\left(\frac{40}{2}\right) \approx 1780 \text{ ft}^2$$

e. Estimate the weight of the skylight in tons.

$$\text{Weight} = 1780 \text{ ft}^2\left(\frac{35 \text{ lb}}{1 \text{ ft}^2}\right)\left(\frac{1 \text{ ton}}{2000 \text{ lb}}\right) \approx 31.15 \text{ tons}$$

63. $a = \text{semimajor axis} = 50$ feet, $b = \text{height} = 30$ feet

$$c^2 = a^2 - b^2$$
$$c^2 = 50^2 - 30^2$$
$$c = \sqrt{1600} = 40$$

The foci are located 40 feet to the right and to the left of center.

65. $2a = 36 \quad 2b = 9$

$\quad\;\; a = 18 \quad\;\; b = \dfrac{9}{2}$

$$c^2 = a^2 - b^2$$
$$c^2 = 18^2 - \left(\frac{9}{2}\right)^2$$
$$c^2 = 324 - \frac{81}{4}$$
$$c^2 = \frac{1215}{4}$$
$$c = \frac{9\sqrt{15}}{2}$$

Since one focus is at $(0, 0)$, the center of the ellipse is at $\left(9\sqrt{15}/2, \; 0\right)$ or $(17.43, 0)$. The equation of the path of Halley's Comet in astronomical units is

$$\frac{\left(x - 9\sqrt{15}/2\right)^2}{324} + \frac{y^2}{81/4} = 1$$

67. $\dfrac{x^2}{75^2} + \dfrac{y^2}{34^2} = 1$

Solve for y, where $x = 55$.

$$\frac{55^2}{75^2} + \frac{y^2}{34^2} = 1$$
$$\frac{y^2}{34^2} = 1 - \frac{55^2}{75^2}$$
$$y^2 = 34^2\left(1 - \frac{55^2}{75^2}\right)$$
$$y = \sqrt{34^2\left(1 - \frac{55^2}{75^2}\right)}$$
$$y \approx 23 \text{ ft}$$
$$h = y + 1 = 23 + 1 = 24 \text{ ft}$$

69. a. $c^2 = a^2 - b^2$
$\qquad c^2 = 4^2 - 3^2$
$\qquad c^2 = 7$
$\qquad\; c = \sqrt{7}$

$\sqrt{7}$ ft to the right and left of O.

b. $2a = 2(4) = 8$ ft

71. $9y^2 + 36y + 16x^2 - 108 = 0$

$$y = \frac{-36 \pm \sqrt{36^2 - 4(9)(16x^2 - 108)}}{2(9)}$$

$$= \frac{-36 \pm \sqrt{1296 - 36(16x^2 - 108)}}{18}$$

$$= \frac{-36 \pm \sqrt{1296 - 576x^2 + 3888}}{18}$$

$$= \frac{-36 \pm \sqrt{-576x^2 + 5184}}{18}$$

$$= \frac{-36 \pm \sqrt{576(-x^2 + 9)}}{18}$$

$$= \frac{-36 \pm 24\sqrt{(-x^2 + 9)}}{18}$$

$$= \frac{-6 \pm 4\sqrt{(-x^2 + 9)}}{3}$$

73. $9y^2 + 18y + 4x^2 + 24x + 44 = 0$

$$y = \frac{-18 \pm \sqrt{18^2 - 4(9)(4x^2 + 24x + 44)}}{2(9)}$$

$$= \frac{-18 \pm \sqrt{324 - 36(4x^2 + 24x + 44)}}{18}$$

$$= \frac{-18 \pm \sqrt{324 - 144x^2 - 864x - 1584}}{18}$$

$$= \frac{-18 \pm \sqrt{-144x^2 - 864x - 1260}}{18}$$

$$= \frac{-18 \pm \sqrt{36(-4x^2 - 24x - 35)}}{18}$$

$$= \frac{-18 \pm 6\sqrt{-4x^2 - 24x - 35}}{18}$$

$$= \frac{-3 \pm \sqrt{-4x^2 - 24x - 35}}{3}$$

75. The sum of the distances between the two foci and a point on the ellipse is $2a$.

$$2a = \sqrt{\left(\frac{9}{2} - 0\right)^2 + (3 - 3)^2} + \sqrt{\left(\frac{9}{2} - 0\right)^2 + (3 + 3)^2}$$

$$= \sqrt{\left(\frac{9}{2}\right)^2} + \sqrt{\frac{225}{4}}$$

$$= \frac{9}{2} + \frac{15}{2}$$

$$= 12$$

$$a = 6$$
$$c = 3$$
$$c^2 = a^2 - b^2$$
$$9 = 36 - b^2$$
$$b^2 = 27$$

$$\frac{x^2}{36} + \frac{y^2}{27} = 1$$

77. Center $(1, -1)$

$$c^2 = a^2 - b^2$$
$$c^2 = 16 - 9$$
$$c^2 = 7$$
$$c = \sqrt{7}$$

The latus rectum is on the graph

of $y = -1 + \sqrt{7}$, or $y = -1 - \sqrt{7}$

$$\frac{(x-1)^2}{9} + \frac{(y+1)^2}{16} = 1$$

$$\frac{(x-1)^2}{9} + \frac{\left(-1 + \sqrt{7} + 1\right)^2}{16} = 1$$

or $\quad \dfrac{(x-1)^2}{9} + \dfrac{\left(-1 - \sqrt{7} + 1\right)^2}{16} = 1$

$$\frac{(x-1)^2}{9} + \frac{7}{16} = 1$$

$$\frac{(x-1)^2}{9} = \frac{9}{16}$$

$$16(x-1)^2 = 81$$

$$(x-1)^2 = \frac{81}{16}$$

$$x - 1 = \pm\sqrt{\frac{81}{16}}$$

$$x - 1 = \pm\frac{9}{4}$$

$$x = \frac{13}{4} \text{ and } -\frac{5}{4}$$

The x-coordinates of the endpoints of the latus rectum

are $\frac{13}{4}$ and $-\frac{5}{4}$.

$$\left|\frac{13}{4}-\left(-\frac{5}{4}\right)\right|=\frac{9}{2}$$

The length of the latus rectum is $\frac{9}{2}$.

79. Let us transform the general equation of an ellipse into

an $x'y'$ - coordinate system where the center is at the

origin by replacing $(x-h)$ by x' and $(y-k)$ by y'.

We have $\dfrac{x'^2}{a^2}+\dfrac{y'^2}{b^2}=1$.

Letting $x'=c$ and solving for y' yields

$$\frac{(c)^2}{a^2}+\frac{y'^2}{b^2}=1$$
$$b^2c^2+a^2y'^2=a^2b^2$$
$$a^2y'^2=a^2b^2-b^2c^2$$
$$a^2y'^2=b^2(a^2-c^2)$$

But since $c^2=a^2-b^2,\ b^2=a^2-c^2$, we can

substitute to obtain

$$a^2y'^2=b^2(b^2)$$
$$y'^2=\frac{b^4}{a^2}$$
$$y'=\pm\sqrt{\frac{b^4}{a^2}}=\pm\frac{b^2}{a}$$

The endpoints of the latus rectum, then, are

$\left(c,\ \dfrac{b^2}{a}\right)$ and $\left(c,-\dfrac{b^2}{a}\right)$.

The distance between these points is $\dfrac{2b^2}{a}$.

81. Kepler was born Dec 17, 1571, and died Nov 15, 1630.

He was among the first strong supporters of the

heliocentric theory. In 1596 Kepler published

Cosmographic Mystery, in which he defended the

Copernican theory.

Tycho Brahe, mathematician at the court of Emperor

Rudolph II, was impressed with the work of Kepler and

invited him to Prague as his assistant. When Brahe died

the following year (1601), Kepler was appointed to the

position held by Brahe. Between 1609 and 1619,

Kepler published his three laws of motion:

1. Each planet moves about the sun in an orbit that is

an ellipse, with the sun at one focus of the ellipse.

2. The straight line joining a planet and the sun sweeps

out equal areas in space in equal intervals of time.

3. The squares of the sidereal periods of the planets are

in direct proportion to the cubes of the semimajor axes

of their orbits. That is, $P^2=ka^3$. The value of k

depends on the units of measurement. If astronomical

units are used, then $k=1$.

Kepler also made contributions to optics and telescope

lenses and gave a physical explanation of nova. His

text *Introduction to Copernican Astronomy* was one of

the most widely read treatises on astronomy.

a. A planet's velocity is greatest when it is at

perihelion. This follows from Kepler's second law.

b. The period of Mars is 1.88 years. This follows from

the third law.

Prepare for Section 8.3

P1. Find the midpoint and length.

$$\frac{4+-2}{2}=1 \quad \text{and} \quad \frac{-3+1}{2}=-1$$

Midpoint: $(1,-1)$

$$\sqrt{(-2-4)^2+(1--3)^2}=\sqrt{52}=2\sqrt{13}$$

Length: $2\sqrt{13}$

P3. Simplify.

$$\frac{4}{\sqrt{8}}=\frac{4\sqrt{8}}{8}=\frac{8\sqrt{2}}{8}=\sqrt{2}$$

P5. Solve.

$$\frac{x^2}{4}-\frac{y^2}{9}=1$$
$$-\frac{y^2}{9}=1-\frac{x^2}{4}$$
$$y^2=\frac{9x^2}{4}-9$$
$$y=\pm\sqrt{\frac{9x^2}{4}-9}$$
$$y=\pm\frac{3}{2}\sqrt{x^2-4}$$

Section 8.3 Exercises

1. Find the vertices of the hyperbola.

   $(h, k+a)$ and $(h, k-a)$

3. Find the eccentricity of the hyperbola.

   $e = \dfrac{c}{a} = \dfrac{\sqrt{a^2 + b^2}}{a}$

5. **a.** iii

   b. ii

   c. i

   d. iv

7. Find the center, vertices, foci, and asymptotes; graph.

   $\dfrac{x^2}{16} - \dfrac{y^2}{25} = 1$

   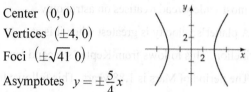

   Center $(0, 0)$

   Vertices $(\pm 4, 0)$

   Foci $(\pm\sqrt{41}\ 0)$

   Asymptotes $y = \pm\dfrac{5}{4}x$

9. Find the center, vertices, foci, and asymptotes; graph.

   $\dfrac{y^2}{4} - \dfrac{x^2}{25} = 1$

   Center $(0, 0)$

   Vertices $(0, \pm 2)$

   Foci $(0, \pm\sqrt{29})$

   Asymptotes $y = \pm\dfrac{2}{5}x$

   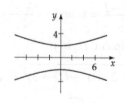

11. Find the center, vertices, foci, and asymptotes; graph.

    $\dfrac{x^2}{7} - \dfrac{y^2}{9} = 1$

    Center $(0, 0)$

    Vertices $(\pm\sqrt{7}, 0)$

    Foci $(\pm 4, 0)$

    Asymptotes $y = \pm\dfrac{3\sqrt{7}}{7}x$

    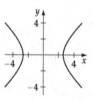

13. Find the center, vertices, foci, and asymptotes; graph.

    $\dfrac{4x^2}{9} - \dfrac{y^2}{16} = 1$

Center $(0, 0)$

Vertices $\left(\pm\dfrac{3}{2}, 0\right)$

Foci $\left(\pm\dfrac{\sqrt{73}}{2}, 0\right)$

Asymptotes $y = \pm\dfrac{8}{3}x$

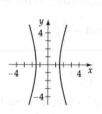

15. Find the center, vertices, foci, and asymptotes; graph.

    $\dfrac{(x-3)^2}{16} - \dfrac{(y+4)^2}{9} = 1$

    Center $(3, -4)$

    Vertices $(3 \pm 4, -4) = (7, -4),\ (-1, -4)$

    Foci $(3 \pm 5, -4) = (8, -4),\ (-2, -4)$

    Asymptotes $y + 4 = \pm\dfrac{3}{4}(x - 3)$

    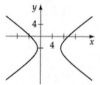

17. Find the center, vertices, foci, and asymptotes; graph.

    $\dfrac{(y+2)^2}{4} - \dfrac{(x-1)^2}{16} = 1$

    Center $(1, -2)$

    Vertices $(1, -2 \pm 2) = (1, 0),\ (1, -4)$

    Foci $(1, -2 \pm 2\sqrt{5})$

    $= (1, -2 + 2\sqrt{5}),\ (1, -2 - 2\sqrt{5})$

    Asymptotes $y + 2 = \pm\dfrac{1}{2}(x - 1)$

    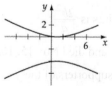

19. Find the center, vertices, foci, and asymptotes; graph.

    $\dfrac{(x+2)^2}{9} - \dfrac{y^2}{25} = 1$

    Center $(-2, 0)$

    Vertices $(-2 \pm 3, 0) = (1, 0),\ (-5, 0)$

    Foci $(-2 \pm\sqrt{34}, 0)$

    Asymptotes $y = \pm\dfrac{5}{3}(x + 2)$

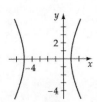

21. Find the center, vertices, foci, and asymptotes; graph.

$$\frac{9(x-1)^2}{16} - \frac{(y+1)^2}{9} = 1$$

$$\frac{(x-1)^2}{16/9} - \frac{(y+1)^2}{9} = 1$$

Center $(1, -1)$

Vertices $\left(1 \pm \frac{4}{3}, -1\right) = \left(\frac{7}{3}, -1\right)$, $\left(-\frac{1}{3}, -1\right)$

Foci $\left(1 \pm \frac{\sqrt{97}}{3}, -1\right)$

Asymptotes $(y+1) = \pm \frac{9}{4}(x-1)$

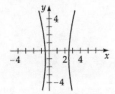

23. Find the center, vertices, foci, and asymptotes; graph.

$$x^2 - y^2 = 9$$

$$\frac{x^2}{9} - \frac{y^2}{9} = 1$$

Center $(0, 0)$

Vertices $(\pm 3, 0)$

Foci $(\pm 3\sqrt{2}, 0)$

Asymptotes $y = \pm x$

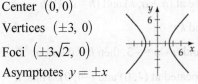

25. Find the center, vertices, foci, and asymptotes; graph.

$$16y^2 - 9x^2 = 144$$

$$\frac{y^2}{9} - \frac{x^2}{16} = 1$$

Center $(0, 0)$

Vertices $(0, \pm 3)$

Foci (0 ± 5)

Asymptotes $y = \pm \frac{3}{4}x$

27. Find the center, vertices, foci, and asymptotes; graph.

$$9y^2 - 36x^2 = 4$$

$$\frac{y^2}{4/9} - \frac{x^2}{1/9} = 1$$

Center $(0, 0)$

Vertices $\left(0, \pm \frac{2}{3}\right)$

Foci $\left(0, \pm \frac{\sqrt{5}}{3}\right)$

Asymptotes $y = \pm 2x$

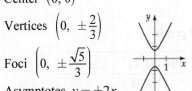

29. Find the center, vertices, foci, and asymptotes; graph.

$$x^2 - y^2 - 6x + 8y - 3 = 0$$

$$x^2 - y^2 - 6x + 8y = 3$$

$$(x^2 - 6x) - (y^2 - 8y) = 3$$

$$(x^2 - 6x + 9) - (y^2 - 8y + 16) = 3 + 9 - 16$$

$$(x-3)^2 - (y-4)^2 = -4$$

$$\frac{(y-4)^2}{4} - \frac{(x-3)^2}{4} = 1$$

Center $(3, 4)$

Vertices $(3, 4 \pm 2) = (3, 6), (3, 2)$

Foci $(3, 4 \pm 2\sqrt{2}) = (3, 4 + 2\sqrt{2}), (3, 4 - 2\sqrt{2})$

Asymptotes $y - 4 = \pm(x-3)$

31. Find the center, vertices, foci, and asymptotes; graph.

$$9x^2 - 4y^2 + 36x - 8y + 68 = 0$$

$$9x^2 + 36x - 4y^2 - 8y = -68$$

$$9(x^2 + 4x) - 4(y^2 + 2y) = -68$$

$$9(x^2 + 4x + 4) - 4(y^2 + 2y + 1) = -68 + 36 - 4$$

$$9(x+2)^2 - 4(y+1)^2 = -36$$

$$\frac{(y+1)^2}{9} - \frac{(x+2)^2}{4} = 1$$

Center $(-2, -1)$

Vertices $(-2, -1 \pm 3) = (-2, 2), (-2, -4)$

Foci $(-2, -1 \pm \sqrt{13})$

$$= (-2, -1 + \sqrt{13}), (-2, -1 - \sqrt{13})$$

Asymptotes $y+1=\pm\frac{3}{2}(x+2)$

33. Solve for y and graph.

$$y=\frac{-6\pm\sqrt{6^2-4(-1)(4x^2+32x+39)}}{2(-1)}$$

$$=\frac{-6\pm\sqrt{36+4(4x^2+32x+39)}}{-2}$$

$$=\frac{-6\pm\sqrt{16x^2+128x+192}}{-2}$$

$$=\frac{-6\pm\sqrt{16(x^2+8x+12)}}{-2}$$

$$=\frac{-6\pm4\sqrt{x^2+8x+12}}{-2}$$

$$=3\pm2\sqrt{x^2+8x+12}$$

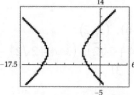

35. Solve for y and graph.

$$y=\frac{64\pm\sqrt{(-64)^2-4(-16)(9x^2-36x+116)}}{2(-16)}$$

$$=\frac{64\pm\sqrt{4096+64(9x^2-36x+116)}}{-32}$$

$$=\frac{64\pm\sqrt{64(9x^2-36x+116+64)}}{-32}$$

$$=\frac{64\pm8\sqrt{(9x^2-36x+180)}}{-32}$$

$$=\frac{64\pm8\sqrt{9(x^2-4x+20)}}{-32}$$

$$=\frac{64\pm24\sqrt{x^2-4x+20}}{-32}$$

$$=\frac{-8\pm3\sqrt{x^2-4x+20}}{4}$$

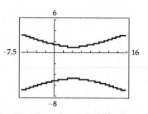

37. Solve for y and graph.

$$y=\frac{18\pm\sqrt{(-18)^2-4(-9)(4x^2+8x-6)}}{2(-9)}$$

$$=\frac{18\pm\sqrt{324+36(4x^2+8x-6)}}{-18}$$

$$=\frac{18\pm\sqrt{36(4x^2+8x-6+9)}}{-18}$$

$$=\frac{18\pm6\sqrt{4x^2+8x+3}}{-18}$$

$$=\frac{-3\pm\sqrt{4x^2+8x+3}}{3}$$

39. Find the equation of the hyperbola in standard form.

vertices $(3, 0)$ and $(-3, 0)$, foci $(4, 0)$ and $(-4, 0)$

Traverse axis is on x-axis. For a standard hyperbola, the vertices are at $(h+a, k)$ and $(h-a, k)$, $h+a=3$, $h-a=-3$, and $k=0$..

If $h+a=3$ and $h-a=-3$, then $h=0$ and $a=3$.

The foci are located at $(4, 0)$ and $(-4, 0)$. Thus, $h=0$ and $c=4$.

Since $c^2=a^2+b^2$, $b^2=c^2-a^2$

$b^2=(4)^2-(3)^2=16-9=7$

$$\frac{(x-h)^2}{a^2}-\frac{(y-k)^2}{b^2}=1$$

$$\frac{(x-0)^2}{(3)^2}-\frac{(y-0)^2}{7}=1$$

$$\frac{x^2}{9}-\frac{y^2}{7}=1$$

41. Find the equation of the hyperbola in standard form.

foci (0, 4) and (0, –4), asymptotes $y = 4x$ and $y = -4x$

Transverse axis is on y-axis. Since foci are at $(h, k + c)$ and $(h, k - c)$, $k + c = 4$, $k - c = -4$, and $h = 0$.

Therefore, $k = 0$ and $c = 4$.

Since one of the asymptotes is $y = \dfrac{a}{b}x$, $\dfrac{a}{b} = 4$ and

$a = 4b$.

$a^2 + b^2 = c^2$; then substituting $a = 4b$ and $c = 4$ yields

$(4b)^2 + b^2 = (4)^2$, or $17b^2 = 16$.

Therefore, $b^2 = 16/17$ and $b = \dfrac{4}{\sqrt{17}}$.

Since $a = 4b$, $a = 4\left(\dfrac{4}{\sqrt{17}}\right) = \dfrac{16}{\sqrt{17}}$

$a^2 = \dfrac{256}{17}$.

$\dfrac{(y-k)^2}{a^2} - \dfrac{(x-h)^2}{b^2} = 1$

$\dfrac{y^2}{256/17} - \dfrac{x^2}{16/17} = 1$

43. Find the equation of the hyperbola in standard form.

vertices (0, 3) and (0, –3), point (2, 4)

The distance between the two vertices is the length of the transverse axis, which is $2a$.

$2a = |3 - (-3)| = 6$ or $a = 3$.

Since the midpoint of the transverse axis is the center of the hyperbola, the center is given by

$\left(\dfrac{0+0}{2}, \dfrac{3+(-3)}{2}\right)$, or (0, 0)

Since both vertices lie on the y-axis, the transverse axis must be on the y-axis.

Taking the standard form of the hyperbola, we have

$\dfrac{y^2}{a^2} - \dfrac{x^2}{b^2} = 1$

Substituting the point (2, 4) for x and y, and 3 for a, we have $\dfrac{16}{9} - \dfrac{4}{b^2} = 1$.

Solving for b^2 yields $b^2 = \dfrac{36}{7}$.

Therefore, the equation is $\dfrac{y^2}{9} - \dfrac{x^2}{36/7} = 1$.

45. Find the equation of the hyperbola in standard form.

vertices (0, 4) and (0, –4),

asymptotes $y = \dfrac{1}{2}x$ and $y = -\dfrac{1}{2}x$.

The length of the transverse axis, or the distance between the vertices, is equal to $2a$.

$2a = 4 - (-4) = 8$, or $a = 4$

The center of the hyperbola, or the midpoint of the line segment joining the vertices, is $\left(\dfrac{0+0}{2}, \dfrac{4+(-4)}{2}\right)$, or

(0, 0).

Since both vertices lie on the y-axis, the transverse axis must lie on the y-axis. Therefore, the asymptotes are

given by $y = \dfrac{a}{b}x$ and $y = -\dfrac{a}{b}x$. One asymptote is

$y = \dfrac{1}{2}x$. Thus $\dfrac{a}{b} = \dfrac{1}{2}$ or $b = 2a$.

Since $b = 2a$ and $a = 4$, $b = 2(4) = 8$.

Thus, the equation is $\dfrac{y^2}{4^2} - \dfrac{x^2}{8^2} = 1$ or $\dfrac{y^2}{16} - \dfrac{x^2}{64} = 1$.

47. Find the equation of the hyperbola in standard form.

vertices (–2, 5) and (12, 5), foci (–5, 5) and (15, 5)

Length of transverse axis = distance between vertices

$2a = |-2 - 12|$

$a = 7$

The center of the hyperbola (h, k) is the midpoint of the line segment joining the vertices, or the

point $\left(\dfrac{-2+12}{2}, \dfrac{5+5}{2}\right)$.

Thus, $h = \dfrac{-2+12}{2}$, or 5, and $k = \dfrac{5+5}{3}$, or 5.

Since both vertices lie on the horizontal line $y = 5$, the transverse axis is parallel to the x-axis. The location of the foci is given by $(h + c, k)$ and $(h - c, k)$, or specifically (–5, 5) and (15, 5). Thus $h + c = -5$,

$h - c = 15$, and $k = 5$. Solving for h and c simultaneously yields $h = 5$ and $c = -10$.

Since $c^2 = a^2 + b^2$, $b^2 = c^2 - a^2$.

Substituting, we have

$b^2 = (-10)^2 - 7^2 = 100 - 49 = 51$.

Substituting $a = 7$, $b^2 = 51$, $h = 5$, and $k = 5$ in the

standard equation $\dfrac{(x-h)^2}{a^2} - \dfrac{(y-k)^2}{b^2} = 1$

yields $\dfrac{(x-5)^2}{49} - \dfrac{(y-5)^2}{51} = 1$.

49. Find the equation of the hyperbola in standard form.

foci $(1, -2)$ and $(7, -2)$, slope of an asymptote $= \dfrac{5}{4}$

Both foci lie on the horizontal line $y = -2$; therefore, the transverse axis is parallel to the x-axis.

The foci are given by $(h + c, k)$ and $(h - c, k)$.

Thus, $h - c = 1$, $h + c = 7$, and $k = -2$. Solving simultaneously for h and c yields $h = 4$ and $c = 3$.

Since $y - k = \dfrac{b}{a}(x - h)$ is the equation for an

asymptote, and the slope of an asymptote is given as

$\dfrac{5}{4}$, $\dfrac{b}{a} = \dfrac{5}{4}$, $b = \dfrac{5a}{4}$, and $b^2 = \dfrac{25a^2}{16}$.

Because $a^2 + b^2 = c^2$, substituting $c = 3$ and

$b^2 = \dfrac{25a^2}{16}$ yields $a^2 = \dfrac{144}{41}$.

Therefore, $b^2 = \dfrac{3600}{656} = \dfrac{225}{41}$.

Substituting in the standard equation for a hyperbola

yields $\dfrac{(x-4)^2}{144/41} - \dfrac{(y+2)^2}{225/41} = 1$

51. Find the equation of the hyperbola in standard form.

Because the transverse axis is parallel to the y-axis and the center is $(7, 2)$, the equation of the hyperbola is

$\dfrac{(y-2)^2}{a^2} - \dfrac{(x-7)^2}{b^2} = 1$.

Because $(9, 4)$ is a point on the hyperbola,

$\dfrac{(4-2)^2}{a^2} - \dfrac{(9-7)^2}{b^2} = 1$.

The slope of the asymptote is $\dfrac{1}{2}$. Therefore

$\dfrac{1}{2} = \dfrac{a}{b}$ or $b = 2a$.

Substituting, we have $\dfrac{4}{a^2} - \dfrac{4}{4a^2} = 1$

$\dfrac{4}{a^2} - \dfrac{1}{a^2} = 1$, or $a^2 = 3$.

Since $b = 2a$, $b^2 = 4a^2$, or $b^2 = 12$. The equation is

$\dfrac{(y-2)^2}{3} - \dfrac{(x-7)^2}{12} = 1$.

53. Find the eccentricity of the hyperbola.

$a^2 = 49$, $b^2 = 25$

$c^2 = a^2 + b^2 = 49 + 25$

$c = \sqrt{74}$

$e = \dfrac{c}{a} = \dfrac{\sqrt{74}}{7} \approx 1.23$

55. Find the eccentricity of the hyperbola.

$a^2 = 9$, $b^2 = 16$

$c^2 = a^2 + b^2 = 9 + 16$

$c = 5$

$e = \dfrac{c}{a} = \dfrac{5}{3}$

57. Find the eccentricity of the hyperbola.

$16y^2 + 32y - 4x^2 - 24x = 84$

$16(y^2 + 2y) - 4(x^2 - 6x) = 84$

$16(y^2 + 2y + 1) - 4(x^2 - 6x + 9) = 84 + 16 - 36$

$16(y + 1)^2 - 4(x - 3)^2 = 64$

$\dfrac{(y+1)^2}{4} - \dfrac{(x-3)^2}{16} = 1$

$a^2 = 4$, $b^2 = 16$

$c^2 = a^2 + b^2 = 4 + 16$

$c = \sqrt{20} = 2\sqrt{5}$

$e = \dfrac{c}{a} = \dfrac{2\sqrt{5}}{2} = \sqrt{5} \approx 2.24$

59. Find the equation of the hyperbola in standard form.

vertices $(1, 6)$ and $(1, 8)$, eccentricity $= 2$

Length of transverse axis = distance between vertices

$$2a = |6 - 8| = 2$$
$$a = 1 \text{ and } a^2 = 1$$

Center (midpoint of transverse axis) is $\left(\dfrac{1+1}{2}, \dfrac{6+8}{2}\right)$,

or $(1, 7)$.

Therefore, $h = 1$ and $k = 7$.

Since both vertices lie on the vertical line $x = 1$, the

transverse axis is parallel to the y-axis.

Since $e = \dfrac{c}{a}$, $c = ae = (1)(2) = 2$.

Because $b^2 = c^2 - a^2$, $b^2 = (2)^2 - (1)^2 = 4 - 1 = 3$.

Substituting h, k, a^2, and b^2 into the standard equation

yields $\dfrac{(y-7)^2}{1} - \dfrac{(x-1)^2}{3} = 1$.

61. Find the equation of the hyperbola in standard form.

foci $(4, 0)$ and $(-4, 0)$, eccentricity $= 2$

Center (midpoint of line segment joining foci) is

$\left(\dfrac{4+(-4)}{2}, \dfrac{0+0}{2}\right)$, or $(0, 0)$

Thus, $h = 0$ and $k = 0$.

Since both foci lie on the horizontal line $y = 0$, the

transverse axis is parallel to the x-axis. The locations of

the foci are given by $(h + c, k)$ and $(h - c, k)$, or

specifically $(4, 0)$ and $(-4, 0)$

Since $h = 0$, $c = 4$.

Because $e = \dfrac{c}{a}$, $a = \dfrac{c}{e} = \dfrac{4}{2} = 2$ and $a^2 = 4$.

Because $b^2 = c^2 - a^2$, $b^2 = 4^2 - 2^2 = 16 - 4 = 12$.

Substituting h, k, a^2 and b^2 into the standard formula

for a hyperbola yields $\dfrac{x^2}{4} - \dfrac{y^2}{12} = 1$.

63. Find the equation of the hyperbola in standard form.

conjugate axis length $= 4$, center $(4, 1)$,

eccentricity $= \dfrac{4}{3}$

$2b =$ conjugate axis length $= 4$

$b = 2$ and $b^2 = 4$

Since $e = \dfrac{c}{a} = \dfrac{4}{3}$, $c = \dfrac{4a}{3}$ and $c^2 = \dfrac{16a^2}{9}$. Since

$a^2 + b^2 = c^2$, substituting $b^2 = 4$ and $c^2 = \dfrac{16a^2}{9}$ and

solving for a^2 yields $a^2 = \dfrac{36}{7}$.

Substituting into the two standard equations of a

hyperbola yields $\dfrac{(x-4)^2}{36/7} - \dfrac{(y-1)^2}{4} = 1$ and

$\dfrac{(y-1)^2}{36/7} - \dfrac{(x-4)^2}{4} = 1$.

65. When the wave hits Earth, $z = 0$.

$$y^2 = x^2 + (z - 10{,}000)^2$$
$$y^2 = x^2 + (0 - 10{,}000)^2$$
$$y^2 - x^2 = 10{,}000^2$$

It is a hyperbola.

67. a. Using the eccentricity, and $a = 2$,

$$\dfrac{c}{2} = \dfrac{\sqrt{17}}{4} \Rightarrow c = \dfrac{\sqrt{17}}{2}$$

Solve for b.

$$a^2 + b^2 = c^2$$
$$b^2 = c^2 - a^2$$
$$b^2 = \left(\dfrac{\sqrt{17}}{2}\right)^2 - 2^2$$
$$b^2 = \dfrac{17}{4} - \dfrac{16}{4}$$
$$b^2 = \dfrac{1}{4}$$
$$b = \dfrac{1}{2} = 0.5$$

$$\dfrac{x^2}{2^2} - \dfrac{y^2}{0.5^2} = 1$$

b. For FG, $y = 0.6$.

$$\frac{x^2}{2^2} - \frac{0.6^2}{0.5^2} = 1$$

$$\frac{x^2}{2^2} = 1 + \frac{0.6^2}{0.5^2}$$

$$x^2 = 2^2\left(1 + \frac{0.6^2}{0.5^2}\right)$$

$$x = \sqrt{2^2\left(1 + \frac{0.6^2}{0.5^2}\right)}$$

$$x \approx 3.1241$$

$$FG = 2x \approx 6.25 \text{ in.}$$

69. Identify the equation and graph.

$$4x^2 + 9y^2 - 16x - 36y + 16 = 0$$

$$4(x^2 - 4x) + 9(y^2 - 4y) = -16$$

$$4(x^2 - 4x + 4) + 9(y^2 - 4y + 4) = -16 + 16 + 36$$

$$4(x - 2)^2 + 9(y - 2)^2 = 36$$

$$\frac{(x - 2)^2}{9} + \frac{(y - 2)^2}{4} = 1$$

ellipse

center $(2, 2)$

vertices $(2 \pm 3, 2) = (5, 2), (-1, 2)$

foci $\left(2 \pm \sqrt{5}, 2\right) = \left(2 + \sqrt{5}, 2\right), \left(2 - \sqrt{5}, 2\right)$

71. Identify the equation and graph.

$$5x - 4y^2 + 24y - 11 = 0$$

$$-4(y^2 - 6y) = -5x + 11$$

$$-4(y^2 - 6y + 9) = -5x + 11 - 36$$

$$-4(y - 3)^2 = -5x - 25$$

$$-4(y - 3)^2 = -5(x + 5)$$

$$(y - 3)^2 = \frac{5}{4}(x + 5)$$

parabola

vertex $(-5, 3)$

focus $\left(-5 + \frac{5}{16}, 3\right) = \left(-\frac{75}{16}, 3\right)$

directrix $x = -5 - \frac{5}{16}$, or $x = \frac{-85}{16}$

73. Identify the equation and graph.

$$x^2 + 2y - 8x = 0$$

$$x^2 - 8x = -2y$$

$$x^2 - 8x + 16 = -2y + 16$$

$$(x - 4)^2 = -2(y - 8)$$

parabola

vertex $(4, 8)$

foci $\left(4, 8 - \frac{1}{2}\right) = \left(4, \frac{15}{2}\right)$

directrix $y = 8 + \frac{1}{2}$, or $y = \frac{17}{2}$

75. Identify the equation and graph.

$$25x^2 + 9y^2 - 50x - 72y - 56 = 0$$

$$25(x^2 - 2x) + 9(y^2 - 8y) = 56$$

$$25(x^2 - 2x + 1) + 9(y^2 - 8y + 16) = 56 + 25 + 144$$

$$25(x - 1)^2 + 9(y - 4)^2 = 225$$

$$\frac{(x - 1)^2}{9} + \frac{(y - 4)^2}{25} = 1$$

ellipse

center $(1, 4)$

vertices $(1, 4 \pm 5) = (1, 9), (1, -1)$

foci $(1, 4 \pm 4) = (1, 8), (1, 0)$

77. Find the equation of the hyperbola in standard form.

foci $F_1(2, 0)$, $F_2(-2, 0)$ passing through $P_1(2, 3)$

$$d(P_1, F_2) - d(P_1, F_1) = \sqrt{(2+2)^2 + 3^2} - \sqrt{(2-2)^2 + 3^2}$$
$$= 5 - 3 = 2$$

Let $P(x, y)$ be any point on the hyperbola. Since the difference between F_1P and F_2P is the same as the difference between F_1P_1 and F_2P_1, we have

$$\sqrt{(x-2)^2 + y^2} - \sqrt{(x+2)^2 + y^2} = 2$$

$$\sqrt{(x-2)^2 + y^2} = 2 + \sqrt{(x+2)^2 + y^2}$$

$$x^2 - 4x + 4 + y^2 = 4 + 4\sqrt{(x+2)^2 + y^2} + x^2 + 4x + 4 + y^2$$

$$-8x - 4 = 4\sqrt{(x+2)^2 + y^2}$$

$$-2x - 1 = \sqrt{(x+2)^2 + y^2}$$

$$4x^2 + 4x + 1 = x^2 + 4x + 4 + y^2$$

$$3x^2 - y^2 = 3$$

$$\frac{x^2}{1} - \frac{y^2}{3} = 1$$

79. Find the equation of the hyperbola in standard form.

foci $(0, 4)$ and $(0, -4)$, point $\left(\frac{7}{3}, 4\right)$

Difference in distances from (x, y) to foci = difference of distances from $\left(\frac{7}{3}, 4\right)$ to foci

$$\sqrt{(x-0)^2 + (y-4)^2} - \sqrt{(x-0)^2 + (y+4)^2}$$
$$= \sqrt{\left(\frac{7}{3} - 0\right)^2 + (4-4)^2} - \sqrt{\left(\frac{7}{3} - 0\right)^2 + (4+4)^2}$$

$$\sqrt{x^2 + y^2 - 8y + 16} - \sqrt{x^2 + y^2 + 8y + 16}$$
$$= \frac{7}{3} - \frac{25}{3} = -6$$

$$\sqrt{x^2 + y^2 - 8y + 16}$$
$$= \sqrt{x^2 + y^2 + 8y + 16} - 6$$

$$\sqrt{(x-0)^2 + (y-4)^2} - \sqrt{(x-0)^2 + (y+4)^2}$$
$$= \sqrt{\left(\frac{7}{3} - 0\right)^2 + (4-4)^2} - \sqrt{\left(\frac{7}{3} - 0\right)^2 + (4+4)^2}$$

$$\sqrt{x^2 + y^2 - 8y + 16} - \sqrt{x^2 + y^2 + 8y + 16}$$
$$= \frac{7}{3} - \frac{25}{3} = -6$$

$$\sqrt{x^2 + y^2 - 8y + 16}$$
$$= \sqrt{x^2 + y^2 + 8y + 16} - 6$$

$$x^2 + y^2 - 8y + 16$$
$$= x^2 + y^2 + 8y + 16 - 12\sqrt{x^2 + y^2 + 8y + 16} + 36$$
$$- 16y - 36$$
$$= -12\sqrt{x^2 + y^2 + 8y + 16}$$

$$4y + 9$$
$$= 3\sqrt{x^2 + y^2 + 8y + 16}$$

$$16y^2 + 72y + 81$$
$$= 9x^2 + 9y^2 + 72y + 144$$

$$7y^2 - 9x^2 = 63$$

Thus, $\dfrac{y^2}{9} - \dfrac{x^2}{7} = 1$

81. Sketch the graph.

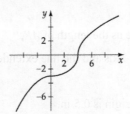

83. a. For the equation of the parabola:

$$x^2 = 4py, \quad p = 16$$
$$x^2 = 64y$$

For the equation of the hyperbola: $V_1(0, -2), V_2(0, 12)$

Halfway between the vertices is the point $(0, 7)$.

$$\frac{(y-7)^2}{a^2} - \frac{x^2}{b^2} = 1$$

a is the distance from the vertex to the center

$$= 12 - 7 = 5.$$

c is the midpoint between the foci $= [16 - (-2)]/2 = 9$.

$$b^2 = c^2 - a^2 = 9^2 - 5^2 = 81 - 25 = 56 = \left(2\sqrt{14}\right)^2$$

$$\frac{(y-7)^2}{5^2} - \frac{x^2}{\left(2\sqrt{14}\right)^2} = 1$$

b. The x-coordinate of D is 4, so set $x = 4$ and solve the parabolic equation for y.

$$(4)^2 = 64y$$
$$y = \frac{64}{16} = 0.25$$

$$D = (4, 0.25)$$

The x-coordinate of P is 1, so set $x = 1$ and solve the hyperbolic equation for y.

$$\frac{(y-7)^2}{5^2} - \frac{1^2}{\left(2\sqrt{14}\right)^2} = 1$$

$$\frac{(y-7)^2}{25} - \frac{1}{56} = 1$$

$$\frac{(y-7)^2}{25} = \frac{57}{56}$$

$$(y-7)^2 = 25\left(\frac{57}{56}\right)$$

$$y - 7 = \sqrt{25\left(\frac{57}{56}\right)} \text{ Reject the negative}$$

$$y = 7 + \sqrt{25\left(\frac{57}{56}\right)} \approx 12.0444$$

$$P = (1, 12.0444)$$

c. The length BE is the same as the length of AR, where R is point where the y-axis and the line extended from EQ intersect.

The distance from A to the origin is 0.5 in.

From part b, we know the distance from the origin to the y-coordinate of P is 12.0444 in.

The distance between the y-coordinates of P and Q is 0.125 in. Therefore the length of

$AR = 0.5 + 12.0444 + 0.125 = 12.6694$ in.

Thus, $BE = 12.6694$ in.

Exploring Concepts with Technology

a. $y = 3$ and $y = \frac{1}{4}x^2 \Rightarrow 3 = \frac{1}{4}x^2 \Rightarrow 12 = x^2 \Rightarrow x = 2\sqrt{3}$.

Thus point A is located at $(2\sqrt{3}, 3)$.

distance from A to $(0, 1)$ = the radius of the circle = 4

distance from A to directrix $(y = -1)$ = distance from

$(2\sqrt{3}, 3)$ to $(2\sqrt{3}, -1)$ is

$$\sqrt{(2\sqrt{3} - 2\sqrt{3})^2 + (3 - (-1))^2} = \sqrt{0 + 4^2} = 4$$

b. $y = 2$ and $y = \frac{1}{4}x^2 \Rightarrow 2 = \frac{1}{4}x^2 \Rightarrow 8 = x^2 \Rightarrow x = 2\sqrt{2}$.

Thus, point B is located at $(2\sqrt{3}, 3)$.

distance from B to $(0, 1)$ = the radius of the circle = 3

distance from B to directrix $(y = -1)$ = distance from

$(2\sqrt{2}, 2)$ to $(2\sqrt{2}, -1)$ is

$$\sqrt{(2\sqrt{2} - 2\sqrt{2})^2 + (2 - (-1))^2} = \sqrt{0 + 3^2} = 3$$

c. For any point on a parabola, the distance from the point to the focus is the same as the distance from the point to the directrix.

Prepare for Section 8.4

P1. Expand.

$$\cos(\alpha + \beta) = \cos\alpha\cos\beta - \sin\alpha\sin\beta$$

P3. Solve.

$$\cot 2\alpha = \frac{\sqrt{3}}{3}$$

$$\tan 2\alpha = \frac{3}{\sqrt{3}}$$

$$2\alpha = \tan^{-1}\left(\frac{3}{\sqrt{3}}\right) = \frac{\pi}{3}$$

$$\alpha = \frac{\pi}{6}$$

P5. Identify the graph.

$$4x^2 - 6y^2 + 9x + 16y - 8 = 0$$
$$A = 4, B = 0, C = -6$$

Since $B^2 - 4AC = 0^2 - 4(4)(-6) = 96 > 0$,

the graph is a hyperbola.

Section 8.4 Exercises

1. Give the equation.

$$Ax^2 + Bxy + Cy^2 + Dx + Ey + F = 0$$

3. Find the coordinates. Round to the nearest tenth.

Use the rotation-of-axes formula (2) for

$P(x, y) = (4, 6)$ and $\alpha = 30°$

$$x' = x\cos\alpha + y\sin\alpha$$

$$= 4\cos 30° + 6\sin 30° = 4\left(\frac{\sqrt{3}}{2}\right) + 6\left(\frac{1}{2}\right)$$

$$= 2\sqrt{3} + 3$$

$y' = y\cos\alpha - x\sin\alpha$

$= 6\cos 30° - 4\sin 30° = 6\left(\dfrac{\sqrt{3}}{2}\right) - 4\left(\dfrac{1}{2}\right)$

$= 3\sqrt{3} - 2$

The $x'y'$-coordinates of P are

$\left(2\sqrt{3}+3,\ 3\sqrt{3}-2\right) \approx (6.5,\ 3.2).$

5. Find the coordinates. Round to the nearest tenth.

Use the rotation-of-axes formula (2) for

$P(x, y) = (-2,\ 4)$ and $\alpha = 60°$

$x' = x\cos\alpha + y\sin\alpha$

$= -2\cos 60° + 4\sin 60° = -2\left(\dfrac{1}{2}\right) + 4\left(\dfrac{\sqrt{3}}{2}\right)$

$= -1 + 2\sqrt{3}$

$y' = y\cos\alpha - x\sin\alpha$

$= 4\cos 60° + 2\sin 60° = 4\left(\dfrac{1}{2}\right) + 2\left(\dfrac{\sqrt{3}}{2}\right)$

$= 2 + \sqrt{3}$

The $x'y'$-coordinates of P are

$\left(2\sqrt{3}-1,\ 2+\sqrt{3}\right) \approx (2.5,\ 3.7).$

7. Find the coordinates. Round to the nearest tenth.

Use the rotation-of-axes formula (2) for

$P(x, y) = (3, -5)$ and $\alpha = 73.5°$

$x' = x\cos\alpha + y\sin\alpha$

$= 3\cos 73.5° - 5\sin 73.5°$

≈ -3.9

$y' = y\cos\alpha - x\sin\alpha$

$= -5\cos 73.5° - 3\sin 73.5°$

≈ -4.3

The $x'y'$-coordinates of P are $\approx (-3.9, -4.3)$.

9. Find α. Round to the nearest 0.1°.

$xy = 3$

$A = 0,\ B = 1,\ C = 0$

$\cot 2\alpha = \dfrac{A-C}{B}$

$\cot 2\alpha = \dfrac{0-0}{1}$

$\cot 2\alpha = 0$

$2\alpha = 90°$

$\alpha = 45°$

11. Find α. Round to the nearest 0.1°.

$5x^2 - 6\sqrt{3}xy - 11y^2 + 4x - 3y + 2 = 0$

$A = 5,\ B = -6\sqrt{3},\ C = -11$

$\cot 2a = \dfrac{A-C}{B}$

$\cot 2a = \dfrac{5-(-11)}{-6\sqrt{3}}$

$\cot 2a = \dfrac{5+11}{-6\sqrt{3}}$

$\cot 2a = \dfrac{16}{-6\sqrt{3}}$

$\cot 2a = \dfrac{16}{-6\sqrt{3}}$

$\cot 2a = -\dfrac{8}{3\sqrt{3}}$

$\cot 2a = -\dfrac{8\sqrt{3}}{9}$

$2\alpha \approx 147°$

$\alpha \approx 73.5°$

13. Find α, then find an equation and graph.

$xy = 4$

$xy - 4 = 0$

$A = 0,\ B = 1,\ C = 0,\ F = -4$

$\cot 2\alpha = \dfrac{A-C}{B} = \dfrac{0-0}{1} = 0$

$\csc^2 2\alpha = \cot^2 2\alpha + 1$

$\csc^2 2\alpha = 0^2 + 1 = 1$

$\csc 2\alpha = +1$ (2α is in the first quadrant.)

$\sin 2\alpha = \dfrac{1}{\csc 2\alpha} = \dfrac{1}{1} = 1$

$\sin^2 2\alpha + \cos^2 2\alpha = 1$

$\cos^2 2\alpha = 1 - \sin^2 2\alpha$

$\cos^2 2\alpha = 1 - (1)^2$

$\cos^2 2\alpha = 0$

$\cos 2\alpha = 0$

$\sin\alpha = \sqrt{\dfrac{1-(0)}{2}} = \dfrac{\sqrt{2}}{2},\qquad \cos\alpha = \sqrt{\dfrac{1+(0)}{2}} = \dfrac{\sqrt{2}}{2}$

$\alpha = 45°$

$A' = A\cos^2\alpha + B\cos\alpha\sin\alpha + C\sin^2\alpha$

$= 0\left(\dfrac{\sqrt{2}}{2}\right)^2 + 1\left(\dfrac{\sqrt{2}}{2}\right)\left(\dfrac{\sqrt{2}}{2}\right) + 0\left(\dfrac{\sqrt{2}}{2}\right)^2 = \dfrac{1}{2}$

$$C' = A\sin^2\alpha - B\cos\alpha\sin\alpha + C\cos^2\alpha$$
$$= 0\left(\frac{\sqrt{2}}{2}\right)^2 - 1\left(\frac{\sqrt{2}}{2}\right)\left(\frac{\sqrt{2}}{2}\right) + 0\left(\frac{\sqrt{2}}{2}\right)^2 = -\frac{1}{2}$$
$$F' = F = -4$$

$$\frac{1}{2}x'^2 - \frac{1}{2}y'^2 - 4 = 0 \text{ or } \frac{x'^2}{8} - \frac{y'^2}{8} = 1$$

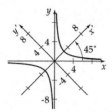

15. Find α, then find an equation and graph.

$$6x^2 - 6xy + 14y^2 - 45 = 0$$
$$A = 6,\, B = -6,\, C = 14,\, F = -45$$

$$\cot 2\alpha = \frac{A-C}{B} = \frac{6-14}{-6} = \frac{4}{3}$$
$$\csc^2 2\alpha = \cot^2 2\alpha + 1$$
$$\csc^2 2\alpha = \left(\frac{4}{3}\right)^2 + 1 = \frac{25}{9}$$
$$\csc 2\alpha = +\sqrt{\frac{25}{9}} = \frac{5}{3} \qquad (2\alpha \text{ is in the first quadrant.})$$

$$\sin 2\alpha = \frac{1}{\csc 2\alpha} = \frac{3}{5}$$
$$\sin^2\alpha + \cos^2 2\alpha = 1$$
$$\cos^2 2\alpha = 1 - \sin^2 2\alpha$$
$$\cos^2 2\alpha = 1 - \left(\frac{3}{5}\right)^2 = \frac{16}{25}$$
$$\cos 2\alpha = +\sqrt{\frac{16}{25}} = \frac{4}{5} \qquad \begin{array}{l}(2\alpha \text{ is in the} \\ \text{first quadrant.})\end{array}$$

$$\sin\alpha = \sqrt{\frac{1-\left(\frac{4}{5}\right)}{2}} = \frac{\sqrt{10}}{10}, \quad \cos\alpha = \sqrt{\frac{1+\left(\frac{4}{5}\right)}{2}} = \frac{3\sqrt{10}}{10}$$

$$\alpha = 18.4°$$
$$A' = A\cos^2\alpha + B\cos\alpha\sin\alpha + C\sin^2\alpha$$
$$= 6\left(\frac{3\sqrt{10}}{10}\right)^2 - 6\left(\frac{3\sqrt{10}}{10}\right)\left(\frac{\sqrt{10}}{10}\right) + 14\left(\frac{\sqrt{10}}{10}\right)^2 = 5$$
$$C' = A\sin^2\alpha - B\cos\alpha\sin\alpha + C\cos^2\alpha$$
$$= 6\left(\frac{\sqrt{10}}{10}\right)^2 + 6\left(\frac{3\sqrt{10}}{10}\right)\left(\frac{\sqrt{10}}{10}\right) + 14\left(\frac{3\sqrt{10}}{10}\right)^2 = 15$$
$$F' = F = -45$$

$$5x'^2 + 15y'^2 - 45 = 0 \text{ or } \frac{(x')^2}{9} + \frac{(y')^2}{3} = 1$$

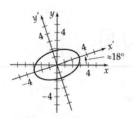

17. Find α, then find an equation and graph.

$$x^2 + 4xy - 2y^2 - 1 = 0$$
$$A = 1,\, B = 4,\, C = -2,\, F = -1$$

$$\cot 2\alpha = \frac{A-C}{B} = \frac{1-(-2)}{4} = \frac{3}{4}$$
$$\csc^2 2\alpha = \cot^2 2\alpha + 1$$
$$\csc^2 2\alpha = \left(\frac{3}{4}\right)^2 + 1 = \frac{25}{16}$$
$$\csc 2\alpha = +\sqrt{\frac{25}{16}} = \frac{5}{4} \qquad (2\alpha \text{ is in the first quadrant.})$$

$$\sin 2\alpha = \frac{1}{\csc 2\alpha} = \frac{4}{5}$$
$$\sin^2\alpha + \cos^2 2\alpha = 1$$
$$\cos^2 2\alpha = 1 - \sin^2\alpha$$
$$\cos^2 2\alpha = 1 - \left(\frac{4}{5}\right)^2 = \frac{9}{25}$$
$$\cos 2\alpha = +\sqrt{\frac{9}{25}} = \frac{3}{5} \qquad \begin{array}{l}(2\alpha \text{ is in the} \\ \text{first quadrant.})\end{array}$$

$$\sin\alpha = \sqrt{\frac{1-\left(\frac{3}{5}\right)}{2}} = \frac{\sqrt{5}}{5}, \quad \cos\alpha = \sqrt{\frac{1+\left(\frac{3}{5}\right)}{2}} = \frac{2\sqrt{5}}{5}$$

$$\alpha \approx 26.6°$$
$$A' = A\cos^2\alpha + B\cos\alpha\sin\alpha + C\sin^2\alpha$$
$$= 1\left(\frac{2\sqrt{5}}{5}\right)^2 + 4\left(\frac{2\sqrt{5}}{5}\right)\left(\frac{\sqrt{5}}{5}\right) - 2\left(\frac{\sqrt{5}}{5}\right)^2 = 2$$
$$C' = A\sin^2\alpha - B\cos\alpha\sin\alpha + C\cos^2\alpha$$
$$= 1\left(\frac{\sqrt{5}}{5}\right)^2 - 4\left(\frac{2\sqrt{5}}{5}\right)\left(\frac{\sqrt{5}}{5}\right) - 2\left(\frac{2\sqrt{5}}{5}\right)^2 = -3$$
$$F' = F = -1$$

$$2(x')^2 - 3(y')^2 = 1$$

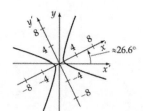

19. Find α, then find an equation and graph.

$3x^2 + 2\sqrt{3}xy + y^2 + 2x - 2\sqrt{3}y + 16 = 0$

$A = 3, B = 2\sqrt{3}, C = 1, D = 2, E = -2\sqrt{3}, F = 16$

$\cot 2\alpha = \dfrac{A - C}{B} = \dfrac{3 - 1}{2\sqrt{3}} = \dfrac{\sqrt{3}}{3}$

$\csc^2 2\alpha = \cot^2 2\alpha + 1$

$\csc^2 2\alpha = \left(\dfrac{\sqrt{3}}{3}\right)^2 + 1 = \dfrac{4}{3}$

$\csc 2\alpha = +\sqrt{\dfrac{4}{3}} = \dfrac{2\sqrt{3}}{3}$ (2α is in the first quadrant.)

$\sin 2\alpha = \dfrac{1}{\csc 2\alpha} = \dfrac{\sqrt{3}}{2}$

$\sin^2 \alpha + \cos^2 2\alpha = 1$

$\quad \cos^2 2\alpha = 1 - \sin^2 2\alpha$

$\quad \cos^2 2\alpha = 1 - \left(\dfrac{\sqrt{3}}{2}\right)^2 = \dfrac{1}{4}$

$\quad\quad \cos 2\alpha = +\sqrt{\dfrac{1}{4}} = \dfrac{1}{2}$ (2α is in the first quadrant.)

$\sin \alpha = \sqrt{\dfrac{1 - \left(\frac{1}{2}\right)}{2}} = \dfrac{1}{2}, \quad \cos \alpha = \sqrt{\dfrac{1 + \left(\frac{1}{2}\right)}{2}} = \dfrac{\sqrt{3}}{2}$

$\alpha = 30°$

$A' = A\cos^2\alpha + B\cos\alpha\sin\alpha + C\sin^2\alpha$

$\quad = 3\left(\dfrac{\sqrt{3}}{2}\right)^2 + 2\sqrt{3}\left(\dfrac{\sqrt{3}}{2}\right)\left(\dfrac{1}{2}\right) + 1\left(\dfrac{1}{2}\right)^2 = 4$

$C' = A\sin^2\alpha - B\cos\alpha\sin\alpha + C\cos^2\alpha$

$\quad = 3\left(\dfrac{1}{2}\right)^2 - 2\sqrt{3}\left(\dfrac{\sqrt{3}}{2}\right)\left(\dfrac{1}{2}\right) + 1\left(\dfrac{\sqrt{3}}{2}\right)^2 = 0$

$D' = D\cos\alpha + E\sin\alpha = 2\left(\dfrac{\sqrt{3}}{2}\right) - 2\sqrt{3}\left(\dfrac{1}{2}\right) = 0$

$E' = -D\sin\alpha + E\cos\alpha = -2\left(\dfrac{1}{2}\right) - 2\sqrt{3}\left(\dfrac{\sqrt{3}}{2}\right) = -4$

$F' = F = 16$

$4(x')^2 - 4y' + 16 = 0$ or $y' = (x')^2 + 4$

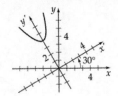

21. Find α, then find an equation and graph.

$9x^2 - 24xy + 16y^2 - 40x - 30y + 100 = 0$

$A = 9, B = -24, C = 16, D = -40, E = -30 \; F = 100$

$\cot 2\alpha = \dfrac{A - C}{B} = \dfrac{9 - 16}{-24} = \dfrac{7}{24}$

$\csc^2 2\alpha = \cot^2 2\alpha + 1$

$\csc^2 2\alpha = \left(\dfrac{7}{24}\right)^2 + 1 = \dfrac{625}{576}$

$\csc 2\alpha = +\sqrt{\dfrac{625}{576}} = \dfrac{25}{24}$ (2α is in the first quadrant.)

$\sin 2\alpha = \dfrac{1}{\csc 2\alpha} = \dfrac{24}{25}$

$\sin^2 \alpha + \cos^2 2\alpha = 1$

$\quad \cos^2 2\alpha = 1 - \sin^2 2\alpha$

$\quad \cos^2 2\alpha = 1 - \left(\dfrac{24}{25}\right)^2 = \dfrac{49}{625}$

$\quad\quad \cos 2\alpha = +\sqrt{\dfrac{49}{625}} = \dfrac{7}{25}$ (2α is in the first quadrant.)

$\sin \alpha = \sqrt{\dfrac{1 - \left(\frac{7}{25}\right)}{2}} = \dfrac{3}{5}, \quad \cos \alpha = \sqrt{\dfrac{1 + \left(\frac{7}{25}\right)}{2}} = \dfrac{4}{5}$

$\alpha \approx 36.9°$

$A' = A\cos^2\alpha + B\cos\alpha\sin\alpha + C\sin^2\alpha$

$\quad = 9\left(\dfrac{4}{5}\right)^2 - 24\left(\dfrac{4}{5}\right)\left(\dfrac{3}{5}\right) + 16\left(\dfrac{3}{5}\right)^2 = 0$

$C' = A\sin^2\alpha - B\cos\alpha\sin\alpha + C\cos^2\alpha$

$\quad = 9\left(\dfrac{3}{5}\right)^2 + 24\left(\dfrac{4}{5}\right)\left(\dfrac{3}{5}\right) + 16\left(\dfrac{4}{5}\right)^2 = 25$

$D' = D\cos\alpha + E\sin\alpha = -40\left(\dfrac{4}{5}\right) - 30\left(\dfrac{3}{5}\right) = -50$

$E' = -D\sin\alpha + E\cos\alpha = 40\left(\dfrac{3}{5}\right) - 30\left(\dfrac{4}{5}\right) = 0$

$F' = F = 100$

$25(y')^2 - 50x' + 100 = 0$ or $(y')^2 = 2(x' - 2)$

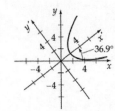

23. Find α, then find an equation and graph.

$$6x^2 + 24xy - y^2 - 12x + 26y + 11 = 0$$

$$A = 6, B = 24, C = -1, D = -12, E = 26 \ F = 11$$

$$\cot 2\alpha = \frac{A-C}{B} = \frac{6-(-1)}{24} = \frac{7}{24}$$

$$\csc^2 2\alpha = \cot^2 2\alpha + 1$$

$$\csc^2 2\alpha = \left(\frac{7}{24}\right)^2 + 1 = \frac{625}{576}$$

$$\csc 2\alpha = +\sqrt{\frac{625}{576}} = \frac{25}{24} \qquad \text{(2α is in the first quadrant.)}$$

$$\sin 2\alpha = \frac{1}{\csc 2\alpha} = \frac{24}{25}$$

$$\sin^2 2\alpha + \cos^2 2\alpha = 1$$

$$\cos^2 2\alpha = 1 - \sin^2 2\alpha$$

$$\cos^2 2\alpha = 1 - \left(\frac{24}{25}\right)^2 = \frac{49}{625}$$

$$\cos 2\alpha = +\sqrt{\frac{49}{625}} = \frac{7}{25} \qquad \text{(2α is in the first quadrant.)}$$

$$\sin \alpha = \sqrt{\frac{1-\left(\frac{7}{25}\right)}{2}} = \frac{3}{5}, \quad \cos \alpha = \sqrt{\frac{1+\left(\frac{7}{25}\right)}{2}} = \frac{4}{5}$$

$$\alpha \approx 36.9°$$

$$A' = A\cos^2 \alpha + B\cos \alpha \sin \alpha + C\sin^2 \alpha$$

$$= 6\left(\frac{4}{5}\right)^2 + 24\left(\frac{4}{5}\right)\left(\frac{3}{5}\right) - 1\left(\frac{3}{5}\right)^2 = 15$$

$$C' = A\sin^2 \alpha - B\cos \alpha \sin \alpha + C\cos^2 \alpha$$

$$= 6\left(\frac{3}{5}\right)^2 - 24\left(\frac{4}{5}\right)\left(\frac{3}{5}\right) - 1\left(\frac{4}{5}\right)^2 = -10$$

$$D' = D\cos \alpha + E\sin \alpha = -12\left(\frac{4}{5}\right) + 26\left(\frac{3}{5}\right) = 6$$

$$E' = -D\sin \alpha + E\cos \alpha = 12\left(\frac{3}{5}\right) + 26\left(\frac{4}{5}\right) = 28$$

$$F' = F = 11$$

$$15(x')^2 - 10(y')^2 + 6x' + 28y' + 11 = 0$$

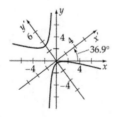

25. Graph the equation using a graphing utility.

$$6x^2 - xy + 2y^2 + 4x - 12y + 7 = 0$$

$$A = 6, B = -1, C = 2, D = 4, E = -12, F = 7$$

Graph

$$y = \frac{-(-x-12) \pm \sqrt{(-x-12)^2 - 8(6x^2 + 4x + 7)}}{4}$$

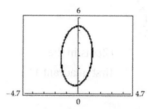

27. Graph the equation using a graphing utility.

$$x^2 - 6xy + y^2 - 2x - 5y + 4 = 0$$

$$A = 1, B = -6, C = 1, D = -2, E = -5, F = 4$$

Graph

$$y = \frac{-(-6x-5) \pm \sqrt{(-6x-5)^2 - 4(x^2 - 2x + 4)}}{2}$$

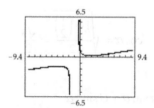

29. Graph the equation using a graphing utility.

$$3x^2 - 6xy + 3y^2 + 10x - 8y - 2 = 0$$

$$A = 3, B = -6, C = 3, D = 10, E = -8, F = -2$$

Graph

$$y = \frac{-(-6x-8) \pm \sqrt{(-6x-8)^2 - 12(3x^2 + 10x - 2)}}{6}$$

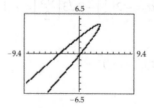

31. Find the equations of the asymptotes.

$$\frac{2x'^2}{1} - \frac{3y'^2}{1} = 1 \qquad \sin\alpha = \frac{\sqrt{5}}{5} \qquad \cos\alpha = \frac{2\sqrt{5}}{5}$$

$$a^2 = \frac{1}{2}; \quad a = \frac{\sqrt{2}}{2}$$

$$b^2 = \frac{1}{3}; \quad b = \frac{\sqrt{3}}{3}$$

Asymptotes: $y' = \pm\dfrac{b}{a}x'$ or $y' = \pm\dfrac{\sqrt{6}}{3}x'$

Using the transformation formulas for x' and y' yields

$$y\cos\alpha - x\sin\alpha = \pm\frac{\sqrt{6}}{3}(x\cos\alpha + y\sin\alpha)$$

$$\frac{2\sqrt{5}}{5}y - \frac{\sqrt{5}}{5}x = \pm\frac{\sqrt{6}}{3}\left(\frac{2\sqrt{5}}{5}x + \frac{\sqrt{5}}{5}y\right)$$

$$\frac{2\sqrt{5}}{5}y - \frac{\sqrt{5}}{5}x = \pm\left(\frac{2\sqrt{30}}{15}x + \frac{\sqrt{30}}{15}y\right)$$

Multiplying both sides of the equation by $\dfrac{15}{\sqrt{5}}$ yields

$$6y - 3x = \pm\left(2\sqrt{6}x + \sqrt{6}y\right)$$

$$6y - 3x = 2\sqrt{6}x + \sqrt{6}y$$

$$6y - \sqrt{6}y = 3x + 2\sqrt{6}x$$

$$y = \frac{3 + 2\sqrt{6}}{6 - \sqrt{6}}x$$

and

$$6y - 3x = -\left(2\sqrt{6}x + \sqrt{6}y\right)$$

$$6y + \sqrt{6}y = 3x - 2\sqrt{6}x$$

$$y = \frac{3 - 2\sqrt{6}}{6 + \sqrt{6}}x$$

Rationalizing the denominators, we obtain

$$y = \frac{2 + \sqrt{6}}{2}x \qquad \text{and} \qquad y = \frac{2 - \sqrt{6}}{2}x$$

33. Find the coordinates of the foci.

From Exercise 11, $\dfrac{x'^2}{9} + \dfrac{y'^2}{3} = 1$, $\sin\alpha = \dfrac{\sqrt{10}}{10}$,

$\cos\alpha = \dfrac{3\sqrt{10}}{10}$, $a^2 = 9$, $b^2 = 3$, $c^2 = 9 - 3 = 6$.

Thus $c = \sqrt{6}$.

Foci in $x'y'$-coordinates are $(\pm\sqrt{6}, 0)$.

Thus $x' = \pm\sqrt{6}$, $y' = 0$.

$$x = x'\cos\alpha - y'\sin\alpha \qquad\qquad y = y'\cos\alpha + x'\sin\alpha$$

$$= \pm\sqrt{6}\left(\frac{3\sqrt{10}}{10}\right) - 0\cdot\frac{\sqrt{10}}{10} \qquad = 0\cdot\cos\alpha \pm\sqrt{6}\left(\frac{\sqrt{10}}{10}\right)$$

$$= \pm\frac{3\sqrt{15}}{5} \qquad\qquad\qquad\qquad = \pm\frac{\sqrt{15}}{5}$$

Foci in the xy-coordinate system are

$$\left(\frac{3\sqrt{15}}{5}, \frac{\sqrt{15}}{5}\right) \text{ and } \left(-\frac{3\sqrt{15}}{5}, -\frac{\sqrt{15}}{5}\right).$$

35. Identify the graph using the Conic Identification Theorem.

$$x^2 + xy - y^2 - 40 = 0$$

$$A = 1, B = 1, C = -1$$

Since $B^2 - 4AC = 1^2 - 4(1)(-1) = 5 > 0$, the graph is a hyperbola.

37. Identify the graph using the Conic Identification Theorem.

$$3x^2 + 2\sqrt{3}xy + y^2 - 3x + 2y + 20 = 0$$

$$A = 3, B = 2\sqrt{3}, C = 1$$

Since $B^2 - 4AC = (2\sqrt{3})^2 - 4(3)(1) = 0$, the graph is a parabola.

39. Identify the graph using the Conic Identification Theorem.

$$4x^2 - 4xy + y^2 - 12y - 20 = 0$$

$$A = 4, B = -4, C = 1$$

Since $B^2 - 4AC = (-4)^2 - 4(4)(1) = 0$, the graph is a parabola.

41. Identify the graph using the Conic Identification Theorem.

$$5x^2 - 6\sqrt{3}xy - 11y^2 + 4x - 3y + 2 = 0$$

$$A = 5, B = -6\sqrt{3}, C = -11$$

Since $B^2 - 4AC = (-6\sqrt{3})^2 - 4(5)(-11) = 328 > 0$, the graph is a hyperbola.

43. Identify the graph using the Conic Identification Theorem.

$$6x^2 + 2\sqrt{3}xy + 5y^2 - 3x + 2y - 20 = 0$$

$$A = 6, B = 2\sqrt{3}, C = 5$$

Since $B^2 - 4AC = (2\sqrt{3})^2 - 4(6)(5) = -108 < 0$, the graph is an ellipse or a circle.

45. Show the statement to be true.

$$x^2 + y^2 = r^2$$

Substitute $x = x'\cos\alpha - y'\sin\alpha$ and $y = y'\cos\alpha + x'\sin\alpha$.

$$r^2 = (x'\cos\alpha - y'\sin\alpha)^2 + (y'\cos\alpha + x'\sin\alpha)^2$$

$$r^2 = x'^2\cos^2\alpha - 2x'y'\cos\alpha\sin\alpha + y'^2\sin^2\alpha$$
$$\quad + y'^2\cos^2\alpha + 2x'y'\cos\alpha\sin\alpha + x'^2\sin^2\alpha$$

$$r^2 = x'^2(\cos^2\alpha + \sin^2\alpha)$$
$$\quad + x'y'(2\cos\alpha\sin\alpha - 2\cos\alpha\sin\alpha)$$
$$\quad + y'^2(\sin^2\alpha + \cos^2\alpha)$$

$$r^2 = x'^2(1) + x'y'(0) + y'^2(1)$$

$$r^2 = x'^2 + y'^2$$

47. Find the equation of the ellipse.

Vertices $(2, 4)$ and $(-2, -4)$ imply that the major axis is on the line $y = 2x$. Consider an $x'y'$-coordinate system rotated an angle α, where $\tan\alpha = 2$. From this equation, using identities,

$$\cos\alpha = \frac{\sqrt{5}}{5} \text{ and } \sin\alpha = \frac{2\sqrt{5}}{5}.$$

The equation of the ellipse in the $x'y'$-coordinate system is

$$\frac{x'^2}{20} + \frac{y'^2}{10} = 1 \qquad (1)$$

This follows from the fact that $(2, 4)$ in xy-coordinates is $(\sqrt{20},\ 0)$ in $x'y'$-coordinates. Therefore, $a = \sqrt{20}$. Also, $(\sqrt{2},\ 2\sqrt{2})$ in xy-coordinates is $(\sqrt{10},\ 0)$ in $x'y'$-coordinates. Therefore, $c = \sqrt{10}$. Thus $b = \sqrt{10}$.

Now let $x' = \frac{\sqrt{5}}{5}x + \frac{2\sqrt{5}}{5}y$ and $y' = \frac{\sqrt{5}}{5}y - \frac{2\sqrt{5}}{5}x$ and substitute into Equation (1):

$$\frac{\left(\frac{\sqrt{5}}{5}x + \frac{2\sqrt{5}}{5}y\right)^2}{20} + \frac{\left(\frac{\sqrt{5}}{5}y - \frac{2\sqrt{5}}{5}x\right)^2}{10} = 1$$

Simplifying, we have

$$\frac{\left(\frac{1}{5}x^2 + \frac{4}{5}xy + \frac{4}{5}y^2\right)}{20} + \frac{\left(\frac{4}{5}x^2 - \frac{4}{5}xy + \frac{1}{5}y^2\right)}{10} = 1$$

$$\frac{\frac{9}{5}x^2 - \frac{4}{5}xy + \frac{6}{5}y^2}{20} = 1$$

$$\frac{9x^2 - 4xy + 6y^2}{100} = 1$$

or $9x^2 - 4xy + 6y^2 = 100$

49. Show that $A' + C' = A + C$.

$$A' + C' = A\cos^2\alpha + B\cos\alpha\sin\alpha + C\sin^2\alpha$$
$$\quad + A\sin^2\alpha - B\cos\alpha\sin\alpha + C\cos^2\alpha$$
$$= A(\cos^2\alpha + \sin^2\alpha) + B(\cos\alpha\sin\alpha - \cos\alpha\sin\alpha)$$
$$\quad + C(\sin^2\alpha + \cos^2\alpha)$$
$$= A + C$$

51. Transform the equation.

We are given $10x^2 + 24xy + 17y^2 - 26 = 0$. Thus, $A = 10$, $B = 24$, $C = 17$, $D = 0$, $E = 0$, and $F = 26$. We seek the equation of the form

$A'(x')^2 + C'(y')^2 - F = 0$. The invariant theorems in Exercises 49 and 50 indicate that $A' + C' = A + C$ and

$$B'^2 - 4A'C' = B^2 - 4AC.$$

$A' + C' = A + C$ implies that $A' + C' = 10 + 17 = 27$.

Hence $C' = 27 - A'$.

$$B'^2 - 4A'C' = B^2 - 4AC = 24^2 - 4(10)(17) = -104$$

In the $x'y'$-coordinate system $B' = 0$, so we have

$$B'^2 - 4A'C' = -104$$
$$-4A'C' = -104$$
$$-4A'(27 - A') = -104 \quad \text{Substitute } C' = 27 - A'.$$
$$4(A')^2 - 108A' + 104 = 0$$
$$(A')^2 - 27A' + 26 = 0$$
$$(A' - 1)(A' - 26) = 0$$

Hence $A' = 1$ or $A' = 26$. If $A' = 1$, then $C' = 26$. If $A' = 26$, then $C' = 1$. Thus we conclude that the conic given by $10x^2 + 24xy + 17y^2 - 26 = 0$ is also represented by

$$26(x')^2 + 1(y')^2 - 26 = 0 \text{ or } 1(x')^2 + 26(y')^2 - 26 = 0$$

in the $x'y'$-coordinate system. An examination of the

following graph shows that we should pick

$(x')^2 + 26(y')^2 - 26 = 0$ if we plan to obtain the

x'-axis by a 53° counterclockwise rotation of the x-

axis and that we should pick $26(x')^2 + (y')^2 - 26 = 0$

if we plan to obtain the x'-axis by a 143°

counterclockwise rotation of the x-axis.

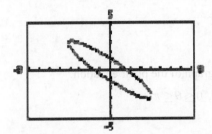

Mid-Chapter 8 Quiz

1. Find the equation of the parabola in standard form.

vertex (6, 2), focus (4, 2)

$(y - k)^2 = 4p(x - h)$

$h = 6,\ k = 2$

Since the focus is $(h + p, k)$, $6 + p = 4$ and $p = -2$.

$(y - 2)^2 = 4(-2)(x - 6)$

$(y - 2)^2 = -8(x - 6)$

3. Find the vertex, focus, and directrix; sketch the graph.

$-2x + 3y^2 + 6y = 3$

$3y^2 + 6y = 2x + 3$

$y^2 + 2y = \dfrac{2}{3}x + 1$

$y^2 + 2y + 1 = \dfrac{2}{3}x + 1 + 1$

$(y + 1)^2 = \dfrac{2}{3}(x + 3)$

vertex $= (-3,\ -1)$

$4p = \dfrac{2}{3},\ \ p = \dfrac{1}{6}$

focus $= \left(-3 + \dfrac{1}{6},\ -1\right) = \left(-\dfrac{17}{6},\ -1\right)$

directrix: $y = -3 - \dfrac{1}{6} = -\dfrac{19}{6}$

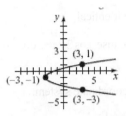

5. Find the center, vertices, foci, and asymptotes.

$4x^2 - 9y^2 - 24x + 18y - 9 = 0$

$4(x^2 - 6x) - 9(y^2 - 2y) = 9$

$4(x^2 - 6x + 9) - 9(y^2 - 2y + 1) = 9 + 36 - 9$

$4(x - 3)^2 - 9(y - 1)^2 = 36$

$\dfrac{(x - 3)^2}{9} - \dfrac{(y - 1)^2}{4} = 1$

Center (3, 1)

Vertices $(3 \pm 3,\ 1) = (0,\ 1),\ (6,\ 1)$

Foci $\left(3 \pm \sqrt{13},\ 1\right) = \left(3 + \sqrt{13},\ 1\right),\ \left(3 - \sqrt{13},\ 1\right)$

Asymptotes $y - 1 = \pm\dfrac{2}{3}(x - 3)$

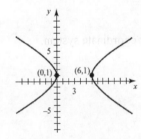

Prepare for Section 8.5

P1. Determine whether $\sin x$ is an even or odd function.

$\sin(-x) = -\sin x$ odd function

P3. Solve.

$\tan \alpha = -\sqrt{3}$

$\alpha = \dfrac{2\pi}{3}, \dfrac{5\pi}{3}$

P5. Write in simplest form.

$(r\cos\theta)^2 + (r\sin\theta)^2 = r^2\cos^2\theta + r^2\sin^2\theta$

$\qquad\qquad\qquad\quad = r^2(\cos^2\theta + \sin^2\theta)$

$\qquad\qquad\qquad\quad = r^2$

Section 8.5 Exercises

1. Determine the quadrant.

The point (–2, 150°) is in quadrant IV.

3. Explain why the graphs are identical.

The graphs are identical because $\cos 2\theta = 2\cos^2 \theta - 1$ is an identity.

5. Plot the point on the polar coordinate system.

7. Plot the point on the polar coordinate system.

9. Plot the point on the polar coordinate system.

11. Plot the point on the polar coordinate system.

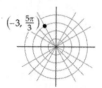

13. Sketch the graph of the polar equation.

$r = 3, \ 0 \le \theta \le 2\pi$

15. Sketch the graph of the polar equation.

$\theta = 2$

17. Sketch the graph of the polar equation.

$r = 6\cos\theta, \ 0 \le \theta \le \pi$

19. Sketch the graph of the polar equation.

$r = 4\cos 2\theta, \ 0 \le \theta \le 2\pi$

21. Sketch the graph of the polar equation.

$r = 2\sin 5\theta, \ 0 \le \theta \le \pi$

23. Sketch the graph of the polar equation.

$r = 2 - 3\sin\theta, \ 0 \le \theta \le 2\pi$

25. Sketch the graph of the polar equation.

$r = 4 + 3\sin\theta, \ 0 \le \theta \le 2\pi$

27. Sketch the graph of the polar equation.

$r^2 = -16\cos 2\theta$

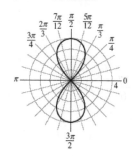

29. Sketch the graph of the polar equation.

$$r = 2[1 + 1.5\sin(-\theta)], \ 0 \le \theta \le 2\pi$$

31. Use a graphing utility to graph the equation.

$$r = 3 + 3\cos\theta \ \text{for} \ 0 \le \theta \le 2\pi.$$

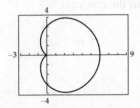

33. Use a graphing utility to graph the equation.

$$r = 4\cos 3\theta \ \text{for} \ 0 \le \theta \le \pi.$$

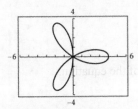

35. Use a graphing utility to graph the equation.

$$r = 3\sec\theta \ \text{for} \ 0 \le \theta \le \pi.$$

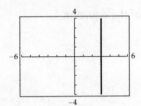

37. Use a graphing utility to graph the equation.

$$r = -5\csc\theta \ \text{for} \ 0 \le \theta \le \pi.$$

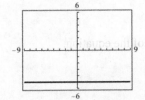

39. Use a graphing utility to graph the equation.

$$r = 4\sin(3.5\theta) \ \text{for} \ 0 \le \theta \le 4\pi.$$

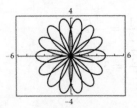

41. Use a graphing utility to graph the equation.

$$r = \theta \ \text{for} \ 0 \le \theta \le 6\pi.$$

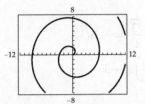

43. Use a graphing utility to graph the equation.

$$r = 2^{\theta} \ \text{for} \ 0 \le \theta \le 2\pi.$$

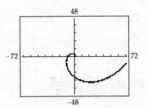

45. Use a graphing utility to graph the equation.

$$r = \frac{6\cos 7\theta + 2\cos 3\theta}{\cos\theta} \ \text{for} \ 0 \le \theta \le 2\pi \ \text{with} \ \theta_{\text{step}} = \frac{\pi}{200}$$

(Some graphing utilities may draw a false asymptote in "connected" mode.)

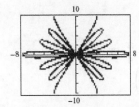

47. Find rectangular coordinates from polar coordinates.

$$\left(4, \ \frac{\pi}{2}\right)$$

$$x = 4\cos\frac{\pi}{2} = 4(0) = 0$$

$$y = 4\sin\frac{\pi}{2} = 4(1) = 4$$

$$(0, \ 4)$$

49. Find rectangular coordinates from polar coordinates.

$$\left(2\sqrt{3},\ \frac{2\pi}{3}\right)$$

$$x = 2\sqrt{3}\cos\frac{2\pi}{3} = 2\sqrt{3}\left(-\frac{1}{2}\right) = -\sqrt{3}$$

$$y = 2\sqrt{3}\sin\frac{2\pi}{3} = 2\sqrt{3}\left(\frac{\sqrt{3}}{2}\right) = 3$$

$$\left(-\sqrt{3},\ 3\right)$$

51. Find rectangular coordinates from polar coordinates.

$$\left(6, -\frac{5\pi}{6}\right)$$

$$x = 6\cos\left(-\frac{5\pi}{6}\right) = 6\left(-\frac{\sqrt{3}}{2}\right) = -3\sqrt{3}$$

$$y = 6\sin\left(-\frac{5\pi}{6}\right) = 6\left(-\frac{1}{2}\right) = -3$$

$$\left(-3\sqrt{3}, -3\right)$$

53. Find 2 pairs of polar coordinates in radians from rectangular coordinates.

$$\left(1, -\sqrt{3}\right) \text{ lies in quadrant IV}$$

$$r^2 = (1)^2 + \left(-\sqrt{3}\right)^2 = 4$$

$$r = 2 \text{ or } r = -2$$

$$\tan\theta = \frac{-\sqrt{3}}{1} = -\sqrt{3} \quad \Rightarrow \quad \theta = \frac{2\pi}{3} \text{ or } \frac{5\pi}{3}$$

$$\left(-2,\ \frac{2\pi}{3}\right) \text{ or } \left(2,\ \frac{5\pi}{3}\right)$$

55. Find 2 pairs of polar coordinates in radians from rectangular coordinates.

$$\left(-3,\ -3\right) \text{ lies in quadrant III}$$

$$r^2 = (-3)^2 + (-3)^2 = 18$$

$$r = 3\sqrt{2} \text{ or } r = -3\sqrt{2}$$

$$\tan\theta = \frac{-3}{-3} = 1 \quad \Rightarrow \quad \theta = \frac{\pi}{4} \text{ or } \frac{5\pi}{4}$$

$$\left(3\sqrt{2},\ \frac{5\pi}{4}\right) \text{ or } \left(-3\sqrt{2},\ \frac{\pi}{4}\right)$$

57. Find 2 pairs of polar coordinates in degrees from rectangular coordinates. Round to the nearest tenth.

$$(2,\ 2) \text{ lies in quadrant I}$$

$$r^2 = (2)^2 + (2)^2 = 8$$

$$r = 2\sqrt{2} \text{ or } r = -2\sqrt{2}$$

$$\tan\theta = \frac{2}{2} = 1 \quad \Rightarrow \quad \theta = 45° \text{ or } 135°$$

$$\left(2\sqrt{2},\ 45°\right) \text{ or } \left(-2\sqrt{2},\ 135°\right)$$

59. Find 2 pairs of polar coordinates in degrees from rectangular coordinates. Round to the nearest tenth.

$$(-8,\ 15) \text{ lies in quadrant II}$$

$$r^2 = (-8)^2 + (15)^2 = 17$$

$$r = 17 \text{ or } r = -17$$

$$\tan\theta = \frac{15}{-8} \quad \Rightarrow \quad \theta = 298.1° \text{ or } 118.1°$$

$$\left(17,\ 118.1°\right) \text{ or } \left(-17,\ 298.1°\right)$$

61. Find the rectangular form of the equation.

$$r = 3\cos\theta$$

$$r - 3\cos\theta = 0$$

$$r^2 - 3r\cos\theta = 0$$

$$x^2 + y^2 - 3x = 0$$

63. Find the rectangular form of the equation.

$$r = 3\sec\theta$$

$$r = \frac{3}{\cos\theta}$$

$$r\cos\theta = 3$$

$$x = 3$$

65. Find the rectangular form of the equation.

$$r = 4$$

$$\sqrt{x^2 + y^2} = 4$$

$$x^2 + y^2 = 16$$

67. Find the rectangular form of the equation.

$$\theta = \frac{\pi}{3}$$

$$\tan\theta = \frac{y}{x}$$

$$\tan\frac{\pi}{3} = \frac{y}{x}$$

$$\frac{\sqrt{3}}{1} = \frac{y}{x}$$

$$y = \sqrt{3}x$$

69. Find the rectangular form of the equation.

$$r = \tan\theta$$

$$r = \frac{\sin\theta}{\cos\theta}$$

$$r\cos\theta = \sin\theta$$

$$r\cos\theta - \sin\theta = 0$$

$$r^2\cos\theta - r\sin\theta = 0$$

$$\sqrt{x^2 + y^2}\,(x) - y = 0$$

$$\sqrt{x^2 + y^2} = \frac{y}{x}$$

$$x^2 + y^2 = \frac{y^2}{x^2}$$

$$x^4 - y^2 + x^2 y^2 = 0$$

71. Find the rectangular form of the equation.

$$r = \frac{2}{1 + \cos\theta}$$

$$r + r\cos\theta = 2$$

$$\sqrt{x^2 + y^2} + x = 2$$

$$\sqrt{x^2 + y^2} = 2 - x$$

$$x^2 + y^2 = 4 - 4x + x^2$$

$$y^2 + 4x - 4 = 0$$

73. Find the rectangular form of the equation.

$$r(\sin\theta - 2\cos\theta) = 6$$

$$r\sin\theta - 2r\cos\theta = 6$$

$$y - 2x = 6$$

$$y = 2x + 6$$

75. Find the polar form of the equation.

$$y = 2$$

$$r\sin\theta = 2$$

$$r = 2\csc\theta$$

77. Find the polar form of the equation.

$$y = x$$

$$r\sin\theta = r\cos\theta$$

$$\tan\theta = 1$$

$$\theta = \frac{\pi}{4}$$

79. Find the polar form of the equation.

$$x = 3$$

$$r\cos\theta = 3$$

$$r = 3\sec\theta$$

81. Find the polar form of the equation.

$$x^2 + y^2 = 4$$

$$r^2 = 4$$

$$r = 2$$

83. Find the polar form of the equation.

$$\left(x^2 + y^2\right)^{3/2} - y = 0$$

$$\left(r^2\cos^2\theta + r^2\sin^2\theta\right)^{3/2} - r\sin\theta = 0$$

$$\left[r^2(\cos^2\theta + \sin^2\theta)\right]^{3/2} = r\sin\theta$$

$$\left[r^2(1)\right]^{3/2} = r\sin\theta$$

$$r^3 = r\sin\theta$$

$$r^2 = \sin\theta$$

85. Find the polar form of the equation.

$$x^2 - y^2 = 25$$

$$r^2\cos^2\theta - r^2\sin^2\theta = 25$$

$$r^2(\cos^2\theta - \sin^2\theta) = 25$$

$$r^2(\cos 2\theta) = 25$$

87. Use a graphing utility to graph the equation.

$$r = 3\cos\left(\theta + \frac{\pi}{4}\right)$$

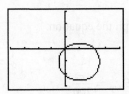

89. Use a graphing utility to graph the equation.

$$r = 2\sin\left(2\theta - \frac{\pi}{3}\right)$$

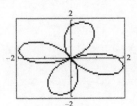

91. Use a graphing utility to graph the equation.

$$r = 2 + 2\sin\left(\theta - \frac{\pi}{6}\right)$$

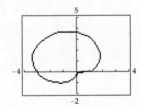

93. Use a graphing utility to graph the equation.

$$r = 1 + 3\cos\left(\theta + \frac{\pi}{3}\right)$$

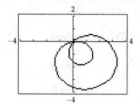

95. Use a graphing utility to graph the equation.

$$r^2 = 4\cos 2\theta$$

Enter as $r = \sqrt{4\cos 2\theta}$ and

$r = -\sqrt{4\cos 2\theta}$ for $0 \le \theta \le 4\pi$.

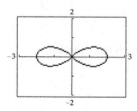

97. Use a graphing utility to graph the equation.

$$r = 2(1 + \sec\theta)$$

Graph for $0 \le \theta \le 2\pi$ with $\theta_{step} = \frac{\pi}{200}$.

(Some graphing utilities may produce a false asymptote in "connected" mode.)

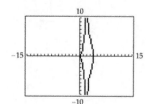

99. Use a graphing utility to graph the equation.

$$r\theta = 2$$

Graph $r = \frac{2}{\theta}$ for $-4\pi < \theta < 4\pi$.

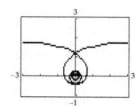

101. Use a graphing utility to graph the equation.

$$r = |\theta|$$

Graph for $-30 \le \theta \le 30$.

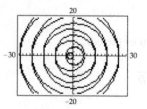

103. Use a graphing utility to graph for each domain.

a. $0 \le \theta \le 5\pi$

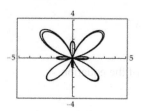

b. $0 \le \theta \le 20\pi$

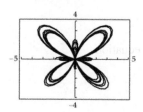

105. a. Verify the graph is a circle.

$$r = a\sin\theta + b\cos\theta$$
$$r^2 = ar\sin\theta + br\cos\theta$$
$$x^2 + y^2 = ay + bx$$
$$x^2 - bx + y^2 - ay = 0$$
$$x^2 - bx + \frac{b^2}{4} + y^2 - ay + \frac{a^2}{4} = \frac{a^2}{4} + \frac{b^2}{4}$$
$$\left(x - \frac{b}{2}\right)^2 + \left(y - \frac{a}{2}\right)^2 = \frac{a^2 + b^2}{4}$$
$$\left(x - \frac{b}{2}\right)^2 + \left(y - \frac{a}{2}\right)^2 = \left(\frac{\sqrt{a^2 + b^2}}{2}\right)^2$$

b. Find the center and radius.

Thus the center of the circle is $\left(\frac{b}{2}, \frac{a}{2}\right)$ and the radius

is $\frac{1}{2}\sqrt{a^2 + b^2}$.

Prepare for Section 8.6

P1. Find the eccentricity of the graph.

$$\frac{x^2}{25} + \frac{y^2}{16} = 1$$

$$a^2 = 25, \ a = 5$$

$$b^2 = 16$$

$$c^2 = a^2 - b^2 = 25 - 16 = 9$$

$$c = 3$$

$$e = \frac{c}{a} = \frac{3}{5}$$

The eccentricity is $\frac{3}{5}$.

P3. Solve for y.

$$y = 2(1 + yx)$$

$$y = 2 + 2yx$$

$$y - 2yx = 2$$

$$y(1 - 2x) = 2$$

$$y = \frac{2}{1 - 2x}$$

P5. Determine which statement is true.

For a hyperbola, $e > 1$.

Section 8.6 Exercises

1. Determine the eccentricity of the ellipse.

$$r = \frac{10}{4 - 2\sin\theta} = \frac{\frac{5}{2}}{1 - \frac{1}{2}\sin\theta}$$

$$e = \frac{1}{2}$$

3. Determine the type of symmetry for $r = \dfrac{6}{1 + \sin\theta}$.

The parabola has a vertical axis of symmetry.

5. Describe and sketch the graph.

$$r = \frac{12}{3 - 6\cos\theta} = \frac{4}{1 - 2\cos\theta}$$

$e = 2$ The graph is a hyperbola.

The transverse axis is on the polar axis.

Let $\theta = 0$.

$$r = \frac{12}{3 - 6\cos 0} = \frac{12}{3 - 6} = -4$$

Let $\theta = \pi$.

$$r = \frac{12}{3 - 6\cos\pi} = \frac{12}{3 + 6} = \frac{4}{3}$$

The vertices are at $(-4, \ 0)$ and $\left(\frac{4}{3}, \ \pi\right)$.

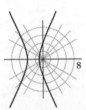

7. Describe and sketch the graph.

$$r = \frac{8}{4 + 3\sin\theta} = \frac{2}{1 + \frac{3}{4}\sin\theta}$$

$e = \frac{3}{4}$ The graph is an ellipse.

The major axis is on the line $\theta = \frac{\pi}{2}$.

Let $\theta = \frac{\pi}{2}$.

$$r = \frac{8}{4 + 3\sin\frac{\pi}{2}} = \frac{8}{4 + 3} = \frac{8}{7}$$

Let $\theta = \frac{3\pi}{2}$.

$$r = \frac{8}{4 + 3\sin\frac{3\pi}{2}} = \frac{8}{4 - 3} = 8$$

Vertices on major axis are at $\left(\frac{8}{7}, \ \frac{\pi}{2}\right)$ and $\left(8, \ \frac{3\pi}{2}\right)$.

Let $\theta = 0$.

$$r = \frac{8}{4 + 3\sin 0} = \frac{8}{4 + 0} = 2$$

Let $\theta = \pi$.

$$r = \frac{8}{4 + 3\sin\pi} = \frac{8}{4 + 0} = 2$$

The curve also goes through $(2, 0)$ and $(2, \pi)$.

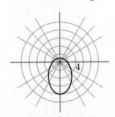

9. Describe and sketch the graph.

$$r = \frac{9}{3 - 3\sin\theta} = \frac{3}{1 - \sin\theta}$$

$e = 1$ The graph is a parabola.

The axis of symmetry is $\theta = \frac{\pi}{2}$.

When $\theta = \frac{\pi}{2}$, r is undefined.

Let $\theta = \frac{3\pi}{2}$.

$$r = \frac{9}{3 - 3\sin\frac{3\pi}{2}} = \frac{9}{3+3} = \frac{3}{2}$$

Vertex is at $\left(\frac{3}{2}, \frac{3\pi}{2}\right)$.

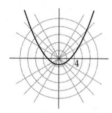

11. Describe and sketch the graph.

$$r = \frac{10}{5 + 6\cos\theta} = \frac{2}{1 + \frac{6}{5}\cos\theta}$$

$e = \frac{6}{5}$ The graph is a hyperbola.

The transverse axis is on the polar axis.

Let $\theta = 0$.

$$r = \frac{10}{5 + 6\cos 0} = \frac{10}{5+6} = \frac{10}{11}$$

Let $\theta = \pi$.

$$r = \frac{10}{5 + 6\cos \pi} = \frac{10}{5-6} = -10$$

The vertices are at $\left(\frac{10}{11}, 0\right)$ and $(-10, \pi)$.

13. Describe and sketch the graph.

$$r = \frac{4\sec\theta}{2\sec\theta - 1}$$

$$= \frac{\dfrac{4}{\cos\theta}}{\dfrac{2}{\cos\theta} - 1} = \frac{4}{2 - \cos\theta}$$

$$= \frac{2}{1 - \frac{1}{2}\cos\theta}$$

$e = \frac{1}{2}$ The graph is an ellipse.

The major axis is on the polar axis.

However, the original equation is undefined at

$\frac{\pi}{2}$ and at $\frac{3\pi}{2}$.

Thus, the ellipse will have holes at those angles.

Let $\theta = 0$.

$$r = \frac{4}{2 - \cos 0} = \frac{4}{2-1} = 4$$

Let $\theta = \pi$.

$$r = \frac{4}{2 - \cos \pi} = \frac{4}{2+1} = \frac{4}{3}$$

Vertices on major axis are at $(4, 0)$ and $\left(\frac{4}{3}, \pi\right)$.

Let $\theta = \frac{\pi}{2}$.

$$r = \frac{4}{2 - \cos\frac{\pi}{2}} = \frac{4}{2-0} = 2$$

Let $\theta = \frac{3\pi}{2}$.

$$r = \frac{4}{2 - \cos\frac{3\pi}{2}} = \frac{4}{2-0} = 2$$

Vertices on minor axis of

$$\frac{2}{1 - \frac{1}{2}\cos\theta} \text{ are at } \left(2, \frac{\pi}{2}\right) \text{ and } \left(2, \frac{3\pi}{2}\right).$$

Thus, the equation $r = \frac{4\sec\theta}{2\sec\theta - 1}$ will have holes at

$\left(2, \frac{\pi}{2}\right)$ and $\left(2, \frac{3\pi}{2}\right)$.

15. Describe and sketch the graph.

$$r = \frac{12\csc\theta}{6\csc\theta - 2} = \frac{\dfrac{12}{\sin\theta}}{\dfrac{6}{\sin\theta} - 2} = \frac{12}{6 - 2\sin\theta} = \frac{2}{1 - \frac{1}{3}\sin\theta}$$

$e = \frac{1}{3}$ The graph is an ellipse.

The major axis is on $\theta = \dfrac{\pi}{2}$.

Let $\theta = \dfrac{\pi}{2}$.

$$r = \dfrac{12}{6 - 2\sin\dfrac{\pi}{2}} = \dfrac{12}{6 - 2} = 3$$

Let $\theta = \dfrac{3\pi}{2}$.

$$r = \dfrac{12}{6 - 2\sin\dfrac{3\pi}{2}} = \dfrac{12}{6 + 2} = \dfrac{3}{2}$$

Vertices on major axis are at $\left(3, \dfrac{\pi}{2}\right)$ and $\left(\dfrac{3}{2}, \dfrac{3\pi}{2}\right)$.

The equation $r = \dfrac{12\csc\theta}{6\csc\theta - 2}$ has holes at $(2, 0)$ and $(2, \pi)$.

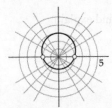

17. Describe and sketch the graph.

$$r = \dfrac{3}{\cos\theta - 1}$$
$$= \dfrac{-3}{1 - \cos\theta}$$

$e = 1$ The graph is a parabola.

The axis of symmetry is the polar axis.

Let $\theta = \pi$.

$$r = \dfrac{-3}{1 - \cos\pi} = \dfrac{-3}{1 - (-1)} = \dfrac{-3}{1 + 1} = -\dfrac{3}{2}$$

Vertex is at $\left(-\dfrac{3}{2}, \pi\right)$.

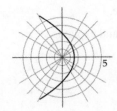

19. Find the rectangular equation.

$$r = \dfrac{12}{3 - 6\cos\theta}$$
$$3r - 6r\cos\theta = 12$$
$$3r = 6r\cos\theta + 12$$
$$r = 2r\cos\theta + 4$$
$$r = 2x + 4$$
$$r^2 = (2x + 4)^2$$
$$r^2 = 4x^2 + 16x + 16$$
$$x^2 + y^2 = 4x^2 + 16x + 16$$
$$3x^2 - y^2 + 16x + 16 = 0$$

21. Find the rectangular equation.

$$r = \dfrac{8}{4 + 3\sin\theta}$$
$$4r + 3r\sin\theta = 8$$
$$4r = -3r\sin\theta + 8$$
$$4r = -3y + 8$$
$$(4r)^2 = (-3y + 8)^2$$
$$16r^2 = 9y^2 - 48y + 64$$
$$16(x^2 + y^2) = 9y^2 - 48y + 64$$
$$16x^2 + 16y^2 = 9y^2 - 48y + 64$$
$$16x^2 + 7y^2 + 48y - 64 = 0$$

23. Find the rectangular equation.

$$r = \dfrac{9}{3 - 3\sin\theta}$$
$$3r - 3r\sin\theta = 9$$
$$3r = 3y + 9$$
$$r = y + 3$$
$$r^2 = (y + 3)^2$$
$$r^2 = y^2 + 6y + 9$$
$$x^2 + y^2 = y^2 + 6y + 9$$
$$x^2 - 6y - 9 = 0$$

25. Find the polar equation, given the eccentricity and directrix.

$$e = \dfrac{3}{2}, \; r\cos\theta = 2, \; d = |2| = 2$$

$$r = \dfrac{ed}{1 + e\cos\theta} = \dfrac{\left(\dfrac{3}{2}\right)(2)}{1 + \left(\dfrac{3}{2}\right)\cos\theta} = \dfrac{3}{1 + \dfrac{3}{2}\cos\theta}$$

$$= \dfrac{6}{2 + 3\cos\theta}$$

27. Find the polar equation, given the eccentricity and directrix.

$e = 1, \ r\sin\theta = 2, \ d = |2| = 2$

$$r = \frac{ed}{1 + e\sin\theta} = \frac{(1)(2)}{1 + (1)\sin\theta}$$

$$= \frac{2}{1 + \sin\theta}$$

29. Find the polar equation, given the eccentricity and directrix.

$e = \dfrac{4}{5}, \ r\sin\theta = -2, \ d = |-2| = 2$

$$r = \frac{ed}{1 - e\sin\theta} = \frac{\left(\frac{4}{5}\right)(2)}{1 - \left(\frac{4}{5}\right)\sin\theta} = \frac{\frac{8}{5}}{1 - \frac{4}{5}\sin\theta}$$

$$= \frac{8}{5 - 4\sin\theta}$$

31. Find the polar equation, given the eccentricity and directrix.

$e = \dfrac{3}{2}, \ r = 2\sec\theta$

$$r = \frac{2}{\cos\theta}$$

$r\cos\theta = 2, \quad d = |2| = 2$

$$r = \frac{ed}{1 + e\cos\theta} = \frac{\left(\frac{3}{2}\right)(2)}{1 + \left(\frac{3}{2}\right)\cos\theta} = \frac{3}{1 + \frac{3}{2}\cos\theta}$$

$$= \frac{6}{2 + 3\cos\theta}$$

33. Find the polar equation.

vertex: $(2, \ \pi)$, curve: parabola

$$r = \frac{ed}{1 - e\cos\theta} \qquad e = 1 \text{ (by definition of a parabola)}$$

When $\theta = \pi, \ r = 2$. Substituting into $r = \dfrac{ed}{1 - e\cos\theta}$, we have

$$2 = \frac{1 \cdot d}{1 - 1 \cdot \cos(\pi)} = \frac{d}{2}$$

Therefore, $d = 4$. Substituting $e = 1$ and $d = 4$ yields

$$r = \frac{(1)(4)}{1 - (1)\cos\theta} \quad \text{or} \quad r = \frac{4}{1 - \cos\theta}$$

35. Find the polar equation.

vertex: $\left(1, \ \dfrac{3\pi}{2}\right), \ e = 2$

$$r = \frac{ed}{1 - e\sin\theta}$$

When $\theta = \dfrac{3\pi}{2}, \ r = 1$. Substituting into $r = \dfrac{ed}{1 - e\sin\theta}$, we have

$$1 = \frac{2d}{1 - 2\sin\left(\frac{3\pi}{2}\right)} = \frac{2d}{3}$$

Therefore $d = \dfrac{3}{2}$. Substituting $e = 2$ and $d = \dfrac{3}{2}$ yields

$$r = \frac{(2)\left(\frac{3}{2}\right)}{1 - (2)\sin\theta} = \frac{3}{1 - 2\sin\theta}$$

37. Use a graphing utility to graph and explain.

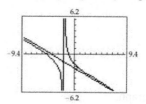

Rotate the graph of Exercise 5 $\dfrac{\pi}{6}$ radian counterclockwise about the pole.

39. Use a graphing utility to graph and explain.

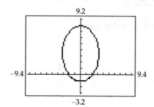

Rotate the graph of Exercise 7 π radians counterclockwise about the pole.

41. Use a graphing utility to graph and explain.

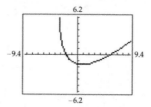

Rotate the graph of Exercise 9 $\dfrac{\pi}{6}$ radian clockwise about the pole.

43. Use a graphing utility to graph and explain.

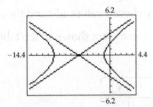

Rotate the graph of Exercise 11 π radians clockwise about the pole.

45. Use a graphing utility to graph $r = \dfrac{3}{3 - \sec\theta}$.

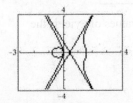

47. Use a graphing utility to graph $r = \dfrac{3}{1 + 2\csc\theta}$.

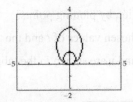

51. Show that $\dfrac{d(P,F)}{d(P,D)} = e$.

Convert the equations for the conic and the directrix to rectangular form.

$d = -r\cos\theta$ (by definition)
$ = -x$
$x = -d$

$$r = \frac{ed}{1 - e\cos\theta}$$
$$r(1 - e\cos\theta) = ed$$
$$r - er\cos\theta = ed$$
$$\sqrt{x^2 + y^2} - ex = ed$$
$$\sqrt{x^2 + y^2} = ex + ed = e(x + d)$$
$$x^2 + y^2 = e^2(x^2 + 2dx + d^2)$$
$$x^2 + y^2 - e^2x^2 - 2e^2dx - e^2d^2 = 0$$
$$(1 - e^2)x^2 + y^2 - (2e^2d)x - (e^2d^2) = 0$$

Solving for y^2 yields

$$y^2 = (e^2 - 1)x^2 + (2e^2d)x + (e^2d^2).$$

Now, with $y^2 = (e^2 - 1)x^2 + (2e^2d)x + (e^2d^2)$ and a

directrix of $x = -d$, let $k = \dfrac{d(P,F)}{d(P,D)}$, where the focus

is at the origin (by definition).

$$k = \frac{d(P,F)}{d(P,D)} = \frac{\sqrt{x^2 + y^2}}{|x + d|}$$

We can substitute

$y^2 = (e^2 - 1)x^2 + (2e^2d)x + (e^2d^2)$ to obtain

$$k = \frac{\sqrt{x^2 + (e^2 - 1)x^2 + (2e^2d)x + (e^2d^2)}}{|x + d|}$$
$$= \frac{\sqrt{e^2x^2 + 2e^2dx + e^2d^2}}{|x + d|}$$

Solving for k^2 gives us

$$k^2 = \frac{e^2x^2 + 2e^2dx + e^2d^2}{x^2 + 2dx + d^2} = \frac{e^2(x^2 + 2dx + d^2)}{x^2 + 2dx + d^2} = e^2.$$

Since $k^2 = e^2$, $k = \pm\sqrt{e^2} = \pm e$.

But, since $k = \dfrac{d(P,F)}{d(P,D)}$, and the ratio of two distances

must be positive, k cannot be negative.

Therefore, $k = e$, or $\dfrac{d(P,F)}{d(P,D)} = e$.

53. Verify.

Use a circle with center (h, k) where $h = a\cos\theta_C$,

$k = a\sin\theta_C$, and radius a.

$$(x - h)^2 + (y - k)^2 = a^2$$
$$(x - a\cos\theta_C)^2 + (y - a\sin\theta_C)^2 = a^2$$

Changing to polar yields

$$a^2 = (r\cos\theta - a\cos\theta_C)^2 + (r\sin\theta - a\sin\theta_C)^2$$
$$a^2 = (r^2\cos^2\theta - 2ra\cos\theta\cos\theta_C + a^2\cos^2\theta_C)$$
$$\qquad + (r^2\sin^2\theta - 2ra\sin\theta\sin\theta_C + a^2\sin^2\theta_C)$$
$$a^2 = r^2(\cos^2\theta + \sin^2\theta)$$
$$\qquad - 2ra(\cos\theta\cos\theta_C + \sin\theta\sin\theta_C)$$
$$\qquad + a^2(\cos^2\theta_C + \sin^2\theta_C)$$

$$a^2 = r^2 - 2ra(\cos\theta\cos\theta_C + \sin\theta\sin\theta_C) + a^2$$

$$0 = r^2 - 2ra[\cos(\theta - \theta_C)]$$

$$0 = r - 2a[\cos(\theta - \theta_C)]$$

$$r = 2a[\cos(\theta - \theta_C)]$$

Prepare for Section 8.7

P1. Complete the square.

$$y^2 + 3y + \left(\frac{3}{2}\right)^2 = \left(y + \frac{3}{2}\right)^2$$

P3. Identify the graph of $\left(\frac{x-2}{3}\right)^2 + \left(\frac{y-3}{2}\right)^2 = 1$.

ellipse

P5. Solve for t.

$$y = \ln t$$

$$e^y = t$$

Section 8.7 Exercises

1. Determine if the statement is true or false.

False

3. Determine if the statement is true or false.

True

5. Find the rectangular coordinates.

For $t = 4$,

$x = 2t - 1 = 2(4) - 1 = 7$

$y = 3 - t = 3 - 4 = -1$

The coordinates are $(7, -1)$.

7. Find the rectangular coordinates.

For $t = \frac{3\pi}{2}$,

$x = 4\cos t = 4\cos\frac{3\pi}{2} = 4(0) = 0$

$y = 6\sin t = 6\sin\frac{3\pi}{2} = 6(-1) = -6$

The coordinates are $(0, -6)$.

9. Find the rectangular coordinates.

For $t = 2$,

$x = |t - 5| = |2 - 5| = 3$

$y = t^3 = 2^3 = 8$

The coordinates are $(3, 8)$.

11. Graph the parametric equation by plotting points. A table of five arbitrarily chosen values of t and the corresponding values of x and y are shown in the table below.

t	$x = 2t$	$y = -t$	(x, y)
-2	-4	2	$(-4, 2)$
-1	-2	1	$(-2, 1)$
0	0	0	$(0, 0)$
1	2	-1	$(2, -1)$
2	4	-2	$(4, -2)$

Plotting points for several values of t yields the following graph.

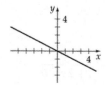

13. Graph the parametric equation by plotting points. A table of five arbitrarily chosen values of t and the corresponding values of x and y are shown in the table below.

t	$x = -t$	$y = t^2 - 1$	(x, y)
-2	2	3	$(2, 3)$
-1	1	0	$(1, 0)$
0	0	-1	$(0, -1)$
1	-1	0	$(-1, 0)$
2	-2	3	$(-2, 3)$

Plotting points for several values of t yields the following graph.

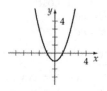

15. Graph the parametric equation by plotting points. A table of five arbitrarily chosen values of t and the corresponding values of x and y are shown in the table below.

t	$x = t^2$	$y = t^3$	(x, y)
-2	4	-8	$(4, -8)$
-1	1	-1	$(1, -1)$
0	0	0	$(0, 0)$
1	1	1	$(1, 1)$
2	4	8	$(4, 8)$

Plotting points for several values of t yields the following graph.

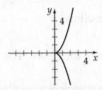

17. Graph the parametric equation by plotting points.

A table of eight arbitrarily chosen values of t and the corresponding values of x and y are shown in the table.

t	$x = 2\cos t$	$y = 3\sin t$	(x, y)
0	2	0	$(2, 0)$
$\pi/4$	$\sqrt{2}$	$3\sqrt{2}/2$	$(\sqrt{2}, 3\sqrt{2}/2)$
$\pi/2$	0	3	$(0, 3)$
$3\pi/4$	$-\sqrt{2}$	$3\sqrt{2}/2$	$(-\sqrt{2}, 3\sqrt{2}/2)$
π	-2	0	$(-2, 0)$
$5\pi/4$	$-\sqrt{2}$	$-3\sqrt{2}/2$	$(-\sqrt{2}, -3\sqrt{2}/2)$
$3\pi/2$	0	-3	$(0, -3)$
$7\pi/4$	$\sqrt{2}$	$-3\sqrt{2}/2$	$(\sqrt{2}, -3\sqrt{2}/2)$

Plotting points for several values of t yields the following graph.

19. Graph the parametric equation by plotting points.

A table of five arbitrarily chosen values of t and the corresponding values of x and y are shown in the table below.

t	$x = 2^t$	$y = 2^{t+1}$	(x, y)
-2	$\dfrac{1}{4}$	$\dfrac{1}{2}$	$\left(\dfrac{1}{4}, \dfrac{1}{2}\right)$
-1	$\dfrac{1}{2}$	1	$\left(\dfrac{1}{2}, 1\right)$
0	1	2	$(1, 2)$
1	2	4	$(2, 4)$
2	4	8	$(4, 8)$

Plotting points for several values of t yields the following graph.

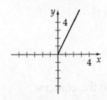

21. Eliminate the parameter and graph the equation.

$$x = \sec t \qquad -\frac{\pi}{2} < t < \frac{\pi}{2}$$
$$y = \tan t$$

$$\tan^2 t + 1 = \sec^2 t$$
$$y^2 + 1 = x^2$$
$$x^2 - y^2 - 1 = 0 \qquad x \geq 1, \ y \in R$$

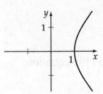

23. Eliminate the parameter and graph the equation.

$$x = 2 - t^2 \qquad t \in R$$
$$y = 3 + 2t^2$$

$$x = 2 - t^2 \rightarrow t^2 = 2 - x$$
$$y = 3 + 2(2 - x)$$
$$y = -2x + 7$$

Because $x = 2 - t^2$ and $t^2 \geq 0$ for all real numbers t, $x \leq 2$ for all t. Similarly, $y \geq 3$ for all t.

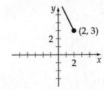

25. Eliminate the parameter and graph the equation.

$$x = \cos^3 t \qquad 0 \le t < 2\pi$$

$$y = \sin^3 t$$

$$\cos^2 t = x^{2/3}$$

$$\sin^2 t = y^{2/3}$$

$$\cos^2 t + \sin^2 t = 1 \qquad -1 \le x \le 1$$

$$x^{2/3} + y^{2/3} = 1 \qquad -1 \le y \le 1$$

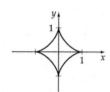

27. Eliminate the parameter and graph the equation.

$$x = \sqrt{t+1} \qquad t \ge -1$$

$$y = t$$

$$x = \sqrt{y+1} \qquad x \ge 0$$

$$y = x^2 - 1 \qquad y \ge -1$$

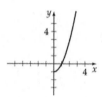

29. Eliminate the parameter and graph the equation.

$$x = t^3 \qquad t > 0,\ x > 0$$

$$y = 3 \ln t \qquad y \in R$$

$$x = t^3 \rightarrow t = x^{1/3}$$

$$y = 3 \ln x^{1/3}$$

$$y = \ln x \text{ for } x > 0 \text{ and } y \in R$$

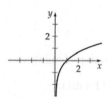

31. Describe the motion as t increases.

$$x = 2 + 3\cos t \qquad y = 3 + 2\sin t$$

$$x - 2 = 3\cos t \qquad y - 3 = 2\sin t$$

$$\frac{x-2}{3} = \cos t \qquad \frac{y-3}{2} = \sin t$$

$$\left(\frac{x-2}{3}\right)^2 + \left(\frac{y-3}{2}\right)^2 = \cos^2 t + \sin^2 t$$

$$\left(\frac{x-2}{3}\right)^2 + \left(\frac{y-3}{2}\right)^2 = 1$$

At $t = 0$,

$$x = 2 + 3\cos 0 = 5 \qquad y = 3 + 2\sin 0 = 3$$

At $t = \pi$,

$$x = 2 + 3\cos \pi = -1 \qquad y = 3 + 2\sin \pi = 3$$

The point traces the top half of the ellipse

$\left(\frac{x-2}{3}\right)^2 + \left(\frac{y-3}{2}\right)^2 = 1$, as shown in the figure. The

point starts at (5, 3) and moves counterclockwise along the ellipse until it reaches the point (–1, 3) at time $t = \pi$.

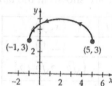

33. Describe the motion as t increases.

$$y = t + 1 \quad \Rightarrow \quad t = y - 1$$

$$x = 2t - 1$$

$$x = 2(y-1) - 1$$

$$x = 2y - 3$$

$$x + 3 = 2y$$

$$y = \frac{1}{2}x + 3$$

At $t = 0$,

$$x = 2(0) - 1 = -1 \qquad y = 0 + 1 = 1$$

At $t = 3$,

$$x = 2(3) - 1 = 5 \qquad y = 3 + 1 = 4$$

The point traces a line segment, as shown in the figure. The point starts at (–1, 1) and moves along the line segment until it reaches the point (5, 4) at time $t = 3$.

35. Describe the motion as t increases.

$$x = \tan\left(\frac{\pi}{4} - t\right) \qquad y = \sec\left(\frac{\pi}{4} - t\right)$$

$$y^2 - x^2 = \sec^2\left(\frac{\pi}{4} - t\right) - \tan^2\left(\frac{\pi}{4} - t\right)$$

$$y^2 - x^2 = 1 \quad \text{Since } 1 + \tan^2\theta = \sec^2\theta$$

At $t = 0$,

$$x = \tan\left(\frac{\pi}{4} - 0\right) = 1 \qquad y = \sec\left(\frac{\pi}{4} - 0\right) = \sqrt{2}$$

At $t = \frac{\pi}{2}$,

$$x = \tan\left(\frac{\pi}{4} - \frac{\pi}{2}\right) = -1 \quad y = \sec\left(\frac{\pi}{4} - \frac{\pi}{2}\right) = \sqrt{2}$$

The point traces a portion of the top branch of the hyperbola $y^2 - x^2 = 1$, as shown in the figure. The point starts at $\left(1, \sqrt{2}\right)$ and moves along the hyperbola until it reaches the point $\left(-1, \sqrt{2}\right)$ at time $t = \frac{\pi}{2}$.

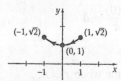

37. Eliminate the parameter and describe the differences in the graphs.

$$C_1 : x = 2 + t^2$$
$$y = 1 - 2t^2$$

$$x = 2 + t^2 \to t^2 = x - 2$$
$$y = 1 - 2(x - 2)$$
$$y = -2x + 5 \quad x \geq 2, \, y \leq 1$$

$$C_2 : x = 2 + t$$
$$y = 1 - 2t$$

$$x = 2 + t \to t = x - 2$$
$$y = 1 - 2(x - 2)$$
$$y = -2x + 5 \quad x \in R, \, y \in R$$

The graph of C_1 is a ray beginning at (2, 1) with slope -2. The graph of C_2 is a line passing through (2, 1) with slope -2.

39. Sketch the graph and then sketch another graph with a different domain.

$$x = \sin t$$
$$y = \csc t$$

$$\csc t = \frac{1}{\sin t}$$

$$y = \frac{1}{x}$$

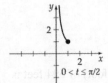
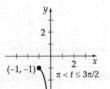

Range for graph 1: $0 < t \leq \frac{\pi}{2}$
$$0 < x \leq 1$$
$$y \geq 1$$

Range for graph 2: $\pi \leq t \leq \frac{3\pi}{2}$
$$-1 \leq x \leq 0$$
$$y \leq -1$$

41. Use a graphing utility to graph the cycloid.

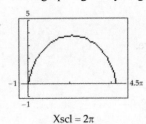

Xscl = 2π

43. a. Find the pair of parametric equations.

For the Hummer,

$$x = 6$$
$$y = 60t \text{ for } t \geq 0$$

b. Determine which vehicle will be first.

Using the graphing calculator in SIMUL and PAR mode, the Hummer is the first to reach the intersection.

45. Find the maximum height and associated value of t.

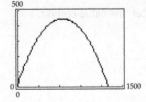

Maximum height (to the nearest foot) of 462 feet is attained when $t \approx 5.38$ seconds.

The projectile has a range (to the nearest foot) of 1295 feet and hits the ground in about 10.75 seconds.

47. Find the maximum height and associated value of t.

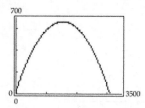

Maximum height (to the nearest foot) of 694 feet is attained when $t \approx 6.59$ seconds.

The projectile has a range (to the nearest foot) of 3084 feet and hits the ground in about 13.17 seconds.

49. Graph the Lissajous figure.

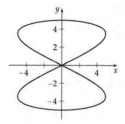

51. Graph the Lissajous figure.

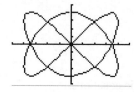

53. Show the statement to be true.

Let $P_1(x_1, y_1)$ and $P_2(x_2, y_2)$ be two distinct points on a line.

If $P(x, y)$ is any other point on the line, then

$\dfrac{y - y_1}{x - x_1} = \dfrac{y_2 - y_1}{x_2 - x_1}$ (Slope is constant along entire line.)

This equation can be rewritten as

$\dfrac{y - y_1}{y_2 - y_1} = \dfrac{x - x_1}{x_2 - x_1}$ Let this value equal t.

Thus, $\dfrac{x - x_1}{x_2 - x_1} = t$ and $\dfrac{y - y_1}{y_2 - y_1} = t$.

Solving for x and y, respectively, we have

$x = x_1 + t(x_2 - x_1)$ and $y = y_1 + t(y_2 - y_1)$

55. Show the statement to be true.

$x = h + a \tan t$ $a > 0, x \in R,\ 0 \le t < 2\pi$
$y = k + b \sec t$ $b > 0, y \in R,\ 0 \le t < 2\pi$

$x = h + a \tan t \rightarrow \tan t = \dfrac{x - h}{a}$

$y = k + b \sec t \rightarrow \sec t = \dfrac{y - k}{b}$

$\tan^2 t + 1 = \sec^2 t$

$\left(\dfrac{x - h}{a}\right)^2 + 1 = \left(\dfrac{y - k}{b}\right)^2$

$\left(\dfrac{y - k}{b}\right)^2 - \left(\dfrac{x - h}{a}\right)^2 = 1$

$\left(\dfrac{y - k}{b}\right)^2 - \left(\dfrac{x - h}{a}\right)^2 = 1$ which is the standard equation

for a hyperbola at (h, k).

57. a. Find the starting altitude.

For $t = 0$, $z = 4 - \dfrac{2t}{\pi} = 4 - \dfrac{2(0)}{\pi} = 4$

The altitude was 4000 ft.

b. Find the location of the plane when $t = \pi$.

$x = 4 \cos t = 4 \cos \pi = 4(-1) = -4$
$y = 4 \sin t = 4 \sin \pi = 4(0) = 0$
$z = 4 - \dfrac{2t}{\pi} = 4 - \dfrac{2\pi}{\pi} = 2$

The plane's location is (–4, 0, 2).

c. Graph the descent using WolframAlpha.

Type "parametric plot (2cost, 8sint, 4-2t/pi), t=0 to 2pi"

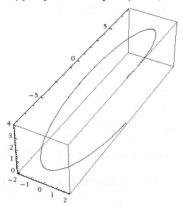

d. Give a reason.

When the plane is about to touch down, the direction of the plane is more closely aligned with the direction of the runway.

Chapter 8 Review Exercises

1. Identify each conic section and parts, then graph. [8.3]

$$x^2 - y^2 = 4$$

$$\frac{x^2}{4} - \frac{y^2}{4} = 1$$

hyperbola

center $(0, 0)$; vertex $(\pm 2, 0)$; foci $(\pm 2\sqrt{2},\ 0)$

asymptotes $y = \pm x$

3. Identify each conic section and parts, then graph. [8.2]

$$x^2 + 4y^2 - 6x + 8y - 3 = 0$$

$$x^2 - 6x + 4(y^2 + 2y) = 3$$

$$(x^2 - 6x + 9) + 4(y^2 + 2y + 1) = 3 + 9 + 4$$

$$(x - 3)^2 + 4(y + 1)^2 = 16$$

$$\frac{(x - 3)^2}{16} + \frac{(y + 1)^2}{4} = 1$$

ellipse

center $(3, -1)$

vertices $(3 \pm 4,\ -1) = (7,\ -1),\ (-1,\ -1)$

foci $(3 \pm 2\sqrt{3},\ -1) = (3 + 2\sqrt{3},\ -1),\ (3 - 2\sqrt{3},\ -1)$

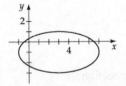

5. Identify each conic section and parts, then graph. [8.1]

$$3x - 4y^2 + 8y + 2 = 0$$

$$-4(y^2 - 2y) = -3x - 2$$

$$-4(y^2 - 2y + 1) = -3x - 2 - 4$$

$$-4(y - 1)^2 = -3(x + 2)$$

$$(y - 1)^2 = \frac{3}{4}(x + 2)$$

parabola

vertex $(-2, 1)$

focus $\left(-2 + \frac{3}{16},\ 1\right) = \left(-\frac{29}{16},\ 1\right)$

directrix $x = -2 - \frac{3}{16}$ or $x = -\frac{35}{16}$

7. Identify each conic section and parts, then graph. [8.2]

$$9x^2 + 4y^2 + 36x - 8y + 4 = 0$$

$$9(x^2 + 4x) + 4(y^2 - 2y) = -4$$

$$9(x^2 + 4x + 4) + 4(y^2 - 2x + 1) = -4 + 36 + 4$$

$$9(x + 2)^2 + 4(y - 1)^2 = 36$$

$$\frac{(x + 2)^2}{4} + \frac{(y - 1)^2}{9} = 1$$

ellipse

center $(-2, 1)$

vertices $(-2,\ 1 \pm 3) = (-2,\ 4),\ (-2,\ -2)$

foci $(-2,\ 1 \pm \sqrt{5}) = (-2,\ 1 + \sqrt{5}),\ (-2,\ 1 - \sqrt{5})$

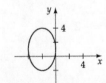

9. Identify each conic section and parts, then graph. [8.3]

$$4x^2 - 9y^2 - 8x + 12y - 144 = 0$$

$$4(x^2 - 2x) - 9\left(y^2 - \frac{4}{3}y\right) = 144$$

$$4(x^2 - 2x + 1) - 9\left(y^2 - \frac{4}{3}y + \frac{4}{9}\right) = 144 + 4 - 4$$

$$4(x-1)^2 - 9\left(y - \frac{2}{3}\right)^2 = 144$$

$$\frac{(x-1)^2}{36} - \frac{(y-2/3)^2}{16} = 1$$

hyperbola

center $\left(1, \frac{2}{3}\right)$

vertices $\left(1 \pm 6, \frac{2}{3}\right) = \left(7, \frac{2}{3}\right), \left(-5, \frac{2}{3}\right)$

foci $\left(1 \pm 2\sqrt{13}, \frac{2}{3}\right) = \left(1 + 2\sqrt{13}, \frac{2}{3}\right), \left(1 - 2\sqrt{13}, \frac{2}{3}\right)$

asymptotes $y - \frac{2}{3} = \pm\frac{2}{3}(x-1)$

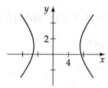

11. Identify each conic section and parts, then graph. [8.1]

$$4x^2 + 28x + 32y + 81 = 0$$

$$4(x^2 + 7x) = -32y - 81$$

$$4\left(x^2 + 7x + \frac{49}{4}\right) = -32y - 81 + 49$$

$$4\left(x + \frac{7}{2}\right)^2 = -32(y+1)$$

$$\left(x + \frac{7}{2}\right)^2 = -8(y+1)$$

parabola

vertex $\left(-\frac{7}{2}, -1\right)$ $4p = -8, p = -2$

focus $\left(-\frac{7}{2}, -3\right)$

directrix $y = 1$

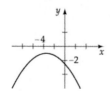

13. Find the eccentricity. [8.2]

$$4x^2 + 49y^2 - 48x - 294y + 389 = 0$$

$$4(x^2 - 12x) + 49(y^2 - 6y) = -389$$

$$4(x^2 - 12x + 36) + 49(y^2 - 6y + 9) = -389 + 144 + 441$$

$$4(x-6)^2 + 49(y-3)^2 = 196$$

$$\frac{(x-6)^2}{49} + \frac{(y-3)^2}{4} = 1$$

$$c^2 = a^2 - b^2 = 49 - 4 = 45$$

$$c = \sqrt{45} = 3\sqrt{5}$$

$$e = \frac{c}{a} = \frac{3\sqrt{5}}{7}$$

15. Find the equation of the conic section. [8.2]

$$2a = |7 - (-3)| = 10$$

$$a = 5$$

$$a^2 = 25$$

$$2b = 8$$

$$b = 4$$

$$b^2 = 16$$

Center (2, 3)

$$\frac{(x-2)^2}{25} + \frac{(y-3)^2}{16} = 1$$

17. Find the equation of the conic section. [8.3]

center $(-2, 2)$, $c = 3$

$$2a = 4$$

$$a = 2$$

$$a^2 = 4$$

$$c^2 = a^2 + b^2$$

$$9 = 4 + b^2$$

$$b^2 = 5$$

$$\frac{(x+2)^2}{4} - \frac{(y-2)^2}{5} = 1$$

19. Find the equation of the conic section. [8.1]

$$(x-h)^2 = 4p(y-k) \qquad (y-k)^2 = 4p(x-h)$$

$$(3-0)^2 = 4p(4+2) \qquad (4+2)^2 = 4p(3-0)$$

$$9 = 4p(6) \qquad\qquad 36 = 4p(3)$$

$$p = \frac{3}{8} \qquad\qquad\qquad p = 3$$

Thus, there are two parabolas that satisfy the given conditions:

$$x^2 = \frac{3}{2}(y+2) \quad \text{or} \quad (y+2)^2 = 12x$$

21. Find the equation of the conic section. [8.2]

$a = 6$ and the transverse axis is on the x-axis.

$$\pm \frac{b}{a} = \pm \frac{1}{9}$$

$$\frac{b}{6} = \frac{1}{9}$$

$$b = \frac{2}{3}$$

$$\frac{x^2}{36} - \frac{y^2}{4/9} = 1$$

23. The equation of the mirror is [8.1]

$$x^2 = 4py \quad -4 \le x \le 4$$

Because $(4, 0.1)$ is a point on the parabola, $(4, 0.1)$ must be a solution of the equation. Thus

$$4^2 = 4p(0.1)$$

$$16 = 0.4p$$

$$40 = p$$

The focus is 40 inches above the vertex.

25. a. Find the distance of the pushpins from O. [8.2]

$$c^2 = a^2 - b^2$$

$$c^2 = 3^2 - \left(\frac{3}{2}\right)^2$$

$$c^2 = \frac{27}{4}$$

$$c = \frac{3\sqrt{3}}{2} \approx 2 \text{ ft } 7.2 \text{ in.}$$

$\frac{3\sqrt{3}}{2}$ ft or approximately 2 ft $7\frac{3}{16}$ in. to the right and left of O.

b. $2a = 2(3) = 6$ ft

27. Explain. [8.1, 8.3]

The parabola was produced by cutting the cup in a plane parallel to line l. The branch of the hyperbola was produced by cutting in a vertical plane.

29. Write the equation and name the graph. [8.4]

$$11x^2 - 6xy + 19y^2 - 40 = 0$$

$$A = 11, \ B = -6, \ C = 19, \ D = 0, \ E = 0, \ F = -40$$

$$\cot 2\alpha = \frac{11-19}{-6} = \frac{-8}{-6} = \frac{4}{3} \quad (2\alpha \text{ is in quadrant I.})$$

$$\csc^2 2\alpha = 1 + \cot^2 2\alpha = 1 + \frac{16}{9} = \frac{25}{9}$$

$$\csc 2\alpha = \frac{5}{3}$$

Thus, $\sin 2\alpha = \frac{3}{5}$ and $\cos 2\alpha = \frac{4}{5}$.

$$\sin \alpha = \sqrt{\frac{1-\frac{4}{5}}{2}} = \frac{\sqrt{10}}{10} \qquad \cos \alpha = \frac{3\sqrt{10}}{10}$$

$$A' = 11\left(\frac{3\sqrt{10}}{10}\right)^2 - 6\left(\frac{\sqrt{10}}{10}\right)\left(\frac{3\sqrt{10}}{10}\right) + 19\left(\frac{\sqrt{10}}{10}\right)^2$$

$$= \frac{99}{10} - \frac{18}{10} + \frac{19}{10} = \frac{100}{10} = 10$$

$$B' = 0$$

$$C' = 11\left(\frac{\sqrt{10}}{10}\right)^2 + 6\left(\frac{\sqrt{10}}{10}\right)\left(\frac{3\sqrt{10}}{10}\right) + 19\left(\frac{3\sqrt{10}}{10}\right)^2$$

$$= \frac{11}{10} + \frac{18}{10} + \frac{171}{10} = \frac{200}{10} = 20$$

$$F' = F$$

$$10(x')^2 + 20(y')^2 - 40 = 0 \text{ or } (x')^2 + 2(y')^2 - 4 = 0$$

The graph is an ellipse.

31. Write the equation and name the graph. [8.4]

$$x^2 + 2\sqrt{3}xy + 3y^2 + 8\sqrt{3}x - 8y + 32 = 0$$

$$A = 1, \ B = 2\sqrt{3}, \ C = 3, \ D = 8\sqrt{3}, \ E = -8, \ F = 32$$

$$\cot 2\alpha = \frac{1-3}{2\sqrt{3}} = -\frac{1}{\sqrt{3}}. \text{ Thus } 90° < 2\alpha < 180°.$$

$$\cot^2 2\alpha + 1 = \csc^2 2\alpha$$

$$\frac{1}{3} + 1 = \csc^2 2\alpha$$

$$\frac{4}{3} = \csc^2 2\alpha, \text{ or } \csc 2\alpha = \frac{2}{\sqrt{3}}$$

Therefore, $\sin 2\alpha = \frac{\sqrt{3}}{2}$ and $\cos 2\alpha = -\frac{1}{2}$.

Since 2α is in quadrant II,

$$\cos \alpha = \sqrt{\frac{1+\left(-\frac{1}{2}\right)}{2}} = \frac{1}{2} \qquad \sin \alpha = \sqrt{\frac{1-\left(-\frac{1}{2}\right)}{2}} = \frac{\sqrt{3}}{2}$$

$$A' = 1\left(\frac{1}{2}\right)^2 + 2\sqrt{3}\left(\frac{1}{2}\right)\left(\frac{\sqrt{3}}{2}\right) + 3\left(\frac{\sqrt{3}}{2}\right)^2 = \frac{1}{4} + \frac{6}{4} + \frac{9}{4} = 4$$

$$B' = 0$$

$$C' = 1\left(\frac{\sqrt{3}}{2}\right)^2 - 2\sqrt{3}\left(\frac{1}{2}\right)\left(\frac{\sqrt{3}}{2}\right) + 3\left(\frac{1}{2}\right)^2 = \frac{3}{4} - \frac{6}{4} + \frac{3}{4} = 0$$

$$D' = 8\sqrt{3}\left(\frac{1}{2}\right) - 8\left(\frac{\sqrt{3}}{2}\right) = 0$$

$$E' = 8\sqrt{3}\left(\frac{\sqrt{3}}{2}\right) - 8\left(\frac{1}{2}\right) = -12 - 4 = -16$$

$$F' = 32$$

$$4(x')^2 - 16y' + 32 = 0 \text{ or } (x')^2 - 4y' + 8 = 0$$

The graph is a parabola.

33. Identify the graph using the Conic Identification Theorem. [8.4]

$$x^2 - 4xy + 3y^2 - 2x + 6y - 24 = 0$$
$$A = 1, B = -4, C = 3$$

Since $B^2 - 4AC = (-4)^2 - 4(1)(3) = 4 > 0$, the graph is a hyperbola.

35. Identify the graph using the Conic Identification Theorem. [8.4]

$$x^2 + 2\sqrt{3}xy + 3y^2 - x + 2y - 20 = 0$$
$$A = 1, B = 2\sqrt{3}, C = 3$$

Since $B^2 - 4AC = (2\sqrt{3})^2 - 4(1)(3) = 0$, the graph is a parabola.

37. Determine whether the graph is symmetric to [8.5]

a. the line $\theta = 0$.

$$r^2 \overset{?}{=} 6\sin 2(-\theta)$$
$$r^2 \neq -6\sin 2\theta$$
No, not symmetric to the line $\theta = 0$.

b. the line $\theta = \frac{\pi}{2}$.

$$(-r)^2 \overset{?}{=} 6\sin 2(-\theta)$$
$$r^2 \neq -6\sin 2\theta$$
No, not symmetric to the line $\theta = \frac{\pi}{2}$.

c. the pole.

$$(-r)^2 \overset{?}{=} 6\sin 2(\pi + \theta)$$
$$r^2 = 6\sin 2\theta$$
Yes, symmetric to the pole.

39. Graph the polar equation $r\sin\theta = 3$. [8.5]

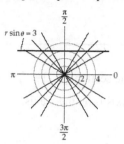

41. Graph the polar equation $r = 4\cos 3\theta$. [8.5]

43. Graph the polar equation $r = 2(1 - 2\sin\theta)$. [8.5]

45. Graph the polar equation $r = 5\sin\theta$. [8.5]

47. Graph the polar equation $r = 4\csc\theta$. [8.5] horizontal line through (0, 4)

49. Graph the polar equation $r = 3 + 2\cos\theta$. [8.5]

51. Find a polar form of the equation. [8.5]

$$y^2 = 16x$$
$$(r\sin\theta)^2 = 16(r\cos\theta)$$
$$r^2\sin^2\theta = 16r\cos\theta$$
$$r\sin^2\theta = 16\cos\theta$$

53. Find a polar form of the equation. [8.5]

$$3x - 2y = 6$$
$$3r\cos\theta - 2r\sin\theta = 6$$

55. Find a rectangular form of the equation. [8.5]

$$r = \frac{4}{1 - \cos\theta}$$
$$r - r\cos\theta = 4$$
$$\sqrt{x^2 + y^2} - x = 4$$
$$\sqrt{x^2 + y^2} = x + 4$$
$$x^2 + y^2 = x^2 + 8x + 16$$
$$y^2 = 8x + 16$$

57. Find a rectangular form of the equation. [8.5]

$$r^2 = \cos 2\theta$$
$$r^2 = \cos^2\theta - \sin^2\theta$$
$$r^4 = r^2\cos^2\theta - r^2\sin^2\theta$$
$$(r^2)^2 = r^2\cos^2\theta - r^2\sin^2\theta$$
$$(x^2 + y^2)^2 = x^2 - y^2$$
$$x^4 + 2x^2y^2 + y^4 = x^2 - y^2$$
$$x^4 + y^4 + 2x^2y^2 - x^2 + y^2 = 0$$

59. Graph the conic. [8.6]

$$r = \frac{4}{3 - 6\sin\theta}$$

61. Graph the conic. [8.6]

$$r = \frac{2}{2 - \cos\theta}$$

63. Eliminate the parameter and graph the curve. [8.7]

$$x = 4t - 2, \; y = 3t + 1, \; t \in R$$

$$4t = x + 2$$
$$t = \frac{x + 2}{4}$$
$$y = 3t + 1$$
$$y = 3\left(\frac{x + 2}{4}\right) + 1$$
$$y = \frac{3}{4}x + \frac{5}{2}$$

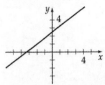

65. Eliminate the parameter and graph the curve. [8.7]

$$x = 4\sin t \qquad y = 3\cos t \qquad 0 \le t < 2\pi$$
$$\frac{1}{4}x = \sin t \qquad \frac{1}{3}y = \cos t$$
$$\frac{1}{16}x^2 = \sin^2 t \qquad \frac{1}{9}y^2 = \cos t^2$$

Using the trigonometric identity $\sin^2 t + \cos^2 t = 1$, we have

$$\frac{1}{16}x^2 + \frac{1}{9}y^2 = 1$$
$$\frac{x^2}{16} + \frac{y^2}{9} = 1$$

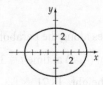

67. Eliminate the parameter and graph the curve. [8.7]

$$x = \frac{1}{t} \qquad y = -\frac{2}{t} \qquad t > 0$$
$$y = -2\left(\frac{1}{t}\right)$$
$$y = -2x, \; x > 0$$

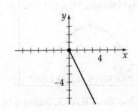

69. Eliminate the parameter and graph the curve. [8.7]

$$x = \sqrt{t}, \ y = 2^{-t}, \ t \geq 0$$

$$t = x^2$$

$$y = 2^{-x^2}, \ x \geq 0$$

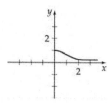

71. Describe the motion of P as t increases. [8.7]

$$x = t + 1 \ \Rightarrow \ t = x - 1, \quad -1 \leq t \leq 3$$

$$y = 2 + \sqrt{t + 1}$$

$$y = 2 + \sqrt{(x-1)+1} \quad \text{Substitute } x - 1 \text{ for } t$$

$$y = 2 + \sqrt{x}, \ 0 \leq x \leq 4$$

The point traces a portion of this parabola.

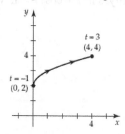

The point starts at (0, 2) and moves along the parabola, as shown, until it reaches the point (4, 4) at time $t = 3$.

73. Use a graphing utility to find the height. [8.7]

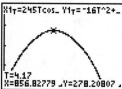

Graph in parametric mode. Use the TRACE feature to determine that the maximum height (to the nearest foot) of 278 feet is attained when $t \approx 4.17$ seconds.

Chapter 8 Test

1. Find the vertex, focus, and directrix. [8.1]

$$y = \frac{1}{8}x^2$$

$$x^2 = 8y$$

$$4p = 8$$

$$p = 2$$

vertex: (0, 0)

focus: (0, 2)

directrix: $y = -2$

3. Find the vertices and foci. [8.2]

$$25x^2 - 150x + 9y^2 + 18y + 9 = 0$$

$$25(x^2 - 6x + 9) + 9(y^2 + 2y + 1) = -9 + 225 + 9$$

$$25(x - 3)^2 + 9(y - 1)^2 = 225$$

$$\frac{(x-3)^2}{9} + \frac{(y+1)^2}{25} = 1$$

$$a = 5 \quad b = 3 \quad c = 4$$

vertices: (3, 4), (3, –6)

foci: (3, 3), (3, –5)

5. The equation of the mirror is [8.1]

$$y^2 = 4px \quad -4 \leq y \leq 4$$

Because (4, 4) is a point on the parabola, (4, 4) must be a solution of the equation. Thus

$$4^2 = 4p(4)$$

$$16 = 16p$$

$$1 = p$$

The focus is 1 inch from the vertex.

7. Graph. [8.3]

$$16y^2 + 32y - 4x^2 - 24x = 84$$

$$16(y^2 + 2y) - 4(x^2 + 6x) = 84$$

$$16(y^2 + 2y + 1) - 4(x^2 + 6x + 9) = 84 + 16 - 36$$

$$16(y + 1)^2 - 4(x + 3)^2 = 64$$

$$\frac{(y+1)^2}{4} - \frac{(x+3)^2}{16} = 1$$

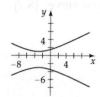

9. Find the distance. [5.1]

$$400 = 76 + c^2$$
$$c^2 = 342$$
$$c = 18$$
$$2c = 36$$

The emitter should be placed 36 cm away.

11. Determine whether the graph is a parabola, ellipse or hyperbola. [8.4]

$A = 8 \quad B = 5 \quad C = 2 \quad D = -10 \quad E = 5 \quad F = 4$

Since $B^2 - 4AC = (5)^2 - 4(8)(2) = -39 < 0$, the graph is an ellipse.

13. Graph. [8.5]

$r = 4 \cos \theta$

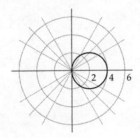

15. Graph. [8.5]

$r = 2 \sin 4\theta$

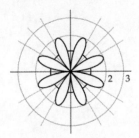

17. Find the rectangular form. [8.5]

$$r - r \cos x = 4$$
$$\sqrt{x^2 + y^2} - x = 4$$
$$\sqrt{x^2 + y^2} = x + 4$$
$$x^2 + y^2 = x^2 + 8x + 16$$
$$y^2 - 8x - 16 = 0$$

19. Eliminate the parameter, and graph the curve. [8.7]

$$x = t - 3$$
$$x + 3 = t$$
$$(x + 3)^2 = t^2$$
$$2(x + 3)^2 = 2t^2$$
$$2(x + 3)^2 = y$$
$$(x + 3)^2 = \frac{1}{2}y$$

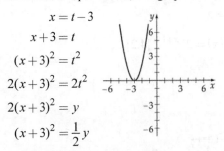

21. Use a graphing utility to graph the cycloid. [8.7]

$$x - 2(t - \sin t), \quad y = 2(1 - \cos t), \quad 0 \le t \le 12\pi$$

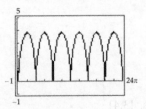

$Xscl = 2\pi$

Cumulative Review Exercises

1. Solve $x^4 - 2x^2 - 8 = 0$. [1.4]

Let $u = x^2$.

$$u^2 - 2u - 8 = 0$$
$$(u - 4)(u + 2) = 0$$

$$u = 4 \quad \text{or} \quad u = -2$$
$$x^2 = 4 \qquad x^2 = -2$$
$$x = \pm 2 \qquad x = \pm i\sqrt{2}$$

The solutions are $2, \ -2, \ i\sqrt{2}, \ -i\sqrt{2}$.

3. Write the difference quotient. [2.6]

$$\frac{f(2+h) - f(2)}{h} = \frac{\left[1 - (2+h)^2\right] - \left[1 - (2)^2\right]}{h}$$

$$= \frac{1 - 4 - 4h - (h)^2 - 1 + 4}{h}$$

$$= \frac{-4h - h^2}{h}$$

$$= -4 - h$$

5. Find the number of complex solutions. [3.4]

By the Linear Factor Theorem, since the polynomial is of degree 6, there are 6 complex number solutions to

$$x^6 + 2x^4 - 3x^3 - x^2 + 5x - 7 = 0.$$

7. Find the equations of the asymptotes. [2.3]

$x = -3, y = 2$

9. Graph $f(x) = 2^{-x+1}$. [4.2]

11. Sketch the graph of $y = -f(x) + 2$. [2.5]

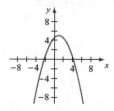

13. Find the remaining zeros. [3.4]

$$
\begin{array}{r|rrrrr}
2i & 1 & 1 & -8 & 4 & -48 \\
& & 2i & -4+2i & -4-24i & 48 \\
\hline
-2i & 1 & 1+2i & -12+2i & -24i & 0 \\
& & -2i & -2i & 24i & \\
\hline
& 1 & 1 & -12 & 0 &
\end{array}
$$

$x^2 + x - 12 = (x-3)(x+4) = 0$

$x = 3, \; x = -4$

The remaining zeros are $-2i$ and -4.

15. Convert $120°$ to radians. [5.1]

$120° = 120° \left(\dfrac{\pi}{180°} \right) = \dfrac{2\pi}{3}$

17. Find the period and amplitude. [5.5]

$f(x) = 3\cos 4x$

period: $\dfrac{\pi}{2}$

amplitude: 3

19. Solve. [6.6]

$\sin x \cos x - \dfrac{1}{2}\cos x = 0$

$\cos x \left(\sin x - \dfrac{1}{2} \right) = 0$

$\cos x = 0 \qquad\qquad \sin x - \dfrac{1}{2} = 0$

$x = \dfrac{\pi}{2}, \dfrac{3\pi}{2} \qquad\qquad \sin x = \dfrac{1}{2}$

$x = \dfrac{\pi}{6}, \dfrac{5\pi}{6}$

Chapter 9 Systems of Equations and Inequalities

Section 9.1 Exercises

1. Determine whether (4, 7) is a solution of the system.

   Equation (1)

   $$2x + 3y = 29$$
   $$2(4) + 3(7) \overset{?}{=} 29$$
   $$8 + 21 \overset{?}{=} 29$$
   $$29 = 29 \quad \text{True}$$

   Equation (2)

   $$-5x + 2y = -6$$
   $$-5(4) + 2(7) \overset{?}{=} -6$$
   $$-20 + 14 \overset{?}{=} -6$$
   $$-6 = -6 \quad \text{True}$$

   Yes, (4, 7) is a solution of the system.

3. Describe the relationship between graphs.

   The graphs are lines that are parallel to each other.

5. Solve by substitution.

   $$\begin{cases} 3x - 2y = 1 \\ y = 4 \end{cases}$$

   $$3x - 2(4) = 1$$
   $$3x = 9$$
   $$x = 3$$

   The solution is (3, 4).

7. Solve by substitution.

   $$\begin{cases} 3x + 4y = 18 \\ y = -2x + 3 \end{cases}$$

   $$3x + 4(-2x + 3) = 18$$
   $$3x - 8x + 12 = 18$$
   $$-5x = 6$$
   $$x = -\frac{6}{5}$$

   $$y = -2\left(-\frac{6}{5}\right) + 3 = \frac{27}{5}$$

   The solution is $\left(-\frac{6}{5}, \frac{27}{5}\right)$.

9. Solve by substitution.

   $$\begin{cases} -2x + 3y = 6 \\ x = 2y - 5 \end{cases}$$

   $$-2(2y - 5) + 3y = 6$$
   $$-4y + 10 + 3y = 6$$
   $$-y = -4$$
   $$y = 4$$

   $$x = 2(4) - 5 = 3$$

   The solution is (3, 4).

11. Solve by substitution.

    $$\begin{cases} 6x + 5y = 1 & (1) \\ x - 3y = 4 & (2) \end{cases}$$

    Solve (2) for x: $x = 3y + 4$

    $$6(3y + 4) + 5y = 1$$
    $$18y + 24 + 5y = 1$$
    $$23y = -23$$
    $$y = -1$$

    $$x = 3(-1) + 4 = 1$$

    The solution is (1, −1).

13. Solve by substitution.

    $$\begin{cases} 7x + 6y = -3 & (1) \\ y = \frac{2}{3}x - 6 & (2) \end{cases}$$

    $$7x + 6\left(\frac{2}{3}x - 6\right) = -3$$
    $$7x + 4x - 36 = -3$$
    $$11x = 33$$
    $$x = 3$$

    $$y = \frac{2}{3}(3) - 6 = -4$$

    The solution is (3, –4).

15. Solve by substitution.

    $$\begin{cases} y = 3x - 5 \\ y = 5x - 7 \end{cases}$$

    $$3x - 5 = 5x - 7$$
    $$2 = 2x$$
    $$1 = x$$

    $$y = 3(1) - 5 = -2$$

    The solution is (1,–2).

855

17. Solve by substitution.

$$\begin{cases} 2y = 3x + 19 \\ x = -2y + 7 \end{cases}$$

$$2y = 3(-2y + 7) + 19$$
$$2y = -6y + 21 + 19$$
$$8y = 40$$
$$y = 5$$

$$x = -2(5) + 7 = -3$$

The solution is (–3, 5).

19. Solve by substitution.

$$\begin{cases} 3x - 4y = 2 & (1) \\ 4x + 3y = 14 & (2) \end{cases}$$

Solve (1) for x and substitute into (2).

$$3x = 4y + 2 \quad \rightarrow \quad x = \frac{4y + 2}{3}$$

$$4\left(\frac{4y + 2}{3}\right) + 3y = 14$$
$$16y + 8 + 9y = 42$$
$$25y = 34$$
$$y = \frac{34}{25}$$

$$x = \frac{4}{3}\left(\frac{34}{25}\right) + \frac{2}{3} = \frac{62}{25}$$

The solution is $\left(\frac{62}{25}, \frac{34}{25}\right)$.

21. Solve by substitution.

$$\begin{cases} 3x - 3y = 5 & (1) \\ 4x - 4y = 9 & (2) \end{cases}$$

Solve (1) for x and substitute into (2).

$$3x - 3y = 5 \quad \rightarrow \quad x = \frac{3y + 5}{3}$$

$$4\left(\frac{3y + 5}{3}\right) - 4y = 9$$
$$12y + 20 - 12y = 27$$
$$20 = 27$$

The system of equations is inconsistent
and has no solution.

23. Solve by substitution.

$$\begin{cases} 4x + 3y = 6 \\ y = -\frac{4}{3}x + 2 \end{cases}$$

$$4x + 3\left(-\frac{4}{3}x + 2\right) = 6$$
$$4x - 4x + 6 = 6$$
$$0 = 0$$

The system of equations is dependent.

Let $x = c$ and $y = -\frac{4}{3}c + 2$.

The solutions are $\left(c, \ -\frac{4}{3}c + 2\right)$.

25. Solve by elimination.

$$\begin{cases} 3x - y = 10 & (1) \\ 4x + 3y = -4 & (2) \end{cases}$$

$$\begin{array}{ll} 9x - 3y = 30 & 3 \text{ times (1)} \\ 4x + 3y = -4 & (2) \\ \hline 13x = 26 \end{array}$$
$$x = 2$$

$$3(2) - y = 10$$
$$6 - y = 10$$
$$y = -4$$

The solution is (2, –4).

27. Solve by elimination.

$$\begin{cases} 4x + 7y = 21 & (1) \\ 5x - 4y = -12 & (2) \end{cases}$$

$$\begin{array}{ll} 20x + 35y = 105 & 5 \text{ times (1)} \\ -20x + 16y = 48 & -4 \text{ times (2)} \\ \hline 51y = 153 \end{array}$$
$$y = 3$$

$$4x + 7(3) = 21$$
$$x = 0$$

The solution is (0, 3).

29. Solve by elimination.

$$\begin{cases} 5x - 3y = 0 & (1) \\ 10x - 6y = 0 & (2) \end{cases}$$

$$-10x + 6y = 0 \quad -2 \text{ times (1)}$$
$$\underline{10x - 6y = 0 \quad (2)}$$
$$0 = 0$$

$$5x - 3c = 0$$
$$x = \frac{3c}{5}$$

The solution is $\left(\dfrac{3c}{5},\ c\right)$.

31. Solve by elimination.

$$\begin{cases} 3x - y = 4 & (1) \\ -6x + 2y = -8 & (2) \end{cases}$$

$$-6x + 2y = -8 \quad -2 \text{ times (1)}$$
$$\underline{-6x + 2y = -8 \quad (2)}$$
$$0 = 0$$

$$3c - y = 4$$
$$-y = -3c + 4$$
$$y = 3c - 4$$

The solution is $(c,\ 3c - 4)$.

33. Solve by elimination.

$$\begin{cases} 3x + 6y = 11 & (1) \\ 2x + 4y = 9 & (2) \end{cases}$$

$$6x + 12y = 22 \quad 2 \text{ times (1)}$$
$$\underline{-6x - 12y = -27 \quad -3 \text{ times (2)}}$$
$$0 = -5$$

The system of equations is
inconsistent and has no solution.

35. Solve by elimination.

$$\begin{cases} \dfrac{5}{6}x - \dfrac{1}{3}y = -6 & (1) \\ \dfrac{1}{6}x + \dfrac{2}{3}y = 1 & (2) \end{cases}$$

$$\dfrac{5}{3}x - \dfrac{2}{3}y = -12 \quad 2 \text{ times (1)}$$
$$\underline{\dfrac{1}{6}x + \dfrac{2}{3}y = 1 \quad\quad (2)}$$
$$\dfrac{11}{6}x \quad\quad = -11$$
$$x = -6$$

$$\frac{1}{6}(-6) + \frac{2}{3}y = 1$$
$$\frac{2}{3}y = 2$$
$$y = 3$$

The solution is $(-6,\ 3)$.

37. Solve by elimination.

$$\begin{cases} \dfrac{3}{4}x + \dfrac{1}{3}y = 1 & (1) \\ \dfrac{1}{2}x + \dfrac{2}{3}y = 0 & (2) \end{cases}$$

$$9x + 4y = 12 \quad 12 \text{ times (1)}$$
$$\underline{-3x - 4y = 0 \quad -6 \text{ times (2)}}$$
$$6x \quad\quad = 12$$
$$x = 2$$

$$3(2) + 4y = 0$$
$$4y = -6$$
$$y = -\frac{3}{2}$$

The solution is $\left(2, -\dfrac{3}{2}\right)$.

39. Solve by elimination.

$$\begin{cases} 2\sqrt{3}x - 3y = 3 & (1) \\ 3\sqrt{3}x + 2y = 24 & (2) \end{cases}$$

$$6\sqrt{3}x - 9y = 9 \quad 3 \text{ times (1)}$$
$$\underline{-6\sqrt{3}x - 4y = -48 \quad -2 \text{ times (2)}}$$
$$-13y = -39$$
$$y = 3$$

$$2\sqrt{3}x - 3(3) = 3$$
$$2\sqrt{3}x = 12$$
$$\sqrt{3}x = 6$$
$$x = 2\sqrt{3}$$

The solution is $(2\sqrt{3},\ 3)$.

41. Solve by elimination.

$$\begin{cases} 3\sqrt{2}x - 4\sqrt{3}y = -6 & (1) \\ 2\sqrt{2}x + 3\sqrt{3}y = 13 & (2) \end{cases}$$

$$6\sqrt{2}x - 8\sqrt{3}y = -12 \qquad \text{2 times (1)}$$
$$\underline{-6\sqrt{2}x - 9\sqrt{3}y = -39} \qquad \text{-3 times (2)}$$
$$-17\sqrt{3}y = -51$$
$$y = \frac{3}{\sqrt{3}} = \sqrt{3}$$

$$9\sqrt{2}x - 12\sqrt{3}y = -18 \qquad \text{3 times (1)}$$
$$\underline{8\sqrt{2}x + 12\sqrt{3}y = 52} \qquad \text{4 times (2)}$$
$$17\sqrt{2}x = 34$$
$$x = \frac{2}{\sqrt{2}}$$
$$x = \sqrt{2}$$

The solution is $\left(\sqrt{2}, \sqrt{3}\right)$.

43. Find the area of the triangle.

Find each vertex by solving a system of equations.

$$\begin{cases} y = 0 & (1) \\ x - y = 1 & (2) \end{cases}$$
$$x - 0 = 1 \Rightarrow x = 1$$

The vertex is $(1, 0)$.

$$\begin{cases} y = 0 & (1) \\ 5x - y = 25 & (3) \end{cases}$$
$$5x - 0 = 25 \quad (3)$$
$$x = 5$$

The vertex is $(5, 0)$.

$$\begin{cases} x - y = 1 & (2) \\ 5x - y = 25 & (3) \end{cases}$$

$$-x + y = -1 \quad \text{-1 times (2)}$$
$$\underline{5x - y = 25 \quad (3)}$$
$$4x = 24$$
$$x = 6$$

$$6 - y = 1 \quad (2)$$
$$y = 5$$

The vertex is $(6, 5)$.

The base of the triangle from $(1, 0)$ to $(5, 0)$ has distance 4.

The height of the triangle from $y = 0$ to $(6, 5)$ has distance 5.

Recall, the formula for the area of a triangle, $A = \frac{1}{2}bh$.

$$A = \frac{1}{2}bh$$
$$A = \frac{1}{2}(4)(5) = 10$$

The area of the triangle is 10 square miles.

45. Find the equilibrium price.

Solve the system by substitution.

$$25p - 3400 = -5p + 2000$$
$$30p = 5400$$
$$p = 180$$

The solution is \$180.

47. Find the rate of the plane in calm air and wind.

Rate of plane with the wind: $r + w$

Rate of plane against the wind: $r - w$

$$r \cdot t = d$$
$$\begin{cases} (r + w) \cdot 3 = 450 & (1) \\ (r - w) \cdot 5 = 450 & (2) \end{cases}$$

$$r + w = 150$$
$$\underline{r - w = 90}$$
$$2r \quad = 240$$
$$r = 120$$

$$120 + w = 150$$
$$w = 30$$

Rate of plane = 120 mph.

Rate of wind = 30mph.

49. Find the rate of the boat in calm water and current.

Rate of boat with the current: $r + w$

Rate of boat against the wind: $r - w$

$$r \cdot t = d$$
$$\begin{cases} (r + w) \cdot 4 = 120 & (1) \\ (r - w) \cdot 6 = 120 & (2) \end{cases}$$

$$r + w = 30$$
$$\underline{r - w = 20}$$
$$2r \quad = 50$$
$$r = 25$$

$$25 + w = 30$$
$$w = 5$$

Rate of boat = 25 mph.

Rate of current = 5 mph.

51. Find the cost for the iron and lead alloys.

$x =$ cost per kilogram of iron alloy

$y =$ cost per kilogram of lead alloy

$$\begin{cases} 30x+45y=1080 & (1) \\ 15x+12y=372 & (2) \end{cases}$$

$$\begin{array}{l} 30x+45y=1080 \quad (1) \\ \underline{-30x-24y=-744} \quad -2 \text{ times } (2) \\ 21y=336 \\ y=16 \end{array}$$

$$\begin{array}{l} 15x+12(16)=372 \\ 15x=180 \\ x=12 \end{array}$$

Cost of iron alloy: \$12 per kilogram

Cost of lead alloy: \$16 per kilogram

53. Find the amount of each alloy.

$x =$ amount of 40% gold

$y =$ amount of 60% gold

$$\begin{cases} x+y=20 & (1) \\ 0.40x+0.60y=(0.52)(20) & (2) \end{cases}$$

$$\begin{array}{l} -0.40x-0.40y=-8 \quad -0.40 \text{ times } (1) \\ \underline{0.40x+0.60y=10.4} \quad (2) \\ 0.20y=2.4 \\ y=12 \end{array}$$

$$\begin{array}{l} x+12=20 \\ x=8 \end{array}$$

Amount of 40% gold: 8 g; Amount of 60% gold: 12 g

55. Find the area.

Sketch a graph to visualize the right triangle.

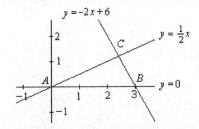

To find the coordinates of point A, solve the system

$$\begin{cases} y=0 \\ y=\dfrac{1}{2}x \end{cases}$$

By substitution, $\dfrac{1}{2}x=0$

$$x=0 \quad \text{Thus } A \text{ is } (0,0).$$

To find the coordinates of point B, solve the system

$$\begin{cases} y=0 \\ y=-2x+6 \end{cases}$$

By substitution, $-2x+6=0$

$$\begin{array}{l} -2x=-6 \\ x=3 \quad \text{Thus } B \text{ is } (3,0). \end{array}$$

To find the coordinates of the point C, solve the system

$$\begin{cases} y=-2x+6 & (1) \\ y=\dfrac{1}{2}x & (2) \end{cases}$$

By substitution, $\dfrac{1}{2}x=-2x+6$

$$\begin{array}{l} \dfrac{5}{2}x=6 \\ x=\dfrac{12}{5} \end{array}$$

Substituting $\dfrac{12}{5}$ for x in Equation (2), we have

$$y=\dfrac{1}{2}\left(\dfrac{12}{5}\right)=\dfrac{6}{5}. \quad \text{Thus } C \text{ is } \left(\dfrac{12}{5},\ \dfrac{6}{5}\right).$$

The base of the triangle is the distance from A to B, or 3 units. The height of the triangle is the vertical distance from the base to C, or $\dfrac{6}{5}$ units.

$$\begin{array}{l} \text{Area} =\dfrac{1}{2}(\text{base})(\text{height}) \\ \phantom{\text{Area}}=\dfrac{1}{2}(3)\left(\dfrac{6}{5}\right)=\dfrac{9}{5} \text{ square units} \end{array}$$

57. Find the largest possible value of Z.

$$\begin{array}{r} 5Z7 \\ +\ \underline{256} \\ XY3 \end{array}$$

Case 1: $Z+5+1\le 9$ Case 2: $Z+5+1>9$

$$\begin{cases} Z+5+1=Y \\ 5+2=X \end{cases} \qquad \begin{cases} Z+5+1=10+Y \\ 5+2+1=X \end{cases}$$

$$\begin{cases} Z+6=Y \\ 7=X \end{cases} \qquad \begin{cases} Z-4=Y \\ 8=X \end{cases}$$

$$\begin{array}{l} X+Y=7+Z+6 \\ X+Y=Z+13 \end{array} \qquad \begin{array}{l} X+Y=8+Z-4 \\ X+Y=Z+4 \end{array}$$

$XY3$ is divisible by $3 \Rightarrow X+Y$ is divisible by 3.

If $Z+13$ is divisible by 3, then $Z=2, 5,$ or 8.

If $Z+4$ is divisible by 3, then $Z=2, 5,$ or 8.

In both cases, the largest digit Z can be is 8.

59. Find the Pythagorean triples.

$14 = c - b$	$18 = c - b$
$126 = c + b$	$98 = c + b$
$140 = 2c$	$116 = 2c$
$70 = c, b = 56$	$58 = c, b = 40$
$294 = c - b$	$2 = c - b$
$6 = c + b$	$882 = c + b$
$300 = 2c$	$884 = 2c$
$150 = c, b = 144$	$442 = c, b = 440$

The Pythagorean triples are: 42, 56, 70; 42, 40, 58; 42, 144, 150; 42, 440, 442.

61. Find how many people like the skin cream.

x = people who like lip balm but do not like skin cream

y = people who like lip balm and skin cream

z = people who do not like lip balm but do like skin cream

w = people who do not like lip balm nor skin cream

$$\begin{cases} x+y+z+w = 100 & (1) \\ 0.80(y+z) = y & (2) \\ 0.50(x+w) = w & (3) \\ x+y = 77 & (4) \end{cases}$$

Rewrite the system by solving eq (2) for z, eq (3) for w, and eq (4) for x.

$$\begin{cases} x+y+z+w = 100 & (1) \\ z = 0.25y & (2) \\ w = x & (3) \\ x = -y+77 & (4) \end{cases}$$

Substitute the values from equations (2), (3), and (4) into equation (1) and solve for y.

$$(-y+77)+y+0.25y+(-y+77) = 100$$
$$-0.75y+154 = 100$$
$$-0.75y = -54$$
$$y = 72$$

$z = 0.25(72) = 18$

$x = -72+77 = 5$

$w = 5$

Find the number of people who like skin cream ($y+z$)

$y+z = 72+18 = 90$ people like the skin cream.

63. Find the pumping rate for each pump.

S = supply pump

A = outlet pump

$$\begin{cases} \frac{1}{2}S - \frac{1}{2}A = 8750 & (1) \\ \frac{3}{4}S - 2\left(\frac{3}{4}A\right) = 11{,}250 & (2) \end{cases}$$

$$\begin{array}{ll} S - A = 17{,}500 & (1) \\ \underline{S - 2A = 15{,}000} & \frac{4}{3} \text{ times (2)} \\ A = 2500 \\ S = 20{,}000 \end{array}$$

The supply pump can pump 20,000 gal/h.

The outlet pump can pump 2500 gal/h.

65. Solve the system of equations.

$$\begin{cases} \frac{2}{x} - \frac{1}{y} = \frac{2}{3} & (1) \\ \frac{5}{x} + \frac{3}{y} = -\frac{1}{6} & (2) \end{cases}$$

Use the substitution, $u = 1/x$ and $v = 1/y$.

$$\begin{cases} 2u - v = \frac{2}{3} & (3) \\ 5u + 3v = -\frac{1}{6} & (4) \end{cases}$$

$$\begin{array}{ll} 36u - 18v = 12 & 18 \text{ times (3)} \\ \underline{30u + 18v = -1} & 6 \text{ times (4)} \\ 66u \quad\quad = 11 \\ \quad u = \frac{1}{6} \end{array}$$

$$2u - v = \frac{2}{3} \quad (3)$$
$$2\left(\frac{1}{6}\right) - v = \frac{2}{3}$$
$$\frac{1}{3} - v = \frac{2}{3}$$
$$v = -\frac{1}{3}$$

Since $u = 1/6$ and $v = -1/3,$ then $x = 6$ and $y = -3$.

The solution is $(6, -3)$.

67. Graph the equations from Exercise 65.

    a. Determine if $(6, -3)$ is a point of intersection.

Yes.

    b. Explain.

The variable terms in the system are undefined

for $x = 0$ and $y = 0$.

Prepare for Section 9.2

P1. Solve.

$$2x - 5y = 15$$
$$-5y = -2x + 15$$
$$y = \frac{2}{5}x - 3$$

P3. Solve by substitution.

$$\begin{cases} 5x - 2y = 10 & (1) \\ 2y = 8 & (2) \end{cases}$$

$$y = 4$$
$$5x - 2(4) = 10$$
$$5x = 18$$
$$x = \frac{18}{5}$$

The solution is $\left(\frac{18}{5}, 4\right)$.

P5. Solve by substitution.

$$\begin{cases} y = 3x - 4 & (1) \\ y = 4x - 2 & (2) \end{cases}$$

$$3x - 4 = 4x - 2$$
$$x = -2$$
$$y = 3(-2) - 4 = -10$$

The solution is $(-2, -10)$.

Section 9.2 Exercises

1. Determine the solutions.

    a. $3x - 2y + z = 6$

          $3(3) - 2(4) + 5 = 6$

                    $6 = 6$ True

    b. $3x - 2y + z = 6$

          $3(5) - 2(3) - 3 = 6$

                    $6 = 6$ True

    c. $3x - 2y + z = 6$

         $3(0) - 2(3) + 11 = 6$

                    $5 = 6$ False

3. Describe the graph of the system.

The graph of the solutions is the line where the graphs

of the equations, which are planes, intersect.

5. Solve the system of equations.

$$\begin{cases} 2x - y + z = 8 & (1) \\ 2y - 3z = -11 & (2) \\ 3y + 2z = 3 & (3) \end{cases}$$

$$\begin{array}{ll} 6y - 9z = -33 & 3 \text{ times (2)} \\ \underline{-6y - 4z = -6} & -2 \text{ times (3)} \\ -13z = -39 \\ z = 3 & (4) \end{array}$$

$$\begin{cases} 2x - y + z = 8 \\ 2y - 3z = -11 \\ z = 3 & (4) \end{cases}$$

$$2y - 3(3) = -11$$
$$y = -1$$
$$2x - (-1) + 3 = 8$$
$$x = 2$$

The solution is $(2, -1, 3)$.

7. Solve the system of equations.

$$\begin{cases} 2x + 2y - 3z = 18 & (1) \\ x - 3y + 4z = -6 & (2) \\ 3x - y + 2z = 10 & (3) \end{cases}$$

$$\begin{array}{ll} 2x + 2y - 3z = 18 & (1) \\ \underline{-2x + 6y - 8z = 12} & -2 \text{ times (2)} \\ 8y - 11z = 30 & (4) \end{array}$$

$$\begin{array}{ll} -3x + 9y - 12z = 18 & -3 \text{ times (2)} \\ \underline{3x - y + 2z = 10} & (3) \\ 8y - 10z = 28 & (5) \end{array}$$

$$\begin{cases} 2x + 2y - 3z = 18 & (1) \\ 8y - 11z = 30 & (4) \\ 8y - 10z = 28 & (5) \end{cases}$$

$$8y - 11z = 30 \quad (4)$$
$$\underline{-8y + 10z = -28} \quad -1 \text{ times } (5)$$
$$-z = 2$$
$$z = -2 \quad (6)$$

$$\begin{cases} 2x + 2y - 3z = 18 & (1) \\ 8y - 11z = 30 & (4) \\ z = -2 & (6) \end{cases}$$

$$8y - 11(-2) = 30 \qquad 2x + 2(1) - 3(-2) = 18$$
$$y = 1 \qquad\qquad x = 5$$

The solution is (5, 1,–2).

9. Solve the system of equations.

$$\begin{cases} 3x + 4y - z = -7 & (1) \\ x - 5y + 2z = 19 & (2) \\ 5x + y - 2z = 5 & (3) \end{cases}$$

$$3x + 4y - z = -7 \quad (1)$$
$$\underline{-3x + 15y - 6z = -57} \quad -3 \text{ times } (2)$$
$$19y - 7z = -64 \quad (4)$$

$$-5x + 25y - 10z = -95 \quad -5 \text{ times } (2)$$
$$\underline{5x + y - 2z = 5} \quad (3)$$
$$26y - 12z = -90 \quad (5)$$

$$\begin{cases} 3x + 4y - z = -7 & (1) \\ 19y - 7z = -64 & (4) \\ 26y - 12z = -90 & (5) \end{cases}$$

$$494y - 182z = -1664 \quad 26 \text{ times } (4)$$
$$\underline{-494y + 228z = 1710} \quad -19 \text{ times } (3)$$
$$46z = 46$$
$$z = 1 \quad (6)$$

$$\begin{cases} 3x + 4y - z = -7 & (1) \\ 19y - 7z = -64 & (4) \\ z = 1 & (6) \end{cases}$$

$$19y - 7(1) = -64$$
$$y = -3$$

$$3x + 4(-3) - 1 = -7$$
$$x = 2$$

The solution is (2, –3, 1).

11. Solve the system of equations.

$$\begin{cases} 3x - 2y + 3z = 8 & (1) \\ 2x + 3y - 4z = -28 & (2) \\ x - y + 2z = 9 & (3) \end{cases}$$

$$3x - 2y + 3z = 8 \quad (1)$$
$$\underline{-3x + 3y - 6z = -27} \quad -3 \text{ times } (3)$$
$$y - 3z = -19 \quad (4)$$

$$2x + 3y - 4z = -28 \quad (2)$$
$$\underline{-2x + 2y - 4z = -18} \quad -2 \text{ times } (3)$$
$$5y - 8z = -46 \quad (5)$$

$$\begin{cases} 3x - 2y + 3z = 8 & (1) \\ y - 3z = -19 & (4) \\ 5y - 8z = -46 & (5) \end{cases}$$

$$-5y + 15z = 95 \quad -5 \text{ times } (4)$$
$$\underline{5y - 8z = -46} \quad (5)$$
$$7z = 49$$
$$z = 7 \quad (6)$$

$$\begin{cases} 3x - 2y + 3z = 8 & (1) \\ y - 3z = -19 & (4) \\ z = 7 & (6) \end{cases}$$

$$y - 3(7) = -19$$
$$y = 2$$

$$3x - 2(2) + 3(7) = 8$$
$$x = -3$$

The solution is (–3, 2, 7).

13. Solve the system of equations.

$$\begin{cases} 3x - 2y + 4z = 16 & (1) \\ 2x + 3y + 2z = 16 & (2) \\ 5x + 2y + 4z = 24 & (3) \end{cases}$$

$$3x - 2y + 4z = 16 \quad (1)$$
$$\underline{-4x - 6y - 4z = -32} \quad -2 \text{ times } (2)$$
$$-x - 8y = -16 \quad (4)$$

$$-3x + 2y - 4z = -16 \quad -1 \text{ times } (1)$$
$$\underline{5x + 2y + 4z = 24} \quad (3)$$
$$2x + 4y = 8 \quad (5)$$

$$\begin{cases} 3x - 2y + 4z = 16 & (1) \\ -x - 8y = -16 & (4) \\ 2x + 4y = 8 & (5) \end{cases}$$

$-2x-16y=-32$ 2 times (4)

$\dfrac{2x+4y\ =8}{}$ (5)

$\qquad -12y=-24$

$\qquad\quad y=2$ (6)

$\begin{cases}3x-2y+4z=16 & (1)\\ -x-8y\ \ \ \ \ \ \ =-16 & (4)\\ \qquad y\ \ \ \ \ =2 & (6)\end{cases}$

$-x-8(2)=-16$

$\qquad x=0$

$3(0)-2(2)+4z=16$

$\qquad\qquad z=5$

The solution is $(0, 2, 5)$.

15. Solve the system of equations.

$\begin{cases}2x-5y+2z=-4 & (1)\\ 3x+2y+3z=13 & (2)\\ 5x-3y-4z=-18 & (3)\end{cases}$

$6x-15y+6z=-12$ 3 times (1)

$\dfrac{-6x-4y-6z=-26}{}$ -2 times (2)

$\qquad -19y\ \ \ \ \ \ =-38$

$\qquad\quad y=2$ (4)

$10x-25y+10z=-20$ 5 times (1)

$\dfrac{-10x+6y+8z=36}{}$ -2 times (3)

$\qquad -19y+18z=16$ (5)

$\begin{cases}2x-5y+2z=-4 & (1)\\ \qquad y=2 & (4)\\ -19y+18z=16 & (5)\end{cases}$

$\quad 19y\ \ \ \ \ =38$ 19 times (4)

$\dfrac{-19y+18z=16}{}$ (5)

$\qquad 18z=54$

$\qquad\quad z=3$ (6)

$\begin{cases}2x-5y+2z=-4 & (1)\\ \qquad y=2 & (4)\\ \qquad z=3 & (6)\end{cases}$

$2x-5(2)+2(3)=-4$

$\qquad x=0$

The solution is $(0, 2, 3)$.

17. Solve the system of equations.

$\begin{cases}2x+y-z=-2 & (1)\\ 3x+2y+3z=21 & (2)\\ 7x+4y+z=17 & (3)\end{cases}$

$6x+3y-3z=-6$ 3 times (1)

$\dfrac{-6x-4y-6z=-42}{}$ -2 times (2)

$\qquad -y-9z=-48$ (4)

$14x+7y-7z=-14$ 7 times (1)

$\dfrac{-14x-8y-2z=-34}{}$ -2 times (3)

$\qquad -y-9z=-48$ (5)

$\begin{cases}2x+y-z=-2 & (1)\\ \quad -y-9z=-48 & (4)\\ \quad -y-9z=-48 & (5)\end{cases}$

$\quad -y-9z=-48$ (4)

$\dfrac{\ \ \ \ y+9z=48}{}$ -1 times (5)

$\qquad\quad 0=0$ (6)

$\begin{cases}2x+y-z=-2 & (1)\\ \quad -y-9z=-48 & (4)\\ \qquad 0=0 & (6)\end{cases}$

The system of equations is dependent.

Let $z=c$. $-y-9c=-48$

$\qquad\qquad\qquad y=48-9c$

$2x+(48-9c)-c=-2$

$\qquad\qquad x=5c-25$

The solution is $(5c-25,\ 48-9c,\ c)$.

19. Solve the system of equations.

$\begin{cases}6x-7y+12z=81 & (1)\\ 4x+3y+2z=7 & (2)\\ 3x-5y+11z=65 & (3)\end{cases}$

$12x-14y+24z=162$ 2 times (1)

$\dfrac{-12x-9y\ -6z\ =-21}{}$ -3 times (2)

$\qquad -23y+18z=141$ (4)

$6x-7y+12z=81$ (1)

$\dfrac{-6x+10y-22z=-130}{}$ -2 times (2)

$\qquad 3y-10z=-49$ (5)

$\begin{cases}6x-7y+12z=81 & (1)\\ \ \ -23y+18z=141 & (4)\\ \ \ \ \ 3y-10z=-49 & (5)\end{cases}$

$$-69y+54z=423 \qquad 3 \text{ times (4)}$$
$$\underline{-69y-230z=-1127 \quad 23 \text{ times (5)}}$$
$$-176z=-704$$
$$z=4 \qquad (6)$$

$$\begin{cases} 6x-7y+12z=81 & (1) \\ -23y+18z=141 & (4) \\ z=4 & (6) \end{cases}$$

$$-23y+18(4)=141 \qquad\qquad 6x-7(-3)+12(4)=81$$
$$y=-3 \qquad\qquad\qquad\qquad x=2$$

The solution is $(2,-3,4)$.

21. Solve the system of equations.

$$\begin{cases} 2x-3y+6z=3 & (1) \\ x+2y-4z=5 & (2) \\ 3x+4y-8z=7 & (3) \end{cases}$$

$$2x-3y+6z=3 \qquad (1)$$
$$\underline{-2x-4y+8z=-10 \quad -2 \text{ times (2)}}$$
$$-7y+14z=-7$$
$$-y+2z=-1 \qquad (4)$$

$$-3x-6y+12z=-15 \quad -3 \text{ times (2)}$$
$$\underline{3x+4y-8z=7 \qquad (3)}$$
$$-2y+4z=-8$$
$$-y+2z=-4 \qquad (5)$$

$$\begin{cases} 2x-3y+6z=3 & (1) \\ -y+2z=-1 & (4) \\ -y+2z=-4 & (5) \end{cases}$$

$$-y+2z=-1 \qquad (4)$$
$$\underline{y-2z=4 \qquad -1 \text{ times (5)}}$$
$$0=3 \qquad (6)$$

$$\begin{cases} 2x-3y+6z=3 & (1) \\ -y+2z=-1 & (4) \\ 0=3 & (6) \end{cases}$$

The system of equations is inconsistent and has no solution.

23. Solve the system of equations.

$$\begin{cases} 2x-3y+5z=14 & (1) \\ x+4y-3z=-2 & (2) \end{cases}$$

$$2x-3y+5z=14 \qquad (1)$$
$$\underline{-2x-8y+6z=4 \qquad -2 \text{ times (2)}}$$
$$-11y+11z=18 \qquad (3)$$

$$\begin{cases} 2x-3y+5z=14 & (1) \\ -11y+11z=18 & (3) \end{cases}$$

Let $z=c$. $\qquad -11y+11c=18$
$$y=\frac{18-11c}{-11}$$
$$y=\frac{11c-18}{11}$$

$$2x-3\left(\frac{11c-18}{11}\right)+5c=14$$
$$2x=14-5c+\frac{33c-54}{11}$$
$$2x=\frac{154-55c+33c-54}{11}$$
$$x=\frac{50-11c}{11}$$

The solution is $\left(\dfrac{50-11c}{11},\ \dfrac{11c-18}{11},\ c\right)$.

25. Solve the system of equations.

$$\begin{cases} 6x-9y+6z=7 & (1) \\ 4x-6y+4z=9 & (2) \end{cases}$$

$$24x-36y+24z=28 \qquad 4 \text{ times (1)}$$
$$\underline{-24x+36y-24z=-54 \quad -6 \text{ times (2)}}$$
$$0=-26 \qquad (3)$$

$$\begin{cases} 6x-9y+6z=7 & (1) \\ 0=-26 & (3) \end{cases}$$

The system of equations is inconsistent and has no solution.

27. Solve the system of equations.

$$\begin{cases} 5x+3y+2z=10 & (1) \\ 3x-4y-4z=-5 & (2) \end{cases}$$

$$15x+9y+6z=30 \qquad 3 \text{ times (1)}$$
$$\underline{-15x+20y+20z=25 \quad -5 \text{ times (2)}}$$
$$29y+26z=55 \qquad (3)$$

$$\begin{cases} 5x+3y+2z=10 & (1) \\ 29y+26z=55 & (3) \end{cases}$$

Let $z=c$. $\qquad 29y+26c=55$
$$y=\frac{55-26c}{29}$$

$$5x + 3\left(\frac{55-26c}{29}\right) + 2c = 10$$

$$5x = 10 - 2c - \frac{165-78c}{29}$$

$$5x = \frac{290 - 58c - 165 + 78c}{29}$$

$$x = \frac{25+4c}{29}$$

The solution is $\left(\dfrac{25+4c}{29},\ \dfrac{55-26c}{29},\ c\right)$.

29. Solve the homogeneous system of equations.

$$\begin{cases} x + 3y - 4z = 0 & (1) \\ 2x + 7y + z = 0 & (2) \\ 3x - 5y - 2z = 0 & (3) \end{cases}$$

$$\begin{array}{ll} -2x - 6y + 8z = 0 & -2 \text{ times } (1) \\ \underline{2x + 7y + z = 0} & (2) \\ y + 9z = 0 & (4) \end{array}$$

$$\begin{array}{ll} -3x - 9y + 12z = 0 & -3 \text{ times } (1) \\ \underline{3x - 5y - 2z = 0} & (3) \\ -14y + 10z = 0 \\ -7y + 5z = 0 & (5) \end{array}$$

$$\begin{cases} x + 3y - 4z = 0 & (1) \\ y + 9z = 0 & (4) \\ -7y + 5z = 0 & (5) \end{cases}$$

$$\begin{array}{ll} 7y + 63z = 0 & 7 \text{ times } (4) \\ \underline{-7y + 5z = 0} & (5) \\ 68z = 0 \\ z = 0 & (6) \end{array}$$

$$\begin{cases} x + 3y - 4z = 0 & (1) \\ y + 9z = 0 & (4) \\ z = 0 & (6) \end{cases}$$

$$\begin{array}{ll} y + 9(0) = 0 & x + 3(0) - 4(0) = 0 \\ y = 0 & x = 0 \end{array}$$

The solution is $(0, 0, 0)$.

31. Solve the homogeneous system of equations.

$$\begin{cases} 2x - 3y + z = 0 & (1) \\ 2x + 4y - 3z = 0 & (2) \\ 6x - 2y - z = 0 & (3) \end{cases}$$

$$\begin{array}{ll} -2x + 3y - z = 0 & -1 \text{ times } (1) \\ \underline{2x + 4y - 3z = 0} & (2) \\ 7y - 4z = 0 & (4) \end{array}$$

$$\begin{array}{ll} -6x + 9y - 3z = 0 & -3 \text{ times } (1) \\ \underline{6x - 2y - z = 0} & (3) \\ 7y - 4z = 0 & (5) \end{array}$$

$$\begin{cases} 2x - 3y + z = 0 & (1) \\ 7y - 4z = 0 & (4) \\ 7y - 4z = 0 & (5) \end{cases}$$

$$\begin{array}{ll} -7y + 4z = 0 & -1 \text{ times } (4) \\ \underline{7y - 4z = 0} & (5) \\ 0 = 0 & (6) \end{array}$$

$$\begin{cases} 2x - 3y + z = 0 & (1) \\ 7y - 4z = 0 & (4) \\ 0 = 0 & (6) \end{cases}$$

Let $z = c$. Then $7y = 4c$ or $y = \dfrac{4}{7}c$. Substitute for y

and z in Eq. (1) and solve for x.

$$2x - 3\left(\frac{4}{7}c\right) + c = 0$$

$$2x = \frac{5}{7}c$$

$$x = \frac{5}{14}c$$

The solution is $\left(\dfrac{5}{14}c,\ \dfrac{4}{7}c,\ c\right)$.

33. Solve the homogeneous system of equations.

$$\begin{cases} 3x - 5y + 3z = 0 & (1) \\ 2x - 3y + 4z = 0 & (2) \\ 7x - 11y + 11z = 0 & (3) \end{cases}$$

$$\begin{array}{ll} -6x + 10y - 6z = 0 & -2 \text{ times } (1) \\ \underline{6x - 9y + 12z = 0} & 3 \text{ times } (2) \\ y + 6z = 0 & (4) \end{array}$$

$$\begin{array}{ll} -21x + 35y - 21z = 0 & -7 \text{ times } (1) \\ \underline{21x - 33y + 33z = 0} & 3 \text{ times } (3) \\ 2y + 12z = 0 & (5) \end{array}$$

$$\begin{cases} 3x - 5y + 3z = 0 & (1) \\ y + 6z = 0 & (4) \\ 2y + 12z = 0 & (5) \end{cases}$$

$$-2y - 12z = 0 \quad -2 \text{ times } (4)$$
$$\underline{2y + 12z = 0 \quad (5)}$$
$$0 = 0 \quad (6)$$

$$\begin{cases} 3x - 5y + 3z = 0 & (1) \\ y + 6z = 0 & (4) \\ 0 = 0 & (6) \end{cases}$$

From Eq. (4), $y = -6z$. Substitute into Eq. (1).

$$3x - 5(-6z) + 3z = 0$$
$$3x = -33z$$
$$x = -11z$$

Let z be any real number c, then the solutions are $(-11c, \ -6c, \ c)$.

35. Solve the homogeneous system of equations.

$$\begin{cases} 4x - 7y - 2z = 0 & (1) \\ 2x + 4y + 3z = 0 & (2) \\ 3x - 2y - 5z = 0 & (3) \end{cases}$$

$$4x - 7y - 2z = 0 \quad (1)$$
$$\underline{-4x - 8y - 6z = 0 \quad -2 \text{ times } (2)}$$
$$-15y - 8z = 0 \quad (4)$$

$$6x + 12y + 9z = 0 \quad 3 \text{ times } (2)$$
$$\underline{-6x + 4y + 10z = 0 \quad -2 \text{ times } (3)}$$
$$16y + 19z = 0 \quad (5)$$

$$\begin{cases} 4x - 7y - 2z = 0 & (1) \\ -15y - 8z = 0 & (4) \\ 16y + 19z = 0 & (5) \end{cases}$$

$$-240y - 128z = 0 \quad 16 \text{ times } (4)$$
$$\underline{240y + 285z = 0 \quad 15 \text{ times } (5)}$$
$$157z = 0$$
$$z = 0 \quad (6)$$

$$\begin{cases} 4x - 7y - 2z = 0 & (1) \\ -15y - 8z = 0 & (4) \\ z = 0 & (6) \end{cases}$$

$z = 0$, $y = 0$, $x = 0$. The solution is $(0, 0, 0)$.

37. Find the equation that passes through the points.

$$y = ax^2 + bx + c$$
$$3 = a(2)^2 + b(2) + c$$
$$7 = a(-2)^2 + b(-2) + c$$
$$-2 = a(1)^2 + b(1) + c$$

$$\begin{cases} 4a + 2b + c = 3 & (1) \\ 4a - 2b + c = 7 & (2) \\ a + b + c = -2 & (3) \end{cases}$$

$$4a + 2b + c = 3 \quad (1)$$
$$\underline{-4a + 2b - c = -7 \quad -1 \text{ times } (2)}$$
$$4b = -4 \quad (4)$$

$$4a + 2b + c = 3 \quad (1)$$
$$\underline{-4a - 4b - 4c = 8 \quad -4 \text{ times } (3)}$$
$$-2b - 3c = 11 \quad (5)$$

$$\begin{cases} 4a + 2b + c = 3 & (1) \\ 4b = -4 & (4) \\ -2b - 3c = 11 & (5) \end{cases}$$

From (4): $4b = -4$
$$b = -1$$

From (5): $-2(-1) - 3c = 11$
$$c = -3$$

From (1): $4a + 2(-1) - 3 = 3$
$$a = 2$$

The equation whose graph passes through the three points is $y = 2x^2 - x - 3$.

39. Find the equation of the circle.

$$x^2 + y^2 + ax + by + c = 0$$
$$5^2 + 3^2 + a(5) + b(3) + c = 0$$
$$(-1)^2 + (-5)^2 + a(-1) + b(-5) + c = 0$$
$$(-2)^2 + 2^2 + a(-2) + b(2) + c = 0$$

$$\begin{cases} 5a + 3b + c = -34 & (1) \\ -a - 5b + c = -26 & (2) \\ -2a + 2b + c = -8 & (3) \end{cases}$$

$$5a + 3b + c = -34 \quad (1)$$
$$\underline{a + 5b - c = 26 \quad -1 \text{ times } (2)}$$
$$6a + 8b = -8$$
$$3a + 4b = -4 \quad (4)$$

$$5a+3b+c=-34 \quad (1)$$
$$2a-2b-c=8 \quad -1 \text{ times (3)}$$
$$\overline{7a+b=-26 \quad (5)}$$

$$\begin{cases} 5a+3b+c=-34 & (1) \\ 3a+4b=-4 & (4) \\ 7a+b=-26 & (5) \end{cases}$$

$$3a+4b=-4 \quad (4)$$
$$-28a-4b=104 \quad -4 \text{ times (5)}$$
$$\overline{-25a =100}$$
$$a=-4 \quad (6)$$

$$\begin{cases} 5a+3b+c=-34 & (1) \\ 3a+4b=-4 & (4) \\ a=-4 & (6) \end{cases}$$

$$3(-4)+4b=-4 \qquad 5(-4)+3(2)+c=-34$$
$$b=2 \qquad\qquad c=-20$$

The equation whose graph passes through the three

points is $x^2+y^2-4x+2y-20=0$.

41. Find the center and radius of the circle.

$$x^2+y^2+ax+by+c=0$$
$$(-2)^2+10^2+a(-2)+b(10)+c=0$$
$$(-12)^2+(-14)^2+a(-12)+b(-14)+c=0$$
$$5^2+3^2+a(5)+b(3)+c=0$$

$$\begin{cases} -2a+10b+c=-104 & (1) \\ -12a-14b+c=-340 & (2) \\ 5a+3b+c=-34 & (3) \end{cases}$$

$$-2a+10b+c=-104 \quad (1)$$
$$12a+14b-c=340 \quad -1 \text{ times (2)}$$
$$\overline{10a+24b=236}$$
$$5a+12b=118 \quad (4)$$

$$-2a+10b+c=-104 \quad (1)$$
$$-5a-3b-c=34 \quad -1 \text{ times (3)}$$
$$\overline{-7a+7b =-70}$$
$$-a+b=-10 \quad (5)$$

$$\begin{cases} -2a+10b+c=-104 & (1) \\ 5a+12b=118 & (4) \\ -a+b=-10 & (5) \end{cases}$$

$$5a+12b=118 \quad (4)$$
$$-5a+5b=-50 \quad 5 \text{ times (5)}$$
$$\overline{17b=68}$$
$$b=4 \quad (6)$$

$$\begin{cases} -2a+10b+c=-104 & (1) \\ 5a+12b=118 & (4) \\ b=4 & (6) \end{cases}$$

$$5a+12(4)=118 \qquad -2(14)+10(4)+c=-104$$
$$a=14 \qquad\qquad c=-116$$

The equation whose graph passes through the three

points is $x^2+y^2+14x+4y-116=0$.

$$(x^2+14x+49)+(y^2+4y+4)=116+49+4$$
$$(x+7)^2+(y+2)^2=169$$

The center is $(-7,\ -2)$ and radius is 13.

43. Find the traffic flow between B and C.

For intersection A, $275+225=x_1+x_2$
$$x_1+x_2=500$$

For intersection B, $x_2+90=x_3+150$
$$x_2-x_3=60$$

For intersection C, $x_1+x_3=240+200$
$$x_1+x_3=440$$

$$\begin{cases} x_1+x_2=500 & (1) \\ x_2-x_3=60 & (2) \\ x_1+x_3=440 & (3) \end{cases}$$

The equations are dependent.

Solve Eq. (2) for x_3 and substitute the inequality for x_2.

$$x_3=x_2-60$$

Because $150 \le x_2 \le 250$, then $90 \le x_3 \le 190$.

The flow between B and C is 90 to 190 cars per hour.

45. Find the estimated traffic flow between C and A,

D and C, and B and D.

For intersection A, $256+x_4=389+x_1$
$$x_1-x_4=-133$$

For intersection B, $437+x_1=x_2+300$
$$x_1-x_2=-137$$

For intersection C, $298 + x_3 = 249 + x_4$

$$x_3 - x_4 = -49$$

For intersection D, $314 + x_2 = 367 + x_3$

$$x_2 - x_3 = 53$$

$$\begin{cases} x_1 - x_4 = -133 & (1) \\ x_1 - x_2 = -137 & (2) \\ x_3 - x_4 = -49 & (3) \\ x_2 - x_3 = 53 & (4) \end{cases}$$

The equations are dependent. Solving the system gives

$$x_1 = x_4 - 133$$

$$x_2 = x_4 + 4$$

$$x_3 = x_4 - 49$$

Because $125 \le x_1 \le 175$, then

$125 \le x_4 - 133 \le 175$ and $258 \le x_2 - 4 \le 308$
$258 \le x_4 \le 308$ $262 \le x_2 \le 312$

and $258 \le x_3 + 49 \le 308$
       $209 \le x_3 \le 259$

The flow between C and A is 258 to 308 cars per hour

The flow between B and D is 262 to 312 cars per hour.

The flow between D and C is 209 to 259 cars per hour.

47. Find the number of coins of each denomination.

Let n = the number of nickels, d = the number of dimes, and q = the number of quarters. The fee is 10% of the total value. So, if the voucher is for $139.50 then the total value is $139.50/90%, or $155.

The total value: $0.05n + 0.10d + 0.25q = 155$

$$5n + 10d + 25q = 15{,}500$$

The total number: $n + d + q = 1025$

Relation between nickels and quarters: $q = 2n$

$$\begin{cases} 5n + 10d + 25q = 15{,}500 & (1) \\ n + d + q = 1025 & (2) \\ q = 2n & (3) \end{cases}$$

Substitute $2n$ for q in Eq. (1) and (2), and simplify.

$5n + 10d + 25(2n) = 15{,}500$

$$55n + 10d = 15{,}500 \quad (4)$$

$n + d + 2n = 1025$

$$3n + d = 1025 \quad (5)$$

Use elimination to solve Eq. (4) and (5).

$$\begin{array}{l} 55n + 10d = 15{,}500 \quad (4) \\ \underline{-30n - 10d = -10{,}250} \quad -10 \text{ times (5)} \\ 25n \qquad\quad = 5250 \\ \qquad\qquad n = 210 \end{array}$$

Substitute the value for n into Eq. (5) and solve.

$$3(210) + d = 1025$$
$$d = 395$$

Substitute the value for n into Eq. (3) and solve.

$$q = 2(210) = 420$$

There are 420 quarters, 395 dimes, and 210 nickels.

49. Find the position of the middle chime.

$w_1 d_1 + w_2 d_2 = w_3 d_3$ and $w_1 = 2$, $w_2 = 6$, $w_3 = 9$

$$2d_1 + 6d_2 = 9d_3 \quad (1)$$

From the words in the exercise,

$$d_1 + d_3 = 13 \quad (2)$$
$$d_2 = \frac{1}{3} d_1 \quad (3)$$

$$\begin{array}{l} 2d_1 + 6d_2 - 9d_3 = 0 \quad (1) \\ \underline{2d_1 - 6d_2 \qquad\quad = 0} \quad 6 \text{ times (3)} \\ 4d_1 \qquad - 9d_3 = 0 \quad (4) \end{array}$$

$$\begin{array}{l} 4d_1 - 9d_3 = 0 \quad (4) \\ \underline{9d_1 + 9d_3 = 117} \quad 9 \text{ times (2)} \\ 13d_1 \qquad = 117 \\ \qquad d_1 = 9 \quad (5) \end{array}$$

Substitute into equation (3) and (2)

$$d_2 = \frac{1}{3}(9) = 3$$

$9 + d_3 = 13$

$d_3 = 4$

Therefore $d_1 = 9$ in., $d_2 = 3$ in., and $d_3 = 4$ in.

$d_2 + d_3 = 3 + 4 = 7$ in.

$d_1 - d_2 = 9 - 3 = 6$ in.

So the middle chime is 7 in. from the 9 ounce chime and 6 in. from the 2 ounce chime.

51. Find the equation of a plane.

$z = ax + by + c$

$$\begin{cases} a - b + c = 5 & (1) \\ 2a - 2b + c = 9 & (2) \\ -3a - b + c = -1 & (3) \end{cases}$$

$$\begin{array}{ll} 3a - 3b + 3c = 15 & \text{3 times (1)} \\ \underline{-3a - b + c = -1} & (3) \\ -4b + 4c = 14 & (4) \end{array}$$

$$\begin{array}{ll} -2a + 2b - 2c = -10 & \text{-2 times (1)} \\ \underline{2a - 2b + c = 9} & (2) \\ -c = -1 \\ \quad c = 1 & (5) \end{array}$$

Substitute 1 for c in Eq (4) and solve.

$-4b + 4(1) = 14$

$b = -\dfrac{5}{2}$

Substitute the value for b in Eq (1) and solve.

$a - \left(-\dfrac{5}{2}\right) + 1 = 5$

$a = \dfrac{3}{2}$

Thus, the equation of the plane is $z = \dfrac{3}{2}x - \dfrac{5}{2}y + 1$ or

$3x - 5y - 2z = -2$.

53-54. These exercises follow the same steps until the end.

$$\begin{cases} x - 3y - 2z = A^2 & (1) \\ 2x - 5y + Az = 9 & (2) \\ 2x - 8y + z = 18 & (3) \end{cases}$$

$$\begin{array}{ll} -2x + 6y + 4z = -2A^2 & \text{-2 times (1)} \\ \underline{2x - 5y + Az = 9} & (2) \\ y + (4 + A)z = -2A^2 + 9 & (4) \end{array}$$

$$\begin{array}{ll} -2x + 6y + 4z = -2A^2 & \text{-2 times (1)} \\ \underline{2x - 8y + z = 18} & (3) \\ -2y + 5z = -2A^2 + 18 & (5) \end{array}$$

$$\begin{cases} x - 3y - 2z = A^2 & (1) \\ y + (4 + A)z = -2A^2 + 9 & (4) \\ -2y + 5z = -2A^2 + 18 & (5) \end{cases}$$

$$\begin{array}{ll} 2y + (8 + 2A)z = -4A^2 + 18 & \text{2 times (4)} \\ \underline{-2y + 5z \qquad = -2A^2 + 18} & (5) \\ (13 + 2A)z = -6A^2 + 36 & (6) \end{array}$$

$$\begin{cases} x - 3y - 2z = A^2 & (1) \\ y + (4 + A)z = -2A^2 + 9 & (4) \\ (13 + 2A)z = -6A^2 + 36 & (6) \end{cases}$$

For Exercise 53, the system of equations has no solution when $2A + 13 = 0$ or $A = -\dfrac{13}{2}$.

55-56. These exercises follow the same steps until the end.

$$\begin{cases} x + 2y + z = A^2 & (1) \\ -2x - 3y + Az = 1 & (2) \\ 7x + 12y + A^2z = 4A^2 - 3 & (3) \end{cases}$$

$$\begin{array}{ll} 2x + 4y + 2z = 2A^2 & \text{2 times (1)} \\ \underline{-2x - 3y + Az = 1} & (2) \\ y + (A + 2)z = 2A^2 + 1 & (4) \end{array}$$

$$\begin{array}{ll} -7x - 14y - 7z = -7A^2 & \text{-7 times (1)} \\ \underline{7x + 12y + A^2z = 4A^2 - 3} & (3) \\ -2y + (A^2 - 7)z = -3A^2 - 3 & (5) \end{array}$$

$$\begin{cases} x + 2y + z = A^2 & (1) \\ y + (A + 2)z = 2A^2 + 1 & (4) \\ -2y + (A^2 - 7)z = -3A^2 - 3 & (5) \end{cases}$$

$$\begin{array}{ll} 2y + (2A + 4)z = 4A^2 + 2 & \text{2 times (4)} \\ \underline{-2y + (A^2 - 7)z = -3A^2 - 3} & (5) \\ (A^2 + 2A - 3)z = A^2 - 1 & (6) \end{array}$$

$$\begin{cases} x + 2y + z = A^2 & (1) \\ y + (A + 2)z = 2A^2 + 1 & (4) \\ (A^2 + 2A - 3)z = A^2 - 1 & (6) \end{cases}$$

In Exercise 55, the system of equations will have a unique solution when $(A^2 + 2A - 3) \neq 0$ in Eq. (6).

Prepare for Section 9.3

P1. Solve for x.

$$x^2 + 2x - 2 = 0$$
$$x^2 + 2x + 1 = 2 + 1$$
$$(x+1)^2 = 3$$
$$x + 1 = \pm\sqrt{3}$$
$$x = -1 \pm \sqrt{3}$$

P3. Name the graph of $(y+3)^2 = 8x$.

parabola

P5. Find the number of times the graphs intersect.

2

Section 9.3 Exercises

1. Determine whether the system is nonlinear.

Yes, $y = 2x^2$ makes the system nonlinear.

3. Determine the number of solutions.

By definition, asymptotes do not interect the graph.

There are 0 solutions to this system.

5. Solve the system of equations.

$$\begin{cases} y = x^2 + 3x & (1) \\ y = 4x + 6 & (2) \end{cases}$$

Set the expressions for y equal to each other.

$$x^2 + 3x = 4x + 6$$
$$x^2 - x - 6 = 0$$
$$(x+2)(x-3) = 0$$

$$x + 2 = 0 \qquad x - 3 = 0$$
$$x = -2 \qquad x = 3$$

Substitute for x in Eq. (1).

When $x = -2$, $y = (-2)^2 + 3(-2) = -2$

When $x = 3$, $y = 3^2 + 3(3) = 18$

The solutions are (–2,–2) and (3, 18).

7. Solve the system of equations.

$$\begin{cases} y = 2x^2 - 3x - 3 & (1) \\ y = x - 4 & (2) \end{cases}$$

Set the expressions for y equal to each other.

$$2x^2 - 3x - 3 = x - 4$$
$$2x^2 - 4x + 1 = 0$$

$$x = \frac{4 \pm \sqrt{16 - 4(2)(1)}}{2 \cdot 2} \quad \text{(Quadratic Formula)}$$
$$= \frac{4 \pm \sqrt{8}}{4} = \frac{4 \pm 2\sqrt{2}}{4} = \frac{2 \pm \sqrt{2}}{2}$$

Substitute for x in (1) and solve for y.

When $x = \frac{2+\sqrt{2}}{2}$, $y = \frac{2+\sqrt{2}}{2} - 4 = \frac{-6+\sqrt{2}}{2}$.

When $x = \frac{2-\sqrt{2}}{2}$, $y = \frac{2-\sqrt{2}}{2} - 4 = \frac{-6-\sqrt{2}}{2}$.

The solutions are

$$\left(\frac{2+\sqrt{2}}{2}, \frac{-6+\sqrt{2}}{2}\right) \text{ and } \left(\frac{2-\sqrt{2}}{2}, \frac{-6-\sqrt{2}}{2}\right).$$

9. Solve the system of equations.

Write the system in standard form.

$$\begin{cases} y = -x^2 + 3x + 5 & (1) \\ y = x^2 - 2x - 7 & (2) \end{cases}$$

Set the expressions for y equal to each other.

$$-x^2 + 3x + 5 = x^2 - 2x - 7$$
$$2x^2 - 5x - 12 = 0$$
$$(2x + 3)(x - 4) = 0$$

$$x = -\frac{3}{2} \text{ or } x = 4$$

Substitute for x in Eq. (2).

$$y = \left(-\frac{3}{2}\right)^2 - 2\left(-\frac{3}{2}\right) - 7 \qquad y = 4^2 - 2(4) - 7$$
$$y = -\frac{7}{4} \qquad\qquad\qquad y = 1$$

The solutions are $\left(-\frac{3}{2}, -\frac{7}{4}\right)$ and (4, 1).

11. Solve the system of equations.

$$\begin{cases} 2x + 3y = 16 & (1) \\ xy = 10 & (2) \end{cases}$$

Substitute y from Eq. (1) into Eq. (2).

$$x\left(\frac{16 - 2x}{3}\right) = 10$$
$$16x - 2x^2 = 30$$

$$0 = 2x^2 - 16x + 30$$
$$0 = 2(x^2 - 8x + 15)$$
$$0 = (x - 3)(x - 5)$$
$$x = 3 \text{ or } x = 5$$

Substitute for x in Eq. (2).

$$(3)y = 10 \qquad\qquad (5)y = 10$$
$$y = \frac{10}{3} \qquad\qquad\qquad y = 2$$

The solutions are $\left(3, \frac{10}{3}\right)$ and $(5, 2)$.

13. Solve the system of equations.

$$\begin{cases} 2x - y = 1 & (1) \\ xy = 6 & (2) \end{cases}$$

Solve Eq. (1) for y.

$$y = 2x - 1 \quad (3)$$

Substitute into Eq. (2).

$$x(2x - 1) = 6$$
$$2x^2 - x = 6$$
$$2x^2 - x - 6 = 0$$
$$(2x + 3)(x - 2) = 0$$
$$2x + 3 = 0, \text{ or } x - 2 = 0$$
$$x = -\frac{3}{2} \quad x = 2$$

Substitute for x in Eq. (3).

When $x = -\frac{3}{2}$, $y = 2\left(-\frac{3}{2}\right) - 1 = -4$.

When $x = 2$, $y = 2(2) - 1 = 3$.

The solutions are $(-3/2, -4)$ and $(2, 3)$.

15. Solve the system of equations.

Write the system in standard form.

$$\begin{cases} 5x^2 - 3y^2 = -7 & (1) \\ y = 3x - 3 & (2) \end{cases}$$

Substitute y from Eq. (2) into Eq. (1).

$$5x^2 - 3(3x - 3)^2 = -7$$
$$5x^2 - 3(9x^2 - 18x + 9) = -7$$
$$5x^2 - 27x^2 + 54x - 27 = -7$$

$$-22x^2 + 54x - 20 = 0$$
$$-2(11x^2 - 27x + 10) = 0$$
$$(11x - 5)(x - 2) = 0$$
$$x = \frac{5}{11} \text{ or } x = 2$$

Substitute for x in Eq. (2).

$$y = 3\left(\frac{5}{11}\right) - 3 \qquad y = 3(2) - 3$$
$$y = -\frac{18}{11} \qquad\qquad y = 3$$

The solutions are $\left(\frac{5}{11}, -\frac{18}{11}\right)$ and $(2, 3)$.

17. Solve the system of equations.

$$\begin{cases} y = x^3 + 4x^2 - 3x - 5 & (1) \\ y = 2x^2 - 2x - 3 & (2) \end{cases}$$

Set the expressions for y equal to each other.

$$x^3 + 4x^2 - 3x - 5 = 2x^2 - 2x - 3$$
$$x^3 + 2x^2 - x - 2 = 0$$
$$x^2(x + 2) - (x + 2) = 0$$
$$(x + 2)(x^2 - 1) = 0$$
$$(x + 2)(x - 1)(x + 1) = 0$$
$$x = -2, \; x = 1, \text{ or } x = -1$$

Substitute for x in Eq. (2).

When $x = -2$, $y = 2(-2)^2 - 2(-2) - 3 = 9$

When $x = 1$, $y = 2(1)^2 - 2(1) - 3 = -3$

When $x = -1$, $y = 2(-1)^2 - 2(-1) - 3 = 1$

The solutions are $(-2, 9)$, $(1, -3)$ and $(-1, 1)$.

19. Solve the system of equations.

$$\begin{cases} 2x^2 + y^2 = 9 & (1) \\ x^2 - y^2 = 3 & (2) \end{cases}$$

$$\begin{array}{ll} 2x^2 + y^2 = 9 & (1) \\ \underline{x^2 - y^2 = 3} & (2) \\ 3x^2 \qquad = 12 \\ x^2 = 4 \\ x = \pm 2 \end{array}$$

When $x = -2$, $(-2)^2 - y^2 = 3$ From Eq. (2)

$$4 - y^2 = 3$$
$$-y^2 = -1$$
$$y^2 = 1$$
$$y = \pm 1$$

When $x = 2$, $(2)^2 - y^2 = 3$

$$4 - y^2 = 3$$
$$-y^2 = -1$$
$$y^2 = 1$$
$$y = \pm 1$$

The solutions are $(-2, 1)$, $(-2, -1)$, $(2, 1)$, and $(2, -1)$.

21. Solve the system of equations.

$$\begin{cases} x^2 - 2y^2 = 8 & (1) \\ x^2 + 3y^2 = 28 & (2) \end{cases}$$

Use the elimination method to eliminate x^2.

$$x^2 - 2y^2 = 8 \qquad (1)$$
$$\underline{-x^2 - 3y^2 = -28} \quad -1 \text{ times (2)}$$
$$-5y^2 = -20$$
$$y^2 = 4$$
$$y = \pm 2$$

Substitute for y in Eq. (1).

$$x^2 - 2(2)^2 = 8 \qquad x^2 - 2(-2)^2 = 8$$
$$x^2 = 16 \qquad\qquad x^2 = 16$$
$$x = \pm 4 \qquad\qquad x = \pm 4$$

The solutions are $(4, 2)$, $(-4, 2)$, $(4, -2)$ and $(-4, -2)$.

23. Solve the system of equations.

$$\begin{cases} 2x^2 + 4y^2 = 5 & (1) \\ 3x^2 + 8y^2 = 14 & (2) \end{cases}$$

Use the elimination method to eliminate y^2.

$$-4x^2 - 8y^2 = -10 \quad -2 \text{ times (1)}$$
$$\underline{3x^2 + 8y^2 = 14} \qquad (2)$$
$$-x^2 = 4$$
$$x^2 = -4$$

$x^2 = -4$ has no real number solutions. The graphs of the equations do not intersect.

25. Solve the system of equations.

$$\begin{cases} x^2 - 2x + y^2 = 1 & (1) \\ 2x + y = 5 & (2) \end{cases}$$

Substitute y from Eq. (2) into Eq. (1).

$$x^2 - 2x + (5 - 2x)^2 = 1$$
$$x^2 - 2x + 25 - 20x + 4x^2 = 1$$
$$5x^2 - 22x + 24 = 0$$
$$(5x - 12)(x - 2) = 0$$
$$x = \frac{12}{5} \text{ or } x = 2$$

Substitute for x in Eq. (2).

$$2\left(\frac{12}{5}\right) + y = 5 \qquad\qquad 2(2) + y = 5$$
$$y = \frac{1}{5} \qquad\qquad\qquad y = 1$$

The solutions are $(12/5, 1/5)$ and $(2, 1)$.

27. Solve the system of equations.

$$\begin{cases} (x - 3)^2 + (y + 1)^2 = 5 & (1) \\ x - 3y = 7 & (2) \end{cases}$$

Substitute x from Eq. (2) into Eq. (1).

$$(3y + 4)^2 + (y + 1)^2 = 5$$
$$9y^2 + 24y + 16 + y^2 + 2y + 1 = 5$$
$$10y^2 + 26y + 12 = 0$$
$$5y^2 + 13y + 6 = 0$$
$$(5y + 3)(y + 2) = 0$$
$$y = -\frac{3}{5} \text{ or } y = -2$$

Substitute for y in Eq. (2).

$$x = 3\left(-\frac{3}{5}\right) + 7 \qquad\qquad x = 3(-2) + 7$$
$$x = \frac{26}{5} \qquad\qquad\qquad x = 1$$

The solutions are $(26/5, -3/5)$ and $(1, -2)$.

29. Solve the system of equations.

$$\begin{cases} x^2 - 3x + y^2 = 4 & (1) \\ 3x + y = 11 & (2) \end{cases}$$

Substitute y from Eq. (2) into Eq. (1).

$$x^2 - 3x + (11 - 3x)^2 = 4$$
$$x^2 - 3x + 121 - 66x + 9x^2 = 4$$
$$10x^2 - 69x + 117 = 0$$
$$(10x - 39)(x - 3) = 0$$

$$x = \frac{39}{10} \text{ or } x = 3$$

Substitute for x in Eq. (2).

$$3\left(\frac{39}{10}\right) + y = 11 \qquad\qquad 3(3) + y = 11$$
$$y = -\frac{7}{10} \qquad\qquad\qquad y = 2$$

The solutions are (39/10, –7/10) and (3, 2).

31. Solve the system of equations.

$$\begin{cases} (x-2)^2 + (y+2)^2 = 160 & (1) \\ (x+3)^2 + (y-1)^2 = 162 & (2) \end{cases}$$

Expand the binomials and then subtract.

$$x^2 - 4x + 4 + y^2 + 4y + 4 = 160 \quad (1)$$
$$\underline{x^2 + 6x + 9 + y^2 - 2y + 1 = 162} \quad (2)$$
$$-10x - 5 \qquad + 6y + 3 = -2$$
$$-10x + 6y = 0$$
$$y = \frac{5}{3}x$$

Substitute for y in Eq. (2).

$$(x+3)^2 + \left(\frac{5}{3}x - 1\right)^2 = 162$$

$$x^2 + 9x + 9 + \frac{25}{9}x^2 + \frac{10}{3}x + 1 = 162$$

$$\frac{34}{9}x^2 + \frac{37}{3}x - 152 = 0$$

$$34x^2 + 111x - 1368 = 0$$

$$(34x + 228)(x - 6) = 0$$

$$x = -\frac{228}{34} = -\frac{114}{17} \quad \text{or} \quad x = 6$$

Substitute for x in $y = \frac{5}{3}x$.

$$y = \frac{5}{3}\left(-\frac{114}{17}\right) = -\frac{190}{17} \qquad y = \frac{5}{3}(6) = 10$$

The solutions are $\left(-\frac{114}{17}, -\frac{190}{17}\right)$ and (6, 10).

33. Solve the system of equations.

$$\begin{cases} (x+3)^2 + (y-2)^2 = 20 & (1) \\ (x-2)^2 + (y-3)^2 = 2 & (2) \end{cases}$$

Expand the binomials and then subtract.

$$x^2 + 6x + 9 + y^2 - 4y + 4 = 20 \qquad (1)$$
$$\underline{x^2 - 4x + 4 + y^2 - 6y + 9 = 2} \qquad (2)$$
$$10x + 5 \qquad + 2y - 5 = 18$$
$$10x + 2y = 18$$
$$y = -5x + 9$$

Find $y - 3$ and substitute in Eq. (2): $y - 3 = -5x + 6$.

$$(x-2)^2 + (-5x+6)^2 = 2$$
$$x^2 - 4x + 4 + 25x^2 - 60x + 36 = 2$$
$$26x^2 - 64x + 38 = 0$$
$$13x^2 - 32x + 19 = 0$$
$$(13x - 19)(x - 1) = 0$$

$$x = \frac{19}{13} \text{ or } x = 1$$

Substitute for x in $y = -5x + 9$.

$$y = -5\left(\frac{19}{13}\right) + 9 \qquad y = -5(1) + 9$$
$$y = \frac{22}{13} \qquad\qquad\qquad y = 4$$

The solutions are (19/13, 22/13) and (1, 4).

35. Solve the system of equations.

$$\begin{cases} (x-1)^2 + (y+1)^2 = 2 & (1) \\ (x+2)^2 + (y-3)^2 = 3 & (2) \end{cases}$$

Expand the binomials and then subtract.

$$x^2 - 2x + 1 + y^2 + 2y + 1 = 2 \qquad (1)$$
$$\underline{x^2 + 4x + 4 + y^2 - 6y + 9 = 3} \qquad (2)$$
$$-6x - 3 \qquad + 8y - 8 = -1$$
$$-6x + 8y = 10$$
$$y = \frac{3x + 5}{4}$$

Find $y + 1$ and substitute in Eq. (1): $y + 1 = \frac{3x + 9}{4}$.

$$(x-1)^2 + \left(\frac{3x+9}{4}\right)^2 = 2$$

$$x^2 - 2x + 1 + \frac{9x^2 + 54x + 81}{16} = 2$$

$$16x^2 - 32x + 16 + 9x^2 + 54x + 81 = 32$$

$$25x^2 + 22x + 65 = 0$$

$$x = \frac{-22 \pm \sqrt{22^2 - 4(25)(65)}}{2(25)} = \frac{-22 \pm \sqrt{-6016}}{50}$$

x is not a real number. There are no real solutions.

The curves do not intersect.

37. Find the width and height.

h = height

w = weight

$$2h + 2w = 25$$

$$wh = 37.5$$

$$w = \frac{37.5}{h}$$

$$2h + 2\left(\frac{37.5}{h}\right) = 25$$

$$2h^2 + 75 = 25h$$

$$2h^2 - 25h + 75 = 0$$

$$(2h - 15)(h - 5) = 0$$

$$2h - 15 = 0 \qquad h - 5 = 0$$

$$h = 7.5 \qquad h = 5$$

Since the height is greater than the width, $h = 7.5$.

$$w = \frac{37.5}{7.5} = 5$$

The width is 5 in. and the height is 7.5 in.

39. Find the dimensions of each carpet.

$$x^2 + y^2 = 865$$

$$\underline{x^2 - y^2 = 703}$$

$$2x^2 = 1568$$

$$x^2 = 784$$

$$x = 28$$

$$28^2 + y^2 = 865$$

$$y^2 = 81$$

$$y = 9$$

The small carpet is 9 ft by 9 ft and the large carpet is 28 ft by 28 ft.

41. Find the radius of each globe.

r = radius of the small globe

R = radius of the large globe

$$V = \frac{4}{3}\pi r^3$$

$$\frac{4}{3}\pi R^3 = 8\left(\frac{4}{3}\pi r^3\right)$$

$$R^3 = 8r^3$$

$$\frac{4}{3}\pi R^3 - \frac{4}{3}\pi r^3 = 15{,}012.62$$

$$-\frac{4}{3}\pi R^3 + \frac{32}{3}\pi r^3 = 0$$

$$\underline{\frac{4}{3}\pi R^3 - \frac{4}{3}\pi r^3 = 15{,}012.62}$$

$$\frac{28}{3}\pi r^3 = 15{,}012.62$$

$$r^3 = \frac{3(15{,}012.62)}{28\pi}$$

$$r = \sqrt[3]{\frac{3(15{,}012.62)}{28\pi}}$$

$$r \approx 8.0$$

$$R^3 = 8r^3$$

$$R^3 = 8(8.0)^3$$

$$R = 16.0$$

The radius of the large globe is 16.0 in. and the radius of the small globe is 8.0 in.

43. Find the perimeter.

$$\begin{cases} x^2 = y & (1) \\ 18x - 22 = 3y + 5 & (2) \end{cases}$$

Substitute for y in Eq. (2).

$$18x - 22 = 3(x^2) + 5$$

$$0 = 3x^2 - 18x + 27$$

$$0 = 3(x^2 - 6x + 9)$$

$$0 = 3(x - 3)^2$$

$$x = 3$$

$$y = 3^2 = 9$$

$$P = x^2 + 3y + 5 + y + 18x - 22$$

$$= 3^2 + 3(9) + 5 + 9 + 18(3) - 22$$

$$= 82 \text{ units}$$

45. Find values of r.

$$\begin{cases} x^2 + y^2 = r^2 & (1) \\ \quad\quad y = 2x + 1 & (2) \end{cases}$$

Substitute for y in Eq. (1)

$$x^2 + (2x+1)^2 = r^2$$
$$x^2 + 4x^2 + 4x + 1 = r^2$$
$$5x^2 + 4x + 1 = r^2$$

Minimize r^2 by completing the square.

$$r^2 = 5x^2 + 4x + 1 = 5\left(x^2 + \frac{4}{5}x + \frac{4}{25}\right) + 1 - \frac{4}{5}$$

$$= 5\left(x + \frac{2}{5}\right)^2 + \frac{1}{5}$$

Thus $\left(-\frac{2}{5}, \frac{1}{5}\right)$ is the point on both $x^2 + y^2 = r^2$ and

$y = 2x + 1$ for which $x^2 + y^2 = r^2$ has the smallest

radius.

Substitute for x in $r^2 = 5x^2 + 4x + 1$

$$r^2 = 5\left(-\frac{2}{5}\right)^2 + 4\left(-\frac{2}{5}\right) + 1 = \frac{1}{5}$$

$$r = \sqrt{\frac{1}{5}} \text{ or } \frac{\sqrt{5}}{5} \text{ is the minimum radius.}$$

Therefore $r \geq \frac{\sqrt{5}}{5}$.

47. Find the numbers.

$$\begin{cases} x + y = 5 & (1) \\ \quad\quad xy = 1 & (2) \end{cases}$$

Solve Eq. (1) for y and substitute $5 - x$ into Eq. (2).

$$x(5 - x) = 1$$
$$5x - x^2 = 1$$
$$x^2 - 5x + 1 = 0$$

$$x = \frac{5 \pm \sqrt{5^2 - 4(1)(1)}}{2(1)}$$

$$x = \frac{5 \pm \sqrt{21}}{2}$$

$$x = \frac{5 + \sqrt{21}}{2} \approx 4.791287847$$

$$x = \frac{5 - \sqrt{21}}{2} \approx 0.2087121525$$

$$y = 5 - \frac{5 + \sqrt{21}}{2} \approx 0.2087121525$$

$$y = 5 - \frac{5 - \sqrt{21}}{2} \approx 4.791287847$$

The numbers are $\frac{5 + \sqrt{21}}{2} \approx 4.791287847$ and

$\frac{5 - \sqrt{21}}{2} \approx 0.2087121525$.

49. Solve the system of equations.

$$\begin{cases} y = 2^x \\ y = x + 1 \end{cases}$$

Using a graphing calculator, graph the two equations
on the same coordinate grid. Using the ZOOM feature,
estimate the coordinates of the points where the graphs
intersect. These coordinates are the solutions of the
system of equations. For this system of equations, the
solutions are $(0, 1)$ and $(1, 2)$.

51. Solve the system of equations.

$$\begin{cases} y = e^{-x} \\ y = x^2 \end{cases}$$

Using a graphing calculator, graph the two equations
on the same coordinate grid. Using the ZOOM feature,
estimate the coordinates of the points where the graphs
intersect. These coordinates are the solutions of the
system of equations. For this system of equations, the
solution is approximately $(0.7035, 0.4949)$.

53. Solve the system of equations.

$$\begin{cases} y = \sqrt{x} \\ y = \dfrac{1}{x - 1} \end{cases}$$

Using a graphing calculator, graph the two equations
on the same coordinate grid. Using the ZOOM feature,
estimate the coordinates of the points where the graphs
intersect. These coordinates are the solutions of the
system of equations. For this system of equations, the
solution is approximately $(1.7549, 1.3247)$.

55. Solve the system of equations for rational-number
ordered pairs.

$$\begin{cases} y = x^2 + 4 \\ x = y^2 - 24 \end{cases}$$

Solve by substitution.

$$x = (x^2 + 4)^2 - 24$$

$$x = x^4 + 8x^2 + 16 - 24$$

$$0 = x^4 + 8x^2 - x - 8$$

$$0 = (x - 1)(x^3 + x^2 + 9x + 8)$$

$x^3 + x^2 + 9x + 8$ is not factorable over the rational

numbers because the Rational Zero Theorem implies

the only rational zeros are $\pm 1, \pm 2, \pm 4, \pm 8$. Thus,

the only rational ordered-pair solution is $(1, 5)$.

57. Solve the system of equations for rational-number

ordered pairs.

$$x^2 - 3xy + y^2 = 5$$
$$x^2 - xy - 2y^2 = 0$$

Factor the second equation.

$$(x - 2y)(x + y) = 0$$

Thus $x = 2y$ or $x = -y$. Substituting each expression into

the first equation, we have

$$(2y)^2 - 3(2y)y + y^2 = 5$$

$$4y^2 - 6y^2 + y^2 = 5$$

$$-y^2 = 5$$

$$y^2 = -5$$

There are no rational solutions.

$$(-y)^2 - 3(-y)y + y^2 = 5$$

$$y^2 + 3y^2 + y^2 = 5$$

$$5y^2 = 5$$

$$y^2 = 1$$

$$y = \pm 1$$

Substituting into $x = -y$, we have $x = -1$ or $x = 1$. The

rational ordered-pair solutions are $(-1, 1)$ and $(1, -1)$.

59. Solve the system of equations for rational-number

ordered pairs.

$$\begin{cases} 2x^2 - 4xy - y^2 = 6 \\ 4x^2 - 3xy - y^2 = 6 \end{cases}$$

Subtract the two equations.

$$-2x^2 - xy = 0$$

$$-x(2x + y) = 0$$

$$x = 0 \text{ or } y = -2x$$

Substituting $x = 0$ into the first equation gives

$-y^2 = 6$ or $y^2 = -6$. There are no rational solutions.

Substituting $y = -2x$ into the first equation gives

$$2x^2 + 8x^2 - 4x^2 = 6$$

$$6x^2 = 6$$

$$x^2 = 1$$

$$x = \pm 1$$

The rational ordered-pair solutions are

$(1, -2)$ and $(-1, 2)$.

61. a. Determine the guard's distance.

Let A represent the starting point for the back of the

parade, B represent the point where the back of the

parade meets the guard, C represent the starting point

of the guard, the starting point for the front of the

parade, and the ending point for the back of the parade,

and D represent the point where the guard meets the

front of the parade.

Recall, that $d = rt$ and $t = \dfrac{d}{r}$.

The time that it takes the guard to reach the back of the

parade is the same time it takes the back of the parade

to reach the guard. Let $x =$ the distance the guard goes

from the front of the parade to the back of the parade,

CB. Let $r =$ the guard's rate. Then $2 - x =$ the distance

from the back of the parade to the guard, AB. The rate

of the parade is 4 mph.

$$\frac{x}{r} = \frac{2 - x}{4}$$

The time that it takes the guard to reach the front of the parade is the same time it takes the end of the parade to reach the point where the guard started. Let $x+2$ = the distance the guard goes from meeting the end of the parade to the front of the parade, BD. Let r = the guard's rate. Then x = the distance from meeting the guard to the end of the parade, BC. The rate of the parade is 4 mph.

$$\frac{x+2}{r} = \frac{x}{4}$$

We have a system of two equation in two variables.

$$\begin{cases} \dfrac{x}{r} = \dfrac{2-x}{4} & (1) \\ \dfrac{x+2}{r} = \dfrac{x}{4} & (2) \end{cases}$$

Clear fractions.

$$\begin{cases} 4x = r(2-x) & (1) \\ 4(x+2) = rx & (2) \end{cases}$$

Solve Eq. (2) for r and substitute $\dfrac{4(x+2)}{x}$ in Eq. (1).

$$4x = \frac{4(x+2)}{x}(2-x)$$
$$4x^2 = -4x^2 + 16$$
$$8x^2 = 16$$
$$x^2 = 2$$
$$x = \sqrt{2} \quad \text{(reject negative)}$$

The guard traveled a total distance of

$x + (x+2) = \sqrt{2} + \sqrt{2} + 2$, or $2 + 2\sqrt{2}$ mi.

b. Determine the guard's rate.

The rate of the parade was 4 mph, and the distance of the parade was 2 mi. Therefore, the time of the parade was 0.5 hr, which is the same time it took the guard to travel $2 + 2\sqrt{2}$ mi.

$$r = \frac{d}{t} = \frac{2+2\sqrt{2}}{0.5 \text{ hr}} = 4 + 4\sqrt{2} \text{ mph}$$

The guard's rate was $4 + 4\sqrt{2}$ mph.

Mid-Chapter 9 Quiz

1. Solve the system of equations.

$$\begin{cases} 2x - 3y = -15 & (1) \\ -3x + 4y = 19 & (2) \end{cases}$$

$$\begin{array}{ll} 6x - 9y = -45 & \text{3 times (1)} \\ \underline{-6x + 8y = 38} & \text{2 times (2)} \\ \quad\; -y = -7 \\ \qquad\;\; y = 7 \end{array}$$

$$2x - 3(7) = -15$$
$$x = 3$$

The solution is (3, 7).

3. Give an example of an inconsistent system of equation in two variables.

Answers will vary.

One example is $\begin{cases} x - 2y = 7 & (1) \\ -5x + 10y = 11 & (2) \end{cases}$

5. Solve the system of equations.

$$\begin{cases} 3x^2 + y^2 = 28 & (1) \\ x^2 - y^2 = 8 & (2) \end{cases}$$

$$\begin{array}{ll} 3x^2 + y^2 = 28 & (1) \\ \underline{x^2 - y^2 = 8} & (2) \\ 4x^2 \qquad\;\; = 36 \\ \quad x^2 = 9 \\ \quad x = \pm 3 \end{array}$$

When $x = -3$, $(-3)^2 - y^2 = 8$
$$-y^2 = -1$$
$$y^2 = 1$$
$$y = \pm 1$$

When $x = 3$, $(3)^2 - y^2 = 8$
$$-y^2 = -1$$
$$y^2 = 1$$
$$y = \pm 1$$

The solutions are (−3, 1), (−3, −1), (3, 1), and (3, −1).

Prepare for Section 9.4

P1. Factor over the real numbers.

$$x^4 + 14x^2 + 49 = (x^2 + 7)^2$$

P3. Simplify.

$$\frac{7}{x} - \frac{6}{x-1} + \frac{10}{(x-1)^2}$$

$$= \frac{(x-1)^2}{(x-1)^2} \cdot \frac{7}{x} - \frac{x(x-1)}{x(x-1)} \cdot \frac{6}{x-1} + \frac{x}{x} \cdot \frac{10}{(x-1)^2}$$

$$= \frac{7x^2 - 14x + 7}{x(x-1)^2} - \frac{6x^2 - 6x}{x(x-1)^2} + \frac{10x}{x(x-1)^2}$$

$$= \frac{x^2 + 2x + 7}{x(x-1)^2}$$

P5. Solve.

$$\begin{cases} 0 = A + B & (1) \\ 3 = -2B + C & (2) \\ 16 = 7A - 2C & (3) \end{cases}$$

Solve Eq (1) for A and substitute into Eq (3).

$$16 = -7B - 2C \quad (4)$$

Multiply Eq (2) by 2 and add to Eq (4).

$$6 = -4B + 2C$$
$$\underline{16 = -7B - 2C}$$
$$22 = -11B$$
$$-2 = B$$

$$A = 2$$

$$C = 2B + 3$$
$$= 2(-2) + 3$$
$$= -1$$

The solution is $(2, -2, -1)$.

Section 9.4 Exercises

1. Give the objective for partial fractions.

   The objective is to write the rational expression as the sum of the two simpler rational expressions.

3. State the first step for the given rational expression.

   Factor the denominator

5. Determine the constants A and B.

$$\frac{-3x + 20}{x(x+5)} = \frac{A}{x} + \frac{B}{x+5}$$

$$-3x + 20 = A(x+5) + Bx$$
$$-3x + 20 = (A+B)x + 5A$$

$$\begin{cases} -3 = A + B \\ 20 = 5A \end{cases}$$

Then $A = 4$.

$$-3 = 4 + B$$
$$B = -7$$

7. Determine the constants A and B.

$$\frac{1}{(2x+3)(x-1)} = \frac{A}{2x+3} + \frac{B}{x-1}$$

$$1 = A(x-1) + B(2x+3)$$
$$1 = Ax - A + 2Bx + 3B$$
$$1 = (A+2B)x + (-A+3B)$$

$$0 = A + 2B$$
$$\underline{1 = -A + 3B}$$
$$1 = \quad 5B$$

$$B = \frac{1}{5} \qquad 0 = A + 2\left(\frac{1}{5}\right)$$

$$A = -\frac{2}{5}$$

9. Determine the constants A, B and C.

$$\frac{x+9}{x(x-3)^2} = \frac{A}{x} + \frac{B}{x-3} + \frac{C}{(x-3)^2}$$

$$x + 9 = A(x-3)^2 + Bx(x-3) + Cx$$
$$x + 9 = Ax^2 - 6Ax + 9A + Bx^2 - 3Bx + Cx$$
$$x + 9 = (A+B)x^2 + (-6A - 3B + C)x + 9A$$

$$\begin{cases} 0 = A + B \\ 1 = -6A - 3B + C \\ 9 = 9A \end{cases}$$

$$A = 1 \qquad A + B = 0 \qquad\qquad -6A - 3B + C = 1$$
$$1 + B = 0 \qquad\qquad -6(1) - 3(-1) + C = 1$$
$$B = -1 \qquad\qquad C = 4$$

11. Determine the constants A, B and C.

$$\frac{4x^2 + 3}{(x-1)(x^2 + x + 5)} = \frac{A}{x-1} + \frac{Bx + C}{x^2 + x + 5}$$

$$4x^2 + 3 = A(x^2 + x + 5) + (Bx + C)(x-1)$$
$$4x^2 + 3 = Ax^2 + Ax + 5A + Bx^2 - Bx + Cx - C$$
$$4x^2 + 3 = (A+B)x^2 + (A - B + C)x + (5A - C)$$

$$\begin{cases} 4 = A + B & (1) \\ 0 = A - B + C & (2) \\ 3 = 5A - C & (3) \end{cases}$$

From (3), $C = 5A - 3$. From (1), $B = 4 - A$.

Substitute C and B into Eq. (2).

$$0 = A - (4 - A) + 5A - 3 \qquad C = 5(1) - 3 \qquad B = 4 - 1$$
$$0 = A - 4 + A + 5A - 3 \qquad C = 2 \qquad B = 3$$
$$7 = 7A$$
$$1 = A$$

13. Determine the constants A, B, C, and D.

$$\frac{x^3 + 2x}{(x^2 + 1)^2} = \frac{Ax + B}{x^2 + 1} + \frac{Cx + D}{(x^2 + 1)^2}$$

$$x^3 + 2x = (Ax + B)(x^2 + 1) + (Cx + D)$$
$$x^3 + 2x = Ax^3 + Ax + Bx^2 + B + Cx + D$$
$$x^3 + 2x = Ax^3 + Bx^2 + (A + C)x + (B + D)$$

$$\begin{cases} 1 = A \\ 0 = B \\ 2 = A + C \\ 0 = B + D \end{cases}$$

$$A = 1 \qquad B = 0 \qquad 1 + C = 2 \qquad 0 + D = 0$$
$$C = 1 \qquad D = 0$$

15. Find the partial fraction decomposition.

$$\frac{5x + 6}{x(x - 3)} = \frac{A}{x} + \frac{B}{x - 3}$$

$$5x + 6 = A(x - 3) + Bx$$
$$5x + 6 = Ax - 3A + Bx$$
$$5x + 6 = (A + B)x - 3A$$

$$\begin{cases} 5 = A + B \\ 6 = -3A \end{cases}$$

$$A = -2$$

$$-2 + B = 5$$
$$B = 7$$

$$\frac{5x + 6}{x(x - 3)} = \frac{-2}{x} + \frac{7}{x - 3}$$

17. Find the partial fraction decomposition.

$$\frac{3x + 50}{x^2 - 7x - 18} = \frac{3x + 50}{(x - 9)(x + 2)} = \frac{A}{x - 9} + \frac{B}{x + 2}$$

$$3x + 50 = A(x + 2) + B(x - 9)$$
$$3x + 50 = Ax + 2A + Bx - 9B$$
$$3x + 50 = (A + B)x + (2A - 9B)$$

$$\begin{cases} 3 = A + B \\ 50 = 2A - 9B \end{cases}$$

$$\begin{array}{r} -2A - 2B = -6 \\ 2A - 9B = 50 \\ \hline -11B = 44 \\ B = -4 \end{array}$$

$$3 = A + (-4)$$
$$7 = A$$

$$\frac{3x + 50}{x^2 - 7x - 18} = \frac{7}{x - 9} + \frac{-4}{x + 2}$$

19. Find the partial fraction decomposition.

$$\frac{16x + 34}{4x^2 + 16x + 15} = \frac{16x + 34}{(2x + 3)(2x + 5)}$$

$$= \frac{A}{2x + 3} + \frac{B}{2x + 5}$$

$$16x + 34 = A(2x + 5) + B(2x + 3)$$
$$16x + 34 = 2Ax + 5A + 2Bx + 3B$$
$$16x + 34 = (2A + 2B)x + (5A + 3B)$$

$$\begin{cases} 16 = 2A + 2B & (1) \\ 34 = 5A + 3B & (2) \end{cases}$$

$$\begin{array}{rl} 6A + 6B = 48 & \text{3 times (1)} \\ -10A - 6B = -68 & \text{-2 times (2)} \\ \hline -4A = -20 \\ A = 5 \end{array}$$

$$2(5) + 2B = 16$$
$$B = 3$$

$$\frac{16x + 34}{4x^2 + 16x + 15} = \frac{5}{2x + 3} + \frac{3}{2x + 5}$$

21. Find the partial fraction decomposition.

$$\frac{x - 4}{(5x + 2)(x - 3)} = \frac{A}{5x + 2} + \frac{B}{x - 3}$$

$$x - 4 = A(x - 3) + B(5x + 2)$$
$$x - 4 = Ax - 3A + 5Bx + 2B$$
$$x - 4 = (A + 5B)x + (-3A + 2B)$$

$$\begin{cases} 1 = A + 5B & (1) \\ -4 = -3A + 2B & (2) \end{cases}$$

$$3A+15B=3 \qquad \text{3 times (1)}$$
$$\underline{-3A+2B=-4 \qquad (2)}$$
$$17B=-1$$
$$B=-\frac{1}{17}$$

$$A+5\left(-\frac{1}{17}\right)=1$$
$$A=\frac{22}{17}$$

$$\frac{x-4}{(5x+2)(x-3)}=\frac{22}{17(5x+2)}+\frac{-1}{17(x-3)}$$

23. Find the partial fraction decomposition.

$$\begin{array}{r} x+3 \\ x^2-4{\overline{\smash{\big)}\,x^3+3x^2-4x-8}} \\ \underline{x^3 \qquad -4x} \\ 3x^2 \qquad -8 \\ \underline{3x^2 \qquad -12} \\ 4 \end{array}$$

$$\frac{x^3+3x^2-4x-8}{x^2-4}=x+3+\frac{4}{(x-2)(x+2)}$$

$$\frac{4}{(x-2)(x+2)}=\frac{A}{x-2}+\frac{B}{x+2}$$

$$4=A(x+2)+B(x-2)$$
$$4=Ax+2A+Bx-2B$$
$$4=(A+B)x+(2A-2B)$$

$$\begin{cases} 0=A+B \qquad (1) \\ 4=2A-2B \qquad (2) \end{cases}$$

$$2A+2B=0 \quad \text{2 times (1)}$$
$$\underline{2A-2B=4 \quad (2)}$$
$$4A \qquad =4$$
$$A=1$$

$$1+B=0$$
$$B=-1$$

$$\frac{x^3+3x^2-4x-8}{x^2-4}=x+3+\frac{1}{x-2}+\frac{-1}{x+2}$$

25. Find the partial fraction decomposition.

$$\frac{3x^2+49}{x(x+7)^2}=\frac{A}{x}+\frac{B}{x+7}+\frac{C}{(x+7)^2}$$

$$3x^2+49=A(x+7)^2+Bx(x+7)+Cx$$
$$3x^2+49=Ax^2+14Ax+49A+Bx^2+7Bx+Cx$$
$$3x^2+49=(A+B)x^2+(14A+7B+C)x+49A$$

$$\begin{cases} 3=A+B \\ 0=14A+7B+C \\ 49=49A \end{cases}$$

$$A=1$$
$$1+B=3$$
$$B=2$$

$$14(1)+7(2)+C=0$$
$$C=-28$$

$$\frac{3x^2+49}{x(x+7)^2}=\frac{1}{x}+\frac{2}{x+7}+\frac{-28}{(x+7)^2}$$

27. Find the partial fraction decomposition.

$$\frac{5x^2-7x+2}{x^3-3x^2+x}=\frac{5x^2-7x+2}{x(x^2-3x+1)}=\frac{A}{x}+\frac{Bx+C}{x^2-3x+1}$$

$$5x^2-7x+2=A(x^2-3x+1)+(Bx+C)x$$
$$5x^2-7x+2=Ax^2-3Ax+A+Bx^2+Cx$$
$$5x^2-7x+2=(A+B)x^2+(-3A+C)x+A$$

$$\begin{cases} 5=A+B \\ -7=-3A+C \\ 2=A \end{cases}$$

$$A=2$$
$$2+B=5$$
$$B=3$$

$$-3(2)+C=-7$$
$$C=-1$$

$$\frac{5x^2-7x+2}{x^3-3x^2+x}=\frac{2}{x}+\frac{3x-1}{x^2-3x+1}$$

29. Find the partial fraction decomposition.

$$\frac{2x^3+9x^2+26x+41}{(x+3)^2(x^2+1)}=\frac{A}{x+3}+\frac{B}{(x+3)^2}+\frac{Cx+D}{x^2+1}$$

$2x^3 + 9x^2 + 26x + 41$

$= A(x+3)(x^2+1) + B(x^2+1) + (Cx+D)(x+3)^2$

$= Ax^3 + Ax + 3Ax^2 + 3A + Bx^2 + B$

$\quad + Cx^3 + 6Cx^2 + 9Cx + Dx^2 + 6Dx + 9D$

$= (A+C)x^3 + (3A+B+6C+D)x^2$

$\quad + (A+9C+6D)x + (3A+B+9D)$

$\begin{cases} 2 = A+C & (1) \\ 9 = 3A+B+6C+D & (2) \\ 26 = A+9C+6D & (3) \\ 41 = 3A+B+9D & (4) \end{cases}$

$\begin{array}{l} 3A+B+6C+D = 9 \quad (2) \\ \underline{-3A-B-9D = -41} \quad -1 \text{ times } (4) \\ \quad\quad 6C - 8D = -32 \quad (5) \end{array}$

$\begin{array}{l} A + 9C + 6D = 26 \quad (3) \\ \underline{-A - C = -2} \quad -1 \text{ times } (1) \\ \quad 8C + 6D = 24 \quad (6) \end{array}$

$\begin{cases} 2 = A+C & (1) \\ 9 = 3A+B+6C+D & (2) \\ -32 = 6C - 8D & (5) \\ 24 = 8C + 6D & (6) \end{cases}$

$\begin{array}{l} 36C - 48D = -192 \quad 6 \text{ times } (5) \\ \underline{64C + 48D = 192} \quad 8 \text{ times } (6) \\ \quad\quad 100C = 0 \end{array}$

$C = 0$
$A + 0 = 2$
$\quad A = 2$
$8(0) + 6D = 24$
$\quad\quad D = 4$
$3(2) + B + 9(4) = 41$
$\quad\quad\quad B = -1$

$\dfrac{2x^3 + 9x^2 + 26x + 41}{(x+3)^2(x^2+1)} = \dfrac{2}{x+3} + \dfrac{-1}{(x+3)^2} + \dfrac{4}{x^2+1}$

31. Find the partial fraction decomposition.

$\dfrac{3x-7}{(x-4)^2} = \dfrac{A}{x-4} + \dfrac{B}{(x-4)^2}$

$3x - 7 = A(x-4) + B$

$3x - 7 = Ax - 4A + B$

$3x - 7 = Ax + (-4A + B)$

$\begin{cases} 3 = A \\ -7 = -4A + B \end{cases}$

$B - 4(3) = -7$

$\quad B = 5$

$\dfrac{3x-7}{(x-4)^2} = \dfrac{3}{x-4} + \dfrac{5}{(x-4)^2}$

33. Find the partial fraction decomposition.

$\dfrac{3x^3 - x^2 + 34x - 10}{(x^2+10)^2} = \dfrac{Ax+B}{x^2+10} + \dfrac{Cx+D}{(x^2+10)^2}$

$3x^3 - x^2 + 34x - 10$

$= (Ax+B)(x^2+10) + Cx + D$

$= Ax^3 + Bx^2 + 10Ax + 10B + Cx + D$

$= Ax^3 + Bx^2 + (10A+C)x + (10B+D)$

$\begin{cases} 3 = A \\ -1 = B \\ 34 = 10A + C \\ -10 = 10B + D \end{cases}$

$\begin{array}{ll} 10(3) + C = 34 & 10(-1) + D = -10 \\ \quad\quad C = 4 & \quad\quad D = 0 \end{array}$

$\dfrac{3x^3 - x^2 + 34x - 10}{(x^2+10)^2} = \dfrac{3x-1}{x^2+10} + \dfrac{4x}{(x^2+10)^2}$

35. Find the partial fraction decomposition.

$\dfrac{1}{k^2 - x^2} = \dfrac{1}{(k-x)(k+x)} = \dfrac{A}{k-x} + \dfrac{B}{k+x}$

$1 = A(k+x) + B(k-x)$

$1 = Ak + Ax + Bk - Bx$

$1 = (A-B)x + (Ak+Bk)$

$\begin{cases} 0 = A - B & (1) \\ 1 = Ak + Bk & (2) \end{cases}$

$\begin{array}{l} Ak - Bk = 0 \quad k \text{ times } (1) \\ \underline{Ak + Bk = 1} \quad (2) \\ \quad\quad 2Ak = 1 \end{array}$

$\quad A = \dfrac{1}{2k}$

$\dfrac{1}{2k} - B = 0$

$\quad B = \dfrac{1}{2k}$

$\dfrac{1}{k^2 - x^2} = \dfrac{1}{2k(k-x)} + \dfrac{1}{2k(k+x)}$

37. Find the partial fraction decomposition.

$$\begin{array}{r} x \\ x^2-x\overline{\smash{\big)}\,x^3-x^2-x-1} \\ \underline{x^3-x^2} \\ -x-1 \end{array}$$

$$\frac{x^3-x^2-x-1}{x^2-x}=x+\frac{-x-1}{x^2-x}$$

$$\frac{-x-1}{x(x-1)}=\frac{A}{x}+\frac{B}{x-1}$$

$$-x-1=Ax-A+Bx$$

$$-x-1=(A+B)x-A$$

$$\begin{cases} -1=A+B \\ -1=-A \end{cases}$$

$$A=1$$

$$1+B=-1$$

$$B=-2$$

$$\frac{x^3-x^2-x-1}{x^2-x}=x+\frac{1}{x}+\frac{-2}{x-1}$$

39. Find the partial fraction decomposition.

$$\begin{array}{r} 2x-2 \\ x^2-x-1\overline{\smash{\big)}\,2x^3-4x^2\quad\ +5} \\ \underline{2x^3-2x^2-2x} \\ -2x^2+2x+5 \\ \underline{-2x^2+2x+2} \\ 3 \end{array}$$

$$\frac{2x^3-4x^2+5}{x^2-x-1}=2x-2+\frac{3}{x^2-x-1}$$

41. Find the partial fraction decomposition.

$$\frac{x^2-1}{(x-1)(x+2)(x-3)}=\frac{(x-1)(x+1)}{(x-1)(x+2)(x-3)}$$

$$=\frac{x+1}{(x+2)(x-3)}=\frac{A}{x+2}+\frac{B}{x-3}$$

$$x+1=A(x-3)+B(x+2)$$

$$x+1=Ax-3A+Bx+2B$$

$$x+1=(A+B)x+(-3A+2B)$$

$$\begin{cases} A+B=1 \quad (1) \\ -3A+2B=1 \quad (2) \end{cases}$$

$$\begin{array}{r} 3A+3B=3 \quad \text{3 times (1)} \\ -3A+2B=1 \quad (2) \\ \hline 5B=4 \end{array}$$

$$B=\frac{4}{5}$$

$$A+\frac{4}{5}=1$$

$$A=\frac{1}{5}$$

$$\frac{x^2-1}{(x-1)(x+2)(x-3)}=\frac{1}{5(x+2)}+\frac{4}{5(x-3)}$$

43. Find the partial fraction decomposition.

$$\frac{-x^4-4x^2+3x-6}{x^4(x-2)}=\frac{A}{x}+\frac{B}{x^2}+\frac{C}{x^3}+\frac{D}{x^4}+\frac{E}{x-2}$$

$$-x^4-4x^2+3x-6$$

$$=Ax^3(x-2)+Bx^2(x-2)+Cx(x-2)$$

$$\quad +D(x-2)+Ex^4$$

$$=Ax^4-2Ax^3+Bx^3-2Bx^2+Cx^2-2Cx$$

$$\quad +Dx-2D+Ex^4$$

$$=(A+E)x^4+(-2A+B)x^3+(-2B+C)x^2$$

$$\quad +(-2C+D)x+(-2D)$$

$$\begin{cases} -1=A+E \quad (1) \\ 0=-2A+B \quad (2) \\ -4=-2B+C \quad (3) \\ 3=-2C+D \quad (4) \\ -6=-2D \quad (5) \end{cases}$$

$$-2D=-6$$

$$D=3$$

$$-2C+3=3$$

$$C=0$$

$$-2B+0=-4$$

$$B=2$$

$$-2A+2=0$$

$$A=1$$

$$1+E=-1$$

$$E=-2$$

$$\frac{-x^4-4x^2+3x-6}{x^4(x-2)}=\frac{1}{x}+\frac{2}{x^2}+\frac{3}{x^4}+\frac{-2}{x-2}$$

45. Find the partial fraction decomposition.

$$\frac{2x^2+3x-1}{(x^3-1)}=\frac{2x^2+3x-1}{(x-1)(x^2+x+1)}=\frac{A}{x-1}+\frac{Bx+C}{x^2+x+1}$$

$$2x^2+3x-1=A(x^2+x+1)+(Bx+C)(x-1)$$

$$2x^2+3x-1=Ax^2+Ax+A+Bx^2-Bx+Cx-C$$

$$2x^2+3x-1=(A+B)x^2+(A-B+C)x+(A-C)$$

$$\begin{cases} 2=A+B & (1) \\ 3=A-B+C & (2) \\ -1=A-C & (3) \end{cases}$$

Solve Eq. (1) for B and Eq. (3) for C and substitute into Eq. (2).

$$A+B=2 \qquad\qquad A-C=-1$$
$$\quad B=2-A \qquad\qquad\quad C=A+1$$

$$A-B+C=3$$
$$A-(2-A)+(A+1)=3$$
$$A-2+A+A+1=3$$
$$3A=4$$
$$A=\frac{4}{3}$$

$$B=2-A \qquad\qquad C=A+1$$
$$B=2-\frac{4}{3} \qquad\qquad C=\frac{4}{3}+1$$
$$B=\frac{2}{3} \qquad\qquad\quad C=\frac{7}{3}$$

$$\frac{2x^2+3x-1}{x^3-1}=\frac{4}{3(x-1)}+\frac{2x+7}{3(x^2+x+1)}$$

47. Show the statement to be true.

$$\frac{1}{(b-a)(p(x)+a)}+\frac{1}{(a-b)(p(x)+b)}$$

$$=\frac{(a-b)(p(x)+b)+(b-a)(p(x)+a)}{(b-a)(a-b)(p(x)+a)(p(x)+b)}$$

$$=\frac{(a-b)p(x)+(a-b)b+(b-a)p(x)+(b-a)a}{(b-a)(a-b)(p(x)+a)(p(x)+b)}$$

$$=\frac{(a-b)p(x)+(a-b)b-(a-b)p(x)-(a-b)a}{(b-a)(a-b)(p(x)+a)(p(x)+b)}$$

$$=\frac{(a-b)b-(a-b)a}{(b-a)(a-b)(p(x)+a)(p(x)+b)}$$

$$=\frac{(a-b)(b-a)}{(b-a)(a-b)(p(x)+a)(p(x)+b)}$$

$$=\frac{1}{(p(x)+a)(p(x)+b)}$$

Prepare for Section 9.5

P1. Graph $y=-2x+3$.

P3. Graph $y=|x|+1$.

P5. Graph $\dfrac{x^2}{16}+\dfrac{y^2}{25}=1$.

Section 9.5 Exercises

1. Define a half-plane.

A half-plane is the set of point on one side of a line in a plane.

3. Determine if $(3, 4)$ is a solution.

$$-2x+5y>11$$
$$-2(3)+5(4)>11$$
$$\qquad\qquad 14>11 \quad\text{True}$$

Yes, $(3, 4)$ is a solution.

5. Determine if $(4, 5)$ is a solution.

$$3x-2y\le 6$$
$$3(4)-2(5)\le 0$$
$$\qquad\quad 2\le 0 \quad\text{False}$$

No, $(4, 5)$ is not a solution.

7. Sketch the graph of the inequality.

$y\le -2$ Use a solid line.

Test point $(0, 0)$: not in region

9. Sketch the graph of the inequality.

$y \geq 2x + 3$ Use a solid line.

Test point (0, 0): not in region

11. Sketch the graph of the inequality.

$2x - 3y < 6$

$y > \frac{2}{3}x - 2$ Use a dashed line.

Test point (0, 0): is in region

13. Sketch the graph of the inequality.

$4x + 3y \leq 12$

$y \leq -\frac{4}{3}x + 4$ Use a solid line.

Test point (0, 0): is in region

15. Sketch the graph of the inequality.

$y < x^2$ Use a dashed curve.

Vertex: (0, 0)

Test point (1, 0): is in region

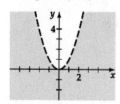

17. Sketch the graph of the inequality.

$y \geq x^2 - 2x - 3$ Use a solid curve.

Vertex: (1, –4)

Test point (0, 0): is in region

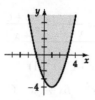

19. Sketch the graph of the inequality.

$(x - 2)^2 + (y - 1)^2 < 16$ Use a dashed graph.

Center: (2, 1), radius: 4

Test point (0, 0): is in region

21. Sketch the graph of the inequality.

$\dfrac{(x - 3)^2}{9} - \dfrac{(y + 1)^2}{16} > 1$ Use a dashed graph.

Center: (3, –1), $a = 3$, $b = 4$

Test point (–1, 0): is in region

23. Sketch the graph of the inequality.

$$4x^2 + 9y^2 - 8x + 18y \geq 23$$
$$4(x^2 - 2x + 1) + 9(y^2 + 2y + 1) \geq 23 + 4 + 9$$
$$4(x - 1)^2 + 9(y + 1)^2 \geq 36$$
$$\frac{(x - 1)^2}{9} + \frac{(y + 1)^2}{4} \geq 1 \quad \text{Use a solid graph.}$$

center (1, –1), $a = 3$, $b = 2$

Test point (0, 0): not in region

25. Sketch the graph of the inequality.

$y \geq |2x - 4|$　Use a solid graph.

Test point (0, 0): not in region

27. Sketch the graph of the inequality.

$y < 2^{x-1}$　Use a dashed curve.

Test point (0, 0): is in region

29. Sketch the graph of the solution set of the system of inequalities.

$$\begin{cases} 1 \leq x < 3 \\ -2 < y \leq 4 \end{cases}$$

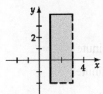

31. Sketch the graph of the solution set of the system of inequalities.

$$\begin{cases} x + y \leq 2 \\ x - y < 2 \end{cases}$$

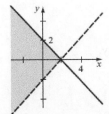

33. Sketch the graph of the solution set of the system of inequalities.

$$\begin{cases} 2x - y \geq -4 \\ 4x - 2y \leq -17 \end{cases}$$

The graphs of each of the two inequalities are shown.

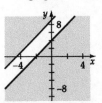

Because the solution sets of the inequalities do not intersect, the system has no solution and cannot be graphed.

35. Sketch the graph of the solution set of the system of inequalities.

$$\begin{cases} 3x + 3y < 6 \\ 3x - 2y \geq -6 \end{cases}$$

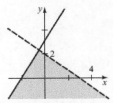

37. Sketch the graph of the solution set of the system of inequalities.

$$\begin{cases} y < 2x + 3 \\ y > 2x - 2 \end{cases}$$

39. Sketch the graph of the solution set of the system of inequalities.

$$\begin{cases} y > x - 1 \\ y \leq -x^2 + 4 \end{cases}$$

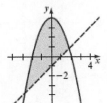

41. Sketch the graph of the solution set of the system of inequalities.

$$\begin{cases} x^2 + y^2 \le 49 \\ 9x^2 + 4y^2 \ge 36 \end{cases}$$

43. Sketch the graph of the solution set of the system of inequalities.

$$\begin{cases} (x-1)^2 + (y+1)^2 \le 16 \\ (x-1)^2 + (y+1)^2 \ge 4 \end{cases}$$

45. Sketch the graph of the solution set of the system of inequalities.

$$\begin{cases} \dfrac{x^2}{4} - \dfrac{y^2}{16} > 1 \\ \dfrac{x^2}{16} + \dfrac{y^2}{4} < 1 \end{cases}$$

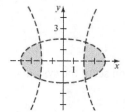

47. Sketch the graph of the solution set of the system of inequalities.

$$\begin{cases} 6x + y \ge 30 \\ x + 4y \ge 40 \\ 2x + 3y \ge 60 \\ x \ge 0,\ y \ge 0 \end{cases}$$

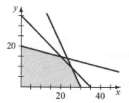

49. Sketch the graph of the solution set of the system of inequalities.

$$\begin{cases} x + 4y \le 80 \\ x + y \le 35 \\ 2x + y \le 60 \\ x \ge 0,\ y \ge 0 \end{cases}$$

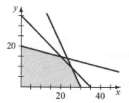

51. Find the heart rate range.

Substitute Ashley's age, 35, in the first inequality to find the minimum value.

$$y \ge 0.55(208 - 0.7x)$$
$$y \ge 0.55(208 - 0.7(35))$$
$$y \ge 100.925$$

Substitute Ashley's age in the second inequality to find the maximum value.

$$y \le 0.75(208 - 0.7x)$$
$$y \le 0.75(208 - 0.7(35))$$
$$y \le 137.625$$

The minimum is 101 beats per minute and maximum is 138 beats per minute.

53. Sketch the graph of the inequality.

$$|y| \ge |x|$$

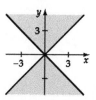

55. Sketch the graph of the inequality.

$$|x + y| \le 1$$

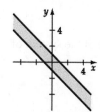

57. Sketch the graph of the inequality.

$$|x|+|y|\leq 1$$

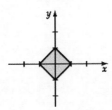

59. Sketch the graph of the inequalities. Explain.

If x is a negative number, then the inequality is reversed when multiplying both sides of the inequality by the negative number $\dfrac{1}{x}$.

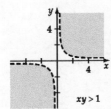   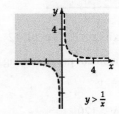

$xy > 1$ $y > \dfrac{1}{x}$

61. a. Sketch the graph for a 25-year old person.

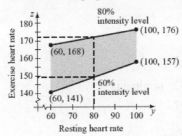

b. Use the graph to estimate Tyler's targeted heart rate.

149 to 172 beats/min

c. Determine your targeted heart rate.

Answes will vary.

Prepare for Section 9.6

P1. Graph $2x+3y\leq 12$.

P3. Evaluate.

$$C = 6x+4y+15$$
$$C(6,\ 20) = 6(0)+4(20)+15 = 95$$
$$C(4,\ 18) = 6(4)+4(18)+15 = 111$$
$$C(10,\ 10) = 6(10)+4(10)+15 = 115$$
$$C(15,\ 0) = 6(15)+4(0)+15 = 105$$

P5. Solve the system of equations.

$$\begin{cases} 300x+100y = 900 & (1) \\ 400x+300y = 2200 & (2) \end{cases}$$

$$-900x-300y = -2700 \quad -3 \text{ times } (1)$$
$$\underline{400x+300y = 2200 \qquad (2)}$$
$$-500x = -500$$
$$x = 1$$

$$300(1)+100y = 900$$
$$100y = 600$$
$$y = 6$$

The solution is (1, 6).

Section 9.6 Exercises

1. Define optimization problem.

It is a problem that seeks a solution that will maximize or minimize a situation.

3. Explain where the set of feasible solutions occur.

It must occur at a vertex of the set of feasible solutions.

5. Find the minimum value and state where it is.

$$C(x,\ y) = 3x+4y$$
$$C(0,\ 5) = 3(0)+4(5) = 20$$
$$C(2,\ 3) = 3(2)+4(3) = 18$$
$$C(6,\ 1) = 3(6)+4(1) = 22$$
$$C(9,\ 0) = 3(9)+4(0) = 27$$

The minimum of 18 occurs at (2, 3).

7. Find the maximum value and state where it is.

$$C(x,\ y) = 2.5x+3y+5$$
$$C(0,\ 20) = 2.5(0)+3(20)+5 = 65$$
$$C(5,\ 19) = 2.5(5)+3(19)+5 = 74.5$$
$$C(20,\ 4) = 2.5(20)+3(4)+5 = 67$$
$$C(22.5,\ 0) = 2.5(22.5)+3(0)+5 = 61.25$$

The maximum of 74.5 occurs at (5, 19).

9. Solve the linear programming problem.

$$C = 4x + 2y$$

$$\begin{cases} x + y \geq 7 \\ 4x + 3y \geq 24 \\ x \leq 10, \ y \leq 10 \\ x \geq 0, \ y \geq 0 \end{cases}$$

$C = 4x + 2y$	
(0, 10)	20
(0, 8)	16 minimum
(3, 4)	20
(7, 0)	28
(10, 0)	40
(10, 10)	60

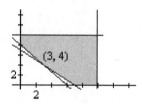

The minimum is 16 at (0, 8).

11. Solve the linear programming problem.

$$C = 6x + 7y$$

$$\begin{cases} x + 2y \leq 16 \\ 5x + 3y \leq 45 \\ x \geq 0, \ y \geq 0 \end{cases}$$

$C = 6x + 7y$	
(0, 8)	56
(0, 0)	0
(9, 0)	54
(6, 5)	71 maximum

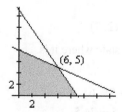

The maximum is 71 at (6, 5).

13. Solve the linear programming problem.

$$C = x + 6y$$

$$\begin{cases} 5x + 8y \leq 120 \\ 7x + 16y \leq 192 \\ x \geq 0, \ y \geq 0 \end{cases}$$

$$C = x + 6y$$

(0, 12)	72 maximum
(0, 0)	0
(24, 0)	24
(16, 5)	46

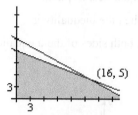

The maximum is 72 at (0, 12).

15. Solve the linear programming problem.

$$C = 4x + y$$

$$\begin{cases} 3x + 5y \geq 120 \\ x + y \geq 32 \\ x \geq 0, \ y \geq 0 \end{cases}$$

$C = 4x + y$	
(40, 0)	160
(0, 32)	32 minimum
(20, 12)	92

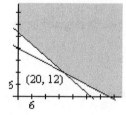

The minimum is 32 at (0, 32).

17. Solve the linear programming problem.

$$C = 2x + 7y$$

$$\begin{cases} x + y \leq 10 \\ x + 2y \leq 16 \\ 2x + y \leq 16 \\ x \geq 0, \ y \geq 0 \end{cases}$$

$C = 2x + 7y$

$(0, 8)$	56	maximum
$(0, 0)$	0	
$(8, 0)$	16	
$(6, 4)$	40	
$(4, 6)$	50	

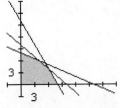

The maximum is 56 at $(0, 8)$.

19. Solve the linear programming problem.

$C = 5x + 2.5y$

$$\begin{cases} 3x + y \geq 12 \\ 2x + 7y \geq 21 \\ x + y \geq 8 \end{cases}$$

$C = 5x + 2.5y$

$(0, 12)$	30	
$(2, 6)$	25	minimum
$(7, 1)$	37.5	
$(10.5, 0)$	52.5	

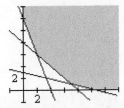

The minimum is 25 at $(2, 6)$.

21. Solve the linear programming problem.

$C = x + 4y$

$$\begin{cases} 2x + y \leq 10 \\ 2x + 3y \leq 18 \\ x - y \leq 2 \end{cases}$$

$C = x + 4y$

$(0, 6)$	24	maximum
$(3, 4)$	19	
$(4, 2)$	12	
$(2, 0)$	2	
$(0, 0)$	0	

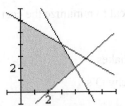

The maximum is 24 at $(0, 6)$.

23. Solve the linear programming problem.

$C = 3x + 2y$

$$\begin{cases} x + 2y \geq 8 \\ 3x + y \geq 9 \\ x + 4y \geq 12 \\ x \geq 0, y \geq 0 \end{cases}$$

$C = 3x + 2y$

$(0, 9)$	18	
$(2, 3)$	12	maximum
$(4, 2)$	16	
$(12, 0)$	36	

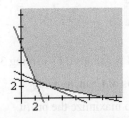

The minimum is 12 at $(2, 3)$.

25. Solve the linear programming problem.

$C = 6x + 7y$

$$\begin{cases} x + 2y \leq 900 \\ x + y \leq 500 \\ 3x + 2y \leq 1200 \\ x \geq 0, y \geq 0 \end{cases}$$

$C = 6x + 7y$

$(0, 450)$	3150	
$(100, 400)$	3400	maximum
$(200, 300)$	3300	
$(400, 0)$	2400	
$(0, 0)$	0	

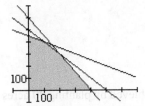

The maximum is 3400 at $(100, 400)$.

27. Find the amount of each cereal to minimize the cost and find the minimum cost.

x = number of cups of Oat Flakes

y = number of cups of Crunchy O's

$C = 0.38x + 0.32y$

Constraints:

$$\begin{cases} 6x + 3y \geq 210 \\ 30x + 40y \geq 1200 \\ x \geq 0 \\ y \geq 0 \end{cases}$$

$C = 0.38x + 0.32y$

(0, 70)	$22.40	
(32, 6)	$14.08	minimum
(40, 0)	$15.20	

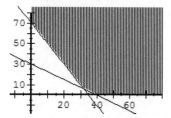

The minimum cost of $14.08 is achieved by mixing 32 cups of Oat Flakes and 6 cups of Crunchy O's.

29. Find the acres of each crop to maximize the profit.

W = acres of wheat to plant; B = acres of barley to plant

$P = 50W + 70B$

Constraints:

$$\begin{cases} 4W + 3B \leq 200 \\ W + 2B \leq 100 \\ W \geq 0, \ B \geq 0 \end{cases}$$

$P = 50W + 70B$

(0, 50)	3500	
(20, 40)	3800	maximum
(50, 0)	2500	
(0, 0)	0	

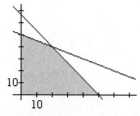

The maximum profit is achieved by planting 20 acres of wheat and 40 acres of barley.

31. Find the amount of each to maximize profit and find the maximum weekly profit.

x = number of economy boards

y = number of superior boards

Profit $= 26x + 42y$

Constraints:

$$\begin{cases} 2x + 2.5y \leq 240 \\ 2x + 4y \leq 312 \\ x \geq 0, \ y \geq 0 \end{cases}$$

Profit $= 26x + 42y$

(0, 78)	3276	
(120, 0)	3120	
(60, 48)	3576	maximum
(0, 0)	0	

The maximized profit of $3576 is achieved by producing 60 economy boards and 48 superior boards.

33. Find the amount of each to minimize cost and find the minimum cost.

A = ounces of food group A

B = ounces of food group B

Cost $= 40A + 10B$

Constraints:

$$\begin{cases} 3A + B \geq 24 \\ A + B \geq 16 \\ A + 3B \geq 30 \\ A \geq 0, \ B \geq 0 \end{cases}$$

Cost $= 40A + 10B$

(0, 24)	240	minimum
(4, 12)	280	
(9, 7)	430	
(30, 0)	1200	

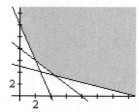

To minimize cost, use 24 ounces of food group B and zero ounces of food group A.

The minimum cost is $2.40.

35. Find the amount of each to maximize profit and find the maximum profit.

x = number of strawberry platters

y = number of crème brulee desserts

Profit $= 3.25x + 5y$

Constraints:

$$\begin{cases} 4x + 6y \leq 1200 \\ 3x + 6y \leq 990 \\ x + 0.75y \leq 300 \\ x \geq 0, \, y \geq 0 \end{cases}$$

Profit $= 3.25x + 5y$

(0, 165)	825
(210, 60)	982.5 maximum
(300, 0)	975
(0, 0)	0

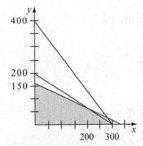

210 strawberry platters and 60 crème brulee desserts produce the maximum daily profit of $982.50.

37. Solve the linear programming problem.

$C = 5x + 3y$

$$\begin{cases} x - y \geq -2 \\ x + 4y \geq 18 \\ 4x + y \leq 42 \end{cases}$$

$C = 5x + 3y$

(2, 4)	22	minimum
(8, 10)	70	maximum
(10, 2)	56	

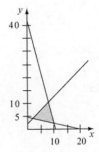

The minimum is 22 at (2, 4).

The maximum is 70 at (8, 10).

Exploring Concepts with Technology

Produce a graph of the set of feasible solutions using Wolframalpha.

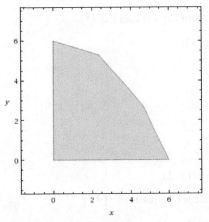

Chapter 9 Review Exercises

1. Solve the system of equations. [9.1]

$$\begin{cases} 5x - 3y = -4 & (1) \\ 2x + 5y = 11 & (2) \end{cases}$$

$$\begin{array}{ll} 10x - 6y = -8 & \text{2 times (1)} \\ -10x - 25y = -55 & -5 \text{ times (2)} \\ \hline -31y = -63 & \end{array}$$

$$y = \frac{63}{31}$$

$$2x + 5\left(\frac{63}{31}\right) = 11$$

$$x = \frac{13}{31}$$

The solution is $\left(\dfrac{13}{31}, \dfrac{63}{31}\right)$.

3. Solve the system of equations. [9.1]

$$\begin{cases} 7x + 2y = 4 & (1) \\ y = \frac{2}{5}x + 3 & (2) \end{cases}$$

$$7x + 2\left(\frac{2}{5}x + 3\right) = 4$$

$$7x + \frac{4}{5}x + 6 = 4$$

$$\frac{39}{5}x = -2$$

$$x = -\frac{10}{39}$$

$$y = \frac{2}{5}\left(-\frac{10}{39}\right) + 3 = \frac{113}{39}$$

The solution is $\left(-\dfrac{10}{39}, \dfrac{113}{39}\right)$.

5. Solve the system of equations. [9.1]

$$\begin{cases} y = 2x - 5 & (1) \\ x = 4y - 1 & (2) \end{cases}$$

Substitute x from Eq. (2) into Eq. (1).

$$y = 2(4y - 1) - 5$$
$$y = 8y - 2 - 5$$
$$-7y = -7$$
$$y = 1$$

$$x = 4(1) - 1$$
$$x = 3$$

The solution is $(3, 1)$.

7. Solve the system of equations. [9.1]

$$\begin{cases} 6x + 9y = 15 & (1) \\ 10x + 15y = 25 & (2) \end{cases}$$

$$2x + 3y = 5 \quad \tfrac{1}{3} \text{ times (1)}$$
$$\underline{2x + 3y = 5} \quad \tfrac{1}{5} \text{ times (2)}$$
$$0 = 0$$

Let $y = c$.

$$2x + 3c = 5$$
$$x = \frac{5 - 3c}{2}$$

The ordered-pair solutions are $\left(\dfrac{5 - 3c}{2},\ c \right)$.

9. Solve the system of equations. [9.2]

$$\begin{cases} 2x - 3y + z = -9 & (1) \\ 2x + 5y - 2z = 18 & (2) \\ 4x - y + 3z = -4 & (3) \end{cases}$$

$$2x - 3y + z = -9 \quad (1)$$
$$\underline{-2x - 5y + 2z = -18} \quad -1 \text{ times (2)}$$
$$8y + 3z = -27 \quad (4)$$

$$-4x + 6y - 2z = 18 \quad -2 \text{ times (1)}$$
$$\underline{4x - y + 3z = -4} \quad (3)$$
$$5y + z = 14 \quad (5)$$

$$\begin{cases} 2x - 3y + z = -9 & (1) \\ -8y + 3z = -27 & (4) \\ 5y + z = 14 & (5) \end{cases}$$

$$-8y + 3z = -27 \quad (4)$$
$$\underline{-15y - 3z = -42} \quad -3 \text{ times (5)}$$
$$-23y = -69$$
$$y = 3 \quad (6)$$

$$\begin{cases} 2x - 3y + z = -9 & (1) \\ -8y + 3z = -27 & (4) \\ y = 3 & (6) \end{cases}$$

$$-8(3) + 3z = -27$$
$$3z = -3$$
$$z = -1$$

$$2x - 3(3) - 1 = -9$$
$$2x = 1$$
$$x = \frac{1}{2}$$

The ordered-triple solution is $\left(\dfrac{1}{2}, 3, -1 \right)$.

11. Solve the system of equations. [9.2]

$$\begin{cases} x + 3y - 5z = -12 & (1) \\ 3x - 2y + z = 7 & (2) \\ 5x + 4y - 9z = -17 & (3) \end{cases}$$

$$-3x - 9y + 15z = 36 \quad -3 \text{ times (1)}$$
$$\underline{3x - 2y + z = 7} \quad (2)$$
$$-11y + 16z = 43 \quad (4)$$

$$-5x - 15y + 25z = 60 \quad -5 \text{ times (1)}$$
$$\underline{5x + 4y - 9z = -17} \quad (3)$$
$$-11y + 16z = 43 \quad (5)$$

$$\begin{cases} x + 3y - 5z = -12 & (1) \\ -11y + 16z = 43 & (4) \\ -11y + 16z = 43 & (5) \end{cases}$$

$$-11y + 16z = 43 \quad (4)$$
$$\underline{11y - 16z = -43} \quad -1 \text{ times (5)}$$
$$0 = 0 \quad (6)$$

The system of equations is dependent.

Let $z = c$.

$$-11y + 16c = 43$$
$$y = \frac{16c - 43}{11}$$

$$x + 3\left(\frac{16c - 43}{11}\right) - 5c = -12$$

$$x = \frac{7c - 3}{11}$$

The solution is $\left(\dfrac{7c-3}{11}, \dfrac{16c-43}{11}, c\right)$.

13. Solve the system of equations. [9.2]

$$\begin{cases} 3x + 4y - 6z = 10 & (1) \\ 2x + 2y - 3z = 6 & (2) \\ x - 6y + 9z = -4 & (3) \end{cases}$$

Rearrange the equations so that the equation with the 1 as the coefficient of x is in the first row.

$$\begin{cases} x - 6y + 9z = -4 & (3) \\ 2x + 2y - 3z = 6 & (2) \\ 3x + 4y - 6z = 10 & (1) \end{cases}$$

$$\begin{aligned} -2x + 12y - 18z &= 8 \quad -2 \text{ times (3)} \\ 2x + 2y - 3z &= 6 \quad (2) \\ \hline 14y - 21z &= 14 \\ 2y - 3z &= 2 \quad (4) \end{aligned}$$

$$\begin{aligned} -3x + 18y - 27z &= 12 \quad -3 \text{ times (3)} \\ 3x + 4y - 6z &= 10 \quad (1) \\ \hline 22y - 33z &= 22 \\ 2y - 3z &= 2 \quad (5) \end{aligned}$$

$$\begin{cases} x - 6y + 9z = -4 & (3) \\ 2y - 3z = 2 & (4) \\ 2y - 3z = 2 & (5) \end{cases}$$

$$\begin{aligned} 2y - 3z &= 2 \quad (4) \\ -2y + 3z &= -2 \quad -1 \text{ times (5)} \\ \hline 0 &= 0 \quad (6) \end{aligned}$$

$$\begin{cases} x - 6y + 9z = -4 & (3) \\ 2y - 3z = 2 & (4) \\ 0 = 0 & (6) \end{cases}$$

The system of equations is dependent.

Let $z = c$.

$$2y - 3c = 2$$

$$y = \frac{3c + 2}{2}$$

$$x - 6\left(\frac{3c + 2}{2}\right) + 9c = -4$$

$$x - 9c - 6 + 9c = -4$$

$$x = 2$$

The solution is $\left(2, \dfrac{3c+2}{2}, c\right)$.

15. Solve the system of equations. [9.2]

$$\begin{cases} 2x + 3y - 2z = 0 & (1) \\ 3x - y - 4z = 0 & (2) \\ 5x + 13y - 4z = 0 & (3) \end{cases}$$

$$\begin{aligned} 6x + 9y - 6z &= 0 \quad 3 \text{ times (1)} \\ -6x + 2y + 8z &= 0 \quad -2 \text{ times (2)} \\ \hline 11y + 2z &= 0 \quad (4) \end{aligned}$$

$$\begin{aligned} 15x - 5y - 20z &= 0 \quad 5 \text{ times (2)} \\ -15x - 39y + 12z &= 0 \quad -3 \text{ times (3)} \\ \hline -44y - 8z &= 0 \\ 11y + 2z &= 0 \quad (5) \end{aligned}$$

$$\begin{cases} 2x + 3y - 2z = 0 & (1) \\ 11y + 2z = 0 & (4) \\ 11y + 2z = 0 & (5) \end{cases}$$

$$\begin{aligned} 11y + 2z &= 0 \quad (4) \\ -11y - 2z &= 0 \quad -1 \text{ times (5)} \\ \hline 0 &= 0 \quad (6) \end{aligned}$$

$$\begin{cases} 2x + 3y - 2z = 0 & (1) \\ 11y + 2z = 0 & (4) \\ 0 = 0 & (6) \end{cases}$$

Let $z = c$ $11y + 2c = 0$

$$y = -\frac{2}{11}c$$

$$2x + 3\left(-\frac{2c}{11}\right) - 2c = 0$$

$$2x = \frac{28c}{11}$$

$$x = \frac{14c}{11}$$

The solution is $\left(\dfrac{14}{11}c, \; -\dfrac{2}{11}c, \; c\right)$.

17. Solve the system of equations. [9.2]

$$\begin{cases} 2x - y + 3z = 6 & (1) \\ x + 2y + 4z = 10 & (2) \end{cases}$$

$$2x - y + 3z = \quad 6 \quad (1)$$
$$\underline{-2x - 4y - 8z = -20 \quad -2 \text{ times (2)}}$$
$$-5y - 5z = -14$$
$$5y + 5z = 14 \quad (3)$$

$$\begin{cases} 2x - y + 3z = 6 & (1) \\ 5y + 5z = 14 & (3) \end{cases}$$

Let $z = c$. $5y + 5c = 14$

$$y = -c + \frac{14}{5}$$

$$2x - \left(-c + \frac{14}{5}\right) + 3c = 6$$

$$2x + 4c - \frac{14}{5} = 6$$

$$x = -2c + \frac{22}{5}$$

The solution is $\left(-2c + \dfrac{22}{5}, \ -c + \dfrac{14}{5}, \ c\right)$.

19. Solve the system of equations. [9.3]

$$\begin{cases} y = x^2 - 2x - 3 \\ y = 2x - 7 \end{cases}$$

$$x^2 - 2x - 3 = 2x - 7$$
$$x^2 - 4x + 4 = 0$$
$$(x - 2)(x - 2) = 0$$
$$x = 2$$

$$y = 2(2) - 7$$
$$y = -3$$

The solution is $(2, -3)$.

21. Solve the system of equations. [9.3]

$$\begin{cases} (x - 1)^2 - y = 4 \\ (x + 2)^2 - y = 9 \end{cases}$$

$$\begin{cases} y = (x - 1)^2 - 4 \\ y = (x + 2)^2 - 9 \end{cases}$$

$$(x - 1)^2 - 4 = (x + 2)^2 - 9$$
$$x^2 - 2x + 1 - 4 = x^2 + 4x + 4 - 9$$
$$-2x - 3 = 4x - 5$$
$$-6x = -2$$
$$x = \frac{1}{3}$$

$$y = \left(\frac{1}{3} + 2\right)^2 - 9$$
$$y = -\frac{32}{9}$$

The solution is $\left(\dfrac{1}{3}, -\dfrac{32}{9}\right)$.

23. Solve the system of equations. [9.3]

$$\begin{cases} (x + 1)^2 + (y - 2)^2 = 4 & (1) \\ 2x + y = 4 & (2) \end{cases}$$

From Eq. (2), $y = -2x + 4$.

Substitute y in Eq. (1).

$$(x + 1)^2 + (-2x + 4 - 2)^2 = 4$$
$$(x + 1)^2 + (-2x + 2)^2 = 4$$
$$x^2 + 2x + 1 + 4x^2 - 8x + 4 = 4$$
$$5x^2 - 6x + 1 = 0$$
$$(5x - 1)(x - 1) = 0$$

$$x = \frac{1}{5} \text{ or } x = 1$$

$$y = -2\left(\frac{1}{5}\right) + 4 = \frac{18}{5}$$

$$y = -2(1) + 4 = 2$$

The solutions are $\left(\dfrac{1}{5}, \ \dfrac{18}{5}\right)$ and $(1, 2)$.

25. Solve the system of equations. [9.3]

$$\begin{cases} (x - 2)^2 + (y - 2)^2 = 1 & (1) \\ (x - 1)^2 + (y + 2)^2 = 16 & (2) \end{cases}$$

Expand the binomials. Then subtract.

$$x^2 - 4x + 4 + y^2 - 4y + 4 = 1$$
$$\underline{x^2 - 2x + 1 + y^2 + 4y + 4 = 16}$$
$$-2x + 3 \qquad - 8y \qquad = -15$$
$$y = -\frac{1}{4}x + \frac{9}{4}$$

Substitute y into Eq. (1).

$$(x - 2)^2 + \left(-\frac{1}{4}x + \frac{9}{4} - 2\right)^2 = 1$$

$$(x - 2)^2 + \left(-\frac{1}{4}x + \frac{1}{4}\right)^2 = 1$$

$$x^2 - 4x + 4 + \frac{1}{16}x^2 - \frac{1}{8}x + \frac{1}{16} = 1$$

$$\frac{17}{16}x^2 - \frac{33}{8}x + \frac{49}{16} = 0$$

$$17x^2 - 66x + 49 = 0$$

$$(x-1)(17x - 49) = 0$$

$$x = 1 \text{ or } x = \frac{49}{17}$$

$$y = -\frac{1}{4}(1) + \frac{9}{4} = 2$$

$$y = -\frac{1}{4}\left(\frac{49}{17}\right) + \frac{9}{4} = \frac{26}{17}$$

The solutions are $(1, 2)$ and $\left(\frac{26}{17}, \frac{49}{17}\right)$.

27. Solve the system of equations. [9.3]

$$\begin{cases} x^2 - 3xy + y^2 = -1 & (1) \\ 3x^2 - 5xy - 2y^2 = 0 & (2) \end{cases}$$

Factor Eq. (2).

$$(3x + y)(x - 2y) = 0$$

$$x = \frac{-y}{3} \text{ or } x = 2y$$

$$x = -\frac{y}{3} \text{ implies } y = -3x. \text{ Substitute for } y \text{ in Eq. (1).}$$

$$x^2 - 3x(-3x) + (-3x)^2 = -1$$

$$x^2 + 9x^2 + 9x^2 = -1$$

$$19x^2 = -1$$

$$x^2 = -\frac{1}{19}$$

This equation yields no real solutions. Substituting $x = 2y$ in Eq. (1) yields

$$(2y)^2 - 3(2y)y + y^2 = -1$$

$$4y^2 - 6y^2 + y^2 = -1$$

$$-y^2 = -1$$

$$y^2 = 1$$

$$y = \pm 1$$

The solutions are $(2, 1)$ and $(-2, -1)$.

29. Solve the system of equations. [9.3]

$$\begin{cases} 2x^2 - 5xy + 2y^2 = 56 & (1) \\ 14x^2 - 3xy - 2y^2 = 56 & (2) \end{cases}$$

Subtract Eq. (1) From Eq. (2)

$$12x^2 + 2xy - 4y^2 = 0$$

Factor $2(3x + 2y)(2x - y) = 0$. Thus

$$3x + 2y = 0 \qquad \text{or} \quad 2x - y = 0$$

$$y = -\frac{3}{2}x \qquad\qquad\qquad y = 2x$$

Substituting $y = -\frac{3}{2}x$ into Eq. (1), we have

$$2x^2 - 5x\left(-\frac{3}{2}x\right) + 2\left(-\frac{3}{2}x\right)^2 = 56$$

$$2x^2 + \frac{15}{2}x^2 + \frac{9}{2}x^2 = 56$$

$$14x^2 = 56$$

$$x^2 = 4$$

$$x = \pm 2$$

When $x = 2, y = -\frac{3}{2}(2) = -3$;

When $x = -2, y = -\frac{3}{2}(-2) = 3.$

Two solutions are $(2, -3)$ and $(-2, 3)$.

Substituting $y = 2x$ into Eq. (1) yields $0 = 56$. Thus the only solutions of the system are $(2, -3)$, and $(-2, 3)$.

31. Find the partial fraction decomposition. [9.4]

$$\frac{5x+1}{x^2+x-6} = \frac{5x+1}{(x+3)(x-2)} = \frac{A}{x+3} + \frac{B}{x-2}$$

$$5x + 1 = A(x-2) + B(x+3)$$

$$5x + 1 = Ax - 2A + Bx + 3B$$

$$5x + 1 = (A+B)x + (-2A + 3B)$$

$$\begin{cases} 5 = A + B \\ 1 = -2A + 3B \end{cases}$$

$$\begin{array}{r} 10 = 2A + 2B \\ 1 = -2A + 3B \\ \hline 11 = 5B \end{array}$$

$$\frac{11}{5} = B$$

$$A + \frac{11}{5} = 5$$

$$A = \frac{14}{5}$$

$$\frac{5x+1}{x^2+x-6} = \frac{14}{5(x+3)} + \frac{11}{5(x-2)}$$

33. Find the partial fraction decomposition. [9.4]

$$\frac{2x-2}{(x^2+1)(x+2)} = \frac{Ax+B}{x^2+1} + \frac{C}{x+2}$$

$$2x-2 = (Ax+B)(x+2) + C(x^2+1)$$
$$2x-2 = Ax^2 + 2Ax + Bx + 2B + Cx^2 + C$$
$$2x-2 = (A+C)x^2 + (2A+B)x + (2B+C)$$

$$\begin{cases} 0 = A+C & (1) \\ 2 = 2A+B & (2) \\ -2 = 2B+C & (3) \end{cases}$$

$$\begin{array}{ll} 0 = A \quad\quad +C & (1) \\ 2 = \quad -2B-C & -1 \text{ times } (3) \\ \hline 2 = A - 2B & (4) \end{array}$$

$$\begin{cases} 0 = A+C & (1) \\ 2 = 2A+B & (2) \\ 2 = A-2B & (4) \end{cases}$$

$$\begin{array}{ll} 4 = 4A+2B & 2 \text{ times } (2) \\ 2 = A-2B & (4) \\ \hline 6 = 5A \end{array}$$

$$\frac{6}{5} = A$$

$$2\left(\frac{6}{5}\right) + B = 2$$

$$B = -\frac{2}{5}$$

$$\frac{6}{5} + C = 0$$

$$C = -\frac{6}{5}$$

$$\frac{2x-2}{(x^2+1)(x-2)} = \frac{6x-2}{5(x^2+1)} + \frac{-6}{5(x+2)}$$

35. Find the partial fraction decomposition. [9.4]

$$\frac{11x^2-x-2}{x^3-x} = \frac{11x^2-x-2}{x(x-1)(x+1)} = \frac{A}{x} + \frac{B}{x-1} + \frac{C}{x+1}$$

$$11x^2-x-2 = A(x-1)(x+1) + Bx(x+1) + Cx(x-1)$$
$$11x^2-x-2 = Ax^2 - A + Bx^2 + Bx + Cx^2 - Cx$$
$$11x^2-x-2 = (A+B+C)x^2 + (B-C)x + (-A)$$

$$\begin{cases} 11 = A+B+C & (1) \\ -1 = B-C & (2) \\ -2 = -A & (3) \end{cases}$$

$$\begin{cases} 11 = A+B+C & (1) \\ -1 = B-C & (2) \\ 2 = A & (4) \end{cases}$$

$$\begin{array}{ll} B+C = 9 & \text{from (1) with } A = 2 \\ \underline{B-C = -1} & (2) \\ 2B \quad\quad = 8 \\ B = 4 \end{array}$$

$$4 - C = -1$$
$$C = 5$$

$$\frac{11x^2-x-2}{x^3-x} = \frac{2}{x} + \frac{4}{x-1} + \frac{5}{x+1}$$

37. Find the partial fraction decomposition. [9.4]

Using long division, we have

$$\frac{x^3-2x^2+5x-3}{x^2+3x} = x-5 + \frac{20x-3}{x^2+3x}$$

Find the partial fraction decomposition of the fraction.

$$\frac{20x-3}{x(x+3)} = \frac{A}{x} + \frac{B}{x+3}$$

$$20x-3 = A(x+3) + Bx$$
$$20x-3 = Ax + 3A + Bx$$
$$20x-3 = (A+B)x + 3A$$

$$\begin{cases} 20 = A+B \\ -3 = 3A \end{cases}$$

$$A = -1$$
$$20 = -1 + B$$
$$B = 21$$

$$\frac{x^3-2x^2+5x-3}{x^2+3x} = x-5 - \frac{1}{x} + \frac{21}{x+3}$$

39. Graph the solution set of the inequality. [9.5]

$$4x - 5y < 20$$

$$y > \frac{4}{5}x - 4 \quad \text{Use a dashed line.}$$

Test point $(0, 0)$: is in region

41. Graph the solution set of the inequality. [9.5]

$$y \geq 2x^2 - x - 1$$

vertex $\left(\dfrac{1}{4}, -\dfrac{9}{8}\right)$

Test point (0, 0): is in region

43. Graph the solution set of the inequality. [9.5]

$$(x-2)^2 + (y-1)^2 > 4$$

center (2, 1), $r = 2$

Test point (2, 0): not in region

45. Graph the solution set of the inequality. [9.5]

$$\dfrac{(x-3)^2}{16} - \dfrac{(y+2)^2}{25} \leq 1$$

Test point (0, 0): is in region

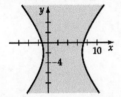

47. Graph the solution set of the inequality. [9.5]

$$(2x - y + 1)(2x - 2y - 2) > 0$$

Test point (0, 0): not in region

49. Graph the solution set of the inequality. [9.5]

$$x^2 y^2 < 1$$

Test point (0, 0): is in region

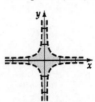

51. Sketch the graph of the solution set of the system of inequalities. [9.5]

$$\begin{cases} 2x - 5y < 9 \\ 3x + 4y \geq 2 \end{cases}$$

53. Sketch the graph of the solution set of the system of inequalities. [9.5]

$$\begin{cases} 2x + 3y > 6 \\ 2x - y > -2 \\ x \leq 4 \end{cases}$$

55. Sketch the graph of the solution set of the system of inequalities. [9.5]

$$\begin{cases} 2x + 3y \leq 18 \\ x + y \leq 7 \\ x \geq 0, \ y \geq 0 \end{cases}$$

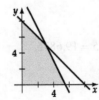

57. Sketch the graph of the solution set of the system of inequalities. [9.5]

$$\begin{cases} 3x+y \geq 6 \\ x+4y \geq 14 \\ 2x+3y \geq 16 \\ x \geq 0,\ y \geq 0 \end{cases}$$

59. Sketch the graph of the solution set of the system of inequalities. [9.5]

$$\begin{cases} y < x^2 - x - 2 \\ y \geq 2x - 4 \end{cases}$$

61. Sketch the graph of the solution set of the system of inequalities. [9.5]

$$\begin{cases} x^2 + y^2 - 2x + 4y > 4 \\ y < 2x^2 - 1 \end{cases}$$

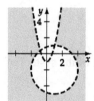

63. Find the minimum value and state where it is. [9.6]

$$C(x,\ y) = \frac{1}{2}x + y$$

$$C(0,\ 30) = \frac{1}{2}(0) + 30 = 30$$

$$C(1.875,\ 18.75) = \frac{1}{2}(1.875) + 18.75 = 19.6875$$

$$C(24,\ 4) = \frac{1}{2}(24) + 4 = 16$$

$$C(40,\ 0) = \frac{1}{2}(40) + 0 = 20$$

The minimum of 16 occurs at (24, 4).

65. Solve the linear programming problem. [9.6]

$$P = 2x + 2y$$

$$\begin{cases} x + 2y \leq 14 \\ 5x + 2y \leq 30 \\ x \leq 0,\ y \leq 0 \end{cases}$$

$P = 2x + 2y$

(0, 7)	14
(6, 0)	12
(4, 5)	18 maximum
(0, 0)	0

The maximum is 18 at (4, 5).

67. Solve the linear programming problem. [9.6]

$$P = 4x + y$$

$$\begin{cases} 5x + 2y \geq 16 \\ x + 2y \geq 8 \\ 0 \leq x \leq 20 \\ 0 \leq y \leq 20 \end{cases}$$

$P = 4x + y$

(0, 8)	8 minimum
(2, 3)	11
(8, 0)	32
(20, 0)	80
(20, 20)	100
(0, 20)	20

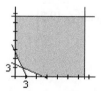

The minimum is 8 at (0, 8).

69. Solve the linear programming problem. [9.6]

$$P = 6x + 3y$$

$$\begin{cases} 5x + 2y \geq 20 \\ x + y \geq 7 \\ x + 2y \geq 10 \\ 0 \leq x \leq 15,\ 0 \leq y \leq 15 \end{cases}$$

$P = 6x + 3y$

(0, 10)	30
(2, 5)	27 minimum
(4, 3)	33
(10, 0)	60
(0, 15)	45
(15, 15)	135
(15, 0)	90

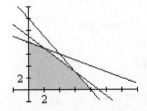

The minimum is 27 at (2, 5).

71. Find the number of each to maximize profit and find the maximum profit. [9.6]

x = number of 4-cylinder engines

y = number of 6-cylinder engines

Profit $= 150x + 250y$

Constraints:

$$\begin{cases} x + y \le 9 \\ 5x + 10y \le 80 \\ 3x + 2y \le 24 \\ x \ge 0, \, y \ge 0 \end{cases}$$

Profit $= 150x + 250y$

(0, 8)	2000
(2, 7)	2050 maximum
(6, 3)	1650
(8, 0)	1200
(0, 0)	0

To achieve the maximum profit of $2050, produce two 4-cylinder engines and seven 6-cylinder engines.

73. Find the equation. [9.2]

$$y = ax^2 + bx + c$$

$$0 = a(1)^2 + b(1) + c$$
$$5 = a(-1)^2 + b(-1) + c$$
$$3 = a(2)^2 + b(2) + c$$

$$\begin{cases} a + b + c = 0 & (1) \\ a - b + c = 5 & (2) \\ 4a + 2b + c = 3 & (3) \end{cases}$$

$$\begin{aligned} a + b + c &= 0 & (1) \\ -a + b - c &= -5 & -1 \text{ times (2)} \\ \hline 2b &= -5 & (4) \end{aligned}$$

$$\begin{aligned} a - b + c &= 5 & (2) \\ -4a - 2b - c &= -3 & -1 \text{ times (3)} \\ \hline -3a - 3b &= 2 & (5) \end{aligned}$$

$$\begin{cases} a + b + c = 0 & (1) \\ 2b = -5 & (4) \\ -3a - 3b = 2 & (5) \end{cases}$$

$$2b = -5$$

$$b = -\frac{5}{2}$$

$$-3a - 3\left(-\frac{5}{2}\right) = 2$$

$$a = \frac{11}{6}$$

$$\frac{11}{6} - \frac{5}{2} + c = 0$$

$$c = \frac{2}{3}$$

The equation of the graph that passes through the three points is $y = \dfrac{11}{6}x^2 - \dfrac{5}{2}x + \dfrac{2}{3}$.

75. Find the equation of the plane. [9.2]

$$z = ax + by + c$$

$$2 = 2a + b + c$$
$$0 = 3a + b + c$$
$$-2 = -2a - 3b + c$$

$$\begin{cases} 2a + b + c = 2 & (1) \\ 3a + b + c = 0 & (2) \\ -2a - 3b + c = -2 & (3) \end{cases}$$

$6a + 3b + 3c = 6$ 3 times (1)
$\underline{-6a - 2b - 2c = 0}$ -2 times (2)
$b + c = 6$ (4)

$2a + b + c = 2$ (1)
$\underline{-2a - 3b + c = -2}$ (3)
$-2b + 2c = 0$
$-b + c = 0$ (5)

$2a + b + c = 2$ (1)
$\underline{-2a - 3b + c = -2}$ (3)
$-2b + 2c = 0$
$-b + c = 0$ (5)

$\begin{cases} 2a + b + c = 2 & (1) \\ b + c = 6 & (4) \\ -b + c = 0 & (5) \end{cases}$

$b + c = 6$ (4)
$\underline{-b + c = 0}$ (5)
$2c = 6$
$c = 3$ (6)

$\begin{cases} 2a + b + c = 2 & (1) \\ b + c = 6 & (4) \\ c = 3 & (6) \end{cases}$

$b + 3 = 6$
$b = 3$

$2a + 3 + 3 = 2$
$2a = -4$
$a = -2$

The equation of the graph that passes through the three points is $z = -2x + 3y + 3$.

77. Find the rate of the wind and plane in calm air. [9.1]

Rate flying with the wind: $r + w$

Rate flying against the wind: $r - w$

$\begin{cases} (r + w)5 = 855 & (1) \\ 5(r - w) = 575 & (2) \end{cases}$

$r + w = 171$ (1)
$\underline{r - w = 115}$ (2)
$2r = 286$
$r = 143$

$143 + w = 171$
$w = 28$

Rate of the wind is 28 mph.

Rate of the plane in calm air is 143 mph.

79. Find the values for a, b, and c. [9.2]

Given (a, b, c) with $ab = c$, $ac = b$.

If $ab = c$, then $b \cdot (ab) = a$ and $b^2 = 1$, $b = \pm 1$, or $a = 0$.

If $bc = a$, then $bc \cdot c = b$ and $c^2 = 1$, $c = \pm 1$, or $b = 0$.

If $ac = b$, then $a \cdot ac = c$ and $a^2 = 1$, $a \pm 1$, or $c = 0$.

The ordered triples are $(1, 1, 1)$, $(1, -1, -1)$, $(-1, -1, 1)$, $(-1, 1, -1)$, $(0, 0, 0)$.

Chapter 9 Test

1. Solve the system of equations. [9.1]

$\begin{cases} 3x + 2y = -5 & (1) \\ 2x - 5y = -16 & (2) \end{cases}$

$15x + 10y = -25$ 5 times (1)
$\underline{4x - 10y = -32}$ 2 times (2)
$19x = -57$
$x = -3$

$3(-3) + 2y = -5$
$2y = 4$
$y = 2$

The solution is $(-3, 2)$.

3. Solve the system of equations. [9.2]

$\begin{cases} 2x - 3y + z = 9 & (1) \\ x - 5y + 3z = 5 & (2) \\ 3x - y - 2z = 8 & (3) \end{cases}$

$-6x + 9y - 3z = -27$ -3 times (1)
$\underline{x - 5y + 3z = 5}$ (2)
$-5x + 4y = -22$ (4)

$4x - 6y + 2z = 18$ 2 times (1)
$\underline{3x - y - 2z = 8}$ (3)
$7x - 7y = 26$ (5)

$\begin{cases} 2x - 3y + z = 9 & (1) \\ -5x + 4y = -22 & (4) \\ 7x - 7y = 26 & (5) \end{cases}$

$-35x + 28y = -154$ 7 times (4)

$\dfrac{28x - 28y = 104}{-7x = -50}$ 4 times (5)

$x = \dfrac{50}{7}$ (6)

$\begin{cases} 2x - 3y + z = 9 & (1) \\ -5x + 4y = -22 & (4) \\ x = \dfrac{50}{7} & (6) \end{cases}$

$-5\left(\dfrac{50}{7}\right) + 4y = -22$

$y = \dfrac{24}{7}$

$2\left(\dfrac{50}{7}\right) - 3\left(\dfrac{24}{7}\right) + z = 9$

$z = 5$

The solution is $\left(\dfrac{50}{7},\ \dfrac{24}{7},\ 5\right)$.

5. Solve the system of equations. [9.2]

$\begin{cases} 2x - 3y + z = -1 & (1) \\ x + 5y - 2z = 5 & (2) \end{cases}$

$2x - 3y + z = -1$ (1)

$\dfrac{-2x - 10y + 4z = -10}{-13y + 5z = -11}$ -2 times (2)

 (3)

$\begin{cases} 2x - 3y + z = -1 & (1) \\ -13y + 5z = -11 & (3) \end{cases}$

The system of equations is dependent.

Let $z = c$.

$-13y + 5(c) = -11$

$y = \dfrac{5c + 11}{13}$

$x + 5\left(\dfrac{5c + 11}{13}\right) - 2c = 5$

$x = \dfrac{c + 10}{13}$

The solution is $\left(\dfrac{c + 10}{13},\ \dfrac{5c + 11}{13},\ c\right)$.

7. Solve the system of equations. [9.3]

$\begin{cases} 8x = 3y + 10 & (1) \\ y = x^2 - 2x - 5 & (2) \end{cases}$

Substitute y from Eq. (2) into Eq. (1).

$8x = 3(x^2 - 2x - 5) + 10$

$8x = 3x^2 - 6x - 15 + 10$

$0 = 3x^2 - 14x - 5$

$0 = (3x + 1)(x - 5)$

$x = -\dfrac{1}{3} \text{ or } x = 5$

Substitute for x in Eq. (2).

$y = \left(-\dfrac{1}{3}\right)^2 - 2\left(-\dfrac{1}{3}\right) - 5$ $y = 5^2 - 2(5) - 5$

$y = -\dfrac{38}{9}$ $y = 10$

The solutions are $\left(-\dfrac{1}{3}, -\dfrac{38}{9}\right)$ and $(5, 10)$.

9. Graph the inequality. [9.5]

$x^2 + 4y^2 \geq 16$

Test point $(0, 0)$: not in region

11. Graph the system of inequalities. [9.5]

$\begin{cases} 2x - 5y \leq 16 \\ x + 3y \geq -3 \end{cases}$

Test point $(0, 0)$: is in region

13. Graph the system of inequalities. [9.5]

$\begin{cases} x + y \geq 8 \\ 2x + y \geq 11 \\ x \geq 0,\ y \geq 0 \end{cases}$

15. Find the partial fraction decomposition. [9.4]

$$\frac{3x-5}{x^2-3x-4}=\frac{3x-5}{(x-4)(x+1)}=\frac{A}{x-4}+\frac{B}{x+1}$$

$$3x-5=A(x+1)+B(x-4)$$

$$3x-5=Ax+A+Bx-4B$$

$$\begin{cases} 3=A+B \\ -5=A-4B \end{cases}$$

$$3=A+B$$
$$\underline{5=-A+4B}$$
$$8=5B$$

$$\frac{8}{5}=B$$

$$3=A+\frac{8}{5}$$
$$\frac{7}{5}=A$$

$$\frac{3x-5}{(x-4)(x+1)}=\frac{7}{5(x-4)}+\frac{8}{5(x+1)}$$

17. Find the partial fraction decomposition. [9.4]

Using long division, we have

$$\frac{x^3+x-2}{x^2+x-12}=x-1+\frac{14x-14}{x^2+x-12}$$

Find the partial fraction decomposition of the fraction.

$$\frac{14x-14}{(x-3)(x+4)}=\frac{A}{x-3}+\frac{B}{x+4}$$

$$14x-14=A(x+4)+B(x-3)$$

$$14x-14=Ax+4A+Bx-3B$$

$$14x-14=(A+B)x+(4A-3B)$$

$$\begin{cases} 14=A+B \\ -14=4A-3B \end{cases}$$

$$-56=-4A-4B$$
$$\underline{-14=\ \ 4A-3B}$$
$$-70=\qquad -7B$$
$$10=B$$

$$14=A+(10)$$
$$4=A$$

$$\frac{x^3+x-2}{x^2+x-12}=x-1+\frac{4}{x-3}+\frac{10}{x+4}$$

19. Find the length and width. [9.1]

$$x=20+y$$
$$2x+\pi y=554.16$$

Using substitution,

$$2(20+y)+\pi y=554.16$$
$$40+2y+\pi y=554.16$$
$$y(2+\pi)=514.16$$
$$y=\frac{514.16}{2+\pi}$$
$$y\approx 100$$

$$x=20+y=20+100=120$$

Length is 120 m and width is 100 m.

21. Find the amount of each to maximize profit. [9.6]

$x=$ Acres of oats

$y=$ Acres of barley

Constraints

$$\begin{cases} x+\ \ y\le 160 \\ 15x+13y\le 2200 \\ 15x+20y\le 2600 \\ x\le 0,\ \ y\ge 0 \end{cases}$$

Profit $=120x+150y$

$(0,130)$	$19,500$
$\left(\dfrac{680}{7},\dfrac{400}{7}\right)$	$20,228.57$ maximum
$\left(146\dfrac{2}{3},0\right)$	$17,600$
$(0,0)$	0

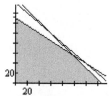

To maximize profit, $\dfrac{680}{7}$ acres of oats and $\dfrac{400}{7}$ acres

of barley must be planted.

Cumulative Review Exercises

1. Find the range. [2.4]

$$f(x) = -x^2 + 2x - 4$$
$$= -(x^2 - 2x + 1) - 4 + 1$$
$$= -(x-1)^2 - 3$$

Vertex (1, –3), parabola opens down.

Range: $\{y \mid y \le -3\}$

3. Find the equation of the parabola. [8.1]

Vertex = (4, 2), point (–1, 1), axis of symmetry $x = 4$

$$(x-4)^2 = 4p(y-2)$$
$$(-1-4)^2 = 4p(1-2)$$
$$25 = 4p(-1)$$
$$p = -\frac{25}{4}$$
$$(x-4)^2 = 4\left(-\frac{25}{4}\right)(y-2)$$
$$(x-4)^2 = -25(y-2)$$

5. Solve. [4.5]

$$\log x - \log(2x-3) = 2$$
$$\log \frac{x}{2x-3} = 2$$
$$\frac{x}{2x-3} = 10^2$$
$$x = 100(2x-3)$$
$$x = 200x - 300$$
$$-199x = -300$$
$$x = \frac{300}{199}$$

7. Find $g\left(-\frac{1}{2}\right)$. [2.2]

$$g\left(-\frac{1}{2}\right) = \frac{-\frac{1}{2} - 2}{-\frac{1}{2}} = \frac{-\frac{5}{2}}{-\frac{1}{2}} = 5$$

9. Find a quadratic regression model. [2.7]

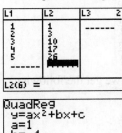

$$y = x^2 + 0.4x - 0.8$$

11. Find the inverse. [4.1]

$$Q(r) = \frac{2}{1-r}$$
$$r = \frac{2}{1-Q}$$
$$r(1-Q) = 2$$
$$r - rQ = 2$$
$$-rQ = 2 - r$$
$$Q = \frac{2-r}{-r}$$
$$Q^{-1}(r) = \frac{r-2}{r}$$

13. Find $g[f(1)]$. [4.2]

$$g[f(1)] = g\left[2^1\right] = g[2] = 3^{2(2)} = 3^4 = 81$$

15. Find the exact value. [5.4]

$$\cos\left(\frac{11\pi}{6}\right) = \frac{\sqrt{3}}{2}$$

17. Find the exact value of $\sin 15°$. [6.2]

$$\sin 15° = \sin(45° - 30°)$$
$$= \sin 45° \cos 30° - \cos 45° \sin 30°$$
$$= \frac{\sqrt{2}}{2} \cdot \frac{\sqrt{3}}{2} - \frac{\sqrt{2}}{2} \cdot \frac{1}{2}$$
$$= \frac{\sqrt{6} - \sqrt{2}}{4}$$

19. Solve. [6.5]

$$\cos^{-1} x + \tan^{-1} x = \frac{\pi}{2}$$
$$\cos\left[\cos^{-1} x + \tan^{-1} x\right] = \cos\left(\frac{\pi}{2}\right)$$
$$\cos(\alpha + \beta) = 0$$
$$\cos\alpha \cos\beta - \sin\alpha \sin\beta = 0$$

$$\cos\alpha = x, \sin\alpha = \sqrt{1+x^2}$$
$$\cos\beta = \sqrt{1-x^2}, \sin\beta = \frac{x}{\sqrt{1+x^2}}$$

$$x\left(\sqrt{1-x^2}\right) - \left(\sqrt{1+x^2}\right) \cdot \frac{x}{\sqrt{1+x^2}} = 0$$
$$x\left(\sqrt{1-x^2}\right) - x = 0$$
$$x\left(\sqrt{1-x^2} - 1\right) = 0$$

$$x = 0 \qquad\qquad \sqrt{1-x^2} - 1 = 0$$
$$\sqrt{1-x^2} = 1$$
$$1 - x^2 = 1$$
$$0 = x^2$$
$$0 = x$$

Chapter 10 Matrices

Section 10.1 Exercises

1. Writing the augmented, coefficient and constant matrices of the given system of equations.

$$\begin{bmatrix} 2 & -3 & 1 & | & 1 \\ 3 & -2 & 3 & | & 0 \\ 1 & 0 & 5 & | & 4 \end{bmatrix}, \quad \begin{bmatrix} 2 & -3 & 1 \\ 3 & -2 & 3 \\ 1 & 0 & 5 \end{bmatrix}, \quad \begin{bmatrix} 1 \\ 0 \\ 4 \end{bmatrix}$$

3. Writing the augmented, coefficient and constant matrices of the given system of equations.

$$\begin{bmatrix} 2 & -3 & -4 & 1 & | & 2 \\ 0 & 2 & 1 & 0 & | & 2 \\ 1 & -1 & 2 & 0 & | & 4 \\ 3 & -3 & -2 & 0 & | & 1 \end{bmatrix}, \quad \begin{bmatrix} 2 & -3 & -4 & 1 \\ 0 & 2 & 1 & 0 \\ 1 & -1 & 2 & 0 \\ 3 & -3 & -2 & 0 \end{bmatrix}, \quad \begin{bmatrix} 2 \\ 2 \\ 4 \\ 1 \end{bmatrix}$$

5. Write the resulting matrix.

$$\begin{bmatrix} 3 & -2 & 6 & 1 \\ 2 & 0 & -1 & 5 \\ 1 & 3 & -6 & 4 \end{bmatrix} \xrightarrow{R_3 \leftrightarrow R_1} \begin{bmatrix} 1 & 3 & -6 & 4 \\ 2 & 0 & -1 & 5 \\ 3 & -2 & 6 & 1 \end{bmatrix}$$

7. Write the resulting matrix.

$$\begin{bmatrix} 1 & 3 & -1 & 5 \\ -3 & -12 & 5 & -6 \\ 2 & -4 & -7 & 8 \end{bmatrix} \xrightarrow{3R_1 + R_2} \begin{bmatrix} 1 & 3 & -1 & 5 \\ 0 & -3 & 2 & 9 \\ 2 & -4 & -7 & 8 \end{bmatrix}$$

9. State whether the matrix is in row echelon form.

No, the matrix is not in row echelon form since the element in row 2 column 2 is not 1.

11. State whether the matrix is in row echelon form.

Yes

The solutions and answers to Exercises 13 – 20 are not unique. The following solutions list the elementary row operations used to produce the row echelon forms shown.

13. Using elementary row operations to achieve row echelon form.

$$\begin{bmatrix} 2 & -1 & 3 & -2 \\ 1 & -1 & 2 & 2 \\ 3 & 2 & -1 & 3 \end{bmatrix} \xrightarrow{R_1 \leftrightarrow R_2} \begin{bmatrix} 1 & -1 & 2 & 2 \\ 2 & -1 & 3 & -2 \\ 3 & 2 & -1 & 3 \end{bmatrix}$$

$$\begin{bmatrix} 1 & -1 & 2 & 2 \\ 2 & -1 & 3 & -2 \\ 3 & 2 & -1 & 3 \end{bmatrix} \xrightarrow[-3R_1 + R_3]{-2R_1 + R_2} \begin{bmatrix} 1 & -1 & 2 & 2 \\ 0 & 1 & -1 & -6 \\ 0 & 5 & -7 & -3 \end{bmatrix}$$

$$\begin{bmatrix} 1 & -1 & 2 & 2 \\ 0 & 1 & -1 & -6 \\ 0 & 5 & -7 & -3 \end{bmatrix} \xrightarrow{-5R_2 + R_3} \begin{bmatrix} 1 & -1 & 2 & 2 \\ 0 & 1 & -1 & -6 \\ 0 & 0 & -2 & 27 \end{bmatrix}$$

$$\begin{bmatrix} 1 & -1 & 2 & 2 \\ 0 & 1 & -1 & -6 \\ 0 & 0 & -2 & 27 \end{bmatrix} \xrightarrow{-\frac{1}{2}R_3} \begin{bmatrix} 1 & -1 & 2 & 2 \\ 0 & 1 & -1 & -6 \\ 0 & 0 & 1 & -\frac{27}{2} \end{bmatrix}$$

Using a graphing calculator, we get the following equivalent answer.

$$\begin{bmatrix} 1 & \frac{2}{3} & -\frac{1}{3} & 1 \\ 0 & 1 & -\frac{11}{7} & \frac{12}{7} \\ 0 & 0 & 1 & -\frac{27}{2} \end{bmatrix}$$

15. Using elementary row operations to achieve row echelon form.

$$\begin{bmatrix} 4 & -5 & -1 & 2 \\ 3 & -4 & 1 & -2 \\ 1 & -2 & -1 & 3 \end{bmatrix} \xrightarrow{R_1 \longleftrightarrow R_3} \begin{bmatrix} 1 & -2 & -1 & 3 \\ 3 & -4 & 1 & -2 \\ 4 & -5 & -1 & 2 \end{bmatrix}$$

$$\xrightarrow[-4R_1 + R_3]{-3R_1 + R_2} \begin{bmatrix} 1 & -2 & -1 & 3 \\ 0 & 2 & 4 & -11 \\ 0 & 3 & 3 & -10 \end{bmatrix}$$

$$\xrightarrow{\frac{1}{2}R_2} \begin{bmatrix} 1 & -2 & -1 & 3 \\ 0 & 1 & 2 & -\frac{11}{2} \\ 0 & 3 & 3 & -10 \end{bmatrix}$$

$$\xrightarrow{-3R_2 + R_3} \begin{bmatrix} 1 & -2 & -1 & 3 \\ 0 & 1 & 2 & -\frac{11}{2} \\ 0 & 0 & -3 & -\frac{13}{2} \end{bmatrix}$$

$$\xrightarrow{-\frac{1}{3}R_3} \begin{bmatrix} 1 & -2 & -1 & 3 \\ 0 & 1 & 2 & -\frac{11}{2} \\ 0 & 0 & 1 & -\frac{13}{6} \end{bmatrix}$$

Using a graphing calculator, we get the following equivalent answer.

$$\begin{bmatrix} 1 & -\frac{5}{4} & -\frac{1}{4} & \frac{1}{2} \\ 0 & 1 & 1 & -\frac{10}{3} \\ 0 & 0 & 1 & -\frac{13}{6} \end{bmatrix}$$

904

17. Using elementary row operations to achieve row echelon form.

$$\begin{bmatrix} 2 & 3 & -1 & 3 \\ 4 & 1 & 1 & -4 \\ 1 & -1 & 1 & 2 \end{bmatrix} \xrightarrow{\ R_1 \longleftrightarrow R_3\ } \begin{bmatrix} 1 & -1 & 1 & 2 \\ 4 & 1 & 1 & -4 \\ 2 & 3 & -1 & 3 \end{bmatrix}$$

$$\xrightarrow[\ -2R_1+R_3\ ]{\ -4R_1+R_2\ } \begin{bmatrix} 1 & -1 & 1 & 2 \\ 0 & 5 & -3 & -12 \\ 0 & 5 & -3 & -1 \end{bmatrix}$$

$$\xrightarrow{\ \frac{1}{5}R_2\ } \begin{bmatrix} 1 & -1 & 1 & 2 \\ 0 & 1 & -\frac{3}{5} & -\frac{12}{5} \\ 0 & 5 & -3 & -1 \end{bmatrix}$$

$$\xrightarrow{\ -5R_2+R_3\ } \begin{bmatrix} 1 & -1 & 1 & 2 \\ 0 & 1 & -\frac{3}{5} & -\frac{12}{5} \\ 0 & 0 & 0 & 11 \end{bmatrix}$$

Using a graphing calculator, we get the following equivalent answer.

$$\begin{bmatrix} 1 & \frac{1}{4} & \frac{1}{4} & -1 \\ 0 & 1 & -\frac{3}{5} & 2 \\ 0 & 0 & 0 & 1 \end{bmatrix}$$

19. Using elementary row operations to achieve row echelon form.

$$\begin{bmatrix} 1 & -3 & 4 & 2 & 1 \\ 2 & -3 & 5 & -2 & -1 \\ -1 & 2 & -3 & 1 & 3 \end{bmatrix}$$

$$\xrightarrow[\ R_1+R_3\ ]{\ -2R_1+R_2\ } \begin{bmatrix} 1 & -3 & 4 & 2 & 1 \\ 0 & 3 & -3 & -6 & -3 \\ 0 & -1 & 1 & 3 & 4 \end{bmatrix}$$

$$\xrightarrow{\ \frac{1}{3}R_2\ } \begin{bmatrix} 1 & -3 & 4 & 2 & 1 \\ 0 & 1 & -1 & -2 & -1 \\ 0 & -1 & 1 & 3 & 4 \end{bmatrix}$$

$$\xrightarrow{\ R_2+R_3\ } \begin{bmatrix} 1 & -3 & 4 & 2 & 1 \\ 0 & 1 & -1 & -2 & -1 \\ 0 & 0 & 0 & 1 & 3 \end{bmatrix}$$

Using a graphing calculator, we get the following equivalent answer.

$$\begin{bmatrix} 1 & -\frac{3}{2} & \frac{5}{2} & -1 & -\frac{1}{2} \\ 0 & 1 & -1 & -2 & -1 \\ 0 & 0 & 0 & 1 & 3 \end{bmatrix}$$

21. Solve the system of equations using the Gaussian elimination method.

$$\begin{cases} x+5y = 16 \\ 4x+3y = 13 \end{cases}$$

$$\begin{bmatrix} 1 & 5 & | & 16 \\ 4 & 3 & | & 13 \end{bmatrix} \xrightarrow{\ -4R_1+R_2\ } \begin{bmatrix} 1 & 5 & | & 16 \\ 0 & -17 & | & -51 \end{bmatrix}$$

$$\xrightarrow{\ -\frac{1}{17}R_2\ } \begin{bmatrix} 1 & 5 & | & 16 \\ 0 & 1 & | & 3 \end{bmatrix}$$

$$\begin{cases} x+5y = 16 \\ \quad\ \ y = 3 \end{cases}$$

$$x+5(3) = 16$$
$$x = 1$$

The solution is (1, 3).

23. Solve the system of equations using the Gaussian elimination method.

$$\begin{cases} 2x-3y = 13 \\ 3x-4y = 18 \end{cases}$$

$$\begin{bmatrix} 2 & -3 & | & 13 \\ 3 & -4 & | & 18 \end{bmatrix} \xrightarrow{\ \frac{1}{2}R_1\ } \begin{bmatrix} 1 & -\frac{3}{2} & | & \frac{13}{2} \\ 3 & -4 & | & 18 \end{bmatrix}$$

$$\xrightarrow{\ -3R_1+R_2\ } \begin{bmatrix} 1 & -\frac{3}{2} & | & \frac{13}{2} \\ 0 & \frac{1}{2} & | & -\frac{3}{2} \end{bmatrix} \xrightarrow{\ 2R_2\ } \begin{bmatrix} 1 & \frac{3}{2} & | & \frac{13}{2} \\ 0 & 1 & | & -3 \end{bmatrix}$$

$$\begin{cases} 2x-3y = 13 \\ \quad\ \ y = -3 \end{cases}$$

$$2x-3(-3) = 13$$
$$2x = 4$$
$$x = 2$$

The solution is (2, –3).

25. Solve the system of equations using the Gaussian elimination method.

$$\begin{cases} x+2y-2z=-2 \\ 5x+9y-4z=-3 \\ 3x+4y-5z=-3 \end{cases}$$

$$\begin{bmatrix} 1 & 2 & -2 & | & -2 \\ 5 & 9 & -4 & | & -3 \\ 3 & 4 & -5 & | & -3 \end{bmatrix} \xrightarrow[\substack{-5R_1+R_2 \\ -3R_1+R_3}]{} \begin{bmatrix} 1 & 2 & -2 & | & -2 \\ 0 & -1 & 6 & | & 7 \\ 0 & -2 & 1 & | & 3 \end{bmatrix}$$

$$\xrightarrow{-1R_2} \begin{bmatrix} 1 & 2 & -2 & | & -2 \\ 0 & 1 & -6 & | & -7 \\ 0 & -2 & 1 & | & 3 \end{bmatrix}$$

$$\xrightarrow{2R_2+R_3} \begin{bmatrix} 1 & 2 & -2 & | & -2 \\ 0 & 1 & -6 & | & -7 \\ 0 & 0 & -11 & | & -11 \end{bmatrix}$$

$$\xrightarrow{-\frac{1}{11}R_3} \begin{bmatrix} 1 & 2 & -2 & | & -2 \\ 0 & 1 & -6 & | & -7 \\ 0 & 0 & 1 & | & 1 \end{bmatrix}$$

$$\begin{cases} x+2y-2z=-2 \\ y-6z=-7 \\ z=1 \end{cases}$$

$$y-6(1)=-7$$
$$y=-1$$

$$x+2(-1)-2(1)=-2$$
$$x=2$$

The solution is $(2, -1, 1)$.

27. Solve the system of equations using the Gaussian elimination method.

$$\begin{cases} 3x+7y-7z=-4 \\ x+2y-3z=0 \\ 5x+6y+z=-8 \end{cases}$$

$$\begin{bmatrix} 3 & 7 & -7 & | & -4 \\ 1 & 2 & -3 & | & 0 \\ 5 & 6 & 1 & | & -8 \end{bmatrix} \xrightarrow{R_1 \longleftrightarrow R_2} \begin{bmatrix} 1 & 2 & -3 & | & 0 \\ 3 & 7 & -7 & | & -4 \\ 5 & 6 & 1 & | & -8 \end{bmatrix}$$

$$\xrightarrow[\substack{-3R_1+R_2 \\ -5R_1+R_3}]{} \begin{bmatrix} 1 & 2 & -3 & | & 0 \\ 0 & 1 & 2 & | & -4 \\ 0 & -4 & 16 & | & -8 \end{bmatrix}$$

$$\xrightarrow{4R_2+R_3} \begin{bmatrix} 1 & 2 & -3 & | & 0 \\ 0 & 1 & 2 & | & -4 \\ 0 & 0 & 24 & | & -24 \end{bmatrix}$$

$$\xrightarrow{\frac{1}{24}R_3} \begin{bmatrix} 1 & 2 & -3 & | & 0 \\ 0 & 1 & 2 & | & -4 \\ 0 & 0 & 1 & | & -1 \end{bmatrix}$$

$$\begin{cases} x+2y-3z=0 \\ y+2z=-4 \\ z=-1 \end{cases}$$

$$y+2(-1)=-4$$
$$y=-2$$

$$x+2(-2)-3(-1)=0$$
$$x=1$$

The solution is $(1, -2, -1)$.

29. Solve the system of equations using the Gaussian elimination method.

$$\begin{cases} x+2y-2z=3 \\ 5x+8y-6z=14 \\ 3x+4y-2z=8 \end{cases}$$

$$\begin{bmatrix} 1 & 2 & -2 & | & 3 \\ 5 & 8 & -6 & | & 14 \\ 3 & 4 & -2 & | & 8 \end{bmatrix} \xrightarrow[\substack{-5R_1+R_2 \\ -3R_1+R_3}]{} \begin{bmatrix} 1 & 2 & -2 & | & 3 \\ 0 & -2 & 4 & | & -1 \\ 0 & -2 & 4 & | & -1 \end{bmatrix}$$

$$\xrightarrow{-\frac{1}{2}R_2} \begin{bmatrix} 1 & 2 & -2 & | & 3 \\ 0 & 1 & -2 & | & \frac{1}{2} \\ 0 & -2 & 4 & | & -1 \end{bmatrix} \xrightarrow{2R_2+R_3} \begin{bmatrix} 1 & 2 & -2 & | & 3 \\ 0 & 1 & -2 & | & \frac{1}{2} \\ 0 & 0 & 0 & | & 0 \end{bmatrix}$$

$$\begin{cases} x+2y-2z=3 \\ y-2z=\frac{1}{2} \end{cases}$$

$$y=2z+\frac{1}{2}$$

$$x+2\left(2z+\frac{1}{2}\right)-2z=3$$

$$x=2-2z$$

Let z be any real number c.

The solution is $\left(2-2c,\ 2c+\frac{1}{2},\ c\right)$.

31. Solve the system of equations using the Gaussian elimination method.

$$\begin{cases} -x+5y-3z=4 \\ 3x+5y-z=3 \\ x+y=2 \end{cases}$$

$$\begin{bmatrix} -1 & 5 & -3 & | & 4 \\ 3 & 5 & -1 & | & 3 \\ 1 & 1 & 0 & | & 2 \end{bmatrix} \xrightarrow{-1R_1} \begin{bmatrix} 1 & -5 & 3 & | & -4 \\ 3 & 5 & -1 & | & 3 \\ 1 & 1 & 0 & | & 2 \end{bmatrix}$$

$$\xrightarrow[\substack{-3R_1+R_2 \\ -1R_1+R_3}]{} \begin{bmatrix} 1 & -5 & 3 & | & -4 \\ 0 & 20 & -10 & | & 15 \\ 0 & 6 & -3 & | & 6 \end{bmatrix}$$

$$\xrightarrow{\frac{1}{20}R_2} \begin{bmatrix} 1 & -5 & 3 & | & -4 \\ 0 & 1 & -\frac{1}{2} & | & \frac{3}{4} \\ 0 & 6 & -3 & | & 6 \end{bmatrix}$$

$$\xrightarrow{-6R_2+R_3} \begin{bmatrix} 1 & -5 & 3 & | & -4 \\ 0 & 1 & -\frac{1}{2} & | & \frac{3}{4} \\ 0 & 0 & 0 & | & \frac{3}{2} \end{bmatrix}$$

$$\begin{cases} x-5y+3z=-4 \\ y-\frac{1}{2}z=\frac{3}{4} \\ 0z=\frac{3}{2} \end{cases}$$

Because $0z=\frac{3}{2}$ has no solution, the system of equations has no solution.

33. Solve the system of equations using the Gaussian elimination method.

$$\begin{cases} x-3y+2z=0 \\ 2x-5y-2z=0 \\ 4x-11y+2z=0 \end{cases}$$

$$\begin{bmatrix} 1 & -3 & 2 & | & 0 \\ 2 & -5 & -2 & | & 0 \\ 4 & -11 & 2 & | & 0 \end{bmatrix} \xrightarrow[\substack{-2R_1+R_2 \\ -4R_1+R_3}]{} \begin{bmatrix} 1 & -3 & 2 & | & 0 \\ 0 & 1 & -6 & | & 0 \\ 0 & 1 & -6 & | & 0 \end{bmatrix}$$

$$\xrightarrow{-1R_2+R_3} \begin{bmatrix} 1 & -3 & 2 & | & 0 \\ 0 & 1 & -6 & | & 0 \\ 0 & 0 & 0 & | & 0 \end{bmatrix}$$

$$\begin{cases} x-3y+2z=0 \\ y-6z=0 \end{cases}$$

$$y=6z$$

$$x-3(6z)+2z=0$$

$$x=16z$$

Let z be any real number c. The solution is $(16c,\ 6c,\ c)$.

35. Solve the system of equations using the Gaussian elimination method.

$$\begin{cases} 2x+y-3z=4 \\ 3x+2y+z=2 \end{cases}$$

$$\begin{bmatrix} 2 & 1 & -3 & | & 4 \\ 3 & 2 & 1 & | & 2 \end{bmatrix} \xrightarrow{-R_1+R_2} \begin{bmatrix} 2 & 1 & -3 & | & 4 \\ 1 & 1 & 4 & | & -2 \end{bmatrix}$$

$$\xrightarrow{R_1 \longleftrightarrow R_2} \begin{bmatrix} 1 & 1 & 4 & | & -2 \\ 2 & 1 & -3 & | & 4 \end{bmatrix}$$

$$\xrightarrow{-2R_1+R_2} \begin{bmatrix} 1 & 1 & 4 & | & -2 \\ 0 & -1 & -11 & | & 8 \end{bmatrix}$$

$$\xrightarrow{-1R_2} \begin{bmatrix} 1 & 1 & 4 & | & -2 \\ 0 & 1 & 11 & | & -8 \end{bmatrix}$$

$$\begin{cases} x+y+4z=-2 \\ y+11z=-8 \end{cases}$$

$$y=-11z-8$$

$$x+(-11z-8)+4z=-2$$

$$x=7z+6$$

Let z be any real number c. The solution is $(7c+6,\ -11c-8,\ c)$.

37. Solve the system of equations using the Gaussian elimination method.

$$\begin{cases} 2x+2y-4z=4 \\ 2x+3y-5z=4 \\ 4x+5y-9z=8 \end{cases}$$

$$\begin{bmatrix} 2 & 2 & -4 & | & 4 \\ 2 & 3 & -5 & | & 4 \\ 4 & 5 & -9 & | & 8 \end{bmatrix} \xrightarrow{\frac{1}{2}R_1} \begin{bmatrix} 1 & 1 & -2 & | & 2 \\ 2 & 3 & -5 & | & 4 \\ 4 & 5 & -9 & | & 8 \end{bmatrix}$$

$$\xrightarrow[\substack{-2R_1+R_2 \\ -4R_1+R_3}]{} \begin{bmatrix} 1 & 1 & -2 & | & 2 \\ 0 & 1 & -1 & | & 0 \\ 0 & 1 & -1 & | & 0 \end{bmatrix}$$

$$\xrightarrow{-1R_2+R_3} \begin{bmatrix} 1 & 1 & -2 & 2 \\ 0 & 1 & -1 & 0 \\ 0 & 0 & 0 & 0 \end{bmatrix}$$

$$\begin{cases} x+y-2z=2 \\ y-z=0 \end{cases}$$

$y=z$

$x+z-2z=2$

$x=z+2$

Let z be any real number c. The solution is

$(c+2,\ c,\ c)$.

39. Solve the system of equations using the Gaussian elimination method.

$$\begin{cases} x+3y+4z=11 \\ 2x+3y+2z=7 \\ 4x+9y+10z=20 \\ 3x-2y+z=1 \end{cases}$$

$$\begin{bmatrix} 1 & 3 & 4 & 11 \\ 2 & 3 & 2 & 7 \\ 4 & 9 & 10 & 20 \\ 3 & -2 & 1 & 1 \end{bmatrix} \xrightarrow[\substack{-2R_1+R_2 \\ -4R_1+R_3 \\ -3R_1+R_4}]{} \begin{bmatrix} 1 & 3 & 4 & 11 \\ 0 & -3 & -6 & -15 \\ 0 & -3 & -6 & -24 \\ 0 & -11 & -11 & -32 \end{bmatrix}$$

$$\xrightarrow{-\frac{1}{3}R_2} \begin{bmatrix} 1 & 3 & 4 & 11 \\ 0 & 1 & 2 & 5 \\ 0 & -3 & -6 & -24 \\ 0 & -11 & -11 & -32 \end{bmatrix}$$

$$\xrightarrow{3R_2+R_3} \begin{bmatrix} 1 & 3 & 4 & 11 \\ 0 & 1 & 2 & 5 \\ 0 & 0 & 0 & -9 \\ 0 & -11 & -11 & -32 \end{bmatrix}$$

$$\xrightarrow{R_3 \longleftrightarrow R_4} \begin{bmatrix} 1 & 3 & 4 & 11 \\ 0 & 1 & 2 & 5 \\ 0 & -11 & -11 & -32 \\ 0 & 0 & 0 & -9 \end{bmatrix}$$

$$\xrightarrow{-\frac{1}{11}R_3} \begin{bmatrix} 1 & 3 & 4 & 11 \\ 0 & 1 & 2 & 5 \\ 0 & 1 & 1 & \frac{32}{11} \\ 0 & 0 & 0 & -9 \end{bmatrix}$$

$$\xrightarrow{-R_2+R_3} \begin{bmatrix} 1 & 3 & 4 & 11 \\ 0 & 1 & 2 & 5 \\ 0 & 0 & -1 & -\frac{23}{11} \\ 0 & 0 & 0 & -9 \end{bmatrix}$$

$$\xrightarrow{-1R_3} \begin{bmatrix} 1 & 3 & 4 & 11 \\ 0 & 1 & 2 & 5 \\ 0 & 0 & 1 & \frac{23}{11} \\ 0 & 0 & 0 & -9 \end{bmatrix}$$

$$\begin{cases} x+3y+4z=11 \\ y+2z=5 \\ z=\dfrac{23}{11} \\ 0z=-9 \end{cases}$$

Because $0z=-9$ has no solutions, the system of equations has no solution.

41. Solve the system of equations using the Gaussian elimination method.

$$\begin{cases} t+2u-3v+w=-7 \\ 3t+5u-8v+5w=-8 \\ 2t+3u-7v+3w=-11 \\ 4t+8u-10v+7w=-10 \end{cases}$$

$$\begin{bmatrix} 1 & 2 & -3 & 1 & -7 \\ 3 & 5 & -8 & 5 & -8 \\ 2 & 3 & -7 & 3 & -11 \\ 4 & 8 & -10 & 7 & -10 \end{bmatrix}$$

$$\xrightarrow[\substack{-3R_1+R_2 \\ -2R_1+R_3 \\ -4R_1+R_4}]{} \begin{bmatrix} 1 & 2 & -3 & 1 & -7 \\ 0 & -1 & 1 & 2 & 13 \\ 0 & -1 & -1 & 1 & 3 \\ 0 & 0 & 2 & 3 & 18 \end{bmatrix}$$

$$\xrightarrow{-1R_2} \begin{bmatrix} 1 & 2 & -3 & 1 & -7 \\ 0 & 1 & -1 & -2 & -13 \\ 0 & -1 & -1 & 1 & 3 \\ 0 & 0 & 2 & 3 & 18 \end{bmatrix}$$

$$\xrightarrow{R_2+R_3} \begin{bmatrix} 1 & 2 & -3 & 1 & -7 \\ 0 & 1 & -1 & -2 & -13 \\ 0 & 0 & -2 & -1 & -10 \\ 0 & 0 & 2 & 3 & 18 \end{bmatrix}$$

$$\xrightarrow{-\frac{1}{2}R_3} \begin{bmatrix} 1 & 2 & -3 & 1 & | & -7 \\ 0 & 1 & -1 & -2 & | & -13 \\ 0 & 0 & 1 & \frac{1}{2} & | & 5 \\ 0 & 0 & 2 & 3 & | & 18 \end{bmatrix}$$

$$\xrightarrow{-2R_3+R_4} \begin{bmatrix} 1 & 2 & -3 & 1 & | & -7 \\ 0 & 1 & -1 & -2 & | & -13 \\ 0 & 0 & 1 & \frac{1}{2} & | & 5 \\ 0 & 0 & 0 & 2 & | & 8 \end{bmatrix}$$

$$\begin{cases} t+ 2u-3v+ w=-7 \\ u- v- 2w=-13 \\ v+\frac{1}{2}w=5 \\ 2w=8 \end{cases}$$

$$v+\frac{1}{2}(4)=5$$
$$v=3$$
$$u-3-2(4)=-13$$
$$u=-2$$
$$t+2(-2)-3(3)+4=-7$$
$$t=2$$

The solution is $(2, -2, 3, 4)$.

43. Solve the system of equations using the Gaussian elimination method.

$$\begin{cases} 2t-u+3v+2w=2 \\ t-u+2v+w=2 \\ 3t -2v-3w=13 \\ 2t+2u -2w=6 \end{cases}$$

$$\begin{bmatrix} 2 & -1 & 3 & 2 & | & 2 \\ 1 & -1 & 2 & 1 & | & 2 \\ 3 & 0 & -2 & -3 & | & 13 \\ 2 & 2 & 0 & -2 & | & 6 \end{bmatrix} \xrightarrow{\frac{1}{2}R_1} \begin{bmatrix} 1 & -\frac{1}{2} & \frac{3}{2} & 1 & | & 1 \\ 1 & -1 & 2 & 1 & | & 2 \\ 3 & 0 & -2 & -3 & | & 13 \\ 2 & 2 & 0 & -2 & | & 6 \end{bmatrix}$$

$$\xrightarrow[\substack{-1R_1+R_2 \\ -3R_1+R_3 \\ -2R_1+R_4}]{} \begin{bmatrix} 1 & -\frac{1}{2} & \frac{3}{2} & 1 & | & 1 \\ 0 & -\frac{1}{2} & \frac{1}{2} & 0 & | & 1 \\ 0 & \frac{3}{2} & -\frac{13}{2} & -6 & | & 10 \\ 0 & 3 & -3 & -4 & | & 4 \end{bmatrix}$$

$$\xrightarrow{-2R_2} \begin{bmatrix} 1 & -\frac{1}{2} & \frac{3}{2} & 1 & | & 1 \\ 0 & 1 & -1 & 0 & | & -2 \\ 0 & \frac{3}{2} & -\frac{13}{2} & -6 & | & 10 \\ 0 & 3 & -3 & -4 & | & 4 \end{bmatrix}$$

$$\xrightarrow[\substack{-\frac{3}{2}R_2+R_3 \\ -3R_2+R_4}]{} \begin{bmatrix} 1 & -\frac{1}{2} & \frac{3}{2} & 1 & | & 1 \\ 0 & 1 & -1 & 0 & | & -2 \\ 0 & 0 & -5 & -6 & | & 13 \\ 0 & 0 & 0 & -4 & | & 10 \end{bmatrix}$$

$$\xrightarrow{-\frac{1}{5}R_3} \begin{bmatrix} 1 & -\frac{1}{2} & \frac{3}{2} & 1 & | & 1 \\ 0 & 1 & -1 & 0 & | & -2 \\ 0 & 0 & 1 & \frac{6}{5} & | & -\frac{13}{5} \\ 0 & 0 & 0 & -4 & | & 10 \end{bmatrix}$$

$$\xrightarrow{-\frac{1}{4}R_4} \begin{bmatrix} 1 & -\frac{1}{2} & \frac{3}{2} & 1 & | & 1 \\ 0 & 1 & -1 & 0 & | & -2 \\ 0 & 0 & 1 & \frac{6}{5} & | & -\frac{13}{5} \\ 0 & 0 & 0 & 1 & | & -\frac{5}{2} \end{bmatrix}$$

$$\begin{cases} t-\frac{1}{2}u+ \frac{3}{2}v+ w=1 \\ u- v =-2 \\ v+\frac{6}{5}w=-\frac{13}{5} \\ w=-\frac{5}{2} \end{cases}$$

$$v+\frac{6}{5}\left(-\frac{5}{2}\right)=-\frac{13}{5}$$
$$v=\frac{2}{5}$$

$$u-\frac{2}{5}=-2$$
$$u= -\frac{8}{5}$$

$$t-\frac{1}{2}\left(-\frac{8}{5}\right)+\frac{3}{2}\left(\frac{2}{5}\right)-\frac{5}{2}=1$$
$$t=\frac{21}{10}$$

The solution is $\left(\frac{21}{10}, -\frac{8}{5}, \frac{2}{5}, -\frac{5}{2}\right)$.

45. Solve the system of equations using the Gaussian elimination method.

$$\begin{cases} 3t+10u+7v-6w=7 \\ 2t+8u+6v-5w=5 \\ t+4u+2v-3w=2 \\ 4t+14u+9v-8w=8 \end{cases}$$

$$\begin{bmatrix} 3 & 10 & 7 & -6 & | & 7 \\ 2 & 8 & 6 & -5 & | & 5 \\ 1 & 4 & 2 & -3 & | & 2 \\ 4 & 14 & 9 & -8 & | & 8 \end{bmatrix} \xrightarrow{R_1 \longleftrightarrow R_3} \begin{bmatrix} 1 & 4 & 2 & -3 & | & 2 \\ 2 & 8 & 6 & -5 & | & 5 \\ 3 & 10 & 7 & -6 & | & 7 \\ 4 & 14 & 9 & -8 & | & 8 \end{bmatrix}$$

$$\xrightarrow[\substack{-2R_1+R_2 \\ -3R_1+R_3 \\ -4R_1+R_4}]{} \begin{bmatrix} 1 & 4 & 2 & -3 & | & 2 \\ 0 & 0 & 2 & 1 & | & 1 \\ 0 & -2 & 1 & 3 & | & 1 \\ 0 & -2 & 1 & 4 & | & 0 \end{bmatrix}$$

$$\xrightarrow[R_2 \longleftrightarrow R_3]{} \begin{bmatrix} 1 & 4 & 2 & -3 & | & 2 \\ 0 & -2 & 1 & 3 & | & 1 \\ 0 & 0 & 2 & 1 & | & 1 \\ 0 & -2 & 1 & 4 & | & 0 \end{bmatrix}$$

$$\xrightarrow[-\frac{1}{2}R_2]{} \begin{bmatrix} 1 & 4 & 2 & -3 & | & 2 \\ 0 & 1 & -\frac{1}{2} & -\frac{3}{2} & | & -\frac{1}{2} \\ 0 & 0 & 2 & 1 & | & 1 \\ 0 & -2 & 1 & 4 & | & 0 \end{bmatrix}$$

$$\xrightarrow[2R_2+R_4]{} \begin{bmatrix} 1 & 4 & 2 & -3 & | & 2 \\ 0 & 1 & -\frac{1}{2} & -\frac{3}{2} & | & -\frac{1}{2} \\ 0 & 0 & 2 & 1 & | & 1 \\ 0 & 0 & 0 & 1 & | & -1 \end{bmatrix}$$

$$\begin{cases} t + 4u + 2v - 3w = 2 \\ u - \frac{1}{2}v - \frac{3}{2}w = -\frac{1}{2} \\ 2v + w = 1 \\ w = -1 \end{cases}$$

$$2v + (-1) = 1$$
$$v = 1$$

$$u - \frac{1}{2}(1) - \frac{3}{2}(-1) = -\frac{1}{2}$$
$$u = -\frac{3}{2}$$

$$t + 4\left(-\frac{3}{2}\right) + 2(1) - 3(-1) = 2$$
$$t = 3$$

The solution is $\left(3, -\frac{3}{2}, 1, -1\right)$.

47. Solve the system of equations using the Gaussian elimination method.

$$\begin{cases} t - u + 2v - 3w = 9 \\ 4t + 11v - 10w = 46 \\ 3t - u + 8v - 6w = 27 \end{cases}$$

$$\begin{bmatrix} 1 & -1 & 2 & -3 & | & 9 \\ 4 & 0 & 11 & -10 & | & 46 \\ 3 & -1 & 8 & -6 & | & 27 \end{bmatrix}$$

$$\xrightarrow[\substack{-4R_1+R_2 \\ -3R_1+R_3}]{} \begin{bmatrix} 1 & -1 & 2 & -3 & | & 9 \\ 0 & 4 & 3 & 2 & | & 10 \\ 0 & 2 & 2 & 3 & | & 0 \end{bmatrix}$$

$$\xrightarrow[R_2 \longleftrightarrow R_3]{} \begin{bmatrix} 1 & -1 & 2 & -3 & | & 9 \\ 0 & 2 & 2 & 3 & | & 0 \\ 0 & 4 & 3 & 2 & | & 10 \end{bmatrix}$$

$$\xrightarrow[-2R_2+R_3]{} \begin{bmatrix} 1 & -1 & 2 & -3 & | & 9 \\ 0 & 2 & 2 & 3 & | & 0 \\ 0 & 0 & -1 & -4 & | & 10 \end{bmatrix}$$

$$\begin{cases} t - u + 2v - 3w = 9 \\ 2u + 2v + 3w = 0 \\ -v - 4w = 10 \end{cases}$$

$$v = -4w - 10$$

$$2u + 2(-4w - 10) + 3w = 0$$
$$u = \frac{5}{2}w + 10$$

$$t - \left(\frac{5}{2}w + 10\right) + 2(-4w - 10) - 3w = 9$$
$$t = \frac{27}{2}w + 39$$

Let w be any real number c. The solution is

$$\left(\frac{27}{2}c + 39, \ \frac{5}{2}c + 10, \ -4c - 10, \ c\right).$$

49. Solve the system of equations using the Gaussian elimination method.

$$\begin{cases} 3t - 4u + v = 2 \\ t + u - 2v + 3w = 1 \end{cases}$$

$$\begin{bmatrix} 3 & -4 & 1 & 0 & | & 2 \\ 1 & 1 & -2 & 3 & | & 1 \end{bmatrix} \xrightarrow[\frac{1}{3}R_1]{} \begin{bmatrix} 1 & -\frac{4}{3} & \frac{1}{3} & 0 & | & \frac{2}{3} \\ 1 & 1 & -2 & 3 & | & 1 \end{bmatrix}$$

$$\xrightarrow[-R_1+R_2]{} \begin{bmatrix} 1 & -\frac{4}{3} & \frac{1}{3} & 0 & | & \frac{2}{3} \\ 0 & \frac{7}{3} & -\frac{7}{3} & 3 & | & \frac{1}{3} \end{bmatrix}$$

$$\xrightarrow[\frac{3}{7}R_2]{} \begin{bmatrix} 1 & -\frac{4}{3} & \frac{1}{3} & 0 & | & \frac{2}{3} \\ 0 & 1 & -1 & \frac{9}{7} & | & \frac{1}{7} \end{bmatrix}$$

$$\begin{cases} t - \frac{4}{3}u + \frac{1}{3}v = \frac{2}{3} \\ u - v + \frac{9}{7}w = \frac{1}{7} \end{cases}$$

$$u = v - \frac{9}{7}w + \frac{1}{7}$$

$$t - \frac{4}{3}\left(v - \frac{9}{7}w + \frac{1}{7}\right) + \frac{1}{3}v = \frac{2}{3}$$

$$t = v - \frac{12}{7}w + \frac{6}{7}$$

Let v be any real number c_1 and w be any real number c_2. The solution is

$$\left(c_1 - \frac{12}{7}c_2 + \frac{6}{7},\ c_1 - \frac{9}{7}c_2 + \frac{1}{7},\ c_1,\ c_2\right).$$

51. Find the polynomial.

Because there are two points, the degree of the interpolating polynomial is at most 1. The form of the polynomial is $p(x) = a_1 x + a_0$

Use this polynomial and the given points to find the system of equations.

$$p(-2) = a_1(-2) + a_0 = -7$$
$$p(1) = a_1(1) + a_0 = -1$$

The system of equations and the associated augmented matrix are

$$\begin{cases} -2a_1 + a_0 = -7 \\ a_1 + a_0 = -1 \end{cases} \qquad \begin{bmatrix} -2 & 1 & | & -7 \\ 1 & 1 & | & -1 \end{bmatrix}$$

The ref (**row echelon form**) feature of a graphing calculator can be used to rewrite the augmented matrix in echelon form.

Consider using the function of your calculator that converts a decimal to a fraction.

The augmented matrix in echelon form and resulting system of equations are

$$\begin{bmatrix} 1 & -\frac{1}{2} & | & \frac{7}{2} \\ 0 & 1 & | & -3 \end{bmatrix} \qquad \begin{cases} a_1 - \frac{1}{2}a_0 = \frac{7}{2} \\ a_0 = -3 \end{cases}$$

Solving by back substitution yields

$$a_0 = -3 \quad \text{and} \quad a_1 = 2.$$

The interpolating polynomial is $p(x) = 2x - 3$.

53. Find the polynomial.

Because there are three points, the degree of the interpolating polynomial is at most 2.

The form of the polynomial is

$$p(x) = a_2 x^2 + a_1 x + a_0.$$

Use this polynomial and the given points to find the

system of equations.

$$p(-1) = a_2(-1)^2 + a_1(-1) + a_0 = 6$$
$$p(1) = a_2(1)^2 + a_1(1) + a_0 = 2$$
$$p(2) = a_2(2)^2 + a_1(2) + a_0 = 3$$

The system of equations and the associated augmented matrix are

$$\begin{cases} a_2 - a_1 + a_0 = 6 \\ a_2 + a_1 + a_0 = 2 \\ 4a_2 + 2a_1 + a_0 = 3 \end{cases} \qquad \begin{bmatrix} 1 & -1 & 1 & | & 6 \\ 1 & 1 & 1 & | & 2 \\ 4 & 2 & 1 & | & 3 \end{bmatrix}$$

The ref (**row echelon form**) feature of a graphing calculator can be used to rewrite the augmented matrix in echelon form.

Consider using the function of your calculator that converts a decimal to a fraction.

The augmented matrix in echelon form and resulting system of equations are

$$\begin{bmatrix} 1 & \frac{1}{2} & \frac{1}{4} & | & \frac{3}{4} \\ 0 & 1 & -\frac{1}{2} & | & -\frac{7}{2} \\ 0 & 0 & 1 & | & 3 \end{bmatrix} \qquad \begin{cases} a_2 + \frac{1}{2}a_1 + \frac{1}{4}a_0 = \frac{3}{4} \\ a_1 - \frac{1}{2}a_0 = -\frac{7}{2} \\ a_0 = 3 \end{cases}$$

Solving by back substitution yields

$$a_0 = 3, \quad a_1 = -2, \quad \text{and} \quad a_2 = 1.$$

The interpolating polynomial is $p(x) = x^2 - 2x + 3$.

55. Find the polynomial.

Because there are four points, the degree of the interpolating polynomial is at most 3.

The form of the polynomial is

$$p(x) = a_3 x^3 + a_2 x^2 + a_1 x + a_0.$$

Use this polynomial and the given points to find the system of equations.

$$p(-2) = a_3(-2)^3 + a_2(-2)^2 + a_1(-2) + a_0 = -12$$
$$p(0) = a_3(0)^3 + a_2(0)^2 + a_1(0) + a_0 = 2$$
$$p(1) = a_3(1)^3 + a_2(1)^2 + a_1(1) + a_0 = 0$$
$$p(3) = a_3(3)^3 + a_2(3)^2 + a_1(3) + a_0 = 8$$

The system of equations and the associated augmented matrix are

$$\begin{cases} -8a_3 + 4a_2 - 2a_1 + a_0 = -12 \\ \qquad\qquad\qquad\quad a_0 = 2 \\ \;\; a_3 + \;\; a_2 + \;\; a_1 + a_0 = 0 \\ 27a_3 + 9a_2 + 3a_1 + a_0 = 8 \end{cases}$$

$$\begin{bmatrix} -8 & 4 & -2 & 1 & -12 \\ 0 & 0 & 0 & 1 & 2 \\ 1 & 1 & 1 & 1 & 0 \\ 27 & 9 & 3 & 1 & 8 \end{bmatrix}$$

The ref (**row echelon form**) feature of a graphing calculator can be used to rewrite the augmented matrix in echelon form.

Consider using the function of your calculator that converts a decimal to a fraction.

The augmented matrix in echelon form and resulting system of equations are

$$\begin{bmatrix} 1 & \frac{1}{3} & \frac{1}{9} & \frac{1}{27} & \frac{8}{27} \\ 0 & 1 & -\frac{1}{6} & \frac{7}{36} & -\frac{13}{9} \\ 0 & 0 & 1 & \frac{5}{6} & \frac{2}{3} \\ 0 & 0 & 0 & 1 & 2 \end{bmatrix}$$

$$\begin{cases} a_3 + \frac{1}{3}a_2 + \frac{1}{9}a_1 + \frac{1}{27}a_0 = \frac{8}{27} \\ \qquad a_2 - \frac{1}{6}a_1 + \frac{7}{36}a_0 = -\frac{13}{9} \\ \qquad\qquad\quad a_1 + \frac{5}{6}a_0 = \frac{2}{3} \\ \qquad\qquad\qquad\quad a_0 = 2 \end{cases}$$

Solving by back substitution yields

$a_0 = 2$, $a_1 = -1$, $a_2 = -2$ and $a_3 = 1$.

The interpolating polynomial is

$p(x) = x^3 - 2x^2 - x + 2$.

57. Find the polynomial.

Because there are three points, the degree of the interpolating polynomial is at most 2.

The form of the polynomial is

$p(x) = a_2 x^2 + a_1 x + a_0$.

Use this polynomial and the given points to find the system of equations.

$$p(-1) = a_2(-1)^2 + a_1(-1) + a_0 = 3$$
$$p(1) = a_2(1)^2 + a_1(1) + a_0 = 7$$
$$p(2) = a_2(2)^2 + a_1(2) + a_0 = 9$$

The system of equations and the associated augmented matrix are

$$\begin{cases} a_2 - \;\; a_1 + a_0 = 3 \\ a_2 + \;\; a_1 + a_0 = 7 \\ 4a_2 + 2a_1 + a_0 = 9 \end{cases}$$

$$\begin{bmatrix} 1 & -1 & 1 & 3 \\ 1 & 1 & 1 & 7 \\ 4 & 2 & 1 & 9 \end{bmatrix}$$

The ref (**row echelon form**) feature of a graphing calculator can be used to rewrite the augmented matrix in echelon form.

Consider using the function of your calculator that converts a decimal to a fraction.

The augmented matrix in echelon form and resulting system of equations are

$$\begin{bmatrix} 1 & \frac{1}{2} & \frac{1}{4} & \frac{9}{4} \\ 0 & 1 & -\frac{1}{2} & -\frac{1}{2} \\ 0 & 0 & 1 & 5 \end{bmatrix}$$

$$\begin{cases} a_2 + \frac{1}{2}a_1 + \frac{1}{4}a_0 = \frac{9}{4} \\ \qquad a_1 - \frac{1}{2}a_0 = -\frac{1}{2} \\ \qquad\qquad a_0 = 5 \end{cases}$$

Solving by back substitution yields

$a_0 = 5$, $a_1 = 2$, and $a_2 = 0$.

The interpolating polynomial is $p(x) = 2x + 5$.

59. Find the equation of the plane.

The form of the polynomial is: $z = ax + by + c$.

Use this polynomial and the given points to find the system of equations.

$$-4 = a(-1) + b(0) + c$$
$$5 = a(2) + b(1) + c$$
$$-1 = a(-1) + b(1) + c$$

The system of equations and the associated augmented matrix are

$$\begin{cases} -a \qquad\;\; + c = -4 \\ 2a + b + c = 5 \\ -a + b + c = -1 \end{cases}$$

$$\begin{bmatrix} -1 & 0 & 1 & -4 \\ 2 & 1 & 1 & 5 \\ -1 & 1 & 1 & -1 \end{bmatrix}$$

The ref (**row echelon form**) feature of a graphing calculator can be used to rewrite the augmented matrix in echelon form.

Consider using the function of your calculator that converts a decimal to a fraction.

The augmented matrix in echelon form and resulting system of equations are

$$\begin{bmatrix} 1 & \frac{1}{2} & \frac{1}{2} & \frac{5}{2} \\ 0 & 1 & 1 & 1 \\ 0 & 0 & 1 & -2 \end{bmatrix} \qquad \begin{cases} a + \frac{1}{2}b + \frac{1}{2}c = \frac{5}{2} \\ \quad\ \ b + c = 1 \\ \qquad\quad\ c = -2 \end{cases}$$

Solving by back substitution yields

$a = 2$, $b = 3$, and $c = -2$.

The interpolating polynomial is $z = 2x + 3y - 2$.

61. Find the equation of the plane.

The form of the polynomial is: $z = ax + by + c$.

Use this polynomial and the given points to find the system of equations.

$3 = a(4) + b(1) + c$

$3 = a(2) + b(2) + c$

$-3 = a(-2) + b(1) + c$

The system of equations and the associated augmented matrix are

$$\begin{cases} 4a + b + c = 3 \\ 2a + 2b + c = 3 \\ -2a + b + c = -3 \end{cases} \qquad \begin{bmatrix} 4 & 1 & 1 & 3 \\ 2 & 2 & 1 & 3 \\ -2 & 1 & 1 & -3 \end{bmatrix}$$

The ref (**row echelon form**) feature of a graphing calculator can be used to rewrite the augmented matrix in echelon form.

Consider using the function of your calculator that converts a decimal to a fraction.

The augmented matrix in echelon form and resulting system of equations are

$$\begin{bmatrix} 1 & \frac{1}{4} & \frac{1}{4} & \frac{3}{4} \\ 0 & 1 & \frac{1}{3} & 1 \\ 0 & 0 & 1 & -3 \end{bmatrix} \qquad \begin{cases} a + \frac{1}{4}b + \frac{1}{4}c = \frac{3}{4} \\ \quad\ \ b + \frac{1}{3}c = 1 \\ \qquad\quad\ c = -3 \end{cases}$$

Solving by back substitution yields

$a = 1$, $b = 2$, and $c = -3$.

The interpolating polynomial is $z = x + 2y - 3$.

63. Find the equation of the circle.

The form of the polynomial is: $x^2 + y^2 + ax + by = c$

Use this polynomial and the given points to find the system of equations.

$(2)^2 + (6)^2 + a(2) + b(6) = c$

$(-4)^2 + (-2)^2 + a(-4) + b(-2) = c$

$(3)^2 + (-1)^2 + a(3) + b(-1) = c$

The system of equations and the associated augmented matrix are

$$\begin{cases} 2a + 6b - c = -40 \\ -4a - 2b - c = -20 \\ 3a - b - c = -10 \end{cases} \qquad \begin{bmatrix} 2 & 6 & -1 & -40 \\ -4 & -2 & -1 & -20 \\ 3 & -1 & -1 & -10 \end{bmatrix}$$

The ref (**row echelon form**) feature of a graphing calculator can be used to rewrite the augmented matrix in echelon form.

Consider using the function of your calculator that converts a decimal to a fraction.

The augmented matrix in echelon form and resulting system of equations are

$$\begin{bmatrix} 1 & \frac{1}{2} & \frac{1}{4} & 5 \\ 0 & 1 & -\frac{3}{10} & -10 \\ 0 & 0 & 1 & 20 \end{bmatrix} \qquad \begin{cases} a - \frac{1}{2}b + \frac{1}{4}c = 5 \\ \quad\ \ b - \frac{3}{10}c = -10 \\ \qquad\quad\ c = 20 \end{cases}$$

Solving by back substitution yields

$a = 2$, $b = -4$, and $c = 20$.

The interpolating polynomial is $x^2 + y^2 + 2x - 4y = 20$.

65. Use a graphing calculator to solve.

$$\begin{bmatrix} 1 & 2 & -1 & 2 & 3 & 11 \\ 1 & -1 & 2 & -1 & 2 & 0 \\ 2 & 1 & -1 & 2 & -1 & 4 \\ 3 & 2 & -1 & 1 & -2 & 2 \\ 2 & 1 & -1 & -2 & 1 & 4 \end{bmatrix}$$

$$\longrightarrow \begin{bmatrix} 1 & 2 & -1 & 2 & 3 & 11 \\ 0 & 1 & -1 & 1 & \frac{1}{3} & \frac{11}{3} \\ 0 & 0 & 1 & -\frac{1}{2} & 3 & \frac{7}{2} \\ 0 & 0 & 0 & 1 & \frac{11}{6} & \frac{14}{3} \\ 0 & 0 & 0 & 0 & 1 & 2 \end{bmatrix}$$

$$\begin{cases} x_1 + 2x_2 - x_3 + 2x_4 + 3x_5 = 11 \\ x_2 - x_3 + x_4 + \frac{1}{3}x_5 = \frac{11}{3} \\ x_3 - \frac{1}{2}x_4 + 3x_5 = \frac{7}{2} \\ x_4 + \frac{11}{6}x_5 = \frac{14}{3} \\ x_5 = 2 \end{cases}$$

The solution is $(1, 0, -2, 1, 2)$.

67. Use a graphing calculator to solve.

$$\begin{bmatrix} 1 & 2 & -3 & -1 & 2 & | & -10 \\ -1 & -3 & 1 & 1 & -1 & | & 4 \\ 2 & 3 & -5 & 2 & 3 & | & -20 \\ 3 & 4 & -7 & 3 & -2 & | & -16 \\ 2 & 1 & -6 & 4 & -3 & | & -12 \end{bmatrix}$$

$$\longrightarrow \begin{bmatrix} 1 & 2 & -3 & -1 & 2 & | & -10 \\ 0 & 1 & 2 & 0 & -1 & | & 6 \\ 0 & 0 & 1 & \frac{4}{3} & -\frac{2}{3} & | & 2 \\ 0 & 0 & 0 & 1 & 3 & | & -7 \\ 0 & 0 & 0 & 0 & 0 & | & 0 \end{bmatrix}$$

$$\begin{cases} x_1 + 2x_2 - 3x_3 - x_4 + 2x_5 = -10 \\ \quad x_2 + 2x_3 \quad - x_5 = 6 \\ \quad\quad x_3 + \frac{4}{3}x_4 - \frac{2}{3}x_5 = 2 \\ \quad\quad\quad x_4 + 3x_5 = -7 \end{cases}$$

$$x_4 = -3x_5 - 7$$

$$x_3 + \frac{4}{3}(-3x_5 - 7) - \frac{2}{3}x_5 = 2$$

$$x_3 = \frac{14}{3}x_5 + \frac{34}{3}$$

$$x_2 + 2\left(\frac{14}{3}x_5 + \frac{34}{3}\right) - x_5 = 6$$

$$x_2 = -\frac{25}{3}x_5 - \frac{50}{3}$$

$$x_1 + 2\left(-\frac{25}{3}x_5 - \frac{50}{3}\right) - 3\left(\frac{14}{3}x_5 + \frac{34}{3}\right) - (-3x_5 - 7) + 2x_5 = -10$$

$$x_1 = \frac{77}{3}x_5 + \frac{151}{3}$$

Let x_5 be any real number c. The solution is

$$\left(\frac{77c + 151}{3}, \frac{-25c - 50}{3}, \frac{14c + 34}{3}, -3c - 7, c\right).$$

69. Find the values for a for a unique solution.

$$\begin{bmatrix} 1 & 3 & -a^2 & | & a^2 \\ 3 & 4 & 2 & | & 3 \\ 2 & 3 & a & | & 2 \end{bmatrix}$$

$$\xrightarrow[\substack{-3R_1 + R_2 \\ -2R_1 + R_3}]{} \begin{bmatrix} 1 & 3 & -a^2 & | & a^2 \\ 0 & -5 & 3a^2 + 2 & | & -3a^2 + 3 \\ 0 & -3 & 2a^2 + a & | & -2a^2 + 2 \end{bmatrix}$$

$$\xrightarrow[\substack{-3R_2 \\ 5R_3}]{} \begin{bmatrix} 1 & 3 & -a^2 & | & a^2 \\ 0 & 15 & -9a^2 - 6 & | & 9a^2 - 9 \\ 0 & -15 & 10a^2 + 5a & | & -10a^2 + 10 \end{bmatrix}$$

$$\xrightarrow[R_2 + R_3]{} \begin{bmatrix} 1 & 3 & -a^2 & | & a^2 \\ 0 & 15 & -9a^2 & -6 & | & 9a^2 - 9 \\ 0 & 0 & a^2 + 5a - 6 & | & -a^2 + 1 \end{bmatrix}$$

$$\begin{cases} x + 3y - a^2 z = a^2 \\ 15y - (9a^2 + 6)z = 9a^2 - 9 \\ (a^2 + 5a - 6)z = -a^2 + 1 \end{cases}$$

For the system of equations to have a unique solution, $a^2 + 5a - 6$ cannot be zero. Thus $a^2 + 5a - 6 \neq 0$, or $a \neq 1$ and $a \neq -6$.

The system of equations has a unique solution for all values of a except 1 and –6.

71. Find the values for a for no solution.

See the solution to exercise 69. For the system of equations to have no solution, $a^2 + 5a - 6$ must be zero and $-a^2 + 1$ must not equal zero. Thus

$$\begin{array}{ll} a^2 + 5a - 6 = 0 \quad \text{and} & -a^2 + 1 \neq 0 \\ (a + 6)(a - 1) = 0 & a^2 \neq 1 \\ a = -6 \text{ or } a = 1 & a \neq 1 \text{ or } a \neq -1 \end{array}$$

The system of equations will have no solution when $a = -6$.

Prepare for Section 10.2

P1. State the additive identity for real numbers.

0

P3. State the multiplicative identity for real numbers.

1

P5. State the order of the matrix.

3×1

Section 10.2 Exercises

1. Decide which statements are true.

(i) $A(BC) = (AB)C$ and (iii) $A(B + C) = AB + AC$ are true statements.

3. Write the 3×3 multiplicative identity matrix.

$$\begin{bmatrix} 1 & 0 & 0 \\ 0 & 1 & 0 \\ 0 & 0 & 1 \end{bmatrix}$$

5. Find the product.

$$\begin{bmatrix} 2 & -4 & 1 \end{bmatrix} \begin{bmatrix} 3 \\ 2 \\ -5 \end{bmatrix} = \begin{bmatrix} 2(3) - 4(2) + 1(-5) \end{bmatrix} = \begin{bmatrix} -7 \end{bmatrix}$$

7. Find $A + B, A - B, 2B,$ and $2A - 3B$.

a. $A + B = \begin{bmatrix} 2 & -1 \\ 3 & 3 \end{bmatrix} + \begin{bmatrix} -1 & 3 \\ 2 & 1 \end{bmatrix} = \begin{bmatrix} 1 & 2 \\ 5 & 4 \end{bmatrix}$

b. $A - B = \begin{bmatrix} 2 & -1 \\ 3 & 3 \end{bmatrix} - \begin{bmatrix} -1 & 3 \\ 2 & 1 \end{bmatrix} = \begin{bmatrix} 3 & -4 \\ 1 & 2 \end{bmatrix}$

c. $2B = 2 \begin{bmatrix} -1 & 3 \\ 2 & 1 \end{bmatrix} = \begin{bmatrix} -2 & 6 \\ 4 & 2 \end{bmatrix}$

d. $2A - 3B = 2 \begin{bmatrix} 2 & -1 \\ 3 & 3 \end{bmatrix} - 3 \begin{bmatrix} -1 & 3 \\ 2 & 1 \end{bmatrix} = \begin{bmatrix} 4 & -2 \\ 6 & 6 \end{bmatrix} - \begin{bmatrix} -3 & 9 \\ 6 & 3 \end{bmatrix}$

$$= \begin{bmatrix} 7 & -11 \\ 0 & 3 \end{bmatrix}$$

9. Find $A + B, A - B, 2B,$ and $2A - 3B$.

a. $A + B = \begin{bmatrix} 0 & -1 & 3 \\ 1 & 0 & -2 \end{bmatrix} + \begin{bmatrix} -3 & 1 & 2 \\ 2 & 5 & -3 \end{bmatrix} = \begin{bmatrix} -3 & 0 & 5 \\ 3 & 5 & -5 \end{bmatrix}$

b. $A - B = \begin{bmatrix} 0 & -1 & 3 \\ 1 & 0 & -2 \end{bmatrix} - \begin{bmatrix} -3 & 1 & 2 \\ 2 & 5 & -3 \end{bmatrix} = \begin{bmatrix} 3 & -2 & 1 \\ -1 & -5 & 1 \end{bmatrix}$

c. $2B = 2 \begin{bmatrix} -3 & 1 & 2 \\ 2 & 5 & -3 \end{bmatrix} = \begin{bmatrix} -6 & 2 & 4 \\ 4 & 10 & -6 \end{bmatrix}$

d. $2A - 3B = 2 \begin{bmatrix} 0 & -1 & 3 \\ 1 & 0 & -2 \end{bmatrix} - 3 \begin{bmatrix} -3 & 1 & 2 \\ 2 & 5 & -3 \end{bmatrix}$

$$= \begin{bmatrix} 0 & -2 & 6 \\ 2 & 0 & -4 \end{bmatrix} - \begin{bmatrix} -9 & 3 & 6 \\ 6 & 15 & -9 \end{bmatrix}$$

$$= \begin{bmatrix} 9 & -5 & 0 \\ -4 & -15 & 5 \end{bmatrix}$$

11. Find $A + B, A - B, 2B,$ and $2A - 3B$.

a. $A + B = \begin{bmatrix} -3 & 4 \\ 2 & -3 \\ -1 & 0 \end{bmatrix} + \begin{bmatrix} 4 & 1 \\ 1 & -2 \\ 3 & -4 \end{bmatrix} = \begin{bmatrix} 1 & 5 \\ 3 & -5 \\ 2 & -4 \end{bmatrix}$

b. $A - B = \begin{bmatrix} -3 & 4 \\ 2 & -3 \\ -1 & 0 \end{bmatrix} - \begin{bmatrix} 4 & 1 \\ 1 & -2 \\ 3 & -4 \end{bmatrix} = \begin{bmatrix} -7 & 3 \\ 1 & -1 \\ -4 & 4 \end{bmatrix}$

c. $2B = 2 \begin{bmatrix} 4 & 1 \\ 1 & -2 \\ 3 & -4 \end{bmatrix} = \begin{bmatrix} 8 & 2 \\ 2 & -4 \\ 6 & -8 \end{bmatrix}$

d. $2A - 3B = 2 \begin{bmatrix} -3 & 4 \\ 2 & -3 \\ -1 & 0 \end{bmatrix} - 3 \begin{bmatrix} 4 & 1 \\ 1 & -2 \\ 3 & -4 \end{bmatrix}$

$$= \begin{bmatrix} -6 & 8 \\ 4 & -6 \\ -2 & 0 \end{bmatrix} - \begin{bmatrix} 12 & 3 \\ 3 & -6 \\ 9 & -12 \end{bmatrix}$$

$$= \begin{bmatrix} -18 & 5 \\ 1 & 0 \\ -11 & 12 \end{bmatrix}$$

13. Find $A + B, A - B, 2B,$ and $2A - 3B$.

a. $A + B = \begin{bmatrix} -2 & 3 & -1 \\ 0 & -1 & 2 \\ -4 & 3 & 3 \end{bmatrix} + \begin{bmatrix} 1 & -2 & 0 \\ 2 & 3 & -1 \\ 3 & -1 & 2 \end{bmatrix}$

$$= \begin{bmatrix} -1 & 1 & -1 \\ 2 & 2 & 1 \\ -1 & 2 & 5 \end{bmatrix}$$

b. $A - B = \begin{bmatrix} -2 & 3 & -1 \\ 0 & -1 & 2 \\ -4 & 3 & 3 \end{bmatrix} - \begin{bmatrix} 1 & -2 & 0 \\ 2 & 3 & -1 \\ 3 & -1 & 2 \end{bmatrix} = \begin{bmatrix} -3 & 5 & -1 \\ -2 & -4 & 3 \\ -7 & 4 & 1 \end{bmatrix}$

c. $2B = 2 \begin{bmatrix} 1 & -2 & 0 \\ 2 & 3 & -1 \\ 3 & -1 & 2 \end{bmatrix} = \begin{bmatrix} 2 & -4 & 0 \\ 4 & 6 & -2 \\ 6 & -2 & 4 \end{bmatrix}$

d. $2A - 3B = 2 \begin{bmatrix} -2 & 3 & -1 \\ 0 & -1 & 2 \\ -4 & 3 & 3 \end{bmatrix} - 3 \begin{bmatrix} 1 & -2 & 0 \\ 2 & 3 & -1 \\ 3 & -1 & 2 \end{bmatrix}$

$$= \begin{bmatrix} -4 & 6 & -2 \\ 0 & -2 & 4 \\ -8 & 6 & 6 \end{bmatrix} - \begin{bmatrix} 3 & -6 & 0 \\ 6 & 9 & -3 \\ 9 & -3 & 6 \end{bmatrix}$$

$$= \begin{bmatrix} -7 & 12 & -2 \\ -6 & -11 & 7 \\ -17 & 9 & 0 \end{bmatrix}$$

15. Find AB and BA, if possible.

$$AB = \begin{bmatrix} 2 & -3 \\ 1 & 4 \end{bmatrix} \begin{bmatrix} -2 & 4 \\ 2 & -3 \end{bmatrix}$$

$$= \begin{bmatrix} (2)(-2)+(-3)(2) & (2)(4)+(-3)(-3) \\ (1)(-2)+(4)(2) & (1)(4)+(4)(-3) \end{bmatrix}$$

$$= \begin{bmatrix} -10 & 17 \\ 6 & -8 \end{bmatrix}$$

$$BA = \begin{bmatrix} -2 & 4 \\ 2 & -3 \end{bmatrix} \begin{bmatrix} 2 & -3 \\ 1 & 4 \end{bmatrix}$$

$$= \begin{bmatrix} (-2)(2)+(4)(1) & (-2)(-3)+(4)(4) \\ (2)(2)+(-3)(1) & (2)(-3)+(-3)(4) \end{bmatrix}$$

$$= \begin{bmatrix} 0 & 22 \\ 1 & -18 \end{bmatrix}$$

17. Find AB and BA, if possible.

$$AB = \begin{bmatrix} 3 & -1 \\ 2 & 3 \end{bmatrix} \begin{bmatrix} 4 & 1 \\ 2 & -3 \end{bmatrix}$$

$$= \begin{bmatrix} (3)(4)+(-1)(2) & (3)(1)+(-1)(-3) \\ (2)(4)+(3)(2) & (2)(1)+(3)(-3) \end{bmatrix}$$

$$= \begin{bmatrix} 10 & 6 \\ 14 & -7 \end{bmatrix}$$

$$BA = \begin{bmatrix} 4 & 1 \\ 2 & -3 \end{bmatrix} \begin{bmatrix} 3 & -1 \\ 2 & -3 \end{bmatrix}$$

$$= \begin{bmatrix} (4)(3)+(1)(2) & (4)(-1)+(1)(3) \\ (2)(3)+(-3)(2) & (2)(-1)+(-3)(3) \end{bmatrix}$$

$$= \begin{bmatrix} 14 & -1 \\ 0 & -11 \end{bmatrix}$$

19. Find AB and BA, if possible.

$$AB = \begin{bmatrix} 2 & -1 \\ 0 & 3 \\ 1 & -2 \end{bmatrix} \begin{bmatrix} 1 & -2 & 3 \\ 2 & 0 & 1 \end{bmatrix}$$

$$= \begin{bmatrix} 2(1)+(-1)(2) & 2(-2)+(-1)(0) & 2(3)+(-1)(1) \\ (0)(1)+(3)(2) & 0(-2)+(3)(0) & 0(3)+(3)(1) \\ 1(1)+(-2)(2) & 1(-2)+(-2)(0) & 1(3)+(-2)(1) \end{bmatrix}$$

$$= \begin{bmatrix} 0 & -4 & 5 \\ 6 & 0 & 3 \\ -3 & -2 & 1 \end{bmatrix}$$

$$BA = \begin{bmatrix} 1 & -2 & 3 \\ 2 & 0 & 1 \end{bmatrix} \begin{bmatrix} 2 & -1 \\ 0 & 3 \\ 1 & -2 \end{bmatrix}$$

$$= \begin{bmatrix} 1(2)+(-2)(0)+3(1) & 1(-1)+(-2)(3)+3(-2) \\ 2(2)+(0)(0)+1(1) & 2(-1)+(0)(3)+(1)(-2) \end{bmatrix}$$

$$= \begin{bmatrix} 5 & -13 \\ 5 & -4 \end{bmatrix}$$

21. Find AB and BA, if possible.

$$AB = \begin{bmatrix} 2 & -1 & 3 \\ 0 & 2 & -1 \\ 0 & 0 & 2 \end{bmatrix} \begin{bmatrix} 2 & 0 & 0 \\ 1 & -1 & 0 \\ 2 & -1 & -2 \end{bmatrix}$$

$$= \begin{bmatrix} 4+(-1)+6 & 0+1+(-3) & 0+0+(-6) \\ 0+2+(-2) & 0+(-2)+1 & 0+0+2 \\ 0+0+4 & 0+0+(-2) & 0+0(0)+(-4) \end{bmatrix}$$

$$= \begin{bmatrix} 9 & -2 & -6 \\ 0 & -1 & 2 \\ 4 & -2 & -4 \end{bmatrix}$$

$$BA = \begin{bmatrix} 2 & 0 & 0 \\ 1 & -1 & 0 \\ 2 & -1 & -2 \end{bmatrix} \begin{bmatrix} 2 & -1 & 3 \\ 0 & 2 & -1 \\ 0 & 0 & 2 \end{bmatrix}$$

$$= \begin{bmatrix} 4+0+0 & -2+0+0 & 6+0+0 \\ 2+0+0 & -1+(-2)+0 & 3+1+0 \\ 4+0+0 & -2+(-2)+0 & 6+1+(-4) \end{bmatrix}$$

$$= \begin{bmatrix} 4 & -2 & 6 \\ 2 & -3 & 4 \\ 4 & -4 & 3 \end{bmatrix}$$

23. Find AB, if possible.

$$AB = \begin{bmatrix} 1 & -2 & 3 \end{bmatrix} \begin{bmatrix} 1 & 0 \\ 2 & -1 \\ 1 & 2 \end{bmatrix}$$

$$= \begin{bmatrix} 1(1)+(-2)(2)+3(1) & 1(0)+(-2)(-1)+3(2) \end{bmatrix}$$

$$= \begin{bmatrix} 0 & 8 \end{bmatrix}$$

25. Find AB, if possible.

The number of columns of the first matrix is not equal to the number of rows of the second matrix. The product is not possible.

27. Find AB, if possible.

$$AB = \begin{bmatrix} 2 & 3 \\ -4 & -6 \end{bmatrix} \begin{bmatrix} 3 & 6 \\ -2 & -4 \end{bmatrix}$$

$$= \begin{bmatrix} (2)(3)+(3)(-2) & (2)(6)+(3)(-4) \\ (-4)(3)+(-6)(-2) & (-4)(6)+(-6)(-4) \end{bmatrix}$$

$$= \begin{bmatrix} 0 & 0 \\ 0 & 0 \end{bmatrix}$$

29. Find AB, if possible.

The number of columns of the first matrix is not equal to the number of rows of the second matrix. The product is not possible.

31. Find the matrix X.

$$3X + A = B$$

$$3X + \begin{bmatrix} -1 & 3 \\ 2 & -1 \\ 3 & 1 \end{bmatrix} = \begin{bmatrix} 0 & -2 \\ 1 & 3 \\ 4 & -3 \end{bmatrix}$$

$$3X = \begin{bmatrix} 0 & -2 \\ 1 & 3 \\ 4 & -3 \end{bmatrix} - \begin{bmatrix} -1 & 3 \\ 2 & -1 \\ 3 & 1 \end{bmatrix}$$

$$3X = \begin{bmatrix} 1 & -5 \\ -1 & 4 \\ 1 & -4 \end{bmatrix}$$

$$X = \begin{bmatrix} \frac{1}{3} & -\frac{5}{3} \\ -\frac{1}{3} & \frac{4}{3} \\ \frac{1}{3} & -\frac{4}{3} \end{bmatrix}$$

33. Find the matrix X.

$$2X - A = X + B$$

$$2X - \begin{bmatrix} -1 & 3 \\ 2 & -1 \\ 3 & 1 \end{bmatrix} = X + \begin{bmatrix} 0 & -2 \\ 1 & 3 \\ 4 & -3 \end{bmatrix}$$

$$X - \begin{bmatrix} -1 & 3 \\ 2 & -1 \\ 3 & 1 \end{bmatrix} = \begin{bmatrix} 0 & -2 \\ 1 & 3 \\ 4 & -3 \end{bmatrix}$$

$$X = \begin{bmatrix} 0 & -2 \\ 1 & 3 \\ 4 & -3 \end{bmatrix} + \begin{bmatrix} -1 & 3 \\ 2 & -1 \\ 3 & 1 \end{bmatrix}$$

$$X = \begin{bmatrix} -1 & 1 \\ 3 & 2 \\ 7 & -2 \end{bmatrix}$$

35. Find A^2.

$$A^2 = A \cdot A = \begin{bmatrix} 2 & -3 \\ 1 & -1 \end{bmatrix} \begin{bmatrix} 2 & -3 \\ 1 & -1 \end{bmatrix}$$

$$= \begin{bmatrix} 2(2)+(-3)(1) & 2(-3)+(-3)(-1) \\ 1(2)+(-1)(1) & 1(-3)+(-1)(-1) \end{bmatrix} = \begin{bmatrix} 1 & -3 \\ 1 & -2 \end{bmatrix}$$

37. Find B^2.

$$B^2 = B \cdot B = \begin{bmatrix} 3 & -1 & 0 \\ 2 & -2 & -1 \\ 1 & 0 & 2 \end{bmatrix} \begin{bmatrix} 3 & -1 & 0 \\ 2 & -2 & -1 \\ 1 & 0 & 2 \end{bmatrix}$$

$$= \begin{bmatrix} 9+(-2)+0 & -3+2+0 & 0+1+0 \\ 6+(-4)+(-1) & -2+4+0 & 0+2+(-2) \\ 3+0+2 & -1+0+0) & 0+0+4 \end{bmatrix}$$

$$= \begin{bmatrix} 7 & -1 & 1 \\ 1 & 2 & 0 \\ 5 & -1 & 4 \end{bmatrix}$$

39. a. 3×4. There are three different fish in four different samples.

b. Fish A was caught in sample number 4.

c. Fish B. There are more 1's in this row than in any other row.

41. a. $H + A = \begin{bmatrix} 13 & 4 \\ 15 & 2 \\ 11 & 6 \end{bmatrix} + \begin{bmatrix} 8 & 9 \\ 6 & 11 \\ 9 & 8 \end{bmatrix} = \begin{bmatrix} 21 & 813 \\ 21 & 13 \\ 20 & 14 \end{bmatrix}$

The matrix represents the total number of wins and losses for each team for the season.

b. $H - A = \begin{bmatrix} 13 & 4 \\ 15 & 2 \\ 11 & 6 \end{bmatrix} - \begin{bmatrix} 8 & 9 \\ 6 & 11 \\ 9 & 8 \end{bmatrix} = \begin{bmatrix} 5 & -5 \\ 9 & -9 \\ 2 & -2 \end{bmatrix}$

The matrix represents the difference between performances at home and performances away.

43. Find and interpret $A - B$.

$$A = \begin{bmatrix} 530 & 650 & 815 \\ 190 & 385 & 715 \\ 485 & 600 & 610 \\ 150 & 210 & 305 \end{bmatrix}, \quad B = \begin{bmatrix} 480 & 500 & 675 \\ 175 & 215 & 345 \\ 400 & 350 & 480 \\ 70 & 95 & 280 \end{bmatrix}$$

$$A - B = \begin{bmatrix} 50 & 150 & 140 \\ 15 & 170 & 370 \\ 85 & 250 & 130 \\ 80 & 115 & 25 \end{bmatrix}$$

$A - B$ is number sold of each item during the week.

45. Find CS and determine which company should be used.

$$C = \begin{bmatrix} 0.04 & 0.06 & 0.05 \\ 0.04 & 0.04 & 0.04 \\ 0.03 & 0.07 & 0.06 \end{bmatrix}, \ S = \begin{bmatrix} 500 & 600 \\ 250 & 450 \\ 600 & 750 \end{bmatrix}$$

$$CS = \begin{bmatrix} 0.04 & 0.06 & 0.05 \\ 0.04 & 0.04 & 0.04 \\ 0.03 & 0.07 & 0.06 \end{bmatrix}\begin{bmatrix} 500 & 600 \\ 250 & 450 \\ 600 & 750 \end{bmatrix}$$

$$= \begin{bmatrix} 65 & 88.5 \\ 54 & 72 \\ 68.5 & 94.5 \end{bmatrix}$$

To minimize commissions costs, customer S_1 should use company T_2.

47. Find the transformation.

$$R_x \cdot \begin{bmatrix} 2 & -3 \\ 5 & 6 \\ 1 & 1 \end{bmatrix} = \begin{bmatrix} 1 & 0 & 0 \\ 0 & -1 & 0 \\ 0 & 0 & 1 \end{bmatrix}\begin{bmatrix} 2 & -3 \\ 5 & 6 \\ 1 & 1 \end{bmatrix}$$

$$= \begin{bmatrix} 2 & -3 \\ -5 & -6 \\ 1 & 1 \end{bmatrix} \Rightarrow \begin{matrix} P'(2,-5) \\ Q'(-3,-6) \end{matrix}$$

49. Find the transformation.

$$R_{xy} \cdot \begin{bmatrix} -3 & -5 \\ 1 & 3 \\ 1 & 1 \end{bmatrix} = \begin{bmatrix} 0 & 1 & 0 \\ 1 & 0 & 0 \\ 0 & 0 & 1 \end{bmatrix}\begin{bmatrix} -3 & -5 \\ 1 & 3 \\ 1 & 1 \end{bmatrix}$$

$$= \begin{bmatrix} 1 & 3 \\ -3 & -5 \\ 1 & 1 \end{bmatrix} \Rightarrow \begin{matrix} P'(1,-3) \\ Q'(3,-5) \end{matrix}$$

51. Find the transformation.

$$R_{xy} \cdot T_{3,-1} \cdot \begin{bmatrix} -1 & 1 & 3 \\ 5 & -2 & 4 \\ 1 & 1 & 1 \end{bmatrix}$$

$$= \begin{bmatrix} 0 & 1 & 0 \\ 1 & 0 & 0 \\ 0 & 0 & 1 \end{bmatrix}\begin{bmatrix} 1 & 0 & 3 \\ 0 & 1 & -1 \\ 0 & 0 & 1 \end{bmatrix}\begin{bmatrix} -1 & 1 & 3 \\ 5 & -2 & 4 \\ 1 & 1 & 1 \end{bmatrix}$$

$$= \begin{bmatrix} 0 & 1 & 0 \\ 1 & 0 & 0 \\ 0 & 0 & 1 \end{bmatrix}\begin{bmatrix} 2 & 4 & 6 \\ 4 & -3 & 3 \\ 1 & 1 & 1 \end{bmatrix}$$

$$= \begin{bmatrix} 4 & -3 & 3 \\ 2 & 4 & 6 \\ 1 & 1 & 1 \end{bmatrix} \Rightarrow \begin{matrix} A'(4,2) \\ B'(-3,4) \\ C'(3,6) \end{matrix}$$

53. Find the transformation.

$$T_{1,-6} \cdot \begin{bmatrix} -1 & -1 & 4 & 4 \\ 2 & 6 & 6 & 2 \\ 1 & 1 & 1 & 1 \end{bmatrix} = \begin{bmatrix} 1 & 0 & 1 \\ 0 & 1 & -6 \\ 0 & 0 & 1 \end{bmatrix}\begin{bmatrix} -1 & -1 & 4 & 4 \\ 2 & 6 & 6 & 2 \\ 1 & 1 & 1 & 1 \end{bmatrix}$$

$$= \begin{bmatrix} 0 & 0 & 5 & 5 \\ -4 & 0 & 0 & -4 \\ 1 & 1 & 1 & 1 \end{bmatrix}$$

$$R_{180} \cdot \begin{bmatrix} 0 & 0 & 5 & 5 \\ -4 & 0 & 0 & -4 \\ 1 & 1 & 1 & 1 \end{bmatrix} = \begin{bmatrix} -1 & 0 & 0 \\ 0 & -1 & 0 \\ 0 & 0 & 1 \end{bmatrix}\begin{bmatrix} 0 & 0 & 5 & 5 \\ -4 & 0 & 0 & -4 \\ 1 & 1 & 1 & 1 \end{bmatrix}$$

$$= \begin{bmatrix} 0 & 0 & -5 & -5 \\ 4 & 0 & 0 & 4 \\ 1 & 1 & 1 & 1 \end{bmatrix}$$

$$T_{-1,6} \cdot \begin{bmatrix} 0 & 0 & -5 & -5 \\ 4 & 0 & 0 & 4 \\ 1 & 1 & 1 & 1 \end{bmatrix}$$

$$= \begin{bmatrix} 1 & 0 & -1 \\ 0 & 1 & 6 \\ 0 & 0 & 1 \end{bmatrix}\begin{bmatrix} 0 & 0 & -5 & -5 \\ 4 & 0 & 0 & 4 \\ 1 & 1 & 1 & 1 \end{bmatrix}$$

$$= \begin{bmatrix} -1 & -1 & -6 & -6 \\ 10 & 6 & 6 & 10 \\ 1 & 1 & 1 & 1 \end{bmatrix} \Rightarrow \begin{matrix} A'(-1,10) & C'(-6,6) \\ B'(-1,6) & D'(-6,10) \end{matrix}$$

55. a. Find the adjacency matrix.

$$A = \begin{bmatrix} 0 & 1 & 1 & 0 \\ 1 & 0 & 1 & 1 \\ 1 & 1 & 0 & 1 \\ 0 & 1 & 1 & 0 \end{bmatrix}$$

b. Find A^2 and the number of walks.

$$A^2 = \begin{bmatrix} 0 & 1 & 1 & 0 \\ 1 & 0 & 1 & 1 \\ 1 & 1 & 0 & 1 \\ 0 & 1 & 1 & 0 \end{bmatrix}\begin{bmatrix} 0 & 1 & 1 & 0 \\ 1 & 0 & 1 & 1 \\ 1 & 1 & 0 & 1 \\ 0 & 1 & 1 & 0 \end{bmatrix} = \begin{bmatrix} 2 & 1 & 1 & 2 \\ 1 & 3 & 2 & 1 \\ 1 & 2 & 3 & 1 \\ 2 & 1 & 1 & 2 \end{bmatrix}$$

A^2 gives the number of walks of length 2.

There are 2 walks of length 2 between vertex 4 and 1.

57. a. Find the adjacency matrix.

$$A = \begin{bmatrix} 0 & 1 & 1 & 0 \\ 1 & 0 & 1 & 1 \\ 1 & 1 & 0 & 0 \\ 0 & 1 & 0 & 0 \end{bmatrix}$$

b. Find A^3 and the number of walks.

$$A^3 = \begin{bmatrix} 0 & 1 & 1 & 0 \\ 1 & 0 & 1 & 1 \\ 1 & 1 & 0 & 0 \\ 0 & 1 & 0 & 0 \end{bmatrix}\begin{bmatrix} 0 & 1 & 1 & 0 \\ 1 & 0 & 1 & 1 \\ 1 & 1 & 0 & 0 \\ 0 & 1 & 0 & 0 \end{bmatrix}\begin{bmatrix} 0 & 1 & 1 & 0 \\ 1 & 0 & 1 & 1 \\ 1 & 1 & 0 & 0 \\ 0 & 1 & 0 & 0 \end{bmatrix}$$

$$= \begin{bmatrix} 2 & 4 & 3 & 1 \\ 4 & 2 & 4 & 3 \\ 3 & 4 & 2 & 1 \\ 1 & 3 & 1 & 0 \end{bmatrix}$$

A^3 gives the number of walks of length 3.

There are 3 walks of length 3 between vertex 2 and 4.

59. a. Find the adjacency matrix.

$$A = \begin{bmatrix} 0 & 1 & 1 & 0 & 0 \\ 1 & 0 & 1 & 1 & 1 \\ 1 & 1 & 0 & 1 & 1 \\ 0 & 1 & 1 & 0 & 0 \\ 0 & 1 & 1 & 0 & 0 \end{bmatrix}$$

b. Find A^3 and the number of walks.

$$A^3 = \begin{bmatrix} 0 & 1 & 1 & 0 & 0 \\ 1 & 0 & 1 & 1 & 1 \\ 1 & 1 & 0 & 1 & 1 \\ 0 & 1 & 1 & 0 & 0 \\ 0 & 1 & 1 & 0 & 0 \end{bmatrix}\begin{bmatrix} 0 & 1 & 1 & 0 & 0 \\ 1 & 0 & 1 & 1 & 1 \\ 1 & 1 & 0 & 1 & 1 \\ 0 & 1 & 1 & 0 & 0 \\ 0 & 1 & 1 & 0 & 0 \end{bmatrix}\begin{bmatrix} 0 & 1 & 1 & 0 & 0 \\ 1 & 0 & 1 & 1 & 1 \\ 1 & 1 & 0 & 1 & 1 \\ 0 & 1 & 1 & 0 & 0 \\ 0 & 1 & 1 & 0 & 0 \end{bmatrix}$$

$$= \begin{bmatrix} 2 & 7 & 7 & 2 & 2 \\ 7 & 6 & 7 & 7 & 7 \\ 7 & 7 & 6 & 7 & 7 \\ 2 & 7 & 7 & 2 & 2 \\ 2 & 7 & 7 & 2 & 2 \end{bmatrix}$$

A^3 gives the number of walks of length 3.

There are 2 walks of length 3 between vertex 1 and 5.

61. a. Find the adjacency matrix.

$$A = \begin{bmatrix} 0 & 0 & 0 & 1 & 1 \\ 0 & 0 & 2 & 0 & 0 \\ 0 & 2 & 0 & 1 & 1 \\ 1 & 0 & 1 & 0 & 0 \\ 1 & 0 & 1 & 0 & 0 \end{bmatrix}$$

b. Find A^4 and the number of walks.

$$A^4 = \begin{bmatrix} 8 & 0 & 16 & 0 & 0 \\ 0 & 24 & 0 & 16 & 16 \\ 16 & 0 & 40 & 0 & 0 \\ 0 & 16 & 0 & 12 & 12 \\ 0 & 16 & 0 & 12 & 12 \end{bmatrix}$$

A^4 gives the number of walks of length 4.

There are 12 walks of length 4 between vertex 5 and 5.

63. a. Find the adjacency matrix.

$$A = \begin{bmatrix} 0 & 1 & 0 & 1 \\ 1 & 0 & 2 & 1 \\ 0 & 2 & 0 & 0 \\ 1 & 1 & 0 & 0 \end{bmatrix}$$

b. Find the matrix.

$$A^3 = \begin{bmatrix} 2 & 7 & 2 & 3 \\ 7 & 2 & 12 & 7 \\ 2 & 12 & 0 & 2 \\ 3 & 7 & 2 & 2 \end{bmatrix}$$

A^3 gives the number of walks of length 3.

65. a. Predict the percent drinking diet soda in 1 year.

For 1 year from now, $n = 2$.

$$\begin{bmatrix} 0.55 & 0.45 \end{bmatrix}\begin{bmatrix} 0.989 & 0.011 \\ 0.007 & 0.993 \end{bmatrix}^2 = \begin{bmatrix} 0.5443 & 0.4557 \end{bmatrix}$$

1 year from now 45.6% of the customers will be drinking diet soda.

b. Predict the percent drinking diet soda in 3 years.

For 3 years from now, $n = 6$.

$$\begin{bmatrix} 0.55 & 0.45 \end{bmatrix}\begin{bmatrix} 0.989 & 0.011 \\ 0.007 & 0.993 \end{bmatrix}^6 = \begin{bmatrix} 0.5334 & 0.4666 \end{bmatrix}$$

3 years from now 46.7% of the customers will be drinking diet soda.

67. a. Predict the percent of customers renting movies online 12 months from now.

For 12 months from now, $n = 12$.

$$\begin{bmatrix} 0.15 & 0.85 \end{bmatrix}\begin{bmatrix} 0.975 & 0.025 \\ 0.014 & 0.986 \end{bmatrix}^{12} = \begin{bmatrix} 0.2293 & 0.7707 \end{bmatrix}$$

12 months from now 22.9% of the customers will be renting movies online.

b. Predict the percent of customers renting movies online 24 months from now.

For 24 months from now, $n = 24$.

$$\begin{bmatrix} 0.15 & 0.85 \end{bmatrix} \begin{bmatrix} 0.975 & 0.025 \\ 0.014 & 0.986 \end{bmatrix}^{24} = \begin{bmatrix} 0.2785 & 0.7215 \end{bmatrix}$$

24 months from now 27.9% of the customers will be renting movies online.

69. Use Exercise 68 to find when store A has 50%.

When $n = 11$, $a > 0.5053$.

$$\begin{bmatrix} 0.25 & 0.75 \end{bmatrix} \begin{bmatrix} 0.98 & 0.02 \\ 0.05 & 0.95 \end{bmatrix}^{11} = \begin{bmatrix} 0.5053 & 0.4947 \end{bmatrix}$$

After 11 months, Store A will have 50% of the town's customers.

71. Using A and B as given and a calculator,

$$AB = \begin{bmatrix} 24 & 21 & -12 & 32 & 0 \\ -7 & -8 & 3 & 21 & 20 \\ 32 & 10 & -32 & 1 & 5 \\ 19 & -15 & -17 & 30 & 20 \\ 29 & 9 & -28 & 13 & -6 \end{bmatrix}$$

73. Using A as given and a calculator,

$$A^3 = \begin{bmatrix} 46 & -100 & 36 & 273 & 93 \\ 82 & -93 & 19 & 27 & 97 \\ 73 & -10 & -23 & 109 & 83 \\ 212 & -189 & 52 & 37 & 156 \\ 68 & -22 & 54 & 221 & 58 \end{bmatrix}$$

75. Using A and B as given and a calculator,

$$A^2 + B^2 = \begin{bmatrix} 76 & -8 & -25 & 30 & 6 \\ 14 & 16 & -10 & 14 & 2 \\ 39 & 0 & -45 & 22 & 27 \\ 0 & -4 & 23 & 83 & -16 \\ 56 & -20 & -22 & 7 & 5 \end{bmatrix}$$

77. a. $R_{90} \cdot \begin{bmatrix} t \\ t+2 \\ 1 \end{bmatrix} = \begin{bmatrix} 0 & -1 & 0 \\ 1 & 0 & 0 \\ 0 & 0 & 1 \end{bmatrix} \begin{bmatrix} t \\ t+2 \\ 1 \end{bmatrix}$

$$= \begin{bmatrix} -t-2 \\ t \\ 1 \end{bmatrix} \Rightarrow (-t-2, \ t)$$

$x = -t-2, \ y = t$

Using substitution, $x = -y - 2$
$$y = -x - 2$$

b. $R_y \cdot \begin{bmatrix} t \\ 3t-1 \\ 1 \end{bmatrix} = \begin{bmatrix} -1 & 0 & 0 \\ 0 & 1 & 0 \\ 0 & 0 & 1 \end{bmatrix} \begin{bmatrix} t \\ 3t-1 \\ 1 \end{bmatrix}$

$$= \begin{bmatrix} -t \\ 3t-1 \\ 1 \end{bmatrix} \Rightarrow (-t, \ 3t-1)$$

$x = -t, \ y = 3t-1$

$t = -x$

Using substitution, $y = 3(-x) - 1$
$$y = -3x - 1$$

c. $T_{-1,-1} \cdot \begin{bmatrix} t \\ \frac{1}{t} \\ 1 \end{bmatrix} = \begin{bmatrix} 1 & 0 & -1 \\ 0 & 1 & -1 \\ 0 & 0 & 1 \end{bmatrix} \begin{bmatrix} t \\ \frac{1}{t} \\ 1 \end{bmatrix} = \begin{bmatrix} t-1 \\ \frac{1}{t}-1 \\ 1 \end{bmatrix}$

$$R_{180} \cdot \begin{bmatrix} t-1 \\ \frac{1}{t}-1 \\ 1 \end{bmatrix} = \begin{bmatrix} -1 & 0 & 0 \\ 0 & -1 & 0 \\ 0 & 0 & 1 \end{bmatrix} \begin{bmatrix} t-1 \\ \frac{1}{t}-1 \\ 1 \end{bmatrix} = \begin{bmatrix} -t+1 \\ -\frac{1}{t}+1 \\ 1 \end{bmatrix}$$

$$T_{1,1} \cdot \begin{bmatrix} -t+1 \\ -\frac{1}{t}+1 \\ 1 \end{bmatrix} = \begin{bmatrix} 1 & 0 & 1 \\ 0 & 1 & 1 \\ 0 & 0 & 1 \end{bmatrix} \begin{bmatrix} -t+1 \\ -\frac{1}{t}+1 \\ 1 \end{bmatrix}$$

$$= \begin{bmatrix} -t+2 \\ -\frac{1}{t}+2 \\ 1 \end{bmatrix} \Rightarrow \left(-t+2, \ -\frac{1}{t}+2\right)$$

$x = -t+2, \ y = -\frac{1}{t}+2$

$t = -x+2$

Using substitution,

$$y = -\frac{1}{-x+2} + 2 = \frac{-1}{-x+2} + \frac{-2x+4}{-x+2} = \frac{-2x+3}{-x+2}$$

$$= \frac{2x-3}{x-2}$$

d. $T_{2,-1} \cdot \begin{bmatrix} t \\ t^2 \\ 1 \end{bmatrix} = \begin{bmatrix} 1 & 0 & 2 \\ 0 & 1 & -1 \\ 0 & 0 & 1 \end{bmatrix} \begin{bmatrix} t \\ t^2 \\ 1 \end{bmatrix}$

$$= \begin{bmatrix} t+2 \\ t^2-1 \\ 1 \end{bmatrix} \Rightarrow (t+2, \ t^2-1)$$

$x = t+2, \ y = t^2-1$

$t = x-2$

Using substitution,

$$y = (x-2)^2 - 1$$
$$y = x^2 - 4x + 3$$

e. $R_{270} \cdot \begin{bmatrix} t \\ t^2 \\ 1 \end{bmatrix} = \begin{bmatrix} 0 & 1 & 0 \\ -1 & 0 & 0 \\ 0 & 0 & 1 \end{bmatrix} \begin{bmatrix} t \\ t^2 \\ 1 \end{bmatrix} = \begin{bmatrix} t^2 \\ -t \\ 1 \end{bmatrix} \Rightarrow (t^2, \; -t)$

$$x = t^2, \quad y = -t$$

Using substitution, $x = y^2$

f. $R_{90} \cdot \begin{bmatrix} t \\ t^2 \\ 1 \end{bmatrix} = \begin{bmatrix} 0 & -1 & 0 \\ 1 & 0 & 0 \\ 0 & 0 & 1 \end{bmatrix} \begin{bmatrix} t \\ t^2 \\ 1 \end{bmatrix} = \begin{bmatrix} -t^2 \\ t \\ 1 \end{bmatrix}$

$T_{-2,\,1} \cdot \begin{bmatrix} -t^2 \\ t \\ 1 \end{bmatrix} = \begin{bmatrix} 1 & 0 & -2 \\ 0 & 1 & 1 \\ 0 & 0 & 1 \end{bmatrix} \begin{bmatrix} -t^2 \\ t \\ 1 \end{bmatrix}$

$= \begin{bmatrix} -t^2 - 2 \\ t+1 \\ 1 \end{bmatrix} \Rightarrow (-t^2 - 2, \; t+1)$

$$x = -t^2 - 2, \quad y = t+1$$
$$t = y - 1$$

Using substitution,

$$x = -(y-1)^2 - 2$$
$$x = -y^2 + 2y - 3$$

Prepare for Section 10.3

P1. Find the multiplicative inverse of $-\dfrac{2}{3}$.

$$-\dfrac{3}{2}$$

P3. State the 3 elementary row operations for matrices.

1. Interchange any two rows.

2. Multiply each elements in a row by the same nonzero constant.

3. Replace a row by the sum of that row and a nonzero multiple of any other row.

P5. Solve for X.

$$AX = B$$
$$A^{-1}AX = A^{-1}B$$
$$X = A^{-1}B$$

Section 10.3 Exercises

1. Determine whether the matrices are inverses.

$$\begin{bmatrix} 3 & -4 \\ -6 & 8 \end{bmatrix} \begin{bmatrix} 4 & 8 \\ -3 & -8 \end{bmatrix} = \begin{bmatrix} 24 & 56 \\ -48 & -112 \end{bmatrix}$$

No, the matrices are not inverses.

3. Write the matrix equation.

$$\begin{bmatrix} 4 & -3 \\ 1 & 2 \end{bmatrix} \begin{bmatrix} x \\ y \end{bmatrix} = \begin{bmatrix} 2 \\ 3 \end{bmatrix}$$

5. Find the inverse of the matrix.

$$\left[\begin{array}{cc|cc} 1 & -3 & 1 & 0 \\ -2 & 5 & 0 & 1 \end{array}\right]$$

$\xrightarrow{\;2R_1 + R_2\;} \left[\begin{array}{cc|cc} 1 & -3 & 1 & 0 \\ 0 & -1 & 2 & 1 \end{array}\right]$

$\xrightarrow{\;-1 \cdot R_2\;} \left[\begin{array}{cc|cc} 1 & -3 & 1 & 0 \\ 0 & 1 & -2 & -1 \end{array}\right]$

$\xrightarrow{\;3R_2 + R_1\;} \left[\begin{array}{cc|cc} 1 & 0 & -5 & -3 \\ 0 & 1 & -2 & -1 \end{array}\right]$

The inverse matrix is $\begin{bmatrix} -5 & -3 \\ -2 & -1 \end{bmatrix}$.

7. Find the inverse of the matrix.

$\left[\begin{array}{cc|cc} 1 & 4 & 1 & 0 \\ 2 & 10 & 0 & 1 \end{array}\right] \xrightarrow{\;-2R_1 + R_2\;} \left[\begin{array}{cc|cc} 1 & 4 & 1 & 0 \\ 0 & 2 & -2 & 1 \end{array}\right]$

$\xrightarrow{\;\frac{1}{2}R_2\;} \left[\begin{array}{cc|cc} 1 & 4 & 1 & 0 \\ 0 & 1 & -1 & \frac{1}{2} \end{array}\right]$

$\xrightarrow{\;-4R_2 + R_1\;} \left[\begin{array}{cc|cc} 1 & 0 & 5 & -2 \\ 0 & 1 & -1 & \frac{1}{2} \end{array}\right]$

The inverse matrix is $\begin{bmatrix} 5 & -2 \\ -1 & \frac{1}{2} \end{bmatrix}$.

9. Find the inverse of the matrix.

$$\left[\begin{array}{ccc|ccc} -1 & 1 & 3 & 1 & 0 & 0 \\ 2 & 1 & -1 & 0 & 1 & 0 \\ 4 & 4 & 1 & 0 & 0 & 1 \end{array}\right]$$

$\xrightarrow{\;-1 \cdot R_1\;} \left[\begin{array}{ccc|ccc} 1 & -1 & -3 & -1 & 0 & 0 \\ 2 & 1 & -1 & 0 & 1 & 0 \\ 4 & 4 & 1 & 0 & 0 & 1 \end{array}\right]$

$$\xrightarrow[\substack{-2R_1+R_2 \\ -4R_1+R_3}]{} \left[\begin{array}{ccc|ccc} 1 & -1 & -3 & -1 & 0 & 0 \\ 0 & 3 & 5 & 2 & 1 & 0 \\ 0 & 8 & 13 & 4 & 0 & 1 \end{array}\right]$$

$$\xrightarrow[\frac{1}{3}R_2]{} \left[\begin{array}{ccc|ccc} 1 & -1 & -3 & -1 & 0 & 0 \\ 0 & 1 & \frac{5}{3} & \frac{2}{3} & \frac{1}{3} & 0 \\ 0 & 8 & 13 & 4 & 0 & 1 \end{array}\right]$$

$$\xrightarrow[-8R_2+R_3]{} \left[\begin{array}{ccc|ccc} 1 & -1 & -3 & -1 & 0 & 0 \\ 0 & 1 & \frac{5}{3} & \frac{2}{3} & \frac{1}{3} & 0 \\ 0 & 0 & -\frac{1}{3} & -\frac{4}{3} & -\frac{8}{3} & 1 \end{array}\right]$$

$$\xrightarrow[-3R_3]{} \left[\begin{array}{ccc|ccc} 1 & -1 & -3 & -1 & 0 & 0 \\ 0 & 1 & \frac{5}{3} & \frac{2}{3} & \frac{1}{3} & 0 \\ 0 & 0 & 1 & 4 & 8 & -3 \end{array}\right]$$

$$\xrightarrow[\substack{3R_3+R_1 \\ -\frac{5}{3}R_3+R_2}]{} \left[\begin{array}{ccc|ccc} 1 & -1 & 0 & 11 & 24 & -9 \\ 0 & 1 & 0 & -6 & -13 & 5 \\ 0 & 0 & 1 & 4 & 8 & -3 \end{array}\right]$$

$$\xrightarrow[R_2+R_1]{} \left[\begin{array}{ccc|ccc} 1 & 0 & 0 & 5 & 11 & -4 \\ 0 & 1 & 0 & -6 & -13 & 5 \\ 0 & 0 & 1 & 4 & 8 & -3 \end{array}\right]$$

The inverse matrix is $\begin{bmatrix} 5 & 11 & -4 \\ -6 & -13 & 5 \\ 4 & 8 & -3 \end{bmatrix}$.

11. Find the inverse of the matrix.

$$\left[\begin{array}{ccc|ccc} 20 & -15 & 17 & 1 & 0 & 0 \\ 1 & -1 & 1 & 0 & 1 & 0 \\ 12 & -9 & 10 & 0 & 0 & 1 \end{array}\right]$$

$$\xrightarrow[R_1\leftrightarrow R_2]{} \left[\begin{array}{ccc|ccc} 1 & -1 & 1 & 0 & 1 & 0 \\ 20 & -15 & 17 & 1 & 0 & 0 \\ 12 & -9 & 10 & 0 & 0 & 1 \end{array}\right]$$

$$\xrightarrow[\substack{-20R_1+R_2 \\ -12R_1+R_3}]{} \left[\begin{array}{ccc|ccc} 1 & -1 & 1 & 0 & 1 & 0 \\ 0 & 5 & -3 & 1 & -20 & 0 \\ 0 & 3 & -2 & 0 & -12 & 1 \end{array}\right]$$

$$\xrightarrow[\frac{1}{5}R_2]{} \left[\begin{array}{ccc|ccc} 1 & -1 & 1 & 0 & 1 & 0 \\ 0 & 1 & -\frac{3}{5} & \frac{1}{5} & -4 & 0 \\ 0 & 3 & -2 & 0 & -12 & 1 \end{array}\right]$$

$$\xrightarrow[-3R_2+R_3]{} \left[\begin{array}{ccc|ccc} 1 & -1 & 1 & 0 & 1 & 0 \\ 0 & 1 & -\frac{3}{5} & \frac{1}{5} & -4 & 0 \\ 0 & 0 & -\frac{1}{5} & -\frac{3}{5} & 0 & 1 \end{array}\right]$$

$$\xrightarrow[-5R_3]{} \left[\begin{array}{ccc|ccc} 1 & -1 & 1 & 0 & 1 & 0 \\ 0 & 1 & -\frac{3}{5} & \frac{1}{5} & -4 & 0 \\ 0 & 0 & 1 & 3 & 0 & -5 \end{array}\right]$$

$$\xrightarrow[\substack{-R_3+R_1 \\ \frac{3}{5}R_3+R_2}]{} \left[\begin{array}{ccc|ccc} 1 & -1 & 0 & -3 & 1 & 5 \\ 0 & 1 & 0 & 2 & -4 & -3 \\ 0 & 0 & 1 & 3 & 0 & -5 \end{array}\right]$$

$$\xrightarrow[R_2+R_1]{} \left[\begin{array}{ccc|ccc} 1 & 0 & 0 & -1 & -3 & 2 \\ 0 & 1 & 0 & 2 & -4 & -3 \\ 0 & 0 & 1 & 3 & 0 & -5 \end{array}\right]$$

The inverse matrix is $\begin{bmatrix} -1 & -3 & 2 \\ 2 & -4 & -3 \\ 3 & 0 & -5 \end{bmatrix}$.

13. Find the inverse of the matrix.

$$\left[\begin{array}{ccc|ccc} -1 & 2 & -5 & 1 & 0 & 0 \\ 2 & -4 & 1 & 0 & 1 & 0 \\ -2 & 4 & 5 & 0 & 0 & 1 \end{array}\right]$$

$$\xrightarrow[-1\cdot R_1]{} \left[\begin{array}{ccc|ccc} 1 & -2 & 5 & -1 & 0 & 0 \\ 2 & -4 & 1 & 0 & 1 & 0 \\ -2 & 4 & 5 & 0 & 0 & 1 \end{array}\right]$$

$$\xrightarrow[\substack{-2R_1+R_2 \\ 2R_1+R_3}]{} \left[\begin{array}{ccc|ccc} 1 & -2 & 5 & -1 & 0 & 0 \\ 0 & 0 & -9 & 2 & 1 & 0 \\ 0 & 0 & 15 & -2 & 0 & 1 \end{array}\right]$$

The matrix does not have an inverse.

15. Use a graphing calculator to find the inverse.

$$\begin{bmatrix} 1 & -1 & 2 & 1 \\ 2 & -1 & 5 & 1 \\ 3 & -3 & 7 & 5 \\ -2 & 3 & -4 & -1 \end{bmatrix}$$

The inverse matrix is $\begin{bmatrix} \frac{19}{2} & -\frac{1}{2} & -\frac{3}{2} & \frac{3}{2} \\ \frac{7}{4} & \frac{1}{4} & -\frac{1}{4} & \frac{3}{4} \\ -\frac{7}{2} & \frac{1}{2} & \frac{1}{2} & -\frac{1}{2} \\ \frac{1}{4} & -\frac{1}{4} & \frac{1}{4} & \frac{1}{4} \end{bmatrix}$.

17. Use a graphing calculator to find the inverse.

$$\begin{bmatrix} -2 & -3 & 1 & 3 \\ 4 & 4 & -1 & 2 \\ -1 & -1 & 4 & -1 \\ -4 & -4 & -5 & -1 \end{bmatrix}$$

The inverse matrix is $\begin{bmatrix} 1 & -31 & -\frac{109}{3} & -\frac{68}{3} \\ -1 & 28 & \frac{98}{3} & \frac{61}{3} \\ 0 & 1 & \frac{4}{3} & \frac{2}{3} \\ 0 & 7 & 8 & 5 \end{bmatrix}$.

19. Use a graphing calculator to find the inverse.

$$\begin{bmatrix} -5 & -5 & -3 & -2 \\ 1 & -5 & 2 & 0 \\ 2 & 4 & -1 & 5 \\ 3 & 1 & 4 & -3 \end{bmatrix}$$

The matrix does not have an inverse.

21. Solve the system using inverses.

$$\begin{bmatrix} 1 & 4 \\ 2 & 7 \end{bmatrix} \begin{bmatrix} x \\ y \end{bmatrix} = \begin{bmatrix} 6 \\ 11 \end{bmatrix} \quad (1)$$

Find the inverse of $\begin{bmatrix} 1 & 4 \\ 2 & 7 \end{bmatrix}$.

$$\begin{bmatrix} 1 & 4 & | & 1 & 0 \\ 2 & 7 & | & 0 & 1 \end{bmatrix} \xrightarrow{-2R_1 + R_2} \begin{bmatrix} 1 & 4 & | & 1 & 0 \\ 0 & -1 & | & -2 & 1 \end{bmatrix}$$

$$\xrightarrow{-1R_2} \begin{bmatrix} 1 & 4 & | & 1 & 0 \\ 0 & 1 & | & 2 & -1 \end{bmatrix}$$

$$\xrightarrow{-4R_2 + R_1} \begin{bmatrix} 1 & 0 & | & -7 & 4 \\ 0 & 1 & | & 2 & -1 \end{bmatrix}$$

The inverse of $\begin{bmatrix} 1 & 4 \\ 2 & 7 \end{bmatrix}$ is $\begin{bmatrix} -7 & 4 \\ 2 & -1 \end{bmatrix}$.

Multiply each side of Eq. (1) by the inverse matrix.

$$\begin{bmatrix} -7 & 4 \\ 2 & -1 \end{bmatrix} \begin{bmatrix} 1 & 4 \\ 2 & 7 \end{bmatrix} \begin{bmatrix} x \\ y \end{bmatrix} = \begin{bmatrix} -7 & 4 \\ 2 & -1 \end{bmatrix} \begin{bmatrix} 6 \\ 11 \end{bmatrix}$$

$$\begin{bmatrix} x \\ y \end{bmatrix} = \begin{bmatrix} 2 \\ 1 \end{bmatrix}$$

The solution is (2, 1).

23. Solve the system using inverses.

$$\begin{bmatrix} 1 & -2 \\ 3 & 2 \end{bmatrix} \begin{bmatrix} x \\ y \end{bmatrix} = \begin{bmatrix} 8 \\ -1 \end{bmatrix} \quad (1)$$

Find the inverse matrix of $\begin{bmatrix} 1 & -2 \\ 3 & 2 \end{bmatrix}$.

$$\begin{bmatrix} 1 & -2 & | & 1 & 0 \\ 3 & 2 & | & 0 & 1 \end{bmatrix} \xrightarrow{-3R_1 + R_2} \begin{bmatrix} 1 & -2 & | & 1 & 0 \\ 0 & 8 & | & -3 & 1 \end{bmatrix}$$

$$\xrightarrow{\frac{1}{8}R_2} \begin{bmatrix} 1 & -2 & | & 1 & 0 \\ 0 & 1 & | & -\frac{3}{8} & \frac{1}{8} \end{bmatrix}$$

$$\xrightarrow{2R_2 + R_1} \begin{bmatrix} 1 & 0 & | & \frac{1}{4} & \frac{1}{4} \\ 0 & 1 & | & -\frac{3}{8} & \frac{1}{8} \end{bmatrix}$$

The inverse matrix is $\begin{bmatrix} \frac{1}{4} & \frac{1}{4} \\ -\frac{3}{8} & \frac{1}{8} \end{bmatrix}$.

Multiply each side of Eq. (1) by the inverse matrix.

$$\begin{bmatrix} \frac{1}{4} & \frac{1}{4} \\ -\frac{3}{8} & \frac{1}{8} \end{bmatrix} \begin{bmatrix} 1 & -2 \\ 3 & 2 \end{bmatrix} \begin{bmatrix} x \\ y \end{bmatrix} = \begin{bmatrix} \frac{1}{4} & \frac{1}{4} \\ -\frac{3}{8} & \frac{1}{8} \end{bmatrix} \begin{bmatrix} 8 \\ -1 \end{bmatrix}$$

$$\begin{bmatrix} x \\ y \end{bmatrix} = \begin{bmatrix} \frac{7}{4} \\ -\frac{25}{8} \end{bmatrix}$$

The solution is $\left(\frac{7}{4}, -\frac{25}{8} \right)$.

25. Solve the system using inverses.

$$\begin{bmatrix} 1 & 1 & 2 \\ 2 & 3 & 3 \\ 3 & 3 & 7 \end{bmatrix} \begin{bmatrix} x \\ y \\ z \end{bmatrix} = \begin{bmatrix} 4 \\ 5 \\ 14 \end{bmatrix} \quad (1)$$

Find the inverse matrix of $\begin{bmatrix} 1 & 1 & 2 \\ 2 & 3 & 3 \\ 3 & 3 & 7 \end{bmatrix}$.

$$\begin{bmatrix} 1 & 1 & 2 & | & 1 & 0 & 0 \\ 2 & 3 & 3 & | & 0 & 1 & 0 \\ 3 & 3 & 7 & | & 0 & 0 & 1 \end{bmatrix} \begin{array}{c} \xrightarrow{-2R_1 + R_2} \\ \xrightarrow{-3R_1 + R_3} \end{array} \begin{bmatrix} 1 & 1 & 2 & | & 1 & 0 & 0 \\ 0 & 1 & -1 & | & -2 & 1 & 0 \\ 0 & 0 & 1 & | & -3 & 0 & 1 \end{bmatrix}$$

$$\xrightarrow{-R_2 + R_1} \begin{bmatrix} 1 & 0 & 3 & | & 3 & -1 & 0 \\ 0 & 1 & -1 & | & -2 & 1 & 0 \\ 0 & 0 & 1 & | & -3 & 0 & 1 \end{bmatrix}$$

$$\begin{array}{c} \xrightarrow{-3R_3 + R_1} \\ \xrightarrow{R_3 + R_2} \end{array} \begin{bmatrix} 1 & 0 & 0 & | & 12 & -1 & -3 \\ 0 & 1 & 0 & | & -5 & 1 & 1 \\ 0 & 0 & 1 & | & -3 & 0 & 1 \end{bmatrix}$$

The inverse matrix is $\begin{bmatrix} 12 & -1 & -3 \\ -5 & 1 & 1 \\ -3 & 0 & 1 \end{bmatrix}$.

Multiply each side of Eq. (1) by the inverse matrix.

$$\begin{bmatrix} 12 & -1 & -3 \\ -5 & 1 & 1 \\ -3 & 0 & 1 \end{bmatrix}\begin{bmatrix} 1 & 1 & 2 \\ 2 & 3 & 3 \\ 3 & 3 & 7 \end{bmatrix}\begin{bmatrix} x \\ y \\ z \end{bmatrix} = \begin{bmatrix} 12 & -1 & -3 \\ -5 & 1 & 1 \\ -3 & 0 & 1 \end{bmatrix}\begin{bmatrix} 4 \\ 5 \\ 14 \end{bmatrix}$$

$$\begin{bmatrix} x \\ y \\ z \end{bmatrix} = \begin{bmatrix} 1 \\ -1 \\ 2 \end{bmatrix}$$

The solution is $(1, -1, 2)$.

27. Solve the system using inverses.

$$\begin{bmatrix} 1 & 2 & 2 \\ -2 & -5 & -2 \\ 2 & 4 & 7 \end{bmatrix}\begin{bmatrix} x \\ y \\ z \end{bmatrix} = \begin{bmatrix} 5 \\ 8 \\ 19 \end{bmatrix} \qquad (1)$$

Find the inverse matrix of $\begin{bmatrix} 1 & 2 & 2 \\ -2 & -5 & -2 \\ 2 & 4 & 7 \end{bmatrix}$.

$$\left[\begin{array}{ccc|ccc} 1 & 2 & 2 & 1 & 0 & 0 \\ -2 & -5 & -2 & 0 & 1 & 0 \\ 2 & 4 & 7 & 0 & 0 & 1 \end{array}\right]$$

$$\xrightarrow[\begin{array}{c} 2R_1 + R_2 \\ -2R_1 + R_3 \end{array}]{} \left[\begin{array}{ccc|ccc} 1 & 2 & 2 & 1 & 0 & 0 \\ 0 & -1 & 2 & 2 & 1 & 0 \\ 0 & 0 & 3 & -2 & 0 & 1 \end{array}\right]$$

$$\xrightarrow{-1R_2} \left[\begin{array}{ccc|ccc} 1 & 2 & 2 & 1 & 0 & 0 \\ 0 & 1 & -2 & -2 & -1 & 0 \\ 0 & 0 & 3 & -2 & 0 & 1 \end{array}\right]$$

$$\xrightarrow{\frac{1}{3}R_3} \left[\begin{array}{ccc|ccc} 1 & 2 & 2 & 1 & 0 & 0 \\ 0 & 1 & -2 & -2 & -1 & 0 \\ 0 & 0 & 1 & -\frac{2}{3} & 0 & \frac{1}{3} \end{array}\right]$$

$$\xrightarrow{-2R_2 + R_1} \left[\begin{array}{ccc|ccc} 1 & 0 & 6 & 5 & 2 & 0 \\ 0 & 1 & -2 & -2 & -1 & 0 \\ 0 & 0 & 1 & -\frac{2}{3} & 0 & \frac{1}{3} \end{array}\right]$$

$$\xrightarrow[\begin{array}{c} 2R_3 + R_2 \\ -6R_3 + R_1 \end{array}]{} \left[\begin{array}{ccc|ccc} 1 & 0 & 0 & 9 & 2 & -2 \\ 0 & 1 & 0 & -\frac{10}{3} & -1 & \frac{2}{3} \\ 0 & 0 & 1 & -\frac{2}{3} & 0 & \frac{1}{3} \end{array}\right]$$

The inverse matrix is $\begin{bmatrix} 9 & 2 & -2 \\ -\frac{10}{3} & -1 & \frac{2}{3} \\ -\frac{2}{3} & 0 & \frac{1}{3} \end{bmatrix}$.

Multiply each side of Eq. (1) by the inverse matrix.

$$\begin{bmatrix} 9 & 2 & -2 \\ -\frac{10}{3} & -1 & \frac{2}{3} \\ -\frac{2}{3} & 0 & \frac{1}{3} \end{bmatrix}\begin{bmatrix} 1 & 2 & 2 \\ -2 & -5 & -2 \\ 2 & 4 & 7 \end{bmatrix}\begin{bmatrix} x \\ y \\ z \end{bmatrix} = \begin{bmatrix} 9 & 2 & -2 \\ -\frac{10}{3} & -1 & \frac{2}{3} \\ -\frac{2}{3} & 0 & \frac{1}{3} \end{bmatrix}\begin{bmatrix} 5 \\ 8 \\ 19 \end{bmatrix}$$

$$\begin{bmatrix} x \\ y \\ z \end{bmatrix} = \begin{bmatrix} 23 \\ -12 \\ 3 \end{bmatrix}$$

The solution is $(23, -12, 3)$.

29. Solve the system using inverses.

$$\begin{bmatrix} 1 & 2 & 0 & 1 \\ 2 & 5 & 1 & 2 \\ 2 & 4 & 1 & 1 \\ 3 & 6 & 0 & 4 \end{bmatrix}\begin{bmatrix} w \\ x \\ y \\ z \end{bmatrix} = \begin{bmatrix} 6 \\ 10 \\ 8 \\ 16 \end{bmatrix} \qquad (1)$$

Find the inverse matrix of $\begin{bmatrix} 1 & 2 & 0 & 1 \\ 2 & 5 & 1 & 2 \\ 2 & 4 & 1 & 1 \\ 3 & 6 & 0 & 4 \end{bmatrix}$.

$$\left[\begin{array}{cccc|cccc} 1 & 2 & 0 & 1 & 1 & 0 & 0 & 0 \\ 2 & 5 & 1 & 2 & 0 & 1 & 0 & 0 \\ 2 & 4 & 1 & 1 & 0 & 0 & 1 & 0 \\ 3 & 6 & 0 & 4 & 0 & 0 & 0 & 1 \end{array}\right]$$

$$\xrightarrow[\begin{array}{c} -2R_1 + R_2 \\ -2R_1 + R_3 \\ -3R_1 + R_4 \end{array}]{} \left[\begin{array}{cccc|cccc} 1 & 2 & 0 & 1 & 1 & 0 & 0 & 0 \\ 0 & 1 & 1 & 0 & -2 & 1 & 0 & 0 \\ 0 & 0 & 1 & -1 & -2 & 0 & 1 & 0 \\ 0 & 0 & 0 & 1 & -3 & 0 & 0 & 1 \end{array}\right]$$

$$\xrightarrow{-2R_2 + R_1} \left[\begin{array}{cccc|cccc} 1 & 0 & -2 & 1 & 5 & -2 & 0 & 0 \\ 0 & 1 & 1 & 0 & -2 & 1 & 0 & 0 \\ 0 & 0 & 1 & -1 & -2 & 0 & 1 & 0 \\ 0 & 0 & 0 & 1 & -3 & 0 & 0 & 1 \end{array}\right]$$

$$\xrightarrow[\begin{array}{c} 2R_3 + R_1 \\ -1R_3 + R_2 \end{array}]{} \left[\begin{array}{cccc|cccc} 1 & 0 & 0 & -1 & 1 & -2 & 2 & 0 \\ 0 & 1 & 0 & 1 & 0 & 1 & -1 & 0 \\ 0 & 0 & 1 & -1 & -2 & 0 & 1 & 0 \\ 0 & 0 & 0 & 1 & -3 & 0 & 0 & 1 \end{array}\right]$$

$$\xrightarrow[\begin{array}{c} R_4 + R_1 \\ -R_4 + R_2 \\ R_4 + R_3 \end{array}]{} \left[\begin{array}{cccc|cccc} 1 & 0 & 0 & 0 & -2 & -2 & 2 & 1 \\ 0 & 1 & 0 & 0 & 3 & 1 & -1 & -1 \\ 0 & 0 & 1 & 0 & -5 & 0 & 1 & 1 \\ 0 & 0 & 0 & 1 & -3 & 0 & 0 & 1 \end{array}\right]$$

The inverse matrix is $\begin{bmatrix} -2 & -2 & 2 & 1 \\ 3 & 1 & -1 & -1 \\ -5 & 0 & 1 & 1 \\ -3 & 0 & 0 & 1 \end{bmatrix}$.

Multiply each side of Eq. (1) by the inverse matrix.

$$\begin{bmatrix} -2 & -2 & 2 & 1 \\ -3 & 1 & -1 & -1 \\ -5 & 0 & 1 & 1 \\ -3 & 0 & 0 & 1 \end{bmatrix}\begin{bmatrix} 1 & 2 & 0 & 1 \\ 2 & 5 & 1 & 2 \\ 2 & 4 & 1 & 1 \\ 3 & 6 & 0 & 4 \end{bmatrix}\begin{bmatrix} w \\ x \\ y \\ z \end{bmatrix} = \begin{bmatrix} -2 & -2 & 2 & 1 \\ 3 & 1 & -1 & -1 \\ -5 & 0 & 1 & 1 \\ -3 & 0 & 0 & 1 \end{bmatrix}\begin{bmatrix} 6 \\ 10 \\ 8 \\ 16 \end{bmatrix}$$

$$\begin{bmatrix} w \\ x \\ y \\ z \end{bmatrix} = \begin{bmatrix} 0 \\ 4 \\ -6 \\ -2 \end{bmatrix}$$

The solution is $(0, 4, -6, -2)$.

31. Find the temperature at x_1 and x_2.

The average temperature for the two points,

$$x_1 = \frac{35 + 50 + x_2 + 60}{4} = \frac{145 + x_2}{4} \text{ or } 4x_1 - x_2 = 145$$

$$x_2 = \frac{x_1 + 50 + 55 + 60}{4} = \frac{165 + x_1}{4} \text{ or } -x_1 + 4x_2 = 165$$

The system of equations and associated matrix equation are

$$\begin{cases} 4x_1 - x_2 = 145 \\ -x_1 + 4x_2 = 165 \end{cases} \qquad \begin{bmatrix} 4 & -1 \\ -1 & 4 \end{bmatrix}\begin{bmatrix} x_1 \\ x_2 \end{bmatrix} = \begin{bmatrix} 145 \\ 165 \end{bmatrix}$$

Solving the matrix equation by using an inverse matrix

gives $\begin{bmatrix} x_1 \\ x_2 \end{bmatrix} = \begin{bmatrix} 49.7 \\ 53.7 \end{bmatrix}$

The temperatures are $x_1 = 49.7°F$, $x_2 = 53.7°F$.

33. Find the temperature at x_1, x_2, x_3, and x_4.

The average temperature for the two points,

$$x_1 = \frac{50 + 60 + x_2 + x_3}{4} = \frac{110 + x_2 + x_3}{4}$$
$$\text{or } 4x_1 - x_2 - x_3 = 110$$

$$x_2 = \frac{x_1 + 60 + 60 + x_4}{4} = \frac{120 + x_1 + x_4}{4}$$
$$\text{or } -x_1 + 4x_2 - x_4 = 120$$

$$x_3 = \frac{50 + x_1 + x_4 + 50}{4} = \frac{100 + x_1 + x_4}{4}$$
$$\text{or } -x_1 + 4x_3 - x_4 = 100$$

$$x_4 = \frac{x_3 + x_2 + 60 + 50}{4} = \frac{110 + x_2 + x_3}{4}$$
$$\text{or } -x_2 - x_3 + 4x_4 = 110$$

The system of equations and associated matrix equation are

$$\begin{cases} 4x_1 - x_2 - x_3 = 110 \\ -x_1 + 4x_2 - x_4 = 120 \\ -x_1 + 4x_3 - x_4 = 100 \\ -x_2 - x_3 + 4x_4 = 110 \end{cases} \qquad \begin{bmatrix} 4 & -1 & -1 & 0 \\ -1 & 4 & 0 & -1 \\ -1 & 0 & 4 & -1 \\ 0 & -1 & -1 & 4 \end{bmatrix}\begin{bmatrix} x_1 \\ x_2 \\ x_3 \\ x_4 \end{bmatrix} = \begin{bmatrix} 110 \\ 120 \\ 100 \\ 110 \end{bmatrix}$$

Solving the matrix equation by using an inverse matrix

gives $\begin{bmatrix} x_1 \\ x_2 \\ x_3 \\ x_4 \end{bmatrix} = \begin{bmatrix} 55 \\ 57.5 \\ 52.5 \\ 55 \end{bmatrix}$

The temperatures are $x_1 = 55°F$, $x_2 = 57.5°F$,

$x_3 = 52.5°F$, $x_4 = 55°F$.

35. Find the number of children and adults.

$A =$ number of adult tickets

$C =$ number of child tickets

Saturday $A + C = 100$

            $20A + 15C = 1900$

$$\begin{bmatrix} 1 & 1 \\ 20 & 15 \end{bmatrix}\begin{bmatrix} A \\ C \end{bmatrix} = \begin{bmatrix} 100 \\ 1900 \end{bmatrix}$$

$$\begin{bmatrix} -3 & \frac{1}{5} \\ 4 & -\frac{1}{5} \end{bmatrix}\begin{bmatrix} 1 & 1 \\ 20 & 15 \end{bmatrix}\begin{bmatrix} A \\ C \end{bmatrix} = \begin{bmatrix} -3 & \frac{1}{5} \\ 4 & -\frac{1}{5} \end{bmatrix}\begin{bmatrix} 100 \\ 1900 \end{bmatrix}$$

$$\begin{bmatrix} A \\ C \end{bmatrix} = \begin{bmatrix} 80 \\ 20 \end{bmatrix}$$

On Saturday, 80 adults and 20 children took the tour.

Sunday $A + C = 120$

            $20A + 15C = 2275$

$$\begin{bmatrix} 1 & 1 \\ 20 & 15 \end{bmatrix}\begin{bmatrix} A \\ C \end{bmatrix} = \begin{bmatrix} 120 \\ 2275 \end{bmatrix}$$

$$\begin{bmatrix} -3 & \frac{1}{5} \\ 4 & -\frac{1}{5} \end{bmatrix}\begin{bmatrix} 1 & 1 \\ 20 & 15 \end{bmatrix}\begin{bmatrix} A \\ C \end{bmatrix} = \begin{bmatrix} -3 & \frac{1}{5} \\ 4 & -\frac{1}{5} \end{bmatrix}\begin{bmatrix} 120 \\ 2275 \end{bmatrix}$$

$$\begin{bmatrix} A \\ C \end{bmatrix} = \begin{bmatrix} 95 \\ 25 \end{bmatrix}$$

On Sunday, 95 adults and 25 children took the tour.

37. Find the amount of additive for each sample.

$x_1 =$ number of 100-gram portions of additive 1

$x_2 =$ number of 100-gram portions of additive 2

$x_3 =$ number of 100-gram portions of additive 3

$$30x_1 + 40x_2 + 50x_3 = 380$$
Sample 1: $$10x_1 + 15x_2 + 5x_3 = 95$$
$$10x_1 + 10x_2 + 5x_3 = 85$$

$$\begin{bmatrix} 30 & 40 & 50 \\ 10 & 15 & 5 \\ 10 & 10 & 5 \end{bmatrix} \begin{bmatrix} x_1 \\ x_2 \\ x_3 \end{bmatrix} = \begin{bmatrix} 380 \\ 95 \\ 85 \end{bmatrix}$$

$$\begin{bmatrix} -\frac{1}{70} & -\frac{6}{35} & \frac{11}{35} \\ 0 & \frac{1}{5} & -\frac{1}{5} \\ \frac{1}{35} & -\frac{2}{35} & -\frac{1}{35} \end{bmatrix} \begin{bmatrix} 30 & 40 & 50 \\ 10 & 15 & 5 \\ 10 & 10 & 5 \end{bmatrix} \begin{bmatrix} x_1 \\ x_2 \\ x_3 \end{bmatrix} = \begin{bmatrix} -\frac{1}{70} & -\frac{6}{35} & \frac{11}{35} \\ 0 & \frac{1}{5} & -\frac{1}{5} \\ \frac{1}{35} & -\frac{2}{35} & -\frac{1}{35} \end{bmatrix} \begin{bmatrix} 380 \\ 95 \\ 85 \end{bmatrix}$$

$$\begin{bmatrix} x_1 \\ x_2 \\ x_3 \end{bmatrix} = \begin{bmatrix} 5 \\ 2 \\ 3 \end{bmatrix}$$

For Sample 1, 500 g of additive 1, 200 g of additive 2, and 300 g of additive 3 are required.

$$30x_1 + 40x_2 + 50x_3 = 380$$
Sample 2: $$10x_1 + 15x_2 + 5x_3 = 110$$
$$10x_1 + 10x_2 + 5x_3 = 90$$

$$\begin{bmatrix} 30 & 40 & 50 \\ 10 & 15 & 5 \\ 10 & 10 & 5 \end{bmatrix} \begin{bmatrix} x_1 \\ x_2 \\ x_3 \end{bmatrix} = \begin{bmatrix} 380 \\ 110 \\ 90 \end{bmatrix}$$

$$\begin{bmatrix} -\frac{1}{70} & -\frac{6}{35} & \frac{11}{35} \\ 0 & \frac{1}{5} & -\frac{1}{5} \\ \frac{1}{35} & -\frac{2}{35} & -\frac{1}{35} \end{bmatrix} \begin{bmatrix} 30 & 40 & 50 \\ 10 & 15 & 5 \\ 10 & 10 & 5 \end{bmatrix} \begin{bmatrix} x_1 \\ x_2 \\ x_3 \end{bmatrix} = \begin{bmatrix} -\frac{1}{70} & -\frac{6}{35} & \frac{11}{35} \\ 0 & \frac{1}{5} & -\frac{1}{5} \\ \frac{1}{35} & -\frac{2}{35} & -\frac{1}{35} \end{bmatrix} \begin{bmatrix} 380 \\ 110 \\ 90 \end{bmatrix}$$

$$\begin{bmatrix} x_1 \\ x_2 \\ x_3 \end{bmatrix} = \begin{bmatrix} 4 \\ 4 \\ 2 \end{bmatrix}$$

For Sample 2, 400 g of additive 1, 400 g of additive 2, and 200 g of additive 3 are required.

39. Find the gross output for the given conditions.

$X = (I - A)^{-1}D$, where X is consumer demand, I is the identity matrix, A is the input-output matrix, and D is the final demand. Thus

$$X = \left(\begin{bmatrix} 1 & 0 & 0 \\ 0 & 1 & 0 \\ 0 & 0 & 1 \end{bmatrix} - \begin{bmatrix} 0.20 & 0.15 & 0.10 \\ 0.10 & 0.30 & 0.25 \\ 0.20 & 0.10 & 0.10 \end{bmatrix} \right)^{-1} \begin{bmatrix} 120 \\ 60 \\ 55 \end{bmatrix}$$

$$= \begin{bmatrix} 0.80 & -0.15 & -0.10 \\ -0.10 & 0.70 & -0.25 \\ -0.20 & -0.10 & 0.90 \end{bmatrix}^{-1} \begin{bmatrix} 120 \\ 60 \\ 55 \end{bmatrix}$$

$$\approx \begin{bmatrix} 194.67 \\ 157.03 \\ 121.82 \end{bmatrix}$$

\$194.67 million worth of manufacturing, \$157.03 million worth of transportation, \$121.82 million worth of services.

41. Find the amount each industry should produce.

The input-output matrix, A, is given by

$$A = \begin{bmatrix} 0.05 & 0.20 & 0.15 \\ 0.02 & 0.03 & 0.25 \\ 0.10 & 0.12 & 0.05 \end{bmatrix}$$

Consumer demand is given by

$$X = (I - A)^{-1}D$$

$$X = \left(\begin{bmatrix} 1 & 0 & 0 \\ 0 & 1 & 0 \\ 0 & 0 & 1 \end{bmatrix} - \begin{bmatrix} 0.05 & 0.20 & 0.15 \\ 0.02 & 0.03 & 0.25 \\ 0.10 & 0.12 & 0.05 \end{bmatrix} \right)^{-1} \begin{bmatrix} 30 \\ 5 \\ 25 \end{bmatrix}$$

$$= \begin{bmatrix} 0.95 & -0.20 & -0.15 \\ -0.02 & 0.97 & -0.25 \\ -0.10 & -0.12 & 0.95 \end{bmatrix}^{-1} \begin{bmatrix} 30 \\ 5 \\ 25 \end{bmatrix}$$

$$\approx \begin{bmatrix} 39.69 \\ 14.30 \\ 32.30 \end{bmatrix}$$

\$39.69 million worth of coal, \$14.30 million worth of iron, \$32.30 million worth of steel.

43. Show that $AB = O$.

$$AB = \begin{bmatrix} 2 & -3 \\ -6 & 9 \end{bmatrix} \begin{bmatrix} -3 & 15 \\ -2 & 10 \end{bmatrix}$$

$$= \begin{bmatrix} 2(-3) + (-3)(-2) & 2(15) + (-3)(10) \\ -6(-3) + 9(-2) & -6(15) + 9(10) \end{bmatrix}$$

$$= \begin{bmatrix} 0 & 0 \\ 0 & 0 \end{bmatrix} = O$$

45. Show $AB = AC$.

$$AB = \begin{bmatrix} 2 & -1 \\ -4 & 2 \end{bmatrix} \begin{bmatrix} 3 & 4 \\ 1 & 5 \end{bmatrix}$$

$$= \begin{bmatrix} 2(3)+(-1)(1) & 2(4)+(-1)(5) \\ -4(3)+2(1) & -4(4)+2(5) \end{bmatrix} = \begin{bmatrix} 5 & 3 \\ -10 & -6 \end{bmatrix}$$

$$AC = \begin{bmatrix} 2 & -1 \\ -4 & 2 \end{bmatrix} \begin{bmatrix} 4 & 7 \\ 3 & 11 \end{bmatrix}$$

$$= \begin{bmatrix} 2(4)+(-1)(3) & 2(7)+(-1)(11) \\ -4(4)+2(3) & -4(7)+2(11) \end{bmatrix} = \begin{bmatrix} 5 & 3 \\ -10 & -6 \end{bmatrix}$$

Mid-Chapter 10 Quiz

1. Solve the system using Gaussian elimination method.

$$\begin{bmatrix} -2 & 1 & -4 & -4 \\ 0 & 1 & -3 & -2 \\ 5 & 1 & -1 & -4 \end{bmatrix} \xrightarrow{-\frac{1}{2}R_1} \begin{bmatrix} 1 & -\frac{1}{2} & 2 & 2 \\ 0 & 1 & -3 & -2 \\ 5 & 1 & -1 & -4 \end{bmatrix}$$

$$\xrightarrow{-5R_1+R_3} \begin{bmatrix} 1 & -\frac{1}{2} & 2 & 2 \\ 0 & 1 & -3 & -2 \\ 0 & \frac{7}{2} & -11 & -14 \end{bmatrix}$$

$$\xrightarrow{-\frac{7}{2}R_2+R_3} \begin{bmatrix} 1 & -\frac{1}{2} & 2 & 2 \\ 0 & 1 & -3 & -2 \\ 0 & 0 & -\frac{1}{2} & -7 \end{bmatrix}$$

$$\begin{cases} x - \dfrac{1}{2}y + 2z = 2 \\ y - 3z = -2 \\ -\dfrac{1}{2}z = -7 \end{cases}$$

$$z = 14$$

$$y - 3(14) = -2$$

$$y = 40$$

$$x - \frac{1}{2}(40) + 2(14) = 2$$

$$x = -6$$

The solution set is $(-6, 40, 14)$

Using a graphing calculator, we get the following equivalent answer.

$$\begin{bmatrix} 1 & \frac{1}{5} & -\frac{1}{5} & -\frac{4}{5} \\ 0 & 1 & -\frac{22}{7} & -4 \\ 0 & 0 & 1 & 14 \end{bmatrix}$$

3. Evaluate.

$$A + C = \begin{bmatrix} 5 & 1 & -4 \\ -3 & 3 & -4 \\ 1 & 5 & -2 \end{bmatrix} + \begin{bmatrix} 1 & -3 & 2 \\ 5 & 0 & -4 \\ 5 & -1 & -3 \end{bmatrix}$$

$$= \begin{bmatrix} 6 & -2 & -2 \\ 2 & 3 & -8 \\ 6 & 4 & -5 \end{bmatrix}$$

5. Evaluate.

$$AB = \begin{bmatrix} 5 & 1 & -4 \\ -3 & 3 & -4 \\ 1 & 5 & -2 \end{bmatrix} \begin{bmatrix} 3 & -4 \\ 3 & 2 \\ 2 & -2 \end{bmatrix} = \begin{bmatrix} 10 & -10 \\ -8 & 26 \\ 14 & 10 \end{bmatrix}$$

Prepare for Section 10.4

P1. Find the order of the matrix.

2

P3. Evaluate.

$$(-1)^{1+1}(-3) + (-1)^{1+2}(-2) + (-1)^{1+3}(5)$$

$$= (-1)^2(-3) + (-1)^3(-2) + (-1)^4(5)$$

$$= -3 + (-1)(-2) + 5$$

$$= -3 + 2 + 5$$

$$= 4$$

P5. Simplify.

$$3\begin{bmatrix} -2 & 1 \\ 3 & -5 \end{bmatrix} = \begin{bmatrix} -6 & 3 \\ 9 & -15 \end{bmatrix}$$

Section 10.4 Exercises

1. Evaluate the determinant.

$$\begin{vmatrix} 2 & -1 \\ 3 & 5 \end{vmatrix} = 2(5) - (-1)(3) = 10 - (-3) = 13$$

3. Evaluate the determinant.

$$\begin{vmatrix} 5 & 0 \\ 2 & -3 \end{vmatrix} = 5(-3) - (2)(0) = -15 - 0 = -15$$

5. Evaluate the determinant.

$$\begin{vmatrix} 4 & 6 \\ 2 & 3 \end{vmatrix} = 4(3) - (2)(6) = 12 - 12 = 0$$

7. Evaluate the minor and cofactor.

$$M_{22} = \begin{vmatrix} 3 & 3 \\ 6 & 3 \end{vmatrix} = 3(3) - 6(3) = -9$$

$$C_{22} = (-1)^{2+2} M_{22} = M_{22} = -9$$

9. Evaluate the minor and cofactor.

$$M_{31} = \begin{vmatrix} -2 & 3 \\ 3 & 0 \end{vmatrix} = -2(0) - 3(3) = -9$$

$$C_{31} = (-1)^{3+1} M_{31} = M_{31} = -9$$

11. Evaluate the determinant by expanding by cofactors.

$$\begin{vmatrix} 2 & -3 & 1 \\ 2 & 0 & 2 \\ 3 & -2 & 4 \end{vmatrix} = -2 \begin{vmatrix} -3 & 1 \\ -2 & 4 \end{vmatrix} + 0 \begin{vmatrix} 2 & 1 \\ 3 & 4 \end{vmatrix} - 2 \begin{vmatrix} 2 & -3 \\ 3 & -2 \end{vmatrix}$$

$$= -2(-10) + 0 - 2(5)$$
$$= 20 - 10$$
$$= 10$$

13. Evaluate the determinant by expanding by cofactors.

$$\begin{vmatrix} -2 & 3 & 2 \\ 1 & 2 & -3 \\ -4 & -2 & 1 \end{vmatrix} = -2 \begin{vmatrix} 2 & -3 \\ -2 & 1 \end{vmatrix} - 3 \begin{vmatrix} 1 & -3 \\ -4 & 1 \end{vmatrix} + 2 \begin{vmatrix} 1 & 2 \\ -4 & -2 \end{vmatrix}$$

$$= -2(-4) - 3(-11) + 2(6)$$
$$= 8 + 33 + 12$$
$$= 53$$

15. Evaluate the determinant by expanding by cofactors.

$$\begin{vmatrix} 2 & -3 & 10 \\ 0 & 2 & -3 \\ 0 & 0 & 5 \end{vmatrix} = 2 \begin{vmatrix} 2 & -3 \\ 0 & 5 \end{vmatrix} - 0 \begin{vmatrix} -3 & 10 \\ 0 & 5 \end{vmatrix} + 0 \begin{vmatrix} -3 & 10 \\ 2 & -3 \end{vmatrix}$$

$$= 2(10) - 0 + 0$$
$$= 20$$

17. Evaluate the determinant by expanding by cofactors.

$$\begin{vmatrix} 0 & -2 & 4 \\ 1 & 0 & -7 \\ 5 & -6 & 0 \end{vmatrix} = 0 \begin{vmatrix} 0 & -7 \\ -6 & 0 \end{vmatrix} - (-2) \begin{vmatrix} 1 & -7 \\ 5 & 0 \end{vmatrix} + 4 \begin{vmatrix} 1 & 0 \\ 5 & -6 \end{vmatrix}$$

$$= 0 + 2(35) + 4(-6)$$
$$= 70 - 24$$
$$= 46$$

19. Evaluate the determinant by expanding by cofactors.

$$\begin{vmatrix} 4 & -3 & 3 \\ 2 & 1 & -4 \\ 6 & -2 & -1 \end{vmatrix} = 4 \begin{vmatrix} 1 & -4 \\ -2 & -1 \end{vmatrix} - (-3) \begin{vmatrix} 2 & -4 \\ 6 & -1 \end{vmatrix} + 3 \begin{vmatrix} 2 & 1 \\ 6 & -2 \end{vmatrix}$$

$$= 4(-9) + 3(22) + 3(-10)$$
$$= -36 + 66 - 30$$
$$= 0$$

21. Give a reason for each equality.

Row 2 consists entirely of zeros. Therefore, the determinant is zero.

23. Give a reason for each equality.

2 was factored from row 2.

25. Give a reason for each equality.

Row 1 was multiplied by –2 and added to row 2.

27. Give a reason for each equality.

2 was factored from column 1.

29. Give a reason for each equality.

The matrix is in triangular form.

The product of the elements on the main diagonal is –12.

Therefore, the value of the determinant is –12.

31. Give a reason for each equality.

Row 1 and row 3 were interchanged.

Therefore, the sign of the determinant was changed.

33. Put the matrix in triangular form and evaluate.

$$\begin{vmatrix} 2 & 4 & 1 \\ 1 & 2 & -1 \\ 1 & 2 & 2 \end{vmatrix} = - \begin{vmatrix} 1 & 2 & -1 \\ 2 & 4 & -1 \\ 1 & 2 & 2 \end{vmatrix} R_1 \leftrightarrow R_2$$

$$= - \begin{vmatrix} 1 & 2 & -1 \\ 0 & 0 & 1 \\ 0 & 0 & 3 \end{vmatrix} \begin{matrix} -2R_1 + R_2 \\ -R_1 + R_3 \end{matrix}$$

$$= -(1)(0)(3) = 0$$

35. Put the matrix in triangular form and evaluate.

$$\begin{vmatrix} 1 & 2 & -1 \\ 2 & 3 & 1 \\ 3 & 4 & 3 \end{vmatrix} = \begin{vmatrix} 1 & 2 & -1 \\ 0 & -1 & 3 \\ 0 & -2 & 6 \end{vmatrix} \begin{matrix} -2R_1 + R_2 \\ -3R_1 + R_3 \end{matrix}$$

$$= \begin{vmatrix} 1 & 2 & -1 \\ 0 & -1 & 3 \\ 0 & 0 & 0 \end{vmatrix} -2R_2 + R_3$$

$$= (-1)(0) = 0$$

37. Put the matrix in triangular form and evaluate.

$$\begin{vmatrix} 0 & -1 & 1 \\ 1 & 0 & -2 \\ 2 & 2 & 0 \end{vmatrix} = - \begin{vmatrix} 1 & 0 & -2 \\ 0 & -1 & 1 \\ 2 & 2 & 0 \end{vmatrix} R_1 \leftrightarrow R_2$$

$$= -\begin{vmatrix} 1 & 0 & -2 \\ 0 & -1 & 1 \\ 0 & 2 & 4 \end{vmatrix} -2R_1 + R_3$$

$$= -\begin{vmatrix} 1 & 0 & -2 \\ 0 & -1 & 1 \\ 0 & 0 & 6 \end{vmatrix} 2R_2 + R_3$$

$$= -(1)(-1)(6) = 6$$

39. Put the matrix in triangular form and evaluate.

$$\begin{vmatrix} 1 & 2 & -1 & 2 \\ 1 & -2 & 0 & 3 \\ 3 & 0 & 1 & 5 \\ -2 & -4 & 1 & 6 \end{vmatrix} = \begin{vmatrix} 1 & 2 & -1 & 2 \\ 0 & -4 & 1 & 1 \\ 0 & -6 & 4 & -1 \\ 0 & 0 & -1 & 10 \end{vmatrix} \begin{matrix} -1R_1 + R_2 \\ -3R_1 + R_3 \\ 2R_1 + R_4 \end{matrix}$$

$$= \begin{vmatrix} 1 & 2 & -1 & 2 \\ 0 & -4 & 1 & 1 \\ 0 & 0 & \frac{5}{2} & -\frac{5}{2} \\ 0 & 0 & -1 & 10 \end{vmatrix} -\frac{3}{2}R_2 + R_3$$

$$= \begin{vmatrix} 1 & 2 & -1 & 2 \\ 0 & -4 & 1 & 1 \\ 0 & 0 & \frac{5}{2} & -\frac{5}{2} \\ 0 & 0 & 0 & 9 \end{vmatrix} \frac{2}{5}R_3 + R_4$$

$$= 1(-4)\left(\frac{5}{2}\right)(9) = -90$$

41. Put the matrix in triangular form and evaluate.

$$\begin{vmatrix} 1 & 2 & 3 & -1 \\ 6 & 5 & 9 & 8 \\ 2 & 4 & 12 & -1 \\ 1 & 2 & 6 & -1 \end{vmatrix} = 3\begin{vmatrix} 1 & 2 & 1 & -1 \\ 6 & 5 & 3 & 8 \\ 2 & 4 & 4 & -1 \\ 1 & 2 & 2 & -1 \end{vmatrix} \quad \text{Factor 3 from } C_3$$

$$= 3\begin{vmatrix} 1 & 2 & 1 & -1 \\ 0 & -7 & -3 & 14 \\ 0 & 0 & 2 & 1 \\ 0 & 0 & 1 & 0 \end{vmatrix} \begin{matrix} -6R_1 + R_2 \\ -2R_1 + R_3 \\ -1R_1 + R_4 \end{matrix}$$

$$= 3\begin{vmatrix} 1 & 2 & 1 & -1 \\ 0 & -7 & -3 & 14 \\ 0 & 0 & 2 & 1 \\ 0 & 0 & 0 & -\frac{1}{2} \end{vmatrix} -\frac{1}{2}R_3 + R_4$$

$$= 3(1)(-7)(2)\left(-\frac{1}{2}\right) = 21$$

43. Use a graphing calculator to find the value of the determinant.

$$\begin{vmatrix} 2 & -2 & 3 & 1 \\ 5 & 2 & -2 & 3 \\ 6 & -1 & 2 & 3 \\ 2 & 3 & -1 & 5 \end{vmatrix} = 3$$

45. Use a graphing calculator to find the value of the determinant.

$$\begin{vmatrix} 3 & -1 & 6 & 2 \\ -1 & 0 & 3 & 6 \\ -1 & -2 & -4 & 6 \\ -6 & 4 & 4 & 6 \end{vmatrix} = 10$$

47. Solve using Cramer's Rule.

$$x = \frac{\begin{vmatrix} -1 & 4 \\ 5 & -6 \end{vmatrix}}{\begin{vmatrix} 5 & 4 \\ 3 & -6 \end{vmatrix}} = \frac{-14}{-42} = \frac{1}{3}$$

$$y = \frac{\begin{vmatrix} 5 & -1 \\ 3 & 5 \end{vmatrix}}{\begin{vmatrix} 5 & 4 \\ 3 & -6 \end{vmatrix}} = \frac{28}{-42} = -\frac{2}{3}$$

The solution is $\left(\frac{1}{3}, -\frac{2}{3}\right)$.

49. Solve using Cramer's Rule.

$$x = \frac{\begin{vmatrix} 0 & 2 \\ -3 & 1 \end{vmatrix}}{\begin{vmatrix} 7 & 2 \\ 2 & 1 \end{vmatrix}} = \frac{6}{3} = 2$$

$$y = \frac{\begin{vmatrix} 7 & 0 \\ 2 & -3 \end{vmatrix}}{\begin{vmatrix} 7 & 2 \\ 2 & 1 \end{vmatrix}} = \frac{-21}{3} = -7$$

The solution is $(2, -7)$.

51. Solve using Cramer's Rule.

$$x = \frac{\begin{vmatrix} 0 & -7 \\ 0 & 4 \end{vmatrix}}{\begin{vmatrix} 3 & -7 \\ 2 & 4 \end{vmatrix}} = \frac{0}{26} = 0 \qquad y = \frac{\begin{vmatrix} 3 & 0 \\ 2 & 0 \end{vmatrix}}{\begin{vmatrix} 3 & -7 \\ 2 & 4 \end{vmatrix}} = \frac{0}{26} = 0$$

The solution is $(0, 0)$.

53. Solve using Cramer's Rule.

$$D = \begin{vmatrix} 3 & -4 & 2 \\ 1 & -1 & 2 \\ 2 & 2 & 3 \end{vmatrix} = -17$$

$$D_x = \begin{vmatrix} 1 & -4 & 2 \\ -2 & -1 & 2 \\ -3 & 2 & 3 \end{vmatrix} = -21$$

$$D_y = \begin{vmatrix} 3 & 1 & 2 \\ 1 & -2 & 2 \\ 2 & -3 & 3 \end{vmatrix} = 3$$

$$D_z = \begin{vmatrix} 3 & -4 & 1 \\ 1 & -1 & -2 \\ 2 & 2 & -3 \end{vmatrix} = 29$$

$$x = \frac{D_x}{D} = \frac{-21}{-17} = \frac{21}{17}$$

$$y = \frac{D_y}{D} = \frac{3}{-17} = -\frac{3}{17}$$

$$z = \frac{D_z}{D} = \frac{29}{-17} = -\frac{29}{17}$$

The solution is $\left(\dfrac{21}{17}, -\dfrac{3}{17}, -\dfrac{29}{17} \right)$.

55. Solve using Cramer's Rule.

$$D = \begin{vmatrix} 1 & 4 & -2 \\ 3 & -2 & 3 \\ 2 & 1 & -3 \end{vmatrix} = 49$$

$$D_x = \begin{vmatrix} 0 & 4 & -2 \\ 4 & -2 & 3 \\ -1 & 1 & -3 \end{vmatrix} = 32$$

$$D_y = \begin{vmatrix} 1 & 0 & -2 \\ 3 & 4 & 3 \\ 2 & -1 & -3 \end{vmatrix} = 13$$

$$D_z = \begin{vmatrix} 1 & 4 & 0 \\ 3 & -2 & 4 \\ 2 & 1 & -1 \end{vmatrix} = 42$$

$$x = \frac{D_x}{D} = \frac{32}{49}$$

$$y = \frac{D_y}{D} = \frac{13}{49}$$

$$z = \frac{D_z}{D} = \frac{42}{49} = \frac{6}{7}$$

The solution is $\left(\dfrac{32}{49}, \dfrac{13}{49}, \dfrac{6}{7} \right)$.

57. Solve using Cramer's Rule.

$$D = \begin{vmatrix} 0 & 2 & -3 \\ 3 & -5 & 1 \\ 4 & 0 & 2 \end{vmatrix} = -64$$

$$D_x = \begin{vmatrix} 1 & 2 & -3 \\ 0 & -5 & 1 \\ -3 & 0 & 2 \end{vmatrix} = 29$$

$$D_y = \begin{vmatrix} 0 & 1 & -3 \\ 3 & 0 & 1 \\ 4 & -3 & 2 \end{vmatrix} = 25$$

$$D_z = \begin{vmatrix} 0 & 2 & 1 \\ 3 & -5 & 0 \\ 4 & 0 & -3 \end{vmatrix} = 38$$

$$x = \frac{D_x}{D} = \frac{29}{-64} = -\frac{29}{64}$$

$$y = \frac{D_y}{D} = \frac{25}{-64} = -\frac{25}{64}$$

$$z = \frac{D_z}{D} = \frac{38}{-64} = -\frac{19}{32}$$

The solution is $\left(-\dfrac{29}{64}, -\dfrac{25}{64}, -\dfrac{19}{32} \right)$.

59. Solve using Cramer's Rule.

$$D = \begin{vmatrix} 4 & -5 & 1 \\ 3 & 1 & 0 \\ 1 & -1 & 3 \end{vmatrix} = 53$$

$$D_x = \begin{vmatrix} -2 & -5 & 1 \\ 4 & 1 & 0 \\ 0 & -1 & 3 \end{vmatrix} = 50$$

$$D_y = \begin{vmatrix} 4 & -2 & 1 \\ 3 & 4 & 0 \\ 1 & 0 & 3 \end{vmatrix} = 62$$

$$D_z = \begin{vmatrix} 4 & -5 & -2 \\ 3 & 1 & 4 \\ 1 & -1 & 0 \end{vmatrix} = 4$$

$$x = \frac{D_x}{D} = \frac{50}{53}$$

$$y = \frac{D_y}{D} = \frac{62}{53}$$

$$z = \frac{D_z}{D} = \frac{4}{53}$$

The solution is $\left(\dfrac{50}{53}, \dfrac{62}{53}, \dfrac{4}{53} \right)$.

61. Find the area of the triangle.

$$\frac{1}{2}\begin{vmatrix} 2 & 3 & 1 \\ -1 & 0 & 1 \\ 4 & 8 & 1 \end{vmatrix} = \frac{1}{2}(-9) = -\frac{9}{2}$$

$$\left|-\frac{9}{2}\right| = \frac{9}{2}$$

The area of the triangle is $\frac{9}{2}$ square units.

63. Find the area of the triangle.

$$\frac{1}{2}\begin{vmatrix} 4 & 9 & 1 \\ 8 & 2 & 1 \\ -3 & -2 & 1 \end{vmatrix} = \frac{1}{2}(-93) = -\frac{93}{2}$$

$$\left|-\frac{93}{2}\right| = \frac{93}{2}$$

The area of the triangle is $\frac{93}{2}$ square units.

65. Find the area of the triangle.

$$\frac{1}{2}\begin{vmatrix} 4 & 3 & 1 \\ 1 & 2 & 1 \\ -2 & 3 & 1 \end{vmatrix} = \frac{1}{2}(-6) = -3$$

$$|-3| = 3$$

The area of the triangle is 3 square units.

67. Find the equation of the line.

$$\begin{vmatrix} x & y & 1 \\ -4 & 1 & 1 \\ 5 & -2 & 1 \end{vmatrix} = 0$$

$$x(1+2) - y(-4-5) + 1(8-5) = 0$$
$$3x + 9y + 3 = 0$$
$$3x + 9y = -3$$
$$3(x + 3y) = 3(-1)$$
$$x + 3y = -1$$

$x + 3y = -1$ is the equation of the line passing through

the points (–4, 1) and (5,–2).

69. Find the equation of the line.

$$\begin{vmatrix} x & y & 1 \\ 2 & 5 & 1 \\ -4 & 5 & 1 \end{vmatrix} = 0$$

$$x(5-5) - y(2+4) + 1(10+20) = 0$$
$$-6y + 30 = 0$$
$$-6y = -30$$
$$y = 5$$

$y = 5$ is the equation of the line passing through the

points (2, 5) and (–4, 5).

71. Find the equation of the line.

$$\begin{vmatrix} x & y & 1 \\ -3 & 4 & 1 \\ 2 & -3 & 1 \end{vmatrix} = 0$$

$$x(4+3) - y(-3-2) + 1(9-8) = 0$$
$$7x + 5y + 1 = 0$$
$$7x + 5y = -1$$

$7x + 5y = -1$ is the equation of the line passing

through the points (–3, 4) and (2,–3).

73. Find the area of the polygon using the surveyor's area

formula.

$$A = \frac{1}{2}\left(\begin{vmatrix} 8 & 25 \\ -4 & 5 \end{vmatrix} + \begin{vmatrix} 25 & 15 \\ 5 & 9 \end{vmatrix} + \begin{vmatrix} 15 & 17 \\ 9 & 20 \end{vmatrix} + \begin{vmatrix} 17 & 0 \\ 20 & 10 \end{vmatrix} + \begin{vmatrix} 0 & 8 \\ 10 & -4 \end{vmatrix}\right)$$

$$= \frac{1}{2}[140 + 150 + 147 + 170 + (-80)]$$

$$= \frac{1}{2}(527) = 263.5 \text{ square units}$$

75. Find k so that the system has a unique solution.

$$D = \begin{vmatrix} 1 & 2 & -3 \\ 2 & k & -4 \\ 1 & -2 & 1 \end{vmatrix} = 4k - 8$$

For the system of equations to have a unique solution,

the determinant of the coefficient matrix cannot be

zero.

$$4k - 8 = 0$$
$$4k = 8$$
$$k = 2$$

The system of equations has a unique solution for all

values of k except $k = 2$.

Exploring Concepts with Technology

$$XT = [0.428 \quad 0.572]$$

$$XT^2 \approx [0.45236 \quad 0.54764]$$

$$XT^3 \approx [0.47355 \quad 0.52645]$$

$$XT^{20} \approx [0.60209 \quad 0.39791]$$

$$XT^{40} \approx [0.61456 \quad 0.38544]$$

$$XT^{60} \approx [0.61533 \quad 0.38467]$$

$$XT^{100} \approx [0.61538 \quad 0.38462]$$

It appears that as the number of weeks increases, Super A will get slightly more than 61.5% of the neighborhood and Super B will get slightly less than 38.5% of the neighborhood.

Changing the market share *does not* affect the result (to 6 decimal places) after 100 weeks.

Three Department Stores: After 100 months, Super A will have 23.87% of the market share, Super B will have 33.65% of the market share, and Super C will have 42.48% of the market share.

Chapter 7 Review Exercises

1. Writing the augmented, coefficient and constant matrices of the given system of equations. [10.1]

$$\begin{bmatrix} 3 & 5 & | & 6 \\ 2 & -7 & | & -1 \end{bmatrix}, \begin{bmatrix} 3 & 5 \\ 2 & -7 \end{bmatrix}, \begin{bmatrix} 6 \\ -1 \end{bmatrix}$$

3. Writing the augmented, coefficient and constant matrices of the given system of equations. [10.1]

$$\begin{bmatrix} 3 & 2 & -3 & 1 & | & 4 \\ 1 & 0 & -4 & 0 & | & -2 \\ -2 & -1 & 0 & -2 & | & 0 \end{bmatrix}, \begin{bmatrix} 3 & 2 & -3 & 1 \\ 1 & 0 & -4 & 0 \\ -2 & -1 & 0 & -2 \end{bmatrix}, \begin{bmatrix} 4 \\ -2 \\ 0 \end{bmatrix}$$

The solutions and answers to Exercises 5 – 12 are not unique. The following solutions list the elementary row operations used to produce the row echelon forms shown.

5. Using elementary row operations to achieve row echelon form. [10.1]

$$\begin{bmatrix} -1 & -2 & -8 \\ 3 & 7 & 3 \end{bmatrix} \xrightarrow{-R_1} \begin{bmatrix} 1 & 2 & 8 \\ 3 & 7 & 3 \end{bmatrix}$$

$$\xrightarrow{-3R_1+R_2} \begin{bmatrix} 1 & 2 & 8 \\ 0 & 1 & -21 \end{bmatrix}$$

Using a graphing calculator, we get the following equivalent answer.

$$\begin{bmatrix} 1 & \frac{7}{3} & 1 \\ 0 & 1 & -21 \end{bmatrix}$$

7. Using elementary row operations to achieve row echelon form. [10.1]

$$\begin{bmatrix} 3 & 3 & -4 \\ 3 & -1 & -4 \end{bmatrix} \xrightarrow{\frac{1}{3}R_1} \begin{bmatrix} 1 & 1 & -\frac{4}{3} \\ 3 & -1 & -4 \end{bmatrix}$$

$$\xrightarrow{-3R_1+R_2} \begin{bmatrix} 1 & 1 & -\frac{4}{3} \\ 0 & -4 & 0 \end{bmatrix} \xrightarrow{-\frac{1}{4}R_2} \begin{bmatrix} 1 & 1 & -\frac{4}{3} \\ 0 & 1 & 0 \end{bmatrix}$$

Using a graphing calculator, we get the same answer.

9. Using elementary row operations to achieve row echelon form. [10.1]

$$\begin{bmatrix} -5 & -3 & 6 & -14 \\ -10 & 7 & -6 & 3 \\ 1 & -1 & 1 & -1 \end{bmatrix}$$

$$\xrightarrow{R_1 \longleftrightarrow R_3} \begin{bmatrix} 1 & -1 & 1 & -1 \\ -10 & 7 & -6 & 3 \\ -5 & -3 & 6 & -14 \end{bmatrix}$$

$$\xrightarrow[5R_1+R_3]{10R_1+R_2} \begin{bmatrix} 1 & -1 & 1 & -1 \\ 0 & -3 & 4 & -7 \\ 0 & -8 & 11 & -19 \end{bmatrix}$$

$$\xrightarrow{-\frac{1}{3}R_2} \begin{bmatrix} 1 & -1 & 1 & -1 \\ 0 & 1 & -\frac{4}{3} & \frac{7}{3} \\ 0 & -8 & 11 & -19 \end{bmatrix}$$

$$\xrightarrow{8R_2+R_3} \begin{bmatrix} 1 & -1 & 1 & -1 \\ 0 & 1 & -\frac{4}{3} & \frac{7}{3} \\ 0 & 0 & \frac{1}{3} & -\frac{1}{3} \end{bmatrix}$$

$$\xrightarrow{3R_3} \begin{bmatrix} 1 & -1 & 1 & -1 \\ 0 & 1 & -\frac{4}{3} & \frac{7}{3} \\ 0 & 0 & 1 & -1 \end{bmatrix}$$

Using a graphing calculator, we get the following equivalent answer.

$$\begin{bmatrix} 1 & -\frac{7}{10} & \frac{3}{5} & -\frac{3}{10} \\ 0 & 1 & -\frac{18}{13} & \frac{31}{13} \\ 0 & 0 & 1 & -1 \end{bmatrix}$$

11. Using elementary row operations to achieve row echelon form. [10.1]

$$\begin{bmatrix} 3 & 3 & 4 & 2 \\ 1 & 2 & 2 & 4 \\ -4 & 4 & 0 & 2 \end{bmatrix} \xrightarrow{R_1 \longleftrightarrow R_2} \begin{bmatrix} 1 & 2 & 2 & 4 \\ 3 & 3 & 4 & 2 \\ -4 & 4 & 0 & 2 \end{bmatrix}$$

$$\xrightarrow[\;4R_1+R_3\;]{-3R_1+R_2} \begin{bmatrix} 1 & 2 & 2 & 4 \\ 0 & -3 & -2 & -10 \\ 0 & 12 & 8 & 18 \end{bmatrix}$$

$$\xrightarrow[\;\frac{1}{2}R_3\;]{-\frac{1}{3}R_2} \begin{bmatrix} 1 & 2 & 2 & 4 \\ 0 & 1 & \frac{2}{3} & \frac{10}{3} \\ 0 & 6 & 4 & 9 \end{bmatrix}$$

$$\xrightarrow{-6R_2+R_3} \begin{bmatrix} 1 & 2 & 2 & 4 \\ 0 & 1 & \frac{2}{3} & \frac{10}{3} \\ 0 & 0 & 0 & -11 \end{bmatrix}$$

Using a graphing calculator, we get the following equivalent answer.

$$\begin{bmatrix} 1 & -1 & 0 & -\frac{1}{2} \\ 0 & 1 & \frac{2}{3} & \frac{7}{12} \\ 0 & 0 & 0 & 1 \end{bmatrix}$$

13. Solve the system using Gaussian elimination. [10.1]

$$\begin{bmatrix} 2 & -3 & 7 \\ 3 & -4 & 10 \end{bmatrix} \xrightarrow{-1R_1+R_2} \begin{bmatrix} 2 & -3 & 7 \\ 1 & -1 & 3 \end{bmatrix}$$

$$\xrightarrow{R_1 \longleftrightarrow R_2} \begin{bmatrix} 1 & -1 & 3 \\ 2 & -3 & 7 \end{bmatrix}$$

$$\xrightarrow{-2R_1+R_2} \begin{bmatrix} 1 & -1 & 3 \\ 0 & -1 & 1 \end{bmatrix}$$

$$\xrightarrow{-1R_2} \begin{bmatrix} 1 & -1 & 3 \\ 0 & 1 & -1 \end{bmatrix}$$

$$\begin{cases} x-y=3 \qquad x-(-1)=3 \\ \quad\; y=-1 \qquad\;\; x=2 \end{cases}$$

The solution is $(2, -1)$.

15. Solve the system using Gaussian elimination. [10.1]

$$\begin{bmatrix} 4 & -5 & 12 \\ 3 & 1 & 9 \end{bmatrix} \xrightarrow{-1R_2+R_1} \begin{bmatrix} 1 & -6 & 3 \\ 3 & 1 & 9 \end{bmatrix}$$

$$\xrightarrow{-3R_1+R_2} \begin{bmatrix} 1 & -6 & 3 \\ 0 & 19 & 0 \end{bmatrix}$$

$$\xrightarrow{\frac{1}{19}R_2} \begin{bmatrix} 1 & -6 & 3 \\ 0 & 1 & 0 \end{bmatrix}$$

$$\begin{cases} x-6y=3 \\ \quad\; y=0 \end{cases}$$

$$x-6(0)=3$$
$$x=3$$

The solution is $(3, 0)$.

17. Solve the system using Gaussian elimination. [10.1]

$$\begin{bmatrix} 1 & 2 & 3 & 5 \\ 3 & 8 & 11 & 17 \\ 2 & 6 & 7 & 12 \end{bmatrix} \xrightarrow[\;-2R_1+R_3\;]{-3R_1+R_2} \begin{bmatrix} 1 & 2 & 3 & 5 \\ 0 & 2 & 2 & 2 \\ 0 & 2 & 1 & 2 \end{bmatrix}$$

$$\xrightarrow{\frac{1}{2}R_2} \begin{bmatrix} 1 & 2 & 3 & 5 \\ 0 & 1 & 1 & 1 \\ 0 & 2 & 1 & 2 \end{bmatrix}$$

$$\xrightarrow{-2R_2+R_3} \begin{bmatrix} 1 & 2 & 3 & 5 \\ 0 & 1 & 1 & 1 \\ 0 & 0 & -1 & 0 \end{bmatrix}$$

$$\xrightarrow{-1R_3} \begin{bmatrix} 1 & 2 & 3 & 5 \\ 0 & 1 & 1 & 1 \\ 0 & 0 & 1 & 0 \end{bmatrix}$$

$$\begin{cases} x+2y+3z=5 \\ \quad\; y+\;z=1 \\ \qquad\quad z=0 \end{cases}$$

$$y+0=1$$
$$y=1$$

$$x+2(1)+3(0)=5$$
$$x=3$$

The solution is $(3, 1, 0)$.

19. Solve the system using Gaussian elimination. [10.1]

$$\begin{bmatrix} 2 & -1 & -1 & 4 \\ 1 & -2 & -2 & 5 \\ 3 & -3 & -8 & 19 \end{bmatrix} \xrightarrow{R_1 \longleftrightarrow R_2} \begin{bmatrix} 1 & -2 & -2 & 5 \\ 2 & -1 & -1 & 4 \\ 3 & -3 & -8 & 19 \end{bmatrix}$$

$$\xrightarrow[\;-3R_1+R_3\;]{-2R_1+R_2} \begin{bmatrix} 1 & -2 & -2 & 5 \\ 0 & 3 & 3 & -6 \\ 0 & 3 & -2 & 4 \end{bmatrix}$$

$$\xrightarrow{\frac{1}{3}R_2} \begin{bmatrix} 1 & -2 & -2 & 5 \\ 0 & 1 & 1 & -2 \\ 0 & 3 & -2 & 4 \end{bmatrix}$$

$$\xrightarrow{-3R_2+R_3} \begin{bmatrix} 1 & -2 & -2 & 5 \\ 0 & 1 & 1 & -2 \\ 0 & 0 & -5 & 10 \end{bmatrix}$$

$$\xrightarrow{-\frac{1}{5}R_3} \begin{bmatrix} 1 & -2 & -2 & | & 5 \\ 0 & 1 & 1 & | & -2 \\ 0 & 0 & 1 & | & -2 \end{bmatrix}$$

$$\begin{cases} x - 2y - 2z = 5 \\ \quad\; y + z = -2 \\ \qquad\quad z = -2 \end{cases}$$

$$y + (-2) = -2$$
$$y = 0$$

$$x - 2(0) - 2(-2) = 5$$
$$x = 1$$

The solution is $(1, 0, -2)$.

21. Solve the system using Gaussian elimination. [10.1]

$$\begin{bmatrix} 4 & -9 & 6 & | & 54 \\ 3 & -8 & 8 & | & 49 \\ 1 & -3 & 2 & | & 17 \end{bmatrix} \xrightarrow{R_1 \longleftrightarrow R_3} \begin{bmatrix} 1 & -3 & 2 & | & 17 \\ 3 & -8 & 8 & | & 49 \\ 4 & -9 & 6 & | & 54 \end{bmatrix}$$

$$\xrightarrow[\;-4R_1 + R_3\;]{-3R_1 + R_2} \begin{bmatrix} 1 & -3 & 2 & | & 17 \\ 0 & 1 & 2 & | & -2 \\ 0 & 3 & -2 & | & -14 \end{bmatrix}$$

$$\xrightarrow{-3R_2 + R_3} \begin{bmatrix} 1 & -3 & 2 & | & 17 \\ 0 & 1 & 2 & | & -2 \\ 0 & 0 & -8 & | & -8 \end{bmatrix}$$

$$\xrightarrow{-\frac{1}{8}R_3} \begin{bmatrix} 1 & -3 & 2 & | & 17 \\ 0 & 1 & 2 & | & -2 \\ 0 & 0 & 1 & | & 1 \end{bmatrix}$$

$$\begin{cases} x - 3y + 2z = 17 \\ \quad\; y + 2z = -2 \\ \qquad\quad z = 1 \end{cases}$$

$$y + 2(1) = -2$$
$$y = -4$$

$$x - 3(-4) + 2(1) = 17$$
$$x = 3$$

The solution is $(3, -4, 1)$.

23. Solve the system using Gaussian elimination. [10.1]

$$\begin{bmatrix} 1 & 4 & -3 & | & 2 \\ -2 & -8 & 6 & | & 1 \\ -1 & 4 & -6 & | & 3 \end{bmatrix} \xrightarrow[\;R_1 + R_3\;]{2R_1 + R_2} \begin{bmatrix} 1 & 4 & -3 & | & 2 \\ 0 & 0 & 0 & | & 5 \\ 0 & 8 & -9 & | & 5 \end{bmatrix}$$

$$\begin{cases} x + 4y - 3z = 2 \\ \qquad\quad 0z = 5 \\ \quad 8y - 9z = 5 \end{cases}$$

Because $0z = 5$ has no solution, the system of equations has no solution.

25. Solve the system using Gaussian elimination. [10.1]

$$\begin{bmatrix} 2 & -5 & 1 & | & -2 \\ 2 & 3 & -3 & | & 2 \\ 2 & -1 & -1 & | & 2 \end{bmatrix} \xrightarrow{\frac{1}{2}R_1} \begin{bmatrix} 1 & -\frac{5}{2} & \frac{1}{2} & | & -1 \\ 2 & 3 & -3 & | & 2 \\ 2 & -1 & -1 & | & 2 \end{bmatrix}$$

$$\xrightarrow[\;-2R_1 + R_3\;]{-2R_1 + R_2} \begin{bmatrix} 1 & -\frac{5}{2} & \frac{1}{2} & | & -1 \\ 0 & 8 & -4 & | & 4 \\ 0 & 4 & -2 & | & 4 \end{bmatrix}$$

$$\xrightarrow{\frac{1}{8}R_2} \begin{bmatrix} 1 & -\frac{5}{2} & \frac{1}{2} & | & -1 \\ 0 & 1 & -\frac{1}{2} & | & \frac{1}{2} \\ 0 & 4 & -2 & | & 4 \end{bmatrix}$$

$$\xrightarrow{-4R_2 + R_3} \begin{bmatrix} 1 & -\frac{5}{2} & \frac{1}{2} & | & -1 \\ 0 & 1 & -\frac{1}{2} & | & \frac{1}{2} \\ 0 & 0 & 0 & | & 2 \end{bmatrix}$$

$$\begin{cases} x - \frac{5}{2}y + \frac{1}{2}z = -1 \\ \qquad\; y - \frac{1}{2}z = \frac{1}{2} \\ \qquad\qquad\; 0z = 2 \end{cases}$$

Because $0z = 2$ has no solution, the system of equations has no solution.

27. Solve the system using Gaussian elimination. [10.1]

$$\begin{bmatrix} 1 & -3 & 1 & | & -3 \\ 2 & 1 & -2 & | & 3 \end{bmatrix} \xrightarrow{-2R_1 + R_2} \begin{bmatrix} 1 & -3 & 1 & | & -3 \\ 0 & 7 & -4 & | & 9 \end{bmatrix}$$

$$\xrightarrow{\frac{1}{7}R_2} \begin{bmatrix} 1 & -3 & 1 & | & -3 \\ 0 & 1 & -\frac{4}{7} & | & \frac{9}{7} \end{bmatrix}$$

$$\begin{cases} x - 3y + z = -3 \\ \quad\; y - \frac{4}{7}z = \frac{9}{7} \end{cases}$$

$$y - \frac{4}{7}z = \frac{9}{7}$$
$$y = \frac{4}{7}z + \frac{9}{7}$$

$$x - 3\left(\frac{4}{7}z + \frac{9}{7}\right) + z = -3$$
$$x = \frac{5}{7}z + \frac{6}{7}$$

Let z be any real number c.

The solution is $\left(\frac{5}{7}c + \frac{6}{7},\; \frac{4}{7}c + \frac{9}{7},\; c\right)$.

29. Solve the system using Gaussian elimination. [10.1]

$$\begin{bmatrix} 1 & 1 & 2 & | & -5 \\ 2 & 3 & 5 & | & -13 \\ 2 & 5 & 7 & | & -19 \end{bmatrix} \xrightarrow[\substack{-2R_1+R_2 \\ -2R_1+R_3}]{} \begin{bmatrix} 1 & 1 & 2 & | & -5 \\ 0 & 1 & 1 & | & -3 \\ 0 & 3 & 3 & | & -9 \end{bmatrix}$$

$$\xrightarrow{-3R_2+R_3} \begin{bmatrix} 1 & 1 & 2 & | & -5 \\ 0 & 1 & 1 & | & -3 \\ 0 & 0 & 0 & | & 0 \end{bmatrix}$$

$$\begin{cases} x+y+2z=-5 \\ \quad y+\ z=-3 \end{cases}$$

$$y=-z-3$$

$$x+(-z-3)+2z=-5$$

$$x=-z-2$$

Let z be any real number c.

The solution is $(-c-2, -c-3, \ c)$.

31. Solve the system using Gaussian elimination. [10.1]

$$\begin{bmatrix} 1 & 2 & -1 & 2 & | & 1 \\ 3 & 8 & 1 & 4 & | & 1 \\ 2 & 7 & 3 & 2 & | & 0 \\ 1 & 3 & -2 & 5 & | & 6 \end{bmatrix} \xrightarrow[\substack{-3R_1+R_2 \\ -2R_1+R_3 \\ -1R_1+R_4}]{} \begin{bmatrix} 1 & 2 & -1 & 2 & | & 1 \\ 0 & 2 & 4 & -2 & | & -2 \\ 0 & 3 & 5 & -2 & | & -2 \\ 0 & 1 & -1 & 3 & | & 5 \end{bmatrix}$$

$$\xrightarrow{-\frac{1}{2}R_2} \begin{bmatrix} 1 & 2 & -1 & 2 & | & 1 \\ 0 & 1 & 2 & -1 & | & -1 \\ 0 & 3 & 5 & -2 & | & -2 \\ 0 & 1 & -1 & 3 & | & 5 \end{bmatrix}$$

$$\xrightarrow[\substack{-3R_2+R_3 \\ -1R_2+R_4}]{} \begin{bmatrix} 1 & 2 & -1 & 2 & | & 1 \\ 0 & 1 & 2 & -1 & | & -1 \\ 0 & 0 & -1 & 1 & | & 1 \\ 0 & 0 & -3 & 4 & | & 6 \end{bmatrix}$$

$$\xrightarrow{-1R_3} \begin{bmatrix} 1 & 2 & -1 & 2 & | & 1 \\ 0 & 1 & 2 & -1 & | & -1 \\ 0 & 0 & 1 & -1 & | & -1 \\ 0 & 0 & -3 & 4 & | & 6 \end{bmatrix}$$

$$\xrightarrow{3R_3+R_4} \begin{bmatrix} 1 & 2 & -1 & 2 & | & 1 \\ 0 & 1 & 2 & -1 & | & -1 \\ 0 & 0 & 1 & -1 & | & -1 \\ 0 & 0 & 0 & 1 & | & 3 \end{bmatrix}$$

$$\begin{cases} w+2x-\ y+2z=\ 1 \\ \quad x+2y-\ z=-1 \\ \qquad\quad y-\ z=-1 \\ \qquad\qquad\ z=\ 3 \end{cases}$$

$$y-3=-1$$

$$y=2$$

$$x+2(2)-3=-1$$

$$x=-2$$

$$w+2(-2)-2+2(3)=1$$

$$w=1$$

The solution is $(1, -2, 2, 3)$.

33. Solve the system using Gaussian elimination. [10.1]

$$\begin{bmatrix} 1 & 3 & 1 & -4 & | & 3 \\ 1 & 4 & 3 & -6 & | & 5 \\ 2 & 8 & 7 & -5 & | & 11 \\ 2 & 5 & 0 & -6 & | & 4 \end{bmatrix} \xrightarrow[\substack{-1R_1+R_2 \\ -2R_1+R_3 \\ -2R_1+R_4}]{} \begin{bmatrix} 1 & 3 & 1 & -4 & | & 3 \\ 0 & 1 & 2 & -2 & | & 2 \\ 0 & 2 & 5 & 3 & | & 5 \\ 0 & -1 & -2 & 2 & | & -2 \end{bmatrix}$$

$$\xrightarrow[\substack{-2R_2+R_3 \\ 1R_2+R_4}]{} \begin{bmatrix} 1 & 3 & 1 & -4 & | & 3 \\ 0 & 1 & 2 & -2 & | & 2 \\ 0 & 0 & 1 & 7 & | & 1 \\ 0 & 0 & 0 & 0 & | & 0 \end{bmatrix}$$

$$\begin{cases} w+3x+\ y-4z=3 \\ \quad x+2y-2z=2 \\ \qquad\quad y+7z=1 \end{cases}$$

$$y=-7z+1$$

$$x+2(-7z+1)-2z=2$$

$$x=16z$$

$$w+3(16z)+(-7z+1)-4z=3$$

$$w=-37z+2$$

Let z be any real number c.

The solution is $(-37c+2, 16c, -7c+1, c)$.

35. Find the polynomial. [10.1]

Because there are three points, the degree of the interpolating polynomial is at most 2.

The form of the polynomial is

$$p(x)=a_2x^2+a_1x+a_0.$$

Use this polynomial and the given points to find the system of equations.

$$p(-1)=a_2(-1)^2+a_1(-1)+a_0=-4$$

$$p(2)=a_2(2)^2+a_1(2)+a_0=8$$

$$p(3)=a_2(3)^2+a_1(3)+a_0=16$$

The system of equations and the associated augmented matrix are

$$\begin{cases} a_2 - a_1 + a_0 = -4 \\ 4a_2 + 2a_1 + a_0 = 8 \\ 9a_2 + 3a_1 + a_0 = 16 \end{cases} \qquad \begin{bmatrix} 1 & -1 & 1 & | & -4 \\ 4 & 2 & 1 & | & 8 \\ 9 & 3 & 1 & | & 16 \end{bmatrix}$$

The ref (**r**ow **e**chelon **f**orm) feature of a graphing calculator can be used to rewrite the augmented matrix in echelon form.

Consider using the function of your calculator that converts a decimal to a fraction.

The augmented matrix in echelon form and resulting system of equations are

$$\begin{bmatrix} 1 & \frac{1}{3} & \frac{1}{9} & | & \frac{16}{9} \\ 0 & 1 & -\frac{2}{3} & | & \frac{13}{3} \\ 0 & 0 & 1 & | & -2 \end{bmatrix} \qquad \begin{cases} a_2 + \frac{1}{3}a_1 + \frac{1}{9}a_0 = \frac{16}{9} \\ a_1 - \frac{2}{3}a_0 = \frac{13}{2} \\ a_0 = -2 \end{cases}$$

Solving by back substitution yields

$$a_0 = -2, \quad a_1 = 3, \quad \text{and} \quad a_2 = 1.$$

The interpolating polynomial is $p(x) = x^2 + 3x - 2$.

37. Find $3A$. [10.2]

$$3A = 3\begin{bmatrix} 2 & -1 & 3 \\ 3 & 2 & -1 \end{bmatrix} = \begin{bmatrix} 6 & -3 & 9 \\ 9 & 6 & -3 \end{bmatrix}$$

39. Find $-A + D$. [10.2]

$$-A + D = -\begin{bmatrix} 2 & -1 & 3 \\ 3 & 2 & -1 \end{bmatrix} + \begin{bmatrix} -3 & 4 & 2 \\ 4 & -2 & 5 \end{bmatrix}$$

$$= \begin{bmatrix} -2 & 1 & -3 \\ -3 & -2 & 1 \end{bmatrix} + \begin{bmatrix} -3 & 4 & 2 \\ 4 & -2 & 5 \end{bmatrix}$$

$$= \begin{bmatrix} -5 & 5 & -1 \\ 1 & -4 & 6 \end{bmatrix}$$

41. Find AB. [10.2]

$$AB = \begin{bmatrix} 2 & -1 & 3 \\ 3 & 2 & -1 \end{bmatrix}\begin{bmatrix} 0 & -2 \\ 4 & 2 \\ 1 & -3 \end{bmatrix} = \begin{bmatrix} -1 & -15 \\ 7 & 1 \end{bmatrix}$$

43. Find BA. [10.2]

$$BA = \begin{bmatrix} 0 & -2 \\ 4 & 2 \\ 1 & -3 \end{bmatrix}\begin{bmatrix} 2 & -1 & 3 \\ 3 & 2 & -1 \end{bmatrix} = \begin{bmatrix} -6 & -4 & 2 \\ 14 & 0 & 10 \\ -7 & -7 & 6 \end{bmatrix}$$

45. Find C^2. [10.2]

$$C^2 = C \cdot C = \begin{bmatrix} 2 & 6 & 1 \\ 1 & 2 & -1 \\ 2 & 4 & -1 \end{bmatrix}\begin{bmatrix} 2 & 6 & 1 \\ 1 & 2 & -1 \\ 2 & 4 & -1 \end{bmatrix} = \begin{bmatrix} 12 & 28 & -5 \\ 2 & 6 & 0 \\ 6 & 16 & -1 \end{bmatrix}$$

47. Find BAC. [10.2]

$$BAC = \begin{bmatrix} 0 & -2 \\ 4 & 2 \\ 1 & -3 \end{bmatrix}\begin{bmatrix} 2 & -1 & 3 \\ 3 & 2 & -1 \end{bmatrix}\begin{bmatrix} 2 & 6 & 1 \\ 1 & 2 & -1 \\ 2 & 4 & -1 \end{bmatrix}$$

$$= \begin{bmatrix} -6 & -4 & 2 \\ 14 & 0 & 10 \\ -7 & -7 & 6 \end{bmatrix}\begin{bmatrix} 2 & 6 & 1 \\ 1 & 2 & -1 \\ 2 & 4 & -1 \end{bmatrix} = \begin{bmatrix} -12 & -36 & -4 \\ 48 & 124 & 4 \\ -9 & -32 & -6 \end{bmatrix}$$

49. Find $AB - BA$. [10.2]

$$AB - BA = \begin{bmatrix} 2 & -1 & 3 \\ 3 & 2 & -1 \end{bmatrix}\begin{bmatrix} 0 & -2 \\ 4 & 2 \\ 1 & -3 \end{bmatrix} - \begin{bmatrix} 0 & -2 \\ 4 & 2 \\ 1 & -3 \end{bmatrix}\begin{bmatrix} 2 & -1 & 3 \\ 3 & 2 & -1 \end{bmatrix}$$

$$= \begin{bmatrix} -1 & -15 \\ 7 & 1 \end{bmatrix} - \begin{bmatrix} -6 & -4 & 2 \\ 14 & 0 & 10 \\ -7 & -7 & 6 \end{bmatrix}$$

Not possible since AB is of order 2×2 and BA is of order 3×3.

51. Find $(A - D)C$. [10.2]

$(A - D)C$

$$= \left(\begin{bmatrix} 2 & -1 & 3 \\ 3 & 2 & -1 \end{bmatrix} - \begin{bmatrix} -3 & 4 & 2 \\ 4 & -2 & 5 \end{bmatrix}\right)\begin{bmatrix} 2 & 6 & 1 \\ 1 & 2 & -1 \\ 2 & 4 & -1 \end{bmatrix}$$

$$= \begin{bmatrix} 5 & -5 & 1 \\ -1 & 4 & -6 \end{bmatrix}\begin{bmatrix} 2 & 6 & 1 \\ 1 & 2 & -1 \\ 2 & 4 & -1 \end{bmatrix} = \begin{bmatrix} 7 & 24 & 9 \\ -10 & -22 & 1 \end{bmatrix}$$

53. Find C^{-1}. [10.3]

$$\begin{bmatrix} 2 & 6 & 1 & | & 1 & 0 & 0 \\ 1 & 2 & -1 & | & 0 & 1 & 0 \\ 2 & 4 & -1 & | & 0 & 0 & 1 \end{bmatrix}$$

$$\xrightarrow{\frac{1}{2}R_1} \begin{bmatrix} 1 & 3 & \frac{1}{2} & | & \frac{1}{2} & 0 & 0 \\ 1 & 2 & -1 & | & 0 & 1 & 0 \\ 2 & 4 & -1 & | & 0 & 0 & 1 \end{bmatrix}$$

$$\xrightarrow[-2R_1+R_3]{-1R_1+R_2} \begin{bmatrix} 1 & 3 & \frac{1}{2} & | & \frac{1}{2} & 0 & 0 \\ 0 & -1 & -\frac{3}{2} & | & -\frac{1}{2} & 1 & 0 \\ 0 & -2 & -2 & | & -1 & 0 & 1 \end{bmatrix}$$

$$\begin{array}{c} -1R_2 \\ -3R_2+R_1 \\ \underline{2R_2+R_3} \end{array} \longrightarrow \left[\begin{array}{ccc|ccc} 1 & 0 & -4 & -1 & 3 & 0 \\ 0 & 1 & \frac{3}{2} & \frac{1}{2} & -1 & 0 \\ 0 & 0 & 1 & 0 & -2 & 1 \end{array}\right]$$

$$\begin{array}{c} 4R_3+R_1 \\ \underline{-\frac{3}{2}R_3+R_2} \end{array} \longrightarrow \left[\begin{array}{ccc|ccc} 1 & 0 & 0 & -1 & -5 & 4 \\ 0 & 1 & 0 & \frac{1}{2} & 2 & -\frac{3}{2} \\ 0 & 0 & 1 & 0 & -2 & 1 \end{array}\right]$$

$$C^{-1} = \left[\begin{array}{ccc} -1 & -5 & 4 \\ \frac{1}{2} & 2 & -\frac{3}{2} \\ 0 & -2 & 1 \end{array}\right]$$

55. Find the adjacency matrix A and A^2. Determine the number of walks of length 2. [10.2]

$$A = \left[\begin{array}{ccccc} 0 & 1 & 0 & 1 & 0 \\ 1 & 0 & 1 & 1 & 1 \\ 0 & 1 & 0 & 0 & 0 \\ 1 & 1 & 0 & 0 & 0 \\ 0 & 1 & 0 & 0 & 0 \end{array}\right], \quad A^2 = \left[\begin{array}{ccccc} 2 & 1 & 1 & 1 & 1 \\ 1 & 4 & 0 & 1 & 0 \\ 1 & 0 & 1 & 1 & 1 \\ 1 & 1 & 1 & 2 & 1 \\ 1 & 0 & 1 & 1 & 1 \end{array}\right]$$

A^2 gives the number of walks of length 2. There is 1 walk of length 2 from vertex 2 to vertex 4.

57. Find the inverse. [10.3]

$$\left[\begin{array}{cc|cc} 2 & -2 & 1 & 0 \\ 3 & -2 & 0 & 1 \end{array}\right] \xrightarrow{\frac{1}{2}R_1} \left[\begin{array}{cc|cc} 1 & -1 & \frac{1}{2} & 0 \\ 3 & -2 & 0 & 1 \end{array}\right]$$

$$\xrightarrow{-3R_1+R_2} \left[\begin{array}{cc|cc} 1 & -1 & \frac{1}{2} & 0 \\ 0 & 1 & -\frac{3}{2} & 1 \end{array}\right]$$

$$\xrightarrow{R_2+R_1} \left[\begin{array}{cc|cc} 1 & 0 & -1 & 1 \\ 0 & 1 & -\frac{3}{2} & 1 \end{array}\right]$$

The inverse matrix is $\left[\begin{array}{cc} -1 & 1 \\ -\frac{3}{2} & 1 \end{array}\right]$.

59. Find the inverse. [10.3]

$$\left[\begin{array}{cc|cc} -2 & 3 & 1 & 0 \\ 2 & 4 & 0 & 1 \end{array}\right] \xrightarrow{-\frac{1}{2}R_1} \left[\begin{array}{cc|cc} 1 & -\frac{3}{2} & -\frac{1}{2} & 0 \\ 2 & 4 & 0 & 1 \end{array}\right]$$

$$\xrightarrow{-2R_1+R_2} \left[\begin{array}{cc|cc} 1 & -\frac{3}{2} & -\frac{1}{2} & 0 \\ 0 & 7 & 1 & 1 \end{array}\right]$$

$$\xrightarrow{\frac{1}{7}R_2} \left[\begin{array}{cc|cc} 1 & -\frac{3}{2} & -\frac{1}{2} & 0 \\ 0 & 1 & \frac{1}{7} & \frac{1}{7} \end{array}\right]$$

$$\xrightarrow{\frac{3}{2}R_2+R_1} \left[\begin{array}{cc|cc} 1 & 0 & -\frac{2}{7} & \frac{3}{14} \\ 0 & 1 & \frac{1}{7} & \frac{1}{7} \end{array}\right]$$

The inverse matrix is $\left[\begin{array}{cc} -\frac{2}{7} & \frac{3}{14} \\ \frac{1}{7} & \frac{1}{7} \end{array}\right]$.

61. Find the inverse. [10.3]

$$\left[\begin{array}{ccc|ccc} 1 & 2 & 1 & 1 & 0 & 0 \\ 2 & 6 & 4 & 0 & 1 & 0 \\ 3 & 8 & 6 & 0 & 0 & 1 \end{array}\right] \begin{array}{c} -2R_1+R_2 \\ \underline{-3R_1+R_3} \end{array} \left[\begin{array}{ccc|ccc} 1 & 2 & 1 & 1 & 0 & 0 \\ 0 & 2 & 2 & -2 & 1 & 0 \\ 0 & 2 & 3 & -3 & 0 & 1 \end{array}\right]$$

$$\xrightarrow{\frac{1}{2}R_2} \left[\begin{array}{ccc|ccc} 1 & 2 & 1 & 1 & 0 & 0 \\ 0 & 1 & 1 & -1 & \frac{1}{2} & 0 \\ 0 & 2 & 3 & -3 & 0 & 1 \end{array}\right]$$

$$\xrightarrow{-2R_2+R_3} \left[\begin{array}{ccc|ccc} 1 & 2 & 1 & 1 & 0 & 0 \\ 0 & 1 & 1 & -1 & \frac{1}{2} & 0 \\ 0 & 0 & 1 & -1 & -1 & 1 \end{array}\right]$$

$$\xrightarrow{-2R_2+R_1} \left[\begin{array}{ccc|ccc} 1 & 0 & -1 & 3 & -1 & 0 \\ 0 & 1 & 1 & -1 & \frac{1}{2} & 0 \\ 0 & 0 & 1 & -1 & -1 & 1 \end{array}\right]$$

$$\begin{array}{c} R_3+R_1 \\ \underline{-R_3+R_2} \end{array} \left[\begin{array}{ccc|ccc} 1 & 0 & 0 & 2 & -2 & 1 \\ 0 & 1 & 0 & 0 & \frac{3}{2} & -1 \\ 0 & 0 & 1 & -1 & -1 & 1 \end{array}\right]$$

The inverse matrix is $\left[\begin{array}{ccc} 2 & -2 & 1 \\ 0 & \frac{3}{2} & -1 \\ -1 & -1 & 1 \end{array}\right]$.

63. Find the inverse. [10.3]

$$\left[\begin{array}{ccc|ccc} 3 & -2 & 7 & 1 & 0 & 0 \\ 2 & -1 & 5 & 0 & 1 & 0 \\ 3 & 0 & 10 & 0 & 0 & 1 \end{array}\right] \xrightarrow{\frac{1}{3}R_1} \left[\begin{array}{ccc|ccc} 1 & -\frac{2}{3} & \frac{7}{3} & \frac{1}{3} & 0 & 0 \\ 2 & -1 & 5 & 0 & 1 & 0 \\ 3 & 0 & 10 & 0 & 0 & 1 \end{array}\right]$$

$$\begin{array}{c} -2R_1+R_2 \\ \underline{-3R_1+R_3} \end{array} \left[\begin{array}{ccc|ccc} 1 & -\frac{2}{3} & \frac{7}{3} & \frac{1}{3} & 0 & 0 \\ 0 & \frac{1}{3} & \frac{1}{3} & -\frac{2}{3} & 1 & 0 \\ 0 & 2 & 3 & -1 & 0 & 1 \end{array}\right]$$

$$\xrightarrow{3R_2} \left[\begin{array}{ccc|ccc} 1 & -\frac{2}{3} & \frac{7}{3} & \frac{1}{3} & 0 & 0 \\ 0 & 1 & 1 & -2 & 3 & 0 \\ 0 & 2 & 3 & -1 & 0 & 1 \end{array}\right]$$

$\xrightarrow{-2R_2+R_3}$ $\begin{bmatrix} 1 & -\frac{2}{3} & \frac{7}{3} & \frac{1}{3} & 0 & 0 \\ 0 & 1 & 1 & -2 & 3 & 0 \\ 0 & 0 & 1 & 3 & -6 & 1 \end{bmatrix}$

$\xrightarrow{\frac{2}{3}R_2+R_1}$ $\begin{bmatrix} 1 & 0 & 3 & -1 & 2 & 0 \\ 0 & 1 & 1 & -2 & 3 & 0 \\ 0 & 0 & 1 & 3 & -6 & 1 \end{bmatrix}$

$\xrightarrow[\,-1R_3+R_2\,]{-3R_3+R_1}$ $\begin{bmatrix} 1 & 0 & 0 & -10 & 20 & -3 \\ 0 & 1 & 0 & -5 & 9 & -1 \\ 0 & 0 & 1 & 3 & -6 & 1 \end{bmatrix}$

The inverse matrix is $\begin{bmatrix} -10 & 20 & -3 \\ -5 & 9 & -1 \\ 3 & -6 & 1 \end{bmatrix}$.

65. Find the inverse. [10.3]

$\begin{bmatrix} 1 & -1 & 2 & 3 & 1 & 0 & 0 & 0 \\ 2 & -1 & 6 & 5 & 0 & 1 & 0 & 0 \\ 3 & -1 & 9 & 6 & 0 & 0 & 1 & 0 \\ 2 & -2 & 4 & 7 & 0 & 0 & 0 & 1 \end{bmatrix}$

$\xrightarrow[\substack{-3R_1+R_3 \\ -2R_1+R_4}]{-2R_1+R_2}$ $\begin{bmatrix} 1 & -1 & 2 & 3 & 1 & 0 & 0 & 0 \\ 0 & 1 & 2 & -1 & -2 & 1 & 0 & 0 \\ 0 & 2 & 3 & -3 & -3 & 0 & 1 & 0 \\ 0 & 0 & 0 & 1 & -2 & 0 & 0 & 1 \end{bmatrix}$

$\xrightarrow{-2R_2+R_3}$ $\begin{bmatrix} 1 & -1 & 2 & 3 & 1 & 0 & 0 & 0 \\ 0 & 1 & 2 & -1 & -2 & 1 & 0 & 0 \\ 0 & 0 & -1 & -1 & 1 & -2 & 1 & 0 \\ 0 & 0 & 0 & 1 & -2 & 0 & 0 & 1 \end{bmatrix}$

$\xrightarrow{-1R_3}$ $\begin{bmatrix} 1 & -1 & 2 & 3 & 1 & 0 & 0 & 0 \\ 0 & 1 & 2 & -1 & -2 & 1 & 0 & 0 \\ 0 & 0 & 1 & 1 & -1 & 2 & -1 & 0 \\ 0 & 0 & 0 & 1 & -2 & 0 & 0 & 1 \end{bmatrix}$

$\xrightarrow{R_2+R_1}$ $\begin{bmatrix} 1 & 0 & 4 & 2 & -1 & 1 & 0 & 0 \\ 0 & 1 & 2 & -1 & -2 & 1 & 0 & 0 \\ 0 & 0 & 1 & 1 & -1 & 2 & -1 & 0 \\ 0 & 0 & 0 & 1 & -2 & 0 & 0 & 1 \end{bmatrix}$

$\xrightarrow[\,-2R_3+R_2\,]{-4R_3+R_1}$ $\begin{bmatrix} 1 & 0 & 0 & -2 & 3 & -7 & 4 & 0 \\ 0 & 1 & 0 & -3 & 0 & -3 & 2 & 0 \\ 0 & 0 & 1 & 1 & -1 & 2 & -1 & 0 \\ 0 & 0 & 0 & 1 & -2 & 0 & 0 & 1 \end{bmatrix}$

$\xrightarrow[\substack{3R_4+R_2 \\ -1R_4+R_3}]{2R_4+R_1}$ $\begin{bmatrix} 1 & 0 & 0 & 0 & -1 & -7 & 4 & 2 \\ 0 & 1 & 0 & 0 & -6 & -3 & 2 & 3 \\ 0 & 0 & 1 & 0 & 1 & 2 & -1 & -1 \\ 0 & 0 & 0 & 1 & -2 & 0 & 0 & 1 \end{bmatrix}$

The inverse matrix is $\begin{bmatrix} -1 & -7 & 4 & 2 \\ -6 & -3 & 2 & 3 \\ 1 & 2 & -1 & -1 \\ -2 & 0 & 0 & 1 \end{bmatrix}$.

67. Find the inverse. [10.3]

$\begin{bmatrix} -1 & 8 & 1 & 1 & 0 & 0 \\ 1 & 3 & -1 & 0 & 1 & 0 \\ 6 & 5 & -6 & 0 & 0 & 1 \end{bmatrix}$ $\xrightarrow{-1R_1}$ $\begin{bmatrix} 1 & -8 & -1 & -1 & 0 & 0 \\ 1 & 3 & -1 & 0 & 1 & 0 \\ 6 & 5 & -6 & 0 & 0 & 1 \end{bmatrix}$

$\xrightarrow[\,-6R_1+R_3\,]{-R_1+R_2}$ $\begin{bmatrix} 1 & -8 & -1 & -1 & 0 & 0 \\ 0 & 11 & 0 & 1 & 1 & 0 \\ 0 & 53 & 0 & 6 & 0 & 1 \end{bmatrix}$

$\xrightarrow{\frac{1}{11}R_2}$ $\begin{bmatrix} 1 & -8 & -1 & -1 & 0 & 0 \\ 0 & 1 & 0 & \frac{1}{11} & \frac{1}{11} & 0 \\ 0 & 53 & 0 & 6 & 0 & 1 \end{bmatrix}$

$\xrightarrow{-53R_2+R_3}$ $\begin{bmatrix} 1 & -8 & -1 & -1 & 0 & 0 \\ 0 & 1 & 0 & \frac{1}{11} & \frac{1}{11} & 0 \\ 0 & 0 & 0 & \frac{13}{11} & -\frac{53}{11} & 1 \end{bmatrix}$

The matrix does not have an inverse.

69. Find the inverse. [10.3]

$\begin{bmatrix} 3 & 7 & -1 & 8 & 1 & 0 & 0 & 0 \\ 2 & 5 & 0 & 5 & 0 & 1 & 0 & 0 \\ 3 & 6 & -4 & 8 & 0 & 0 & 1 & 0 \\ 2 & 4 & -4 & 4 & 0 & 0 & 0 & 1 \end{bmatrix}$

$\xrightarrow{\frac{1}{3}R_1}$ $\begin{bmatrix} 1 & \frac{7}{3} & -\frac{1}{3} & \frac{8}{3} & \frac{1}{3} & 0 & 0 & 0 \\ 2 & 5 & 0 & 5 & 0 & 1 & 0 & 0 \\ 3 & 6 & -4 & 8 & 0 & 0 & 1 & 0 \\ 2 & 4 & -4 & 4 & 0 & 0 & 0 & 1 \end{bmatrix}$

$\xrightarrow[\substack{-3R_1+R_3 \\ -2R_1+R_4}]{-2R_1+R_2}$ $\begin{bmatrix} 1 & \frac{7}{3} & -\frac{1}{3} & \frac{8}{3} & \frac{1}{3} & 0 & 0 & 0 \\ 0 & \frac{1}{3} & \frac{2}{3} & -\frac{1}{3} & -\frac{2}{3} & 1 & 0 & 0 \\ 0 & -1 & -3 & 0 & -1 & 0 & 1 & 0 \\ 0 & -\frac{2}{3} & -\frac{10}{3} & -\frac{4}{3} & -\frac{2}{3} & 0 & 0 & 1 \end{bmatrix}$

$$\xrightarrow{3R_2}\begin{bmatrix}1 & \frac{7}{3} & -\frac{1}{3} & \frac{8}{3} & \frac{1}{3} & 0 & 0 & 0 \\ 0 & 1 & 2 & -1 & -2 & 3 & 0 & 0 \\ 0 & -1 & -3 & 0 & -1 & 0 & 1 & 0 \\ 0 & -\frac{2}{3} & -\frac{10}{3} & -\frac{4}{3} & -\frac{2}{3} & 0 & 0 & 1 \end{bmatrix}$$

$$\xrightarrow[\frac{2}{3}R_2+R_4]{1R_2+R_3}\begin{bmatrix}1 & \frac{7}{3} & -\frac{1}{3} & \frac{8}{3} & \frac{1}{3} & 0 & 0 & 0 \\ 0 & 1 & 2 & -1 & -2 & 3 & 0 & 0 \\ 0 & 0 & -1 & -1 & -3 & 3 & 1 & 0 \\ 0 & 0 & -2 & -2 & -2 & 2 & 0 & 1 \end{bmatrix}$$

$$\xrightarrow{-1R_3}\begin{bmatrix}1 & \frac{7}{3} & -\frac{1}{3} & \frac{8}{3} & \frac{1}{3} & 0 & 0 & 0 \\ 0 & 1 & 2 & -1 & -2 & 3 & 0 & 0 \\ 0 & 0 & 1 & 1 & 3 & -3 & -1 & 0 \\ 0 & 0 & -2 & -2 & -2 & 2 & 0 & 1 \end{bmatrix}$$

$$\xrightarrow{2R_3+R_4}\begin{bmatrix}1 & \frac{7}{3} & -\frac{1}{3} & \frac{8}{3} & \frac{1}{3} & 0 & 0 & 0 \\ 0 & 1 & 2 & -1 & -2 & 3 & 0 & 0 \\ 0 & 0 & 1 & 1 & 3 & -3 & -1 & 0 \\ 0 & 0 & 0 & 0 & 4 & -4 & -2 & 1 \end{bmatrix}$$

The matrix does not have an inverse.

71. Solve the system for the set of constants. [10.3]

a. $b_1 = 2,\ b_2 = -3$

$$\begin{bmatrix}3 & 4 \\ 2 & 3\end{bmatrix}\begin{bmatrix}x \\ y\end{bmatrix}=\begin{bmatrix}2 \\ -3\end{bmatrix}$$

$$\begin{bmatrix}3 & -4 \\ -2 & 3\end{bmatrix}\begin{bmatrix}3 & 4 \\ 2 & 3\end{bmatrix}\begin{bmatrix}x \\ y\end{bmatrix}=\begin{bmatrix}3 & -4 \\ -2 & 3\end{bmatrix}\begin{bmatrix}2 \\ -3\end{bmatrix}$$

$$\begin{bmatrix}x \\ y\end{bmatrix}=\begin{bmatrix}18 \\ -13\end{bmatrix}$$

The solution is $(18, -13)$.

b. $b_1 = -2,\ b_2 = 4$

$$\begin{bmatrix}3 & 4 \\ 2 & 3\end{bmatrix}\begin{bmatrix}x \\ y\end{bmatrix}=\begin{bmatrix}-2 \\ 4\end{bmatrix}$$

$$\begin{bmatrix}3 & -4 \\ -2 & 3\end{bmatrix}\begin{bmatrix}3 & 4 \\ 2 & 3\end{bmatrix}\begin{bmatrix}x \\ y\end{bmatrix}=\begin{bmatrix}3 & -4 \\ -2 & 3\end{bmatrix}\begin{bmatrix}-2 \\ 4\end{bmatrix}$$

$$\begin{bmatrix}x \\ y\end{bmatrix}=\begin{bmatrix}-22 \\ 16\end{bmatrix}$$

The solution is $(-22, 16)$.

73. Solve the system for the set of constants. [10.3]

a. $b_1 = -1,\ b_2 = 2,\ b_3 = 4$

$$\begin{bmatrix}2 & 1 & -1 \\ 4 & 4 & 1 \\ 2 & 2 & -3\end{bmatrix}\begin{bmatrix}x \\ y \\ z\end{bmatrix}=\begin{bmatrix}-1 \\ 2 \\ 4\end{bmatrix}$$

$$\begin{bmatrix}1 & -\frac{1}{14} & -\frac{5}{14} \\ -1 & \frac{2}{7} & \frac{3}{7} \\ 0 & \frac{1}{7} & -\frac{2}{7}\end{bmatrix}\begin{bmatrix}2 & 1 & -1 \\ 4 & 4 & 1 \\ 2 & 2 & -3\end{bmatrix}\begin{bmatrix}x \\ y \\ z\end{bmatrix}=\begin{bmatrix}1 & -\frac{1}{14} & -\frac{5}{14} \\ -1 & \frac{2}{7} & \frac{3}{7} \\ 0 & \frac{1}{7} & -\frac{2}{7}\end{bmatrix}\begin{bmatrix}-1 \\ 2 \\ 4\end{bmatrix}$$

$$\begin{bmatrix}x \\ y \\ z\end{bmatrix}=\begin{bmatrix}-\frac{18}{7} \\ \frac{23}{7} \\ -\frac{6}{7}\end{bmatrix}$$

The solution is $\left(-\dfrac{18}{7},\ \dfrac{23}{7},\ -\dfrac{6}{7}\right)$.

b. $b_1 = -2,\ b_2 = 3,\ b_3 = 0$

$$\begin{bmatrix}2 & 1 & -1 \\ 4 & 4 & 1 \\ 2 & 2 & -3\end{bmatrix}\begin{bmatrix}x \\ y \\ z\end{bmatrix}=\begin{bmatrix}-2 \\ 3 \\ 0\end{bmatrix}$$

$$\begin{bmatrix}1 & -\frac{1}{14} & -\frac{5}{14} \\ -1 & \frac{2}{7} & \frac{3}{7} \\ 0 & \frac{1}{7} & -\frac{2}{7}\end{bmatrix}\begin{bmatrix}2 & 1 & -1 \\ 4 & 4 & 1 \\ 2 & 2 & -3\end{bmatrix}\begin{bmatrix}x \\ y \\ z\end{bmatrix}=\begin{bmatrix}1 & -\frac{1}{14} & -\frac{5}{14} \\ -1 & \frac{2}{7} & \frac{3}{7} \\ 0 & \frac{1}{7} & -\frac{2}{7}\end{bmatrix}\begin{bmatrix}-2 \\ 3 \\ 0\end{bmatrix}$$

$$\begin{bmatrix}x \\ y \\ z\end{bmatrix}=\begin{bmatrix}-\frac{31}{14} \\ \frac{20}{7} \\ \frac{3}{7}\end{bmatrix}$$

The solution is $\left(-\dfrac{31}{14},\ \dfrac{20}{7},\ \dfrac{3}{7}\right)$.

75. Evaluate the determinant. [10.4]

$$\begin{vmatrix}-2 & 3 \\ -5 & 1\end{vmatrix}=-2(1)-(-5)(3)=-2+15=13$$

77. Find M_{11} and C_{11}. [10.4]

$$M_{11}=\begin{vmatrix}0 & 3 \\ 2 & 5\end{vmatrix}=0(5)-2(3)=-6$$

$$C_{11}=(-1)^{1+1}M_{11}=(1)(-6)=-6$$

79. Find M_{12} and C_{12}. [10.4]

$$M_{12}=\begin{vmatrix}1 & 3 \\ -2 & 5\end{vmatrix}=1(5)-(-2)(3)=11$$

$$C_{12}=(-1)^{1+2}M_{12}=(-1)(11)=-11$$

81. Evaluate the determinant by cofactors. [10.4]

$$\begin{vmatrix} 4 & 1 & 3 \\ -2 & 3 & -8 \\ -1 & 4 & -5 \end{vmatrix} = 4\begin{vmatrix} 3 & -8 \\ 4 & -5 \end{vmatrix} - 1\begin{vmatrix} -2 & -8 \\ -1 & -5 \end{vmatrix} + 3\begin{vmatrix} -2 & 3 \\ -1 & 4 \end{vmatrix}$$

$$= 4(17) - 1(2) + 3(-5)$$
$$= 51$$

83. Evaluate the determinant by cofactors. [10.4]

$$\begin{vmatrix} 8 & -3 & 0 \\ 10 & -2 & -5 \\ 3 & -4 & 0 \end{vmatrix} = 0\begin{vmatrix} 10 & -2 \\ 3 & -4 \end{vmatrix} - (-5)\begin{vmatrix} 8 & -3 \\ 3 & -4 \end{vmatrix} + 0\begin{vmatrix} 8 & -3 \\ 10 & -2 \end{vmatrix}$$

$$= 0 + 5(-23) + 0$$
$$= -115$$

85. Evaluate the determinant by cofactors. [10.4]

$$\begin{vmatrix} 1 & -1 & -4 & 2 \\ 0 & 1 & 5 & 0 \\ -2 & 4 & -4 & 0 \\ 3 & 4 & 5 & 0 \end{vmatrix} = -2\begin{vmatrix} 0 & 1 & 5 \\ -2 & 4 & -4 \\ 3 & 4 & 5 \end{vmatrix} + 0 - 0 + 0$$

$$= -2\left[0\begin{vmatrix} 4 & -4 \\ 4 & 5 \end{vmatrix} - 1\begin{vmatrix} -2 & -4 \\ 3 & 5 \end{vmatrix} + 5\begin{vmatrix} -2 & 4 \\ 3 & 4 \end{vmatrix}\right]$$

$$= -2[0 - 1(2) + 5(-20)]$$
$$= 204$$

87. Evaluate the determinant using elementary row or column operations. [10.4]

$$\begin{vmatrix} 2 & 6 & 4 \\ 1 & 2 & 1 \\ 3 & 8 & 6 \end{vmatrix} \xrightarrow{\text{Factor 2 from row 1}} 2\begin{vmatrix} 1 & 3 & 2 \\ 1 & 2 & 1 \\ 3 & 8 & 6 \end{vmatrix}$$

$$\xrightarrow[-3R_1 + R_3]{-1R_1 + R_2} 2\begin{vmatrix} 1 & 3 & 2 \\ 0 & -1 & -1 \\ 0 & -1 & 0 \end{vmatrix}$$

$$\xrightarrow{-R_2 + R_3} 2\begin{vmatrix} 1 & 3 & 2 \\ 0 & -1 & -1 \\ 0 & 0 & 1 \end{vmatrix}$$

$$= 2(1)(-1)(1) = -2$$

89. Evaluate the determinant using elementary row or column operations. [10.4]

$$\begin{vmatrix} 3 & -8 & 7 \\ 2 & -3 & 6 \\ 1 & -3 & 2 \end{vmatrix} \xrightarrow[-\frac{1}{3}R_1 + R_3]{-\frac{2}{3}R_1 + R_2} \begin{vmatrix} 3 & -8 & 7 \\ 0 & \frac{7}{3} & \frac{4}{3} \\ 0 & -\frac{1}{3} & -\frac{1}{3} \end{vmatrix}$$

$$\xrightarrow{\frac{1}{7}R_2 + R_3} \begin{vmatrix} 3 & -8 & 7 \\ 0 & \frac{7}{3} & \frac{4}{3} \\ 0 & 0 & -\frac{1}{7} \end{vmatrix}$$

$$= 3\frac{7}{3}\left(-\frac{1}{7}\right) = -1$$

91. Evaluate the determinant using elementary row or column operations. [10.4]

$$\begin{vmatrix} 1 & -1 & 2 & 1 \\ 2 & -1 & 6 & 3 \\ 3 & -1 & 8 & 7 \\ 3 & 0 & 9 & 9 \end{vmatrix} \xrightarrow{\text{Factor 3 from row 4}} 3\begin{vmatrix} 1 & -1 & 2 & 1 \\ 2 & -1 & 6 & 3 \\ 3 & -1 & 8 & 7 \\ 1 & 0 & 3 & 3 \end{vmatrix}$$

$$\xrightarrow[\substack{-3R_1 + R_3 \\ -1R_1 + R_4}]{-2R_1 + R_2} 3\begin{vmatrix} 1 & -1 & 2 & 1 \\ 0 & 1 & 2 & 1 \\ 0 & 2 & 2 & 4 \\ 0 & 1 & 1 & 2 \end{vmatrix}$$

$$\xrightarrow[-1R_2 + R_4]{-2R_2 + R_3} 3\begin{vmatrix} 1 & -1 & 2 & 1 \\ 0 & 1 & 2 & 1 \\ 0 & 0 & -2 & 2 \\ 0 & 0 & -1 & 1 \end{vmatrix}$$

$$\xrightarrow[R_3 + R_4]{\text{Factor } -2 \text{ from row 3}} -6\begin{vmatrix} 1 & -1 & 2 & 1 \\ 0 & 1 & 2 & 1 \\ 0 & 0 & 1 & -1 \\ 0 & 0 & 0 & 0 \end{vmatrix}$$

$$= -6(1)(1)(1)(0) = 0$$

93. Evaluate the determinant using elementary row or column operations. [10.4]

$$\begin{vmatrix} 1 & 2 & -2 & 1 \\ 2 & 5 & -3 & 1 \\ 2 & 0 & -10 & 1 \\ 3 & 8 & -4 & 1 \end{vmatrix} \xrightarrow[\substack{-R_4 + R_2 \\ -R_4 + R_3}]{-R_4 + R_1} \begin{vmatrix} -2 & -6 & 2 & 0 \\ -1 & -3 & 1 & 0 \\ -1 & -8 & -6 & 0 \\ 3 & 8 & -4 & 1 \end{vmatrix}$$

$$\xrightarrow[\frac{1}{3}R_3 + R_1]{\frac{7}{6}R_3 + R_2} \begin{vmatrix} -\frac{7}{3} & -\frac{26}{3} & 0 & 0 \\ -\frac{7}{6} & -\frac{13}{3} & 0 & 0 \\ -1 & -8 & -6 & 0 \\ 3 & 8 & -4 & 1 \end{vmatrix}$$

$$\xrightarrow{-2R_2 + R_1} \begin{vmatrix} 0 & 0 & 0 & 0 \\ -\frac{7}{6} & -\frac{13}{3} & 0 & 0 \\ -1 & -8 & -6 & 0 \\ 3 & 8 & -4 & 1 \end{vmatrix}$$

$$= 0\left(-\frac{13}{3}\right)(-6)(1) = 0$$

95. Solve the system using Cramer's Rule. [10.4]

$$x = \dfrac{\begin{vmatrix} -1 & 3 \\ 1 & -1 \end{vmatrix}}{\begin{vmatrix} 5 & 3 \\ -2 & -1 \end{vmatrix}} = \dfrac{-2}{1} = -2 \qquad y = \dfrac{\begin{vmatrix} 5 & -1 \\ -2 & 1 \end{vmatrix}}{\begin{vmatrix} 5 & 3 \\ -2 & -1 \end{vmatrix}} = \dfrac{3}{1} = 3$$

The solution is (–2, 3).

97. Solve the system using Cramer's Rule. [10.4]

$$x = \dfrac{\begin{vmatrix} 19 & -2 \\ 8 & -1 \end{vmatrix}}{\begin{vmatrix} 5 & -2 \\ 2 & -1 \end{vmatrix}} = \dfrac{-3}{-1} = 3 \qquad y = \dfrac{\begin{vmatrix} 5 & 19 \\ 2 & 8 \end{vmatrix}}{\begin{vmatrix} 5 & -2 \\ 2 & -1 \end{vmatrix}} = \dfrac{2}{-1} = -2$$

The solution is (3,–2).

99. Solve the system using Cramer's Rule. [10.4]

$$D = \begin{vmatrix} 0 & 2 & 4 \\ 1 & 4 & 4 \\ 2 & 4 & 1 \end{vmatrix} = -2, \quad D_x = \begin{vmatrix} 14 & 2 & 4 \\ 11 & 4 & 4 \\ -3 & 4 & 1 \end{vmatrix} = 10$$

$$D_y = \begin{vmatrix} 0 & 14 & 4 \\ 1 & 11 & 4 \\ 2 & -3 & 1 \end{vmatrix} = -2, \quad D_z = \begin{vmatrix} 0 & 2 & 14 \\ 1 & 4 & 11 \\ 2 & 4 & -3 \end{vmatrix} = -6$$

$$x = \dfrac{D_x}{D} = \dfrac{10}{-2} = -5$$

$$y = \dfrac{D_y}{D} = \dfrac{-2}{-2} = 1$$

$$z = \dfrac{D_z}{D} = \dfrac{-6}{-2} = 3$$

The solution is (–5, 1, 3).

101. Find the area of the triangle. [10.4]

$$\dfrac{1}{2} \begin{vmatrix} 0 & -1 & 1 \\ 3 & -5 & 1 \\ -1 & 3 & 1 \end{vmatrix} = \dfrac{1}{2}(8) = 4$$

The area of the triangle is 4 square units.

103. Find the area of the triangle. [10.4]

$$\dfrac{1}{2} \begin{vmatrix} 2 & 4 & 1 \\ 2 & 2 & 1 \\ -3 & 2 & 1 \end{vmatrix} = \dfrac{1}{2}(-10) = -5$$

$$|-5| = 5$$

The area of the triangle is 5 square units.

105. Find the equation of the line. [10.4]

$$\begin{vmatrix} x & y & 1 \\ 5 & 5 & 1 \\ 4 & 2 & 1 \end{vmatrix} = 0$$

$$x(5-2) - y(5-4) + 1(10-20) = 0$$
$$3x - y - 10 = 0$$
$$3x - y = 10$$

$3x - y = 10$ is the equation of the line passing through the points (5, 5) and (4, 2).

107. Find the equation of the line. [10.4]

$$\begin{vmatrix} x & y & 1 \\ 1 & -5 & 1 \\ -3 & 5 & 1 \end{vmatrix} = 0$$

$$x(-5-5) - y(1+3) + 1(5-15) = 0$$
$$-10x - 4y - 10 = 0$$
$$-10x - 4y = 10$$
$$-2(5x + 2y) = -2(-5)$$
$$5x + 2y = -5$$

$5x + 2y = -5$ is the equation of the line passing through the points (1,–5) and (–3, 5).

109. Use transformations to find the endpoints. [10.2]

$$T_{3,-1} \cdot \begin{bmatrix} -5 & 4 \\ 3 & -2 \\ 1 & 1 \end{bmatrix} = \begin{bmatrix} 1 & 0 & 3 \\ 0 & 1 & -1 \\ 0 & 0 & 1 \end{bmatrix} \begin{bmatrix} -5 & 4 \\ 3 & -2 \\ 1 & 1 \end{bmatrix}$$

$$= \begin{bmatrix} -2 & 7 \\ 2 & -3 \\ 1 & 1 \end{bmatrix}$$

$$R_{xy} \cdot \begin{bmatrix} -2 & 7 \\ 2 & -3 \\ 1 & 1 \end{bmatrix} = \begin{bmatrix} 0 & 1 & 0 \\ 1 & 0 & 0 \\ 0 & 0 & 1 \end{bmatrix} \begin{bmatrix} -2 & 7 \\ 2 & -3 \\ 1 & 1 \end{bmatrix}$$

$$= \begin{bmatrix} 2 & -3 \\ -2 & 7 \\ 1 & 1 \end{bmatrix} \Rightarrow \begin{matrix} A'(2, -2) \\ B'(-3, 7) \end{matrix}$$

111. Find the populations in 5 years. [10.2]

$$\begin{bmatrix} 250,000 & 400,000 \end{bmatrix} \begin{bmatrix} 0.90 & 0.10 \\ 0.15 & 0.85 \end{bmatrix}^5$$

$$\approx \begin{bmatrix} 356,777 & 293,223 \end{bmatrix}$$

The city population will be about 357,000.

The suburb population will be about 293,000.

113. Find the temperatures at x_1 and x_2. [10.3]

The average temperature for the two points,

$$x_1 = \frac{60+40+x_2+50}{4} = \frac{150+x_2}{4} \text{ or } 4x_1 - x_2 = 150$$

$$x_2 = \frac{60+70+x_1+50}{4} = \frac{180+x_1}{4} \text{ or } -x_1 + 4x_2 = 180$$

The system of equations and associated matrix equation are

$$\begin{cases} 4x_1 - x_2 = 150 \\ -x_1 + 4x_2 = 180 \end{cases} \qquad \begin{bmatrix} 4 & -1 \\ -1 & 4 \end{bmatrix}\begin{bmatrix} x_1 \\ x_2 \end{bmatrix} = \begin{bmatrix} 150 \\ 180 \end{bmatrix}$$

Solving the matrix equation by using an inverse matrix

gives $\begin{bmatrix} x_1 \\ x_2 \end{bmatrix} = \begin{bmatrix} 52 \\ 58 \end{bmatrix}$

The temperatures are $x_1 = 52°F$, $x_2 = 58°F$.

115. The input-output matrix A is given by

$$A = \begin{bmatrix} 0.05 & 0.06 & 0.08 \\ 0.02 & 0.04 & 0.04 \\ 0.03 & 0.03 & 0.05 \end{bmatrix}$$

Consumer demand X is given by $X = (I - A)^{-1}D$.

$$X = \left(\begin{bmatrix} 1 & 0 & 0 \\ 0 & 1 & 0 \\ 0 & 0 & 1 \end{bmatrix} - \begin{bmatrix} 0.05 & 0.06 & 0.08 \\ 0.02 & 0.04 & 0.04 \\ 0.03 & 0.03 & 0.05 \end{bmatrix} \right)^{-1} \begin{bmatrix} 30 \\ 12 \\ 21 \end{bmatrix}$$

$$= \begin{bmatrix} 0.95 & -0.06 & -0.08 \\ -0.02 & 0.96 & -0.04 \\ -0.03 & -0.03 & 0.95 \end{bmatrix}^{-1} \begin{bmatrix} 30 \\ 12 \\ 21 \end{bmatrix}$$

$$\approx \begin{bmatrix} 34.47 \\ 14.20 \\ 23.64 \end{bmatrix}$$

$34.47 million computer division, $14.20 million monitor division, $23.64 million disk drive division.

Chapter 10 Test

1. Write the augmented matrix, coefficient matrix and constant matrix. [10.1]

$$\begin{bmatrix} 2 & 3 & -3 & | & 4 \\ 3 & 0 & 2 & | & -1 \\ 4 & -4 & 2 & | & 3 \end{bmatrix}, \begin{bmatrix} 2 & 3 & -3 \\ 3 & 0 & 2 \\ 4 & -4 & 2 \end{bmatrix}, \begin{bmatrix} 4 \\ -1 \\ 3 \end{bmatrix}$$

3. Solve the system using Gaussian elimination. [10.1]

$$\begin{bmatrix} 1 & -2 & 3 & | & 10 \\ 2 & -3 & 8 & | & 23 \\ -1 & 3 & -2 & | & -9 \end{bmatrix} \xrightarrow[R_1+R_3]{-2R_1+R_2} \begin{bmatrix} 1 & -2 & 3 & | & 10 \\ 0 & 1 & 2 & | & 3 \\ 0 & 1 & 1 & | & 1 \end{bmatrix}$$

$$\xrightarrow{-R_2+R_3} \begin{bmatrix} 1 & -2 & 3 & | & 10 \\ 0 & 1 & 2 & | & 3 \\ 0 & 0 & -1 & | & -2 \end{bmatrix} \xrightarrow{-R_3} \begin{bmatrix} 1 & -2 & 3 & | & 10 \\ 0 & 1 & 2 & | & 3 \\ 0 & 0 & 1 & | & 2 \end{bmatrix}$$

$$\begin{cases} x - 2y + 3z = 10 \\ y + 2z = 3 \\ z = 2 \end{cases}$$

$$y + 2(2) = 3$$
$$y = -1$$

$$x - 2(-1) + 3(2) = 10$$
$$x = 2$$

The solution is $(2, -1, 2)$.

5. Solve the system using Gaussian elimination. [10.1]

$$\begin{bmatrix} 4 & 3 & 5 & | & 3 \\ 2 & -3 & 1 & | & -1 \\ 2 & 0 & 2 & | & 3 \end{bmatrix} \xrightarrow{\frac{1}{4}R_1} \begin{bmatrix} 1 & \frac{3}{4} & \frac{5}{4} & | & \frac{3}{4} \\ 2 & -3 & 1 & | & -1 \\ 2 & 0 & 2 & | & 3 \end{bmatrix}$$

$$\xrightarrow[-2R_1+R_3]{-2R_1+R_2} \begin{bmatrix} 1 & \frac{3}{4} & \frac{5}{4} & | & \frac{3}{4} \\ 0 & -\frac{9}{2} & -\frac{3}{2} & | & -\frac{5}{2} \\ 0 & -\frac{3}{2} & -\frac{1}{2} & | & \frac{3}{2} \end{bmatrix}$$

$$\xrightarrow{-\frac{2}{9}R_2} \begin{bmatrix} 1 & \frac{3}{4} & \frac{5}{4} & | & \frac{3}{4} \\ 0 & 1 & \frac{1}{3} & | & \frac{5}{9} \\ 0 & -\frac{3}{2} & -\frac{1}{2} & | & \frac{3}{2} \end{bmatrix} \xrightarrow{\frac{3}{2}R_2+R_3} \begin{bmatrix} 1 & \frac{3}{4} & \frac{5}{4} & | & \frac{3}{4} \\ 0 & 1 & \frac{1}{3} & | & \frac{5}{9} \\ 0 & 0 & 0 & | & \frac{7}{3} \end{bmatrix}$$

$$\begin{cases} x + \frac{3}{4}y + \frac{5}{4}z = \frac{3}{4} \\ y + \frac{1}{3}z = \frac{5}{9} \\ 0z = \frac{7}{3} \end{cases}$$

Because $0z = \frac{7}{3}$ has no solution, the system of equations has no solution.

7. Find $-3A$. [10.2]

$$-3A = -3\begin{bmatrix} -1 & 3 & 2 \\ 1 & 4 & -1 \end{bmatrix} = \begin{bmatrix} 3 & -9 & -6 \\ -3 & -12 & 3 \end{bmatrix}$$

9. Find $3B - 2C$. [10.2]

$$3B - 2C = 3\begin{bmatrix} 2 & -1 & 3 \\ 4 & -2 & -1 \\ 3 & 2 & 2 \end{bmatrix} - 2\begin{bmatrix} 1 & -2 & 3 \\ 2 & -3 & 8 \\ -1 & 3 & -2 \end{bmatrix}$$

$$= \begin{bmatrix} 6 & -3 & 9 \\ 12 & -6 & -3 \\ 9 & 6 & 6 \end{bmatrix} - \begin{bmatrix} 2 & -4 & 6 \\ 4 & -6 & 16 \\ -2 & 6 & -4 \end{bmatrix} = \begin{bmatrix} 4 & 1 & 3 \\ 8 & 0 & -19 \\ 11 & 0 & 10 \end{bmatrix}$$

11. Find CA. [10.2]

$$CA = \begin{bmatrix} 1 & -2 & 3 \\ 2 & -3 & 8 \\ -1 & 3 & -2 \end{bmatrix}\begin{bmatrix} -1 & 3 & 2 \\ 1 & 4 & -1 \end{bmatrix}$$

The number of columns of the first matrix is not equal to the number of rows of the second matrix. The product is not possible.

13. Find B^2. [10.2]

$$B^2 = \begin{bmatrix} 2 & -1 & 3 \\ 4 & -2 & -1 \\ 3 & 2 & 2 \end{bmatrix}\begin{bmatrix} 2 & -1 & 3 \\ 4 & -2 & -1 \\ 3 & 2 & 2 \end{bmatrix} = \begin{bmatrix} 9 & 6 & 13 \\ -3 & -2 & 12 \\ 20 & -3 & 11 \end{bmatrix}$$

15. Find the minor and cofactor of b_{21}. [10.4]

$$M_{21} = \begin{vmatrix} -1 & 3 \\ 2 & 2 \end{vmatrix} = -2 - 6 = -8$$

$$C_{21} = (-1)^{2+1} M_{21} = -(-8) = 8$$

17. Find the determinant using elementary row operations. [10.4]

$$\begin{vmatrix} 1 & -2 & 3 \\ 2 & -3 & 8 \\ -1 & 3 & -2 \end{vmatrix} \begin{matrix} -2R_1 + R_2 \\ R_1 + R_3 \end{matrix} = \begin{vmatrix} 1 & -2 & 3 \\ 0 & 1 & 2 \\ 0 & 1 & 1 \end{vmatrix}$$

$$= -R_2 + R_3 \begin{vmatrix} 1 & -2 & 3 \\ 0 & 1 & 2 \\ 0 & 0 & -1 \end{vmatrix}$$

$$= (1)(1)(-1)$$

$$= -1$$

19. Find the area of the triangle. [10.4]

$$\frac{1}{2}\begin{vmatrix} 7 & 1 & 1 \\ -2 & 2 & 1 \\ -1 & 3 & 1 \end{vmatrix} = \frac{1}{2}(-10) = -5$$

$$|-5| = 5$$

The area of the triangle is 5 square units.

21. Use the inverse to solve the system of equations. [10.3]

$$\begin{bmatrix} 1 & -3 & 0 \\ -5 & -5 & 3 \\ 3 & 4 & -2 \end{bmatrix}\begin{bmatrix} x \\ y \\ z \end{bmatrix} = \begin{bmatrix} 1 \\ -2 \\ 0 \end{bmatrix}$$

$$\begin{bmatrix} -2 & -6 & -9 \\ -1 & -2 & -3 \\ -5 & -13 & -20 \end{bmatrix}\begin{bmatrix} 1 \\ -2 \\ 0 \end{bmatrix} = \begin{bmatrix} 10 \\ 3 \\ 21 \end{bmatrix} = \begin{bmatrix} x \\ y \\ z \end{bmatrix}$$

The solution is $(10, 3, 21)$.

23. Find the vertices. [10.2]

$$\begin{bmatrix} 0 & -1 & 0 \\ 1 & 0 & 0 \\ 0 & 0 & 1 \end{bmatrix}\begin{bmatrix} -3 & 1 & 4 \\ -2 & 3 & 1 \\ 1 & 1 & 1 \end{bmatrix} = \begin{bmatrix} 2 & -3 & -1 \\ -3 & 1 & 4 \\ 1 & 1 & 1 \end{bmatrix}$$

$A^1(2, -3)$, $B^1(-3, 1)$, $C^1(-1, 4)$

25. Set up the matrix equation. [10.3]

$$X = (I - A)^{-1} D$$

$$\left(\begin{bmatrix} 1 & 0 & 0 \\ 0 & 1 & 0 \\ 0 & 0 & 1 \end{bmatrix} - \begin{bmatrix} 0.15 & 0.23 & 0.11 \\ 0.08 & 0.10 & 0.05 \\ 0.16 & 0.11 & 0.07 \end{bmatrix}\right)^{-1}\begin{bmatrix} 50 \\ 32 \\ 8 \end{bmatrix}$$

Cumulative Review Exercises

1. Find the equation of the line. [2.3]

$$y - 5 = -\frac{1}{2}(x - (-4))$$

$$y - 5 = -\frac{1}{2}x - 2$$

$$y = -\frac{1}{2}x + 3$$

3. Find $h[k(0)]$. [4.2]

$$h[k(0)] = h[3^0]$$

$$= h[1]$$

$$= e^{-1}$$

$$\approx 0.3679$$

5. Solve the system of equations. [9.1]

$$\begin{cases} 3x - 4y = 4 & (1) \\ 2x - 3y = 1 & (2) \end{cases}$$

Solve (2) for x and substitute into (1).

$$2x - 3y = 1$$

$$2x = 3y + 1$$

$$x = \frac{3y + 1}{2}$$

$$3\left(\frac{3y+1}{2}\right) - 4y = 4$$

$$9y + 3 - 8y = 8$$

$$y = 5$$

$$x = \frac{3(5)+1}{2} = 8$$

The solution is (8, 5).

7. Find the vertical asymptotes. [3.5]

$$x^2 + 4x - 5 = 0$$

$$(x+5)(x-1) = 0$$

$$x + 5 = 0 \qquad x - 1 = 0$$

$$x = -5 \qquad x = 1$$

Vertical asymptotes: $x = -5$, $x = 1$

9. Solve and express answer in interval notation. [1.5]

$$\frac{x+2}{x+1} > 0$$

The quotient $\frac{x+2}{x+1}$ is positive.

The critical values are –2 and –1.

$\frac{x+2}{x+1}$

```
          + + + + + + + + - -  - + + + + + +
        ──┼───┼───┼───┼───┼───┼──
         -4  -3  -2  -1   0   1
```

$$(-\infty, -2) \cup (-1, \infty)$$

11. Find the difference quotient. [2.6]

$$\frac{f(x+h) - f(x)}{h}$$

$$= \frac{(x+h)^2 - 3(x+h) + 2 - (x^2 - 3x + 2)}{h}$$

$$= \frac{x^2 + 2xh + h^2 - 3x - 3h + 2 - x^2 + 3x - 2}{h}$$

$$= \frac{h^2 + 2xh - 3h}{h}$$

$$= 2x - 3 + h$$

13. Find the partial fraction decomposition. [9.4]

$$\frac{x-2}{x^2 - 5x - 6} = \frac{x-2}{(x+1)(x-6)}$$

$$\frac{x-2}{(x-6)(x+1)} = \frac{A}{x-6} + \frac{B}{x+1}$$

$$x - 2 = A(x+1) + B(x-6)$$

$$x - 2 = Ax + A + Bx - 6B$$

$$x - 2 = (A+B)x + (A - 6B)$$

$$\begin{cases} 1 = A + B & (1) \\ -2 = A - 6B & (2) \end{cases}$$

$$\begin{array}{ll} -1 = -A - B & -1 \text{ times } (1) \\ \underline{-2 = A - 6B} & (2) \\ -3 = -7B \end{array}$$

$$\frac{3}{7} = B$$

$$1 = A + \frac{3}{7} \quad \Rightarrow \quad A = \frac{4}{7}$$

$$\frac{x-2}{x^2 - 5x - 6} = \frac{4}{7(x-6)} + \frac{3}{7(x+1)}$$

15. Find the exact value. [5.2]

$$\cos 30° \sin 30° + \sec 45° \tan 60°$$

$$= \cos 30° \sin 30° + \frac{1}{\sin 45°} \tan 60°$$

$$= \frac{\sqrt{3}}{2} \cdot \frac{1}{2} + \frac{2}{\sqrt{2}} \cdot \sqrt{3}$$

$$= \frac{\sqrt{3}}{4} + \frac{2\sqrt{6}}{2}$$

$$= \frac{\sqrt{3} + 4\sqrt{6}}{4}$$

17. Find the area of the triangle. [7.2]

$$K = \frac{1}{2} bc \sin A$$

$$= \frac{1}{2}(11)(4) \sin 65° = 22 \sin 65°$$

$$\approx 20 \text{ m}^2$$

19. Write in trigonometric form. [7.4]

$$z = 2 - 2i\sqrt{3}$$

$$r = \sqrt{2^2 + (-2\sqrt{3})^2} = \sqrt{4 + 12} = \sqrt{16} = 4$$

$$\alpha = \tan^{-1}\left|\frac{b}{a}\right| = \tan^{-1}\left|\frac{-2\sqrt{3}}{2}\right| = \tan^{-1}|-\sqrt{3}| = 60°$$

z is in the fourth quadrant $270° < \theta < 360°$,

$$\theta = 360° - 60° = 300°$$

$$z = r \text{cis} \theta$$

$$= 4(\cos 300° + i \sin 300°)$$

Chapter 11 Sequences, Series, and Probability

Section 11.1 Exercises

1. Find terms of the sequence.

   a. fourth term: 8

   b. fourth partial sum: $S_4 = 1 + 2 + 4 + 8 = 15$

3. Define alternating sequence.

   A sequence in which the terms alternate signs

5. Evaluate.

   $7! = 7 \cdot 6 \cdot 5 \cdot 4 \cdot 3 \cdot 2 \cdot 1 = 5040$

7. Find terms 1, 2, 3, and 8 of the sequence.

   $a_1 = 1(1-1) = 1 \cdot 0 = 0$
   $a_2 = 2(2-1) = 2 \cdot 1 = 2$
   $a_3 = 3(3-1) = 3 \cdot 2 = 6$
   $a_8 = 8(8-1) = 8 \cdot 7 = 56$

9. Find terms 1, 2, 3, and 8 of the sequence.

   $a_1 = 1 - \dfrac{1}{1} = 0$

   $a_2 = 1 - \dfrac{1}{2} = \dfrac{1}{2}$

   $a_3 = 1 - \dfrac{1}{3} = \dfrac{2}{3}$

   $a_8 = 1 - \dfrac{1}{8} = \dfrac{7}{8}$

11. Find terms 1, 2, 3, and 8 of the sequence.

    $a_1 = \dfrac{(-1)^{1+1}}{1^2} = \dfrac{(-1)^2}{1} = 1$

    $a_2 = \dfrac{(-1)^{2+1}}{2^2} = \dfrac{(-1)^3}{4} = -\dfrac{1}{4}$

    $a_3 = \dfrac{(-1)^{3+1}}{3^2} = \dfrac{(-1)^4}{9} = \dfrac{1}{9}$

    $a_8 = \dfrac{(-1)^{8+1}}{8^2} = \dfrac{(-1)^9}{64} = -\dfrac{1}{64}$

13. Find terms 1, 2, 3, and 8 of the sequence.

    $a_1 = \dfrac{(-1)^{2 \cdot 1 - 1}}{3 \cdot 1} = \dfrac{(-1)^{2-1}}{3} = -\dfrac{1}{3}$

    $a_2 = \dfrac{(-1)^{2 \cdot 2 - 1}}{3 \cdot 2} = \dfrac{(-1)^{4-1}}{6} = -\dfrac{1}{6}$

    $a_3 = \dfrac{(-1)^{2 \cdot 3 - 1}}{3 \cdot 3} = \dfrac{(-1)^{6-1}}{9} = -\dfrac{1}{9}$

    $a_8 = \dfrac{(-1)^{2 \cdot 8 - 1}}{3 \cdot 8} = \dfrac{(-1)^{16-1}}{24} = -\dfrac{1}{24}$

15. Find terms 1, 2, 3, and 8 of the sequence.

    $a_1 = \left(\dfrac{2}{3}\right)^1 = \dfrac{2}{3}$

    $a_2 = \left(\dfrac{2}{3}\right)^2 = \dfrac{4}{9}$

    $a_3 = \left(\dfrac{2}{3}\right)^3 = \dfrac{8}{27}$

    $a_8 = \left(\dfrac{2}{3}\right)^8 = \dfrac{256}{6561}$

17. Find terms 1, 2, 3, and 8 of the sequence.

    $a_1 = 1 + (-1)^1 = 1 + (-1) = 0$
    $a_2 = 1 + (-1)^2 = 1 + 1 = 2$
    $a_3 = 1 + (-1)^3 = 1 + (-1) = 0$
    $a_8 = 1 + (-1)^8 = 1 + 1 = 2$

19. Find terms 1, 2, 3, and 8 of the sequence.

    $a_1 = (1.1)^1 = 1.1$
    $a_2 = (1.1)^2 = 1.21$
    $a_3 = (1.1)^3 = 1.331$
    $a_8 = (1.1)^8 = 2.14358881$

21. Find terms 1, 2, 3, and 8 of the sequence.

    $a_1 = \dfrac{(-1)^{1+1}}{\sqrt{1}} = \dfrac{(-1)^2}{1} = 1$

    $a_2 = \dfrac{(-1)^{2+1}}{\sqrt{2}} = \dfrac{(-1)^3}{\sqrt{2}} = -\dfrac{1}{\sqrt{2}} = -\dfrac{\sqrt{2}}{2}$

    $a_3 = \dfrac{(-1)^{3+1}}{\sqrt{3}} = \dfrac{(-1)^4}{\sqrt{3}} = \dfrac{1}{\sqrt{3}} = \dfrac{\sqrt{3}}{3}$

    $a_8 = \dfrac{(-1)^{8+1}}{\sqrt{8}} = \dfrac{(-1)^9}{2\sqrt{2}} = -\dfrac{1}{2\sqrt{2}} = -\dfrac{\sqrt{2}}{4}$

23. Find terms 1, 2, 3, and 8 of the sequence.

    $a_1 = 1! = 1$
    $a_2 = 2! = 2 \cdot 1 = 2$
    $a_3 = 3! = 3 \cdot 2 \cdot 1 = 6$
    $a_8 = 8! = 8 \cdot 7 \cdot 6 \cdot 5 \cdot 4 \cdot 3 \cdot 2 \cdot 1 = 40,320$

945

25. Find terms 1, 2, 3, and 8 of the sequence.

$$a_1 = \log 1 = 0$$
$$a_2 = \log 2 \approx 0.3010$$
$$a_3 = \log 3 \approx 0.4771$$
$$a_8 = \log 8 \approx 0.9031$$

27. Find terms 1, 2, 3, and 8 of the sequence.

$$a_1 = 1 \qquad \frac{1}{7} = 0.\overline{142857}$$
$$a_2 = 4$$
$$a_3 = 2$$
$$a_8 = 4$$

29. Find terms 1, 2, 3, and 8 of the sequence.

$$a_1 = 3$$
$$a_2 = 3$$
$$a_3 = 3$$
$$a_8 = 3$$

31. Find the first three terms of the sequence.

$$a_1 = 5$$
$$a_2 = 2 \cdot a_1 = 2 \cdot 5 = 10$$
$$a_3 = 2 \cdot a_2 = 2 \cdot 10 = 20$$

33. Find the first three terms of the sequence.

$$a_1 = 2$$
$$a_2 = 2 \cdot a_1 = 2 \cdot 2 = 4$$
$$a_3 = 3 \cdot a_2 = 3 \cdot 4 = 12$$

35. Find the first three terms of the sequence.

$$a_1 = 2$$
$$a_2 = (a_1)^2 = (2)^2 = 4$$
$$a_3 = (a_2)^2 = (4)^2 = 16$$

37. Find the first three terms of the sequence.

$$a_1 = 2$$
$$a_2 = 2 \cdot 2 \cdot a_1 = 4 \cdot 2 = 8$$
$$a_3 = 2 \cdot 3 \cdot a_2 = 6 \cdot 8 = 48$$

39. Find the first three terms of the sequence.

$$a_1 = 3$$
$$a_2 = (a_1)^{1/2} = (3)^{1/2} = \sqrt{3}$$
$$a_3 = (a_2)^{1/3} = \left(3^{1/2}\right)^{1/3} = 3^{1/6} = \sqrt[6]{3}$$

41. Find a_3, a_4, and a_5.

$$a_1 = 1$$
$$a_2 = 3$$
$$a_3 = \frac{1}{2}(a_2 + a_1) = \frac{1}{2}(3 + 1) = \frac{1}{2}(4) = 2$$
$$a_4 = \frac{1}{2}(a_3 + a_2) = \frac{1}{2}(2 + 3) = \frac{1}{2}(5) = \frac{5}{2}$$
$$a_5 = \frac{1}{2}(a_4 + a_3) = \frac{1}{2}\left(\frac{5}{2} + 2\right) = \frac{1}{2}\left(\frac{9}{2}\right) = \frac{9}{4}$$

43. Evaluate the factorial expression.

$$7! - 6! = 7 \cdot 6! - 6! = 6!(7 - 1)$$
$$= 6! \cdot 6 = 6 \cdot 5 \cdot 4 \cdot 3 \cdot 2 \cdot 1 \cdot 6$$
$$= 4320$$

45. Evaluate the factorial expression.

$$\frac{9!}{7!} = \frac{9 \cdot 8 \cdot 7!}{7!} = 72$$

47. Evaluate the factorial expression.

$$\frac{8!}{3!5!} = \frac{8 \cdot 7 \cdot 6 \cdot 5!}{3 \cdot 2 \cdot 1 \cdot 5!} = 56$$

49. Evaluate the factorial expression.

$$\frac{100!}{99!} = \frac{100 \cdot 99!}{99!} = 100$$

51. Evaluate the series.

$$\sum_{i=1}^{5} i = 1 + 2 + 3 + 4 + 5 = 15$$

53. Evaluate the series.

$$\sum_{i=1}^{5} i(i-1)$$
$$= 1(1-1) + 2(2-1) + 3(3-1) + 4(4-1) + 5(5-1)$$
$$= 1 \cdot 0 + 2 \cdot 1 + 3 \cdot 2 + 4 \cdot 3 + 5 \cdot 4$$
$$= 0 + 2 + 6 + 12 + 20 = 40$$

55. Evaluate the series.

$$\sum_{k=1}^{4} \frac{1}{k} = \frac{1}{1} + \frac{1}{2} + \frac{1}{3} + \frac{1}{4} = \frac{12}{12} + \frac{6}{12} + \frac{4}{12} + \frac{3}{12} = \frac{25}{12}$$

57. Evaluate the series.

$$\sum_{j=1}^{8} 2j = 2\sum_{j=1}^{8} j$$
$$= 2(1 + 2 + 3 + 4 + 5 + 6 + 7 + 8)$$
$$= 2(36) = 72$$

59. Evaluate the series.

$$\sum_{i=1}^{5}(-1)^{i-1}2^{i}$$

$$=(-1)^{0}2^{1}+(-1)^{1}2^{2}+(-1)^{2}2^{3}+(-1)^{3}2^{4}+(-1)^{4}2^{5}$$

$$=2-4+8-16+32$$

$$=22$$

61. Evaluate the series.

$$\sum_{n=1}^{7}\log\left(\frac{n+1}{n}\right)$$

$$=\log\left(\frac{1+1}{1}\right)+\log\left(\frac{2+1}{2}\right)+\log\left(\frac{3+1}{3}\right)+\log\left(\frac{4+1}{4}\right)$$

$$+\log\left(\frac{5+1}{5}\right)+\log\left(\frac{6+1}{6}\right)+\log\left(\frac{7+1}{7}\right)$$

$$=\log\left(2\cdot\frac{3}{2}\cdot\frac{4}{3}\cdot\frac{5}{4}\cdot\frac{6}{5}\cdot\frac{7}{6}\cdot\frac{8}{7}\right)=\log 8$$

$$=3\log 2\approx 0.9031$$

63. Evaluate the series.

$$\sum_{k=0}^{8}\frac{8!}{k!(8-k)!}$$

$$=\frac{8!}{0!(8-0)!}+\frac{8!}{1!(8-1)!}+\frac{8!}{2!(8-2)!}+\frac{8!}{3!(8-3)!}$$

$$+\frac{8!}{4!(8-4)!}+\frac{8!}{5!(8-5)!}+\frac{8!}{6!(8-6)!}$$

$$+\frac{8!}{7!(8-7)!}+\frac{8!}{8!(8-8)!}$$

$$=\frac{8!}{0!8!}+\frac{8!}{1!7!}+\frac{8!}{2!6!}+\frac{8!}{3!5!}+\frac{8!}{4!4!}+\frac{8!}{5!3!}+\frac{8!}{6!2!}$$

$$+\frac{8!}{7!1!}+\frac{8!}{8!0!}$$

$$=1+8+\frac{8\cdot 7}{2}+\frac{8\cdot 7\cdot 6}{3\cdot 2}+\frac{8\cdot 7\cdot 6\cdot 5}{4\cdot 3\cdot 2}+\frac{8\cdot 7\cdot 6}{3\cdot 2\cdot 1}+\frac{8\cdot 7}{2}$$

$$+8+1$$

$$=1+8+28+56+70+56+28+8+1$$

$$=256$$

65. Write the series in summation notation.

$$\frac{1}{1}+\frac{1}{4}+\frac{1}{9}+\frac{1}{16}+\frac{1}{25}+\frac{1}{36}$$

$$=\frac{1}{1^{2}}+\frac{1}{2^{2}}+\frac{1}{3^{2}}+\frac{1}{4^{2}}+\frac{1}{5^{2}}+\frac{1}{6^{2}}$$

$$=\sum_{i=1}^{6}\frac{1}{i^{2}}$$

67. Write the series in summation notation.

$$2-4+8-16+32-64+128$$

$$=2^{1}(-1)^{1+1}+2^{2}(-1)^{2+1}+2^{3}(-1)^{3+1}+2^{4}(-1)^{4+1}$$

$$+2^{5}(-1)^{5+1}+2^{6}(-1)^{6+1}+2^{7}(-1)^{7+1}$$

$$=\sum_{i=1}^{7}2^{i}(-1)^{i+1}$$

69. Write the series in summation notation.

$$7+10+13+16+19$$

$$=7+(7+3)+(7+3\cdot 2)+(7+3\cdot 3)+(7+3\cdot 4)$$

$$=\sum_{i=0}^{4}(7+3i)$$

71. Write the series in summation notation.

$$\frac{1}{2}+\frac{1}{4}+\frac{1}{8}+\frac{1}{16}=\frac{1}{2}+\frac{1}{2^{2}}+\frac{1}{2^{3}}+\frac{1}{2^{4}}=\sum_{i=1}^{4}\frac{1}{2^{i}}$$

73. Approximate $\sqrt{7}$ by computing a_4 and compare to calculator value.

Let $N=7$.

$$a_{1}=\frac{7}{2}=3.5$$

$$a_{2}=\frac{1}{2}\left(3.5+\frac{7}{3.5}\right)=2.75$$

$$a_{3}=\frac{1}{2}\left(2.75+\frac{7}{2.75}\right)\approx 2.6477273$$

$$a_{4}\approx\frac{1}{2}\left(2.6477273+\frac{7}{2.6477273}\right)\approx 2.6457520$$

75. Determine the sum of the first 10 terms of the Fibonacci sequence.

$$1+1=2$$

$$1+1+2=4$$

$$1+1+2+3=7$$

The sum of the first n terms of a Fibonacci sequence equals the $(n+2)$ term – 1.

Therefore the sum of the first ten terms is

(10+2) term – 1, i.e. 12th term – 1.

$$144-1=143$$

77. Find F_{10} and F_{15}.

$$F_{10} = \frac{\left(\frac{1+\sqrt{5}}{2}\right)^{10} - \left(\frac{1-\sqrt{5}}{2}\right)^{10}}{\sqrt{5}} = 55$$

$$F_{15} = \frac{\left(\frac{1+\sqrt{5}}{2}\right)^{15} - \left(\frac{1-\sqrt{5}}{2}\right)^{15}}{\sqrt{5}} = 610$$

79. Find the next two terms of the RATS sequence.

$$\begin{array}{r} 668 \\ +\ 866 \\ \hline 1534 \end{array} \qquad \begin{array}{r} 1345 \\ +\ 5431 \\ \hline 6776 \end{array}$$

Sorting 1534 gives   Sorting 6776 gives

1345   6677

Prepare for Section 11.2

P1. Solve for d.

$$-3 = 25 + (15-1)d$$
$$-3 = 25 + 14d$$
$$-28 = 14d$$
$$-2 = d$$

P3. Evaluate.

$$S = \frac{50\left[2(2) + (50-1)\frac{5}{4}\right]}{2} = \frac{50\left[4 + \frac{245}{4}\right]}{2}$$

$$= \frac{50\left[\frac{261}{4}\right]}{2} = \frac{6525}{4}$$

P5. Find the twentieth term.

$$a_{20} = 52 + (20-1)(-3) = -5$$

Section 11.2 Exercises

1. Find the common difference.

$$d = 8 - 3 = 5$$

3. Find the third partial sum.

$$S_3 = 1 + 4 + 7 = 12$$

5. Determine whether the sequence is arithmetic.

$$\begin{aligned} a_{i+1} - a_i &= [7(i+1)+3] - [7i+3] \\ &= [7i+7-3] - [7i+3] \\ &= 7 \end{aligned}$$

Yes, since the difference between successive terms is a constant, $a_n = 7n+3$ is an arithmetic sequence.

7. Determine whether the sequence is arithmetic.

$$\begin{aligned} a_{i+1} - a_i &= [4-(i+1)] - [4-i] \\ &= [4-i-1] - [4-i] \\ &= -1 \end{aligned}$$

Yes, since the difference between successive terms is a constant, $a_n = 4-n$ is an arithmetic sequence.

9. Determine whether the sequence is arithmetic.

$$\begin{aligned} a_{i+1} - a_i &= [2^{i+1}] - [2^i] \\ &= 2^i(2-1) \\ &= 2^i \end{aligned}$$

No, since the difference between successive terms is not a constant, $a_n = 2^n$ is not an arithmetic sequence.

11. Determine whether the sequence is arithmetic.

$$a_{i+1} - a_i = [3] - [3] = 0$$

Yes, since the difference between successive terms is a constant, $a_n = 3$ is an arithmetic sequence.

13. Find the 9th, 24th, and nth terms.

$$d = 10 - 6 = 4$$
$$a_n = 6 + (n-1)4 = 6 + 4n - 4 = 4n + 2$$
$$a_9 = 4 \cdot 9 + 2 = 36 + 2 = 38$$
$$a_{24} = 4 \cdot 24 + 2 = 98$$

15. Find the 9th, 24th, and nth terms.

$$d = 4 - 6 = -2$$
$$a_n = 6 + (n-1)(-2) = 6 - 2n + 2 = 8 - 2n$$
$$a_9 = 8 - 2 \cdot 9 = 8 - 18 = -10$$
$$a_{24} = 8 - 2 \cdot 24 = 8 - 48 = -40$$

17. Find the 9th, 24th, and nth terms.

$$d = -5 - (-8) = 3$$
$$\begin{aligned} a_n &= -8 + (n-1)3 = -8 + 3n - 3 \\ &= 3n - 11 \end{aligned}$$
$$a_9 = 3 \cdot 9 - 11 = 27 - 11 = 16$$
$$a_{24} = 3 \cdot 24 - 11 = 72 - 11 = 61$$

19. Find the 9th, 24th, and nth terms.

$$d = 4 - 1 = 3$$
$$a_n = 1 + (n-1)3 = 1 + 3n - 3 = 3n - 2$$
$$a_9 = 3 \cdot 9 - 2 = 27 - 2 = 25$$
$$a_{24} = 3 \cdot 24 - 2 = 72 - 2 = 70$$

21. Find the 9th, 24th, and nth terms.

$$d = (a+2) - a = 2$$

$$a_n = a + (n-1)2 = a + 2n - 2$$
$$a_9 = a + 2 \cdot 9 - 2 = a + 18 - 2 = a + 16$$
$$a_{24} = a + 2 \cdot 24 - 2 = a + 48 - 2 = a + 46$$

23. Find the 9th, 24th, and nth terms.

$$d = \log 14 - \log 7 = \log \frac{14}{7} = \log 2$$

$$a_n = \log 7 + (n-1)\log 2$$
$$a_9 = \log 7 + 8\log 2$$
$$a_{24} = \log 7 + 23\log 2$$

25. Find the 9th, 24th, and nth terms.

$$d = \log a^2 - \log a = \log \frac{a^2}{a} = \log a$$

$$a_n = \log a + (n-1)\log a$$
$$= (1 + n - 1)\log a = n\log a$$
$$a_9 = 9\log a$$
$$a_{24} = 24\log a$$

27. Find the 20th term.

$$d = 15 - 13 = 2$$

$$a_4 = a_1 + (4-1)2 = 13$$
$$a_1 + 6 = 13$$
$$a_1 = 7$$
$$a_{20} = 7 + (20-1)2 = 7 + (19)2 = 7 + 38 = 45$$

29. Find the 17th term.

First find d, then find a_1. $a_5 = -19,\ a_7 = -29$

$$a_7 = a_5 + 2d$$
$$d = \frac{a_7 - a_5}{2}$$
$$d = \frac{-29 - (-19)}{2}$$
$$d = -5$$
$$a_n = a_1 + (n-1)d$$
$$a_5 = a_1 + (5-1)(-5)$$
$$-19 = a_1 + (-20)$$
$$a_1 = 1$$
$$a_{17} = 1 + (17-1)(-5) = 1 + 16(-5) = 1 - 80 = -79$$

31. Find the nth partial sum.

$$a_1 = 3(1) + 2 = 5$$
$$a_{10} = 3(10) + 2 = 32$$

$$S_{10} = \frac{10}{2}(a_1 + a_{10}) = 5(5 + 32) = 185$$

33. Find the nth partial sum.

$$a_1 = 3 - 5(1) = -2$$
$$a_{15} = 3 - 5(15) = -72$$

$$S_{15} = \frac{15}{2}(a_1 + a_{15}) = \frac{15}{2}(-2 + (-72)) = \frac{15}{2}(-74)$$
$$= -555$$

35. Find the nth partial sum.

$$a_1 = 6(1) = 6$$
$$a_{12} = 6(12) = 72$$

$$S_{12} = \frac{12}{2}(a_1 + a_{12}) = 6(6 + 72) = 6(78) = 468$$

37. Find the nth partial sum.

$$a_1 = 1 + 8 = 9$$
$$a_{25} = 25 + 8 = 33$$

$$S_{25} = \frac{25}{2}(a_1 + a_{25}) = \frac{25}{2}(9 + 33) = \frac{25}{2}(42) = 525$$

39. Find the nth partial sum.

$$a_1 = -1$$
$$a_{30} = -30$$

$$S_{30} = \frac{30}{2}(a_1 + a_{30}) = 15(-1 + (-30))$$
$$= 15(-31)$$
$$= -465$$

41. Find the nth partial sum.

$$a_1 = 1 + x$$
$$a_{12} = 12 + x$$

$$S_{12} = \frac{12}{2}(a_1 + a_{12})$$
$$= 6(1 + x + 12 + x)$$
$$= 78 + 12x$$

43. Find the nth partial sum.

$$a_1 = (1)x = x$$
$$a_{20} = (20)x = 20x$$

$$S_{20} = \frac{20}{2}(a_1 + a_{20}) = 10(x + 20x) = 210x$$

45. Insert k arithmetic means between the numbers.

$-1,\ c_2,\ c_3,\ c_4,\ c_5,\ c_6,\ 23$

$a_1 = -1$

$a_7 = a_1 + (n-1)d$

$23 = -1 + (7-1)d$

$23 = -1 + 6d$

$24 = 6d$

$d = 4$

$c_2 = -1 + (2-1)4 = -1 + 4 = 3$

$c_3 = -1 + (3-1)4 = -1 + (2)4 = 7$

$c_4 = -1 + (4-1)4 = -1 + (3)4 = 11$

$c_5 = -1 + (5-1)4 = -1 + (4)4 = 15$

$c_6 = -1 + (6-1)4 = -1 + (5)4 = 19$

47. Insert k arithmetic means between the numbers.

$3,\ c_2,\ c_3,\ c_4,\ c_5,\ \dfrac{1}{2}$

$a_1 = 3$

$a_6 = a_1 + (n-1)d$

$\dfrac{1}{2} = 3 + (6-1)d$

$-\dfrac{5}{2} = 5d$

$d = -\dfrac{1}{2}$

$c_2 = 3 + (2-1)\left(-\dfrac{1}{2}\right) = 3 - \dfrac{1}{2} = \dfrac{5}{2}$

$c_3 = 3 + (3-1)\left(-\dfrac{1}{2}\right) = 3 - 1 = 2$

$c_4 = 3 + (4-1)\left(-\dfrac{1}{2}\right) = 3 - \dfrac{3}{2} = \dfrac{3}{2}$

$c_5 = 3 + (5-1)\left(-\dfrac{1}{2}\right) = 3 - 2 = 1$

49. Show the sum.

$a_1 = 1,\ d = 2$

$S_n = \dfrac{n[2(1) + (n-1)2]}{2} = \dfrac{n}{2}[2n] = n^2$

51. Write the formula. Find the number of weeks.

$a_1 = 15$

$a_2 = 15 + 8 = 23$

$d = 8$

$a_n = 8n + 7$

When $a_n = 55$, find n.

$55 = 8n + 7$

$48 = 8n$

$6 = n$

In 6 weeks, the client will be walking 55 min a day.

53. Write the formula. Find the year for the value $425,000.

$a_1 = 600,000,\ d = -25,000$

$V = 600,000 - 25,000n$

When $V = \$425,000$, find n.

$425,000 = 600,000 - 25,000n$

$-175,000 = -25,000n$

$7 = n$

In 7 years, the earthmover will have a value of $425,000.

55. Write the formula for handshakes.

$a_1 = 1,\ S_1 = 0$

$a_2 = 2,\ S_2 = 0 + 1$

$a_3 = 3,\ S_3 = 0 + 1 + 2$

$a_4 = 4,\ S_4 = 0 + 1 + 2 + 3$

$a_5 = 5,\ S_5 = 0 + 1 + 2 + 3 + 4$

Using the sum of an arithmetic sequence, the sum of n people shaking hands can be found.

$S_n = \dfrac{n}{2}[a_1 + a_n] = \dfrac{n}{2}[0 + (n-1)]$

$ = \dfrac{n^2 - n}{2}$

57. Find the number of seats in the tenth row and the total number of logs for the section.

$a_1 = 27,\ d = 2$

$a_{10} = 27 + (10-1)(2) = 27 + 18 = 45$

$S_{10} = \dfrac{10}{2}(27 + 45) = 5(72) = 360$

45 seats in the tenth row; 360 seats in ten rows

59. Find the distance.

$a_1 = 16,\ d = 32$

$S_7 = \dfrac{7}{2}[2(16) + (7-1)32] = \dfrac{7}{2}[32 + 192] = \dfrac{7}{2}[224]$

$ = 784$

The total distance the object falls is 784 ft.

61. Find the number of cross threads.

$$a_n = a_1 + (n-1)d$$
$$46.9 = 3.5 + (n-1)1.4$$
$$43.4 = (n-1)1.4$$
$$31 = n-1$$
$$32 = n$$

There are 32 cross threads.

63. Show $f(n)$ is an arithmetic sequence.

If $f(x)$ is a linear function, then $f(x) = mx + b$. To show that $f(n)$, where n is a positive integer, is an arithmetic sequence, we must show that $f(n+1) - f(n)$ is a constant. We have

$$f(n+1) - f(n) = (m(n+1) + b) - (m(n) + b)$$
$$= mn + m + b - mn - b$$
$$= m$$

Thus, the difference between any two successive terms is m, the slope of the linear function.

65. Find the formula for a_n in terms of a_1.

$$a_1 = 4, \ a_n = a_{n-1} - 3$$

Rewriting $a_n = a_{n-1} - 3$ as $a_n - a_{n-1} = -3$, we find that the difference between successive terms is the same constant -3.

Thus the sequence is an arithmetic sequence with $a_1 = 4$ and $d = -3$.

$$a_n = a_1 + (n-1)d$$

Substituting,

$$a_n = 4 + (n-1)(-3) = 4 - 3n + 3 = 7 - 3n$$

67. Show that a_n and b_n are arithmetic sequences and find a_{50}.

$$a_1 = 1, a_n = b_{n-1} + 7; b_1 = -2, b_n = a_{n-1} + 1$$

To show that a_n is an arithmetic sequence, we must show that $a_n - a_{n-1} = d$, where d is a constant. We begin by finding a relationship between a_n and a_{n-2}.

$$a_n = b_{n-1} + 7 = a_{n-2} + 1 + 7$$
$$a_n = a_{n-2} + 8$$

This establishes a relationship between *alternate* successive terms. We now examine some terms of a_n.

$$a_1 = 1$$
$$a_2 = b_1 + 7 = -2 + 7 = 5$$
$$a_3 = a_1 + 8 \hspace{2cm} (a_n = a_{n-2} + 8)$$
$$a_4 = a_2 + 8$$
$$a_5 = a_3 + 8 = (a_1 + 8) + 8 = a_1 + 2(8)$$
$$a_6 = a_4 + 8 = (a_2 + 8) + 8 = a_2 + 2(8)$$
$$a_7 = a_5 + 8 = (a_1 + 2(8)) + 8 = a_1 + 3(8)$$
$$a_8 = a_6 + 8 = (a_2 + 2(8)) + 8 = a_2 + 3(8)$$

Now consider two cases. First, n is an even integer, $n = 2k$.

$$a_{2k} = a_2 + (k-1)8 \hspace{1cm} k \geq 2$$

When n is an odd integer, $n = 2k - 1$.

$$a_{2k-1} = a_1 + (k-1)8 \hspace{1cm} k \geq 2$$

Thus $a_{2k} - a_{2k-1} = (a_2 + (k-1)8) - (a_1 + (k-1)8)$
$$= a_2 - a_1 = 5 - 1 = 4$$

Therefore, the difference between successive terms is the constant. To find a_{50}, use $a_n = a_1 + (n-1)d$.

$$a_{50} = 1 + (49)(4) = 197$$

To show that b_n is an arithmetic sequence, we have

$$b_n - b_{n-1} = (a_{n-1} + 1) - (a_{n-2} + 1) = a_{n-1} - a_{n-2}$$

Because a_n is an arithmetic sequence, $a_{n-1} - a_{n-2}$ is a constant. Thus b_n is an arithmetic sequence.

Prepare for Section 11.3

P1. Find the ratio of any two successive terms.

$$\frac{4}{2} = 2, \ \frac{8}{4} = 2$$

The ratio is 2.

P3. Evaluate.

$$S = \frac{3(1 - (-2)^5)}{1 - (-2)} = 33$$

P5. Write the first three terms of the sequence.

$$a_1 = 3\left(-\frac{1}{2}\right)^1 = -\frac{3}{2}$$

$$a_2 = 3\left(-\frac{1}{2}\right)^2 = \frac{3}{4}$$

$$a_3 = 3\left(-\frac{1}{2}\right)^3 = -\frac{3}{8}$$

Section 11.3 Exercises

1. Find the common ratio.

$$r = \frac{64}{16} = 4$$

3. Determine whether the sequence is geometric.

$$\frac{a_{n+1}}{a_n} = \frac{3^{n+1}}{3^n} = 3^{n+1-n} = 3$$

Yes, since the ratio is a constant, $a_n = 3^n$ is a

geometric sequence.

5. Determine whether the sequence is geometric.

$$\frac{a_{n+1}}{a_n} = \frac{\dfrac{4}{5^{n+1}}}{\dfrac{4}{5^n}} = \frac{4}{5^{n+1}} \cdot \frac{5^n}{4} = 5^{n-(n+1)} = 5^{-1} = \frac{1}{5}$$

Yes, since the ratio is a constant, $a_n = \dfrac{4}{5^n}$ is a

geometric sequence.

7. Determine whether the sequence is geometric.

$$\frac{a_{n+1}}{a_n} = \frac{\dfrac{1}{(n+1)!}}{\dfrac{1}{n!}} = \frac{n!}{(n+1)!} = \frac{n!}{n!(n+1)} = \frac{1}{n+1}$$

No, since the ratio is not a constant, $a_n = \dfrac{1}{n!}$ is not a

geometric sequence.

9. Determine whether the sequence is geometric.

$$\frac{a_{n+1}}{a_n} = \frac{2(0.35)^{n+1}}{2(0.35)^n} = 0.35$$

Yes, since the ratio is a constant, $a_n = 2(0.35)^n$ is a

geometric sequence.

11. Find the nth term of the geometric sequence.

$$r = \frac{8}{2} = 4$$

$$a_n = 2 \cdot 4^{n-1} = 2 \cdot 2^{2(n-1)} = 2^{2n-1}$$

13. Find the nth term of the geometric sequence.

$$r = \frac{12}{-4} = -3$$

$$a_n = -4(-3)^{n-1}$$

15. Find the nth term of the geometric sequence.

$$r = \frac{4}{6} = \frac{2}{3}$$

$$a_n = 6\left(\frac{2}{3}\right)^{n-1}$$

17. Find the nth term of the geometric sequence.

$$r = -\frac{5}{6}$$

$$a_n = -6\left(-\frac{5}{6}\right)^{n-1}$$

19. Find the nth term of the geometric sequence.

$$r = \frac{-3}{9} = -\frac{1}{3}$$

$$a_n = 9\left(-\frac{1}{3}\right)^{n-1} = \left(-\frac{1}{3}\right)^{n-3}$$

21. Find the nth term of the geometric sequence.

$$r = \frac{-x}{1} = -x$$

$$a_n = 1(-x)^{n-1} = (-x)^{n-1}$$

23. Find the nth term of the geometric sequence.

$$r = \frac{c^5}{c^2} = c^3$$

$$a_n = c^2\left(c^3\right)^{n-1} = c^2 c^{3n-3} = c^{3n-1}$$

25. Find the nth term of the geometric sequence.

$$r = \frac{\dfrac{3}{10,000}}{\dfrac{3}{100}} = \frac{1}{100}$$

$$a_n = \frac{3}{100}\left(\frac{1}{100}\right)^{n-1} = 3\left(\frac{1}{100}\right)^n$$

27. Find the nth term of the geometric sequence.

$$r = \frac{0.05}{0.5} = 0.1$$

$$a_n = 0.5(0.1)^{n-1} = 5(0.1)^n$$

29. Find the nth term of the geometric sequence.

$$r = \frac{0.0045}{0.45} = 0.01$$

$$a_n = 0.45(0.01)^{n-1} = 45(0.01)^n$$

31. Find the requested term.

$$a_1 = 2, \qquad a_5 = 162, \qquad a_n = a_1 r^{n-1}$$

$$162 = 2r^{5-1}$$

$$r^4 = 81$$

$$r = 3$$

$$a_3 = 2(3)^{3-1} = 2 \cdot 9 = 18$$

33. Find the requested term.

$$a_3 = \frac{4}{3}, \qquad a_6 = -\frac{32}{81}$$

$$\frac{\frac{4}{3}}{-\frac{32}{81}} = \frac{a_1(r)^{3-1}}{a_1(r)^{6-1}}$$

$$\frac{-27}{8} = \frac{1}{r^3}$$

$$r^3 = -\frac{8}{27}$$

$$r = \frac{-2}{3}$$

$$\frac{4}{3} = a_1 \left(-\frac{2}{3}\right)^{3-1}$$

$$a_1 = \frac{4}{3}\left(\frac{9}{4}\right) = 3$$

$$a_2 = 3\left(\frac{-2}{3}\right) = -2$$

35. Classify the sequence $a_n = \dfrac{1}{n^2}$ as arithmetic, geometric or neither.

An arithmetic sequence has a common difference.

$$\frac{1}{1^2}, \; \frac{1}{2^2}, \; \frac{1}{3^2}, \; \frac{1}{4^2}, \ldots = 1, \; \frac{1}{2}, \; \frac{1}{9}, \; \frac{1}{16}, \ldots$$

No common difference; the sequence is not arithmetic.

A geometric sequence has a common ratio.

$$\frac{\frac{1}{2}}{1}, \; \frac{\frac{1}{9}}{\frac{1}{2}}, \; \frac{\frac{1}{16}}{\frac{1}{9}}, \ldots = \frac{1}{2}, \; \frac{2}{9}, \; \frac{9}{16}, \ldots$$

No common ratio; the sequence is not geometric.

Neither.

37. Classify the sequence $a_n = 2n - 7$ as arithmetic, geometric or neither.

An arithmetic sequence has a common difference.

$$2(1) - 7, \; 2(2) - 7, \; 2(3) - 7, \; 2(4) - 7, \ldots = -5, -3, -1, 1, \ldots$$

The common difference is $d = 2$.

Arithmetic

39. Classify the sequence $a_n = \left(-\dfrac{6}{5}\right)^n$ as arithmetic, geometric or neither.

A geometric sequence has a common ratio.

$$\text{common ratio: } r = \frac{\left(\frac{6}{5}\right)^{i+1}}{\left(-\frac{6}{5}\right)^i} = -\frac{6}{5}$$

Geometric

41. Classify the sequence $a_n = \dfrac{n}{2^n}$ as arithmetic, geometric or neither.

An arithmetic sequence has a common difference.

$$\frac{1}{2^1}, \; \frac{2}{2^2}, \; \frac{3}{2^3}, \; \frac{4}{2^4}, \ldots = \frac{1}{2}, \; \frac{1}{2}, \; \frac{3}{8}, \; \frac{1}{4}, \ldots$$

No common difference; the sequence is not arithmetic.

A geometric sequence has a common ratio.

$$\frac{\frac{1}{2}}{\frac{1}{2}}, \; \frac{\frac{3}{8}}{\frac{1}{2}}, \; \frac{\frac{1}{4}}{\frac{3}{8}}, \ldots = 1, \; \frac{3}{4}, \; \frac{2}{3}, \ldots$$

No common ratio; the sequence is not geometric.

Neither.

43. Classify the sequence $a_n = -3n$ as arithmetic, geometric or neither.

An arithmetic sequence has a common difference.

$$-3(1), \; -3(2), \; -3(3), \; -3(4), \ldots = -3, -6, -9, -12, \ldots$$

The common difference is $d = -3$.

Arithmetic

45. Classify the sequence $a_n = \dfrac{3^n}{2}$ as arithmetic,

geometric or neither.

A geometric sequence has a common ratio.

common ratio: $r = \dfrac{\frac{3^{i+1}}{2}}{\frac{3^i}{2}} = 3$

Geometric

47. Find the sum of the finite geometric series.

$a_1 = 3, \qquad a_2 = 9, \qquad r = \dfrac{9}{3} = 3$

$S_5 = \dfrac{3(1-3^5)}{1-3} = \dfrac{3(-242)}{-2} = 363$

49. Find the sum of the finite geometric series.

$a_1 = \dfrac{2}{3}, \qquad a_2 = \dfrac{4}{9}, \qquad r = \dfrac{\frac{4}{9}}{\frac{2}{3}} = \dfrac{2}{3}$

$S_6 = \dfrac{\frac{2}{3}\left[1-\left(\frac{2}{3}\right)^6\right]}{1-\frac{2}{3}} = \dfrac{\frac{2}{3}\left(\frac{665}{729}\right)}{\frac{1}{3}} = \dfrac{1330}{729}$

51. Find the sum of the finite geometric series.

$a_1 = 1, \ a_2 = -\dfrac{2}{5}, \ r = -\dfrac{2}{5}$

$S_9 = \dfrac{1\left[1-\left(-\frac{2}{5}\right)^9\right]}{1-\left(-\frac{2}{5}\right)} = \dfrac{\frac{1,953,637}{1,953,125}}{\frac{7}{5}} = \dfrac{279,091}{390,625}$

53. Find the sum of the finite geometric series.

$a_1 = 1, \ a_2 = -2, \ r = -2$

$S_{10} = \dfrac{1[1-(-2)^{10}]}{1-(-2)} = -341$

55. Find the sum of the finite geometric series.

$a_1 = 5, \ a_2 = 15, \ r = 3$

$S_{10} = \dfrac{5[1-3^{10}]}{1-3} = 147,620$

57. Find the sum of the infinite geometric series.

$a_1 = \dfrac{1}{3}, \ a_2 = \dfrac{1}{9}, \ r = \dfrac{1}{3}$

$S = \dfrac{\frac{1}{3}}{1-\frac{1}{3}} = \dfrac{\frac{1}{3}}{\frac{2}{3}} = \dfrac{1}{2}$

59. Find the sum of the infinite geometric series.

$a_1 = -\dfrac{2}{3}, \ r = -\dfrac{2}{3}$

$S = \dfrac{-\frac{2}{3}}{1-\left(-\frac{2}{3}\right)} = \dfrac{-\frac{2}{3}}{\frac{5}{3}} = -\dfrac{2}{5}$

61. Find the sum of the infinite geometric series.

$a_1 = \dfrac{9}{100}, \ r = \dfrac{9}{100}$

$S = \dfrac{\frac{9}{100}}{1-\frac{9}{100}} = \dfrac{\frac{9}{100}}{\frac{91}{100}} = \dfrac{9}{91}$

63. Find the sum of the infinite geometric series.

$a_1 = 0.1, \ r = 0.1$

$S = \dfrac{0.1}{1-0.1} = \dfrac{0.1}{0.9} = \dfrac{1}{9}$

65. Find the sum of the infinite geometric series.

$a_1 = 1, \ r = -0.4$

$S = \dfrac{1}{1-(-0.4)} = \dfrac{1}{1.4} = \dfrac{5}{7}$

67. Write as a quotient of two integers in simplest form.

$0.\overline{3} = \dfrac{3}{10} + \dfrac{3}{100} + \dfrac{3}{1000} + \cdots$

$a_1 = \dfrac{3}{10}, \ r = \dfrac{\frac{3}{100}}{\frac{3}{10}} = \dfrac{1}{10}$

$0.\overline{3} = \dfrac{\frac{3}{10}}{1-\frac{1}{10}} = \dfrac{\frac{3}{10}}{\frac{9}{10}} = \dfrac{1}{3}$

69. Write as a quotient of two integers in simplest form.

$0.\overline{45} = \dfrac{45}{100} + \dfrac{45}{10,000} + \dfrac{45}{1,000,000} + \cdots$

$a_1 = \dfrac{45}{100}, \ r = \dfrac{\frac{45}{10,000}}{\frac{45}{100}} = \dfrac{1}{100}$

$0.\overline{45} = \dfrac{\frac{45}{100}}{1-\frac{1}{100}} = \dfrac{\frac{45}{100}}{\frac{99}{100}} = \dfrac{5}{11}$

71. Write as a quotient of two integers in simplest form.

$$0.\overline{123} = \frac{123}{1000} + \frac{123}{1,000,000} + \frac{123}{1,000,000,000} + \cdots$$

$$a_1 = \frac{123}{1000}, \; r = \frac{1}{1000}$$

$$0.\overline{123} = \frac{\frac{123}{1000}}{1 - \frac{1}{1000}} = \frac{123}{999} = \frac{41}{333}$$

73. Write as a quotient of two integers in simplest form.

$$0.\overline{422} = \frac{422}{1000} + \frac{422}{1,000,000} + \frac{422}{1,000,000,000} + \cdots$$

$$a_1 = \frac{422}{1000}, \; r = \frac{1}{1000}$$

$$0.\overline{422} = \frac{\frac{422}{1000}}{1 - \frac{1}{1000}} = \frac{422}{999}$$

75. Write as a quotient of two integers in simplest form.

$$0.25\overline{4} = \frac{25}{100} + \left[\frac{4}{1000} + \frac{4}{10,000} + \frac{4}{100,000} + \cdots\right]$$

$$a_1 = \frac{4}{1000}, \; r = \frac{1}{10}$$

$$0.25\overline{4} = \frac{25}{100} + \frac{\frac{4}{1000}}{1 - \frac{1}{10}} = \frac{25}{100} + \frac{4}{900} = \frac{229}{900}$$

77. Write as a quotient of two integers in simplest form.

$$1.20\overline{84} = 1 + \frac{2}{10} + \left[\frac{84}{10,000} + \frac{84}{1,000,000} + \frac{84}{100,000,000} + \cdots\right]$$

$$a_1 = \frac{84}{10,000}, \; r = \frac{1}{100}$$

$$1.20\overline{84} = 1 + \frac{2}{10} + \frac{\frac{84}{10,000}}{1 - \frac{1}{100}} = \frac{12}{10} + \frac{7}{825} = \frac{1994}{1650} = \frac{997}{825}$$

79. Find the future value in 8 years.

$$P = 100, \; i = 0.09, \; n = 2, \; t = 8$$

$$r = \frac{i}{n} = \frac{0.09}{2} = 0.045, \; m = nt = 2 \cdot 8 = 16$$

$$A = 100 \frac{\left[(1 + 0.045)^{16} - 1\right]}{0.045} \approx 2271.93367$$

The future value is $2271.93.

81. Find the dividend growth rate.

$$\text{Stock value} = \frac{D(1+g)}{i-g}$$

$$67 = \frac{1.32(1+g)}{0.20 - g} \quad \text{Solve for } g.$$

$$67(0.20 - g) = 1.32(1 + g)$$

$$13.4 - 67g = 1.32 + 1.32g$$

$$12.08 = 68.32g$$

$$0.1768 = g$$

The dividend growth rate is 17.68%.

83. Find the price per share.

$$\text{Stock value (no dividend growth)} = \frac{D}{i} = \frac{2.94}{0.15} = \$19.60$$

85. Explain.

If g was **not** less than i in the Gordon model of stock valuation, the common ratio of the geometric sequence would be greater than 1 and the sum of the infinite geometric series would not be defined.

87. Find the net effect.

Using the multiplier effect,

$$\frac{25}{1 - 0.75} = 100$$

The net effect of $25 million is $100 million.

89. Find the amount each recipient would receive.

When a name was removed from the top of the list, the letter had been sent to

$$5(5^5) = 15,625 \text{ people}$$

who sent 10 cents each for a total of

$$0.10(15,625) = \$1562.50 \text{ for each recipient.}$$

91. Find the concentration to the nearest hundredth.

$$A = 0.5, \; n = 4, \; k = -0.876, \; t = 4$$

$$A + Ae^{kt} + Ae^{2kt} + Ae^{3kt}$$

$$= 0.5 + 0.5e^{-0.867(4)} + 0.5e^{2(-0.867)(4)} + 0.5e^{3(-0.867)(4)}$$

$$\approx 0.52 \text{ mg}$$

Or $A = 0.5, \; r = e^{kt}, \; n = 4, \; k = -0.876, \; t = 4$

$$S_4 = \frac{0.5\left(1 - e^{-.867(4)(3)}\right)}{1 - e^{-.867(4)}} \approx 0.52 \text{ mg}$$

93. Find the number of grandparents.

The nth generation has $a_n = 2^n$ grandparents. Since this is a geometric sequence, the sum can be found by a formula.

$$S_n = \frac{a_1(1-r^n)}{1-r}$$

$$S_{10} = \frac{2(1-2^{10})}{1-2} = \frac{2(1-1024)}{-1} = 2046$$

When $n = 1$, $a_n = 2$ and there are no grandparents.

Therefore there are $2046 - 2 = 2044$ grandparents by the 10th generation.

Mid-Chapter 11 Quiz

1. Find the fourth and eighth terms.

$$a_4 = \frac{4}{2^4} = \frac{4}{16} = \frac{1}{4}$$

$$a_8 = \frac{8}{2^8} = \frac{8}{256} = \frac{1}{32}$$

3. Evaluate.

$$\sum_{k=1}^{5} \frac{(-1)^{k-1}}{k^2} = \frac{(-1)^0}{1^2} + \frac{(-1)^1}{2^2} + \frac{(-1)^2}{3^2} + \frac{(-1)^3}{4^2} + \frac{(-1)^4}{5^2}$$

$$= 1 - \frac{1}{4} + \frac{1}{9} - \frac{1}{16} + \frac{1}{25}$$

$$= \frac{3019}{3600}$$

5. Find the sum of the first 25 terms.

$$a_n = 5 - n$$

$$a_1 = 4, \ a_2 = 3, \ a_3 = 2$$

$$d = -1$$

$$S_{25} = \frac{25}{2}[2(4) + (25-1)(-1)] = -200$$

7. Find the sum of the first eight terms.

$$a_n = (-3)^n$$

$$a_1 = (-3)^1 = -3$$

$$a_2 = (-3)^2 = 9$$

$$r = \frac{9}{-3} = -3$$

$$S_8 = \frac{-3(1-(-3)^8)}{1-(-3)} = 4920$$

Prepare for Section 11.4

P1. Show the statement is true for $n = 4$.

$$\sum_{i=1}^{4} \frac{1}{i(i+1)} = \frac{1}{1(1+1)} + \frac{1}{2(2+1)} + \frac{1}{3(3+1)} + \frac{1}{4(4+1)}$$

$$= \frac{1}{2} + \frac{1}{6} + \frac{1}{12} + \frac{1}{20} = \frac{4}{5} = \frac{4}{4+1}$$

P3. Simplify.

$$\frac{k}{k+1} + \frac{1}{(k+1)(k+2)} = \frac{k+2}{k+2} \cdot \frac{k}{k+1} + \frac{1}{(k+1)(k+2)}$$

$$= \frac{k^2 + 2k + 1}{(k+1)(k+2)}$$

$$= \frac{(k+1)(k+1)}{(k+1)(k+2)}$$

$$= \frac{k+1}{k+2}$$

P5. Write in simplest form.

$$S_n + a_{n+1} = \frac{n(n+1)}{2} + n + 1$$

$$= \frac{n^2 + n}{2} + \frac{2n+2}{2} = \frac{n^2 + 3n + 2}{2}$$

$$= \frac{(n+1)(n+2)}{2}$$

Section 11.4 Exercises

1. Write the two steps in a proof by mathematical induction.

Step 1: Prove the statement is true for $n = 1$.

Step 2: Assume that the statement is true for $n = k$ and then prove it is true for $n = k + 1$.

3. Prove the statement by mathematical induction.

1. Let $n = 1$. $S_1 = 3 \cdot 1 - 2 = 1 = \frac{1(3 \cdot 1 - 1)}{2}$

2. Assume the statement is true for some positive integer k.

$$S_k = 1 + 4 + 7 + \cdots + (3k - 2) = \frac{k(3k-1)}{2}$$

(Induction Hypothesis)

Verify that the statement is true when $n = k + 1$

$$S_{k+1} = \frac{(k+1)(3k+2)}{2}.$$

$$a_k = 3k - 2, \ a_{k+1} = 3k + 1$$

$$S_{k+1} = S_k + a_{k+1}$$
$$= \frac{k(3k-1)}{2} + 3k + 1 = \frac{3k^2 - k}{2} + \frac{6k+2}{2}$$
$$= \frac{3k^2 + 5k + 2}{2} = \frac{(k+1)(3k+2)}{2}$$

By the Principle of Mathematical Induction, the statement is true for all positive integers.

5. Prove the statement by mathematical induction.

1. Let $n = 1$. $S_1 = 1^3 = 1 = \dfrac{1^2(1+1)^2}{4}$

2. Assume $S_k = 1 + 8 + 27 + \cdots + k^3 = \dfrac{k^2(k+1)^2}{4}$ is true for some positive integer k.(Induction Hypothesis).

Verify that $S_{k+1} = \dfrac{(k+1)^2(k+2)^2}{4}$.

$$a_k = k^3, \quad a_{k+1} = (k+1)^3$$

$$S_{k+1} = S_k + a_{k+1} = \frac{k^2(k+1)^2}{4} + (k+1)^3$$
$$= \frac{k^2(k+1)^2 + 4(k+1)^3}{4}$$
$$= \frac{(k+1)^2(k^2 + 4k + 4)}{4} = \frac{(k+1)^2(k+2)^2}{4}$$

By the Principle of Mathematical Induction, the statement is true for all positive integers.

7. Prove the statement by mathematical induction.

1. Let $n = 1$. $S_1 = 4 \cdot 1 - 1 = 3 = 1(2 \cdot 1 + 1)$

2. Assume that $S_k = 3 + 7 + 11 + \cdots + (4k-1) = k(2k+1)$ is true for some positive integer k (Induction Hypothesis).
   Verify that $S_{k+1} + (k+1)(2k+3)$.

$$a_k = 4k - 1, \quad a_{k+1} = 4k + 3$$

$$S_{k+1} = S_k + a_{k+1}$$
$$= k(2k+1) + 4k + 3$$
$$= 2k^2 + 5k + 3 = (k+1)(2k+3)$$

By the Principle of Mathematical Induction, the statement is true for all positive integers.

9. Prove the statement by mathematical induction.

1. Let $n = 1$. $S_1 = (2 \cdot 1 - 1)^3 = 1 = 1^2(2 \cdot 1^2 - 1)$

2. Assume that
$$S_k = 1 + 27 + 125 + \cdots + (2k-1)^3 = k^2(2k^2 - 1)$$
is true for some positive integer k (Induction Hypothesis).

Verify that $S_{k+1} = (k+1)^2(2k^2 + 4k + 1)$.

$$a_k = (2k-1)^3, \quad a_{k+1} = (2k+1)^3$$

$$S_{k+1} = S_k + a_{k+1} = k^2(2k^2 - 1) + (2k+1)^3$$
$$= 2k^4 - k^2 + 8k^3 + 12k^2 + 6k + 1$$
$$= 2k^4 + 8k^3 + 11k^2 + 6k + 1$$
$$= (k+1)(2k^3 + 6k^2 + 5k + 1)$$
$$= (k+1)^2(2k^2 + 4k + 1)$$

By the Principle of Mathematical Induction, the statement is true for all positive integers.

11. Prove the statement by mathematical induction.

1. Let $n = 1$. $S_1 = \dfrac{1}{(2 \cdot 1 - 1)(1 \cdot 1 + 1)} = \dfrac{1}{3} = \dfrac{1}{2 \cdot 1 + 1}$

2. Assume that
$$S_k = \frac{1}{1 \cdot 3} + \frac{1}{3 \cdot 5} + \frac{1}{5 \cdot 7} + \cdots + \frac{1}{(2k-1)(2k+1)}$$
$$= \frac{k}{2k+1}$$
for some positive integer k (Induction Hypothesis).

Verify that $S_{k+1} = \dfrac{k+1}{2k+3}$.

$$a_k = \frac{1}{(2k-1)(2k+1)}, \quad a_{k+1} = \frac{1}{(2k+1)(2k+3)}$$

$$S_{k+1} = \frac{k}{2k+1} + \frac{1}{(2k+1)(2k+3)}$$
$$= \frac{2k^2 + 3k + 1}{(2k+1)(2k+3)}$$
$$= \frac{(2k+1)(k+1)}{(2k+1)(2k+3)} = \frac{k+1}{2k+3}$$

By the Principle of Mathematical Induction, the statement is true for all positive integers.

13. Prove the statement by mathematical induction.

1. Let $n = 1$.

$$S_1 = 1^4 = 1$$

$$= \frac{1(1+1)(2 \cdot 1+1)(3 \cdot 1^2 + 3 \cdot 1 - 1)}{30} = \frac{2 \cdot 3 \cdot 5}{30} = 1$$

2. Assume that

$$S_k = 1 + 16 + 81 + \cdots + k^4$$

$$= \frac{k(k+1)(2k+1)(3k^2 + 3k - 1)}{30}$$

for some positive integer k (Induction Hypothesis).

Verify that

$$S_{k+1} = \frac{(k+1)(k+2)(2k+3)(3k^2 + 9k + 5)}{30}.$$

$$a_k = k^4, \quad a_{k+1} = (k+1)^4$$

$$S_{k+1} = \frac{k(k+1)(2k+1)(3k^2 + 3k - 1)}{30} + (k+1)^4$$

$$= \frac{(k+1)[k(2k+1)(3k^2 + 3k - 1) + 30(k+1)^3]}{30}$$

$$= \frac{(k+1)[6k^4 + 39k^3 + 91k^2 + 89k + 30]}{30}$$

$$= \frac{(k+1)(k+2)(6k^3 + 27k^2 + 37k + 15)}{30}$$

$$= \frac{(k+1)(k+2)(2k+3)(3k^2 + 9k + 5)}{30}$$

By the Principle of Mathematical Induction, the statement is true for all positive integers.

15. Prove the inequality by mathematical induction.

1. Let $n = 4$. Then $\left(\frac{3}{2}\right)^4 = \frac{81}{16} = 5\frac{1}{16}$; $4 + 1 = 5$

Thus, $\left(\frac{3}{2}\right)^n > n + 1$ for $n = 4$.

2. Assume $\left(\frac{3}{2}\right)^k > k + 1$ is true for some positive

integer $k \geq 4$ (Induction Hypothesis).

Verify that $\left(\frac{3}{2}\right)^{k+1} > k + 2$.

$$\left(\frac{3}{2}\right)^{k+1} = \left(\frac{3}{2}\right)^k \left(\frac{3}{2}\right)$$

$$> (k+1)\left(\frac{3}{2}\right) = \frac{1}{2}(3k+3) = \frac{1}{2}(2k + k + 3)$$

$$> \frac{1}{2}(2k + 1 + 3) = k + 2$$

Thus $\left(\frac{3}{2}\right)^{k+1} > k + 2$. By the Principle of

Mathematical Induction, $\left(\frac{3}{2}\right)^n > n + 1$ for all $n \geq 4$.

17. Prove the inequality by mathematical induction.

1. Let $n = 1$.

$$0 < a < 1$$

$$0 < a \cdot a < a \cdot 1$$

$$a^{1+1} = a^2 < a$$

Thus, if $0 < a < 1$, then $a^{1+1} < a^n$ for $n = 1$.

2. Assume $a^{k+1} < a^k$ is true for some positive integer

k, if $0 < a < 1$ (Induction Hypothesis).

Verify $a^{k+2} < a^{k+1}$.

$$0 < a < 1$$

$$0 < a \cdot a^{k+1} < 1 \cdot a^{k+1}$$

$$a^{k+2} < a^{k+1}$$

By the Principle of Mathematical Induction, if

$0 < a < 1$, then $a^{n+1} < a^n$ for all positive integers.

19. Prove the inequality by mathematical induction.

1. Let $n = 4$. $1 \cdot 2 \cdot 3 \cdot 4 = 24$, $2^4 = 16$

Thus, $1 \cdot 2 \cdot 3 \cdot 4 > 2^n$ for $n = 4$.

2. Assume $1 \cdot 2 \cdot 3 \cdot 4 \cdot \cdots \cdot k > 2^k$ is true for some

positive integer $k \geq 4$ (Induction Hypothesis).

Verify $1 \cdot 2 \cdot 3 \cdot \cdots \cdot k \cdot (k+1) > 2^{k+1}$.

$$1 \cdot 2 \cdot 3 \cdot \cdots \cdot k \cdot (k+1) > 2^k(k+1) > 2^k \cdot 2 = 2^{k+1}$$

Thus, $1 \cdot 2 \cdot 3 \cdot \cdots \cdot k \cdot (k+1) > 2^{k+1}$. By the Principle of

Mathematical Induction, $1 \cdot 2 \cdot 3 \cdot \cdots \cdot n > 2^n$ for all

$n \geq 4$.

21. Prove the inequality by mathematical induction.

1. Let $n = 1$ and $a > 0$. $(1+a)^1 = 1+a = 1+1 \cdot a$

    Thus $(1+a)^n \geq 1+na$ for $n = 1$.

2. Assume $(1+a)^k \geq 1+ka$ is true for some positive integer k (Induction Hypotheses).

    Verify $(1+a)^{k+1} \geq 1+(k+1)a$.

    $$(1+a)^{k+1} = (1+a)^k (1+a)$$
    $$\geq (1+ka)(1+a) = 1+(k+1)a+ka^2$$
    $$\geq 1+(k+1)a$$

    Thus $(1+a)^{k+1} \geq 1+(k+1)a$. By the Principle of Mathematical Induction, $(1+a)^n \geq 1+na$ for all positive integers n.

23. Prove the statement by mathematical induction.

1. Let $n = 1$. $1^2 +1 = 2$, $2 = 2 \cdot 1$

    Thus 2 is a factor of $n^2 +n$ for $n = 1$.

2. Assume 2 is a factor of $k^2 +k$ for some positive integer k (Induction Hypothesis).

    Verify 2 is a factor of $(k+1)^2 +k+1$.

    $$(k+1)^2 +k+1 = (k+1)(k+1+1)$$
    $$= (k+1)(k+2)$$

    Since $k^2 +k = k(k+1)$, 2 is a factor of k or $k+1$.

    If 2 is a factor of $k + 1$, then 2 is a factor of $(k+1)(k+2)$.

    If 2 is a factor of k, then 2 is a factor of $k + 2$.

    Thus, 2 is a factor of $(k+1)^2 +k+1$. By the Principle of Mathematical Induction, 2 is a factor of $n^2 +n$ for all positive integers.

25. Prove the statement by mathematical induction.

1. Let $n = 1$. $5^1 -1 = 4$, $4 = 4 \cdot 1$

    Thus, 4 is a factor of $5^n -1$ for $n = 1$.

2. Assume 4 is a factor of $5^k -1$ for some positive

integer k (Induction Hypothesis).

Verify 4 is a factor of $5^{k+1} -1$.

Now $5^{k+1} -1 = 5 \cdot 5^k -5+4 = 5(5^k -1)+4$

which is the sum of two multiples of 4.

Thus, 4 is a factor of $5^{k+1} -1$. By the Principle of Mathematical Induction, 4 is a factor of $5^n -1$ for all positive integers.

27. Prove the statement by mathematical induction.

1. Let $n = 1$. $(xy)^1 = xy = x^1 y^1$

    Thus, $(xy)^n = x^n y^n$ for $n = 1$.

2. Assume $(xy)^k = x^k y^k$ is true for some positive integer k (Induction Hypothesis).

    Verify $(xy)^{k+1} = x^{k+1} y^{k+1}$.

    $$(xy)^{k+1} = (xy)^k (xy)^1 = x^k y^k \cdot xy = x^{k+1} y^{k+1}$$

    Thus $(xy)^{k+1} = x^{k+1} y^{k+1}$. By the Principle of Mathematical Induction, $(xy)^n = x^n y^n$ for all positive integers.

29. Prove the statement by mathematical induction.

1. Let $n = 1$. $a^1 -b^1 = a-b$

    Thus $a-b$ is a factor of $a^n -b^n$ for $n = 1$.

2. Assume $a-b$ is a factor of $a^k -b^k$ for some positive integer k (Induction Hypothesis).

    Verify $a-b$ is a factor of $a^{k+1} -b^{k+1}$.

    $$a^{k+1} -b^{k+1} = (a \cdot a^k -ab^k)+(ab^k -b \cdot b^k)$$
    $$= a(a^k -b^k)+b^k (a-b)$$

    The sum of two multiples of $a-b$ is a multiple of $a-b$. Thus, $a-b$ is a factor of $a^{k+1} -b^{k+1}$.

By the Principle of Mathematical Induction, $a-b$ is a factor of $a^n -b^n$ for all positive integers.

31. Prove the statement by mathematical induction.

1. Let $n = 1$. $\quad ar^{1-1} = a \cdot 1 = a = \dfrac{a(1-r^1)}{1-r}$

   Thus, the statement is true for $n = 1$.

2. Assume $\displaystyle\sum_{k=1}^{j} ar^{k-1} = \dfrac{a(1-r^j)}{1-r}$ is true for some

positive integer j.

Verify $\displaystyle\sum_{k=1}^{j+1} ar^{k-1} = \dfrac{a(1-r^{j+1})}{1-r}$ is true for

$n = j+1$.

$$\sum_{k=1}^{j+1} ar^{k-1} = \sum_{k=1}^{j}(ar^{k+1}-1) + ar^j = \dfrac{a(1-r^j)}{1-r} + ar^j$$

$$= \dfrac{a(1-r^j) + ar^j(1-r)}{1-r}$$

$$= \dfrac{a[1-r^j + r^j - r^{j+1}]}{1-r}$$

$$= \dfrac{a(1-r^{j+1})}{1-r}$$

By the Principle of Mathematical Induction,

$$\sum_{k=1}^{n} ar^{k-1} = \dfrac{a(1-r^n)}{1-r}$$

33. Prove the statement by mathematical induction.

1. If $N = 25$, then $\log 25! \approx 25.19 > 25$.

2. Assume $\log k! > k$ for $k > 25$ (Induction

   Hypothesis).

   Prove $\log(k+1)! > k+1$.

   $\log(k+1)! = \log[(k+1)k!]$

   $\qquad\qquad = \log(k+1) + \log k!$

   $\qquad\qquad > \log(k+1) + k$

   Because $k > 25$, $\log(k+1) > 1$.

   Thus, $\log(k+1)! > k+1$.

   Therefore, $\log n! > n$ for all $n > 25$.

35. Prove the statement by mathematical induction.

1. When $n = 1$, we have $(x^m)^1 = x^m$ and $x^{m \cdot 1} = x^m$.

   Therefore, the statement is true for $n = 1$.

2. Assume the statement is true for $n = k$. That is,

assume $(x^m)^k = x^{mk}$ (Induction Hypothesis).

Prove the statement is true for $n = k + 1$.

$$x^{m(k+1)} = x^{mk+m} = x^{mk} \cdot x^m = (x^m)^k \cdot x^m$$

$$= (x^m)^{k+1}$$

Thus, the statement is true for all positive integers n and m.

37. Prove the statement by mathematical induction.

1. When $n = 3$, we have $\left(\dfrac{3+1}{3}\right)^3 = \left(\dfrac{4}{3}\right)^3 = \dfrac{64}{27} < 3$.

   Thus the statement is true for $n = 3$.

2. Assume the statement is true for $n = k$. That is,

   assume $\left(\dfrac{k+1}{k}\right)^k < k$ (Induction Hypothesis).

   Prove the statement is true for $n = k + 1$. That is,

   prove $\left(\dfrac{k+2}{k+1}\right)^{k+1} < k+1$.

   We begin by noting that $\left(\dfrac{k+2}{k+1}\right) < \dfrac{k+1}{k}$.

   Therefore

   $$\left(\dfrac{k+2}{k+1}\right)^{k+1} < \left(\dfrac{k+1}{k}\right)^{k+1} = \left(\dfrac{k+1}{k}\right)^k \left(\dfrac{k+1}{k}\right)$$

   By the Induction Hypothesis, $\left(\dfrac{k+1}{k}\right)^k < k$; thus

   $$\left(\dfrac{k+1}{k}\right)^k \left(\dfrac{k+1}{k}\right) < k\left(\dfrac{k+1}{k}\right) = k+1$$

   We now have $\left(\dfrac{k+2}{k+1}\right)^{k+1} < k+1$. The induction

   is complete.

   Thus $\left(\dfrac{n+1}{n}\right)^n < n$ is true for all $n \geq 3$.

Prepare for Section 11.5

P1. Expand.

$$(a+b)^3 = (a+b)(a+b)(a+b)$$

$$= a^3 + 3a^2b + 3ab^2 + b^3$$

P3. Evaluate.

$$0! = 1$$

P5. Evaluate.

$$\frac{7!}{3!(7-3)!} = \frac{5040}{6(24)} = 35$$

Section 11.5 Exercises

1. Write the degree of each term.

After expanding $(a+b)^n$, the degree of each term is n.

3. Evaluate the binomial coefficient.

$$\binom{7}{4} = \frac{7!}{4!(7-4)!} = \frac{7 \cdot 6 \cdot 5 \cdot 4!}{4!(3 \cdot 2 \cdot 1)} = 35$$

5. Evaluate the binomial coefficient.

$$\binom{9}{2} = \frac{9!}{2!(9-2)!} = \frac{9 \cdot 8 \cdot 7!}{2 \cdot 1 \cdot 7!} = 36$$

7. Evaluate the binomial coefficient.

$$\binom{12}{9} = \frac{12!}{9!(12-9)!} = \frac{12 \cdot 11 \cdot 10 \cdot 9!}{9! \cdot 3 \cdot 2 \cdot 1} = 220$$

9. Evaluate the binomial coefficient.

$$\binom{11}{0} = \frac{11!}{0!(11-0)!} = \frac{11!}{1 \cdot 11!} = 1$$

11. Expand the binomial.

$$(x+y)^5 = \binom{5}{0}x^5 + \binom{5}{1}x^4 \cdot y + \binom{5}{2}x^3 \cdot y^2 + \binom{5}{3}x^2 \cdot y^3$$

$$+ \binom{5}{4}x \cdot y^4 + \binom{5}{5}y^5$$

$$= x^5 + 5x^4y + 10x^3y^2 + 10x^2y^3 + 5xy^4 + y^5$$

13. Expand the binomial.

$$(a-b)^4 = \binom{4}{0}a^4 + \binom{4}{1}a^3(-b) + \binom{4}{2}a^2(-b)^2$$

$$+ \binom{4}{3}a(-b)^3 + \binom{4}{4}(-b)^4$$

$$= a^4 - 4a^3b + 6a^2b^2 - 4ab^3 + b^4$$

15. Expand the binomial.

$$(x+5)^4 = \binom{4}{0}x^4 + \binom{4}{1}x^3 5 + \binom{4}{2}x^2 5^2 + \binom{4}{3}x5^3 + \binom{4}{4}5^4$$

$$= x^4 + 20x^3 + 150x^2 + 500x + 625$$

17. Expand the binomial.

$$(a-3)^5 = \binom{5}{0}a^5 + \binom{5}{1}a^4(-3) + \binom{5}{2}a^3(-3)^2$$

$$+ \binom{5}{3}a^2(-3)^3 + \binom{5}{4}a(-3)^4 + \binom{5}{5}(-3)^5$$

$$= a^5 - 15a^4 + 90a^3 - 270a^2 + 405ab^4 - 243$$

19. Expand the binomial.

$$(2x-4)^7 = \binom{7}{0}(2x)^7 + \binom{7}{1}(2x)^6(-4) + \binom{7}{2}(2x)^5(-4)^2$$

$$+ \binom{7}{3}(2x)^4(-4)^3 + \binom{7}{4}(2x)^3(-4)^4$$

$$+ \binom{7}{5}(2x)^2(-4)^5 + \binom{7}{6}(2x)(-4)^6 + \binom{7}{7}(-4)^7$$

$$= 128x^7 - 1792x^6 + 10,752x^5 - 35,840x^4$$

$$+ 71,680x^3 - 86,016x^2 + 57,344x - 16,384$$

21. Expand the binomial.

$$(x+3y)^6 = \binom{6}{0}x^6 + \binom{6}{1}x^5(3y) + \binom{6}{2}x^4(3y)^2$$

$$+ \binom{6}{3}x^3(3y)^3 + \binom{6}{4}x^2(3y)^4$$

$$+ \binom{6}{5}x(3y)^5 + \binom{6}{6}(3y)^6$$

$$= x^6 + 18x^5y + 135x^4y^2 + 540x^3y^3$$

$$+ 1215x^2y^4 + 1458xy^5 + 729y^6$$

23. Expand the binomial.

$$(2x-5y)^4$$

$$= \binom{4}{0}(2x)^4 + \binom{4}{1}(2x)^3(-5y) + \binom{4}{2}(2x)^2(-5y)^2$$

$$+ \binom{4}{3}(2x)(-5y)^3 + \binom{4}{4}(-5y)^4$$

$$= 16x^4 - 160x^3y + 600x^2y^2 - 1000xy^3 + 625y^4$$

25. Expand the binomial.

$$(x^2-4)^7$$

$$= \binom{7}{0}(x^2)^7 + \binom{7}{1}(x^2)^6(-4) + \binom{7}{2}(x^2)^5(-4)^2$$

$$+ \binom{7}{3}(x^2)^4(-4)^3 + \binom{7}{4}(x^2)^3(-4)^4$$

$$+ \binom{7}{5}(x^2)^2(-4)^5 + \binom{7}{6}(x^2)(-4)^6 + \binom{7}{7}(-4)^7$$

$$= x^{14} - 28x^{12} + 336x^{10} - 2240x^8 + 8960x^6$$

$$- 21504x^4 + 28672x^2 - 16384$$

27. Expand the binomial.

$$\left(2x^2 + y^3\right)^5 = \binom{5}{0}\left(2x^2\right)^5 + \binom{5}{1}\left(2x^2\right)^4\left(y^3\right)$$

$$+ \binom{5}{2}\left(2x^2\right)^3\left(y^3\right)^2 + \binom{5}{3}\left(2x^2\right)^2\left(y^3\right)^3$$

$$+ \binom{5}{4}\left(2x^2\right)\left(y^3\right)^4 + \binom{5}{5}\left(y^3\right)^5$$

$$= 32x^{10} + 80x^8 y^3 + 80x^6 y^6 + 40x^4 y^9$$

$$+ 10x^2 y^{12} + y^{15}$$

29. Expand the binomial.

$$\left(x + \sqrt{y}\right)^5 = \binom{5}{0}x^5 + \binom{5}{1}x^4 \cdot \sqrt{y} + \binom{5}{2}x^3 \cdot \left(\sqrt{y}\right)^2$$

$$+ \binom{5}{3}x^2 \cdot \left(\sqrt{y}\right)^3 + \binom{5}{4}x \cdot \left(\sqrt{y}\right)^4 + \binom{5}{5}\left(\sqrt{y}\right)^5$$

$$= x^5 + 5x^4 y^{1/2} + 10x^3 y$$

$$+ 10x^2 y^{3/2} + 5xy^2 + y^{5/2}$$

31. Expand the binomial.

$$\left(\frac{2}{x} - \frac{x}{2}\right)^4 = \binom{4}{0}\left(\frac{2}{x}\right)^4 + \binom{4}{1}\left(\frac{2}{x}\right)^3\left(-\frac{x}{2}\right) + \binom{4}{2}\left(\frac{2}{x}\right)^2\left(-\frac{x}{2}\right)^2$$

$$+ \binom{4}{3}\left(\frac{2}{x}\right)\left(-\frac{x}{2}\right)^3 + \binom{4}{4}\left(-\frac{x}{2}\right)^4$$

$$= \frac{16}{x^4} - \frac{16}{x^2} + 6 - x^2 + \frac{x^4}{16}$$

33. Expand the binomial.

$$\left(s^{-2} + s^2\right)^6$$

$$= \binom{6}{0}\left(s^{-2}\right)^6 + \binom{6}{1}\left(s^{-2}\right)^5\left(s^2\right) + \binom{6}{2}\left(s^{-2}\right)^4\left(s^2\right)^2$$

$$+ \binom{6}{3}\left(s^{-2}\right)^3\left(s^2\right)^3 + \binom{6}{4}\left(s^{-2}\right)^2\left(s^2\right)^4 + \binom{6}{5}\left(s^{-2}\right)\left(s^2\right)^5$$

$$+ \binom{6}{6}\left(s^2\right)^6$$

$$= s^{-12} + 6s^{-8} + 15s^{-4} + 20 + 15s^4 + 6s^8 + s^{12}$$

35. Find the term without expanding.

eighth term is $\binom{10}{7}(3x)^3(-y)^7 = -3240x^3 y^7$

37. Find the term without expanding.

third term is $\binom{12}{2}x^{10}(4y)^2 = 1056x^{10}y^2$

39. Find the term without expanding.

fifth term is $\binom{9}{4}\left(\sqrt{x}\right)^5\left(-\sqrt{y}\right)^4 = 126x^2 y^2 \sqrt{x}$

41. Find the term without expanding.

ninth term is $\binom{11}{8}\left(\frac{a}{b}\right)^3\left(\frac{b}{a}\right)^8 = \frac{165b^5}{a^5}$

43. Find the term in the expansion.

$$\binom{n}{i-1}a^{n-(i-1)}b^{i-1}, \text{ if } b^{i-1} = b^8, \text{ then } i = 9.$$

ninth term is $\binom{10}{8}(2a)^2(-b)^8 = 180a^2 b^8$

45. Find the term in the expansion.

$$\left(y^2\right)^{i-1} = y^8, \text{ if } 2i - 2 = 8, \text{ then } 2i = 10 \text{ or } i = 5.$$

fifth term is $\binom{6}{4}(2x)^2\left(y^2\right)^4 = 60x^2 y^8$

47. Find the term in the expansion.

sixth term is $\binom{10}{5}(3a)^5(-b)^5 = -61,236a^5 b^5$

49. Find the term in the expansion.

fifth term is $\binom{9}{4}\left(s^{-1}\right)^5(s)^4 = 126s^{-1}$

sixth term is $\binom{9}{5}\left(s^{-1}\right)^4(s)^5 = 126s$

51. Simplify the power of the complex number.

$$(2 - i)^4 = \binom{4}{0}\left(2^4\right) + \binom{4}{1}(2)^3(-i)^1 + \binom{4}{2}(2)^2(-i)^2$$

$$+ \binom{4}{3}2(-i)^3 + \binom{4}{4}(-i)^4$$

$$= 16 + 32(-i) + 24(-1) + 8(-i)^3 + 1$$

$$= 16 - 32i - 24 + 8i + 1$$

$$= -7 - 24i$$

53. Simplify the power of the complex number.

$$(1 + 2i)^5 = \binom{5}{0}(1)^5 + \binom{5}{1}(1)^4(2i)^1 + \binom{5}{2}(1)^3(2i)^2$$

$$+ \binom{5}{3}(1)^2(2i)^3 + \binom{5}{4}(1)(2i)^4 + \binom{5}{5}(2i)^5$$

$$= 1 + 10i - 40 - 80i + 80 + 32i$$

$$= 41 - 38i$$

55. Simplify the power of the complex number.

$$\left(\frac{\sqrt{2}}{2}+i\frac{\sqrt{2}}{2}\right)^8$$

$$=\left(\frac{\sqrt{2}}{2}\right)^8(1+i)^8$$

$$\quad+\frac{1}{16}(1+8i-28-56i+70+56i-28-8i+1)$$

$$=\frac{1}{16}(16)=1$$

Prepare for Section 11.6

P1. Evaluate.

$$7!=7\cdot6\cdot5\cdot4\cdot3\cdot2\cdot1=5040$$

P3. Evaluate.

$$\binom{7}{1}=\frac{7!}{1!(7-1)!}=\frac{7\cdot6!}{1(6!)}=7$$

P5. Evaluate.

$$\frac{10!}{(10-2)!}=\frac{10\cdot9\cdot8!}{8!}=90$$

Section 11.6 Exercises

1. Define the Fundamental Counting Principle.

The number of ways a sequence can occur is the product of the number of ways each of the n events can occur.

3. Classify as a permutation or combination.

Since the order of six M&Ms does not matter, it is a combination.

5. Classify as a permutation or combination.

Since the order of the telephone numbers matters, it is a permutation.

7. Evaluate.

$$P(6,\ 2)=\frac{6!}{(6-2)!}=\frac{6\cdot5\cdot4!}{4!}=30$$

9. Evaluate.

$$C(8,\ 4)=\frac{8!}{4!(8-4)!}=\frac{8\cdot7\cdot6\cdot5\cdot4!}{4!\cdot4\cdot3\cdot2\cdot1}=70$$

11. Evaluate.

$$P(8,\ 0)=\frac{8!}{(8-0)!}=\frac{8!}{8!}=1$$

13. Evaluate.

$$C(7,\ 7)=\frac{7!}{7!(7-7)!}=\frac{7!}{7!\ \cdot1}=1$$

15. Evaluate.

$$C(10,\ 4)=\frac{10!}{4!(10-4)!}=\frac{10\cdot9\cdot8\cdot7\cdot6!}{4\cdot3\cdot2\cdot1\cdot6!}=210$$

17. Find the number of ways.

Use the counting principle.

$$3\cdot2\cdot2=12$$

There are 12 different possible computer systems.

19. a. Find the number of ways.

There are a total of 10 dogs.

$$10!=3,628,800$$

b. Find the number of orders.

$$4!\cdot3!\cdot3!\cdot3!=24\cdot6\cdot6\cdot6=5184$$

21. Find the number of ways.

$$P(6,6)=6!=720$$

23. Find the number of ways.

Select 2 A's from 9 tiles and select 7 tiles from the remaining 91 tiles.

$$C(9,\ 2)\cdot C(91,\ 7)=36(46,504,458)=1,674,160,488$$

25. Find the number of tests.

Use the combination formula with $n=25$, $r=5$.

$$C(25,\ 5)=\frac{25!}{5!(20)!}=53,130$$

27. Explain.

There are 676 ways to arrange 26 letters taken two at a time $(26\cdot26=676)$. Now if there are more than 676 employees, then at least two employees have the same first and last initials.

29. Find the number of ways.

$$C(6,\ 3)\cdot C(8,\ 3)=1120$$

31. Find the number of ways.

$$2^{10}=1024$$

33. Find the number of ways.

Select 5 from 56 numbers and 1 from 46 numbers.

$$C(56,\ 5)\cdot C(46,\ 1)=175,711,536$$

35. a. Find the number of ways for 0 defective computers.

$C(7, 5) = 21$

b. Find the number of ways for 1 defective computer.

$C(7, 4) \cdot C(3, 1) = 35 \cdot 3 = 105$

c. Find the number of ways for 3 defective computers.

$C(7, 2) \cdot C(3, 3) = 21 \cdot 1 = 21$

37. Find the possible number of serial numbers.

$3 \cdot 12 \cdot 5 \cdot 10^7 = 1.8 \times 10^9$

39. Find the number of ways to select at most 1 defective drive.

$C(10, 3) - C(8, 1) = 120 - 8 = 112$

41. Find the number of ways.

Select 1 from 4 professionals and 3 from 12 amateurs.

$C(4, 1) \cdot C(12, 3) = 880$

43. Find the number of lines.

$C(7, 2) = 21$

45. Find the number of ways.

$\dfrac{16 \cdot 14}{2} = 112$

47. Find the number of ways.

$C(20, 10) = 184{,}756$

49. Find the number of ways.

$C(20, 12) \cdot C(12, 4) = 125{,}970 \cdot 495 = 62{,}355{,}150$

51. Find the number of possible cones.

A triple-decker cone could have all one flavor ice cream, or two different flavors with one scoop of the first flavor and two scoops of the second flavor (such as one scoop of vanilla and two scoops of chocolate), or two different flavors with two scoops of the first flavor and one scoop of the second flavor (such as two scoops of vanilla and one scoop of chocolate), or three different flavors.

$C(31, 1) + C(31, 2) + C(31, 2) + C(31, 3)$

$= \dfrac{31!}{1!30!} + \dfrac{31!}{2!29!} + \dfrac{31!}{2!29!} + \dfrac{31!}{3!28!}$

$= 31 + 465 + 465 + 4495$

$= 5456$

53. Find the number of arrangements.

$19!$

55. Find the number of lines.

There are n ways to choose the first point and $n - 1$ ways to choose the second point. Thus there are $n(n - 1)$ ways to choose both points. Since the direction of the line is not important, the number of lines is

$\dfrac{n(n-1)}{2}$ or $\dbinom{n}{2}$.

57. Find the number of combinations.

To return to the original spot, the tourist must toss an equal number of heads and tails. This is $C(10, 5) = 252$. There are 252 different toss combinations that return the tourist to the origin.

Prepare for Section 11.7

P1. Define the Fundamental Counting Principle.

See Section 8.6.

P3. Evaluate.

$P(7,2) = \dfrac{7!}{(7-2)!} = \dfrac{7 \cdot 6 \cdot 5!}{5!} = 42$

P5. Evaluate.

$\dbinom{8}{5}\left(\dfrac{1}{4}\right)^5\left(\dfrac{3}{4}\right)^{8-5} = \dfrac{8!}{5!(8-5)!}\left(\dfrac{1}{4}\right)^5\left(\dfrac{3}{4}\right)^3 = \dfrac{189}{8192}$

Section 11.7 Exercises

1. State the difference between a sample space and an event.

A sample space consists of all possible outcomes from an experiment. An even is a subset of the sample space.

3. Determine whether the events are mutually exclusive.

No, the events are not mutually exclusive, for example, any odd natural number greater than 4, such as $\{5\}$.

5. List the elements in the sample space.

Label senators S_1, S_2 and representatives R_1, R_2, R_3.

The sample space is $S = \{S_1R_1,\ S_1R_2,\ S_1R_3,\ S_2R_1,$

$S_2R_2,\ S_2R_3,\ R_1R_2,\ R_1R_3,\ R_2R_3,\ S_1S_2\}$

7. List the elements in the sample space.

Label coin H, T and integers 1, 2, 3, 4.

$S = \{H1, H2, H3, H4, T1, T2, T3, T4\}$

9. List the elements in the sample space.

Let the three cans be represented by A, B, and C and (x, y) represent the cans that balls 1 and 2 are placed in; e.g., (A, B) means ball 1 is in can A and ball 2 is in can B.

$S = \{(A, A), (A, B), (A, C), (B, A), (B, B), (B, C), (C, A), (C, B), (C, C)\}$

11. Express the event as a subset of the sample space.

$E = \{HHHH\}$

13. Express the event as a subset of the sample space.

$E = \{TTTT, HTTT, THTT, TTHT, TTTH, TTHH,$
$\quad THTH, HTHT, THHT, HTTH, HHTT\}$

The sample space S for the events in Exercises 15—20 is

$S = \{(1,1),\ (1,2),\ (1,3),\ (1,4),\ (1,5),\ (1,6),\ (2,1),\ (2,2),$
$\quad (2,3),\ (2,4),\ (2,5),\ (2,6),\ (3,1),\ (3,2),\ (3,3),\ (3,4),$
$\quad (3,5),\ (3,6),\ (4,1),\ (4,2),\ (4,3),\ (4,4),\ (4,5),\ (4,6),$
$\quad (5,1),\ (5,2),\ (5,3),\ (5,4),\ (5,5),\ (5,6),\ (6,1),\ (6,2),$
$\quad (6,3),\ (6,4),\ (6,5),\ (6,6)\}$

15. Determine the number of elements in the sample space.

36

17. Find the probability: sum of the numbers is 7.

$E = \{(1,6),(2,5),(3,4),(4,3),(5,2),(6,1)\}$

$P(E) = \dfrac{6}{36} = \dfrac{1}{6}$

19. Find the probability: sum of the two numbers is greater than 3.

$S = \{(1,3),\ (1,4),\ (1,5),\ (1,6),\ (2,2),\ (2,3),\ (2,4),$
$\quad (2,5),\ (2,6),\ (3,1),\ (3,2),\ (3,3),\ (3,4),\ (3,5),$
$\quad (3,6),\ (4,1),\ (4,2),\ (4,3),\ (4,4),\ (4,5),\ (4,6),$
$\quad (5,1),\ (5,2),\ (5,3),\ (5,4),\ (5,5),\ (5,6),\ (6,1),$
$\quad (6,2),\ (6,3),\ (6,4),\ (6,5),\ (6,6)\}$

$P(E) = \dfrac{33}{36} = \dfrac{11}{12}$

21. a. Find the probability: a king.

$P(\text{king}) = \dfrac{4}{52} = \dfrac{1}{13}$

b. Find the probability: a spade.

$P(\text{spade}) = \dfrac{13}{52} = \dfrac{1}{4}$

23. Find the probabilities.

The sample space is

$S = \{HHHH, THHH, HTHH, HHHT, TTHH, THHT,$
$\quad HTTH, HHTT, THTH, HTHT, HHTH, HTTT, TTTH,$
$\quad THTT, TTHT, TTTT\}$

a. $E = \{TTTT\}$

$P(\text{all } T) = \dfrac{1}{16}$

b. $E = \{HHHH, THHH, HTHH, HHHT, TTHH,$
$\quad THHT, HTTH, HHTT, THTH, HTHT, HHTH, HTTT,$
$\quad TTTH, THTT, TTHT\}$

$P(\text{at least } 1\ H) = \dfrac{15}{16}$

25. a. Yes, the events are mutually exclusive since $E_1 \cap E_2 = \varnothing$. That is, no two digit numbers that are perfect squares with the first digit the number 5.

b. $E_1 = \{50,\ 51,\ 52,\ 53,\ 54,\ 55,\ 56,\ 57,\ 58,\ 59\}$

$E_2 = \{16,\ 25,\ 36,\ 49,\ 64,\ 81\}$

$P(E_1) = \dfrac{10}{90}$ and $P(E_2) = \dfrac{6}{90}$

$P(E_1 \cup E_2) = \dfrac{10}{90} + \dfrac{6}{90} = \dfrac{16}{90} = \dfrac{8}{45}$

27. a. No, the events are not mutually exclusive, since the numbers 35 and 70 are divisible by both 5 and 7.

b. $E_1 = \{5,10,15,20,25,30,35,40,45,50,55,$
$\quad 60,65,70,75,80,85,90,95,100\}$

$E_2 = \{7,14,21,28,35,42,49,56,63,70,77,84,91,98\}$

$P(E_1) = \dfrac{20}{100},\ P(E_2) = \dfrac{14}{100}$ and $P(E_1 \cap E_2) = \dfrac{2}{100}$

$P(E_1 \cup E_2) = \dfrac{20}{100} + \dfrac{14}{100} - \dfrac{2}{100} = \dfrac{32}{100} = \dfrac{8}{25}$

29. Find the probability.

$P(\text{increase GNP}) + P(\text{increase inflation})$
$\quad - P(\text{increase GNP and inflation})$
$= 0.64 + 0.55 - 0.22 = 0.97$

31. Find the probability: at least one contract.

$$P(\text{1st}) + P(\text{2nd}) - P(\text{1st and 2nd})$$

$$= \frac{1}{2} + \frac{1}{5} - \frac{1}{10} = \frac{6}{10} = \frac{3}{5}$$

33. Find the probability: 0 does not occur.

Because sampling is with replacement, the events are independent. On one trial, the probability of not selecting a 0 is $\frac{9}{10}$. Therefore, the probability of not selecting a 0 on five trials is $\left(\frac{9}{10}\right)^5 = 0.59$.

35. Find the probability: at least $50.

To receive at least $50, an envelope with $50 in cash or $100 in cash must be selected. The probability is

$$\frac{75}{500} + \frac{50}{500} = \frac{125}{500} = \frac{1}{4} = 0.25 \,.$$

37. Find the probability: alternating boys and girls.

There are $P(6, 6) = 720$ seating arrangements for the 6 children. There are $2 \cdot 3! \cdot 3!$ ways to have boys and girls alternate. Therefore the probability of boys and girls alternating is $\frac{2 \cdot 3! \cdot 3!}{720} = \frac{72}{720} = \frac{1}{10} = 0.1$.

39. Find the probability: 2 cards correct.

The subject can select $C(5, 2) = 10$ different sets of 2 cards. The magician must name the set the subject has drawn. Therefore, the probability that a magician can guess the answers is $\frac{1}{10}$ or 0.1.

41. Find the probability of rolling 10 in 2 rolls of 2 dice.

Here, order does not matter. Rolling 2 dice 2 times can give $(6 \cdot 6) \cdot (6 \cdot 6) = 1296$ different results.

For the dice to add to 10, the rolls can be:
1126, or 1135, or, 1144 or 1225, or 1234, or 1333, or 2224, or 2233.

The roll 1234 can be rolled 24 different ways.

The rolls 1126, 1135, and 1125 can be rolled 12 different ways.

The rolls 1144, 1333, and 2224 can be rolled 6 different ways.

The rolls 1333, and 2224 can be rolled 4 different ways.

Probability $= \dfrac{12 + 12 + 6 + 12 + 24 + 4 + 4 + 6}{1296} = \dfrac{5}{81}$

The probability is $\dfrac{5}{81}$.

43. Find the probability: at least 1 unprofitable well.

The probability of at least one unprofitable

$$= 1 - \text{ probability of all profitable}$$
$$= 1 - [(0.10)(0.10)(0.10)(0.10)]$$
$$= 1 - 0.0001 = 0.9999$$

45. Find the probability: at most 3 have a part-time job.

This is a binomial probability with

$$p = 0.80, \ q = 0.20, \ n = 7, \text{ and } k = 0, \ 1, \ 2, \text{ and } 3.$$

$$P = \binom{7}{0}(0.2)^7 + \binom{7}{1}(0.8)^1 (0.2)^6$$
$$+ \binom{7}{2}(0.8)^2 (0.2)^5 + \binom{7}{3}(0.8)^3 (0.2)^4$$

The probability is 0.033344.

47. Find the probability: at least one defective.

The probability of at least one defective equals 1 minus the probability of no defectives. The probability of no defectives is $(0.95)^5$. Therefore the probability of at least one defective is $1 - (0.95)^5 \approx 0.2262$

49. Find the probability 21 or more people show up.

This is a binomial probability with $p = \dfrac{3}{4}$, $q = \dfrac{1}{4}$, $n = 25$, and $k = 21, \ 22, \ 23, \ 24, \text{ and } 25.$

$$P = \binom{25}{21}\left(\frac{3}{4}\right)^{21}\left(\frac{1}{4}\right)^4 + \binom{25}{22}\left(\frac{3}{4}\right)^{22}\left(\frac{1}{4}\right)^3$$
$$+ \binom{25}{23}\left(\frac{3}{4}\right)^{23}\left(\frac{1}{4}\right)^2 + \binom{25}{24}\left(\frac{3}{4}\right)^{24}\left(\frac{1}{4}\right)^1 + \binom{25}{25}\left(\frac{3}{4}\right)^{25}$$

The probability is approximately 0.2137.

51. Find the probability the rumor told 3 times without returning to originator.

Let the originator tell a member of the club. (Rumor told once.) This person repeats it to any of 7 remaining members with probability $\frac{7}{8}$. (Rumor told twice.) That person repeats it to any of 7 members with probability $\frac{7}{8}$. Probability of both events is $\left(\frac{7}{8}\right)^2 = \frac{49}{64}$.

53. Find the probability of winning a game presently tied. Recall from Section 11.3 the Sum of a Geometric Series

$$S = \frac{a_1}{1-r}$$

$$S = \frac{p^2}{1-2p(1-p)} \qquad a_1 = p^2,\ r = 2p(1-p)$$

$$S = \frac{(0.55)^2}{1-2(0.55)(1-0.55)}$$

$$S \approx 0.599$$

The probability is 0.599.

Exploring Concepts with Technology

Casino 1

Mark	Catch	Win	Expectation
6	4	$8	$0.23
6	5	$176	$0.54
6	6	$2960	$0.38
			$1.15

Casino 2

Mark	Catch	Win	Expectation
6	4	$6	$0.17
6	5	$160	$0.50
6	6	$3900	$0.50
			$1.17

Casino 3

Mark	Catch	Win	Expectation
6	4	$8	$0.23
6	5	$180	$0.56
6	6	$3000	$0.39
			$1.18

Casino 4

Mark	Catch	Win	Expectation
6	4	$6	$0.17
6	5	$176	$0.54
6	6	$3000	$0.39
			$1.10

Each mathematical expectation was determined using the formula

$$\text{Mathematical expectation} = \frac{C(20,\ c)\cdot C(60,\ 6-c)}{C(80,\ 6)}W$$

W is the number of dollars you win for c catches. Thus, casino 3 offers the greatest mathematical expectation. For each $2 bet at casino 3, the gambler has a mathematical expectation of winning $1.18.

Chapter 11 Review Exercises

1. Find the third and seventh terms. [11.1]

$$a_n = 3n+1$$
$$a_3 = 3(3)+1 = 10$$
$$a_7 = 3(7)+1 = 22$$

3. Find the third and seventh terms. [11.1]

$$a_n = n^2$$
$$a_3 = 3^2 = 9$$
$$a_7 = 7^2 = 49$$

5. Find the third and seventh terms. [11.1]

$$a_n = \frac{1}{n}$$
$$a_3 = \frac{1}{3}$$
$$a_7 = \frac{1}{7}$$

7. Find the third and seventh terms. [11.1]

$$a_n = 2^n$$
$$a_3 = 2^3 = 8$$
$$a_7 = 2^7 = 128$$

9. Find the third and seventh terms. [11.1]

$$a_n = \left(\frac{2}{3}\right)^n$$
$$a_3 = \left(\frac{2}{3}\right)^3 = \frac{8}{27}$$
$$a_7 = \left(\frac{2}{3}\right)^7 = \frac{128}{2187}$$

11. Find the third and seventh terms. [11.1]

$a_1 = 2, \ a_n = 3a_{n-1}$

$a_2 = 3a_1 = 3 \cdot 2 = 6$

$a_3 = 3a_2 = 3 \cdot 6 = 18$ •

$a_4 = 3a_3 = 3 \cdot 18 = 54$

$a_5 = 3a_4 = 3 \cdot 54 = 162$

$a_6 = 3a_5 = 3 \cdot 162 = 486$

$a_7 = 3a_6 = 3 \cdot 486 = 1458$ •

13. Find the third and seventh terms. [11.1]

$a_1 = 1, \ a_n = -na_{n-1}$

$a_2 = -2a_1 = -2 \cdot 1 = -2$

$a_3 = -3a_2 = -3 \cdot (-2) = 6$ •

$a_4 = -4a_3 = -4 \cdot 6 = -24$

$a_5 = -5a_4 = -5 \cdot (-24) = 120$

$a_6 = -6a_5 = -6 \cdot 120 = -720$

$a_7 = -7a_6 = -7 \cdot (-720) = 5040$ •

15. Find the third and seventh terms. [11.1]

$a_1 = 1, \ a_2 = 2, \ a_n = a_{n-1}a_{n-2}$

$a_3 = a_2 \cdot a_1 = 2 \cdot 1 = 2$ •

$a_4 = a_3 \cdot a_2 = 2 \cdot 2 = 4$

$a_5 = a_4 \cdot a_3 = 4 \cdot 2 = 8$

$a_6 = a_5 \cdot a_4 = 8 \cdot 4 = 32$

$a_7 = a_6 \cdot a_5 = 32 \cdot 8 = 256$ •

17. Find the third and seventh terms. [11.1]

$a_1 = 1, \ a_2 = 2, \ a_n = 2a_{n-2} - a_{n-1}$

$a_3 = 2a_1 - a_2 = 2 \cdot 1 - 2 = 0$ •

$a_4 = 2a_2 - a_3 = 2 \cdot 2 - 0 = 4$

$a_5 = 2a_3 - a_4 = 2 \cdot 0 - 4 = -4$

$a_6 = 2a_4 - a_5 = 2 \cdot 4 - (-4) = 12$

$a_7 = 2a_5 - a_6 = 2 \cdot (-4) - 12 = -20$ •

19. Evaluate. [11.1]

$5! + 3! = 5 \cdot 4 \cdot 3 \cdot 2 \cdot 1 + 3 \cdot 2 \cdot 1 = 120 + 6 = 126$

21. Evaluate. [11.1]

$\dfrac{10!}{6!} = \dfrac{10 \cdot 9 \cdot 8 \cdot 7 \cdot 6!}{6!} = 5040$

23. Evaluate. [11117]

$\dbinom{12}{3} = \dfrac{12!}{3!(12-3)!} = \dfrac{12 \cdot 11 \cdot 10 \cdot 9!}{3 \cdot 2 \cdot 1 \cdot 9!} = 220$

25. Evaluate. [11.1]

$\displaystyle\sum_{k=1}^{5} k^2 = 1^2 + 2^2 + 3^2 + 4^2 + 5^2$

$= 1 + 4 + 9 + 16 + 25 = 55$

27. Find the term for the arithmetic sequence. [11.2]

$a_1 = 3, \ a_2 = 7, \ a_3 = 11, \ d = 7 - 3 = 4$

$a_n = 3 + (n-1)(4) = 3 + 4n - 4 = 4n - 1$

$a_{25} = 4(25) - 1 = 99$

29. Find the term for the arithmetic sequence. [11.2]

$a_1 = -2, \ a_{10} = 25$

$25 = -2 + (10-1)d$

$27 = 9d$

$d = 3$

$a_{15} = -2 + (15-1)(3) = 40$

31. Find the term for the arithmetic sequence. [11.2]

$a_n = 3n - 4$

$a_1 = 3(1) - 4 = -1$

$a_{20} = 3(20) - 4 = 56$

$S_{20} = \dfrac{20}{2}(-1 + 56) = 550$

33. Find the sum for the arithmetic sequence. [11.2]

$a_1 = 6, \ a_2 = 8, \ a_3 = 10, \ d = 8 - 6 = 2$

$S_{100} = \dfrac{100}{2}[2(6) + (100-1)(2)] = 10,500$

35. Find the arithmetic means. [11.2]

$13, \ c_2, \ c_3, \ c_4, \ c_5, \ 28$

$a_1 = 13$

$a_6 = a_1 + (n-1)d$

$28 = 13 + (6-1)d$

$15 = 5d$

$d = 3$

$c_2 = 13 + (2-1)(3) = 13 + 3 = 16$

$c_3 = 13 + (3-1)(3) = 13 + 6 = 19$

$c_4 = 13 + (4-1)(3) = 13 + 9 = 22$

$c_5 = 13 + (5-1)(3) = 13 + 12 = 25$

37. Find the term for the geometric sequence. [11.3]

$r = \dfrac{-2}{4} = -\dfrac{1}{2}$

$a_n = 4\left(-\dfrac{1}{2}\right)^{n-1}$

39. Find the term for the geometric sequence. [11.3]

$$r = \frac{\frac{15}{4}}{5} = \frac{3}{4}$$

$$a_n = 5\left(\frac{3}{4}\right)^{n-1}$$

41. Find the sum for the geometric sequence. [11.3]

$a_1 = 1,\ a_2 = 2,\ r = 2$

$$S_8 = \frac{1(1-2^8)}{1-2} = 255$$

43. Find the sum for the geometric sequence. [11.3]

$a_1 = 1,\ r = -\dfrac{1}{2}$

$$S = \frac{1}{1-\left(-\frac{1}{2}\right)} = \frac{1}{\frac{3}{2}} = \frac{2}{3}$$

45. Evaluate the series. [11.1]

$$\sum_{k=1}^{5} 2\left(\frac{1}{5}\right)^{k-1} = 2\sum_{k=1}^{5}\left(\frac{1}{5}\right)^{k-1}$$

$$= 2\left(1 + \frac{1}{5} + \frac{1}{25} + \frac{1}{125} + \frac{1}{625}\right)$$

$$= 2\left(\frac{781}{625}\right) = \frac{1562}{625}$$

47. Evaluate the series. [11.3]

$a_1 = 1,\ a_2 = -\dfrac{5}{6},\ r = -\dfrac{5}{6}$

$$S = \frac{1}{1-\left(-\frac{5}{6}\right)} = \frac{1}{\frac{11}{6}} = \frac{6}{11}$$

49. Find the ratio of integers in lowest form. [11.3]

$$0.2\overline{3} = \frac{2}{10} + \left[\frac{3}{100} + \frac{3}{1000} + \frac{3}{10,000}\cdots\right]$$

$$a_1 = \frac{3}{100},\ r = \frac{1}{10}$$

$$0.2\overline{3} = \frac{2}{10} + \frac{\frac{3}{100}}{1-\frac{1}{10}} = \frac{2}{10} + \frac{1}{30} = \frac{7}{30}$$

51. Classify the sequence $a_n = n^2$ as arithmetic, geometric or neither. [11.3]

An arithmetic sequence has a common difference.

$1^2,\ 2^2,\ 3^2,\ 4^2,... = 1,\ 4,\ 9,\ 16,...$

No common difference; the sequence is not arithmetic.

A geometric sequence has a common ratio.

$\dfrac{4}{1},\ \dfrac{9}{4},\ \dfrac{16}{9},...$

No common ratio; the sequence is not geometric.

Neither.

53. Classify the sequence $a_n = (-2)^n$ as arithmetic, geometric or neither. [11.3]

A geometric sequence has a common ratio.

common ratio: $r = \dfrac{(-2)^{i+1}}{(-2)^i} = -2$

Geometric

55. Classify the sequence $a_n = \dfrac{n}{2} + 1$ as arithmetic, geometric or neither. [11.3]

An arithmetic sequence has a common difference.

$\dfrac{1}{2}+1,\ \dfrac{2}{2}+1,\ \dfrac{3}{2}+1,\ \dfrac{4}{2}+1,... = \dfrac{3}{2},\ 2,\ \dfrac{5}{2},\ 3,...$

The common difference is $d = \dfrac{1}{2}$.

Arithmetic

57. Classify the sequence $a_n = n2^n$ as arithmetic, geometric or neither. [11.3]

An arithmetic sequence has a common difference.

$1(2)^1,\ 2(2)^2,\ 3(2)^3,\ 4(2)^4,... = 2,\ 8,\ 24,\ 64,...$

No common difference; the sequence is not arithmetic.

A geometric sequence has a common ratio.

$\dfrac{8}{2},\ \dfrac{24}{8},\ \dfrac{64}{24},...$

No common ratio; the sequence is not geometric.

Neither.

59. Prove the statement $\displaystyle\sum_{i=1}^{n}(5i+1) = \dfrac{n(5n+7)}{2}$ by mathematical induction. [11.4]

1. For $n = 1$, we have $\displaystyle\sum_{i=1}^{1}(5i+1) = 6$ and

$$\frac{1(5+7)}{2} = 6.$$

Therefore that statement is true for $n = 1$.

2.   Assume the statement is true for $n = k$.

$$\sum_{i=1}^{k}(5i+1) = \frac{k(5k+7)}{2}$$ Induction Hypothesis

Prove the statement is true for $n = k + 1$.

$$\sum_{i=1}^{k+1}(5i+1)$$

$$= \sum_{i=1}^{k}(5i+1)+5(k+1)+1 = \sum_{i=1}^{k}(5i+1)+5k+6$$

$$= \frac{k(5k+7)}{2}+5k+6 = \frac{k(5k+7)+10k+12}{2}$$

$$= \frac{5k^2+7k+10k+12}{2} = \frac{5k^2+17k+12}{2}$$

$$= \frac{(k+1)(5k+12)}{2}$$

Thus the statement is true for $n = k+1$. By the Induction Axiom, the statement is true for all positive integers.

61.   Prove the statement $\displaystyle\sum_{i=0}^{n}\left(-\frac{1}{2}\right)^i = \frac{2\left(1-\left(-\frac{1}{2}\right)^{n+1}\right)}{3}$ by mathematical induction. [11.4]

1.   This induction begins with $n = 0$.

$$\sum_{i=0}^{0}\left(-\frac{1}{2}\right)^i = \left(-\frac{1}{2}\right)^0 = 1 \text{ and } \frac{2\left(1-\left(-\frac{1}{2}\right)^{1}\right)}{3} = 1$$

Thus the statement is true when $n = 0$.

2.   Assume the statement is true for $n = k$.

$$\sum_{i=0}^{k}\left(-\frac{1}{2}\right)^i = \frac{2\left(1-\left(-\frac{1}{2}\right)^{k+1}\right)}{3}$$ Induction Hypothesis

Prove the statement is true for $n = k + 1$.

$$\sum_{i=0}^{k+1}\left(-\frac{1}{2}\right)^i = \sum_{i=0}^{k}\left(-\frac{1}{2}\right)^i + \left(-\frac{1}{2}\right)^{k+1}$$

$$= \frac{2\left(1-\left(-\frac{1}{2}\right)^{k+1}\right)}{3}+\left(-\frac{1}{2}\right)^{k+1}$$

$$= \frac{2\left(1-\left(-\frac{1}{2}\right)^{k+1}\right)+3\left(-\frac{1}{2}\right)^{k+1}}{3}$$

$$= \frac{2-2\left(-\frac{1}{2}\right)^{k+1}+3\left(-\frac{1}{2}\right)^{k+1}}{3}$$

$$= \frac{2+\left(-\frac{1}{2}\right)^{k+1}}{3}$$

Thus the statement is true for all integers $n \geq 0$.

63.   Prove the statement $n^n \geq n!$ by mathematical induction. [11.4]

1.   When $n = 1$, $1^1 = 1$ and $n! = 1$. The statement is true for $n = 1$.

2.   Assume the statement is true for $n = k$.

$k^k \geq k!$ Induction Hypothesis

Prove the statement is true for $n = k + 1$. That is,

$(k+1)^{k+1} \geq (k+1)!$
$(k+1)^{k+1} = (k+1)(k+1)^k > (k+1)k^k$
By Induction Hypothesis $k^k \geq k!$. We have
$(k+1)k^k \geq (k+1)k! = (k+1)!$
Therefore $(k+1)^{k+1} \geq (k+1)!$

Therefore the statement is true for all integers $n \geq 1$.

65.   Prove the statement 3 is a factor of $n^3 + 2n$ for all positive integers by mathematical induction. [11.4]

1. When $n = 1$, we have $1^3 + 2(1) = 3$. . Since 3 is a factor of 3, the statement is true for $n = 1$.

2.   Assume the statement is true for $n = k$.

3 is a factor of $k^3 + 2k$ Induction Hypothesis

Prove the statement is true for $n = k + 1$. That is,

3 is a factor of $(k+1)^3 + 2(k+1)$.

$$(k+1)^3 + 2(k+1) = k^3 + 3k^2 + 3k + 1 + 2k + 2$$
$$= (k^3 + 2k) + 3(k^2 + k + 1)$$

By Induction Hypothesis, 3 is a factor of $k^3 + 2k$. Three is also a factor of $3(k^2 + k + 1)$. Thus 3 is a factor of $(k+1)^3 + 2(k+1)$.

The statement is true for all positive integers n.

67. Expand the binomial. [11.5]

$$(4a-b)^5 = \binom{5}{0}(4a)^5 + \binom{5}{1}(4a)^4(-b)^1$$

$$+ \binom{5}{2}(4a)^3(-b)^2 + \binom{5}{3}(4a)^2(-b)^3$$

$$+ \binom{5}{4}(4a)(-b)^4 + \binom{5}{5}(-b)^5$$

$$= 1024a^5 - 1280a^4b + 640a^3b^2 - 160a^2b^3$$

$$+ 20ab^4 - b^5$$

69. Expand the binomial. [11.5]

$$(a-b)^7 = \binom{7}{0}a^7 + \binom{7}{1}a^6b + \binom{7}{2}a^5b^2 + \binom{7}{3}a^4b^3$$

$$+ \binom{7}{4}a^3b^4 + \binom{7}{5}a^2b^5 + \binom{7}{6}ab^6 + \binom{7}{7}b^7$$

$$= a^7 - 7a^6b + 21a^5b^2 - 35a^4b^3$$

$$+ 35a^3b^4 - 21a^2b^5 + 7ab^6 - b^7$$

71. Find the fifth term. [11.5]

$$\binom{7}{4}(3x)^3(-4y)^4 = 35(27x^3)(256y^4) = 241,920x^3y^4$$

73. Find the number of color selections. [11.6]

$12 \cdot 8 = 96$

75. Find the number of ways. [11.6]

There are 26 choices for each letter. By the

Fundamental Counting Principle, there are 26^8

possible passwords.

77. Find the number of ways. [11.6]

This is a permutation with $n = 15$ and $r = 3$.

$$P(15,3) = \frac{15!}{(15-3)!} = \frac{15!}{12!} = 2730$$

79. Find the number of ways. [11.6]

There are $\binom{4}{1}$ ways to choose a supervisor and $\binom{12}{3}$

ways to choose 3 regular employees.

$$\binom{4}{1}\binom{12}{3} = 4 \cdot 220 = 880 \text{ shifts have 1 supervisor.}$$

81. Find the number of ways. [11.6]

$$C(52,4) = \frac{52!}{4!(52-4)!} = \frac{52 \cdot 51 \cdot 50 \cdot 49 \cdot 48!}{4 \cdot 3 \cdot 2 \cdot 1 \cdot 48!} = 270,725$$

83. a. Yes, the number 431 is in the sample space. [11.7]

b. Yes, with replacement, the number 313 is in the

sample space.

85. List the elements: upward faces total 10. [11.7]

$\{(4, 6), (5, 5), (6, 4)\}$

87. a. Determine if mutually exclusive. [11.7]

The events are not mutually exclusive. For example, 61

is a prime number that is above 50.

b. Find the probability.

$$P(E_1) = \frac{25}{100}, \ P(E_2) = \frac{50}{100}, \ P(E_1 \cap E_2) = \frac{10}{100}$$

$$P(E_1 \cup E_2) = P(E_1) + P(E_2) - P(E_1 \cap E_2)$$

$$= \frac{25}{100} + \frac{50}{100} - \frac{10}{100} = \frac{13}{20}$$

89. Find which one has the greater probability. [11.7]

The probability of drawing an ace and a 10 card from

one regular deck of playing cards is

$$\frac{\binom{4}{1}\binom{16}{1}}{\binom{52}{2}} = \frac{4 \cdot 16}{\frac{52 \cdot 51}{2}} \approx 0.0483.$$

The probability of drawing an ace and a 10 card from

two regular decks of playing cards is

$$\frac{\binom{8}{1}\binom{32}{1}}{\binom{104}{2}} = \frac{8 \cdot 32}{\frac{104 \cdot 103}{2}} \approx 0.0478.$$

Drawing an ace and a 10 card from *one* deck has the

greater probability.

91. Find the probability. [11.7]

$$P = \binom{10}{8}(0.9)^8(0.1)^2 \approx 0.19$$

93. Find the probability. [11.7]

There are $\binom{12}{3}$ ways of choosing 3 people from 12.

There are $\binom{11}{2} \cdot 1$ ways of choosing 2 people and the person with badge number 6.

$$\text{Probability} = \frac{\binom{11}{2} \cdot 1}{\binom{12}{3}} = \frac{\frac{11 \cdot 10}{2}}{\frac{12 \cdot 11 \cdot 10}{3 \cdot 2}} = \frac{1}{4}$$

95. Find the net effect. [11.3]

Using the multiplier effect,

$$\frac{15}{1 - 0.80} = 75$$

The net effect of $15 million is $75 million.

Chapter 11 Test

1. Find the third and fifth terms. [11.1]

$$a_3 = \frac{2^3}{3!} = \frac{8}{6} = \frac{4}{3}$$

$$a_5 = \frac{2^5}{5!} = \frac{32}{120} = \frac{4}{15}$$

3. Classify the sequence as arithmetic, geometric or neither. [11.2]

$$a_{n+1} - a_n = [-2(n+1) + 3] - (-2n + 3)$$
$$= -2n - 2 + 3 + 2n - 3$$
$$= -2 = \text{constant}$$

arithmetic

5. Classify the sequence as arithmetic, geometric or neither. [11.2]

$$\frac{a_{n+1}}{a_n} = \frac{\frac{(-1)^{n+1-1}}{3^{n+1}}}{\frac{(-1)^{n-1}}{3^n}} = \frac{-1}{3} = \text{constant}$$

geometric

7. Find the sum of the series. [11.3]

$$\sum_{j=1}^{10} \frac{1}{2j} = \frac{1}{2} + \frac{1}{4} + \frac{1}{8} + \frac{1}{16} + \cdots \frac{1}{1024}$$

$$= \frac{\frac{1}{2}\left(1 - \left(\frac{1}{2}\right)^{10}\right)}{1 - \frac{1}{2}} = 1 - \left(\frac{1}{2}\right)^{10}$$

$$= 1 - \frac{1}{1024} = \frac{1023}{1024}$$

9. Find the twentieth term. [11.2]

$$a_3 = a_1 + (3-1)d = 7,$$
$$a_8 = a_1 + (8-1)d = 22$$

$$a_1 + 2d = 7$$
$$\underline{a_1 + 7d = 22}$$
$$-5d = -15$$
$$d = 3$$

$$a_1 = a_3 - 2(3)$$
$$= 7 - 6$$
$$= 1$$

$$a_{20} = a_1 + (20-1)d$$
$$= 1 + (19)(3)$$
$$= 58$$

11. Write as a quotient of integers in simplest form. [11.3]

$$0.\overline{15} = 0.15 + 0.0015 + 0.000015 + \cdots$$

$$= \frac{0.15}{1 - 0.01} = \frac{0.15}{0.99} = \frac{15}{99} = \frac{5}{33}$$

13. Prove by mathematical induction. [11.4]

1. Let $n = 7$

$$7! = 5040 \quad 3^7 = 2187$$

Thus $n! > 3^n$ for $n = 7$.

2. Assume $k! > 3^k$

Verify $(k+1)! > 3^{k+1}$
$$k! > 3^k$$
$$k + 1 > 3$$
$$(k+1)k! > 3 \cdot 3^k$$
$$(k+1)! > 3^{k+1}$$

Thus the formula has been established by the extended principle of mathematical induction.

15. Write the binomial expansion. [11.5]

$$\left(x + \frac{1}{x}\right)^6 = x^6 + 6(x)^5\left(\frac{1}{x}\right) + 15(x)^4\left(\frac{1}{x}\right)^2 + 20(x)^3\left(\frac{1}{x}\right)^3$$
$$+ 15(x)^2\left(\frac{1}{x}\right)^4 + 6x\left(\frac{1}{x}\right)^5 + \left(\frac{1}{x}\right)^6$$
$$= x^6 + 6x^4 + 15x^2 + 20 + \frac{15}{x^2} + \frac{6}{x^4} + \frac{1}{x^6}$$

17. Find the number of ways. [11.6]

$$52 \cdot 51 \cdot 50 = 132,600$$

19. Find the probability. [11.7]

$$\frac{C(8,\,3)C(10,\,2)}{C(18,\,5)}=\frac{56\cdot 45}{8568}=\frac{5}{17}\approx 0.294118$$

Cumulative Review Exercises

1. Solve, write in interval notation. [1.5]

$$|3-5x|\le 4$$
$$-4\le 3-5x\le 4$$
$$-7\le -5x\le 1$$
$$\frac{7}{5}\ge x\ge -\frac{1}{5}$$

The solution set is $\left[-\frac{1}{5},\frac{7}{5}\right]$.

3. Graph $y=x^2-x-2$. [2.1]

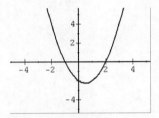

5. Divide. [3.1]

$$
\begin{array}{r}
x^2-x+1 \\
x+1\overline{\smash{\big)}\,x^3\qquad\;-1} \\
\underline{x^3+x^2} \\
-x^2 \\
\underline{-x^2-x} \\
x-1 \\
\underline{x+1} \\
-2
\end{array}
$$

$$\frac{x^3-1}{x+1}=x^2-x+1-\frac{2}{x+1}$$

7. Expand. [4.4]

$$\log_b\left(\frac{xy^2}{z^3}\right)=\log_b x+\log_b y^2-\log_b z^3$$

$$=\log_b x+2\log_b y-3\log_b z$$

9. Solve the system of equations. [9.1]

$$\begin{cases}2x-3y=8 & (1)\\ x+4y=-7 & (2)\end{cases}$$

Solve (2) for x and substitute into (1).

$$x=-4y-7$$

$$2(-4y-7)-3y=8$$
$$-8y-14-3y=8$$
$$-11y=22$$
$$y=-2$$

$$x=-4(-2)-7=1$$

The solution is $(1, -2)$.

11. Simplify. [P.2]

$$-2\sqrt[4]{80}+3\sqrt[4]{405}=-4\sqrt[4]{5}+9\sqrt[4]{5}=5\sqrt[4]{5}$$

13. Find the coordinates of the vertex. [2.4]

$$-\frac{b}{2a}=-\frac{5}{2(-2)}=\frac{5}{4}$$

$$F\left(\frac{5}{4}\right)=-2\left(\frac{5}{4}\right)^2+5\left(\frac{5}{4}\right)-2=\frac{9}{8}$$

Vertex: $\left(\frac{5}{4},\frac{9}{8}\right)$

15. Find the horizontal asymptote. [3.5]

$$y=0$$

17. Solve. Round to the nearest tenth. [4.5]

$$4^{2x+1}=3^{x-2}$$
$$\ln 4^{2x+1}=\ln 3^{x-2}$$
$$(2x+1)\ln 4=(x-2)\ln 3$$
$$2x\ln 4+\ln 4=x\ln 3-2\ln 3$$
$$x\ln 4^2-x\ln 3=-2\ln 3-\ln 4$$
$$x(\ln 16-\ln 3)=-2\ln 3-\ln 4$$
$$x=\frac{-2\ln 3-\ln 4}{\ln 16-\ln 3}\approx -2.1$$

19. Find the product. [10.2]

$$\begin{bmatrix}3 & 2\\ -2 & 1\\ 1 & -4\end{bmatrix}\begin{bmatrix}2 & 3 & 1 & 1\\ -2 & 0 & 4 & -3\end{bmatrix}=\begin{bmatrix}2 & 9 & 11 & -3\\ -6 & -6 & 2 & -5\\ 10 & 3 & -15 & 13\end{bmatrix}$$